12/5/73

Analysis of
Electric Circuits

PRENTICE-HALL SERIES IN ELECTRONIC TECHNOLOGY

IRVING L. KOSOW, editor

Charles M. Thomson, Joseph J. Gershon and Joseph A. Labok, consulting editors

Analysis of
Electric Circuits

FREDERICK F. DRISCOLL

Wentworth Institute
Boston, Massachusetts

PRENTICE-HALL, INC.   *Englewood Cliffs, New Jersey*

©1973
PRENTICE-HALL, INC.
Englewood Cliffs, N. J.

10 9 8 7 6 5 4 3 2 1

ISBN: 13-032912-6

Library of Congress Catalog Card Number: 72-3691

Printed in the United States of America

PRENTICE-HALL INTERNATIONAL, INC., *London*
PRENTICE-HALL OF AUSTRALIA PTY. LTD., *Sydney*
PRENTICE-HALL OF CANADA, LTD., *Toronto*
PRENTICE-HALL OF INDIA, PRIVATE LIMITED, *New Delhi*
PRENTICE-HALL OF JAPAN, INC., *Tokyo*

for Jean

contents

APPENDICES *430*

BIBLIOGRAPHY *447*

INDEX *449*

preface

The electronic technician must be familiar not only with the symbols and terminology used in his field, but also must be acquainted with as many different circuits and systems as possible, so that he may perform his work with skill.

Keeping this in mind, I have attempted to write a book which will fulfill both of these functions. This text may be used as the basis for a two semester course in electronic fundamentals for technical institute students majoring in electronic technology or electronic engineering technology.

Unlike most other fundamental texts, this one introduces both ac and dc waveforms together in Chapter 2. This technique permits the earlier introduction of the impedance concept, enabling the student to begin to analyze circuits containing all of the passive elements sooner in the course. It is my contention that this will foster a greater understanding of the principles involved.

To utilize the concepts presented throughout the book, the reader needs only a background in high school mathematics—algebra and an introduction to trigonometry are helpful. Calculus is not stressed and is used only where defining equations allows no alternative method.

Each chapter contains at least one example of each new concept. Examples progress from the simple to the complex, and the problems at the end of each chapter are based on the examples and arranged by degree of difficulty.

As aids in the preparation of this text, I am indebted to Dean Charles M. Thomson of Wentworth Institute, who took the time from his busy schedule to read the manuscript and offer suggestions; to Mr. Robert F. Coughlin, who helped me to work through various questions and issues; to Mr. Lawrence A. Coppola, with whom I began this text as a collaborative effort and whose conception of the joint ac and dc approach is a central theme of the book; to my wife Jean, who read the manuscript for style and form and typed all of the drafts; and especially to Dr. Irving L. Kosow, who edited the chapters as they were written and who provided innumerable suggestions for improving the work.

Frederick F. Driscoll

Analysis of
Electric Circuits

CHAPTER 1

fundamental concepts

Modern man lives in a world of necessities, conveniences, and gadgets, many of which depend upon electricity. Although electricity has not been adequately defined even today, it does conform to definite rules and laws.

Our understanding of electricity can be traced back to such men as Faraday, Lenz, Ohm, and Kirchhoff, who formulated many of the theories still in use today. A thorough knowledge of these theories involves more than the simple application of formulas to solve a particular problem. Rather, it necessitates a basic understanding of the concepts behind the theories and the ability to apply correctly the knowledge acquired.

Certain building blocks are essential in relating abstract electronic theory to practical use. Among these are atomic structure, the nature of electricity, and the language that scientists and technicians use in this field.

This chapter provides a framework of knowledge which enables the reader to understand the basic concepts involved, for example, in Fig. 1-1.

The first four sections discuss the MKS system, scientific notation, metric prefixes, and symbols. A knowledge of these provides a base from which to present atomic structure and the movement of electrons, how electricity travels through a conductor, what distinguishes conductors from insulators, and what the meters shown in Fig. 1-1 actually measure.

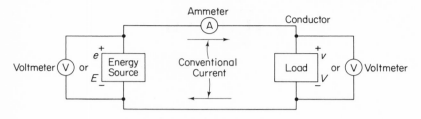

Fig. 1-1. A simplified circuit.

1-1. THE MKS SYSTEM

Certain basic definitions used by scientists, engineers, and technicians in the course of their work form a standardized "technical language." In measuring basic quantities of length, mass, and time, for example, the following units are standard: length is measured in meters, mass is measured in kilograms, and time is measured in seconds. The meter (m), kilogram (kg), and second (s) form the basis for the MKS, or *metric, system.* This system was first suggested by G. Giorgi in 1901 and has become conventional in scientific and engineering work.

The MKS system is widely used because the basic quantities are multiples of 10. Thus they are simpler to use in calculations than the British engineering system, for example, which has for its basic units the foot, pound, and second. While the meter equals 100 centimeters (cm) and the kilogram equals 1,000 grams (g), the foot equals 12 inches (in.) and the pound equals 16 ounces (oz).

1-2. THE POWER-OF-10 NOTATION

A method for expressing large or small numbers as whole numbers and multipliers is by the *power-of-10, or scientific notation.* For example, 13,800 m may be expressed as 1.38×10^4 m. This procedure, also called *standard notation*, greatly simplifies mathematical operations.

To write a number in standard notation, move the decimal point after the first significant digit to obtain a number between 1 and 10. Then count the number of decimal places the original decimal was moved — this number becomes the *exponent* of the multiplier. If the decimal point is moved to the *left*, the exponent is *positive*. If the decimal point is moved to the *right*, the exponent is *negative*.

Examples of expressing large whole numbers in standard notation are

$$100 = 1.0 \times 10^2 \qquad 10,000 = 1.0 \times 10^4$$
$$720 = 7.2 \times 10^2 \qquad 15,300 = 1.53 \times 10^4$$
$$1,000 = 1.0 \times 10^3 \qquad 60,000 = 6.0 \times 10^4$$
$$6,400 = 6.4 \times 10^3 \qquad 100,000 = 1.0 \times 10^5$$
$$3,240 = 3.24 \times 10^3 \qquad 450,000 = 4.5 \times 10^5$$

Examples of expressing decimal numbers in standard notation are

$$0.01 = 1.0 \times 10^{-2} \qquad 0.0001 = 1.0 \times 10^{-4}$$
$$0.53 = 5.3 \times 10^{-1} \qquad 0.0022 = 2.2 \times 10^{-3}$$
$$0.001 = 1.0 \times 10^{-3} \qquad 0.00001 = 1.0 \times 10^{-5}$$
$$0.007 = 7.0 \times 10^{-3} \qquad 0.00025 = 2.5 \times 10^{-4}$$
$$0.085 = 8.5 \times 10^{-2} \qquad 0.0056 = 5.6 \times 10^{-3}$$

In the processes of addition and subtraction, the exponents of the multipliers must be equal. If they are not, it is necessary to *scale* one of the numbers. The following examples illustrate this point.

Example 1-1: Add 4.5×10^5 and 73.2×10^4.

Solution: First scale one number so that the powers of 10 (exponents of the multipliers) are equal.

$$4.5 \times 10^5 = \quad 45.0 \times 10^4$$
$$+ \quad 73.2 \times 10^4$$
$$\overline{118.2 \times 10^4} = 1.182 \times 10^6$$

Alternatively,

$$73.2 \times 10^4 = \quad 7.32 \times 10^5$$
$$+ \quad 4.5 \times 10^5$$
$$\overline{11.82 \times 10^5} = 1.182 \times 10^6$$

Remember that in the process of addition, any number may be scaled to bring about equal powers of 10.

Other examples of addition are

a. $41 \times 10^2 + 6 \times 10^3 = 10.1 \times 10^3 = 1.01 \times 10^4$
b. $2.8 \times 10^4 + 0.37 \times 10^5 = 6.5 \times 10^4$
c. $0.56 \times 10^5 + 144 \times 10^3 = 200 \times 10^3 = 2 \times 10^5$
d. $80 \times 10^6 + 1.8 \times 10^7 = 98 \times 10^6 = 9.8 \times 10^7$
e. $0.256 \times 10^{-3} + 1.7 \times 10^{-4} = 4.26 \times 10^{-4}$

The same method is used in subtraction.

Example 1-2: Solve $3.2 \times 10^7 - 20.8 \times 10^6$.

Solution: Again scale one of the numbers so that the powers of 10 are equal.

$$3.2 \times 10^7 = \begin{array}{r} 32.0 \times 10^6 \\ - \; 20.8 \times 10^6 \\ \hline 11.2 \times 10^6 = 1.12 \times 10^7 \end{array}$$

Alternatively,

$$-20.8 \times 10^6 = \begin{array}{r} 3.2 \;\; \times 10^7 \\ - \; 2.08 \times 10^7 \\ \hline 1.12 \times 10^7 \end{array}$$

In subtraction as in addition, one or more numbers may be scaled to obtain the same exponents or powers of 10. The answer is expressed in the same power of 10 as the numbers being subtracted.

Other examples of subtraction are

a. $32.5 \;\; \times 10^4 \;\; - 0.155 \times 10^6 \;\; = \;\; 17.0 \;\; \times 10^4 = 1.7 \times 10^5$
b. $\;\; 0.18 \times 10^6 \;\; - 0.022 \times 10^7 \;\; = -0.04 \times 10^6 = -4 \times 10^4$
c. $13.5 \;\; \times 10^{-5} - 115 \;\;\;\; \times 10^{-6} = \;\; 2.0 \;\; \times 10^{-5}$
d. $40.3 \;\; \times 10^{-3} - 153 \;\;\;\; \times 10^{-4} = 250 \;\;\;\; \times 10^{-4}$
e. $\;\; 5.7 \;\; \times 10^7 \;\; - 120 \;\;\;\; \times 10^5 \;\; = \;\; 4.5 \;\; \times 10^7$

In performing multiplication and division, scientific notation greatly simplifies our work. To multiply or divide, it is not necessary to have the same powers of 10.

To multiply two numbers expressed in powers of 10, we *multiply* the coefficients (whole numbers) and *add* the exponents of the multiplier.

Example 1-3: Solve $3.5 \times 10^{-6} \times 7.0 \times 10^3$.

Solution: Multiply the coefficients or whole numbers:

$$3.5 \times 7.0 = 24.5$$

Add the exponents:

$$10^{-6} \times 10^3 = 10^{-3}$$

Therefore,

$$(3.5 \times 10^{-6})(7.0 \times 10^3) = 24.5 \times 10^{-3} = 2.45 \times 10^{-2}$$

Other examples of multiplication are

a. $(1.5 \times 10^3) \;\;\;\; (2.5 \times 10^2) \;\; = \;\; 3.75 \times 10^5$
b. $(0.4 \times 10^{-6}) \;\; (1.8 \times 10^6) \;\; = \;\; 0.72 \times 10^0 \;\; = 7.2 \times 10^{-1}$
c. $(5.5 \times 10^{-3}) \;\; (2.0 \times 10^6) \;\; = 11.0 \;\; \times 10^3 \;\; = 1.1 \times 10^4$
d. $(6.0 \times 10^{-9}) \;\; (12.0 \times 10^6) = 72 \;\;\;\; \times 10^{-3} = 7.2 \times 10^{-2}$
e. $(8.2 \times 10^9) \;\;\;\; (0.5 \times 10^{-3}) = \;\; 4.1 \;\;\;\; \times 10^6$

Note in example b that $10^0 = 1$, because *any* number raised to the zero power equals 1.

To divide, we use another special property of exponents. *Any number raised to a power may be moved from the denominator to the numerator or vice versa, provided that the sign of the exponent is changed.* We divide coefficients and add exponents algebraically.

Example 1-4: Solve

$$\frac{6.6 \times 10^{-12}}{4.0 \times 10^{-9}}$$

Solution: Divide the coefficients, move the exponents to the numerator, and add them algebraically:

$$\frac{6.6 \times 10^{-12}}{4.0 \times 10^{-9}} = \frac{6.6 \times 10^{-12} \times 10^9}{4.0} = 1.65 \times 10^{-3}$$

Other examples of division are

a. $\dfrac{4.5 \times 10^6}{2.5 \times 10^{-3}} = \dfrac{4.5 \times 10^6 \times 10^3}{2.5} = 1.8 \times 10^9$

b. $\dfrac{8.4 \times 10^3}{4.2 \times 10^9} = \dfrac{8.4 \times 10^3 \times 10^{-9}}{4.2} = 2.0 \times 10^{-6}$

c. $\dfrac{6.0 \times 10^5}{2.5 \times 10^{-2}} = \dfrac{6.0 \times 10^5 \times 10^2}{2.5} = 2.4 \times 10^7$

d. $\dfrac{4.0 \times 10^3}{5.0 \times 10^{-3}} = \dfrac{4.0 \times 10^3 \times 10^3}{5.0} = 0.8 \times 10^6 = 8 \times 10^5$

e. $\dfrac{75 \times 10^4}{5.0 \times 10^3} = \dfrac{75 \times 10^4 \times 10^{-3}}{5.0} = 15 \times 10^1 = 1.5 \times 10^2$

1-3. METRIC PREFIXES

Although most of the above examples have been written as whole numbers with one significant decimal place times the power of 10, it is not necessary to express all numbers this way. Engineers and technicians use a system of *metric prefixes*. Table 1-1 lists the prefixes, symbols, and values that are most often encountered.

TABLE 1-1

METRIC PREFIXES

Prefix	Prefix Symbol	Value
pico	p	10^{-12}
nano	n	10^{-9}
micro	μ	10^{-6}
milli	m	10^{-3}
kilo	k	10^{3}
mega	M	10^{6}
giga	G	10^{9}
tera	T	10^{12}

Since a prefix is a letter, word, or syllable placed at the beginning of a word, the prefixes listed in Table 1-1 should not be used without an accompanying fundamental unit. Some examples are

a. 40 ms = 40 milliseconds = 40 $\times 10^{-3}$ s
b. 0.1 μs = 0.1 microsecond = 0.1 $\times 10^{-6}$ s
c. 25 ns = 25 nanoseconds = 25 $\times 10^{-9}$ s
d. 100 ns = 100 nanoseconds = 100 $\times 10^{-9}$ s or 0.1 μs

1-4. SYMBOLS

In electronics, as in other sciences, we use symbols to represent a given *physical quantity* or variable. These *quantity* symbols should not be confused with the prefix symbols in Table 1-1. The most common symbols used are the capital and lowercase letters of the English alphabet (usually in *italic)* and the Greek alphabet (see Table 1-2). Standard symbols are used throughout this text and are defined as they are introduced.

A note of caution before we proceed: Even with both alphabets available, it sometimes becomes necessary to use the *same* symbol to represent unrelated quantities. For example, Q is used to represent both electric charge and quality factor.

In the expression $V = 50$ mV, V is the symbol for the physical quantity (voltage) and mV is the symbol for a unit of that quantity (millivolts).

TABLE 1-2

GREEK ALPHABET

Name	Capital	Lower case	Name	Capital	Lower case
alpha	A	α	nu	N	ν
beta	B	β	xi	Ξ	ξ
gamma	Γ	γ	omicron	O	o
delta	Δ	δ, ∂	pi	Π	π
epsilon	E	ε	rho	P	ρ
zeta	Z	ζ	sigma	Σ	σ
eta	H	η	tau	T	τ
theta	Θ	θ	upsilon	Υ	v
iota	I	ι	phi	Φ	ϕ
kappa	K	κ	chi	X	χ
lambda	Λ	λ	psi	Ψ	ψ
mu	M	μ	omega	Ω	ω

Some frequently used symbols, their meanings, and examples are given in Table 1-3.

TABLE 1-3

FREQUENTLY USED SYMBOLS AND THEIR MEANINGS

Symbol	Meaning	Examples
$\neq$	not equal to	$4.85 \neq 4.79$
$\approx$	approximately equal to	$7.14324 \approx 7.14$
$>$	greater than	$7.9 > 7.7$
$<$	less than	$10^4 < 10^5$
$\geqslant$	greater than or equal to	$a \geqslant b$ if $b = 4$, then $a > 4$ or $a = 4$
$\leqslant$	less than or equal to	$a \leqslant b$ if $b = 4$, then $a < 4$ or $a = 4$
$\lessgtr$	greater than but less than	$2 < t < 4$ $t > 2$ but $t < 4$
$\dfrac{d}{dt}$	a change with respect to time	$\dfrac{dy}{dt}$ the value of y changes with time
$\rightarrow$	approaches	$9.999 \rightarrow 10.0$

1-5. ATOMIC STRUCTURE

We may interpret our physical world in terms of *matter* and *energy*, both of which exist in a variety of forms.   Matter undergoes change; energy either causes or is a result of the change.

Matter has been defined as anything that occupies space and possesses mass.   Energy, on the other hand, is the ability to do work.   Although energy itself cannot be measured, its results can.   The basic unit for energy in the MKS system is the joule, named in honor of James Prescott Joule (1818–1889).

All matter exists in a solid, a liquid, or a gaseous state.   Since most electronic components are solids, our study of circuit analysis begins with the physical, chemical, and electrical properties of a solid.

The idea that all matter is composed of atoms was first introduced by the Greeks more than 2,000 years ago.   Although the atom is the smallest particle of an element, it, too, has many subdivisions.   Atoms are constructed like miniature solar systems, with a central core and orbiting particles.   The central core of an atom is called the *nucleus* and the orbiting particles are called *electrons.*

The unique property of the electron is its negative electrical charge.   The charge of the electron is the smallest electrical charge that can exist in nature. Because of this, it is more practical to group a large number of electrons together for a more workable quantity.   Hence the basic quantity of electrical charge is the *coulomb* represented by the symbol C.   One coulomb (1C) equals the accumulated charge of 6.24×10^{18} electrons.   The coulomb is named in honor of Charles A. de Coulomb (1736–1806) and is represented by the symbol Q.

We know that elements in their normal state are electrically neutral. Since the electron is negatively charged, there is a positive charge in the atom to counteract the charge of the electron.   This positive charge is known as the *proton* and is located in the nucleus.   The mass of the proton is 1837 times the mass of the electron.   Since the proton is so much heavier than the electron, the nucleus contains most of the mass of the atom and is the predominating factor in holding the atom together.   The nucleus of most atoms is actually heavier than just the mass of the protons.   To account for this additional weight, the nucleus contains another element.   This element is called the *neutron*, and, as the name implies, it is electrically neutral.   The number of protons equal the atomic number.   The sum of the protons and neutrons equals the atomic weight of an atom.

The electrons closest to the nucleus do not contain as much energy as the electrons farther away and thus are more tightly bound to the nucleus. Since the amount of energy an electron has determines the orbit of the electron, we may refer to a particular orbit of an electron as a distinct *energy level.* The electrons closest to the nucleus are said to be in the lowest energy level.   Electrons with approximately the same energy are grouped into electron shells or subshells.   Actually, no two electrons in the same atom may have exactly the same energy, spin direction, position of orbit, or orbital shape.   This is known as *Pauli's exclusion principle.*   Fig. 1-2 shows a simplified two-dimensional model of an atom.   Shells are designated by the symbols K, L, M, N, O, P, and Q.   The first four shells are shown in Fig. 1-2.

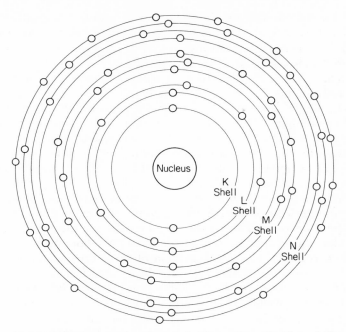

Fig. 1-2. Two-dimensional model of an atom showing shells and subshells.

The maximum number of electrons that a shell may contain is given by $2n^2$, where n is the number of the shell, counting from the nucleus. For example, the K shell is the first shell from the nucleus, then $n = 1$, and the maximum number of electrons in this shell is 2. For the L shell it is 8. For the M shell it is 18, and so forth.

Electrons for a particular element do not arbitrarily exist in any shell. They do, on the other hand, fill the shells in an orderly arrangement, from the lightest element, hydrogen, to the heaviest natural element, uranium. The number of electrons in the outermost shell of an atom is of particular importance because it dictates whether or not a material will be a good conductor of electric current. The electrons in the outermost shell of an atom are called *valence electrons*.

1-6.   CONDUCTORS, INSULATORS, AND SEMICONDUCTORS

Why is one material a better conductor than another? Most metals, which are good conductors of electric current, are composed of atoms with only one valence electron. In a metal, the valence electrons are not tightly bound to the nucleus and are said to be *free* or *mobile*. A good conductor contains high concentrations of these free or mobile charge carriers. The greater the number of free electrons per unit of volume in a metal, the better the material behaves, as an *electric conductor*.

Some materials with high concentrations of free electrons are silver, copper, and aluminum. It has been estimated that copper has approximately 8.55×10^{22} free electrons per cubic centimeter. Silver has approximately 5 per cent more free electrons per unit volume, and aluminum has only 60 per cent of the free electrons of copper per unit volume. Although silver has more free electrons per unit volume, its high cost prohibits its extensive use. Copper and aluminum, on the other hand, are less expensive and easily manufactured into wire.

Materials such as air, glass, mica, and rubber have considerably fewer free electrons per unit of volume than materials that make good conductors. In these poor conducting materials, the outer shell of the atom has several valence electrons and thus they are more tightly bound to the nucleus. Materials that *lack* a high concentration of free electrons are classified as *insulators*. Because no material has absolutely zero free electrons, even an insulator can be made to conduct. This, of course, would take a considerable amount of energy, usually more energy than would prove economical.

A semiconductor is a solid material that contains fewer free electrons than a conductor, but more than an insulator. Normally, at room temperature, a semiconductor is neither a good conductor nor a good insulator. However, at extremely low temperatures, semiconductors are insulators, whereas at very high temperatures they are reasonably good conductors. Materials most commonly used in the manufacture of semiconductors are germanium and silicon, both of which have four valence electrons. Germanium and silicon are the two elements mainly used in the fabrication of diodes and transistors.

1-7. CURRENT

To understand the electrical conduction process (or how an electric current flows), let us consider a portion of a conductor (the shaded volume) as shown in Fig. 1-3. We shall first consider the case where no outside

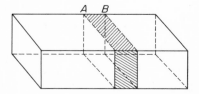

Fig. 1-3. A conductor.

energy, except thermal energy, is applied to the conductor. Thermal energy is always present because we are unable to achieve a temperature of absolute zero. Under normal conditions, the free electrons, or mobile charge carriers, are said to be *uniformly distributed* and to have a *random motion.*

By *uniformly distributed* we mean that the free electrons will not be concentrated in one particular location. They will, however, permeate throughout the metal, like a gas diffuses through a closed container. Thus at any instant, *any* incremental volume of the *same* size as that of the shaded portion of Fig. 1-3 would contain the *same* number of free electrons.

By *random motion* we mean that no specific direction can be assigned to the free electrons. Fig. 1-4 shows a two-dimensional example of random motion.

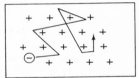

Fig. 1-4. Two-dimensional example of random motion.

Mathematically, the probability of finding a number of free electrons moving in one direction is the same as finding an equal number of free electrons moving in the opposite direction. Thus the net motion in any one direction is zero. In order for the mobile charge carriers to have a net direction, some energy other than thermal energy must be applied to the conductor.

Let us consider the same conductor with one end connected to a positive charge while the other end is connected to a negative charge (see Fig. 1-5).

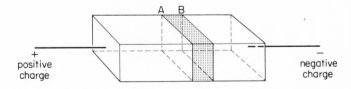

Fig. 1-5. Conductor with one end connected to a positive charge while the other end is connected to a negative charge.

Since like charges repel and unlike charges attract, electrons (negative charges) will travel toward the positive end of the conductor.

Even though a difference of charge has been applied to a conductor and the free electrons do move toward the positive charge, they do not move in a straight path. Instead, they follow a zigzag pattern similar to random motion. Unlike true random motion, however, there is now a definite drift toward the positive charge, as shown in Fig. 1-6. This net movement of free electrons in one direction constitutes an electric current, known as *drift current.*

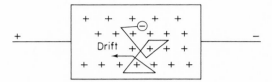

Fig. 1-6. Two-dimensional example of random motion when a difference of charge is applied to a conductor.

The actual value of drift current depends on the *rate* at which the free electrons move. The term "rate" implies any quantity divided by a specific time interval. A note of caution: Current is a *rate-of-flow* term and not a *velocity* term. Velocity is a rate that is measured in terms of length per unit time, such as feet per second or inches per second. Rate of flow, on the other hand, is a measure of the quantity in motion per unit time, such as gallons per minute, coulombs per second, or electrons per second.

The actual velocity of an electron is only a few thousands of an inch per second. However, current (rate of charge flow) is *not* dependent on the *speed* of the electron but rather on the net number of a large quantity of free electrons in motion. For this reason the effect of current is realized instantaneously.

Consider the cross section of a conductor, such as plane A of Fig. 1-5. The amount of current that flows depends on the quantity of charge, ΔQ, that crosses this plane in a given time interval, Δt. Mathematically, the current or rate of charge flow is

$$I = \frac{\Delta Q}{\Delta t} \frac{\text{coulombs}}{\text{seconds}} \qquad (1\text{-}1)$$

where I is the conventional symbol for current, ΔQ is the charge in coulombs, and Δt is the time in seconds.

Equation (1-1) gives the value of current in the fundamental units – coulombs and seconds. However, in honor of one of the pioneers of science, Andre Marie Ampère, the basic unit of current is called the *ampere*. A definition of the ampere is: *If 1 coulomb of charge passes through a cross-sectional area of a conductor in 1 second, 1 ampere of current flows,* or

$$1 \text{ ampere} = \frac{1 \text{ coulomb}}{1 \text{ second}}$$

Example 1-5: What is the current in an electrical conductor if 0.012 C of charge passes through a cross section of the conductor in 1 min?

Solution: $I = \dfrac{\Delta Q}{\Delta t} = \dfrac{0.012 \text{ C}}{60 \text{ s}} = 0.0002 \text{ A} = 0.2 \text{ mA}$

Note in Example 1-5 that the time (in minutes) first had to be converted into seconds. The answer is also best expressed using the metric prefixes.

If the number of free electrons that traverse the cross section in a given time interval is not constant, we may then calculate the current at a particular instant by letting $\Delta t \to 0$. This may be expressed mathematically as

$$i = \frac{dq}{dt} \tag{1-2}$$

where i is the instantaneous current in amperes and dq/dt is rate of change of charge with respect to time in coulombs per second.

Comparing Eq. (1-1) to Eq. (1-2), we note that the units for both are the same. However, Eq. (1-1) is expressed in uppercase letters while Eq. (1-2) is expressed in lowercase letters. It is conventional to use uppercase letters to represent constant values – values that do not vary with time – and lowercase letters to represent instantaneous values – values that are not constant for all time.

Conventional Current Flow. Since an electron has a negative charge, it drifts from a negative terminal to a positive one. However, the early experimenters thought that current flowed in solids from the positive terminal to the negative terminal. As a result, they formulated their rules on the direction of current accordingly. Fortunately, however, to solve an electronic problem mathematically, it makes no difference which direction is chosen as long as the same direction is used thoughout the solution.

In this text we shall assign the *direction* of current to be from positive to negative in solid materials. This is known as *conventional current flow*. The actual direction of the free electrons in solid conductors is from negative to positive, and this is termed *electron flow*. Both alternatives are shown in Fig. 1-7.

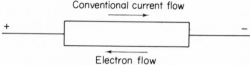

Conventional current flow

Electron flow

Fig. 1-7. Distinction between electron and conventional current flow.

Ammeters. Equations (1-1) and (1-2) are used mostly for defining current in fundamental units — charge and time — rather than calculating current. Current is a measurable quantity. The instrument commonly used for measuring current is called an *ammeter*. Examples of connecting an ammeter in a circuit are shown in Fig. 1-8a and b. Note that an ammeter is connected in *series* with an element. Therefore, all the current that flows through the circuit element must flow through the ammeter. In Fig. 1-8a the ammeter measures the current through the single element. In Fig. 1-8b,

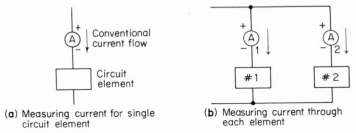

(a) Measuring current for single   (b) Measuring current through
     circuit element   each element

Fig. 1-8. Connection of an ammeter in a circuit.

ammeter 1 measures the current only through element 1, and ammeter 2 measures the current only through element 2.

In electronic work the values of current are small, on the order of milliamperes or microamperes. Therefore, the meters used are either the milliammeter or the microammeter. At this time we are not concerned with the internal structure of the ammeter but with its external operation.

The face of the ammeter has a calibrated scale to indicate the magnitude of the current flowing through it. Two of the most common ammeter scales are shown in Fig. 1-9a and b. A conventional ammeter with a

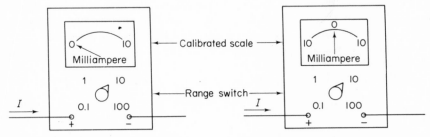

(a) Conventional scale: measures current   (b) Zero-center scale: measures (steady)
    only when the current through it   current regardless of the direction
    flows in the proper direction

Fig. 1-9. Front panel of two ammeters.

scale similar to that of Fig. 1-9a reads the magnitude of current only when current flows through it in the proper direction. If the current should be reversed, the pointer would deflect below zero. A zero-center ammeter with a scale as shown in Fig. 1-9b may be connected into the circuit regardless of the direction of current flow. The magnitude of the current may be read on either side of zero.

In addition to the calibrated scale, the ammeter also has a *range switch.* The value at which this is set dictates the *largest* value of current that the meter will read. For example, if the range switch is set on 10 mA, the ammeter can measure any value of current from 0 to 10 mA and the scale is calibrated proportionally, as shown in Fig. 1-9a and b.

Ammeters have a positive and negative terminal. Sometimes these terminals are color-coded. For instance, the positive terminal may be red and the negative terminal black. The path of conventional current is into the positive terminal, through the meter, and out the negative terminal.

1-8. VOLTAGE

In Section 1-7 we saw that for an electric current to flow in a conductor it is necessary to have the ends of the conductor connected to a charge difference. Only then is it possible to have a movement of free electrons in a specific direction.

In this section we shall introduce the concept of *potential difference,* or *voltage.* A voltage exists between two points in a circuit whenever one point has a more positive charge with respect to the other. It is this un-balance of charge or *potential energy difference* that causes current flow. In present-day circuit analysis, this potential-energy difference is known as voltage.

Voltage is referred to as either a potential rise or fall. A potential *rise* occurs when work is done *on* the charge by an element. An example is an energy source. A potential *fall* exists when work is done *by* the charge on an element. An example is a load.

The distinction between potential rise and potential fall is maintained by polarity markings across an element. Fig. 1-10 shows polarity mark-

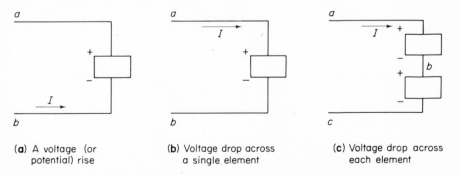

(**a**) A voltage (or potential) rise

(**b**) Voltage drop across a single element

(**c**) Voltage drop across each element

Fig. 1-10. Examples of polarity markings, and potential rise and drop.

ings for voltages. In Fig. 1-10a the direction of conventional current flow is from negative (−) to positive (+). This signifies a voltage rise or potential rise. In Fig. 1-10b and c the direction of conventional current flow is from positive (+) to negative (−). This signifies a voltage drop or potential fall.

Voltage is represented by the symbols e, E and v, V. The symbols e and E usually represent a potential rise while v and V represent a potential fall. As in the discussion of current, the lowercase letters represent instantaneous values (values that change with time) and uppercase letters represent constant values (values that do not change with time).

The *basic unit of voltage*, whether it is a potential rise or fall, is the *volt*, named in honor of Alessandro Volta. In terms of the fundamental units, voltage can be defined as follows: *If 1 joule of energy is expended in moving 1 coulomb of charge from one point to another in an electric circuit, the voltage is 1 volt.* This statement may be expressed in terms of work per unit charge as

$$1 \text{ volt} = \frac{1 \text{ joule}}{1 \text{ coulomb}}$$

Using the conventional symbols,

$$E = \frac{W}{Q} \tag{1-3a}$$

$$V = \frac{W}{Q} \tag{1-3b}$$

where E and V are constant values of voltage in volts, W the energy in joules, and Q the charge in coulombs. The instantaneous value of voltage may also be written in equation form using conventional symbols:

$$e = \frac{dw}{dq} \tag{1-4a}$$

$$v = \frac{dw}{dq} \tag{1-4b}$$

where e and v are the instantaneous values of voltage in volts and dw/dq is the change in energy with respect to change in charge.

Voltmeters. Equations (1-3) and (1-4) are used mostly for defining voltage in fundamental quantities (energy and charge) rather than calculating voltage. Voltage, like current, is a measurable quantity. The instrument for measuring voltage is the *voltmeter*, which measures the potential difference between two points in a circuit in volts.

Examples of connecting a voltmeter are shown in Fig. 1-11. Note that a voltmeter is connected *across* (in parallel with) an element or elements. In Fig. 1-11a the voltmeter measures the voltage across the element. In Fig. 1-11b voltmeter 1 measures the voltage across element 1 and voltmeter

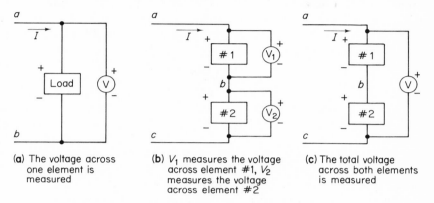

(a) The voltage across one element is measured

(b) V_1 measures the voltage across element #1, V_2 measures the voltage across element #2

(c) The total voltage across both elements is measured

Fig. 1-11. Examples of connecting voltmeter into a circuit.

2 measures the voltage across element 2. Between terminals *a* and *c* of Fig. 1-11c, the voltmeter measures the total voltage across both elements 1 and 2.

The voltmeter, like the ammeter, has a calibrated scale and a range switch. The calibrated scale is from zero to the largest value of voltage determined by the range switch, as shown in Fig. 1-12. The magnitude

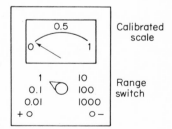

Fig. 1-12. Front panel of voltmeter.

of the voltage is indicated by the reading on the meter face. The voltmeter does not read negative values. It always indicates positive, or upscale, readings. Therefore, to ensure an upscale reading, the meter must be connected into the circuit with the correct polarity. That is, the positive terminal of the meter is connected to the positive side of the circuit and the negative terminal of the meter is connected to the negative side of the circuit, as shown in Fig. 1-11.

1-9. POWER

The goal of electrical circuits is to transfer energy from one point to another to accomplish useful work. Electric circuits are composed of (1) sources and (2) loads as shown in Fig. 1-13. The source supplies the

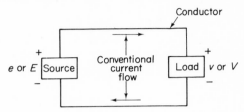

Fig. 1-13. A simplified circuit.

electrical energy while the load converts this energy into work. *The rate at which energy is transferred from the source to the load is called* power.

Assuming ideal conductors, all the energy produced by the source must be delivered to the load. Thus the power of the source must equal the power of the load. Power is defined as work per unit time or energy per unit time. Power may be calculated in terms of other measurable electrical quantities — current and voltage. The letter symbol for power is P or p and the basic unit is the *watt*, named in honor of James Watt.

We may define: *1 watt is the transfer of 1 joule of electrical energy in 1 second*. This statement written in equation form is

$$1 \text{ watt} = \frac{1 \text{ joule}}{1 \text{ second}}$$

In electric circuits, power is equal to the product of voltage times current. If the voltage and current are constant values, then

$$P = EI \qquad (1\text{-}5a)$$
$$P = VI \qquad (1\text{-}5b)$$

where P is the power in watts, E and V the voltages in volts across a source and a load, respectively, and I the current in amperes through the element. If the voltage and current vary with time, the instantaneous power is

$$p = ei \qquad (1\text{-}6a)$$
$$p = vi \qquad (1\text{-}6b)$$

where p, e or v, and i represent the instantaneous values of power, voltage, and current, respectively.

Since power is the product of voltage times current, the fundamental units may be derived as follows:

$$\text{power} = \text{voltage} \times \text{current} = \frac{\text{joule}}{\text{coulomb}} \times \frac{\text{coulomb}}{\text{second}} = \frac{\text{joule}}{\text{second}} = \text{watt}$$

which corresponds to our definition of power.

In the previous two sections we discussed the fact that current and voltage are measurable quantities. Although power may also be measured directly (using a meter known as a *wattmeter*), in electronic circuits of low power it is usually calculated as the product of voltage times current.

Example 1-6: What is the output power of a 24 V source with 5 mA of current flowing through it?

Solution: $P = EI = (24\text{ V})(5\text{ mA}) = 120\text{ mW}$

Example 1-7: The voltage across a load is measured as 10 V, the current through it is 10 mA. What is the power?

Solution: $P = VI = (10\text{ V})(10\text{ mA}) = 100\text{ mW}$

Example 1-8: A 12 V source is capable of delivering 600 mW of power. What is the largest value of current this source can produce?

Solution: Rearranging Eq. (1-5),

$$I = \frac{P}{E} = \frac{600\text{ mW}}{12\text{ V}} = 50\text{ mA}$$

The three examples above have dealt with power as the product of voltage times current.

Example 1-9: Consider the circuit in Fig. 1-14 and from the measured values calculate (a) the power delivered by the source, (b) the power of load 1, and (c) the power of load 2.

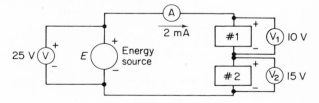

Fig. 1-14. Circuit for Ex. 1-9.

Solution: (a) $P\ \ = EI\ = (25\text{ V})(2\text{ mA}) = 50\text{ mW}$
(b) $P_1 = V_1 I = (10\text{ V})(2\text{ mA}) = 20\text{ mW}$
(c) $P_2 = V_2 I = (15\text{ V})(2\text{ mA}) = 30\text{ mW}$

Note that the power delivered by the source (50 mW) equals the sum of the load powers (20 mW + 30 mW).

SUMMARY

This chapter has presented the basic concepts with which any student of electronic engineering technology must become familiar (recall Fig. 1-1). The reader should be able to explain why current travels through the conducting material, what the ammeter and voltmeter measure in this circuit, and how to calculate power. The chapters to follow will build on this fundamental beginning.

PROBLEMS

1-1. Express the following numbers in scientific notation.

    (a) 14700   (f) 0.0004
    (b) 0.0135   (g) 0.102
    (c) 71.3   (h) 438.5
    (d) 0.003   (i) 482
    (e) 2580   (j) 0.0014

1-2. Add the following numbers as powers of 10 and express the answer in scientific notation.

    (a) $13.2 \times 10^3 + 62 \times 10^2$
    (b) $0.82 \times 10^2 + 0.15 \times 10^1$
    (c) $45 \times 10^{-3} + 580 \times 10^{-4}$
    (d) $0.012 \times 10^3 + 20 \times 10^0$
    (e) $8.6 \times 10^{-2} + 720 \times 10^{-4}$

1-3. Subtract the following numbers as powers of 10 and express the answer in scientific notation.

    (a) $37.5 \times 10^4 - 0.185 \times 10^6$
    (b) $10.7 \times 10^7 - 120 \times 10^5$
    (c) $13.3 \times 10^{-3} - 123 \times 10^{-4}$
    (d) $43.5 \times 10^{-5} - 95 \times 10^{-6}$
    (e) $0.22 \times 10^6 - 0.018 \times 10^7$

1-4. Multiply the following numbers as powers of 10 and express the answer in scientific notation.

    (a) $(10.2 \times 10^8)(0.2 \times 10^{-2})$
    (b) $(8.0 \times 10^{-9})(4.0 \times 10^6)$
    (c) $(3.5 \times 10^3)(6.0 \times 10^{-3})$
    (d) $(0.3 \times 10^4)(1.5 \times 10^4)$
    (e) $(12 \times 10^{12})(5.0 \times 10^{-6})$

1-5. Divide the following numbers as power of 10 and express the answer in scientific notation.

$$\text{(a)} \quad (12.0 \times 10^{12}) \div (5.0 \times 10^6)$$
$$\text{(b)} \quad (4.0 \times 10^6) \div (8.0 \times 10^6)$$
$$\text{(c)} \quad (4.0 \times 10^6) \div (8.0 \times 10^{-6})$$
$$\text{(d)} \quad (3.5 \times 10^9) \div (0.7 \times 10^3)$$
$$\text{(e)} \quad (6.0 \times 10^3) \div (3.0 \times 10^6)$$

1-6. Express the following quantities using metric prefixes.

$$\text{(a)} \quad 400 \times 10^{-6}\,\text{s}$$
$$\text{(b)} \quad 10 \times 10^3\,\Omega$$
$$\text{(c)} \quad 5 \times 10^5\,\Omega$$
$$\text{(d)} \quad 2.5 \times 10^{-8}\,\text{s}$$
$$\text{(e)} \quad 3 \times 10^{-4}\,\text{A}$$

1-7. Perform the necessary operations:

(a) $\dfrac{(4 \times 10^3)(8 \times 10^3)}{4 \times 10^3 + 8 \times 10^3}$

(b) $\dfrac{(2.5 \times 10^3)(2.5 \times 10^3)}{2.5 \times 10^3 + 2.5 \times 10^3}$

(c) $\dfrac{10 \times 10^6 + 50 \times 10^3}{40 \times 10^3 + 6 \times 10^6}$

(d) $\dfrac{(6 \times 10^{-3})(8 \times 10^{-3})}{(2 \times 10^{-6})(4 \times 10^{-6})}$

1-8. What is the current in an electrical conductor if 64 μC of charge passes through a cross section of the conductor in 0.8 ms?

1-9. Calculate the current in a conductor if 90 mC of charge passes through a cross section of a conductor in 1.5 min.

1-10. How many coulombs of charge flow through a cross section of a conductor in 3 min if there is a steady current of 18 mA?

1-11. How long will it take 25 mC of charge to pass through a cross section of a conductor if there is a steady current of 10 mA?

1-12. What is the voltage drop between two points in an electric circuit if it takes 4 J of energy to move 10 C of charge from one point to another?

1-13. If 2 J of energy is expended in moving 0.5 C of charge from one point to another, what is the potential difference between the points?

1-14. Find the charge, Q, required to move 50 J of energy through a potential difference of 15 V.

1-15. A conductor carrying a current of 20 mA converts 10 J of energy to heat in 1 min. What is the voltage drop across the conductor?

1-16. What is the output power of a 10 V source if 250 mA of current flows through it?

1-17. What is the power dissipated by a load if the voltage across it is 2.5 V and 10 mA of current flows through it?

1-18. A 15 V source is capable of delivering 1.5 W. What is the largest value of current this source can produce?

1-19. A load dissipates 50 mW of power and the voltage across it is 10 V. What is the current through the load?

1-20. A load dissipates 200 mW of power when 25 mA of current flows through it. What is the voltage across the load?

1-21. For the circuit of Fig. 1-15, calculate

    (a) the power delivered by the source,
    (b) the power of the load.

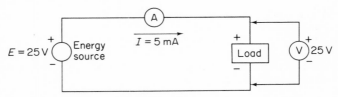

Fig. 1-15. Circuit diagram for Prob. 21.

1-22. For the circuit of Fig. 1-16, calculate

    (a) the power delivered by the source,
    (b) the power of load 1,
    (c) the power of load 2.

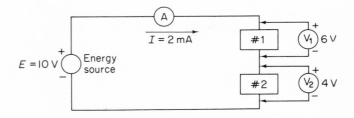

Fig. 1-16. Circuit diagram for Prob. 22.

1-23. In the circuit of Fig. 1-17, the power of load 1 is 50 mW. With the other measured values shown on the diagram, calculate

    (a)   the voltage across load 1,

    (b)   the power of load 2,

    (c)   the power delivered by the source.

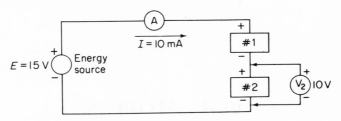

Fig. 1-17. Circuit diagram for Prob. 23.

CHAPTER 2

waveforms

In Chapter 1 we discussed some of the fundamental concepts involved in electric circuits. This chapter will concentrate on two of these concepts, current and voltage, as waveforms.

Section 2-1 introduces the different *types* of waveforms. Section 2-2 develops the sine wave from a rotating vector, and Section 2-3 is concerned with phase shift in waveforms. The last two sections, 2-4 and 2-5, discuss direct current, average values, and effective or rms values, respectively.

2-1. CLASSIFICATION OF WAVEFORMS

A waveform is a graph that shows how a quantity varies as a function of time. In circuit analysis, we deal with waveforms of current and voltages. Although in electronic applications many different waveforms are encountered, they all fall into at least one of the following five categories.

Ac Waveform. An *ac,* or alternating-current, *waveform* is a waveform whose average value is zero. That is, the area of the positive cycle equals the area of the negative cycle. Examples are shown in Fig. 2-1a and b.

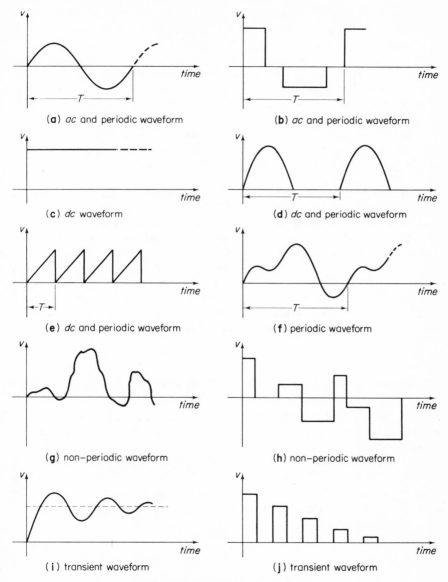

Fig. 2-1. Examples of waveforms.

Dc Waveform. A *dc,* or *direct-current, waveform* is any waveform whose average value is not zero. This average value may be either positive or negative. The most common dc waveform is one whose value is constant for all time as shown in Fig. 2-1c. Examples of waveforms containing a dc component are shown in Fig. 2-1d, e, f, g, h, i, and j.

Periodic Waveform. A periodic waveform is a waveform that has a repeated cycle. The time required for a cycle is called the *period* and is represented by *T*. See Fig. 2-1a, b, d, e and f.

Nonperiodic Waveform. A nonperiodic waveform is a waveform that does not have a repeated cycle. This type of waveform is also called an *arbitrary waveform.* Examples of this type are shown in Fig. 2-1g and h.

Transient Waveform. A transient waveform is one whose amplitude approaches a constant value with an increase of time. This constant value may be positive, negative, or zero. See Fig. 2-1i and j for examples.

Note that it is possible for a waveform to fall into more than one category. For example, a sine wave is not only an ac waveform but also a periodic waveform. Another example is the waveform shown in Fig. 2-1d. It is both a dc and a periodic waveform. The reader will find other such examples in further study of electronics.

Since electronic applications exhibit an infinite number of waveforms, we must restrict ourselves to those which occur most often. These are either a sinusoidal waveform – sine wave or cosine wave – or a constant dc waveform. We will study the development and characteristics of each of these waveforms in the next two sections.

2-2. DEVELOPMENT OF THE SINE WAVE

Some physical phenomena may best be represented by a phasor whose magnitude is constant but whose direction varies. Sinusoidal waveforms can be generated by such a phasor.

This development is based on using the *unit circle* (a circle whose radius is equal to unity), as shown in Fig. 2-2. Note that because the radius is equal to one unit of length, the circumference of this circle is equal to 2π units of length. Let the phasor $\overline{OP}$ (the radius of the unit circle) be rotated

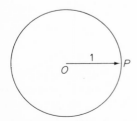

Fig. 2-2. Unit circle.

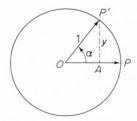

Fig. 2-3. Unit phasor rotated through an angle α.

counterclockwise through an angle, α (alpha), which is measured from the horizontal axis, to a position P' as shown in Fig. 2-3. We are now able to apply the rules of trigonometry to the triangle OAP' of Fig. 2-3.

$$\sin \alpha = \frac{y}{|\overline{OP}|} = \frac{y}{1} = y \tag{2-1}$$

where y is the perpendicular distance AP' and $\overline{OP}$ is the magnitude of the rotating vector (in this case, 1). Equation (2-1) shows that the vertical component (y) of the phasor OP is a function of the sine of the angle α.

If the phasor is constantly rotated, Eq. (2-1) can be used to compute a table for the different values of the angle α. Table 2-1 is tabulated for

TABLE 2-1

RELATIONSHIP BETWEEN THE ANGLE OF ROTATION AND THE VERTICAL COMPONENT OF A PHASOR

α (in degrees)	$\sin \alpha = y$
0	0.000
30	0.500
60	0.866
90	1.000
120	0.866
150	0.500
180	0.000
210	-0.500
240	-0.866
270	-1.000
300	-0.866
330	-0.500
360	0.000

intervals of 30° for one complete cycle of rotation. A graphical representation of Table 2-1 can be obtained by plotting α along the horizontal axis (abscissa) and y along the vertical axis (ordinate). The waveform shown in Fig. 2-4 is called a *sine* (or *sinusoidal*) *wave*. Table 2-1 could be continued by rotating the phasor at a constant rate for more than one cycle. This results in a *periodic wave*. That is, the value of *y* would continue to change in a repetitive manner, as shown in Fig. 2-5.

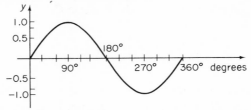

Fig. 2-4. Graphical representation of Table 2-1.

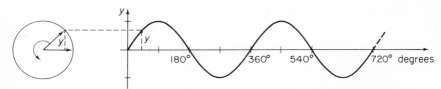

Fig. 2-5. A periodic sine wave generated by rotating a phasor at a constant rate.

From the above analysis we may conclude that the vertical component of a rotating phasor is directly proportional to the sine of the angle of rotation. For simplicity, the maximum value of the vertical axis was chosen as unity and the horizontal axis expressed in degrees. The scale of both these axes may be expressed differently. For example, we used the "degree" as our unit for measuring an angle. However, it is sometimes convenient to use another unit of angular measure, the *radian*. Referring to Fig. 2-3, a radian is defined as the value of the angle α when the length of the arc PP' equals the radius $|\overline{OP}|$. The term "radian" is a combination of the words "radius" and "angle."

The rotation of the phasor through one complete revolution (360° or 2π radians) is one *cycle*. Thus the relationship between radians and degrees is given by the following:

$$1 \text{ radian} = \frac{360°}{2\pi} = \frac{180°}{\pi} = \frac{180°}{3.14} = 57.3°$$

Example 2-1: Convert (a) 25° to radians and (b) 1.2π radians to degrees.

Solution: (a)

$$25° \times \frac{\pi \text{ rad}}{180 \text{ deg}} = 0.139\pi \text{ rad} = 0.437 \text{ rad}$$

(b)

$$1.2\pi \text{ rad} \times \frac{180 \text{ deg}}{\pi \text{ rad}} = 216°$$

Figure 2-6 shows a sine wave with the abscissa (horizontal axis) plotted as radians.

The time for one cycle of a periodic wave is called the *period* and is designated by the letter T. The reciprocal of the period has special sig-

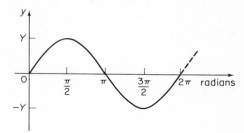

Fig. 2-6. A sine wave plotted in radians.

nificance in circuit analysis and is called the *frequency*. It is represented by the symbol f. Frequency is defined as the number of cycles per unit time. In equation form,

$$f = \frac{1}{T} \qquad (2\text{-}2)$$

The units for frequency are *cycles per second* (c/s), now more commonly referred to as *hertz* (Hz).

Example 2-2: Determine the frequencies of the waveforms in Fig. 2-7.

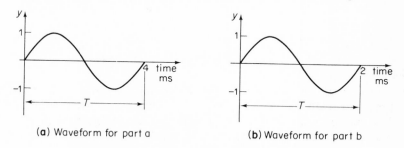

(a) Waveform for part a   (b) Waveform for part b

Fig. 2-7. Periodic waveforms for Ex. 2-2.

Solution: Waveform (a)

$$f = \frac{1}{T} = \frac{1}{4 \text{ ms}} = 250 \text{ Hz}$$

Waveform (b):

$$f = \frac{1}{T} = \frac{1}{2 \text{ ms}} = 500 \text{ Hz}$$

Note that as the duration of the period decreases, the frequency increases. In this example, the period was halved and the frequency doubled.

The previous discussion dealt with only a rotating phasor of unit magnitude. This results in an amplitude, or peak value, of the sine wave equal to unity. However, if the magnitude of the rotating vector is changed, the amplitude of the sine wave will also change by the same amount. A general mathematical form for a sine wave is

$$y = Y \sin \alpha \qquad (2\text{-}3)$$

where y is the instantaneous value, Y is the amplitude or peak value of the sine wave, and α is the angle of rotation (see Fig. 2-8).

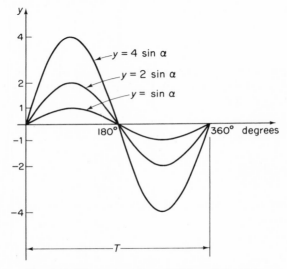

Fig. 2-8. Illustrating the principles of Eq. (2-3).

In circuit analysis we are usually more concerned with frequency and time than with the angle of the sine wave. Since α is a fractional part of $360°$ or 2π radians, let us express the corresponding time as t, which is a fractional part of the period, T, as shown in Fig. 2-9.

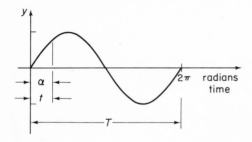

Fig. 2-9. Showing the relationships of α, t, 2π and T.

The following ratio shows the relationship among α, t and T:

$$\frac{\alpha}{2\pi} = \frac{t}{T} \tag{2-4}$$

Solving for α we obtain

$$\alpha = \frac{2\pi t}{T} \tag{2-5}$$

Substituting Eq. (2-2) into Eq. (2-5) results in an expression for the angle, α, in the most common form:

$$\alpha = 2\pi f t \tag{2-6a}$$

and

$$\frac{\alpha}{t} = 2\pi f \tag{2-6b}$$

Because the factor $2\pi f$ recurs so often, it is sometimes designated by the Greek letter ω (omega):

$$\omega = 2\pi f \tag{2-7}$$

ω is termed the angular frequency, the rate at which the phasor is rotating, expressed in radians per second (i, e., 2π rad/s $\times f$ Hz). Substituting Eqs. (2-6) and (2-7) into Eq. (2-3) yields a general form for the instantaneous value of y:

$$y = Y \sin \omega t \tag{2-8}$$

where y is the instantaneous value, Y the peak value, and ωt the angle of rotation in radians. The coefficient of the sine term, Y, is always the amplitude, or peak value, of the wave.

The general expression for instantaneous voltage and current is given by

$$e = E_m \sin \omega t \tag{2-9a}$$

or

$$v = V_m \sin \omega t \tag{2-9b}$$

and

$$i = I_m \sin \omega t \tag{2-10}$$

where e and i are the instantaneous values of voltage and current, respectively, and E_m and I_m are the peak values.

Example 2-3: For the sine wave of voltage shown in Fig. 2-10 determine (a) the peak value, (b) the frequency, and (c) the instantaneous value of voltage at $t = 0.5 \, \mu s$.

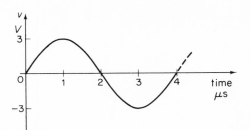

Fig. 2-10. Sine wave of voltage for Ex. 2-3.

Solution:

(a) $V_m = 3\,\text{V}$

(b) $f = \dfrac{1}{T} = \dfrac{1}{4\,\mu s} = 0.25 \times 10^6\,\text{Hz} = 250\,\text{kHz}$

(c) $v = V_m \sin \omega t$

$$\omega = 2\pi f = (2)(3.14)(0.25 \times 10^6) = 1.57 \times 10^6\,\text{rad/s}$$

$$v = 3 \sin (1.57 \times 10^6\,\text{rad/s})(0.5 \times 10^{-6}\,\text{s})$$

$$v = 3 \sin (0.785\,\text{rad}) = 3 \sin \left(0.785\,\text{rad} \times \frac{180°}{\pi\,\text{rad}} \right)$$

or

$$v = 3 \sin 45° = (3)(0.707) = 2.121\,\text{V}$$

Note that 0.785 rad was converted to degrees, as shown in Example 2-1. The sin 45° may be found either from a set of trigonometry tables or by using a slide rule.

Example 2-4: Write the general expression and draw the waveform for the instantaneous current if the frequency is 10 kHz and the peak value is 5 mA.

Solution:

$$i = I_m \sin \omega t$$

$$I_m = 5\,\text{mA}$$

$$\omega = 2\pi f = (2)(3.14)(10 \times 10^3\,\text{Hz}) = 6.28 \times 10^4\,\text{rad/s}$$

$$i = 5 \sin (6.28 \times 10^4 t)\,\text{mA}$$

To draw the current waveform we first must solve for the period, T. $T = = 1/f = 1/10\,\text{kHz} = 0.1\,\text{ms}$ (Fig. 2-11).

Note that the period, T, is calculated from the reciprocal of frequency, f, *not* the reciprocal of ω.

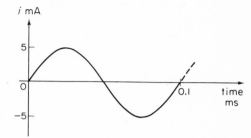

Fig. 2-11. The sine wave of current for Ex. 2-4.

2-3. PHASE SHIFT

Figure 2-12 shows two sinusoidal waveforms having the same maximum value but *shifted* from one another in time or phase. That is, at a specific time they do *not* have the same amplitude. The angle by which they are shifted is known as the *phase angle*. By convention, the phase angle is usually measured from 0° or 0 rad as a reference. Thus a sine wave has a phase shift of 0° or 0 rad. Figure 2-12 shows that both waveforms are "out of phase" with a (reference) sine wave by $\pi/6$ rad, or 30°. Waveform y_1 has a phase angle of $\pi/6$ rad and waveform y_2 has a phase angle of $-\pi/6$ rad. These two waveforms are then "out of phase" with each other by $\pi/3$ rad, or 60°.

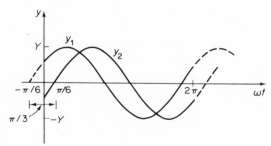

Fig. 2-12. An example of phase shift.

Let us now consider the generation and mathematical expression of a shifted waveform. The rotating phasor of Fig. 2-13a begins at an angle of $+ 30°$, thereby creating the waveform of Fig. 2-13b. When a phasor rotates in a *counterclockwise* direction, *positive* time is measured. Therefore, the rotating phasor of Fig. 2-13a, which is initially at $+ 30°$, is also referred to as being 30° *earlier* in *time* or *phase*. For this reason the waveform of Fig. 2-13b is said to *lead* a sine wave by 30°.

Similarly, a rotating vector starting at an angle of $-30°$ in phase as shown in Fig. 2-13c creates the waveform shown in Fig. 2-13d.

Since positive time is measured in a counterclockwise direction, a negative angle represents a delay in time or phase. For this case a $-30°$ means 30° later in time or phase. The waveform of Fig. 2-13d is said to *lag* a sine wave by 30°

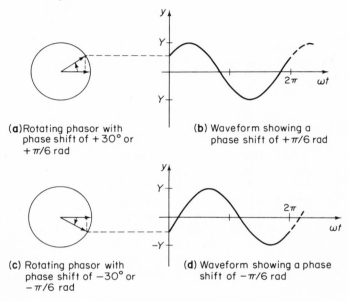

(a)Rotating phasor with phase shift of $+30°$ or $+\pi/6$ rad

(b) Waveform showing a phase shift of $+\pi/6$ rad

(c) Rotating phasor with phase shift of $-30°$ or $-\pi/6$ rad

(d) Waveform showing a phase shift of $-\pi/6$ rad

Fig. 2-13. Examples of phase shift.

Expressing these waveforms in equation form only requires writing the equation of a sine wave but including the phase shift. The general equation of a sine wave, Eq. (2-8), is written in terms of radians rather than degrees. In order for all elements of the equation to have the same units, we must first express the phase shift in radians. For Fig. 2-13b the phase shift is $+30°$, or $+\pi/6$ rad. The phase shift of Fig. 2-13d is $-30°$, or $-\pi/6$ rad.

Therefore, the mathematical expression in terms of a sine wave for Figs. 2-13b and 2-12 is

$$y = Y \sin\left(\omega t + \frac{\pi}{6}\right)$$

Similarly, the mathematical expression in terms of a sine wave for Figs. 2-13d and 2-12 is

$$y = Y \sin\left(\omega t - \frac{\pi}{6}\right)$$

where y is the instantaneous value and Y is the peak value of the sinusoidal waveform.

Frequently in waveform analysis a phase shift of $+90°$ or $-90°$ occurs. A waveform whose phase shift is $+90°$, or $+\pi/2$ rad, is shown in Fig. 2-14b.

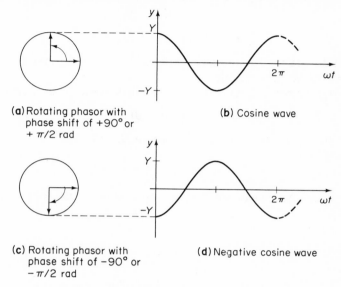

(a) Rotating phasor with phase shift of $+90°$ or $+\pi/2$ rad

(b) Cosine wave

(c) Rotating phasor with phase shift of $-90°$ or $-\pi/2$ rad

(d) Negative cosine wave

Fig. 2-14. The generation of a cosine and negative cosine wave.

This waveform is known as a *cosine wave*. The equation for this waveform is expressed in either of the following forms:

$$y = Y \sin\left(\omega t + \frac{\pi}{2}\right) \qquad \text{or} \qquad y = Y \cos \omega t$$

Likewise, a phase shift of $-90°$, or $\pi/2$ rad, results in the sinusoidal waveform of Fig. 2-14d. This waveform is a *negative cosine wave* and may be expressed as

$$y = Y \sin\left(\omega t - \frac{\pi}{2}\right) \qquad \text{or} \qquad y = -Y \cos \omega t$$

A general expression for any sinusoidal wave may now be written as

$$y = Y \sin(\omega t + \theta) \qquad\qquad (2\text{-}11)$$

where y is the instantaneous value, Y the peak or maximum value, and θ the phase angle. The phase angle may either be a positive or negative angle. Remember that since ωt is expressed in radians in Eq. (2-11), θ must also be expressed in radians.

Equation (2-11) may be expressed in terms of either voltage or current. The general expression for the instantaneous voltage is

$$e = E_m \sin (\omega t + \theta) \qquad (2\text{-}12a)$$

or

$$v = V_m \sin (\omega t + \theta) \qquad (2\text{-}12b)$$

where e and v are the instantaneous values of voltage, E_m and V_m the peak values, and θ the phase angle.

Similarly, the general expression for the instantaneous current is

$$i = I_m \sin (\omega t + \theta) \qquad (2\text{-}13)$$

where i is the instantaneous value of current, I_m the peak value, and θ the phase angle.

Example 2-5: Draw the waveform for each of the following expressions:

(a)

$$i = 3 \sin \left(\omega t + \frac{\pi}{4} \right) \text{mA}$$

(b)

$$v = 4 \cos \left(\omega t + \frac{\pi}{6} \right) \text{V}$$

(c)

$$v = -6 \sin \omega t \text{ V}$$

(d)

$$i = -8 \cos \left(\omega t - \frac{\pi}{6} \right) \text{mA}$$

Solution:

(a) $i = 3 \sin (\omega t + \pi/4)$ mA. From this equation the phase shift is $\pi/4$ rad (or 45°). This relationship is shown in Fig. 2-15.

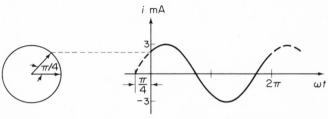

(a) Rotating phasor showing   (b) Current waveform showing phase
the proper phase shift   shift of $\pi/4$ rad (45°)
at $t = 0$

Fig. 2-15. Solution for part 1 of Ex. 2-5.

(b) $v = 4 \cos (\omega t + \pi/6)$ V.   This waveform may be expressed as a sine function $v = 4 \sin (\omega t + 2\pi/3)$ V.   Thus from 0° we have a phase shift of $2\pi/3$ rad (or 120°).   This relationship is shown in Fig. 2-16.

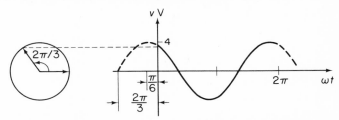

(**a**) Rotating phasor showing the proper phase shift   (**b**) Voltage waveform showing phase shift of $2\pi/3$ rad (120°)

Fig. 2-16. Solution for part 2 of Ex. 2-5.

(c) $v = -6 \sin \omega t$ V.   This waveform may be expressed as $v = 6 \sin (\omega t + \pi)$ V.   Therefore the phase shift is π rad (or 180°).   This relationship is shown in Fig. 2-17.

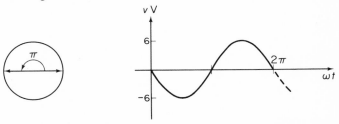

(**a**) Rotating phasor showing the proper phase shift   (**b**) Voltage waveform showing phase shift of π rad (180°)

Fig. 2-17. Solution for part 3 of Ex. 2-5.

(d) $i = -8 \cos (\omega t - \pi/6)$ mA.   This waveform may be expressed as $i = 8 \sin (\omega t + 4\pi/3)$ mA.   Therefore, from 0° we have a phase shift of $4\pi/3$ rad (or 240°).   This relationship is shown in Fig. 2-18.

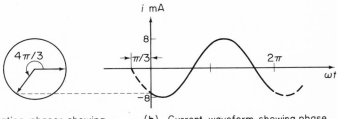

(**a**) Rotating phasor showing the proper phase shift   (**b**) Current waveform showing phase shift of $4\pi/3$ rad (240°)

Fig. 2-18. Solution for part 4 of Ex. 2-5.

The above examples show that it is always possible to express a sinusoidal waveform as either a sine wave, cosine wave, negative sine wave, or negative cosine wave along with the proper phase angle.

Example 2-6: Write the expression for each of the sinusoidal waveforms shown in Fig. 2-19.

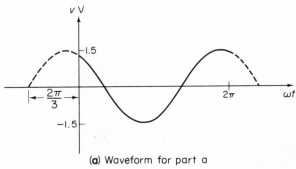

(a) Waveform for part a

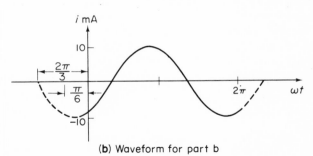

(b) Waveform for part b

Fig. 2-19. Sinusoidal waveforms for Ex. 2-6.

Solution: Waveform (a): In terms of a sine function

$$v = 1.5 \sin \left(\omega t + \frac{2\pi}{3} \right) \text{V}$$

or, in terms of a cosine function,

$$v = 1.5 \cos \left(\omega t + \frac{\pi}{6} \right) \text{V}$$

As seen in Fig. 2-19a, this waveform is generated by adding $2\pi/3$ rad (120°) to the sine wave or by adding $\pi/6$ rad (30°) to the cosine wave.

Waveform (b): In terms of a negative cosine function,

$$i = -10 \cos \left(\omega t + \frac{\pi}{6} \right) \text{mA}$$

or, in terms of a negative sine function,

$$i = -10 \sin\left(\omega t + \frac{2\pi}{3}\right) \text{mA}$$

As in waveform (a), we see that waveform (b) may be expressed in more than one way. Waveform (b) may be written as either a sine or a cosine function.

2-4. DIRECT CURRENT AND AVERAGE VALUES

In Section 2-1, dc waveforms were defined. However, the general definition and the waveforms shown in Fig. 2-1 include unidirectional dc signals that are constant for all time (Fig. 2-1c), pulsating unidirectional dc signals (Fig. 2-1d), and other waveforms whose average values are not zero (Fig. 2-1e, f,g,h,i, and j). The constant unidirectional waveform (Fig. 2-1c) is so common that it is simply referred to as dc. Although dc stands for *direct current*, we may also refer to a *voltage* that is unidirectional and constant for all time as a dc voltage. Unless otherwise specified, a dc energy source is one that produces a constant unidirectional value for all time. The generation of a dc voltage waveform is shown in Fig. 2-20.

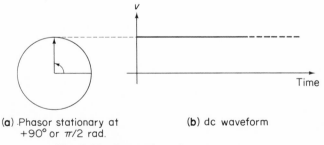

(a) Phasor stationary at
  +90° or π/2 rad.

(b) dc waveform

Fig. 2-20. Generation of a dc waveform.

One may produce this waveform in a manner similar to that of a sine wave. Although a dc waveform may be generated at any angle, for our example at time $t = 0$, the vector is at $+90°$, as shown in Fig. 2-20. This is the same *initial* position as that for a cosine wave. At this position there is a maximum value. To be able to produce this constant value for all time, the frequency of such a phasor must equal zero. In other words, the phasor must be stationary and cannot rotate. Since $f = 0$ and $\omega = 2\pi f = 0$ for all positive time, $v = V_m \cos \omega t = V_m$. Therefore, the voltage at any time after 0 s is equal to the voltage at $t = 0$.

In the ac condition, during the positive half-cycle current flows in one direction, and then during the next half-cycle current flows in the opposite direction. However, in the dc case there is no polarity reversal. Therefore, current is always unidirectional; i.e., it flows in one direction.

Two of the more common dc voltage sources are batteries and dc regulated power supplies. **Note:** *We should not think of dc and ac as being two completely separate entities but rather both as being waveforms. The dc waveform is simply a special case of the ac waveform where the frequency is equal to zero and the average and maximum values are the same.*

We have seen that a periodic waveform may contain an *average value.* This average is known as a *dc component* or a *dc level.* This average value or dc component could be either zero or a positive or a negative value, as shown in Fig. 2-21. Figure 2-21a shows that if the dc component equals

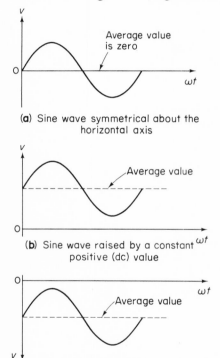

(a) Sine wave symmetrical about the horizontal axis

(b) Sine wave raised by a constant positive (dc) value

(c) Sine wave lowered by a constant negative (dc) value

Fig. 2-21. Examples showing average values.

zero, the periodic wave is an ac waveform. The average value or dc component of a periodic waveform is calculated by

$$\text{average value} = \frac{A}{T} \qquad (2\text{-}14)$$

where A is the area under the waveform for one complete cycle and T the period, is the duration of one cycle.

Example 2-7: Determine the average value and the frequency for each of the periodic waveforms shown in Fig. 2-22.

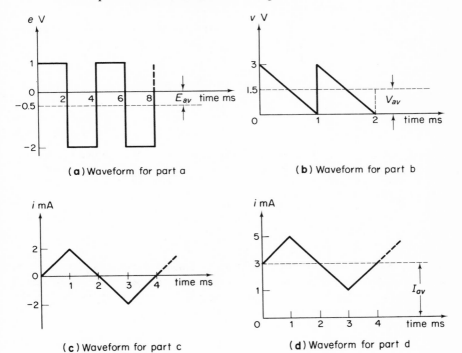

(a) Waveform for part a

(b) Waveform for part b

(c) Waveform for part c

(d) Waveform for part d

Fig. 2-22. Waveforms for Ex. 2-7.

Solution: Waveform (a): To determine the average value we need the area under the waveform for one complete cycle:

$$\text{area} = (1 \text{ V}) \cdot (2 \text{ ms}) + (-2 \text{ V}) \cdot (2 \text{ ms}) = -2 \text{ V} \cdot \text{ms}$$

Therefore,

$$E_{\text{av}} = \frac{A}{T} = \frac{\text{area}}{\text{period}} = \frac{-2 \text{ V} \cdot \text{ms}}{4 \text{ ms}} = -0.5 \text{ V}$$

The frequency

$$f = \frac{1}{T} = \frac{1}{4 \text{ ms}} = 250 \text{ Hz}$$

Waveform (b): The area under the waveform for one complete cycle is

$$\text{area} = (3 \text{ V})(\tfrac{1}{2})(1 \text{ ms}) = \tfrac{3}{2} \text{ V} \cdot \text{ms}$$

Therefore,

$$V_{\text{av}} = \frac{A}{T} = \frac{\text{area}}{\text{period}} = \frac{\tfrac{3}{2} \text{ V} \cdot \text{ms}}{1 \text{ ms}} = 1.5 \text{ V}$$

The frequency

$$f = \frac{1}{T} = \frac{1}{1\ \text{ms}} = 1\ \text{kHz}$$

Waveform (c):   The area under the waveform for one complete cycle is

$$\text{area} = (2\ \text{mA})\,(\tfrac{1}{2})\,(1\ \text{ms})\,(2) + (-2\ \text{mA})\,(\tfrac{1}{2})\,(1\ \text{ms})\,(2) = 0$$

Therefore,

$$I_{\text{av}} = \frac{A}{T} = \frac{\text{area}}{\text{period}} = \frac{0}{4\ \text{ms}} = 0$$

The frequency

$$f = \frac{1}{T} = \frac{1}{4\ \text{ms}} = 250\ \text{Hz}$$

Waveform (d):   This triangular waveform is similar to that of waveform (c) but is raised to a dc level of 3 mA.   The area above the dc level equals the area below the dc level.   Therefore, the average value of the entire waveform is the dc level:

$$I_{\text{av}} = 3\ \text{mA}$$

The frequency

$$f = \frac{1}{T} = \frac{1}{4\ \text{ms}} = 250\ \text{Hz}$$

This example shows that periodic waveforms have a dc component which may be either zero or a positive or a negative number.   The frequency is equal to the reciprocal of the period and is in no way dependent on the average value.

2-5. EFFECTIVE OR RMS VALUES

Is there any relationship between ac voltage and current and dc voltage and current?   The best way to analyze this problem is to compare the amount of power delivered to a load by *both* sources.   Consider the circuit of Fig. 2-23.

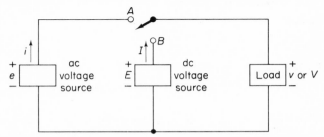

Fig. 2-23. Circuit diagram to illustrate the concept of effective values.

We wish to have the same power delivered to the load whether the switch is in position A or position B.   The average power delivered by the ac voltage source must equal the average power delivered by the dc voltage source. The instantaneous power developed by the ac source is

$$p = ei \tag{2-15}$$

if

$$e = E_m \sin \omega t \quad \text{and} \quad i = I_m \sin \omega t$$

Then, since power is the product of voltage and current,

$$p = (E_m \sin \omega t)(I_m \sin \omega t) \tag{2-16}$$

or

$$p = E_m I_m \sin^2 \omega t \tag{2-17}$$

According to trigonometric identities,

$$\sin^2 \omega t = \frac{1}{2}(1 - \cos 2\omega t)$$

Substituting the trigonometric identity into Eq. (2-17), we obtain

$$p = \frac{E_m I_m}{2}(1 - \cos 2\omega t) \tag{2-18}$$

Figure 2-24c is a plot of Eq. (2-18). Note that the value of power is never negative (never goes below the horizontal axis). Also, because of the term cos $2\omega t$, the frequency of the power is twice the frequency of either the voltage or current.

The average value of the ac source is just the first term of Eq. (2-18), because the average value of the cos $2\omega t$ term is zero. Therefore, the average power delivered by the ac source to the load is

$$P_{av} = \frac{1}{2} E_m I_m \tag{2-19}$$

Since the average power is constant (refer to Fig. 2-24c), the possibility exists that a dc voltage source can supply the same amount of power to the load. The equation for the dc power, which also is the average power, is

$$P = EI \tag{2-20}$$

In order to obtain the same power at the load whether the ac or the dc source is connected, Eq. (2-19) must equal Eq. (2-20):

$$P_{av} = P$$

or

$$\frac{1}{2} E_m I_m = EI \tag{2-21}$$

where E_m and I_m are the peak or maximum values of voltage and current, respectively, and E and I are the dc values of voltage and current, respectively.

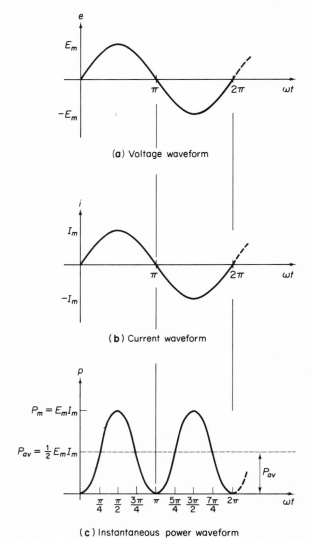

Fig. 2-24. Waveforms illustrating the concept of instantaneous power when voltage and current are in phase.

In other words, Eq. (2-21) gives what the average value of power due to an ac sine wave of voltage and current must be in order to deliver the same power to a load as a dc source.

Solution: (a) To determine the average value, we need the area under the waveform of Fig. 2-27a:

$$\text{area} = (4\text{ V})(2\text{ s}) = 8\text{ V} \cdot \text{s}$$

Therefore,

$$V_{av} = \frac{\text{area}}{\text{period}} = \frac{8\text{ V} \cdot \text{s}}{5\text{ s}} = 1.6\text{ V}$$

(b) To determine the rms value, we first need the area under the square of the voltage waveform, as shown in Fig. 2-27b. For one cycle,

$$\text{area} = (16\text{ V})(2\text{ s}) = 32\text{ V}^2 \cdot \text{s}$$

Using Eq. (2-24),

$$\hat{V} = \sqrt{\frac{\text{area}}{\text{period}}} = \sqrt{\frac{32\text{ V}^2 \cdot \text{s}}{5\text{ s}}} = 2.53\text{ V}$$

Note that the period of the waveform is 5 s and not just the time that current flows. In this example, the dc level is a positive value. The rms value of any periodic waveform is always a positive number.

Example 2-10: Consider the current waveform of Fig. 2-28a and determine (a) the average value and (b) the rms value.

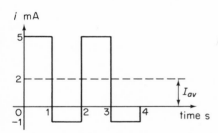

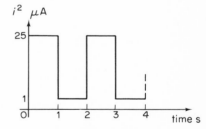

(a) Rectangular current waveform and its average value

(b) Waveform for calculating rms value

Fig. 2-28. Periodic waveforms for Ex. 2-10.

Solution: (a) To determine the average value we need the area under the waveform of Fig. 2-28a:

$$\text{area} = (5\text{ mA})(1\text{ s}) + (-1\text{ mA})(1\text{ s}) = 4\text{ mA} \cdot \text{s}$$

Then

$$I_{av} = \frac{\text{area}}{\text{period}} = \frac{4\text{ mA} \cdot \text{s}}{2\text{ s}} = 2\text{ mA}$$

(b) To determine the rms value, we need the area under the *square* of the current waveform as shown in Fig. 2-28b. For one complete cycle,

$$\text{area} = (25\ \mu\text{A}^2) \cdot (1\text{ s}) + (1\ \mu\text{A}^2) \cdot (1\text{ s}) = 26\ \mu\text{A}^2 \cdot \text{s}$$

Using Eq. (2-24),

$$I = \sqrt{\frac{\text{area}}{\text{period}}} = \sqrt{\frac{26\,\mu\text{A}^2 \cdot \text{s}}{2\,\text{s}}} = 3.6\,\text{mA}$$

In this example the average value of the periodic waveform is positive and once again the rms value of any periodic waveform is always a positive number. Note also that when a waveform is squared to obtain an rms value, only the amplitude is squared, never the frequency.

At the beginning of our discussion on effective values, we assumed that both the voltage and current are sine waves:

$$e = E_m \sin \omega t \qquad \text{and} \qquad i = I_m \cos \omega t$$

This was not by coincidence. If one waveform is a sine wave and the other waveform is a cosine wave, no usable power is delivered to the load. Let us analyze this case:

$$i = I_m \cos \omega t \qquad \text{and} \qquad e = E_m \sin \omega t$$

According to Eq. (2-15),

$$p = ei$$

Then

$$p = (E_m \sin \omega t)\,(I_m \cos \omega t) \qquad \text{or} \qquad p = E_m I_m \sin \omega t \cos \omega t$$

Using the trigonometric identity,

$$\sin \omega t \cos \omega t = \frac{1}{2} \sin 2\omega t$$

We obtain

$$p = \frac{1}{2} E_m I_m \sin 2\omega t \qquad\qquad (2\text{-}25)$$

Figure 2-29 shows a plot of the voltage and current waveforms and the instantaneous power waveform, which is a plot of Eq. (2-25). It can be seen from Eq. (2-25) and Fig. 2-29c that the average value of the instantaneous power is zero. Therefore, there can be no usable power delivered to the load under such a condition, i.e., voltage and current 90° out of phase.

Also consider the situation in which both the voltage and current are cosine waves. The effective or rms value of voltage and current should be the same as those given in Eqs. (2-22) and (2-23).

If the load is made up of *more* than one element, only that element for which the voltage and current are in phase is capable of dissipating energy (doing work). Any other element in the load for which the voltage and current are not in phase is not capable of dissipating energy. We shall study these various elements in detail in Chapter 3.

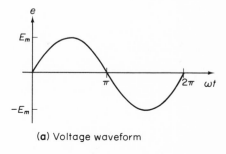

(a) Voltage waveform

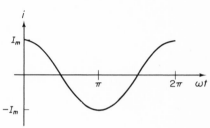

(b) Current waveform

Fig. 2-29. Waveforms illustrating the concept of instantaneous power when voltage and current waves are $\pi/2$ (or $90°$) out of phase.

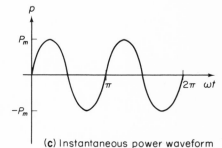

(c) Instantaneous power waveform

SUMMARY

In circuit analysis, currents and voltages are dealt with as waveforms. This chapter has presented fundamental information about the five categories of waveforms. We have developed a sine wave from a rotating phasor and discussed the concept of phase shift. The reader should now also be able to calculate average values, and effective (or rms) values.

Along with this knowledge is the fundamental concept that ac and dc are *not* separate entities — both are waveforms, with dc a special case of ac, in which the average, effective, and maximum values are the same and the frequency is zero.

PROBLEMS

2-1. Classify each of the waveforms in Fig. 2-30. (Note that some waveforms fall into more than one category.)

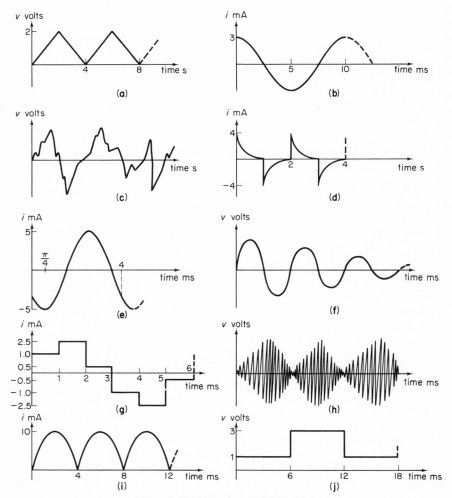

Fig. 2-30. Waveforms for Prob. 1.

2-2. Convert the following angles in degrees to radians.

    (a) 90°   (c) 120°   (e) −60°   (g) −90°
    (b) 45°   (d) −30°   (f) 180°   (h) 50°

2-3. Convert the following angles in radians to degrees.

    (a) π rad   (c) 0.785 rad   (e) 2.4 rad   (g) $\dfrac{3\pi}{4}$ rad

    (b) $-\dfrac{\pi}{2}$ rad   (d) −1.2 rad   (f) $-\dfrac{\pi}{4}$ rad   (h) $\dfrac{4\pi}{3}$ rad

2-4. Refer to Fig. 2-30 and calculate the frequency for the sinusoidal waveforms b and e.

2-5. Refer to waveform b of Fig. 2-30 and calculate the value of current at the following instants:

    (a) $t = 1$ ms   (b) $t = 2.5$ ms   (c) $t = 6$ ms

2-6. Convert the following frequencies in cycles per second to angular frequency — ω.

    (a) $f = 60$ Hz   (c) $f = 10$ kHz   (e) $f = 25$ kHz
    (b) $f = 0.5$ kHz   (d) $f = 1.5$ MHz   (f) $f = 3.5$ MHz

2-7. Convert the following angular frequencies to frequencies in hertz.

    (a) $\omega = 100$ rad/s   (c) $\omega = 250$ rad/s   (e) $\omega = 1.0$ Mrad/s
    (b) $\omega = 7.5$ krad/s   (d) $\omega = 10$ krad/s   (f) $\omega = 8.5$ Mrad/s

2-8. Write the general expression and draw the waveform for the instantaneous current if the frequency is 1.5 kHz and the peak value is 20 mA.

2-9. Write the general expression and draw the waveform for the instantaneous voltage if the frequency is 10 kHz and the peak value is 8 V.

2-10. Draw the waveforms for each of the following mathematical expressions.

    (a) $e = 10 \sin(\omega t + \pi/4)$ V
    (b) $v = 1.5 \cos(\omega t + \pi/4)$ V
    (c) $i = 12 \sin(\omega t - \pi/6)$ mA
    (d) $v = 6 \cos(\omega t - \pi/3)$ V
    (e) $i = -5 \sin(\omega t + \pi/2)$ mA

2-11. Write the mathematical expression for the waveforms b and e of Fig. 2-30.

2-12. Write the mathematical expressions for each of the sinusoidal waveforms in Fig. 2-31.

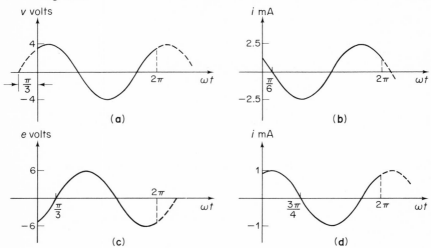

Fig. 2-31. Waveforms for Prob. 12.

2-13. Determine the phase shift by which the voltage either leads or lags the current for each of the following cases:

(a) $v = 4 \sin(\omega t + \pi/6)$ V, $i = 2 \sin \omega t$ mA
(b) $v = 7.5 \sin(\omega t + \pi/4)$ V, $i = 3 \sin(\omega t + \pi/6)$ mA
(c) $e = 8 \sin(\omega t + \pi/5)$ V, $i = 1.5 \sin(\omega t - \pi/3)$ mA
(d) $v = 10 \sin(\omega t - \pi/6)$ V, $i = 4 \cos \omega t$ mA
(e) $v = 15 \cos(\omega t + \pi/4)$ V, $i = -2 \sin(\omega t + \pi/4)$ mA

2-14. Calculate the angle in degrees for each of the following times. Let $f = 20$ Hz.

(a) $t = 1$ ms   (c) $t = 7.5$ ms   (e) $t = 20\ \mu s$
(b) $t = 3.5$ ms   (d) $t = 9$ ms   (f) $t = 450\ \mu s$

2-15. Evaluate $i = 4 \sin(\omega t + 3\pi/2)$ mA for values of ωt of 0 rad, $\pi/4$ rad, $\pi/6$ rad, $-3\pi/2$ rad, and $-\pi/2$ rad. Plot the current waveform as a function of ωt.

2-16. Write the following sine function as a cosine function:

$$v = -1 \sin(\omega t + \pi/6) \text{ V}$$

2-17. Write the following cosine function as a sine function:

$$i = 5 \cos(\omega t - \pi/3) \text{ mA}$$

2.18. Draw the waveforms for each of the following mathematical expressions.

(a) $i = 10 + 2 \sin \omega t$ mA   (c) $v = -5 + 3 \sin \omega t$ V
(b) $v = 4 + 1 \cos \omega t$ V   (d) $i = 3 + 4 \sin \omega t$ mA

2-19.   Determine the average value for the periodic wave forms of Fig. 2-32.

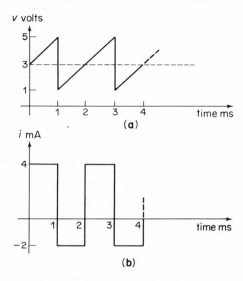

Fig. 2-32. Waveforms for Prob. 19.

2-20.   The following are peak values of a sinusoidal waveform.   Calculate the rms value.

(a)   4 V   (c)   3 mA   (e)   60 mA   (g)   2 mA
(b)   2 V   (d)   0.5 V   (f)   15 V   (h)   10 μA

2-21.   The following are rms values of a sinusoidal waveform.   Calculate the peak-to-peak value.

(a)   10 V   (c)   2 V   (e)   1 mA   (g)   4 V
(b)   16 mA   (d)   4.5 V   (f)   14 μA   (h)   5 mA

2-22.   Calculate the rms values of waveforms b and e of Fig. 2-30.

2-23.   Calculate the rms values of the waveforms of Fig. 2-31.

2-24.   Determine the average and the rms values of waveforms g and j of Fig. 2-30.

2-25.   Determine the rms value of waveform b of Fig. 2-32.

CHAPTER 3

elements, terminology, and laws

As the title of this chapter suggests, we shall be considering three basic areas. The first two sections deal with elements in a circuit, both passive and active. Section 3-3 covers network terminology, and the last section introduces Kirchhoff's voltage and current laws.

3-1. PASSIVE ELEMENTS

A *passive element* is only capable of storing or dissipating energy. The passive elements are resistance, inductance (self and mutual), and capacitance. Resistance converts electrical energy into other energy forms. Self-inductance stores energy in a magnetic field, and mutual inductance transfers energy from one part of a circuit to another through a magnetic field. Capacitance stores energy in an electric field. Let us now analyze the unique characteristics of each element separately.

Resistance. In Section 1-7 current was discussed. When a potential energy is applied to a conductor, a net movement of free electrons in one direction exists. Even under the influence of the potential energy the path of a single electron is not straight but a zigzag pattern similar to random motion. A cause of the zigzag pattern is collisions of the electron with other atoms.

Since energy cannot be destroyed, the externally applied potential energy is transformed to kinetic energy of the electron. When the electron collides with an atom, heat is produced. The heat produced as a result of these collisions increases the thermal vibrations in the material, which in turn causes more collisions. This thermodynamically *irreversible* process is called *Joule heating*. The collisions, which simultaneously produce heat, also constitute the opposition to the flow of current. Opposition to current flow is *resistance*. For circuit analysis, the primary characteristic of resistance is energy conversion (dissipation of heat or other forms of energy).

Every material used as a conductor has some degree of resistance. Some materials conduct electric charge flow more efficiently, i.e., with less resistance and production of heat than others. Copper, for example, is an excellent conductor, while steel is a good conductor but not an excellent one.

Resistors[1] may be classified as either fixed or variable. A fixed resistor is one whose value is constant while a variable resistor is one whose value can be changed. The circuit symbols for both a fixed and a variable resistance are shown in Fig. 3-1.

Fig. 3-1. Circuit symbols for resistors. **(a)** Fixed resistor   **(b)** Variable resistor

In 1827, Georg Simon Ohm first discovered that the ratio of voltage across a resistance to the current through it is equal to the value of the resistance. In equation form,

$$\frac{V_R}{I_R} = R \qquad \text{or} \qquad \frac{v_R}{i_R} = R \qquad (3\text{-}1)$$

where V_R or v_R is the voltage across the resistor in volts, I_R or i_R the current through the resistor in amperes, and R the value of the resistance. This fundamental relation is known as *Ohm's law of constant proportionality* or simply as *Ohm's law*.

As can be seen by Eq. (3-1), resistance is directly proportional to the voltage across the resistor and inversely proportional to the current through the resistor. Equation (3-1) may be considered as a definition of resistance.

---

[1] A resistor is a physical device having the properties of resistance.

According to Ohm's law, the units for resistance are volts per ampere. However, in honor of Georg Simon Ohm, the basic unit of resistance is named the *ohm*, represented by the Greek letter Ω (omega). The symbol for resistance is R, as defined in Eq. (3-1).

In Section 2-5 we saw that an element dissipates energy only when the voltage and current waveforms are in phase. For example, if the voltage and current waveforms for an element are both sine waves, the average power dissipated by the element is found by Eq. (2-19). These waveforms differ only in their peak value. The maximum value of the current is

$$I_m = \frac{E_m}{R} \tag{3-2}$$

where I_m and E_m are the peak values of the current and voltage, respectively, and R is the value of resistance.

Ohm's law does have one limitation: The ambient temperature (the temperature surrounding the resistor) must be kept constant, in order to satisfy the law. Although it is seldom necessary to consider the surrounding temperature when calculations involving resistance are made, the reader should keep in mind that resistance values usually vary with temperature. In many materials, resistance usually increases with an increase of temperature. Other materials, such as germanium, silicon and carbon, are the exceptions. For such materials, resistance decreases with an increase of temperature.

The major cause for destruction of a resistor is an excess of heat. The heat created by the flow of current through a resistor may increase faster than the resistor can possibly dissipate it. The end result is that the resistor either melts or burns. We must be able to calculate the power being dissipated by a resistor to ensure that the power-dissipation rating of the resistor is not exceeded.

In Section 2-5 we saw that the average power dissipated by an element is given by

$$P = VI \quad \text{(in dc circuits)} \tag{3-3a}$$

$$P = \hat{V}\hat{I} \quad \text{(in ac circuits)} \tag{3-3b}$$

Since resistance is the only element that dissipates power,

$$P = V_R I_R \quad \text{(in dc circuits)} \tag{3-4a}$$

$$P = \hat{V}_R \hat{I}_R \quad \text{(in ac circuits)} \tag{3-4b}$$

where P in both equations is the average power in watts, V_R and $\hat{V}_R$ are the dc and rms voltages, respectively, across the resistance, and I_R and $\hat{I}_R$ are the dc and rms currents in amperes, respectively, through the resistance. According to Ohm's law,

$$V_R = I_R R \tag{3-5a}$$

$$\hat{V}_R = \hat{I}_R R \tag{3-5b}$$

Substituting Eq. (3-5) into Eq. (3-4) yields

$$P = I_R^2 R \text{ (in dc circuits)} \tag{3-6a}$$

$$P = \hat{I}_R^2 R \text{ (in ac circuits)} \tag{3-6b}$$

Equation (3-6) gives the average power, P, dissipated by a resistance, R, when a current, I_R or $\hat{I}_R$, flows through it.   By rearranging Ohm's law such that

$$I_R = \frac{V_R}{R} \tag{3-7a}$$

$$\hat{I}_R = \frac{\hat{V}_R}{R} \tag{3-7b}$$

and then substituting Eq. (3-7) into Eq. (3-4), we obtain

$$P = \frac{V_R^2}{R} \text{ (in dc circuits)} \tag{3-8a}$$

$$P = \frac{\hat{V}_R^2}{R} \text{ (in ac circuits)} \tag{3-8b}$$

Equation (3-8) yields the average power, P, dissipated by a resistor in terms of the voltage across the resistor and the value of resistor.

Example 3-1: A dc voltage of 20 V appears across a 5 kΩ resistor.   Determine the current through the resistor.

Solution: Rearranging Ohm's law,

$$I_R = \frac{V_R}{R} = \frac{20 \text{ V}}{5 \text{ k}\Omega} = 4 \text{ mA}$$

This information may be shown schematically as in Fig. 3-2.

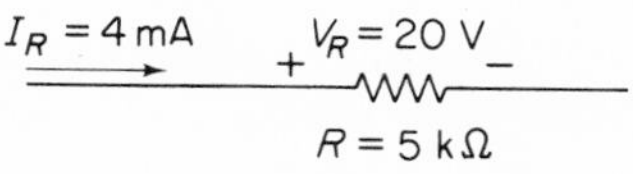

$I_R = 4\,\text{mA} \qquad V_R = 20\,\text{V}$
$+\!\!\!-\!\!\!\Lambda\!\Lambda\!\Lambda\!-$
$R = 5\,\text{k}\Omega$

Fig. 3-2. Fixed resistor of Ex. 3-1.

Note that the direction of conventional current has determined the polarity marking for voltage across the resistor.

Example 3-2: The voltage across a resistor is 10 V rms and the current through it is 2.5 mA. What is the value of the resistor?

Solution: According to Ohm's law.

$$R = \frac{\hat{V}_R}{\hat{I}_R} = \frac{10\ \text{V}}{2.5\ \text{mA}} = 4\ \text{k}\Omega$$

This information is shown schematically in Fig. 3-3.

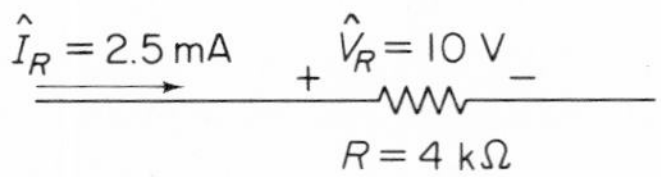

$$\hat{I}_R = 2.5\ \text{mA} \qquad +\ \hat{V}_R = 10\ \text{V}\ -$$

$$R = 4\ \text{k}\Omega$$

Fig. 3-3. Fixed resistor of Ex. 3-2.

As long as the voltage across and the current through a resistor are known, the value of resistor can always be determined.

Example 3-3: A 10 kΩ resistor is capable of dissipating 0.25 W. What is the largest rms value of current that can flow through the resistor?

Solution: Since $P = \hat{I}_R^2 R$, then

$$\hat{I}_R = \sqrt{\frac{P}{R}} = \sqrt{\frac{0.25\ \text{W}}{10\ \text{k}\Omega}} = 5\ \text{mA}$$

Note that although the rms value of current was asked for, 5 mA is also the largest dc value of current that can flow through the resistor without exceeding 0.25 W. Thus 5 mA of ac has the same heating effect as 5 mA of dc.

Example 3-4: The periodic waveform shown in Fig. 3-4 is the voltage across a 25 kΩ resistor. Draw the current waveform.

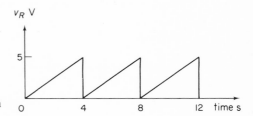

Fig. 3-4. Voltage waveform for Ex. 3-4.

Solution: Since the voltage and current waveforms for a resistor are in phase, the current waveform will be similar to Fig. 3-4. However, the maximum value of current will be determined from Eq. (3-2):

$$I_m = \frac{V_m}{R} = \frac{5\ \text{V}}{25\ \text{k}\Omega} = 0.2\ \text{mA}$$

Then we get the result shown in Fig. 3-5.

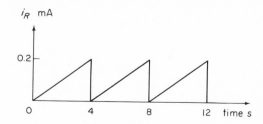

Fig. 3-5. Current waveform for Ex. 3-4.

Thus the voltage and current waveforms for a resistor differ only in their vertical scale.

Example 3-5: The current waveform through a 2 kΩ resistor is shown in Fig. 3-6a. What is the total power dissipated by this resistor?

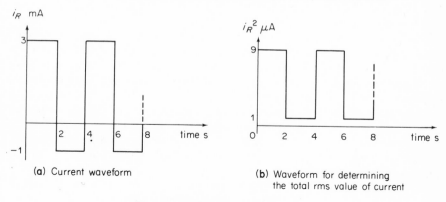

(a) Current waveform

(b) Waveform for determining the total rms value of current

Fig. 3-6. Waveforms for Ex. 3-5.

Solution: The equation for power dissipated by a resistance for a periodic waveform is

$$P = \hat{I}_R^2 R$$

To find $\hat{I}_R$ the total rms value of current, from Eq. (2-24),

$$\hat{I}_R = \sqrt{\frac{A}{T}}$$

We must first determine the area under the waveform of Fig. 3-6b for one complete cycle:

$$\text{area} = (9 \ \mu A^2) \cdot (2 \ s) + (1 \ \mu A^2) \cdot (2 \ s) = 20 \ \mu A^2 \cdot s$$

Then

$$\hat{I}_R = \sqrt{\frac{20\ \mu A^2 \cdot s}{4\ s}} = 2.24\ mA$$

Therefore,

$$P = \hat{I}_R^2 R = (2.24\ mA)^2\ (2\ k\Omega) = 10\ mW$$

In conclusion, we must determine the total rms value of current for a periodic waveform in order to calculate the total power.

Self-inductance.   The second passive element is inductance.   The unique property of this element is that it stores energy in a *magnetic field*.   Let us, then, examine some fundamental properties of a magnetic field.

Before 1819, electricity and magnetism were considered independent natural phenomena.   However, in that year Hans Christian Oersted (1777-1851) observed an interaction between electricity and magnetism.   Thus a new science of electromagnetism was born.

Oersted's experiment showed that an electric current caused a magnetic compass needle to move.   Because the wire through which the electric current flowed did not come into direct contact with the compass needle, he concluded that the interaction must have involved the surrounding space.

Whenever there is an interaction between two points in space with no visible contact between them, the surrounding space is called a *field*.   The two important properties of a field are intensity (magnitude) and direction. It is because of a field that the flow of current had an effect on the magnetic compass.

It was later shown that whenever a current flows in a conductor, a magnetic field is created.   The flow of current dictates both the intensity and the direction of the magnetic field.   The larger the current, the greater the intensity of the magnetic field.   The direction of the magnetic field is determined by the *right-hand rule*.   That is, if the thumb of the right hand points in the direction of conventional current, the fingers of the right hand will curl around the conductor in the direction of the magnetic field.

Because it is a human trait to want to visualize every concept, it is convenient to have some way of depicting a field.   Michael Faraday, a pioneer in the science of electromagnetism, suggested adopting "lines of force" to represent a magnetic field.   Although these lines are imaginary, they are the best way to visualize a magnetic field.   For a current-carrying conductor, Fig. 3-7 shows magnetic lines of force and their direction.   The direction of the magnetic field is always perpendicular to the direction of current and must always form a complete loop.

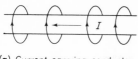

(a) Current carrying conductor

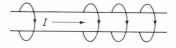

(b) Current carrying conductor

(c) Cross–section view of the
magnetic field

(d) Cross–sectional view of the
magnetic field

Fig. 3-7. "Lines of force" concept to visualize a magnetic field.

If a current - carrying conductor is wound in the form of a coil, not only will the magnetic lines of force encircle each infinitesimal portion of the conductor, but they will add to produce a magnetic field surrounding the entire coil, as shown in Fig. 3-8.

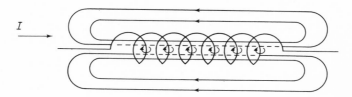

Fig. 3-8. Magnetic field for a current carrying conductor wound in a coil.

Up to this point, our discussion of the interaction between electricity and magnetism has dealt with the fact that the flow of current in a conductor produces a magnetic field about the conductor. We now wish to consider another interaction between electricity and magnetism. That is, a changing magnetic field surrounding a conductor induces a voltage in the conductor. This induced voltage is called a *self-induced voltage* because it is induced in the *same* conductor that is carrying the current. Note that it is a voltage that is induced into the conductor and not a current. The flow of current depends upon whether or not there is a completed circuit.

If the circuit is complete (a continuous path for current to flow), then the direction of the self-induced voltage is given by Lenz's law: *The direction of a self-induced voltage is such as to oppose the change in current.* For example, suppose that the current in a coil tends to increase; then the magnetic field about the coil increases and the direction of the self-induced voltage according to Lenz's law is such as to oppose any change, thus delaying or prolonging the increase as shown in Fig. 3-9.

Fig. 3-9. Direction of the self-induced voltage for an increase in current.

When the current in a circuit tends to decrease, the magnetic field also decreases and the direction of the self-induced voltage tends to keep the current from changing, thus prolonging the decay (see Fig. 3-10).   In the latter case, when the magnetic field is collapsing, it gives energy back to the circuit, temporarily acting as an energy source.   This opposition to change in current is an effect of electromagnetic induction.

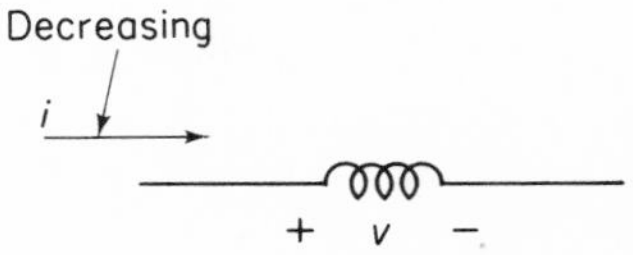

Fig. 3-10. Direction of the self-induced voltage for a decrease in current.

The circuit element that has the property of opposing any change in the current through that circuit by storing energy in a magnetic field is called *self-inductance*.   The basic unit for inductance is the *henry*, H, in honor of Joseph Henry.   Inductance is represented by the symbol *L*.   *Inductors*[2] may be classified as fixed or variable.   The circuit symbols for both are shown in Fig. 3-11.   A fixed inductor has a nonadjustable value.

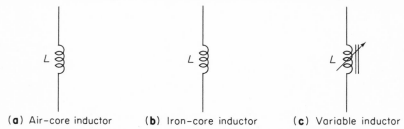

(**a**) Air–core inductor   (**b**) Iron–core inductor   (**c**) Variable inductor

Fig. 3-11. Circuit symbols for inductors.

Fixed inductors have either an air core or an iron core.   Variable inductors have an iron core that can be moved within the coil to vary the magnetic field and thereby vary its inductance.

---

[2] Inductance is the electrical property of storing energy in a magnetic field.   An inductor is a physical device.

Theoretically, any current-carrying conductor has an inductance associated with it. However, only at extremely high frequencies (100 MHz or greater) will the inductance of a straight piece of wire have to be considered. Since our discussion of electric circuit analysis does not involve such high frequencies, we consider only the inductance of a physical coil.

Inductance depends on the physical characteristics of a coil, such as the material of the core, the cross-sectional area of the core, the length of the core, the number of turns of the coil, and how tightly the turns are wound. Although the value of inductance may be calculated by the physical characteristics of a coil, we are primarily interested in defining inductance in terms of voltage and current. The inductance of an inductor is equal to the ratio of the induced voltage to the rate of change of current with respect to time. Or,

$$L = \frac{v_L}{di_L/dt} \tag{3-9}$$

where L is the inductance in henries, v_L the self-induced voltage in volts, and di_L/dt the rate of change of current with respect to time in amperes per second. Equation (3-9) may be taken as a defining equation for inductance regardless of the size or shape and whether or not the core is air or a magnetic material.

We have previously mentioned that the basic unit for inductance is the henry. The value of inductance, however, varies over a wide range in electronic applications depending on the frequency and use. In ac power supplies where the frequencies are low (20 Hz to 2 kHz), the inductance values are on the order of several henries and the core material is usually iron laminations. Circuits subject to higher frequencies (10 to 30 kHz) usually have inductance values in the order of millihenries ($mH = 10^{-3}$ henry). At still higher frequencies (300 kHz to 3 MHz), inductance values are on the order of microhenries ($\mu H = 10^{-6}$ henry). The core material for the last two cases usually is air. If the frequency is increased much higher, the inductance of a straight piece of wire may have to be considered.

Equation (3-9) is not only a defining equation for inductance, but may be rearranged to express the voltage-current relationship:

$$v_L = L \frac{di_L}{dt} \tag{3-10}$$

In other words, Eq. (3-10) states that the voltage across an inductance is directly proportional to the value of inductance and also the rate of change of current with respect to time. Note that the voltage depends on the *rate of change* of current and not on the current itself. Thus it is possible to have a steady current through an inductance but no voltage across it. Alternatively, the current at a particular instant may be zero but the voltage across the inductance does not necessarily have to be zero.   In the examples that follow, pay particular attention to this point.

Unlike resistance, where the voltage and current waveforms are similar, or in phase, the voltage and current waveforms for an inductance are not similar because of the term di_L/dt.

We have discussed the concept of inductance storing energy in a magnetic field.   This energy must come from a source.   However, all the energy delivered by a source is not stored in the magnetic field.   It can be shown by calculus that only one half of the total energy delivered by the source is stored by the inductance.   At this time, however, the derivation is not of critical importance.   The average energy stored by an inductance is

$$W_m = \tfrac{1}{2} L I_{av}^2 \text{ joules} \tag{3-11}$$

where W_m is the average energy in joules stored in a magnetic field, L the inductance in henries, and I_{av} the average current in amperes through the inductance.

Note that Eq. (3-11) is written in terms of average values.   If an ac current flows through an inductance, the average energy stored over one complete cycle is zero.   Therefore, average values and not instantaneous values are of primary concern.

Let us study some examples to see the differences in waveforms.

Example 3-6:   The periodic waveform shown in Fig. 3-12 is the current through a 20 mH inductor.   Draw the voltage waveform.

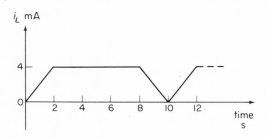

Fig. 3-12.   Current waveform for Ex. 3-6.

Solution: We shall treat the problem in three steps, separating the time axis between 0 and 2 s, 2 and 8 s, and 8 and 10 s, in other words, between times when the slope of the current waveform is not changing.   Since the slope between a time interval does not change instantaneously we may write

$$\frac{di}{dt} = \frac{\Delta i}{\Delta t}$$

With this expression, let us now analyze the three time intervals.

$0\,\text{s} < t < 2\,\text{s}$:

$$\frac{di_L}{dt} = \frac{\Delta i_L}{\Delta t} = \frac{4\,\text{mA} - 0}{2\,\text{s} - 0} = 2\,\text{mA/s}$$

$$v_L = L\,\frac{di_L}{dt} = (20\,\text{mH})\,(2\,\text{mA}) = 40\,\mu\text{V}$$

$2\,\text{s} < t < 8\,\text{s}$:

$$\frac{di_L}{dt} = \frac{\Delta i_L}{\Delta t} = \frac{4\,\text{mA} - 4\,\text{mA}}{8\,\text{s} - 2\,\text{s}} = 0$$

$$v_L = L\,\frac{di_L}{dt} = (20\,\text{mH})\,(0\,\text{A}) = 0$$

Note that there is no change in the magnitude of current between 2 and 8 s. Therefore, the voltage across the inductor is 0.

$8\,\text{s} < t < 10\,\text{s}$:

$$\frac{di_L}{dt} = \frac{\Delta i_L}{\Delta t} = \frac{0 - 4\,\text{mA}}{10\,\text{s} - 8\,\text{s}} = -2\,\text{mA/s}$$

$$v_L = L\,\frac{di_L}{dt} = (20\,\text{mH})\,(-2\,\text{mA/s}) = -40\,\mu\text{V}$$

Now that the voltage across the inductor has been calculated between the time intervals, we are able to draw the voltage waveform, as shown in Fig. 3-13.

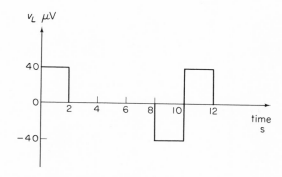

Fig. 3-13. Voltage waveform for Ex. 3-6.

Note that a voltage is produced across an inductance only when the magnitude of current changes with time. If the current increases with time, there is a positive voltage across the inductor. If the current decreases with time, the voltage across the inductor is negative.

Example 3-7: Draw the voltage waveform across a 10 mH inductor for the current waveform shown in Fig. 3-14.

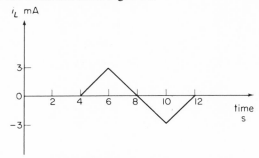

Fig. 3-14. Current waveform for Ex. 3-7.

Solution: As in Example 3-6, we shall analyze the current waveform of Fig. 3-14 during specific time intervals.

$0\,\text{s} < t < 4\,\text{s}$:

$$\frac{di_L}{dt} = \frac{\Delta i_L}{\Delta t} = \frac{0 - 0}{4\,\text{s} - 0\,\text{s}} = 0\,\text{A/s}$$

$$v_L = L\,\frac{di_L}{dt} = (10\,\text{mH})\,(0\,\text{A/s}) = 0\,\text{V}$$

$4\,\text{s} < t < 6\,\text{s}$:

$$\frac{di_L}{dt} = \frac{\Delta i_L}{\Delta t} = \frac{3\,\text{mA} - 0}{6\,\text{s} - 4\,\text{s}} = 1.5\,\text{mA/s}$$

$$v_L = L\,\frac{di_L}{dt} = (10\,\text{mH})\,(1.5\,\text{mA/s}) = 15\,\mu\text{V}$$

$6\,\text{s} < t < 10\,\text{s}$:

$$\frac{di_L}{dt} = \frac{\Delta i_L}{\Delta t} = \frac{-3\,\text{mA} - 3\,\text{mA}}{10\,\text{s} - 6\,\text{s}} = -1.5\,\text{mA/s}$$

$$v_L = L\,\frac{di_L}{dt} = (10\,\text{mH})\,(-1.5\,\text{mA/s}) = -15\,\mu\text{V}$$

$10\,\text{s} < t < 12\,\text{s}$:

$$\frac{di_L}{dt} = \frac{\Delta i_L}{\Delta t} = \frac{0 - (-3\,\text{mA})}{12\,\text{s} - 10\,\text{s}} = 1.5\,\text{mA/s}$$

$$v_L = L\,\frac{di_L}{dt} = (10\,\text{mH})\,(1.5\,\text{mA/s}) = 15\,\mu\text{V}$$

$t > 12\,\text{s}$:

$$\frac{di_L}{dt} = \frac{\Delta i_L}{\Delta t} = 0\,\text{A/s}$$

$$v_L = L\,\frac{di_L}{dt} = (10\,\text{mH})\,(0\,\text{A/s}) = 0\,\text{V}$$

From these calculated values of voltage for each time interval, we can draw the voltage waveform, as shown in Fig. 3-15.

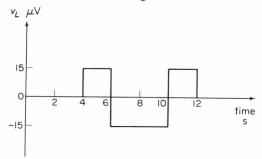

Fig. 3-15. Voltage waveform for Ex. 3-7.

Note that when current equals zero (at $t = 8$ s) the rate of change of current is -1.5 mA/s. The voltage then equals -15 mV and *not* zero.

Example 3-8: The average value of current flowing through a 3 mH inductor is measured to be 6 mA. What is the energy stored by this element?

Solution: According to Eq. (3-11),

$$W_m = \tfrac{1}{2} L I_{av}^2 = \tfrac{1}{2}(3 \text{ mH})(6 \text{ mA}) = 9 \ \mu\text{J}$$

The average energy stored must be calculated from the average value of current.

Example 3-9: The current waveform flowing through a 10 mH inductor is shown in Fig. 3-16. What is the energy stored by the inductor?

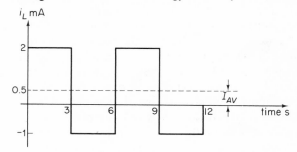

Fig. 3-16. Current waveform for Ex. 3-9.

Solution: The average value of the current waveform is

$$I_{av} = \frac{A}{T}$$

$$\text{area} = (2 \text{ mA})(3 \text{ s}) - (1 \text{ mA})(3 \text{ s}) = 3 \text{ mA} \cdot \text{s}$$

Then
$$I_{av} = \frac{A}{T} = \frac{3\,\text{mA}\cdot\text{s}}{6\,\text{s}} = 0.5\,\text{mA}$$

Therefore,
$$W_m = \tfrac{1}{2} LI_{av}^2 = \tfrac{1}{2}(10\,\text{mH})(0.5\,\text{mA}) = 2.5\,\mu\text{J}$$

Thus the procedure developed in Chapter 2 for determining the average value is the first step in solving for the average energy stored in an inductor.

Mutual Inductance. In circuit analysis, *mutual inductance* is that element which accounts for the transfer of energy, through a magnetic field, from one circuit to another. The resistive, inductive, and capacitive cases relate voltage and current for a single element. Mutual inductance, on the other hand, is the effect of voltage at one point because of a changing current at a different point in the circuit.

The transformer is a device that utilizes the principle of mutual inductance. For this reason we shall treat mutual inductance in the chapter on transformers, Chapter 11.

Capacitance. The third passive element is *capacitance*. Its properties are as unique as those of resistance and inductance. A *capacitor* stores energy in an *electric field*. In Section 1-8 we saw that whenever there is a difference of charge, a voltage (or potential difference) exists. Although not mentioned at the time, the space between a positive and negative charge is known as an *electric field*.

As in the case of a magnetic field, the "lines-of-force" concept may be used to visualize an electric field. Unlike the lines of force of a magnetic field, which from *closed* loops, the lines of force of an electric field always *begin and end* on a charged surface.

Similar to the lines of force of a magnetic field, the lines of force of an electric field indicate both strength and direction. The greater the number of lines, the more intense the electric field. The strength of an electric field is directly proportional to the charge and indirectly proportional to the distance squared. Thus the strength of an electric field is greatest closest to the charge and decreases as one moves away from it.

The direction of the electric field is the path a positive charge would travel if placed in the field. Since a basic phenomenon of charge is that like charges repel and unlike charges attract, then the *direction* of the lines of force is always *from the positive charge to the negative charge*, as shown in Fig. 3-17.

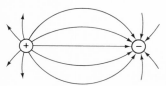

Fig. 3-17. "Lines of force" concept to visualize an electric field.

An electric field can be created in a circuit by placing two conducting plates in parallel and having one plate more positive than the other, as shown in Fig. 3-18a. The material between the two plates is nonconducting, or insulating. This insulating material is also called a *dielectric*. Examples of some dielectrics used are air, bakelite, ceramic, formica, glass, kraft paper, polyethylene, and Teflon.

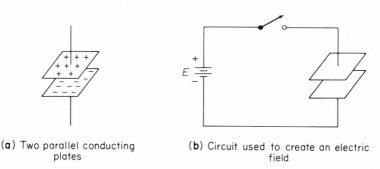

(a) Two parallel conducting   (b) Circuit used to create an electric
         plates   field

Fig. 3-18. Circuit diagrams used to illustrate the concept of electric field and capacitance.

Capacitance is the property of storing energy in an electric field. The energy is stored by the capacitance between parallel plates. A *capacitor* is a physical device having the property of capacitance.

We shall now examine how one plate of a capacitor can be made positive with respect to the other plate. Let us consider the circuit of Fig. 3-18b. A capacitor is connected to a battery through a switch. When the switch is closed, electrons from the negative terminal of the battery will collect on the bottom plate. To maintain equilibrium, an equal number of electrons will leave the top plate and drift toward the positive terminal of the battery. While this is happening the capacitor is said to be *charging*. This flow of charges will continue until the voltage across the capacitor equals the source voltage. When the capacitor voltage and applied voltage are equal, the capacitor is said to be *charged*. The capacitor will remain in this charged condition even after the source is removed. In order for the capacitor to return to a neutral or uncharged state, an external path must be provided for the flow of current. This external path may range from a short circuit (zero resistance) to a complicated network. While the capacitor is returning to a neutral (or uncharged) state, it is said to be *discharging*.

Although a capacitor supplies energy to a circuit while it is discharging, it is not classified as an energy source because a capacitance cannot *transform* some other form of energy into electrical energy; it can only *store* energy.

In Section 1-3 we saw that a movement of charges in one direction con-stitutes an electric current.   The current associated with a capacitor occurs only when the capacitor is either charging or discharging, for only at these times is there a movement of electrons.   The reader should be aware of the fact that a current does *not* flow *between* the plates of a capacitor but only in the external circuit.

Capacitors may be classified as either fixed or variable.   The circuit and letter symbols are shown in Fig. 3-19.   A fixed capacitor has a non-adjustable value.   A variable capacitor has one set of plates that are movable, thereby changing the value of the capacitance.

(**a**) Fixed capacitor   (**b**) Variable capacitor

Fig. 3-19. Circuit symbols for capacitors.

Both fixed and variable capacitors have maximum voltage ratings specified by the manufacturer.   If the voltage across the plates of a ca-pacitor exceeds the maximum rating, the dielectric will break down and permit current to flow between the plates.

The value of capacitance depends on a number of things, such as the insulating material, the distance between the parallel plates, the area of the plates, and the number of plates.   However, in terms of electrical quan-tities (charge, current, voltage, etc.) capacitance is defined as the ratio of charge on the plates to the voltage across the plates.   This statement in equation form is

$$\frac{Q}{V} = C \qquad\qquad (3\text{-}12)$$

where Q is the charge on the plates, in coulombs, V the voltage across the plates in volts, and C the value of the capacitance.

The basic unit of capacitance is the *farad*, named in honor of Michael Faraday.   However, the farad is such a large unit that its use is impractical. For this reason, capacitor values are usually in the order of microfarads (μF $= 10^{-6}$ farad) or picofarads (pF $= 10^{-12}$ farad).

Equation (3-12) defines capacitance in terms of capital letters Q and V. According to convention, capital letters designate dc values.   Since dc conditions are not the only conditions that can exist in a circuit, let us in-vestigate what happens when the voltage across the capacitance changes with time.   If the voltage is changing with respect to time, so will the charge on the plates.   The notation of change with respect to time is d/dt.   Thus

$$\frac{\dfrac{dq}{dt}}{\dfrac{dv}{dt}} = C$$

or

$$\frac{dq}{dt} = C\frac{dv}{dt} \tag{3-13}$$

From Eq. (1-2),

$$i = \frac{dq}{dt}$$

Then

$$i_C = C\frac{dv_C}{dt} \tag{3-14}$$

where i_C is the capacitance current in amperes, C the value of the capacitance in farads, and dv_C/dt the change of voltage across the capacitance with respect to time in volts per second.   Capacitance has the property of opposing any change in voltage across it.

At this time the author wishes to stress the importance of the fact that the current through a capacitance[3] depends on the *time rate of change of voltage* (rate at which the voltage is either increasing or decreasing) and not on the voltage itself.   Thus it is possible to have a large constant value of voltage across a capacitor without a flow of current through the capacitor.   Alternatively, stating that the voltage across a capacitor is zero does not necessarily imply that the current through the capacitor is zero.   The following examples illustrate this point.

We have discussed the concept that capacitance stores energy in an electric field.   This energy must come from a source.   However, all the energy delivered by the source is not stored in the electric field.   It can be shown by calculus that only one half of the total energy delivered by the source is stored by the capacitance.   At this time, however, the derivation is not of critical importance.   In equation form the average energy stored by a capacitance is

$$W_E = \frac{1}{2}CV_{av}^2 \text{ joules} \tag{3-15}$$

where W_E represents the average energy stored in the electric field in joules, C the value of the capacitance in farads, and V_{av} an average value of voltage across the capacitance in volts.

---

[3] As we have mentioned, free electrons do *not* flow between the plates of a capacitor. However, the expression "current *through* a capacitance" is used as standard terminology.   It simply means that charges flow onto one plate and off the other, giving the *effect* of a current flowing *between* the plates.

Note that Eq. (3-15) is written in terms of average values. This is because it is the average value and not an instantaneous value that is of primary concern. If an ac voltage waveform is impressed across a capacitor, the average energy stored over one complete cycle is zero.

Example 3-10: The triangular waveform shown in Fig. 3-20 is the voltage waveform across a $2\ \mu F$ capacitor. Draw the current waveform.

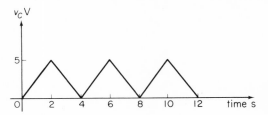

Fig. 3-20. Voltage waveform for Ex. 3-10.

Solution: As in previous examples, analyze the waveform during time intervals when the slope is constant. The slope is calculated by letting

$$\frac{dv_C}{dt} = \frac{\Delta v_C}{\Delta t}$$

$0\ s < t < 2\ s$:

$$\frac{dv_C}{dt} = \frac{\Delta v_C}{\Delta t} = \frac{5\ V - 0}{2\ s - 0} = 2.5\ V/s$$

$$i_C = C\frac{dv_C}{dt} = (2\ \mu F)(2.5\ V/s) = 5\ \mu A$$

$2\ s < t < 4\ s$:

$$\frac{dv_C}{dt} = \frac{\Delta v_C}{\Delta t} = \frac{0 - 5\ V}{4\ s - 2\ s} = -2.5\ V/s$$

$$i_C = C\frac{dv_C}{dt} = (2\ \mu F)(-2.5\ V/s) = -5\ \mu A$$

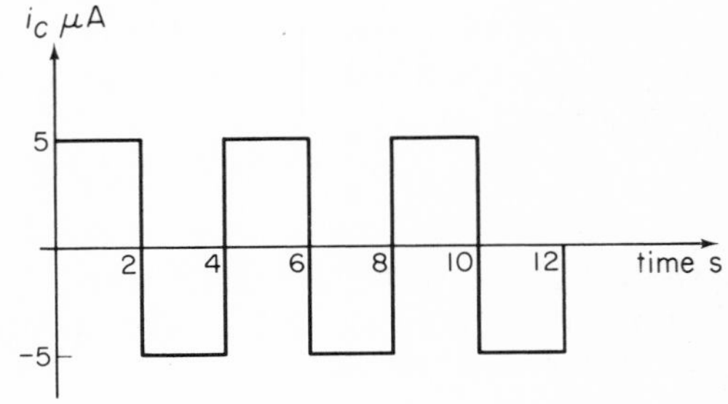

Fig. 3-21. Current waveform for Ex. 3-10.

It is not necessary to calculate the current for any other time intervals because the voltage waveform is periodic every 4 sec. Therefore, the current waveform will also be periodic, as shown in Fig. 3-21.

Comparing the current waveform, Fig. 3-20, to the voltage waveform, Fig. 3-21, we note that as the voltage increases across the capacitor, a positive current flows through it. Similarly, as the voltage decreases across the capacitor, the current is negative. Note that whenever the slope of the voltage waveform changes instantaneously, such as at $t = 2$ s, $t = 4$ s, etc., the value of current also changes instantaneously.

Example 3-11: The voltage waveform across a 10 μF capacitor is shown in Fig. 3-22. Draw the current waveform.

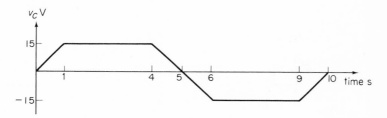

Fig. 3-22. Voltage waveform for Ex. 3-11.

Solution:

0 s $< t < 1$ s:

$$\frac{dv_C}{dt} = \frac{\Delta v_C}{\Delta t} = \frac{15 \text{ V} - 0}{1 \text{ s} - 0} = 15 \text{ V/s}$$

$$i_C = C\frac{dv_C}{dt} = (10 \ \mu\text{F})(15 \text{ V/s}) = 150 \ \mu\text{A} = 0.15 \text{ mA}$$

1 s $< t < 4$ s:

$$\frac{dv_C}{dt} = \frac{\Delta v_C}{\Delta t} = \frac{0 \text{ V}}{4 \text{ s} - 1 \text{ s}} = 0 \text{ V/s}$$

$$i_C = C\frac{dv_C}{dt}(10 \ \mu\text{F})(0 \text{ V/s}) = 0 \text{ A}$$

4 s $< t < 6$ s:

$$\frac{dv_C}{dt} = \frac{\Delta v_C}{\Delta t} = \frac{-15 \text{ V} - 15 \text{ V}}{6 \text{ s} - 4 \text{ s}} = -15 \text{ V/s}$$

$$i_C = C\frac{dv_C}{dt} = (10 \ \mu\text{F})(-15 \text{ V/s}) = -150 \ \mu\text{A} = -0.15 \text{ mA}$$

$6\,s < t < 9\,s$:

$$\frac{dv_C}{dt} = \frac{\Delta v_C}{\Delta t} = \frac{0\text{ V}}{9\text{ s} - 6\text{ s}} = 0\text{ V/s}$$

$$i_C = C\frac{dv_C}{dt} = (10\,\mu\text{F})\,(0\text{ V/s}) = 0\text{ A}$$

$9\,s < t < 10\,s$:

$$\frac{dv_C}{dt} = \frac{\Delta v_C}{\Delta t} = \frac{0\text{ V} - (-15\text{ V})}{10\text{ s} - 9\text{ s}} = 15\text{ V/s}$$

$$i_C = C\frac{dv_C}{dt} = (10\,\mu\text{F})\,(15\text{ V/s}) = 150\,\mu\text{A} = 0.15\text{ mA}$$

See Fig. 3-23.

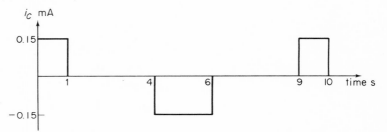

Fig. 3-23. Current waveform for Ex. 3-11.

Note that between time intervals 1 to 4 s and 6 to 9 s, the voltage across the capacitor is 15 V and − 15 V, respectively. Since the voltage remains constant, the rate of change of voltage is zero and the current is zero. Thus it is possible to have a large constant value of voltage across a capacitor *without* a flow of current through it. Now check $t = 5\,s$ from Fig. 3-22. The voltage across the capacitor is zero. However, the rate of change of voltage is *not* zero, it is − 15 V/s. Therefore, the current at this instant is not equal to zero.

Example 3-12: Calculate the energy stored by a 10 μF capacitor if 50 V dc is applied across its terminals.

Solution: Using Eq. (3-15),

$$W_E = \tfrac{1}{2}CV_{av}^2 = \tfrac{1}{2}(10\,\mu\text{F})\,(50\text{ V})^2 = 12.5\text{ mJ}$$

Since a dc value is an average value, we may use it in Eq. (3-15).

Example 3-13: Calculate the charge on the plates of the capacitor of Example 3-12.

Solution: Rearranging Eq. (3-12),

$$Q = CV = (10\ \mu\text{F})(50\ \text{V}) = 500\ \mu\text{C} = 0.5\ \text{mC}$$

Example 3-14: If the voltage waveform of Fig. 3-20 is applied across a 20-μF capacitor, what is the stored energy?

Solution: To determine the average value of a periodic waveform use Eq. (2-14):

$$V_{\text{av}} = \frac{A}{T}$$

The area under the waveform for one complete cycle is

$$A = \text{area} = \tfrac{1}{2}(5\ \text{V})(2\ \text{s})(2) = 10\ \text{V} \cdot \text{s} \qquad V_{\text{av}} = \frac{10\ \text{V} \cdot \text{s}}{4\ \text{s}} = 2.5\ \text{V}$$

Then

$$W_E = \tfrac{1}{2}CV_{\text{av}}^2 = \tfrac{1}{2}(20\ \mu\text{F})(2.5\ \text{V})^2 = 62.5\ \mu\text{J}$$

In conclusion, we see that when a periodic waveform whose average value is *not* zero is applied across a capacitor, the capacitor stores energy in its electric field.

3-2. ACTIVE ELEMENTS

Active elements are capable of supplying energy to a circuit. A rotating generator produces electrical energy from the mechanical energy of rotation. The battery produces electrical energy from a chemical reaction. Both the rotating generator and the battery are active elements. Active elements are classified as either *voltage sources* or *current sources*. In circuit analysis, we treat energy sources or active elements as if they were ideal sources.

Ideal sources are capable of delivering energy *without limit*. An ideal source will deliver the *same* magnitude of voltage or current regardless of the load connected to its terminals. In circuit theory two types of ideal sources are defined: (1) the ideal voltage source and (2) the ideal current source.

Ideal Voltage Source. An ideal voltage source exists if the *voltage* waveform generated by the source is *independent* of the network connected to its terminals. The ideal voltage source, therefore, has no internal losses. Thus the voltage produced internally by the ideal voltage source is delivered directly to its output terminals. In practice, the ideal voltage source does not exist. However, the concept of an ideal source is used to an advantage in circuit analysis — problems can be greatly simplified in this way. Unless otherwise stated, all voltage sources in this text may be considered ideal voltage sources. Chapter 5 deals with practical sources, where we shall see how the practical voltage source differs from the ideal voltage source.

Two of the basic waveforms used in electronic applications are the ac sine wave and the constant dc waveform. Since the sine wave is the most common ac waveform, a sine wave of voltage is referred to as an ac voltage. The circuit symbol for an ideal ac voltage source is shown in Fig. 3-24a. A constant dc waveform of voltage is simply referred to as a dc voltage. The circuit symbol for an ideal dc voltage source is shown in Fig. 3-24b.

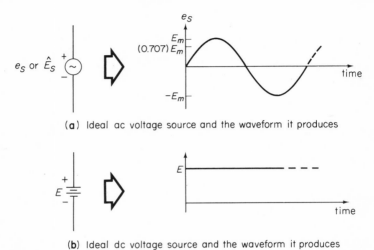

(a) Ideal ac voltage source and the waveform it produces

(b) Ideal dc voltage source and the waveform it produces

Fig. 3-24. Circuit symbols for ideal sources.

Ideal Current Source. An ideal current source exists if the *current* waveform generated by the source is *independent* of the network connected to its terminals. The ideal current source, like the ideal voltage source, has no internal losses. Therefore, any current produced by the ideal current source is delivered directly to its output terminals. Although the concept of a current source with no internal losses does not exist in practice, the ideal current source nevertheless is used in circuit analysis. Unless otherwise stated, all current sources in this text may be considered ideal current sources. Chapter 5 deals not only with practical voltage sources but also with practical current sources. Thus in Chapter 5 we will see the difference between ideal and practical current sources.

Since the sine wave is the most common ac waveform used in electronic applications, a sine wave of current is simply referred to as an ac current. The circuit symbol for an ideal current source is a circle with an arrow indicating the direction of conventional current. The circuit symbol for an ideal ac current source is shown in Fig. 3-25a. A constant dc waveform of current is referred to as a dc current.   The circuit symbol for an ideal dc current source is shown in Fig. 3-25b.

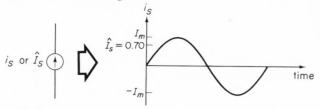

(a) Ideal ac current source and the waveform it produces

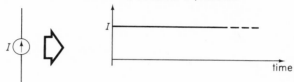

(b) Ideal dc current source and the waveform it produces

Fig. 3-25. Circuit symbols for ideal current sources.

3-3.   NETWORK TERMINOLOGY

In science and other disciplines, one must become familiar with certain defining terms and phrases.   Electronics as a specialized branch of physical science is no exception.

Electric circuits are combinations of active and passive elements.   When these elements are connected together, new terminology is needed to express the resultant conditions.

The circuit of Fig. 3-26 is used to define our new expressions.   Re-

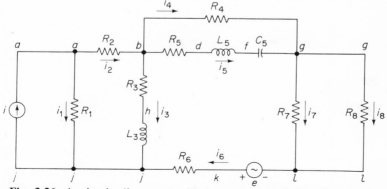

Fig. 3-26. A circuit diagram to illustrate the principles of network terminology.

member, however, that in some texts these expressions may be defined slightly differently.

Series Connection.   A *series connection* is two or more elements (either active or passive) connected end to end so that the current that flows through one element must *also* flow through each other element.   Elements that are series connected carry the *same* current.   Consider the circuit diagram of Fig. 3-26. R_3 and L_3 are in series.   Other series connections are R_5, L_5, and C_5. R_6 and the voltage source, e, are also in series.

Parallel Connection.   A *parallel conection* is two or more elements (either active or passive) which share *two* common *nodes* (see definition below). Elements that are connected in parallel have the *same* voltage across them. Referring to the circuit diagram of Fig. 3-26 we see that the current source, i, and the resistance, R_1, are in parallel.   Resistances R_7 and R_8 are in parallel.   The series combination of R_5, L_5, and C_5 is in parallel with R_4.

Nodes.   A *node* (also referred to as a *junction*) is a common point between two or more circuit elements.   In the circuit diagram of Fig. 3-26, common points a, b, d, f, g, h, j, k, and l are nodes.

Branch.   A *branch* is a single path between two nodes.   In Fig. 3-26, branches are aj, ab, bj, bg, jl, and gl.   Note that there are two branches aj — the branch consisting of the current source, i, and the resistive branch containing R_1.   The branch ab consists of R_2.   The branch bj consists of the series combination of R_3 and L_3.   There are two branches bg — the resistive branch R_4 and the branch consisting of the series combination of R_5, L_5, and C_5.   The branch jl contains the series combination of the resistance R_6 and the voltage source, e.   There are two branches labeled gl.   Both are resistive, one containing R_7, the other R_8.

Branch Current.   A *branch current* is the current that flows through an element or elements of a single branch.   In the circuit diagram of Fig. 3-26, currents $i_1, i_2, i_3, i_4, i_5, i_6, i_7$, and i_8 are all branch currents.   The source current, i, is also a branch current.

Loop.   A *loop* is any sequence of circuit elements forming a complete or closed path in a circuit which does not pass through any node more than once.   We may trace out the following loops in Fig. 3-26:   Loop aja, consisting of resistance R_1 and the current source, i; loop *abja*, consisting of R_2, R_3, L_3, and R_1 or R_2, R_3, L_3, and the current source, i; loop *bdfglkjb*, consisting of R_5, L_5, C_5, R_7, the voltage source, e, R_6, L_3, and R_3, loop *bgfdb*, consisting of R_4, C_5, L_5, and R_5; loop *glg*, consisting of R_8 and R_7. These are not all the loops that can be identified in the circuit diagram of Fig. 3-26. The reader should try to trace out the remaining loops. Note that it is possible for a branch to be used more than once when tracing out different loops of a network. Branch *aj* was used in the loop *aja* and also loop *abja*. Branch *bdfg* was used in tracing out loop *bdfglkjb* and also loop *bgfdb*.

3-4. KIRCHHOFF'S LAWS

The law of conservation of energy states that energy can be neither created nor destroyed. When we connect active and passive elements to form a network we automatically must conform to this law. This places certain restrictions on electric circuits. Elements that are connected in *series*, for example, must have the *same current* flowing through them. If two or more branches are connected in *parallel*, as the branches *dg* of Fig. 3-27, the *same voltage* must be across them.

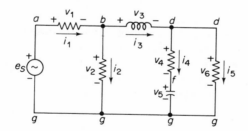

Fig. 3-27. Circuit diagram illustrating the principle of Kirchhoff's voltage law.

These and other conditions were incorporated into two basic laws of circuit analysis by Gustav Robert Kirchhoff (1824-1887). These laws are simply the law of conservation of energy as applied to electric circuits. They are Kirchhoff's voltage law (KVL) and Kirchhoff's current law (KCL).

Kirchhoff's Voltage Law. *At any instant of time, the algebraic sum of potential rises around any closed loop equals the algebraic sum of the potential drops.*

Let us consider the circuit of Fig. 3-27 and apply Kirchhoff's voltage law to the loop *abga*:

$$\text{potential rise} = \text{potential drop}$$

$$e_s = v_1 + v_2$$

or

$$e_s - (v_1 + v_2) = 0$$

Similarly, consider the loop *bdfgb*.

$$\text{potential rise} = \text{potential drop}$$

$$v_2 = v_3 + v_4 + v_5$$

or

$$v_2 - (v_3 + v_4 + v_5) = 0$$

Around the loop *dgfd*,

$$\text{potential rise} = \text{potential drop}$$

$$v_4 + v_5 = v_6$$

or

$$v_4 + v_5 - v_6 = 0$$

Kirchhoff's Current Law. *At any instant of time, the sum of the currents entering a node equals the sum of the currents leaving the node.*

The words *entering* and *leaving* refer to the direction of the arrow of current. If the head of the arrow points toward a node, the current is said to be entering that node. If the head of the arrow points away from a node, the current is said to be leaving that node. Consider Fig. 3-28. Currents i_1 and i_2 are entering the node while currents i_3, i_4, and i_5 are

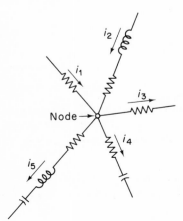

Fig. 3-28. Circuit diagram illustrating the principle of Kirchhoff's current law.

leaving the node.   Applying Kirchhoff's current law to Fig. 3-28, we obtain

currents entering = currents leaving

$$i_1 + i_2 = i_3 + i_4 + i_5 \quad \text{or} \quad i_1 + i_2 - (i_3 + i_4 + i_5) = 0$$

For further illustrations of Kirchhoff's current law, again consider the circuit of Fig. 3-27.

At node b: $i_1 = i_2 + i_3$ or $i_1 - (i_2 + i_3) = 0$

At node d: $i_3 = i_4 + i_5$ or $i_3 - (i_4 + i_5) = 0$

At node g: $i_1 = i_2 + i_4 + i_5$ or $i_1 - (i_2 + i_4 + i_5) = 0$

SUMMARY

Chapter 3 introduces new concepts in electric circuits.   These include the two types of elements in a circuit, passive and active.   The passive elements, resistors, inductors, and capacitors, either store or dissipate energy. The active elements, voltage sources and current sources, deliver energy to a circuit.

The section on network terminology presents definitions of expressions used in further study.   The last section introduced Kirchhoff's voltage and current laws, which will be used in analysis of circuits.

PROBLEMS

3-1.   A dc voltage of 75 V appears across a 10 kΩ resistor.   Calculate the current through the resistance.

3-2.   If the current through a 20 MΩ resistor is 4 μA, calculate the voltage across it.

3-3.   Calculate the value of resistance if the current through it is 10 mA and the voltage across it is 8.5 V.

3-4.   A 5 kΩ resistor dissipates 25 mW.   Calculate the current through the resistor.

3-5.   Calculate the current through each resistor for the diagrams shown in Fig. 3-29.

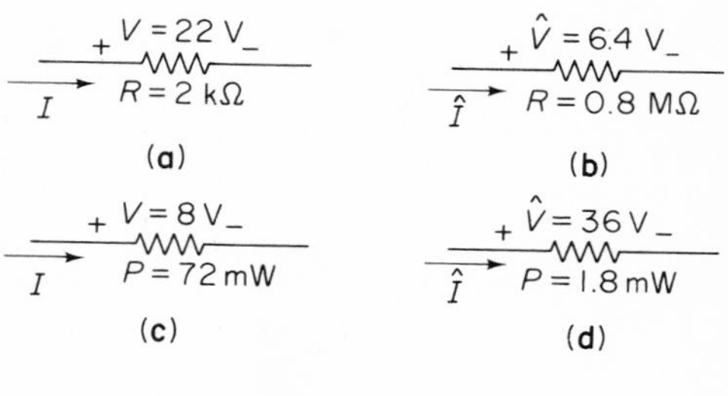

Fig. 3-29. Circuit diagrams for Prob. 5.

(a) $+\ V = 22\ V\ -$, $R = 2\ k\Omega$, I

(b) $+\ \hat{V} = 6.4\ V\ -$, $R = 0.8\ M\Omega$, $\hat{i}$

(c) $+\ V = 8\ V\ -$, $P = 72\ mW$, I

(d) $+\ \hat{V} = 36\ V\ -$, $P = 1.8\ mW$, $\hat{i}$

3-6. Calculate the value of each resistor in the diagrams shown in Fig. 3-30.

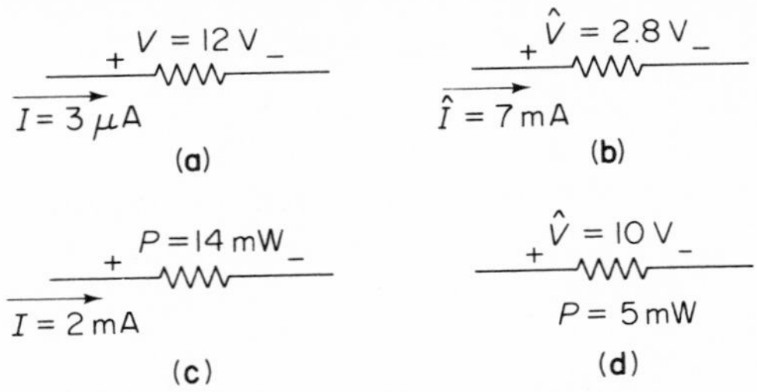

$+\ V = 12\ V\ -$
$I = 3\ \mu A$

(a)

$+\ \hat{V} = 2.8\ V\ -$
$\hat{I} = 7\ mA$

(b)

$+\ P = 14\ mW\ -$
$I = 2\ mA$

(c)

$+\ \hat{V} = 10\ V\ -$
$P = 5\ mW$

(d)

Fig. 3-30. Circuit diagrams for Prob. 6.

3-7. For each of the resistors shown in Fig. 3-31, calculate the voltage drop.

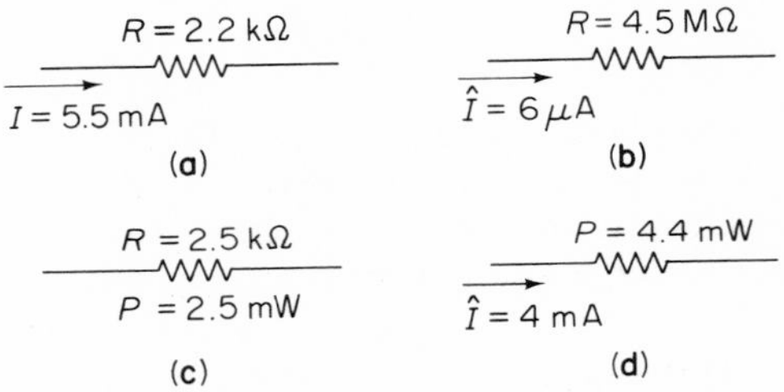

$R = 2.2\ k\Omega$
$I = 5.5\ mA$

(a)

$R = 4.5\ M\Omega$
$\hat{I} = 6\ \mu A$

(b)

$R = 2.5\ k\Omega$
$P = 2.5\ mW$

(c)

$P = 4.4\ mW$
$\hat{I} = 4\ mA$

(d)

Fig. 3-31. Circuit diagrams for Prob. 7.

3-8. The periodic waveform shown in Fig. 3-32 is the voltage across a 12 kΩ resistor. Draw the current waveform.

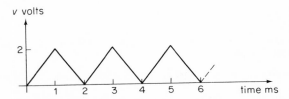

Fig. 3-32. Voltage waveform for Prob. 8.

3-9. The periodic waveform shown in Fig. 3-33 is the current through a 2.5 kΩ resistor. Draw the voltage waveform and calculate the frequency.

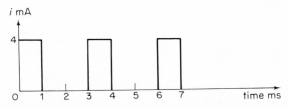

Fig. 3-33. Current waveform for Prob. 9.

3-10. Calculate the total power dissipated by the resistor of Problem 3-9.

3-11. Draw the waveform for the voltage across a 100 mH coil if the current waveform is as shown in Fig. 3-34.

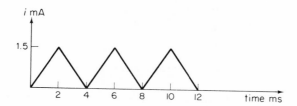

Fig. 3-34. Current waveform for Prob. 11.

3-12. Draw the waveform for the voltage across a 40 μH coil if the current waveform is as shown in Fig. 3-35.

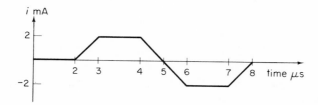

Fig. 3-35. Current waveform for Prob. 12.

3-13. Repeat Problem 3-12 for a 20 μH coil and a 80 μH coil. Compare the voltage waveforms.

3-14. The average value of current flowing through a 5 mH inductor is 12 mA. Calculate the energy stored by this element.

3-15. A dc current of 2.5 mA flows through a 4.2 H coil. Calculate the energy stored by this element.

3-16. The energy stored by a 40 mH inductor is 20 μJ. Calculate the value of current flowing through it.

3-17. If the current waveform of Fig. 3-33 is applied to a 10 mH inductor, calculate the stored energy.

3-18. Calculate the energy stored by the coil of Problem 3-11.

3-19. Figure 3-36 shows the voltage waveform across a 10 μF capacitor. Draw the current waveform.

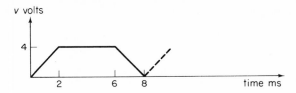

Fig. 3-36. Voltage waveform for Prob. 19.

3-20. Figure 3-37 shows the voltage waveform across a 30 μF capacitor. Draw the current waveform.

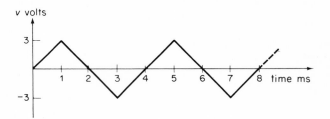

Fig. 3-37. Voltage waveform for Prob. 20.

3-21. Repeat Problem 3-20 for a 15 μF capacitor and then for a 60 μF capacitor. Compare current waveforms.

3-22. Calculate the energy stored by a 2.5 μF capacitor if 10 V dc is applied across its terminals.

3-23. The average value of voltage across a 40 μF capacitor is 30 V. Calculate the energy stored by this element.

3-24. The energy stored by a 50 μF capacitor is 100 mJ. Calculate the voltage across the capacitor.

3-25. Calculate the charge on the plates of the capacitor of Problem 3-22.

3-26. Calculate the charge on the plates of the capacitor of Problem 3-23.

3-27. If a 15 μF capacitor has a charge of 200 μC, calculate the voltage across it.

3-28. Calculate the charge on the plates of the capacitor of Problem 3-24.

3-29. If the voltage waveform of Fig. 3-32 is applied across a 25 μF capacitor, calculate (a) the stored energy and (b) the charge on the plates.

CHAPTER 4

series and parallel circuits

Beginning with this chapter, we will be using some of the fundamental concepts previously discussed in dealing with more complex circuits. Sections 4-1 and 4-2 deal with series and parallel circuits, respectively, and Section 4-3 is concerned with series-parallel circuits. In all three sections, resistors, inductors, and capacitors are the elements used to demonstrate the properties of the various circuit connections.

4-1. SERIES CIRCUITS

In Chapter 3 we defined series-connected elements and showed some examples. Now we shall deal with three special conditions: (1) resistors in series, (2) inductors in series, and (3) capacitors in series. The discussion of different elements such as a resistor and an inductor, a resistor and a capacitor, and a resistor, inductor, and capacitor connected in series will be left to Chapter 7.

Resistors in Series. Figure 4-1 shows three resistors connected in series. Using Kirchhoff's voltage law, a procedure can be developed to determine the total resistance. Although Fig. 4-1 contains only three resistors in series, the results can be generalized to any number of series resistors. Applying Kirchhoff's voltage law to Fig. 4-1, we obtain

$$v_T = v_1 + v_2 + v_3 \tag{4-1}$$

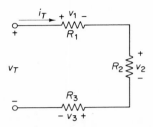

Fig. 4-1. Three resistors connected in series.

By definition, the *same* current has to flow through all elements connected in series. Thus i_T flows through each resistance of Fig. 4-1. Dividing both sides of Eq. (4-1) by i_T we obtain

$$\frac{v_T}{i_T} = \frac{v_1}{i_T} + \frac{v_2}{i_T} + \frac{v_3}{i_T} \tag{4-2}$$

Analyzing Eq. (4-2), we find that the left-hand side of the equation is the total voltage divided by the total current. This ratio always yields the total resistance, R_T, of the circuit. Using Ohm's law, the terms on the right-hand side of Eq. (4-2) yield

$$\frac{v_1}{i_T} = R_1 \qquad \frac{v_2}{i_T} = R_2 \qquad \frac{v_3}{i_T} = R_3$$

Therefore,

$$R_T = R_1 + R_2 + R_3 \tag{4-3}$$

Equation (4-3) states that when resistors are connected in series the total resistance is the sum of the individual resistances. This rule may be applied to *any* number of resistors in series.

If a number of *equal* resistors are connected in series, we may use

$$R_T = NR \tag{4-4}$$

where N is the number of equal resistors and R is the value of (any) one of the resistors.

Let us consider some series resistive circuits.

Example 4-1: Calculate the total resistance for each of the circuits in Fig. 4-2.

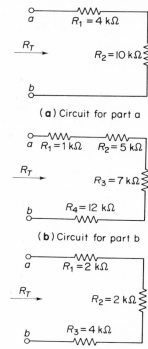

(**a**) Circuit for part a

(**b**) Circuit for part b

(**c**) Circuit for part c

Fig. 4-2. Circuit diagrams for Ex. 4-1.

Solution: Circuit a:

$$R_T = R_1 + R_2$$
$$R_T = 4\,\text{k}\Omega + 10\,\text{k}\Omega$$
$$R_T = 14\,\text{k}\Omega$$

Circuit b:

$$R_T = R_1 + R_2 + R_3 + R_4$$
$$R_T = 1\,\text{k}\Omega + 5\,\text{k}\Omega + 7\,\text{k}\Omega + 12\,\text{k}\Omega$$
$$R_T = 25\,\text{k}\Omega$$

Circuit c:

$$R_T = NR + R_3$$
$$R_T = (2) \cdot (2\,k\Omega) + 4\,\text{k}\Omega$$
$$R_T = 8\,\text{k}\Omega$$

Note that each circuit of Fig. 4-2 may be represented by an equivalent resistor, R_T. This total or equivalent resistor is capable of dissipating the same amount of energy as the sum of the individual resistors. Consider the circuit in Fig. 4-2a. The equivalent resistance, 14 kΩ, dissipates the same energy in the same amount of time as the series combination of 4 kΩ and 10 kΩ. The voltage across an equivalent resistance equals the sum of the voltages across the series combination. Although the calculation for the total or equivalent resistor of series resistors is not difficult, it is usually only the first step in a problem, as shown in Example 4-2.

Example 4-2:   For the series circuit of Fig. 4-3a, calculate
(a)   the total equivalent resistance,
(b)   the current waveform,
(c)   the rms value of the current,
(d)   the power dissipated by each resistor,
(e)   the total power dissipated by the circuit.

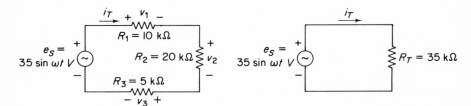

(**a**) Circuit for Ex. 4-2   (**b**) Equivalent circuit of (a)
Fig. 4-3. Circuit diagrams of Ex. 4-2.

Solution: (a)
$$R_T = R_1 + R_2 + R_3$$
$$R_T = 10\,\text{k}\Omega + 20\,\text{k}\Omega + 5\,\text{k}\Omega$$
$$R_T = 35\,\text{k}\Omega$$

The equivalent circuit is shown in Fig. 4-3b.

(b)   According to Ohm's law,

$$i_T = \frac{e_S}{R_T} = \frac{35\sin\omega t\ \text{V}}{35\,\text{k}\Omega} = 1\sin\omega t\ \text{mA}$$

A plot of the voltage and current waveforms are shown in Fig. 4-4a and and b.

(c)

$$\hat{I}_T = \frac{I_m}{\sqrt{2}} = \frac{1\,\text{mA}}{\sqrt{2}} = 0.707\,\text{mA}$$

(d)   From the answer of part (c) and the given values of the resistors, we may calculate the power dissipated by each resistor using the following method.

$$P_1 = (\hat{I}_T)^2 R_1 = (0.707\,\text{mA})^2 \cdot (10\,\text{k}\Omega) = \quad 5 \quad \text{mW}$$
$$P_2 = (\hat{I}_T)^2 R_2 = (0.707\,\text{mA})^2 \cdot (20\,\text{k}\Omega) = 10 \quad \text{mW}$$
$$P_3 = (\hat{I}_T)^2 R_3 = (0.707\,\text{mA})^2 \cdot (5\,\text{k}\Omega) = \quad 2.5\,\text{mW}$$

(e)   One method of solving for the total power dissipated by a circuit is by finding the product of the rms values of voltage and current:

$$P_T = \hat{V}_T \hat{I}_T = \frac{V_m}{\sqrt{2}} \times \frac{I_m}{\sqrt{2}} = \frac{35\text{ V}}{\sqrt{2}} \times \frac{1\text{ mA}}{\sqrt{2}} = 17.5\text{ mW}$$

An alternative procedure for obtaining total power dissipated by a circuit is addition of the powers dissipated by the individual resistors.   Note that this yields the same answer as part (e):

$$P_T = P_1 + P_2 + P_3$$
$$P_T = 5\text{ mW} + 10\text{ mW} + 2.5\text{ mW}$$
$$P_T = 17.5\text{ mW}$$

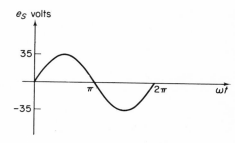

(a) Voltage waveform

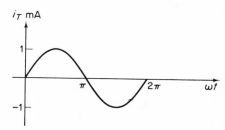

Fig. 4-4. Waveforms for Ex. 4-2.   (b) Current waveform

In conclusion, the reader should note that the voltage and current waveforms for a resistor differ only in magnitude and not in frequency.   In part (e) of this example, note that the total power dissipated equals the power dissipated by the series combination.   The power dissipated by any resistor or the power dissipated by the total circuit also could have been obtained by any variation of the formula for power $P = \hat{V}\hat{I}$, $P = \hat{I}^2 R$, *or* $P = \hat{V}^2/R$. In determining the power dissipated by a single resistor and using one of the formulas containing voltage, use only the voltage across that resistor and not the total voltage.

Example 4-3: Consider the circuit of Fig. 4-5 and determine the voltages V_1 and V_2 in terms of the resistances, R_1 and R_2, and the source voltage E_S.

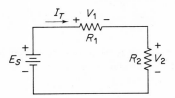

Fig. 4-5. Circuit diagram for Ex. 4-3.

Solution: According to Ohm's law,

$$V_1 = I_T R_1$$

However,

$$I_T = \frac{E_S}{R_T} = \frac{E_S}{R_1 + R_2}$$

Therefore,

$$V_1 = \frac{R_1}{R_1 + R_2} E_S \qquad (4\text{-}5)$$

By the same procedure,

$$V_2 = I_T R_2$$

and

$$I_T = \frac{E_S}{R_1 + R_2}$$

Then

$$V_2 = \frac{R_2}{R_1 + R_2} E_S \qquad (4\text{-}6)$$

Equations (4-5) and (4-6) demonstrate a principle known as the *voltage division formula*. This formula may be generalized to include any number of series resistors and is stated as: *The voltage across a single resistor equals the ratio of that resistor to the total resistance times the applied voltage.* The voltage may be either ac or dc.

Using voltage division method, we eliminate solving for the current. For a numerical example let $R_1 = R_2 = 10$ kΩ and $E_S = 5$ V. Then

$$V_1 = \frac{10 \text{ k}\Omega}{10 \text{ k}\Omega + 10 \text{ k}\Omega} (5 \text{ V}) = 2.5 \text{ V}$$

and

$$V_2 = \frac{10 \text{ k}\Omega}{10 \text{ k}\Omega + 10 \text{ k}\Omega} (5 \text{ V}) = 2.5 \text{ V}$$

Note that because R_1 and R_2 are equal, the voltages across them are also equal.

Inductors in Series. The method for determining the total inductance of Fig. 4-6 will be similar to that used for resistors in series. According to Kirchhoff's voltage law, we may write the following equation for Fig. 4-6:

$$v_T = v_1 + v_2 + v_3 \tag{4-7}$$

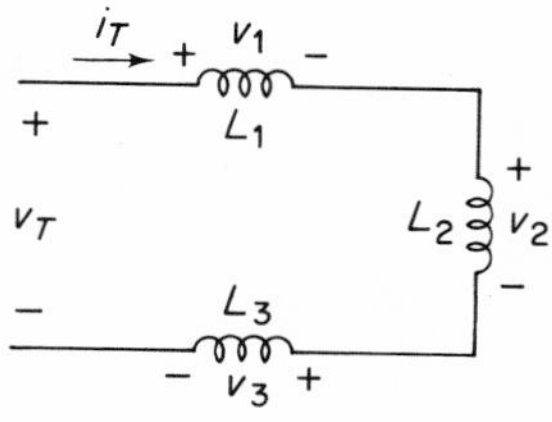

Fig. 4-6. Three inductors connected in series.

The current flowing in the circuit of Fig. 4-6 is i_T. However, in the discussion on inductances in Chapter 3 we saw that the voltage across an inductor was proportional to the rate of change of current and not the current itself. The rate of change of the current flowing in the circuit of Fig. 4-6 is di_T/dt. Dividing both sides of Eq. (4.7) by di_T/dt we obtain the following expression:

$$\frac{v_T}{di_T/dt} = \frac{v_1}{di_T/dt} + \frac{v_2}{di_T/dt} + \frac{v_3}{di_T/dt} \tag{4-8}$$

Let us analyze both sides of Eq. (4-8) to determine how inductors in series combine. The left-hand side of Eq. (4-8) is the total voltage divided by the rate of change of current. This term gives the total inductance, L_T. Each term on the right-hand side of Eq. (4-8) gives the value of an individual inductance:

$$\frac{v_1}{di_T/dt} = L_1, \quad \frac{v_2}{di_T/dt} = L_2, \quad \frac{v_3}{di_T/dt} = L_3$$

Therefore, the total inductance of Fig. 4-6 is

$$L_T = L_1 + L_2 + L_3 \tag{4-9}$$

In other words, Eq. (4-9) states that when inductors are connected in series the total inductance is the sum of the individual inductances. As in the case of resistors, this rule may be generalized to include any number of inductors in series.

If there are two or more equal inductors in series, the total inductance may be found by

$$L_T = NL \tag{4-10}$$

where N is the number of equal inductors and L the value of one of the equal inductors.

Now with the theory as a background, let us proceed to solve some problems.

Example 4-4: Consider the circuit diagrams of Fig. 4-7 and determine the total inductance between terminals *ab*.

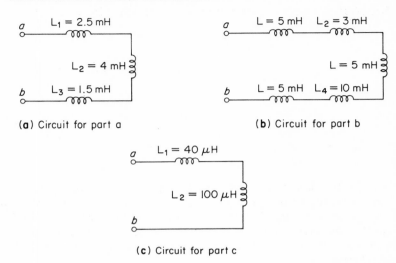

(**a**) Circuit for part a　　　　　　　　　　　(**b**) Circuit for part b

(**c**) Circuit for part c

Fig. 4-7. Circuit diagrams for Ex. 4-4.

Solution:　Circuit a:

$$L_T = L_1 + L_2 + L_3$$
$$L_T = 2.5\,\text{mH} + 4\,\text{mH} + 1.5\,\text{mH}$$
$$L_T = 8\,\text{mH}$$

Circuit b:

$$L_T = NL + L_2 + L_4$$
$$L_T = (3) \cdot (5\,\text{mH}) + 3\,\text{mH} + 10\,\text{mH}$$
$$L_T = 28\,\text{mH}$$

Circuit c:

$$L_T = L_1 + L_2$$
$$L_T = 40\,\mu\text{H} + 100\,\mu\text{H}$$
$$L_T = 140\,\mu\text{H}$$

The total inductance between terminals *ab* is an equivalent inductance. That is, it is capable of performing the same functions as the combination of series inductors. The energy stored in the magnetic field of an equivalent inductance is equal to the sum of the energy stored by each series inductor. The voltage across an equivalent inductance is equal to the sum of the voltages across the series combination.

Example 4-5: Consider the circuit of Fig. 4-8 and determine the total inductance when the switch is in position A and when it is in position B.

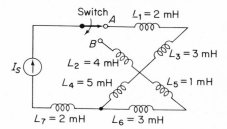

Fig. 4-8. Circuit diagram for Ex. 4-5.

Solution: With the switch in position A the total inductance is

$$L_T = L_1 + L_3 + L_4 + L_7$$
$$L_T = 2\,\text{mH} + 3\,\text{mH} + 5\,\text{mH} + 2\,\text{mH} = 12\,\text{mH}$$

With the switch in position B the total inductance seen by the current source is

$$L_T = L_2 + L_5 + L_6 + L_7$$
$$L_T = 4\,\text{mH} + 1\,\text{mH} + 3\,\text{mH} + 2\,\text{mH} = 10\,\text{mH}$$

Thus only the inductances that are connected together to form a closed path are involved in the calculation of the total inductance.

Example 4-6: In the circuit of Fig. 4-9, the current source supplies energy to the inductors connected in series. Show that the energy stored by a total equivalent inductance equals the energy stored by the series combination.

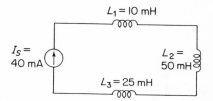

Fig. 4-9. Circuit diagram for Ex. 4-6.

Solution: The total inductance is

$$L_T = L_1 + L_2 + L_3$$
$$L_T = 10\,\text{mH} + 50\,\text{mH} + 25\,\text{mH}$$
$$L_T = 85\,\text{mH}$$

The energy stored by this total inductance is

$$W_M = \tfrac{1}{2} L_T I_S^2$$
$$W_M = \tfrac{1}{2}(85\,\text{mH})(40\,\text{mA})^2$$
$$W_M = 68\,\mu\text{J}$$

The energy stored by the series combination of inductances is

$$W_M = \tfrac{1}{2}L_1 I_S^2 + \tfrac{1}{2}L_2 I_S^2 + \tfrac{1}{2}L_3 I_S^2$$
$$W_M = \tfrac{1}{2}(10 \text{ mH})(40 \text{ mA})^2 + \tfrac{1}{2}(50 \text{ mH})(40 \text{ mA})^2 + \tfrac{1}{2}(25 \text{ mH})(40 \text{ mA})^2$$
$$W_M = 8 \; \mu\text{J} + 40 \; \mu\text{J} + 20 \; \mu\text{J}$$
$$W_M = 68 \; \mu\text{J}$$

Thus we see that the energy stored by an equivalent inductance does equal the energy stored by the series combination.

Capacitors in Series.   In a manner similar to that used for resistors and inductors, let us apply Kirchhoff's voltage law to Fig. 4-10 as the initial step in determining the total capacitance.   Using Kirchhoff's voltage law, we obtain

$$v_T = v_1 + v_2 + v_3 \tag{4-11}$$

In the discussion on capacitances in Chapter 3 we showed that the current "through" a capacitor was proportional to the rate of change of voltage

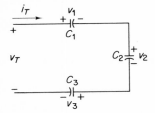

Fig. 4-10. Three capacitors connected in series.

across the capacitor.   Therefore, it is necessary to express both sides of Eq. (4-11) as a rate of change of voltage.

$$\frac{dv_T}{dt} = \frac{d}{dt}(v_1 + v_2 + v_3) \tag{4-12}$$

or

$$\frac{dv_T}{dt} = \frac{dv_1}{dt} + \frac{dv_2}{dt} + \frac{dv_3}{dt} \tag{4-13}$$

Since the current flowing through the capacitor is given by the equation $i_C = C(dv_C/dt)$, we may rewrite Eq. (4-13) as

$$\frac{i_T}{C_T} = \frac{i_T}{C_1} + \frac{i_T}{C_2} + \frac{i_T}{C_3} \tag{4-14}$$

Dividing both sides of Eq. (4-14) by i_T yields

$$\frac{1}{C_T} = \frac{1}{C_1} + \frac{1}{C_2} + \frac{1}{C_3} \tag{4-15}$$

Note that the total capacitance for two or more capacitors in series is calculated quite differently from that of series resistors and series inductors. Equation (4-15) states that the reciprocal of the total capacitance is equal to the sum of the reciprocals of the individual capacitors.

In the case when only two capacitors are connected in series,

$$\frac{1}{C_T} = \frac{1}{C_1} + \frac{1}{C_2}$$

Solving the above expression for C_T, the equivalent capacitance, we obtain

$$C_T = \frac{C_1 C_2}{C_1 + C_2} \tag{4-16}$$

Equation (4-16) is known as the *product over the sum rule*. It may be used *only* when *two* capacitors are in series. If more than two capacitors are connected in series, the total capacitance must be found in a manner similar to that of Eq. (4-15).

In the special case when two or more series capacitors are equal, the following shortcut may be used:

$$C_T = \frac{C}{N} \tag{4-17}$$

where C is the value of one of the equal capacitors in farads and N is the number of equal capacitors.

Let us now apply the formulas concerning capacitances to the following examples.

Example 4-7: Calculate the total capacitance between terminals *ab* for each of the circuits of Fig. 4-11.

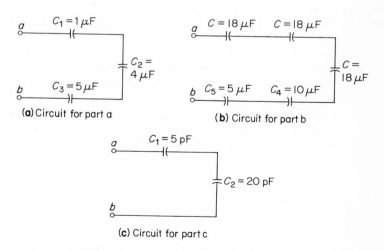

(a) Circuit for part a

(b) Circuit for part b

(c) Circuit for part c

Fig. 4-11. Circuit diagrams for Ex. 4-7.

Solution: Circuit a:

$$\frac{1}{C_T} = \frac{1}{C_1} + \frac{1}{C_2} + \frac{1}{C_3}, \qquad \frac{1}{C_T} = \frac{1}{1\ \mu F} + \frac{1}{4\ \mu F} + \frac{1}{5\ \mu F}$$

$$\frac{1}{C_T} = 1.0 \times 10^{+6} + 0.25 \times 10^{+6} + 0.2 \times 10^{+6} = 1.45 \times 10^{+6}$$

Therefore,

$$C_T = \frac{1}{1.45 \times 10^{+6}} = 0.69\ \mu F$$

Circuit b: The equivalent capacitance of the three equal capacitors is

$$C_{eq} = \frac{C}{N} = \frac{18\ \mu F}{3} = 6\ \mu F$$

The total capacitance can then be found by

$$\frac{1}{C_T} = \frac{1}{6\ \mu F} + \frac{1}{10\ \mu F} + \frac{1}{5\ \mu F}$$

$$\frac{1}{C_T} = 0.167 \times 10^{+6} + 0.1 \times 10^{+6} + 0.2 \times 10^{+6} = 0.467 \times 10^{+6}$$

Therefore,

$$C_T = \frac{1}{0.467 \times 10^{+6}} = 2.14\ \mu F$$

Circuit c: Using the product over the sum rule may be the easiest method of solving for the total capacitance:

$$C_T = \frac{C_1 C_2}{C_1 + C_2} = \frac{(5\ pF) \cdot (20\ pF)}{5\ pF + 20\ pF} = 4\ pF$$

Now that we have solved for the total capacitance for a series circuit, let us investigate some of the properties of this circuit. Note that the total capacitance of a series combination is less than the smallest value of capacitance in the circuit. This is because in determining the total capacitance we must use the reciprocals. The total capacitance is the equivalent capacitance between the terminals *ab*. An equivalent capacitance is capable of storing the same amount of energy in the same time as the series combination. The voltage across the equivalent capacitance is equal to the sum of voltages across the individual capacitances.

Example 4-8: In the circuit of Fig. 4-12, the voltage across each capacitor is specified. Show that the energy stored in the total capacitance is equal to the energy stored in the series combination.

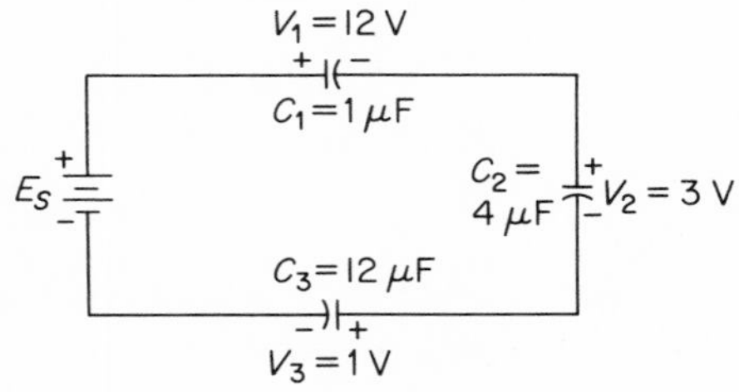

Fig. 4-12. Circuit diagram of Ex. 4-8.

Solution: The sum of the energies stored by the individual capacitors is

$$W_E = \tfrac{1}{2} C_1 V_1^2 + \tfrac{1}{2} C_2 V_2^2 + \tfrac{1}{2} C_3 V_3^2$$
$$W_E = \tfrac{1}{2}(1\ \mu\text{F}) \cdot (12\ \text{V})^2 + \tfrac{1}{2}(4\ \mu\text{F}) \cdot (3\ \text{V})^2 + \tfrac{1}{2}\ (12\ \mu\text{F}) \cdot (1\ \text{V})^2$$
$$W_E = 72\ \mu\text{J} + 18\ \mu\text{J} + 6\ \mu\text{J}$$
$$W_E = 96\ \mu\text{J}$$

The total capacitance may be found by

$$\frac{1}{C_T} = \frac{1}{C_1} + \frac{1}{C_2} + \frac{1}{C_3}$$

$$\frac{1}{C_T} = \frac{1}{1\ \mu\text{F}} + \frac{1}{4\ \mu\text{F}} + \frac{1}{12\ \mu\text{F}}$$

$$\frac{1}{C_T} = 1 \times 10^{+6} + 0.25 \times 10^{+6} + 0.083 \times 10^{+6} = 1.333 \times 10^{+6}$$

Therefore,

$$C_T = \frac{1}{1.333 \times 10^{+6}} = 0.75\ \mu\text{F}$$

According to Kirchhoff's voltage law,

$$E_S = V_1 + V_2 + V_3$$

$$E_S = 12\ \text{V} + 3\ \text{V} + 1\ \text{V} = 16\ \text{V}$$

The energy stored by the total equivalent capacitance is

$$W_E = \tfrac{1}{2} C_T E_S^2 = \tfrac{1}{2}(0.75\ \mu\text{F}) \cdot (16\ \text{V})^2 = 96\ \mu\text{J}$$

Thus the energy stored by the equivalent capacitance does equal the energy stored by the series combination.

Example 4-9: For the circuit of Fig. 4-12, calculate the charge on each capacitor and the total charge in microcoulombs (μC).

Solution: From Eq. (3-12)

$$C = \frac{Q}{V}$$

or

$$Q = CV$$

Then

$$Q_1 = C_1 V_1 = (1\ \mu\text{F}) \cdot (12\ \text{V}) = 12\ \mu\text{C}$$

$$Q_2 = C_2 V_2 = (4\ \mu\text{F}) \cdot (3\ \text{V}) = 12\ \mu\text{C}$$

$$Q_3 = C_3 V_3 = (12\ \mu\text{F}) \cdot (1\ \text{V}) = 12\ \mu\text{C}$$

and

$$Q_T = C_T E_S = (0.75\ \mu\text{F}) \cdot (16\ \text{V}) = 12\ \mu\text{C}$$

From the above results, we conclude that when capacitors are connected in series, the charge on each capacitance and the total charge are equal. In effect, this is saying that the same current flows through each element in a series circuit.

4-2. PARALLEL CIRCUITS

A second basic type of electric circuit is one in which the terminals of the elements are connected to common nodes.   This type of circuit is defined as a parallel circuit.   In general, *two or more elements are considered in parallel if they have the same voltage across them.*

In this section we shall deal with three special parallel circuits: (1) resistors in parallel, (2) inductors in parallel, and (3) capacitors in parallel.   As in the case of series circuits, the discussion of different elements in parallel such as a resistor and an inductor, a resistor and a capacitor, and a resistor, inductor, and capacitor is left to Chapter 7.

Resistors in Parallel.   Figure 4-13 shows three resistors connected in parallel.   While in series circuits, Kirchhoff's voltage law is the initial step in the procedure for determining total resistance, for parallel circuits

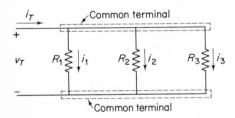

Fig. 4-13. Three resistors connected in parallel.

Kirchhoff's current law is used.   Although Fig. 4-13 contains only three resistors in parallel, the principle can be generalized to any number of parallel resistors.   Applying Kirchhoff's current law to the circuit of Figure 4-13, we obtain

$$i_T = i_1 + i_2 + i_3 \tag{4-18}$$

Since v_T is the voltage across each of the parallel resistors, we may divide Eq. (4-18) by v_T:

$$\frac{i_T}{v_T} = \frac{i_1}{v_T} + \frac{i_2}{v_T} + \frac{i_3}{v_T} \tag{4-19}$$

Analyzing both sides of Eq. (4-19) we see that the terms are current divided by voltage.   Current divided by voltage is the reciprocal of resistance.   The left-hand side of Eq. (4-19) is the total current divided by the total voltage.   This ratio yields the reciprocal of the total resistance ($i_T/v_T = 1/R_T$).   The terms on the right-hand side of Eq. (4-19) yield

$$\frac{i_1}{v_T} = \frac{1}{R_1}, \quad \frac{i_2}{v_T} = \frac{1}{R_2}, \quad \frac{i_3}{v_T} = \frac{1}{R_3}$$

Thus

$$\frac{1}{R_T} = \frac{1}{R_1} + \frac{1}{R_2} + \frac{1}{R_3} \qquad (4\text{-}20)$$

Equation (4-20) states that the reciprocal of the total resistance is equal to the sum of the reciprocals of each resistance in parallel.

Comparing Eqs. (4-15) and (4-20) we note that resistors in parallel combine in the same manner as capacitors in series. Therefore, the same concepts learned in the discussion of capacitors in series may now be applied to resistors in parallel. For any two resistors in parallel we may use the product over the sum rule to determine the total resistance:

$$R_T = \frac{R_1 R_2}{R_1 + R_2} \qquad (4\text{-}21)$$

where R_T is the total equivalent resitance for R_1 in parallel with R_2.

We may also apply the following concept. If two or more parallel resistors are equal, the total resistance is equal to the value of one of the resistors divided by the number of equal resistors:

$$R_T = \frac{R}{N} \qquad (4\text{-}22)$$

where R is the value of one of the parallel resistors in ohms and N is the number of equal resistors.

When determining the equivalent resistance of a parallel circuit, it becomes necessary to use the reciprocal of resistance as shown by Eq. (4-20). The reciprocal of resistance is defined as *conductance:*

$$G = \frac{1}{R} \quad \text{or} \quad R = \frac{1}{G} \qquad (4\text{-}23)$$

The letter quantity symbol G is used to represent conductance. The units of conductance are *mhos.* A conventional unit symbol for the mho is ℧.

In circuits that have resistors in parallel, it may be more convenient to work with conductances rather than resistances. Equation (4-20) expressed in terms of conductances is

$$G_T = G_1 + G_2 + G_3 \qquad (4\text{-}24)$$

In general, *the total conductance of a parallel circuit equals the sum of the conductances of the individual branches.*

Let us now apply the above rules and formulas to circuits.

Example 4-10: Calculate the total resistance between terminals *ab* for each of the circuits in Fig. 4-14.

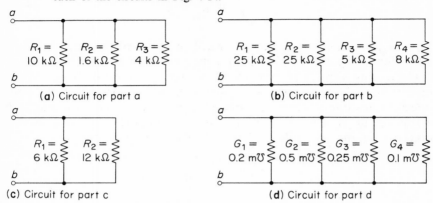

Fig. 4-14. Circuit diagrams for Ex. 4-10.

Solution: Circuit a:

$$\frac{1}{R_T} = \frac{1}{10 \text{ k}\Omega} + \frac{1}{1.6 \text{ k}\Omega} + \frac{1}{4 \text{ k}\Omega}$$

$$\frac{1}{R_T} = 0.1 \text{ m}\mho + 0.625 \text{ m}\mho + 0.25 \text{ m}\mho = 0.975 \text{ m}\mho$$

Therefore,

$$R_T = \frac{1}{0.925 \text{ m}\mho} = 1.08 \text{ k}\Omega$$

Circuit b:

$$\frac{1}{R_T} = \frac{1}{25 \text{ k}\Omega} + \frac{1}{25 \text{ k}\Omega} + \frac{1}{5 \text{ k}\Omega} + \frac{1}{8 \text{ k}\Omega}$$

$$\frac{1}{R_T} = 0.04 \text{ m}\mho + 0.04 \text{ m}\mho + 0.2 \text{ m}\mho + 0.125 \text{ m}\mho = 0.405 \text{ m}\mho$$

Therefore,

$$R_T = \frac{1}{0.405 \text{ m}\mho} = 2.47 \text{ k}\Omega$$

Circuit c: Since there are only two resistors in parallel, use the product over the sum rule:

$$R_T = \frac{R_1 R_2}{R_1 + R_2} = \frac{(6 \text{ k}\Omega)(12 \text{ k}\Omega)}{6 \text{ k}\Omega + 12 \text{ k}\Omega} = 4 \text{ k}\Omega$$

Circuit d: First find the total conductance:

$$G_T = G_1 + G_2 + G_3 + G_4$$

$$G_T = 0.2 \text{ m}\mho + 0.5 \text{ m}\mho + 0.25 \text{ m}\mho + 0.1 \text{ m}\mho$$

$$G_T = 1.05 \text{ m}\mho$$

Then

$$R_T = \frac{1}{G_T} = \frac{1}{1.05\,\text{m}\mho} = 952\,\Omega$$

Example 4-11: For the parallel resistive circuit of Fig. 4-15, calculate (a) the current through each resistor and the total current and (b) the power dissipated by each resistor and the total power.

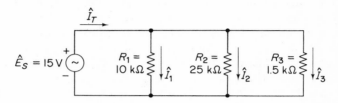

Fig. 4-15. Circuit diagram for Ex. 4-11.

Solution: (a) Since the three resistors are connected in parallel with the voltage source, then

$$\hat{E}_S = \hat{V}_1 = \hat{V}_2 = \hat{V}_3 = 15\,\text{V}$$

Using Ohm's law,

$$\hat{I}_1 = \frac{\hat{V}_1}{R_1} = \frac{15\,\text{V}}{10\,\text{k}\Omega} = 1.5\,\text{mA}$$

$$\hat{I}_2 = \frac{\hat{V}_2}{R_2} = \frac{15\,\text{V}}{25\,\text{k}\Omega} = 0.6\,\text{mA}$$

$$\hat{I}_3 = \frac{\hat{V}_3}{R_3} = \frac{15\,\text{V}}{1.5\,\text{k}\Omega} = 10\,\text{mA}$$

Applying Kirchhoff's current law, we may determine the total current:

$$\hat{I}_T = \hat{I}_1 + \hat{I}_2 + \hat{I}_3$$

$$\hat{I}_T = 1.5\,\text{mA} + 0.6\,\text{mA} + 10\,\text{mA}$$

$$\hat{I}_T = 12.1\,\text{mA}$$

(b) Using the formula $P = \hat{I}^2 R$,

$$P_1 = \hat{I}_1^2 R_1 = (1.5\,\text{mA})^2\,(10\,\text{k}\Omega) = 22.5\,\text{mW}$$

$$P_2 = \hat{I}_2^2 R_2 = (0.6\,\text{mA})^2\,(25\,\text{k}\Omega) = 9.0\,\text{mW}$$

$$P_3 = \hat{I}_3^2 R_3 = (10\,\text{mA})^2\,(1.5\,\text{k}\Omega) = 150\,\text{mW}$$

The total power may be found by

$$P_T = \hat{E}_S \hat{I}_T = (15\,\text{V})\,(12.1\,\text{mA}) = 181.5\,\text{mW}$$

or

$$P_T = P_1 + P_2 + P_3$$

$$P_T = 22.5\,\text{mW} + 9.0\,\text{mW} + 150\,\text{mW}$$

$$P_T = 181.5\,\text{mW}$$

Therefore, knowing the voltage across resistors connected in parallel and the value of the resistor, we may apply Ohm's law to find the branch currents. The total current is determined by Kirchhoff's current law. In part (b) we found the power dissipated by using one of the three formulas for calculating power. The reader should check the above answers by applying the remaining two formulas for calculating power. Note that the total power delivered by the energy source equals the sum of the power dissipated by the resistors. This is the same result we saw when resistors are connected in series. Thus power delivered equals power dissipated regardless of the circuit.

Example 4-12: For the circuit of Fig. 4-16, calculate $\hat{I}_S$.

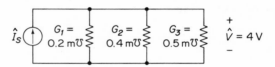

Fig. 4-16. Circuit diagram for Ex. 4-12.

Solution: Begin by determining the total conductance of the circuit:

$$G_T = G_1 + G_2 + G_3$$

$$G_T = 0.2\,\text{m}\mho + 0.4\,\text{m}\mho + 0.5\,\text{m}\mho$$

$$G_T = 1.1\,\text{m}\mho$$

From Ohm's law

$$\hat{V} = \hat{I}R$$

and from Eq. (4-23)

$$R = \frac{1}{G}$$

we obtain

$$\hat{V} = \hat{I}\frac{1}{G}$$

or

$$\hat{I} = \hat{V}G \tag{4-25}$$

Therefore,

$$\hat{I}_S = (4\,\text{V})(1.1\,\text{m}\mho) = 4.4\,\text{mA}$$

According to Eq. (4-25), if the conductance value and the voltage across it are known, the current may be determined. This problem may also have been solved by first determining the current through each branch and then applying Kirchhoff's current law.

Example 4-13: Consider the circuit of Fig. 4-17 and determine I_1 and I_2 in terms of the total current and the resistors.

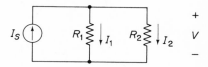

Fig. 4-17. Circuit diagram for Ex. 4-13.

Solution: Since R_1 is in parallel with R_2,

$$V_1 = V_2 = V \quad \text{and} \quad V = I_S R_T$$

where

$$R_T = \frac{R_1 R_2}{R_1 + R_2}$$

Applying Kirchhoff's current law and solving for I_1, we obtain

$$I_1 = I_S - I_2$$

$$I_1 = I_S - \frac{V}{R_2}$$

$$I_1 = I_S - \frac{I_S R_T}{R_2}$$

$$I_1 = \left(1 - \frac{R_T}{R_2}\right) I_S$$

or

$$I_1 = \left(\frac{R_2}{R_1 + R_2}\right) I_S \qquad (4\text{-}26)$$

Using the same procedure to solve for I_2,

$$I_2 = I_S - I_1$$

$$I_2 = I_S - \frac{V}{R_1}$$

$$I_2 = I_S - \frac{I_S R_T}{R_1}$$

$$I_2 = \left(1 - \frac{R_T}{R_1}\right) I_S$$

or

$$I_2 = \left(\frac{R_1}{R_1 + R_2}\right) I_S \qquad (4\text{-}27)$$

Equations (4-26) and (4-27) are known as the *current division formula.*
This formula may be stated as: *The current in one branch of a two branch
parallel circuit is equal to the value of resistance in the other branch divided
by the sum of the two resistances times the total current entering the node.*

Refer to the circuit of Fig. 4-17 and let $I_S = 6$ mA, $R_1 = 6$ kΩ, and
$R_2 = 3$ kΩ. Then

$$I_1 = \frac{R_2}{R_1 + R_2} I_S$$

$$I_1 = \frac{3 \text{ k}\Omega}{6 \text{ k}\Omega + 3 \text{ k}\Omega} \times 6 \text{ mA} = 2 \text{ mA}$$

$$I_2 = \frac{R_1}{R_1 + R_2} I_S$$

$$I_2 = \frac{6 \text{ k}\Omega}{6 \text{ k}\Omega + 3 \text{ k}\Omega} \times 6 \text{ mA} = 4 \text{ mA}$$

Thus the current division formula eliminates the need of solving first for
the total resistance and then applying Ohm's law. In the numerical problem, since R_1 is twice R_2, the current through it is one half of that through R_2.

Inductors in Parallel. Applying Kirchhoff's current law to Fig. 4-18, we
can determine how inductors in parallel combine:

$$i_T = i_1 + i_2 + i_3 \tag{4-28}$$

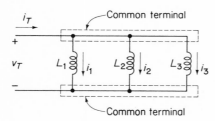

Fig. 4-18. Inductors connected in parallel.

However, in the discussion on inductance in Chapter 3 we expressed the
voltage across an inductor in terms of the rate of change of current. Therefore, it is necessary to express both sides of Eq. (4-28) as a rate of change
of current:

$$\frac{di_T}{dt} = \frac{d}{dt}(i_1 + i_2 + i_3) \tag{4-29}$$

or

$$\frac{di_T}{dt} = \frac{di_1}{dt} + \frac{di_2}{dt} + \frac{di_3}{dt} \tag{4-30}$$

Since the voltage across an inductor is given by $v_L = L(di_L/dt)$ and also since v_T is the voltage across the parallel inductance, we may rewrite Eq. (4-30) as

$$\frac{v_T}{L_T} = \frac{v_T}{L_1} + \frac{v_T}{L_2} + \frac{v_T}{L_3} \qquad (4\text{-}31)$$

Dividing both sides of Eq. (4-31) by v_T gives us a form for combining inductors in parallel:

$$\frac{1}{L_T} = \frac{1}{L_1} + \frac{1}{L_2} + \frac{1}{L_3} \qquad (4\text{-}32)$$

In other words, Eq. (4-32) states that the reciprocal of the total inductance is equal to the sum of the reciprocals of the individual inductances connected in parallel.

As in the case of two resistors in parallel or two capacitors in series, we may use the product over the sum rule to determine the total inductance for two inductors in parallel:

$$L_T = \frac{L_1 L_2}{L_1 + L_2} \qquad (4\text{-}33)$$

where L_T is the total inductance and L_1 and L_2 are the parallel inductors.

If two or more parallel inductors are equal, the total inductance may be determined by dividing the value of one of the inductors by the number of equal inductors:

$$L_T = \frac{L}{N} \qquad (4\text{-}34)$$

where L is the value of one of the equal inductors and N is the number of equal inductors.

Example 4-14: Calculate the total inductance between terminals *ab* for each of the circuits shown in Fig. 4-19.

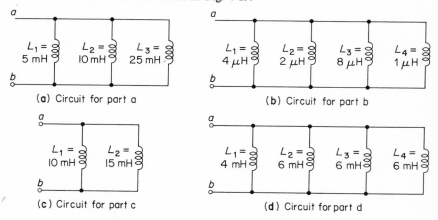

(a) Circuit for part a

(b) Circuit for part b

(c) Circuit for part c

(d) Circuit for part d

Fig. 4-19. Circuit diagrams for Ex. 4-14.

Solution: Circuit a:

$$\frac{1}{L_T} = \frac{1}{L_1} + \frac{1}{L_2} + \frac{1}{L_3}$$

$$\frac{1}{L_T} = \frac{1}{5\,\text{mH}} + \frac{1}{10\,\text{mH}} + \frac{1}{25\,\text{mH}}$$

$$\frac{1}{L_T} = 0.2 \times 10^3 + 0.1 \times 10^3 + 0.04 \times 10^3 = 0.34 \times 10^3$$

Therefore,

$$L_T = \frac{1}{0.34 \times 10^3} = 2.94\,\text{mH}$$

Circuit b:

$$\frac{1}{L_T} = \frac{1}{L_1} + \frac{1}{L_2} + \frac{1}{L_3} + \frac{1}{L_4}$$

$$\frac{1}{L_T} = \frac{1}{4\,\mu\text{H}} + \frac{1}{2\,\mu\text{H}} + \frac{1}{8\,\mu\text{H}} + \frac{1}{1\,\mu\text{H}}$$

$$\frac{1}{L_T} = 0.25 \times 10^6 + 0.5 \times 10^6 + 0.125 \times 10^6 + 1 \times 10^6 = 1.875 \times 10^6$$

Therefore,

$$L_T = \frac{1}{1.875 \times 10^6} = 0.533\,\mu\text{H}$$

Circuit c: Using the product over the sum rule,

$$L_T = \frac{L_1 L_2}{L_1 + L_2}$$

$$L_T = \frac{(10\,\text{mH})\,(15\,\text{mH})}{10\,\text{mH} + 15\,\text{mH}} = 6\,\text{mH}$$

Circuit d: Since three of the inductors are equal, an equivalent inductance is

$$L_{\text{eq}} = \frac{L}{N} = \frac{6\,\text{mH}}{3} = 2\,\text{mH}$$

The total inductance is then

$$L_T = \frac{L_1\,L_{\text{eq}}}{L_1 + L_{\text{eq}}}$$

$$L_T = \frac{(4\,\text{mH})\,(2\,\text{mH})}{4\,\text{mH} + 2\,\text{mH}} = 1.33\,\text{mH}$$

A total or equivalent inductance is capable of storing the same amount of energy as the sum of the inductors connected in parallel and also draws an equal amount of current from an energy source.

Example 4-15: For the circuit of Fig. 4-20, show that the energy stored by a total equivalent inductance is the sum of the energy stored by the individual inductors.

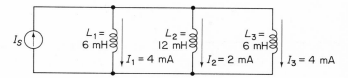

Fig. 4-20. Circuit diagram for Ex. 4-15.

Solution: The energy stored by each inductance is given by Eq. (3-11):

$$W_{M1} = \tfrac{1}{2}L_1 I_1^2 = \tfrac{1}{2}(6\,\text{mH})(4\,\text{mA})^2 = 48\,\text{nJ}$$

$$W_{M2} = \tfrac{1}{2}L_2 I_2^2 = \tfrac{1}{2}(12\,\text{mH})(2\,\text{mA})^2 = 24\,\text{nJ}$$

$$W_{M3} = \tfrac{1}{2}L_3 I_3^2 = \tfrac{1}{2}(6\,\text{mH})(4\,\text{mA})^2 = 48\,\text{nJ}$$

Thus the total energy is

$$W_T = W_{M1} + W_{M2} + W_{M3}$$

$$W_T = 48\,\text{nJ} + 24\,\text{nJ} + 48\,\text{nJ} = 120\,\text{nJ}$$

The total equivalent inductance is

$$\frac{1}{L_T} = \frac{1}{L_1} + \frac{1}{L_2} + \frac{1}{L_3}$$

$$\frac{1}{L_T} = \frac{1}{6\,\text{mH}} + \frac{1}{12\,\text{mH}} + \frac{1}{6\,\text{mH}} = 0.415 \times 10^3$$

Therefore,

$$L_T = \frac{1}{0.415 \times 10^3} = 2.4\,\text{mH}$$

According to Kirchhoff's current law,

$$I_S = I_1 + I_2 + I_3$$

$$I_S = 4\,\text{mA} + 2\,\text{mA} + 4\,\text{mA} = 10\,\text{mA}$$

Thus the total energy is

$$W_T = \tfrac{1}{2}L_T I_T^2$$

$$W_T = \tfrac{1}{2}(2.4\,\text{mH})(10\,\text{mA})^2 = 120\,\text{nJ}$$

Thus an equivalent inductance does store the same amount of energy as the individual inductors.

Capacitors in Parallel. Fig. 4-21 shows three capacitors connected in parallel. Applying Kirchhoff's current law to this circuit we obtain.

$$i_T = i_1 + i_2 + i_3 \qquad\qquad (4\text{-}35)$$

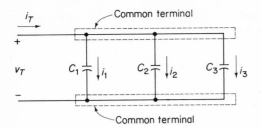

Fig. 4-21. Three capacitors connected in parallel.

From our knowledge of the voltage-current relationship of capacitance as developed in Chapter 3, we know that $i_C = C(dv_C/dt)$. Using the fact that v_T is the voltage across the parallel capacitors and knowing the voltage-current relationship for capacitance we may rewrite Eq. (4-25) as

$$C_T \frac{dv_T}{dt} = C_1 \frac{dv_T}{dt} + C_2 \frac{dv_T}{dt} + C_3 \frac{dv_T}{dt} \tag{4-36}$$

Equation (4-36) reduces to

$$C_T = C_1 + C_2 + C_3 \tag{4-37}$$

In other words, Eq. (4-37) states that when capacitors are connected in parallel, the total capacitance is equal to the sum of the individual capacitors. Unlike parallel resistors or parallel inductors, which are added only by their reciprocals, parallel capacitors are combined like series resistors or series inductors.

Example 4-16: Calculate the total capacitance between terminals *ab* for each of the circuits of Fig. 4-22.

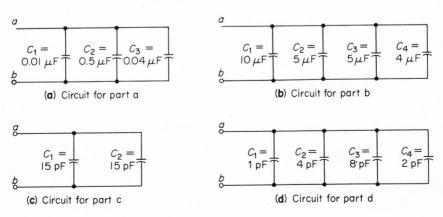

Fig. 4-22. Circuit diagrams for Ex. 4-16.

Solution: Circuit a:

$$C_T = C_1 + C_2 + C_3$$
$$C_T = 0.01\ \mu\text{F} + 0.5\ \mu\text{F} + 0.04\ \mu\text{F}$$
$$C_T = 0.55\ \mu\text{F}$$

Circuit b:

$$C_T = C_1 + C_2 + C_3 + C_4$$
$$C_T = 10\ \mu\text{F} + 5\ \mu\text{F} + 5\ \mu\text{F} + 4\ \mu\text{F}$$
$$C_T = 24\ \mu\text{F}$$

Circuit c:

$$C_T = C_1 + C_2$$
$$C_T = 15\ \text{pF} + 15\ \text{pF}$$
$$C_T = 30\ \text{pF}$$

Circuit d:

$$C_T = C_1 + C_2 + C_3 + C_4$$
$$C_T = 1\ \text{pF} + 4\ \text{pF} + 8\ \text{pF} + 2\ \text{pF}$$
$$C_T = 15\ \text{pF}$$

These total capacitances are an equivalent capacitance of the parallel elements. The following examples demonstrate this equivalence.

Example 4-17: For the circuit of Fig. 4-23, show that the sum of the energies stored by each capacitor equals the energy stored by an equivalent capacitor.

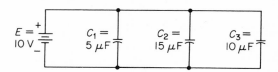

Fig. 4-23. Circuit diagram for Ex. 4-17.

Solution: Since we are analyzing a parallel circuit,

$$E = V_1 = V_2 = V_3 = 10\ \text{V}$$

According to Eq. (3-15),

$$W_{E1} = \tfrac{1}{2} C_1 V_1^2 = \tfrac{1}{2}(\ 5\ \mu\text{F})(10\ \text{V})^2 = 250\ \mu\text{J}$$
$$W_{E2} = \tfrac{1}{2} C_2 V_2^2 = \tfrac{1}{2}(15\ \mu\text{F})(10\ \text{V})^2 = 750\ \mu\text{J}$$
$$W_{E3} = \tfrac{1}{2} C_3 V_3^2 = \tfrac{1}{2}(10\ \mu\text{F})(10\ \text{V})^2 = 500\ \mu\text{J}$$

Therefore, the sum of the energies is

$$W_T = W_{E1} + W_{E2} + W_{E3}$$

$$W_T = 250\ \mu\text{J} + 750\ \mu\text{J} + 500\ \mu\text{J}$$

$$W_T = 1500\ \mu\text{J or 1.5 mJ}$$

A total equivalent capacitance of the circuit is

$$C_T = C_1 + C_2 + C_3$$

$$C_T = 5\ \mu\text{F} + 15\ \mu\text{F} + 10\ \mu\text{F}$$

$$C_T = 30\ \mu\text{F}$$

The energy stored by a total equivalent capacitance is

$$W_T = \tfrac{1}{2} C_T E^2 = \tfrac{1}{2}(30\ \mu\text{F})(10\ \text{V})^2 = 1.5\ \text{mJ}$$

Thus the energy stored by an equivalent capacitor equals the sum of the energies stored by individual capacitors.

Example 4-18: Again consider the circuit diagram of Fig. 4-23 and calculate the charge stored by each capacitor and the total charge.

Solution: Since we are analyzing a parallel circuit,

$$E_S = V_1 = V_2 = V_3 = 10\ \text{V}$$

According to Eq. (3-12),

$$Q_1 = C_1 V_1 = (5\ \ \mu\text{F})(10\ \text{V}) = \ \ 50\ \mu\text{C}$$

$$Q_2 = C_2 V_2 = (15\ \mu\text{F})(10\ \text{V}) = 150\ \mu\text{C}$$

$$Q_3 = C_3 V_3 = (10\ \mu\text{F})(10\ \text{V}) = 100\ \mu\text{C}$$

and

$$Q_T = C_T E_S = (30\ \mu\text{F})(10\ \text{V}) = 300\ \mu\text{C}$$

Alternatively, the total charge equals

$$Q_T = Q_1 + Q_2 + Q_3$$

$$Q_T = 50\ \mu\text{C} + 150\ \mu\text{C} + 100\ \mu\text{C} = 300\ \mu\text{C}$$

From the above results, we may conclude that when capacitors are connected in parallel the total charge equals the sum of the charges – in effect this is Kirchhoff's current law. Note how this differs from the series circuit of Example 4-9.

4-3. SERIES-PARALLEL CIRCUITS

The two previous sections dealt with series and parallel circuits as separate entities. However, it is quite conceivable and practical to have a circuit containing *both* series and parallel elements. This type of circuit is a *series-parallel circuit*. The analysis of series-parallel circuits requires a firm understanding of the basic principles of both series and parallel circuits. At this time we shall treat series-parallel combinations containing only one type of passive element.

Resistors in Series-Parallel. The most direct method for determining total resistance of a series-parallel circuit is to obtain an equivalent resistance of each branch independently and then to reduce the remaining circuit to a total resistance. This concept is shown in the following examples.

Example 4-19: For the circuit of Fig. 4-24a, calculate the resistance between terminals *ab*.

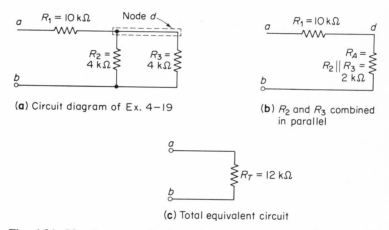

(a) Circuit diagram of Ex. 4−19

(b) R_2 and R_3 combined in parallel

(c) Total equivalent circuit

Fig. 4-24. Step by step reduction of a series-parallel resistive circuit.

Solution: Let resistance R_A equal to the parallel combination of R_2 and R_3 Since the values of R_2 and R_3 are equal, we may calculate R_A by the following:

$$R_A = \frac{4\,\text{k}\Omega}{2} = 2\,\text{k}\Omega$$

The equivalent circuit showing R_A is Fig. 4-24b. The total resistance, R_T, between terminals *ab* is

$$R_T = R_1 + R_A = 10\,\text{k}\Omega + 2\,\text{k}\Omega = 12\,\text{k}\Omega$$

Thus the resistance between terminals *ab* may be represented by a single resistance, R_T, as shown in Fig. 4-24c.

Note that in drawing the equivalent circuit, node *d* was eliminated. Since we are only interested in finding the input resistance between nodes *a* and *b*, the elimination of node *d* is not of importance.

Example 4-20: For the circuit diagram of Fig. 4-25a, calculate the input resistance between terminals *ab*.

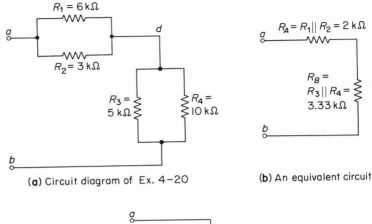

(a) Circuit diagram of Ex. 4-20   (b) An equivalent circuit

(c) Total equivalent resistance

Fig. 4-25. Reduction of a series-parallel circuit.

Solution: Let $R_A = R_1 \parallel R_2$ and $R_B = R_3 \parallel R_4$. Using the product over the sum rule we obtain

$$R_A = \frac{R_1 R_2}{R_1 + R_2} = \frac{(6\,\text{k}\Omega)\,(3\,\text{k}\Omega)}{6\,\text{k}\Omega + 3\,\text{k}\Omega} = 2\,\text{k}\Omega$$

and

$$R_B = \frac{R_3 R_4}{R_3 + R_4} = \frac{(5\,\text{k}\Omega)\,(10\,\text{k}\Omega)}{5\,\text{k}\Omega + 10\,\text{k}\Omega} = 3.33\,\text{k}\Omega$$

An equivalent circuit diagram for Fig. 4-25a is Fig. 4-25b. The total input resistance, R_T, is found by

$$R_T = R_A + R_B = 2\,\text{k}\Omega + 3.33\,\text{k}\Omega = 5.33\,\text{k}\Omega$$

Therefore, the resistance between terminals *ab* may be represented by a single resistance, R_T, as shown in Fig. 4-25c.

Example 4-21: Calculate the resistance between terminals ab for the series-parallel resistive circuit shown in Fig. 4-26a.

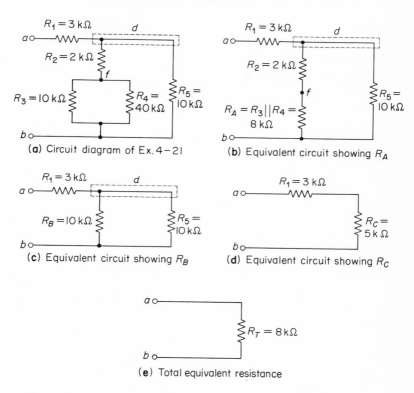

(a) Circuit diagram of Ex. 4-21

(b) Equivalent circuit showing R_A

(c) Equivalent circuit showing R_B

(d) Equivalent circuit showing R_C

(e) Total equivalent resistance

Fig. 4-26. Step by step reduction of a series-parallel resistive circuit.

Solution: Letting $R_A = R_3 \| R_4$ and using the product over the sum rule yields

$$R_A = \frac{R_3 R_4}{R + R_4} = \frac{(10\,k\Omega)\,(40\,k\Omega)}{10\,k\Omega + 40\,k\Omega} = 8\,k\Omega$$

The equivalent circuit is shown in Fig. 4-26b. Referring to Fig. 4-26b, let

$$R_B = R_2 + R_A = 2\,k\Omega + 8\,k\Omega = 10\,k\Omega$$

Figure 4-26c shows the result of combining R_2 with R_A. From the circuit diagram of Fig. 4-26c we see that R_B is in parallel with R_5. Let $R_C = R_B \| R_5$. Since both resistors are equal,

$$R_C = \frac{10\,k\Omega}{2} = 5\,k\Omega$$

The equivalent circuit after combining R_B and R_5 is shown in Fig. 4-26d.

The last step involves only the addition of two resistors in series, resulting in the total input resistance:

$$R_T = R_1 + R_C = 3\,k\Omega + 5\,k\Omega = 8\,k\Omega$$

Therefore, the resistance between terminals *ab* once again may be represented by a single resistance (see Fig. 4-26e).

We may conclude from the previous examples that the resistance between any two terminals can be found by logically reducing the original network. The steps used in reducing a network involve the concepts learned in the sections on series and parallel circuits.

Example 4-22: For the circuit diagram of Fig. 4-27a, calculate
- (a) total resistance, R_T,
- (b) branch current, $\hat{I}_3$,
- (c) voltage, $\hat{V}$,
- (d) total power dissipated, P_T.

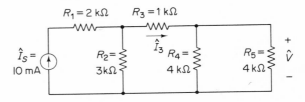

(a) Circuit diagram for Ex. 4-22

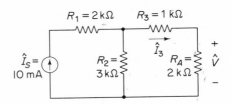

(b) Equivalent circuit for Fig. 4-27a

Fig. 4-27. Circuit diagrams for Ex. 4-22.

Solution: (a)   Let $R_A = R_4 \| R_5$.   Since both resistances are equal,

$$R_A = \frac{R}{N} = \frac{4\,k\Omega}{2} = 2\,k\Omega$$

Figure 4-27b shows that R_3 and R_A are in series.   Let

$$R_B = R_3 + R_A$$
$$R_B = 1\,k\Omega + 2\,k\Omega = 3\,k\Omega$$

R_B is in parallel with R_2. Let R_C be the parallel combination of R_2 and R_B. Since both resistors are equal to 3 kΩ,

$$R_C = \frac{R}{N} = \frac{3 \text{ k}\Omega}{2} = 1.5 \text{ k}\Omega$$

The total resistance is now the series combination of R_1 and R_C, then

$$R_T = R_1 + R_C$$
$$R_T = 2 \text{ k}\Omega + 1.5 \text{ k}\Omega = 3.5 \text{ k}\Omega$$

(b) Using the current division law,

$$\hat{I}_3 = \frac{R_2}{R_2 + R_3 + R_A} \times \hat{I}_S$$

$$\hat{I}_3 = \frac{3 \text{ k}\Omega}{3 \text{ k}\Omega + 1 \text{ k}\Omega + 2 \text{ k}\Omega} \times 10 \text{ mA}$$

$$\hat{I}_3 = 5 \text{ mA}$$

(c) Figure 4-27b shows that the voltage $\hat{V}$ is across R_A. Since R_3 and R_A are in series, then $\hat{I}_3$ also flows through R_A:

$$\hat{V} = \hat{I}_3 R_A$$
$$\hat{V} = (5 \text{ mA})(2 \text{ k}\Omega) = 10 \text{ V}$$

(d) Using Eq. (3-6a),

$$P_T = I_S^2 R_T$$
$$P_T = (10 \text{ mA})^2 (3.5 \text{ k}\Omega)$$
$$P_T = 350 \text{ mW} \text{ or } 0.35 \text{ W}$$

Thus the total resistance is only one step in the solution of the problem. Note that in using the current division law resistance R_A had to be included. In the calculation of $\hat{V}$ the equivalent resistance (R_A) and the current I_3 through it was used. An alternative method would be first to solve for either the current through R_4 or through R_5 and then to apply Ohm's law. This is left as an exercise for the reader. Also, the reader should check to see that the total power is the sum of the individual powers.

Inductors in Series-Parallel. In this section we reduce inductive series-parallel circuits in steps using the rules learned in both the sections on series and parallel circuits.

Example 4-23: For the circuit of Fig. 4-28, calculate the inductance between the terminals *ab*.

Solution: Let $L_A = L_1 \| L_2$. Using the product over the sum rule,

$$L_A = \frac{L_1 L_2}{L_1 + L_2} = \frac{(2 \text{ mH})(6 \text{ mH})}{2 \text{ mH} + 6 \text{ mH}} = 1.5 \text{ mH}$$

(see Fig. 4-28b). The total inductance is then

$$L_T = L_A + L_3 = 1.5 \text{ mH} + 2.5 \text{ mH} = 4 \text{ mH}$$

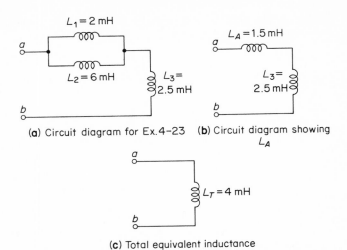

(a) Circuit diagram for Ex. 4-23   (b) Circuit diagram showing L_A

(c) Total equivalent inductance

Fig. 4-28. Step by step reduction of inductors connected in series-parallel.

The inductance L_T is the equivalent of the three inductances L_1, L_2, and L_3. Any voltage between the terminals of Fig. 4-28c is the same as between the terminals of Fig. 4-28a. The energy stored by an equivalent inductor such as L_T equals the sum of the energy stored by the individual inductors.

Example 4-24: For the circuit of Fig. 4-29a, calculate equivalent or total inductance between terminals *ab*.

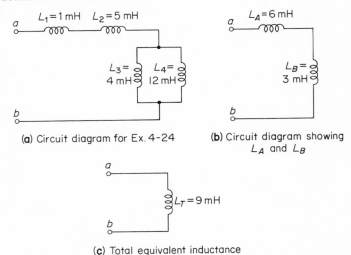

(a) Circuit diagram for Ex. 4-24   (b) Circuit diagram showing L_A and L_B

(c) Total equivalent inductance

Fig 4-29. Circuit diagrams showing reduction of inductors connected in series-parallel.

Solution: Let $L_A = L_1 + L_2$ and $L_B = L_3 \| L_4$. Then

$$L_A = 1\,\text{mH} + 5\,\text{mH} = 6\,\text{mH}$$

and

$$L_B = \frac{L_3 L_4}{L_3 + L_4} = \frac{(4\,\text{mH})(12\,\text{mH})}{4\,\text{mH} + 12\,\text{mH}} = 3\,\text{mH}$$

The equivalent circuit containing L_A and L_B is shown in Fig. 4-29b. The total inductance, L_T, is

$$L_T = L_A + L_B = 6\,\text{mH} + 3\,\text{mH} = 9\,\text{mH}$$

The total or equivalent inductor is shown in Fig. 4-29c.

Example 4-25: The total of two inductors connected in series is 8 mH. When the same (two) inductors are connected in parallel, the total inductance is 2 mH. Calculate the values of the individual inductors.

Solution: Let L_1 and L_2 represent the two inductors whose values are to be found. Then

$$L_1 + L_2 = 8\,\text{mH} \tag{4-38}$$

and

$$\frac{L_1 L_2}{L_1 + L_2} = 2\,\text{mH} \tag{4-39}$$

Substituting Eq. (4-38) into Eq. (4-39), we obtain

$$\frac{L_1 L_2}{8\,\text{mH}} = 2\,\text{mH}$$

or

$$L_1 L_2 = 16\,\text{mH}$$

$$L_2 = \frac{16\,\text{mH}}{L_1} \tag{4-40}$$

The substitution of Eq. (4-40) into Eq. (4-38) yields

$$L_1 + \frac{16\,\text{mH}}{L_1} = 8\,\text{mH}$$

Rearranging this expression into the form of a binomial equation, we get

$$L_1^2 - (8\,\text{mH})\,L_1 + 16\,\text{mH} = 0$$

The value of L_1 may be solved by means of the quadratic formula $(-b \pm \sqrt{b^2 - 4ac})/2a$, which yields

$$L_1 = 4\,\text{mH}$$

Therefore,

$$L_2 = 4\,\text{mH}$$

This example shows that the values of the individual inductances can be calculated knowing how they are connected together.

Capacitors in Series-Parallel.   Capacitors connected in a series-parallel configuration may also be analyzed in a way similar to that of resistors and inductors.

Example 4-26: Determine the total capacitance between terminals *ab* for the series-parallel circuit of Fig. 4-30a.

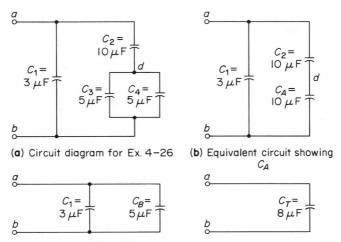

(**a**) Circuit diagram for Ex. 4–26

(**b**) Equivalent circuit showing C_A

(**c**) Equivalent circuit showing C_B

(**d**) Total equivalent capacitance

Fig. 4-30. Circuit diagram showing the reduction of capacitors connected in series-parallel.

Solution: Let $C_A = C_3 \| C_4$. Then

$$C_A = C_3 + C_4 = 5\,\mu\text{F} + 5\,\mu\text{F} = 10\,\mu\text{F}$$

as shown in Fig. 4-30b.   Now let C_B be equal to the series combination of C_2 and C_A. Since both capacitors are equal,

$$C_B = \frac{10\,\mu\text{F}}{2} = 5\,\mu\text{F}$$

The equivalent circuit of Fig. 4-30c shows that node *d* has been eliminated. The total equivalent capacitance, C_T, is

$$C_T = C_1 + C_B = 3\,\mu\text{F} + 5\,\mu\text{F} = 8\,\mu\text{F}$$

As far as the terminals *ab* are concerned, the equivalent capacitance, C_T, of Fig. 4-30d and the circuit of Fig. 4-30a perform the same functions.   That is the voltage across the terminals *ab* of Fig. 4-30d is equal to the voltage across the terminals *ab* of Fig. 4-30a.   The energy stored by C_T is equal to the sum of the energy stored by the individual capacitances of Fig. 4-30a.

Example 4-27: For the circuit of Fig. 4-31, calculate the total capacitance and the energy stored in it.

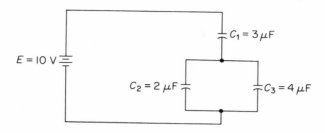

Fig. 4-31. Circuit diagram for Ex. 4-27.

Solution: Let $C_A = C_2 \| C_3$. Then

$$C_A = C_2 + C_3 = 2\,\mu F + 4\,\mu F = 6\,\mu F$$

The total capacitance is then

$$C_T = \frac{C_1 C_A}{C_1 + C_A}$$

$$C_T = \frac{(3\,\mu F)(6\,\mu F)}{3\,\mu F + 6\,\mu F} = 2\,\mu F$$

Using Eq. (3-15) to solve for the total capacitance yields

$$W_T = \tfrac{1}{2} C_T E^2$$

$$W_T = \tfrac{1}{2}(2\,\mu F)(10\,V)^2 = 100\,\mu J$$

Note that the total energy stored in the circuit is easily found after the total capacitance is known.

SUMMARY

The three sections of this chapter provide an introduction to the three types of electric circuits. For each type of circuit — series, parallel and their combination, series-parallel — the procedures are given for determining the equivalent resistance, inductance, and capacitance. Once these quantities are known, they may be used in solving for other circuit variables.

In analyzing any circuit, remember an important distinction brought out in the preceding pages: Two or more elements are in series if they have the *same current* through them; two or more elements are in parallel if they have the *same voltage* across them.

PROBLEMS

4-1. Calculate the total resistance for each of the circuits of Fig. 4-32.

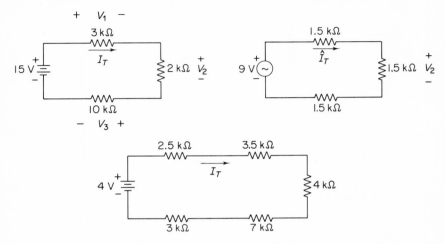

Fig. 4-32. Circuit diagrams for Prob. 1.

4-2. For each of the series circuits of Fig. 4-32, calculate

    (a) the total current,
    (b) the power dissipated by each resistor,
    (c) the total power disspated by the circuit.

4-3. In each of the circuits in Fig. 4-33, find the source voltage.

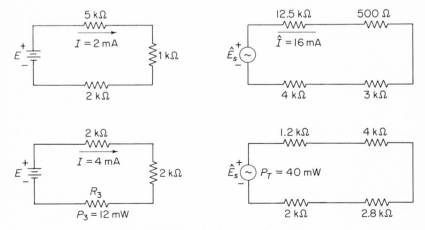

Fig. 4-33. Circuit diagrams for Prob. 3.

4-4. Using the voltage division law, calculate V_1, V_2, and V_3 in the circuit of Fig. 4-32a.

4-5. Using the voltage division law, calculate the voltage across R_2 (500 Ω) in the circuit of Fig. 4-33b.

4-6. In each of the circuits of Fig. 4-34, calculate the unknown quantities.

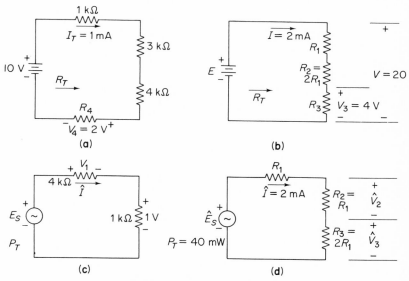

Fig. 4-34. Circuit diagrams for Prob. 6.

4-7. Calculate the total inductance between terminals *ab* for the circuits of Fig. 4-35.

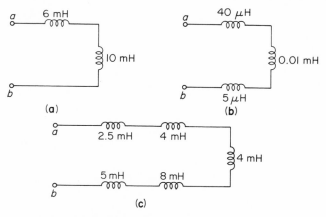

Fig. 4-35. Circuit diagrams for Prob. 7.

4-8. For the circuit of Fig. 4-36, determine the total inductance when the switch is in position A and when it is in position B.

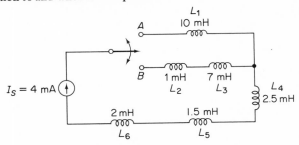

Fig. 4-36. Circuit diagram for Prob. 8.

4-9. For the circuit of Fig. 4-36, calculate the energy stored by each inductor and the total energy stored by the circuit:

(a)   when the switch is in position A,

(b)   when the switch is in position B.

4-10. If the current waveform of Fig. 4-37a is applied to the circuit of Fig. 4-37b, calculate the total energy stored by the circuit.

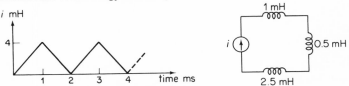

Fig. 4-37. Diagrams for Prob. 10.

4-11. Calculate the total capacitance between terminals *ab* for each of the circuits of Fig. 4-38.

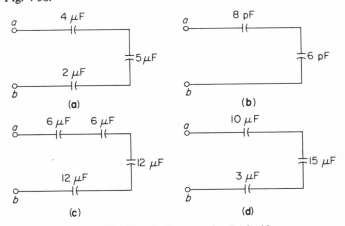

Fig. 4-38. Circuit diagrams for Prob. 11.

4-12. For the circuit of Fig. 4-39, calculate the energy stored by each capacitance and the total energy stored by the circuit.

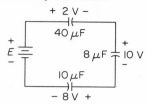

Fig. 4-39. Circuit diagram for Prob. 12.

4-13. If the voltage waveform of Fig. 4-40a is applied to the circuit of Fig. 4-40b, calculate the total energy stored by the circuit.

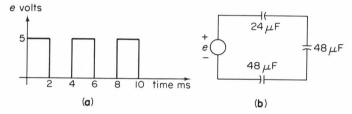

Fig. 4-40. Diagrams for Prob. 13.

4-14. For the circuit of Fig. 4-39, calculate the charge on each capacitor and the total charge.

4-15. For the circuit of Fig. 4-40, calculate the total charge.

4-16. Calculate the total resistance between terminals *ab* for each of the circuits of Fig. 4-41.

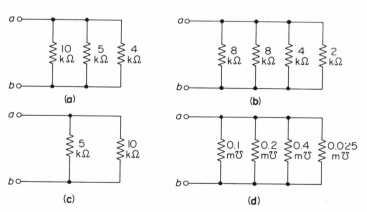

Fig. 4-41. Circuit diagrams for Prob. 16.

4-17. For the parallel resistive circuits of Fig. 4-42, calculate

    (a) the current through each resistor and the total current,

    (b) the power dissipated by each resistor and the total power.

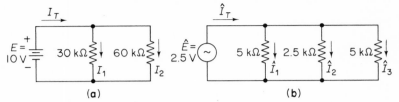

Fig. 4-42. Circuit diagrams for Prob. 17.

4-18. For the circuits of Fig. 4-43, calculate the source current.

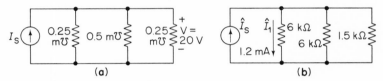

Fig. 4-43. Circuit diagrams for Prob. 18.

4-19. Using the current division law, calculate the current through R_1 for each of the circuits of Fig. 4-44.

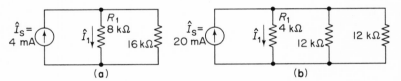

Fig. 4-44. Circuit diagrams for Prob. 19.

4-20. For each of the circuits of Fig. 4-45, calculate the values of the unknown quantities.

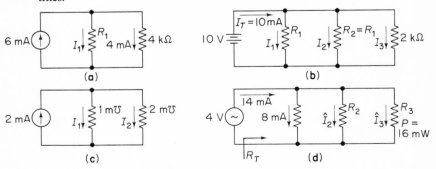

Fig. 4-45. Circuit diagrams for Prob. 20.

4-21. If the voltage waveform of Fig. 4-46a is applied to the circuit of Fig. 4-46b, calculate the total power dissipated by the circuit.

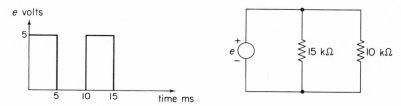

Fig. 4-46. Diagrams for Prob. 21.

4-22. Calculate the total inductance between terminals *ab* for each circuit shown in Fig. 4-47.

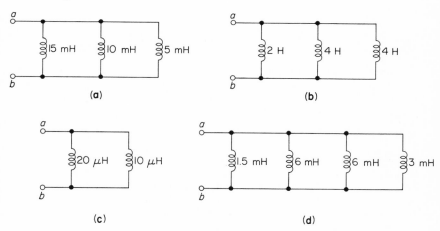

Fig. 4-47. Circuit diagrams for Prob. 22.

4-23. For the circuit of Fig. 4-48, calculate the energy stored by each inductor and the total energy stored by the circuit.

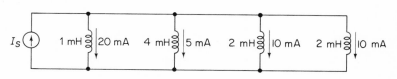

Fig. 4-48. Circuit diagram for Prob. 23.

4-24. If the current waveform of Fig. 4-37a is applied to the circuit of Fig. 4-47a, calculate the total energy stored by the circuit.

4-25. Calculate the total capacitance between terminals *ab* for each of the circuits shown in Fig. 4-49.

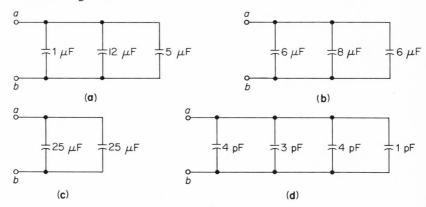

Fig. 4-49. Circuit diagrams for Prob. 25.

4-26. For the circuits of Fig. 4-50, calculate the energy stored by each capacitor and the total energy stored by the circuit.

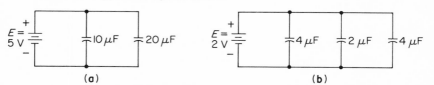

Fig. 4-50. Circuit diagrams for Prob. 26.

4-27. For the circuits of Fig. 4-50, calculate the charge stored by each capacitor and the total charge.

4-28. Calculate the total resistance of Fig. 4-24a if

$$R_1 = 20 \text{ k}\Omega, \qquad R_2 = 8 \text{ k}\Omega, \qquad R_3 = 8 \text{ k}\Omega$$

4-29. Calculate the total resistance of Fig. 4-25a if

$$R_1 = 10 \text{ k}\Omega, \qquad R_2 = 5 \text{ k}\Omega, \qquad R_3 = 1 \text{ k}\Omega, \qquad R_4 = 3 \text{ k}\Omega$$

4-30. Calculate the total resistance of Fig. 4-26a if

$$R_1 = 1 \text{ k}\Omega, \qquad R_2 = 4 \text{ k}\Omega, \qquad R_3 = 2 \text{ k}\Omega, \qquad R_4 = 2 \text{ k}\Omega, \qquad R_5 = 0.5 \text{ k}\Omega$$

4-31. For the circuits of Fig. 4-51, calculate

(a)   total resistance, R_T,
(b)   voltage, $\hat{V}$,
(c)   branch current, $\hat{I}_5$,
(d)   total power dissipated, P_T.

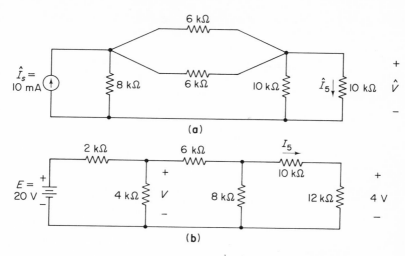

Fig. 4-51. Circuit diagrams for Prob. 31.

4-32. Refer to the circuit of Fig. 4-24a. If 24 V dc is applied between terminals *ab*, calculate the current through each resistor.

4-33. Refer to the circuit of Fig. 4-25a. If a 25-V dc source is connected between terminals *ab*, calculate the voltage across the parallel combination of R_1 and R_2 and also across the parallel combination of R_3 and R_4.

4-34. An ac voltage source is connected between terminals *ab* of Fig. 4-26a. If the value of the source is 32 V (rms), calculate the current through each resistor.

4-35. Refer to Fig. 4-28a and calculate the total inductance if

$$L_1 = 4\,mH, \qquad L_2 = 10\,mH, \qquad L_3 = 4.5\,mH$$

4-36. Refer to Fig. 4-29a and calculate the total inductance if

$$L_1 = 5\,\mu H, \qquad L_2 = 5\,\mu H, \qquad L_3 = 2\,\mu H, \qquad L_4 = 6\,\mu H$$

4-37. If the current waveform of Fig. 4-37a is applied to the circuit of Fig. 4-28a, calculate the total energy stored by the circuit.

4-38. If the current waveform of Fig. 4-37a is applied to the circuit of Fig. 4-29a, calculate the total energy stored by the circuit.

4-39. Refer to Fig. 4-30a and calculate the total capacitance if

(a) $C_1 = 5\,\mu F$
    $C_2 = 50\,\mu F$
    $C_3 = 25\,\mu F$
    $C_4 = 25\,\mu F$

(b) $C_1 = 3\,\mu F$
    $C_2 = 6\,\mu F$
    $C_3 = 2\,\mu F$
    $C_4 = 1\,\mu F$

4-40. If the voltage waveform of Fig. 4-40a is applied to the circuit of Fig. 4-30a, calculate

(a) the total energy stored,
(b) the total charge.

4-41. For the circuit of Fig. 4-52, calculate

(a) the total capacitance,
(b) the total energy stored,
(c) the total charge.

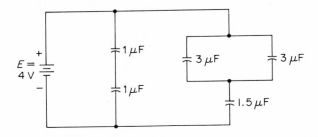

Fig. 4-52. Circuit diagram for Prob. 41.

CHAPTER 5

resistive circuits

Recall that in Chapter 3 two active elements were defined—the ideal voltage source and the ideal current source. The *ideal voltage source*, either ac or dc, has the property of maintaining a *constant voltage* regardless of any network that may be connected to its terminals. Similarly, the ideal *current source*, either ac or dc, has the property of maintaining a *constant current* regardless of any network that may be connected to its terminals.

Up to this point we have not considered any effect the network may have on the energy source. However, the terminal voltage of a practical source, in some instances, may be affected significantly by the loading network, because the transfer of energy from the source to the network is never 100 per cent efficient.

This lack of perfect efficiency is the result of internal passive element or elements within the source which either dissipate or store energy. The passive elements in many, but not necessarily all, sources are resistances (referred to either as *source resistances* or *internal resistances*). For our analysis, we may represent a *practical* source as the combination of an ideal source and resistor.

Figures 5-1a and b show practical dc and ac voltage sources as represented by an ideal voltage source in *series* with a resistor. Practical current sources are represented by an ideal current source in *parallel* with a resistor in Fig. 5-1c and d.

This chapter is divided into two main parts. Sections 5-1, 5-2, and 5-3 discuss practical sources, and sections 5-4, 5-5, 5-6, 5-7, and 5-8 discuss five methods for analyzing a network.

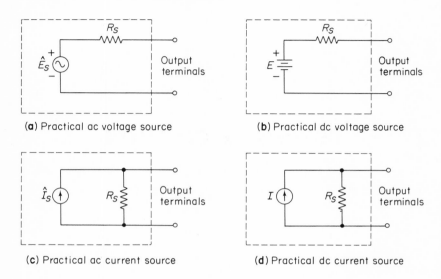

(**a**) Practical ac voltage source   (**b**) Practical dc voltage source

(**c**) Practical ac current source   (**d**) Practical dc current source

Fig. 5-1. Equivalent circuits for practical sources.

5-1. PRACTICAL VOLTAGE SOURCES

Consider a practical dc voltage source to which a load resistor, R_L, has been connected to the output terminals as shown in Fig. 5-2. Load

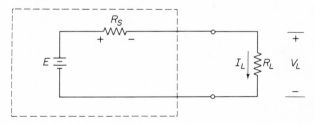

Fig. 5-2. Circuit diagram for a practical dc voltage source.

resistor R_L has now completed the circuit, thereby providing a path for current to flow. This flow of current produces a voltage drop across the internal resistance, R_S. Applying Kirchhoff's voltage law to this equivalent circuit, we may determine the effect that the voltage drop across the internal resistor, R_S, has on the terminal load voltage, V_L:

$$E = I_L R_S + I_L R_L \qquad (5\text{-}1)$$

or

$$V_L = E - I_L R_S = I_L R_L \qquad (5\text{-}2)$$

Equation (5-2) shows that the voltage drop across the internal resistor decreases the available voltage V_L at the output terminals (the voltage across R_L). Note in Eq. (5-2) that when $R_S = 0$, a practical source becomes an ideal source and the voltage at the output terminals is always equal to E.

We have now established that a practical voltage source is not 100 per cent efficient and that there is an internal voltage drop due to the internal source resistor. The next step in the treatment of practical voltage sources is to formulate a procedure for determining the value of the internal resistor.

This is done by determining both the voltage across and the current through the internal resistor, and, applying Ohm's law,

$$R_S = \frac{I_L R_S}{I_L} = \frac{V_{\text{int}}}{I_{\text{int}}} = \frac{E - V_L}{I_L} \qquad (5\text{-}3)$$

These values of voltage and current may be determined by the following measurements and calculations:

1. Remove the load resistor, R_L, and measure the open-circuit voltage E_{oc} between the output terminals. If this voltage is measured without drawing any appreciable current ($I_{\text{int}} \simeq 0$), then the voltage drop across R_S will be equal to zero ($V_{\text{int}} = 0$). Under these conditions the measured value of open-circuit voltage is equal to the voltage of the ideal source, $E_{\text{oc}} = E$. The term open-circuit implies that the load resistance is removed. See Fig. 5-3a.

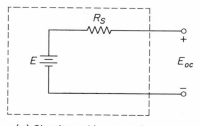
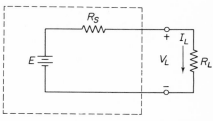

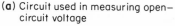
(**a**) Circuit used in measuring open–
circuit voltage   (**b**) Circuit used in measuring load voltage.

Fig. 5-3. Circuit diagrams for practical voltage source.

2. Connect the load resistor, R_L, back to the output terminals and measure the voltage across the output terminals, V_L. This voltage, V_L,

across R_L must be less than E_{oc} because we have now completed the circuit and provided a path for the current to flow. The flow of current produces a voltage drop across the internal resistor, which in turn lowers the available voltage at the output terminals [see Fig. 5-3b and Eq. (5.2)].

Applying Kirchhoff's voltage law to the equivalent circuit of Fig. 5-3b, we obtain

$$E = V_{int} + V_L \tag{5-4}$$

or

$$V_{int} = E - V_L \tag{5-5}$$

Since both E and V_L are known from the measured values, V_{int} is also known. The circuit of Fig. 5-3b is a series circuit; therefore,

$$I_{int} = I_L = \frac{V_L}{R_L} \tag{5-6}$$

The remaining step in this procedure is to use Ohm's law to solve for the internal resistance:

$$R_S = \frac{V_{int}}{I_{int}} \tag{5-7}$$

The following examples illustrate this procedure.

Example 5-1: A voltmeter that draws negligible current is connected to the output terminals of a practical voltage source. The open-circuit voltage reading is 10 V. When a load resistor of 1.5 kΩ is connected to the output terminals, the voltmeter reading drops to 7.5 V. Determine the internal resistance of the source and draw the equivalent circuit with the proper values.

Solution: Since the open-circuit voltage is 10 V, the ideal voltage source must be equal to 10 V ($E = E_{oc} = 10$ V). The load voltage, V_L, equals 7.5 V.

Applying Kirchhoff's voltage law, we get

$$E = V_{int} + V_L$$

or

$$V_{int} = E - V_L = 10 \text{ V} - 7.5 \text{ V} = 2.5 \text{ V}$$

From Ohm's law the current in the circuit is determined:

$$I_{int} = I_L = \frac{V_L}{R_L} = \frac{7.5 \text{ V}}{1.5 \text{ k}\Omega} = 5 \text{ mA}$$

Therefore,

$$R_S = \frac{V_{int}}{I_{int}} = \frac{2.5 \text{ V}}{5 \text{ mA}} = 500 \text{ }\Omega$$

The equivalent circuit is shown in Fig. 5-4

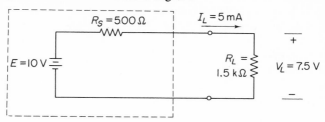

Fig. 5-4. Circuit diagram for Ex. 5-1.

Example 5-2: The open-circuit voltage of a practical ac voltage source is measured as 12 V rms. When a load resistor of 5 kΩ is connected to the output terminals, then load current is measured as 2 mA. Determine the value of the internal resistance and draw the equivalent circuit with the proper values.

Solution: The load voltage is

$$\hat{V}_L = \hat{I}_L R_L = (2 \text{ mA})(5 \text{ k}\Omega) = 10 \text{ V}$$

From Kirchhoff's voltage law

$$\hat{V}_{int} = \hat{E}_S - \hat{V}_L = 12 \text{ V} - 10 \text{ V} = 2 \text{ V}$$

Applying Ohm's law to find the internal resistance,

$$R_S = \frac{\hat{V}_{int}}{\hat{I}_L} = \frac{2 \text{ V}}{2 \text{ mA}} = 1 \text{ k}\Omega$$

See Fig. 5.5.

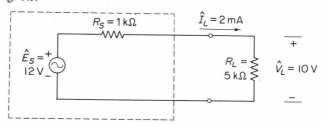

Fig. 5-5. Circuit diagram for Ex. 5-2.

Although we have developed a method for finding the internal resistance, our discussion of practical voltage sources is not complete. It is also necessary to be able to analyze the circuit when different values of load resistance are connected to the output terminals. The reason for doing this is so that we may interpret the effects that R_S and R_L have on the circuit. We also may wish to determine that value of R_L at which maximum power is delivered (transfered) to the load.

For a numerical analysis of a practical source, let the ideal voltage source equal 10 V dc and its internal resistance equal 500 Ω, as shown in the circuit diagram of Fig. 5-6. The range of possible load resistance R_L

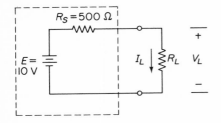

Fig. 5-6. Circuit diagram of a practical dc voltage source.

varies from zero (short circuit) to infinity (open circuit). Table 5-1 shows all pertinent information for this circuit for different values of R_L.

TABLE 5-1

EFFECT OF VARYING THE LOAD ON OTHER CIRCUIT PARAMETERS OF A PRACTICAL VOLTAGE SOURCE

R_L (ohms)	I_L (mA)	V_L (volts)	P_L (mW)
0	20	0	0
100	16.6	1.66	27.6
200	14.3	2.86	40.9
400	11.1	4.45	49.6
500	10	5.0	50.0
600	9.1	5.46	49.7
800	7.7	6.16	47.5
1 k	6.67	6.67	44.5
2 k	4	8.0	32.0
5 k	1.82	9.1	16.6
10 k	0.95	9.5	9.0
100 k	0.099	9.9	0.98
∞	0	10.0	0.0

The data of Table 5-1 may also be represented in graphical form. Figure 5-7a is a plot of load voltage, V_L, versus load current, I_L. Figure 5-7b is a plot of the power dissipated in the load versus the load resistance. Let us analyze each graph separately, beginning with Fig. 5-7a. This graph shows that as the load current increases, the load voltage decreases. The points of primary importance are the intercepts of the vertical axis (10 V) and the horizontal axis (20 mA).

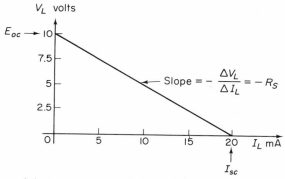

(a) Voltage – current characteristics for the circuit of Fig. 5–6

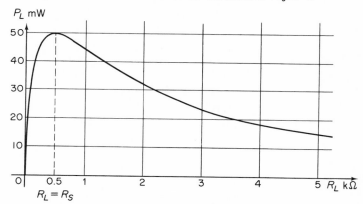

(b) Power – transfer characteristics for the circuit of Fig. 5–6

Fig. 5-7. Characteristics for a practical voltage source.

The maximum load voltage is the open-circuit voltage.   In this example, $E_{oc} = 10$ V.   The open-circuit voltage refers to the condition of infinite load resistance ($R_L = \infty$) and no current flow, $I_L = 0$.   Thus removing the load gives the vertical axis intercept ($V_L = 10$ V and $I_L = 0$ A).

The horizontal-axis intercept is the other extreme condition, that is, when the load resistance is a short circuit ($R_L = 0$ Ω).   The voltage across any short circuit is always zero.   This does *not* mean that no current flows in the circuit.   On the contrary, *maximum* current flows.   This maximum current is referred to as the *short-circuit current*, I_{sc}.   In this example, where R_L is zero, $I_{sc} = 20$ mA and $V_L = 0$ V.   These values represent the horizontal-axis intercept.

Because the circuit elements are linear, a straight line having a negative slope can be constructed between the vertical and horizontal intercepts. From this graph (Fig. 5-7a), all values of load voltage and load current can be found. By applying Ohm's law, the load resistance can be determined for a particular value of voltage and current. Consider the particular case where the load voltage is one half of the open-circuit voltage:

$$V_L = \frac{1}{2} E_{oc} = \frac{1}{2}(10 \text{ V}) = 5 \text{ V}$$

From the graph of Fig. 5-7a when $V_L = 5$ V, $I_L = 10$ mA. Therefore,

$$R_L = \frac{V_L}{I_L} = \frac{5 \text{ V}}{10 \text{ mA}} = 500 \ \Omega$$

Note that this value of load resistance is equal to the internal resistance of the source. This is not just a coincidence. If the load voltage is one half of the open-circuit voltage, then (and only then) is the load resistance equal to the internal resistance. Further, if the load current is one half of the short-circuit current, then (and only then) is the load resistance equal to the internal resistance.

The reader should keep in mind that the internal source resistance is assumed constant. This constant value of internal resistance may also be determined by using the graph of Fig. 5-7a. The slope of the line plotted on Fig. 5-7a is the negative value of the internal resistance:

$$R_S = -\text{ slope} = -\frac{\Delta V_L}{\Delta I_L} = -\frac{(0 - 10) \text{ V}}{(20 - 0) \text{ mA}} = 500 \ \Omega$$

From this graph, then, we have obtained the same information contained in the first three columns of Table 5-1.

In electronic networks, a primary concern is the amount of power that a source may deliver to a load. Therefore, we shall also attempt to determine what value of load resistance results in maximum power transferred to the load. A graphical interpretation of the data contained in Table 5-1 has been plotted in Fig. 5-7b. This graph is a plot of the power dissipated in the load, P_L, versus the load resistance, R_L. As the load changes from a short circuit ($R_L = 0$) to an open circuit ($R_L = \infty$), the power delivered to the load increases rapidly until a maximum is reached and then begins to decrease. The graph shows that the load receives the maximum amount of power from the source when the load resistance is equal to the internal resistance ($R_L = R_S$). For this example, $R_L = R_S = 500 \ \Omega$.

The graphical information may also be stated as a theorem, known as the *maximum power transfer theorem: Maximum power is delivered by a source to a load when the load resistance is equal to the internal resistance of the practical source.*

Note that even though power is being delivered to the load, there must also be power dissipated by the internal resistance. When $R_L = R_S$, the power dissipated by the internal resistance *equals* the power delivered to the load.

At quick glance it would seem that a practical voltage source is not very efficient. Efficiency for electronic networks is defined as the ratio of the power in the load resistance, P_L, to the total power developed by the source, P_T. The symbol for efficiency is the Greek letter eta, η:

$$\eta = \frac{P_L}{P_T}$$

For the circuit under discussion, the efficiency for the maximum power output is

$$\eta = \frac{P_L}{P_T} = \frac{50 \text{ mW}}{100 \text{ mW}} = 0.5 \quad \text{or, in per cent,} \quad \eta = 50 \text{ %}$$

Therefore, when maximum power is being delivered to a load, a practical voltage source has an efficiency of only 50 per cent. This rather low efficiency is tolerated in electronic applications such as audio, communications, and microwave systems to obtain maximum power transfer. On the other hand, commercial power distribution systems must avoid this low efficiency because of high line losses. Their object is high efficiency, not maximum power transfer.

It is clear, therefore, that maximum power does *not* mean maximum efficiency. We may also note from Table 5-1 that as R_L *increases*, the efficiency *increases*. For the above example, consider when $R_L = 100$ kΩ. From Table 5-1, $I_L = 0.099$ mA, $V_L = 9.9$ V, and $P_L = 0.98$ mW. The total power delivered by the source under these conditions is

$$P_T = I_L E = (0.099 \text{ mA})(10 \text{ V}) = 0.99 \text{ mW}$$

Since

$$\eta = \frac{P_L}{P_T} \quad \text{then} \quad \eta = \frac{0.98 \text{ mW}}{0.99 \text{ mW}} = 0.99$$

or, in terms of per cent, $\eta = 99$ per cent.

5-2. PRACTICAL CURRENT SOURCES

Since practical current sources find extensive use in electronic applications, their characteristics also should be considered. For this text, *practical* current sources are represented by an ideal current source in *parallel* with a resistance; see Fig. 5-1c and d. The analysis for the practical current source is similar to that of a practical voltage source.

Let us begin by substituting a single equivalent load resistor, R_L, for the external network connected to the output terminals, as shown in Fig. 5-8.

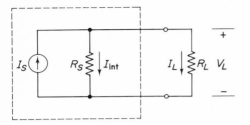

Fig. 5-8. Equivalent circuit for a practical dc current source supplying energy to a resistive load.

By applying Kirchhoff's current law to this equivalent circuit, we may establish the effect of internal resistance when current is drawn from the ideal source:

$$I_S = I_{int} + I_L \tag{5-8}$$

Rearranging, we obtain

$$I_L = I_S - I_{int} \tag{5-9}$$

As seen by Eq. (5-9), the load current equals the ideal source current only when $I_{int} = 0$. This can happen only when $R_S = \infty$. The internal resistance accounts for the fact that a practical source is not 100 per cent efficient in supplying energy to a load.

To calculate the value of R_S, it is first necessary to determine the voltage across and the current through the internal resistance. These values may be found by the following procedure.

1. Remove R_L and connect a low-resistance ($R \approx 0$) ammeter between the output terminals. This causes a short circuit between the terminals

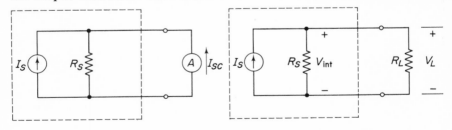

(a) Circuit used in measuring
short–circuit current

(b) Circuit used in measuring
load voltage

Fig. 5-9. Circuit diagrams for a practical current source.

because the resistance of the meter is negligible. Since R_S is now in parallel with a short circuit, all the current supplied by the ideal source will flow through the ammeter. The value of current measured is referred to as the short-circuit current, I_{sc} (see Fig. 5-9a).

2.   Replace R_L in lieu of the ammeter and measure the voltage between the output terminals, V_L In this case R_S and R_L are in parallel.   Therefore, V_L equals V_{int} (the voltage across R_S) (see Fig. 5-9b).

The internal resistance can be found by applying Ohm's law:

$$R_S = \frac{V_{int}}{I_{int}} \qquad (5\text{-}10)$$

where

$$I_{int} = I_{sc} - I_L \quad \text{or} \quad I_{int} = I_{sc} - V_L/R_L$$

Example 5-3:   An ammeter is connected between the output terminals of a practical current source.   The short-circuit current reading is 3 mA. When a load resistor of 2 kΩ is connected to the output terminals, a voltmeter reads 2 V.   Determine the internal resistance of the source and draw the equivalent circuit.

Solution:

$$V_{int} = V_L = 2 \text{ V}$$

$$I_{int} = I_{sc} - I_L$$

$$I_{int} = I_{sc} - \frac{V_L}{R_L} = 3 \text{ mA} - \frac{2 \text{ V}}{2 \text{ k}\Omega} = 2 \text{ mA}$$

$$R_S = \frac{2 \text{ V}}{2 \text{ mA}} = 1 \text{ k}\Omega$$

The equivalent circuit is shown in Fig. 5-10.

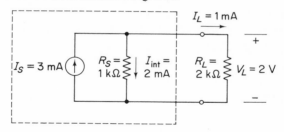

Fig. 5-10. Circuit diagram for Ex. 5-3.

Example 5-4:   A practical current source has a short-circuit current equal to 5 mA and an internal resistor of 9.5 kΩ.   What power will this source deliver to a 500-Ω load?

Solution: Since the power in the load may be found by the formula $P_L = I_L^2 R_L$, it is first necessary to find the load current, I_L, using the current division law:

$$I_L = \frac{R_S}{R_S + R_L} (I_S)$$

$$I_L = \frac{9.5 \text{ k}\Omega}{9.5 \text{ k}\Omega + 0.5 \text{ k}\Omega} \times (5 \text{ mA}) = 4.75 \text{ mA}$$

Therefore,

$$P_L = I_L^2 R_L = (4.75 \text{ mA})^2 (0.5 \text{ k}\Omega) = 11.3 \text{ mW}$$

To further develop our knowledge of practical current sources, let us vary the load from a short circuit ($R_L = 0$) to an open circuit ($R_L = \infty$). Again we wish to know what effect R_S and R_L have on the equivalent circuit. For a numerical analysis, let the ideal dc current source equal 3 mA and the internal resistor equal 2 kΩ, as shown in Fig. 5-11. Table 5-2 has been formed to show the relationships among current, voltage, and power for different values of R_L.

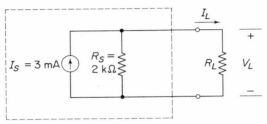

Fig. 5-11. Circuit diagram of a practical ac current source.

TABLE 5-2

EFFECT OF VARYING THE LOAD ON OTHER CIRCUIT PARAMETERS OF A PRACTICAL CURRENT SOURCE

R_L (kΩ)	I_L (mA)	V_L (volts)	P_L (mW)
0	3.0	0.0	0.0
0.5	2.4	1.2	2.88
1.0	2.0	2.0	4.0
1.5	1.71	2.56	4.38
1.8	1.58	2.85	4.48
2.0	1.5	3.0	4.5
2.2	1.43	3.14	4.48
2.5	1.33	3.33	4.43
3.0	1.2	3.6	4.32
4.0	1.0	4.0	4.0
5.0	0.857	4.29	3.68
10.0	0.5	5.0	2.5
100.0	0.059	5.9	0.35
∞	0.0	6.0	0.0

Note that when $R_L = \infty$, there is a voltage measured between the output terminals (in this example, 6 V). This voltage, as described in the previous section, is called the open-circuit voltage and is caused by the voltage drop across R_S.

The effect of varying R_L may also be represented in graphical form. Figure 5-12a is a plot of load current versus load voltage and Fig. 5-12b is

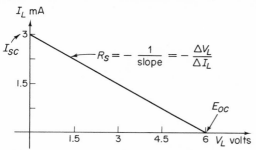

(a) Current–voltage characteristic for the circuit of Fig. 5–11

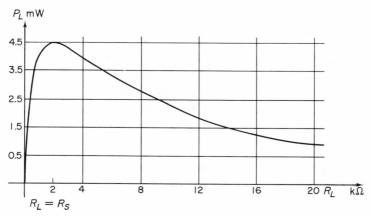

(b) Power–transfer characteristic for the circuit of Fig. 5–11

Fig. 5-12. Characteristics for a practical current source. (Note that Fig. 5-12(a) is a plot of I_L versus V_L, whereas Fig. 5-7(b) is a plot of V_L versus I_L. The axes have been interchanged so that you may become familiar with the various ways of depicting load voltage and current. This will become useful in applications such as transistor characteristics.)

a plot of power dissipated by the load versus the load resistance. Let us analyze each graph. Points of primary interest in Fig. 5-12a are the vertical and horizontal intercepts. The vertical intercept is the short-circuit current (in this example, 3 mA). The horizontal intercept is the open-circuit voltage (in this example, 6 V). Between these extreme points a straight line can be constructed, on which all possible values fall. Thus this graph contains the data of the first three columns of Table 5-2. It also contains other information, such as the value of internal resistance. From the graph of Fig. 5-12a, the internal resistance is equal to the negative reciprocal of the slope:

$$R_S = -\frac{1}{\text{slope}} = \frac{1}{\Delta I/-\Delta V} = \frac{-\Delta V}{\Delta I} \tag{5-11}$$

For this example,

$$R_S = \frac{-\Delta V}{\Delta I} = \frac{-(0-6)\text{ V}}{(3-0)\text{ mA}} = 2\text{ k}\Omega$$

Consider the case where $I_L = \frac{1}{2} E_{oc}$ and we wish to determine the load resistance. From Fig. 5-12a,

$$V_L = \tfrac{1}{2} E_{oc}$$

Since

$$R_L = \frac{V_L}{I_L}$$

then

$$R_L = \frac{\tfrac{1}{2}E_{oc}}{\tfrac{1}{2}I_{sc}} \quad \frac{6\text{ V}}{3\text{ mA}} = 2\text{ k}\Omega$$

In conclusion, when the load current is one half of the short-circuit current, then, and only then, $R_L = R_S$.

Since power delivered to a load is always a primary concern in electronic networks, it must be discussed in connection with practical current sources. The graph of Fig. 5-12b is similar to that of Fig. 5-7b; that is, the maximum power delivered to a resistive load occurs when $R_L = R_S$. The maximum power transfer theorem stated in the previous section is equally applicable to practical current sources.

The efficiency for the maximum power output of a practical current source is also the same as that of a practical voltage source, 50 per cent:

$$\eta = \frac{P_L}{P_T} = \frac{4.5\text{ mW}}{9.0\text{ mW}} = 50\,\%$$

As discussed previously, this low efficiency is tolerated for electronic applications.

5-3. EQUIVALENT PRACTICAL SOURCES

In the development of practical voltage sources and practical current sources, many similarities should have been noticed — such as internal resistance, voltage and current characteristics, and maximum power transfer. These similarities lead one to ask if it is possible to *interchange* one practical source for the other and retain the same results at the load. Not only is it possible, but at times it is a very useful technique in the simplification of a network. Determining the equivalent values of these sources is the next step.

If the *same* value of load resistance, R_L, is connected to the circuits of Fig. 5-13, the load current flowing in circuit (a) is

$$\hat{I}_L = \frac{\hat{E}_S}{R_S + R_L} \qquad (5\text{-}12)$$

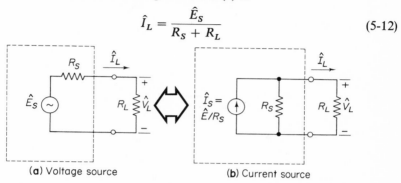

(a) Voltage source   (b) Current source

Fig. 5-13. Two practical sources which yield the same output characteristics.

and the load current flowing in circuit (b) is

$$\hat{I}_L = \frac{R_S}{R_S + R_L} \hat{I}_S \qquad (5\text{-}13)$$

To produce the same effects at the load, Eqs. (5-12) and (5-13) must be equal:

$$\frac{\hat{E}_S}{R_S + R_L} = \frac{R_S}{R_S + R_L} \hat{I}_S$$

Thus establishing the result that

$$\hat{E}_S = \hat{I}_S R_S \quad \text{or} \quad \hat{I}_S = \frac{\hat{E}_S}{R_S} \qquad (5\text{-}14)$$

The value of internal resistance for both practical current and voltage sources does *not* change. Although these equations have been derived for ac conditions, they are also equally applicable to dc circuits.

Note that the same values of voltage, current, and power may be obtained at the output terminals by either a practical voltage or a practical current source. Note, however, there is *no* equivalence of voltage, current, or power *inside* the practical sources. This point may be seen by considering the case where $R_L = \infty$. Inside the practical voltage source, there is no closed path $\hat{I}_{int} = 0$. However, for the practical current source, there is a closed path and $\hat{I}_{int} = \hat{I}_S$. On the other hand, the voltage between the output terminals, when $R_L = \infty$, for both sources is E_{oc}.

A final comment on equivalent practical sources concerns internal resistance. The internal resistance for the practical voltage source has been represented by a single resistance, R_S. But R_S could have been the equi-

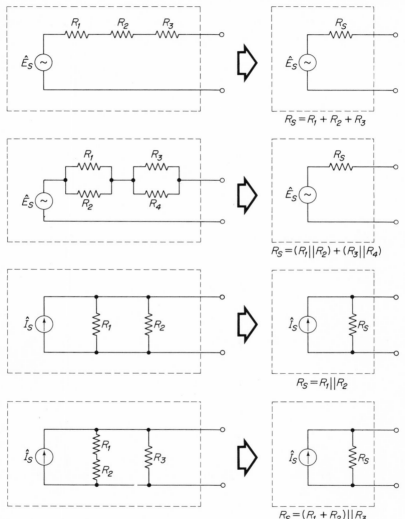

Fig. 5-14. Examples of equivalent sources.

valent resistance of two (or more) series or parallel resistors. Similarly, the internal resistance for the practical current source could have been the equivalent resistance of two or more series or parallel resistors.

The circuit diagrams shown in Fig. 5-14 show these relationships.

Example 5-5: Consider the network of Fig. 5-15. (a) For which value of R_L will the practical voltage source deliver maximum power? (b) Draw the current equivalent circuit.

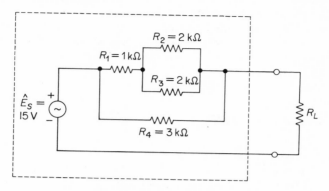

Fig. 5-15. Circuit diagram for Ex. 5-5.

Solution: (a) Let

$$R_A = R_2 \| R_3 = 2\,\text{k}\Omega \| 2\,\text{k}\Omega = 1\,\text{k}\Omega$$

and

$$R_B = R_1 + R_A = 1\,\text{k}\Omega + 1\,\text{k}\Omega = 2\,\text{k}\Omega$$

Then

$$R_S = R_B \| R_4 = 2\,\text{k}\Omega \| 3\,\text{k}\Omega$$

$$R_S = \frac{(2\,\text{k}\Omega)\,(3\,\text{k}\Omega)}{2\,\text{k}\Omega + 3\,\text{k}\Omega} = 1.2\,\text{k}\Omega$$

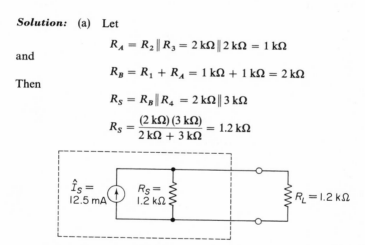

Fig. 5-16. The current source equivalent circuit of Fig. 5-15.

To deliver maximum power to the load,

$$R_L = R_S = 1.2\,\text{k}\Omega$$

(b) The value of the current source

$$\hat{I}_s = \frac{\hat{E}_s}{R_S} = \frac{15\ \text{V}}{1.2\,\text{k}\Omega} = 12.5\,\text{mA}$$

The current equivalent circuit is shown in Fig. 5-16.

The remaining five sections discuss five methods for analyzing a network, using loop analysis (Section 5-4), nodal analysis (Section 5-5), the superposition theorem (Section 5-6), Thévenin's theorem (Section 5-7), and Norton's theorem (Section 5-8).

There is no one simple method that is best for analyzing every network. The five approaches to follow can be selectively applied to different circuits depending on the nature of the circuit and the variables to be determined.

5-4. LOOP ANALYSIS

In this section we show that a set of simultaneous linear equations can be written which, for all practical purposes, completely describes the network.   This set of equations depends on a choice of currents (not branch currents) as variables, used in connection with Kirchhoff's voltage law. This allows us to analyze networks that have two or more loops.   The circuits shown in Fig. 5-17a and b have two and three loops, respectively.

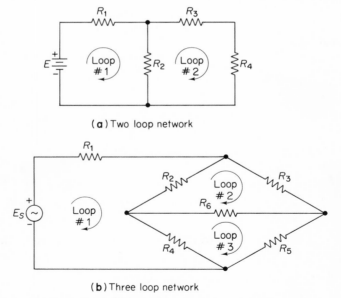

(**a**) Two loop network

(**b**) Three loop network

Fig. 5-17. Circuit diagrams illustrating the principles of loops.

We can assign currents to represent each loop, such as I_A to represent loop 1, I_B to represent loop 2, I_C to represent loop 3, etc.   These currents − I_A, I_B, and I_C − are called *loop currents*.   The term "current" is somewhat misleading because these are not measurable quantities. The loop currents do not actually flow through the elements of a network. The measurable quantities are the branch currents. Loop currents, however, provide us with a mathematical solution for the branch currents and thus the voltage drops in a network.

A general procedure for analyzing a network by loop analysis is as follows:

1. Assign a loop current for each independent closed path. The number of independent loops in a network is given by the expression

$$T - N + 1 \qquad\qquad (5\text{-}15)$$

where T is the *total* of elements in the network (both *active* and *passive*) and N the number of nodes in the network. Although the direction of loop currents is arbitrary, we shall only assign clockwise directions for the sake of uniformity.

2. Indicate the polarity markings across each *passive element* due to the direction of the loop current in that closed path. The polarities of voltage sources are determined by their own positive and negative terminals.

3. Apply Kirchhoff's voltage law to each closed path. If we are applying Kirchhoff's voltage law to one loop, and a *passive element* has two or more loop currents flowing through it, the total voltage drop equals the sum of the voltage drops due to each loop current separately. If the loop currents flow in the same direction, the voltage drops add. However, if the loop currents flow in opposite directions, the voltage drops subtract from one another.

4. Solve the set of simultaneous linear equations for the loop currents.

Let us now apply the above procedure in analyzing some examples.

Example 5-6: Consider the network of Fig. 5-18 and determine the current through each resistance.

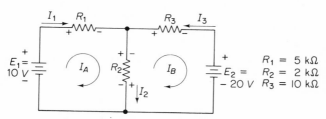

Fig. 5-18. Circuit diagram for Ex. 5-6.

Solution: Steps 1 and 2 are shown in the circuit of Fig. 5-18. Applying Kirchhoff's voltage law to each loop,

Loop A:

$$I_A R_1 + I_A R_2 - I_B R_2 = E_1 \qquad \text{or} \qquad I_A (R_1 + R_2) - I_B R_2 = E_1$$

Substituting numerical values,

$$I_A \, 7 \, k\Omega - I_B \, 2 \, k\Omega = 10 \, V$$

Loop B:

$$I_B R_2 - I_A R_2 + I_B R_3 = -E_2$$

or

$$-I_A R_2 + I_B(R_2 + R_3) = -E_2$$

Substituting numerical values,

$$-I_A \, 2 \, k\Omega + I_B \, 12 \, k\Omega = -20 \, V$$

Thus the set of simultaneous equations is

$$I_A \, 7 \, k\Omega - I_B \, 2 \quad k\Omega = 10 \, V$$

$$-I_A \, 2 \, k\Omega + I_B \, 12 \, k\Omega = -20 \, V$$

Different techniques have been developed to solve a set of simultaneous linear equations. The examples in this book will be solved using the method of determinants. The basics of determinants are given in the Appendix for the reader who is not familiar with this principle.

$$I_A = \frac{\begin{vmatrix} 10 \, V & -2 \times 10^3 \, \Omega \\ -20 \, V & 12 \times 10^3 \, \Omega \end{vmatrix}}{\begin{vmatrix} 7 \times 10^3 \, \Omega & -2 \times 10^3 \, \Omega \\ -2 \times 10^3 \, \Omega & 12 \times 10^3 \, \Omega \end{vmatrix}} = \frac{(120 \times 10^3 - 40 \times 10^3) \, V \cdot \Omega}{(84 \times 10^6 - 4 \times 10^6) \Omega^2} =$$

$$= \frac{80 \times 10^3 \, V \cdot \Omega}{80 \times 10^6 \Omega^2}$$

Therefore,

$$I_A = 1 \, mA$$

$$I_B = \frac{\begin{vmatrix} 7 \times 10^3 \, \Omega & 10 \, V \\ -2 \times 10^3 \, \Omega & -20 \, V \end{vmatrix}}{80 \times 10^6 \, \Omega^2} = \frac{(-140 \times 10^3 + 20 \times 10^3) \, V \; \Omega}{80 \times 10^6 \, \Omega^2} =$$

$$= \frac{-120 \times 10^3 \, V \cdot \Omega}{80 \times 10^6 \, \Omega^2}$$

Therefore,

$$I_B = -1.5 \, mA$$

We may now solve for the branch currents in terms of the loop currents:

$$I_1 = I_A = 1 \, mA$$

$$I_2 = I_A - I_B = 1 \, mA - (-1.5 \, mA) = 2.5 \, mA$$

$$I_3 = -I_B = -(-1.5 \, mA) = 1.5 \, mA$$

Note that in order to find the branch currents — I_1, I_2, and I_3 — the loop currents — I_A and I_B — first had to be solved.

The directions of the loop currents and the branch currents are arbitrary. A minus sign preceding a value of current only means that the direction has been chosen incorrectly. However, the magnitude of the answer is correct.

Example 5-7: For the circuit of Fig. 5-19, calculate the branch current, $\hat{I}$.

Solution: Steps 1 and 2 are shown in the circuit of Fig. 5-19. Applying Kirchhoff's voltage law to each loop,

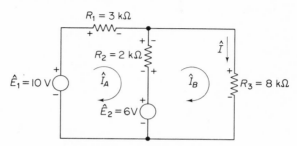

Fig. 5-19. Circuit diagram for Ex. 5-7.

Loop A:

$$\hat{I}_A R_1 + \hat{I}_A R_2 - \hat{I}_B R_2 = \hat{E}_1 - \hat{E}_2$$

or

$$\hat{I}_A (R_1 + R_2) - \hat{I}_B R_2 = \hat{E}_1 - \hat{E}_2$$

Substituting numerical values,

$$\hat{I}_A\, 5\,\mathrm{k\Omega} - \hat{I}_B\, 2\,\mathrm{k\Omega} = 4\,\mathrm{V}$$

Loop B:

$$-\hat{I}_A R_2 + \hat{I}_B R_2 + \hat{I}_B R_3 = \hat{E}_2$$

or

$$-\hat{I}_A R_2 + \hat{I}_B (R_2 + R_3) = \hat{E}_2$$

Substituting numerical values,

$$-\hat{I}_A\, 2\,\mathrm{k\Omega} + \hat{I}_B\, 10\,\mathrm{k\Omega} = 6\,\mathrm{V}$$

Thus the set of simultaneous equation is

$$\hat{I}_A\, 5\,\mathrm{k\Omega} - \hat{I}_B\ \ 2\,\mathrm{k\Omega} = 4\,\mathrm{V}$$

$$-\hat{I}_A\, 2\,\mathrm{k\Omega} + \hat{I}_B\, 10\,\mathrm{k\Omega} = 6\,\mathrm{V}$$

Since $\hat{I} = \hat{I}_B$, it is necessary only to solve for loop current $\hat{I}_B$:

$$\hat{I}_B = \frac{\begin{vmatrix} 5\,\mathrm{k\Omega} & 4\,\mathrm{V} \\ -2\,\mathrm{k\Omega} & 6\,\mathrm{V} \end{vmatrix}}{\begin{vmatrix} 5\,\mathrm{k\Omega} & -\ 2\,\mathrm{k\Omega} \\ -2\,\mathrm{k\Omega} & 10\,\mathrm{k\Omega} \end{vmatrix}}$$

$$\hat{I}_B = \frac{(30 \times 10^3 + 8 \times 10^3)\,\mathrm{V \cdot \Omega}}{(50 \times 10^6 - 4 \times 10^6)\,\Omega^2} = \frac{38 \times 10^3\ \mathrm{V \cdot \Omega}}{46 \times 10^6\ \Omega^2}$$

$$\hat{I} = \hat{I}_B = 0.826\,\mathrm{mA}$$

Note that in the circuit of Fig. 5-19 the voltage source $\hat{E}_2$ is common to both loops. Therefore, it must be included in both loop equations. When writing the equation for loop A, $\hat{E}_2$ appears as a potential drop (plus to minus). However, when writing the equation for loop B, $\hat{E}_2$ appears as a potential rise (minus to plus).

Example 5-8: The circuit of Fig. 5-20 is called a *bridge circuit*. Determine the current through R_6.

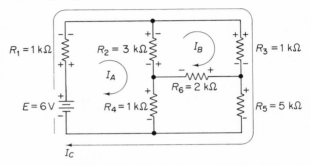

Fig. 5-20. Circuit diagram for Ex. 5-8.

Solution: The loop currents have been chosen so that only loop current I_B flows through R_6. Therefore, only loop current I_B needs to be solved for. Steps 1 and 2 are shown in Fig. 5-20. Applying Kirchhoff's voltage law to each loop,

Loop A:

$$I_A R_1 + I_C R_1 + I_A R_2 - I_B R_2 + I_A R_4 = E$$

or

$$I_A(R_1 + R_2 + R_4) - I_B R_2 + I_C R_1 = E$$

Substituting numerical values,

$$I_A\,5\,k\Omega - I_B\,3\,k\Omega + I_C\,1\,k\Omega = 6\,V$$

Loop B:

$$+ I_B R_2 - I_A R_2 + I_B R_3 + I_B R_6 + I_C R_3 = 0$$

or

$$- I_A R_2 + I_B(R_2 + R_3 + R_6) + I_C R_3 = 0$$

Substituting numerical values,

$$- I_A\,3\,k\Omega + I_B\,6\,k\Omega + I_C\,1\,k\Omega = 0$$

Loop C:

$$I_C R_1 + I_A R_1 + I_C R_3 + I_B R_3 + I_C R_5 = E$$

or

$$I_A R_1 + I_B R_3 + I_C(R_1 + R_3 + R_5) = E$$

Substituting numerical values,

$$I_A\,1\,k\Omega + I_B\,1\,k\Omega + I_C\,7\,k\Omega = 6\,V$$

Thus the set of simultaneous linear equations is

$$I_A\,5\,\text{k}\Omega - I_B\,3\,\text{k}\Omega + I_C\,1\,\text{k}\Omega = 6\,\text{V}$$

$$-I_A\,3\,\text{k}\Omega + I_B\,6\,\text{k}\Omega + I_C\,1\,\text{k}\Omega = 0$$

$$I_A\,1\,\text{k}\Omega + I_B\,1\,\text{k}\Omega + I_C\,7\,\text{k}\Omega = 6\,\text{V}$$

Using the method of determinants,

$$I_B = \frac{\begin{vmatrix} 5 \times 10^3\,\Omega & 6\,\text{V} & 1 \times 10^3\,\Omega \\ -3 \times 10^3\,\Omega & 0 & 1 \times 10^3\,\Omega \\ 1 \times 10^3\,\Omega & 6\,\text{V} & 7 \times 10^3\,\Omega \end{vmatrix}}{\begin{vmatrix} 5 \times 10^3\,\Omega & -3 \times 10^3\,\Omega & 1 \times 10^3\,\Omega \\ -3 \times 10^3\,\Omega & 6 \times 10^3\,\Omega & 1 \times 10^3\,\Omega \\ 1 \times 10^3\,\Omega & 1 \times 10^3\,\Omega & 7 \times 10^3\,\Omega \end{vmatrix}}$$

$$I_B = \frac{[(5 \times 10^3)(-6 \times 10^3) - 6(-21 \times 10^6 - 1 \times 10^6) + \\ (1 \times 10^3)(-18 \times 10^3)]\,\text{V}\cdot\Omega^2}{[(5 \times 10^3)(42 \times 10^6 - 1 \times 10^6) - (-3 \times 10^3)(-21 \times 10^6 - 1 \times 10^6) + \\ 1 \times 10^3(-3 \times 10^3 - 6 \times 10^3)]\,\Omega^3}$$

$$I_B = \frac{84 \times 10^6\,\text{V}\cdot\Omega^2}{130 \times 10^9\,\Omega^3}$$

Therefore,

$$I_B \simeq 0.646\,\text{mA}$$

Thus the current through R_6 is 0.646 mA and flows through it in the direction of I_B. Again note that by choosing the loop currents as we did, only I_B had to be solved for.

Example 5-9: Write the set of simultaneous linear equations for the circuit shown in Fig. 5-21. Do not solve.

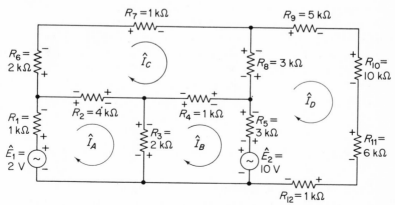

Fig. 5-21. Circuit diagram for Ex. 5-9.

Solution: Applying Kirchhoff's voltage law to each loop,

Loop A:

$$\hat{I}_A(R_1 + R_2 + R_3) - \hat{I}_B R_3 - \hat{I}_C R_2 = \hat{E}_1$$

Then

$$\hat{I}_A 7\,k\Omega - \hat{I}_B 2\,k\Omega - \hat{I}_C 4\,k\Omega = 2\text{ V}$$

Loop B:

$$-\hat{I}_A R_3 + \hat{I}_B(R_3 + R_4 + R_5) - \hat{I}_C R_4 - \hat{I}_D R_5 = -\hat{E}_2$$

Then

$$-\hat{I}_A 2\,k\Omega + \hat{I}_B 6\,k\Omega - \hat{I}_C 1\,k\Omega - \hat{I}_D 3\,k\Omega = -10\text{ V}$$

Loop C:

$$-\hat{I}_A R_2 - \hat{I}_B R_4 + \hat{I}_C(R_2 + R_4 + R_6 + R_7 + R_8) - \hat{I}_D R_8 = 0$$

Then

$$-\hat{I}_A 4\,k\Omega - \hat{I}_B 1\,k\Omega + \hat{I}_C 11\,k\Omega - \hat{I}_D 3\,k\Omega = 0$$

Loop D:

$$-\hat{I}_B R_5 - \hat{I}_C R_8 + \hat{I}_D(R_5 + R_8 + R_9 + R_{10} + R_{11} + R_{12}) = \hat{E}_2$$

Then

$$-\hat{I}_B 3\,k\Omega - \hat{I}_C 3\,k\Omega + \hat{I}_D 28\,k\Omega = 10\text{ V}$$

Thus the set of simultaneous linear equations is

$$\hat{I}_A 7\,k\Omega - \hat{I}_B 2\,k\Omega - \hat{I}_C\ 4\,k\Omega + 0 = 2\text{V}$$
$$-\hat{I}_A 2\,k\Omega + \hat{I}_B 6\,k\Omega - \hat{I}_C\ 1\,k\Omega - \hat{I}_D\ 3\,k\Omega = -10\text{ V}$$
$$-\hat{I}_A 4\,k\Omega - \hat{I}_B 1\,k\Omega + \hat{I}_C 11\,k\Omega - \hat{I}_D\ 3\,k\Omega = 0$$
$$0 - \hat{I}_B 3\,k\Omega - \hat{I}_C\ 3\,k\Omega + \hat{I}_D 28\,k\Omega = 10\text{ V}$$

Note that it is possible for two loop currents not to share a common element. For example, loop currents $\hat{I}_A$ and $\hat{I}_D$ do not flow together through any element. Also, it is not necessary for a voltage source to be included in the choice of a path for a loop current, e.g., loop $\hat{I}_C$.

It is important to understand that loop analysis can be applied to either ac resistive circuits or dc resistive circuits. In any type of network, however, the number of linear equations is dependent on the number of loop currents.

5-5. NODAL ANALYSIS

The object of this section is also to obtain a set of simultaneous linear equations that uniquely describe the network. However, unlike the loop-analysis method, the procedure developed in this section depends on the choice of certain voltages used as variables in conjunction with Kirchhoff's current law. This method of solving problems is called *nodal analysis.* Nodal analysis offers not only an alternative, but in some cases the easiest approach to the solution of the problem.

Since this method uses Kirchhoff's current law, it will be more convenient to treat problems containing only current sources. If a circuit should contain a voltage source, we may obtain the equivalent current source from the procedure developed in Section 5-3.

A general procedure for analyzing a network by nodal analysis is as follows:

1. Convert all voltage sources to current sources.

2. Indicate one node as a reference node. This reference node is used as the common point in the network from which the node voltages $-V_a$, V_b, etc.—are measured. The number of node voltages and thus the number of linear equations is given by

$$N - 1 \qquad\qquad (5\text{-}16)$$

where N is the total number of nodes in the network (see Fig. 5-22).

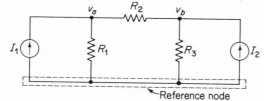

(a) Circuit illustrating two node voltages

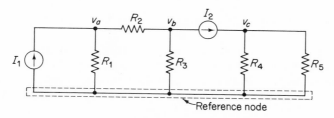

(b) Circuit illustrating three node voltages

Fig. 5-22. Circuit diagrams illustrating the principles of nodal analysis.

3. At each node, except the reference node, apply Kirchhoff's current law.

4. Solve the set of simultaneous linear equations for the node voltages $-V_a$, V_b, etc.

Let us now apply the above procedure to numerical examples.

Example 5-10: Consider the circuit diagram of Fig. 5-23 and determine the branch currents I_1, I_2, and I_3.

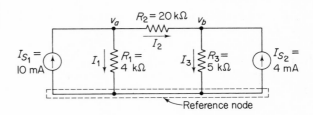

Fig. 5-23. Circuit diagram for Ex. 5-10.

Solution: Steps 1 and 2 are shown in Fig. 5-23. Applying Kirchhoff's current law at each node,

Node *a:*

$$I_{S_1} = I_1 + I_2$$

However, in terms of node voltages,

$$I_1 = \frac{V_a}{R_1} \quad \text{and} \quad I_2 = \frac{V_a - V_b}{R_2}$$

Therefore,

$$I_{S_1} = \frac{V_a}{R_1} + \frac{V_a - V_b}{R_2}$$

or

$$I_{S_1} = V_a\left(\frac{1}{R_1} + \frac{1}{R_2}\right) - V_b\frac{1}{R_2}$$

Substituting numerical values,

$$10\,\text{mA} = V_a\left(\frac{1}{4\,\text{k}\Omega} + \frac{1}{20\,\text{k}\Omega}\right) - V_b\frac{1}{20\,\text{k}\Omega}$$

or

$$10\,\text{mA} = V_a\,0.3\,\text{m}\mho - V_b\,0.05\,\text{m}\mho$$

Node *b:*

$$I_{S_2} + I_2 = I_3$$

or

$$I_{S_2} = -I_2 + I_3$$

In terms of node voltages,

$$I_{S_2} = -\frac{V_a - V_b}{R_2} + \frac{V_b}{R_3}$$

or

$$I_{S_2} = -V_a\frac{1}{R_2} + V_b\left(\frac{1}{R_2} + \frac{1}{R_3}\right)$$

Substituting numerical values,

$$4\,\text{mA} = -V_a\frac{1}{20\,\text{k}\Omega} + V_b\left(\frac{1}{20\,\text{k}\Omega} + \frac{1}{5\,\text{k}\Omega}\right)$$

or

$$4\,\text{mA} = -V_a\,0.05\,\text{m}\mho + V_b\,0.25\,\text{m}\mho$$

Thus the set of simultaneous linear equations is

$$10\,\text{mA} = V_a\,0.3\,\text{m}\mho - V_b\,0.05\,\text{m}\mho$$

$$4\,\text{mA} = -V_a\,0.05\,\text{m}\mho + V_b\,0.25\,\text{m}\mho$$

Using the method of determinants,

$$V_a = \frac{\begin{vmatrix} 10\,\text{mA} & -0.05\,\text{m}\mho \\ 4\,\text{mA} & 0.25\,\text{m}\mho \end{vmatrix}}{\begin{vmatrix} 0.3\,\text{m}\mho & -0.05\,\text{m}\mho \\ -0.05\,\text{m}\mho & 0.25\,\text{m}\mho \end{vmatrix}}$$

$$V_a = \frac{2.7 \times 10^{-6}\,\text{A}\cdot\mho}{0.0725 \times 10^{-6}\,\mho^2}$$

$$V_a = 37.2\,\text{V}$$

$$V_b = \frac{\begin{vmatrix} 0.3\;\text{m}\mho & 10\,\text{mA} \\ -0.05\,\text{m}\mho & 4\,\text{mA} \end{vmatrix}}{0.0725 \times 10^{-6}\,\mho^2}$$

$$V_b = \frac{1.7 \times 10^{-6}\,\text{A}\cdot\mho}{0.0725 \times 10^{-6}\,\mho^2}$$

$$V_b = 23.4\,\text{V}$$

The branch currents then are

$$I_1 = \frac{V_a}{R_1} = \frac{37.2\,\text{V}}{4\,\text{k}\Omega} = 9.3\,\text{mA}$$

$$I_2 = \frac{V_a - V_b}{R_2} = \frac{37.2\,\text{V} - 23.4\,\text{V}}{20\,\text{k}\Omega} = 0.69\,\text{mA}$$

$$I_3 = \frac{V_b}{R_3} = \frac{23.4\,\text{V}}{5\,\text{k}\Omega} = 4.68\,\text{mA}$$

Thus branch currents may be solved in terms of node voltages. Note that an element that has one terminal connected to the reference node has only one node voltage associated with it, for example R_1 and R_3. However, an element connected as R_2 required the difference of two node voltages to find the voltage across it.

Example 5-11: For the circuit of Fig. 5-24, calculate the node voltages $\hat{V}_a$ and $\hat{V}_b$.

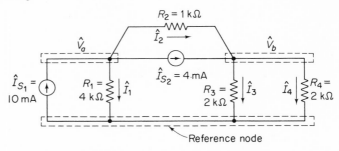

Fig. 5-24. Circuit diagram for Ex. 5-11.

Solution: Applying Kirchhoff's current law to each node,

Node *a:*

$$\hat{I}_{S1} - \hat{I}_{S2} = \hat{I}_1 + \hat{I}_2$$

where

$$\hat{I}_1 = \frac{\hat{V}_a}{R_1} \quad \text{and} \quad \hat{I}_2 = \frac{\hat{V}_a - \hat{V}_b}{R_2}$$

Then

$$\hat{I}_{S_1} - \hat{I}_{S_2} = \frac{\hat{V}_a}{R_1} + \frac{\hat{V}_a - \hat{V}_b}{R_2} = \hat{V}_a\left(\frac{1}{R_1} + \frac{1}{R_2}\right) - \hat{V}_b\frac{1}{R_2}$$

substituting numerical values,

$$10\,\text{mA} - 4\,\text{mA} = \hat{V}_a\left(\frac{1}{4\,\text{k}\Omega} + \frac{1}{1\,\text{k}\Omega}\right) - \hat{V}_b\frac{1}{1\,\text{k}\Omega}$$

$$6\,\text{mA} = \hat{V}_a\,1.25\,\text{m}\mho - \hat{V}_b\,1\,\text{m}\mho$$

Node *b:*

$$\hat{I}_{S_2} + \hat{I}_2 = \hat{I}_3 + \hat{I}_4$$

or

$$I_{S_2} = -\hat{I}_2 + \hat{I}_3 + \hat{I}_4$$

In terms of node voltages,

$$\hat{I}_{S_2} = -\left(\frac{\hat{V}_a - \hat{V}_b}{R_2}\right) + \frac{\hat{V}_b}{R_3} + \frac{V_b}{R_4}$$

or

$$\hat{I}_{S_2} = -\hat{V}_a\frac{1}{R_2} + \hat{V}_b\left(\frac{1}{R_2} + \frac{1}{R_3} + \frac{1}{R_4}\right)$$

Substituting numerical values,

$$4\,\text{mA} = -\hat{V}_a\frac{1}{1\,\text{k}\Omega} + \hat{V}_b\left(\frac{1}{1\,\text{k}\Omega} + \frac{1}{2\,\text{k}\Omega} + \frac{1}{2\,\text{k}\Omega}\right)$$

or

$$4\,\text{mA} = -\hat{V}_a\,1\text{m}\mho + \hat{V}_b\,2\,\text{m}\mho$$

Thus the set of simultaneous equations is

$$6 \text{ mA} = \hat{V}_a \, 1.25 \text{ m}\mho - \hat{V}_b \, 1 \text{ m}\mho$$

$$4 \text{ mA} = -\hat{V}_a \, 1 \text{ m}\mho + \hat{V}_b \, 2 \text{ m}\mho$$

Using the method of determinants,

$$\hat{V}_a = \frac{\begin{vmatrix} 6 \text{ mA} & -1 \text{ m}\mho \\ 4 \text{ mA} & 2 \text{ m}\mho \end{vmatrix}}{\begin{vmatrix} 1.25 \text{ m}\mho & -1 \text{ m}\mho \\ -1 \text{ m}\mho & 2 \text{ m}\mho \end{vmatrix}}$$

$$\hat{V}_a = \frac{[(6 \times 10^{-3})(2 \times 10^{-3}) - (-1 \times 10^{-3})(4 \times 10^{-3})] \, \text{A} \cdot \mho}{[(1.25 \times 10^{-3})(2 \times 10^{-3}) - (-1 \times 10^{-3})(-1 \times 10^{-3})] \mho^2}$$

$$\hat{V}_a = \frac{16 \times 10^{-6} \, \text{A} \cdot \mho}{1.5 \times 10^{-6} \, \mho^2} = 10.6 \text{ V}$$

$$\hat{V}_b = \frac{\begin{vmatrix} 1.25 \text{ m}\mho & 6 \text{ mA} \\ -1 \text{ m}\mho & 4 \text{ mA} \end{vmatrix}}{1.5 \times 10^{-6} \, \mho^2}$$

$$\hat{V}_b = \frac{(5 \times 10^{-6} + 6 \times 10^{-6}) \, \text{A} \cdot \mho}{1.5 \times 10^{-6} \, \mho^2} = 7.33 \text{ V}$$

Thus a current source $(\hat{I}_{s_2})$ that is connected between node a and node b must appear in both node equations. The polarity is determined by whether the current enters or leaves the node. At node a it leaves, therefore, $-\hat{I}_{s_2}$; at node b it enters, thus $+\hat{I}_{s_2}$.

Example 5-12: Consider the network of Fig. 5-25 and determine the branch current $\hat{I}_3$ using nodal analysis.

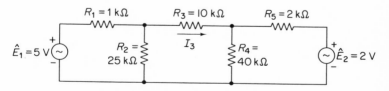

Fig. 5-25. Circuit diagram for Ex. 5-12.

Solution: Using the procedure developed in Section 5-3, we may convert the voltage sources to current sources as shown in Fig. 5-26. To simplify the circuit of Fig. 5-26a, combine the parallel resistors R_1 and R_2 and R_4 and R_5. Let

$$R_A = \frac{R_1 R_2}{R_1 + R_2} = \frac{(1 \text{ k}\Omega)(25 \text{ k}\Omega)}{1 \text{ k}\Omega + 25 \text{ k}\Omega} \approx 1 \text{ k}\Omega$$

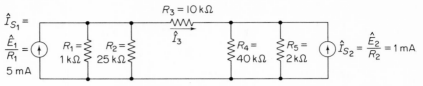

(a) Circuit diagram showing the conversion from voltage
sources to current sources

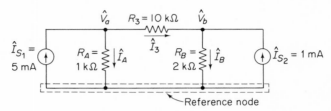

(b) Equivalent circuit showing R_A and R_B

Fig. 5-26. Circuit diagrams used in the solution of Ex. 5-12.

and

$$R_B = \frac{R_4 R_5}{R_4 + R_5} = \frac{(40 \text{ k}\Omega)(2 \text{ k}\Omega)}{40 \text{ k}\Omega + 2 \text{ k}\Omega} \approx 2 \text{ k}\Omega$$

The equivalent and more familiar circuit is shown in Fig. 5-26b. Now apply Kirchhoff's current law to this circuit.

Node *a*:

$$\hat{I}_{S_1} = \hat{I}_A + \hat{I}_3$$

In terms of node voltages,

$$\hat{I}_A = \frac{\hat{V}_a}{R_A}$$

and

$$\hat{I}_3 = \frac{\hat{V}_a - \hat{V}_b}{R_3}$$

Therefore,

$$\hat{I}_{S_1} = \frac{\hat{V}_a}{R_A} + \frac{\hat{V}_a - \hat{V}_b}{R_3}$$

or

$$\hat{I}_{S_1} = \hat{V}_a \left(\frac{1}{R_A} + \frac{1}{R_3} \right) - \hat{V}_b \frac{1}{R_3}$$

Substituting numerical values,

$$5 \text{ mA} = \hat{V}_a \left(\frac{1}{1 \text{ k}\Omega} + \frac{1}{10 \text{ k}\Omega} \right) - \hat{V}_b \frac{1}{10 \text{ k}\Omega}$$

$$5 \text{ mA} = \hat{V}_a 1.1 \text{ m}\mho - \hat{V}_b 0.1 \text{ m}\mho$$

Node b:

$$\hat{I}_{S_2} + \hat{I}_3 = \hat{I}_B$$

or

$$\hat{I}_{S_2} = \hat{I}_B - \hat{I}_3$$

In terms of node voltages,

$$\hat{I}_{S_2} = \frac{\hat{V}_b}{R_B} - \frac{\hat{V}_a - \hat{V}_b}{R_3}$$

$$\hat{I}_{S_2} = -\hat{V}_a \frac{1}{R_3} + \hat{V}_b \left(\frac{1}{R_B} + \frac{1}{R_3} \right)$$

Substituting numerical values,

$$1 \text{ mA} = -\hat{V}_a \frac{1}{10 \text{ k}\Omega} + \hat{V}_b \left(\frac{1}{2 \text{ k}\Omega} + \frac{1}{10 \text{ k}\Omega} \right)$$

$$1 \text{ mA} = -\hat{V}_a \, 0.1 \text{ m}\mho + \hat{V}_b \, 0.6 \text{ m}\mho$$

Thus the set of simultaneous linear equations is

$$5 \text{ mA} = \hat{V}_a \, 1.1 \text{ m}\mho - \hat{V}_b \, 0.1 \text{ m}\mho$$

$$1 \text{ mA} = -\hat{V}_a \, 0.1 \text{ m}\mho + \hat{V}_b \, 0.6 \text{ m}\mho$$

Using the method of determinants,

$$\hat{V}_a = \frac{\begin{vmatrix} 5 \text{ mA} & -0.1 \text{ m}\mho \\ 1 \text{ mA} & 0.6 \text{ m}\mho \end{vmatrix}}{\begin{vmatrix} 1.1 \text{ m}\mho & -0.1 \text{ m}\mho \\ -0.1 \text{ m}\mho & 0.6 \text{ m}\mho \end{vmatrix}}$$

$$\hat{V}_a = \frac{3.1 \times 10^{-6} \text{ A} \cdot \mho}{0.65 \times 10^{-6} \, \mho^2} = 4.78 \text{ V}$$

$$\hat{V}_b = \frac{\begin{vmatrix} 1.1 \text{ m}\mho & 5 \text{ mA} \\ -0.1 \text{ m}\mho & 1 \text{ mA} \end{vmatrix}}{0.65 \times 10^{-6} \, \mho^2}$$

$$\hat{V}_b = \frac{1.6 \times 10^{-6} \text{ A} \cdot \mho}{0.65 \times 10^{-6} \, \mho^2} = 2.46 \text{ V}$$

Since

$$\hat{I}_3 = \frac{\hat{V}_a - \hat{V}_b}{R_3}$$

then

$$\hat{I}_3 = \frac{4.78 \text{ V} - 2.46 \text{ V}}{10 \text{ k}\Omega} = 0.232 \text{ mA}$$

The equivalent network of Fig. 5-26b was obtained so that the procedure for nodal analysis could easily be applied. Although only branch current $\hat{I}_3$ has to be found, it depended on node voltages $\hat{V}_a$ and $\hat{V}_b$ and thus both of these variables first had to be determined. Since $R_2 > 10 R_1$ then their parallel combination is approximately R_1. Similarly $R_4 > 10 R_5$ and their parallel combination is approximately R_5.

Example 5-13: Consider the network of Fig. 5-27 and write (do not solve) the set of simultaneous linear equations.

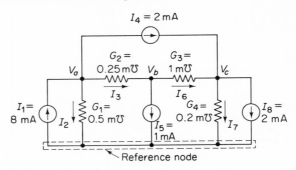

Fig. 5-27. Circuit diagram for Ex. 5-13.

Solution: Steps 1 and 2 are shown in Fig. 5-27. Applying Kirchhoff's current law to each node,

Node *a:*

$$I_1 = I_2 + I_3 + I_4$$

or

$$I_1 - I_4 = I_2 + I_3$$

In terms of node voltages,

$$I_1 - I_4 = V_a G_1 + (V_a - V_b) G_2$$

Substituting numerical values and rearranging,

$$6 \text{ mA} = V_a 0.75 \text{ m}\mho \ - V_b 0.25 \text{ m}\mho$$

Node *b:*

$$I_3 = I_5 + I_6$$

or

$$I_5 = I_3 - I_6$$

In terms of node voltages,

$$I_5 = (V_a - V_b)G_2 - (V_b - V_c)G_3$$

Substituting and rearranging,

$$1 \text{ mA} = V_a 0.25 \text{ m}\mho \ - V_b 1.25 \text{ m}\mho \ + V_c 1 \text{ m}\mho$$

Node *c:*

$$I_4 + I_6 = I_7 + I_8$$

or

$$I_4 - I_8 = I_7 - I_6$$

In terms of node voltages,

$$I_4 - I_8 = V_c G_4 - (V_b - V_c)G_3$$

Substituting and rearranging,

$$0 = -V_{b_1}\, \text{m}\mho + V_c\, 1.2\, \text{m}\mho$$

Thus the set of simultaneous linear equations is

$$6\,\text{mA} = V_a\, 0.75\,\text{m}\mho - V_b\, 0.25\,\text{m}\mho + 0$$

$$1\,\text{mA} = V_a\, 0.25\,\text{m}\mho - V_b\, 1.25\,\text{m}\mho + V_c\, 1\ \text{m}\mho$$

$$0 = \quad 0 \quad - V_{b_1}\ \text{m}\mho + V_c\, 1.2\,\text{m}\mho$$

Although the circuit diagram of Fig. 5-27 has a current source connected to every node, this is not a requirement of nodal analysis. Also note that one branch current is not dependent on every node voltage. For example, I_3 is independent of V_c.

5-6. SUPERPOSITION THEOREM

When dealing with networks of more than one active source, it is possible to use a method that avoids solving a set of simultaneous linear equations. This is accomplished by obtaining the effects of each source separately. Consider the circuits of Fig. 5-28.

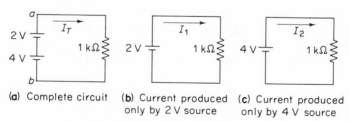

(a) Complete circuit   (b) Current produced only by 2 V source   (c) Current produced only by 4 V source

Fig. 5-28. Circuit diagrams illustrating the method of super-position.

The circuit of Fig. 5-28a is a combination of the voltage sources of circuit b and circuit c. The total voltage between terminals ab of circuit a is 6 V. Therefore, the total current flowing in circuit a is $I_T = 6\ \text{V}/1\ \text{k}\Omega = 6\ \text{mA}$. The current flowing in circuit b and circuit c is $I_1 = 2\ \text{V}/1\ \text{k}\Omega = 2\ \text{mA}$ and $I_2 = 4\ \text{V}/1\ \text{k}\Omega = 4\ \text{mA}$, respectively. The addition of I_1 and I_2 also yields 6 mA. The result is not coincidental.

This technique of analyzing a network with only one source operating at a time is known as the method of *superposition*. The above example is not a proof of the principle of superposition; it only offers a simple demonstration of the method. It is not within the scope of this book to provide a rigorous proof for all theorems. Nevertheless, we still may apply the principles to solve network problems.

This procedure can be applied to *any* passive network containing *any* number of voltage sources, current sources, or a combination of both. These sources do not necessarily have to be in series (or parallel in the case of current sources) and in more elaborate networks they are usually arbitrarily distributed. The *superposition theorem* may be stated as follows: *In a network containing more than one source (of voltage or current), the total current in any branch is the algebraic sum of the individual currents produced by each source acting alone. All other sources are replaced by their internal resistance.*

In obtaining the algebraic sum, if the currents flow in the same direction they add. However, if the currents flow in opposite directions, they subtract from one another. For a numerical analysis, consider the following examples.

Example 5-14: Using the method of superposition, determine the current through the 5 kΩ resistor for the circuit of Fig. 5-29.

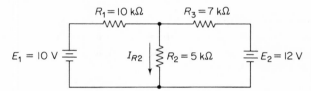

Fig. 5-29. Circuit diagram for Ex. 5-14.

Solution: Consider the circuit of Fig. 5-30a. Let

$$R_A = R_2 \| R_3$$

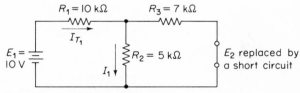

(a) Considering only the effects of E_1

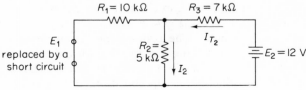

(b) Considering only the effects of E_2

Fig. 5-30. Circuit diagrams used in analyzing Fig. 5-29 by the method of superposition.

Then

$$R_A = \frac{R_2 R_3}{R_2 + R_3} = \frac{(5\,k\Omega)\,(7\,k\Omega)}{5\,k\Omega + 7\,k\Omega} = 2.92\,k\Omega$$

and the total resistance is

$$R_{T_1} = R_1 + R_A = 10\,k\Omega + 2.92\,k\Omega = 12.92\,k\Omega$$

The current drawn from E_1 is

$$I_{T_1} = \frac{E_1}{R_{T_1}} = \frac{10\,V}{12.92\,k\Omega} = 0.775\,mA$$

Using the current division law,

$$I_1 = \frac{R_3}{R_2 + R_3} \times I_{T1}$$

$$I_1 = \frac{7\,k\Omega}{5\,k\Omega + 7\,k\Omega}\,(0.775\,mA) = 0.45\,mA$$

Now consider the circuit of Fig. 5-30b. Let

$$R_B = R_1 \| R_2$$

Then

$$R_B = \frac{R_1 R_2}{R_1 + R_2} = \frac{(10\,k\Omega)\,(5\,k\Omega)}{10\,k\Omega + 5\,k\Omega} = 3.33\,k\Omega$$

and the total resistance is

$$R_{T_2} = R_3 + R_B = 7\,k\Omega + 3.33\,k\Omega = 10.33\,k\Omega$$

The current drawn from E_2 is

$$I_{T_2} = \frac{E_2}{R_{T_2}} = \frac{12\,V}{10.33\,k\Omega} = 1.16\,mA$$

Using the current division law,

$$I_2 = \frac{R_1}{R_1 + R_2} \times I_{T_2}$$

$$I_2 = \frac{10\,k\Omega}{10\,k\Omega + 5\,k\Omega}\,(1.16\,mA) = 0.775\,mA$$

Therefore, the total current through the $5\,k\Omega$ resistance is

$$I_{R_2} = I_1 + I_2$$

$$I_{R_2} = 0.45\,mA + 0.775\,mA = 1.225\,mA$$

Note that the currents I_1 and I_2 flow through the resistance R_2 in the same direction. Therefore, they add to give I_{R_2}. This answer may be checked using either loop analysis or nodal analysis.

Example 5-15: Using the method of superposition, calculate the branch current I_2 in the circuit of Fig. 5-23.

Solution: With the current source I_{S_2} replaced by an open circuit as shown in Fig. 5-31a, R_2 and R_3 are then in series and the current I_A flows through them. Using the current division law,

$$I_A = \frac{R_1}{R_1 + R_2 + R_3} \times I_{S_1}$$

$$I_A = \frac{4 \text{ k}\Omega}{4 \text{ k}\Omega + 20 \text{ k}\Omega + 5 \text{ k}\Omega}(10 \text{ mA}) = 1.38 \text{ mA}$$

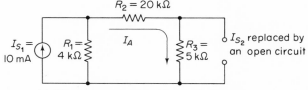

(a) Circuit considering only the effects of I_{S_1}

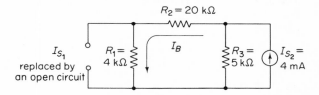

(b) Circuit considering only the effects of I_{S_2}

Fig. 5-31. Circuit diagrams used to analyze Fig. 5-23 by super-position.

With the current source I_{S_1} replaced by an open circuit as shown in Figure 5-31b, R_1 and R_2 are then in series and the current I_B flows through them. Using the current division law,

$$I_B = \frac{R_3}{R_1 + R_2 + R_3} \times I_{S_2}$$

$$I_B = \frac{5 \text{ k}\Omega}{4 \text{ k}\Omega + 20 \text{ k}\Omega + 5 \text{ k}\Omega}(4 \text{ mA}) = 0.69 \text{ mA}$$

The branch current I_2 is now the algebraic sum of I_A and I_B. Since I_B opposes the direction of I_2, then

$$I_2 = I_A - I_B$$

$$I_2 = 1.38 \text{ mA} - 0.69 \text{ mA} = 0.69 \text{ mA}$$

This answer checks with Example 5-10. Similarly, branch currents I_1 and I_3 may be calculated using the method of superposition. This is left as an exercise for the reader.

Example 5-16: Consider the circuit of Fig. 5-32 and using the method of superposition, determine the branch current $\hat{I}_{R_2}$.

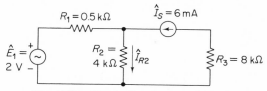

Fig. 5-32. Circuit diagram for Ex. 5-16.

Solution: With the current source replaced by an open circuit as shown in Fig. 5-33a then the current $\hat{I}_1$ is the current drawn from the source E. Thus

$$\hat{I}_1 = \frac{\hat{E}_1}{R_1 + R_2} = \frac{2\,V}{0.5\,k\Omega + 4\,k\Omega} = 0.445\,mA$$

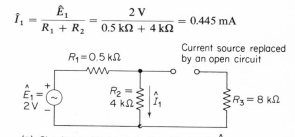

(a) Circuit considering only the effects of $\hat{E}_1$

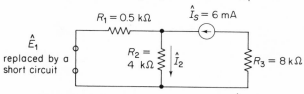

(b) Circuit considering only the effects of $\hat{I}_S$

Fig. 5-33. Circuit diagrams used to **analyze** Fig. 5-32 by superposition.

Considering the circuit of Fig. 5-33b and using the current division law,

$$\hat{I}_2 = \frac{R_1}{R_1 + R_2} \times \hat{I}_S, \qquad \hat{I}_2 = \frac{0.5\,k\Omega}{0.5\,k\Omega + 4\,k\Omega}(6\,mA) = 0.666\,mA$$

Therefore, the branch current I_{R_2} is

$$\hat{I}_{R_2} = \hat{I}_1 + \hat{I}_2, \qquad \hat{I}_{R_2} = 0.445\,mA + 0.666\,mA = 1.111\,mA$$

The branch current $\hat{I}_{R_2}$ is easily found using the method of superposition. However, both loop analysis and nodal analysis would first involve a source transformation. The reader should note that a loop equation cannot be written around any closed path containing an ideal current source because the voltage across an ideal current source is undefined.

Example 5-17: Determine the current through and the power dissipated by R_2 in the circuit of Fig. 5-34.

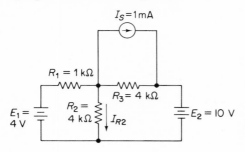

Fig. 5-34. Circuit diagram for Ex. 5-17.

Solution: Referring to the circuit of Fig. 5-35a we see that

$$I_{T1} = \frac{E_1}{R_{T1}}$$

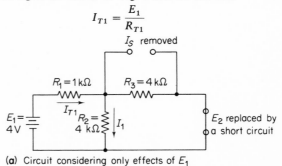

(a) Circuit considering only effects of E_1

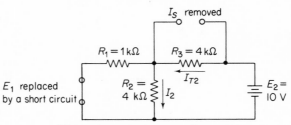

(b) Circuit considering only effects of E_2

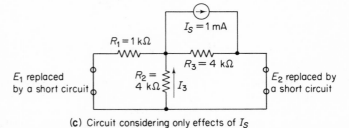

(c) Circuit considering only effects of I_S

Fig. 5-35. Circuit diagrams used to analyze Fig. 5-34 by superposition.

where

$$R_{T_1} = R_1 + \frac{R_2 R_3}{R_2 + R_3}$$

$$R_{T_1} = 1\,\text{k}\Omega + \frac{(4\,\text{k}\Omega)\,(4\,\text{k}\Omega)}{4\,\text{k}\Omega + 4\,\text{k}\Omega} = 3\,\text{k}\Omega$$

Then

$$I_{T_1} = \frac{4\,\text{V}}{3\,\text{k}\Omega} = 1.33\,\text{mA}$$

Using the current division law,

$$I_1 = \frac{R_3}{R_2 + R_3} \times I_{T_1}$$

$$I_1 = \frac{4\,\text{k}\Omega}{4\,\text{k}\Omega + 4\,\text{k}\Omega}\,(1.33\,\text{mA}) = 0.665\,\text{mA}$$

Analyzing the circuit of Fig. 5-35b,

$$I_{T_2} = \frac{E_2}{R_{T_2}}$$

where

$$R_{T_2} = R_3 + \frac{R_1 R_2}{R_1 + R_2}$$

$$R_{T_2} = 4\,\text{k}\Omega + \frac{(1\,\text{k}\Omega)\,(4\,\text{k}\Omega)}{1\,\text{k}\Omega + 4\,\text{k}\Omega} = 4.8\,\text{k}\Omega$$

Then

$$I_{T_2} = \frac{10\,\text{V}}{4.8\,\text{k}\Omega} = 2.08\,\text{mA}$$

Using the current division law,

$$I_2 = \frac{R_1}{R_1 + R_2} \times I_{T2}$$

$$I_2 = \frac{1\,\text{k}\Omega}{1\,\text{k}\Omega + 4\,\text{k}\Omega}\,(2.08\,\text{mA}) = 0.416\,\text{mA}$$

The circuit of Fig. 5-35c shows that R_1, R_2, and R_3 are in parallel. Let

$$R_A = \frac{R_1 R_3}{R_1 + R_3}$$

Then

$$R_A = \frac{(1\,\text{k}\Omega)\,(4\,\text{k}\Omega)}{1\,\text{k}\Omega + 4\,\text{k}\Omega} = 0.8\,\text{k}\Omega$$

Using the current division law,

$$I_3 = \frac{R_A}{R_A + R_2} \times I_S$$

$$I_3 = \frac{0.8\,\text{k}\Omega}{0.8\,\text{k}\Omega + 4\,\text{k}\Omega}\,(1\,\text{mA}) = 0.166\,\text{mA}$$

Therefore, the total current through R_2 with all sources activated is

$$I_{R_2} = I_1 + I_2 - I_3$$

$$I_{R_2} = 0.665 \text{ mA} + 0.416 \text{ mA} - 0.166 \text{ mA}$$

$$I_{R_2} = 0.915 \text{ mA}$$

The power dissipated by R_2 is

$$P_2 = I_{R_2}^2 R_2$$

$$P_2 = (0.915 \text{ mA})^2 \, (4 \text{ k}\Omega)$$

$$P_2 = 3.35 \text{ mW}$$

Since the direction of current I_3 is opposite to that of I_1 and I_2, I_3 is subtracted.

Note that to calculate the power dissipated by a resistance we first must know the total current through the element. We *cannot* calculate the power due to each individual current and then add.

5-7. THEVENIN'S THEOREM

Let us consider the problem where an entire resistive network except for a pair of terminals is sealed in a box as shown in Fig. 5-36a.

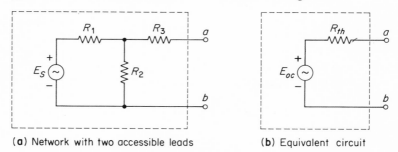

(**a**) Network with two accessible leads   (**b**) Equivalent circuit

Fig. 5-36. Circuit diagrams illustrating Thévenin's theorem.

Having no previous knowledge of the contents of the box, we may assume that it contains an arbitrary number of active and passive elements. By experiment, though, we may establish what effect the network produces at terminals *ab*. One stipulation is that any load connected between terminals *ab* must first be removed. Only after this has been done may the following two tests be performed.

1. An ideal voltmeter — one that draws negligible current — is connected to terminals *ab*. A measured value of voltage indicates that the network inside the box contains at least one energy source. This measured value is the open-circuit voltage, E_{oc}.

2.   An ideal ammeter — negligible resistance — connected between terminals *ab* has the effect of a short circuit.   A measured value of current indicates that the voltage source, E_{oc}, produces a current when the terminals are shorted.   This value of current is I_{sc}.

Since a finite value of current flows when the terminals *ab* are shorted, we may conclude that for all practical purposes a voltage source in series with a resistance exists in the box.   The value of the resistor can be calculated from Ohm's law:

$$R_{Th} = \frac{E_{oc}}{I_{sc}}$$

where R_{Th} is the series resistance in ohms, E_{oc} is the open-circuit voltage in volts, and I_{sc} is the short-circuit current in amperes.   The equivalent circuit is shown in Fig. 5-36b and is similar to that of a practical voltage source.

Therefore, any elaborate network may be reduced to a voltage source in series with a resistor.   Once this equivalent circuit is developed, no matter what load is connected between terminals *ab*, it will have exactly the same current and voltage relationship as it would have if connected to the original circuit.   Otherwise, it could be extremely laborious to solve for the current in the original circuit by loop analysis, nodal analysis, or superposition.

Rather than leave this concept of an equivalent voltage source in series with an equivalent resistance as only an experiment, M. L. Thévenin stated the procedure in the form of a theorem: *Any two terminals of an active network may be represented by a series combination of a voltage source and a resistor.   The value of this equivalent source is the voltage appearing at the open-circuited terminals.   The value of the equivalent resistance is equal to that resistance seen looking back into the network with all energy sources replaced by their respective internal resistance.*

Therefore, using Thévenin's theorem we are able to mathematically analyze any network to obtain an equivalent circuit between any two terminals of the network.

Example 5-18:   Consider the circuit of Fig. 5-37 and determine the Thévenin equivalent circuit for the load resistance, R_L, then determine I_L if R_L is 4, 8, and 10 kΩ.

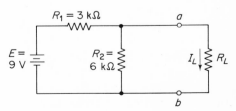

Fig. 5-37. Circuit diagram for Ex. 5-18.

Solution: From Fig. 5-38a we see that E_{Th} is the voltage across R_2.   Using the voltage division law,

$$E_{Th} = \frac{R_2}{R_1 + R_2} \times E, \qquad E_{Th} = \frac{6\,k\Omega}{3\,k\Omega + 6\,k\Omega}(9\,V) = 6\,V$$

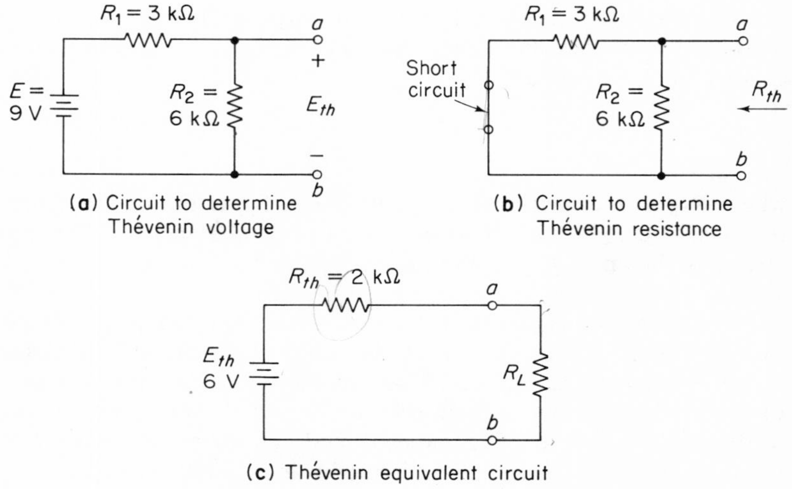

(a) Circuit to determine
Thévenin voltage

(b) Circuit to determine
Thévenin resistance

(c) Thévenin equivalent circuit

Fig. 5-38. Circuit diagrams for the solution of Ex. 5-18.

The Thévenin equivalent resistance is determined using Fig. 5-38b.   Using the product over the sum,

$$R_{Th} = \frac{R_1 R_2}{R_1 + R_2} = \frac{(3\,k\Omega)(6\,k\Omega)}{3\,k\Omega + 6\,k\Omega} = 2\,k\Omega$$

Using the Thévenin equivalent circuit shown in Fig. 5-38c we may determine I_L for different values of R_L:

If $R_L = 4$ kΩ, then

$$I_L = \frac{E_{Th}}{R_{Th} + R_L} = \frac{6\,V}{2\,k\Omega + 4\,k\Omega} = 1\,mA$$

If $R_L = 8$ kΩ, then

$$I_L = \frac{E_{Th}}{R_{Th} + R_L} = \frac{6\,V}{2\,k\Omega + 8\,k\Omega} = 0.6\,mA$$

If $R_L = 10$ kΩ, then

$$I_L = \frac{E_{Th}}{R_{Th} + R_L} = \frac{6\,V}{2\,k\Omega + 10\,k\Omega} = 0.5\,mA$$

A Thévenin circuit delivers the same amount of power to a load as would the original network.   Therefore, when comparison of different loads is desired, a Thévenin circuit is usually easier to analyze than the original network, particularly if the original network has to be analyzed by loop analysis, nodal analysis, or superposition.

Example 5-19: Obtain the Thévenin equivalent circuit of Fig. 5-39.

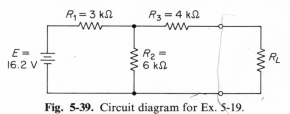

Fig. 5-39. Circuit diagram for Ex. 5-19.

Solution: From the circuit of Fig. 5-40a, E_{Th} is the voltage across R_2 because the voltage across and the current through R_3 is zero. Using the voltage division law,

$$E_{Th} = \frac{R_2}{R_1 + R_2} \times E, \qquad E_{Th} = \frac{6\ k\Omega}{3\ k\Omega + 6\ k\Omega}(16.2\ \text{V}) = \underline{\underline{10.8\ \text{V}}}$$

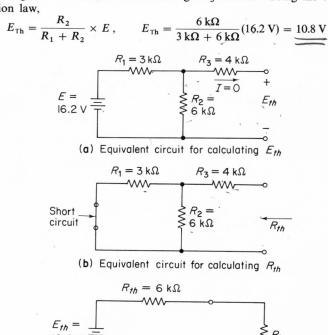

(a) Equivalent circuit for calculating E_{th}

(b) Equivalent circuit for calculating R_{th}

(c) Thévenin equivalent circuit of Fig. 5-39

Fig. 5-40. Circuit diagrams used in the solution of Ex. 5-19.

From the circuit of Fig. 5-40b,

$$R_{Th} = R_3 + R_1 \| R_2$$

$$R_{Th} = R_3 + \frac{R_1 R_2}{R_1 + R_2}$$

$$R_{Th} = 4\,k\Omega + \frac{(3\,k\Omega)(6\,k\Omega)}{3\,k\Omega + 6\,k\Omega} = 6\,k\Omega$$

The Thévenin equivalent circuit is shown in Fig. 5-40c.   Note although R_3 is not used calculating E_{Th}, it is used in calculating R_{Th}.

Example 5-20:   Consider the circuit of Fig. 5-41 and find the Thévenin equivalent circuit for the 2 kΩ resistor.

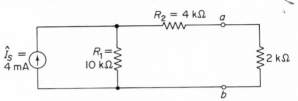

Fig. 5-41. Circuit diagram for Ex. 5-20.

Solution: From Fig. 5-42a note that terminals *ab* are open circuited.   Therefore, the current through and the voltage across the 4 kΩ resistor is zero.

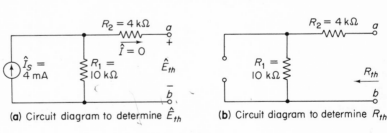

(a) Circuit diagram to determine $\hat{E}_{th}$ **(b)** Circuit diagram to determine R_{th}

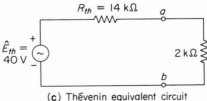

(c) Thévenin equivalent circuit

Fig. 5-42. Circuit diagrams for the solution of Ex. 5-20.

E_{Th} is then equal to the voltage across the 10 kΩ resistor:

$$\hat{E}_{Th} = \hat{I}_s R_1 = (4\,mA)(10\,k\Omega) = 40\,V$$

From Fig. 5-42b,

$$R_{Th} = R_1 + R_2 = 10\,k\Omega + 4\,k\Omega = 14\,k\Omega$$

The Thévenin equivalent circuit is shown in Fig. 5-42c. Note again that the voltage across R_2 in Fig. 5-42a is zero because it is in series with an open circuit and no current flows through an open circuit.

Example 5-21: Consider the circuit of Fig. 5-43 and determine (a) the Thévenin equivalent circuit between terminals *ab* and (b) the current through the 1 kΩ resistor.

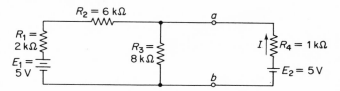

Fig. 5-43. Circuit diagram for Ex. 5-21.

Solution: (a) Applying the voltage division law to Fig. 5-44a yields

$$E_{Th} = \frac{R_3}{R_1 + R_2 + R_3} \times E$$

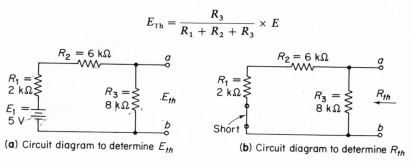

(a) Circuit diagram to determine E_{th} **(b)** Circuit diagram to determine R_{th}

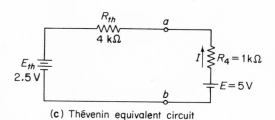

(c) Thévenin equivalent circuit

Fig. 5-44. Circuit diagrams for the solution of Ex. 5-21.

Then

$$E_{Th} = \frac{8\,k\Omega}{2\,k\Omega + 6\,k\Omega + 8\,k\Omega}(5\,V) = 2.5\,V$$

From Fig. 5-44b we see that

$$R_{Th} = (R_1 + R_2) \parallel R_3$$

Using the product over the sum rule,

$$R_{Th} = \frac{(R_1 + R_2) R_3}{R_1 + R_2 + R_3}$$

$$R_{Th} = \frac{(2\,k\Omega + 6\,k\Omega)(8\,k\Omega)}{2\,k\Omega + 6\,k\Omega + 8\,k\Omega} = 4\,k\Omega$$

The Thévenin equivalent circuit is shown in Fig. 5-44c.

(b) Applying Kirchhoff's voltage law to the circuit of Fig. 5-44c we obtain

$$E_{Th} = IR_{Th} + IR_4 + E_2$$

Solving for I yields

$$I = \frac{E_{Th} - E_2}{R_{Th} + R_4}$$

Substituting numerical values yields

$$I = \frac{5\,V - 2.5\,V}{4\,k\Omega + 1\,k\Omega} = 0.5\,mA$$

Thus we have used a Thévenin equivalent circuit and Kirchhoff's voltage law in place of loop analysis, nodal analysis, or superposition.

Example 5-22: Consider the circuit of Fig. 5-45 and calculate (a) the Thévenin equivalent circuit between terminals ab and (b) the current through the 5 kΩ resistor, I_L.

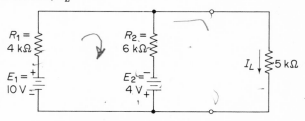

Fig. 5-45. Circuit diagram for Ex. 5-22.

Solution: (a) From the circuit of Fig. 5-46a,

$$E_1 + E_2 = IR_1 + IR_2$$

Rearranging,

$$I = \frac{E_1 + E_2}{R_1 + R_2}$$

$$I = \frac{10\,V + 4\,V}{4\,k\Omega + 6\,k\Omega} = 1.4\,mA$$

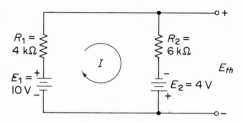

(a) Equivalent circuit for calculating E_{th}

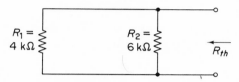

(b) Equivalent circuit for calculating R_{th}

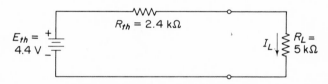

(c) Thévenin equivalent circuit of Fig. 5-45

Fig. 5-46. Circuit diagrams used in the analysis of Fig. 5-45.

Then

$$E_{Th} + E_2 = IR_2$$

or

$$E_{Th} = IR_2 - E_2$$

$$E_{Th} = (1.4 \text{ mA}) (6 \text{ k}\Omega) - 4\text{V} = 4.4 \text{ V}$$

From the circuit of Fig. 5-46b,

$$R_{Th} = \frac{R_1 R_2}{R_1 + R_2}$$

$$R_{Th} = \frac{(4 \text{ k}\Omega) (6 \text{ k}\Omega)}{4 \text{ k}\Omega + 6 \text{ k}\Omega} = 2.4 \text{ k}\Omega$$

The Thévenin equivalent circuit is shown in Fig. 5-46c.

(b) From Fig. 5-46c,

$$I_L = \frac{E_{Th}}{R_{Th} + R_L}$$

$$I_L = \frac{4.4 \text{ k}\Omega}{2.4 \text{ k}\Omega + 5 \text{ k}\Omega} = 0.595 \text{ mA}$$

Thus a Thévenin equivalent circuit can be obtained from a circuit containing more than one active element. The loop-analysis method was used to calculate E_{Th}.

Example 5-23: For the circuit of Fig. 5-47, obtain the Thévenin equivalent circuit.

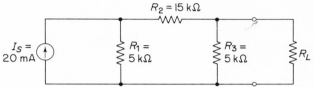

Fig. 5-47. Circuit diagram for Ex. 5-23.

Solution: When R_L is removed, R_2 and R_3 are then in series and E_{Th} is the voltage across R_3, as shown in Fig. 5-48a. Using the current division law,

$$I = \frac{R_1}{R_1 + R_2 + R_3} \times I_s$$

$$I = \frac{5\,k\Omega}{5\,k\Omega + 15\,k\Omega + 5\,k\Omega}\,(20\,mA) = 4\,mA$$

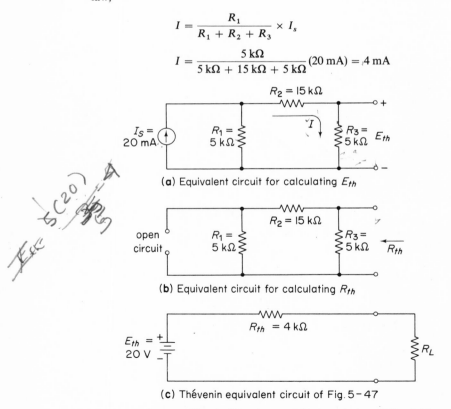

(**a**) Equivalent circuit for calculating E_{th}

(**b**) Equivalent circuit for calculating R_{th}

(**c**) Thévenin equivalent circuit of Fig. 5-47

Fig. 5-48. Circuit diagrams used in the solution of Ex. 5-23.

Then

$$E_{Th} = IR_3$$

$$E_{Th} = (4\,mA)(5\,k\Omega) = 20\,V$$

From the circuit of Fig. 5-48b when I_s is open circuited, R_1 and R_2 are in series. Then

$$R_{Th} = (R_1 + R_2)\|R_3$$

$$R_{Th} = \frac{(R_1 + R_2)R_3}{R_1 + R_2 + R_3}$$

$$R_{Th} = \frac{(5\,k\Omega + 15\,k\Omega)5\,k\Omega}{5\,k\Omega + 15\,k\Omega + 5\,k\Omega} = 4\,k\Omega$$

The Thévenin equivalent circuit is shown in Fig. 5-48c.

Thus it is possible to obtain a Thévenin equivalent circuit from a circuit containing a current source. This problem also may be analyzed by first converting the current source in parallel with R_1 to a voltage source ($E = I_S R_1$) in series with R_1 and then obtaining the Thévenin equivalent circuit.

5-8. NORTON'S THEOREM

In Section 5-3 we saw that it is possible to represent every series combination of a voltage source and a resistor as a parallel combination of a current source and a resistor. This parallel combination is known as a *Norton's equivalent circuit*, shown in Fig. 5-49c.

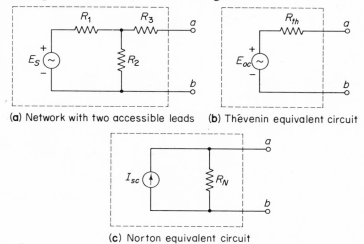

(a) Network with two accessible leads **(b)** Thévenin equivalent circuit

(c) Norton equivalent circuit

Fig. 5-49. Circuit diagrams illustrating Norton's theorem.

The values for a Norton's equivalent circuit are found by using the same two tests as that for the Thévenin's equivalent circuit.

1.   An ideal voltmeter connected between terminals *ab* of Fig. 5-49a measures the open-circuit voltage, E_{oc}.

2.   An ideal ammeter connected between terminals *ab* of Fig. 5-49a measures the short-circuit current, I_{sc}.   This value is the value of the current source in the Norton equivalent circuit.

The internal resistance is calculated using Ohm's law,

$$R_N = \frac{E_{oc}}{I_{sc}}$$

where R_N is the parallel resistor in ohms, E_{oc} is the open-circuit voltage in volts and I_{sc} is the short-circuit current in amperes.

Note that the Thévenin equivalent resistance and the Norton equivalent resistance are equal ($R_{Th} = R_N$).

Norton summed the above results as a theorem: *Any two terminals of an active resistive network may be represented by a parallel combination of a current source and a resistor.   The value of the current source is the short-circuit current at the terminals.   The value of the equivalent resistance is equal to that resistance seen looking back into the network with all energy sources replaced by their respective internal resistance.*

Using Norton's theorem we are able to mathematically analyze any network to obtain an equivalent circuit between any two terminals of the network.

Example 5-24:   Consider the circuit of Fig. 5-50 and find the Norton equivalent circuit for the load resistance, R_L.

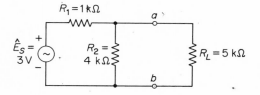

Fig. 5-50. Circuit diagram for Ex. 5-24.

Solution: From Fig. 5-51a, note that when terminals *ab* are connected together, R_2 is short-circuited.   Then

$$\hat{I}_N = \frac{\hat{E}_S}{R_1} = \frac{3\text{ V}}{1\text{ k}\Omega} = 3\text{ mA}$$

From Fig. 5-51b,

$$R_N = \frac{R_1 R_2}{R_1 + R_2} = \frac{(1\text{ k}\Omega)(4\text{ k}\Omega)}{1\text{ k}\Omega + 4\text{ k}\Omega} = 0.8\text{ k}\Omega$$

The Norton equivalent circuit is shown in Fig. 5-51c.

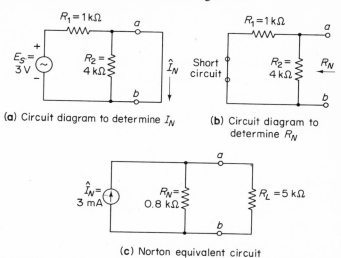

(a) Circuit diagram to determine I_N (b) Circuit diagram to determine R_N

(c) Norton equivalent circuit

Fig. 5-51. Circuit diagrams used in the solution of Ex. 5-24.

Thus a Norton equivalent circuit can be obtained for a network driven by a voltage source. In some problems it may be easier first to obtain the Thévenin equivalent circuit and from this circuit obtain the Norton equivalent circuit.

Example 5-25: Find the Norton equivalent circuit for the 1 kΩ load resistor in the circuit of Fig. 5-52. If the load resistor is changed to 6 kΩ, determine the current through it.

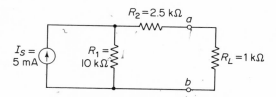

Fig. 5-52. Circuit diagram for Ex. 5-25.

Solution: From Fig. 5-53a we see that when terminals *ab* are shorted, R_1 is in parallel with R_2. Using the current division law,

$$I_N = \frac{R_1}{R_1 + R_2} \times I_S$$

$$I_N = \frac{10 \text{ k}\Omega}{10 \text{ k}\Omega + 2.5 \text{ k}\Omega}(5 \text{ mA}) = 4 \text{ mA}$$

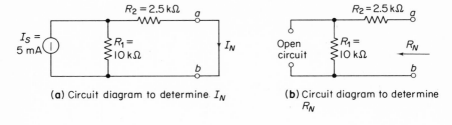

(a) Circuit diagram to determine. I_N (b) Circuit diagram to determine R_N

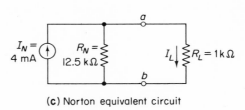

(c) Norton equivalent circuit

Fig. 5-53. Circuit diagrams used in the solution of Ex. 5-25.

Figure 5-53b shows that

$$R_N = R_1 + R_2$$
$$R_N = 10\,\text{k}\Omega + 2.5\,\text{k}\Omega = 12.5\,\text{k}\Omega$$

The Norton equivalent circuit is shown in Fig. 5-53c.

Using the current division law,

$$I_L = \frac{R_N}{R_L + R_N} \times I_S$$

If $R_L = 1\,\text{k}\Omega$, then

$$I_L = \frac{12.5\,\text{k}\Omega}{1\,\text{k}\Omega + 12.5\,\text{k}\Omega}(4\,\text{mA}) = 3.7\,\text{mA}$$

If $R_L = 6\,\text{k}\Omega$, then

$$I_L = \frac{12.5\,\text{k}\Omega}{6\,\text{k}\Omega + 12.5\,\text{k}\Omega}(4\,\text{mA}) = 2.7\,\text{mA}$$

This example, like Example 5-18, shows that an equivalent circuit may be used to determine the load current for different values of R_L.

Example 5-26: For the circuit of Fig. 5-39, obtain the Norton equivalent circuit between terminals *ab*.

Solution: To calculate I_N it is first necessary to find the total current (I_T) drawn from the voltage source in Fig. 5-54a:

$$I_T = \frac{E}{R_T}$$

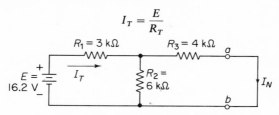

(a) Equivalent current for calculating I_N

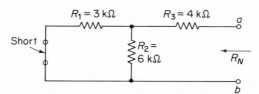

(b) Equivalent circuit for calculating R_N

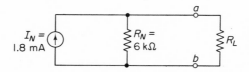

(c) Norton equivalent circuit of Fig. 5-39 .

Fig. 5-54. Circuit diagrams used in the solution of Ex. 5-26.

where

$$R_T = R_1 + R_2 \| R_3$$

$$R_T = R_1 + \frac{R_2 R_3}{R_2 + R_3}$$

$$R_T = 3\,k\Omega + \frac{(6\,k\Omega)\,(4\,k\Omega)}{6\,k\Omega + 4\,k\Omega} = 5.4\,k\Omega$$

Then

$$I_T = \frac{16.2\,V}{5.4\,k\Omega} = 3\,mA$$

Using the current division law,

$$I_N = \frac{R_2}{R_2 + R_3} \times I_T$$

$$I_N = \frac{6\,k\Omega}{6\,k\Omega + 4\,k\Omega}(3\,mA) = 1.8\,mA$$

From the circuit of Fig. 5-54b,

$$R_N = R_3 + R_1 \| R_2.$$

$$R_N = R_3 + \frac{R_1 R_2}{R_1 + R_2}$$

$$R_N = 4\,k\Omega + \frac{(3\,k\Omega)(6\,k\Omega)}{3\,k\Omega + 6\,k\Omega} = 6\,k\Omega$$

The Norton equivalent circuit is shown in Fig. 5-54c.

The Norton equivalent circuit of Fig. 5-54c may be obtained from the Thévenin equivalent circuit of Fig. 5-40c by simply converting the voltage source (E_{Th}) in series with the resistor (R_{Th}) to a current source in parallel with a resistor. The Norton equivalent current is

$$I_N = \frac{E_{Th}}{R_{Th}} = \frac{10.8\,V}{6\,k\Omega} = 1.8\,mA$$

and

$$R_N = R_{Th} = 6\,k\Omega$$

This is the same procedure used in Section 5-3. Note that the Norton equivalent resistance equals the Thévenin equivalent resistance because both circuits from which they are calculated are the same.

SUMMARY

In this chapter we have discussed the effect of load resistor on ideal sources. Because of this, the transfer of energy from the source to the network is never 100 per cent efficient, and these sources are termed practical as opposed to ideal sources.

Section 5-1 considered practical voltage sources and showed that there is an internal voltage drop when a network contains an internal resistor or resistors. By determining the voltage across and the current through the internal resistance, and applying Ohm's law, we were able to determine the value of the internal resistance. A procedure was developed to measure the values of voltage and current. Further discussion centered on the determination of maximum power transfer and the development of the *maximum power transfer theorem*.

Section 5-2 was a consideration of practical current sources and their analysis. A procedure was developed to determine the voltage across and the current through the internal resistance. The *maximum power transfer theorem* was also applied to practical current sources.

Section 5-3 showed how to interchange one practical source for the other and retain the same results at the load. This procedure was made use of in the following sections on analyzing networks.

The remaining five sections discussed different methods for analyzing networks. Using loop analysis, currents were chosen as variables and used in connection with Kirchhoff's voltage law. This procedure provided us with a set of simultaneous linear equations which described the network. Nodal analysis, which also produced a set of simultaneous linear equations, utilized voltages in conjunction with Kirchhoff's current law. When networks contained more than one source, superposition — determining the effects of each source separately — was another method described.

Both Thévenin's theorem and Norton's theorem were used to mathematically analyze a network to obtain an equivalent circuit between any two terminals of the network. Thévenin's theorem represented the equivalent circuit by a series combination of a voltage source and a resistor, while the Norton equivalent circuit was composed of a parallel combination, a current source and a resistor.

PROBLEMS

5-1. A voltmeter that draws negligible current is connected to the output terminals of a practical voltage source. The open-circuit voltage reading is 18 V. When a load resistor of 2 kΩ is connected to the output terminals, the voltmeter reading drops to 16 V. Calculate the internal resistance of the source and draw the equivalent circuit with the proper values.

5-2. An ideal voltmeter measures the open-circuit voltage to be 15 V. When a 1 kΩ load is connected to the output terminals, the voltmeter reading drops to 7.5 V. Calculate the internal resistance.

5-3. The open-circuit voltage of a practical voltage source is measured as 8 V. When a load resistor of 0.8 kΩ is connected to the output terminals, the load current is measured as 2 mA. Calculate the internal resistance and draw the equivalent circuit with the proper values.

5-4. Calculate the value of the internal resistance, R_S, in the circuits of Fig. 5-55.

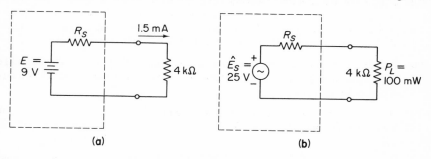

(a)   **(b)**

Fig. 5-55. Circuit diagrams for Prob. 5-4.

5-5. The voltage-current characteristics of a practical dc voltage source are shown in Fig. 5-56. Calculate the values of the ideal voltage source and the internal resistance.

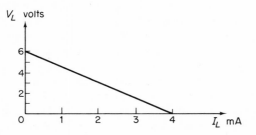

Fig. 5-56. Voltage - current characteristics of Prob. 5-5.

5-6. If a load resistor of 2.5 kΩ is connected to the practical voltage source developed in Problem 5-5, calculate

    (a)   load current, I_L,
    (b)   load voltage, V_L,
    (c)   load power, P_L.

5-7. The power-transfer characteristic of a practical ac voltage source is shown in Fig. 5-57. Determine the values of $\hat{E}_{oc}$, R_S, R_L, and I_L for maximum power.

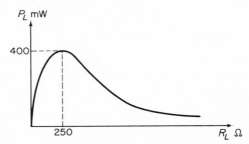

Fig. 5-57. Power-transfer characteristics for Prob. 5-7.

5-8. An ammeter is connected between the output terminals of a practical current source. The short-circuit current reading is 12 mA. When a load resistor of 4 kΩ is connected to the output terminals, a voltmeter reads 4 V. Determine the internal resistance of the source and draw the equivalent circuit.

5-9. An ideal ammeter measures the short-circuit current to be 6 mA. When a 10 kΩ load is connected to the output terminals, a voltmeter reading is 25 V. Calculate the internal resistance.

5-10. A practical current source has a short-circuit current equal to 8 mA and an internal resistance of 15 kΩ. What power will this source deliver to a 1 kΩ load?

5-11. For a practical current source, the open-circuit voltage is 10 V and the short-circuit current is 2 mA. Calculate I_S and R_S and draw the equivalent circuit.

5-12. Calculate the value of the internal resistance, R_S, in the circuits of Fig. 5-58.

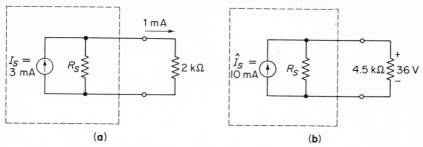

(a)   **(b)**

Fig. 5-58. Circuit diagrams for Prob. 5-12.

5-13. The current-voltage characteristic of a practical current source is shown in Fig. 5-59. Calculate the values of the ideal current source and the internal resistance.

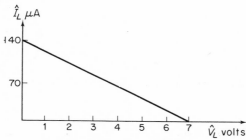

Fig. 5-59. Diagram for Prob. 5-13.

5-14. If a load resistor of 10 kΩ is connected to the output terminals of the practical current source developed in Problem 5-13, calculate

(a) load current, $\hat{I}_L$,
(b) load voltage, $\hat{V}_L$,
(c) load power, P_L.

5-15. The power-transfer characteristic for a practical dc current source is shown in Fig. 5-60. Calculate the values of I_S, R_S, R_L, and I_L for maximum power.

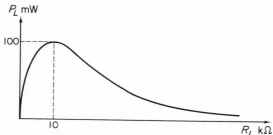

Fig. 5-60. Diagram for Prob. 5-15.

5-16. For each of the practical voltage sources shown in Fig. 5-61, draw the current equivalent circuits.

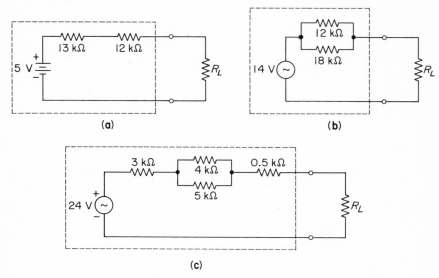

(a)   (b)

(c)

Fig. 5-61. Circuit diagrams for Prob. 5-16.

5-17. For each of the practical current sources shown in Fig. 5-62, draw the voltage equivalent circuit.

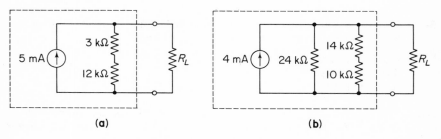

(a)   (b)

Fig. 5-62. Circuit diagrams for Prob. 5-17.

5-18. Refer to the circuit of Fig. 5-18 and calculate the branch currents, I_1, I_2, and I_3 for the following circuit quantities:

(a) $R_1 = 10\,k\Omega$ (b) $R_1 = 1\,k\Omega$ (c) $R_1 = 4\,k\Omega$
    $R_2 = 2\,k\Omega$ $R_2 = 1\,k\Omega$ $R_2 = 2\,k\Omega$
    $R_3 = 5\,k\Omega$ $R_3 = 1\,k\Omega$ $R_3 = 3\,k\Omega$
    $E_1 = 20\,V$ $E_1 = 4\,V$ $E_1 = 5\,V$
    $E_2 = 10\,V$ $E_2 = 2\,V$ $E_2 = 10\,V$

5-19. In the circuit of Fig. 5-20, if both R_2 and R_5 are changed to 1 kΩ, calculate the current through R_6.

5-20. Using loop analysis, solve for the unknown quantities in each of the circuits of Fig. 5-63.

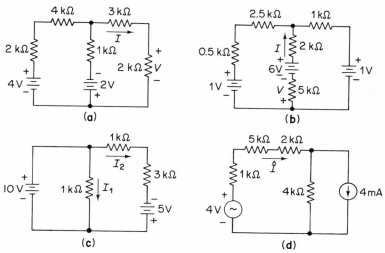

Fig. **5-63.** Circuit diagrams for Prob. 5-20.

5-21. Calculate the current through R_2 in the circuit of Fig. 5-20 using loop analysis.

5-22. If the polarity of the 2-V source in circuit a of Fig. 5-63a is reversed, calculate I using loop analysis.

5-23. Draw a corresponding circuit for each of the following sets of loop equations:

(a) $10\,\text{V} = 5\,\text{k}\Omega\,I_A - 1\,\text{k}\Omega\,I_B$
$5\,\text{V} = -1\,\text{k}\Omega\,I_A + 4\,\text{k}\Omega\,I_B$

(b) $2\,\text{V} = 8\,\text{k}\Omega\,\hat{I}_A - 3\,\text{k}\Omega\,\hat{I}_B$
$4\,\text{V} = -3\,\text{k}\Omega\,\hat{I}_A + 6\,\text{k}\Omega\,\hat{I}$

(c) $0 = 3\,\text{k}\Omega\,I_A - 2\,\text{k}\Omega\,I_B + 0\,I_C$
$-8\,\text{V} = -2\,\text{k}\Omega\,I_A + 6\,\text{k}\Omega\,I_B + 0\,I_C$
$8\,\text{V} = 0\,I_A + 0\,I_B + 6\,\text{k}\Omega\,I_C$

5-24. Refer to the circuit of Fig. 5-23 and calculate the branch currents I_1, I_2, and I_3 using nodal analysis.

(a) $R_1 = 1\,\text{k}\Omega$
$R_2 = 2\,\text{k}\Omega$
$R_3 = 1\,\text{k}\Omega$
$I_{S_1} = 5\,\text{mA}$
$I_{S_2} = 10\,\text{mA}$

(b) $R_1 = 5\,\text{k}\Omega$
$R_2 = 4\,\text{k}\Omega$
$R_3 = 4\,\text{k}\Omega$
$I_{S_1} = 2\,\text{mA}$
$I_{S_2} = 2\,\text{mA}$

(c) $G_1 = 1\,\text{m}\mho$
$G_2 = 4\,\text{m}\mho$
$G_3 = 2\,\text{m}\mho$
$I_{S_1} = 4\,\text{mA}$
$I_{S_2} = 1\,\text{mA}$

5-25. Refer to the circuit of Fig. 5-25 and using nodal analysis calculate the branch current I_3 for the following circuit quantities:

(a) $R_1 = 10\,k\Omega$
     $R_2 = 10\,k\Omega$
     $R_3 = 4\,k\Omega$
     $R_4 = 20\,k\Omega$
     $R_5 = 20\,k\Omega$
     $\hat{E}_1 = 20\,V$
     $\hat{E}_2 = 40\,V$

(b) $R_1 = 3\,k\Omega$
     $R_2 = 6\,k\Omega$
     $R_3 = 1\,k\Omega$
     $R_4 = 1.5\,k\Omega$
     $R_5 = 3\,k\Omega$
     $\hat{E}_1 = 3\,V$
     $\hat{E}_2 = 6\,V$

5-26. Write the nodal equations for each of the circuits of Fig. 5-64.

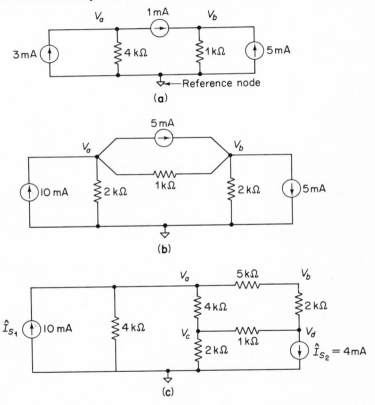

Fig. 5-64. Circuit diagrams for Prob. 5-26.

5-27. Solve for the current through and the voltage across the 1 kΩ resistor in the circuit of Fig. 5-64a and b.

5-28. Using nodal analysis, calculate the current I in the circuit of Fig. 5-63a. (*Hint:* Convert both voltage sources to current sources.)

5-29. Solve for the branch current I_3 in the circuit of Fig. 5-27 using nodal analysis.

5-30. Draw a corresponding circuit for each of the following sets of nodal equations:

(a) $6\,\text{mA} = 1.5\,\text{m}\mho\ V_a - 0.5\,\text{m}\mho\ V_b$
     $6\,\text{mA} = -0.5\,\text{m}\mho\ V_a + 1\,\text{m}\mho\ V_b$

(b) $-2\,\text{mA} = 2\text{m}\mho\ \hat{V}_a - 0.5\,\text{m}\mho\ \hat{V}_b$
     $4\,\text{mA} = -0.5\,\text{m}\mho\ \hat{V}_a + 4\,\text{m}\mho\ \hat{V}_b$

(c) $10\,\text{mA} = 6\,\text{m}\mho\ \hat{V}_a - 4\,\text{m}\mho\ \hat{V}_b + 0\,\hat{V}_c$
     $4\,\text{mA} = -4\,\text{m}\mho\ \hat{V}_a + 7\,\text{m}\mho\ \hat{V}_b - 2\,\text{m}\mho\ \hat{V}_c$
     $-1\,\text{mA} = 0\,\hat{V}_a - 2\,\text{m}\mho\ \hat{V}_b + 10\,\text{m}\mho\ \hat{V}_c$

5-31. Refer to the circuit of Fig. 5-29 and using the method of superposition solve for I_{R_2} if the circuit quantities are

(a) $R_1 = 6\,\text{k}\Omega$ (b) $R_1 = 4\,\text{k}\Omega$
     $R_2 = 6\,\text{k}\Omega$ $R_2 = 4\,\text{k}\Omega$
     $R_3 = 3\,\text{k}\Omega$ $R_3 = 4\,\text{k}\Omega$
     $E_1 = 4\,\text{V}$ $E_1 = 2\,\text{V}$
     $E_2 = 10\,\text{V}$ $E_2 = 4\,\text{V}$

5-32. If the polarity of the 2 V source in the circuit of Fig. 5-32 is reversed, calculate the value of $\hat{I}_{R_2}$.

5-33. Refer to Fig. 5-32. If the direction of the 6 mA current source is reversed, calculate the value of $\hat{I}_{R_2}$.

5-34. What is the value of I_{R_2} and V_{R_2} in the circuit of Fig. 5-34 if

$$R_1 = 4\,\text{k}\Omega \qquad E_1 = 10\,\text{V}$$
$$R_2 = 10\,\text{k}\Omega \qquad E_2 = 10\,\text{V}$$
$$R_3 = 4\,\text{k}\Omega \qquad I_S = 5\,\text{mA}$$

5-35. If the direction of the current source, I_S, in Fig. 5-34 is reversed, calculate I_{R_2}.

5-36. Use the method of superposition and calculate the current $\hat{I}_3$ in the circuit of Fig. 5-24.

5-37. Refer to the circuit of Fig. 5-37 and obtain the Thévenin equivalent circuit between terminals *ab* if the circuit values are

(a) $R_1 = 4\,\text{k}\Omega$ (b) $R_1 = 4\,\text{k}\Omega$ (c) $R_1 = 1.5\,\text{k}\Omega$
     $R_2 = 4\,\text{k}\Omega$ $R_2 = 2\,\text{k}\Omega$ $R_2 = 3\,\text{k}\Omega$
     $E = 10\,\text{V}$ $E = 10\,\text{V}$ $E = 4.5\,\text{V}$

5-38. What is the Thévenin equivalent circuit between terminals *ab* of Fig. 5-41 if

(a) $R_1 = 15\,\text{k}\Omega$ (b) $R_1 = 10\,\text{k}\Omega$
     $R_2 = 3\,\text{k}\Omega$ $R_2 = 10\,\text{k}\Omega$
     $\hat{I}_S = 1.5\,\text{mA}$ $\hat{I}_S = 4\,\text{mA}$

5-39. For each of the circuits of Fig. 5-65, calculate

  (a) the Thévenin equivalent circuit between terminals *ab*,

  (b) the current *I*.

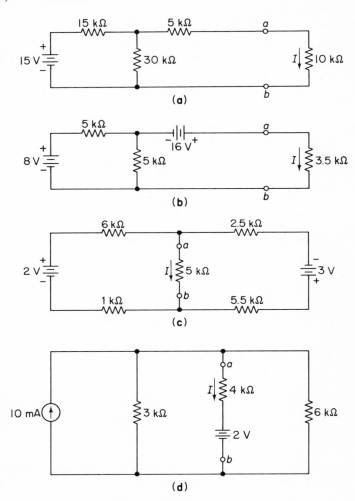

Fig. 5-65. Circuit diagrams for Prob. 5-39.

5-40. Obtain the Norton equivalent circuit for the circuit of Fig. 5-50 if the values are

  (a) $R_1 = 5\,k\Omega$ (b) $R_1 = 3\,k\Omega$ (c) $R_1 = 4\,k\Omega$
       $R_2 = 5\,k\Omega$ $R_2 = 6\,k\Omega$ $R_2 = 1\,k\Omega$
       $E_S = 10\,V$ $E_S = 18\,V$ $E_S = 2\,V$

5-41. Obtain the Norton equivalent circuit for the circuit of Fig. 5-52 if the values are

(a) $R_1 = 3\,k\Omega$ (b) $R_1 = 6\,k\Omega$ (c) $R_1 = 10\,k\Omega$
$\quad\;\; R_2 = 7\,k\Omega$ $R_2 = 6\,k\Omega$ $R_2 = 5\,k\Omega$
$\quad\;\; I_S = 25\,mA$ $I_S = 12\,mA$ $I_S = 30\,mA$

5-42. For each of the circuits of Fig. 5-65 calculate

(a) the Norton equivalent circuit between terminals *ab*,
(b) the current *I*.

CHAPTER 6

addition of sinusoidal waves and complex numbers

In Chapter 4 circuit currents and voltages were found by simple addition and subtraction. However, these circuits contained only one type of circuit element – resistance, inductance, or capacitance. When two different elements appear in a circuit, the currents and voltages can no longer be added or subtracted by simple algebra. In this chapter we shall investigate the special properties of these circuits.

The sections of this chapter deal with (1) ac voltage and current relationships of an inductor, (2) ac voltage and current relationships of a capacitor, (3) addition of waveforms, (4) an introduction to complex numbers, and (5) complex numbers and their use in electric circuits.

6-1. AC VOLTAGE AND CURRENT RELATIONSHIPS OF AN INDUCTOR

A unique property of an inductor is that it opposes any change of current. In this section we will show what happens if an ac waveform of current flows through an inductor. (We will consider dc conditions in Chapter 9.)

The analysis is best approached from a mathematical standpoint: If the reader does not completely understand the derivative operation, he may consider Eq. (6-5) as a definition and proceed.

To determine the relationship of voltage and current, let the instantaneous current through the inductor be a sine wave of the form

$$i_L = I_m \sin \omega t \tag{6-1}$$

We can express the instantaneous value of voltage across the inductor as given in Chapter 3:

$$v_L = L \frac{di_L}{dt} \tag{6-2}$$

where L is the value of inductance and is a constant (does not vary with time).   Substituting Eq. (6-1) into Eq. (6-2),

$$v_L = L \frac{d}{dt}(I_m \sin \omega t) \tag{6-3}$$

Since the maximum value of current, I_m, is also constant,

$$v_L = LI_m \frac{d}{dt}(\sin \omega t) \tag{6-4}$$

By performing the differentiating operation in Eq. (6-4) we obtain a mathematical expression for the voltage across the inductor at any instant of time when an alternating current flows.   Therefore,

$$v_L = LI_m \omega \cos \omega t \tag{6-5}$$

or

$$v_L = V_m \cos \omega t \tag{6-6}$$

where

$$V_m = LI_m \omega$$

L is the value of the inductor in henries, I_m the peak value of current in amperes, and ω the angular velocity in radians per second.

Plotting the sine wave of Eq. (6-1) and the cosine wave of Eq. (6-6) versus time, Fig. 6-1a is obtained.   The phasor diagram is shown in Fig. 6-1b.

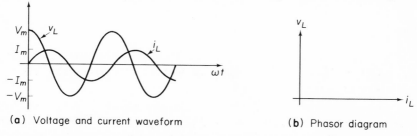

(**a**) Voltage and current waveform   (**b**) Phasor diagram

Fig. 6-1. Voltage and current diagrams for an inductance.

A cosine wave *leads* a sine wave by 90° or $\pi/2$ rad $[\cos \omega t = \sin (\omega t + \pi/2)]$. Therefore, the voltage across a pure inductor, v_L, is commonly referred to as *leading* the current through a pure inductor, i_L, by 90° or $\pi/2$ rad.

Looking at it in another way, a sine wave *lags* a cosine wave by 90° or $\pi/2$ rad $[\sin \omega t = \cos (\omega t - \pi/2)]$. Thus we may state that the current through a pure inductor *lags* the voltage across a pure inductor by 90 or $\pi/2$ rad.

Example 6-1: If a cosine wave of current flows through an inductor, draw the voltage waveform.

Solution: The current wave is of the form

$$i_L = I_m \cos \omega t$$

For an inductor, the voltage waveform *leads* the current waveform by 90° or $\pi/2$ rad. Then

$$v_L = V_m \cos (\omega t + \tfrac{\pi}{2})$$

or

$$v_L = -V_m \sin \omega t$$

Figure 6-2a shows a plot of i_L and v_L versus ωt while Fig. 6-2b shows the phasor diagram. Note that the voltage *leads* the current or the current *lags* the voltage. This property remains unchanged regardless of the sinusoidal current waveform impressed on the inductor.

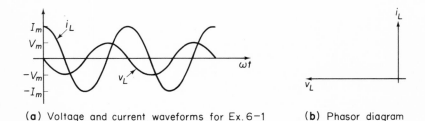

(**a**) Voltage and current waveforms for Ex. 6-1   (**b**) Phasor diagram

Fig. 6-2. Diagrams for Ex. 6-1.

6-2. AC VOLTAGE AND CURRENT RELATIONSHIPS OF A CAPACITOR

A unique property of a capacitor is that it opposes any change of voltage. Let us consider what happens if an ac voltage is applied across the capacitor. In this case we know the voltage waveform and wish to determine the resulting current waveform. Equation (6-11) may be considered as a definition for those who are unfamiliar with the derivative operation.

Let the voltage across the capacitor be a sine wave of the form

$$v_C = V_m \sin \omega t \tag{6-7}$$

Chapter 3 also gave the expression of instantaneous current through the capacitor as

$$i_C = C \frac{dv_C}{dt} \tag{6-8}$$

where C is the value of the capacitance and is independent of time. Substituting Eq. (6-7) into Eq. (6-8),

$$i_C = C \frac{d}{dt}(V_m \sin \omega t) \tag{6-9}$$

Since the maximum value of voltage, V_m, in Eq. (6-9) also does not vary with time,

$$i_C = CV_m \frac{d}{dt}(\sin \omega t) \tag{6-10}$$

Then

$$i_C = CV_m \omega \cos \omega t \tag{6-11}$$

or

$$i_C = I_m \cos \omega t \tag{6-12}$$

where

$$I_m = CV_m \omega$$

C is the value of capacitance in farads, V_m the peak value of voltage in volts, and ω the angular velocity in radians per second. Equation (6-12) yields an expression for the cosine wave of current through a capacitor at any instant of time when a sine wave of voltage is across it. Plotting Eqs. (6-7) and (6-12) versus time, we obtain Fig. 6-3a. The phasor diagram is shown in Fig. 6-3b.

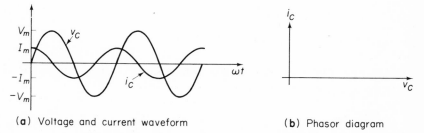

(**a**) Voltage and current waveform   (**b**) Phasor diagram

Fig. 6-3. Voltage and current diagrams for a capacitor.

In the present discussion we have a sine wave of applied voltage, v_C, and a cosine wave of current, i_C. A sine wave *lags* a cosine wave by 90° or $\pi/2$ rad [$\sin \omega t = \cos (\omega t - \pi/2)$]. Thus the voltage across a pure capacitor *lags* the current through a pure capacitor by 90° or $\pi/2$ rad.

Looking at it in another way, a cosine wave *leads* a sine wave by 90° or $\pi/2$ rad [$\cos \omega t = \sin (\omega t + \pi/2)$]. Therefore, we may state that the current through a pure capacitor *leads* the voltage across a pure capacitor by 90° or $\pi/2$ rad.

Example 6-2: If a sine wave of current flows through a capacitor, draw the voltage waveform.

Solution: The current wave is of the form

$$i_C = I_m \sin \omega t$$

For a capacitor the voltage waveform *lags* the current waveform by 90° or $\pi/2$ rad. Then

$$v_C = V_m \sin (\omega t - \tfrac{\pi}{2})$$

or

$$v_C = -V_m \cos \omega t$$

Figure 6-4a shows a plot of i_C and v_C versus ωt while Fig. 6-4b shows the phasor diagram. Note that the current *leads* the voltage or the voltage *lags* the current. This property of a capacitor remains unchanged regardless of the sinusoidal voltage waveform across a capacitance.

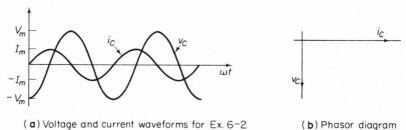

(**a**) Voltage and current waveforms for Ex. 6-2   (**b**) Phasor diagram

Fig. 6-4. Diagrams for Ex. 6-2.

6-3. ADDITION OF WAVEFORMS

In Chapter 5 we used Kirchhoff's law to solve both dc and ac circuits containing only one type of passive element (resistance). Thus we did not involve ourselves directly with time although it was included in the statement of the law. Kirchhoff's voltage law: *At any instant of time, the algebraic sum of potential rises around a closed loop equals the algebraic sum of the potential drops.*

Let us use Kirchhoff's voltage law to solve for the total value of voltage across two different elements — resistance and inductance $(R\text{-}L)$ and resistance and capacitance $(R\text{-}C)$. The $R\text{-}L$ circuit of Fig. 6-5a is a typical

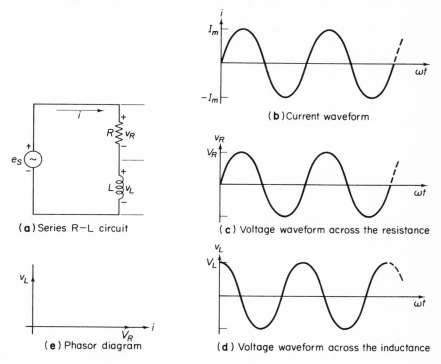

(**b**) Current waveform

(**a**) Series R—L circuit

(**c**) Voltage waveform across the resistance

(**e**) Phasor diagram

(**d**) Voltage waveform across the inductance

Fig. 6-5. Waveforms in a series $R\text{-}L$ circuit.

example of the problems encountered. The voltage source produces alternating current flow in the circuit. Since this is a series circuit, the instantaneous current, i, must be the *same* in all parts of the circuit. (*Note:* The current in the inductor must have the same magnitude at every instant of time as the current in the resistor.)

The alternating current through the resistor causes an ac voltage drop across it. The voltage waveform is exactly in phase with the current waveform. However, from Sec. 6-1 we know that an alternating current through an inductor *lags* the voltage across it by 90°. Or, to look at it another way, the alternating voltage across an inductor *must lead* the alternating current through it by 90°. Therefore, the voltage across the resistor and the voltage across the inductor are 90° out of phase. The peak value of the instantaneous voltage across the inductor will occur 90° earlier than the peak value of the instantaneous voltage across the resistance. This relationship is shown in Fig. 6-5c and d.   Figure 6-5e shows the phasor diagram.

In the case of one element (R, L, or C), where the maximum values and the minimum values of voltage occur at the same instant of time, the calculation of the total value of voltage is obtained by simple addition.   However, our two-element case (R-L circuit) involves the sum of a sine wave and a cosine wave.   The addition of two waves of the same frequency but different amplitude and phase is common in circuit analysis.   Referring to the R-L circuit and using Kirchhoff's voltage law, we obtain.

$$e_S = v_R + v_L \tag{6-13}$$

where the voltage across the resistor is

$$v_R = V_R \sin \omega t \tag{6-14}$$

and the voltage across the inductor is

$$v_L = V_L \cos \omega t \tag{6-15}$$

where V_R and V_L are the maximum voltages across the resistor and inductor, respectively.

For a numerical analysis let

$$V_R = V_L = 1 \text{ V}$$

Then

$$v_R = 1 \sin \omega t$$

and

$$v_L = 1 \cos \omega t$$

The waveform of e_S is obtained by adding the instantaneous values of the sinusoidal waves of Fig. 6-5b and c.   If intervals of $\pi/36$ rad (5°) are taken, a fairly accurate plot of the resultant waveform can be drawn.   Figure 6-6a shows the waveforms of v_R and v_L along with the resultant waveform, e_S.

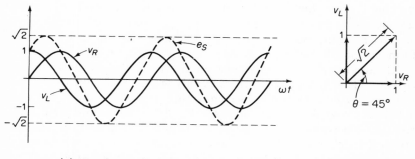

(a) Waveforms of v_R, v_L and e_S (b) Phasor diagram

Fig. 6-6. Addition of a sine and cosine wave of equal amplitude.

From this discussion of a resistor in series with an inductor, we may conclude that when voltages are 90° ($\pi/2$ rad) out of phase, the total voltage is *not* simply the algebraic addition of the individual voltages (either rms or peak). However, the resultant waveform may be obtained by adding the instantaneous values.

From the waveform of Fig. 6-6a note the addition of a sine wave of voltage and a cosine wave of voltage results in a sinusoidal waveform whose peak value occurs neither at the peak value of v_R nor at the peak value of v_L, but at some time in between these peak values. The total or resultant waveform may be represented by a sine wave shifted a certain number of radians (represented by the symbol θ). The resultant wave will also have a greater amplitude than either v_R or v_L.

With this in mind we are able to express the total voltage of Fig. 6-6a in the form of an equation:

$$e_S = E_m \sin(\omega t + \theta) \tag{6-16}$$

where E_m is the maximum value of the resultant waveform and θ is the phase angle. For our present example, E_m can be read from Fig. 6-6a. $E_m = \sqrt{2}$ and $\theta = \pi/4$ rad. Therefore,

$$e_S = \sqrt{2} \sin(\omega t + \tfrac{\pi}{4} \text{ rad})$$

Also note that the addition of two or more sinusoidal waves of the same frequency results in a sinusoidal wave of the same frequency. The phasor diagram is shown in Fig. 6-6b.

Example 6-3: Consider the series *R-L* circuit of Fig. 6-3a. The current waveform is a sine wave. The maximum values of voltage are $V_R = 4$ V and $V_L = 2.3$ V. Draw the waveforms of v_R and v_L along with the resultant waveform.

Solution: Since the voltage across a resistance is in phase with the current, then

$$v_R = 4 \sin \omega t$$

The voltage across an inductance *leads* the current by 90° ($\pi/2$ rad). Then

$$v_L = 2.3 \sin(\omega t + \tfrac{\pi}{2})$$

or

$$v_L = 2.3 \cos \omega t$$

The resultant waveform, e_S, is obtained by adding the instantaneous values of v_R and v_L. The waveforms are shown in Fig. 6-7a and the phasor diagram is shown in Fig. 6-7b. From Fig. 6-7 the expression of e_S is

$$e_S = 4.63 \sin(\omega t + \pi/6)$$

Note that the maximum value of e_S in general does not occur at the inter-sections of v_R and v_L because the maximum values of the latter are unequal.

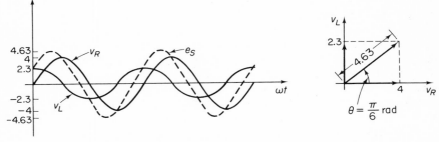

(a) Waveforms of v_R, v_L and e_S
for Ex. 6-3

(b) Phasor diagram

Fig. 6-7. Diagrams for Ex. 6-3.

Example 6-4: A sine wave of current flows through the series R-C circuit of Fig. 6-8. The rms values of voltage across the resistor and capacitor are 0.707 and 1.22 V, respectively. Draw the waveforms of v_R and v_C and the resultant waveform.

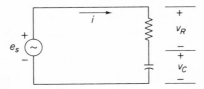

Fig. 6-8. Circuit diagram Ex. 6-4.

Solution: The instantaneous voltage across the resistor is

$$v_R = \sqrt{2}(0.707) \sin \omega t = 1 \sin \omega t$$

The current waveform is a sine wave and since the voltage waveform across a capacitor lags the current through the capacitor by $90°$ ($\pi/2$ rad), then

$$v_C = \sqrt{2}(1.22) \sin (\omega t - \tfrac{\pi}{2})$$
$$v_C = 1.73 \sin (\omega t - \tfrac{\pi}{2})$$

or

$$v_C = -1.73 \cos \omega t$$

The resultant waveform, e_S, is obtained by adding the instantaneous value of v_R and v_C. Figure 6-9a shows a plot of v_R and v_C along with e_S. Figure 6-9b shows the phasor diagram. From Fig. 6-9a the expression of e_S is

$$e_S = 2 \sin (\omega t - \tfrac{\pi}{3})$$

Regardless of whether it is a series *R-L* or a series *R-C* circuit, the rms values of voltage must be first converted to peak values and then plotted.

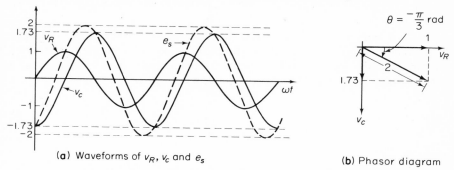

(a) Waveforms of v_R, v_c and e_s (b) Phasor diagram

Fig. 6-9. Addition of a sine and a negative cosine wave of different amplitude for a series R-C circuit.

6-4.   INTRODUCTION TO COMPLEX NUMBERS

In Section 6-3 we found that the addition of a sine wave and a cosine wave yields a waveform of the same frequency but different amplitude. Such waveforms appear frequently in circuit analysis. The total voltage may be represented in a form other than a wave. However, to be able to express these other forms we shall need additional background in mathematics. Until now, the reader has been mainly involved with performing certain mathematical operations — addition, multiplication, division, squares, etc. — with a set of real numbers. The set of real numbers includes

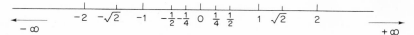

Fig. 6-10. Set of real numbers.

all numbers from $-\infty$ to $+\infty$ (see Fig. 6-10). However, consider the quadratic equation

$$x^2 + 1 = 0$$

Then

$$x = \pm \sqrt{-1}$$

This equation cannot be solved for x because there is no number in the set of real numbers whose square is equal to -1. The laws of algebra state that a positive number squared is a positive number $(+2)^2 = +4$. Similarly, a negative number squared is also a positive number $(-2)^2 = +4$. Therefore, the solution for x can be neither a positive nor a negative number.

Under these conditions, the square root of any negative number (-3, -100, etc.) is impossible to calculate. To provide solutions to equations such as that given above and to complete the number set, mathematicians have introduced a set of imaginary numbers.

-1 is defined as the *unit imaginary number* and given the letter symbol j. (The reader should note that mathematicians denote the unit imaginary number by the letter i. However, to prevent confusion, engineers use j because the symbol i is used for instantaneous current.) Since

$$j = \sqrt{-1}$$

then

$$j^2 = -1$$

$$j^3 = j^2 \times j = -1 \times j = -j$$

$$j^4 = j^2 \times j^2 = -1 \times -1 = 1$$

This procedure may be carried on for any powers of j with the result that they will repeat $-j^5 = j$, etc.

Example 6-5: Solve the equation $x^2 + 36 = 0$ for x.

Solution:

$$x^2 + 36 = 0$$

$$x^2 = -36$$

$$x = \pm \sqrt{-36}$$

$$x = \pm \sqrt{36 \times -1}$$

$$x = \pm 6 \sqrt{-1}$$

$$x = \pm j6$$

A geometrical interpretation of real and imaginary numbers can be achieved by a set of perpendicular axes. Let one of these axes be taken as the axis of real numbers and the other the axis of imaginary numbers (see Fig. 6-11). Therefore, geometrically, we may regard $+j$ as an operator

Fig. 6-11. Real and imaginary axes.

whose effect when applied to a positive real number results in a number on the positive imaginary axis, a rotation of $+90°$ or $\pi/2$ rad (see Fig. 6-12a).

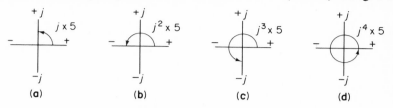

Fig. 6-12. Effect of the $j =$ operator.

If a positive real number is multiplied by the operator j^2, the result is a number on the negative real axis, a rotation of 180° or π rad (see Fig. 6-12b). Figure 6-12c and d show the results of multiplying a positive real number by j^3 (or $-j$) and j^4, a rotation of $+270°$ or $3\pi/2$ ($-90°$ or $-\pi/2$ rad) and 360° or 2π rad, respectively. [*Note:* The multiplication of $+j$ results in a 90° (or $\pi/2$ rad) lead, whereas the multiplication of $-j$ results in a lag of 90° (or $\pi/2$ rad).]

Certain algebraic equations have a solution that is neither a real number nor an imaginary number. This problem is overcome by introducing another class of numbers, called complex numbers, of the form $a + jb$. Both real and imaginary numbers are actually only a special case of complex numbers. Consider the following two cases:

Case 1:
$$b = 0$$

Then $a + jb = a + j0 = a$, a real number.

Case 2:
$$a = 0$$

Then $a + jb = 0 + jb$, an imaginary number.

Complex numbers are made up of two parts, the real part a and the imaginary part jb. Therefore, the addition of a real number and an imaginary number is a complex number which can be plotted on a graph similar to Fig. 6-11. A complex number may be expressed in either rectangular form or polar form.

Rectangular Form. The rectangular form of a complex number is

$$a + jb \qquad (6\text{-}17)$$

Examples of plotting complex numbers in rectangular form are shown in Fig. 6-13.

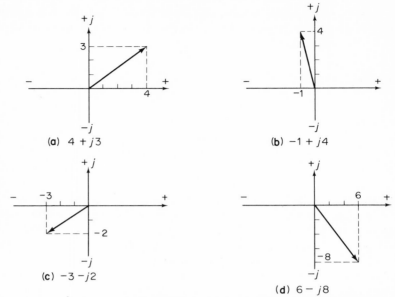

Fig. 6-13. Examples of graphical representation of the rectangular form of a complex number.

Polar Form.   The polar form of a complex number is

$$A \underline{/\theta} \tag{6-18}$$

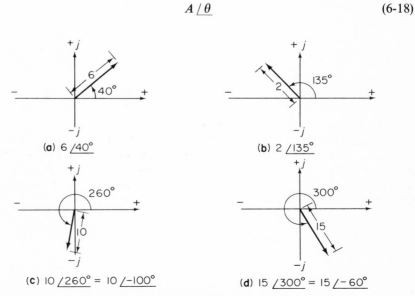

Fig. 6-14. Examples of plotting the polar form of a complex number.

where A is the magnitude and θ is the phase angle. θ is measured from the positive real axis. Examples of plotting a complex number in polar form are shown in Fig. 6-14.

Converting from Rectangular to Polar. The polar form is

$$A \underline{/\theta}$$

where in terms of the rectangular values

$$A = \sqrt{a^2 + b^2} \tag{6-19}$$

and

$$\theta = \underline{/\tan^{-1} \frac{b}{a}} \tag{6-20}$$

Example 6-6: Convert the following complex numbers from rectangular form to polar form:

(a)	$4 + j3$	(c)	$-2 - j6$
(b)	$-1 + j2$	(d)	$6 - j8$

Solution: (a) $4 + j3$

$$A = \sqrt{4^2 + 3^2} = \sqrt{25} = 5$$

$$\theta = \underline{/\tan^{-1} \frac{3}{4}} = 36.9°$$

Therefore,

$$4 + j3 = 5 \underline{/36.9°}$$

A graphical representation is shown in Fig. 6-15a.

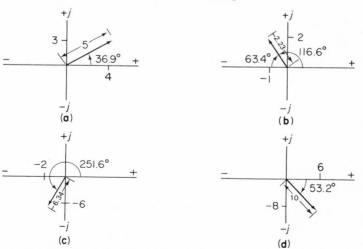

Fig. 6-15. Graphical representation of the complex numbers of Ex. 6-6.

(b) $-1 + j2$

$$A = \sqrt{(-1)^2 + 2^2} = \sqrt{5} = 2.24$$

$$\beta = \underline{/\tan^{-1}\frac{2}{-1}} = -63.4°$$

$$\theta = 180° - 63.4° = 116.6°$$

Therefore,

$$-1 + j2 = 2.24\underline{/116.6°}$$

A graphical representation is shown in Fig. 6-15b.

(c) $-2 - j6$

$$A = \sqrt{(-2)^2 + (-6)^2} = \sqrt{40} = 6.34$$

$$\beta = \underline{/\tan^{-1}\frac{-6}{-2}} = 71.6°$$

$$\theta = 180° + 71.6° = 251.6°$$

Therefore,

$$-2 - j6 = 6.34\underline{/251.6°}$$

A graphical representation is shown in Fig. 6-15c.

(d) $6 - j8$

$$A = \sqrt{6^2 + (-8)^2} = \sqrt{100} = 10$$

$$\theta = \underline{/\tan^{-1}\frac{-8}{6}} = -53.1°$$

Therefore,

$$6 - j8 = 10\underline{/-53.1°}$$

A graphical representation is shown in Fig. 6-15d.

Converting from Polar to Rectangular. The rectangular form is

$$a + jb$$

where in terms of the polar form

$$a = A\cos\theta \qquad\qquad (6\text{-}21)$$

and

$$b = A\sin\theta \qquad\qquad (6\text{-}22)$$

Example 6-7: Convert the following complex numbers from polar form to rectangular form.

  (a) $5\underline{/45°}$ (c) $4\underline{/210°}$

  (b) $8\underline{/120°}$ (d) $1\underline{/-50°}$

Solution: (a) $5\,/\,45°$

$$a = 5\cos(45°) = (5)(0.707) = 3.54$$
$$b = 5\sin(45°) = (5)(0.707) = 3.54$$

Therefore,

$$5\,/\,45° = 3.54 + j3.54$$

Figure 6-16a shows a graphical representation.

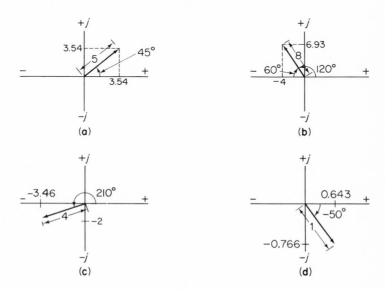

Fig. 6-16. Graphical representation of complex numbers of Ex. 6-7.

(b) $8\,/\,120°$

$$a = 8\cos(180° - 120°) = 8\cos(60°) = (8)(0.5) = 4$$
$$b = 8\sin(60°) = (8)(0.866) \approx 6.93$$

Therefore,

$$8\,/\,120° = -4 + j6.93$$

Figure 6-16b shows a graphical representation.   Note that the phase angle (120°) is in the second quadrant; therefore, the sign of the real part is negative.

(c) $4\,/\,210°$

$$a = 4\cos(210° - 180°) = 4\cos(30°) = (4)(0.866) = 3.46$$
$$b = 4\sin(30°) = (4)(0.5) = 2$$

Therefore,

$$4\underline{/210°} = -3.46 - j2$$

Figure 6-16c shows a graphical representation. Note that the phase angle (210°) is in the third quadrant; therefore, both the real and imaginary parts are preceded by a negative sign.

(d) $1\underline{/-50°}$

$$a = 1 \cos(+50°) = (1)(0.643) = 0.643$$
$$b = 1 \sin(-50°) = (1)(-0.766) = -0.766$$

Therefore,

$$1\underline{/-50°} = 0.643 - j0.766$$

Figure 6-16d shows a graphical representation.

Addition of Complex Numbers.   Whenever it is necessary to add complex numbers, the rectangular form must be used. Using this form, the real parts and the imaginary parts are added separately. The final answer may then be converted to polar form if desired.

Example 6-8:   Add the following complex numbers.

(a) $4 + j3$ and $3 + j4$

(b) $5 + j6$ and $-1 - j2$

(c) $5 - j3$ and $5 + j3$

(d) $1 - j7$ and $3 - j1$

(e) $2\underline{/60°}$ and $1\underline{/-30°}$

Solution: (a)

$$\begin{array}{r} 4 + j3 \\ 3 + j4 \\ \hline 7 + j7 \end{array} = 9.9\underline{/45°}$$

(b)

$$\begin{array}{r} 5 + j6 \\ -1 - j2 \\ \hline 4 + j4 \end{array} = 5.66\underline{/45°}$$

(c)

$$\begin{array}{r} 5 - j3 \\ 5 + j3 \\ \hline 10 + j0 \end{array} = 10\underline{/0°}$$

(d)

$$\begin{array}{r} 1 - j7 \\ 3 - j1 \\ \hline 4 - j8 \end{array} = 8.94 \underline{/-63.4°}$$

(e)

$$\begin{array}{r} 2\underline{/60°} = 1.0 + j1.73 \\ 1\underline{/-30°} = 0.87 - j0.5 \\ \hline 1.87 + j1.23 = 2.24\underline{/33.4°} \end{array}$$

Note that the rules of algebra apply to both the real and imaginary terms. In part (c) the imaginary term equals zero, and the answer is then simply a real number. In part (e) the polar forms first had to be converted to rectangular form and then added. The result then may be converted back to polar form.

Subtraction of Complex Numbers.   Subtraction, like addition, must be performed using the rectangular form. The real and imaginary parts are subtracted separately. The final answer may be expressed in polar form.

Example 6-9: Subtract the following complex numbers.

(a) $(7 + j6) - (3 + j3)$

(b) $(4 + j2) - (1 - j2)$

(c) $(5 + j5) - (2 + j5)$

(d) $(6 - j8) - (6 + j8)$

(e) $(5\underline{/45°}) - (2\underline{/60°})$

Solution: (a)

$$\begin{array}{r} 7 + j6 \\ -3 - j3 \\ \hline 4 + j3 = 5\underline{/36.9°} \end{array}$$

(b)

$$\begin{array}{r} 4 + j2 \\ -1 + j2 \\ \hline 3 + j4 = 5\underline{/53.1°} \end{array}$$

(c)

$$\begin{array}{r} 5 + j5 \\ -2 - j5 \\ \hline 3 + j0 = 3\underline{/0°} \end{array}$$

(d)

$$\begin{array}{r} 6 - j8 \\ -6 - j8 \\ \hline 0 - j16 = 16\underline{/-90°} \end{array}$$

(e)

$$\begin{array}{r} 5\underline{/45°} = 3.54 + j3.54 \\ -2\underline{/60°} = -1.0 - j1.73 \\ \hline 2.54 + j1.81 = 3.12\underline{/35.5°} \end{array}$$

Note that the sign of both the real and imaginary parts of the subtrahend are changed. In part (c) the imaginary part is zero, and the answer is then simply a real number. In part (d) the real part equals zero and the answer is then an imaginary number (note the phase angle). In part (e) the polar forms first had to be converted to rectangular form. The answer may then be converted to polar form.

Multiplication of Complex Numbers. Unlike addition or subtraction, the multiplication of complex numbers may be performed in either the rectangular or the polar form.

When multiplying two complex numbers in rectangular form, treat each number as a binomial. In this way, the regular rules of algebra may be applied.

The product of two complex numbers in rectangular form will contain a j^2 term. According to the properties of complex numbers, whenever a j^2 term appears, it can be replaced by the number -1:

$$(a + jb)(c + jd) = ac + jad + jbc + j^2bd \qquad (6\text{-}23a)$$

Since

$$j^2 = -1$$

then

$$(a + jb)(c + jd) = (ac - bd) + j(ad + bc) \qquad (6\text{-}23b)$$

Therefore, the product of two complex numbers in rectangular form is another complex number in rectangular form.

The multiplication of two complex numbers also can be carried out in polar form.

$$(A_1 \underline{/\theta_1})(A_2 \underline{/\theta_2}) = |A_1||A_2| \underline{/\theta_1 + \theta_2} \qquad (6\text{-}24)$$

The product of two complex numbers in polar form is obtained, as shown in Eq. (6-24), by multiplying the magnitudes and adding the angles.

Example 6-10: Multiply the following complex numbers.

(a) $(1 + j1)(3 + j2)$

(b) $(6 + j8)(2 - j1)$

(c) $(3 - j4)(3 + j4)$

(d) $(4 \underline{/45°})(1 \underline{/30°})$

(e) $(5 \underline{/-80°})(2 \underline{/50°})$

		Rectangular Form	*Polar Form*

Solution:

(a)

$$1 + j1$$
$$\underline{3 + j2}$$
$$3 + j3$$
$$\underline{+\, j2 + j^2\, 2}$$
$$3 + j5 - 2$$

or

$$1 + j5 = 5.12\,\underline{/78.7°}$$

Polar Form (a):

$$1.41\,\underline{/45°}$$
$$\underline{3.5\ \ \underline{/33.7°}}$$
$$5.12\,\underline{/78.7°}$$

(b)

$$6 + j8$$
$$\underline{2 - j1}$$
$$12 + j16$$
$$\underline{-\, j6 - j^2\, 8}$$
$$12 + j10 + 8$$

or

$$20 + j10 = 22.35\,\underline{/26.6°}$$

Polar Form (b):

$$10\ \ \ \ \underline{/53.1°}$$
$$\underline{2.235\,\underline{/-26.5°}}$$
$$22.35\ \ \underline{/26.6°}$$

(c)

$$3 - j4$$
$$\underline{3 + j4}$$
$$9 - j12$$
$$\underline{+\, j12 - j^2\, 16}$$
$$9 + j0\ \ + 16$$

or

$$25 + j0\ \ = 25\,\underline{/0°}$$

Polar Form (c):

$$5\,\underline{/-53.2°}$$
$$\underline{5\,\underline{/53.2°}}$$
$$25\,\underline{/0°}$$

(d)

$$2.83 + j2.83$$
$$\underline{0.87 + j0.5}$$
$$2.45 + j2.45$$
$$\underline{+\, j1.42 + j^2\, 1.42}$$
$$2.45 + j3.87\ \ \ - 1.42$$

or

$$1.03 + j3.87 \approx 4\,\underline{/75°}$$

Polar Form (d):

$$4\ \underline{/45°}$$
$$\underline{1\ \underline{/30°}}$$
$$4\ \underline{/75°}$$

(e)

$$0.866 - j4.93$$
$$\underline{1.28 + j1.53}$$
$$1.11 - j6.33$$
$$\underline{+\, j1.33 - j^2\, 7.55}$$
$$1.11 - j4.97 - 7.55$$

or

$$8.66 - j5.00 = 10\,\underline{/-30°}$$

Polar Form (e):

$$5\ \underline{/-80°}$$
$$\underline{2\ \underline{/50°}}$$
$$10\ \underline{/-30°}$$

This example shows that multiplication of complex numbers may be performed using either the rectangular or the polar form.

Division of Complex Numbers. Like multiplication, the division of complex numbers may also be performed in either the rectangular or the polar form.   However, to divide two complex numbers in rectangular form, it is necessary to eliminate (rationalize) the imaginary part (the j term) from the denominator.   Otherwise, division in this form is impossible.   This elimination is done by multiplying the numerator and the denominator by a term called the *conjugate* of the denominator.   The conjugate of any complex number is obtained by changing only the sign of the imaginary part (the sign preceding the j term).   There is no change in sign for the real part.

As a general expression for the division of two complex numbers, consider

$$\frac{a + jb}{c + jd} \tag{6-25}$$

The conjugate of the denominator $(c + jd)$ is $c - jd$.   Only the sign in front of the j term has been changed.   Multiplying the numerator and denominator by the conjugate, we obtain

$$\frac{a + jb}{c + jd} \times \frac{c - jd}{c - jd} = \frac{(ac + bd) + j(bc - ad)}{c^2 + d^2} \tag{6-26a}$$

$$= \frac{ac + bd}{c^2 + d^2} + j\frac{bc - ad}{c^2 + d^2} \tag{6-26b}$$

Therefore, the division of two complex numbers actually involves the multiplication of two complex numbers, thereby reducing the denominator to a real number.   This method is called "rationalization."   [*Note:* Eq. (6-26b) is a complex number.   Thus the ratio of two complex numbers in rectangular form is a third complex number of the same form.]

The operation of division in polar form may be performed by using the formula

$$\frac{A_1 \,/\, \theta_1}{A_2 \,/\, \theta_2} = \frac{A_1}{A_2} \,/\, \theta_1 - \theta_2 \tag{6-27}$$

Therefore, the ratio of any two complex numbers in polar form is obtained by dividing the magnitudes and subtracting the angle of the denominator from the angle of the numerator.

Example 6-11:   Divide the following complex numbers.

(a) $\dfrac{4 + j3}{1 + j2}$

(b) $\dfrac{6 - j6}{1 - j1}$

(c) $\dfrac{5 - j0}{0 + j2}$

(d) $\dfrac{10 \,/\, -40°}{5 \,/\, 20°}$

(e) $\dfrac{14 \,/\, 30°}{7 \,/\, -10°}$

Solution: (a)

$$\frac{4 + j3}{1 + j2} = \frac{4 + j3}{1 + j2} \times \frac{1 - j2}{1 - j2} = \frac{10 - j5}{5} = 2 - j1 = 2.24 \underline{/-26.6°}$$

or

$$\frac{4 + j3}{1 + j2} = \frac{5 \underline{/36.8°}}{2.24 \underline{/63.4°}} \approx 2.24 \underline{/-26.6°}$$

(b)

$$\frac{6 - j6}{1 - j1} = \frac{6 - j6}{1 - j1} \times \frac{1 + j1}{1 + j1} = \frac{12 + j0}{2} = 6 + j0 = 6 \underline{/0°}$$

or

$$\frac{6 - j6}{1 - j1} = \frac{8.5 \underline{/-45°}}{1.41 \underline{/-45°}} = 6 \underline{/0°}$$

(c)

$$\frac{5 - j0}{0 + j2} = \frac{5 - j0}{0 + j2} \times \frac{0 - j2}{0 - j2} = \frac{0 - j10}{4} = 0 - j2.5 = 2.5 \underline{/-90°}$$

or

$$\frac{5 - j0}{0 + j2} = \frac{5 \underline{/0°}}{2 \underline{/90°}} = 2.5 \underline{/-90°}$$

(d)

$$\frac{10 \underline{/-40°}}{15 \underline{/20°}} = \frac{2}{3} \underline{/-60°}$$

(e)

$$\frac{14 \underline{/30°}}{7 \underline{/-10°}} = 2 \underline{/40°}$$

6-5. COMPLEX NUMBERS AND THEIR USE IN ELECTRIC CIRCUITS

Consider again the series R-L circuit of Fig. 6-5a and let us apply complex notation to this example.   Let the rms values of voltage equal

$$\hat{V}_R = 2 \text{ V}$$

$$\hat{V}_L = 2 \text{ V}$$

As mentioned previously, $\hat{V}_R$ is in phase with the current, but $\hat{V}_L$ leads the current by 90° or $\pi/2$ rad.   Since the voltage leads the current by $\pi/2$ rad, it can be represented by the imaginary number, $j\hat{V}_L$.   Therefore, the voltage across the resistor and the voltage across the inductor can be represented as being 90° out of phase by use of the j notation.

If the above information is superimposed on Fig. 6-11, the result will be the phasor diagram, as shown in Fig. 6-17.   Since the current is common to both elements, it is used as a reference in Fig. 6-17.   Therefore, the total value of voltage expressed in rectangular form is

$$\bar{E}_S = \hat{V}_R + j\hat{V}_L$$

$$\bar{E}_S = (2 + j2) \text{ V}$$

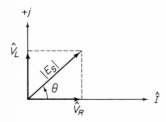

Fig. 6-17. Phasor diagram of voltage for a series R-L circuit.

In polar form,

$$\bar{E}_S = |E_S|\,\underline{/\theta}$$

$$\bar{E}_S = \sqrt{(2 \text{ V})^2 + (2 \text{ V})^2}\,\underline{/\tan^{-1} \frac{2 \text{ V}}{2 \text{ V}}}$$

$$\bar{E}_S = 2.83\,\underline{/45°} \text{ V}$$

$$\bar{E}_S = 2.83\,\underline{/\pi /4} \text{ V}$$

The above formulas and procedures are for a series *R-L* circuit.   Let us now consider a series *R-C* circuit.

> **Example 6-12:**   In the *R-C* circuit of Example 6-4, the values of voltage across the resistor and capacitor are 0.707 and 1.22 V, respectively.   Express the total voltage both in rectangular and polar form.
>
> *Solution:*   The current is taken as the reference because the circuit is a series circuit.   The total voltage for the *R-C* circuit is of the form
>
> $$\bar{E}_S = \hat{V}_R - j\hat{V}_C$$
>
> The rectangular form is
>
> $$\bar{E}_S = (0.707 - j1.22) \text{ V}$$
>
> In polar form,
>
> $$\bar{E}_S = |E_S|\,\underline{/\theta}$$
>
> $$\bar{E}_S = \sqrt{(0.707 \text{ V})^2 + (-1.22 \text{ V})^2}\,\underline{/\tan^{-1} \frac{-1.22 \text{ V}}{0.707 \text{ V}}}$$
>
> $$E_S = 1.414\,\underline{/-60°} \text{ V}$$

The phasor diagram is shown in Fig. 6-18.

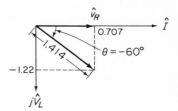

Fig. 6-18. Phasor diagram for the
R-C circuit of Ex. 6-12.

Since the voltage across a capacitor lags the current through a capacitor
by 90°, the $\hat{V}_C$ term is preceded by a minus sign as shown in Eq. (6-20).

Example 6-13: For the network of Fig. 6-19, determine the voltage across
the capacitor both in rectangular and polar form.

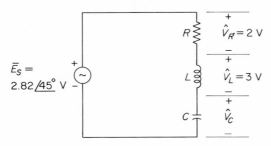

Fig. 6-19. Circuit diagram for Ex. 6-13.

Solution: The rms value of source voltage is 2.82 V and leads the current
by 45°. In rectangular form

$$\bar{E}_S = 2 \underline{/45°} \text{ V} = (2 + j2) \text{ V}$$

Since the current through all elements is the same in this circuit, we take the
phase angle of the current as being zero. According to Kirchhoff's voltage law, the source voltage in the circuit of Fig. 6-19 must be the algebraic
sum of the voltage drops using j notation:

$$\bar{E}_S = \hat{V}_R + j\hat{V}_L - j\hat{V}_C$$

Rearranging the equation to solve for voltage across the capacitor yields

$$-j\hat{V}_C = \bar{E}_S - \hat{V}_R - j\hat{V}_L$$
$$-j\hat{V}_C = (2 + j2 - 2 - j3) \text{ V}$$
$$-j\hat{V}_C = -j\,1 \text{ V} = 1 \underline{/-90°} \text{ V}$$

Therefore, the rms value of voltage across the capacitor is 1 V and lags the
current through it by 90°.

Note that it is necessary to add or subtract ac currents and voltages in rectangular form. The real parts and imaginary parts are added (or subtracted) separately. The final answer may then be converted to polar form.

Example 6-14: For the circuit of Fig. 6-20, calculate the source current in both rectangular and polar form.

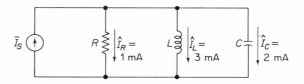

Fig. 6-20. Circuit diagram for Ex. 6-13.

Solution: Since the voltage is the same across all elements in this circuit, we take the phase angle of the voltage as being equal to zero. According to Kirchhoff's current law, the source current in the circuit of Fig. 6-20 must be the algebraic sum of the branch currents. Using the j notation:

$$\bar{I}_S = \hat{I}_R - j\hat{V}_L + j\hat{V}_C$$

$$\bar{I}_S = (1 - j3 + j2)\,\text{mA}$$

or

$$\bar{I}_S = (1 - j1)\,\text{mA} = 1.41\,\underline{/-45°}\,\text{mA}$$

Therefore, the rms value of source current is 1.41 mA and lagging the voltage by 45°.

SUMMARY

In Chapter 6 we have learned the properties of voltage and current for *R-L*, *R-C*, and *R-L-C* circuits driven by an ac energy source.

The first section discussed inductance from an ac voltage and current standpoint. We saw what happens when an ac waveform of current flows through an inductor, which opposes any change of current. This section also introduced the concepts of lead and lag as applied to voltage and current.

Section 6-2 was concerned with capacitance, which opposes any change of voltage. We showed the resultant current wave through a capacitor when an ac voltage wave is applied. The concept of lead and lag also applies to a capacitor.

To represent mathematically the phase angle of 90° ($\frac{\pi}{2}$ rad) between voltage and current, complex numbers were introduced in Section 6-3.

The final section involved the addition and subtraction of complex currents and voltages, using the concepts learned in the preceding section.

PROBLEMS

6-1. The following are sinusoidal expressions of current through an inductor. Draw the current and voltage waveforms and also the phasor diagrams.

(a) $i_L = -I_m \sin \omega t$ (c) $i_L = I_m \sin(\omega t + \pi/4)$
(b) $i_L = -I_m \cos \omega t$ (d) $i_L = I_m \sin(\omega t - \pi/3)$

6-2. The following are sinusoidal expressions of voltage across an inductor. Draw the voltage and current waveforms and also the phasor diagrams.

(a) $v_L = V_m \sin \omega t$ (c) $v_L = V_m \sin(\omega t - \pi/4)$
(b) $v_L = V_m \cos \omega t$ (d) $v_L = V_m \sin(\omega t + \pi/4)$

6-3. The current through a 10-mH inductor is given by the following expressions. Calculate the sinusoidal expression of voltage across the inductor in each case: (Hint: remember to add the $\frac{\pi}{2}$ rad $= 90°$ phase shift.)

(a) $i_L = 10 \sin 50t$ mA   (c) $i_L = 2 \sin(10t + \pi/4)$ mA
(b) $i_L = 4 \cos 200t$ mA   (d) $i_L = -5 \sin 40t$ mA

6-4. The following are sinusoidal expressions of current through a capacitor. Draw the current and voltage waveforms and also the phasor diagrams.

(a) $i_C = I_m \cos \omega t$ (c) $i_C = -I_m \cos \omega t$
(b) $i_C = -I_m \sin \omega t$ (d) $i_C = I_m \sin(\omega t + \pi/6)$

6-5. The following are sinusoidal expressions of voltage across a capacitor. Draw the voltage and current waveforms and also the phasor diagrams.

(a) $v_C = V_m \cos \omega t$ (c) $v_C = V_m \sin(\omega t + \pi/4)$
(b) $v_C = -V_m \sin \omega t$ (d) $v_C = V_m \sin(\omega t - \pi/6)$

6-6. The voltage across a 10 μF capacitor is given by the following expressions. Calculate the sinusoidal expression of current "through" the capacitor in each case. (Hint: remember to add the $\frac{\pi}{2}$ rad $= 90°$ phase shift.)

(a) $v_C = 10 \sin 50t$ V   (c) $v_C = -50 \sin 10t$ V
(b) $v_C = 20 \cos 30t$ V   (d) $v_C = 4 \cos(100t + \pi/6)$ V

6-7. A sine wave of current flows through the R-L circuit of Fig. 6-5a. The maximum values are $V_R = 2$ V and $V_L = 1$ V. Draw the waveforms of v_R and v_L and the resultant waveform, e_S. Write the expression of e_S.

6-8. For the series R-L circuit of Fig. 6-5a, let the current waveform be a cosine wave. The maximum values of voltage $V_R = 1$ V and $V_L = 1$ V. Draw the waveforms of v_R and v_L and the resultant waveform, e_S. Write the expression of e_S.

6-9. A sine wave of current flows through the series R-C circuit of Fig. 6-8. The maximum values are $V_R = 1$ V and $V_C = 1$ V. Draw the waveforms of v_R and v_C and the resultant waveform, e_S. Write the expression of e_S.

6-10. A cosine wave of current flows through the series R-C circuit of Fig. 6-8. The rms values of voltage are $\hat{V}_R = 3$ V and $\hat{V}_C = 6$ V. Draw the waveforms of v_R and v_C and the resultant waveform, e_S. Write the expression of e_S.

6-11. Solve the following equations for x.

(a) $x^2 + 25 = 0$ (d) $x^2 = -3.6 \times 10^5$
(b) $x^2 + 132 = 0$ (e) $x^2 + 144 = 0$
(c) $x^2 + 16 \times 10^4 = 0$ (f) $x^2 + 90 \times 10^{-3} = 0$

6-12. Convert the following complex numbers from rectangular form to polar form.

(a) $1 + j2$ (c) $-5 - j5$ (e) $0 + j4$
(b) $-1 + j2$ (d) $-3 + j1$ (f) $4 + j0$

6-13. Convert the following complex numbers from polar form to rectangular form.

(a) $4\,\underline{/20°}$ (c) $10\,\underline{/120°}$ (e) $2\,\underline{/90°}$
(b) $4\,\underline{/-20°}$ (d) $5\,\underline{/210°}$ (f) $1\,\underline{/-90°}$

6-14. Perform the necessary operations and express the answers in both rectangular and polar form.

(a) $(3 + j4) + (4 + j3)$ (h) $(6 - j8) \div (3 + j4)$
(b) $(1 - j2) + (-1 - j2)$
(c) $(-2 + j2) - (-2 + j2)$ (i) $\dfrac{(4 + j4)(1 + j2)}{(4 + j4) + (1 + j2)}$
(d) $(4 + j1) - (1 + j4)$
(e) $(1 + j5) \times (4 + j2)$
(f) $(-1 - j1) \times (-1 + j1)$ (j) $\dfrac{(2 - j3)(4 + j2)}{(2 - j3) + (4 + j2)}$
(g) $(10 + j12) \div (5 + j5)$

6-15. For the series R-L circuit of Fig. 6-5a, calculate the total value of voltage both in rectangular and polar form for the following values of voltage.

(a) $\hat{V}_R = 1$ V, $\hat{V}_L = 2$ V
(b) $\hat{V}_R = 4$ V, $\hat{V}_L = 3$ V
(c) $\hat{V}_R = 1$ V, $\hat{V}_L = 1$ V
(d) $\hat{V}_R = 6$ V, $\hat{V}_L = 8$ V
(e) $\hat{V}_R = 6$ V, $\hat{V}_L = 3$ V

6-16. For the R-C circuit of Fig. 6-8, calculate the total value of voltage both in rectangular and polar form for the following values of voltage.

(a) $\hat{V}_R = 1$ V, $\hat{V}_C = 4$ V
(b) $\hat{V}_R = 4$ V, $\hat{V}_C = 3$ V
(c) $\hat{V}_R = 12$ V, $\hat{V}_C = 16$ V
(d) $\hat{V}_R = 2$ V, $\hat{V}_C = 2$ V
(e) $\hat{V}_R = 1$ V, $\hat{V}_C = 2$ V

6-17. For the circuit of Fig. 6-20, let $|\hat{I}_R| = 2\,\text{mA}$, $|\hat{I}_L| = 4\,\text{mA}$, and $|\hat{I}_C| = 6\,\text{mA}$. Calculate the source current both in rectangular and polar form.

6-18. For each of the circuits of Fig. 6-21, calculate the value of the unknown current.

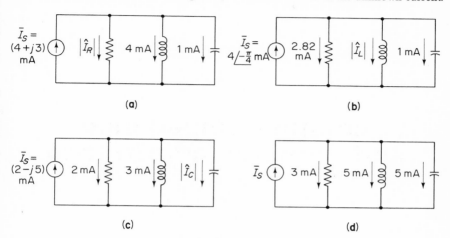

(a)   (b)

(c)   (d)

Fig. 6-21. Circuit diagrams of Prob. 6-18.

6-19. For the circuit of Fig. 6-19, let $\hat{E}_S = (1 + j2)$ V, $|\hat{V}_R| = 1$ V, and $|\hat{V}_L| = 5$ V. Calculate the voltage across the capacitor.

6-20. For each of the circuits of Fig. 6-22, calculate the value of the unknown voltage.

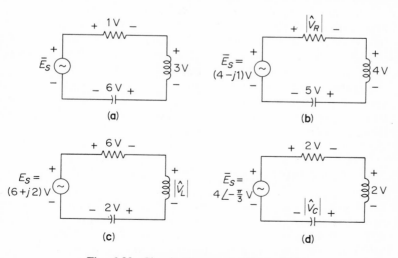

(a)   (b)

(c)   (d)

Fig. 6-22. Circuit diagrams of Prob. 6-20.

CHAPTER 7

complex impedance
and complex admittance

This chapter, like Chapter 6, deals with *R-L*, *R-C*, and *R-L-C* circuits. In Chapter 6 the main concern was determining the total value of voltage or current. Chapter 7, however, is concerned with determining the total value of impedance and admittance as well as voltage and current.

The concepts learned in this chapter will be carried over to later chapters on impedance network theory. Therefore, it is necessary to obtain a thorough working knowledge of complex impedance, complex admittance, and the methods used to determine their values.

In Chapter 6 only the addition and subtraction of complex numbers were needed to solve for the values of voltage or current. However, as we shall see, some complex impedance and admittance problems involve multiplication and division of complex numbers as well.

7-1. COMPLEX IMPEDANCE

Complex impedance is defined as the total opposition to alternating current. It is the ratio of total voltage to total current. In this section we shall apply ac energy sources to series *R-L*, *R-C*, and *R-L-C* circuits.

The values we will obtain for impedance are in the form of complex numbers, with a real and an imaginary part. The *real part* is the value of the *resistance*, and the *imaginary part* is called *reactance*. Like voltage and current, we are also able to express impedance in either rectangular or polar form.

Inductive Reactance.   Consider the *R-L* circuit of Fig. 7-1. The total impedance of the circuit may be determined by the following procedure.

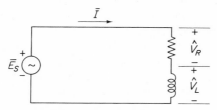

Fig. 7-1. Series *R-L* circuit.

First, let us find the total value of voltage. From Chapter 6 it is known that

$$\bar{E}_S = \hat{V}_R + j\hat{V}_L \tag{7-1}$$

Since this is a series circuit, the current, $\bar{I}$, must be the same through both elements. For this type of circuit it is more convenient to use the current as the reference axis when drawing the phasor diagram. Let the phase angle associated with the current be $0°$. The reader should refer to the phasor diagram of voltage, Fig. 6-9, and note that the current is along the *real* axis.

Dividing Eq. (7-1) by $\bar{I}$ yields

$$\frac{\bar{E}_S}{\bar{I}} = \frac{\hat{V}_R}{\bar{I}} + j\frac{\hat{V}_L}{\bar{I}} \tag{7-2}$$

The left side of Eq. (7-2) is the total voltage, $\bar{E}_S$, divided by the total current, $\bar{I}$. This ratio is called the *total impedance* of the circuit:

$$\frac{\bar{E}_S}{\bar{I}} = \bar{Z}_T \tag{7-3}$$

where $\bar{E}_S$ is the applied voltage in volts, $\bar{I}$ the current in amperes, and $\bar{Z}_T$ the total impedance in ohms.

Now consider each term on the right side of Eq. (7-2) separately. The first term, $\hat{V}_R/\bar{I}$, is the voltage across the resistor divided by the current through the resistor. This quotient is equal to the actual value of resistance in the circuit:

$$\frac{\hat{V}_R}{\bar{I}} = R \tag{7-4}$$

In Chapters 4 and 5 we used this relationship to solve resistive circuits.

One might expect that the second term, $j(\hat{V}_L/\bar{I})$, is the actual value of the inductance in the circuit. However, this is not the case. First consider the $\hat{V}_L/\bar{I}$ term. This is the value of voltage across the inductor divided by the value of current through the inductor. Since voltage is expressed in volts and current in amperes, the resultant units must be ohms. However, the units of inductance are henries. Thus $\hat{V}_L/\bar{I}$ cannot be the actual value of the inductor. It is, in fact, a new concept.

When the unit of ohms is associated with an element that has no phase shift between voltage and current, it is termed *resistance*, as in Eq. (7-4). However, we know that the $+j$ in the term $+j(\hat{V}_L/\bar{I})$ means that the voltage across the inductor leads the current through it by 90°. Therefore, let us define the magnitude of voltage across an element divided by the current through the element, where the voltage and current are 90° (or $\pi/2$ rad) out of phase as *reactance* and give it the symbol X.

In this case, we are only concerned about an inductor. Therefore, our new term for opposition to current will be defined as *inductive reactance*, X_L:

$$\left|\frac{\hat{V}_L}{\bar{I}}\right| = X_L \tag{7-5}$$

Substituting Eqs. (7-3), (7-4), and (7-5) into Eq. (7-2) yields

$$\bar{Z}_T = R + jX_L \tag{7-6}$$

where $\bar{Z}_T$ is the total impedance, R the resistance, and X_L the inductive reactance.

Example 7-1: Let us consider the circuit of Fig. 7-1 and determine the impedance. Let $|\hat{V}_R| = 6$ V, $|\hat{V}_L| = 8$ V, and $\bar{I} = 2$ mA.

Solution:

$$\bar{E}_S = (6 + j8)\text{ V}$$

$$\frac{\bar{E}_S}{\bar{I}} = \frac{6\text{ V}}{2\text{ mA}} + j\frac{8\text{ V}}{2\text{ mA}}$$

$$\bar{Z}_T = (3 + j4)\text{ k}\Omega$$

Thus the value of the resistor equals 3 kΩ and the value of inductive reactance equals 4 kΩ. (*Note:* It is the value of inductive reactance and not the value of the inductor that is equal to 4 kΩ.)

The total voltage can be expressed in both rectangular form and the polar form; so can the total impedance. In polar form,

$$\bar{Z}_T = |Z_T|\underline{/\theta} \tag{7-7}$$

For the above example,

$$\bar{Z}_T = \sqrt{(3\text{ k}\Omega)^2 + (4\text{ k}\Omega)^2}\,\underline{/\tan^{-1}\frac{4\text{ k}\Omega}{3\text{ k}\Omega}}$$

$$\bar{Z}_T = 5\,\underline{/53.1°}\text{ k}\Omega$$

The phasor diagram is shown in Fig. 7-2.

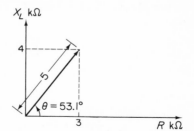

Fig. 7-2. Phasor diagram of a series
R-L circuit.

Until now we have discussed and defined inductive reactance as the magnitude of voltage divided by current, where the voltage and current are 90° out of phase. However, it would also seem advantageous if we were able to express the value of reactance in terms of the actual value of the element.

The following is a derivation for the inductive reactance in terms of usually known quantities (inductance and frequency) as opposed to measurable values (voltage and current). In Chapter 6 the instantaneous values of voltage and current for the inductor were given as

$$v_L = LI_m \omega \cos \omega t \tag{7-8a}$$

or

$$v_L = LI_m \omega \sin \left(\omega t + \tfrac{\pi}{2}\right) \tag{7-8b}$$

and

$$i_L = I_m \sin \omega t \tag{7-9}$$

Inductive reactance is defined as the magnitude of voltage across the inductor divided by the current through the inductor. This definition will not change, but the reader should realize that the definition did not require that the values of voltage or current be rms, peak, or instantaneous. However, it should be noted that if the voltage is rms, the current must also be rms. Similarly, if the voltage is an instantaneous value, the current will also have to be an instantaneous value. In other words, voltage and current must be expressed in the same terms.

Dividing Eq. (7-8b) by Eq. (7-9) we obtain

$$\frac{v_L}{i_L} = \frac{LI_m \sin (\omega t + \pi/2)}{I_m \sin \omega t}$$

$$\frac{v_L}{i_L} = L\omega \underline{/ +90°}$$

or

$$\frac{v_L}{i_L} = +j\omega L \tag{7-10}$$

Once again, the $+j$ term accounts for the fact that the voltage and current are not in phase. The magnitude of Eq. (7-10) gives the value of inductive reactance:

$$\left|\frac{v_L}{i_L}\right| = \omega L$$

or

$$X_L = \omega L \qquad (7\text{-}11)$$

where $\omega = 2\pi f$, f is the frequency in hertz, and L is the value of the inductor in henries. Equation (7-11) shows that inductive reactance depends on the values of the inductor, frequency, and the constant 2π.

Example 7-2: Consider the circuit of Fig. 7-3a and determine the total impedance in (a) rectangular form and (b) polar form.

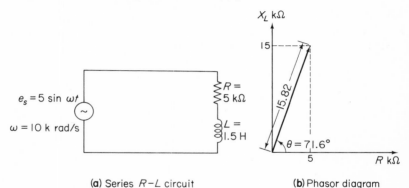

(a) Series R–L circuit   (b) Phasor diagram

Fig. 7-3. Diagrams for Ex. 7-2.

Solution: (a)

$$\bar{Z}_T = R + j\omega L$$
$$\bar{Z}_T = 5\,\text{k}\Omega + j(10\,\text{krad/s})\,(1.5\,\text{H})$$
$$\bar{Z}_T = (5 + j15)\,\text{k}\Omega$$

(b)

$$\bar{Z}_T = |\bar{Z}_T|\,\underline{/\theta}$$
$$\bar{Z}_T = \sqrt{(5\,\text{k}\Omega)^2 + (15\,\text{k}\Omega)^2}\,\underline{/\tan^{-1}\frac{15\,\text{k}\Omega}{5\,\text{k}\Omega}}$$
$$\bar{Z}_T = 15.82\,\underline{/71.6°}\,\text{k}\Omega$$

The phasor diagram is shown in Fig. 7-3b.

Although the instantaneous value of source voltage was given, it was not needed in the direct calculation of total impedance in Example 7-2.

Note that the angular velocity, ω, was given in radians per second and not in hertz. Therefore, the constant, 2π, was included in the calculation of inductive reactance.

Example 7-3: For the circuit of Fig. 7-4a, calculate the total impedance in (a) rectangular form and (b) polar form.

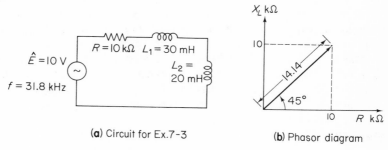

             (a) Circuit for Ex.7-3   **(b)** Phasor diagram

Fig. 7-4. Diagrams for Ex. 7-3.

Solution: $\omega = 2\pi f = (6.28)(31.8 \text{ kHz}) = 200 \text{ krad/s}$.

(a)

$$\bar{Z}_T = 10 \text{ k}\Omega + j(200 \text{ krad/s})(30 \text{ mH}) + j(200 \text{ krad/s})(20 \text{ mH})$$

$$\bar{Z}_T = 10 \text{ k}\Omega + j6 \text{ k}\Omega + j4 \text{ k}\Omega$$

or

$$\bar{Z}_T = (10 + j10) \text{ k}\Omega$$

(b)

$$\bar{Z}_T = \sqrt{(10 \text{ k}\Omega)^2 + (10 \text{ k}\Omega)^2} \, \underline{/\tan^{-1} \dfrac{10 \text{ k}\Omega}{10 \text{ k}\Omega}}$$

$$\bar{Z}_T = 14.14 \, \underline{/45°} \text{ k}\Omega$$

The phasor diagram is shown in Fig. 7-4b.

This problem could be solved by first adding L_1 and L_2 and then calculating the inductive reactance. This is left as an exercise for the reader.

The value of inductive reactance according to Eq. (7-11) is directly proportional to only two parameters, frequency and inductance. Therefore, if we double the frequency without changing the source voltage or the inductance, the value of inductive reactance is doubled. Likewise, if we doubled the value of inductance, leaving both the source voltage and frequency constant, the value of inductive reactance is also doubled. However, doubling only the value of the source voltage while leaving the frequency and inductance constant does not change the inductive reactance. The reason for this is that if the source voltage is doubled, the total value of current is also doubled. Consequently the ratio of the voltage across the inductor divided by the current through the inductor remains constant. Therefore, unlike the two previous cases, changing the source voltage does not change the total impedance of the circuit.

Capacitive Reactance.   The procedure for determining the complex impedance of a series *R-C* circuit is similar to that of a series *R-L* circuit.   Remember, however, that the voltage across a capacitor lags the current through the capacitor by 90° ($\pi/2$ rad).   This lag, in complex notation, is represented by a $-j$ term.

The voltage across the inductor divided by the current through the inductor is defined as inductive reactance.   Likewise, the magnitude of voltage across the capacitor divided by the current through the capacitor is defined as *capacitive reactance*, represented by the symbol X_C.   The following example will hopefully clarify any problems dealing with voltage and current relationships in series *R-C* circuits.

Example 7-4:   Determine the complex impedance both in rectangular and polar form for the circuit of Fig. 7-5a.

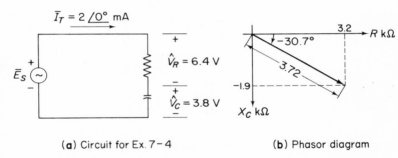

(a) Circuit for Ex. 7-4   (b) Phasor diagram

Fig. 7-5. Diagrams for Ex. 7-4.

Solution: (a)   By Kirchhoff's voltage law,

$$\bar{E}_S = \hat{V}_R - j\hat{V}_C \qquad (7\text{-}12)$$

Dividing both sides of Eq. (7-12) by the current, $\bar{I}$, we obtain

$$\frac{\bar{E}_S}{\bar{I}} = \frac{\hat{V}_R}{\bar{I}} - j\frac{\hat{V}_C}{\bar{I}}$$

$$\bar{Z}_T = R - jX_C \qquad (7\text{-}13)$$

where $\hat{V}_R/\bar{I} = R$ and $|\hat{V}_C/\bar{I}| = X_C$.   Thus

$$\frac{\bar{E}_S}{\bar{I}} = \frac{6.4\text{ V}}{2\text{ mA}} - j\frac{3.8\text{ V}}{2\text{ mA}}$$

$$\bar{Z}_T = (3.2 - j1.9)\text{ k}\Omega$$

(b)   In polar form,

$$\bar{Z}_T = |Z_T|\underline{/\theta}$$

$$\bar{Z}_T = \sqrt{(3.2\ k\Omega)^2 + (1.9\ k\Omega)^2}\ \underline{/\tan^{-1}\dfrac{-1.9\ k\Omega}{3.2\ k\Omega}}$$

$$\bar{Z}_T = 3.72\ \underline{/-30.7^\circ}\ k\Omega$$

The phasor diagram is shown in Fig. 7-5b. Note that it is the value of capacitive *reactance* that is equal to 1.9 kΩ and not the value of the capacitor.

The following procedure shows that capacitive reactance may be expressed in terms of circuit quantities (capacitance and frequency) rather than measurable quantities (voltage and current).

In Chapter 6 we showed that when an alternating voltage appears across a capacitor, the instantaneous current waveform leads the voltage waveform by 90°:

$$v_C = E_m \sin \omega t \tag{7-14}$$

and

$$i_C = CE_m \omega \cos \omega t \tag{7-15a}$$

or

$$i_C = CE_m \omega \sin (\omega t + \pi/2) \tag{7-15b}$$

Then dividing Eq. (7-14) by Eq. (7-15b) we obtain

$$\frac{v_C}{i_C} = \frac{E_m \sin \omega t}{CE_m\ \omega \sin (\omega t + \pi/2)}$$

$$\frac{v_C}{i_C} = \frac{1}{\omega C \underline{/90^\circ}}$$

$$\frac{v_C}{i_C} = \frac{1}{j\omega C} \tag{7-16}$$

According to the definition of capacitive reactance, which is the magnitude of voltage across the capacitor divided by the current through the capacitor,

$$\left|\frac{v_C}{i_C}\right| = \frac{1}{\omega C}$$

or

$$X_C = \frac{1}{\omega C} \tag{7-17}$$

where $\omega = 2\pi f$, f is the frequency in hertz, and C is the value of capacitance in farads.

We are now able to express the impedance of a series *R-C* circuit in terms of circuit parameters. This expression is obtained by substituting Eq. (7-17) into Eq. (7-13). Therefore,

$$\bar{Z}_T = R - j\frac{1}{\omega C} \tag{7-18}$$

Example 7-5: For the circuit of Fig. 7-6a, calculate the total impedance in (a) rectangular form and (b) polar form.

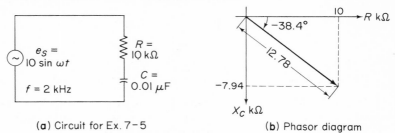

(a) Circuit for Ex. 7-5   (b) Phasor diagram

Fig. 7-6. Diagrams for Ex. 7-5.

Solution: (a)

$$\bar{Z}_T = R - j\frac{1}{\omega C}$$

$$\bar{Z}_T = R - j\frac{1}{2\pi f C}$$

$$\bar{Z}_T = 10\,\text{k}\Omega - j\frac{1}{(6.28)\,(2\,\text{kHz})\,(0.01\,\mu\text{F})}$$

$$\bar{Z}_T = 10\,\text{k}\Omega - j\frac{1}{0.126\,\text{m}\mho}$$

$$\bar{Z}_T = (10 - j7.94)\,\text{k}\Omega$$

(b)

$$\bar{Z}_T = |Z_T|\,\underline{/\theta}$$

$$\bar{Z}_T = \sqrt{(10\,\text{k}\Omega)^2 + (7.94\,\text{k}\Omega)^2}\,\underline{/\tan^{-1}\frac{-7.94\,\text{k}\Omega}{10\,\text{k}\Omega}}$$

$$\bar{Z}_T = 12.78\,\underline{/-38.4°}\,\text{k}\Omega$$

The phasor diagram is shown in Fig. 7-6b.

Example 7-6: For the circuit of Fig. 7-7a, calculate the total impedance in (a) rectangular form and (b) polar form.

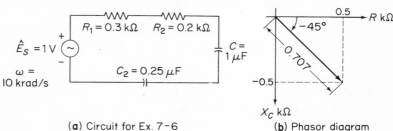

(a) Circuit for Ex. 7-6   (b) Phasor diagram

Fig. 7-7. Diagrams for Ex. 7-6.

Solution: (a)

$$\bar{Z}_T = R_1 + R_2 - j\frac{1}{\omega C_1} - j\frac{1}{\omega C_2}$$

$$\bar{Z}_T = 0.3 \text{ k}\Omega + 0.2 \text{ k}\Omega - j\frac{1}{(10 \text{ krad/s})(1 \text{ }\mu\text{F})} - j\frac{1}{(10 \text{ krad/s})(0.25 \text{ }\mu\text{F})}$$

$$\bar{Z}_T = 0.3 \text{ k}\Omega + 0.2 \text{ k}\Omega - j0.1 \text{ k}\Omega - j0.4 \text{ k}\Omega$$

or

$$\bar{Z}_T = (0.5 - j0.5) \text{ k}\Omega$$

(b)

$$\bar{Z}_T = \sqrt{(0.5 \text{ k}\Omega)^2 + (0.5 \text{ k}\Omega)^2} / \tan^{-1}\frac{-0.5 \text{ k}\Omega}{0.5 \text{ k}\Omega}$$

$$\bar{Z}_T = 0.707 \underline{/-45°} \text{ k}\Omega$$

The phasor diagram is shown in Fig. 7-7b. The reader should check these results by first adding R_1 and R_2 and C_1 and C_2.

Equation (7-17) shows that capacitive reactance is independent of voltage and current, although it can be calculated from these values. It can also be calculated from the circuit quantities of frequency, capacitance, and the constant, 2π.

Now that we have established Eq. (7-17) as a method of calculating capacitive reactance, let us again consider the effects of varying the amplitude and the frequency of the source voltage. Let us double the frequency without changing the peak value of the source voltage or the value of capacitance. Then according to Eq. (7-17) the capacitive reactance is halved. Likewise, doubling the capacitance without altering the peak value of the source voltage or the frequency also results in one half the original capacitive reactance value.

However, doubling the value of the voltage source without changing the other parameters has no effect on calculating the value of capacitive reactance.

Therefore, capacitive reactance is inversely proportional to only the frequency and the capacitance. The amplitude of the source voltage has no effect on the value of capacitive reactance.

Series *R-L-C* Circuits. Up to this point we have discussed series *R-L* and series *R-C* circuits. We shall now examine the series combination of all three passive elements and determine the total impedance. This type of circuit will not involve new concepts but can be solved by simple addition of complex numbers.

Example 7-7: Determine the complex impedance in (a) rectangular form and (b) polar form for the network of Fig. 7-8 and draw the equivalent circuit.

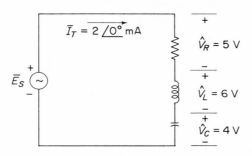

$\hat{V}_R = 5\ \text{V}$

$\hat{V}_L = 6\ \text{V}$

$\hat{V}_C = 4\ \text{V}$ **Fig. 7-8.** Circuit diagram for Ex. 7-7.

Solution: (a) According to Kirchhoff's voltage law,

$$\bar{E}_S = \hat{V}_R + j\hat{V}_L - j\hat{V}_C$$

Then

$$\bar{Z}_T = \frac{\bar{E}_S}{\bar{I}} = \frac{5\ \text{V}}{2\ \text{mA}} + j\frac{6\ \text{V}}{2\ \text{mA}} - j\frac{4\ \text{V}}{2\ \text{mA}}$$

$$\bar{Z}_T = (2.5 + j3 - j2)\ \text{k}\Omega$$

or

$$\bar{Z}_T = (2.5 + j1)\ \text{k}\Omega$$

(b)

$$\bar{Z}_T = 2.69\ \underline{/21.8°}\ \text{k}\Omega$$

Note that the series R-L-C circuit of Fig. 7-8 has the following circuit values: $R = 2.5$ kΩ, $X_L = 3$ kΩ, and $X_C = 2$ kΩ, and draws a current of $2\ \underline{/0°}$ mA from the voltage source. An equivalent circuit must always draw the same amount of current from the same source.

In this case the equivalent circuit is a resistance in series with an inductance, because the inductive reactance is greater than the capacitive reactance. The equivalent circuit is shown in Fig. 7-9a and the phasor diagram is shown in Fig. 7-9b.

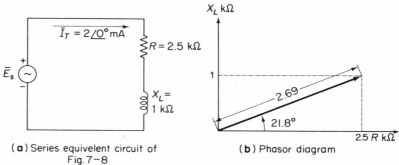

(**a**) Series equivelent circuit of Fig. 7-8

(**b**) Phasor diagram

Fig. 7-9. Diagrams of the solution of Ex. 7-7.

Example 7-8: The series R-L-C circuit of Fig. 7-10 has the following circuit parameters: $R = 12$ kΩ, $X_L = 6$ kΩ, and $X_C = 18$ kΩ. Calculate the total impedance in (a) rectangular form and (b) polar form and draw the equivalent circuit.

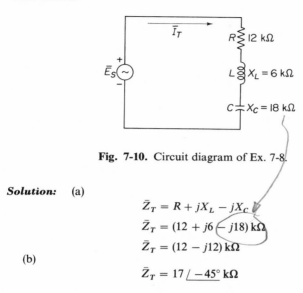

Fig. 7-10. Circuit diagram of Ex. 7-8.

Solution: (a)

$$\bar{Z}_T = R + jX_L - jX_C$$
$$\bar{Z}_T = (12 + j6 - j18)\,\text{k}\Omega$$
$$\bar{Z}_T = (12 - j12)\,\text{k}\Omega$$

(b)

$$\bar{Z}_T = 17\,\underline{/-45°}\;\text{k}\Omega$$

The equivalent of the circuit of Fig. 7-10 is shown in Fig. 7-11a and the phasor diagram is shown in Fig. 7-11b. The equivalent circuit is a resistor in series with a capacitor because the capacitive reactance is greater than the inductive reactance.

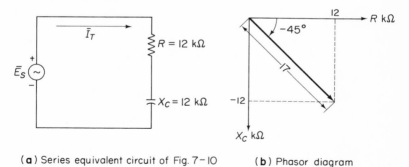

(**a**) Series equivalent circuit of Fig. 7-10 (**b**) Phasor diagram

Fig. 7-11. Diagrams for the solution of Ex. 7-8.

Example 7-9: For the circuit of Fig. 7-12, calculate the total impedance in (a) rectangular form and (b) polar form. Also draw the equivalent circuit and phasor diagram.

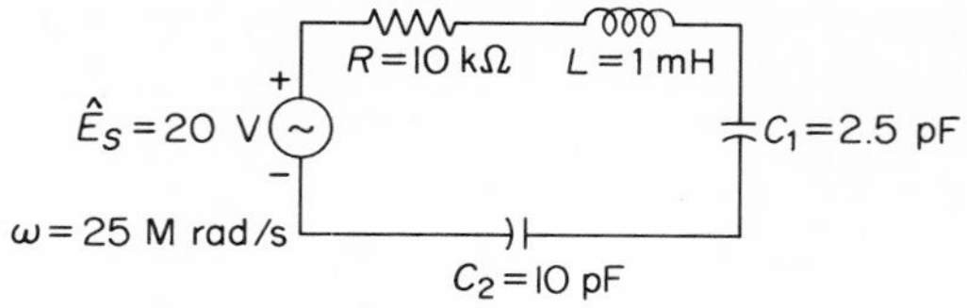

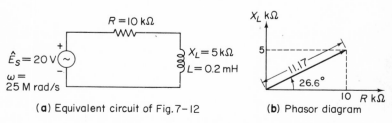

Fig. 7-12. Circuit diagram for Ex. 7-9.

Solution: (a)   For the circuit of Fig. 7-12,

$$j\omega L = j\,(25\ \text{Mrad/s})\,(1\ \text{mH}) = 25\ \text{k}\Omega$$

$$-j\frac{1}{\omega C_1} = -j\frac{1}{(25\ \text{Mrad/s})\,(2.5\ \text{pF})} = -j16\ \text{k}\Omega$$

$$-j\frac{1}{\omega C_2} = -j\frac{1}{(25\ \text{Mrad/s})\,(10\ \text{pF})} = -j4\ \text{k}\Omega$$

Then

$$\bar{Z}_T = R + j\omega L - j\frac{1}{\omega C_1} - j\frac{1}{\omega C_2}$$

$$\bar{Z}_T = (10 + j25 - j16 - j4)\ \text{k}\Omega$$

or

$$\bar{Z}_T = (10 + j5)\ \text{k}\Omega$$

(b)

$$\bar{Z}_T = \sqrt{(10\ \text{k}\Omega)^2 + (5\ \text{k}\Omega)^2}\,\underline{/\tan^{-1}\frac{5\ \text{k}\Omega}{10\ \text{k}\Omega}}$$

$$\bar{Z}_T = 11.17\,\underline{/26.6°}\ \text{k}\Omega$$

The equivalent circuit for Fig. 7-12 is shown in Fig. 7-13a and the phasor diagram is shown in Fig. 7-13b. Note that the equivalent inductance must be calculated from the imaginary part of the total impedance. For this example,

$$L = X_L/\omega = \frac{5\ \text{k}\Omega}{25\ \text{Mrad/s}} = 0.2\ \text{mH}$$

(a) Equivalent circuit of Fig. 7-12   **(b)** Phasor diagram

Fig. 7-13. Diagrams for the solution of Ex. 7-9.

7-2. COMPLEX ADMITTANCE

Admittance is the reciprocal of impedance and is represented by the symbol Y:

$$\bar{Y} = \frac{1}{\bar{Z}} \tag{7-19}$$

In Chapter 4 we encountered a special form of admittance — conductance. Conductance is the reciprocal of resistance and is used when determining the equivalent resistance for a parallel combination. This section will involve a resistor in parallel with other passive elements (inductors and/or capacitors). The combination of two or more different elements in ac circuits involves our complex notation.

Inductive Susceptance.   Consider the parallel R-L circuit of Fig. 7-14. By Kirchhoff's current law,

$$\bar{I}_T = \hat{I}_R - j\hat{I}_L \tag{7-20}$$

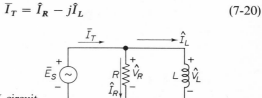

Fig. 7-14. Parallel R-L circuit.

Since we are treating a parallel circuit, the source voltage, $\bar{E}_S$, is common to both elements. For this type of circuit it is more convenient to use the voltage as the reference axis when drawing the phasor diagram. For our present analysis, let us consider the phase angle of the voltage source to be $0°$. Then

$$\bar{E}_S = \hat{V}_R = \hat{V}_L \tag{7-21}$$

Dividing Eq. (7-20) by Eq. (7-21) yields

$$\frac{\bar{I}_T}{\bar{E}_S} = \frac{\hat{I}_R}{\hat{V}_R} - j\frac{\hat{I}_L}{\hat{V}_L} \tag{7-22}$$

The left side of Eq. (7-22) is the total value of current divided by the total value of voltage; thus it must be the total admittance of the circuit:

$$\frac{\bar{I}_T}{\bar{E}_S} = \bar{Y}_T \tag{7-23}$$

As in the previous section, let us treat each term on the right side of Eq. (7-22) separately. In Chapter 4 the current through a resistor divided by the voltage across a resistor was defined as conductance, and represented by the symbol G:

$$\frac{\hat{I}_R}{\hat{V}_R} = G \tag{7-24}$$

Remember that conductance is the reciprocal of resistance:

$$G = \frac{1}{R}$$

The units of Eq. (7-24) are mhos (amperes divided by volts). The units of the second term, $\hat{I}_L/\hat{V}_L$, are also mhos. However, the term conductance has been used for the element that has no phase shift between current and voltage. Therefore, let us define a new term, *susceptance*, which is the magnitude of current through the element divided by the voltage across the element when the current and voltage are 90° (or $\pi/2$ rad) out of phase. The symbol B is associated with susceptance.

For the circuit of Fig. 7-14, the magnitude of current through the inductance divided by the voltage across the inductance is defined as *inductive susceptance*, B_L:

$$\left|\frac{\hat{I}_L}{\hat{V}_L}\right| = B_L \tag{7-25}$$

Substituting Eqs. (7-23), (7-24), and (7-25) into Eq. (7-22) yields the total admittance for the circuit of Fig. 7-14:

$$\overline{Y}_T = G - jB_L \tag{7-26}$$

Note that the coefficient of an inductive reactance term is $+ j$, whereas the coefficient of an inductive susceptance term is $-j$. Why?

Just as conductance is the reciprocal of resistance, so susceptance is the reciprocal of reactance, or

$$B = \frac{1}{X} \tag{7-27}$$

Example 7-10: For the parallel *R-L* circuit of Fig. 7-15a, determine the complex admittance in both (a) rectangular form and (b) polar form.

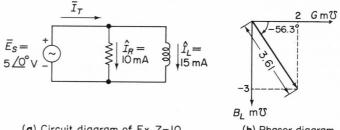

(a) Circuit diagram of Ex. 7-10   (b) Phasor diagram

Fig. 7-15. Diagrams for Ex. 7-10.

Solution: According to Kirchhoff's current law,

$$\bar{I}_T = \hat{I}_R - j\hat{I}_L$$

$$\bar{I}_T = (10 - j15)\,\text{mA}$$

$$\frac{\bar{I}_T}{\bar{E}_S} = \frac{10\,\text{mA}}{5\,\text{V}} - j\frac{15\,\text{mA}}{5\,\text{V}}$$

$$\bar{Y}_T = (2 - j3)\,\text{m}\mho$$

Complex admittance, like any other complex number, also may be expressed in polar form. For the parallel *R-L* circuit,

$$\bar{Y}_T = |Y_T|\,\underline{/\phi} \tag{7-28}$$

where

$$|Y_T| = \sqrt{G^2 + (-B_L)^2}$$

$$\phi = \underline{/\tan}^{-1}\frac{-B_L}{G}$$

For our example,

$$\bar{Y}_T = \sqrt{(2\,\text{m}\mho)^2 + (-3\,\text{m}\mho)^2}\,\underline{/\tan}^{-1}\frac{-3\,\text{m}\mho}{2\,\text{m}\mho}$$

$$\bar{Y}_T = 3.61\,\underline{/-56.3°}\,\text{m}\mho$$

The phasor diagram for a parallel *R-L* circuit is shown in Fig. 7-15b. Note that the angle θ is reserved for impedance expressed in polar form, whereas ϕ is the angle for the polar form of complex admittance. This is the conventional notation set up by the Institute of Electrical and Electronic Engineers (IEEE). Both angles, θ and ϕ, are always measured from the positive real axis.

It is useful to be able to express the inductive susceptance term utilizing circuit quantities (frequency and inductance) in the same way as for inductive reactance.

The inductive reactance term was determined from the instantaneous values of voltage and current. In a like manner, susceptance may be determined by obtaining the magnitude of the instantaneous current divided by the instantaneous voltage. This method is left for the reader to solve and yields the same result as the following procedure.

In Chapter 4, conductance was defined as the reciprocal of resistance. Similarly, inductive susceptance can be defined as the reciprocal of inductive reactance.

$$B_L = \frac{1}{X_L} \tag{7-29}$$

Since

$$X_L = \omega L$$

Then

$$B_L = \frac{1}{\omega L} \tag{7-30}$$

where $\omega = 2\pi f$, f is the frequency in hertz, and L is the value of inductor in henries.

An admittance expression can be obtained for a resistor and inductor connected in parallel by substituting Eq. (7-30) into Eq. (7-26).

$$\overline{Y}_T = G - j\frac{1}{\omega L} \tag{7-31}$$

Example 7-11: Determine the total admittance in (a) rectangular form and (b) polar form for the circuit of Fig. 7-16a.

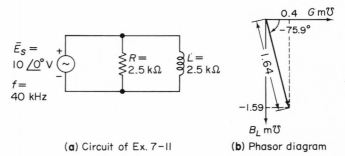

(a) Circuit of Ex. 7-11   (b) Phasor diagram

Fig. 7-16. Diagram for Ex. 7-11.

Solution: (a)

$$\overline{Y}_T = G - jB_L$$

$$G = \frac{1}{R} = \frac{1}{2.5\,\text{k}\Omega} = 0.4\,\text{m}\mho$$

$$B_L = \frac{1}{\omega L} = \frac{1}{(6.28)\,(40\,\text{kHz})\,(2.5\text{mH})} = 1.59\,\text{m}\mho$$

$$\overline{Y}_T = (0.4 - j1.59)\,\text{m}\mho$$

(b)

$$\overline{Y}_T = |Y_T|\underline{/\phi}$$

$$\overline{Y}_T = \sqrt{(0.4\,\text{m}\mho)^2 + (1.59\,\text{m}\mho)^2}\,\underline{/\tan^{-1}\dfrac{-1.59\,\text{m}\mho}{0.4\,\text{m}\mho}}$$

$$Y_T = 1.64\,\underline{/-75.9°}\,\text{m}\mho$$

The phasor diagram is shown in Fig. 7-16b. Note that for a resistor in parallel with an inductor the angle is negative.

Example 7-12: For the parallel circuit shown in Fig. 7-17a, draw the equivalent circuit with the circuit values.

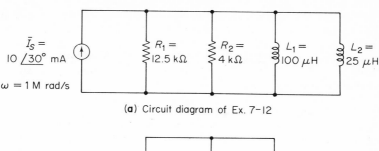

(a) Circuit diagram of Ex. 7-12

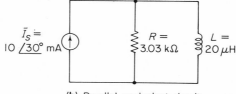

(b) Parallel equivalent circuit

Fig. 7-17. Diagrams for Ex. 7-12.

Solution: In order to draw the equivalent circuit, we must first obtain the total admittance in rectangular form:

$$G_1 = \frac{1}{R_1} = \frac{1}{12.5\text{ k}\Omega} = 0.08 \text{ m}\mho$$

$$G_2 = \frac{1}{R_2} = \frac{1}{4\text{ k}\Omega} = 0.25 \text{ m}\mho$$

$$B_{L_1} = \frac{1}{\omega L_1} = \frac{1}{(1\text{ Mrad/s})(100\ \mu\text{H})} = 10 \text{ m}\mho$$

$$B_{L_2} = \frac{1}{\omega L_2} = \frac{1}{(1\text{ Mrad/s})(25\ \mu\text{H})} = 40 \text{ m}\mho$$

Since

$$\bar{Y}_T = G_1 + G_2 - jB_{L_1} - jB_{L_2}$$

then

$$\bar{Y}_T = (0.08 + 0.25 - j10 - j40)\text{ m}\mho$$

$$\bar{Y}_T = (0.33 - j50)\text{ m}\mho$$

Therefore, the conductance value of the equivalent circuit equals 0.33 m$\mho$ and the inductive susceptance value is 50 m$\mho$. The circuit values of the equivalent network will then be

$$R = \frac{1}{G} = \frac{1}{0.33\text{ m}\mho} = 3.03 \text{ k}\Omega$$

$$L = \frac{1}{B_L \omega} = \frac{1}{(50\text{ m}\mho)(1\text{ Mrad/s})} = 20\ \mu\text{H}$$

The equivalent circuit is shown in Fig. 7-17b. Note that the values for the equivalent circuit are calculated only from the total admittance.

Capacitive Susceptance. The development of a parallel R-C circuit is similar to that of a parallel R-L circuit. The following example will help to demonstrate the results.

> **Example 7-13:** The parallel R-C circuit of Fig. 7-18a is energized by a current source. Determine the admittance seen by the current source in (a) rectangular form and (b) polar form and draw the phasor diagram.

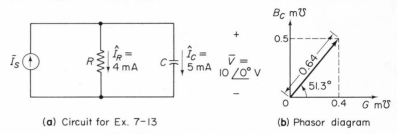

(a) Circuit for Ex. 7-13   (b) Phasor diagram

Fig. 7-18. Diagrams for Ex. 7-13.

Solution: (a)

$$\hat{I}_S = \hat{I}_R + j\hat{I}_C$$

$$\bar{Y}_T = \frac{\bar{I}_S}{\bar{V}} = \frac{4\text{ mA}}{10\text{ V}} + j\frac{5\text{ mA}}{10\text{ V}}$$

$$\bar{Y}_T = (0.4 + j0.5)\text{ m}\mho$$

(b)

$$\bar{Y}_T = |\bar{Y}_T|\underline{/\phi}$$

$$\bar{Y}_T = 0.64\underline{/51.3°}\text{ m}\mho$$

The phasor diagram is shown in Fig. 7-18b. Note that for a resistor in parallel with a capacitor the angle is positive.

A capacitor connected in parallel with a resistor is the fourth combination of either R-L or R-C networks. Like inductive susceptance, capacitive susceptance can also be defined in two ways: (1) the magnitude of current through the capacitor divided by the voltage across the capacitor or (2) the reciprocal of capacitive reactance. The first case is left to the reader, starting with the instantaneous expressions of current and voltage—Eqs. (7-14) and (7-15b), respectively. The results would be the same as those obtained by the second case.

Case 2:

$$B_C = \frac{1}{X_C} \tag{7-32}$$

Since

$$X_C = \frac{1}{\omega C}$$

then

$$B_C = \frac{1}{1/\omega C}$$

Therefore,

$$B_C = \omega C \tag{7-33}$$

where $\omega = 2\pi f$, f is the frequency in hertz, and C is the capacitance in farads.

Equation (7-33), therefore, allows us to express capacitive susceptance in terms of circuit quantities (frequency, capacitance, and the constant, 2π).

Since the expression for total admittance of a parallel R-C circuit is

$$\bar{Y}_T = G + jB_C$$

then in terms of circuit quantities,

$$\bar{Y}_T = G + j\omega C \tag{7-34}$$

Again note the $+j$ term. Since we are involved with a parallel circuit, it accounts for the fact that the current through the capacitor leads the voltage across the capacitor by 90° (or $\pi/2$ radians).

Let us apply the above results to the following problem.

Example 7-14: For the circuit of Fig. 7-19a driven by a current source, I_S, calculate the total admittance, $\bar{Y}_T$, in (a) rectangular form and (b) polar form and draw the phasor diagram.

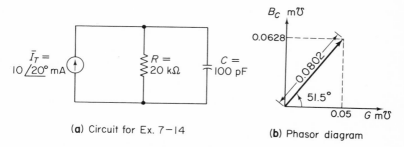

(a) Circuit for Ex. 7-14   (b) Phasor diagram

Fig. 7-19. Diagrams for Ex. 7-14.

Solution: (a)

$$\bar{Y}_T = G + jB_C$$

$$G = \frac{1}{R} = \frac{1}{20\,k\mho} = 0.05\,m\mho$$

$$B_C = j\omega C = (6.28)\,(100\,kHz)\,(100\,pF) = 0.0628\,m\mho$$

Therefore,
$$\bar{Y}_T = (0.05 + j0.0628) \text{ m}\mho$$

(b)

$$\bar{Y}_T = |Y_T| \underline{/\phi}$$

$$\bar{Y}_T = \sqrt{(0.05 \text{ m}\mho)^2 + (0.0628 \text{ m}\mho)^2} \underline{/\tan^{-1} \dfrac{0.0628 \text{ m}\mho}{0.05 \text{ m}\mho}}$$

$$\bar{Y}_T = 0.0802 \underline{/51.5°} \text{ m}\mho$$

The phasor diagram is shown in Fig. 7-19b. Note that when a resistor is connected in parallel with a capacitor, the phase angle is positive because the voltage has been taken as the reference axis.

Parallel R-L-C Circuits.   Determining the total admittance of a parallel R-L-C circuit involves the addition of complex numbers and the concepts learned for the parallel *R-L* and parallel *R-C* circuits. The following examples illustrate this procedure.

Example 7-15:   For the parallel circuit of Fig. 7-20a, calculate the total admittance of the circuit in (a) rectangular form and (b) polar form and draw the phasor diagram.

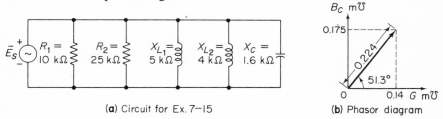

(a) Circuit for Ex. 7–15   (b) Phasor diagram

Fig. 7-20. Diagrams for Ex. 7-15.

Solution: Since resistance and reactance values are given, it is only necessary to obtain the conductance and susceptance values to determine the total admittance.   Therefore,

$$G_1 = \frac{1}{R_1} = \frac{1}{10 \text{ k}\Omega} = 0.1 \text{ m}\mho$$

$$G_2 = \frac{1}{R_2} = \frac{1}{25 \text{ k}\Omega} = 0.04 \text{ m}\mho$$

$$B_{L_1} = \frac{1}{X_{L_1}} = \frac{1}{5 \text{ k}\Omega} = 0.2 \text{ m}\mho$$

$$B_{L_2} = \frac{1}{X_{L_2}} = \frac{1}{4 \text{ k}\Omega} = 0.25 \text{ m}\mho$$

$$B_C = \frac{1}{X_C} = \frac{1}{1.6 \text{ k}\Omega} = 0.625 \text{ m}\mho$$

(a)

$$\bar{Y}_T = G_1 + G_2 - jB_{L_1} - jB_{L_2} + jB_C$$

$$\bar{Y}_T = (0.1 + 0.04 - j0.2 - j0.25 + j0.625)\,\text{m}\mho$$

$$\bar{Y}_T = (0.14 + j0.175)\,\text{m}\mho$$

(b)

$$\bar{Y}_T = |Y_T|\underline{/\phi}$$

$$\bar{Y}_T = \sqrt{(0.14\,\text{m}\mho)^2 + (0.175\,\text{m}\mho)^2}\,\underline{/\tan^{-1}\frac{0.6175\,\text{m}\mho}{0.14\,\text{m}\mho}}$$

$$\bar{Y}_T = 0.224\,\underline{/51.3°}\,\text{m}\mho$$

The phasor diagram is shown in Fig. 7-20b. From the total admittance, we see that an equivalent circuit would be a resistor in parallel with a capacitor.

Example 7-16: For the parallel circuit of Fig. 7-21, draw the parallel equivalent circuit and the phasor diagram.

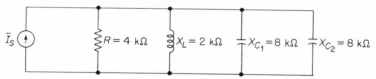

Fig. 7-21. Circuit diagram for Ex. 7-16.

Solution: To draw the equivalent circuit we must first determine the total admittance in rectangular form. This means finding the conductance and susceptance values.

$$G = \frac{1}{R} = \frac{1}{4\,\text{k}\Omega} = 0.25\,\text{m}\mho$$

$$B_L = \frac{1}{X_L} = \frac{1}{2\,\text{k}\Omega} = 0.5\,\text{m}\mho$$

$$B_{C_1} = \frac{1}{X_{C_1}} = \frac{1}{8\,\text{k}\Omega} = 0.125\,\text{m}\mho$$

$$B_{C_2} = \frac{1}{X_{C_2}} = \frac{1}{8\,\text{k}\Omega} = 0.125\,\text{m}\mho$$

Then

$$\bar{Y}_T = G - jB_L + jB_{C_1} + jB_{C_2}$$

$$\bar{Y}_T = (0.25 - j0.5 + j0.125 + j0.125)\,\text{m}\mho$$

$$\bar{Y}_T = (0.25 - j0.25)\,\text{m}\mho$$

Therefore, the values of the equivalent circuit are

$$G = 0.25\,m\Omega \qquad \text{and} \qquad B_L = 0.25\,\text{m}\Omega$$

The parallel equivalent circuit is shown in Fig. 7-22a and the phasor diagram is shown in Fig. 7-22b. As with the previous example, the reactance terms must first be converted to susceptance terms and then the total admittance may be found. The equivalent circuit is a resistor in parallel with an inductor because the inductive susceptance is larger than the capacitive susceptance.

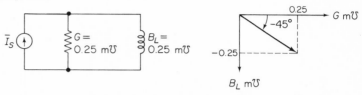

(a) Parallel equivalent circuit of Fig 7-21   (b) Phasor diagram

Fig. 7-22. Diagrams of the solution of Ex. 7-16.

Example 7-17: For the circuit of Fig. 7-23, draw the equivalent circuit with the circuit values and the phasor diagram.

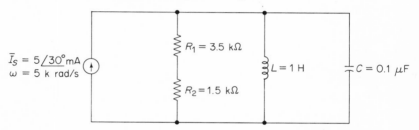

Fig. 7-23. Circuit diagram for Ex. 7-17.

Solution: In order to determine the equivalent circuit, it is again necessary to determine the total admittance in rectangular form.

$$G = \frac{1}{R_1 + R_2} = \frac{1}{3.5 \, \text{k}\Omega + 1.5 \, \text{k}\Omega} = 0.2 \, \text{m}\mho$$

$$B_L = \frac{1}{\omega L} = \frac{1}{(5 \, \text{krad/s})(1 \, \text{H})} = 0.2 \, \text{m}\mho$$

$$B_C = \omega C = (5 \, \text{krad/s})(0.1 \, \mu\text{F}) = 0.5 \, \text{m}\mho$$

Since

$$\bar{Y}_T = G - jB_L + jB_C$$

$$\bar{Y}_T = (0.2 - j0.2 + j0.5) \, \text{m}\mho$$

Therefore,

$$\bar{Y}_T = (0.2 + j0.3) \, \text{m}\mho$$

This equation shows that the equivalent circuit is a resistor in parallel with a capacitor. The circuit values for the equivalent circuit now can be determined.

$$R = \frac{1}{G} = \frac{1}{0.2 \, m\mho} = 5 \, k\Omega$$

$$C = \frac{B_C}{\omega} = \frac{0.3 \, m\mho}{5 \, krad/s} = 0.06 \, \mu F$$

The equivalent circuit is shown in Fig. 7-24a and the phasor diagram is shown in Fig. 7-24b. Note that resistances R_1 and R_2 are first added and the conductance of the total resistance is found. The value of the equivalent capacitance is found from the expression of the total admittance. The equivalent resistance equals the sum of R_1 and R_2 because there are no other resistors in Fig. 7-23. However, the equivalent capacitance must be calculated from the total admittance.

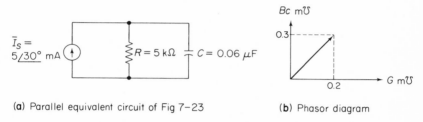

(a) Parallel equivalent circuit of Fig 7-23   (b) Phasor diagram

Fig. 7-24. Diagrams of the solution of Ex. 7-17.

7-3. COMPLEX NETWORKS

To analyze circuits containing either complex impedance or complex admittance, we rely on manipulation of complex numbers. In Chapter 6 the addition, subtraction, multiplication, and division of complex numbers were studied. The following examples analyze complex circuits.

Example 7-18: Determine the source voltage in polar form for the series circuit shown in Fig. 7-25.

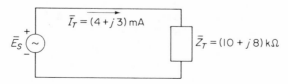

Fig. 7-25. Circuit diagram for Ex. 7-18.

Solution: Using Ohm's law,

$$\bar{E}_S = \bar{I}_T \bar{Z}_T$$

$$\bar{E}_S = (4 + j3)\,\text{mA}\,(10 + j8)\,\text{k}\Omega$$

$$\bar{E}_S = (4)(10) + j(4)(8) + j(3)(8) + j^2(3)(8)$$

$$\bar{E}_S = (16 + j62)\,\text{V}$$

In polar form,

$$\bar{E}_S = \sqrt{(16\,\text{V})^2 + (62\,\text{V})^2}\,\underline{/\tan^{-1}\frac{62\,\text{V}}{16\,\text{V}}}$$

$$\bar{E}_S = 64.2\,\underline{/75.5°}\,\text{V}$$

This problem could also be solved by first converting the rectangular form of current and impedance to polar form:

$$\bar{I}_T = (4 + j3)\,\text{mA} = 5\,\underline{/36.9°}\,\text{mA}$$

$$\bar{Z}_T = (10 + j8)\,\text{k}\Omega = 12.8\,\underline{/38.7°}\,\text{k}\Omega$$

Then

$$\bar{E}_S = \bar{I}_T \bar{Z}_T$$

$$\bar{E}_S = (5\,\underline{/36.9°}\,\text{mA})(12.8\,\underline{/38.7°}\,\text{k}\Omega)$$

$$\bar{E}_S = 64\,\underline{/75.6°}\,\text{V}$$

The reader should be able to answer the following questions: (1) Could the load be represented by a resistor and an inductor or a resistor and capacitor? (2) What would the voltage phasor diagram be? Remember the voltage across a resistor is always in phase with the current.

Example 7-19: For the circuit of Fig. 7-26, determine the total current.

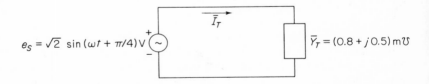

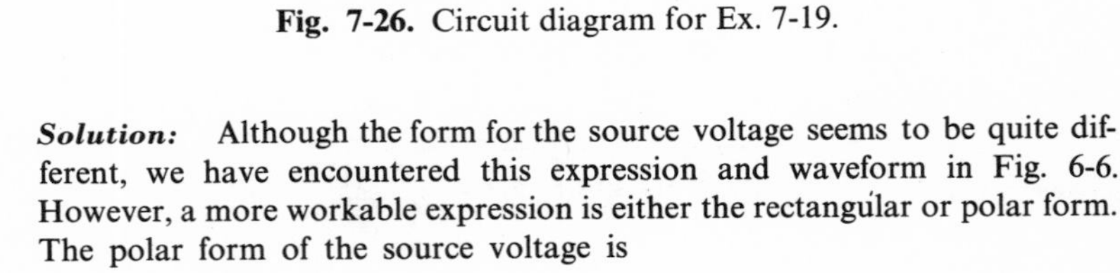

$$e_S = \sqrt{2}\,\sin(\omega t + \pi/4)\,\text{V}$$ $$\bar{Y}_T = (0.8 + j\,0.5)\,\text{m}\mho$$

Fig. 7-26. Circuit diagram for Ex. 7-19.

Solution: Although the form for the source voltage seems to be quite different, we have encountered this expression and waveform in Fig. 6-6. However, a more workable expression is either the rectangular or polar form. The polar form of the source voltage is

$$e_S = \sqrt{2}\,\sin(\omega t + \pi/4)\,\text{V}$$

$$\bar{E}_S = \frac{\sqrt{2}}{\sqrt{2}}\,\underline{/\pi/4} = 1\,\underline{/45°}\,\text{V}$$

Converting $\bar{Y}_T$ to polar form, we obtain

$$\bar{Y}_T = (0.8 + j0.5)\,\text{m}\mho = 0.945\,\underline{/32°}\,\text{m}\mho$$

Knowing the voltage across an admittance, we can calculate the current through it by using

$$\bar{I}_T = \bar{E}_S\bar{Y}_T \qquad\qquad (7\text{-}35)$$

Then

$$\bar{I}_T = (1\,\underline{/45°}\,\text{V})(0.945\,\underline{/32°}\,\text{mA})$$

$$\bar{I}_T = (0.945\,\underline{/77°})\,\text{mA}$$

or

$$\bar{I}_T = (0.212 + j0.92)\,\text{mA}$$

This example could also have been solved by using the rectangular form for both the source voltage and the total admittance. This is left as an exercise for the reader. Note, however, that the source voltage first had to be converted to a fixed value. Equation (7-35) allows the calculation of current given the voltage across an admittance.

Example 7-20: For the parallel circuit of Fig. 7-27, determine the total impedance.

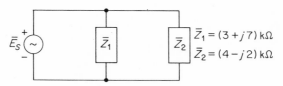

Fig. 7-27. Circuit diagram for Ex. 7-20.

Solution: This problem can be solved most easily by using the product over the sum rule. For complex impedance this rule is

$$\bar{Z}_T = \frac{\bar{Z}_1\bar{Z}_2}{\bar{Z}_1 + \bar{Z}_2} \qquad\qquad (7\text{-}36)$$

$$\bar{Z}_1 = (3 + j7)\,\text{k}\Omega = (7.62\,\underline{/66.8°})\,\text{k}\Omega$$

$$\bar{Z}_2 = (4 - j2)\,\text{k}\Omega = (4.47\,\underline{/-26.6°})\,\text{k}\Omega$$

$$\bar{Z}_T = \frac{(7.62\,\underline{/66.8°}\,\text{k}\Omega)(4.47\,\underline{/-26.6°}\,\text{k}\Omega)}{(3 + j7)\,\text{k}\Omega + (4 - j2)\,\text{k}\Omega}$$

$$\bar{Z}_T = \frac{34.1\,\underline{/40.2°}\,\text{M}\Omega}{(7 + j5)\,\text{k}\Omega}$$

$$\bar{Z}_T = \frac{34.1\,\underline{/40.2°}\,\text{M}\Omega}{8.59\,\underline{/35.6°}\,\text{k}\Omega}$$

$$\bar{Z}_T = 3.97\,\underline{/4.6°}\,\text{k}\Omega$$

In rectangular form,

$$\bar{Z}_T = (3.93 + j0.328) \text{ k}\Omega$$

If asked, an equivalent circuit could be drawn from the rectangular form expression. The circuit would be a resistor in series with an inductor. The value of the resistor is 3.93 kΩ; the value of the inductive reactance is 328 Ω. The specific frequency would have to be known to find the value of the inductor. [*Note:* Since the numerator of Ex. 7-20 involves multiplication, the polar form is used. However, in the denominator the addition has to be done in rectangular form. It is then converted to the polar form in preparation for the final step. The final step (the division of two complex numbers) is done in polar form.]

This example could also have been done by first converting $\bar{Z}_1$ and $\bar{Z}_2$ to admittance values which in turn would be added to give the total admittance of the circuit. The total admittance may then be converted to give the total impedance. The reader should check the answer by this method.

It is the author's contention that the operation of multiplication and division of complex numbers is more easily performed in the polar form.

Example 7-21: Consider the series-parallel circuit of Fig. 7-28a. Determine the total impedance, $\bar{Z}_T$, and the total current, $\bar{I}_T$.

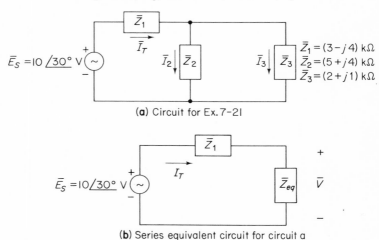

(a) Circuit for Ex. 7-21

(b) Series equivalent circuit for circuit a

Fig. 7-28. Circuit diagrams for Ex. 7-21.

Solution: (1) For the above network, the impedances $\bar{Z}_2$ and $\bar{Z}_3$ are in parallel and that combination is in series with $\bar{Z}_1$. Thus, to determine the total impedance, first obtain the equivalent impedance of $\bar{Z}_2$ and $\bar{Z}_3$.

$$\bar{Z}_{eq} = \frac{\bar{Z}_2 \bar{Z}_3}{\bar{Z}_2 + \bar{Z}_3}$$

$$\bar{Z}_{eq} = \frac{(6.4 \,\underline{/\,38.7^\circ}\, k\Omega)(2.23 \,\underline{/\,26.6^\circ}\, k\Omega)}{(5 + j4)\, k\Omega + (2 + j1)\, k\Omega}$$

$$\bar{Z}_{eq} = \frac{(14.3 \,\underline{/\,65.3^\circ})\, M\Omega}{(7 + j5)\, k\Omega}$$

$$\bar{Z}_{eq} = \frac{(14.3 \,\underline{/\,65.3^\circ})\, M\Omega}{(8.59 \,\underline{/\,35.6^\circ})\, k\Omega}$$

$$\bar{Z}_{eq} = 1.665 \,\underline{/\,29.7^\circ}\, k\Omega$$

$$\bar{Z}_{eq} = (1.43 + j0.825)\, k\Omega$$

Now we are able to reduce the original circuit to the equivalent circuit shown in Fig. 7-28b. Since $\bar{Z}_1$ and $\bar{Z}_{eq}$ are in series, the total impedance is the addition of $\bar{Z}_1$ and $\bar{Z}_{eq}$:

$$Z_T = \bar{Z}_1 + \bar{Z}_{eq}$$

$$\bar{Z}_T = (3 - j4)\, k\Omega + (1.43 + j0.825)\, k\Omega$$

$$\bar{Z}_T = (4.43 - j3.175)\, k\Omega$$

or

$$\bar{Z}_T = 5.45 \,\underline{/\,-35.6^\circ}\, k\Omega$$

(2)

$$\bar{I}_T = \frac{\bar{E}_T}{\bar{Z}_T}$$

$$\bar{I}_T = \frac{10 \,\underline{/\,30^\circ}\, V}{(5.45 \,\underline{/\,-35.6^\circ})\, k\Omega}$$

$$\bar{I}_T = 1.835 \,\underline{/\,65.6^\circ}\, mA$$

or

$$\bar{I}_T = (0.76 + j1.67)\, mA$$

This example shows that the rules developed in Chapter 4 for resistive circuits are applicable to impedance networks.

Example 7-22: Again consider the network of Fig. 7-28a, this time to determine the branch currents $\bar{I}_2$ and $\bar{I}_3$.

Solution: This problem may be approached in two ways:

1. Knowing the total value of current and the impedance values, the current division law can be applied to determine each branch current.

$$\bar{I}_2 = \frac{\bar{Z}_3}{\bar{Z}_2 + \bar{Z}_3} \times \bar{I}_T$$

$$\bar{I}_2 = \frac{(2.23 \,\underline{/\,26.6^\circ})\, k\Omega}{(5 + j4)\, k\Omega + (2 + j1)\, k\Omega} \times (1.835 \,\underline{/\,65.6^\circ})\, mA$$

$$\bar{I}_2 = \frac{4.1 \,\underline{/\,92.2^\circ}\, V}{8.59 \,\underline{/\,35.6^\circ}\, k\Omega}$$

$$\bar{I}_2 = 0.48 \,\underline{/\,56.6^\circ}\, mA$$

Similarly, the branch current, I_3, can be determined:

$$\bar{I}_3 = \frac{\bar{Z}_2}{\bar{Z}_2 + \bar{Z}_3} \times \bar{I}_T$$

$$\bar{I}_3 = \frac{(6.4 \underline{/38.7°}) \, k\Omega}{(5 + j4) \, k\Omega + (2 + j1) \, k\Omega} \times (1.835 \underline{/65.6°}) \, mA$$

$$\bar{I}_3 = \frac{11.75 \underline{/104.3°} \, V}{(8.59 \underline{/35.6°}) \, k\Omega}$$

$$\bar{I}_3 = 1.365 \underline{/68.7°} \, mA$$

2.   An alternative solution is first to obtain the value of voltage, $\bar{V}$, across the parallel combination of $\bar{Z}_2$ and $\bar{Z}_3$ and then to apply Ohm's law to each element to find the branch currents:

$$\bar{V} = \bar{I}_T \bar{Z}_{eq}$$

$$\bar{V} = (1.835 \underline{/65.6°} \, mA)(1.66 \underline{/29.8°} \, k\Omega)$$

$$\bar{V} = 3.05 \underline{/95.4°} \, V$$

Then

$$\bar{I}_2 = \frac{\bar{V}}{\bar{Z}_2} = \frac{3.05 \underline{/95.4°} \, V}{(6.4 \underline{/38.7°}) \, k\Omega} = 0.48 \underline{/56.6°} \, mA$$

and

$$\bar{I}_3 = \frac{\bar{V}}{\bar{Z}_2} = \frac{3.05 \underline{/95.4°} \, V}{(2.23 \underline{/26.6°}) \, k\Omega} = 1.365 \underline{/68.7°} \, mA$$

Note that the alternative (second) solution involves less work.

Example 7-23:   Consider each of the networks a and b in Fig. 7-29 and draw the parallel equivalent circuits.

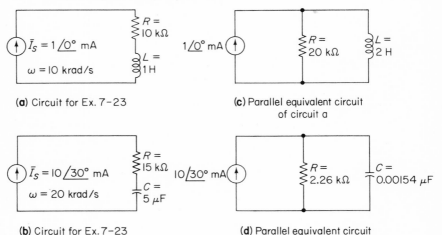

(**a**) Circuit for Ex. 7-23

(**c**) Parallel equivalent circuit of circuit a

(**b**) Circuit for Ex. 7-23

(**d**) Parallel equivalent circuit of circuit b

Fig. 7-29. Circuit diagrams for Ex. 7-23.

Solution: Circuit (a): The total impedance is

$$\bar{Z}_T = R + j\omega L$$

$$\bar{Z}_T = 10 \text{ k}\Omega + j(10 \text{ krad/s}) (1 \text{ H})$$

$$\bar{Z}_T = (10 + j10) \text{ k}\Omega$$

$$\bar{Z}_T = (14.14 \,\underline{/\,45°}) \text{ k}\Omega$$

The total admittance is

$$\bar{Y}_T = \frac{1}{Z_T} = \frac{1}{(14.14 \,\underline{/\,45°}) \text{ k}\Omega} = (0.0707 \,\underline{/\,-45°}) \text{ m}\mho$$

$$\bar{Y}_T = (0.05 - j0.05) \text{ m}\mho$$

From this expression we know that the equivalent circuit is a resistor in parallel with an inductor. The values of the equivalent circuit can be found by

$$R = \frac{1}{G} = \frac{1}{0.05 \text{ m}\mho} = 20 \text{ k}\Omega$$

$$L = \frac{1}{\omega B_L} = \frac{1}{(10 \text{ krad/s}) (0.05 \text{ m}\mho)} = 2 \text{ H}$$

The equivalent circuit is shown in Fig. 7-29c. The reader should take special note that in order to determine the values for the equivalent circuit, one must first calculate the total admittance in rectangular form. Then from this expression the value for the parallel resistor and parallel inductor can be found.

It is *not* correct to calculate the parallel resistor and parallel inductor from the impedance form but rather from the admittance form:

Circuit (b): The total impedance is

$$\bar{Z}_T = R - j\frac{1}{\omega C}$$

$$\bar{Z}_T = 15 \text{ k}\Omega - j\frac{1}{(20 \text{ krad/s}) (5 \text{ }\mu\text{F})}$$

$$\bar{Z}_T = (15 - j10) \text{ k}\Omega$$

$$\bar{Z}_T = (18 \,\underline{/\,-33.7°}) \text{ k}\Omega$$

The total admittance is

$$\bar{Y}_T = \frac{1}{Z_T} = \frac{1}{(18 \,\underline{/\,-33.7°}) \text{ k}\Omega} = (0.0555 \,/\, 33.7) \text{ m}\mho$$

or

$$\bar{Y}_T = (0.0462 + j0.0308) \text{ m}\mho$$

From the rectangular form of the total admittance, the equivalent circuit is a resistor in parallel with a capacitor. The circuit values for the equivalent circuit are

$$R = \frac{1}{G} = \frac{1}{0.0462 \text{ m}\mho} = 21.65 \text{ k}\Omega$$

$$C = \frac{B_C}{\omega} = \frac{0.0308 \text{ m }\mho}{20 \text{ krad/s}} = 0.00154 \text{ }\mu\text{F}$$

The equivalent circuit is shown in Fig. 7-29d. Note that in Example 7-23, the values for the parallel R-C circuit, like that of the parallel R-L circuit, must be calculated from the *admittance* form and not from the impedance form.

7-4. DUALITY

Compare the equation of complex impedance for a series R-L-C circuit,

$$\overline{Z}_T = R + jX_L - jX_C$$

with that of the equation of complex admittance for a parallel G-C-L circuit,

$$\overline{Y}_T = G + jB_C - jB_L$$

Note that the equations are of the same form (not necessarily the same letter symbols). The similarity between these equations is based on a circuit principle known as *duality*. The dual of a circuit is another circuit whose equations are of the same form as the original circuit. For example, the voltage-current relationship of impedance is $\overline{V} = \overline{I}\overline{Z}$ and the relationship of admittance is $\overline{I} = \overline{V}\overline{Y}$. We note between these two expressions the voltage and current variables are interchanged. Therefore, the two circuits for which these equations are written are duals of one another.

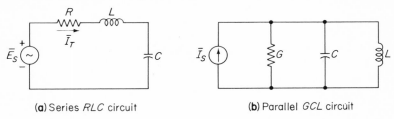

(a) Series *RLC* circuit   (b) Parallel *GCL* circuit

Fig. 7-30. Dual networks.

Figure 7-30 shows dual circuits. Table 7-1 lists circuit variables of a series circuit and its corresponding dual in the parallel circuit.

If the principle of duality had been used in the previous sections after having derived the expressions for a series circuit, it would not have been necessary to derive the expressions for a parallel circuit.

TABLE 7-1

DUAL-CIRCUIT QUANTITIES

Series	Parallel
$\bar{E}_S$	$\bar{I}_S$
$\hat{V}$	$\hat{I}$
$\hat{I}$	$\hat{V}$
$\Sigma \hat{V} = 0$	$\Sigma \hat{I} = 0$
R	G
$X_L = \omega L$	$B_C = \omega C$
$X_C = \dfrac{1}{\omega C}$	$B_L = \dfrac{1}{\omega L}$
$\theta = \underline{/\tan}^{-1} \dfrac{X}{R}$	$\phi = \underline{/\tan}^{-1} \dfrac{B}{G}$
$\bar{Z}$	$\bar{Y}$
$\bar{V} = \bar{I}\bar{Z}$	$\bar{I} = \bar{V}\bar{Z}$

SUMMARY

Chapter 7, like Chapter 6, has dealt with R-L, R-C, and R-L-C circuits driven by an ac source. However, Chapter 7 has been concerned with determining the total value of impedance and admittance.

Section 7-1 defined complex impedance as the total opposition to current. Procedures were developed for determining the total impedance of series R-L, R-C, and R-L-C circuits. This section also dealt with inductive and capacitive reactance. A main concern was to be able to express the value of reactance in terms of circuit parameters (the value of the element and frequency).

Section 7-2 showed that when analyzing parallel circuits, it is sometimes more convenient to use admittance (which is the reciprocal of impedance). The real part of complex admittance is conductance (G) and the imaginary part is susceptance (B). Procedures were developed to express both inductive and capacitive susceptance in terms of circuit quantities.

Section 7-3 showed that rules and laws such as the voltage division law, current division law, and product over the sum rule apply to complex circuits as they do to resistive circuits.

Section 7-4 showed the similarity between equations of a series circuit and those of a parallel circuit. The similarities are based on a circuit concept known as *duality*.

PROBLEMS

7-1. Calculate the impedance of the series R-L circuit of Fig. 7-1 in rectangular and polar form if

(a) $|\hat{V}_R| = 12$ V   (b) $|\hat{V}_R| = 6$ V   (c) $|\hat{V}_R| = 12$ V
$\,|\hat{V}_L| = 8$ V   \,$|\hat{V}_L| = 4$ V   \,$|\hat{V}_L| = 8$ V
$\,\,\,\bar{I} = 2$ mA   \,\,\,\,$\bar{I} = 2$ mA   \,\,\,\,$\bar{I} = 4$ mA

7-2. For each of the circuits of Fig. 7-31, calculate the impedance in rectangular and polar form.

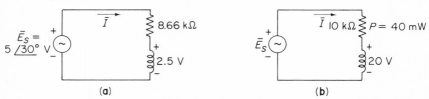

(a)   (b)

Fig. 7-31. Circuit diagrams for Prob. 7-2.

7-3. Calculate the value of inductive reactance for a 100 mH inductor if the frequency is

(a) 0 Hz (dc)   (c) 10 kHz   (e) 60 Hz
(b) 5 kHz   (d) 10 krad/s   (f) 100 krad/s

7-4. What is the value of inductor for each of the following values of inductive reactance. $f = 7$ kHz.

(a) 20 kΩ   (c) 200 Ω   (e) 200 kΩ
(b) 50 kΩ   (d) 100 kΩ   (f) 4 kΩ

7-5. Calculate the frequency at which a 4 H inductor has the following values of inductive reactance

(a) 400 Ω   (c) 4 kΩ   (e) 40 kΩ
(b) 1 kΩ   (d) 500 Ω   (f) 30 MΩ

7-6. If the frequency of the circuit of Fig. 7-4a is doubled, calculate the total impedance in rectangular and polar form.

7-7. For each of the circuits of Fig. 7-32, calculate the value of impedance in rectangular and polar form.

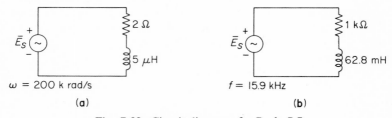

$\omega = 200$ k rad/s $f = 15.9$ kHz

(a)   (b)

Fig. 7-32. Circuit diagrams for Prob. 7-7.

7-8. At a frequency of 10 kHz, the total impedance of a series R-L circuit is $2\,\underline{/\,30°}$ kΩ. Calculate the value of inductance.

7-9. At a frequency of 20 krad/s, the total impedance of a series R-L circuit is $(4 + j4)$ kΩ. What is the value of the inductor.

7-10. Calculate the complex impedance of the R-C circuit of Fig. 7-5a in rectangular and polar form if

(a) $|\hat{V}_R| = 12.8$ V   (b) $|\hat{V}_R| = 3.2$ V   (c) $|\hat{V}_R| = 6.4$ V
$|\hat{V}_C| = 3.8$ V $|\hat{V}_C| = 7.6$ V $|\hat{V}_C| = 3.8$ V
$\bar{I} = 2$ mA $\bar{I} = 2$ mA $\bar{I} = 1$ mA

7-11. Determine the impedance in rectangular and polar form for each of the circuits of Fig. 7-33.

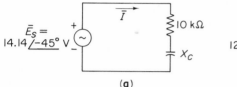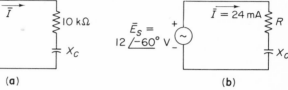

Fig. 7-33. Circuit diagrams for Prob. 7-11.

7-12. Calculate the value of capacitive reactance for a 5 μF capacitor if the frequency is

(a) 0 Hz (dc)   (c) 4 kHz   (e) 40 kHz
(b) 2 kHz   (d) 10 krad/s   (f) 100 krad/s

7-13. What is the value of the capacitor for each of the following values of capacitive reactance. $\omega = 25$ krad/s.

(a) 10 Ω (c) 100 Ω (e) 1 kΩ
(b) 25 kΩ (d) 50 kΩ (f) 100 kΩ

7-14. Calculate the frequency at which a 0.1 μF capacitor has the following values of capacitive reactance:

(a) 25 Ω (c) 100 Ω (e) 25 kΩ
(b) 1 kΩ (d) 5 kΩ (f) 10 kΩ

7-15. For each of the circuits of Fig. 7-34, calculate the value of impedance in rectangular and polar form.

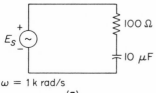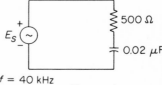

Fig. 7-34. Circuit diagrams for Prob. 7-15.

7-16. At a frequency of 25 kHz, the total impedance of a series R-C circuit is $10\underline{/-30°}$ kΩ. Calculate the value of the resistor and capacitor.

7-17. At a frequency of 50 krad/s, the total impedance of a series R-C circuit is $70.7\underline{/-45°}$ kΩ. Calculate the value of the resistor and capacitor.

7-18. For each of the series R-L-C circuits of Fig. 7-35, calculate (a) the impedance, $\bar{Z}_T$ and (b) the series equivalent circuit.

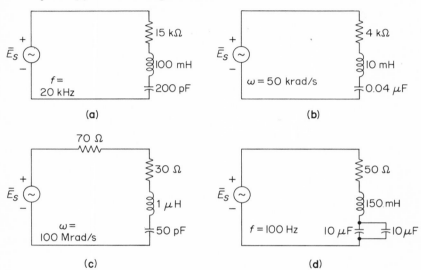

Fig. 7-35. Circuit diagrams for Prob. 7-18.

7-19 Calculate the admittance of the parallel R-L circuit of Fig. 7-15a in rectangular and polar form if

(a) $|\hat{I}_R| = 10$ mA   (b) $|\hat{I}_R| = 5$ mA   (c) $|\hat{I}_R| = 10$ mA

     $|\hat{I}_L| = 30$ mA $|\hat{I}_L| = 15$ mA $|\hat{I}_L| = 15$ mA

     $\bar{E} = 5$ V $\bar{E} = 5$ V $\bar{E} = 10$ V

7-20. For each of the circuits of Fig. 7-36, calculate the admittance in rectangular and polar form.

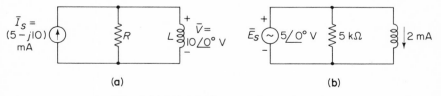

Fig. 7-36. Circuit diagrams for Prob. 7-20.

7-21. Calculate the value of inductive susceptance for a 5-mH inductor if the frequency is

(a) 0 Hz (dc)   (c) 200 kHz   (e) 2 kHz
(b) 60 Hz   (d) 10 krad/s   (f) 20 krad/s

7-22. Calculate the inductive susceptance value from each of the following inductive reactance values.

(a) $X_L = 1\,k\Omega$ (b) $X_L = 5\,\Omega$ (c) $X_L = 250\,\Omega$

7-23. Calculate the frequency at which a 2 H inductor has the following values of inductive susceptance.

(a) 0.1 m℧   (c) 0.01 ℧   (e) 1 m℧
(b) 0.02 ℧   (d) 0.04 ℧   (f) 0.2 μ℧

7-24. For each of the circuits of Fig. 7-37, calculate the admittance impedance in rectangular and polar form.

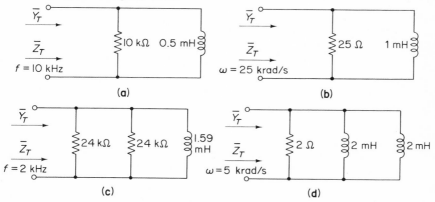

Fig. 7-37. Circuit diagrams for Prob. 7-24.

7-25. For each of the circuits of Fig. 7-32, calculate the parallel equivalent circuit.

7-26. At a frequency of 100 krad/s the total admittance of a parallel *R-L* circuit is $10\,/\!-45°$ m℧.   Calculate the value of the resistor and inductor.

7-27. At a frequency of 5 kHz, the total admittance of a parallel *R-L* circuit is $(2 - j1)$ m℧.   Calculate the value of the resistor and inductor.

7-28. Calculate the complex admittance of the parallel *R-C* circuit of Fig. 7-18a in rectangular and polar form if

(a) $|\hat{I}_R| = 4\,mA$ (b) $|\hat{I}_R| = 8\,mA$ (c) $|\hat{I}_R| = 4\,mA$
     $|\hat{I}_C| = 10\,mA$ $|\hat{I}_C| = 5\,mA$ $|\hat{I}_C| = 5\,mA$
     $\bar{V} = 10\,V$ $\bar{V} = 10\,V$ $\bar{V} = 20\,V$

7-29. Determine the admittance in rectangular and polar form for each of the circuits of Fig. 7-38.

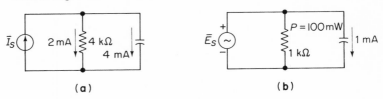

(a)   (b)

Fig. 7-38. Circuit diagrams for Prob. 7-29.

7-30. Calculate the value of capacitive susceptance for a 1 μF capacitor if the frequency is

(a)   0 Hz (dc)   (c)   200 Hz   (e)   2 kHz
(b)   60 Hz   (d)   4 krad/s   (f)   100 krad/s

7-31. Calculate the capacitive susceptance value from each of the following capacitive reactance values:

(a) $X_C = 4\,k\Omega$ (b) $X_C = 10\,\Omega$ (c) $X_C = 250\,\Omega$

7-32. Calculate the frequency at which, 4 μF capacitor has the following values of capacitive susceptance.

(a)   0.1 ℧   (c)   0.01 ℧   (e)   1 m℧
(b)   0.04 m℧   (d)   4 ℧   (f)   20 m℧

7-33. For each of the circuits of Fig. 7-39, calculate the admittance and impedance in rectangular and polar form.

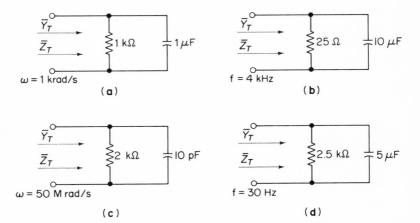

(a)   (b)

(c)   (d)

Fig. 7-39. Circuit diagrams for Prob. 7-33.

7-34. At a frequency of 100 krad/s, the total admittance of a parallel *R-C* circuit is $50 \underline{/45°}$ m℧. Calculate the value of the resistor and capacitor.

7-35. At a frequency of 250 krad/s, the total admittance of a parallel *R-C* circuit is $(5 + j6)$ m℧. Calculate the value of the resistor and capacitor.

7-36. For each of the parallel *R-L-C* circuits of Fig. 7-40, calculate (a) the admittance, $\overline{Y}_T$, (b) the parallel equivalent circuit, and (c) the series equivalent circuit.

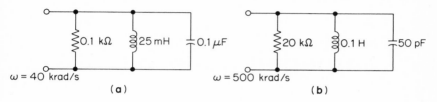

$\omega = 40$ krad/s $\omega = 500$ krad/s

(a)   (b)

Fig. 7-40. Circuit diagrams for Prob. 7-36.

CHAPTER 8

impedance networks

In Chapter 5 we discussed five methods for analyzing resistive networks: loop analysis, nodal analysis, superposition, Thévenin's theorem, and Norton's theorem.

In this chapter we shall again use these methods but will apply them to impedance networks (those containing resistance, inductance, and capacitance).

8-1. LOOP ANALYSIS

This section again shows that a set of simultaneous linear equations can be written which describes the network. This set of equations depends on a choice of loop currents used in connection with Kirchhoff's voltage law.

The general procedure for analyzing a network by loop analysis is again presented here so that the reader does not have to refer back to Chapter 5.

1. Assign a loop current for each independent closed path. The number of independent loops in a network is given by the expression

$$T - N + 1 \qquad (8\text{-}1)$$

where T is the *total* number of elements in the network (both active and passive) and N is the number of nodes in the network.

Although the direction of loop currents is arbitrary, in this book we shall only assign clockwise directions, for the sake of uniformity.

2. Indicate the polarity markings across each passive element due to the direction of the loop current in that closed path. The polarities of voltage sources are determined by their own positive and negative terminals.

3. Apply Kirchhoff's voltage law to each closed path.

If we are applying Kirchhoff's voltage law to one loop, and a passive element has two or more loop currents flowing through it, then the total voltage drop equals the sum of the voltage drops due to each loop current separately. If the loop currents flow in the same direction, the voltage drops add. However, if the loop currents flow in opposite directions, the voltage drops subtract from one another.

4. Solve the set of simultaneous linear equations for the loop currents.

Let us now apply the above procedure in analyzing some examples.

Example 8-1:   Calculate the branch current, $\bar{I}$, in the circuit of Fig. 8-1a.

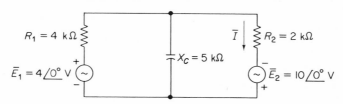

(a) Circuit diagram for Ex. 8-1

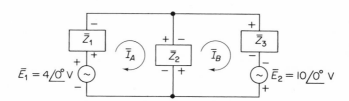

(b) Impedance diagram of circuit a

Fig. 8-1. Circuit diagrams for Ex. 8-1.

Solution: The impedance values for circuit b of Fig. 8-1 are

$$\bar{Z}_1 = R_1 = 4\,\text{k}\Omega$$

$$\bar{Z}_2 = -jX_c = -j5\,\text{k}\Omega$$

$$\bar{Z}_3 = R_2 = 2\text{k}\Omega$$

Steps 1 and 2 of the loop-analysis method are shown in Fig. 8-1b. Applying Kirchhoff's voltage law to each loop:

Loop A:

$$\bar{E}_1 = \bar{I}_A \bar{Z}_1 + \bar{I}_A \bar{Z}_2 - \bar{I}_B \bar{Z}_2$$

or

$$\bar{E}_1 = \bar{I}_A (\bar{Z}_1 + \bar{Z}_2) - \bar{I}_B \bar{Z}_2$$

Substituting numerical values,

$$4 \underline{/0^\circ} \text{ V} = \bar{I}_A (4 - j5) \text{ k}\Omega - \bar{I}_B (-j5) \text{ k}\Omega$$

Loop B:

$$\bar{E}_2 = -\bar{I}_A \bar{Z}_2 + \bar{I}_B \bar{Z}_2 + \bar{I}_B \bar{Z}_3$$

or

$$\bar{E}_2 = -\bar{I}_A \bar{Z}_2 + \bar{I}_B (\bar{Z}_2 + \bar{Z}_3)$$

Substituting numerical values,

$$10 \underline{/0^\circ} \text{ V} = -\bar{I}_A (-j5) \text{ k}\Omega + \bar{I}_B (2 - j5) \text{ k}\Omega$$

The set of simultaneous linear equations is

$$4 \text{ V} = \bar{I}_A (4 - j5) \text{ k}\Omega + \bar{I}_B (j5) \text{ k}\Omega$$

$$10 \text{ V} = \bar{I}_A (j5) \text{ k}\Omega + \bar{I}_B (2 - j5) \text{ k}\Omega$$

For this example, only the loop current $\bar{I}_B$ has to be solved, because $\bar{I} = \bar{I}_B$. Using the method of determinants,

$$\bar{I}_B = \frac{\begin{vmatrix} (4 - j5) \text{ k}\Omega & 4 \text{ V} \\ j5 \text{ k}\Omega & 10 \text{ V} \end{vmatrix}}{\begin{vmatrix} (4 - j5) \text{ k}\Omega & j5 \text{k}\Omega \\ j5 \text{ k}\Omega & (2 - j5) \text{ k}\Omega \end{vmatrix}}$$

$$\bar{I}_B = \frac{[10 (4 - j5) \times 10^3 - 4 (j5) \times 10^3] \text{ V} \cdot \Omega}{[(4 - j5)(2 - j5) \times 10^6 - (j5)(j5) \times 10^6] \Omega^2}$$

$$\bar{I}_B = \frac{(40 - j70) \times 10^3 \text{ V} \cdot \Omega}{(8 - j30) \times 10^6 \Omega^2} = \frac{80.6 \underline{/-60.3^\circ} \times 10^3 \text{ V} \cdot \Omega}{31.1 \underline{/-75.1^\circ} \times 10^6 \Omega^2}$$

$$\bar{I}_B = I = 2.59 \underline{/14.8^\circ} \text{ mA}$$

Again note that only loop current $\bar{I}_B$ needed to be solved because only loop current $\bar{I}_B$ flowed through $\bar{Z}_3$.

Example 8-2: Using the method of loop analysis, calculate the branch current, $\bar{I}$, in the circuit of Fig. 8-2a.

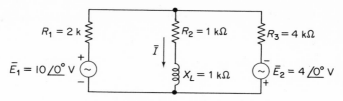

(a) Circuit diagram for Ex. 8-2

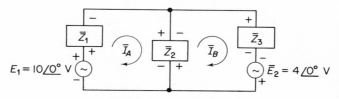

(b) Impedance diagram of circuit a

Fig. 8-2. Circuit diagrams for Ex. 8-2.

Solution: The impedance values are

$$\bar{Z}_1 = R_1 = 2\,\text{k}\Omega$$

$$\bar{Z}_2 = R_2 + jX_L = (1 + j1)\,\text{k}\Omega$$

$$\bar{Z}_3 = R_3 = 4\,\text{k}\Omega$$

Steps 1 and 2 of the loop-analysis method are shown in Fig. 8-2b. Applying Kirchhoff's voltage law to each loop yields.

Loop A:

$$\bar{E}_1 = \bar{I}_A\bar{Z}_1 + \bar{I}_A\bar{Z}_2 - \bar{I}_B\bar{Z}_2$$

or

$$\bar{E}_1 = \bar{I}_A(\bar{Z}_1 + \bar{Z}_2) - \bar{I}_B\bar{Z}_2$$

Substituting numerical values,

$$10\,\underline{/0°}\,\text{V} = \bar{I}_A(3 + j1)\,\text{k}\Omega - \bar{I}_B(1 + j1)\,\text{k}\Omega$$

Loop B:

$$\bar{E}_2 = -\bar{I}_A\bar{Z}_2 + \bar{I}_B\bar{Z}_2 + \bar{I}_B\bar{Z}_3$$

or

$$\bar{E}_2 = -\bar{I}_A\bar{Z}_2 + \bar{I}_B(\bar{Z}_2 + \bar{Z}_3)$$

Substituting numerical values,

$$4\,\underline{/0°}\,\text{V} = -\bar{I}_A(1 + j1)\,\text{k}\Omega + \bar{I}_B(5 + j1)\,\text{k}\Omega$$

Thus the set of simultaneous equations for the circuit of Fig. 8-2 is

$$10\,\text{V} = \bar{I}_A(3 + j1)\,\text{k}\Omega - \bar{I}_B(1 + j1)\,\text{k}\Omega$$

$$4\,\text{V} = -\bar{I}_A(1 + j1)\,\text{k}\Omega + \bar{I}_B(5 + j1)\,\text{k}\Omega$$

Using the method of determinants,

$$\bar{I}_A = \frac{\begin{vmatrix} 10\,\text{V} & -(1+j1)\,\text{k}\Omega \\ 4\,\text{V} & (5+j1)\,\text{k}\Omega \\ (3+j1)\,\text{k}\Omega & -(1+j1)\,\text{k}\Omega \\ -(1+j1)\,\text{k}\Omega & (5+j1)\,\text{k}\Omega \end{vmatrix}}{}$$

$$\bar{I}_A = \frac{[10(5+j1) \times 10^3 + 4(1+j1) \times 10^3]\,\text{V} \cdot \Omega}{[(3+j1)(5+j1) \times 10^6 - (1+j1)(1+j1) \times 10^6]\,\Omega^2}$$

$$\bar{I}_A = \frac{(54+j14) \times 10^3\,\text{V} \cdot \Omega}{(14+j6) \times 10^6\,\Omega^2} = \frac{56\,\underline{/14.5^\circ} \times 10^3\,\text{V} \cdot \Omega}{15.2\,\underline{/23.2^\circ} \times 10^6\,\Omega^2}$$

$$\bar{I}_A = 3.68\,\underline{/-9.7^\circ}\,\text{mA} = (3.63 - j0.62)\,\text{mA}$$

$$\bar{I}_B = \frac{\begin{vmatrix} (3+j1)\,\text{k}\Omega & 10\,\text{V} \\ -(1+j1)\,\text{k}\Omega & 4\,\text{V} \end{vmatrix}}{15.2\,\underline{/23.2^\circ} \times 10^6\,\Omega^2}$$

$$\bar{I}_B = \frac{[4(3+j1) \times 10^3 + 10(1+j1) \times 10^3]\,\text{V} \cdot \Omega}{15.2\,\underline{/23.2^\circ} \times 10^6\,\Omega^2}$$

$$\bar{I}_B = \frac{(22+j14) \times 10^3\,\text{V} \cdot \Omega}{15.2\,\underline{/23.2^\circ} \times 10^6\,\Omega^2} = \frac{26.1\,\underline{/32.5^\circ} \times 10^3\,\text{V} \cdot \Omega}{15.2\,\underline{/23.2^\circ} \times 10^6\,\Omega^2}$$

$$\bar{I}_B = 1.72\,\underline{/9.3^\circ}\,\text{mA} = (1.7 + j0.278)\,\text{mA}$$

Since

$$\bar{I} = \bar{I}_A - \bar{I}_B$$

then

$$\bar{I} = (3.63 - j0.62)\,\text{mA} - (1.7 + j0.278)\,\text{mA}$$

$$\bar{I} = (1.93 - j0.898)\,\text{mA}$$

In this example, both loop currents had to be solved because both loop currents flowed through $\bar{Z}_2$.

Example 8-3: Consider the circuit of Fig. 8-3 and determine the branch current $\bar{I}_2$.

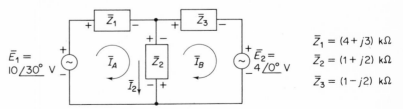

$\bar{Z}_1 = (4+j3)\,\text{k}\Omega$

$\bar{Z}_2 = (1+j2)\,\text{k}\Omega$

$\bar{Z}_3 = (1-j2)\,\text{k}\Omega$

Fig. 8-3. Circuit diagram for Ex. 8-3.

Solution: Steps 1 and 2 are shown in the diagram of Fig. 8-3. Applying Kirchhoff's voltage law to each loop:

Loop A:

$$\bar{I}_A \bar{Z}_1 + \bar{I}_A \bar{Z}_2 - \bar{I}_B \bar{Z}_2 = \bar{E}_1$$

or

$$\bar{I}_A(\bar{Z}_1 + \bar{Z}_2) - \bar{I}_B \bar{Z}_2 = \bar{E}_1$$

Substituting numerical values,

$$\bar{I}_A (5 + j5)\,k\Omega - \bar{I}_B(1 + j2)\,k\Omega = 10\,\underline{/30°}\,V$$

or

$$\bar{I}_A (7.07\,\underline{/45°})\,k\Omega - \bar{I}_B (2.24\,\underline{/63.5°})\,k\Omega = 10\,\underline{/30°}\,V$$

Loop B:

$$\bar{I}_B \bar{Z}_2 - \bar{I}_A \bar{Z}_2 + \bar{I}_B \bar{Z}_3 = -\bar{E}_2$$

or

$$-\bar{I}_A \bar{Z}_2 + \bar{I}_B(\bar{Z}_2 + \bar{Z}_3) = -\bar{E}_2$$

Substituting numerical values,

$$-\bar{I}_A(1 + j2)\,k\Omega + \bar{I}_B(2 + j0)\,k\Omega = -4\,\underline{/0°}\,V$$

or

$$-\bar{I}_A(2.24\,\underline{/63.5°})\,k\Omega + \bar{I}_B(2\,\underline{/0°})\,k\Omega = -4\,\underline{/0°}\,V$$

Thus the set of simultaneous linear equations is

$$\bar{I}_A (7.07\,\underline{/45°})\,k\Omega - \bar{I}_B (2.24\,\underline{/63.5°})\,k\Omega = 10\,\underline{/30°}\,V$$

$$-\bar{I}_A (2.24\,\underline{/63.5°})\,k\Omega + \bar{I}_B (2\,\underline{/0°})\,k\Omega = -4\,\underline{/0°}\,V$$

Using the method of determinants,

$$\bar{I}_A = \frac{\begin{vmatrix} 10\,\underline{/30°}\,V & -2.24\,\underline{/63.5°}\,k\Omega \\ -4\,\underline{/0°}\,V & 2\,\underline{/0°}\,k\Omega \end{vmatrix}}{\begin{vmatrix} 7.07\,\underline{/45°}\,k\Omega & -2.24\,\underline{/63.5°}\,k\Omega \\ -2.24\,\underline{/63.5°}\,k\Omega & 2\,\underline{/0°}\,k\Omega \end{vmatrix}}$$

$$\bar{I}_A = \frac{[(10\,\underline{/30°})(2\,\underline{/0°} \times 10^3) - (-2.24\,\underline{/63.5°} \times 10^3)(-4\,\underline{/0°})]\,V \cdot \Omega}{[(7.07\underline{/45°} \times 10^3)(2\underline{/0°} \times 10^3) - (-2.24\underline{/63.5°} \times 10^3)(-2.24\underline{/63.5°} \times 10^3)]\Omega^2}$$

$$\bar{I}_A = \frac{13.5\,\underline{/8.5°} \times 10^3\,V \cdot \Omega}{14.2\,\underline{/24.8°} \times 10^6\,\Omega^2}$$

$$\bar{I}_A = 0.95\,\underline{/-16.3°}\,mA = 0.95 - j0.266)\,mA$$

$$\bar{I}_B = \frac{\begin{vmatrix} 7.07\,\underline{/45°}\,k\Omega & 10\,\underline{/30°}\,V \\ -2.24\,\underline{/63.5°}\,k\Omega & -4\,\underline{/0°}\,V \end{vmatrix}}{14.2\,\underline{/24.8°} \times 10^6\,\Omega^2}$$

$$\bar{I}_B = \frac{[(7.07\,\underline{/45°} \times 10^3)(-4\,\underline{/0°}) - (10\,\underline{/30°})(-2.24\,\underline{/63.5°} \times 10^3)]\,V \cdot \Omega}{14.2\,\underline{/24.8°} \times 10^6\,\Omega^2}$$

$$\bar{I}_B = \frac{21.4\,\underline{/174°} \times 10^3}{14.2\,\underline{/24.8°} \times 10^6}$$

$$\bar{I}_B = 1.5\,\underline{/149.2°}\,mA = (-1.28 + j0.77)\,mA$$

From the circuit of Fig. 8-3,

$$\bar{I}_2 = \bar{I}_A - \bar{I}_B$$

$$\bar{I}_2 = [0.92 - j0.266 - (-1.28 + j0.77)]\,\text{mA}$$

Therefore,

$$\bar{I}_2 = (2.20 - j1.036)\,\text{mA}$$

Example 8-4: For the circuit of Fig. 8-4, obtain a set of simultaneous equations. Do not solve. $\omega = 12.5$ krad/s.

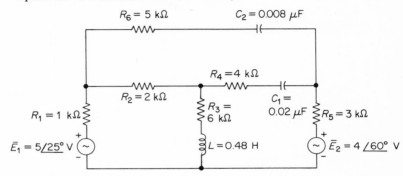

(a) Circuit diagram for Ex. 8-4

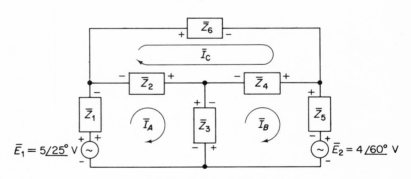

(b) Impedance diagram of circuit a

Fig. 8-4. Circuit diagrams for Ex. 8-4.

Solution: Converting the capacitor and inductor values to capacitive reactance and inductive reactance yields

$$X_L = \omega L = (12.5\ \text{krad/s})(0.48\ \text{H}) = 6\ \text{k}\Omega$$

$$X_{C1} = \frac{1}{\omega C_1} = \frac{1}{(12.5\ \text{krad/s})(0.02\ \mu\text{F})} = 4\ \text{k}\Omega$$

$$X_{C2} = \frac{1}{\omega C_2} = \frac{1}{(12.5\ \text{krad/s})(0.008\ \mu\text{F})} = 10\ \text{k}\Omega$$

The impedance values in the circuit of Fig. 8-4b are

$$\bar{Z}_1 = R_1 = 1 \text{ k}\Omega$$

$$\bar{Z}_2 = R_2 = 2 \text{ k}\Omega$$

$$\bar{Z}_3 = R_3 + jX_L = (6 + j6) \text{ k}\Omega$$

$$\bar{Z}_4 = R_4 - jX_{C1} = (4 - j4) \text{ k}\Omega$$

$$\bar{Z}_5 = R_5 = 3 \text{ k}\Omega$$

$$\bar{Z}_6 = R_6 - jX_{C_2} = (5 - j10) \text{ k}\Omega$$

Steps 1 and 2 of the loop-analysis method are shown in Fig. 8-4b. Applying Kirchhoff's voltage law to each loop,

Loop A:

$$\bar{E}_1 = \bar{I}_A(\bar{Z}_1 + \bar{Z}_2 + \bar{Z}_3) - \bar{I}_B\bar{Z}_3 - \bar{I}_C\bar{Z}_2$$

$$5 \underline{/25°} \text{ V} = \bar{I}_A(9 + j6) \text{ k}\Omega - \bar{I}_B(6 + j6) \text{ k}\Omega - \bar{I}_C 2 \text{ k}\Omega$$

Loop B:

$$-\bar{E}_2 = -\bar{I}_A\bar{Z}_3 + \bar{I}_B(\bar{Z}_3 + \bar{Z}_4 + \bar{Z}_5) - \bar{I}_C\bar{Z}_4$$

$$-4 \underline{/60°} \text{ V} = -\bar{I}_A(6 + j6) \text{ k}\Omega + \bar{I}_B(13 + j2) \text{ k}\Omega - \bar{I}_C(4 - j4) \text{ k}\Omega$$

Loop C:

$$0 = -\bar{I}_A\bar{Z}_2 - \bar{I}_B\bar{Z}_4 + \bar{I}_C(\bar{Z}_2 + \bar{Z}_4 + \bar{Z}_6)$$

$$0 = -\bar{I}_A 2 \text{ k}\Omega - \bar{I}_B(4 - j4) \text{ k}\Omega + \bar{I}_C(11 - j14) \text{ k}\Omega$$

A set of simultaneous linear equations for the circuit of Fig. 8-4 is

$$5 \underline{/25°} \text{ V} = \bar{I}_A(9 + j6) \text{ k}\Omega - \bar{I}_B(6 + j6) \text{ k}\Omega - \bar{I}_C 2 \text{ k}\Omega$$

$$-4 \underline{/60°} \text{ V} = -\bar{I}_A(6 + j6) \text{ k}\Omega + \bar{I}_B(13 + j2) \text{ k}\Omega - \bar{I}_C(4 - j4) \text{ k}\Omega$$

$$0 = -\bar{I}_A 2 \text{ k}\Omega - \bar{I}_B(4 - j4) \text{ k}\Omega + \bar{I}_C(11 - j14) \text{ k}\Omega$$

Note that the inductive and capacitive values first had to be converted to reactance values before the loop equations could be written. The angular frequency of both voltage sources is 12.5 krad/s. Chapter 12 treats circuits that have two sources set at different frequencies.

8-2. NODAL ANALYSIS

The object of this section is also to obtain a set of simultaneous linear equations that uniquely describe the network. However, unlike the loop analysis method, the procedure developed in this section depends on the choice of certain voltages used as variables in conjunction with Kirchhoff's current law.

The general procedure for anlyzing a network by nodal analysis is presented here so that the reader does not have to refer back to Chapter 5.

1.   Convert all voltage sources to current sources,
2.   Indicate one node as a reference node.   This reference node is used as the common point in the network from which the node voltages — $\bar{V}_a$, $\bar{V}_b$, etc. — are measured.   The number of node voltages and thus the number of linear equations is given by

$$N - 1 \qquad (8\text{-}2)$$

where N is the total number of nodes in the network.
3.   At each node, except the reference node, apply Kirchhoff's current law.
4.   Solve the set of simultaneous linear equations for the node voltages — $\bar{V}_a$, $\bar{V}_b$, etc.

Let us now apply the above procedure to numerical examples.

> **Example 8-5:** Using nodal analysis, calculate the branch current, $\bar{I}$, in the circuit of Fig. 8-2.
>
> *Solution:*   The circuit of Fig. 8-5 shows the voltage sources of Fig. 8-2b converted to current sources:
>
> $$\bar{I}_{S_1} = \frac{\bar{E}_1}{\bar{Z}_1} = \frac{10\,\underline{/0°}\text{ V}}{2\text{ k}\Omega} = 5\text{ mA}$$
>
> $$\bar{I}_{S_2} = \frac{\bar{E}_2}{\bar{Z}_3} = \frac{4\,\underline{/0°}\text{ V}}{4\text{ k}\Omega} = 1\text{ mA}$$

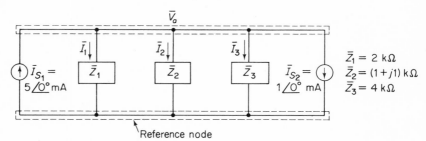

Fig. 8-5. Current source equivalent circuit of Fig. 8-2b.

Figure 8-5 shows that there is only one node voltage and therefore only one equation. Applying Kirchhoff's current law,

$$\bar{I}_{S_1} = \bar{I}_1 + \bar{I}_2 + \bar{I}_3 + \bar{I}_{S_2}$$

where

$$\bar{I}_1 = \frac{\bar{V}_a}{\bar{Z}_1}, \qquad \bar{I}_2 = \frac{\bar{V}_a}{\bar{Z}_2}, \qquad \bar{I}_3 = \frac{\bar{V}_a}{\bar{Z}_3}$$

Then

$$\bar{I}_{S_1} - \bar{I}_{S_2} = \bar{V}_a \left(\frac{1}{\bar{Z}_1} + \frac{1}{\bar{Z}_2} + \frac{1}{\bar{Z}_3} \right)$$

Substituting numerical values,

$$5\,\text{mA} - 1\,\text{mA} = \bar{V}_a \left(\frac{1}{2\,\text{k}\Omega} + \frac{1}{\sqrt{2}\,\underline{/45°}\,\text{k}\Omega} + \frac{1}{4\,\text{k}\Omega} \right)$$

$$4\,\text{mA} = \bar{V}_a \,(0.5\,\text{m}\mho + 0.707\,\underline{/-45°}\,\text{m}\mho + 0.25\,\text{m}\mho)$$

$$4\,\text{mA} = \bar{V}_a \,[0.5\,\text{m}\mho + (0.5 - j0.5)\,\text{m}\mho + 0.25\,\text{m}\mho]$$

$$4\,\text{mA} = \bar{V}_a (1.25 - j0.5)\,\text{m}\mho$$

$$\bar{V}_a = \frac{4\,\text{mA}}{(1.25 - j0.5)\,\text{m}\mho} = \frac{4\,\text{mA}}{1.345\,\underline{/-21.8°}\,\text{m}\mho}$$

$$\bar{V}_a = 2.98\,\underline{/21.8°}\,\text{V}$$

Therefore,

$$\bar{I} = \bar{I}_2 = \frac{\bar{V}_a}{\bar{Z}_2} = \frac{2.98\,\underline{/21.8°}\,\text{V}}{\sqrt{2}\,\underline{/45°}\,\text{k}\Omega}$$

$$\bar{I} = 2.1\,\underline{/-23.2°}\,\text{mA} = (1.93 - j0.83)\,\text{mA}$$

From the circuit of Fig. 8-5, there is only one node voltage; therefore, there is only one equation to be solved. This method involved less work than that of Example 8-2.

Example 8-6: Using the method of nodal analysis, calculate the value of current I_2 in the circuit of Fig. 8-6.

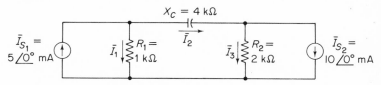

(a) Circuit diagram for Ex. 8-6

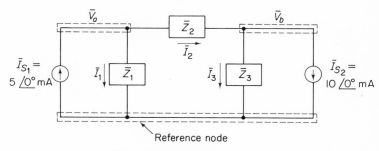

(b) Impedance diagram of circuit a

Fig. 8-6. Circuit diagrams for Ex. 8-6.

Solution: The impedance values of circuit b are

$$\bar{Z}_1 = R_1 = 1\,\text{k}\Omega$$

$$\bar{Z}_2 = -jX_c = -j4\,\text{k}\Omega$$

$$\bar{Z}_3 = R_2 = 2\,\text{k}\Omega$$

Applying Kirchhoff's current law to each node:

Node a:

$$\bar{I}_{S_1} = \bar{I}_1 + \bar{I}_2$$

where

$$\bar{I}_1 = \frac{\bar{V}_a}{\bar{Z}_1} \quad \text{and} \quad \bar{I}_2 = \frac{\bar{V}_a - \bar{V}_b}{\bar{Z}_2}$$

Then

$$\bar{I}_{S_1} = \frac{\bar{V}_a}{\bar{Z}_1} + \frac{\bar{V}_a - \bar{V}_b}{\bar{Z}_2}$$

$$\bar{I}_{S_1} = \bar{V}_a\left(\frac{1}{\bar{Z}_1} + \frac{1}{\bar{Z}_2}\right) - \bar{V}_b\frac{1}{\bar{Z}_2}$$

Substituting numerical values,

$$5\,\text{mA} = \bar{V}_a\left(\frac{1}{1\,\text{k}\Omega} + \frac{1}{-j4\,\text{k}\Omega}\right) - \bar{V}_b\frac{1}{-j4\,\text{k}\Omega}$$

$$5\,\text{mA} = \bar{V}_a(1 + j0.25)\,\text{m}\mho - \bar{V}_b(j0.25)\,\text{m}\mho$$

Node b:

$$\bar{I}_2 = \bar{I}_3 + \bar{I}_{S_2}$$

Rearranging,

$$-\bar{I}_{S_2} = -\bar{I}_2 + \bar{I}_3$$

where

$$\bar{I}_3 = \frac{\bar{V}_b}{\bar{Z}_3}$$

Then

$$-\bar{I}_{S_2} = -\frac{\bar{V}_a - \bar{V}_b}{\bar{Z}_2} + \frac{\bar{V}_b}{\bar{Z}_3}$$

or

$$-\bar{I}_{S_2} = -\bar{V}_a\frac{1}{Z_2} + \bar{V}_b\left(\frac{1}{\bar{Z}_2} + \frac{1}{\bar{Z}_3}\right)$$

Substituting numerical values,

$$-10\,\text{mA} = -\bar{V}_a\frac{1}{-j4\,\text{k}\Omega} + \bar{V}_b\left(\frac{1}{-j4\,\text{k}\Omega} + \frac{1}{2\,\text{k}\Omega}\right)$$

$$-10\,\text{mA} = -\bar{V}_a(j0.25)\,\text{m}\mho + \bar{V}_b(0.5 + j0.25)\,\text{m}\mho$$

The set of simultaneous equations is

$$5\,\text{mA} = \bar{V}_a(1 + j0.25)\,\text{m}\mho - \bar{V}_b(j0.25)\,\text{m}\mho$$

$$-10\,\text{mA} = -\bar{V}_a(j0.25)\,\text{m}\mho + \bar{V}_b(0.5 + j0.25)\,\text{m}\mho$$

Using the method of determinants,

$$\bar{V}_a = \frac{\begin{vmatrix} 5\text{ mA} & -j0.25\text{ mU} \\ -10\text{ mA} & (0.5 + j0.25)\text{ mU} \end{vmatrix}}{\begin{vmatrix} (1 + j0.25)\text{ mU} & -j0.25\text{ mU} \\ -j0.25\text{ mU} & (0.5 + j0.25)\text{ mU} \end{vmatrix}}$$

$$\bar{V}_a = \frac{[5(0.5 + j0.25) \times 10^{-6} - (-j0.25)(-10) \times 10^{-6}]\,\text{A}\cdot\text{U}}{[(1 + j0.25)(0.5 + j0.25) \times 10^{-6} - (-j0.25)(-j0.25) \times 10^{-6}]\,\text{U}^2}$$

$$\bar{V}_a = \frac{(2.5 - j1.25) \times 10^{-6}\,\text{A}\cdot\text{U}}{(0.5 + j0.375) \times 10^{-6}\,\text{U}^2} = \frac{2.79\,\underline{/-26.6°} \times 10^{-6}\,\text{A}\cdot\text{U}}{0.625\,\underline{/36.9°} \times 10^{-6}\,\text{U}^2}$$

$$\bar{V}_a = 4.47\,\underline{/-63.5°}\,\text{V} = (2 - j4)\,\text{V}$$

$$\bar{V}_b = \frac{\begin{vmatrix} (1 + j0.25)\text{ mU} & 5\text{ mA} \\ -j0.25\text{ mU} & -10\text{ mA} \end{vmatrix}}{0.625\,\underline{/36.9°} \times 10^{-6}\,\text{U}^2}$$

$$\bar{V}_b = \frac{[-10(1 + j0.25) \times 10^{-6} - (5)(-j0.25) \times 10^{-6}]\,\text{A}\cdot\text{U}}{0.625\,\underline{/36.9°} \times 10^{-6}\,\text{U}^2}$$

$$\bar{V}_b = \frac{(-10 - j1.25) \times 10^{-6}\,\text{A}\cdot\text{U}}{0.625\,\underline{/36.9°} \times 10^{-6}\,\text{U}^2} = \frac{-10.05\,\underline{/7.14} \times 10^{-6}\,\text{A}\cdot\text{U}}{0.625\,\underline{/36.9°} \times 10^{-6}\,\text{U}^2}$$

$$\bar{V}_b = -16.1\,\underline{/-29.76°}\,\text{V} = (-14 + j8)\,\text{V}$$

Since

$$\bar{I}_2 = \frac{\bar{V}_a - \bar{V}_b}{\bar{Z}_2}$$

then

$$\bar{I}_2 = \frac{(2 - j4)\,\text{V} - (-14 + j8)\,\text{V}}{-j4\,\text{k}\Omega}$$

$$\bar{I}_2 = \frac{(16 - j12)\,\text{V}}{-j4\,\text{k}\Omega} = \frac{20\,\underline{/-36.9°}\,\text{V}}{4\,\underline{/-90°}\,\text{k}\Omega}$$

$$\bar{I}_2 = 5\,\underline{/53.1°}\,\text{mA} = (3 + j4)\,\text{mA}$$

This example required the solution of two equations because $\bar{I}_2$ is associated with two node voltages.

Example 8-7: Consider the circuit of Fig. 8-7 and determine the branch currents $\bar{I}_1$, $\bar{I}_2$, and $\bar{I}_3$.

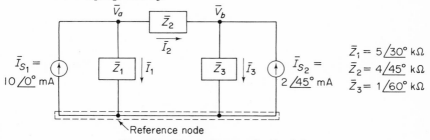

Fig. 8-7. Circuit diagram for Ex. 8-7.

Solution: Steps 1 and 2 are shown in Fig. 8-7. Applying Kirchhoff's current law:

Node *a:*

$$\bar{I}_{S1} = \bar{I}_1 + \bar{I}_2$$

In terms of node voltages

$$\bar{I}_1 = \frac{\bar{V}_a}{\bar{Z}_1} \quad \text{and} \quad \bar{I}_2 = \frac{\bar{V}_a - \bar{V}_b}{\bar{Z}_2}$$

Therefore,

$$\bar{I}_{S1} = \frac{\bar{V}_a}{\bar{Z}_1} + \frac{\bar{V}_a - \bar{V}_b}{\bar{Z}_2}$$

$$\bar{I}_{S1} = \bar{V}_a \left(\frac{1}{\bar{Z}_1} + \frac{1}{\bar{Z}_2} \right) - \bar{V}_b \frac{1}{\bar{Z}_2}$$

Substituting numerical values,

$$10 \underline{/0°} \text{ mA} = \bar{V}_a \left(\frac{1}{5 \underline{/30°} \text{ k}\Omega} + \frac{1}{4 \underline{/45°} \text{ k}\Omega} \right) - \bar{V}_b \frac{1}{4 \underline{/45°} \text{ k}\Omega}$$

$$10 \underline{/0°} \text{ mA} = \bar{V}_a \, 0.445 \underline{/-38.5°} \text{ m}\mho - \bar{V}_b \, 0.25 \underline{/-45°} \text{ m}\mho$$

Node *b:*

$$\bar{I}_{S2} + \bar{I}_2 = \bar{I}_3$$

or

$$\bar{I}_{S2} = \bar{I}_3 - \bar{I}_2$$

Therefore, in terms of node voltages,

$$\bar{I}_{S2} = \frac{\bar{V}_b}{\bar{Z}_3} - \left(\frac{\bar{V}_a - \bar{V}_b}{\bar{Z}_2} \right)$$

or

$$\bar{I}_{S2} = -\bar{V}_a \frac{1}{\bar{Z}_2} + \bar{V}_b \left(\frac{1}{\bar{Z}_2} + \frac{1}{\bar{Z}_3} \right)$$

Substituting numerical values,

$$2 \underline{/45°} \text{ mA} = -\bar{V}_a \frac{1}{4 \underline{/45°} \text{ k}\Omega} + \bar{V}_b \left(\frac{1}{4 \underline{/45°} \text{ k}\Omega} + \frac{1}{1 \underline{/60°} \text{ k}\Omega} \right)$$

$$2 \underline{/45°} \text{ mA} = -\bar{V}_a \, 0.25 \underline{/-45°} \text{m}\mho + \bar{V}_b \, 1.24 \underline{/-57°} \text{ m}\mho$$

Thus the set of simultaneous linear equations is

$$10 \underline{/0°} \text{ mA} = \bar{V}_a \, 0.445 \underline{/-38.4°} \text{ m}\mho - \bar{V}_b \, 0.25 \underline{/-45°} \text{ m}\mho$$

$$2 \underline{/45°} \text{ mA} = -\bar{V}_a \, 0.25 \underline{/-45°} \text{ m}\mho + \bar{V}_b \, 1.24 \underline{/-57°} \text{ m}\mho$$

Using the method of determinants,

$$\bar{V}_a = \frac{\begin{vmatrix} 10 \underline{/0°} \text{ mA} & -0.25 \underline{/-45°} \text{ m}\mho \\ 2 \underline{/45°} \text{ mA} & 1.24 \underline{/-57°} \text{ m}\mho \end{vmatrix}}{\begin{vmatrix} 0.445 \underline{/-38.4°} \text{ m}\mho & -0.25 \underline{/-45°} \text{ m}\mho \\ -0.25 \underline{/-45°} \text{ m}\mho & 1.24 \underline{/-57°} \text{ m}\mho \end{vmatrix}}$$

$$\bar{V}_a = \frac{(12.4\,\underline{/-57°} + 0.5\,\underline{/0°}) \times 10^{-6}\,A \cdot \mho}{(0.555\,\underline{/-95.4°} - 0.0625\,\underline{/-90°}) \times 10^{-6}\,\mho^2}$$

$$\bar{V}_a = \frac{12.6\,\underline{/-55°} \times 10^{-6}\,A \cdot \mho}{0.49\,\underline{/-95.5°} \times 10^{-6}\,\mho^2}$$

$$\bar{V}_a = 25.7\,\underline{/40°}\,V$$

or

$$\bar{V}_a = (19.4 + j16.5)\,V$$

$$\bar{V}_b = \frac{\begin{vmatrix} 0.445\,\underline{/-38.4°}\,m\mho & 10\,\underline{/0°}\,mA \\ -0.25\,\underline{/-45°}\,m\mho & 2\,\underline{/45°}\,mA \end{vmatrix}}{0.49\,\underline{/-95.5°} \times 10^{-6}\,\mho^2}$$

$$\bar{V}_b = \frac{(0.89\,\underline{/6.5°} + 2.5\,\underline{/-45°}) \times 10^{-6}\,A \cdot \mho}{0.49\,\underline{/-95.5°} \times 10^{-6}\,\mho^2}$$

$$\bar{V}_b = \frac{3.12\,\underline{/-32.2°} \times 10^{-6}\,A \cdot \mho}{0.49\,\underline{/-95.5°} \times 10^{-6}\,\mho^2}$$

$$\bar{V}_b = 6.4\,\underline{/63.2°}\,V$$

or

$$\bar{V}_b = (2.88 + j5.7)\,V$$

With the values of the node voltages known, we now may solve for the branch currents:

$$\bar{I}_1 = \frac{\bar{V}_a}{\bar{Z}_1} = \frac{25.7\,\underline{/40°}\,V}{5\,\underline{/30°}\,k\Omega} = 5.14\,\underline{/10°}\,mA$$

$$\bar{I}_2 = \frac{\bar{V}_a - \bar{V}_b}{\bar{Z}_2} = \frac{19.8\,\underline{/33.2°}\,V}{4\,\underline{/45°}\,k\Omega} = 4.95\,\underline{/-11.8}\,mA$$

$$\bar{I}_3 = \frac{\bar{V}_b}{\bar{Z}_3} = \frac{6.4\,\underline{/63.2°}\,V}{1\,\underline{/60°}\,k\Omega} = 6.4\,\underline{/3.2°}\,mA$$

Thus a complex impedance network is analyzed in the same manner as that of resistive networks. As a check the reader should apply Kirchhoff's current law at node *a* and at node *b*.

Example 8-8: Write the set of simultaneous linear equations for the circuit shown in Fig. 8-8. Do not solve.

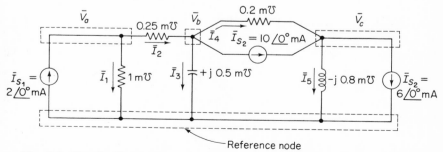

Fig. 8-8. Circuit diagram for Ex. 8-8. (Note admittance values are given.)

Solution: Applying Kirchhoff's current law to each node:

Node *a:*

$$\bar{I}_{S_1} = \bar{I}_1 + \bar{I}_2$$

where

$$\bar{I}_1 = \bar{V}_a \, 1 \, \text{m℧}$$

$$\bar{I}_2 = (\bar{V}_a - \bar{V}_b) \, 0.25 \, \text{m℧}$$

Then

$$2 \underline{/30°} \, \text{mA} = \bar{V}_a (1.25 \, \text{m ℧}) - \bar{V}_b (0.25 \, \text{m℧})$$

Node *b:*

$$\bar{I}_2 = \bar{I}_3 + \bar{I}_4 + \bar{I}_{S_2} \qquad \text{or} \qquad -\bar{I}_{S_2} = -\bar{I}_2 + \bar{I}_3 + \bar{I}_4$$

where

$$\bar{I}_3 = \bar{V}_b \, (j0.5 \, \text{m℧})$$

$$\bar{I}_4 = (\bar{V}_b - \bar{V}_c) \, 0.2 \, \text{m℧}$$

$$-10 \underline{/0°} \, \text{mA} = -(\bar{V}_a - \bar{V}_b) \, 0.25 \, \text{m℧} + \bar{V}_b \, (j0.5 \, \text{m℧}) + (\bar{V}_b - \bar{V}_c) \, 0.2 \, \text{m℧}$$

$$-10 \underline{/0°} \, \text{mA} = -\bar{V}_a (0.25 \, \text{m℧}) + \bar{V}_b (0.45 + j0.5) \, \text{m℧} - \bar{V}_c (0.2 \, \text{m℧})$$

Node *c:*

$$\bar{I}_{S_2} + \bar{I}_4 = \bar{I}_5 + \bar{I}_{S_3}$$

$$\bar{I}_{S_2} - \bar{I}_{S_3} = -\bar{I}_4 + \bar{I}_5$$

where

$$\bar{I}_5 = \bar{V}_c \, (-j0.8 \, \text{m℧})$$

Then

$$10 \underline{/0°} \, \text{mA} - 6 \underline{/0°} \, \text{mA} = -(\bar{V}_b - \bar{V}_c) \, 0.2 \, \text{m℧} + \bar{V}_c \, (-j0.8 \, \text{m℧})$$

$$4 \underline{/0°} \, \text{mA} = -\bar{V}_b (0.2 \, \text{m℧}) + \bar{V}_c (0.2 - j0.8) \, \text{m℧}$$

Thus the set of simultaneous equations is

$$2 \underline{/30°} \, \text{mA} = \bar{V}_a (1.25 \, \text{m℧}) - \bar{V}_b (0.25 \, \text{m℧}) \qquad\qquad + \; 0$$

$$-10 \underline{/0°} \, \text{mA} = -\bar{V}_a (0.25 \, \text{m℧}) + \bar{V}_b (0.45 + j0.5) \, \text{m℧} - \bar{V}_c (0.2 \, \text{m℧})$$

$$4 \underline{/0°} \, \text{mA} = \quad 0 \qquad\qquad\quad - \bar{V}_b (0.2 \, \text{m℧}) \qquad + \bar{V}_c (0.2 - j0.8) \, \text{m℧}$$

8-3. SUPERPOSITION THEOREM

The method of superposition to analyze an impedance network is the same as it is to analyze a resistive circuit. Superposition avoids solving a set of simultaneous linear equations. This is accomplished by obtaining the effects of each energy source separately. The superposition theorem may be stated as follows: *In a network containing more than one source (of voltage or current), the total current in any branch is the algebraic sum of the individual currents produced by each source acting alone. All other sources are replaced by their internal impedance.*

In obtaining the algebraic sum, if the currents flow in the same direction they add. If the currents flow in opposite directions they subtract from one another. For a numerical analysis, consider the following examples.

Example 8-9: Consider again the circuit of Fig. 8-6 and determine the branch current $\bar{I}_2$. This time, however, use the method of superposition.

Solution: From the circuit of Fig. 8-9a when $\bar{I}_{S_2}$ is open circuited, $\bar{Z}_2$ and $\bar{Z}_3$ are in series and $\bar{I}_A$ flows through them. Using the current division law:

$$\bar{I}_A = \frac{\bar{Z}_1}{\bar{Z}_1 + \bar{Z}_2 + \bar{Z}_3} \times I_{S_1}$$

$$\bar{I}_A = \frac{1\,\text{k}\Omega}{1\,\text{k}\Omega - j4\,\text{k}\Omega + 2\,\text{k}\Omega}(5\,\text{mA})$$

$$\bar{I}_A = \frac{5\,\text{mA}}{(3 - j4)\,\text{k}\Omega} = \frac{5\,\text{mA}}{5\,\underline{/-53.1°}\,\text{k}\Omega}$$

$$\bar{I}_A = 1\,\underline{/53.1°}\,\text{mA} = (0.6 + j0.8)\,\text{mA}$$

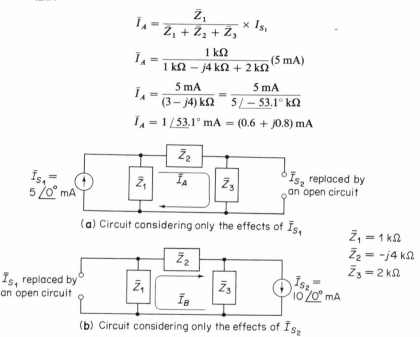

(a) Circuit considering only the effects of $\bar{I}_{S_1}$

(b) Circuit considering only the effects of $\bar{I}_{S_2}$

$$\bar{Z}_1 = 1\,\text{k}\Omega$$
$$\bar{Z}_2 = -j4\,\text{k}\Omega$$
$$\bar{Z}_3 = 2\,\text{k}\Omega$$

Fig. 8-9. Circuit diagrams to analyze Fig. 8-6 by superposition.

From the circuit of Fig. 8-9b when $\bar{I}_{S_1}$ is open circuited, $\bar{Z}_1$ and $\bar{Z}_2$ are in series and I_B flows through them. Using the current division law

$$\bar{I}_B = \frac{\bar{Z}_3}{\bar{Z}_1 + \bar{Z}_2 + \bar{Z}_3} \times \bar{I}_{S_2}$$

$$\bar{I}_B = \frac{2\,\text{k}\Omega}{1\,\text{k}\Omega - j4\,\text{k}\Omega + 2\,\text{k}\Omega}(10\,\text{mA})$$

$$\bar{I}_B = \frac{20\,\text{mA}}{(3 - j4\,\text{k}\Omega} = \frac{20\,\text{mA}}{5\,\underline{/-53.1°}\,\text{k}\Omega}$$

$$\bar{I}_B = 4\,\underline{/53.1°}\,\text{mA} = (2.4 + j3.2)\,\text{mA}$$

Since both $\bar{I}_A$ and $\bar{I}_B$ flow through $\bar{Z}_2$ in the same direction, then

$$\bar{I}_2 = \bar{I}_A + \bar{I}_B$$
$$\bar{I}_2 = (0.6 + j0.8)\,\text{mA} + (2.4 + j3.2)\,\text{mA}$$
$$\bar{I}_2 = (3 + j4)\,\text{mA}$$

This answer checks with the answer of Example 8-6 and does not involve as much work.

Example 8-10: Using the method of superposition, calculate the current $\bar{I}$ in the circuit of Fig. 8-10.

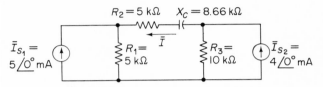

Fig. 8-10. Circuit diagram for Ex. 8-10.

Solution: From circuit a of Fig. 8-11 when $\bar{I}_{S_2}$ is open circuited, $\bar{Z}_2$ is in series with $\bar{Z}_3$ and $\bar{I}_A$ flows through them. Using the current division law,

$$I_A = \frac{\bar{Z}_1}{\bar{Z}_1 + \bar{Z}_2 + \bar{Z}_3} \times \bar{I}_{S_1}$$

$$\bar{I}_A = \frac{5\,\text{k}\Omega}{(5 + 5 - j8.66 + 10)\,\text{k}\Omega}(5\,\underline{/0^\circ}\,\text{mA})$$

$$\bar{I}_A = \frac{25\,\text{mA}}{(20 - j8.66)\,\text{k}\Omega} = \frac{25\,\text{mA}}{21.8\,\underline{/-23.4^\circ}\,\text{k}\Omega}$$

$$\bar{I}_A = 1.145\,\underline{/23.4^\circ}\,\text{mA} = (1.05 + j0.445)\,\text{mA}$$

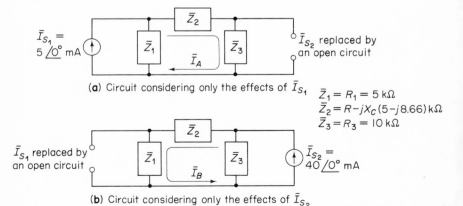

(a) Circuit considering only the effects of $\bar{I}_{S_1}$

$\bar{Z}_1 = R_1 = 5\,\text{k}\Omega$
$\bar{Z}_2 = R - jX_C\,(5 - j8.66)\,\text{k}\Omega$
$\bar{Z}_3 = R_3 = 10\,\text{k}\Omega$

(b) Circuit considering only the effects of $\bar{I}_{S_2}$

Fig. 8-11. Circuits diagrams used to analyze Fig. 8-10 by superposition.

From circuit b of Fig. 8-11 when $\bar{I}_{S_1}$ is open circuited, Z_1 is in series with $\bar{Z}_2$ and $\bar{I}_B$ flows through them. Using the current division law,

$$\bar{I}_B = \frac{\bar{Z}_3}{\bar{Z}_1 + \bar{Z}_2 + \bar{Z}_3} \times \bar{I}_{S_2}$$

$$\bar{I}_B = \frac{10 \text{ k}\Omega}{(5 + 5 - j8.66 + 10) \text{ k}\Omega} (4 \underline{/0°} \text{ mA})$$

$$\bar{I}_B = \frac{40 \text{ mA}}{(20 - j8.66) \text{ k}\Omega} = \frac{40 \text{ mA}}{21.8 \underline{/-23.4°} \text{ k}\Omega}$$

$$\bar{I}_B = 1.835 \underline{/23.4°} \text{ mA} = (1.68 + j0.728) \text{ mA}$$

Since $\bar{I}_A$ and $\bar{I}_B$ flow through $\bar{Z}_2$ in opposite directions, then

$$\bar{I} = \bar{I}_B - \bar{I}_A$$

$$\bar{I} = (1.68 + j0.728) \text{ mA} - (1.05 + j0.445) \text{ mA}$$

$$I = (0.63 + j0.283) \text{ mA}$$

Note that in calculating I it is I_B minus I_A because I and I_B flow in the same direction while $\bar{I}$ and $\bar{I}_A$ flow in opposite directions.

Example 8-11: Consider again the circuit of Fig. 8-3 and determine the branch current through $\bar{Z}_2$. This time, however, use the method of superposition.

Solution: Consider the circuit of Fig. 8-12a. Let

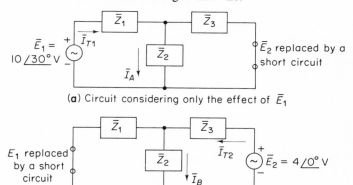

(a) Circuit considering only the effect of $\bar{E}_1$

(b) Circuit considering only the effect of $\bar{E}_2$

Fig. 8-12. Circuit diagrams to analyze Fig. 8-3 by superposition.

$$\bar{Z}_A = \bar{Z}_2 \| \bar{Z}_3 = \frac{\bar{Z}_2 \bar{Z}_3}{\bar{Z}_2 + \bar{Z}_3}$$

Then

$$\bar{Z}_A = \frac{(2.24 \underline{/63.5°} \text{ k}\Omega)(2.24 \underline{/-63.5°} \text{k}\Omega)}{(1 + j2) \text{ k}\Omega + (1 - j2) \text{ k}\Omega}$$

$$\bar{Z}_A = 2.5 \underline{/0°} \text{ k}\Omega = (2.5 + j0) \text{ k}\Omega$$

Let
$$\bar{Z}_{T1} = \bar{Z}_1 + \bar{Z}_A$$

Then
$$\bar{Z}_{T1} = (4 + j3)\,k\Omega + (2.5 + j0)\,k\Omega$$

$$\bar{Z}_{T1} = (6.5 + j3)\,k\Omega = 7.15\,\underline{/24.8°}\,k\Omega$$

The current drawn from E_1 is
$$\bar{I}_{T1} = \frac{\bar{E}_1}{\bar{Z}_{T1}} = \frac{10\,\underline{/30°}\,\text{V}}{7.14\,\underline{/24.8°}\,k\Omega} = 1.4\,\underline{/5.2°}\,\text{mA}$$

Using the current division law,
$$\bar{I}_A = \frac{\bar{Z}_3}{\bar{Z}_2 + \bar{Z}_3} \times I_{T1}$$

$$\bar{I}_A = \frac{2.24\,\underline{/-63.5°}\,k\Omega}{2\,\underline{/0°}\,k\Omega}(1.4\,\underline{/5.2°}\,\text{mA})$$

$$\bar{I}_A = 1.57\,\underline{/-58.3°}\,\text{mA} = (0.825 - j1.33)\,\text{mA}$$

Now consider the circuit of Fig. 8-12b. Let
$$\bar{Z}_B = \bar{Z}_1 \| \bar{Z}_2 = \frac{\bar{Z}_1 \bar{Z}_2}{\bar{Z}_1 + \bar{Z}_2}$$

Then
$$\bar{Z}_B = \frac{(5\,\underline{/37°}\,k\Omega)\,(2.24\,\underline{/63.5°}\,k\Omega)}{(4 + j3)\,k\Omega + (1 + j2)\,k\Omega}$$

$$\bar{Z}_B = 1.58\,\underline{/55.5°}\,k\Omega = (0.89 + j1.3)\,k\Omega$$

Let
$$\bar{Z}_{T2} = \bar{Z}_B + \bar{Z}_3$$

$$\bar{Z}_{T2} = (0.89 + j1.3)\,k\Omega + (1 - j2)\,k\Omega$$

$$\bar{Z}_{T2} = (1.89 - j0.7)\,k\Omega = 2\,\underline{/-20.4°}\,k\Omega$$

The current drawn from E_2 is
$$\bar{I}_{T2} = \frac{\bar{E}_2}{\bar{Z}_{T2}} = \frac{4\,\underline{/0°}\,\text{V}}{2\,\underline{/-20.4°}\,k\Omega} = 2\,\underline{/20.4°}\,\text{mA}$$

Using the current division law,
$$\bar{I}_B = \frac{\bar{Z}_1}{\bar{Z}_1 + \bar{Z}_2} \times \bar{I}_{T2}$$

$$\bar{I}_B = \frac{5\,\underline{/37°}\,k\Omega}{7.07\,\underline{/45°}\,k\Omega}(2\,\underline{/20.4°}\,\text{mA})$$

$$\bar{I}_B = 1.41\,\underline{/12.4°}\,\text{mA} = (1.38 + j0.34)\,\text{mA}$$

Therefore, the total branch current through $\bar{Z}_2$ is
$$\bar{I}_2 = \bar{I}_A + \bar{I}_B$$

$$\bar{I}_2 = (0.825 - j1.33)\,\text{mA}$$

$$\bar{I}_2 = (2.2 - j1.0)\,\text{mA}$$

Thus the method of analysis of a complex impedance network is similar to that of resistive networks. Note that this answer checks with that of Example 8-3.

Example 8-12: Use the method of superposition and calculate the current $\bar{I}$ in the circuit of Fig. 8-13.

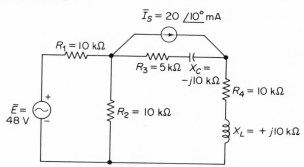

Fig. 8-13. Circuit diagram for Ex. 8-12.

Solution: Considering circuit a of Fig. 8-14, let

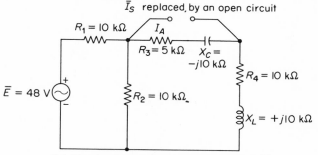

(a) Circuit considering only the effects of $\bar{E}$

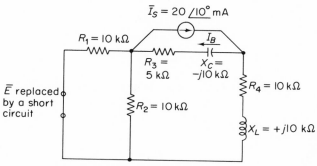

(b) Circuit considering only the effects of $\bar{I}_S$

Fig. 8-14. Circuit diagrams used to analyze Fig. 8-13 by superposition.

$$\bar{Z}_A = R_3 - jX_c + R_4 + jX_L$$

$$\bar{Z}_A = (5 - j10) \, k\Omega + (10 + j10) \, k\Omega = 15 \, k\Omega$$

and

$$\bar{Z}_B = \bar{Z}_A \| R_2 = \frac{(15 \, k\Omega)(10 \, k\Omega)}{15 \, k\Omega + 10 \, k\Omega} = 6 \, k\Omega$$

Then

$$\bar{Z}_T = R_1 + \bar{Z}_B = 10 \, k\Omega + 6 \, k\Omega = 16 \, k\Omega$$

$$\bar{I}_T = \frac{\bar{E}}{\bar{Z}_T} = \frac{48 \, V}{16 \, k\Omega} = 3 \, mA$$

Using the current division law,

$$\bar{I}_A = \frac{R_2}{R_2 + \bar{Z}_A} \times \bar{I}_T$$

$$\bar{I}_A = \frac{10 \, k\Omega}{10 \, k\Omega + 15 \, k\Omega}(3 \, mA) = 1.2 \, mA$$

Considering circuit b of Fig. 8-14, let

$$\bar{Z}_C = R_1 \| R_2 = \frac{10 \, k\Omega}{2} = 5 \, k\Omega$$

and

$$\bar{Z}_D = \bar{Z}_C + R_4 + jX_L$$

$$\bar{Z}_D = 5 \, k\Omega + (10 + j10) \, k\Omega = (15 + j10) \, k\Omega$$

Using the current divison law,

$$\bar{I}_B = \frac{R_3 - jX_C}{\bar{Z}_D + R_3 - jX_C} \times \bar{I}_S$$

$$\bar{I}_B = \frac{(5 - j10) \, k\Omega}{(15 + j10) \, k\Omega + (5 - j10) \, k\Omega}(20 \, \underline{/10°} \, mA)$$

$$\bar{I}_B = \frac{10.75 \, \underline{/-26.6°} \, k\Omega}{20 \, k\Omega}(20 \, \underline{/10°} \, mA)$$

$$\bar{I}_B = 10.75 \, \underline{/-16.6°} \, mA = (10.3 - j3.08) \, mA$$

Since $\bar{I}_A$ and $\bar{I}_B$ flow in the same direction, then

$$\bar{I} = \bar{I}_A + \bar{I}_B$$

$$\bar{I} = 1.2 \, mA + (10.3 - j3.08) \, mA$$

$$\bar{I} = (11.5 - j3.08) \, mA$$

The superposition method may be used for circuits containing both voltage sources and current sources.

8-4. THÉVENIN'S THEOREM

Restating Thévenin's theorem for impedance networks: *Any two terminals of an active network may be represented by a series combination of a voltage source and an impedance. The value of this equivalent source is the voltage appearing at the open-circuited terminals. The value of the equivalent impedance is equal to that impedance seen looking back into the network with all energy sources replaced by their respective internal impedance.*

Using Thévenin's theorem, we are able to mathematically analyze any network to obtain an equivalent circuit between any two terminals of the network. Therefore, any elaborate network may be reduced to a voltage source in series with an impedance. Once this equivalent circuit is developed, no matter what impedance is connected between the output terminals, it will have exactly the same current and voltage relationship as it would have if it were connected to the original circuit.

Example 8-13: Obtain the Thévenin equivalent circuit for the 1 kΩ resistor in the circuit of Fig. 8-15.

Fig. 8-15. Circuit diagrams for Ex. 8-13.

Solution: The impedance values in Fig. 8-16 are

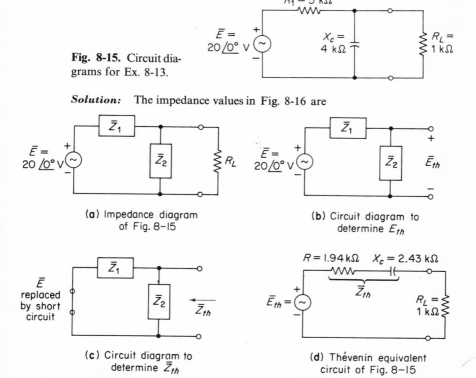

(a) Impedance diagram of Fig. 8-15

(b) Circuit diagram to determine E_{th}

(c) Circuit diagram to determine $\bar{Z}_{th}$

(d) Thévenin equivalent circuit of Fig. 8-15

Fig. 8-16. Circuit diagrams used in the solution of Ex. 8-13.

$$\bar{Z}_1 = R_1 = 5\,\text{k}\Omega$$

$$\bar{Z}_2 = -jX_C = -j4\,\text{k}\Omega$$

Applying the voltage division law to circuit b of Fig. 8-16,

$$\bar{E}_{\text{Th}} = \frac{\bar{Z}_2}{\bar{Z}_1 + \bar{Z}_2} \times \bar{E}$$

$$\bar{E}_{\text{Th}} = \frac{-j4\,\text{k}\Omega}{(5 - j4)\,\text{k}\Omega}(20\,\underline{/\,0°}\,\text{V})$$

$$\bar{E}_{\text{Th}} = \frac{80\,\underline{/-90°}\,\text{V}}{6.4\,\underline{/-38.6°}} = 12.5\,\underline{/-51.4°}\,\text{V}$$

From circuit c of Fig. 8-16,

$$\bar{Z}_{\text{Th}} = \frac{\bar{Z}_1\bar{Z}_2}{\bar{Z}_1 + \bar{Z}_2}$$

$$\bar{Z}_{\text{Th}} = \frac{(5\,\text{k}\Omega)(-j4\,\text{k}\Omega)}{(5 - j4)\,\text{k}\Omega}$$

$$\bar{Z}_{\text{Th}} = \frac{20\,\underline{/-90°} \times 10^6\,\Omega^2}{6.44\,\underline{/-38.6°} \times 10^3\,\Omega}$$

$$\bar{Z}_{\text{Th}} = 3.11\,\underline{/-51.4°}\,\text{k}\Omega = (1.94 - j2.43)\,\text{k}\Omega$$

The Thévenin equivalent circuit is shown in circuit d of Fig. 8-16. Note that if the frequency of the voltage source is known, then the value of the capacitor in the equivalent circuit may be calculated ($X_c = 1/\omega C$).

Example 8-14: Consider again circuit a of Fig. 8-1. This time solve for $\bar{I}$ from a Thévenin equivalent circuit.

Solution: Considering $\bar{Z}_3$ and $\bar{E}_2$ as the load and removing them, circuit a of Fig. 8-17 is left. Applying the voltage division law to circuit a,

$$\bar{E}_{\text{Th}} = \frac{\bar{Z}_2}{\bar{Z}_1 + \bar{Z}_2} \times \bar{E}_1$$

$$\bar{E}_{\text{Th}} = \frac{-j5\,\text{k}\Omega}{(4 - j5)\,\text{k}\Omega}(4\,\text{V})$$

$$E_{\text{Th}} = \frac{20\,\underline{/-90°}\,\text{V}}{6.4\,\underline{/-51.35°}}$$

$$\bar{E}_{\text{Th}} = 3.13\,\underline{/-38.65°}\,\text{V} = (2.44 - j1.95)\,\text{V}$$

From circuit b of Fig. 8-17,

$$\bar{Z}_{\text{Th}} = \frac{\bar{Z}_1\bar{Z}_2}{\bar{Z}_1 + \bar{Z}_2}$$

$$\bar{Z}_{\text{Th}} = \frac{(4\,\text{k}\Omega)(-j5\,\text{k}\Omega)}{(4 - j5)\,\text{k}\Omega}$$

$$\bar{Z}_{\text{Th}} = \frac{20\,\underline{/-90°} \times 10^6\,\Omega^2}{6.4\,\underline{/-51.35°} \times 10^3\,\Omega}$$

$$\bar{Z}_{\text{Th}} = 3.13\,\underline{/-38.65°}\,\text{k}\Omega = (2.44 - j1.95)\,\text{k}\Omega$$

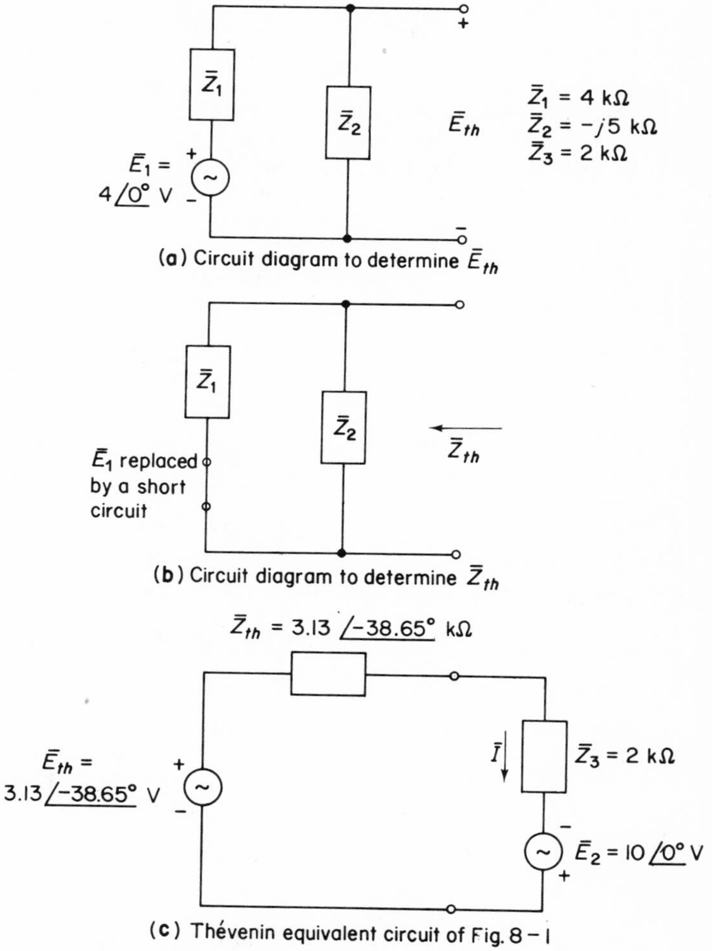

(a) Circuit diagram to determine $\bar{E}_{th}$

$\bar{Z}_1 = 4\,\text{k}\Omega$
$\bar{Z}_2 = -j5\,\text{k}\Omega$
$\bar{Z}_3 = 2\,\text{k}\Omega$

(b) Circuit diagram to determine $\bar{Z}_{th}$

(c) Thévenin equivalent circuit of Fig. 8 – 1

Fig. 8-17. Circuit diagrams used in the solution of Ex. 8-14.

The Thévenin equivalent circuit is shown in circuit c of Fig. 8-17, and from this circuit

$$\bar{E}_{\text{Th}} + \bar{E}_2 = \bar{I}\bar{Z}_{\text{Th}} + \bar{I}\bar{Z}_2$$

or

$$\bar{I} = \frac{\bar{E}_{\text{Th}} + \bar{E}_2}{\bar{Z}_{\text{Th}} + \bar{Z}_2}$$

$$\bar{I} = \frac{(2.44 - j1.95)\,\text{V} + 10\,\text{V}}{(2.44 - j1.95)\,\text{k}\Omega + 2\,\text{k}\Omega}$$

$$\bar{I} = \frac{(12.44 - j1.95)\,\text{V}}{(4.44 - j1.95)\,\text{k}\Omega} = \frac{12.6\,/\!-8.9^\circ\,\text{V}}{4.85\,/\!-23.7^\circ\,\text{k}\Omega}$$

$$\bar{I} = 2.59\,/\underline{14.8^\circ}\,\text{mA}$$

This answer checks with that of Example 8-1.

Example 8-15: Obtain the Thévenin equivalent circuit for the load impedance, Z_L, in the circuit of Fig. 8-18.

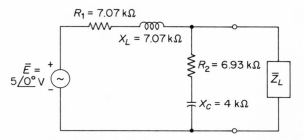

Fig. 8-18. Circuit diagram for Ex. 8-15.

Solution: The impedance values in the circuits of Fig. 8-19 are

$$\bar{Z}_1 = R_1 + jX_L = (7.07 + j7.07)\,\text{k}\Omega = 10\,/\underline{45^\circ}\,\text{k}\Omega$$

$$\bar{Z}_2 = R_2 - jX_C = (6.93 - j4)\,\text{k}\Omega = 8\,/\!-30^\circ\,\text{k}\Omega$$

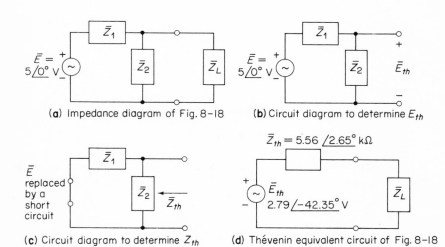

(a) Impedance diagram of Fig. 8-18   (b) Circuit diagram to determine E_{th}

(c) Circuit diagram to determine Z_{th} (d) Thévenin equivalent circuit of Fig. 8-18

Fig. 8-19. Circuit diagrams used in the solution of Ex. 8-15.

Applying the voltage division law to circuit b of Fig. 8-19,

$$\bar{E}_{Th} = \frac{\bar{Z}_2}{\bar{Z}_1 + \bar{Z}_2} \times \bar{E}$$

$$\bar{E}_{Th} = \frac{(8 \,/\!-30°\ k\Omega)\,(5 \,/\!0°\ V)}{(7.07 + j7.07)\ k\Omega + (6.93 - j4)\ k\Omega}$$

$$E_{Th} = \frac{40 \,/\!-30°\ V}{14.35 \,/\!12.35°\ k\Omega} = 2.79 \,/\!-42.35°\ V$$

From circuit c of Fig. 8-18,

$$\bar{Z}_{Th} = \frac{\bar{Z}_1\bar{Z}_2}{\bar{Z}_1 + \bar{Z}_2}$$

$$\bar{Z}_{Th} = \frac{(10 \,/\!45°\ k\Omega)\,(8 \,/\!-30°\ k\Omega)}{14.35 \,/\!12.35°\ k\Omega}$$

$$\bar{Z}_{Th} = \frac{80 \,/\!15°\ k\Omega}{14.35 \,/\!12.35°}$$

$$\bar{Z}_{Th} = 5.56 \,/\!2.65°\ k\Omega = (5.55 + j0.257)\ k\Omega$$

The Thévenin equivalent circuit is shown in circuit d of Fig. 8-19. If the frequency of the voltage source is known, the value of the inductor may be calculated $(X_L = \omega L)$. Note that for different circuit values in Fig. 8-18 the Thévenin equivalent impedance could have been a resistor in series with a capacitor.

Example 8-16: Consider the circuit of Fig. 8-20 and determine the Thévenin equivalent circuit for $\bar{Z}_L$.

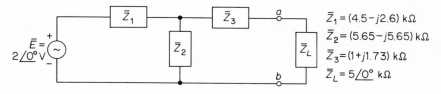

$\bar{Z}_1 = (4.5 - j2.6)\ k\Omega$
$\bar{Z}_2 = (5.65 - j5.65)\ k\Omega$
$\bar{Z}_3 = (1 + j1.73)\ k\Omega$
$\bar{Z}_L = 5 \,/\!0°\ k\Omega$

Fig. 8-20. Circuit diagram for Ex. 8-16.

Solution: From Fig. 8-21a E_{Th} is the voltage across $\bar{Z}_2$. Using the *voltage division law,*

$$\bar{E}_{Th} = \frac{\bar{Z}_2}{\bar{Z}_1 + \bar{Z}_2} \times \bar{E}$$

$$\bar{E}_{Th} = \frac{(5.65 - j5.65)\ k\Omega}{(4.5 - j2.6)\ k\Omega + (5.65 - j5.65)\ k\Omega}(2\ V)$$

$$\bar{E}_{Th} = 1.28 \,/\!-9.5°\ V$$

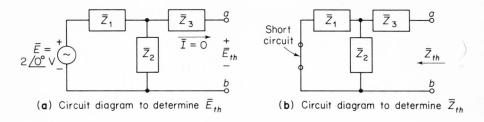

(a) Circuit diagram to determine $\bar{E}_{th}$ (b) Circuit diagram to determine $\bar{Z}_{th}$

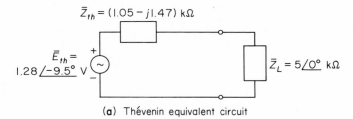

(a) Thévenin equivalent circuit

Fig. 8-21. Circuit diagrams for the solution of Ex. 8-16.

From Fig. 8-21b,

$$\bar{Z}_{Th} = \frac{\bar{Z}_1 \bar{Z}_2}{\bar{Z}_1 + \bar{Z}_2} + \bar{Z}_3$$

$$\bar{Z}_{Th} = \frac{(5\,\underline{/30°}\ k\Omega)(8\,\underline{/-45°}\ k\Omega)}{5\,\underline{/30°}\ k\Omega + 8\,\underline{/-45°}\ k\Omega} + 2\,\underline{/60°}\ k\Omega$$

$$\bar{Z}_{Th} = 3.78\,\underline{/-58.3°}\ k\Omega + 2\,\underline{/60°}\ k\Omega$$

$$\bar{Z}_{Th} = (1.05 - j1.47)\ k\Omega$$

The Thévenin equivalent circuit is shown in Fig. 8-21 c, note that the Thévenin equivalent voltage is the voltage across $\bar{Z}_2$. The voltage across $\bar{Z}_3$ is zero because the current through it is zero.

8-5. NORTON'S THEOREM

In Chapter 5 we saw that it is possible to represent every series combination of a voltage source and a resistance as a parallel combination of a current source and a resistance. This parallel combination is known as a *Norton's equivalent circuit*. A general circuit that shows impedance other than just resistance is Fig. 8-22c.

A statement of Norton's theorem is: *Any two terminals of an active network may be represented by a parallel combination of a current source and an impedance. The value of the current source is the short-circuit current at the terminals. The value of the equivalent impedance is equal to that impedance seen looking back into the network with all energy sources replaced by their respective internal impedance.*

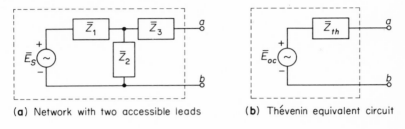

(**a**) Network with two accessible leads   (**b**) Thévenin equivalent circuit

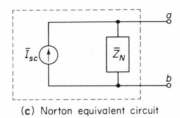

(**c**) Norton equivalent circuit

Fig. 8-22. Circuit diagrams illustrating Norton's theorem.

Using Norton's theorem we are able to mathematically analyze any network to obtain an equivalent circuit between any two terminals of the network.

Example 8-17: For the circuit of Fig. 8-23, (a) obtain the Norton equivalent circuit and (b) calculate $\bar{I}$ from the equivalent circuit.

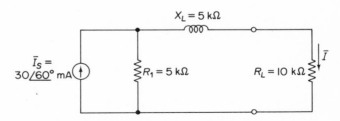

Fig. 8-23. Circuit diagram for Ex. 8-17.

Solution: (a) The impedance values in Fig. 8-24 are

$$\bar{Z}_1 = R_1 = 5\,\text{k}\Omega$$

$$\bar{Z}_2 = jX_L = j5\,\text{k}\Omega$$

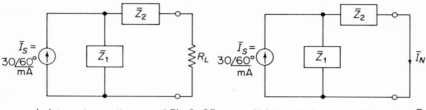

(**a**) Impedance diagram of Fig. 8-23   (**b**)Circuit diagram to determine $\bar{I}_N$

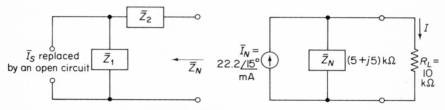

(**c**)Circuit diagram to determine $\bar{Z}_N$ (**d**)Norton equivalent circuit of Fig. 8-23

Fig. 8-24. Circuit diagrams used in the solution of Ex. 8-17.

The equivalent current, $\bar{I}_N$, is determined from Fig. 8-24b. Using the current division law,

$$\bar{I}_N = \frac{\bar{Z}_1}{\bar{Z}_1 + \bar{Z}_2} \times \bar{I}_S$$

$$\bar{I}_N = \frac{5\,\text{k}\Omega}{(5 + j5)\,\text{k}\Omega}\,(30\,\underline{/60°}\,\text{mA})$$

$$\bar{I}_N = \frac{150\,\underline{/60°}\,\text{mA}}{7.07\,\underline{/45°}} \approx 22.2\,\underline{/15°}\,\text{mA}$$

From Fig. 8-24c,

$$\bar{Z}_N = \bar{Z}_1 + \bar{Z}_2 = (5 + j5)\,\text{k}\Omega$$

The Norton equivalent circuit is shown in Fig. 8-24d.

(b)   Using the current division law in circuit d of Fig. 8-24,

$$I = \frac{\bar{Z}_N}{\bar{Z}_N + R_L} \times I_N$$

$$\bar{I} = \frac{(5 + j5)\,\text{k}\Omega}{(5 + j5)\,\text{k}\Omega + 10\,\text{k}\Omega}\,(22.2\underline{/15°}\,\text{mA})$$

$$\bar{I} = \frac{7.07\,\underline{/45°}}{15.8\,\underline{/18.4°}}\,(22.2\underline{/15°}\,\text{mA})$$

$$\bar{I} = 9.94\,\underline{/41.6°}\,\text{mA}$$

Example 8-18:   Consider the circuit of Fig. 8-25 and determine the Norton equivalent circuit for $\bar{Z}_L$.

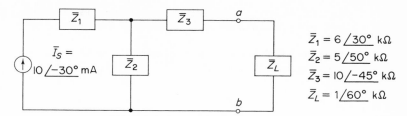

Fig. 8-25. Circuit diagram for Ex. 8-18.

Solution: The equivalent current, $\bar{I}_N$, is determined from Fig. 8-26a. Using the current division law,

$$\bar{I}_N = \frac{\bar{Z}_2}{\bar{Z}_2 + \bar{Z}_3} \times \bar{I}_S$$

$$\bar{I}_N = \frac{4\,\underline{/50°}\text{ k}\Omega}{4\,\underline{/50°}\text{ k}\Omega + 10\,\underline{/-45°}\text{ k}\Omega} \times 10\,\underline{/-30°}\text{ mA}$$

$$\bar{I}_N = 3.87\,\underline{/42.7°}\text{ mA}$$

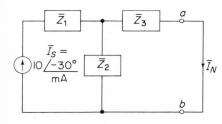

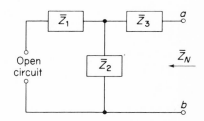

(a) Circuit diagram to determine $\bar{I}_N$

(b) Circuit diagram to determine $\bar{Z}_N$

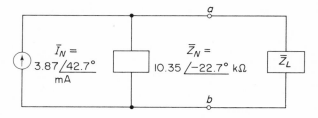

(c) Norton equivalent circuit

Fig. 8-26. Circuit diagrams used in the solution of Ex. 8-18.

From Fig. 8-26b we see that

$$\bar{Z}_N = \bar{Z}_2 + \bar{Z}_3$$
$$\bar{Z}_N = (2.57 + j3.06)\text{ k}\Omega + (7.07 - j7.07)\text{ k}\Omega$$
$$\bar{Z}_N = (9.57 - j4)\text{ k}\Omega = 10.35\,\underline{/-22.7°}\text{ k}\Omega$$

The Norton equivalent circuit is shown in Fig. 8-26c.

Note that $\bar{Z}_1$ is not included in determining the Norton equivalent im-

pedance because of the open circuit as shown in Fig. 8-26b. Complex
impedance networks are analyzed the same as resistive circuits.

Example 8-19: For the circuit of Fig. 8-27, obtain a Norton equivalent
circuit between terminals *ab*.

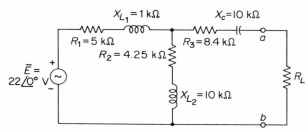

Fig. 8-27. Circuit diagram for Ex. 8-19.

Solution: The impedance values in Fig. 8-28 are

$$\bar{Z}_1 = R_1 + jX_{L_1} = (5 + j1)\,k\Omega \qquad = 5.11\,\underline{/11.3°}\,k\Omega$$

$$\bar{Z}_2 = R_2 + jX_{L_2} = (4.25 + j10)\,k\Omega = 10.9\,\underline{/67°}\,k\Omega$$

$$\bar{Z}_3 = R_3 - jX_C = (8.4 - j10)\,k\Omega\ = 13.1\,\underline{/-50°}\,k\Omega$$

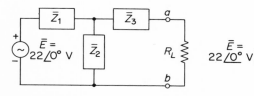

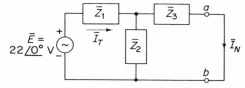

(a) Impedence diagram of Fig 8-27   (b) Circuit diagram to determine $\bar{I}_N$

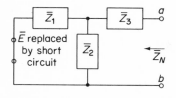

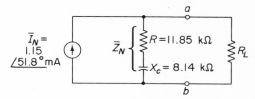

(c) Circuit diagram to determine $\bar{Z}_N$ (d) Norton equivalent circuit of Fig 8-27

Fig. 8-28. Circuit diagrams used in the solution of Ex. 8-19.

From circuit b of Fig. 8-28, let

$$\bar{Z}_T = \bar{Z}_1 + \bar{Z}_2\|\bar{Z}_3 = \bar{Z}_1 + \frac{\bar{Z}_2\bar{Z}_3}{\bar{Z}_2 + \bar{Z}_3}$$

$$\bar{Z}_T = (5 + j1)\,k\Omega + \frac{(10.9\,\underline{/67°}\,k\Omega)(13.1\,\underline{/-50°}\,k\Omega)}{(4.25 - j10)\,k\Omega + (8.4 - j10)\,k\Omega}$$

$$\bar{Z}_T = (5 + j1)\,\text{k}\Omega + \frac{143\,\underline{/\,17^\circ}\,\text{k}\Omega}{12.65\,\text{k}\Omega}$$

$$\bar{Z}_T = (5 + j1)\,\text{k}\Omega + (10.8 + j3.3)\,\text{k}\Omega$$

$$Z_T = (15.8 + 4.3)\,\text{k}\Omega = 16.4\,\underline{/\,15.2^\circ}\,\text{k}\Omega$$

and

$$I_T = \frac{\bar{E}}{\bar{Z}_T} = \frac{22\,\underline{/\,0^\circ}\,\text{V}}{16.4\,\underline{/\,15.2^\circ}\,\text{k}\Omega} = 1.34\,\underline{/\,-15.2}\,\text{mA}$$

Using the current division law,

$$\bar{I}_N = \frac{\bar{Z}_2}{\bar{Z}_2 + \bar{Z}_3} \times \bar{I}_T$$

$$\bar{I}_N = \frac{10.9\,\underline{/\,67^\circ}\,\text{k}\Omega}{(4.25 + j10)\,\text{k}\Omega + (8.4 - j10)\,\text{k}\Omega}(1.34\,\underline{/\,-15.2^\circ}\,\text{mA})$$

$$\bar{I}_N = \frac{14.6\,\underline{/\,51.8^\circ}\,\text{mA}}{12.65\,\text{k}\Omega} = 1.15\,\underline{/\,51.8^\circ}\,\text{mA}$$

From circuit c of Fig. 8-28,

$$\bar{Z}_N = \bar{Z}_1\|\bar{Z}_2 + \bar{Z}_3 = \frac{\bar{Z}_1\bar{Z}_2}{Z_1 + Z_2} + Z_3$$

$$\bar{Z}_N = \frac{(5.11\,\underline{/\,11.3^\circ}\,\text{k}\Omega)\,(10.9\,\underline{/\,67^\circ}\,\text{k}\Omega)}{(5 + j1)\,\text{k}\Omega + (4.25 + j10)\,\text{k}\Omega} + (8.4 - j10)\,\text{k}\Omega$$

$$\bar{Z}_N = \frac{55.6\,\underline{/\,78.3^\circ}\,\text{k}\Omega}{(9.25 + j11)\text{k}\Omega} + (8.4 - j10)\,\text{k}\Omega$$

$$\bar{Z}_N = \frac{55.6\,\underline{/\,78.3^\circ}\,\text{k}\Omega}{14.2\,\underline{/\,50^\circ}\,\text{k}\Omega} + (8.4 - j10)\,\text{k}\Omega$$

$$\bar{Z}_N = 3.92\,\underline{/\,28.3^\circ}\,\text{k}\Omega + (8.4 - j10)\,\text{k}\Omega$$

$$\bar{Z}_N = (3.45 + j1.86)\,\text{k}\Omega + (8.4 - j10)\,\text{k}\Omega$$

$$\bar{Z}_N = (11.85 - j8.14)\,\text{k}\Omega$$

The Norton equivalent circuit is shown in Fig. 8-28d.

SUMMARY

Five methods for analyzing a network—loop analysis, nodal analysis, superposition, Thévenin's theorem, and Norton's theorem—introduced in Chapter 5 are again presented in this chapter. In Chapter 5 they were applied only to resistive circuits. This chapter shows that they are also applicable to impedance networks.

PROBLEMS

8-1. If the polarity of voltage source $\bar{E}_2$ of Fig. 8-1a is reversed, calculate the branch current $\bar{I}$.

8-2. If the polarity of voltage source $\bar{E}_1$ of Fig. 8-2a is reversed, calculate the branch current $\bar{I}$.

8-3. Refer to the circuit of Fig. 8-3 and calculate the branch current, $\bar{I}_2$, for the following circuit quantities.

(a) $\bar{Z}_1 = (1 + j1)\,\text{k}\Omega$
    $\bar{Z}_2 = (5 + j5)\,\text{k}\Omega$
    $\bar{Z}_3 = (4 - j4)\,\text{k}\Omega$
    $\bar{E}_1 = 2\,\underline{/0^\circ}\,\text{V}$
    $\bar{E}_2 = 4\,\underline{/0^\circ}\,\text{V}$

(b) $\bar{Z}_1 = 8\,\underline{/30^\circ}\,\text{k}\Omega$
    $\bar{Z}_2 = 10\,\underline{/45^\circ}\,\text{k}\Omega$
    $\bar{Z}_3 = 1\,\underline{/0^\circ}\,\text{k}\Omega$
    $\bar{E}_1 = 10\,\underline{/30^\circ}\,\text{V}$
    $\bar{E}_2 = 2\,\underline{/0^\circ}\,\text{V}$

8-4. For each of the circuits of Fig. 8-29, find the unknown quantities. Use the method of loop analysis.

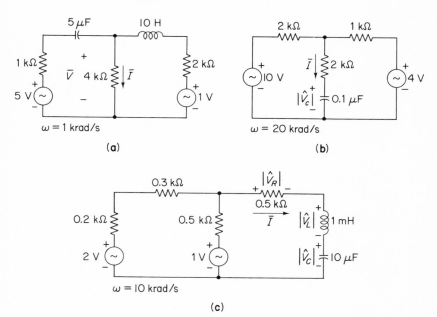

Fig. 8-29. Circuit diagrams for Prob. 8-4.

8-5. In the circuit of Fig. 8-6a, if the direction of current source $\bar{I}_{S_2}$ is reversed, calculate the branch current $\bar{I}_2$.

8-6. If the direction of the current source $\bar{I}_{S_1}$ in the circuit of Fig. 8-7 is reversed, calculate the branch currents $\bar{I}_1$, $\bar{I}_2$, and $\bar{I}_3$.

8-7. Refer to the circuit of Fig. 8-7 and solve for the branch currents $\bar{I}_1$, $\bar{I}_2$, and $\bar{I}_3$ if the values of the circuit parameters are

(a) $\bar{Y}_1 = (1 + j1)\,\text{m}\mho$ $\qquad$ (b) $\bar{Y}_1 = (4 + j2)\,\text{m}\mho$
$\quad\ \bar{Y}_2 = (4 - j4)\,\text{m}\mho$ $\qquad\qquad\ \bar{Y}_2 = (3 - j3)\,\text{m}\mho$
$\quad\ \bar{Y}_3 = (3 + j4)\,\text{m}\mho$ $\qquad\qquad\ \bar{Y}_3 = (1 + j2)\,\text{m}\mho$
$\quad\ \bar{I}_{S_1} = 10\,\underline{/0^\circ}\,\text{mA}$ $\qquad\qquad\ \bar{I}_{S_1} = 5\,\underline{/0^\circ}\,\text{mA}$
$\quad\ \bar{I}_{S_2} = 4\,\underline{/30^\circ}\,\text{mA}$ $\qquad\qquad\ \bar{I}_{S_2} = 1\,\underline{/0^\circ}\,\text{mA}$

8-8. For circuits a and b of Fig. 8-29, use nodal analysis to find the unknown quantities. (Compare answers with those of Problem 8-4.)

8-9. Solve for the branch current $\bar{I}_5$ in the circuit of Fig. 8-8. Note that the set of simultaneous linear equations is given in Example 8-8.

8-10. Use the method of superposition and solve for the current, $\bar{I}$, in the circuit of Fig. 8-1.

8-11. Use the method of superposition and solve for the current, $\bar{I}$, in the circuit of Fig. 8-2.

8-12. Using the method of superposition solve for I in the circuit of Fig. 8-30.

8-13. Refer to the circuits of Fig. 8-29a and use the method of superposition to solve for the branch currents $\bar{I}$ in each circuit. (Compare the answers to those of Problems 8-4 and 8-8.)

8-14. Refer to the circuit of Fig. 8-20 and obtain the Thévenin circuit between terminals ab if

(a) $\bar{Z}_1 = (5 - j5)\,\text{k}\Omega$ $\qquad$ (b) $\bar{Z}_1 = 4\,\underline{/30^\circ}\,\text{k}\Omega$
$\quad\ \bar{Z}_2 = (5 + j5)\,\text{k}\Omega$ $\qquad\qquad\ \bar{Z}_2 = 1\,\underline{/-30^\circ}\,\text{k}\Omega$
$\quad\ \bar{Z}_3 = (2 + j1)\,\text{k}\Omega$ $\qquad\qquad\ \bar{Z}_3 = 5\,\underline{/45^\circ}\,\text{k}\Omega$
$\quad\ \bar{E} = 2\,\underline{/45^\circ}\,\text{V}$ $\qquad\qquad\quad\ \bar{E} = 3\,\underline{/0^\circ}\,\text{V}$

8-15. Solve for the current $\bar{I}$ in the circuit of Fig. 8-30 by a Thévenin equivalent circuit.

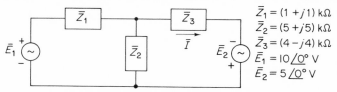

$\bar{Z}_1 = (1 + j1)\,\text{k}\Omega$
$\bar{Z}_2 = (5 + j5)\,\text{k}\Omega$
$\bar{Z}_3 = (4 - j4)\,\text{k}\Omega$
$\bar{E}_1 = 10\,\underline{/0^\circ}\,\text{V}$
$\bar{E}_2 = 5\,\underline{/0^\circ}\,\text{V}$

Fig. 8-30. Circuit diagram for Prob. 8-12.

8-16. Refer to the circuit of Fig. 8-2 and solve for $\bar{I}$ by a Thévenin equivalent circuit. *(Hint:* Let R_2 and X_L be considered the load.)

8-17. Obtain the Thévenin equivalent circuit for the capacitor in the circuit of Fig. 8-6. *(Hint:* Convert the current sources to voltage sources.)

8-18. Refer to the circuit of Fig. 8-25 and obtain the Norton equivalent circuit between terminals *ab* if

    (a) $\bar{Z}_1 = 8\,\underline{/45°}\ \text{k}\Omega$ (b) $\bar{Z}_1 = 4\,\underline{/30°}\ \text{k}\Omega$

           $\bar{Z}_2 = 5\,\underline{/30°}\ \text{k}\Omega$ $\bar{Z}_2 = 1\,\underline{/-30°}\ \text{k}\Omega$

           $\bar{Z}_3 = 6\,\underline{/60°}\ \text{k}\Omega$ $\bar{Z}_3 = 5\,\underline{/45°}\ \text{k}\Omega$

           $\bar{I}_S = 4\,\underline{/0°}\ \text{mA}$ $\bar{I}_S = 3\,\underline{/0°}\ \text{mA}$

CHAPTER 9

transient response

The three sections of this chapter consider circuits with resistance-inductance (R-L), resistance-capacitance (R-C), and resistance-inductance-capacitance (R-L-C) combinations. The circuits analyzed are energized by dc voltage or current sources. Therefore, we are concerned with both the steady-state and transient values of voltage or current. Both series and parallel combinations of the passive elements will be treated in each section.

9-1. THE R-L CIRCUIT

At time $t = 0$, the switch in the circuit of Fig. 9-1 is closed. At the instant of switch closure ($t = 0$), a current tends to flow. However, the

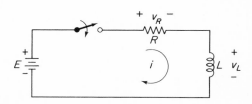

Fig. 9-1. Series R-L circuit.

characteristic of an inductor is to oppose any instantaneous change of current. This property stems from the voltage-current relationship of an inductor $v_L = L(di_L/dt)$. From this expression, as noted in Chapter 3, the voltage across an inductor depends on the rate of change of current and not on the current itself. It is because of this unique property that the current through an inductor cannot change instantaneously. (There is no restriction, however, on how the voltage across an inductor can change.) In other words, it is the property of the inductor to keep the value of current the same as it was before the switch was closed. Prior to closure, $i = 0$. Therefore, the current must be zero at $t = 0$. Only after the switch has been closed for a sufficiently long period of time is the current able to build up to a steady-state value. Because it takes time for the current to reach the constant value, it is best to analyze the circuit of Fig. 9-1 at three different times:

1. the instant that the switch is closed ($t = 0$),
2. after the switch has been closed for a long period of time,
3. the interval between the two extremes above.

Let us determine the initial rate of change of current at time 1 ($\pm = 0$): Applying Kirchhoff's voltage law, we obtain

$$E = v_R + v_L \tag{9-1}$$

From Ohm's law,

$$v_R = Ri \tag{9-2}$$

and from Eq. (3-10),

$$v_L = L\frac{di}{dt} \tag{9-3}$$

Substituting Eqs. (9-2) and (9-3) into Eq. (9-1) yields

$$E = Ri + L\frac{di}{dt} \tag{9-4}$$

At $t = 0$ we have stated that $i = 0$. Thus

$$E = R(0) + L\frac{di}{dt}$$

or

$$E = L\frac{di}{dt} \tag{9-5}$$

and

$$\frac{di}{dt} = \frac{E}{L} \tag{9-6}$$

where di/dt is the *initial* rate of rise of current, E the value of the dc voltage source in volts, and L the value of the inductor in henries. Equation (9-6) shows that the initial rate of rise of current through the inductor depends directly on the source voltage and inversely on the inductor. (It does not depend on the resistance.) A plot of Eq. (9-6) is shown in Fig. 9-2. Note

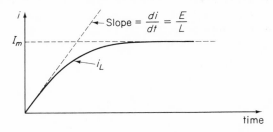

Fig. 9-2. Rise of current in the circuit of Fig. 9-1.

at the instant of switch closure, that although $i = 0$, this does not mean that di/dt is zero.

Now consider time 2, when the switch has been closed for sufficient time so that the current has been able to build up to a (final) steady-state value. The steady-state value, unlike the initial rate of rise, is limited by the resistor of the circuit because it is not possible for the voltage across the resistor to be greater than the applied voltage. When a constant current flows in the circuit of Fig. 9-1, $di/dt = 0$. Thus Eq. (9-4) reduces to

$$E = Ri + L(0)$$

or

$$i = I_m = \frac{E}{R} \tag{9-7}$$

where I_m is the maximum value of current that can flow in the circuit, shown in Fig. (9-2). As may be seen from Eq. (9-7), the final value of current, I_m, does *not* depend on the inductor but only the source voltage, E, and the resistor, R.

Note that for the steady-state condition, although the rate of change of current, di/dt, is equal to zero, this does not mean that the current, i, is zero, as shown by Eq. (9-7).

Obtaining the instantaneous values of current between the two time limits above involves solving Eq. (9-4) for i. Although solving for i from this differential equation is beyond the scope of this text, we are able, nevertheless, to use the solution of this equation to analyze the type of electric circuit shown in Fig. 9-1. The solution is

$$i = I_m (1 - e^{-t/\tau}) \tag{9-8}$$

where $I_m = E/R$, $e = 2.718$ (the base of the natural logarithms), and $\tau = L/R$. A plot of Eq. (9-8) is an *exponential curve*, as shown in Fig. 9-2.

Therefore, after the switch is closed, the current in the circuit of Fig. 9-1 rises initially according to Eq. (9-6), then levels off and approaches the final value asymptotically (reaches the value only at $t = \infty$).

To solve the expression $e^{-t/\tau}$, one or more of the following methods is usually used: slide rule, tables or a graph. Most engineering slide rules have scales for determining $e^{-t/\tau}$. A set of tables has been included in Appendix E. The graph of Fig. 9-3 is a universal plot of $e^{-t/\tau}$ and

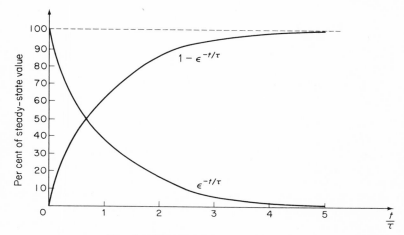

Fig. 9-3. Universal exponential curves.

$1 - e^{-t/\tau}$. It is possible to plot universal curves for $e^{-t/\tau}$ and $1 - e^{-t/\tau}$ regardless of the values of source voltage, resistance, and inductance, because the initial rate of rise will always be E/L and the steady-state value will always be E/R. So that the graph of Fig. 9-3 will be applicable to any circuit similar to that of Fig. 9-1, the vertical axis is calibrated as a percentage of the final value and the horizontal axis is calibrated in terms of t/τ (where $\tau = L/R$).

> **Example 9-1:** Consider the circuit of Fig. 9-4 and determine (a) the steady-state value of current and (b) the current 1 μs after the switch is closed.

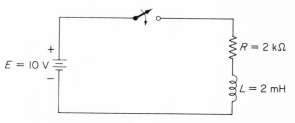

Fig. 9-4. Circuit diagram for Ex. 9-1.

Solution: (a) The steady-state or maximum value of current is

$$I_m = \frac{E}{R} = \frac{10\,V}{2\,k\Omega} = 5\,mA$$

(b) According to Eq. (9-8),

$$i = I_m(1 - e^{-t/\tau})$$

where

$$\tau = \frac{L}{R} = \frac{2\,mH}{2\,k\Omega} = 1\,\mu s$$

$$\frac{t}{\tau} = \frac{1\,\mu s}{1\,\mu s} = 1$$

Therefore,

$$i = 5(1 - e^{-1})\,mA$$

$$i = 5(1 - 0.37)\,mA$$

$$i = 5(0.63)\,mA = 3.15\,mA$$

Until the steady-state value is reached, Eq. (9-8) has to be used to determine the current.

Time constant. After the switch of the circuit in Fig. 9-1 is closed, we have shown that the current rises exponentially. The rate of rise depends on the exponent $-t/\tau$. In terms of circuit quantities, $t/\tau = t/(L/R)$. The denominator (L/R) of this expression has a special significance in circuit analysis. It is called the *time constant* of an *R-L* circuit and is represented by the Greek letter τ (tau). Then

$$\tau = \frac{L}{R} \qquad (9\text{-}9)$$

Then Eq. (9-8) may be written as

$$i_L = I_m(I - e^{-tR/L}) \qquad (9\text{-}10)$$

where L is the value of the inductor in henries, R the value of the resistor in ohms, and I_m the final or steady-state value of current.

Whenever the time constant, τ, equals the time, t, then $t/\tau = 1$. Under this condition, refer to Fig. 9-3 and note that the current has risen to 63 per cent of its final value. Thus the time constant for an *R-L* circuit may be defined as *the time it takes the instantaneous current to reach 63 per cent of its final value.*

Note that the time constant, τ, is *directly* proportional to the *inductance*. If we double the inductance, it would take twice as long for the current to reach 63 per cent of its final value and thus twice as long to approach the steady-state value. Likewise, the time constant, τ, is *inversely* proportional to the *resistance*. Therefore, if we were to double the resistance, the current will reach 63 per cent of its final value in one half the time and thus it will approach the steady-state value in one half the time.

We have previously mentioned that the current rises exponentially but never reaches the final value until $t = \infty$. We must therefore consider some time after which, for all practical purposes, a constant current flows. When the current has risen to 98 per cent of its final value, then for all practical purposes we may say the current is constant. The current reaches 98 per cent of its final value in five time constants, 5τ. Thus

$$5\tau = 5\frac{L}{R} \text{ seconds} \tag{9-11}$$

For this reason, it is not necessary to have the universal exponential curves extend farther than five time constants.

Example 9-2: Calculate (a) the time constant for the circuit of Fig. 9-4 and (b) the time for the current to reach the steady-state value.

Solution: (a) The time constant for an R-L circuit is

(a) $\tau = \dfrac{L}{R} = \dfrac{2 \text{ mH}}{2 \text{ k}\Omega} = 1 \ \mu\text{s}$

Then

(b) $5\tau = (5)(1 \ \mu\text{s}) = 5 \ \mu\text{s}$

(b) For all practical purposes, the current in the inductor reaches its steady-state value in 5 μs.

Example 9-3: At $t = 0$ the switch of Fig. 9-5 is closed. Calculate (a) the time constant and (b) value of current at $t = 9$ ms.

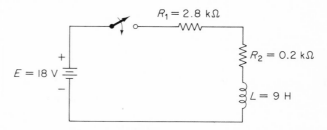

Fig. 9-5. Circuit diagram for Ex. 9-3.

Solution: (a) The time constant is

$$\tau = \frac{L}{R_1 + R_2} = \frac{9 \text{ H}}{2.8 \text{ k}\Omega + 0.2 \text{ k}\Omega} = 3 \text{ ms}$$

(b) The time to reach the steady-state value is

$$5\tau = (5)(3 \text{ ms}) = 15 \text{ ms}$$

Since we are interested in a value of current between 0 and 5τ, Eq. (9-8) must be used:

$$i = I_m (1 - e^{-t/\tau})$$

where

$$I_m = \frac{E}{R_1 + R_2} = \frac{18\ V}{2.8\ k\Omega + 0.2\ k\Omega} = 6\ mA$$

and

$$\frac{t}{\tau} = \frac{9\ ms}{3\ ms} = 3$$

Therefore,

$$i = 6(1 - e^{-3})\ mA$$

$$i = 6(1 - 0.05)\ mA$$
$$i = 6(0.95)\ mA = 5.7\ mA$$

Therefore, between 0 and 5τ, Eq. (9-8) is used. At a time greater than 5τ, the steady-state value is reached given by Eq. (9-7).

Deenergizing an R-L Circuit. We analyzed an *R-L* circuit when a dc voltage is applied. We now analyze what happens in a circuit containing an inductor when the switch is opened. Consider the circuit of Fig. 9-6a.

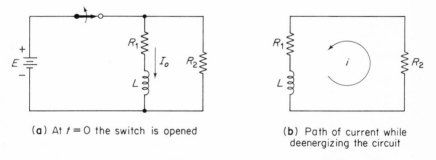

(**a**) At $t = 0$ the switch is opened   (**b**) Path of current while deenergizing the circuit

Fig. 9-6. Circuit diagrams illustrating deenergizing an *R-L* circuit.

The reason for the resistor in parallel with the series *R-L* combination is to provide a path for the current to flow after the switch is opened. If a suitable path is not provided, most of the energy stored in the inductor is dissipated at the switch contacts, creating an extremely large arc at the switch contacts and possible permanent damage to the circuit elements.

We have stated that a characteristic of an inductor is that the current through it cannot change instantaneously. In other words, an inductor opposes any change of current. Since the voltage source was connected when the switch was closed, a current had to flow. Let this current be I_o. Then the instant after the switch is opened, the current in the inductor is still I_o.

I_o is referred to as an *initial condition current*. If the switch was closed for a sufficient length of time (at least five time constants),

$$I_o = I_m = \frac{E}{R_1}$$

where I_m is the maximum value of current, E the voltage source, and R_1 the resistor in series with the inductor. After the switch is opened, R_1 and R_2 are in series, as shown in Fig. 9-6b. The total resistance, R_T, then equals

$$R_T = R_1 + R_2$$

The property of a resistor is to dissipate energy when a current flows through it. Therefore, I_o decays toward zero as the energy stored by the inductor is converted into heat by the resistors. Let us analyze how I_o decays toward zero. We begin by applying Kirchhoff's voltage law to the circuit of Fig. 9-6b:

$$v_{R1} + v_{R2} + v_L = 0$$

Then

$$R_1 i + R_2 i + L \frac{di}{dt} = 0$$

or

$$R_T i + L \frac{di}{dt} = 0 \tag{9-12}$$

The solution of Eq. (9-12) is

$$i = I_o e^{-t/\tau} \tag{9-13}$$

where I_o is the current through the inductance the instant the switch is opened and $\tau = L/R$. A plot of $e^{-t/\tau}$ is an exponential decay curve, as shown in Fig. 9-3.

The rate at which the current decays depends on the exponent t/τ, where $t/\tau = t/(L/R_T)$. The denominator (L/R_T) of this expression is the decaying time constant, τ:

$$\tau = \frac{L}{R_T} \tag{9-14}$$

Equation (9-13) then may be written as

$$i = I_o e^{-t R_T /L} \tag{9-15}$$

where L is the value of the inductor in henries and R_T is the value of the total resistance in ohms.

When the time constant, τ, equals the time, t, then $t/\tau = 1$. Under this condition, let us refer to Fig. 9-3 and note that the current has decayed by 63 per cent of its initial value. Another way of stating this is to say that after one time constant, only 37 per cent of I_0 remains flowing in the circuit.

As in the case where the switch was closed, it takes five time constants for the current to reach its final value after the switch is opened:

$$5\tau = 5\frac{L}{R_T} \text{ seconds} \qquad (9\text{-}16)$$

For all practical purposes, almost all the electrical energy stored in the inductor at $t = 0$ is converted to heat (and light) after 5τ.

Note that the time constant is the ratio of the inductor to the total resistance "seen" by the inductor. Consider the circuit of Fig. 9-6a. When the switch is closed, the voltage source, E, is connected directly across R_1 and L. Therefore the equivalent (or Thévenin) resistance seen by the inductor is R_1. When the switch is open, however, the inductor sees R_1 and R_2 in series. Therefore, the reader is cautioned that the time constant for energizing an R-L circuit is not always the same as for deenergizing the circuit.

Example 9-4: For the circuit of Fig. 9-7, at $t = 0$, the switch is opened. Calculate

(a) the current through the inductor at $t = 0$,
(b) the time constant for deenergizing the inductor,
(c) the current through the inductor at $t = 10\ \mu s$,
(d) the current through the inductor at $t = 3\ \mu s$,
(e) the voltage across R_1 at $t = 3\ \mu s$.

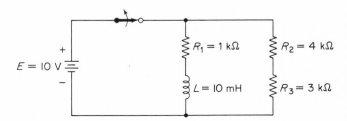

Fig. 9-7. Circuit diagram for Ex. 9-4.

Solution: (a) If the switch is closed for a sufficient length of time, the current through the inductor will be a maximum value.

$$I_m = \frac{E}{R_1} = \frac{10\ V}{1\ k\Omega} = 10\ mA$$

Thus at $t = 0$,

$$I_0 = I_m = 10\ mA$$

(b)

$$\tau = \frac{L}{R_T}$$

where

$$R_T = R_1 + R_2 + R_3$$

$$R_T = 1\,\text{k}\Omega + 4\,\text{k}\Omega + 3\,\text{k}\Omega = 8\,\text{k}\Omega$$

Therefore,

$$\tau = \frac{10\,\text{mH}}{8\,\text{k}\Omega} = 1.25\,\mu\text{s}$$

(c) The current through the inductance reaches the steady-state value in 5τ:

$$5\tau = 5(1.25\,\mu\text{s}) = 6.25\,\mu\text{s}$$

Therefore, at $t = 10\,\mu\text{s}$,

$$i_L = 0$$

(d) At $t = 3\,\mu\text{s}$,

$$i_L = I_0 e^{-t/\tau}$$

where

$$I_o = I_m = 10\,\text{mA}$$

and

$$\frac{t}{\tau} = \frac{3\,\mu\text{s}}{1.25\,\mu\text{s}} = 2.4$$

Therefore,

$$i_L = 10e^{-2.4}\,\text{mA}$$

$$i_L = 10(0.091)\,\text{mA}$$

$$i_L = 0.91\,\text{mA}$$

(e) R_1 and L are series; therefore, the voltage across R_1 at $t = 3\,\mu\text{s}$ is

$$v_{R1} = R_1 i = (1\,\text{k}\Omega)(0.91\,\text{mA}) = 0.91\,\text{V}$$

Note that the answer to part (c) is 0 because the time is greater than 5τ. However in part (d), Eq. (9-13) had to be used because the time was between 0 and 5τ.

Example 9-5: The waveform of Fig. 9-8a is applied to the circuit of Figure 9-8b. Draw the following waveforms:

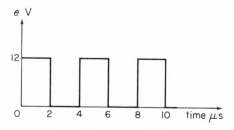

(a) Source voltage waveform for Ex. 9-5

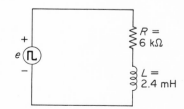

(b) Circuit diagram for Ex. 9-5

Fig. 9-8. Diagrams for Ex. 9-5.

(a) the current through the inductor,
(b) the voltage across the resistor,
(c) the voltage across the inductor.

Solution: (a) In this circuit the time constant to energize and deenergize the circuit is the same:

$$\tau = \frac{L}{R} = \frac{2.4\,\text{mH}}{6\,\text{k}\Omega} = 0.4\,\mu\text{s}$$

and

$$5\tau = 5(0.4\,\mu\text{s}) = 2\,\mu\text{s}$$

Therefore, the current through the inductor is able to reach the steady-state value at 2 μs and at 4μs. The steady-state value of current at 2 μs is

$$I_m = \frac{E}{R} = \frac{12\,\text{V}}{6\,\text{k}\Omega} = 2\,\text{mA}$$

The steady-state value of current at 4 μs is zero. The current waveform is shown in Fig. 9-9a.

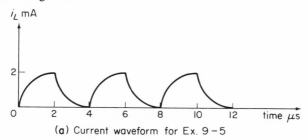

(a) Current waveform for Ex. 9 − 5

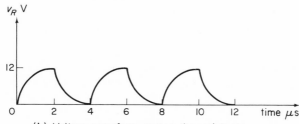

(b) Voltage waveform across the resistance

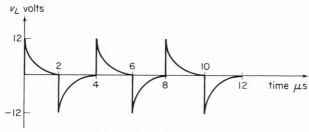

(c) Voltage waveform across the inductance

Fig. 9-9. Waveforms for Ex. 9-5.

(b)   A property of resistance is that the current and voltage are in phase. Therefore, the voltage waveform across the resistor is similar to the current waveform (Fig. 9-9a) and

$$V_m = I_m R = (2 \text{ mA})(6 \text{ k}\Omega) = 12 \text{ V}$$

This waveform is shown in Fig. 9-9b.

(c)   According to Kirchhoff's voltage law,

$$E = v_R + v_L$$

Therefore, while the voltage across the resistor is an exponential rise, the voltage across the inductor then must be an exponential decay.   The voltage waveform across the inductor is shown in Fig. 9-9c.

Since the voltage across and the current through a resistor are in phase, Figs. 9-9a and 9-9b are similar.   Analyzing waveforms of Figs. 9-9a and 9-9c show that as the current through the inductance rises exponentially, the voltage across the inductance decays exponentially.

Parallel R-L Circuit.   Let us consider the circuit of Fig. 9-10 and note the similarities between series and parallel *R-L* circuits.   At $t = 0$ the switch

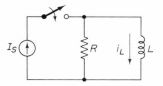

Fig. 9-10. Circuit diagram to illustrate principles of parallel *R-L* circuits.

is closed.   The properties of the inductance are the same whether it is connected in series or parallel.   Therefore, the current through the inductance must rise exponentially from zero to the steady-state value.   Thus at $t = 0$, $i_L = 0$.   After the switch has been closed for a sufficiently long time (5τ),

$$i_L = I_m = I_S \tag{9-17}$$

The value of current between these two extremes is given by

$$i_L = I_m(1 - e^{-t/\tau}) \tag{9-18a}$$

where

$$I_m = I_S$$

and

$$\tau = \frac{L}{R}$$

Then

$$I_L = I_m(1 - e^{-tR/L}) \tag{9-18b}$$

Thus the expression for current for a parallel R-L circuit is the same as that for a series circuit. Also, the time constant for the parallel circuit is the same as that for the series circuit. For all practical purposes the current through the inductor reaches the steady-state value in 5τ.

Example 9-6: Consider the circuit of Fig. 9-11. At $t = 0$ the switch is closed. Determine

(a) the current through the inductor at $t = 0$,
(b) the steady-state value of current,
(c) the time to reach the steady-state value,
(d) the current through the inductor at $t = 2$ μs.

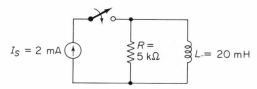

$I_S = 2$ mA $R = 5$ kΩ $L = 20$ mH

Fig. 9-11. Circuit diagram for Ex. 9-9.

Solution: (a) At $t = 0$ the inductance appears as an open circuit; therefore, $i_L = 0$.

(b) The steady-state or maximum value of current is

$$I_m = I_S = 2 \text{ mA}$$

(c) Since

$$\tau = \frac{L}{R} = \frac{20 \text{ mH}}{5 \text{ k}\Omega} = 4 \text{ } \mu s$$

the time to reach the steady-state current value is

$$5\tau = (5)(4 \text{ } \mu s) = 20 \text{ } \mu s$$

(d)

$$i = I_m(1 - e^{-t/\tau})$$

where

$$I_m = 2 \text{ mA}$$

and

$$\frac{t}{\tau} = \frac{2 \text{ } \mu s}{4 \text{ } \mu s} = 0.5$$

Then

$$i = 2(1 - e^{-0.5}) \text{ mA}$$
$$i = 2(1 - 0.607) \text{ mA}$$
$$i = 0.786 \text{ mA}$$

Note that because the energy source is a current source, (1) the time constant is determined by the inductance and the parallel resistance, and

(2) other formulas developed for the series circuit are applicable to the parallel circuit.

Example 9-7: Again consider the circuit of Fig. 9-11. At $t = 0$ the switch is opened. Calculate

(a)   the current through the inductor at $t = 0$,
(b)   the steady-state value of current,
(c)   the time to reach the steady-state value,
(d)   the current through the inductor at $t = 2 \, \mu s$.

Solution: (a)   If the switch has been closed for a sufficient length of time, the current through the inductor has reached a maximum value. Therefore, at $t = 0$,

$$i_L = I_S = 2 \, \text{mA}$$

(b)   After the switch is opened, the current in the inductor decays exponentially toward zero. Thus the steady-state value is zero.

(c)   Since

$$\tau = \frac{L}{R} = \frac{20 \, \text{mH}}{5 \, \text{k}\Omega} = 4 \, \mu s$$

the time to reach the steady-state current value is

$$5\tau = 5(4 \, \mu s) = 20 \, \mu s$$

(d)   Since the current is decaying exponentially, Eq. (9-13) is applicable:

$$i_L = I_0 e^{-t/\tau}$$

where

$$I_0 = I_m = 2 \, \text{mA}$$

and

$$\frac{t}{\tau} = \frac{2 \, \mu s}{4 \, \mu s} = 0.5$$

Then

$$i_L = 2e^{-0.5} \, \text{mA}$$

$$i_L = 2(0.607) \, \text{mA} = 1.214 \, \text{mA}$$

Note that when the switch of circuit of Fig. 9-11 is opened, the resistance provides a path for deenergizing the circuit. In part (d) the value of current between 0 and 5τ is desired; therefore, Eq. (9-13) has to be used.

9-2. THE R-C CIRCUIT

In this section we notice certain similarities between the analysis for R-C circuits and that previously discussed for the R-L circuit, such as exponential curves, time constant, and initial conditions. Consider the R-C circuit of Fig. 9-12. Let us determine what happens after the switch is closed.

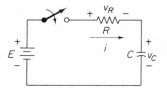

Fig. 9-12. Series *R-C* circuit.

Before performing the initial step in the analysis. let us restate the properties of a capacitor. In Chapter 3 the voltage-current relationship of a capacitor was given as $i_C = C(dv_C/dt)$. This expression states that the current is proportional to the rate of change of voltage and not the voltage itself. It is because of this unique property that the voltage across a capacitor cannot change instantaneously. Thus a capacitor opposes any change in voltage or, to put it another way, the capacitor in the circuit of Fig. 9-12 will have the same voltage across it the instant after the switch is closed as it had the instant before the switch was closed. (There is no restriction, however, on how the current "through" a capacitor can change.) The instant before the switch was closed $v_C = 0$. Therefore, at $t = 0$, v_C must equal zero. Only after the switch has been closed for a sufficiently long period of time is the voltage across the capacitor able to build up to a steady-state value. For this reason, let us analyze the circuit of Fig. 9-12 at three different times:

1. the instant that the switch is closed $(t = 0)$,
2. after the switch has been closed for a long period of time,
3. the interval between the above two extremes.

Applying Kirchhoff's voltage law to the circuit of Fig. 9-12, we obtain

$$E = v_R + v_C \tag{9-19}$$

From Ohm's law,

$$I_R = Ri$$

Then

$$E = Ri + v_C \tag{9-20}$$

Since the current, i, is the same through each element of a series circuit and the current through a capacitor is given by the expression

$$v_C = C\frac{dv_C}{dt}$$

then

$$E = RC\frac{dv_C}{dt} + v_C \tag{9-21}$$

At $t = 0$, $v_C = 0$. Thus the initial rate of rise of voltage across the capacitor is

$$E = RC\frac{dv_C}{dt} + 0$$

or

$$\frac{dv_C}{dt} = \frac{E}{RC} \tag{9-22}$$

where dv_C/dt is the initial rate of rise of voltage, E the value of the dc voltage source in volts, R the resistor value in ohms, and C the capacitor value in farads.

Equation (9-22) shows that the initial rate of rise of voltage across the capacitor depends not only on the source voltage and capacitor but also on the resistor. The initial rate of rise is shown in Fig. 9-13.

The final value of voltage across the capacitance is reached when $dv_C/dt = 0$. When this happens it means that there is no longer a rate of change of voltage across the capacitance. Equation (9-21) is now written as

$$E = RC(0) + v_C$$

or

$$E = v_C \tag{9-23}$$

Equation (9-23) states that after the switch of Fig. 9-12 has been closed for a long period of time, the voltage across the capacitance has built up to the source voltage. This final value of voltage is also shown in Fig. 9-13.

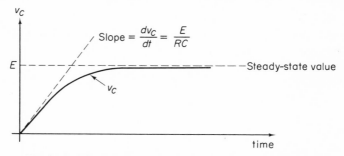

Fig. 9-13. Rise of voltage across the capacitance of Fig. 9-2.

To determine the voltage across the capacitance at any instant of time it is necessary to solve the differential equation of Eq. (9-21), the solution of which is

$$v_C = E(1 - e^{-t/\tau}) \tag{9-24a}$$

where E is the source voltage, $e = 2.718$ (the base of the natural logarithms), and $\tau = RC$. A plot of Eq. (9-24) is an exponential curve, as shown in Fig.

9-13. Therefore, the rise of voltage across a capacitance is similar to the rise of current through an inductance.

The value of the expression $e^{-t/\tau}$ may be determined not only by slide rule or tables but also by the universal exponential curves of Fig. 9-3. When using Fig. 9-3 to solve for $(1 - e^{-t/\tau})$ the vertical axis is now calibrated as the percentage of the final value of voltage across the capacitance and the horizontal axis is calibrated in terms of t/τ $(t/\tau = t/RC)$. Thus Eq. (9-24a) may be written as

$$v_c = E_m(1 - e^{-t/RC}) \qquad (9\text{-}24b)$$

Example 9-8: Consider the circuit of Fig. 9-14. At $t = 0$ the switch is closed. Determine

    (a)   the voltage across the capacitor at $t = 0$,
    (b)   the voltage across the resistor at $t = 0$,
    (c)   the steady-state value of voltage, across the capacitor,
    (d)   the voltage across the capacitor at $t = 3$ s.

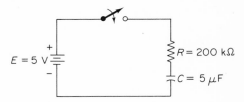

Fig. 9-14. Circuit diagram for Ex. 9-8.

Solution: (a) At $t = 0$, $v_C = 0$.
    (b) From Kirchhoff's voltage law,

$$E = v_R + v_C$$

Since at $t = 0$, $v_C = 0$,

$$v_R = E = 5 \text{ V}$$

    (c) The steady-state value of voltage across the capacitor is

$$v_C = E = 5 \text{ V}$$

    (d) According to Eq. (9-24a),

$$v_C = E(1 - e^{-t/\tau})$$

Since

$$\frac{t}{\tau} = \frac{t}{RC} = \frac{3 \text{ s}}{(200 \text{ k}\Omega)(5 \text{ }\mu\text{F})} = 3$$

then

$$v_C = 5(1 - e^{-3}) \text{ V}$$
$$v_C = 5(1 - 0.05) \text{ V}$$
$$v_C = 4.75 \text{ V}$$

Note that Kirchhoff's voltage law must apply at any instant of time. For example, at $t = 0$, $v_C = 0$ and $v_R = 5$ V.

Time Constant.   The R-C circuit, like the R-L circuit, has a time constant, τ, associated with it.   The time constant for the R-C circuit is

$$\tau = RC \qquad (9\text{-}25)$$

Whenever the time constant, τ, equals the time, t, then $t/\tau = 1$.   According to the universal exponential curves of Fig. 9-3, when $t/\tau = 1$, the voltage across the capacitance has risen to 63 per cent of its final value.   The time constant of an R-C circuit may be defined as *the time it takes for the instantaneous voltage across the capacitor to reach 63 per cent of its steady-state value.*

Unlike the R-L circuit, the time constant for the R-C circuit is directly proportional to the values of both circuit elements (resistance and capacitance).   For example, if the resistor is doubled while the capacitor remains the same, the time constant also doubles.   The result is that it takes twice as long for the voltage across the capacitor to reach 63 per cent of its steady-state value and thus twice as long to approach its final value.   The result is the same when the resistor remains constant and the capacitor is doubled.

We have shown that the voltage across the capacitor rises exponentially and therefore the steady-state value is not reached until $t = \infty$. (However, we know that it is impossible to wait until $t = \infty$ and thus a time limit must be set.)   Therefore, for all practical purposes, the voltage across the capacitor after $t = 5\tau$ may be considered constant.

$$5\tau = 5RC \text{ seconds} \qquad (9\text{-}26)$$

Example 9-9:   Determine the time constant for the circuit of Fig. 9-14 and the time for the capacitor to charge up to the steady-state value.

Solution: The time constant for an R-C circuit is

$$\tau = RC = (200 \text{ k}\Omega)(5 \text{ }\mu\text{F}) = 1 \text{ s}$$

Then

$$5\tau = 5(1 \text{ s}) = 5 \text{ s}$$

In 5τ the capacitor, for all practical purposes, has charged up to the steady-state value.

Example 9-10:   Consider the circuit of Fig. 9-15.   At $t = 0$ the switch is closed.   Determine the time constant and the value of voltage across the capacitor at $t = 0.9$ ms and $t = 4$ ms.

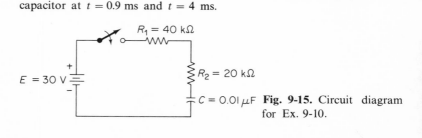

$R_1 = 40$ kΩ

$E = 30$ V

$R_2 = 20$ kΩ

$C = 0.01$ μF   **Fig. 9-15.** Circuit diagram for Ex. 9-10.

Solution:

In this circuit

$$\tau = RC$$

Then

$$R = R_1 + R_2 = 40\,k\Omega + 20\,k\Omega = 60\,k\Omega$$

$$\tau = (60\,k\Omega)(0.01\,\mu F) = 0.6\,ms$$

The voltage across the capacitance at $t = 0.9$ ms is found by

$$v_C = E(1 - e^{-t/\tau})$$

where

$$\frac{t}{\tau} = \frac{0.9\,ms}{0.6\,ms} = 1.5$$

Then

$$v_C = 30\,(1 - e^{-1.5})\,V$$
$$v_C = 30\,(1 - 0.224)\,V$$
$$v_C = 23.28\,V$$

In this example $\tau = 0.6$ ms. Thus the voltage across the capacitor has reached the steady-state value in (5) (0.6 ms) = 3 ms. Therefore, at $t = 4$ ms

$$v_C = E = 30\,V$$

Note that Eq. (9-24) has to be used if the time is between 0 and 5τ. After 5τ, $v_C = E$.

Deenergizing an R-C Circuit. Up to this point we have shown how a dc voltage builds up across a capacitor, thus storing energy. We have also shown that after five time constants the steady-state value of voltage is reached. We shall now analyze what happens when the switch is opened and the capacitor gives up its stored energy. The appropriate circuits are shown in Fig. 9-16a and b. Although the circuit of Fig. 9-16b is similar

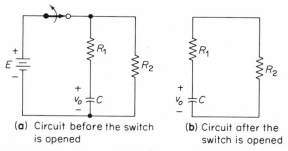

(a) Circuit before the switch is opened (b) Circuit after the switch is opened

Fig. 9-16. Circuit diagrams to illustrate the principles for deenergizing an *R-C* circuit.

to the one considered in the *R-L* case, it is not necessary in the *R-C* case to provide a path for current to flow the instant the switch is opened. Unlike the inductor, the capacitor would continue to hold its stored energy until such time that a suitable path is provided.

We have stated that the characteristic of a capacitor is that the voltage across it cannot change instantaneously. Thus any voltage across the capacitor the instant before the switch is opened must equal the voltage across the capacitor the instant after the switch is opened. Let this voltage be represented by V_o. The instant after the switch is opened, the capacitor will appear to be a voltage source whose initial value is V_o. V_o is called an *initial condition.* If the switch is closed for a sufficient length of time (at least five time constants), then

$$V_o = E$$

After the switch is opened, V_o decays toward zero as the energy stored in the capacitor is converted into heat by the resistors R_1 and R_2. Let us now analyze how the voltage across the capacitor decays toward zero. We begin by applying Kirchhoff's voltage law to the circuit of Fig. 9-16b:

$$v_{R1} + v_{R2} + v_C = 0 \tag{9-27}$$

$$R_1 i + R_2 i + v_C = 0 \tag{9-28}$$

or

$$R_T i + v_C = 0 \tag{9-29}$$

where

$$R_T = R_1 + R_2$$

Since the circuit of Fig. 9-16b is a series circuit and we know that the equation of current for a capacitor is $i_C = C(dv_C/dt)$,

$$R_T C \frac{dv_C}{dt} + v_C = 0 \tag{9-30}$$

or

$$\frac{dv_C}{dt} + \frac{1}{R_T C} v_C = 0 \tag{9-31}$$

The solution of Eq. (9-31) is

$$v_C = V_o e^{-t/\tau} \tag{9-32}$$

where V_o is the voltage across the capacitor the instant the switch is opened and $\tau = R_T C$. A plot of $e^{-t/\tau}$ is an exponential decaying curve as shown in Fig. 9-3.

The rate at which the voltage across the capacitor decays depends on the expression $t/R_T C$. The denominator, $R_T C$, of the term is the time constant of decay:

$$\tau = R_T C \tag{9-33}$$

Equation (9-32) may be written as

$$v_C = V_o e^{-t/R_T C} \tag{9-34}$$

Refer to Fig. 9-3 and note that at one time constant ($t/\tau = 1$) 37 per cent of V_o remains across the capacitor.

As in the case when the switch is closed, it also takes five time constants for the voltage to reach its steady-state value when the switch is opened. Thus for all practical purposes, all the energy stored by the capacitor will be converted into heat by the total resistance of the circuit in five time constants:

$$5\tau = 5R_TC \text{ seconds}$$

The reader is cautioned that the time constant for energizing an R-C circuit is not always the same as that for deenergizing the circuit. For example, consider the circuit of Fig. 9-16a. If the switch is closed, the time constant is determined only by R_1 and C. However, when the switch is opened, the time constant is determined by R_1, C, and R_2.

Example 9-11: At $t = 0$ the switch in the circuit of Fig. 9-17 is opened. Determine

(a)   the initial voltage across the capacitor,
(b)   the time constant for deenergizing the circuit,
(c)   the voltage across the capacitor at $t = 20$ ms.

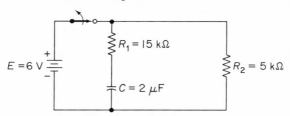

Fig. 9-17. Circuit diagram for Ex. 9-11.

Solution: (a)   If the switch is closed for a sufficient length of time, the capacitor will charge up to a voltage equal to the source voltage. Thus

$$V_0 = E = 6 \text{ V}$$

(b)

$$\tau = R_TC$$

where $R_T = R_1 + R_2$ for deenergizing the circuit:

$$R_T = 15 \text{ k}\Omega + 5 \text{ k}\Omega = 20 \text{ k}\Omega$$

Then

$$\tau = (20 \text{ k}\Omega)(2 \text{ }\mu\text{F}) = 40 \text{ ms}$$

(c) According to Eq. (9-32),

$$v_C = V_0 e^{-t/\tau}$$

where

$$\frac{t}{\tau} = \frac{20 \text{ ms}}{40 \text{ ms}} = 0.5$$

Then

$$v_C = 6e^{-0.5} \text{ V}$$
$$v_C = 6(0.607) \text{ V}$$
$$v_C = 3.642 \text{ V}$$

Note that Eq. (9-32) has to be used to solve for voltage across the capacitor. After 5τ the voltage would be decayed to zero.

Example 9-12: The waveform of Fig. 9-18a is applied to the circuit of Fig. 9-18b. Draw

    (a)   the voltage waveform across the capacitor,
    (b)   the voltage waveform across the resistor,
    (c)   the current waveform.

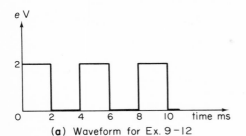
(**a**) Waveform for Ex. 9-12

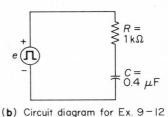
(**b**) Circuit diagram for Ex. 9-12

Fig. 9-18. Diagrams for Ex. 9-12.

Solution: (a) The time constant for both energizing and deenergizing the circuit is

$$\tau = RC = (1 \text{ k}\Omega)(0.4 \text{ }\mu\text{F}) = 0.4 \text{ ms}$$

and

$$5\tau = (5)(0.4 \text{ ms}) = 2 \text{ ms}$$

Therefore, the capacitor is able to charge and discharge to its steady-state value. The steady-state value at 2 ms is

$$v_C = E = 2 \text{ V}$$

The steady-state value at 4 ms is zero. The voltage waveform across the capacitance is shown in Fig. 9-19a.

    (b)   According to Kirchhoff's voltage law,

$$E = v_R + v_C$$

Therefore, while the voltage across the capacitor is an exponential rise, the voltage across the resistor is an exponential decay. This waveform is shown in Fig. 9-19b.

    (c)   Since the voltage and current in a resistor are in phase, the current waveform is similar to the voltage waveform, as show in Fig. 9-19c. The pulse waveform of Fig. 9-18a has the same effect as opening and closing

a switch. Thus the waveforms of Fig. 9-19 repeat every 4 ms. Note that
the maximum value of current is

$$I_m = \frac{V_m}{R} = \frac{2\,V}{1\,k\Omega} = 2\,mA$$

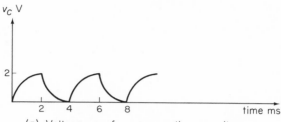

(a) Voltage waveform across the capacitance

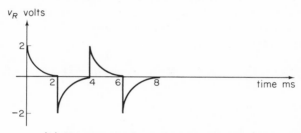

(b) Voltage waveform across the resistance

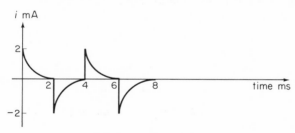

(c) Current waveform

Fig. 9-19. Waveform diagrams for the circuit of Fig. 9-18 b.

Parallel R-C Circuit. Let us consider the circuit of Fig. 9-20 and see the
similarities between the series *R-C* circuit and the parallel circuit. At

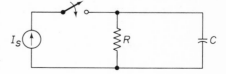

Fig. 9-20. Circuit diagram to illustrate the principles of parallel
R-C circuit.

$t = 0$ the switch is closed. The properties of the capacitor are the same
whether it is connected in series or parallel. Therefore, the voltage across

the capacitor must rise exponentially from zero to the steady-state value. Thus at $t = 0$, $v_C = 0$.

If the switch has been closed for a sufficient length of time (5τ),

$$v_C = V_m = I_S R \tag{9-35}$$

The value of current between these two extremes is given by

$$v_C = V_m(I - e^{-t/\tau}) \tag{9-36}$$

where

$$V_m = I_S R$$

and

$$\tau = RC$$

Thus the expression for voltage across the capacitor for a parallel R-C circuit is the same as that for a series circuit. Note also that the expression for the time constant for both circuits is the same. As in the previous cases, the steady-state value is reached in 5τ.

> **Example 9-13:** Consider the circuit of Fig. 9-21. At $t = 0$ the switch is closed. Determine
>
> (a) the voltage across the capacitor at $t = 0$,
> (b) the steady-state value of voltage,
> (c) the time to reach the steady-state value,
> (d) the voltage across the capacitor at $t = 15$ ms.

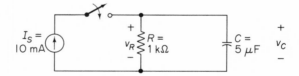

Fig. 9-21. Circuit diagram for Ex. 9-13.

Solution: (a) At $t = 0$ the capacitor appears as a short circuit. Therefore,

$$v_C = 0$$

(b) Since R and C are in parallel, then

$$v_C = v_R$$

Therefore, the steady-state value of voltage across the capacitance is

$$V_m = I_S R = (10 \text{ mA})(1 \text{ k}\Omega) = 10 \text{ V}$$

(c) Since

$$\tau = RC = (1 \text{ k}\Omega)(5 \text{ }\mu\text{F}) = 5 \text{ ms}$$

the time to reach the steady-state value is

$$5\tau = (5)(5 \text{ ms}) = 25 \text{ ms}$$

(d)
$$v_C = V_m(1 - e^{-t/\tau})$$

where
$$V_m = 10 \text{ V}$$

and
$$\frac{t}{\tau} = \frac{15 \text{ ms}}{25 \text{ ms}} = 0.6$$

Then
$$v_C = 10(1 - e^{-0.6}) \text{ V}$$
$$v_C = 10(1 - 0.55) \text{ V} = 4.5 \text{ V}$$

Note that because the energy source is a current source, the time constant is determined by the capacitor and the parallel resistor and that the other formulas developed for the series circuit are applicable to the parallel circuit.

Example 9-14: Again consider the circuit of Fig. 9-21. At $t = 0$, the switch is opened. Determine
   (a) the voltage across the capacitor at $t = 0$,
   (b) the steady-state value of voltage,
   (c) the time to reach the steady-state value,
   (d) the voltage across the capacitor at $t = 15$ ms.

Solution: (a) If the switch has been closed for a sufficient length of time, the voltage across the capacitor would be a maximum value. Therefore, at $t = 0$,
$$v_C = V_m = 10 \text{ V}$$

(b) After the switch is opened, the voltage across the capacitance decays exponentially toward zero. Thus the steady-state value is zero.

(c) Since
$$\tau = RC = (1 \text{ k}\Omega)(5 \text{ }\mu\text{F}) = 5 \text{ ms}$$

the time to reach the steady-state value is
$$5\tau = 5(5 \text{ ms}) = 25 \text{ ms}$$

(d) Equation (9-32) is applicable to a parallel circuit,
$$v_C = V_0 e^{-t/\tau}$$

where for this example
$$V_0 = V_m = 10 \text{ V}$$

and
$$\frac{t}{\tau} = \frac{15 \text{ ms}}{25 \text{ ms}} = 0.6$$

Then
$$v_C = 10e^{-0.6} \text{ V}$$
$$v_C = 10(0.55) \text{ V} = 5.5 \text{ V}$$

Note that the charging time constant of Example 9-13 equals the decaying time constant of this example. However, the value of v_C at $t = 15$ ms of Example 9-13 does not equal the value of v_C at $t = 15$ ms of this example.

SUMMARY

In Section 9-1 we analyzed a dc circuit containing a resistance and an inductance, first connected in series and then in parallel. The concept of a time constant of an R-L circuit was developed and defined as the time it takes the instantaneous current to reach 63 per cent of its steady-state value. Concluding our analysis, we determined the process of deenergizing the R-L circuit.

The circuit combination of a resistance and a capacitance, both in series and in parallel, was the topic of Section 9-2. As in Section 9-1, we analyzed the energizing of the circuit and the time constant, in this case defined as the time it takes for the instantaneous voltage across the capacitance to reach 63 per cent of its steady-state value. Unlike the R-L circuit, it was not necessary to provide a path for current to flow the instant the switch is opened in the R-C circuit.

PROBLEMS

9-1. Refer to the circuit of Fig. 9-4 and determine
  (a) the steady-state value of current,
  (b) the current 1 μs after the switch is closed for each of the following conditions.

  (1) $R = 4\,k\Omega$ (2) $R = 2\,k\Omega$ (3) $R = 2\,k\Omega$
        $L = 2\,mH$ $L = 4\,mH$ $L = 2\,mH$
        $E = 10\,V$ $E = 10\,V$ $E = 20\,V$

9-2. For each of the conditions of Problem 9-1, calculate
  (a) the time constant,
  (b) the time to reach the steady-state value.

9-3. If the inductor in the circuit of Fig. 9-5 is changed to 4.5 H, is the time to reach the steady-state value increased or decreased?

9-4. If the voltage source in the circuit of Fig. 9-5 is changed to 9 V, is the time constant changed? What is the value of current at $t = 6$ ms?

9-5. For each of the circuits of Fig. 9-22, calculate
  (a) time constant,
  (b) steady-state value,
  (c) current through the inductor at $t = 5\ \mu$s.

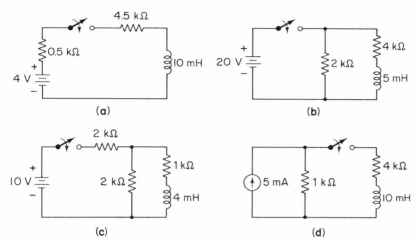

Fig. 9-22. Circuit diagrams for Prob. 9-5.

9-6. Refer to the circuit of Fig. 9-7 and calculate
   (a) the time constant to energize the inductor,
   (b) the time constant to deenergize the inductor for each of the following conditions.

(1) $R_1 = 2\,\text{k}\Omega$ (2) $R_1 = 1\,\text{k}\Omega$ (3) $R_1 = 1\,\text{k}\Omega$
   $R_2 = 4\,\text{k}\Omega$ $R_2 = 4\,\text{k}\Omega$ $R_2 = 5\,\text{k}\Omega$
   $R_3 = 3\,\text{k}\Omega$ $R_3 = 3\,\text{k}\Omega$ $R_3 = 4\,\text{k}\Omega$
   $L = 10\,\text{mH}$ $L = 20\,\text{mH}$ $L = 10\,\text{mH}$
   $E = 10\,\text{V}$ $E = 5\,\text{V}$ $E = 10\,\text{V}$

9-7. For each of the conditions of Problem 9-6, calculate
   (a) the current through the inductor at $t = 0$,
   (b) the current through the inductor at $t = 50\ \mu\text{s}$,
   (c) the current through the inductor at $t = 3\ \mu\text{s}$,
   (d) the voltage across R_1 at $t = 3\ \mu\text{s}$.
   Consider at $t = 0$ the switch is closed.

9-8. Refer to Fig. 9-8 and draw the waveform for
   (a) current through the inductor,
   (b) voltage across the resistor,
   (c) voltage across the inductor,
   for $L = 4.8$ mH and $L = 1.2$ mH.

9-9. Repeat Example 9-6 with the following circuit values: $I_S = 10\,\text{mA}$, $R = 2.5$ k, $L = 10$ mH.

9-10. Repeat Example 9-7 with the following circuit values: $I_S = 10$ mA, $R = 2.5\,\text{k}\Omega$, $L = 10$ mH.

9-11. For each of the circuits of Fig. 9-23, calculate
   (a) the voltage across the capacitor at $t = 0$,
   (b) the voltage across R at $t = 0$,
   (c) the steady-state value of voltage across the capacitor.

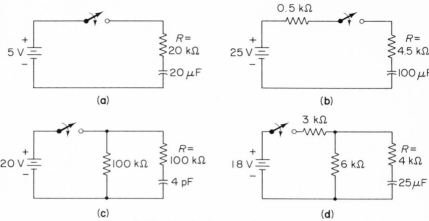

Fig. 9-23. Circuit diagrams for Prob. 9-11.

9-12. Calculate the charging time constant for each of the circuits of Fig. 9-23.

9-13. For circuits a and b of Fig. 9-23, calculate the voltage across the capacitor at $t = 0.8$ s.

9-14. For the circuit of Fig. 9-23d, calculate the voltage across the capacitor and resistor, R, at $t = 0.6$ s.

9-15. Repeat Example 9-11 with the following circuit quantities.

    (a) $R_1 = 30\,k\Omega$ (b) $R_1 = 15\,k\Omega$ (c) $R_1 = 15\,k\Omega$
          $R_2 = 5\,k\Omega$ $R_2 = 10\,k\Omega$ $R_2 = 5\,k\Omega$
          $C = 2\,\mu F$ $C = 2\,\mu F$ $C = 2\,\mu F$
          $E = 6\,V$ $E = 6\,V$ $E = 12\,V$

9-16. For circuits c and d of Fig. 9-23, the switch is opened after the capacitor has charged up to its final value. Calculate
    (a) the initial value of voltage,
    (b) the discharge time constant,
    (c) the voltage across the capacitor at $t = 0.6$ s.
    (d) the voltage across R at $t = 0.6$ s.

9-17. Repeat Example 9-12 for capacitor values of $C = 0.8\ \mu F$ and $C = 0.2\ \mu F$.

9-18. For the circuit of Fig. 9-24, calculate V and I at the instant the switch is closed and the final values.

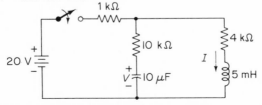

Fig. 9-24. Circuit diagram for Prob. 9-18.

9-19. Repeat Example 9-13 with the following circuit values: $I_S = 15$ mA, $R = 5\,k\Omega$, $C = 10\ \mu F$.

9-20. Repeat Example 9-14 with the following circuit values: $I_S = 15$ mA, $R = 5\,k\Omega$, $C = 10\ \mu F$.

CHAPTER 10

tuned circuits

A tuned circuit is a *filter* possessing the properties of selection and rejection. A filter is a circuit designed to pass only a specific band of frequencies while rejecting all other frequencies. The band of frequencies which the filter is allowed to pass is called the *bandwidth*.

A tuned circuit is designed so that the bandwidth is centered about one frequency. This center frequency is called the *resonant frequency*. When a circuit is tuned to the resonant frequency, the condition is known as *resonance*.

An *ideal filter* passes a desired band of frequencies with no change in either the amplitude or phase angle. For a practical tuned circuit to approach the ideal case, the bandwidth must be small with respect to the resonant frequency. That is, the bandwidth must be less than one tenth of the resonant frequency. Under this (ideal) restriction, the circuit is known as a *narrowband tuned circuit*. In this text we are concerned only with narrowband tuned circuits.

Tuned circuits are either a series or parallel combination of resistance, inductance, and capacitance. Although it is possible to have both series and parallel tuned circuits, the majority of electronic applications employ only the parallel case. Systems that use such tuned circuits are transmitters, receivers, and measuring devices.

Section 10-1 treats the theoretical or idealized parallel-tuned circuits, Section 10-2 deals with practical parallel-tuned circuits, Section 10-3 is concerned with series-tuned circuits, and in Section 10-4 other filters are discussed.

10-1. PARALLEL-TUNED CIRCUITS

The circuit shown in Fig. 10-1a contains a parallel combination of a high resistor, a pure inductor and a capacitor. This circuit was analyzed in Chapter 7 from the standpoint of impedance and admittance. It is called a parallel-tuned or parallel-resonant circuit.

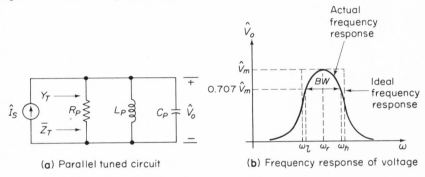

(a) Parallel tuned circuit   (b) Frequency response of voltage

Fig. 10-1. Diagrams that describe the parallel-tuned circuit.

Applications of parallel-tuned circuits are usually in conjunction with either transistors or vacuum-tube pentodes, both of which are represented by an equivalent circuit containing a constant current source. For this reason we shall analyze the parallel-tuned circuit driven by an ideal current source.

The properties of selection and rejection are accomplished in a tuned circuit because the complex admittance (or impedance) varies with frequency. The frequency-dependent terms of complex admittance are inductive susceptance ($B_L = 1/\omega L$) and capacitive susceptance ($B_C = \omega C$).

The important properties of a parallel-tuned circuit are *input admittance, resonant frequency, bandwidth*, and the *quality factor*. The remaining part of this section describes each property and provides examples.

Input Admittance. The input admittance for the circuit of Fig. 10-1a is given by

$$\bar{Y} = G - jB_L + jB_C \tag{10-1}$$

where Y_T is the total input admittance in mhos, $G = 1/R_p$, $B_L = 1/\omega L_p$, and $B_C = \omega C_p$.

Since B_L and B_C are frequency dependent, let us consider different frequency ranges and the effect on the input admittance.

At low values of frequency (frequencies below resonance) the inductive susceptance $(B_L = 1/\omega L_p)$ is large and the capacitive susceptance $(B_C = \omega C_p)$ is small. Therefore, at frequencies below resonance, $B_L > B_C$ and an equivalent circuit of Fig. 10-1a is a resistor in parallel with an inductor.

At high frequencies (frequencies above resonance) it is the capacitive susceptance value that is large; the inductive susceptance value is small. Thus at frequencies above resonance, $B_C > B_L$ and an equivalent circuit of Fig. 10-1a is a resistor in parallel with a capacitor.

We have considered an equivalent circuit for frequencies above and below the resonant frequency. At resonance $B_L = B_C$ or $B_L - B_C = 0$. Since the susceptance term equals zero, Eq. (10-1) reduces to

$$\overline{Y}_T = G_p \tag{10-2a}$$

or

$$\overline{Z}_T = R_p \tag{10-2b}$$

where R_p is the total parallel resistance.

This large value of input impedance (resistance) at resonance is the major advantage of the parallel-tuned circuit. This large resistance produces the maximum output voltage, which is usually a primary concern with transistor and vacuum-tube pentode circuits. A plot of voltage versus frequency for a parallel-tuned circuit is shown in Fig. 10-1b (solid line). The dotted curve is a plot of the ideal frequency response.

Example 10-1: For the circuit of Fig. 10-2, calculate the input admittance for the following angular frequencies:

  (a) $\omega = 6$ Mrad/s
  (b) $\omega = 11.2$ Mrad/s
  (c) $\omega = 20$ Mrad/s

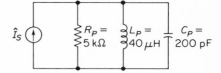

Fig. 10-2. Circuit diagram for Ex. 10-1.

Solution: (a) At $\omega = 6$ Mrad/s,

$$B_L = \frac{1}{\omega L_p} = \frac{1}{(6 \times 10^6 \text{ rad/s})(40 \times 10^{-6} \text{ H})} = 4.16 \text{ m}\mho$$

and

$$B_C = \omega C_p = (6 \times 10^6 \text{ rad/s})(200 \times 10^{-12} \text{ F}) = 1.2 \text{ m}\mho$$

and

$$G = \frac{1}{R_p} = \frac{1}{5 \text{ k}\Omega} = 0.2 \text{ m}\mho$$

Therefore,

$$\bar{Y}_T = (0.2 - j4.16 + j1.2)\ \text{m}\mho$$

or

$$\bar{Y}_T = (0.2 - j2.96)\ \text{m}\mho$$

(b)   At $\omega = 11.2$ Mrad/s,

$$B_L = \frac{1}{\omega L_p} = \frac{1}{(11.2 \times 10^6\ \text{rad/s})\,(40 \times 10^{-6}\ \text{H})} = 2.23\ \text{m}\mho$$

and

$$B_C = \omega C_p = (11.2 \times 10^6\ \text{rad/s})\,(200 \times 10^{-12}\ \text{F}) = 2.23\ \text{m}\mho$$

and

$$G = \frac{1}{R_p} = \frac{1}{5\ \text{k}\Omega} = 0.2\ \text{m}\mho$$

Therefore,

$$\bar{Y}_T = (0.2 - j2.23 + j2.23)\ \text{m}\mho$$

$$\bar{Y}_T = 0.2\ \text{m}\mho$$

or

$$\bar{Z}_T = \frac{1}{\bar{Y}_T} = \frac{1}{0.2\ \text{m}\mho} = 5\ \text{k}\Omega$$

(c)   At $\omega = 20$ Mrad/s,

$$B_L = \frac{1}{\omega L_p} = \frac{1}{(20 \times 10^6\ \text{rad/s})\,(40 \times 10^{-6}\ \text{H})} = 1.25\ \text{m}\mho$$

and

$$B_C = \omega C_p = (20 \times 10^6\ \text{rad/s})\,(200 \times 10^{-12}\ \text{F}) = 4\ \text{m}\mho$$

and

$$G = \frac{1}{R_p} = \frac{1}{5\ \text{k}\Omega} = 0.2\ \text{m}\mho$$

Therefore,

$$\bar{Y}_T = (0.2 - j1.25 + j4)\ \text{m}\mho$$

or

$$\bar{Y}_T = (0.2 + j2.75)\ \text{m}\mho$$

Part (b) shows that $B_L = B_C$. Therefore, 11.2 Mrad/s is the angular resonant frequency. At frequencies below resonance (such as 6 Mrad/s) the value of the total admittance shows that an equivalent circuit is a resistor in parallel with an inductor. Part (c) shows that at frequencies above resonance an equivalent circuit is a resistor in parallel with a capacitor.

Resonant Frequency.   Resonant frequency for a parallel-tuned circuit is defined by one of the following conditions:

1. when the input impedance (Z_T) is maximum (Y_T is a minimum),
2. when the input impedance (Z_T) is resistive (Y_T is conductive),
3. when the inductive reactance (X_L) equals the capacitive reactance (X_C).

For the theoretical or idealized parallel-tuned circuit as shown in Fig. 10-1a, all three conditions occur at the same frequency. Since

$$X_L = X_C$$

then

$$\omega_r L_p = \frac{1}{\omega_r C_p}$$

Solving for ω_r, we obtain

$$\omega_r = \frac{1}{\sqrt{L_p C_p}} \tag{10-3a}$$

or

$$f_r = \frac{1}{2\pi \sqrt{L_p C_p}} \tag{10-3b}$$

where ω_r is the resonant frequency in radians per second, f_r the resonant frequency in hertz, L_p the total parallel inductance in henries, and C_p the total parallel capacitance in farads.

Example 10-2: For the circuit of Fig. 10-2, calculate the resonant frequency.

Solution: Using Eq. (10.3a),

$$\omega_r = \frac{1}{\sqrt{L_p C_p}} = \frac{1}{\sqrt{(40 \times 10^{-6}\,\mathrm{H})(200 \times 10^{-12}\,\mathrm{F})}} = 11.2\ \mathrm{Mrad/s}$$

or

$$f_r = \frac{\omega_r}{2\pi} = \frac{11.2\ \mathrm{Mrad/s}}{6.28} = 1.79\ \mathrm{mHz}$$

Checking with part (b) of Example 10-1, note that the value of $\omega_r = 11.2$ Mrad/s is the angular frequency that resulted in $B_L = B_C$ and therefore maximum impedance.

Example 10-3: For the circuit of Fig. 10-3, calculate the resonant frequency.

Fig. 10-3. Circuit diagram for Ex. 10-3.

Solution: To calculate the resonant frequency the total parallel capacitance is needed. Therefore,

$$C_p = C_1 + C_2$$

$$C_p = 0.1\ \mu\mathrm{F} + 0.4\ \mu\mathrm{F} = 0.5\ \mu\mathrm{F}$$

Then

$$\omega_r = \frac{1}{\sqrt{L_p C_p}} = \frac{1}{\sqrt{(5 \times 10^{-3} \, \text{H})(0.5 \times 10^{-6} \, \text{F})}} = 20 \, \text{krad/s}$$

or

$$f_r = \frac{\omega_r}{2\pi} = \frac{20 \, \text{krad/s}}{6.28} = 3.18 \, \text{kHz}$$

Equation (10-3a) or (10-3b) is applicable only after the total parallel inductance and total parallel capacitance are calculated. Note that the resonant frequency for a parallel-tuned circuit is independent of the resistance and the source current.

Bandwidth. At resonance the voltage is a maximum. Therefore, the power dissipated is also a maximum. If the frequency is varied off resonance, the output voltage decreases as shown by Fig. 10-1b. A decrease in output voltage causes a decrease in output power. There are, then, two frequencies, one above and one below the resonant frequency, at which the power dissipated is one half of that at resonance. The frequency above resonance is called the *upper half-power frequency* (ω_h); the frequency below resonance is called the *lower half-power frequency* (ω_l). These frequencies are also known as the *upper* and *lower cutoff frequencies*.

In terms of voltage, the cutoff frequencies occur at $0.707V_m$, as shown in Fig. 10-1b. Since $P = V^2/R$, the relationship between voltage and power at the cutoff frequencies is shown by

$$P = \frac{(0.707V_m)^2}{R} = \frac{1}{2} P_m$$

As the voltage drops to $0.707V_m$, the admittance must increase by $1/0.707$, or 1.414 times the admittance at resonance. Therefore, if the admittance at resonance is G_p, the magnitude of the admittance at the upper and lower cutoff frequencies equals $1.414G_p$. The increase in admittance is caused by the frequency-dependent terms — inductive susceptance and the capacitive susceptance. At the lower cutoff frequency

$$B_L - B_C = G_p$$

and at the upper cutoff frequency

$$B_C - B_L = G_p$$

Figure 10-4 shows the phasor diagrams for admittance at the cutoff frequencies.

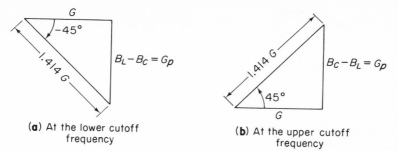

(a) At the lower cutoff frequency   **(b)** At the upper cutoff frequency

Fig. 10-4. Phasor diagrams of admittance (parallel-tuned circuit) at the cut off frequencies.

The bandwidth, BW, is defined as the difference between the upper cutoff frequency and the lower cutoff frequency:

$$\text{BW} = \omega_h - \omega_l \quad \text{rad/s} \tag{10-4a}$$

or

$$\text{BW} = f_h - f_l \quad \text{Hz} \tag{10-4b}$$

In terms of circuit parameters

$$\text{BW} = \frac{1}{R_p C_p} \quad \text{rad/s} \tag{10-5a}$$

or

$$\text{BW} = \frac{1}{2\pi R_p C_p} \quad \text{Hz} \tag{10-5b}$$

where BW is the bandwidth, R_p the total parallel resistance in ohms, and C_p the total parallel capacitance in farads. Equation (10-5) shows that the bandwidth is inversely proportional to both the parallel resistance, R_p, and the parallel capacitance, C_p. Therefore, an increase of either the resistance or the capacitance results in a decrease of the bandwidth. Conversely, a decrease of either the resistance or the capacitance results in an increase of the bandwidth.

In conclusion, we have developed a formula for expressing the bandwidth in terms of the circuit parameters. The reader should note that the parallel resistance, R_p, was a pure resistor (not frequency dependent). The next section will show that although R_p may be frequency dependent, the formula for bandwidth will not change. Once again, these formulas apply to a narrowband parallel-tuned circuit.

For narrowband tuned circuits (BW $\leq 1/10\omega_r$), the bandwidth is symmetrical about the resonant frequency and thus

$$\omega_l = \omega_r - \frac{\text{BW}}{2} \tag{10-6a}$$

$$\omega_h = \omega_r + \frac{\text{BW}}{2} \tag{10-6b}$$

Example 10-4: Calculate the bandwidth for the circuit of Fig. 10-2 and for the circuit of Fig. 10-3.

Solution: For the circuit of Fig. 10-2,

$$\text{BW} = \frac{1}{R_p C_p} = \frac{1}{(5\text{ k}\Omega)(200\text{ pF})} = 1\text{ Mrad/s}$$

In terms of hertz,

$$\text{BW} = \frac{1\text{ Mrad/s}}{6.28} = 159\text{ kHz}$$

For the circuit of Fig. 10-3,

$$\text{BW} = \frac{1}{R_p C_p} = \frac{1}{(10\text{ krad/s})(0.5\text{ }\mu\text{F})} = 0.2\text{ krad/s}$$

In terms of hertz,

$$\text{BW} = \frac{0.2\text{ krad/s}}{6.28} = 31.8\text{ Hz}$$

Note that each of these circuits is a narrowband tuned circuit — $\text{BW} \leq \frac{1}{10}\omega_r$. For a circuit similar to that of Fig. 10-3, the total parallel capacitance must first be calculated.

Example 10-5: The circuit of Fig. 10-5 is to be tuned to a resonant frequency of 455 kHz and have a bandwidth of 10kHz. Calculate

    (a) the capacitor to give the resonant frequency,
    (b) the resistor to give the desired bandwidth.

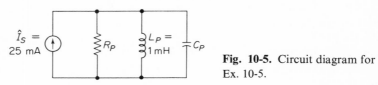

Fig. 10-5. Circuit diagram for Ex. 10-5.

Solution: (a) Rearranging Eq. (10-3b),

$$C_p = \frac{1}{(2\pi f_r)^2 L_p} = \frac{1}{[(6.28)(455 \times 10^3\text{ Hz})]^2 (1 \times 10^{-3})\text{ H}} = 122\text{ pF}$$

(b) Rearranging Eq. (10-5b) yields

$$R_p = \frac{1}{2\pi(\text{BW})C_p} = \frac{1}{(6.28)(10 \times 10^3)\text{ Hz}(122 \times 10^{-12}\text{ F})} = 131\text{ k}\Omega$$

This example shows how a parallel-tuned circuit may be designed to meet specific requirements.

Quality Factor. The symbol Q is frequently encountered in tuned circuits. It is called *quality factor*. (Note that the symbol Q used in tuned circuits should not be confused with the symbol used for charge.) The Q

of a component or a circuit is a measure of its energy-storing abilities. The two energy-storing elements are the inductor and capacitor. Thus Q is associated with these elements or circuits in which they are contained. Some of this stored energy is converted into heat by resistance. Q may be defined as

$$Q = \frac{\text{reactive power}}{\text{real power}} \tag{10-7}$$

where reactive power is the power delivered to the inductor or a capacitor and real power is the power delivered to the resistor.

For the parallel-tuned circuit of Fig. 10-1a the value of Q expressed in terms of R_p and L_p is

$$Q = \frac{\text{reactive power}}{\text{real power}} = \frac{V^2/X_L}{V^2/R_p}$$

or

$$Q = \frac{R_p}{X_L} = \frac{R_p}{\omega_r L_p} \tag{10-8}$$

where R_p is the total parallel resistance in ohms, L_p the total parallel inductance in henries, and ω_r the resonant angular frequency in radians per second.

The value of Q also may be expressed in terms of R_p and C_p:

$$Q = \frac{\text{reactive power}}{\text{real power}} = \frac{V^2/X_C}{V^2/R_p}$$

Then

$$Q = \frac{R}{X_C} = \omega_r R_p C_p \tag{10-9}$$

where R_p is the total parallel resistance in ohms, C_p the total parallel capacitance in farads, and ω_r the resonant angular frequency in radians per second.

The Q of a circuit is also defined as the ratio of resonant frequency to bandwidth:

$$\frac{\text{resonant frequency}}{\text{bandwidth}} = \frac{\omega_r}{\text{BW}} = \frac{\omega_r}{1/R_p C_p} = \omega_r R_p C_p = Q \tag{10-10}$$

where ω_r, R_p, and C_p have the same values as those of Eq. (10-9). Rearranging Eq. (10-10) yields

$$\text{BW} = \frac{\omega_r}{Q} \quad \text{rad/s} \tag{10-11a}$$

or

$$\text{BW} = \frac{f_r}{Q} \quad \text{Hz} \tag{10-11b}$$

Equations (10-11) show that for a particular resonant frequency, if the Q is increased, the bandwidth decreases; if the Q decreases, the bandwidth increases.

In audio circuits, Q determines the selectivity of a circuit. The term "selectivity" arises from the ability of a resonant circuit to choose only one band of frequencies. For example, several radio stations broadcast simultaneously with different carrier frequencies. The radio antenna may be represented by an R-L-C circuit. When the circuit is tuned to the desired frequency (station) and the Q of the circuit is large ($Q > 10$), all other adjacent frequencies are rejected. Thus only one frequency (station) has been selected. It should be noted that the larger the value of Q, the better the selectivity of the circuit. For our analysis, a narrowband tuned circuit has to have a Q value of at least 10.

Example 10-6: For the circuit of Fig. 10-2, calculate the value of Q.

Solution: Using Eq. (10-9),

$$Q = \omega_r R_p C_p = (11.2 \text{ Mrad/s}) (5 \text{ k}\Omega) (200 \text{ pF}) = 11.2$$

Using Eq. (10-10) as a check yields

$$Q = \frac{\omega_r}{\text{BW}} = \frac{11.2 \text{ Mrad/s}}{1 \text{ Mrad/s}} = 11.2$$

The value of the bandwidth was calculated in Example 10-4. Since $Q > 10$, the circuit is a narrowband tuned circuit.

Example 10-7: Calculate the circuit Q for Fig. 10-3.

Solution: Using Eq. (10-9),

$$Q = \omega_r R_p C_p$$

For this circuit

$$C_p = C_1 + C_2 = 0.5 \,\mu\text{F}$$

Therefore,

$$Q = (20 \text{ krad/s}) (10 \text{ k}\Omega) (0.5 \,\mu\text{F}) = 100$$

Using Eq. (10-10) as a check yields

$$Q = \frac{\omega_r}{\text{BW}} = \frac{20 \text{ krad/s}}{0.2 \text{ krad/s}} = 100$$

Since the Q of the circuit of Fig. 10-3 is greater than the Q of the circuit of Fig. 10-2, it is more selective.

Plotting the Resonant Curve.   As previously stated, the resonant curve for a narrowband tuned circuit is symmetrical about ω_r. Figure 10-6a shows an exact plot of a resonant curve compared to the ideal response. For all practical purposes, the exact resonant curve may be approximated by straight lines as shown in Fig. 10-6b.

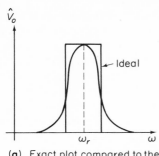

(a) Exact plot compared to the ideal response

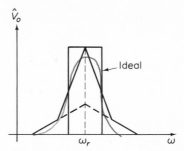

(b) Straight line approximation compared to the ideal response

Fig. 10-6. Plots of the resonant curve of voltage for a narrowband parallel-tuned circuit.

The following procedure shows how the straight-line approximation may be plotted without calculating and plotting the magnitude of $\hat{V}_0$ for numerous values of ω.

1. Calculate the resonant frequency:

$$\omega_r = \frac{1}{\sqrt{L_p C_p}}$$

At the resonant frequency the magnitude of the voltage is a maximum. Therefore,

$$|\hat{V}_0| = |\hat{V}_{\max}| = \hat{I}_s R_p$$

2. Scale the horizontal and vertical axes for frequency and voltage, respectively. The horizontal axis should include at least $10\omega_r$, while $|\hat{V}_{\max}|$ must be able to be plotted on the vertical axis. Therefore, log-log graph paper is the best to use. Locate the intersection of ω_r and $|\hat{V}_{\max}|$ —point (1) in Fig. 10-7.

3. Constructing the upper portion of the resonant curve. First calculate ω_l and ω_h:

$$\omega_l = \omega_r - \frac{BW}{2}$$

$$\omega_h = \omega_r + \frac{BW}{2}$$

At both ω_l and ω_h

$$|\hat{V}_0| = 0.707 |\hat{V}_{\max}|$$

Referring to Fig. 10-7, point (2) is the intersection of ω_l and $|\hat{V}_0|$. Point (3) is the intersection of ω_h and $|\hat{V}_0|$. Line (a) is drawn from point (1) through point (2). Line (b) is drawn from point (1) through point (3).

4. Constructing the lower portion (or skirts) of the resonant curve. Again refer to Fig. 10-6b and note that if the lower portions are extended, they intersect at ω_r. However, the magnitude of voltage is not $\hat{V}_{max}$. At ω_r, $X_L = X_C$ and the magnitude of $\hat{V}_0$ at the intersection of the lower portion is

$$|\hat{V}_0| = \hat{I}_s X_L \quad \text{or} \quad |\hat{V}_0| = \hat{I}_s X_C$$

On Fig. 10-7, this is point (4). If the frequency decreases by a decade, the $|\hat{V}_0|$ decreases by a decade. This locates point (5), and line (c) is drawn from (4) through (5). If the frequency increases by a decade, the $|\hat{V}_0|$ again decreases by a decade. This locates point (6) on Fig. 10-7 and line (d) is drawn from (4) through (6).

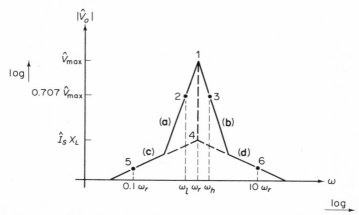

Fig. 10-7. Straight line approximation illustrating frequency response plot for a parallel-tuned circuit.

The solid line in Fig. 10-7 is the straight-line approximation for the frequency-response curve. Now let us apply the above procedure to a numerical example.

Example 10-8: For the circuit of Fig. 10-8, draw the resonant curve using the straight-line approximation.

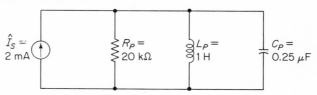

Fig. 10-8. Circuit diagram for Ex. 10-8.

Solution: 1.

$$\omega_r = \frac{1}{\sqrt{L_p C_p}} = \frac{1}{\sqrt{(1\ \text{H})(0.25\ \mu\text{F})}} = 2\ \text{krad/s}$$

At ω_r, the voltage is a maximum for a parallel-tuned circuit:

$$|\hat{V}_{max}| = \hat{I}_s R_p = (2 \text{ mA})(20 \text{ k}\Omega) = 40 \text{ V}$$

2. Figure 10-9a shows the horizontal and vertical axes scaled. Figure 10-9b shows the intersection of ω_r and $|\hat{V}_{max}|$.

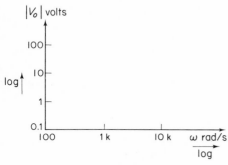

(a) Desired range for horizontal and vertical axes

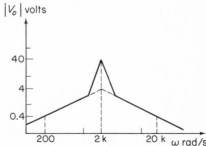

(b) Intersection of ω_r and $|V_{max}|$

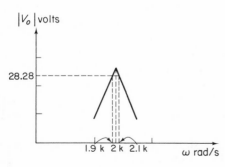

(c) Plot of the upper portion

(d) Straight line approximation

Fig. 10-9. Diagrams showing procedure for drawing a resonant curve.

3. $$\omega_l = \omega_r - \frac{BW}{2} \quad \text{and} \quad \omega_h = \omega_r + \frac{BW}{2}$$

where

$$BW = \frac{1}{R_p C_p} = \frac{1}{(20 \text{ k}\Omega)(0.25 \text{ } \mu\text{F})} = 200 \text{ rad/s}$$

Therefore,

$$\omega_l = 2 \text{ krad/s} - \frac{200 \text{ rad/s}}{2} = 1.9 \text{ krad/s}$$

and

$$\omega_h = 2 \text{ krad/s} + \frac{200 \text{ rad/s}}{2} = 2.1 \text{ krad/s}$$

At both of these frequencies

$$|\hat{V}_0| = (0.707) \ |\hat{V}_{max}|$$

$$|\hat{V}_0| = (0.707)(40 \text{ V}) = 28.28 \text{ V}$$

Figure 10-9c shows the top portion of the resonant curve constructed by straight lines.

4. According to step 4 of the procedure, the lower portion, if extended, intersects at $|\hat{V}_0| = \hat{I}_s X_L$ or $|\hat{V}_0| = \hat{I}_s X_C$. Therefore,

$$|\hat{V}_0| = (2 \text{ mA})(2 \text{ krad/s})(1 \text{ H}) = 4 \text{ V}$$

Then at frequencies of $\frac{1}{10}\omega_r$ (200 rad/s) and $10\omega_r$ (20 krad/s), the $|\hat{V}_0|$ is $\frac{1}{10}$ (4 V) = 0.4 V. Figure 10-9d shows the straight-line approximation of the resonant curve.

10-2. PRACTICAL PARALLEL TUNED CIRCUIT

The circuit of Fig. 10-1a is sometimes referred to as the idealized or theoretical parallel-tuned circuit. This is because the resistor (which is not frequency dependent) is in parallel with a *pure* inductor and capacitor. Physically, however, a pure inductor is not possible because of internal losses. To account for these losses in a circuit diagram, the convention is to represent a resistance in series with the inductance. It is possible, however, to manufacture a capacitor that closely approximates an ideal capacitance — that is, a capacitor having negligible losses in the frequency range of operation. Thus a practical capacitor need not be represented with any series resistive element.

The circuit diagram of Fig. 10-10a shows a parallel-tuned circuit accounting for the losses in the coil. The circuit of Fig. 10-10a is transformed

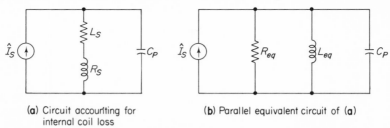

(a) Circuit accounting for   (b) Parallel equivalent circuit of (a)
internal coil loss

Fig. 10-10. Circuit diagrams of a practical parallel tuned circuit.

into the parallel equivalent circuit of Fig. 10-10b.   The circuit of Fig. 10-10b under a valid approximation has the same formula for resonant frequency and bandwidth as the theoretical circuit.   The values of R_{eq} and L_{eq} however, are now frequency dependent.

We shall begin our analysis by transforming the series R-L branch of

Fig. 10-10a to an equivalent parallel R-L combination. The impedance of the series R-L branch is

$$\bar{Z} = R_s + j\omega L_s$$

The admittance of this expression is

$$\bar{Y} = \frac{1}{\bar{Z}} = \frac{1}{R_s + j\omega L_s} = \frac{R_s}{R_s^2 + (\omega L_s)^2} - j\frac{\omega L_s}{R_s^2 + (\omega L_s)^2}$$

This is in the form

$$Y = G - jB_L$$

where

$$G = \frac{1}{R_{eq}} = \frac{R_s}{R_s^2 + (\omega L_s)^2}$$

and

$$B_L = \frac{1}{\omega L_{eq}} = \frac{\omega L_s}{R_s^2 + (\omega L_s)^2}$$

Thus the parallel equivalent resistance is

$$R_{eq} = \frac{R_s^2 + (\omega L_s)^2}{R_s} \tag{10-12}$$

and the parallel inductance is

$$L_{eq} = \frac{R_S^2 + (\omega L_s)^2}{\omega L_S} \tag{10-13}$$

Note that both R_{eq} and L_{eq} are frequency dependent.

The total input admittance of the circuit of Fig. 10-10b may be expressed as

$$Y_T = G + j(B_C - B_L)$$

or

$$Y_T = \frac{R_S}{R_S^2 + (\omega L_S)^2} + j\omega \left(C_p - \frac{L_S}{R_S^2 + (\omega L_S)^2} \right) \tag{10-14}$$

The resonant frequency of the parallel-tuned circuit is not found as easily as the resonant frequency of the ideal parallel-tuned circuit. The reason is that the parallel-tuned circuit has three frequency-dependent parameters — conductance, inductive susceptance, and capacitive susceptance. Because of the interaction between these three frequency-dependent terms, there exist not one but three resonant frequencies for a parallel-tuned circuit. These frequencies occur

1. when the total impedance ($\bar{Z}_T$) is resistive ($\bar{Y}_T$ is conductive). This frequency is the lowest resonant frequency.

2. when the total impedance ($\bar{Z}_T$) is a maximum value ($\bar{Y}_T$ is a minimum value). This frequency is the middle resonant frequency.

3. when the inductive reactance (X_L) equals the capacitive reactance (X_C). This frequency is the highest resonant frequency and is also the same as the resonant frequency of the ideal tuned circuit ($\omega_r = 1/\sqrt{LC}$).

Lowest Resonant Frequency. The parallel-tuned circuit of Fig. 10-10b appears as a resistive circuit when

$$B_C = B_L \quad \text{or} \quad \omega C = \frac{\omega L_S}{R_S^2 + (\omega L_S)^2}$$

Solving for ω yields

$$\omega_{r1} = \frac{1}{\sqrt{L_S C_p}} \sqrt{1 - \frac{C_p R_S^2}{L_S}} \tag{10-15}$$

where L_S is the value of the inductor, as shown in Fig. 10-10a; C_p the total parallel capacitance; and R_S the resistance in series with L_S, as shown in Fig. 10-10a.

Equation (10-15) differs from Eq. (10-5a) by the term $\sqrt{1 - C_p R_S^2/L_S}$. If $C_p R_S^2/L_S > 1$, there is no resonant frequency for that circuit. In fact, in most practical applications $1 > C_p R_S^2/L_S$, at least by a factor of 10 for a narrowband tuned circuit, thereby reducing Eq. (10-15) to

$$\omega_r = \frac{1}{\sqrt{L_S C_p}} \tag{10-16a}$$

or

$$f_r = \frac{1}{2\pi \sqrt{L_S C_p}} \tag{10-16b}$$

where L_S is the value of the inductor and C_p is the total parallel capacitor. Equations (10-16a) and (10-16b) show that it is not necessary to calculate the equivalent parallel inductor. In fact, we will show that if the Q of the coil is greater than 10, $L_p \approx L_S$.

Middle Resonant Frequency. For the ideal parallel-tuned circuit the frequency at which $B_C = B_L$ is the same frequency at which the total impedance is a maximum. For the practical-tuned circuit, however, the resistance is also frequency dependent. Therefore, to obtain the frequency at which the impedance is a maximum (or admittance is a minimum), take the derivative of Eq. (10-14) with respect to ω and set the expression equal to zero. The frequency obtained equals

$$\omega_{r2} = \frac{1}{\sqrt{L_S C_p}} \sqrt{1 - \frac{C_p R_S^2}{4L_S}} \tag{10-17}$$

For narrowband tuned circuits $1 > C_p R_s^2 / 4L_s$, at least by a factor of 10, thereby reducing Eq. (10-17) to Eq. (10-16).

Highest Resonant Frequency. Although there are, theoretically, three resonant frequencies for the practical-parallel tuned circuit, Eq. (10-16) is applicable to most applications. This is because we concern ourselves primarily with narrowband tuned circuits—$BW \leq 1/10\ \omega_r$, or, to express it another way, $Q > 10$.

For the narrowband tuned circuit the bandwidth for the practical and theoretical circuits is given by Eq. (10-5):

$$BW = \frac{1}{R_p C_p}\ \text{rad/s}$$

or

$$BW = \frac{1}{2\pi R_p C_p}\ \text{Hz}$$

Example 10-9: For the circuit of Fig. 10-11, calculate (a) the highest angular resonant frequency and (b) the bandwidth.

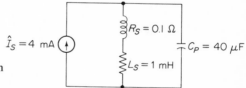

Fig. 10-11. Circuit diagram for Ex. 10-9.

Solution: (a)   Using Eq. (10-16a),

$$\omega_r = \frac{1}{\sqrt{L_s C_p}} = \frac{1}{\sqrt{(1\ \text{mH})(40\ \mu\text{F})}} = 5\ \text{krad/s}$$

(b)   According to Eq. (10-5a),

$$BW = \frac{1}{R_p C_p}$$

where

$$R_p = R_{eq} = \frac{R_s^2 + (\omega_r L_s)^2}{R_s}$$

$$R_p = R_{eq} = \frac{(0.1\ \Omega)^2 + [(5\ \text{krad/s})(1\ \text{mH})]^2}{0.1\ \Omega}$$

$$R_p = 250\ \Omega$$

Therefore,

$$BW = \frac{1}{(250\ \Omega)(40\ \mu\text{F})} = 100\ \text{rad/s}$$

Note that the series resistor, R_s, has to be transformed to the parallel resistor, R_p, to calculate bandwidth.   For narrowband tuned circuits, Eq. (10-16a) is used instead of Eq. (10-15) or Eq. (10-17).   Figure 10-11 is a narrowband tuned circuit because BW $\leq 0.1\omega_r$.

Example 10-10: For the circuit of Fig. 10-12, calculate (a) the highest angular resonant frequency and (b) the bandwidth.

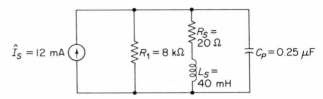

Fig. 10-12. Circuit diagram for Ex. 10-10.

Solution: (a)   Using Eq. (10-16a),

$$\omega_r = \frac{1}{\sqrt{L_s C_p}} = \frac{1}{\sqrt{(40 \text{ mH})(0.25 \ \mu\text{F})}} = 10 \text{ krad/s}$$

(b)   According to Eq. (10-5a),

$$\text{BW} = \frac{1}{R_p C_p}$$

where R_p is the total parallel resistance.

$$R_{eq} = \frac{R_s^2 + (\omega_r L_s)^2}{R_s}$$

$$R_{eq} = \frac{(20 \ \Omega)^2 + [(10 \text{ krad/s})(40 \text{ mH})]^2}{20 \ \Omega} = 8 \text{ k}\Omega$$

Now

$$R_p = R_1 \parallel R_{eq}$$

Since

$$R_1 = R_{eq} = 8 \text{ k}\Omega$$

then

$$R_p = \frac{8 \text{ k}\Omega}{2} = 4 \text{ k}\Omega$$

and

$$\text{BW} = \frac{1}{(4 \text{ k}\Omega)(0.25 \ \mu\text{F})} = 1 \text{ krad/s}$$

Resistance, R_1, may be either the internal resistance of a practical current source or some other external resistance in the circuit.   In either case, however, it must be considered in calculating the total parallel resistance of the circuit because it affects the bandwidth.

In Section 10-1 we discussed and defined (Eq. 10-7) quality factor, Q. It was noted that either a component or a circuit may have energy-storing abilities and losses. At the time, however, we dealt only with the quality factor of a circuit. The circuit of Fig. 10-1a, used as an example, contained both a pure inductor and a pure capacitor and there are no losses associated with these components. Since a capacitor for all practical purposes has negligible internal losses, the Q of this element is very large. An inductor, on the other hand, has internal losses in addition to its energy-storing abilities. Therefore, a Q of a coil is obtainable. Losses of an inductor may be measured by either an impedance bridge or a Q meter. This measurement can represent the inductor as either a parallel combination of resistor and inductor, $R_p - L_p$, or a series combination of resistor and inductor, $R_s - L_s$.

If the bridge reads a *parallel combination*, the recorded measurements are Q_p, ω_r, and L_{eq}. From Eq. (10-7),

$$Q = \frac{\text{reactive power}}{\text{real power}}$$

Then

$$Q_p = \frac{V^2/X_L}{V^2/R_{eq}} = \frac{R_{eq}}{\omega_r L_{eq}} \tag{10-18}$$

where R_{eq} corresponds to the parallel equivalent resistor in the circuit of Fig. 10-10b, Q_p is the quality factor of the coil for the parallel combination, ω_r is the angular resonant frequency at which the measurement is being made, and L_{eq} corresponds to the parallel inductance in the circuit of Figure 10-10b.

If the bridge reads a *series combination*, the recorded measurements are Q_s, ω_r, and L_s. From Eq. (10-7),

$$Q = \frac{\text{reactive power}}{\text{real power}}$$

Then

$$Q_s = \frac{I^2 X_{L_s}}{I^2 R_s} = \frac{\omega_r L_s}{R_s} \tag{10-19}$$

where R_s is the series resistor as shown in Fig. 10-10a, ω_r the angular resonant frequency, L_s the series inductor as shown in Fig. 10-10a, and Q_s, the quality factor of the coil for the series combination. For coils with quality factors greater than 10,

$$Q_p \approx Q_s \tag{10-20}$$

Therefore, Eqs. (10-12) and (10-13) reduce to

$$R_{eq} \approx Q^2 R_s \qquad (10\text{-}21)$$

and

$$L_{eq} \approx L_s \qquad (10\text{-}22)$$

It is not only possible, but very common, for the Q of the circuit to be different from the Q of the coil. (Remember that the Q of a capacitance has been disregarded because of its large value.) In this situation, another symbol must be used to differentiate the two. By convention Q_L is called the *circuit Q* or the *loaded Q* to signify external *loading* by the circuit while Q_p and Q_s represent the parallel and series Q's of the coil. For a parallel circuit, the expression for the loaded Q is the same as Eq. (10-8):

$$Q_L = \frac{R_p}{\omega_r L_p}$$

where Q_L is the quality factor of the circuit, R_p the total parallel resistance in ohms, ω_r the angular resonant frequency of the circuit in radians per second, and L_p the total parallel inductance in henries.

Example 10-11: Calculate the Q of the coil and the loaded Q for the circuits of (a) Fig. 10-11 and (b) Fig. 10-12.

Solution: (a) For Fig. 10-11, the Q of the coil is

$$Q_s = \frac{\omega_r L_s}{R_s} = \frac{(5 \text{ krad/s})(1 \text{ mH})}{0.1 \ \Omega} = 50$$

The loaded Q is

$$Q_L = \frac{R_p}{\omega_r L_p}$$

where

$$L_p \approx L_s = 1 \text{ mH}$$

from Example 10-9,

$$R_p = R_{eq} = 250 \ \Omega$$

Therefore,

$$Q_L = \frac{250 \ \Omega}{(5 \text{ krad/s})(1 \text{ mH})} = 50$$

(b) For Fig. 10-12, the Q of the coil is

$$Q_s = \frac{\omega_r L_s}{R_s} = \frac{(10 \text{ krad/s})(40 \text{ mH})}{20 \ \Omega} = 20$$

The loaded Q is

$$Q_L = \frac{R_p}{\omega_r L_p}$$

where

$$L_p \approx L_s = 40\,\text{mH}$$

From Example 10-10,

$$R_p = R_1 \parallel R_{eq} = \frac{(8\,\text{k}\Omega)(8\,\text{k}\Omega)}{8\,\text{k}\Omega + 8\,\text{k}\Omega} = 4\,\text{k}\Omega$$

Therefore,

$$Q_L = \frac{4\,\text{k}\Omega}{(10\,\text{krad/s})(40\,\text{mH})} = 10$$

Thus in the circuit of Fig. 10-11, $Q_s = Q_L$ because the only resistance in the circuit is the resistance of the coil, R_s. In the circuit of Fig. 10-12, resistance R_1 affects only Q_L and not Q_s.

Example 10-12: A Q meter is used to take measurements on an inductor. The measuring frequency is 200 kHz. The value of the inductance is 5 mH and a Q of 100. Calculate the losses of the coil and whether the value represents R_{eq} or R_s.

Solution: Assume that 5 mH represents L_p and $Q_p = 100$. Then, rearranging Eq. (10-18),

$$R_{eq} = Q_p \omega_r L_q$$

$$R_{eq} = (100)(6.28)(200\,\text{kHz})(5\,\text{mH}) = 628\,\text{k}\Omega$$

Now let us assume that 5 mH represents L_s and $Q_s = 100$. Then, rearranging Eq. (10-19),

$$R_s = \frac{\omega_r L_s}{Q_s}$$

$$R_s = \frac{(6.28)(200\,\text{krad/s})(5\,\text{mH})}{100} = 62.8\,\Omega$$

Since $Q > 10$, R_s may be transformed into R_{eq} by using Eq. (10-21):

$$R_{eq} = Q_s^2 R_s = (100)^2(62.8\,\Omega) = 628\,\text{k}\Omega$$

This example shows that as long as the quality factor is greater than 10, either the series combination or parallel combination yields the same result.

10-3. SERIES-TUNED CIRCUITS

Figure 10-13a shows a series combination of a resistor, inductor, and capacitor. This circuit is called a *series-tuned* or *series-resonant* circuit. The resistance, R, represents either the internal losses of the inductor, the

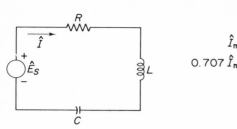

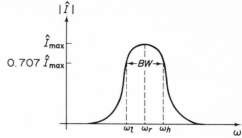

(**a**) Series–tuned circuit   (**b**) Frequency response of current

Fig. 10-13. Diagrams that describe the series-tuned circuit.

internal resistance of a practical voltage source, an external resistance added in series with the inductor and capacitor, or any combination of the three.

Selection and rejection are accomplished in a series-tuned circuit because the complex impedance varies with frequency. The frequency-dependent variables are inductive reactance ($X_L = \omega L$) and capacitive reactance ($X_C = 1/\omega C$). To show the selection and rejection properties, the series-tuned circuit will be analyzed in terms of input impedance, resonant frequency, bandwidth, and quality factor.

Input Impedance.   The equation for complex impedance for the circuit of Fig. 10-13a is

$$\overline{Z}_T = R + jX_L - jX_C \qquad (10\text{-}23)$$

where $\overline{Z}_T$ is the total input impedance, R is the value of resistance, $X_L = \omega L$ and $X_C = 1/\omega C$. Since both X_L and X_C are frequency dependent, the input impedance ($\overline{Z}_T$) varies as the frequency is varied.

At low frequencies (frequencies below resonance), the inductive reactance is small while the capacitive reactance is large. Therefore, $X_L < X_C$ and the circuit of Fig. 10-13a would appear to be a resistor in series with a capacitor.

At high frequencies (frequencies above resonance), the inductive reactance is large while the capacitive reactance is small. Since $X_L > X_C$, the circuit of Fig. 10-13a would appear to be a resistor in series with an inductor.

At the resonant frequency, $X_L = X_C$ or $X_L - X_C = 0$. Therefore, at resonance Eq. (10-23) reduces to

$$\overline{Z}_T = R \qquad (10\text{-}24\text{a})$$

or, in terms of admittance,

$$\overline{Y}_T = \frac{1}{R} \qquad (10\text{-}24\text{b})$$

The small value of input impedance is an advantage of the series-tuned circuit because a large value of current flows. A plot of current versus frequency for a series-tuned circuit is shown in Fig. 10-13b.

Example 10-13: For the circuit of Fig. 10-14, calculate the input impedance for the following angular frequencies.

    (a) $\omega = 5\,\text{krad/s}$,

    (b) $\omega = 10\,\text{krad/s}$,

    (c) $\omega = 20\,\text{krad/s}$.

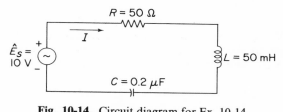

Fig. 10-14. Circuit diagram for Ex. 10-14.

Solution: (a)   At $\omega = 5\,\text{krad/s}$,

$$X_L = \omega L = (5\,\text{krad/s})(50\,\text{mH}) = 250\,\Omega$$

$$X_C = \frac{1}{\omega C} = \frac{1}{(5\,\text{krad/s})(0.2\,\mu\text{F})} = 1000\,\Omega$$

Therefore,

$$\bar{Z}_T = (50 + j250 - j1000)\,\Omega$$

or

$$\bar{Z}_T = (50 - j750)\,\Omega$$

    (b)   At $\omega = 10\,\text{krad/s}$,

$$X_L = \omega L = (10\,\text{krad/s})(50\,\text{mH}) = 500\,\Omega$$

$$X_C = \frac{1}{\omega C} = \frac{1}{(10\,\text{krad/s})(0.2\,\mu\text{F})} = 500\,\Omega$$

Therefore,

$$\bar{Z}_T = (50 + j500 - j500)\,\Omega$$

or

$$\bar{Z}_T = 50\,\Omega$$

    (c)   At $\omega = 20\,\text{krad/s}$,

$$X_L = \omega L = (20\,\text{krad/s})(50\,\text{mH}) = 1000\,\Omega$$

$$X_C = \frac{1}{\omega C} = \frac{1}{(20\,\text{krad/s})(0.2\,\mu\text{F})} = 250\,\Omega$$

Therefore,

$$\bar{Z}_T = (50 + j1000 - j250)\,\Omega$$

or

$$\bar{Z}_T = (50 + j750)\,\Omega$$

Note that in part (b) of this example $X_L = X_C = 500\,\Omega$. Therefore, 10 krad/s is the angular resonant frequency. At frequencies below resonance (such as 5 krad/s), the value of the total impedance shows that an equivalent circuit is a resistor in series with a capacitor. Part (c) shows that at frequencies above resonance ($\omega = 20$ krad/s), an equivalent circuit is a resistor in series with an inductor.

Resonant Frequency. The resonant frequency for a series-tuned circuit is defined as that frequency at which the input impedance is a minimum in terms of circuit parameters. At resonance

$$X_L = X_C$$

Then

$$\omega_r L = \frac{1}{\omega_r C}$$

Solving for ω_r yields

$$\omega_r = \frac{1}{\sqrt{LC}} \tag{10-25a}$$

or

$$f_r = \frac{1}{2\pi\sqrt{LC}} \tag{10-25b}$$

where ω_r is the angular frequency in radians per second, f_r the frequency in hertz, L the value of the total series inductance in henries, and C the value of the total series capacitance in farads. Note that Eq. (10-3) for the narrowband parallel-tuned circuit and Eq. (10-25) for the series-tuned circuit are the same.

Example 10-14: For the circuit of Fig. 10-14, calculate the resonant frequency.

Solution: Using Eq. (10-25a),

$$\omega_r = \frac{1}{\sqrt{LC}} = \frac{1}{\sqrt{(50\ \text{mH})(0.2\ \mu\text{F})}} = 10\ \text{krad/s}$$

or

$$f_r = \frac{\omega_r}{2\pi} = \frac{10\ \text{krad/s}}{6.28} = 1.59\ \text{kHz}$$

Checking with part (b) of Example 10-13, note that the value of $\omega_r = 10$ krad/s is the angular frequency that resulted in $X_L = X_C$ and therefore minimum impedance.

Example 10-15:   For the circuit of Fig. 10-15, calculate the resonant frequency.

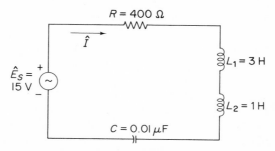

Fig. 10-15. Circuit diagram for Ex. 10-15.

Solution: To calculate the resonant frequency the total series inductance is needed.   Therefore,

$$L_T = L_1 + L_2 = 3\text{ H} + 1\text{ H} = 4\text{ H}$$

Then

$$\omega_r = \frac{1}{\sqrt{L_T C}} = \frac{1}{\sqrt{(4\text{ H})(0.01\text{ }\mu\text{F})}} = 5\text{ krad/s}$$

or

$$f_r = \frac{\omega_r}{2\pi} = \frac{5\text{ krad/s}}{6.28} = 796\text{ Hz}$$

Equation (10-25) is applicable only after the total series inductance and total series capacitance are calculated.   Note that the resonant frequency for a series-tuned circuit is independent of the resistor and the voltage source.

Bandwidth.   The bandwidth, for the series-tuned circuit, as with the parallel-tuned circuit, is the band of frequencies between the upper and lower cutoff frequencies:

$$\text{BW} = \omega_h - \omega_l \qquad \text{or} \qquad \text{BW} = f_h - f_l$$

In terms of circuit parameters, the bandwidth is expressed as

$$\text{BW} = \frac{R}{L} \quad \text{rad/s} \tag{10-26a}$$

or

$$\text{BW} = \frac{R}{2\pi L} \quad \text{Hz} \tag{10-26b}$$

where BW is the bandwidth, R the series resistor value in ohms, and L the inductor value in henries.

For a narrowband series-tuned circuit the bandwidth is symmetrical about the resonant frequency and therefore the lower and upper cutoff frequencies may be expressed as given by Eq. (10-6):

$$\omega_l = \omega_r - \frac{BW}{2}$$

$$\omega_h = \omega_r + \frac{BW}{2}$$

where ω_l and ω_h are the lower and upper cutoff frequencies in radians per second, respectively, ω_r is the angular resonant frequency in radians per second, and BW is the bandwidth in radians per second and determined by Eq. (10-26a).

> **Example 10-16:** Calculate the bandwidth for the circuit of Figs. 10-14 and 10-15.
>
> *Solution:* For the circuit of Fig. 10-14,
>
> $$BW = \frac{R}{L} = \frac{50\,\Omega}{50\,\text{mH}} = 1\,\text{krad/s}$$
>
> In terms of hertz,
>
> $$BW = \frac{1\,\text{krad/s}}{6.28} = 159\,\text{Hz}$$
>
> For the circuit of Fig. 10-15,
>
> $$BW = \frac{R}{L} = \frac{400\,\Omega}{4\,\text{H}} = 100\,\text{rad/s}$$
>
> In terms of hertz,
>
> $$BW = \frac{100\,\text{rad/s}}{6.28} = 15.9\,\text{Hz}$$

Note that each of these circuits is a narrowband tuned circuit — BW $\leq 1/10\,\omega_r$. To determine the bandwidth for the circuit of Fig. 10-15 the total series inductance first had to be calculated. Unlike the narrowband parallel-tuned circuit, the bandwidth of the narrowband series-tuned circuit depends only on resistance and inductance and is independent of capacitance and the voltage source.

> **Example 10-17:** The circuit of Fig. 10-16 is to be tuned to an angular resonant frequency of 200 krad/s and have a bandwidth of 10 krad/s. Calculate:

(a) the capacitance to give the resonant frequency,

(b) the resistance to give the desired bandwidth,

(c) the lower and upper cutoff frequencies.

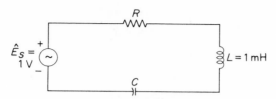

Fig. 10-16. Circuit diagram for Ex. 10-17.

Solution: (a) Rearranging Eq. (10-25a),

$$C = \frac{1}{\omega_r^2 L} = \frac{1}{(200 \text{ krad/s})^2 (1 \text{ mH})} = 0.025 \ \mu\text{F}$$

(b) Rearranging Eq. (10-26a),

$$R = (\text{BW})(L) = (10 \text{ krad/s})(1 \text{ mH}) = 10 \ \Omega$$

(c) Using Eq. (10-6),

$$\omega_l = \omega_r - \frac{\text{BW}}{2} = 200 \text{ krad/s} - \frac{10 \text{ krad/s}}{2} = 195 \text{ krad/s}$$

and

$$\omega_h = \omega_r + \frac{\text{BW}}{2} = 200 \text{ krad/s} + \frac{10 \text{ krad/s}}{2} = 205 \text{ krad/s}$$

This example shows how a series-tuned circuit may be designed to meet specific requirements.

Quality Factor. The quality factor *(Q)* for a series-tuned circuit at resonance is defined the same as that for a parallel-tuned circuit at resonance. That is, at resonance,

$$Q = \frac{\text{reactive power}}{\text{real power}}$$

where reactive power is the power delivered to the inductor or capacitor and real power is the power delivered to the resistor.

For the series-tuned circuit of Fig. 10-13a the value of *Q* expressed in terms of *R* and *L* is

$$Q = \frac{\text{reactive power}}{\text{real power}} = \frac{\hat{I}^2 X_L}{\hat{I}^2 R}$$

Then

$$Q = \frac{X_L}{R} = \frac{\omega_r L}{R} \qquad (10\text{-}27)$$

where R is the total series resistance in ohms, L the total series inductance in henries, and ω_r the resonant angular frequency in radians per second. Note that Eqs. (10-27) and (10-19) are the same.

The quality factor also may be expressed in terms of R and C:

$$Q = \frac{\text{reactive power}}{\text{real power}} = \frac{I^2 X_C}{I^2 R}$$

Then

$$Q = \frac{X_C}{R} = \frac{1}{\omega_r R C} \qquad (10\text{-}28)$$

where R is the total series resistance in ohms, C the total series capacitance in farads, and ω_r the resonant angular frequency in radians per second.

The value of Q for a series-tuned circuit also may be calculated from the ratio of resonant frequency to bandwidth:

$$\frac{\text{resonant frequency}}{\text{bandwidth}} = \frac{\omega_r}{\text{BW}} = \frac{\omega_r}{R/L} = \frac{\omega_r L}{R} = Q \qquad (10\text{-}29)$$

where ω_r, L, and R are the same values as those of Eq. (10-27). Equations (10-27), (10-28), and (10-29) all yield the same value of Q for a particular circuit.

Rearranging Eq. (10-29), the bandwidth may be expressed in terms of Q:

$$\text{BW} = \frac{\omega_r}{Q}$$

Thus as the value of Q increases, the bandwidth decreases and the circuit is said to be more selective.

Example 10-18: For the circuit of Fig. 10-14, calculate the value of Q.

Solution: From Eq. (10-27),

$$Q = \frac{\omega_r L}{R} = \frac{(10 \text{ krad/s})(50 \text{ mH})}{50 \text{ }\Omega} = 10$$

Using Eq. (10-29) as a check yields

$$Q = \frac{\omega_r}{\text{BW}} = \frac{10 \text{ krad/s}}{1 \text{ krad/s}} = 10$$

The value of the bandwidth was calculated in Example 10-16. The circuit is a narrowband tuned circuit since $Q \geq 10$.

Example 10-19: For the circuit of Fig. 10-15, calculate the value of Q.

Solution: From Eq. (10-27),

$$Q = \frac{\omega_r L}{R} = \frac{(5 \text{ krad/s})(4 \text{ H})}{400 \text{ }\Omega} = 50$$

Using Eq. (10-29) as a check yields

$$Q = \frac{\omega_r}{BW} = \frac{5\,\text{krad/s}}{100\,\text{krad/s}} = 50$$

Since the Q of the circuit of Fig. 10-15 is greater than the Q of the circuit of Fig. 10-14, it is more selective.

Plotting the Resonant Curve.   The frequency response of current for a narrowband series-tuned circuit may be plotted in the same manner as the frequency response of voltage for a narrowband parallel-tuned circuit. That is, the exact response curve as shown in Fig. 10-17a may be approximated by straight lines as shown in Fig. 10-17b.

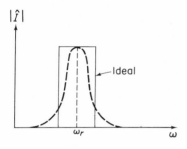

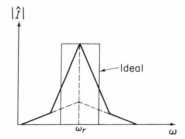

(a) Exact plot compared to the ideal response

(b) Straight line approximation compared to the ideal response

Fig. 10-17. Plots of the resonant curve of current for a narrowband series-tuned circuit.

The following example shows that the procedure developed for plotting the resonant curve of voltage for the narrowband parallel-tuned circuit is applicable to plotting the resonant curve of current for the narrowband series-tuned circuit.

Example 10-20:   For the circuit of Fig. 10-18, draw the resonant curve using the straight-line approximation.

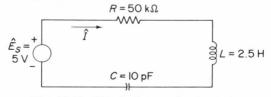

Fig. 10-18. Circuit diagram for Ex. 10-20.

Solution: 1.

$$\omega_r = \frac{1}{\sqrt{LC}} = \frac{1}{\sqrt{(2.5\,\text{H})(10\,\text{pF})}} = 200\,\text{krad/s}$$

At ω_r the current is a maximum for a series-tuned circuit:

$$|\hat{I}_{\text{max}}| = \frac{\hat{E}_s}{R} = \frac{5\,\text{V}}{50\,\text{k}\Omega} = 100\,\mu\text{A}$$

2. Figure 10-19a shows the horizontal and vertical axes scaled. Figure 10-19b shows the intersection of ω_r and $|\hat{I}_{\text{max}}|$.

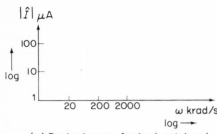

(**a**) Desired range for horizontal and vertical axes

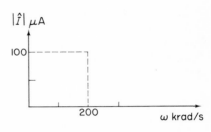

(**b**) Intersection of ω_r and $|\hat{I}_{\text{max}}|$

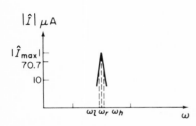

(**c**) Plot of upper portion

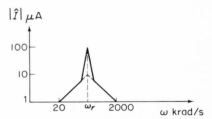

(**d**) Straight line approximation

Fig. 10-19. Diagrams showing procedure for drawing the resonant curve.

3. $$\omega_l = \omega_r - \frac{\text{BW}}{2} \quad \text{and} \quad \omega_h = \omega_r + \frac{\text{BW}}{2}$$

where

$$\text{BW} = \frac{R}{L} = \frac{50\,\text{k}\Omega}{2.5\,\text{H}} = 20\,\text{krad/s}$$

Therefore,

$$\omega_l = 200\,\text{krad/s} - \frac{20\,\text{krad/s}}{2} = 190\,\text{krad/s}$$

and

$$\omega_h = 200\,\text{krad/s} + \frac{20\,\text{krad/s}}{2} = 210\,\text{krad/s}$$

At both of these frequencies

$$|\hat{I}| = (0.707)\,|\hat{I}_{\text{max}}|$$

Therefore,

$$|\hat{I}| = (0.707)(100\,\mu\text{A}) = 70.7\,\mu\text{A}$$

Figure 10-19c shows the top portion of the resonant curve constructed by straight lines.

4. The lower portion of the resonant curve intersects at ω_r and

$$|\hat{I}| = \frac{\hat{E}_s}{X_L} \quad \text{or} \quad |\hat{I}| = \frac{\hat{E}_s}{X_C}$$

$$|\hat{I}| = \frac{5\ \text{V}}{(200\ \text{krad/s})(2.5\ \text{H})} = 10\ \mu\text{A}$$

Then at frequencies of $\frac{1}{10}\omega_r$ (20 krad/s) and $10\omega_r$ (2000 krad/s) the $|\hat{I}|$ is $\frac{1}{10}$ (10 μA) = 1 μA. Figure 10-19d shows the straight-line approximation of the resonant curve.

10-4. FILTERS

A filter is a network designed to pass a specified band of frequencies while attenuating all signals outside this band. The tuned circuits previously discussed are one class of filters — *bandpass filters.*

Filters are either active or passive. An *active filter* is one that includes an amplifying device. Modern active filters are usually designed with operational amplifiers and *R-C* circuits. In this type of filter inductors are not used because of their size, weight, and losses. Capacitors, on the other hand, more closely approach an ideal element.

Passive filters contain resistors, capacitors, and/or inductors. They do not contain an amplifying device. Three types of passive filters are

1. *R-C* filters, designed only with resistors and capacitors,
2. *R-L-C* filters, designed with resistors, inductors, and capacitors,
3. Crystal filters, designed with resistors, inductors, capacitors, and a piezoelectric resonant element.

Piezoelectricity is an electric phenomenon resulting from pressure upon the material. Quartz crystal is one type of material used most often because of its stability and its extremely high value of Q ($Q > 1000$), a value that

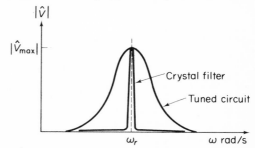

Fig. 11-20. Comparison of the best tuned circuit with that of a crystal filter.

is not obtainable with discrete components. Figure 10-20 shows a comparison between the best *R-L-C* tuned circuit and a crystal filter.

Filters fall into one of the following four categories.

Low-pass Filters. A low-pass filter is a circuit that has a constant output voltage up to a critical frequency, ω_c. As the frequency increases above ω_c, the output voltage is attenuated. This frequency (ω_c) is referred to as the *corner*, or *break*, *frequency*.

The solid line in Fig. 10-21 shows the frequency response for an ideal low-pass filter. The dotted lines in Fig. 10-21 show the frequency response

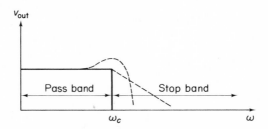

Fig. 10-21. Frequency response of a low-pass filter.

for practical low-pass filters. The range of frequencies that is *transmitted* is known as the *pass band*. The range of frequencies that is *attenuated* is referred to as the *stop band*.

High-pass Filters. A high-pass filter is a circuit that attenuates the output voltage for all frequencies below a critical frequency, ω_c. Above ω_c the

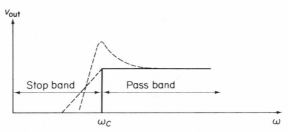

Fig. 10-22. Frequency response of a high-pass filter.

voltage output is constant. The solid line in Fig. 10-22 shows the frequency response for an ideal high-pass filter. The dotted lines show the frequency response for practical high-pass filters.

Bandpass Filters. The purpose of a bandpass filter is to select (pass) a band of frequencies with little or no attenuation while rejecting all other frequencies outside this band. Thus the purpose of the bandpass filter is the same as that of a narrowband tuned circuit. In general, however, one may describe all narrowband tuned circuits as bandpass filters, but

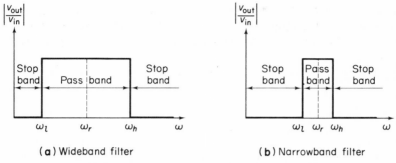

(a) Wideband filter   (b) Narrowband filter

Fig. 10-23. Ideal frequency response of bandpass filters.

not all bandpass filters are narrowband tuned circuits. Bandpass filters may be either wideband or narrowband. Fig. 10-23 shows the ideal frequency response of both types.

Band-elimination Filters. The purpose of a band elimination filter is to reject a band of frequencies while passing all other frequencies outside this band with little or no attenuation. Thus, the band-elimination performs the opposite function of that of the bandpass filter. Band-elimination filters may be either wideband or narrowband. Fig. 10-24 shows the ideal frequency response of both.

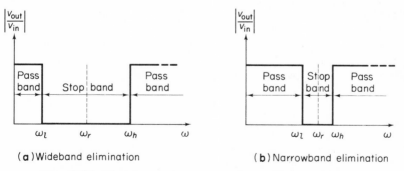

(a) Wideband elimination   (b) Narrowband elimination

Fig. 10-24. Ideal frequency response of band elimination filters.

From the previous discussion, the reader may realize that filter networks may become complex and difficult to analyze. Since we do not wish to become deeply involved with filter design in this book, we shall consider only the low-pass R-C and the high-pass R-C filter. These have been chosen because (1) they are a basic building block of filter networks, and

(2) at low frequencies and at high frequencies the input circuit of a transistor is represented by such networks.

Low-pass R-C Filter.   The circuit of Fig. 10-25a is a low-pass filter.   If the frequency of the input voltage is below the break frequency, the capacitive reactance is large and the capacitor for an ideal filter appears to be an

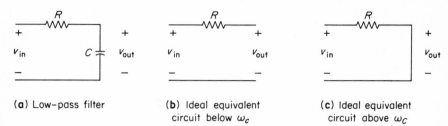

(a) Low-pass filter   (b) Ideal equivalent   (c) Ideal equivalent
                                  circuit below ω_c circuit above ω_c

Fig. 10-25. Circuit diagrams for an ideal low-pass filter.

open circuit, as shown in Fig. 10-25b.   Thus the output voltage is approximately equal to the input voltage.   As the frequency is increased, the capacitive reactance decreases, causing the output voltage to decrease. Figure 10-25c shows the ideal equivalent circuit at all frequencies above the break frequency.   To analyze the circuit of Fig. 10-25a mathematically, apply the voltage division law to obtain v_{out}:

$$v_{out} = \frac{1/j\omega C}{R + 1/j\omega C} \times v_{in} \qquad (10\text{-}30)$$

Dividing both sides of Eq. (10-30) by v_{in}, we have what is known as the *voltage transfer function:*

$$\frac{v_{out}}{v_{in}} = \frac{1/j\omega C}{R + 1/j\omega C} \qquad (10\text{-}31)$$

Rearranging the terms of Eq. (10-31) yields

$$\frac{v_{out}}{v_{in}} = \frac{1}{1 + j\omega RC} \qquad (10\text{-}32)$$

or

$$\frac{v_{out}}{v_{in}} = \frac{1}{1 + j\omega\tau} \qquad (10\text{-}33)$$

where $\tau = RC$, the time constant of the circuit.   Equation (10-33) is a general expression for circuits similar to that of Fig. 10-25a.

Let us consider how the magnitude of the transfer function changes as the frequency is varied.   At low frequencies — $\omega \ll 1$ —

$$\left| \frac{v_{out}}{v_{in}} \right| = 1 \qquad (10\text{-}34)$$

Therefore, a plot of the transfer function at these low frequencies is a horizontal line — a line with a slope of $0°$.

At high frequencies — $\omega \gg 1$ —

$$\left|\frac{v_{out}}{v_{in}}\right| = \left|\frac{1}{\omega\tau}\right| \tag{10-35}$$

From this expression we see that as the frequency increases, the magnitude of the transfer function decreases. Plotting Eq. (10-35) on log-log graph paper results in the following: If ω increases by a decade, $|v_{out}/v_{in}|$ decreases by a decade. This plot is a line with a slope of $-45°$.

The frequency at which the slope changes from $0°$ to $-45°$ is the reciprocal of the circuit's time constant. Thus the break frequency is

$$\omega_c = \frac{1}{\tau} = \frac{1}{RC} \tag{10-36}$$

A plot of the magnitude of the voltage transfer function on log-log graph paper is shown in Fig. 10-26a.

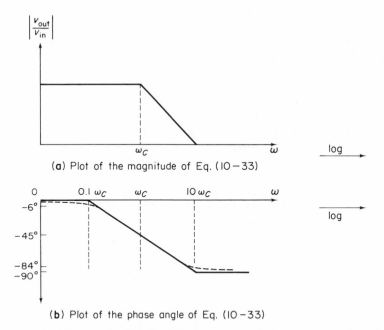

(a) Plot of the magnitude of Eq. (10-33)

(b) Plot of the phase angle of Eq. (10-33)

Fig. 10-26. Plots of the magnitude and phase angle for a low-pass filter.

An exact plot of the transfer function differs from the straight-line approximation in the vicinity of the break frequency. Substituting Eq. (10-36) into Eq. (10-33), we obtain the actual magnitude at the break frequency:

$$\left|\frac{v_{out}}{v_{in}}\right| = \left|\frac{1}{1 + j(1/\tau)\tau}\right| = \left|\frac{1}{1 + j1}\right| = \frac{1}{\sqrt{2}} = 0.707$$

For greater accuracy of $|1/1 + j\omega\tau|$ in the range of a decade above and below the break frequency, Table 10-1 may be used.

TABLE 10-1

FOR GREATER ACCURACY OF $\left|\dfrac{1}{1 + j\omega\tau}\right|$

ω	$\left|v_{out}/v_{in}\right|$
$\frac{1}{10}\,\omega_c$	1.0
$\frac{1}{4}\,\omega_c$	0.97
$\frac{1}{2}\,\omega_c$	0.89
ω_c	0.707
$2\omega_c$	0.445
$4\omega_c$	0.25
$10\omega_c$	0.1

As frequency varies, both the magnitude and the phase angle vary. A plot of the phase angle versus frequency for the transfer function of Eq. (10-33) is shown in Fig. 10-26b. At zero frequency the phase angle is $0°$, at the break frequency the phase angle is $-45°$, and at infinite frequency the phase angle is $-90°$. The plot of the phase angle is symmetrical about the break frequency when the frequency axis is a log scale. A smooth curve drawn through the points given in Table 10-2 results in a fairly accurate plot of the phase angle.

TABLE 10-2

VALUES OF PHASE ANGLE
FOR A LOW-PASS FILTER

ω	Phase angle
$\frac{1}{10}\,\omega_c$	$-5.7°$
$\frac{1}{2}\,\omega_c$	$-26.6°$
ω_c	$-45°$
$2\omega_c$	$-63.4°$
$10\omega_c$	$-84.3°$

As shown in Fig. 10-26b, the phase-angle curve is approximately linear within a decade above and below the break frequency.

Example 10-21: For the circuit of Fig. 10-27 plot the magnitude of the voltage transfer function and the phase-angle characteristics.

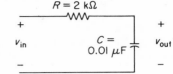

Fig. 10-27. Circuit diagram for Ex. 10-21.

Solution: According to Eq. (10-33),

$$\frac{v_{out}}{v_{in}} = \frac{1}{1 + j\omega\tau}$$

where $\tau = RC = (2\ k\Omega)(0.01\ \mu F) = 0.02$ ms. From Eq. (10-36),

$$\omega_c = \frac{1}{\tau} = \frac{1}{0.02\ ms} = 50\ krad/s$$

Therefore, at frequencies below 50 krad/s, $|v_{out}/v_{in}| = 1$; above 50 krad/s the $|v_{out}/v_{in}|$ has a slope of $-45°$. The magnitude of the voltage transfer function is shown in Fig. 10-28a. Using Table 10-1 a more exact plot is

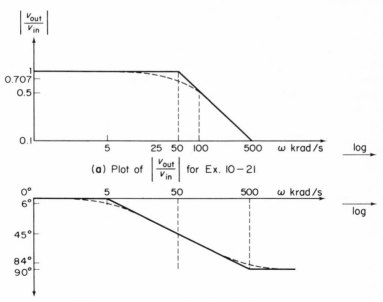

(a) Plot of $\left|\dfrac{v_{out}}{v_{in}}\right|$ for Ex. 10-21

(b) Plot of phase angle for Ex. 10-21

Fig. 10-28. Diagrams for Ex. 10-21.

obtained. The phase-angle curve may be plotted by using Table 10-2. This plot is shown in Fig. 10-28b.

This example shows the ease with which the voltage transfer function is plotted for a low-pass filter using the straight-line approximation and Tables 10-1 and 10-2.

Example 10-22: Again consider the circuit of Fig. 10-27 and calculate the break frequency if

(a) $R = 2 \text{ k}\Omega$ and $C = 0.02 \ \mu\text{F}$,

(b) $R = 1 \text{ k}\Omega$ and $C = 0.01 \ \mu\text{F}$,

(c) $R = 4 \text{ k}\Omega$ and $C = 0.01 \ \mu\text{F}$,

(d) $R = 1 \text{ k}\Omega$ and $C = 0.02 \ \mu\text{F}$.

Solution:

(a) $\omega_c = \dfrac{1}{RC} = \dfrac{1}{(2 \text{ k}\Omega)(0.02 \ \mu\text{F})} = 25 \text{ krad/s}$

(b) $\omega_c = \dfrac{1}{RC} = \dfrac{1}{(1 \text{ k}\Omega)(0.01 \ \mu\text{F})} = 100 \text{ krad/s}$

(c) $\omega_c = \dfrac{1}{RC} = \dfrac{1}{(4 \text{ k}\Omega)(0.01 \ \mu\text{F})} = 25 \text{ krad/s}$

(d) $\omega_c = \dfrac{1}{RC} = \dfrac{1}{(1 \text{ k}\Omega)(0.02 \ \mu\text{F})} = 50 \text{ krad/s}$

This example shows that if R and C are varied, only the break frequency of the voltage transfer function varies. Part (a) shows that if C increases, ω_c decreases. Part (b) shows that if R decreases, ω_c increases. Comparing the answer of Example 10-21 and the answer of part (a) with the answers of parts (c) and (d) shows that there is not a unique solution to obtain a particular break frequency.

High-pass R-C Filter. The circuit of Fig. 10-29a is a high-pass filter. At frequencies below ω_c the capacitive reactance $(X_C = 1/\omega_c)$ is large and the

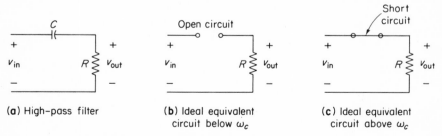

(a) High-pass filter   (b) Ideal equivalent circuit below ω_c (c) Ideal equivalent circuit above ω_c

Fig. 10-29. Circuit diagrams for an ideal high-pass filter.

capacitor for an ideal filter appears to be an open circuit, as shown in Fig. 10-29b. The output voltage, therefore, equals zero. At frequencies above ω_c, the capacitive reactance is small and the capacitor for an ideal filter is a short circuit, as shown in Fig. 10-29c. The output voltage for this circuit equals the input voltage.

To analyze the circuit of Fig. 10-29a mathematically, write the voltage transfer function:

$$\frac{v_{out}}{v_{in}} = \frac{R}{R + 1/j\omega c} \tag{10-37}$$

Rearranging terms we obtain

$$\frac{v_{out}}{v_{in}} = \frac{1}{1 + 1/j\omega\tau} \tag{10-38}$$

where $\tau = RC$, the time constant of the circuit. Equation (10-38) is a general equation for circuits similar to that of Fig. 10-29a.

As with low-pass filters, we are interested in how the magnitude of the transfer function varies as frequency varies. At high frequencies, $\omega \gg 1$,

$$\left|\frac{v_{out}}{v_{in}}\right| = 1 \tag{10-39}$$

At low frequencies, $\omega \ll 1$, Eq. (10-38) becomes

$$\left|\frac{v_{out}}{v_{in}}\right| = |j\omega\tau| \tag{10-40}$$

Equation (10-40) shows that as the frequency is decreased below ω_c, the magnitude of the transfer function is decreased. Plotting Eq. (10-40) on log-log graph paper results in the following: If ω decreases by a decade, the $|v_{out}/v_{in}|$ decreases by a decade. The break frequency is

$$\omega_c = \frac{1}{\tau} = \frac{1}{RC} \tag{10-41}$$

A plot of the magnitude of the voltage transfer function on log-log scales is shown in Fig. 10-30a.

An exact plot of the transfer function differs from the straight-line approximation in the vicinity of the break frequency. Substituting Eq. (10-41) into Eq. (10-38) we obtain the actual magnitude at the break frequency:

$$\left|\frac{v_{out}}{v_{in}}\right| = \left|\frac{1}{1 + \dfrac{1}{j(1/\tau)\tau}}\right| = \left|\frac{1}{1 + j1}\right| = \frac{1}{\sqrt{2}} = 0.707$$

For greater accuracy of $|1/(1 + 1/j\omega\tau)|$ in the range of a decade above and below the break frequency, Table 10-3 may be used.

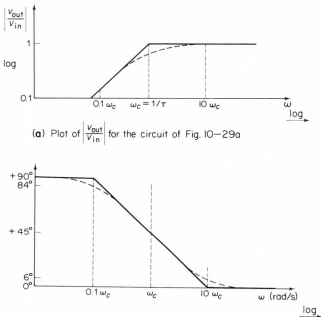

(**a**) Plot of $\left|\dfrac{v_{out}}{v_{in}}\right|$ for the circuit of Fig. 10—29a

(**b**) Plot of phase angle for the circuit of Fig 10—29b

Fig. 10-30. Diagrams for the high-pass filter.

TABLE 10-3

FOR GREATER ACCURACY OF $\left|\dfrac{1}{1 + 1/j\omega\tau}\right|$

ω	$\left|v_{out}/v_{in}\right|$
$\frac{1}{10}\omega_c$	0.1
$\frac{1}{4}\omega_c$	0.25
$\frac{1}{2}\omega_c$	0.445
ω_c	0.707
$2\omega_c$	0.89
$4\omega_c$	0.97
$10\omega_c$	1.0

As frequency varies, both the magnitude and the phase angle vary. At zero frequency the phase angle is $+90°$, at the break frequency the phase angle is $+45°$, and at infinite frequency the phase angle is $0°$. The plot of the phase angle is symmetrical about the break frequency when the frequency axis is a log scale. A smooth curve drawn through the points given in Table 10-4 results in a fairly accurate plot of the phase angle.

TABLE 10-4

**VALUES OF PHASE ANGLE
FOR A HIGH-PASS FILTER**

ω	Phase angle
$\frac{1}{10}\,\omega_c$	84.3°
$\frac{1}{4}\,\omega_c$	63.4°
ω_c	45°
$4\omega_c$	26.6°
$10\omega_c$	5.7°

As shown in Fig. 10-30b, the phase-angle curve is approximately linear within a decade above and below the break frequency.

Example 10-23: For the circuit of Fig. 10-31, plot the magnitude of the voltage transfer function and the phase characteristics.

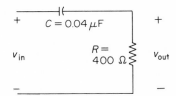

Fig. 10-31. Circuit diagram for Ex. 10-23.

Solution: According to Eq. (10-38),

$$\frac{v_{\text{out}}}{v_{\text{in}}} = \frac{1}{1 + 1/j\omega\tau}$$

where $\tau = RC = (400\,\Omega)(0.04\,\mu F) = 16\,\mu s$. The break frequency is then

$$\omega_c = \frac{1}{\tau} = \frac{1}{16\,\mu s} = 62.5\text{ krad/s}$$

Therefore, at frequencies above 62.5 krad/s the $|v_{\text{out}}/v_{\text{in}}| = 1$; below ω_c the $|v_{\text{out}}/v_{\text{in}}|$ decreases with a slope of 45°.

The magnitude of the voltage transfer function is shown in Fig. 10-32a. A more exact plot is obtained by using Table 10-3, as shown by the dotted curve.

The phase angle is approximately linear from a decade below the break frequency ($0.1\omega_c$) to a decade above the break frequency ($10\omega_c$). The dotted curve (a more exact plot) is obtained by using Table 10-4.

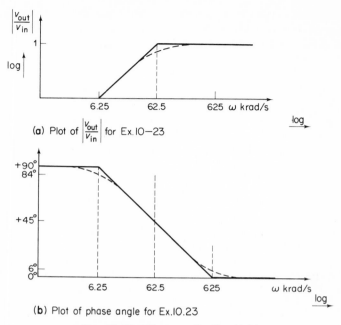

(a) Plot of $\left|\dfrac{V_{out}}{V_{in}}\right|$ for Ex. 10—23

(b) Plot of phase angle for Ex.10.23

Fig. 10-32. Diagrams for Ex. 10-23.

As in the case of the low-pass filter, by varying the values of resistance and/or capacitance of the high-pass filter, only the value of the break frequency (ω_c) is changed.

SUMMARY

The first three sections of this chapter have dealt with tuned circuits, which are designed to pass only a specific band of frequencies while rejecting all other frequencies. We have discussed parallel-tuned circuits in relation to their important properties — input admittance, resonant frequency, bandwidth, and quality factor in Section 10-1.

The second section treated practical parallel resonant circuits and developed the techniques for analyzing them for the same four properties. Section 10-3 was developed along the same lines for a series-tuned circuit. In this type of circuit, selection and rejection are accomplished because the complex impedance varies with frequency.

All the above sections dealt with tuned circuits, a specific type of filter. The last section discussed the four types of filters: low-pass, high-pass, bandpass, and band-elimination filters with emphasis on the low-pass and high-pass R-C filter.

PROBLEMS

10-1. For each of the circuits in Fig. 10-33, calculate
   (a)   resonant frequency,
   (b)   bandwidth.

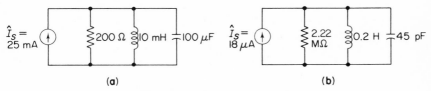

(a)   (b)

Fig. 10-33. Circuit diagrams for Prob. 10-1.

10-2. Are both the circuits of Fig. 10-33 narrowband tuned circuits?   Explain.

10-3. Calculate the Q of both the circuits of Fig. 10-33.   Which circuit is more selective?

10-4. For the circuit of Fig. 10-2, calculate
   (a)   the bandwidth,
   (b)   the quality factor.

10-5. What is the largest bandwidth obtainable from the circuit of Fig. 10-2 which may still be classified as a narrowband tuned circuit?   Calculate the resistance to give this bandwidth.

10-6. If the bandwidth of Example 10-5 is to be doubled but the resonant frequency remains unchanged, calculate the new circuit values.

10-7. Repeat Example 10-5 for an inductor value of 2 mH.

10-8. For both circuits of Fig. 10-33, draw the resonant curve using the straight-line approximation.

10-9. For each of the circuits of Fig. 10-34, calculate
   (a)   resonant frequency,
   (b)   bandwidth,
   (c)   quality factor.

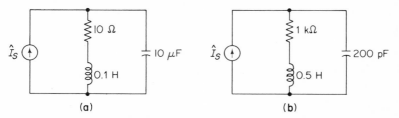

(a)   (b)

Fig. 10-34. Circuit diagrams for Prob. 10-9.

10-10.   Which circuit of Fig. 10-34 is more selective? Explain.

10-11.   For the circuits of Fig. 10-34, is the Q of the coil equal to the Q of the circuit? Explain.

10-12.   If a 2.5 MΩ resistor is connected in parallel with the capacitor in the circuit of Fig. 10-34b, calculate the bandwidth and the Q of the circuit.

10-13.   A Q meter is used to make measurements on an inductor. The measuring frequency is 455 kHz. The value of $C_p = 200$ pF and $Q = 200$. Calculate
(a)   the value of the inductor,
(b)   the losses of the inductor.

10-14.   A 200 μH coil is in parallel with a 50 pF capacitor to form a resonant circuit. The Q of the coil at the resonant frequency is 20. Calculate the frequencies at the half-power points.

10-15.   For each of the circuits of Fig. 10-35, calculate
(a)   resonant frequency,
(b)   bandwidth.

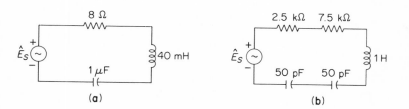

Fig. 10-35. Circuit diagrams for Prob. 10-15.

10-16.   For each of the circuits of Fig. 10-35, calculate ω_l and ω_h.

10-17.   A series-tuned circuit is to be designed to have a resonant frequency of 31.8 kHz and a bandwidth of 2 kHz. If the inductor value is 10 mH, calculate
(a)   the capacitor value,
(b)   the resistor value.

10-18.   If a 10 V source is connected to the circuit of Problem 10-17, calculate the maximum value of current.

10-19.   Calculate the Q of each circuit of Fig. 10-35. Which circuit is more selective?

10-20.   For each circuit of Fig. 10-35, draw the resonant curve of current using the straight-line approximation. $E_S = 1$ V.

10-21. For each low-pass filter of Fig. 10-36, determine
    (a) the break frequency;
    (b) plot the magnitude of the voltage transfer and phase angle characteristics.

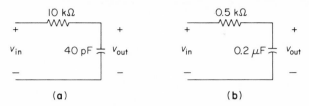

Fig. 10-36. Circuit diagrams for Prob. 10-21.

10-22. If the resistance in the circuit of Fig. 10-36b is changed to 5 kΩ, calculate the new break frequency.

10-23. Let v_{in} in the circuit of Fig. 10-36b be 10 V. Determine v_{out} at
    (a) $0.1\omega_C$,
    (b) ω_C,
    (c) $10\omega_C$.

10-24. For each high-pass filter of Fig. 10-37 determine
    (a) the break frequency,
    (b) plot the magnitude of the voltage transfer and phase angle characteristics.

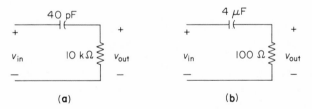

Fig. 10-37. Circuit diagrams for Prob. 10-24.

10-25. If the resistance of the circuit of Fig. 10-37a is changed to 5 kΩ, calculate the new break frequency.

10-26. Let v_{in} in the circuit of Fig. 10-37b be 2 V. Calculate v_{out} at
    (a) $0.1\omega_C$,
    (b) ω_C,
    (c) $10\omega_C$.

CHAPTER 11

transformers

Thus far in our study of networks there has always been an electrical connection between two elements. This is known as *conduction*. The joining or coupling of the networks may be either *direct* or *indirect*. For example, two loops are coupled directly if they share the same element. Two loops are coupled indirectly if they share an element with a common third loop. Consider the network of Fig. 11-1. Loops 1 and 2 share the same element,

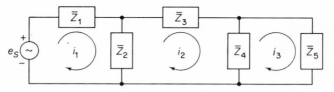

Fig. 11-1. Circuit diagram illustrating the principle of conduction.

$\overline{Z}_2$; loops 2 and 3 share a common element, $\overline{Z}_4$. Therefore, loops 1 and 2 and loops 2 and 3 are directly coupled through conduction. However, loops 1 and 3 do not share a common element but nevertheless do share a common loop. Thus loops 1 and 3 are coupled indirectly through conduction.

In this chapter we shall be concerned with the transfer of energy from one circuit to another through a magnetic field. In network analysis this transfer of energy is represented by an element called *mutual inductance*. If two circuits are coupled through mutual inductance, they are said to be coupled through *induction*. In contrast to conduction, induction does not require either a direct or indirect electrical path, but it does require a magnetic field for energy transfer.

Section 11-1 deals in a general way with the concept of mutual inductance, shows how it is expressed in terms of voltage and current, and explains two important concepts of induction—the coefficient of coupling and the dot notation.

In the second section, discussion centers around the measurement of mutual inductance using an impedance bridge.

The following five sections explain transformers, which are basically two coils connected (or coupled) by a magnetic field. Section 11-3 gives an overview of transformers in general and Section 11-4 deals with ideal transformers. In Section 11-5, a special type of transformer, an autotransformer, with only one coil is explained. Sections 11-6 and 11-7 deal with the efficiency of transformers and tuned transformer circuits, respectively.

11-1. MUTUAL INDUCTANCE

In Chapter 3 it is stated that any current-carrying conductor has a magnetic field associated with it. If the current is ac, the strength of the magnetic field is constantly increasing and decreasing. Likewise, if a changing magnetic field exists about the conductor, an ac voltage is induced in it. (A current flows in the conductor provided a closed path exists.) The magnetic field is increased by winding the conductor in the form of a coil, thus creating the passive element self-inductance.

If two coils are placed so that a changing magnetic field of one induces a voltage in the other, the two coils are connected through mutual induction. The mutual inductance between two coils is directly proportional to the number of magnetic lines of force that are common to both coils. The closer the coils are to one another, the larger is the value of mutual inductance. Mutual inductance is increased further if both coils are wound on a common iron core. If, on the other hand, the coils are separated, the mutual inductance is reduced. Mutual inductance is represented by the symbol M and has the same units as self-inductance, henries.

Mutual inductance may exist between two coils connected in series as shown in Fig. 11-2a or two coils connected in parallel as shown in Fig. 11-2b. It is not necessary, however, to have a direct electrical connection between the coils to have mutual inductance. Figure 11-2c shows such an example.

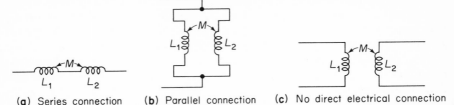

(a) Series connection   (b) Parallel connection   (c) No direct electrical connection

Fig. 11-2. Examples of mutual inductance existing between two coils.

The other passive elements — resistance, self-inductance, and capacitance — have all been expressed in terms of voltage and current. Mutual inductance may also be expressed by such a relationship:

$$v = M \frac{di}{dt} \tag{11-1}$$

where v is the value of induced voltage, M the value of mutual inductance, and di/dt the time rate of change of current. Note that if i is a constant value, $di/dt = 0$ and the induced voltage is zero. Therefore, to induce a voltage in one coil requires that an alternating current exist in the other.

An induced voltage is represented in a circuit diagram by a *dependent voltage source*. The term "dependent voltage source" is used because the voltage induced into one coil is a result of (depends on) current flowing in the other. To illustrate this point, consider the circuit of Fig. 11-3a,

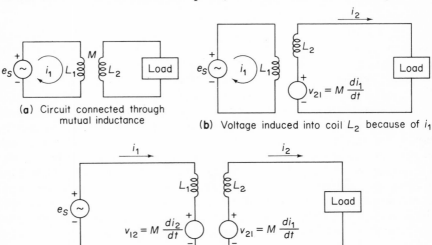

(a) Circuit connected through mutual inductance

(b) Voltage induced into coil L_2 because of i_1

(c) Dependent source associated with both L_1 and L_2

Fig. 11-3. Circuit diagrams illustrating induced voltage represented by dependent sources.

which is connected by mutual inductance. The circuit of Fig. 11-3a shows a current, i_1, flowing through a coil, L_1. Since L_1 and L_2 are connected through mutual inductance, a voltage is induced in coil 2. This induced voltage is shown as v_{21} in Fig. 11-3b. Figure 11-3b also shows that the dependent voltage source causes current i_2 to flow. This current flows through L_2. Therefore, a voltage is induced in coil L_1. Figure 11-3c shows both coils and their respective dependent sources. Later in this section we will show how the polarity markings are chosen for the dependent sources.

Coefficient of Coupling. An expression that relates the mutual inductance to the self-inductances is

$$M = k\sqrt{L_1 L_2} \tag{11-2}$$

where M is the value of the mutual inductance in henries, k is the *coefficient of coupling* and is dimensionless, $|k| \le 1$, and L_1 and L_2 are the values of the self-inductances in henries.

The mutual inductance and thus the coefficient of coupling depend on the core material, the spacing between the coils, and their orientation to each other. If k is approximately equal to 1, the mutual inductance is said to be *closely*, or *tightly*, coupled. If, on the other hand, the value of k is small, the mutual inductance is said to be *loosely coupled*. Maximum coupling is achieved if one coil is wound over the other and the core material has a high permeability. Although the ideal condition, $k = 1$, is never quite possible, it is closely approached if the core material is highly magnetic. Coupling is small when the coils are spaced far apart and/or the coils are placed at right angles to each other.

Dot Notation. At this point we would like to be able to write a set of loop equations for inductively coupled circuits similar to that of conductively coupled circuits in Chapter 5 and 8. With the solution of these equations, all other network relationships can readily be solved. Unlike conductively connected circuits, however, for inductively coupled circuits we must know something about the geometry of the coils — that is, something about the relationship of the windings of coil 1 to that of coil 2. This information is necessary to determine whether an induced voltage will aid or oppose the direction of the loop current. The conventional method of denoting the winding geometry is the *dot notation*. A dot is placed at the ends of the coil showing induced voltages, produced by the mutual flux, which are in phase. The dots have nothing to do with either the direction of the loop current or branch current. The way in which the coils are wound are the only factors that designate the polarity of the induced voltage and the position of the dot. The dotted end of the coils must be known before the loop equations are written.

The circuit of Fig. 11-4 is a test circuit to determine the ends of the coils that are in phase. A dc voltage applied to coil 1 induces a voltage in coil 2 only when the switch is closed or opened. When the switch is closed

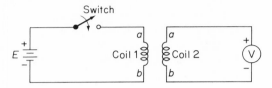

Fig. 11-4. Circuit to test phase windings.

the *a* terminal of coil 1 is driven positive. If at the instant of closure the voltmeter reads upscale, the induced voltage at the *a* terminal for coil 2 must also be positive. Therefore, dots would be placed at *both a* terminals.

If, on the other hand, the *a* terminal of coil 1 is driven positive and the voltmeter (connected to coil 2) reads downscale, the induced voltages at the *a* terminal of coil 1 and the *b* terminal of coil 2 are in phase and the dots would be placed at these terminals. Note that the voltmeter will read either upscale or downscale and then return to zero. Note, too, that we are not interested in the value of voltage but only in the direction of the needle. Therefore, *E* need not be large ($E < 1$ V).

Equation (11-1) is the expression for the induced voltage in terms of mutual inductance and the time rate of change of current. However, the polarity of the induced voltage depends on the location of the dots and the direction of loop currents. Consider the circuit of Fig. 11-5a. To determine the induced voltage in L_2:

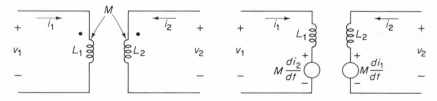

(a) Two circuits connected through mutual inductance

(b) Equivalent circuit showing dependent voltage sources

Fig. 11-5. Circuit diagrams of mutual inductance.

1. Let current i_2 be zero.
2. Note that current i_1 flows into the dotted terminal of L_1. Since the dots designate the ends of the coil that are in phase, the dotted end of L_2 is positive. The value of the dependent source in series with L_2 is

$$v_{21} = M \frac{di_1}{dt}$$

where v_{21} is the voltage in L_2 due to a change in current in L_1, M the mutual inductance, and di_1/dt the time rate of change of current i_1 (see Fig. 11-5b).

To determine the induced voltage in L_1:

1. Let current i_1 be zero.
2. Current i_2 flows into the dotted end of L_2. Therefore, the dotted end of L_1 is driven positive and the value of the induced voltage is

$$v_{12} = M \frac{di_2}{dt}$$

where v_{12} is the voltage induced in L_1 due to a current in L_2, M the value of mutual inductance, and di_2/dt the time rate of change of current i_2 (see Figure 11-5b).

For another example, consider the circuit of Fig. 11-6.

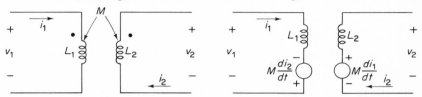

(a) An inductively coupled circuit (b) Equivalent circuit showing dependent voltage sources

Fig. 11-6. Circuit diagram showing mutual inductance.

To determine the induced voltage in L_2:

1. Let current i_2 be zero.
2. Current i_1 flows into the dotted end of L_1. Therefore, the dotted end of L_2 is driven positive as shown in Fig. 11-6b. The value of v_{21} is

$$v_{21} = M \frac{di_1}{dt}$$

To determine the induced voltage in L_1:

1. Let current i_1 be zero.
2. Current i_2 flows into the undotted end of L_2. Then the dotted end of L_1 is driven negative as shown in Fig. 11-6b. The value is

$$v_{12} = -M \frac{di_1}{dt}$$

(The negative sign is accounted for in the circuit of Fig. 11-6b by a potential rise.)

The other two possible circuits containing the location of dots and the direction of current are shown in Fig. 11-7.

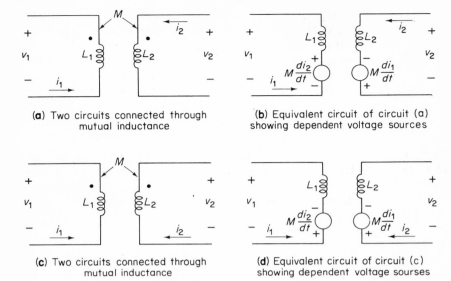

(a) Two circuits connected through mutual inductance

(b) Equivalent circuit of circuit (a) showing dependent voltage sources

(c) Two circuits connected through mutual inductance

(d) Equivalent circuit of circuit (c) showing dependent voltage sourses

Fig. 11-7. Other possible location of dots and directions of currents.

11-2. MEASUREMENT OF MUTUAL INDUCTANCE

The mutual inductance between two coils may be determined by taking two measurements on an impedance bridge. Before these measurements are taken, however, the location of the dots should be known. (If they are not known, use the circuit of Fig. 11-4 to determine their location.)

Connect terminals b and d of Fig. 11-8a together as shown in Fig. 11-8b. Using an impedance bridge, measure the total inductance between terminals a and c. This net value is

$$L_A = L_1 + L_2 - 2M \qquad (11-3)$$

where L_A is the total or net value of inductance between terminals a and c, L_1 the self-inductance of coil 1 (measured with terminals cd open), L_2 the self-inductance of coil 2 (measured with terminals ab open), and M the mutual inductance between the coils.

Proof of Eq. (11-3) may be shown by the following: With terminals b and d connected, the induced voltages are represented by the dependent sources having polarities, as shown in Fig. 11-8c. Using the circuit of Fig. 11-8c to write a loop equation yields

$$v_{ac} = L_1 \frac{di}{dt} - M \frac{di}{dt} - M \frac{di}{dt} + L_2 \frac{di}{dt} \qquad (11-4)$$

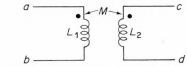

(a) Two coils connected through mutual inductance

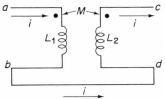

(b) Undotted terminals connected together

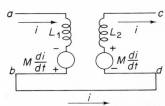

(c) Circuit (b) drawn with dependent sources

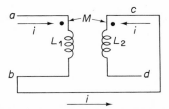

(d) Terminals *b* and *c* connected together

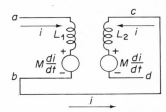

(e) Circuit (d) drawn with dependent sources

Fig. 11-8. Circuit diagrams to measure mutual inductance.

Dividing both sides of Eq. (11-4) by di/dt yields

$$\frac{v_{ac}}{di/dt} = L_A = L_1 + L_2 - 2M$$

For the second measurement connect terminals *b* and *c* together as shown in Fig. 11-8d. An impedance bridge measurement between terminals *a* and *d* yields

$$L_B = L_1 + L_2 + 2M \tag{11-5}$$

where L_B is the total inductance between terminals *a* and *d*, L_1 the self-inductance of coil 1 (as measured previously), L_2 the self-inductance of coil 2 (as measured previously), and *M* the mutual inductance between the two coils.

Proof of Eq. (11-5) may be shown by the following: With terminals *b* and *c* connected, the induced voltages are now represented by the dependent sources having polarities as shown in Fig. 11-8e. Writing a loop equation around the circuit of Fig. 11-8e yields

$$v_{ad} = L_1 \frac{di}{dt} + M \frac{di}{dt} + M \frac{di}{dt} + L_2 \frac{di}{dt} \tag{11-6}$$

Dividing Eq. (11-6) by di/dt we obtain

$$\frac{v_{ad}}{di/dt} = L_B = L_1 + L_2 + 2M$$

Subtracting Eq. (11-3) from Eq. (11-5) results in

$$L_B - L_A = 4M$$

or

$$M = \frac{L_B - L_A}{4} \tag{11-7}$$

where M is the mutual inductance between the coils, L_B the total inductance between terminals a and d, and L_A the total inductance between terminals a and c.

> **Example 11-1:** For the two coils shown in Fig. 11-9, L_A was measured as 2 mH and L_B as 20 mH. Calculate (a) the mutual inductance and (b) the coefficient of coupling.

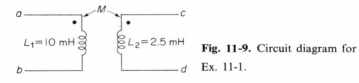

Fig. 11-9. Circuit diagram for Ex. 11-1.

> *Solution:* (a) Using Eq. (11-7),
>
> $$M = \frac{L_B - L_A}{4} = \frac{20\,\text{mH} - 2\,\text{mH}}{4} = 4.5\,\text{mH}$$
>
> (b) According to Eq. (11-2),
>
> $$k = \frac{M}{\sqrt{L_1 L_2}} = \frac{4.5\,\text{mH}}{\sqrt{(10\,\text{mH})(2.5\,\text{mH})}} = 0.9$$

Although this example gave the values of L_1 and L_2, they also could be measured by an impedance bridge. When measuring the self-inductance of one coil the other should be left open.

> **Example 11-2:** The coil of Fig. 11-10 has a center tap. The self-inductance from terminal a to the center tap (CT) is 1 mH and the self-inductance from

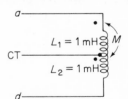

Fig. 11-10. Circuit diagram for Ex. 11-2.

the center tap to terminal d is 1 mH.   If the coefficient of coupling equals one $(k = 1)$, calculate the total inductance between terminals a and b.

Solution: Since

$$M = k \sqrt{L_1 L_2}$$

then

$$M = \sqrt{(1 \text{ mH})(1 \text{ mH})} = 1 \text{ mH}$$

The total inductance between terminals a and d is

$$L_T = L_1 + L_2 + 2M$$

$$L_T = 1 \text{ mH} + 1 \text{ mH} + 2(1 \text{ mH}) = 4 \text{ mH}$$

Note that the total inductance for the circuit of Fig. 11-10 is given by Eq. (11-5) because of the location of the dots.   If the area around the center tap is expanded, then the circuit of Fig. 11-10 is the same as circuit d of Fig. 11-8.   The total inductance is not the sum of L_1 and L_2 when a mutual inductance exists between the coils.   It may be concluded from this example that if the coil is center tapped $(L_1 = L_2)$, the total inductance is $4L_1$ or $4L_2$.

11-3.   TRANSFORMERS

A transformer is two or more coils connected by a mutual (common) magnetic field.   Thus mutual inductance is the phenomenon basic to the operation of the transformer.   The transformer has two (or more) pairs of terminals — an input pair and an output pair.   The input terminal pair — usually connected to an energy source — is called the *primary*.   The output terminal pair — usually connected to the load — is called the *secondary* (see Fig. 11-11a).

The usefulness of the transformer is that electrical energy can be transferred from one circuit to another without a direct electrical connection. Basic applications of a transformer include the ability of changing voltage and current from one magnitude to another, impedance matching devices, and isolating one circuit from another.

To induce a voltage into the secondary requires a changing magnetic field in the primary.   Therefore, a transformer is generally used only when alternating current is applied.   If a direct current is applied to the primary of a transformer, a voltage is induced in the secondary only at the instants when the source is applied and when it is removed, i.e., when the magnetic field is changing.

The transformer circuit of Fig. 11-11a is to be analyzed by loop analysis. Figure 11-11b shows the transformer redrawn with the dependent sources.

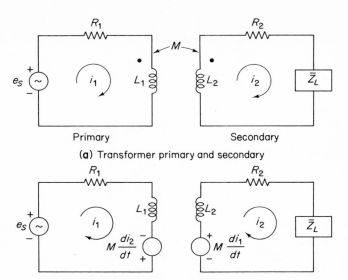

(a) Transformer primary and secondary

(b) Transformer redrawn with dependent sources

Fig. 11-11. Circuit diagrams for a transformer.

The resistor R_1 represents the internal resistance of L_1. R_2 represents the internal resistance of L_2.

Loop 1:

$$e_s = R_1 i_1 + L\frac{di_1}{dt} - M\frac{di_2}{dt} \tag{11-8}$$

Loop 2:

$$0 = - M\frac{di_1}{dt} + R_2 i_2 + L\frac{di_2}{dt} + Z_L i_2 \tag{11-9}$$

In terms of the steady state and not the instantaneous value $- L(di/dt)$ becomes $j\omega L\hat{I}$ and $M(di/dt)$ becomes $j\omega M\hat{I}$ — Eqs. (11-8) and (11-9) are written as

$$\hat{E}_S = (R + j\omega L_1)\hat{I}_1 - j\omega M\hat{I}_2 \tag{11-10}$$

$$0 = -j\omega M\hat{I}_1 + (R_2 + j\omega L_2 + \overline{Z}_L)\hat{I}_2 \tag{11-11}$$

or simply

$$E_S = \overline{Z}_{11}\hat{I}_1 + \overline{Z}_{12}\hat{I}_2 \tag{11-12}$$

$$0 = \overline{Z}_{12}\hat{I}_2 + \overline{Z}_{22}\hat{I}_2 \tag{11-13}$$

where

$$\overline{Z}_{11} = R_1 + j\omega L$$

$$\overline{Z}_{12} = - j\omega M$$

$$\overline{Z}_{22} = R_2 + j\omega L_2 + \overline{Z}_L$$

Using the method of determinants to solve for $\hat{I}_1$ yields

$$\hat{I}_1 = \frac{\begin{vmatrix} \hat{E}_S & \overline{Z}_{12} \\ 0 & \overline{Z}_{22} \end{vmatrix}}{\begin{vmatrix} \overline{Z}_{11} & \overline{Z}_{12} \\ \overline{Z}_{12} & \overline{Z}_{22} \end{vmatrix}} = \frac{\hat{E}_S \overline{Z}_{22}}{\overline{Z}_{11}\overline{Z}_{22} - \overline{Z}_{12}^2}$$

Dividing numerator and denominator by Z_{22},

$$\hat{I}_1 = \frac{\hat{E}_S}{\overline{Z}_{11} - \overline{Z}_{12}^2/\overline{Z}_{22}} = \frac{\hat{E}_S}{\overline{Z}_{\text{in}}} \qquad (11\text{-}14)$$

Thus the input impedance seen by the voltage source is

$$\overline{Z}_{\text{in}} = \overline{Z}_{11} - \frac{\overline{Z}_{12}^2}{\overline{Z}_{22}} \qquad (11\text{-}15)$$

where $\overline{Z}_{11}$ is the impedance of the primary by itself and $-\overline{Z}_{12}^2/\overline{Z}_{22}$ is the impedance of the secondary reflected into the primary. In terms of circuit parameters,

$$\overline{Z}_{\text{in}} = R_1 + j\omega L_1 + \frac{(\omega M)^2}{R_2 + j\omega L_2 + \overline{Z}_L} \qquad (11\text{-}16)$$

Equation (11-16) shows that the input impedance of the primary is increased by the impedance of the secondary reflected (or coupled) into the primary. This increase of impedance causes the primary current to decrease. The reflected impedance varies as M varies. According to Eq. (11-2), the mutual inductance, M, depends on the coefficient of coupling, k. Therefore, if k is small, the reflected impedance is small and vice versa. Also note that the reflected impedance is dependent on ω (the angular frequency). In Section 11-7 we will see what effect varying the frequency has on a frequency-response plot.

Example 11-3: For the circuit of Fig. 11-12, calculate the input impedance.

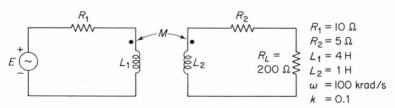

Fig. 11-12. Circuit diagram for Ex. 11-3.

Solution: According to Eq. (11-16),

$$\overline{Z}_{\text{in}} = R_1 + j\omega L_1 + \frac{(\omega M)^2}{R_2 + j\omega L_2 + R_L}$$

Then
$$M = k\sqrt{L_1 L_2} = 0.1\ \sqrt{(4\text{ H})(1\text{ H})} = 0.2\text{ H}$$

and
$$j\omega L_1 = j(100\text{ krad/s})(4\text{ H}) = j4 \times 10^5\ \Omega$$
$$j\omega L_2 = j(100\text{ krad/s})(1\text{ H}) = j1 \times 10^5\ \Omega$$

Since
$$j\omega L_1 \gg R_1 \quad \text{and} \quad j\omega L_2 \gg R_2$$

then
$$\bar{Z}_{\text{in}} \approx j4 \times 10^5 + \frac{[(100 \times 10^3\text{ rad/s})(0.2\text{ H})]^2}{(200 + j1 \times 10^5)\,\Omega}$$

Obtaining a common denominator yields
$$\bar{Z}_{\text{in}} = \frac{j4 \times 10^5\,(200 + j1 \times 10^5) + 0.04 \times 10^{10}}{200 + j1 \times 10^5}$$

Since
$$200 + j1 \times 10^5 \approx j1 \times 10^5$$

and
$$j^2 = -1$$

then
$$\bar{Z}_{\text{in}} = \frac{-4 \times 10^{10} + j0.8 \times 10^8 + 0.04 \times 10^{10}}{j1 \times 10^5}$$

or
$$\bar{Z}_{\text{in}} = (0.8 \times 10^3 + j3.96 \times 10^5)\,\Omega$$

Note that the coefficient of coupling for this example was small ($k = 0.1$).

Now let us examine a transformer circuit for the coefficient of coupling equal to 1.

Example 11-4: For the circuit of Fig. 11-12, calculate the input impedance when $k = 1$.

Solution:
$$M = k\ \sqrt{L_1 L_2} = 1\ \sqrt{(4\text{ H})(1\text{ H})} = 2\text{ H}$$
$$j\omega L_1 = j4 \times 10^5\ \Omega$$

and
$$j\omega L_2 = j1 \times 10^5\ \Omega$$

Since
$$j\omega L_1 \gg R_1 \quad \text{and} \quad j\omega L_2 \gg R_2$$

then
$$\bar{Z}_{\text{in}} \approx +j4 \times 10^5\ \Omega + \frac{[(100\text{ krad/s})(2\text{ H})]^2}{(200 + j1 \times 10^5)\,\Omega}$$

Obtaining the common denominator,
$$\bar{Z}_{\text{in}} = \frac{+j4 \times 10^5\,(200 + j1 \times 10^5) + 4 \times 10^{10}}{200 + j1 \times 10^5}\ \Omega$$

Since

$$200 + j1 \times 10^5 \approx j1 \times 10^5$$

and

$$j^2 = -1$$

then

$$\bar{Z}_{in} = \frac{-4 \times 10^{10} + j0.8 \times 10^8 + 4 \times 10^{10}}{+j1 \times 10^5} \, \Omega$$

$$\bar{Z}_{in} = 0.8 \times 10^3 \, \Omega = 800 \, \Omega$$

Note that when the transformer is considered ideal ($k = 1$) the input impedance is a real number, not a complex number. The next section shows that the input impedance for an ideal transformer may be found without using Eq. (11-16).

The value of the secondary current and voltage is also obtained from Eqs. (11-12) and (11-13):

$$\hat{I}_2 = \frac{\begin{vmatrix} \bar{Z}_{11} & \hat{E}_S \\ \bar{Z}_{12} & 0 \end{vmatrix}}{\begin{vmatrix} \bar{Z}_{11} & \bar{Z}_{12} \\ \bar{Z}_{12} & \bar{Z}_{22} \end{vmatrix}} = \frac{-\hat{E}_S\bar{Z}_{12}}{\bar{Z}_{11}\bar{Z}_{22} - \bar{Z}_{12}^2} \tag{11-17}$$

The ratio of secondary current to primary current is found by dividing Eq. (11-17) by Eq. (11-14):

$$\frac{\hat{I}_2}{\hat{I}_1} = \frac{-\hat{E}_S\bar{Z}_{12}(\bar{Z}_{11}\bar{Z}_{22} - \bar{Z}_{12}^2)}{\hat{E}_S\bar{Z}_{22}(\bar{Z}_{11}\bar{Z}_{22} - \bar{Z}_{12}^2)} = \frac{-\bar{Z}_{12}}{\bar{Z}_{22}} \tag{11-18}$$

or

$$\hat{I}_2 = -\frac{\bar{Z}_{12}}{\bar{Z}_{22}} \hat{I}_1 \tag{11-19}$$

The numerator of Eq. (11-19) is $-\bar{Z}_{12}\hat{I}_1 = j\omega M\hat{I}$. This term is the voltage induced in the secondary because of a current flowing in the primary. The purpose of finding Eq. (11-19) is not so much to show the value of the secondary current written in terms of the steady-state conditions; rather it is to be used in the next section for obtaining the current ratios for the ideal transformer.

11-4. IDEAL TRANSFORMERS

Section 11-3 provided a general discussion of transformers. At times the general equations must be used. However, any approximation or simplification is always appreciated in solving problems. This section, then, deals with ideal transformers. The iron-core transformers used in power and audio systems closely approximate the ideal. For the ideal transformer, the reflected impedance, current ratio, and voltage ratio are quickly and easily calculated in terms of the number of turns of the primary and secondary coil.

Figure 11-13 shows the schematic for an ideal transformer. In Fig. 11-13

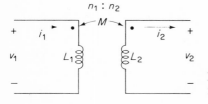

Fig. 11-13. Circuit diagram of an ideal transformer.

voltages v_1 and v_2 are positive at the dotted ends of the coils and the primary current, i_1, enters the dotted terminal while the secondary current, i_2, leaves the dotted terminal. These directions correspond to the condition where the primary is connected to an energy source and the secondary is connected to a passive circuit. The following analysis is written with these conditions in mind.

An ideal transformer has the following properties:

1. The reactance of the primary coil $(j\omega L_1)$ and the reactance of the secondary coil $(j\omega L_2)$ are very much larger than any impedance connected to the secondary terminals.

$$j\omega L_1 \gg \bar{Z}_L \quad \text{and} \quad j\omega L_2 \gg \bar{Z}_L$$

2. The ratio of the square root of the primary to the secondary inductance is

$$\sqrt{\frac{L_1}{L_2}} = \frac{n_1}{n_2} \tag{11-20}$$

where n_1/n_2 is the ratio of the number of turns of the primary coil to the number of turns of the secondary coil.

3. 100 per cent coupling. All the lines of force produced by one coil links the other coil $(k = 1)$. Then $M = \sqrt{L_1 L_2}$.

4. Negligible ac and dc resistance in both the primary and secondary coils. Therefore, the ideal transformer dissipates no energy:

$$\hat{V}_1 \hat{I}_1 = \hat{V}_2 \hat{I}_2 \tag{11-21}$$

Current Ratio. The general expression for the current ratio is given by Eq. (11-18):

$$\frac{\hat{I}_2}{\hat{I}_1} = \frac{-\bar{Z}_{12}}{\bar{Z}_{22}} = \frac{j\omega M}{R_2 + j\omega L_2 + \bar{Z}_L}$$

Since R_2 is only the winding resistance of the secondary coil, then $j\omega L_2 \gg R_2 + Z_L$ and

$$\frac{\hat{I}_2}{\hat{I}_1} = \frac{j\omega\sqrt{L_1 L_2}}{j\omega L_2} = \sqrt{\frac{L_1}{L_2}}$$

or

$$\frac{\hat{I}_2}{\hat{I}_1} = \frac{n_1}{n_2} \tag{11-22}$$

Equation (11-22) states that the ratio of secondary current to primary current is inversely proportional to the turns ratio.

Example 11-5: For the circuit of Fig. 11-14, calculate the primary current, $\hat{I}_1$.

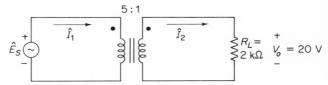

Fig. 11-14. Circuit diagram for Ex. 11-5.

Solution:

$$\hat{I}_2 = \frac{\hat{V}_0}{R_L} = \frac{20\,\text{V}}{2\,\text{k}\Omega} = 10\,\text{mA}$$

Rearranging Eq. (11-22) yields

$$\hat{I}_1 = \frac{n_2}{n_1} \times \hat{I}_2$$

$$\hat{I}_1 = \tfrac{1}{5}(10\,\text{mA}) = 2\,\text{mA}$$

Voltage Ratio. Rearranging Eq. (11-21) we obtain

$$\frac{\hat{V}_2}{\hat{V}_1} = \frac{\hat{I}_1}{\hat{I}_2}$$

and since Eq. (11-22) gives the turns ratio for current, then

$$\frac{\hat{V}_2}{\hat{V}_1} = \frac{n_2}{n_1} \tag{11-23}$$

Equation (11-23) states that the ratio of secondary voltage to primary voltage is directly proportional to the turns ratio.

Example 11-6: For the circuit of Fig. 11-14, calculate the primary voltage.

Solution: From Eq. (11-23),

$$\hat{V}_1 = \frac{n_1}{n_2} \times \hat{V}_2 = \tfrac{5}{1}(20\,\text{V}) = 100\,\text{V}$$

Transformers are classified as either *step-up* or *step-down*. A step-up transformer is one where the secondary voltage is *greater* than the primary voltage. A step-down transformer is one where the secondary voltage is *less* than the primary voltage. The circuit of Fig. 11-14 is a step-down transformer. It should be noted that whenever the voltage is stepped up, the current is stepped down by the same ratio. Likewise, whenever the voltage is stepped down the current is stepped up by the same ratio.

Reflected impedance. The presence of a transformer in a circuit modifies the impedance seen looking into either the primary or secondary. Equation (11-16) is the general expression for reflected impedance.

$$\bar{Z}_{in} = R_1 + j\omega L_1 + \frac{\omega^2 M^2}{R_2 + j\omega L_2 + \bar{Z}_L}$$

Since $j\omega L_1 \gg R_1$ and $j\omega L_2 \gg R_2$, then

$$\bar{Z}_{in} \approx j\omega L_1 + \frac{\omega^2 M^2}{j\omega L_2 + \bar{Z}_L}$$

Obtaining a common denominator,

$$\bar{Z}_{in} = \frac{j\omega L_1 \bar{Z}_L + j^2 \omega^2 L_1 L_2 + \omega^2 M^2}{j\omega L_2 + \bar{Z}_L}$$

Now from the properties of an ideal transformer,

$$j\omega L_2 \gg \bar{Z}_L$$

and

$$k = 1$$

Then

$$M^2 = L_1 L_2$$

$$j^2 = -1$$

Then

$$\bar{Z}_{in} = \bar{Z}_R = \frac{j\omega L_1 \bar{Z}_L}{j\omega L_2} = \sqrt{\frac{L_1}{L_2}}\, \bar{Z}_L$$

or

$$\bar{Z}_R = \left(\frac{n_1}{n_2}\right)^2 \bar{Z}_L \tag{11-24}$$

where Z_R is the impedance of the secondary reflected into the primary $(n_1/n_2)^2$ the turns ratio squared, and $\bar{Z}_L$ the load impedance.

Figure 11-15 shows a circuit with an ideal transformer. Note that the reflected impedance is in series with the source resistance, R_S.

A review of Example 11-4 shows that Eq. (11-24) yields the same results.

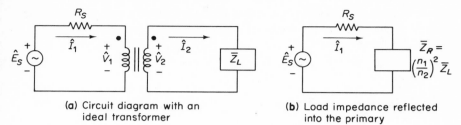

(a) Circuit diagram with an
ideal transformer

(b) Load impedance reflected
into the primary

Fig. 11-15. Circuit diagrams for an ideal transformer.

Example 11-7: For the circuit of Fig. 11-16a calculate (a) the reflected impedance, (b) the primary current, $\hat{I}_1$, and (c) the secondary voltage, $\hat{V}_2$.

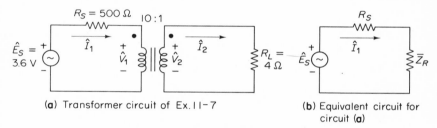

(a) Transformer circuit of Ex. 11-7

(b) Equivalent circuit for
circuit (a)

Fig. 11-16. Circuit diagram for Ex. 11-7.

Solution: (a)   Using Eq. (11-24),

$$\bar{Z}_R = \left(\frac{n_1}{n_2}\right)^2 \bar{Z}_L = \left(\frac{10}{1}\right)^2 (4\,\Omega) = 400\,\Omega$$

(b)   The circuit of Fig. 11-16b shows the reflected impedance in series with R_S.   Therefore,

$$\hat{I}_1 = \frac{\hat{E}_S}{R_S + \bar{Z}_R} = \frac{3.6\,\text{V}}{500\,\Omega + 400\,\Omega} = 4\,\text{mA}$$

(c)   Rearranging Eq. (11-23),

$$\hat{V}_2 = \frac{n_2}{n_1} \times \hat{V}_1$$

where

$$\hat{V}_1 = \hat{I}_1 \bar{Z}_R = (4\,\text{mA})(400\,\Omega) = 1.6\,\text{V}$$

Therefore,

$$\bar{V}_2 = (\tfrac{1}{10})(1.6\,\text{V}) = 0.16\,\text{V} = 160\,\text{mV}$$

Note that the secondary voltage, $\hat{V}_2$, is calculated from the turns ratio and the primary voltage—not the source voltage. In this example the voltage, V_2, equals the voltage across R_L.

Example 11-8: For the circuit of Fig. 11-17a calculate (a) the reflected impedance, (b) the primary voltage, (c) the secondary voltage, and (d) the secondary current.

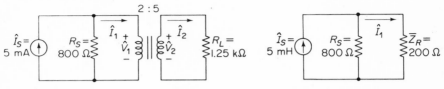

(a) Ideal transformer driven by a practical current source

(b) Primary equivalent circuit

Fig. 11-17. Circuit diagrams for Ex. 11-8.

Solution: (a)

$$\bar{Z}_R = \left(\frac{n_1}{n_2}\right)^2 \bar{Z}_L = (\tfrac{2}{5})^2 \, (1.25 \text{ k}\Omega) = 200 \, \Omega$$

This equivalent circuit is shown in Fig. 11-17b.

(b)

$$\hat{V}_1 = \hat{I}_1 \bar{Z}_R$$

Applying the current division law to Fig. 11-17b,

$$\hat{I}_1 = \frac{R_S}{R_S + \bar{Z}_R} \, \hat{I}_S = \frac{800 \, \Omega}{800 \, \Omega + 200 \, \Omega} (5 \text{ mA}) = 4 \text{ mA}$$

Therefore,

$$\hat{V}_1 = (4 \text{ mA}) \, (200 \, \Omega) = 0.8 \text{ V}$$

(c) From Eq. (11-23),

$$\hat{V}_2 = \frac{n_2}{n_1} \times \hat{V}_1 = (\tfrac{5}{2}) \, (0.8 \text{ V}) = 2 \text{ V}$$

(d)

$$\hat{I}_2 = \frac{\hat{V}_2}{R_L} = \frac{2 \text{ V}}{1.25 \text{ k}\Omega} = 1.6 \text{ mA}$$

The secondary current could also have been found by using Eq. (11-22),

$$\hat{I}_2 = \frac{n_1}{n_2} \times \hat{I}_1 = (\tfrac{2}{5}) \, (4 \text{ mA}) = 1.6 \text{ mA}$$

In this example the source current does not equal the primary current. Therefore, when using the turns ratio to find the secondary current, be sure to use $\hat{I}_1$ and not $\hat{I}_S$.

Example 11-9: The load on an ideal transformer is a 0.01-μF capacitor, as shown in Fig. 11-18. Calculate the reflected capacitance on the primary side.

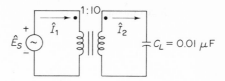

Fig. 11-18. Circuit diagram for Ex. 11-9.

Solution: According to Eq. (11-23),

$$\bar{Z}_R = \left(\frac{n_1}{n_2}\right)^2 \bar{Z}_L$$

But

$$\bar{Z}_L = \frac{1}{\omega C_L}$$

The load capacitor reflected in the primary of an ideal transformer appears as a capacitance and $\bar{Z}_R = 1/\omega C_R$. Therefore,

$$\frac{1}{\omega C_R} = \left(\frac{n_1}{n_2}\right)^2 \frac{1}{\omega C_L}$$

or

$$C_R = \left(\frac{n_2}{n_1}\right)^2 C_L \qquad (11\text{-}25)$$

For this example,

$$C_R = (\tfrac{10}{1})^2 \,(0.01 \ \mu\text{F}) = 1 \ \mu\text{F}$$

Note that capacitance is not reflected the same as resistance.

Example 11-10: For the circuit of Fig. 11-19 obtain (a) the equivalent circuit between terminals b and c and (b) the equivalent circuit between terminals d and e. The inductance between terminals a and c is 10 mH.

$$\frac{n_1}{n_2} = \tfrac{5}{2} \qquad \frac{n_1}{n_3} = \tfrac{2}{1}$$

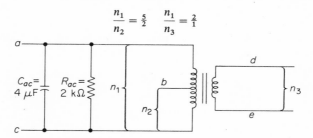

Fig. 11-19. Circuit diagram for Ex. 11-10.

Solution: (a) Reflecting C_{ac}, R_{ac}, and L_{ac} between terminals b and c,

$$C_{bc} = \left(\frac{n_1}{n_2}\right)^2 C_{ac} = (\tfrac{5}{2})^2 \,(4 \ \mu\text{F}) = 25 \ \mu\text{F}$$

$$R_{bc} = \left(\frac{n_2}{n_1}\right)^2 R_{ac} = (\tfrac{2}{5})^2 \,(2 \ \text{k}\Omega) = 320 \ \Omega$$

Rearranging Eq. (11-20),

$$L_{bc} = \left(\frac{n_2}{n_1}\right)^2 L_{ac} = (\tfrac{2}{5})^2 \,(10\text{ mH}) = 1.6\text{ mH}$$

The equivalent circuit between terminals b and c is shown in Fig. 11-20a.

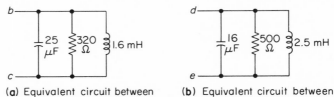

(a) Equivalent circuit between terminals b and c (b) Equivalent circuit between terminals d and e

Fig. 11-20. Circuit diagrams showing the reflected impedance for Ex. 11-10.

(b) Reflecting C_{ac}, R_{ac}, and L_{ac} between terminals d and e,

$$C_{de} = \left(\frac{n_1}{n_3}\right)^2 C_{ac} = (\tfrac{2}{1})^2 \,(4\ \mu\text{F}) = 16\ \mu\text{F}$$

$$R_{de} = \left(\frac{n_3}{n_1}\right)^2 R_{ac} = (\tfrac{1}{2})^2 \,(2\text{ k}\Omega) = 500\ \Omega$$

$$L_{de} = \left(\frac{n_3}{n_1}\right)^2 L_{ac} = (\tfrac{1}{2})^2 \,(10\text{ mH}) = 2.5\text{ mH}$$

The equivalent circuit between terminals d and e is shown in Fig. 11-20b.

This example shows that capacitance is reflected differently than either resistance or inductance. The expression for reflected inductance is obtained by rearranging Eq. (11-20). The equivalent circuits as shown in Fig. 11-20 are parallel-tuned circuits. In Section 11-7 we shall analyze these tuned circuits in the same way the parallel-tuned circuits of Chapter 10 were analyzed.

A transformer with a tapped primary (or secondary) provides several different combinations to obtain the proper equivalent circuit. For example, a desired equivalent circuit between terminals bc could be obtained by connecting a capacitor between terminals ac while a resistor is connected between terminals de.

11-5. AUTOTRANSFORMERS

A transformer that has only one coil is called an *autotransformer*. Figure 11-21 shows the schematic diagrams of fixed and variable autotransformers. As shown, autotransformers do not have electrical isolation

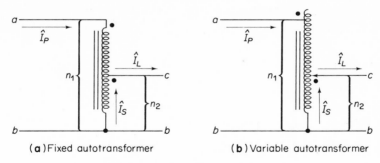

(a) Fixed autotransformer   (b) Variable autotransformer

Fig. 11-21. Types of transformers.

between the input circuit (terminals *ab*) and the output circuit (terminals *cb*). This lack of isolation is their main disadvantage. Their major advantage is that a larger load voltage is obtainable than would be possible if the transformer were operated as an isolation transformer. With the variable autotransformer, the secondary voltage may be varied so as to obtain voltages between 0 and V_{max} (which usually is 5 per cent more than the input voltage).

As in the previous sections, the energy source is connected to the primary and the output circuit is connected to the secondary. The number of turns on the primary is n_1 and the number of turns on the secondary is n_2. Note that the dots in Fig. 11-21 show the parts of the coil where the voltage is in phase. The following examples show how an autotransformer may be analyzed.

Example 11-11: For the circuit of Fig. 11-22, calculate (a) $\hat{V}_L$, (b) $\hat{I}_L$, (c) $\hat{I}_P$, and (d) $\hat{I}_S$. Let $n_1/n_2 = \frac{5}{1}$.

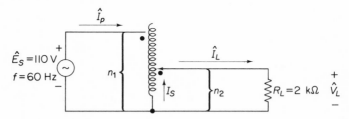

Fig. 11-22. Circuit diagram for Ex. 11-11.

Solution: (a) In terms of the turns ratio,

$$\frac{\hat{E}_S}{\hat{V}_L} = \frac{n_1}{n_2}$$

or

$$\hat{V}_L = \frac{n_2}{n_1} \hat{E}_S = (\tfrac{1}{5})(110 \text{ V}) = 22 \text{ V}$$

(b)

$$\hat{I}_L = \frac{\hat{V}_L}{R_L} = \frac{22 \text{ V}}{2 \text{ k}\Omega} = 11 \text{ mA}$$

(c) In terms of the turns ratio,

$$\frac{\hat{I}_P}{\hat{I}_L} = \frac{n_2}{n_1}$$

or

$$\hat{I}_P = \frac{n_2}{n_1} \hat{I}_L = (\tfrac{1}{5})(11 \text{ mA}) = 2.2 \text{ mA}$$

(d) Applying Kirchhoff's current law,

$$\hat{I}_S = \hat{I}_L - \hat{I}_P = 11 \text{ mA} - 2.2 \text{ mA} = 8.8 \text{ mA}$$

Note that the polarity of $\hat{V}_L$ is determined by the dot notation. This in turn determines the direction of $\hat{I}_S$. The value of $\hat{I}_S$ may also be calculated from Fig. 11-23. From Fig. 11-23, $n_1 = n_2 + n_3$. In terms of the turns ratio,

$$\frac{\hat{I}_P}{\hat{I}_S} = \frac{n_2}{n_3}$$

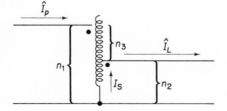

Fig. 11-23. Circuit diagram showing the turns ratios.

and then

$$\hat{I}_S = \frac{n_3}{n_2} \hat{I}_P = (\tfrac{4}{1})(2.2 \text{ mA}) = 8.8 \text{ mA}$$

If we are discussing an autotransformer with 100 per cent efficiency, then the power developed by the source must be delivered to the load. As a check,

$$\hat{E}_S \hat{I}_P = (100 \text{ V})(2.2 \text{ mA}) = 242 \text{ mW}$$

and

$$\hat{V}_L \hat{I}_L = (22 \text{ V})(11 \text{ mA}) = 242 \text{ mW}$$

An alternative method of calculating $\hat{I}_P$ is first to reflect R_L to the primary side and then apply Ohm's law:

$$R_R = \left(\frac{n_1}{n_2}\right)^2 R_L = \left(\frac{5}{1}\right)^2 (2\,k\Omega) = 50\,k\Omega$$

and then

$$\hat{I}_P = \frac{\hat{E}_S}{R_R} = \frac{110\,V}{50\,k\Omega} = 2.2\,mA$$

Example 11-12: If R_L in the circuit of Fig. 11-22 is replaced with a 10 μF capacitor, what is the value of capacitance on the primary side of the auto-transformer?

Solution:

$$C_P = \left(\frac{n_2}{n_1}\right)^2 C_S = (\tfrac{1}{5})^2 (10\,\mu F) = 0.4\,\mu F$$

Therefore, a capacitor connected to the secondary of an autotransformer is reflected to the primary in the same manner as the isolation transformer.

11-6. EFFICIENCY

The efficiency of a transformer may be expressed as

$$\text{effciency} = \frac{\text{output}}{\text{input}} = \frac{\text{output}}{\text{output} + \text{losses}}$$

where the output is the product of the secondary voltage and current. The losses in iron-core transformers are hysteresis losses, eddy-current losses, and $\hat{I}^2 R$ losses. Hysteresis and eddy-current are due to properties of the iron. The hysteresis losses may be minimized by the proper choice of the alloy. The eddy-current losses may be minimized by laminating the core in thin sheets and separating them from one another by a coating of lacquer or varnish which provides an insulator. The $\hat{I}^2 R$ losses are due to the resistance of the windings.

Even with these losses the efficiency of the transformer usually is quite high—80 per cent or better. If the transformer efficiency is high, then for practical purposes Eq. (11-24) (the equation for reflected impedance) developed for the ideal transformer may still be used.

Example 11-13: The transformer of Fig. 11-24 is 80 per cent efficient. The power delivered to the 8 Ω speaker is 4 W. Calculate (a) the primary voltage and (b) the primary current.

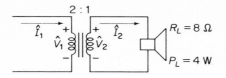

Fig. 11-24. Circuit diagram for Ex. 11-13.

Solution: (a) Since the secondary power is 80 per cent (0.8) of the primary power,

$$P_L = (0.8)\, P_P$$

then

$$P_P = \frac{P_L}{0.8} = \frac{4\,\text{W}}{0.8} = 5\,\text{W}$$

The 8 Ω of the secondary reflected into the primary is

$$R_R = \left(\frac{n_1}{n_2}\right)^2 R_L = (\tfrac{2}{1})^2\,(8\,\Omega) = 32\,\Omega$$

$$P_P = \frac{\hat{V}_1^2}{R_R}$$

or

$$\hat{V}_1 = \sqrt{P_P R_R} = \sqrt{(5\,\text{W})(32\,\Omega)} = 12.7\,\text{V}$$

(b) Since

$$P_P = \hat{V}_1 \hat{I}_1$$

then

$$\hat{I}_1 = \frac{P_P}{\hat{V}_1} = \frac{5\,\text{W}}{12.7\,\text{V}} = 0.394\,\text{A}$$

11-7. TUNED TRANSFORMER CIRCUITS

In Chapter 10 we learned how to analyze the frequency response of parallel-tuned circuits. Section 11-5 has shown that the ideal transformer, because of the turns ratio, allows greater flexibility in choosing resistors and capacitors. Since it is the values of resistance and capacitance that determine the resonant frequency (ω_r) and bandwidth (**BW**) of a tuned circuit, we should have flexibility over ω_r and **BW**.

When a capacitor is connected to either the secondary terminals, primary terminals or both, the circuit is called a *tuned transformer circuit*. Three commonly used transformer circuits are the single-tuned, double-tuned, and stagger-tuned circuits. For our analysis these circuits are to be driven by a current source. The reason for this is that in applications these circuits are primarily used as a load on a transistor or vacuum tube, both of which have an output equivalent circuit represented by a current source.

Single-tuned Circuit. The circuit of Fig. 11-25a with a capacitor connected across the output terminals is called a *single-tuned circuit*. For our analysis the transformer is considered ideal. Not only does this assumption reduce the work involved, but it yields answers that are acceptable in practical applications.

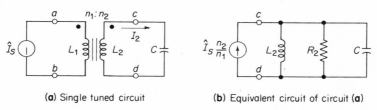

(a) Single tuned circuit   (b) Equivalent circuit of circuit (a)

Fig. 11-25. Circuit diagrams of a single tuned transformer.

To obtain a parallel-tuned circuit, either reflect the capacitor connected to the secondary into the primary or the current source in the primary to the secondary. The latter is shown in Fig. 11-25b, where R_2 is the parallel resistor determined from Eq. (10-8), $R_P = Q_P \omega_r L_P$. The circuit of Figure 11-25b is a parallel-tuned circuit and may be analyzed according to the principles developed in Chapter 10.

Example 11-14: For the single tuned circuit of Fig. 11-26, (a) show the equivalent circuit as seen by the primary, (b) calculate the resonant frequency, and (c) calculate the bandwidth.

$$L_1 = 0.1 \text{ mH} \qquad \frac{n_1}{n_2} = \frac{2}{1}$$

$$Q_1 = 50$$

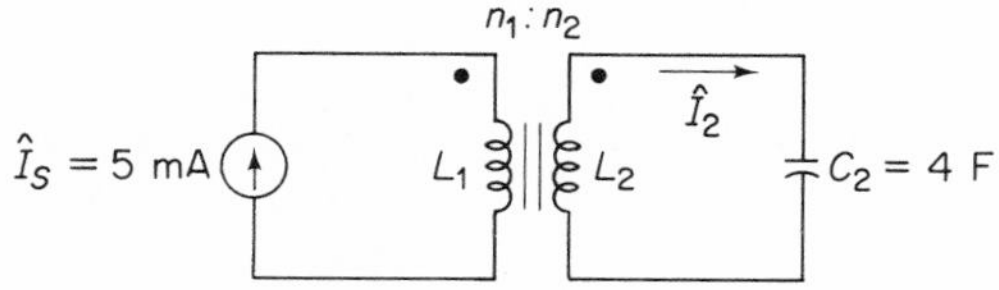

Fig. 11-26. Circuit diagram for Ex. 11-14.

Solution: (a) Reflecting the capacitor of the secondary into the primary is

$$C_1 = \left(\frac{n_2}{n_1}\right)^2 C_2 = (\tfrac{1}{2})^2 (4 \text{ } \mu\text{F}) = 1 \text{ } \mu\text{F}$$

(b) From Eq. (10-17a),

$$\omega_r = \frac{1}{\sqrt{L_1 C_1}} = \frac{1}{\sqrt{(0.1 \text{ mH})(1 \text{ } \mu\text{F})}} = 100 \text{ krad/s}$$

(c) From Eq. (10-5a),

$$BW = \frac{1}{R_1 C_1}$$

where R_1 is the parallel equivalent resistance of the primary and is determined from Eq. (10-8):

$$R_1 = Q_1 \omega_r L_1 = (50)(100 \text{ krad/s})(0.1 \text{ mH}) = 500 \,\Omega$$

Therefore,

$$BW = \frac{1}{(500 \,\Omega)(1 \,\mu F)} = 2 \text{ krad/s}$$

The parallel equivalent circuit is shown in Fig. 11-27.

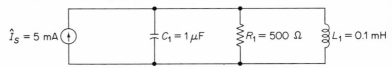

$\hat{I}_S = 5 \text{ mA}$ $C_1 = 1 \,\mu F$ $R_1 = 500 \,\Omega$ $L_1 = 0.1 \text{ mH}$

Fig. 11-27. Equivalent circuit on the primary side of the transformer.

We are able to use the equation developed in Chapter 10 because the circuit of Fig. 11-26 is a narrowband tuned circuit ($BW \leq 0.1\omega_r$). As in any parallel-tuned circuit, the bandwidth may be increased by connecting a resistor in parallel with R_1. The values of L_1 and Q_2 should be measured on a Q meter at the resonant frequency with the secondary terminals left open.

Example 11-15: Given the following parameters for the single-tuned circuit of Fig. 11-28,

$$Q_2 = 23, \quad L_2 = 760 \,\mu H, \quad \text{and} \quad C = 160 \text{ pF}$$

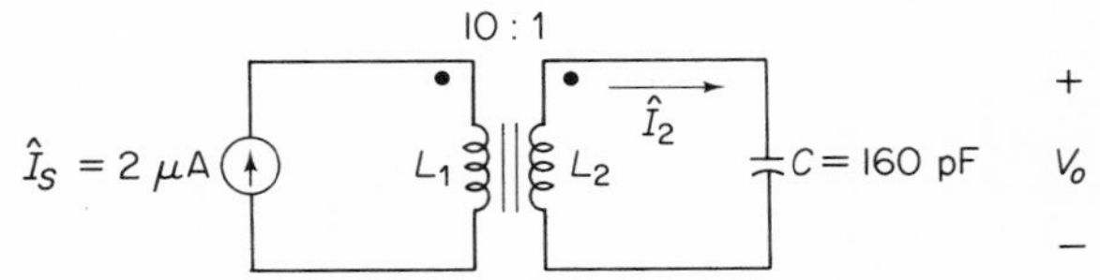

10 : 1

$\hat{I}_S = 2 \,\mu A$ L_1 L_2 $\hat{I}_2$ $C = 160 \text{ pF}$ $+$ V_o $-$

Fig. 11-28. Circuit diagram for Ex. 11-15.

Calculate
  (a) resonant frequency,
  (b) bandwidth,
  (c) frequency response plot of $\hat{V}_0$ versus ω.

Solution: (a) From Eq. (10-3a),

$$\omega_r = \frac{1}{\sqrt{L_2 C}} = \frac{1}{\sqrt{(760 \,\mu H)(160 \text{ pF})}} = 2.86 \text{ Mrad/s}$$

or

$$f_r = \frac{2.86 \text{ Mrad/s}}{6.28} = 455 \text{ kHz}$$

(b) According to Eq. (10-5a),

$$BW = \frac{1}{R_2 C}$$

Since

$$Q_2 = \frac{R_2}{\omega_r L_2}$$

then

$$R_2 = Q_2 \omega_r L_2 = (23)(2.86 \text{ Mrad/s})(760 \ \mu\text{H}) = 50 \text{ k}\Omega$$

Therefore,

$$BW = \frac{1}{(50 \text{ k}\Omega)(160 \text{ pF})} = 125 \text{ krad/s}$$

or

$$BW = \frac{125 \text{ krad/s}}{6.28} = 20 \text{ kHz}$$

Reflecting the source current, $\hat{I}_s$ into the secondary is

$$\hat{I}_2 = \frac{n_1}{n_2} \times \hat{I}_s = (\tfrac{10}{1})(2 \ \mu\text{A}) = 20 \ \mu\text{A}$$

The parallel equivalent circuit is shown in Fig. 11-29a.

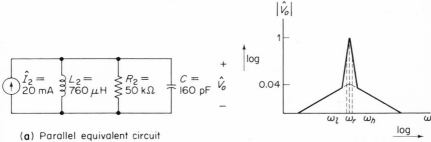

(a) Parallel equivalent circuit
of Fig. 11-28

(b) Frequency response plot

Fig. 11-29. Diagrams for the single tuned circuit of Ex. 11-15.

(c) The frequency response may be plotted by using the straight-line approximation method developed in Section 10-1.

$$\hat{V}_{max} = \hat{I}_2 R_2 = (20 \ \mu\text{A})(50 \text{ k}\Omega) = 1 \text{ V}$$

$$\omega_r = 2.86 \text{ Mrad/s}$$

$$\omega_l = \omega_r - \frac{BW}{2} = 2.86 \text{ Mrad/s} - \frac{125 \text{ krad/s}}{2} = 2.8 \text{ Mrad/s}$$

$$\omega_h = \omega_r + \frac{BW}{1} = 2.86 \text{ Mrad/s} + \frac{125 \text{ krad/s}}{2} = 2.92 \text{ Mrad/s}$$

The intersection of the lower skirts is

$$|\hat{V}_0| = \hat{I}_2 X_L \quad \text{or} \quad |\hat{V}_0| = \hat{I}_2 X_C$$

Then

$$|\hat{V}_0| = \hat{I}_2 \omega_r L = (20 \ \mu A)(2.86 \ \text{Mrad/s})(760 \ \mu H) = 0.04315 \ V$$

The frequency-response plot is shown in Fig. 11-29b.

Example 11-16: *Q*-meter measurements between terminals *ac* of the transformer of Fig. 11-30 are $Q = 120$, $f_r = 500$ kHz, and $C_{ac} = 100$ pF. From voltage measurements, it is determined that the turns ratios are

$$\frac{n_{ac}}{n_{bc}} = \frac{2}{1}, \quad \frac{n_{ac}}{n_{de}} = \frac{2}{1}, \quad \frac{b_{bc}}{n_{de}} = \frac{1}{1}$$

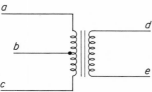

Fig. 11-30. Circuit diagram for Ex. 11-16.

(a) Find the parallel equivalent circuit between terminals *bc*.

(b) Calculate the bandwidth.

(c) If a current source whose value is 0.1 mA is connected between terminals *bc*, plot the frequency response.

Solution: (a) According to Eq. (10-3b),

$$f_r = \frac{1}{2\pi \sqrt{L_{ac} C_{ac}}}$$

Then

$$L_{ac} = \frac{1}{(2\pi f_r)^2 C_{ac}} = \frac{1}{[(6.28)(500 \ \text{kHz})]^2 (100 \ \text{pF})} \approx 1 \ \text{mH}$$

Rearranging Eq. (10-8),

$$R_{ac} = Q \omega_r L_{ac}$$

where

$$\omega_r = 2\pi f_r = (6.28)(500 \ \text{kHz}) = 3140 \ \text{krad/s}$$

Then

$$R_{ac} = (120)(3.16 \ \text{Mrad/s})(1 \ \text{mH}) = 377 \ \text{k}\Omega$$

A parallel equivalent circuit between terminals *ac* is shown in Fig. 11-31a.

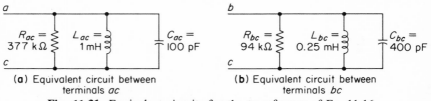

(a) Equivalent circuit between terminals *ac*

(b) Equivalent circuit between terminals *bc*

Fig. 11-31. Equivalent circuits for the transformer of Ex. 11-16.

To obtain the equivalent circuit between terminals bc, each element must be reflected.   Let

$$R_{bc} = R_{ac} \left(\frac{n_{bc}}{n_{ac}} \right)^2 = (377\text{ k}\Omega)\left(\tfrac{1}{2}\right)^2 = 94\text{ k}\Omega$$

$$L_{bc} = L_{ac} \left(\frac{n_{bc}}{n_{ac}} \right)^2 = (1\text{ mH})\left(\tfrac{1}{2}\right)^2 = 0.25\text{ mH}$$

$$C_{bc} = C_{ac} \left(\frac{n_{ac}}{n_{bc}} \right)^2 = (100\text{ pF})\left(\tfrac{2}{1}\right)^2 = 400\text{ pF}$$

The equivalent circuit between terminals bc is shown in Fig. 11-31b.

(b)   The bandwidth is

$$\text{BW} = \frac{1}{R_{bc}C_{bc}} = \frac{1}{(94\text{ k}\Omega)(400\text{ pF})} = 26.6\text{ krad/s}$$

In terms of cycles per second,

$$\text{BW} = \frac{26.6\text{ krad/s}}{6.28} = 4.24\text{ kHz}$$

Note that the bandwidth could have been calculated from the values of Fig. 11-31a and we would obtain BW = 26.6 krad/s (or 4.24 kHz).

(c)   Using the straight-line approximation developed in Section 10-1,

$$\hat{V}_{\max} = I_S R_{bc} = (0.1\text{ mA})(94\text{ k}\Omega) = 9.4\text{ V}$$

$$\omega_r = 3.16\text{ Mrad/s}$$

$$\omega_l = \omega_r - \frac{\text{BW}}{2} = 3.14\text{ Mrad/s} - \frac{26.6\text{ krad/s}}{2} = 3.127\text{ Mrad/s}$$

$$\omega_h = \omega_r + \frac{\text{BW}}{2} = 3.14\text{ Mrad/s} + \frac{26.6\text{ krad/s}}{2} = 3.153\text{ Mrad/s}$$

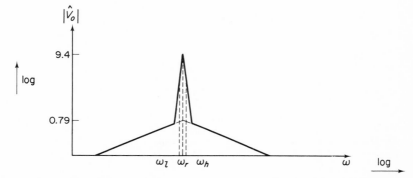

Fig. 11-32. Frequency response plot for the single tuned circuit of Fig. 11-31b.

The intersection of the lower skirts is

$$|\hat{V}_0| = \hat{I}_S X_L \quad \text{or} \quad |\hat{V}_0| = \hat{I}_S X_C$$

Then

$$|\hat{V}_0| = \hat{I}_S \omega_r L_{bc} = (0.1 \text{ mA})(3.14 \text{ Mrad/s})(0.25 \text{ mH})$$

$$|\hat{V}_0| \approx 0.79 \text{ V}$$

The plot of the frequency is shown in Fig. 11-32.

Note that the Q-meter measurements do not read out the value of inductance directly. However, it is calculated from Eq. (10-8). Once the turns ratio and the values of the elements are known between one pair of terminals, these values are easily reflected to any other set of terminals.

Double-tuned Circuit. A transformer coupled circuit that has a tuned primary and a tuned secondary as shown in Fig. 11-33 is called a *double-tuned circuit*. Our analysis of this network will be based on three assumptions:

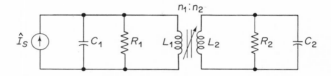

Fig. 11-33. Circuit diagram of a double-tuned circuit.

1. The value of Q for both the primary and secondary is greater than 10 and equal. $Q_1 = Q_2$.
2. Both the primary and secondary are tuned to the same resonant frequency:

$$\omega_r = 2\pi f_r = \frac{1}{\sqrt{L_1 C_1}} = \frac{1}{\sqrt{L_2 C_2}}$$

3. The coefficient of coupling is

$$k = \frac{M}{\sqrt{L_1 L_2}}$$

An advantage of this circuit (over the single-tuned circuit) is that a flatter frequency response about the resonant frequency is obtainable for a given Q. A double-tuned circuit always has a wider bandwidth than a single-tuned circuit.

A frequency response — whether it be a plot of primary current versus frequency, secondary current versus frequency, or output voltage versus frequency — is dependent on k. As the coefficient of coupling increases, the peak value of the output voltage increases until a maximum value is reached. This condition is called *critical coupling.*

If k is increased beyond the critical value, the curve of the secondary current and voltage begins to display a decrease at ω_r and an increase at frequencies on either side of ω_r. (see Fig. 11-34.) The reason for the

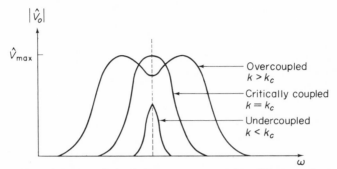

Fig. 11-34. Plot of the frequency response for different values of k.

decrease is that $k < 1$, and now the impedance reflected into the primary $(\omega M)^2/Z_{22}$ is frequency dependent. This affects the primary current, which in turn affects the induced voltage into the secondary. At low frequencies the coupled impedance is a maximum value and is resistive. At high frequencies the coupled impedance is capacitive. It is this coupled impedance that depends on the value of k.

The general expression for calculating the value of the critical coefficient of coupling is

$$k_c = \frac{1}{\sqrt{Q_1 Q_2}} \tag{11-26}$$

where Q_1 is the quality factor of the primary and Q_2 is the quality factor of the secondary. If $Q_1 = Q_2 = Q$, then Eq. (11-26) may be rewritten as

$$k_c = \frac{1}{Q} \tag{11-27}$$

Coupling is classified as either *undercoupled* (or *loosely coupled*) when $k < k_c$, critically coupled $k = k_c$, or overcoupled (or *tightly coupled*) $k > k_c$. In the undercoupled situation the curve of the output voltage is that of just a single-tuned circuit, as shown in Fig. 11-34. As the coefficient of coupling is increased to the critical value, the curve of the output voltage reaches its maximum value and begins to become flat or broader at the maximum value, as shown in Fig. 11-34. If the coefficient of coupling is increased further, the curve of the output voltage begins to display two peaks. If the Q's are equal, both peaks will have maximum values equal to the maximum value at critical coupling.

As previously mentioned, an advantage of the double-tuned circuit is an increase in the bandwidth. When the tuned circuits are critically coupled the bandwidth has increased by the $\sqrt{2}$ over the bandwidth of a single-tuned circuit:

$$\text{BW} = \frac{\sqrt{2}}{RC} = \sqrt{2}\frac{\omega_r}{Q} \qquad (11\text{-}28)$$

where

$$R = \sqrt{R_1 R_2} \quad \text{and} \quad C = \sqrt{C_1 C_2}$$

Therefore,

$$k_c = \frac{1}{Q} = \frac{\text{BW}}{\sqrt{2}\,\omega_r} \qquad (11\text{-}29)$$

The maximum value of voltage for critical coupling is

$$\hat{V}_{\text{max}} = \hat{I}_s\frac{R}{2} \qquad (11\text{-}30)$$

where

$$R = \sqrt{R_1 R_2}$$

The maximum value occurs at ω_r.

A comparison of bandwidths of the single-tuned circuit with that of a circuit with critical coupling is shown in Fig. 11-35.

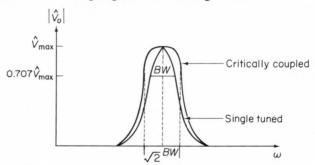

Fig. 11-35. Frequency response plot comparing the bandwidths of the critical coupling with that of the single tuned.

When $k > k_c$, double peaks occur in the output waveforms. The frequencies at which these peaks occur are approximately

$$\omega_1 = \omega_r - \frac{k\omega_r}{2} \qquad (11\text{-}31a)$$

and

$$\omega_2 = \omega_r + \frac{k\omega_r}{2} \qquad (11\text{-}31b)$$

Figure 11-36 shows the relationship between the band of frequencies between the peaks ($\Delta\omega = \omega_2 - \omega_1 = k\omega_r$) and $\sqrt{2}\,\Delta\omega$. A ratio of $\hat{V}_{min}$ to $\hat{V}_{max}$ is usually specified for a particular problem. This ratio is the

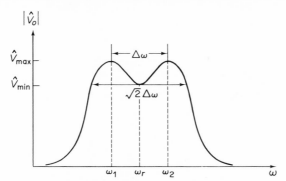

Fig. 11-36. Relationship between the band of frequencies at $\hat{V}_{max}$ and the band of frequencies at $\hat{V}_{min}$.

amount of deviation permitted from the maximum value. The ratio is usually between 0.8 and 1.0. The ratio 1.0 corresponds to the case of critical coupling. With this information, the response of an overcoupled double-tuned circuit can be plotted. The maximum value is obtained from the critical coupling case, Eq. (11-29). Note that the $\sqrt{2}\,\Delta\omega$ is not the bandwidth. The bandwidth is still defined to be the band of frequencies at $0.707\hat{V}_{max}$.

Example 11-17: For the double-tuned circuit of Fig. 11-37, calculate the following for critical coupling.

    (a) coefficient of coupling,
    (b) resonant frequency,
    (c) bandwidth,
    (d) maximum voltage.

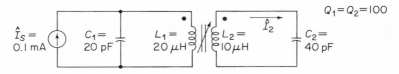

Fig. 11-37. Circuit diagram for Ex. 11-17.

Solution   (a) $k_c = 1/Q = \frac{1}{100} = 0.01$.

        (b) Since

$$\omega_r = \frac{1}{\sqrt{L_1 C_1}} \quad \text{or} \quad \omega_r = \frac{1}{\sqrt{L_2 C_2}}$$

then

$$\omega_r = \frac{1}{\sqrt{L_1 C_1}} = \frac{1}{\sqrt{(20 \ \mu\text{H})(20 \ \text{pF})}} = 50 \ \text{Mrad/s}$$

or

$$f_r = \frac{\omega_r}{2\pi} = \frac{50 \ \text{Mrad/s}}{6.28} = 7.96 \ \text{MHz}$$

(c)   Rearranging Eq. (11-29),

$$\text{BW} = \sqrt{2} \, k_c \, \omega_r$$

$$\text{BW} = \sqrt{2} (0.01)(50 \ \text{Mrad/s}) = 0.707 \ \text{Mrad/s}$$

or

$$\text{BW} = \frac{0.707 \ \text{Mrad/s}}{6.28} = 0.113 \ \text{MHz} = 113 \ \text{kHz}$$

(d)   According to Eq. (11-30),

$$\hat{V}_{\text{max}} = \hat{I}_s \frac{R}{2}$$

where

$$R = \sqrt{R_1 R_2}$$

From Eq. (10-8),

$$R_1 = Q_1 \omega_r L_1 = (100)(50 \ \text{Mrad/s})(20 \ \mu\text{H}) = 100 \ \text{k}\Omega$$

and

$$R_2 = Q_2 \omega_r L_2 = (100)(50 \ \text{Mrad/s})(10 \ \mu\text{H}) = 50 \ \text{k}\Omega$$

$$R = \sqrt{(100 \ \text{k}\Omega)(50 \ \text{k}\Omega)} = 70.7 \ \text{k}\Omega$$

Therefore,

$$\hat{V}_{\text{max}} = (0.1 \ \text{mA}) \frac{70.7 \ \text{k}\Omega}{2} = 3.54 \ \text{V}$$

Example 11-18:   Again consider the circuit of Fig. 11-37.   The coefficient of coupling is increased to 0.02.   At this value of coupling the ratio of $\hat{V}_{\text{min}}$ to $\hat{V}_{\text{max}}$ equals 0.8.   Plot the frequency response.

Solution:

$$\frac{\hat{V}_{\text{min}}}{\hat{V}_{\text{max}}} = 0.8$$

From Example 11-17,

$$\hat{V}_{\text{max}} = 3.54 \ \text{V}$$

Therefore,

$$\hat{V}_{\text{min}} = (0.8)(3.54 \ \text{V}) \approx 2.83 \ \text{V}$$

Since $\omega_r = 50$ Mrad/s,

$$\omega_1 = \omega_r - k\frac{\omega_r}{2}$$

$$\omega_1 = 50 \text{ Mrad/s} - (0.02)\frac{50 \text{ Mrad/s}}{2} = 49.5 \text{ Mrad/s}$$

and

$$\omega_2 = \omega_r + k\frac{\omega_r}{2}$$

$$\omega_2 = 50 \text{ Mrad/s} + (0.02)\frac{50 \text{ Mrad/s}}{2} = 50.5 \text{ Mrad/s}$$

The other band of frequencies of interest centered about ω_r is

$$\sqrt{2} \ (\Delta\omega)$$

where

$$\Delta\omega = \omega_2 - \omega_1 = 50.5 \text{ Mrad/s} - 49.5 \text{ Mrad/s} = 1 \text{ Mrad/s}$$

Therefore,

$$\sqrt{2}\,(1 \text{ Mrad/s}) = 1.414 \text{ Mrad/s}$$

The frequency-response plot is shown in Fig. 11-38.

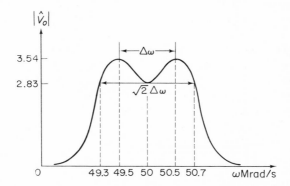

Fig. 11-38. Frequency response plot for Ex. 11-18.

Adjusting the Value of Q. Equation (11-26) shows that the coefficient of coupling is dependent on the value of Q. Since Q is given by Eq. (10-8),

$$Q = \frac{R_p}{\omega_r L_p}$$

where R_p is the total parallel resistance and L_p is the total parallel inductance, then by varying R_p we are able to control k_c. Thus either the Q of the primary or the secondary or both may be adjusted.

Stagger Tuning. The objective of *stagger tuning* is to obtain a uniform or constant output voltage over the entire bandwidth, as shown in Fig. 11-39

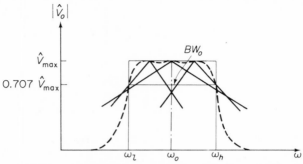

Fig. 11-39. Frequency response plot of three staggered stages. The dotted line shows how stagger tuning approaches the ideal response.

(the ideal frequency response). Stagger tuning employs the use of several stages to give the desired output. In the discussion on double-tuned circuits, we saw that it is possible to achieve a flat output if the coefficient of coupling equals the critical coefficient of coupling ($k = k_c$). As the coefficient of coupling increased, however, double peaks were prominent in the output curve.

Stagger tuning allows us to use the principles learned in the single-tuned circuits and achieve the goal of a uniform output. The transformers are considered ideal ($k = 1$) and are isolated from one another by an amplifier, as shown in Fig. 11-40. Since the coefficient of coupling equals 1 ($k = 1$), an impedance is easily reflected from one side of the transformer to the other $\bar{Z}_p = (n_1/n_2)^2 \bar{Z}_s$. The resonant frequency and the bandwidth for each transformer is the same as that for a parallel-tuned circuit:

$$\omega_r = 2\pi f_r = \frac{1}{\sqrt{L_T C_T}} \text{ rad/s}$$

and

$$\text{BW} = \frac{1}{R_T C_T} \text{ rad/s}$$

where R_T is the total parallel resistance in ohms, L_T the total parallel inductance in henries, and C_T the total parallel capacitance in farads. Although it is possible to stagger-tune double-tuned circuits, most applications employ single-tuned transformer circuits because they are easier to work with.

Stagger tuning uses connected stages in series as shown in Fig. 11-40. These stages are not tuned to the same frequency. They are, however,

tuned to frequencies above and below the *overall center frequency*.   The
bandwidths of each stage are not necessarily equal.   To obtain a uniform

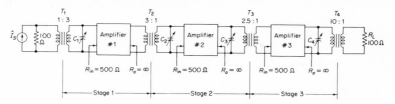

Fig. 11-40. Staggered tuned circuit for Ex. 11-19.

output using a number of staggered stages, we must know at what frequency
and bandwidth to set each stage.   This information is given in Table 11-1
for two to five stages.   This table is derived for an overall center frequency
of ω_0 and an overall bandwidth of BW_0.

TABLE 11-1

DESIGN OF STAGGERED STAGES

Number of stages	Known as	Resonant frequency of each stage	Bandwidth of each stage
2	Staggered pair	$\omega_1 = \omega_0 - 0.35\mathrm{BW}_0$	$\mathrm{BW}_1 = 0.71\mathrm{BW}_0$
		$\omega_2 = \omega_0 + 0.35\mathrm{BW}_0$	$\mathrm{BW}_2 = 0.71\mathrm{BW}_0$
3	Staggered triple	$\omega_1 = \omega_0 - 0.43\mathrm{BW}_0$	$\mathrm{BW}_1 = 0.5\mathrm{BW}_0$
		$\omega_2 = \omega_0 + 0.43\mathrm{BW}_0$	$\mathrm{BW}_2 = 0.5\mathrm{BW}_0$
		$\omega_3 = \omega_0$	$\mathrm{BW}_3 = \mathrm{BW}_0$
4	Staggered quadruple	$\omega_1 = \omega_0 - 0.19\mathrm{BW}_0$	$\mathrm{BW}_1 = 0.92\mathrm{BW}_0$
		$\omega_2 = \omega_0 + 0.19\mathrm{BW}_0$	$\mathrm{BW}_2 = 0.92\mathrm{BW}_0$
		$\omega_3 = \omega_0 - 0.46\mathrm{BW}_0$	$\mathrm{BW}_3 = 0.38\mathrm{BW}_0$
		$\omega_4 = \omega_0 + 0.46\mathrm{BW}_0$	$\mathrm{BW}_4 = 0.38\mathrm{BW}_0$
5	Staggered quintuple	$\omega_1 = \omega_0 - 0.29\mathrm{BW}_0$	$\mathrm{BW}_1 = 0.81\mathrm{BW}_0$
		$\omega_2 = \omega_0 + 0.29\mathrm{BW}_0$	$\mathrm{BW}_2 = 0.81\mathrm{BW}_0$
		$\omega_3 = \omega_0 - 0.48\mathrm{BW}_0$	$\mathrm{BW}_3 = 0.31\mathrm{BW}_0$
		$\omega_4 = \omega_0 + 0.48\mathrm{BW}_0$	$\mathrm{BW}_4 = 0.31\mathrm{BW}_0$
		$\omega_5 = \omega_0$	$\mathrm{BW}_5 = \mathrm{BW}_0$

The following example shows how Table 11-1 is used.

Example 11-19: For the stagger-tuned circuit of Fig. 11-40, calculate the resonant frequency and the bandwidth for each stage. The overall center frequency (f_0) is 455 kHz and the overall bandwidth (BW_0) is 20 kHz.

Solution:
$$\omega_0 = 2\pi f_0 = (6.28)(455\ \text{kHz}) = 2.86\ \text{Mrad/s}$$

$$BW_{0\text{rad/s}} = 2\pi BW_{0\text{Hz}} = (6.28)(20\ \text{kHz}) = 125\ \text{krad/s}$$

Using Table 11-1 for three stages,
    Stage 1:
$$\omega_1 = \omega_0 - 0.43\ BW_0$$

$$\omega_1 = 2.86\ \text{Mrad/s} - 0.43\ (125\ \text{krad/s}) \approx 2.806\ \text{Mrad/s}$$
and
$$BW_1 = 0.5\ BW_0 = 0.5\ (125\ \text{krad/s}) = 62.5\ \text{krad/s}$$

    Stage 2:
$$\omega_2 = \omega_0 = 2.86\ \text{Mrad/s}$$
and
$$BW_2 = BW_0 = 125\ \text{krad/s}$$

    Stage 3:
$$\omega_3 = \omega_0 + 0.43\ BW_0$$

$$\omega_3 = (2.86\ \text{Mrad/s}) + 0.43\ (125\ \text{krad/s}) = 2.914\ \text{Mrad/s}$$
and
$$BW_3 = 0.5\ BW_0 = 0.5\ (125\ \text{krad/s}) = 62.5\ \text{krad/s}$$

Note that in an odd number of stages it is usually the middle stage that has its resonant frequency and bandwidth equal to the overall center frequency and bandwidth. This example shows that it is the number of isolating amplifiers that determine the number of stages and not the number of transformers.

Figure 11-40 shows that each amplifier has two tuned circuits connected to it. In practice, then, the bandwidth of each stage will decrease from the calculated value. This decrease is called *bandwidth shrinkage*. One method to compensate for this shrinkage is to increase the bandwidth of stage 2 while maintaining the same resonant frequency. Therefore, the bandwidths of transformers T_2 and T_3 must be increased. An equation[1] to determine the new bandwidths is

$$\text{stage bandwidth} = \frac{\text{new bandwidth}}{1.2\sqrt{n}} \tag{11-32}$$

where n is the number of tuned circuits involved. In this example $n = 2$ because there are two single-tuned circuits (T_2 and T_3) connected to stage 2. (The stage number has nothing to do with n.) Therefore, instead of having

---

[1] Equation (11-32) is an approximate formula. The exact expression is $BW = \text{stage bandwidth}\ (\sqrt{2^{1/n} - 1})$

a bandwidth of 125 krad/s, T_2 and T_3 should have new bandwidths to compensate for shrinkage. These bandwidths are

$$\text{new bandwidth} = (125 \text{ krad/s})(1.2\sqrt{2}) = 212 \text{ krad/s}$$

The bandwidths of stage 1 and stage 3 have remained unchanged. All compensation was done on stage 2.

Example 11-20: For the stagger-tuned circuit of Fig. 11-40, calculate the value of (a) C_1 and (b) C_3.

Solution: The passive network of just T_1 is drawn in Fig. 11-41a. Reflecting the 100 Ω into the secondary yields

$$R_2 = \left(\frac{3}{1}\right)^2 (100 \,\Omega) = 900 \,\Omega$$

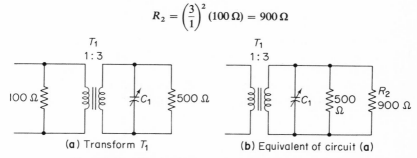

(a) Transform T_1 (b) Equivalent of circuit (a)

Fig. 11-41. Circuit diagrams for part (a) of Ex. 11-20.

The equivalent circuit is shown in Fig. 11-41b, from which the bandwidth is

$$\text{BW}_1 = \frac{1}{R_p C_1}$$

where

$$R_p = \frac{(500 \,\Omega)(900 \,\Omega)}{500 \,\Omega + 900 \,\Omega} = 321 \,\Omega$$

and then

$$C_1 = \frac{1}{\text{BW}_1 R_p} = \frac{1}{(62.5 \text{ krad/s})(900 \,\Omega)} = 0.0178 \,\mu\text{F}$$

(b) The passive network of just T_3 is shown in Fig. 11-42a.

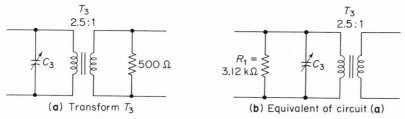

(a) Transform T_3 (b) Equivalent of circuit (a)

Fig. 11-42. Circuit diagrams for part (b) of Ex. 11-20.

Reflecting the 500 Ω resistance into the primary yields

$$R_1 = \left(\frac{2.5}{1}\right)^2 (500 \, \Omega) = 3.12 \, \text{k}\Omega$$

The equivalent circuit is shown in Fig. 11-42b, from which the bandwidth is

$$\text{BW} = \frac{1}{R_p C_3}$$

Remember, however, that the bandwidths of T_2 and T_3 have been increased to account for overall shrinkage. The new bandwidth equals 212 krad/s. Therefore,

$$C_3 = \frac{1}{\text{BW} \cdot R_p} = \frac{1}{(212 \, \text{krad/s})(3.12 \, \text{k}\Omega)} = 0.0015 \, \mu\text{F}$$

In conclusion, if the transformers of a stagger-tuned network are studied individually, they are treated as single-tuned circuits. The capacitors in the stagger-tuned network are shown as variable capacitors. The reason for this is that in an actual network capacitors that are varied to obtain the correct resonant frequency and/or bandwidth.

SUMMARY

The fundamental concept of this chapter was the transfer of energy from one circuit to another through a magnetic field. Section 11-1 introduced mutual inductance along with its voltage – current relationships, showed how an induced voltage is repesented in a circuit diagram, and explained coefficient of coupling and dot notation. Section 11-2 showed how mutual inductance is calculated from two measured values of self-inductance.

Since mutual inductance is the phenomenon basic to the operation of the transformer, we were then led to Section 11-3, which began a general discussion on transformers. Iron-core transformers used in power and audio systems approach what is known as an ideal transformer. Section 11-4 listed the properties and conditions under which a transformer may be considered ideal. Section 11-5 dealt with an autotransformer – a one-coil transformer – its advantages and disadvantages. Since no device is perfectly ideal, it becomes necessary to account for losses, that is, to determine the efficiency of a device. Section 11-6 treated the efficiency of a transformer. If a capacitor is connected across the terminals of a transformer, a tuned circuit is created. Section 11-7 dealt with three commonly used transformer tuned circuits – single tuned, double tuned, and stagger tuned.

PROBLEMS

11-1. For the two coils of Fig. 11-8a, L_A is measured as 5 mH, L_B as 25 mH, and
L_1 as 2 mH. Calculate
(a) the mutual inductance,
(b) the self-inductance, L_2,
(c) the coefficient of coupling.

11-2. For each of the coils of Fig. 11-43, calculate
(a) mutual inductance,
(b) coefficient of coupling.

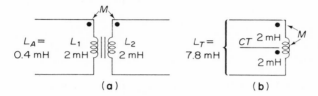

Fig. 11-43. Circuit diagrams for Prob. 11-2.

11-3. Calculate the input impedance for the circuit of Fig. 11-12 for $k = 0.4$.

11-4. For the circuit of Fig. 11-12, remove R_L and replace it with a 10-μF capacitor
and calculate the input impedance.

11-5. Repeat Problem 11-4 with $k = 1$.

11-6. If the load resistor in the circuit of Fig. 11-14 is increased to 4 kΩ, calculate
the primary current and voltage. $\hat{V}_0 = 20$ V.

11-7. For each of the ideal transformer circuits of Fig. 11-44, calculate

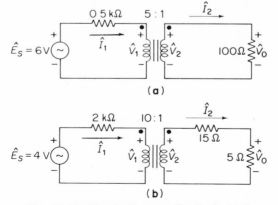

Fig. 11-44. Circuit diagrams for Prob. 11-7.

    (a)   reflected resistance, R_R,
    (b)   input resistance, R_{in},
    (c)   primary current, $\hat{I}_1$,
    (d)   primary voltage, $\hat{V}_1$,
    (e)   secondary current, $\hat{I}_2$,
    (f)   secondary voltage, $\hat{V}_2$,
    (g)   output voltage, $\hat{V}_0$.

11-8. Refer to the circuit of Fig. 11-18. Calculate the reflected capacitance if the turns ratio $n_1/n_2 = 10/1$.

11-9. Repeat Example 11-10 for the following turns ratios:
$$\frac{n_1}{n_2} = \frac{6}{1} \qquad \frac{n_1}{n_3} = \frac{4}{1}$$

11-10. Repeat Example 11-11 for a turns ratio $n_1/n_2 = 2/1$.

11-11. For the autotransformer circuit of Fig. 11-45, calculate
    (a) $\hat{I}_P$ (c) $\hat{V}_L$ (e) $\hat{I}_S$

    (b) $\hat{V}_1$ (d) $\hat{I}_L$ (f) $\dfrac{n_1}{n_3}$

    Let $n_1/n_2 = 4/1$.

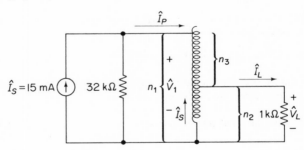

Fig. 11-45. Circuit diagram for Prob. 11-11.

11-12. If the inductance between terminals bc of Fig. 11-46 is 5 mH, determine the parallel equivalent circuit between terminals ac. Let $n_1/n_2 = 5/1$ and $Q = 50$.

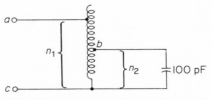

Fig. 11-46. Circuit diagram for Prob. 11-12.

11-13. Repeat Example 11-13 for a transformer efficiency of 90 per cent.

11-14. The transformer in the circuit of Fig. 11-47 has an efficiency of 95 per cent. Calculate

(a) $\hat{I}_1$ (c) $\hat{I}_2$.   (e) $\hat{V}_0$
(b) $\hat{V}_1$ (d) $\hat{V}_2$.

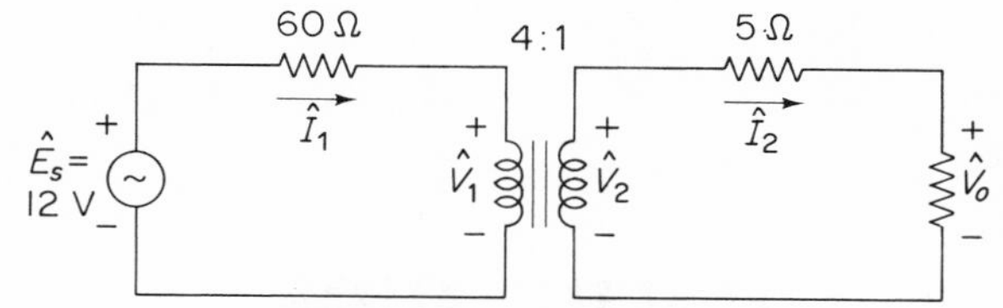

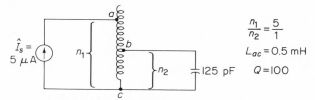

Fig. 11-47. Circuit diagram for Prob. 11-14.

11-15. Repeat Example 11-14 for $Q_1 = 100$.

11-16. For the single-tuned circuit of Fig. 11-28, the bandwidth is to be increased to 100 krad/s about the same resonant frequency. What compensation may be made

(a) on the secondary side of the transformer?,
(b) on the primary side of the transformer?

11-17. For the single-tuned circuit of Fig. 11-48, determine

(a) resonant frequency,
(b) bandwidth,
(c) frequency response plot of $\hat{V}_0$ versus ω.

Fig. 11-48. Circuit diagram for Prob. 11-17.

11-18. Repeat Example 11-16 for $Q = 60$.

11-19. For the double-tuned circuit of Fig. 11-37, let $Q_1 = Q_2 = 50$ and repeat Example 11-17.

11-20. Repeat Example 11-18 for the quality factors used in Problem 11-19 and the ratio of $\hat{V}_{min}$ to $\hat{V}_{max}$ equals 0.9.

11-21. The overall bandwidth (BW_0) for the staggered-tuned circuit of Fig. 11-40 is to be 10 kHz. Calculate the resonant frequency and the bandwidth for each stage.

11-22. For the overall bandwidth used in Problem 11-21, calculate C_1 and C_3.

CHAPTER 12

harmonics
and fourier analysis

In Chapter 2 we defined different types of waveforms. Our analysis of circuits, thus far, has dealt only with sinusoidal waves. However, we have found that it is possible to have a periodic waveform that is not sinusoidal.

In Chapter 9 we analyzed networks with more than one energy source. We restricted this analysis so that all ac sources in a particular network had the *same* frequency. However, in a network having two audio oscillators, it is possible to have one at 1 kHz while the other is generating at 3 kHz. In this chapter, we shall treat such problems.

Section 12-1 shows how a periodic nonsinusoidal waveform is generated, and the following section describes how to determine the effective values of such waveforms. Sections 12-3 and 12-4 deal with the Fourier series. The first of these two sections gives a general explanation for this solution for a periodic waveform. The next section focuses on a graphic analysis of Fourier coefficients. The procedure for calculating the impedance circuit for each harmonic is described in Section 12-5 and the last section shows how to add and subtract nonsinusoidal waveforms.

12-1. FUNDAMENTAL AND HARMONICS

To analyze a periodic nonsinusoidal waveform, it is helpful to know how one is generated.   The circuit of Fig. 12-1a shows two audio oscilla-

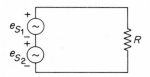

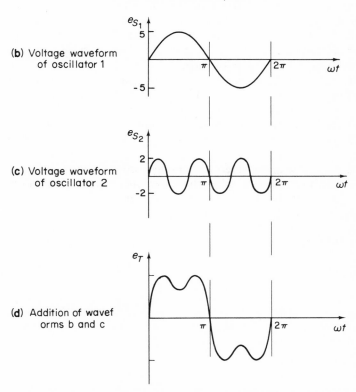

(a) Circuit having two voltage sources
set at different frequencies

(b) Voltage waveform
of oscillator 1

(c) Voltage waveform
of oscillator 2

(d) Addition of wavef
orms b and c

Fig. 12-1. Generation of a periodic nonsinusoidal waveform.

tors set at different amplitudes and frequencies.   One oscillator has a peak voltage of 5 V and a frequency of 1 kHz (6.28 or 2π krad/s).   The other os-cillator has a peak voltage of 2 V and a frequency of 3 kHz (18.84 or 6π krad/s).   The waveforms are shown in Fig. 12-1b and c.   The resultant waveform shown in Fig. 12-1d is plotted by adding the instantaneous values of Fig. 12-1b and c. In equation form,

$$e_{s1} = 5 \sin 2\pi kt$$

and

$$e_{s2} = 2 \sin 6\pi kt$$

The total voltage is

$$e_T = e_{s1} + e_{s2}$$

or

$$e_T = 5 \sin 2\pi kt + 2 \sin 6\pi kt$$

Analyzing the nonsinusoidal waveform of Fig. 12-1d shows that it has a fundamental a frequency ($f = 1/T$) of 1 kHz. This is the *minimum* frequency of the nonsinusoidal waveform and is called the *fundamental frequency*. All other higher-order frequencies that are integers (2, 3, 4, 5, etc.) of the fundamental are called *harmonics*. In this example, one other frequency is present and it has a frequency three times the fundamental. It is called the third harmonic.

Note that the harmonic terms are integers (whole-number multiples) of the fundamental frequency. The sin $5\omega t$ is the fifth harmonic of the sin ωt. However, the sin $3.5\omega t$ is not a harmonic of the sin ωt. Rather the sin $3.5\omega t$ and the sin ωt may be harmonics of the sin $0.5\omega t$. The sin $3.5\omega t$ is the seventh harmonic and the sin ωt is the second harmonic of the sin $0.5\omega t$.

In general, the amplitude of the higher-order harmonics is smaller than that of either the fundamental or the lower-order harmonics. For example, we should expect that the amplitude of the ninth harmonic is smaller than the amplitude of the fifth harmonic.

Example 12-1: For the circuit of Fig. 12-2, plot the total voltage.

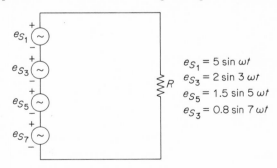

$$e_{S_1} = 5 \sin \omega t$$
$$e_{S_3} = 2 \sin 3 \omega t$$
$$e_{S_5} = 1.5 \sin 5 \omega t$$
$$e_{S_3} = 0.8 \sin 7 \omega t$$

Fig. 12-2. Circuit diagram for Ex. 12-1.

Solution: The numbers 1, 3, 5, and 7 designate the fundamental, third, fifth, and seventh harmonics, respectively. Figure 12-3 is a plot of the fundamental and the harmonics and their addition. Since the harmonics do not lead or lag the fundamental, they are *in phase* and begin at the origin. Note that in Fig. 12-3, the number of peaks of the resultant waveform equals the number of waves added. The resultant waveform has a frequency equal to the fundamental.

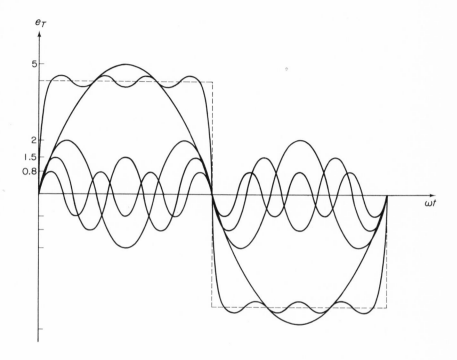

Fig. 12-3. Addition of a fundamental and three harmonics to approximate a square wave.

Both Figs. 12-1d and 12-3 show that the fundamental and the harmonics are *in phase*. That is, each wave (including the fundamental) starts at zero and goes positive. It is *not* required, however, that the harmonics be in phase. For example, the fundamental may be $4 \sin \omega t$ while the *third* harmonic is $-2 \sin 3\omega t$, lagging the fundamental by 180°, as shown in Fig. 12-4a. Another example is shown in Fig. 12-4b — the fundamental is $3 \sin \omega t$ and the second harmonic is $-1 \cos 2\omega t$. Therefore, the harmonic terms may lead or lag the fundamental by a specified phase angle. Note that the phase angle of the harmonic term is based on the fact that one cycle of the harmonic is 360° (2π rad). The dotted lines in Fig. 12-4a and b are the resultant waveforms.

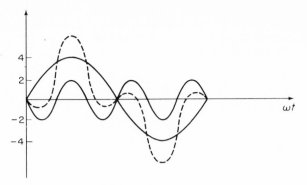

(**a**) Addition of $4 \sin \omega t$ and $-2 \sin 3\omega t$

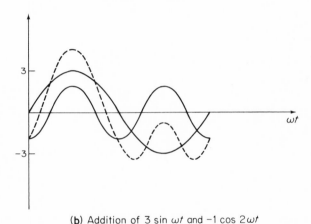

(**b**) Addition of $3 \sin \omega t$ and $-1 \cos 2\omega t$

Fig. 12-4. Examples of fundamental and harmonics not in phase.

12-2.   EFFECTIVE VALUE OF A NONSINUSOIDAL WAVEFORM

In Chapter 2 and Chapter 3 we defined and determined the effective (rms) value of a sine wave. The expressions for the rms voltage and rms current are, respectively,

$$\hat{E} = \sqrt{PR}$$

and

$$\hat{I} = \sqrt{\frac{P}{R}}$$

Normally we cannot add powers because they are not a linear function (refer to Example 5-17). However, because of a unique property of harmonic circuits, we are able to add the powers delivered to a load by each harmonic acting independently. This is similar to the method of superposition we applied in Chapter 5 and Chapter 9 to find current. The circuit of Fig. 12-5 contains both a dc source and ac sources of different frequencies.

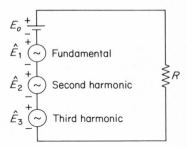

Fig. 12-5. Circuit diagram used to illustrate effective value.

Using the unique property of harmonic circuits, the total power delivered to the resistor is

$$P_T = \frac{E_0^2}{R} + \frac{\hat{E}_1^2}{R} + \frac{\hat{E}_2^2}{R} + \frac{\hat{E}_3^2}{R}$$

Therefore, the total rms value of voltage is

$$\hat{E}_T = \sqrt{PR} = \sqrt{E_0^2 + \hat{E}_1^2 + \hat{E}_2^2 + \hat{E}_3^2} \qquad (12\text{-}1)$$

where $\hat{E}_T$ is the total rms value; E_0 the dc value; and $\hat{E}_1$, $\hat{E}_2$, and E_3 the rms values of the fundamental, second, and third harmonics, respectively.

Similarly, the expression for the effective value of current for a nonsinusoidal waveform may be written

$$P_T = I_0^2 R + \hat{I}_1^2 R + \hat{I}_2^2 R + \hat{I}_3^2 R$$

$$\hat{I}_T = \sqrt{\frac{P_T}{R}} = \sqrt{I_0^2 + \hat{I}_1^2 + \hat{I}_2^2 + \hat{I}_3^2} \qquad (12\text{-}2)$$

where $\hat{I}_T$ is the total rms value; I_0 the dc value; and $\hat{I}_1$, $\hat{I}_2$, and $\hat{I}_3$ are the rms values of the fundamental, second, and third harmonics, respectively.

Equations (12-1) and (12-2) may be generalized to any number of harmonic terms by adding the square of the rms values to the other terms under the radical.

Example 12-2: The circuit of Fig. 12-5 has the values $E_0 = 2$ V, $\hat{E}_1 = 5$ V, $\hat{E}_2 = 3$ V, $\hat{E}_3 = 1$ V, and $R = 2$ kΩ. Calculate the effective values of voltage and current.

Solution: According to Eq. (12-1),

$$\hat{E}_T = \sqrt{(2 \text{ V})^2 + (5 \text{ V})^2 + (3 \text{ V})^2 + (1 \text{ V})^2}$$

$$\hat{E}_T = \sqrt{39 \text{ V}^2} = 6.26 \text{ V}$$

and

$$\hat{I}_T = \frac{\hat{E}_T}{R} = \frac{6.26 \text{ V}}{2 \text{ k}\Omega} = 3.13 \text{ mA}$$

Example 12-3: For the circuit of Fig. 12-2, calculate the total effective value of voltage.

Solution: Before applying Eq. (12-1), the rms value of each voltage source must be calculated:

$$\hat{E}_1 = \frac{5 \text{ V}}{\sqrt{2}} = 3.53 \text{ V}$$

$$\hat{E}_3 = \frac{2 \text{ V}}{\sqrt{2}} = 1.41 \text{ V}$$

$$\hat{E}_5 = \frac{1.5 \text{ V}}{\sqrt{2}} = 1.06$$

$$\hat{E}_7 = \frac{0.8 \text{ V}}{\sqrt{2}} = 0.566 \text{ V}$$

Therefore,

$$\hat{E}_T = \sqrt{(3.43 \text{ V})^2 + (1.41 \text{ V})^2 + (1.06 \text{ V})^2 + (0.566 \text{ V})^2}$$

$$\hat{E}_T = \sqrt{15.94 \text{ V}^2} \approx 4 \text{ V}$$

12-3. FOURIER SERIES

Sections 12-1 and 12-2 have established that a constant (a dc component), a fundamental, and a number of harmonics produce a periodic waveform. A French mathematician, Jean Baptiste Fourier, showed that the reverse is also true. Any periodic waveform is composed of a constant, a fundamental, and harmonics. The general solution of a periodic waveform is called a *Fourier series* and is expressed as

$$y = A_0 + A_1 \sin \omega t + A_2 \sin 2 \omega t + A_3 \sin 3 \omega t \ldots$$

$$+ B_1 \cos \omega t + B_2 \cos 2 \omega t + B_3 \cos 3 \omega t \ldots \quad (12\text{-}3)$$

where y may represent voltage, current, or power; A_0 is the constant or dc component [it is the average value for one complete cycle and is given by Eq. (2-14)]; A_1, A_2, and A_3 represent the coefficients of the sine terms of the fundamental, second, and third harmonic terms, respectively; and B_1, B_2, and B_3 represent the coefficients of the cosine terms of the fundamental, second, and third harmonic terms, respectively. Depending on the type of periodic waveform, many of the coefficients in Eq. (12-3) may equal zero. An examination of the waveform quickly shows which terms will be zero.

If the waveform has as much area above the x axis as below it, the average value is zero ($A_0 = 0$). The waveforms of Fig. 12-6a, b, c, and d all have

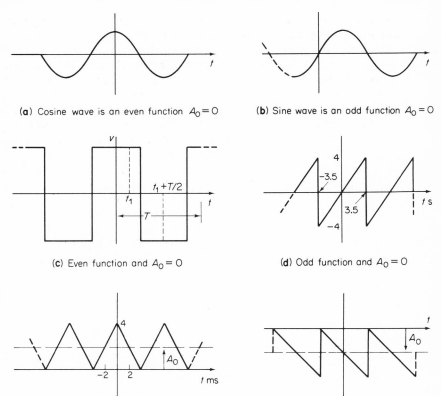

(a) Cosine wave is an even function $A_0 = 0$ (b) Sine wave is an odd function $A_0 = 0$

(c) Even function and $A_0 = 0$ (d) Odd function and $A_0 = 0$

(e) Even function and $A_0 \neq 0$ (f) Odd function and $A_0 \neq 0$

Fig. 12-6. Periodic waveforms illustrating even and odd functions.

$A_0 = 0$; the waveforms of Fig. 12-6e and f have average values other than zero and A_0 can be calculated from Eq. (2-14).

A waveform that is symmetrical about the y axis is called an *even function*. Examples of even functions are shown in Fig. 12-6a, c, and e. Note that a cosine wave is an even function. For all even functions, the coefficients of the sine terms (A_1, A_2, A_3, etc.) are equal to zero. A waveform is an even function if the amplitude at $+t$ (or $+\omega t$) equals that at $-t$ (or

$-\omega t$). For example, the amplitude of the waveform of Fig. 12-6e at $+2$ ms equals the amplitude of that at -2 ms. This waveform is, therefore, an even function.

The waveform of Fig. 12-6c shows that

$$v(t_1) = -v\left(t_1 + \frac{T}{2}\right) \tag{12-4}$$

This waveform is said to have half-wave or mirror symmetry. The even harmonics of both the sine and cosine terms are zero. Or to put it another way, such waveforms contain only odd harmonics.

A waveform that is symmetrical about an origin is called an *odd function*. Examples of odd functions are shown in Fig. 12-6b, d, and f. Note that a sine wave is an odd function. For all odd functions the coefficients of the cosine terms $(B_1, B_2, B_3,$ etc.) equal zero. A waveform is an odd function if the amplitude at $+t$ (or $+\omega t$) is the negative of the amplitude at $-t$ (ot $-\omega t$). For example, the amplitude of the waveform of Fig. 12-6d at $+3.5$ s ($+4$ V) is the negative of the amplitude at -3.5 s (-4 V).

12-4.   GRAPHICAL ANALYSIS OF FOURIER COEFFICIENTS

A general method of evaluating the coefficients of a Fourier series is by graphical analysis. This method requires finding the area under the curve of *each* harmonic. This is done by (1) dividing each harmonic into m equal parts as shown in Fig. 12-7, (2) determining the area of each rec-

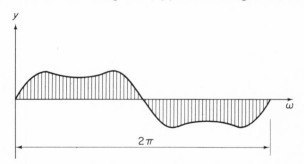

Fig. 12-7. Periodic waveform divided into a number of equal parts.

tangle, (3) summing the areas, and (4) dividing the summation by m. In equation form,

$$A_n = \frac{1}{m}\sum_0^m y_\theta \sin_n \theta \tag{12-5}$$

$$B_n = \frac{1}{m}\sum_0^m y_\theta \cos_n \theta \tag{12-6}$$

where A_n and B_n are the coefficients of the sine and cosine terms, respectively; n equals 1, 2, 3, etc., which correspond to the fundamental, second harmonic, third harmonic terms respectively; and y_θ is the amplitude at the angle θ. Angles of every $10°$ should be accurate for most examples. If, however, greater accuracy is desired, choose divisions of every $5°$.

The average value or dc component is determined by Eq. (2-14):

$$A_0 = \text{average value} = \frac{\text{area}}{\text{base}}$$

The discussion of Section 12-3 showed that if the waveform is an even function, all the coefficients of the sine terms (A_1, A_2, A_3, etc.) are zero. Similarly, if the waveform is an odd function, all the coefficients of the cosine terms (B_1, B_2, B_3, etc.) are zero. Also, if the positive area equals the negative area for one cycle, the Fourier series contains only odd harmonics (A_1, B_1, A_3, B_3, A_5, B_5, etc.)

For a waveform that has symmetrical positive and negative areas, it is necessary to calculate only the area under the curve for the first $180°$ and multiply it by 2. The reason for this is that the area under the curve for the second $180°$ equals the area of the first $180°$.

Example 12-4: Obtain the Fourier series for the waveform shown in Figure 12-8.

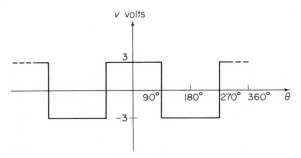

Fig. 12-8. Square wave of Ex. 12-4.

Solution: The waveform of Fig. 12-8 has as much positive area as negative area for one complete cycle. Therefore, the waveform will be analyzed over the first $180°$ and multiplied by 2. Also, $A_o = 0$. Since the waveform is symmetrical about the y axis, it is an even function and thus the coefficients of the sine terms equal zero ($A_n = 0$).

Dividing the wave at every $10°$ yields

$$m = \frac{180°}{10°} = 18$$

Fundamental — B_1

θ	$\cos \theta$	v (V)	$v \cos \theta$
10°	0.985	3	2.95
20°	0.940	3	2.82
30°	0.866	3	2.60
40°	0.766	3	2.30
50°	0.643	3	1.93
60°	0.500	3	1.50
70°	0.342	3	1.03
80°	0.174	3	0.52
90°	0.000	3	0.00
100°	−0.174	−3	0.52
110°	−0.342	−3	1.03
120°	−0.500	−3	1.50
130°	−0.643	−3	1.93
140°	−0.766	−3	2.30
150°	−0.866	−3	2.60
160°	−0.940	−3	2.82
170°	−0.985	−3	2.95
180°	−1.000	−3	3.00
			34.30

Therefore,

$$B_1 = \frac{34.30}{18} \times 2 = 3.80$$

Third Harmonic - B_3

θ	$\cos 3\theta$	v (V)	$v \cos 3\theta$
10°	0.866	3	2.60
20°	0.500	3	1.50
30°	0.000	3	0.00
40°	−0.500	3	−1.50
50°	−0.866	3	−2.60
60°	−1.000	3	−3.00
70°	−0.866	3	−2.60
80°	−0.500	3	−1.50
90°	−0.000	3	−0.00
100°	0.500	−3	−1.50
110°	0.866	−3	−2.60
120°	1.000	−3	−3.00

(Continued)

θ	$\cos \theta$	v (V)	$v \cos 3\theta$	
130°	0.866	−3		−2.60
140°	0.500	−3		−1.50
150°	0.000	−3		−0.00
160°	−0.500	−3	1.50	
170°	−0.866	−3	2.60	
180°	−1.000	−3	3.00	
			11.20	−22.40

Therefore,

$$B_3 = \frac{11.2 - 22.4}{18} \times 2 = -1.25$$

Fifth Harmonic—B_5

θ	$\cos 5\theta$	v (V)	$v \cos 5\theta$	
10°	0.643	3	1.93	
20°	−0.174	3		−0.52
30°	−0.866	3		−2.60
40°	−0.940	3		−2.82
50°	−0.342	3		−1.03
60°	0.500	3	1.50	
70°	0.985	3	2.95	
80°	0.766	3	2.30	
90°	0.000	3	0.00	
100°	−0.766	−3	2.30	
110°	−0.985	−3	2.95	
120°	−0.500	−3	1.50	
130°	0.342	−3		−1.03
140°	0.940	−3		−2.82
150°	0.866	−3		−2.60
160°	0.174	−3		−0.52
170°	−0.643	−3	1.93	
180°	−1.000	−3	3.00	
			20.36	−13.94

Therefore,

$$B_5 = \frac{20.36 - 13.94}{18} \times 2 = 0.715$$

The Fourier series for the waveform of Fig. 12-8 is

$$v = 3.8 \cos \omega t - 1.25 \cos 3\omega t + 0.715 \cos 5\omega t$$

Example 12-5:   For the waveform of Fig. 12-9 determine the Fourier series.

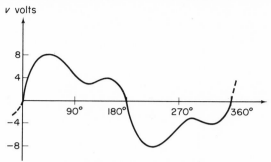

Fig. 12-9. Waveform of Ex. 12-5.

Solution: The waveform of Fig. 12-9 has as much positive area as negative area; therefore, $A_0 = 0$.   The waveform is an odd function; therefore, all the coefficients of the cosine terms equal zero ($B_n = 0$):

$$m = \frac{180°}{10°} = 18$$

Fundamental—A_1

θ	$\sin \theta$	v (V)	$v \sin \theta$
0°	0.0	0	0.0
10°	0.173	4.0	0.68
20°	0.342	6.2	2.12
30°	0.500	7.3	3.65
40°	0.643	7.9	5.07
50°	0.766	8.0	6.14
60°	0.866	7.8	6.75
70°	0.940	6.5	6.11
80°	0.985	5.4	5.31
90°	1.000	4.5	4.50
100°	0.985	3.7	3.64
110°	0.940	3.0	2.82
120°	0.866	3.2	2.78
130°	0.766	3.6	2.76
140°	0.643	4.0	2.57
150°	0.500	4.2	2.10
160°	0.342	3.5	1.20
170°	0.173	2.5	0.43
180°	0.0	0.0	0.0

58.63

Therefore,

$$A_1 = \frac{58.63}{18} \times 2 = 6.53$$

Third Harmonic—A_3

θ	$\sin 3\theta$	v (V)	$v \sin 3\theta$	
0°	0	0	0.00	
10°	0.500	4.0	2.00	
20°	0.866	6.2	5.38	
30°	1.000	7.3	7.30	
40°	0.866	7.9	6.85	
50°	0.500	8.0	4.00	
60°	0.000	7.8	0.00	
70°	−0.500	6.5		−3.25
80°	−0.866	5.4		−4.68
90°	−1.000	4.5		−4.50
100°	−0.866	3.7		−3.21
110°	−0.500	3.0		−1.50
120°	0.000	3.2		0.00
130°	0.500	3.6	1.80	
140°	0.866	4.0	3.46	
150°	1.000	4.2	4.20	
160°	0.866	3.5	3.03	
170°	0.500	2.5	1.25	
180°	0.000	0.0	0.00	
			39.27	−17.14

Therefore,

$$A_3 = \frac{39.27 - 17.14}{18} \times 2 = 2.46$$

Fifth Harmonic $-A_5$

θ	$\sin 5\theta$	v (V)	$v \sin 5\theta$	
0°	0	0	0.0	
10°	0.766	4.0	3.17	
20°	0.985	6.2	6.10	
30°	0.500	7.3	3.67	
40°	−0.342	7.9		−3.70
50°	−0.940	8.0		−7.50
60°	−0.866	7.8		−6.75

(Continued)

θ	$\sin 5\theta$	v (V)	$v \sin 5\theta$	
70°	−0.174	6.5		−1.13
80°	0.643	5.4	3.37	
90°	1.000	4.5	4.50	
100°	0.643	3.7	2.37	
110°	−0.174	3.0		−0.52
120°	−0.866	3.2		−2.77
130°	−0.940	3.6		−3.38
140°	−0.342	4.0		−1.37
150°	0.500	4.2	2.10	
160°	0.985	3.5	3.45	
170°	0.766	2.5	1.92	
180°	0.000	0.0	0.0	
			30.65	−27.12

Therefore,

$$A_5 = \frac{30.65 - 27.12}{18} \times 2 = 0.392$$

The Fourier series for the waveform of Fig. 12-9 is

$$v = 6.53 \sin \omega t + 2.46 \sin 3\omega t + 0.392 \sin 5\omega t$$

12-5. CIRCUIT ANALYSIS FOR A NONSINUSOIDAL INPUT

In the impedance networks analyzed in Chapter 9 neither X_L ($X_L = \omega L$) nor X_C ($X_C = 1/\omega C$) changed, because the frequency of each energy source was the same. In this chapter we have seen that a nonsinusoidal energy source may be represented by a number of energy sources, each of a different frequency. Since both X_L and X_C are frequency dependent, the impedance must be calculated for each frequency (harmonic). This procedure is shown in the following example.

Example 12-6: For the circuit of Fig. 12-10, determine the expression for the current, i. The voltage

$$e = 10 \sin \omega t + 17 \sin 3\omega t + 8 \sin (5\omega t + 30°)$$

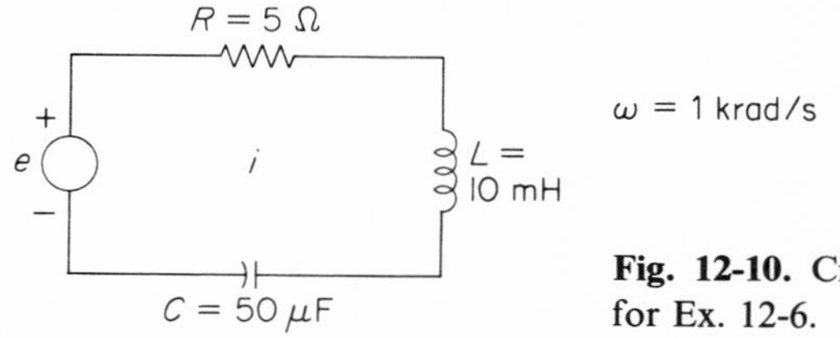

$\omega = 1$ krad/s

Fig. 12-10. Circuit diagram for Ex. 12-6.

Solution: Since both inductive reactance and capacitive reactance are frequency dependent, the fundamental and each harmonic must be analyzed separately. The subscripts 1, 3, and 5 correspond to the fundamental, third, and fifth harmonics, respectively.

Fundamental

$$\hat{E}_1 = \frac{10 \text{ V}}{\sqrt{2}} = 7.07 \text{ V}$$

$$R_1 = 5 \,\Omega$$

$$X_{L_1} = \omega L_1 = (1 \text{ krad/s}) (10 \text{ mH}) = 10 \,\Omega$$

$$X_{C_1} = \frac{1}{\omega C_1} = \frac{1}{(1 \text{ krad/s}) (50 \,\mu\text{F})} = 20 \,\Omega$$

$$\bar{Z}_1 = (5 + j10 - j20) \,\Omega$$

$$\bar{Z}_1 = (5 - j10) \,\Omega = 10.9 \,\underline{/63.8°} \,\Omega$$

Then

$$\hat{I}_1 = \frac{\hat{E}_1}{\bar{Z}_1} = \frac{7.07 \text{ V}}{10.9 \,\underline{/63.8°} \,\Omega} = 0.65 \,\underline{/-63.8°} \text{ A}$$

The expression of current due to the fundamental component is

$$i_1 = \sqrt{2} \,(0.65) \sin (\omega t - 63.8°) \text{ A}$$

$$i_1 = 0.92 \sin (\omega t - 63.8°) \text{ A}$$

Third Harmonic

$$\hat{E}_3 = \frac{17}{\sqrt{2}} = 12 \text{ V}$$

$$R_3 = 5 \,\Omega$$

$$X_{L_3} = 3X_{L_1} = 3(10 \,\Omega) = 30 \,\Omega$$

$$X_{C_3} = \frac{X_{C_1}}{3} = \frac{20 \,\Omega'}{3} = 6.66 \,\Omega$$

$$\bar{Z}_3 = (5 + j30 - j6.67) \,\Omega$$

$$\bar{Z}_3 = (5 + j23.33) \,\Omega = 26 \,\underline{/77.8°} \,\Omega$$

Then

$$\hat{I}_3 = \frac{\hat{E}_3}{\bar{Z}_3} = \frac{12 \text{ V}}{26 \,\underline{/77.8°} \,\Omega} = 0.46 \,\underline{/-77.8°} \text{ A}$$

The expression of current due to the third harmonic is

$$i_3 = \sqrt{2} \,(0.46) \sin (3\omega t - 77.8°) \text{ A}$$

$$i_3 = 0.65 \sin (3\omega t - 77.8°) \text{ A}$$

Fifth Harmonic

$$\hat{E}_5 = \frac{8\text{ V}}{\sqrt{2}} = 5.65\text{ V}$$

$$R_5 = 5\,\Omega$$

$$X_{L_5} = 5X_{L_1} = (5)(10\,\Omega) = 50\,\Omega$$

$$X_{C_5} = \frac{X_{C_5}}{5} = \frac{20\,\Omega}{5} = 4\,\Omega$$

$$\bar{Z}_5 = (5 + j50 - j4)\,\Omega$$

$$\bar{Z}_5 = (5 + j46)\,\Omega = 46.4\,\underline{/83.8°}\,\Omega$$

$$I_5 = \frac{\hat{E}_5}{\bar{Z}_5} = \frac{5.65\text{ V}}{46.4\,\underline{/83.8°}\,\Omega} = 0.122\,\underline{/-83.8°}\,\text{A}$$

The expression of current due to the fifth harmonic is

$$i_5 = \sqrt{2}\,(0.122)\sin(5\,\omega t + 30° - 83.8°)\,\text{A}$$

$$i_5 = 0.173\sin(5\omega t - 53.8°)\,\text{A}$$

The expression for the total current is

$$i_T = 0.92\sin(\omega t - 63.8°) + 0.65\sin(3\omega t - 77.8°) + \\ + 0.173\sin(5\omega t - 53.8°)\,\text{A}$$

Thus the total current is determined by using superposition. The difference between this example and the examples that were analyzed in Chapter 9 is that the inductive and capacitive reactance values have to be calculated for each frequency.

12-6. ADDITION AND SUBTRACTION OF NONSINUSOIDAL WAVEFORMS

To either add or subtract nonsinusoidal waveforms, they first must be expressed in terms of a Fourier series. In this form the addition (or subtraction) is accomplished by adding (or subtracting) the terms with the same frequency.

Example 12-7: Obtain the sum of v_1 and v_2:

$$v_1 = 5 + 10\sin\omega t + 4\sin 2\omega t + 1\sin 3\omega t + 0.7\sin 5\omega t$$

$$v_2 = 6 + 8\sin\omega t + 2.5\sin 3\omega t + 2\sin(5\omega t + \tfrac{\pi}{4})$$

Solution: The addition of the dc terms

$$v_{T_0} = 5\text{ V} + 6\text{ V} = 11\text{ V}$$

The fundamentals

$$v_{T_1} = 10 \sin \omega t + 8 \sin \omega t = 18 \sin \omega t$$

Only the peak values are added and the frequency remains unchanged. Since v_2 does not contain a second harmonic term,

$$v_{T_2} = 4 \sin 2\omega t$$

The third harmonic

$$v_{T_3} = 1 \sin 3\omega t + 2.5 \sin 3\omega t = 3.5 \sin 3\omega t$$

The fifth harmonic:

Since the phase angles are not the same, it is necessary to convert both terms to rectangular form and then add:

$$0.7 \sin 5\omega t = \frac{0.7}{\sqrt{2}} \underline{/0} \text{ rad} = 0.495 + j0$$

and

$$2 \sin (5\omega t + \tfrac{\pi}{4}) = \frac{2}{\sqrt{2}} \underline{/\tfrac{\pi}{4}} \text{ rad} = 1.414 \underline{/\tfrac{\pi}{4}} \text{ rad} = 1 + j1$$

$$\hat{V}_{T_5} = 0.495 + j0 + 1 + j1 = 1.495 + j1 = 1.8 \underline{/33.8°}$$

or

$$\hat{V}_{T_5} = 1.8 \underline{/0.59} \text{ rad}$$

Therefore,

$$v_{T_5} = \sqrt{2} (1.8) (0.59 \text{ rad})$$

$$v_{T_5} = 2.55 \sin (5\omega t + 0.59 \text{ rad})$$

Then

$$v_T = 11 + 18 \sin \omega t + 4 \sin 2\omega t + 3.5 \sin 3\omega t + 2.55 \sin (5\omega t + 0.59 \text{ rad})$$

SUMMARY

In this chapter we introduced the concept of networks in which the sources were not of the same frequency. The first section began with a description of a periodic nonsinusoidal waveform, which results from the addition of the waveforms of two such sources. Section 12-2 showed the method for determining the effective value of nonsinusoidal waveforms.

The generation of the waveform is the first part of the problem. Then we must be able to analyze it. Section 12-3 gave the basic equation for a nonsinusoidal periodic waveform. This analysis is known as Fourier analysis of a waveform. A graphical method of solving for the coefficients of a Fourier equation was given in Section 12-4. Since both inductive and capacitive reactance are frequency dependent, Section 12-5 showed that both of these values had to be calculated for each harmonic. Section 12-6 showed how to add and subtract nonsinusoidal waveforms.

PROBLEMS

12-1. Plot the following fundamental, second, and fourth harmonics and their resultant:

$$e_{S_1} = 4 \cos \omega t$$

$$e_{S_2} = 2 \cos 2\omega t$$

$$e_{S_4} = 1 \cos 4\omega t$$

12-2. Calculate the effective value for the voltages of Problems 12-1.

12-3. For the circuit of Fig. 12-11, calculate the effective value of voltage and current.

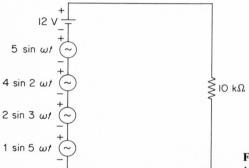

Fig. 12-11. Circuit diagram for Prob. 12-3.

12-4. Classify each of the waveforms of Fig. 12-12 as either even or odd functions and calculate the average value.

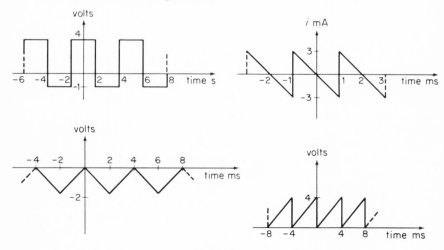

Fig. 12-12. Waveforms for Prob. 12-4.

12-5. Refer to the waveform of Fig. 12-8. If the maximum values are $+10$ V and -10 V, obtain the first three terms of the Fourier series.

12-6. Obtain the first three terms of the Fourier series for the waveform of Figune 12-6d.

12-7. Obtain the first three terms of the Fourier series for the waveform of Figure 12-6e.

12-8. Refer to the circuit of Fig. 12-10 and determine the expression of current if the voltage is

$$e = 4 \sin \omega t + 2 \sin 2 \omega t + 1 \sin 4 \omega t$$

12-9. Repeat Example 12-6 for $L = 20$ mH.

12-10. Obtain the sum of v_1 and v_2:

$$v_1 = 10 + 8 \sin \omega t + 6 \sin 2\omega t + 4 \sin 5\omega t$$

$$v_2 = 4 \sin \omega t + 2 \sin 2\omega t + 1.5 \sin (5 \omega t + \pi/6) + 0.5 \sin 7\omega t$$

12-11. For the values of voltage of Problems 12-10, obtain the result of $v_1 - v_2$.

appendices

A. WYE–DELTA TRANSFORMATION

Figure A-1 shows two types of configurations — the wye, Y, and the delta,

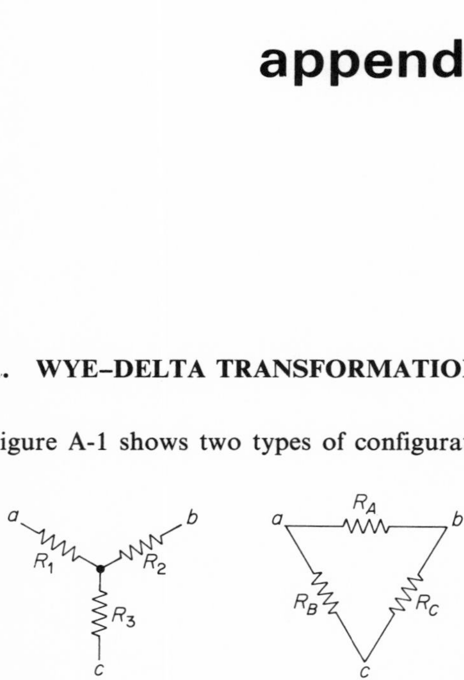

Fig. A-1. Circuit diagrams of the Y and Δ.

Δ — encountered in circuit analysis. Converting from one configuration to another often simplifies the problem. Converting from Y to Δ,

$$R_A = \frac{R_1 R_2 + R_1 R_3 + R_2 R_3}{R_3}$$

$$R_B = \frac{R_1 R_2 + R_1 R_3 + R_2 R_3}{R_2}$$

$$R_C = \frac{R_1 R_2 + R_1 R_3 + R_2 R_3}{R_1}$$

Converting from Δ to Y,

$$R_1 = \frac{R_A R_B}{R_A + R_B + R_C}$$

$$R_2 = \frac{R_A R_C}{R_A + R_B + R_C}$$

$$R_3 = \frac{R_B R_C}{R_A + R_B + R_C}$$

For impedance networks the conversion formulas are the same except that the R's are replaced by Z's.

Example A-1: For the circuit of Fig. A-2a, calculate the total resistance.

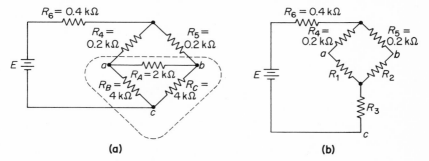

Fig. A-2. Circuit diagrams for Ex. A-1.

Solution: Using the Δ-to-Y transformation,

$$R_1 = \frac{R_A R_B}{R_A + R_B + R_C} = \frac{(2\,k\Omega)(4\,k\Omega)}{2\,k\Omega + 4\,k\Omega + 4\,k\Omega} = 0.8\,k\Omega$$

$$R_2 = \frac{R_A R_C}{R_A + R_B + R_C} = \frac{(2\,k\Omega)(4\,k\Omega)}{10\,k\Omega} = 0.8\,k\Omega$$

$$R_3 = \frac{R_B R_C}{R_A + R_B + R_C} = \frac{(4\,k\Omega)(4\,k\Omega)}{10\,k\Omega} = 1.6\,k\Omega$$

The Y transformation is shown in the circuit of Fig. A-2b.
From Fig. A-2b Let

$$R_X = R_4 + R_1 = 0.2\,k\Omega + 0.8\,k\Omega = 1\,k\Omega$$

and

$$R_Y = R_5 + R_2 = 0.2\,k\Omega + 0.8\,k\Omega = 1\,k\Omega$$

Let

$$R = R_X \| R_Y = \frac{1\,k\Omega}{2} = 0.5\,k\Omega$$

Then

$$R_T = R_6 + R + R_3$$
$$R_T = 0.4\,k\Omega + 0.5\,k\Omega + 1.6\,k\Omega = 2.5\,k\Omega$$

B. SLIDE–RULE TECHNIQUES

The conversion from rectangular form to polar form or vice versa can be performed using a slide rule. From Section 6-4,

$$a + jb = A \,\underline{/\theta}$$

$$A = \sqrt{a^2 + b^2}$$

$$\theta = \underline{/\tan^{-1}\frac{b}{a}}$$

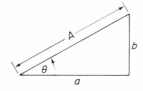

Fig. B-1. Graphical representation of a complex number.

Converting from Rectangular Form to Polar Form. When $0.1 \leq b/a \leq 10$ or $0.1 \leq a/b \leq 10$, the following procedure is applicable:

1. Set the index (1) of the C scale over the larger number (a or b) on the D scale.
2. Set the hairline over the smaller number on the D scale.
3. The angle θ is read on the T scale. If $a > b$, the angle is less than $45°$ (black numbers). If $b < a$, the angle is greater than $45°$ (red numbers).
4. Keep the hairline set. Using the S scale, move the slide until the angle θ determined in step 3 (and of the same color) is under the hairline.
5. The magnitude, A, is now read on the D scale under the index of the C scale.

If $a > 10b$ or $a > 100b$ or $a > 1000b$, the magnitude, A, is approximately a. The angle, θ, may be determined from the following:

When $a > 10b$, the angle is read on the ST scale instead of the T scale.

When $a > 100b$, the angle is read on the ST scale but the decimal point of the angle is moved one place to the left.

When $a > 1000b$, the angle is read on the ST scale but the decimal point of the angle is moved two places to the left.

If $b > 10a$ or $b > 100a$ or $b > 1000a$, the magnitude, A, is approximately b. The angle, θ, may be determined from the following:

When $b > 10a$, the angle read on the ST scale is subtracted from $90°$.

When $b > 100a$, the decimal point of the angle that is read on the ST scale is moved one place to the left. The result is subtracted from $90°$.

When $b > 1000a$, the decimal point of the angle that is read on the ST scale is moved two places to the left. The result is subtracted from $90°$.

Converting from Polar Form to Rectangular Form. For $5.7 > \theta > 84.3°$:

1. Set the index of the C scale over the magnitude, A, on the D scale.
2. To read a, set the hairline over the cos θ (angles in the red numbers) on the S scale.
3. To read b, set the hairline over the sin θ (angles in the black numbers) on the S scale.

If the hairline cannot be set over either cos θ or sin θ, move the slide so that the other index of the C scale is over the magnitude, A, on the D scale. Now set the hairline.
For $\theta < 5.73°$,

$$a \approx A$$

1. Set the hairline over the angle θ on the ST scale and read the value b on the D scale.
   For $\theta > 84,3°$,

$$b \approx A$$

1. Since cos θ = sin $(90° - \theta)$, set the hairline over sin $(90° - \theta)$ on the ST scale and read the value a on the D scale.

Powers of e. The constant e is the base of the natural logarithms. Its value is approximately 2.718. Slide rules containing log-log (LL) scales may be used to determine the values of e^x and e^{-x}.
To find values of e^x:

1. Set the hairline to x on the D scale.
2. Read the value of e^x on the proper LL scale. Guides to indicate the proper LL scale are on the right end of the slide rule.

To find values of e^{-x}:

1. Set the hairline to x on the D scale.
2. Read the values of e^{-x} on the proper LLO scale. Guides to indicate the proper LLO scale are on the right end of the slide rule.

Example B-1: Determine the value of (a) e^4 and (b) e^{-4}. Note that in terms of the exponential expressions in Chapter 9

$$e^{-t/\tau} = e^{-x}$$

and

$$1 - e^{-t/\tau} = 1 - e^{-\tau}$$

Solution: (a) Set the hairline to 4 on the D scale; at the hairline on the LL3 scale read 50.45:

$$e^4 = 50.45$$

(b) Set the hairline to 4 on the D scale; at the hairline on the LL03 scale read 0.0184:

$$e^{-4} = 0.0184$$

Example B-2: Determine the value of (a) $e^{0.5}$ and (b) $e^{-0.5}$

Solution: (a) Set the hairline to 5 on the D scale: at the hairline on the LL2 scale read 1.65:

$$e^{0.5} = 1.65$$

(b) Set the hairline to 5 on the D scale; at the hairline on the LL02 scale read 0.607:

$$e^{-5} = 0.607$$

C. DETERMINANTS

A determinant is defined as a *number* obtained from evaluating a square array of numbers. The determinant of a 2 by 2 square array is

$$\text{row} \longrightarrow \begin{vmatrix} a_{11} & a_{12} \\ a_{21} & a_{22} \end{vmatrix} = a_{11}a_{22} - a_{12}a_{21} \qquad \text{(C-1)}$$
$$\text{columns}$$

Elements in horizontal lines form rows; elements in vertical lines form columns. A square array has as many rows as columns. The first subscript of the *a* terms designate the rows. The second subscript designates the columns. For example, in a_{12} the 1 stands for the first row and the subscript 2 refers to the second column. The schematic array of elements within the brackets of Eq. (C-1) is called a determinant of order 2.

Example C-1: Calculate the determinant for each of the following square arrays:

(a) $\begin{vmatrix} 1 & 2 \\ 3 & 4 \end{vmatrix}$ (b) $\begin{vmatrix} 1 & -1 \\ 4 & 5 \end{vmatrix}$

(c) $\begin{vmatrix} 3 & 1 \\ 8 & 0 \end{vmatrix}$ (d) $\begin{vmatrix} 10 & 5 \\ 4 & 2 \end{vmatrix}$

Solution: (a) $\begin{vmatrix} 1 & 2 \\ 3 & 4 \end{vmatrix} = (1)(4) - (2)(3) = 4 - 6 = -2$

(b) $\begin{vmatrix} 1 & -1 \\ 4 & 5 \end{vmatrix} = (1)(5) - (-1)(4) = 5 - (-4) = 9$

(c) $\begin{vmatrix} 3 & 1 \\ 8 & 0 \end{vmatrix} = (3)(0) - (1)(8) = 0 - 8 = -8$

(d) $\begin{vmatrix} 10 & 5 \\ 4 & 2 \end{vmatrix} = (10)(2) - (5)(4) = 20 - 20 = 0$

Thus the determinant may be a positive number, a negative number, or zero.

A 3 by 3 square array (determinant of order 3) may be evaluated as

$$
\begin{vmatrix} a_{11} & a_{12} & a_{13} \\ a_{21} & a_{22} & a_{23} \\ a_{31} & a_{32} & a_{33} \end{vmatrix} = a_{11} \begin{vmatrix} a_{22} & a_{23} \\ a_{32} & a_{33} \end{vmatrix} - a_{12} \begin{vmatrix} a_{21} & a_{23} \\ a_{31} & a_{33} \end{vmatrix} + a_{13} \begin{vmatrix} a_{21} & a_{22} \\ a_{31} & a_{32} \end{vmatrix} \tag{C-2}
$$

The expansion (C-2) on the elements of the first row is called a *Laplace development.*

Example C-2: Calculate the determinant for the following square array:

$$
\begin{vmatrix} 2 & -1 & 0 \\ 4 & 5 & 1 \\ -2 & 0 & 3 \end{vmatrix}
$$

Solution:

$$
\begin{vmatrix} 2 & -1 & 0 \\ 4 & 5 & 1 \\ -2 & 0 & 3 \end{vmatrix} = 2 \begin{vmatrix} 5 & 1 \\ 0 & 3 \end{vmatrix} - (-1) \begin{vmatrix} 4 & 1 \\ -2 & 3 \end{vmatrix} + 0 \begin{vmatrix} 4 & 5 \\ -2 & 0 \end{vmatrix}
$$

$$
= 2\,[(5)(3) - (1)(0)] + 1\,[(4)(3) - (1)(-2)] + 0\,[(4)(0) - (5)(-2)]
$$

$$
= 2(15) + (14) + 0(10) = 44
$$

In circuit analysis we often obtain a set of simultaneous linear equations. In such a set, there are as many equations as there are unknown values. The equations are of the form

$$
\begin{aligned} a_{11}x_1 + a_{12}x_2 &= b_1 \\ a_{21}x_1 + a_{22}x_2 &= b_2 \end{aligned} \tag{C-3}
$$

where the a's and b's are known quantities and the x's unknown values. To solve for the unknown terms, x_1 and x_2, we may determine the ratio of two second-order determinants:

$$
x_1 = \frac{\begin{vmatrix} b_1 & a_{12} \\ b_2 & a_{22} \end{vmatrix}}{\begin{vmatrix} a_{11} & a_{12} \\ a_{21} & a_{22} \end{vmatrix}} \qquad x_2 = \frac{\begin{vmatrix} a_{11} & b_1 \\ a_{21} & b_2 \end{vmatrix}}{\begin{vmatrix} a_{11} & a_{12} \\ a_{21} & a_{22} \end{vmatrix}} \tag{C-4}
$$

The denominators for both unknown terms are the coefficients of x_1 and x_2. The numerator of x_1 is obtained by replacing the coefficients of x_1 by b_1 and b_2. Similarly, the numerator of x_2 is obtained by replacing the coefficients of x_2 by b_1 and b_2. This method is known as *Cramer's rule*.

Example C-3: Calculate x_1 and x_2:

$$2x_1 + 3x_2 = 4$$

$$3x_1 + 4x_2 = 10$$

Solution:

$$x_1 = \frac{\begin{vmatrix} 4 & 3 \\ 10 & 4 \end{vmatrix}}{\begin{vmatrix} 2 & 3 \\ 3 & 4 \end{vmatrix}} = \frac{(4)(4) - (3)(10)}{(2)(4) - (3)(3)} = \frac{-14}{-1} = 14$$

$$x_2 = \frac{\begin{vmatrix} 2 & 4 \\ 3 & 10 \end{vmatrix}}{-1} = \frac{(2)(10) - (4)(3)}{-1} = \frac{8}{-1} = -8$$

Therefore, $x_1 = 14$ and $x_2 = -8$. Cramer's rule is not restricted to only a set of two equations.

Example C-4: Calculate x_1, x_2, and x_3.

$$4x_1 - 1x_2 - 2x_3 = 4$$

$$-1x_1 + 5x_2 + 0x_3 = 2$$

$$-2x_1 + 0x_2 + 3x_3 = 0$$

Solution:

$$x_1 = \frac{\begin{vmatrix} 4 & -1 & -2 \\ 2 & 5 & 0 \\ 0 & 0 & 3 \end{vmatrix}}{\begin{vmatrix} 4 & -1 & -2 \\ -1 & 5 & 0 \\ -2 & 0 & 3 \end{vmatrix}} = \frac{4\begin{vmatrix} 5 & 0 \\ 0 & 3 \end{vmatrix} - (-1)\begin{vmatrix} 2 & 0 \\ 0 & 3 \end{vmatrix} - 2\begin{vmatrix} 2 & 5 \\ 0 & 0 \end{vmatrix}}{4\begin{vmatrix} 5 & 0 \\ 0 & 3 \end{vmatrix} - 1(-1)\begin{vmatrix} -1 & 0 \\ -2 & 3 \end{vmatrix} - 2\begin{vmatrix} -1 & 5 \\ -2 & 0 \end{vmatrix}}$$

$$= \frac{4(15) + 1(6) - 2(0)}{4(15) + 1(-3) - 2(10)} = \frac{66}{37} \approx 1.78$$

$$x_2 = \frac{\begin{vmatrix} 4 & 4 & -2 \\ -1 & 2 & 0 \\ 0 & 0 & 3 \end{vmatrix}}{37} = \frac{4\begin{vmatrix} 2 & 0 \\ 0 & 3 \end{vmatrix} - (4)\begin{vmatrix} -1 & 0 \\ 0 & 3 \end{vmatrix} - 2\begin{vmatrix} -1 & 2 \\ 0 & 0 \end{vmatrix}}{37}$$

$$= \frac{4(6) - 4(-3) - 2(0)}{37} = \frac{12}{37} \approx 0.324$$

$$x_3 = \frac{\begin{vmatrix} 4 & -1 & 4 \\ -1 & 5 & 2 \\ -2 & 0 & 0 \end{vmatrix}}{37} = \frac{4 \begin{vmatrix} 5 & 2 \\ 0 & 0 \end{vmatrix} - (-1) \begin{vmatrix} -1 & 2 \\ -2 & 0 \end{vmatrix} + 4 \begin{vmatrix} -1 & 5 \\ -2 & 0 \end{vmatrix}}{37}$$

$$= \frac{4(0) + 1(4) + 4(10)}{37} = \frac{44}{37} \approx 1.19$$

Therefore, $x_1 = 1.78$, $x_2 = 0.324$ and $x_3 = 1.19$. When using loop analysis in Chapters 5 and 9 the x's are the loop currents. When using nodal analysis the x's are node voltages.

D. NATURAL TRIGONOMETRIC FUNCTIONS

Natural Trigonometric Functions: Sine

Angles from 0°.0 to 44°.9

	.0	.1	.2	.3	.4	.5	.6	.7	.8	.9
0°	.0000	.0017	.0035	.0052	.0070	.0087	.0105	.0122	.0140	.0157
1°	.0175	.0192	.0209	.0227	.0244	.0262	.0279	.0297	.0314	.0332
2°	.0349	.0366	.0384	.0401	.0419	.0436	.0454	.0471	.0488	.0506
3°	.0523	.0541	.0558	.0576	.0593	.0610	.0628	.0645	.0663	.0680
4°	.0698	.0715	.0732	.0750	.0767	.0785	.0802	.0819	.0837	.0854
5°	.0872	.0889	.0906	.0924	.0941	.0958	.0976	.0993	.1011	.1028
6°	.1045	.1063	.1080	.1097	.1115	.1132	.1149	.1167	.1184	.1201
7°	.1219	.1236	.1253	.1271	.1288	.1305	.1323	.1340	.1357	.1374
8°	.1392	.1409	.1426	.1444	.1461	.1478	.1495	.1513	.1530	.1547
9°	.1564	.1582	.1599	.1616	.1633	.1650	.1668	.1685	.1702	.1719
10°	.1736	.1754	.1771	.1788	.1805	.1822	.1840	.1857	.1874	.1891
11°	.1908	.1925	.1942	.1959	.1977	.1994	.2011	.2028	.2045	.2062
12°	.2079	.2096	.2113	.2130	.2147	.2164	.2181	.2198	.2215	.2233
13°	.2250	.2267	.2284	.2300	.2317	.2334	.2351	.2368	.2385	.2402
14°	.2419	.2436	.2453	.2470	.2487	.2504	.2521	.2538	.2554	.2571
15°	.2588	.2605	.2622	.2639	.2656	.2672	.2689	.2706	.2723	.2740
16°	.2756	.2773	.2790	.2807	.2823	.2840	.2857	.2874	.2890	.2907
17°	.2924	.2940	.2957	.2974	.2990	.3007	.3024	.3040	.3057	.3074
18°	.3090	.3107	.3123	.3140	.3156	.3173	.3190	.3206	.3223	.3239
19°	.3256	.3272	.3289	.3305	.3322	.3338	.3355	.3371	.3387	.3404
20°	.3420	.3437	.3453	.3469	.3486	.3502	.3518	.3535	.3551	.3567
21°	.3584	.3600	.3616	.3633	.3649	.3665	.3681	.3697	.3714	.3730
22°	.3746	.3762	.3778	.3795	.3811	.3827	.3843	.3859	.3875	.3891
23°	.3907	.3923	.3939	.3955	.3971	.3987	.4003	.4019	.4035	.4051
24°	.4067	.4083	.4099	.4115	.4131	.4147	.4163	.4179	.4195	.4210
25°	.4226	.4242	.4258	.4274	.4289	.4305	.4321	.4337	.4352	.4368
26°	.4384	.4399	.4415	.4431	.4446	.4462	.4478	.4493	.4509	.4524
27°	.4540	.4555	.4571	.4586	.4602	.4617	.4633	.4648	.4664	.4679
28°	.4695	.4710	.4726	.4741	.4756	.4772	.4787	.4802	.4818	.4833
29°	.4848	.4863	.4879	.4894	.4909	.4924	.4939	.4955	.4970	.4985
30°	.5000	.5015	.5030	.5045	.5060	.5075	.5090	.5105	.5120	.5135
31°	.5150	.5165	.5180	.5195	.5210	.5225	.5240	.5255	.5270	.5284
32°	.5299	.5314	.5329	.5344	.5358	.5373	.5388	.5402	.5417	.5432
33°	.5446	.5461	.5476	.5490	.5505	.5519	.5534	.5548	.5563	.5577
34°	.5592	.5606	.5621	.5635	.5650	.5664	.5678	.5693	.5707	.5721
35°	.5736	.5750	.5764	.5779	.5793	.5807	.5821	.5835	.5850	.5864
36°	.5878	.5892	.5906	.5920	.5934	.5948	.5962	.5976	.5990	.6004
37°	.6018	.6032	.6046	.6060	.6074	.6088	.6101	.6115	.6129	.6143
38°	.6157	.6170	.6184	.6198	.6211	.6225	.6239	.6252	.6266	.6280
39°	.6293	.6307	.6320	.6334	.6347	.6361	.6374	.6388	.6401	.6414
40°	.6428	.6441	.6455	.6468	.6481	.6494	.6508	.6521	.6534	.6547
41°	.6561	.6574	.6587	.6600	.6613	.6626	.6639	.6652	.6665	.6678
42°	.6691	.6704	.6717	.6730	.6743	.6756	.6769	.6782	.6794	.6807
43°	.6820	.6833	.6845	.6858	.6871	.6884	.6896	.6909	.6921	.6934
44°	.6947	.6959	.6972	.6984	.6997	.7009	.7022	.7034	.7046	.7059

Natural Trigonometric Functions: Sine

Angles from 45°.0 to 89°.9

	.0	.1	.2	.3	.4	.5	.6	.7	.8	.9
45°	.7071	.7083	.7096	.7108	.7120	.7133	.7145	.7157	.7169	.7181
46°	.7193	.7206	.7218	.7230	.7242	.7254	.7266	.7278	.7290	.7302
47°	.7314	.7325	.7337	.7349	.7361	.7373	.7385	.7396	.7408	.7420
48°	.7431	.7443	.7455	.7466	.7478	.7490	.7501	.7513	.7524	.7536
49°	.7547	.7559	.7570	.7581	.7593	.7604	.7615	.7627	.7638	.7649
50°	.7660	.7672	.7683	.7694	.7705	.7716	.7727	.7738	.7749	.7760
51°	.7771	.7782	.7793	.7804	.7815	.7826	.7837	.7848	.7859	.7869
52°	.7880	.7891	.7902	.7912	.7923	.7934	.7944	.7955	.7965	.7976
53°	.7986	.7997	.8007	.8018	.8028	.8039	.8049	.8059	.8070	.8080
54°	.8090	.8100	.8111	.8121	.8131	.8141	.8151	.8161	.8171	.8181
55°	.8192	.8202	.8211	.8221	.8231	.8241	.8251	.8261	.8271	.8281
56°	.8290	.8300	.8310	.8320	.8329	.8339	.8348	.8358	.8368	.8377
57°	.8387	.8396	.8406	.8415	.8425	.8434	.8443	.8453	.8462	.8471
58°	.8480	.8490	.8499	.8508	.8517	.8526	.8536	.8545	.8554	.8563
59°	.8572	.8581	.8590	.8599	.8607	.8616	.8625	.8634	.8643	.8652
60°	.8660	.8669	.8678	.8686	.8695	.8704	.8712	.8721	.8729	.8738
61°	.8746	.8755	.8763	.8771	.8780	.8788	.8796	.8805	.8813	.8821
62°	.8829	.8838	.8846	.8854	.8862	.8870	.8878	.8886	.8894	.8902
63°	.8910	.8918	.8926	.8934	.8942	.8949	.8957	.8965	.8973	.8980
64°	.8988	.8996	.9003	.9011	.9018	.9026	.9033	.9041	.9048	.9056
65°	.9063	.9070	.9078	.9085	.9092	.9100	.9107	.9114	.9121	.9128
66°	.9135	.9143	.9150	.9157	.9164	.9171	.9178	.9184	.9191	.9198
67°	.9205	.9212	.9219	.9225	.9232	.9239	.9245	.9252	.9259	.9265
68°	.9272	.9278	.9285	.9291	.9298	.9304	.9311	.9317	.9323	.9330
69°	.9336	.9342	.9348	.9354	.9361	.9367	.9373	.9379	.9385	.9391
70°	.9397	.9403	.9409	.9415	.9421	.9426	.9432	.9438	.9444	.9449
71°	.9455	.9461	.9466	.9472	.9478	.9483	.9489	.9494	.9500	.9505
72°	.9511	.9516	.9521	.9527	.9532	.9537	.9542	.9548	.9553	.9558
73°	.9563	.9568	.9573	.9578	.9583	.9588	.9593	.9598	.9603	.9608
74°	.9613	.9617	.9622	.9627	.9632	.9636	.9641	.9646	.9650	.9655
75°	.9659	.9664	.9668	.9673	.9677	.9681	.9686	.9690	.9694	.9699
76°	.9703	.9707	.9711	.9715	.9720	.9724	.9728	.9732	.9736	.9740
77°	.9744	.9748	.9751	.9755	.9759	.9763	.9767	.9770	.9774	.9778
78°	.9781	.9785	.9789	.9792	.9796	.9799	.9803	.9806	.9810	.9813
79°	.9816	.9820	.9823	.9826	.9829	.9833	.9836	.9839	.9842	.9845
80°	.9848	.9851	.9854	.9857	.9860	.9863	.9866	.9869	.9871	.9874
81°	.9877	.9880	.9882	.9885	.9888	.9890	.9893	.9895	.9898	.9900
82°	.9903	.9905	.9907	.9910	.9912	.9914	.9917	.9919	.9921	.9923
83°	.9925	.9928	.9930	.9932	.9934	.9936	.9938	.9940	.9942	.9943
84°	.9945	.9947	.9949	.9951	.9952	.9954	.9956	.9957	.9959	.9960
85°	.9962	.9963	.9965	.9966	.9968	.9969	.9971	.9972	.9973	.9974
86°	.9976	.9977	.9978	.9979	.9980	.9981	.9982	.9983	.9984	.9985
87°	.9986	.9987	.9988	.9989	.9990	.9990	.9991	.9992	.9993	.9993
88°	.9994	.9995	.9995	.9996	.9996	.9997	.9997	.9997	.9998	.9998
89°	.9998	.9999	.9999	.9999	.9999	1.000	1.000	1.000	1.000	1.000

Angles from 0.0 to 44.9

	.0	.1	.2	.3	.4	.5	.6	.7	.8	.9
0°	1.0000	1.0000	1.0000	1.0000	1.0000	1.0000	.9999	.9999	.9999	.9999
1°	.9998	.9998	.9998	.9997	.9997	.9997	.9996	.9996	.9995	.9995
2°	.9994	.9993	.9993	.9992	.9991	.9990	.9990	.9989	.9988	.9987
3°	.9986	.9985	.9984	.9983	.9982	.9981	.9980	.9979	.9978	.9977
4°	.9976	.9974	.9973	.9972	.9971	.9969	.9968	.9966	.9965	.9963
5°	.9962	.9960	.9959	.9957	.9956	.9954	.9952	.9951	.9949	.9947
6°	.9945	.9943	.9942	.9940	.9938	.9936	.9934	.9932	.9930	.9928
7°	.9925	.9923	.9921	.9919	.9917	.9914	.9912	.9910	.9907	.9905
8°	.9903	.9900	.9898	.9895	.9893	.9890	.9888	.9885	.9882	.9880
9°	.9877	.9874	.9871	.9869	.9866	.9863	.9860	.9857	.9854	.9851
10°	.9848	.9845	.9842	.9839	.9836	.9833	.9829	.9826	.9823	.9820
11°	.9816	.9813	.9810	.9806	.9803	.9799	.9796	.9792	.9789	.9785
12°	.9781	.9778	.9774	.9770	.9767	.9763	.9759	.9755	.9751	.9748
13°	.9744	.9740	.9736	.9732	.9728	.9724	.9720	.9715	.9711	.9707
14°	.9703	.9699	.9694	.9690	.9686	.9681	.9677	.9673	.9668	.9664
15°	.9659	.9655	.9650	.9646	.9641	.9636	.9632	.9627	.9622	.9617
16°	.9613	.9608	.9603	.9598	.9593	.9588	.9583	.9578	.9573	.9568
17°	.9563	.9558	.9553	.9548	.9542	.9537	.9532	.9527	.9521	.9516
18°	.9511	.9505	.9500	.9494	.9489	.9483	.9478	.9472	.9466	.9461
19°	.9455	.9449	.9444	.9438	.9432	.9426	.9421	.9415	.9409	.9403
20°	.9397	.9391	.9385	.9379	.9373	.9367	.9361	.9354	.9348	.9342
21°	.9336	.9330	.9323	.9317	.9311	.9304	.9298	.9291	.9285	.9278
22°	.9272	.9265	.9259	.9252	.9245	.9239	.9232	.9225	.9219	.9212
23°	.9205	.9198	.9191	.9184	.9178	.9171	.9164	.9157	.9150	.9143
24°	.9135	.9128	.9121	.9114	.9107	.9100	.9092	.9085	.9078	.9070
25°	.9063	.9056	.9048	.9041	.9033	.9026	.9018	.9011	.9003	.8996
26°	.8988	.8980	.8973	.8965	.8957	.8949	.8942	.8934	.8926	.8918
27°	.8910	.8902	.8894	.8886	.8878	.8870	.8862	.8854	.8846	.8838
28°	.8829	.8821	.8813	.8805	.8796	.8788	.8780	.8771	.8763	.8755
29°	.8746	.8738	.8729	.8721	.8712	.8704	.8695	.8686	.8678	.8669
30°	.8660	.8652	.8643	.8634	.8625	.8616	.8607	.8599	.8590	.8581
31°	.8572	.8563	.8554	.8545	.8536	.8526	.8517	.8508	.8499	.8490
32°	.8480	.8471	.8462	.8453	.8443	.8434	.8425	.8415	.8406	.8396
33°	.8387	.8377	.8368	.8358	.8348	.8339	.8329	.8320	.8310	.8300
34°	.8290	.8281	.8271	.8261	.8251	.8241	.8231	.8221	.8211	.8202
35°	.8192	.8181	.8171	.8161	.8151	.8141	.8131	.8121	.8111	.8100
36°	.8090	.8080	.8070	.8059	.8049	.8039	.8028	.8018	.8007	.7997
37°	.7986	.7976	.7965	.7955	.7944	.7934	.7923	.7912	.7902	.7891
38°	.7880	.7869	.7859	.7848	.7837	.7826	.7815	.7804	.7793	.7782
39°	.7771	.7760	.7749	.7738	.7727	.7716	.7705	.7694	.7683	.7672
40°	.7660	.7649	.7638	.7627	.7615	.7604	.7593	.7581	.7570	.7559
41°	.7547	.7536	.7524	.7513	.7501	.7490	.7478	.7466	.7455	.7443
42°	.7431	.7420	.7408	.7396	.7385	.7373	.7361	.7349	.7337	.7325
43°	.7314	.7302	.7290	.7278	.7266	.7254	.7242	.7230	.7218	.7206
44°	.7193	.7181	.7169	.7157	.7145	.7133	.7120	.7108	.7096	.7083

Natural Trigonometric Functions: Cosine

Angles from 45°.0 to 89°.9

	.0	.1	.2	.3	.4	.5	.6	.7	.8	.9
45°	.7071	.7059	.7046	.7034	.7022	.7009	.6997	.6984	.6972	.6959
46°	.6947	.6934	.6921	.6909	.6896	.6884	.6871	.6858	.6845	.6833
47°	.6820	.6807	.6794	.6782	.6769	.6756	.6743	.6730	.6717	.6704
48°	.6691	.6678	.6665	.6652	.6639	.6626	.6613	.6600	.6587	.6574
49°	.6561	.6547	.6534	.6521	.6508	.6494	.6481	.6468	.6455	.6441
50°	.6428	.6414	.6401	.6388	.6374	.6361	.6347	.6334	.6320	.6307
51°	.6293	.6280	.6266	.6252	.6239	.6225	.6211	.6198	.6184	.6170
52°	.6157	.6143	.6129	.6115	.6101	.6088	.6074	.6060	.6046	.6032
53°	.6018	.6004	.5990	.5976	.5962	.5948	.5934	.5920	.5906	.5892
54°	.5878	.5864	.5850	.5835	.5821	.5807	.5793	.5779	.5764	.5750
55°	.5736	.5721	.5707	.5693	.5678	.5664	.5650	.5635	.5621	.5606
56°	.5592	.5577	.5563	.5548	.5534	.5519	.5505	.5490	.5476	.5461
57°	.5446	.5432	.5417	.5402	.5388	.5373	.5358	.5344	.5329	.5314
58°	.5299	.5284	.5270	.5255	.5240	.5225	.5210	.5195	.5180	.5165
59°	.5150	.5135	.5120	.5105	.5090	.5075	.5060	.5045	.5030	.5015
60°	.5000	.4985	.4970	.4955	.4939	.4924	.4909	.4894	.4879	.4863
61°	.4848	.4833	.4818	.4802	.4787	.4772	.4756	.4741	.4726	.4710
62°	.4695	.4679	.4664	.4648	.4633	.4617	.4602	.4586	.4571	.4555
63°	.4540	.4524	.4509	.4493	.4478	.4462	.4446	.4431	.4415	.4399
64°	.4384	.4368	.4352	.4337	.4321	.4305	.4289	.4274	.4258	.4242
65°	.4226	.4210	.4195	.4179	.4163	.4147	.4131	.4115	.4099	.4083
66°	.4067	.4051	.4035	.4019	.4003	.3987	.3971	.3955	.3939	.3923
67°	.3907	.3891	.3875	.3859	.3843	.3827	.3811	.3795	.3778	.3762
68°	.3746	.3730	.3714	.3697	.3681	.3665	.3649	.3633	.3616	.3600
69°	.3584	.3567	.3551	.3535	.3518	.3502	.3486	.3469	.3453	.3437
70°	.3420	.3404	.3387	.3371	.3355	.3338	.3322	.3305	.3289	.3272
71°	.3256	.3239	.3223	.3206	.3190	.3173	.3156	.3140	.3123	.3107
72°	.3090	.3074	.3057	.3040	.3024	.3007	.2990	.2974	.2957	.2940
73°	.2924	.2907	.2890	.2874	.2857	.2840	.2823	.2807	.2790	.2773
74°	.2756	.2740	.2723	.2706	.2689	.2672	.2656	.2639	.2622	.2605
75°	.2588	.2571	.2554	.2538	.2521	.2504	.2487	.2470	.2453	.2436
76°	.2419	.2402	.2385	.2368	.2351	.2334	.2317	.2300	.2284	.2267
77°	.2250	.2233	.2215	.2198	.2181	.2164	.2147	.2130	.2113	.2096
78°	.2079	.2062	.2045	.2028	.2011	.1994	.1977	.1959	.1942	.1925
79°	.1908	.1891	.1874	.1857	.1840	.1822	.1805	.1788	.1771	.1754
80°	.1736	.1719	.1702	.1685	.1668	.1650	.1633	.1616	.1599	.1582
81°	.1564	.1547	.1530	.1513	.1495	.1478	.1461	.1444	.1426	.1409
82°	.1392	.1374	.1357	.1340	.1323	.1305	.1288	.1271	.1253	.1236
83°	.1219	.1201	.1184	.1167	.1149	.1132	.1115	.1097	.1080	.1063
84°	.1045	.1028	.1011	.0993	.0976	.0958	.0941	.0924	.0906	.0889
85°	.0872	.0854	.0837	.0819	.0802	.0785	.0767	.0750	.0732	.0715
86°	.0698	.0680	.0663	.0645	.0628	.0610	.0593	.0576	.0558	.0541
87°	.0523	.0506	.0488	.0471	.0454	.0436	.0419	.0401	.0384	.0366
88°	.0349	.0332	.0314	.0297	.0279	.0262	.0244	.0227	.0209	.0192
89°	.0175	.0157	.0140	.0122	.0105	.0087	.0070	.0052	.0035	.0017

Natural Trigonometric Functions: Tangent

Angles from 0°.0 to 44°.9

	.0	.1	.2	.3	.4	.5	.6	.7	.8	.9
0°	.0000	.0017	.0035	.0052	.0070	.0087	.0105	.0122	.0140	.0157
1°	.0175	.0192	.0209	.0227	.0244	.0262	.0279	.0297	.0314	.0332
2°	.0349	.0367	.0384	.0402	.0419	.0437	.0454	.0472	.0489	.0507
3°	.0524	.0542	.0559	.0577	.0594	.0612	.0629	.0647	.0664	.0682
4°	.0699	.0717	.0734	.0752	.0769	.0787	.0805	.0822	.0840	.0857
5°	.0875	.0892	.0910	.0928	.0945	.0963	.0981	.0998	.1016	.1033
6°	.1051	.1069	.1086	.1104	.1122	.1139	.1157	.1175	.1192	.1210
7°	.1228	.1246	.1263	.1281	.1299	.1317	.1334	.1352	.1370	.1388
8°	.1405	.1423	.1441	.1459	.1477	.1495	.1512	.1530	.1548	.1566
9°	.1584	.1602	.1620	.1638	.1655	.1673	.1681	.1709	.1727	.1745
10°	.1763	.1781	.1799	.1817	.1835	.1853	.1871	.1890	.1908	.1926
11°	.1944	.1962	.1980	.1998	.2016	.2035	.2053	.2071	.2089	.2107
12°	.2126	.2144	.2162	.2180	.2199	.2217	.2235	.2254	.2272	.2290
13°	.2309	.2327	.2345	.2364	.2382	.2401	.2419	.2438	.2456	.2475
14°	.2493	.2512	.2530	.2549	.2568	.2586	.2605	.2623	.2642	.2661
15°	.2679	.2698	.2717	.2736	.2754	.2773	.2792	.2811	.2830	.2849
16°	.2867	.2886	.2905	.2924	.2943	.2962	.2981	.3000	.3019	.3038
17°	.3057	.3076	.3096	.3115	.3134	.3153	.3172	.3191	.3211	3230
18°	.3249	.3269	.3288	.3307	.3327	.3346	.3365	.3385	.3404	.3424
19°	.3443	.3463	.3482	.3502	.3522	.3541	.3561	.3581	.3600	.3620
20°	.3640	.3659	.3679	.3699	.3719	.3739	.3759	.3779	.3799	.3819
21°	.3839	.3859	.3879	.3899	.3919	.3939	.3959	.3979	.4000	.4020
22°	.4040	.4061	.4081	.4101	.4122	.4142	.4163	.4183	.4204	.4224
23°	.4245	.4265	.4286	.4307	.4327	.4348	.4369	.4390	.4411	.4431
24°	.4452	.4473	.4494	.4515	.4536	.4557	.4578	.4599	.4621	.4642
25°	.4663	.4684	.4706	.4727	.4748	.4770	.4791	.4813	.4834	.4856
26°	.4877	.4899	.4921	.4942	.4964	.4986	.5008	.5029	.5051	.5073
27°	.5095	.5117	.5139	.5161	.5184	.5206	.5228	.5250	.5272	.5295
28°	.5317	.5340	.5362	.5384	.5407	.5430	.5452	.5475	.5498	.5520
29°	.5543	.5566	.5589	.5612	.5635	.5658	.5681	.5704	.5727	.5750
30°	.5774	.5797	.5820	.5844	.5867	.5890	.5914	.5938	.5961	.5985
31°	.6009	.6032	.6056	.6080	.6104	.6128	.6152	.6176	.6200	.6224
32°	.6249	.6273	.6297	.6322	.6346	.6371	.6395	.6420	.6445	.6469
33°	.6494	.6519	.6544	.6569	.6594	.6619	.6644	.6669	.6694	.6720
34°	.6745	.6771	.6796	.6822	.6847	.6873	.6899	.6924	.6950	.6976
35°	.7002	.7028	.7054	.7080	.7107	.7133	.7159	.7186	.7212	.7239
36°	.7265	.7292	.7319	.7346	.7373	.7400	.7427	.7454	.7481	.7508
37°	.7536	.7563	.7590	.7618	.7646	.7673	.7701	.7729	.7757	.7785
38°	.7813	.7841	.7869	.7898	.7926	.7954	.7983	.8012	.8040	.8069
39°	.8098	.8127	.8156	.8185	.8214	.8243	.8273	.8302	.8332	.8361
40°	.8391	.8421	.8451	.8481	.8511	.8541	.8571	.8601	.8632	.8662
41°	.8693	.8724	.8754	.8785	.8816	.8847	.8878	.8910	.8941	.8972
42°	.9004	.9036	.9067	.9099	.9131	.9163	.9195	.9228	.9260	.9293
43°	.9325	.9358	.9391	.9424	.9457	.9490	.9523	.9556	.9590	.9623
44°	.9657	.9691	.9725	.9759	.9793	.9827	.9861	.9896	.9930	.9965

Natural Trigonometric Functions: Tangent

Angles from 45°.0 to 89°.9

	.0	.1	.2	.3	.4	.5	.6	.7	.8	.9
45°	1.000	1.003	1.007	1.011	1.014	1.018	1.021	1.025	1.028	1.032
46°	1.036	1.039	1.043	1.046	1.050	1.054	1.057	1.061	1.065	1.069
47°	1.072	1.076	1.080	1.084	1.087	1.091	1.095	1.099	1.103	1.107
48°	1.111	1.115	1.118	1.122	1.126	1.130	1.134	1.138	1.142	1.146
49°	1.150	1.154	1.159	1.163	1.167	1.171	1.175	1.179	1.183	1.188
50°	1.192	1.196	1.200	1.205	1.209	1.213	1.217	1.222	1.226	1.230
51°	1.235	1.239	1.244	1.248	1.253	1.257	1.262	1.266	1.271	1.275
52°	1.280	1.285	1.289	1.294	1.299	1.303	1.308	1.313	1.317	1.322
53°	1.327	1.332	1.337	1.342	1.347	1.351	1.356	1.361	1.366	1.371
54°	1.376	1.381	1.387	1.392	1.397	1.402	1.407	1.412	1.418	1.423
55°	1.428	1.433	1.439	1.444	1.450	1.455	1.460	1.466	1.471	1.477
56°	1.483	1.488	1.494	1.499	1.505	1.511	1.517	1.522	1.528	1.534
57°	1.540	1.546	1.552	1.558	1.564	1.570	1.576	1.582	1.588	1.594
58°	1.600	1.607	1.613	1.619	1.625	1.632	1.638	1.645	1.651	1.658
59°	1.664	1.671	1.678	1.684	1.691	1.698	1.704	1.711	1.718	1.725
60°	1.732	1.739	1.746	1.753	1.760	1.767	1.775	1.782	1.789	1.797
61°	1.804	1.811	1.819	1.827	1.834	1.842	1.849	1.857	1.865	1.873
62°	1.881	1.889	1.897	1.905	1.913	1.921	1.929	1.937	1.946	1.954
63°	1.963	1.971	1.980	1.988	1.997	2.006	2.014	2.023	2.032	2.041
64°	2.050	2.059	2.069	2.078	2.087	2.097	2.106	2.116	2.125	2.135
65°	2.145	2.154	2.164	2.174	2.184	2.194	2.204	2.215	2.225	2.236
66°	2.246	2.257	2.267	2.278	2.289	2.300	2.311	2.322	2.333	2.344
67°	2.356	2.367	2.379	2.391	2.402	2.414	2.426	2.438	2.450	2.463
68°	2.475	2.488	2.500	2.513	2.526	2.539	2.552	2.565	2.578	2.592
69°	2.605	2.619	2.633	2.646	2.660	2.675	2.689	2.703	2.718	2.733
70°	2.727	2.762	2.778	2.793	2.808	2.824	2.840	2.856	2.872	2.888
71°	2.904	2.921	2.937	2.954	2.971	2.989	3.006	3.024	3.042	3.060
72°	3.078	3.096	3.115	3.133	3.152	3.172	3.191	3.211	3.230	3.251
73°	3.271	3.291	3.312	3.333	3.354	3.376	3.398	3.420	3.442	3.465
74°	3.487	3.511	3.534	3.558	3.582	3.606	3.630	3.655	3.681	3.706
75°	3.732	3.758	3.785	3.812	3.839	3.867	3.895	3.923	3.952	3.981
76°	4.011	4.041	4.071	4.102	4.134	4.165	4.198	4.230	4.264	4.297
77°	4.331	4.366	4.402	4.437	4.474	4.511	4.548	4.586	4.625	4.665
78°	4.705	4.745	4.787	4.829	4.872	4.915	4.959	5.005	5.050	5.097
79°	5.145	5.193	5.242	5.292	5.343	5.396	5.449	5.503	5.558	5.614
80°	5.671	5.730	5.789	5.850	5.912	5.976	6.041	6.107	6.174	6.243
81°	6.314	6.386	6.460	6.535	6.612	6.691	6.772	6.855	6.940	7.026
82°	7.115	7.207	7.300	7.396	7.495	7.596	7.700	7.806	7.916	8.028
83°	8.144	8.264	8.386	8.513	8.643	8.777	8.915	9.058	9.205	9.357
84°	9.514	9.677	9.845	10.02	10.20	10.39	10.58	10.78	10.99	11,20
85°	11.43	11.66	11.91	12.16	12.43	12.71	13.00	13.30	13.62	13.95
86°	14.30	14.67	15.06	15.46	15.89	16.35	16.83	17.34	17.89	18.46
87°	19.08	19.74	20.45	21.20	22.02	22.90	23.86	24.90	26.03	27.27
88°	28.64	30.14	31.82	33.69	35.80	38.19	40.92	44.07	47.74	52.08
89°	57.29	63.66	71.62	81.85	95.49	114.6	143.2	191.0	286.5	573.0

E. TABLE OF EXPONENTIAL FUNCTIONS

x	e^x	e^{-x}	x	e^x	e^{-x}
0.00	1.0000	1.000000	0.74	2.0959	.477114
0.02	1.0202	.980199	0.76	2.1383	.467666
0.04	1.0408	.960789	0.78	2.1815	.458406
0.06	1.0618	.941765			
0.08	1.0833	.923116	**0.80**	2.2255	0.449329
			0.82	2.2705	.440432
0.10	1.1052	0.904837	0.84	2.3164	.431711
0.12	1.1275	.886920	0.86	2.3632	.423162
0.14	1.1503	.869358	0.88	2.4109	.414783
0.16	1.1735	.852144			
0.18	1.1972	.835270	**0.90**	2.4596	0.406570
			0.92	2.5093	.398519
0.20	1.2214	0.818731	0.94	2.5600	.390628
0.22	1.2461	.802519	0.96	2.6117	.382893
0.24	1.2712	.786628	0.98	2.6645	.375311
0.26	1.2969	.771052			
0.28	1.3231	.755784	**1.00**	2.7183	0.367879
			1.05	2.8577	.349938
0.30	1.3499	0.740818	1.10	3.0042	.332871
0.32	1.3771	.726149	1.15	3.1582	.316637
0.34	1.4049	.711770	1.20	3.3201	.301194
0.36	1.4333	.697676			
0.38	1.4623	.683861	**1.25**	3.4903	0.286505
			1.30	3.6693	.272532
0.40	1.4918	0.670320	1.35	3.8574	.259240
0.42	1.5220	.657047	1.40	4.0552	.246597
0.44	1.5527	.644036	1.45	4.2631	.234570
0.46	1.5841	.631284			
0.48	1.6161	.618783	**1.50**	4.4817	0.223130
			1.55	4.7115	.212248
0.50	1.6487	0.606531	1.60	4.9530	.201897
0.52	1.6820	.594521	1.65	5.2070	.192050
0.54	1.7160	.582748	1.70	5.4739	.182684
0.56	1.7507	.571209			
0.58	1.7860	.559898	**1.75**	5.7546	0.173774
			1.80	6.0496	.165299
0.60	1.8221	0.548812	1.85	6.3598	.157237
0.62	1.8589	.537944	1.90	6.6859	.149569
0.64	1.8965	.527292	1.95	7.0287	.142274
0.66	1.9348	.516851			
0.68	1.9739	.506617	**2.00**	7.3891	0.135335
			2.05	7.7679	.128735
0.70	2.0138	0.496585	2.10	8.1662	.122456
0.72	2.0544	.486752	2.15	8.5849	.116484

TABLE OF EXPONENTIAL FUNCTIONS (continued)

x	e^x	e^{-x}	x	e^x	e^{-x}
2.20	9.0250	.110803	**5.00**	148.41	0.006738
			5.05	156.02	.006409
2.25	9.4877	0.105399	5.10	164.02	.006097
2.30	9.9742	.100259	5.15	172.43	.005799
2.35	10.486	.095369	5.20	181.27	.005517
2.40	11.023	.090718			
2.45	11.588	.086294	**5.25**	190.57	0.005248
			5.30	200.34	.004992
2.50	12.182	0.082085	5.35	210.61	.004748
2.55	12.807	.078082	5.40	221.41	.004517
2.60	13.464	.074274	5.45	232.76	.004296
2.65	14.154	.070651			
2.70	14.880	.067206	**5.50**	244.69	0.0040868
			5.55	257.24	.0038875
2.75	15.643	0.063928	5.60	270.43	.0036979
2.80	16.445	.060810	5.65	284.29	.0035175
2.85	17.288	.057844	5.70	298.87	.0033460
2.90	18.174	.055023			
2.95	19.106	.052340	**5.75**	314.19	0.0031828
			5.80	330.30	.0030276
3.00	20.086	0.049787	5.85	347.23	.0028799
3.05	21.115	.047359	5.90	365.04	.0027394
3.10	22.198	.045049	5.95	383.75	.0026058
3.15	23.336	.042852			
3.20	24.533	.040762	**6.00**	403.43	0.0024788
			6.05	424.11	.0023579
3.25	25.790	0.038774	6.10	445.86	.0022429
3.30	27.113	.036883	6.15	468.72	.0021335
3.35	28.503	.035084	6.20	492.75	.0020294
3.40	29.964	.033373			
3.45	31.500	.031746	**6.25**	518.01	0.0019305
			6.30	544.57	.0018363
3.50	33.115	0.030197	6.35	572.49	.0017467
3.55	34.813	.028725	6.40	601.85	.0016616
3.60	36.598	.027324	6.45	632.70	.0015805
3.65	38.475	.025991			
3.70	40.447	.024724	**6.50**	665.14	0.0015034
			6.55	699.24	.0014301
3.75	42.521	0.023518	6.60	735.10	.0013604
3.80	44.701	.022371	6.65	772.78	.0012940
3.85	46.993	.021280	6.70	812.41	.0012309
3.90	49.402	.020242			
3.95	51.935	.019255	**6.75**	854.06	0.0011709
			6.80	897.85	.0011138
4.00	54.598	0.018316	6.85	943.88	.0010595
4.05	57.397	.017422	6.90	992.27	.0010078
4.10	60.340	.016573	6.95	1043.1	.0009586
4.15	63.434	.015764			
4.20	66.686	.014996	**7.00**	1096.6	0.0009119
			7.05	1152.9	.0008674
4.25	70.105	0.014264	7.10	1212.0	.0008251
4.30	73.700	.013569	7.15	1274.1	.0007849
4.35	77.478	.012907	7.20	1339.4	.0007466
4.40	81.451	.012277			
4.45	85.627	.011679	**7.25**	1408.1	0.0007102
			7.30	1480.3	.0006755
4.50	90.017	0.011109	7.35	1556.2	.0006426
4.55	94.632	.010567	7.40	1636.0	.0006113
4.60	99.484	.010052	7.45	1719.9	.0005814
4.65	104.58	.009562			
4.70	109.95	.009095	**7.50**	1808.0	.0005531
			7.55	1900.7	.0005261
4.75	115.58	0.008652	7.60	1998.2	.0005005
4.80	121.51	.008230	7.65	2100.6	.0004760
4.85	127.74	.007828	7.70	2208.3	.0004528
4.90	134.29	.007447			
4.95	141.17	.007083	**7.75**	2321.6	0.0004307

TABLE OF EXPONENTIAL FUNCTIONS (continued)

x	e^x	e^{-x}	x	e^x	e^{-x}
7.80	2440.6	.0004097	8.95	7707.9	.0001297
7.85	2565.7	.0003898			
7.90	2697.3	.0003707	9.00	8103.1	0.0001234
7.95	2835.6	.0003527	9.05	8518.5	.0001174
			9.10	8955.3	.0001117
8.00	2981.0	0.0003355	9.15	9414.4	.0001062
8.05	3133.8	.0003191	9.20	9897.1	.0001010
8.10	3294.5	.0003035			
8.15	3463.4	.0002887	9.25	10405	0.0000961
8.20	3641.0	.0002747	9.30	10938	.0000914
			9.35	11499	.0000870
8.25	3827.6	0.0002613	9.40	12088	.0000827
8.30	4023.9	.0002485	9.45	12708	.0000787
8.35	4230.2	.0002364			
8.40	4447.1	.0002249	9.50	13360	0.0000749
8.45	4675.1	.0002139	9.55	14045	.0000712
			9.60	14765	.0000677
8.50	4914.8	0.0002035	9.65	15522	.0000644
8.55	5166.8	.0001935	9.70	16318	.0000613
8.60	5431.7	.0001841			
8.65	5710.1	.0001751	9.75	17154	0.0000583
8.70	6002.9	.0001666	9.80	18034	.0000555
			9.85	18958	.0000527
8.75	6310.7	0.0001585	9.90	19930	.0000502
8.80	6634.2	.0001507	9.95	20952	.0000477
8.85	6974.4	.0001434			
8.90	7332.0	.0001364	10.00	22026	0.0000454

bibliography

Adler, R. B., A. C. Smith and R. L. Longini, *Introduction to Semiconductor Physics*. New York: John Wiley & Sons, Inc., 1964.

Babb, Daniel S., *Resistive Circuits*. Scranton, Penn.: International Textbook Co., 1968.

Boylestad, Robert L., *Introductory Circuit Analysis*. Columbus, Ohio: Charles E. Merrill Publishing Co., 1968.

Brenner, Egon and Mansour Javid, *Analysis of Electric Circuits*. New York: McGraw-Hill Book Co., Inc., 1959.

Carter, Robert C., *Introduction to Electrical Circuit Analysis*. New York: Holt, Rinehart and Winston, 1966.

Eisberg, Robert Martin, *Fundamentals of Modern Physics*. New York: John Wiley & Sons, Inc., 1961.

Ghausi, Mohammed Shuaib, *Principles and Design of Linear Active Circuits*. New York: McGraw-Hill Book Co., 1964.

Gillie, Angelo C., *Electrical Principles of Electronics* (2nd ed.). New York: McGraw-Hill Book Co., 1969.

Gray, Alexander and G. A. Wallace, *Principles and Practice of Electrical Engineering* (8th ed.). New York: McGraw-Hill Book Co., Inc., 1962.

Halliday, David and Robert Resnick, *Physics for Students of Science and Engineering*. New York: John Wiley & Sons, Inc., 1960.

Headquarters Staff of the American Radio Relay League, *The Radio Amateur's Handbook* (43rd ed.), ed. Byron Goodman, Newington, Conn.: American Radio Relay League, 1966.

Jackson, Herbert W., *Introduction to Electric Circuits* (2nd ed.). Englewood Cliffs, N.J.: Prentice-Hall, Inc., 1965.

Jackson, Herbert W., *Introduction to Electric Circuits* (3rd ed.). Englewood Cliffs, N.J.: Prentice-Hall, Inc., 1970.

Kerchner, Russell M. and George F. Corcoran, *Alternating-Current Circuits* (4th ed.). New York: John Wiley & Sons, Inc., 1960.

Lenert, Louis H., *Semiconductor Physics, Devices, and Circuits*. Columbus, Ohio: Charles E. Merrill Publishing Co., 1968.

Middleton, Robert and Milton Goldstein, *Basic Electricity for Electronics*. New York: Holt, Rinehart and Winston, Inc., 1966.

Romanowitz, H. Alex, *Introduction to Electric Circuits*. New York: John Wiley & Sons, Inc., 1971.

Terman, Frederick Emmons et al., *Electronic and Radio Engineering* (4th ed.). New York: McGraw-Hill Book Co., 1955.

Van Valkenburg, M. E., *Network Analysis*. Englewood Cliffs, N.J.: Prentice-Hall, Inc., 1955.

Wallman, Henry, "Synchronous and Staggered Single-Tuned High-Frequency Bandpass Amplifiers," in *Vacuum Tube Amplifiers*, eds. George E. Valley Jr. and Henry Wallman, *M.I.T. Radiation Laboratory Series* (No. 18). Lexington, Mass.: Boston Technical Publishers, Inc., 1964.

index

Italian Art

A   Duccio (Sienese): *The Annunciation* from the *Maestà*
(*London, National Gallery*)

ITALIAN ART

by

ANDRÉ CHASTEL

TRANSLATED BY

PETER AND LINDA MURRAY

NEW YORK
THOMAS YOSELOFF

*First published in France in 1956
under the title 'L'Art Italien'
First American Edition 1963
Published by Thomas Yoseloff Inc.
New York
Printed in Great Britain*

Contents

CONTENTS

PUBLISHER'S NOTE

The publishers wish to acknowledge with
gratitude editorial help from Miss Albinia
de la Mare and Mr. R. L. Evans

vi

ILLUSTRATIONS

COLOUR PLATES

MONOCHROME PLATES

between pages 64 and 65

between pages 80 and 81

ILLUSTRATIONS

between pages 96 *and* 97

between pages 112 *and* 113

between pages 128 *and* 129

ILLUSTRATIONS

ILLUSTRATIONS

ILLUSTRATIONS

between pages 272 and 273

ILLUSTRATIONS

Text Figures

TEXT FIGURES

MAPS

ABBREVIATIONS

A.B.	*The Art Bulletin* (New York)
B.A.	*Bollettino d'arte* (Rome)
B.M.	*Burlington Magazine* (London)
C.A.	*Cahiers archéologiques* (Paris)
Com.	*Commentari* (Rome)
Cr.A.	*Critica d'arte* (Florence)
G.B.A.	*Gazette des beaux-arts* (Paris, later New York)
J.B.	*Jahrbuch der preussischen Kunstsammlungen* (Berlin)
J.W.	*Jahrbuch der kunsthistorischen Sammlungen in Wien* (Vienna)
J.W.C.I.	*Journal of the Warburg and Courtauld Institutes* (London)
M.J.	*Marburger Jahrbuch für Kunstwissenschaft* (Marburg a. Lahn)
Pa.	*Paragone* (Florence)
Pr.	*Proporzioni* (Florence)
R.A.	*Rassegna d'arte* (Rome)
R.d.A.	*Revue des arts* (Paris)
Rep.KW.	*Repertorium für Kunstwissenschaft* (Munich)
Riv.A.	*Rivista d'arte* (Florence)
R.J.	*Römisches Jahrbuch für Kunstgeschichte* (Rome)
W.J.K.G.	*Wiener Jahrbuch für Kunstgeschichte* (Vienna)
Z.B.K.	*Zeitschrift für bildende Kunst* (Leipzig)
Z.K.G.	*Zeitschrift für Kunstgeschichte* (Stuttgart)

Introduction

The Country and its Schools of Art

Stendhal, who greatly admired Italian art, explained it by Italian intensity of feeling and liveliness of manners, and accounted for these characteristics in their turn by reference to the soil, the rhythm of work and life, and by the prevailing climate.

'The southerner,' he wrote in his *Histoire de la Peinture en Italie* (1817), 'lives on little and in a fertile country; the northerner consumes much in a sterile land. One seeks repose and the other movement. The southerner, through his physical inactivity, finds himself continually drawn towards meditation. For him a pinprick is more cruel than a sabre cut for the other. Artistic expression, therefore, was bound to be born in the South'; and later, 'In Italy, the climate engenders stronger passions; governments there do not weigh heavily on the passions; there is no capital city. There is, therefore, more originality, more natural genius.' It is thus in nature itself, in the geographical physiognomy of the peninsula, and in the diversity of its provinces that the most interesting source of Italian art must be sought. This interpretation by temperament and by manners has been favoured by foreign travellers both in the classical period and in the last century. All Taine had to do was to reduce it to a system.

Geographic Diversity. When they arrived in Piedmont or in Lombardy through the alpine valleys, the 'Barbarians' were always dazzled by the land before them. Villages, sober in form, with colour-washed walls and frescoed gables, crowned the tops of the hills or clustered on the borders of lakes. As far as the Adriatic stretched a plain covered mile upon mile with olive trees and vines alternating with maize and mulberries. Dotted about the landscape were villas and chapels; on the millenary roads from Milan to Padua, Milan to Bologna, lay one interesting town after another. Stormy skies and a brilliant sun that quickly dries the earth and turns it to clouds of dust alternate with a disconcerting speed that stimulates liveliness of perception. There is a striking contrast with the South, so soft in winter, so torrid in summer, so often a desert of cactus, with fig trees and valleys of almond trees between the plateaux of clay and chalk. Between the Alps and Africa, Italy contains a full range of habitat and culture, but the landscape still owes its variety to the long oblique barrier of the Apennines which multiplies picturesque valleys, harsh crests, and wildernesses, close to the most fertile plains. From Lombardy to Apulia, there is no region of Italy which is not close to mountains. Hence contrasts, and the separation of one area from another.

Almost every aspect of Italian landscape has figured in its art. Early Christian mosaics contain typical landscapes of purely Italian character, and so, later, are the gardens of Fra Angelico. The foothills of Cadore were celebrated by Titian, the Euganean Hills by Mantegna, the vine-studded Tuscan hillsides by Piero or by Gozzoli. The banks of the Brenta inspired Guardi, the black crests of the Apennines Salvator Rosa and Magnasco, the Lagoons all the moderns.

The only thing lacking in Italy is the balance given by big, deep-flowing rivers. Italian rivers are torrential, unsuited to navigation, uncertain for irrigation. Everywhere nature requires to be controlled, either to struggle against marshes, lagoons or floods, or to exploit fully the fecundity of the soil with the olive, the mulberry or the vine. Few minerals, but an abundance of different clays permitted almost everywhere the manufacture of bricks and work in terracotta. There are also magnificent quarries of marble and stone on the edge of the mountain chains, in the region of Luni and Carrara in Tuscany, and in Apulia, and in many districts, such as the neighbourhood of Verona and in Tuscany, coloured limestone has for a long time suggested the use of polychromy in architecture.[1]

From all this emerge two facts with a bearing on the arts; first, the strength of local schools and the isolation of provinces, but also, as a result of certain common factors in the taste for exteriors and for decoration, the continual mixture of techniques in order to create unified effects in which it is no longer possible to distinguish between structure and ornament.

The Schools. '. . . and so it is . . . that he who changes country and place, it seems also that he changes nature, abilities, manners, and personal customs, so much so that he no longer seems the same person, but a different one, and all disoriented and bewildered.'

In these words Vasari explains Rosso's complete inability to work in Rome, and the rapid departure of Fra Bartolommeo and Andrea del Sarto, who 'fled without doing anything', but who, when back in Florence again, were once more able to pursue their careers. The town in which one was trained has a certain quality which can make other places unendurable. Donatello explained this amusingly by saying that he had to leave Padua, where he was praised and where he did nothing well, for Florence, where, endlessly criticized, he was stimulated and able to do better work.

The taste for spoken criticism, for public argument was always lively. This curiosity about art, this love of beauty, often mixed up with familiar works, is a fairly general characteristic among the people. Everyone thinks himself an artist. When, at the beginning of the sixteenth century, work was proceeding on the building of S. Petronio in Bologna, the architects in charge complained that 'The moment work was begun on the pilasters, there rose up so many architects that one would never have believed that so many existed in the world. From everywhere, priests, friars, artisans, peasants, school-teachers, cleaners, ostlers, metal-founders, baggage-carriers, and even the water-carriers discovered that they were architects and gave their advice'.

Italy remained, from Antiquity onwards, very much an urban country; from the eleventh century onwards the Communes were extremely important. Until the middle of the fifteenth century the three most interesting cities in Tuscany, Pisa, Siena and Florence, lived in rivalry, in art as in politics. The same applies to Padua and Verona. The slow regional unification, brought about by the 'great States' of the fifteenth century, Lombardy, the Republic of Venice, Florence, the Papal States, and the Kingdom of Naples, never broke up this division. Something quite new could appear in one city without having any effect elsewhere. Towards 1450, the idea of 'treatises' was in the air; several were written in various centres, in Milan, in Padua, in Florence, but little contact was brought about through them, and their effect was purely local. At the end of the fifteenth century, the printed book and engraving became such important means of communication that the circulation of ideas and styles was accelerated. But for all that, uniformity was not established; local academies maintained their local characteristics, and one can sense a considerable distance between Parma and Lecce, or from Bologna to Naples or Palermo.

This natural state of affairs was reinforced by the fact that each region had its own cultural resources, in its history and in its contacts with outside. Lombardy, rich in Early Christian heritage and a country of builders, collected at its end of the valleys the novelties which arrived from France and the Rhineland. Its links with the Gothic, in painting as in architecture, were very close around 1400. This evolution explains why the province was to play a larger part in Mannerism than in the development of classic art, and was to furnish innumerable artisans to the Baroque. The position of the South is somewhat similar. With Byzantium and Islam, it experienced the rich current of Mediterranean art that helped to make up Romanesque. The coming of the Normans, followed by the Angevins, introduced Gothic, which was treated very broadly, and Neapolitan and Palermitan taste inclined always towards a Baroque richness and colour, with Naples maintaining, moreover, a link with Aragonese art through Colantonio during the fifteenth century, and Ribera during the seventeenth.

The two extremities of Italy were specially active during the twelfth and the end of the seventeenth centuries. Venice on the east and Tuscany in the centre followed very different rhythms which filled the intervals—the fourteenth, fifteenth and sixteenth centuries. Possessing very sharply defined characteristics, the one through its island site and its position as a trading post facing the Balkan world, the other by its soil, its culture, and later by its industry and banking activity, both established themselves, by the beginning of the eleventh century, through a creativeness which gave them an enduring and outstanding position. They developed during the fifteenth century their contrasting qualities, the one of intellectual refinement and abstract thought, the other of colour and atmosphere. In the sixteenth century, they were clearly competing with each other, and their duality became necessary to Italy, even in the twentieth century. Rome, which was very important during the eleventh and twelfth centuries, was curiously inactive and ineffectual during the whole Gothic period; it was as if her landscape of ruins was dead. Her late artistic rebirth marks the central phase of the Renaissance, when Antique art seemed to rise again from the ground in new schemes and effects, with the Imperial munificence of the Church, which died down during the eighteenth century, and was extinguished in the nineteenth.

It also seems possible to carve out a further artistic province, at least for medieval Italy, by associating all the maritime cities. They had close contacts, either of trade or of rivalry, Pisa fighting against Amalfi, or Ancona envious of Bari, but they also had common obsessions, and the same taste for oriental splendours, which made them accumulate stores of exoticism important for the future, and erect unusual monuments, such as the cathedral of Bousketos in Pisa, or the Paradise cloister; the last of them—Venice—was to prolong this aspect of Italian culture.

It is therefore impossible to treat each province as a constant centre of energy, worth examining from its origins up to the present day, and all Italian art as a team of schools with parallel careers, as they were presented by Félibien in the seventeenth century, and Abbé Lanzi in 1800. It would then be necessary to maintain—as is sometimes done with a certain complacency— the originality of productions of a very limited local interest, such as the Tuscan crucifixes of the thirteenth century, or the Neapolitan cribs of the sixteenth and seventeenth centuries. What is most striking in the panorama of Italian art is rather the antagonisms and rivalries, and the compensatory value of the results. In speculating on the importance of the links between Tuscany and Rome during the twelfth century, or between Umbria, Urbino, and Rome during the fifteenth century, it may be said that the predominance of the central zone indicates an increase in classical depth and fullness. Conversely, the most exaggerated period of the Baroque is dominated by a parallelism of the most northerly and most southerly regions, Piedmont and Sicily, which recalls the situation during the eleventh and twelfth centuries, when peripheral centres such as Lombardy and the Two Sicilies predominated. But it is also necessary to underline the role of intermediary, or lateral, centres, like Apulia during the thirteenth century, Urbino and the Marches during the fifteenth, Bologna during the sixteenth, Bergamo during the seventeenth, and Genoa during the eighteenth century, when the most attractive and often the most unexpected work was produced in them. The idea of the school is adequate to demonstrate that at each period the synchronization of styles is never perfect, and to indicate the bounds of the traditional periods of the history of styles, which here are less forcefully applied than elsewhere.

The coincidence of the political unification by a bourgeois monarchy with the most extreme form of cultural provincialism explains, in great measure, the decadence of the nineteenth century.

The Unity of the Arts. Other characteristics are linked with those of the soil, the climate, and the country. The ancestral tradition of decorating interiors with tiles, with ornamental painting, stuccoes, mosaics, meant that a special class of artisan always existed, from which could spring an innovating artist—a Cimabue, or a Borromini, for instance. But the communal life in the open, so characteristic of the South, obliged every township, however small, to have its shady forum, and this led to a more lively interest in decorative architecture which, ultimately, consists of treating an open space and the façades which surround it—that is, the exterior rather than the building—as a monumental unit. This was the case not only in Venice, which is an exceptional example with the Piazza S. Marco as no more than a kind of huge quadriporticus with the basilica closing one side, but all through medieval Italy, in Lombardy and in Tuscany where the façade of a palace or a church almost always forms part of a larger composition, and in Baroque Italy involves vast town-planning schemes. Hence the interest in a return to galleried streets with arcades, a form created during the Hellenistic period, and to decorative façades once the street was again considered, during the eleventh or twelfth centuries, as an entity in itself.[2] The same preoccupation survives from Bonnano to Maderna, from the continuous arcades of Pisa to the colonnades of St. Peter's in Rome.

The artist easily develops, therefore, into a kind of scenic designer of social life, and in no other way can the almost incredible prestige and power enjoyed in their day by Giotto, Raphael, Bernini or Tiepolo be explained. This feature, almost unknown elsewhere, is linked with the appearance of a 'universal' artist who determines a whole style, and this last is a characteristic phenomenon of Italian art. At a certain level, universality is the custom; long before the extraordinary Leonardo, who was painter, sculptor, architect, engraver, scene designer, musician, poet, there had been Giotto, who ended his career as Controller of Fine Arts, architect and town-planning expert; Verrocchio, painter, decorator and sculptor, was already a kind of Haephaistos Polytechnos, like the romantic Francesco di Giorgio of Siena. Michelangelo was a master of all techniques, and in all arts; the same can be said of Vasari, and even more of Bernini, the last of these great universal figures.

If it is not a single man, then it is sometimes a group of powerful personalities that determine upon similar innovations in all the arts. At the beginning of the fifteenth century, Brunelleschi's friendship with Donatello and with Masaccio illustrates the similarity of outlook between the architect, the sculptor and the painter at a certain point in the evolution of style, and this is repeated somewhat in the accord between Veronese, Palladio and Vittoria in the middle of the sixteenth century.

Even where it is impossible to establish these parallels on a personal plane, the arts obey the same rhythm and tend towards integration. Both large cloisters and small aedicules inside Roman churches of the eleventh and twelfth centuries were richly decorated with marble inlays and reliefs, which sometimes extended to the pavements and walls, creating an unusually harmonious effect. Tradition ascribes this work to dynasties of artists—the Cosmati and Vassaletti families. This is the first powerful manifestation of a demand which reappeared periodically, though with the substitution of other decorative techniques.

A fair number of the most important Italian paintings are wall decorations; these are not to be considered as series of individual paintings, but as compositions arranged in imaginary architecture. Into this category come the Giottos in the Arena Chapel, the Pieros in Arezzo, and the Tiepolos in palaces and villas. Mural decoration at first replaced stained glass, then tapestry, then leather and other hangings. It derived from a general tendency—and basically a Mediterranean one—to cover the interior of buildings with colour and varied ornament. Before fresco, stucco, mosaic and inlays were used, and all these techniques continued to be used. Stuccoes introduced the effects of coloured sculpture; from Brunelleschi's time onwards the vogue for painted ceilings and for elaborately carved and gilded coffering eliminated vaulting; finally, in certain cases such as the cathedral of Siena, the pavement with an historiated decoration transformed the building into a sort of precious casket in which all the varied techniques made their impact simultaneously. On the lower plane of furniture, it is noticeable that in contrast to the northern taste for a fine decoration in carved wood, Italian taste preferred combinations of wood, stucco and painting, in coffers and cabinets as in the buildings themselves.

This sometimes resulted in insipidity and oversweet elaboration, as occasionally occurred in the fourteenth century; but it also formed the basis of an outstanding success like Venice. A feeling for scenic coherence is never lacking in Italy; it is to be found in the care taken to give a building set in a city, or a picture or a piece of sculpture set in a building, an appropriate and carefully thought out position. Pictures, fragments of altarpieces, reliefs or frescoes that one admires in isolation benefit from a setting which is worth reconstruction. How could one interpret the tomb by Rossellino if it were removed from the chapel in S. Miniato, or the Carpaccios away from the Scuola di S. Giorgio degli Schiavoni, or the Heavenly visions of Padre Pozzo away from the churches of Rome? There is little doubt that to no art has the modern dissociation of techniques done so much harm. It would be quite wrong to break this partnership systematically, and to discuss the architecture of S. Vitale or S. Apollinare separately from their mosaics, or the baptistery in Parma apart from its sculpture, or S. Ignazio without its ceiling.

It is therefore necessary to begin by according to Italian art basic ideas quite different from those obtaining for, say, French art, where architecture and decoration are not so freely blended together. In Italy, the purest monumental values are probably expressed in painting, and the finest picturesque values in sculpture and architecture. The wall was never a distinct unit; either it was treated as an arrangement of partitions, with galleries, as in Lombard art, or with coloured revetments, as in Tuscany; it served as a support for mosaics, paintings, stuccoes, and became subservient to the decoration on it. If the stress was on mass, then the wall became a rampart which delimited the exterior of a block, giving unexpected glimpses into fresh, light-filled courtyards, which by contrast emphasized its importance. The Pitti Palace, so subtly pierced by Ammannati, demonstrates this. The scenographic effect which will increase the impact of the decoration is never far away. When, from Brunelleschi's time onwards, the articulation of space and harmonious relationships became the objects of new concern, the architect started from a privileged standpoint, giving equal attention to the perspective composition and the proportions of the building, as if it were a painted architecture.[3] When Noto in Sicily was rebuilt at the beginning of the eighteenth century, the sharp perspectives of theatrical scenery were wonderfully adapted by the town-planners to the slope of the streets on the hillside.

In generalizing this feeling for picturesque composition, the care taken everywhere over sites and the special features of the landscape, a particular value must be accorded to the art of designing gardens. The antique tradition of topiary art furnishes a thread of continuity through the amusing conceits of the sixteenth century and the sumptuous follies of Baroque parks; in a carefully deliberate multiplicity of screens and perspective vistas the architecture appears and disappears with the ease of an actor playing on a stage of bewildering freshness and variety. The entirely southern love of shade and living water, the innate sense of theatrical effects, the natural association of all the arts, have always made the Italian garden an unparalleled success.

Art and Civilization

Italian art has for long been the special object of literary attention, which has sometimes added to its enchantment, and sometimes obscured its meaning. But Browning on Lippi, Walter Pater on Giorgione, Suarès on Donatello, Barrès on the 'magic four-leafed clover' of Pisa, Goethe on Roman palaces, Rilke on fountains, Aldous Huxley on Baroque tombs, Sitwell on Baroque towns, are sufficient proof that the main force of Italian art is to be found in its complete and very powerful hold on the spectator's sensibilities. And these poetic values can influence the historian, faced with the contradictory interpretations advanced on the basic principles of these creations and their bearing on civilization.

INTRODUCTION

Religious Art and Secular Art. Sometimes Italy appears to be the natural cradle of a profoundly secular and lay art, long constrained to dissimulate, or at least to come to terms with the Church, in a pleasurably Machiavellian way: such is Stendhal's interpretation. Sometimes, on the contrary, the grandeur or the seductiveness of works of art has been confused with their Christian content in thought and feeling: this is Ruskin's interpretation. The first seems not unlikely with masters such as Pinturicchio or Sodoma, whose cynicism is obvious, but is less certain for a Caravaggio; there can be no question of it with the Middle Ages or even with the Baroque; it takes too little account of the fact that passionate feeling in the South cannot be separated from many popular superstitions, and from surges of religious emotion in artists.

The second argument is a naïve one; it is inspired by contrasting the 'healthy' art of Giotto or of the Venetian Trecento with the 'corrupt' art of Tintoretto or Raphael; it is as weak as the attempt to explain the 'true' Renaissance—Giotto's—in terms of Franciscan spirituality. It fails to take into account the need to find an artistic expression for every aspect of piety.

There is even less occasion to consider the two principles as succeeding each other, one determining the Middle Ages and the other provoking the Renaissance. This error of perspective is common, but it involves a misconception of the unique character of Italian spirituality. This appeared clearly when the popes, protesting against the iconoclasm of the Greek bishops, supported the cult of images, gave a home to exiled artists, and gathered works of art into the churches of Rome. There followed a period when the glory of every city required that it should create a luxurious sanctuary, which should proclaim the greatness of the Christian world in the face of Islam, and that of the Roman world in face of the Barbarians. Then monastic houses and cathedral chapters rivalled one another in zeal, and the cult of saints led to gigantic undertakings, like those at Assisi for S. Francis, or at Padua for S. Antony. When magnates like the Medici or the Sforza multiplied pious foundations, it was natural for the clergy to honour the name of their benefactor, and tombs and chapels thus developed more and more into personal monuments. The popes of the Baroque period were eager to anticipate their eternal glory by acquiring the splendours of worldly glory. In this way, the Roman Church never ceased, until very recently, to encourage an artistic activity which constitutes, to a certain degree, its pagan aspect, but which was nowhere so deeply rooted nor so strong as in Italy. Iconoclasm never reared its head there, and reforming movements, like that of the Franciscans or the Jesuits, who began by advocating austerity, rapidly came to terms with, and finally used more energetically than ever, the resources of art.

In the thirteenth century, the Cistercian reform was spread in Italy by S. Bernard's journeys to Rome. It introduced a new type of monastic architecture which was soon taken over by the Mendicant Orders. But a comparison between the austerity of a Cistercian interior, and the abundant decoration

8

and innumerable pictures inspired by the Franciscans and Dominicans, demonstrates that rules for self-control and discretion where images were concerned were not valid for Italy.

The crisis during the sixteenth century is more eloquent still. To the iconoclasts of Geneva and Wittenberg the Roman Church opposed its undying fidelity to the cult of images; the Catholic reformers who advocated more severity, and sought to eliminate from Christian art, as S. Bernard had done, all sorts of secular absurdities and licence, were satisfied in principle, while there soon followed a riot of ornament and picturesque iconography. With its cenotaphs, its private altars, its statues and altarpieces, the Italian church tended to the same apparently useless diversity as did the wealthy private house, the public square, or the palace.

A constant interpenetration of sacred and secular seems to be one of the hallmarks of Italian art. The habit of introducing sarcophagi, Antique objects and exotic forms into the church and to treat the church as a kind of treasure-house or museum, is clear from the thirteenth century onwards. This usage occasioned some rather disconcerting developments during the fifteenth and sixteenth centuries, when tombs decorated with allegories, and scenes of increasing freedom and worldliness penetrated into the church from all quarters; it is difficult to conceive of the Tempio Malatestiano in Rimini as a church. The same tendencies prevailed during the sixteenth and above all during the seventeenth and eighteenth centuries: the Baroque church is an edifice in which, together with the monuments and tombs, were accumulated not only all the marvels of the goldsmith's art, but even curios.

This is because the church became the repository of every human sentiment. Preachers sometimes tried to resist the transports of the faithful, who introduced all their own preoccupations into the sanctuary, fashioned saints for all their needs, miracles for all their dramatic moments, and demanded saints that could be prayed to for the first, and buildings that should commemorate the second. During a whole half century, the greatest architects of the Renaissance built an enormous basilica around the Santa Casa at Loreto, which was believed to be the Virgin's house brought there by angels on the 10th December, 1294, and which is in fact a small Romanesque church without foundations.

Religious art, therefore, tends to incorporate all the variety found in secular art. The Virgin becomes a young girl frightened by, or attentive to, the arrival of the Messenger. The Madonna laughs or cries with her Son, sometimes offering Him with trepidation the flower, the fruit, or the bird which symbolizes His cruel fate. The Nativity is a popular or family fête; the Lamentation the explosive grief of a family faced with death. Familiarity with the saints, or with the Madonna and Child, is truly without bounds. It may be asked what religion loses and what it gains from this kind of human generalization, but it is fundamental none the less. Every formal ceremonial designed to give distance and dignity to sacred scenes is matched by popular reaction;

9

to an archaising stylization is opposed the tenderness of the Mendicant Friars; after the chill formalities of Mannerism followed the warmth of Caravaggio.

Only thus may be explained the constant analogies which were established between a purely profane art, represented by the Antique, and Christian art. Already, by the fourth and fifth centuries, Imperial pomp and the apotheosis of princes took the same form as the Divine celebration. In the thirteenth century, the Ghibelline sculptors of Pisa used the same style for both purposes. But the most interesting evolution was during the fifteenth century, when ancient texts, abundant and descriptive, and more detailed than the scenes known in art, stirred the imagination of artists who ended by re-creating the pagan gods in the likeness of Christian images—Botticelli treating Venus like a nude Madonna, Signorelli evoking the court of Pan in the guise of a *Sacra Conversazione*. Somewhat later, the balance is re-established; for Raphael and Michelangelo, the Virgin borrows her reserve from the chaste goddesses of mythology, her beauty from Antique statues, and Christ derives His fulminating force from the image of Jupiter. But the discrepancy between the text and the image explains the multiplicity of treatises on mythology[4] and their often amazing fantasy. Incorrect inventions, resulting from confusion in the use of the orders of being, or from familiarity with objects of piety, were always possible. Rosso, for instance, on the suggestion of one of his friends who was a canon, represented in one of the churches of Arezzo Adam and Eve attached to the tree of the first sin, with the Virgin extracting the apple from their throats while Diana and Phoebus cavorted in the sky.

These intrusions of fable into Christian imagery have, in Italy, something of the same function as the realist detail and the earthy scene in Northern art. Condemned by the Council of Trent, they reappeared irresistibly in decorative themes and became part of the content of a style, but under the theological and moral cloak of allegory.

The Power of the Masterpiece. A great deal of significance attaches to the writing of the history of art in Italy. For a time it followed the lines of humanist literary history, introducing the essential names into the lists of 'Illustrious Men' indispensable to every major city. But, almost at once, there developed the literature of the connoisseurs characterized by inventories, lists of works worthy of admiration, and, finally, guides for visitors, based on the model of the pilgrims' guides, which, from the sixteenth century up to the present day, often combine the two types. Italian artistic criticism, no less precocious, has, since the fifteenth century, continued to justify the importance thus attributed to works of art; it has elevated them to the rank of the Liberal Arts (Landino, 1481), it has underlined their philosophical and humane purpose (Alberti, 1436), and their poetic value (Leonardo).

These commentaries and historical perspectives explain in part why the great interpretations of art considered as revealing the nature of a race and a society (Taine), as an autonomous force (Wölfflin), as a creation of the spirit (Dvořak), have been formulated on the basis of Italian art. But the Italians themselves have successively elaborated the basic types of the history of art in terms of their own experience, as an exposition of the theory of great style working in irresistible cycles (Vasari, 1st ed. 1550; Toesca, 1927 et seq.), as a framework of local schools (Abbé Lanzi, 1789; Adolfo Venturi, 1901–41), as a sequence of personalities (Vasari; Bellori, 1672; Longhi: *Piero della Francesca*, 1927; *Caravaggio* 1951), as the observations of a connoisseur (Marcantonio Michiel, *c.* 1522; Morelli, 1886–97).

The most revealing aspect of this 'artistic mentality' is perhaps to be found in the importance attached to the masterpiece as such. The famous procession in Siena in 1311, which greeted Duccio's *Maestà*, must not be used to colour a misconception. As the anonymous contemporary chronicler records: 'many prayers were recited . . . and the people besought Christ and His Mother to protect the city against all ill'. In making the installation of the picture in the cathedral into a great feast day, it was primarily the Virgin, Protectress of the city, who was being honoured. But, in fact, the pride of possessing a masterpiece was not infrequently blended with devotion to a cult; it is often quite clear that the sometimes rather idolatrous veneration of an image slipped easily into the veneration of a masterpiece. Leonardo's *Last Supper*, Raphael's and Titian's Madonnas are instances.

A whole popular literature of anecdotes, witticisms, affecting or terrible stories surrounds a celebrated work—Milan cathedral, for example, or the famous Sicilian and Veronese tombs. The figures on the walls of the Carmine in Florence, or the Campo Santo in Pisa, or in the Vatican frescoes, are identified. No work exists without its legend. The instinctive and good natured attachment of the Italian for the celebrities of his province or township, the '*campanilismo*', or parish-pump attitude, so marked in Florence or Venice, play an important part in this, but the fanatical love felt for churches and statues helps to surround the work of art with a climate of affection and contemplation; even if rather uninformed, it still explains the facility with which Italy has been able to acquire and conserve the beauty of a great natural museum. Nowhere else perhaps is glory so sweet to savour. It is the masterpiece which regulates, dominates and finally explains the evolution of the arts. This law, which seems to have a universal application, has never been more clearly manifested than here, and it explains, in all probability, the instinctive mistrust of Italians towards theories which tend to explain the course of art by mechanisms outside the history or internal dialectic of styles. They know by experience that the first essential is to reckon with the power of the creative act; the work of art constitutes in itself a new event; its greatness is better understood by the results which it provokes than by the forces that have brought it to birth.

Certain careers impose a new direction; thus Giotto fulfilled the evolution of Romanesque, assimilated the monumentality of Western Gothic, and established, in fact, the first stage of the Renaissance. Neither was Arnolfo any more easily foreseen, nor the extraordinary vitality which makes every one of Donatello's works an event, at first of local import and then of more general significance in the world of art. The altar of the Santo in Padua is the source of a whole art, working through Mantegna. There was a moment when all Michelangelo's acts were spied upon, his drawings circulated, his compositions copied, even by Tintoretto. The Tenebrist revolution which had so much effect upon all the arts during the seventeenth century, and which had a precursor in certain aspects of Northern Mannerism, derives, properly speaking, from a few canvases by Caravaggio.

These are instances of powerful personalities surrounded by a curiosity made more lively by the romantic fervour of the Italians. But sometimes the effect exercised by an outstanding work is of particular importance; the altarpiece painted by Antonello da Messina in Venice in 1475, the unfinished frescoes in the Carmine in Florence, Vignola's Gesù in Rome, are among those works which act immediately upon a whole generation. Connoisseurs and artists mingled together in the enthusiastic crowd which gazed at Leonardo's *S. Anne,* and at his *Battle of Anghiari.*

Concern with art is fundamental in Italy. Of all the characteristics of this ancient civilization, the most significant is certainly its ability to enjoy the spectacle of the world and life. A whole philosophy has been derived from it, as well as a clear recognition of the way in which to meet difficulties in theory and technique. To the French master-mason of scholastic leanings who reminded the Milanese chapter in 1401 that *ars sine scientia nihil est,* the Italians replied unanimously *scientia sine arte nihil.*

Italy, the Antique, and the West

Italian art, like French art, presents a history so long and full that it seems to contain all the problems of art. Right up to the eighteenth century, when its fecundity weakened, and the nineteenth century, when it comes almost inexplicably to a halt, its creations succeeded each other with a richness and diversity so great that the museums of the world are filled with them, and yet the traveller has never finished exploring the resources of the large and the small towns of the peninsula. During the last three centuries, the masterpieces of Raphael, Michelangelo, Correggio, Titian, have held, sometimes to an obsessive degree, the attention of the whole Western world. It was in Italy, finally, that there appeared during the very special period known as the Renaissance the most impressive and complete geniuses of the world of art.

If an effort is made to see all these developments as a whole, one is tempted to discern their origins in the privileged position of Italy, and in its inheritance from Rome, which for centuries was the axis of the Ancient World; for architecture, sculpture, and ornamental arts present, from the fifth century onwards, an uninterrupted development extending over some fifteen centuries. In the cities and towns of Roman, Imperial or, at the latest, Carolingian foundation, the use of arcades of semi-circular arches, or coloured inlays, was never abandoned; brick and marble were always used for building; sculptured medallions were always introduced into façades, or used in jewelled plaques and on tombs. In other countries, the art often seems to start from scratch; from Milan to Palermo, it derives from the most abundant and strongly established corpus of works—those of the Roman Empire.

To what degree can one speak of an Italian art before the twelfth century, or even before the fourteenth? Early Christian architecture and decoration represent the last phase of Antique art; the works of the Middle Ages, when they are not derived from Byzantine centres, are the result of Germanic invasions. There is, apparently, no unity in Italy before the age of the Baroque. Is it possible to find in this history a single nexus which is not either artificial or external? This problem must be considered, for it is closely connected with that of the 'Renaissances'.

The 'Renaissances'. It is the fourth-century city, already Christian and still intact, which has for ever fixed the legendary image of *Roma perennis*; the gigantic baths, the colonnades, the temples from which all the altars had not yet vanished, the *fora* with their rich shops—all the beauties of Greece absorbed into a sumptuously cosmopolitan setting were to leave in the minds of the invaders, as in those of the Italic people themselves, ineradicable memories. About A.D. 500 Fulgentius wrote: 'What must be the beauty of the celestial Jerusalem, if such is the brilliance of Rome upon this earth?' Dante never hesitated to compare the dazzlement of the gates of Paradise to the emotion of the Northern Barbarians who 'discovered Rome and her great works, with stupefaction. . . .' (Par. XXXI, 34–36.) Every society interested in architecture and art was to turn inevitably towards Rome, and when abandon, war, the erosion of the hills, had slowly buried marbles and bronzes in the dust, the image of Rome still worked like an irreplaceable myth on artists, just as the memory of the Empire impressed itself indelibly on all political life.

One negative proof of this is in the excesses of vandalism which the city has always provoked. In 410, the Visigoths of Alaric pillaged Rome for three days; in 455, the Vandals of Genseric sacked it for two weeks *sine ferro et igne* (without fire and sword) according to the promise made to Leo I. The Ostrogoths of Totila broke into the city in 544 and again in 547. Thus began its disintegration and the dispersal of its treasures. A reduced and fearful population built shelters for itself within the Colosseum and the Theatre of Marcellus, which served as quarries until the High Renaissance. The ports silted up . . . , but pagan works were not the only ones to suffer. In 846, under Pope Sergius II, an invasion of Saracens coming from Corsica ended in a frightful sack and the complete desecration of the basilica of St. Peter's.[5] The city continued to attract periodically such wild excesses, for example those of the

Normans in the twelfth century. The pillage of the Lutheran mercenaries and the Imperial troops under the Constable of Bourbon in 1527 was still within memory at the period when there commenced another more knowledgeable and respectful pillage, though one which often completed the ruin of ancient sites—that of the collectors. This movement began tentatively in the four-teenth century with minor works and coins; it developed in the sixteenth century and soon furnished the fine classical galleries of the princes, and finally the museums of the whole world.

In this way, Italian art grew up in a country covered with familiar ruins, and where innumerable legends and powerful superstitions assured, during the whole of the Middle Ages, the prestige attaching to the *mirabilia urbis*, until the moment came when the buried remains of arches, *fora*, temples, were examined in a spirit of emulation which has never before been equalled anywhere. Thus, there were several small 'Renaissances' in Italy before the Renaissance; the appropriate state of mind was always latent. In the monastic workshop most alien to classical culture, in the most rustic of mason's yards, there sometimes crops up, during the ninth or the twelfth centuries, an Antique motif used with the greatest care. In peripheral regions where art, and above all architecture, obeys standards imposed from outside, the same phenomenon can always arise. Nothing more clearly demonstrates the arti-san's fidelity to his models and, at the same time, the disintegration of culture.

There is a temptation to attach a more general significance to the successive restorations of the Antique order that characterize the Middle Ages, from the Carolingians to the Ottonians and the Swabian Emperors. Their ideology is entirely Roman, and their cultural range must not be underestimated;[6] these movements had, at least in certain provinces, marked repercussions on activity in the arts. But it is not so much a question of 'renaissance' as a revival, a *renovatio*, of the world, inspired by political and religious considerations, and emanating from centres outside Italy, and these movements did not entail a complete transformation of art. More significant was that which, in Rome itself towards the end of the thirteenth century, deliberately returned to the basilicas and mosaics of the Early Christian epoch.[7] New cities then in full growth, like Florence, could not conceive any other patent of nobility than that of their Roman origin; if the chroniclers were to be believed, the city was the favourite daughter of the Empire, confirmed in its privileges by Charle-magne, and Dante even affirms that it was built on the model of Rome, *ad imaginem suam atque similitudinem*.[8] It is often thought that in the city of Dante, Arnolfo and Giotto, about 1300, one is already in the presence of the Renaissance. But if this movement is placed in its historical context it is more reasonable to see it as a moment of national self-discovery strongly marked by truly medieval characteristics and thus an original but still provincial episode falling between the Proto-Renaissance of the South (twelfth century) and the Renaissance of the fifteenth century. Its orientation was opposed in other provinces by very different forms of art, more clearly linked to Western Gothic and, finally, this Neoclassical interest was to encounter a sharp check during the second half of the fourteenth century.

It was not until after 1400 that the cult of Rome, matured by many trials, supported by literary circles, and developed precociously in northern Italy, finally oriented a new culture which ended by sweeping all before it. Curiosity was rampant; ruins were closely investigated and measured. Great finds, like the *Three Graces* (Sienta), the *Laocoon*, the *Torso* and the *Apollo Belvedere* (Vatican), were discovered so as to become, at exactly the right moment, long-awaited models. But this orientation, which powerfully concentrated interest on Rome during the sixteenth century, was not a deeply archaeological one; rather is it another phase of the eternal dialogue between Italian art and its ancestors. Forms so rich, so developed, and sometimes so unexpected, are born from it that they overshadow the models by which they are inspired just as their revival is being attempted.

Italian art then took complete possession of its inheritance, but with the new feeling of 'historical distance' which invited artists to use the Antique as a source of inspiration, so as to equal, if not even surpass it.[9] The models furnished by Roman reliefs and bronzes either nourished an enthusiasm which encouraged ordered, controlled, 'Apollonian' forms, or, on the contrary, confirmed more fiery temperaments in a feeling for the pathetic, and for movement, which would eventually lead to the Baroque. The crisis of Mannerism, which broke out at the heart of the Renaissance after 1530, is sufficient demonstration that the classic order, both in its grand style and its minor phases, was inadequate to satisfy all minds. It soon became clear that for the modern artist Antique art was no more than a precious fiction, and that it was possible to use its resources freely, and even wilfully. To the academic purists, who considered Antiquity, now that it was better known, as the ideal of taste, the Baroque masters retorted by considering it as no more than a repertory of expressive forms. Finally, during the eighteenth century, Italy was to become the natural home of Neoclassicism; through excavations and collections of engraved plates, she furnished all manner of elements which she did not always herself exploit, except for the great Piranesi, who expressed and summed up all the links between Italy and her past.

This general framework sets out at the same time the main branches and the continuity of Italian art, but it must be modified in two ways. The inheritance from the Antique cannot delimit a style; it contains several, and each of its aspects has been given currency in turn by a predilection which remains to be interpreted. In the second place, the evolution of Italian art obviously cannot be explained without taking account of the continual exchanges between the East and the West; these provoked in the South, as in the North, increasingly sharp spasms of innovation which forced artists to interpret their tradition more fully, so that each time they should re-assert their own originality, and, finally, their prestige.

INTRODUCTION

Aspects of Tradition. Many kinds of very different art were grouped under the general heading of Antiquity; among the marble or bronze statues which could always be seen in the fourth century in Rome and in other Italian towns, and which ten centuries later were to be dug up in Tuscany and Lombardy, there were innumerable Roman copies of much older Greek works—Venuses, youths, graceful reliefs, delicate ornamental motifs in marble and bronze. Roman taste tended to the multiplication of more solemn types and less charming proportions, but a thread of Hellenism ran through all the works of the Imperial world and, with or without the resurrection of the Antique, this taste had a fair chance of coming once more to the fore. This can clearly be seen in Tuscany from 1300 onwards, and it was the mission of Florence to rediscover and redeploy it during the fifteenth century, when a refined sensibility was emerging. This 'Greek' taste, which can be recognized in the deliberate search for tension, intelligence and purity, triumphed with Brunelleschi and with sculptors like Desiderio. Its expression found unexpected support, and it in fact played a part in every kind of style, from Ghiberti to Giovanni Bologna, and later Canova, in the classic serenity of Raphael and the tenderness of Correggio. All that is most subtle in Antique mythology, of most exquisite in Christian thought, was to be given valid form through this source of inspiration, from which were derived some of the finest and loveliest works produced in the West.

To an art of elegance and harmony was contrasted the more immediate and massive art of the ancient Romans, founded on monumental grandeur, addicted to state portraiture and to historical scenes; this is the art rediscovered in the thirteenth century by the extraordinary despot of Palermo and Capua, Frederick II of Hohenstaufen. Perhaps it is to this that is due the narrative sense, the note of authority, which appears so clearly in Giotto's compositions, and certainly in Dante. During the sixteenth century, a remarkable historical event restored the sense of the Imperial past and the feeling for the monumental grandeur of ancient Rome to the seat of the Catholic Church; power and prosperity, eloquence and strength were renewed during the time of Julius II and Innocent X. The craving for this return to the past also revived, more stormily, during the time of Victor-Emmanuel II, and more cynically during the years of Mussolini.

But it is not enough to distinguish only these two antagonistic elements in the heritage of Antiquity, as it was understood in Italy. It contains beauties at once more fundamental and more recondite. The grand treatment of volumes, so knowledgeably handled that they penetrate and dissolve space itself, walls painted or picked out in stucco so that they dazzle the eyes, a richness of effect which bemuses the spectator—this also is Antiquity, and it is perhaps in these things that it is more fully expressed than in refined Hellenism or pure Roman solemnity. From the fourth century onwards, great undertakings such as the Baths and the least of private buildings alike were treated in this spirit. Greece was plastic; Rome was monumental; the Late Empire and the Early

Christian Empire concentrated on colour and on a continuous and vibrant treatment of interior space.

It is not necessary to go so far as to maintain that Byzantine art was only a higher development of this aspect of Roman art[10] and that Byzantine influences only restored to Italy its own inheritance; nevertheless, Italy remained permeated by this love of diffused colour and this architecture of outward appearance, better understood nowadays.[11] Ravenna and Sicilian Romanesque art must not be considered as accidents of history, Eastern enclaves on Roman territory. They reveal the lasting implantation of a particular taste; they add to Italy's prestige with one of the major features of the Mediterranean world; and if it is easy to take as points of reference Florence, so eminently sensitive to the Hellenistic element, and Rome with its entirely Roman character, Venice must inevitably be felt to uphold and amplify that which for simplicity's sake may be continued to be called Byzantine. It is the root of all 'pictorial' vision, alive to the effect of the patch of colour, to enveloping light; its metamorphoses may be followed from Titian to Guardi, even to de Pisis; and with it must be associated an architecture, rich in effects of light and scenic qualities, which has always been current in certain parts of Italy.

The measure of the varied resources of this powerful tradition is given by the rapid development of these three centres which, around 1500, established from north to south a fruitful competition. Italy's difficulty in establishing its unity derives from many causes, among which were an unbalanced economy and the parcelling up of power, but to them may be added the fragmentation which created such strongly differentiated modes of feeling. The link with Antiquity is thus far from being simple, but it is this fundamental polarity which explains why, for all its perpetual renewals, Italian art presents no breaks.

Italy and the West. From time to time, Italy was able to dominate the West, or at least to serve as a stimulant to it. For several centuries, the churches of the fourth and fifth centuries provided models for Western Christendom; during the eleventh century, Lombard master builders shared in the development of Romanesque art, and the precocity of Lombard pointed vaulting must not be overlooked. In 1400, Italy was intimately connected, through Siena and Lombardy, with the development of the most elaborate Gothic style. After 1500, master builders, decorators and painters came from as far afield as Castile, the Loire district of France, and the Rhineland, to learn their art in Rome. In 1600, Rome, which was still attracting amateurs and artists from all over Europe, was at the same time the home of new artistic conventions, based on the Accademia di S. Luca, and the centre of new and revolutionary ideas, stimulated by Caravaggio. In 1750, every educated European was interested in Pompeii, and a whole decorative art was inspired by the albums of engravings compiled in Rome after Antique remains.

But, conversely, Italian art was continually affected by major styles evolved outside Italy. The Byzantium of the Golden Age (tenth to eleventh centuries) conquered several provinces; French Gothic of the thirteenth and fourteenth centuries dominated others, and spread its pointed arches, its quatrefoils and its slender figures throughout the peninsula. It is necessary to place these actions and reactions on record so as not to falsify a history the most attractive episodes of which are far from being only those where Florence, Rome, or Venice legislated for the whole of Europe. These exchanges did not come to an end with the Renaissance; on the contrary, they increased with the impact of Flemish painters of 1420–30 on Antonello and Piero, and a century later when Dürer's prints so strongly affected Mannerist artists like Pontormo and Lotto. And if the Baroque decorators owed nothing to anyone else, the France of Louis XIV and Louis XV which replaced Italy in her classicizing role, achieved in architecture, in painting, and even in sculpture, a brilliance which, in turn, had powerful repercussions in the peninsula. Italian art must therefore be considered in a definite perspective so that the whole of Western art is seen in conjunction with the Mediterranean world, to which the peninsula serves as a gateway.

The great phases of Italian art do not fit into the general framework of Western art in the way that a species is part of a genus. In the development of the Latin world, Italy was preoccupied with certain things that after serving as a brake in the twelfth century furthered her advance in the fourteenth, explain her triumph in the seventeenth and, perhaps, her decline in the nineteenth. These things were concerned with the very basis of visual presentation, and the history of Italian art is linked with a 'history of vision'.[12] The 'classic vision', in separate units and with a limited space, had already given way in the fourth century to a vision in which brilliance of colour, with its levelling effect on voids and solids, encouraged a unified representation where figures no longer told as masses but as arabesques. It is in Western Romanesque art that the full results of this attitude can be seen in its application to sculpture, which encloses all forms according to an ornamental type within a closed framework. But Italy was not completely dominated by this attitude; she was more intimately connected with Byzantium where painting, and even the relief arts, maintained the elements of a spatial construction throughout all its imagery. The breach became sharper in the thirteenth century with the success of Gothic in France, Germany, and even in Italy, since the new style of construction was linked with an articulation of space, and a search for volume, which liberated the statue from a decorative system while integrating it with the decoration. This was the genesis of that creative upsurge from which, about 1300, a galaxy of geniuses emerged in Rome, Pisa and Florence. By adding to the plastic feeling of Gothic the scenographic qualities preserved in Byzantine art, and by correcting the formalism of the latter by a new feeling for mass and spatial coherence, Italian artists achieved in the fourteenth century a decisive move towards the elaboration of perspective space, for which the Flemings of the fifteenth century were indebted to them.[13] The late development of art in Italy was partly due to her long ambiguous position between the two worlds of north and south: eventually all was grist to her mill.

The activity of the seventeenth and eighteenth centuries in Italy must, in the same way, be considered in its European context. For it was at the moment when the Baroque was sweeping all before it that, as a result of certain easily identifiable northern influences, there appeared in Rome, Bologna, and Venice, a passionate interest in all forms of 'pure painting', whose qualities of the fantastic, the intimate, and the poetic, compensate to a certain degree for the wholly external character of the decorative arts. Something essential would have been lacking in Italy without Caravaggio, Magnasco, Pannini, and finally, without Venetians like Guardi, in whom was paramount that lively sensibility for the most diverse accidents of light and atmosphere which has stimulated the whole of Western painting. If, during the nineteenth century, Italy rested on her laurels—rather too much, in fact—it is perhaps because between 1650 and 1750 all the newest ideas had multiplied there, and all future developments had been adumbrated.

Uses and Abuses of Tradition. If, in Italian art, it is possible to speak of the abuse of tradition, it is to explain the tendency, apparently stronger there than elsewhere, to work always in the same places and on the same buildings. They were preserved by being completed, sometimes by adding to them inappropriate and incongruous embellishments, such as the unfortunate seventeenth-century paintings in S. Vitale in Ravenna.

There were notable and irreparable destructions: Old St. Peter's in Rome, and S. Francesco, the largest church in Milan; in the nineteenth century, the centre of Florence and the old parts of Milan. But from the earliest times the Italians have exercised their outstanding technical ability in raising fallen walls and the discreet restoration of paintings. Restoration is with them a permanent branch of art; in the twentieth, as in the sixteenth century, restorers do not hesitate to go to the lengths of pastiche, and the result is often disconcerting to the expert. The Tuscan frescoes and the Ravenna mosaics could not have survived without repeated 'rejuvenations'. Prodigies of patience have recently been performed in the reconstruction of the Mantegnas in the Eremitani chapel, and to restore Leonardo's unhappy *Last Supper* and recover what survives of the original work.

The feeling for the continuity of art is generally stronger than the opposition of various 'manners'. Thus later masters may be found completing unfinished works, and adapting themselves to a style that had already run its course, and was apparently far distant from their own. When Michelangelo inserted into the *tepidarium* of the Baths of Diocletian the church of S. Maria degli Angeli (1563), which Vanvitelli finished in the Baroque taste, it seems as if he intended to demonstrate that his art could approximate to the Antique, and give it renewed authority. When still a young man, and a refugee in Bologna, he was commissioned to carve two statues and an angel candelabrum for the *arca* of S. Dominic, begun by Niccolo Pisano (1264), almost completed by Niccolo da Bari, and then recently surmounted by a further storey (1473). In the same way, Filippino Lippi had the honour of completing in the Carmine in Florence the fresco cycle by Masaccio and Masolino, left unfinished in 1427.

The most lively intellect of the fifteenth century, Brunelleschi, had his first great success by surmounting the cathedral, S. Maria del Fiore, with a dome foreshadowed in the plans by Arnolfo, a Gothic master, a hundred and twenty years earlier; and Brunelleschi also used in his churches ideas derived from the Romanesque church of SS. Apostoli, which had conserved in Florence the spirit of Early Christian basilicas and was believed to have been founded by Charlemagne. Alberti, the theorist of what may be called 'humanist' art, did not hesitate to complete the old façade of S. Maria Novella in a style consistent with the Romanesque structure. In a broader sense still, the example of Venice, like that of Rome, demonstrates the natural and continuous growth of the modern on the old, and even within the old. This adaptability is the sign of an enlightened historical conscience, able to interpret the past. In Venice, the Piazza S. Marco evolved during the sixteenth and seventeenth centuries into a unified design which included a Gothic palace of the fourteenth century and cathedral of the eleventh century, Early Christian in style.

A building which may be considered as typical of this kind of evolution, to which numerous and considerable alterations have been made over the centuries, is the basilica of S. Maria Maggiore in Rome. The kernel of the foundation by Pope Liberius (352–66) survives under the fifth-century restorations, the additions made during the thirteenth century (the transept), the embellishments of the sixteenth century (Michelangelo's Sforza chapel and Fontana's chapel of Sixtus V), and Fuga's eighteenth-century façade; it is a kind of compendium of all styles, and the finest example of their continuity. Here the modernizations can easily be distinguished, but this is not always the case. The dating of the plan and parts of some Florentine buildings, such as the Baptistery or S. Miniato, continuously reworked and completed from the fourth to the sixteenth centuries, is the more difficult in that the edifice as a whole remains homogeneous and very pure in appearance. Sometimes it was not until the nineteenth century that the façades of medieval churches were finally built, their incomplete state having always been an embarrassment. Hence, to say nothing of Milan cathedral, finished in 1813, both at Sta Croce and the cathedral in Florence there are unfortunate pastiches born of this concern for the continuation and perfecting of the works of the past, which is one of the fundamental traits of Italian art. The historian has to bear it in mind.

Part One

THE MIDDLE AGES

I

Roman Art, Greek Art and the Art of the Early Middle Ages

(Fifth to Tenth Centuries)

Introduction: The End of the Antique World

It is the custom to start a new era with the beginning of the fifth century, at the moment when the structure of the Roman Empire was crumbling in the West. But the term Middle Ages, invented by historians of classical antiquity, and intended to cover what was thought to be a cultural and artistic void lasting for ten centuries from 400 to 1400, lacks the necessary subdivisions. It should properly be reserved for the earliest period of Western history, the period which opens after the year 1000; the predominance of the Benedictines of Monte Cassino and later of Cluny, which created a new framework in the West, the quarrel between the Papacy and the Empire, which led to a hierarchy of powers different from that of the Byzantine world, the success of the Norman invasions of England (1066) and Sicily (1060–72), and finally the Crusades (1096), correspond to the emergence of a new consciousness in Latin Christendom.

If the true Middle Ages only began in Byzantium after the Iconoclastic crisis of the eighth century, then there should be no question of there being a Middle Ages in Italy until after the breakdown of responsibility and the progressive loss of control which followed the end of Carolingian rule—until, that is, after the void of the tenth century. Until then, it was still the end of the Antique world, a long agony which presented occasional respites, but during which no radically new culture appeared.

In 311 the Edict of Galerius brought the persecutions to an end and opened the road to the official appearance of Christianity which became the State religion under Theodosius (380); in 330 Constantinople was founded and displaced the centre of gravity of the Empire; in 410 Rome was conquered by Alaric; in 476 the very institution of the Western Empire was abolished. But each of these dates marks less a new era than a stage in an irresistible evolution. The dispersal of the capitals was something which had already occurred in the third century under Diocletian, and this is a fact which the art historian must take into account if he is not to be paralysed by a problem now considered out-of-date—the fallacious dilemma of 'Rome or the Middle East?' There was, in fact, to be an art specifically of the great cities, and similar buildings can be met with in places as far apart as Antioch and Milan.

Since the culture of the third and fourth centuries was entirely composite, it is as impossible to explain everything through Italy as it is to understand anything without her, and the activity of the period which followed Constantine was extraordinary in every domain. The crumbling of the Empire, at the end of the fifth century, was followed by efforts towards its restoration which the Barbarians, obsessed by its past greatness, ceaselessly attempted for their own benefit. This was because for a very long time there was no other conceivable political framework than that of the Roman Empire. S. Augustine, who wrote his *City of God* at a moment when the position of the Latin world was clearly compromised, nevertheless inscribed the destinies of the Christian Church within those of the Empire.

Its ideal of authority impressed itself upon the Goths of Theodoric in 500, on the Lombards of Liutprand in 600, and above all on the Franks of Charlemagne in 800. Each time, the divergence from the original is more perceptible, the artifice more evident in the use made of formulae and titles. After this whole series of recastings, it only requires one last crisis to open a new era. This is also observable in the history of art, and particularly in that of architecture, where the typical creations of the fourth century were to be more and more transformed by those who took them over.

The Evolution of Roman Art. On the basis provided by Hellenistic formulae, the fusion of the arts of the Mediterranean area had already taken place during the first century. From the Flavian Emperors onwards, a spectacular and composite grand style spread from Syria to Spain. By its monumental undertakings, the technique of concrete vaulting, organic plans, the use of arcading and niches, and, finally, by its coloured mosaic revetments, this architecture no longer has anything in common with ancient Greek art in which the edifice has primarily a plastic value; it owes to the East the detail of certain forms rather than the system. It is characterized by a search for large unified volumes, or, to express it differently, spatial blocks, either on a long axis as in the rectangularly planned *aula* closed either end by porticoes, or in the centrally planned type with a massive vaulting stretched over a void, as in the Pantheon. A cosmopolitan and luxurious art, tending to illusionistic effects, appeared with the Golden House of Nero (64–68) and reached its highest point in the Baths of Caracalla (211–17), crowned by a dome over 100 feet in diameter. Before the building of the Holy Sepulchre in Jerusalem in 335 by Constantine's mother the Empress Helena—it may be questioned whether this rotunda of the Anastasis was not by origin hypaethral, that is, open at the top—there were to all intents and purposes no buildings of this type in the East.[14] This fact has considerable bearing on the architecture of the High Middle Ages.

It is necessary to distinguish official, or 'aulic', art, that of Rome, Milan, Ravenna—and Constantinople—from provincial art, freer with regard to Antique models and often full of new ideas. In the Roman provinces, art was left much more to its own devices than in the East, where it was dominated by Byzantine autarchy, and bit by bit it evolved, not only in Gaul, in the Rhineland, and on the Danube, but in Italy itself—in Lombardy and Tuscany— what is known as Romanesque art. With this was reached the final stage of the decline of the figure in favour of the decorative effect, and the abandon of the monumental in favour of the ornamental and the precious. Sometimes this art had a direct effect on grand official art: for instance, as early as the reliefs on the Arch of Constantine (312–15) the stiff and monotonous arrangement of the figures surrounding the Emperor contrasts with that in the re-used fragments of second-century sculpture.

In the famous porphyry *Tetrarchs* (*c.* 300) built into the corner of the wall of the Doge's palace in Venice very little indeed survives of Roman. Their stature, their garments, their swords with eagle-head pommels, their heavy gestures, can be explained by the new style of portraiture deriving from the sculptors of Palmyra.[15] In any case, they prove that the masters of the world were already turning away from the canonical styles, and it is this profound hesitation of Roman taste between its academic forms and its provincial forms that is the prelude to the hieratic attitudes and frozen masks of the Middle Ages.

During the Constantinian period, marble busts, still numerous, seem to have been picked out in colour, and sarcophagi were given a polychrome and gilded decoration which sometimes makes them resemble Byzantine enamels.[16] From the Italy of the fourth century, the one the Barbarians discovered, there survived in the domain of sculpture two virtually opposing obsessions: that of the majesty of forms carved or cast in noble and permanent materials, marble and bronze; that of scintillating and glittering polychromed surfaces.

Finally, the pagan world displayed a yearning for subjective, irrational expression, which it has been usual to explain by the rise of Christianity. In particular, the presentation of the image according to an objective order based on perspective which permitted the gradation of figures in space was abandoned and supplanted by a system which superimposed forms by distributing them across the surface so that appropriate distortion should enhance the spiritual content of the image.[17]

The Art of the Catacombs. While Antique art was in this manner achieving its own metamorphosis within a new civilization, the Christians were using it secretly for the purposes of their faith. The catacombs were burial grounds requiring only the most rudimentary architecture, situated outside the city, along the main roads; for example, in Rome there was the Domitilla on the

Via Ardeatina, the S. Agnes on the Via Nomentana, the Pretextatus on the Via Appia, and in other cities, the S. Januarius outside Naples, or the S. Antiochus in Sardinia. During the fourth century, they became true sanctuaries, but it is solely from the point of view of iconography and decoration that this hidden period of Christian art stands at the beginning of Western painting and sculpture. Nothing distinguishes it from pagan art, except the dignity and gravity of the images, which adapt the repertory of funerary art, and chiefly that of marble sarcophagi, to the exaltation of the new faith. It is mythology turned against itself, so that Prometheus and Orpheus represent God the Father and Christ. Pastoral motives abound. The Good Shepherd, transposed from Mercury Criophoros, appears amid praying figures on the Ram Sarcophagus, and in the midst of grape-harvesting putti (both Lateran Mus.). In painting there is the History of Psyche, an image of the soul (Catacomb of Domitilla), banqueting scenes that conceal under a profane appearance the eucharistic agapes (Catacomb of Priscilla), praying figures and beautiful emblematical forms such as the fish—symbol of Christ—the dove of eternal life, the olive branch, the basket of loaves, grouped sometimes in still-life arrangements.

The style, above all in the frescoes, is often poor and summary; the execution is limited to a few touches; but some pieces of fine quality are met with, like the *Apostles* in the Hypogaeum of the Aurelians, classical in handling, and the medallion of *Daniel* in the catacomb of Pretextatus. Only a later composition like the *Orant Madonna and Child* (Ostia cemetery, c. 350) which authoritatively restates a similar motive in the Catacomb of Priscilla (second century), appears as a direct forerunner of monumental art.[18] In sculpture, the transition is more easily perceived from the plastic style, where the figure is simply treated in relief (*Ram Sarcophagus*), to the picturesque effect which sets it among arabesques and subjects it to the play of light and shadow by the variations of the material (*Jonas* and *Good Shepherd* sarcophagi, Lateran Mus.). Of greater significance still, the later sarcophagus of Junius Bassus (d. 352) presents with more animation figures in two rows, the upper ones being between columns and the lower ones under pediments and arches. In the few free-standing figures, such as the *Good Shepherd* (Lateran) and the *Christ teaching* (Mus. delle Terme), the graceful modelling is of pure Hellenistic type, but already under the serenity, the eloquence of the gesture, the tender and thoughtful melancholy, which were to flower during the next two centuries in ivories,[19] are more perceptible.

The Constantinian Period

Once it had been officially recognized, the Church created sanctuaries everywhere; the fourth century is thus a period rich in architectural experiments, and it is a period when—since there were as yet no fixed types—no anomaly can surprise. Only rarely was there a change of use in buildings already constructed; even less was there any substitution of an architecture created from scratch by the bishops for pagan temples and porticoes. In fact, an interesting series of researches can be discerned, some of them without any future, designed to adapt familiar elements of religious and secular architecture to new functions.[20] A fair number of these fourth-century buildings, hastily erected, were destroyed or had to be rebuilt during succeeding centuries, which, of course, does not make the process of reconstruction easier. First, the basilicas must be examined.

The great colonnaded portico of Hellenistic cities (*basilike stoa*) had evolved in Rome into a series of covered galleries borne on columns—a quadriporticus, in fact—surrounding an open space, sometimes enclosing a rectangle similar to a cloister, sometimes on a definite axis with an absidal end. This type of palace courtyard, as in Hadrian's Villa, became the edifice open to the sky which was designed as a setting for Imperial ceremonies, as in Diocletian's palace at Spalato, and later, that of Theodoric in Ravenna. Covered with a timber roof, the central space became a 'nave'. The arrangement of this *domus* to suit the needs of a Christian community seems to have been attempted at the end of the third century, for instance, by Bishop Theodorus at Aquileia. But it was under Constantine and his mother Helena that this architectural setting, which had the double significance of a meeting place and a place of celebration, came to be used in a spectacular fashion as a church. The term 'basilica', despite its Greek origin, was only to become current in the West as a designation for the new Constantinian Christian buildings;[21] it was the *aula Dei*.

St. Peter's was built on the Vatican Hill (324–44) on the site of the Apostle's tomb, lost in a pagan and Christian cemetery on the Via Cornelia, recently explored. The basilica was, in fact, a double building: one part, the quadriporticus, formed the courtyard (the *atrium*), with the gallery at the base of the façade forming the narthex; then came the hall covered with a timber roof, into which access was by five doors, with a nave and four aisles each lit by its own clerestory. The body of the church gave on to a transverse section which projected slightly on the outside—the transept—separated from the nave by a round-headed triumphal arch, and having, on the same axis, a fairly deep niche covered with an apsidal dome. In front of this apse, which contained at the end the papal throne, was an altar covered by a tabernacle borne on columns. This arrangement of the richly decorated altar was completed by a setting of marble with columns, with lamps and the images of saints (*iconostasis*), and ambones or lecterns, which cut off the part reserved to the clergy (Fig. I).

This gigantic type was reproduced at S. Paolo-fuori-le-mura, rebuilt from 386 onwards, and surviving until the fire in 1823, after which it was again rebuilt in an approximative reconstruction; at S. John Lateran, reworked in the tenth century, and completely transformed in the seventeenth century by

Borromini; and, finally, at S. Maria Maggiore, somewhat later in that it dated
from the pontificate of Liberius (352–66), also frequently reworked. S. Paolo-
fuori is, however, with its nave and four aisles, the building which best affords
an idea of what the great Early Christian churches looked like. The Corinthian
columns of the nave bear an entablature which creates a striking rectilinear
perspective, reinforced by the two rows of columns of the aisles; the whole
building converges on the central altar, the place of sacrifice, which was also in
St. Peter's and S. Paolo-fuori the point of 'confession' where the tomb of the
saint was venerated.

The architectural effect aimed therefore at creating in the spectator a
sense of communion with a mystic assembly. On the walls were mosaic
decorations and paintings which induced an ecstatic state of mind, and which

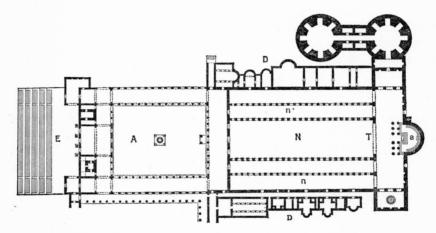

FIG. I. *Old St. Peter's, Rome (fourth cent.). Plan (after Dehio and Bezold):
E, Entrance stairway; A, Atrium and quadriporticus; N, Nave; n n, Aisles and
arcades; T, Triumphal arch; D, Dependencies; a, Apse.*

were conceived in a vision free from precise definition, from exact limitation,
and entirely subordinated to light, colour, and a deliberate suggestion of the
indefinite. In this way was created the powerful illusion of space which
haunted the world of Rome. And it comes as no surprise to find in contemporary writings that the church, and particularly the apse, scintillating with
mosaics, was treated as a 'paradise' where, in the exalted state induced by the
gold and the light, the pious soul achieved its mystical union with the Divine.
The inscriptions often placed, in the sixth and seventh centuries, round the
mosaics, extol in this fashion 'the glory of the Temple of God'; in S. Agnese-
fuori-le-mura one can even read that the colours are those of the captive dawn
and similar to the resplendent peacock that symbolizes eternity.

But near to these vast edifices there existed in Rome other smaller buildings with only a single nave: S. Agnese, and SS. Nereo ed Achilleo are instances. S. Sebastiano was even a basilica with an ambulatory, and Sta Croce in Gerusalemme a church with transverse arches. It must not be imagined, therefore, that the typical formula represented by Old St. Peter's and S. Maria Maggiore was evolved immediately. Archaeological investigation or reconstruction undertaken on their façades demonstrates that at S. Clemente, SS. Giovanni e Paolo, and S. Maria Maggiore itself, the primitive basilicas were always 'open', that is that they terminated not in a façade pierced with doorways, as at St. Peter's, but in colonnades, sometimes two-storeyed ones, opening into three or five open spaces, as may be seen in sarcophagus reliefs.

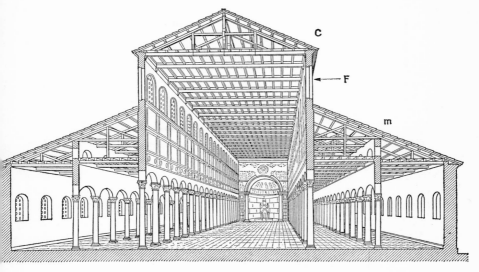

FIG. 2. *S. Paolo-fuori-le-mura, Rome (fourth cent.). Section (after Dehio and Bezold): C, Open timber roof; F, Windows; m, Wall separating the aisles, with passage under the roof.*

Not less important is the variant offered by the delicate little church of S. Sabina, which dates from the pontificate of Celestine I (422–33). Founded on the site of a pagan temple, and contsructed with a nave and two aisles, its lovely proportions and its open apse are splendidly set off by the re-used Antique fluted columns, surmounted not by an entablature but by arches. These effects of emphasis and light, and the purity of its relationships, give the building a particular value. The carved wooden doors, composed of panels in which appear for the first time motifs derived from Syria and Palestine (the Crucifixion, Elijah, etc.) and a kind of concordance of the two Testaments, were made by an assortment of artisans from Rome, the East, and probably also from the North, very uneven in quality, some panels being very heavy, others much lighter in design; they are also unique.[22]

If occasionally columns and decorative elements from pagan temples were re-used, it was nevertheless from the secular and official buildings of the Empire that the basilica was evolved. There was, however, another important source for Christian art: the *martyrium* or the circular commemorative temple raised on the site sanctified by the memory of a saint, or more simply over his tomb. This type of building, which was extremely widespread in the East where it led directly to the Byzantine central plan, was also current in the West from very early times. An enclosed space open to the sky, and originally open all round, these buildings rapidly adapted the form of mausolea or Antique *heroa* to the need of the Christian cult. This transfer of a pagan structure to a Christian use probably explains their character, but it does not seem an exaggeration to say that this kind of building inherited from Antique funerary and religious art played a part as important as that of the basilica.[23]

Here again, it is fruitless to inquire whether the initiative came from Italy or the East. The great martyria of Asia arose in the sixth century; it would clarify many problems if we knew whether the quatrefoil plan of S. Lorenzo in Milan, which will be discussed more fully, corresponded—always admitting that it goes back as far as the end of the fourth century—to the function of a martyrium, but there is apparently no early mention of relics.

Certain characteristics of Old St. Peter's can be explained by this role of martyrium; the crypt, in particular, was probably inspired by a Roman mausoleum. In any case, the type was to be fully developed in the West and during the seventh, eighth and ninth centuries, was widely used, first in Italy. With the mausoleum of S. Helena (*c.* 350) and that of S. Costanza (d. 354), the daughter of Constantine, which is one of the most astonishing monuments in Rome, the fourth century saw the creation in Italy of important architectonic models based on the central plan. The first building, called, after its medieval alterations, the *Tor pignattara*, from the '*pignatte*' or hollow amphorae with which the dome was made, is an octagon with niches in the walls; the second is built like a double ring, preceded by a small atrium with a double apse. Its coffered dome covers an enormous space, brilliantly lit by the high windows in the drum; the whole, to which must be added the mosaics in the attic, is borne by twelve coupled columns, the entablatures of which give an impression of converging towards the centre, and these columns are linked by arches. The barrel-vaulted ring of the ambulatory balances the mass of the dome, and the circular exterior wall is pierced by niches. The result is an impressive contrast of voids and solids, emphasized by the relation of the columns to the walls they support, at the point at which the light ceases to penetrate. All the vaulting was once covered with mosaics, and from the surviving areas of mosaic in the ambulatory it is still possible to form an impression of the magical transformation effected by this light and scintillating decoration. The branches entwined against a light ground, and the scenes of grape-harvesting, have given the impression that this was once a temple of Bacchus; in fact, this is one of the cases when a monumental use was made of themes derived from the art of the catacombs, by way of the luxurious technique of Hellenistic and Roman decorators. The same forms stylized, harder, treated in low relief, may be found on the great porphyry sarcophagus (now in the Vatican) which once stood in the centre of the mausoleum and contained the remains of the saint. All the arts collaborated to create a unique effect of contrast and continuous movement.

These ideas are also to be found in the baptisteries, of which the most

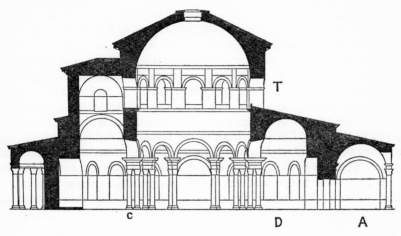

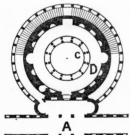

FIG. 3. *S. Costanza, Rome (fourth cent.); Section and plan (after Dehio and Bezold): A, Atrium; D, Circular ambulatory; T, Drum, pierced with windows; c, Coupled Corinthian columns.*

outstanding was that of S. John Lateran. Founded by Constantine, it was re-worked in octagonal form in the following century. From the original construction there survive the lovely, slender columns of black porphyry which carry an entablature—not arches—and are surmounted by a storey which serves as a gallery, articulated by smaller porphyry columns. The baptismal pool was in the centre of this entirely linear design in which the polychromed revetments played the major decorative role. The building was reworked in the sixteenth century under Paul III, and this must be taken into consideration when making any reconstruction of its original appearance, though this is another instance of the continued interest taken in Roman buildings by artists of later date.

The central plan was not limited to buildings which were specifically mausolea and baptisteries—baptisteries being dedicated always to S. John the Baptist. S. Stefano Rotondo was built during the pontificate of Simplicius (468–83) at the moment when the last Roman Emperor was disappearing from the scene. It consists of a double circular nave with re-used heavy Antique Ionic columns, marking the position of the original dome. The church has exactly the same dimensions as the Holy Sepulchre in Jerusalem—60 feet high and 72 feet diameter—of which it is a fair copy. The fact of this imitation is interesting: not only was the type of domed baptistery to remain associated, in Romanesque art, with the memory of the rotunda of the Anastasis in Jerusalem, as is the case in Pisa (1153) and all through Tuscany, but the symbolic theme of the Holy Sepulchre figures also in church plans. This can clearly be seen in the group of three churches dedicated to S. Stephen in Bologna which is a 'copy' of the group in Jerusalem going back at least to the ninth century and in its present state to the eleventh and twelfth centuries.[24]

These 'ideal' forms were therefore highly prized, but it is also necessary to restore to the early Christian temples the beauty of the new rites and the ardour of the faithful who decorated them, coloured them and lit them in all parts. S. Paulinus of Nola, a patrician who became a hermit, evoked in these words in 400 the radiance of the church on great feast days: 'Others . . . bring, to hang them before the doors, magnificent veils shining with the splendour of linen, or embroidered with richly coloured figures; others make and polish silver plaques and cover the holy threshold by fixing there their ex-votos; others light painted candles and hang from the coffers of the ceiling lamps with many tongues of flame, quivering in the motion of the cords from which they hang. . . .'

The Art of the Barbarian and Greek Capitals: Fifth to Sixth Centuries

The division of the Empire into two halves by Theodosius (395) was swiftly followed by the decline of Rome; Syriac or Greek popes and insignificant emperors succeeded each other until the moment when Odoacer, King of the Heruli, drove out the unworthy Imperial representatives who returned to Byzantium. At the same time, the Arian heresy condemned at the Council of Nicea (325) but adopted by powerful barbarians like the Visigoths was a serious challenge to the primacy of the Holy See in Rome. Until the advent of Gregory the Great (600), the ancient capital of the world was to be no more than a celebrated provincial city, exposed to the ravages of the wandering hordes that roamed Italy from north to south.

Culture took refuge in the north. Milan became the capital of the Empire under Theodosius; with S. Ambrose, S. Augustine, and Stilicho, it became by the end of the fourth century a centre of considerable importance. But in 409 Honorius, the conqueror of the West, established the seat of the Empire in Ravenna, which, with a port on the Adriatic at Classe, and defended on the landward side by lagoons, was much safer than Rome. The city owed its brilliance to Galla Placidia, the sister of Honorius, who was first a hostage among the Vandals in Aquitaine, where she married Ataulph, and later was a refugee in Constantinople, before she returned in triumph to Ravenna and governed the Empire from 425 to 450 in the name of Valentinian III without sacrificing anything of Roman greatness and dignity. This quality of originality of the fifth century, springing up outside the now fallen Rome, but continuing the same trend as the Constantinian age, must now be examined.

Milan during the Fourth Century. From the fourth century onwards, Milan contained, as did Rome, a great diversity of Christian buildings, and some which, though rebuilt or even vanished altogether, are of the very greatest interest, and often difficult to date.

Between 378 and 386, S. Ambrose built, in honour of the martyrs SS. Gervasius and Protasius, the basilica of the martyrs, with nave and two aisles, an apse and arcades. The altar was placed under a ciborium with four porphyry columns, which today are virtually all that survives of it, since the whole was completely reconstructed during the Carolingian period. For other buildings, the projecting transept provided an example of a cruciform plan with the altar at the crossing: of this type are the basilicas of SS. Apostoli (or S. Nazzaro Maggiore), the *basilica virginum* (S. Simplicianus), and the basilica of S. Tecla which also had an octagonal baptistery with eight niches. S. Ambrose himself seems to have recommended this, because of the symbolic quality of the figure eight.

The capital of the Empire under Theodosius, Milan was more concerned than Rome with the provinces on the periphery. There was apparent in the city, and in particular in the Ambrosian liturgy, a truly 'Oecumenical' need which conduced to a culture embracing all traditions, those of Rome, the East, and principally those of Syria. It was on the basis of this Hellenized latinity, more concerned with colour and the picturesque, that was to develop, first in Milan, and then from there in Ravenna—and not in the reverse direction as has for so long been thought—the new Italian style.

But the most remarkable monument is certainly S. Lorenzo, the form of which has been clarified by recent archaeological investigation;[25] the basilica consisted of a central square with four exedrae, contained within an ambulatory, quatrefoil in shape, with galleries, four angle towers, a huge atrium, and a quadriporticus with sixteen columns. The whole was covered with paintings and mosaics. It would be very interesting to know what kind of roofing was used on the ambulatory and galleries (a flat ceiling, or barrel-vaulting?) on the one hand, and over the central space (ribbed vaulting or dome?) on the other. The reworkings done in the twelfth and sixteenth centuries make this an open

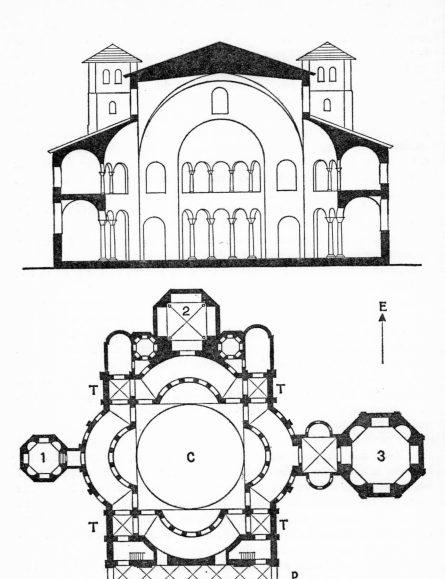

FIG. 4. *S. Lorenzo, Milan (fourth cent.). Section and plan (after A. Calderini, G. Chierici and C. Cecchelli): C, Quatrefoil building; P, Portico; T, Towers. Annexes: 1, S. Sisto; 2, S. Ippolito; 3, S. Aquilino (not in the section).*

34

question. This church built outside the Porta Ticinese had no Christian or pagan precedents, if in fact it goes back as far, in its first form, as the Arian Bishop Auxentius (355–72). Similar buildings were numerous from the Balkans to Mesopotamia during the sixth century. The examples which come closest to it are found around Antioch, and this coincidence suggests very strongly that since the architecture of the Roman Empire, from the fourth to the sixth century, provided a kind of common basis for all innovations, parallel developments, inspired by the desire to erect grandiose churches, may well have occurred in cities containing Imperial residences.

At all events, the group in Milan, to which must also be added the octagonal chapel of S. Aquilino (at S. Lorenzo) and the martyrium of S. Victor (which remains to be identified with certainty), seems to represent a typically Roman character; it is sufficiently early for there to be no question of its being derived from Ravenna. It exercised, in the Po valley, an influence which was powerful enough, during the seventh and eighth centuries, after the Lombard conquest, to eliminate all useless hypotheses about the genius of the Barbarians as builders, since they merely carried on a great style of architecture.

The most important examples of Early Christian architecture in Italy are equally ambiguous. It is possible that the baptistery in Florence, decorated in the thirteenth century, goes back to a fifth-century building, and that the peculiar rotunda of S. Salvatore in Spoleto, where the tendency to separate the presbytery from the body of the church—that is, to establish a choir—first appears, must be dated about 400. The end part is, in effect, covered by a dome, and this composite building demonstrates the importance of vaulting and of the circular plan in a period when the basilica is far from being a unique model.[26]

Ravenna during the Fifth Century. The unusual beauty of Honorius's capital, which was to fascinate in turn the Romanized Goths and the Byzantines, was celebrated by Sidonius Apollinaris (who died about 489) in terms which stress its strangeness and its air of sumptuous decadence: 'The living die of thirst, and the dead float in the water . . . , the priests practise usury, and the Syrians sing the Psalms . . . , the eunuchs learn the art of war, and the barbarian mercenaries study literature'.

Apart from the cathedral—a basilica with nave and four aisles, without transept, built by Bishop Orso about 390—the first group of buildings appeared in Ravenna during the fifth century. It consists of the church of S. Giovanni Evangelista (425) where, with a lower apse, is to be found something of the pure rhythm of S. Sabina; the mosaics of the Baptistery of the Orthodox; and above all the Mausoleum of Galla Placidia, the name of which only goes back to the fourteenth century. This lovely building is in the form of a Greek cross with arms of equal length; these are barrel-vaulted, with, at the crossing, a dome on pendentives, the weight of which is reduced by its being constructed of amphorae. On the outside, the brick walls are enlivened by large blind arches, a motive which was to have a great future, and by pediments, punctuated by small consoles.

The interior, which still has its facing of mosaics, is one of the loveliest creations of late Roman art. The dome is dark blue, like the night sky shimmering with golden stars, and with a great cross at the top; on the vaults, also blue, vines twine among lilies and roses, and in the lunettes are the golden figures of apostles, and deer facing the Fountain of Life; in the entrance tympanum, the Good Shepherd sits in a rocky landscape, and in the end one there is an enigmatic scene of martyrdom or a Last Judgement.[27] This profound and serene decoration, of a completely classical authority in an atmosphere of quiet sumptuousness, with nothing funereal about it, would be sufficient to justify the glory of Ravenna and the legend of Galla Placidia—legend, in that there is nothing to prove that the three sarcophagi in the chapel are those of the Empress, her brother and her husband, or even that this is in fact her tomb.

The Baptistery of the Orthodox, built about 450–52, by Bishop Neon, made the same adaptation of Roman decorations, which so frequently were illusionistic and picturesque. The use of light materials permitted the raising of a dome in eight sections, resting on arches which suggest, in a rather *trompe-l'oeil* fashion, decorative loggias. Mosaics enliven the successive rings of the dome; Christ in the Jordan in the centre, the apostles in the second zone, separated by acanthus leaves, solemn and rigid above abstract architecture, thrones and altars. The storey with the windows contains large reliefs in stucco, originally polychromed, in a much more feeble style. The association of architecture and colour, of strict composition and image, encouraged certainly by Hellenistic taste and Eastern fashions, corresponds to what had become from the third century onwards Roman practice. This elaborate form reappeared from time to time during the Middle Ages and the Baroque period, but it was in Ravenna first that, in the course of the next century, the Gothic, and then the Greek, conquerors were to push it to its highest point of refinement.

Ravenna during the Sixth Century. The barbarian chieftains who completed the ruin of the Empire, the Herulian Odoacer and the Ostrogoth Theodoric did not contest each other's possession of Rome, but of Ravenna; it was Theodoric, adroitly deflected by Byzantine policy from the Danube towards the Po, who finally won. From 493 to 526, a highly cultured Imperial court of Arian Goths, headed by Cassiodorus, was established in the capital where many spectacular edifices were erected. But soon afterwards, the province became an Exarchate and the object of rivalry for the domination of Italy between Goths and Byzantines. With his generals Belisarius and Narses, Justinian defeated Totila, the heir of Theodoric, and for the first time Ravenna became a Byzantine capital (540–65). Until the moment when, two centuries later, the withdrawal of the Greeks became certain, Ravenna enjoyed an exceptional prestige in northern Italy, where it affirmed with brilliance the twin greatness of Rome and Byzantium.

Under Theodoric, more ardent in his Latinity than the Romans themselves, churches with nave and two aisles of the Constantinian type were not lacking: S. Teodoro (or Sto Spirito), the cathedral of the Arians, and above all S. Apollinare Nuovo (519). The increasing of its height, the disappearance of the apse, and the opening of chapels during the sixteenth century have altered its design, but on the walls of the nave are the most complete set of mosaics. It has also lost the splendid ceiling which once earned it the name of

Fig. 5. *S. Vitale in Ravenna. Section on the south-east axis (after Dehio and Bezold): A, Apse; T, Tribunes; F, Windows.*

S. Martino al Ciel d'Oro. The superb row of Corinthian columns with marble dosserets is echoed in the upper storeys by the rows of scintillating figures; the framings, the friezes, and all the multicoloured decoration, are made of the same material as the saints under the golden baldaquins which frame the windows; above, there are twenty-six small panels representing the Life of Christ. Everywhere, golden grounds replace the blue grounds of the earlier period. This prodigality in light gives a supernatural dignity to all the motifs, and even to the architectural symbols like the façade of Theodoric's palace, with its intercolumniations decorated with garlands and hangings, and the depiction of the port of Classe. The procession of virgins and saints in the lowest storey is an addition of the Justinian period (under Agnellus, 556–69);

37

the drawing is poorer, there are pieces of mother-of-pearl among the coloured tesserae, and white and gold are used in the costumes.

In the field of the central plan, the period is less fortunate; the Baptistery of the Arians, a domed octagon with, originally, an ambulatory, is a rather poor imitation, above all in the mosaics in the dome, of the Baptistery of the Orthodox. Only a few vestiges remain of Theodoric's palace, which can be reconstructed from the projection in flattened perspective in the mosaic in S. Apollinare Nuovo; an echo of it appeared later in the façade, with its arches and ceremonial balcony, of the palace of the Exarchate,[28] which was incorporated in the ninth century into the atrium of S. Salvatore a Calchi. At a short distance outside the city, the two-storey tomb of Theodoric, which inside consists of a cruciform *cella* surmounted by a circular hall, retains on the outside, with its enormous monolithic dome and its powerful articulations, a Roman sense of massiveness; only the decorative frieze on the dome betrays a barbarian source of inspiration, since it has arabesques similar to those on the breastplate of the Gothic prince.

When, twenty years later, the Byzantines were masters of the Exarchate, and Justinian held Ravenna, Constantinople, which had never known the vicissitudes of the Italian cities, was in the full flood of its artistic effort. With this new stimulus, the art of Ravenna, swinging to the other extreme of its former activity, produced the masterpiece of the end of Antique art, the church of S. Vitale. This church was consecrated in 547 by Bishop Maximianus, but the decision to build it had been taken twenty years before by the banker Julius Argentarius, and there is no question of the project deriving from S. Sophia in Constantinople, which was begun in 532. These two exactly contemporary works were the last effort of this Imperial art, which expressed its love of the luxurious and the marvellous by seeking to enhance the animation, the colour and the size of the interior space of its buildings. S. Lorenzo in Milan had already advanced a long way on this road, while retaining a kind of geometrical rigidity. In S. Vitale the proportions are more elegant, the rhythm more lively, the colour and the transitions from light to shadow of the most strange and perfect poetry. To the eight sections of the octagon correspond eight fluted pilasters which support the dome and its high drum pierced with windows. Two remarkable inventions stress the effect of the interior: the main arches are broken up into two storeys, each divided into three bays, but these exedrae are concave, and introduce a delicate softening effect into the structure, extended by the unexpected depth of the apse. This creates a kind of irradiation, or centrifugal movement, which enchants and captivates eye and mind. The grand sweep of the ambulatory adds a contrast of light and shadow, brilliantly accentuated by the sculptural decoration. The curious forms of the pilasters multiply the projecting ridges, and the capitals are covered with lacy motifs whose delicacy is increased by their black background. Every part of the apsidal portion is covered with mosaics, to which the atrium, placed obliquely in relation to the axis of the church, directed the spectator's enchanted concentration: the spheroid portion consists of a Theophany centred upon a beardless Christ, seated upon a blue universe in the midst of a visionary landscape. At the sides are S. Vitale and the Bishop Ecclesius, and principally the two great scenes of the donors: the group of Justinian, in an ample toga, accompanied by dignitaries and Bishop Maximianus, contains striking portraits, powerfully conceived and inserted into the frontally presented figures; on the other side is Theodora, with a halo, covered with jewels, draped in her purple mantle and surrounded by her suite. These are intense and hieratic images, in which the forms are virtually mummified in the fluid gold and the essences of the East. The two

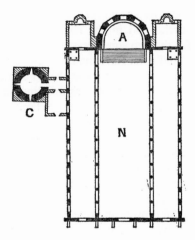

FIG. 6. *S. Apollinare in Classe. Plan (after Dehio and Bezold): N, Nave; A, Raised apse; C, Circular campanile.*

sovereigns bring the eucharistic offering; these official effigies, which are in harmony with the dedication of the church, appear like 'consecration pictures'. In any event, the supreme homage is paid as much to the theocrats as to the Godhead, and this betrays the Constantinopolitan origin of the work.[29]

On the outside are to be found the framing bands, supplemented later by buttresses. The original quadriporticus has disappeared, but nothing in the rather drab design of the exterior and the articulation of the apses with the rest of the building, gives any idea of the splendours of the interior. This is characteristic of nearly all the architecture discussed up to now; it is not so with the architecture of the Middle Ages. The last great building in Ravenna was the church of S. Apollinare in Classe, dedicated in 549. Stripped of its marble revetment during the fifteenth century, the basilica, with its twenty-four columns of Greek marble on heavy pedestals, and with curiously carved capitals, still gives an impression of magnificence; never was the unity of the decoration and the architectonic space more completely expressed, and this is particularly so in the splendour of the immense apse where, in a mosaic of extraordinarily symbolic and visionary iconography, a huge jewelled cross hovers above the praying saint who stands with white lambs on either side of him in a strange, unreal and glittering landscape. Outside, in the flat country-side of Classe, the volumes of the architecture stand out clearly, but without delicacy, and the surfaces are outlined rather tentatively with bands. The cylindrical campanile, built in the seventh century at the earliest, adds an essential element to the outside of the church, as is also the case at S. Vitale.[30]

The art of Ravenna is the end of an age. It affected both Lombardy and the east coast of the Adriatic, for example, in the basilica at Poreč (Yugoslavia, formerly Parenzo, 530–60), decorated with mosaics and with an octagonal baptistery adjoining—an important precursor in the development of the centrally planned church—and in the slightly later nexus of buildings at Grado (c. 580). But for the two considerable problems which were to occupy the minds of medieval builders, Ravenna not only provided no solutions, but left the problems untouched. The first was the grouping round the cathedral, either actually joined on to it or treated as buildings entirely separated from the liturgical ones, of such ancillary constructions as the baptistery, the sanctuary, the campanile, which was soon to become the general rule, and even the funerary chapels which were no longer to be separate entities. The mausoleum said to be that of Galla Placidia communicated, in fact, with the portico of the basilica (which has now disappeared) of Sta Croce; the atrium of S. Vitale was placed obliquely the better to integrate four small earlier chapels. The unification and the combination of these structures was to be-come more and more the central monumental focus of Italian cities.

But a second more subtle problem still was to arise as the basilical plan of the Constantinian type was modified and transformed. The central plan—apart from baptisteries—was appropriate at the start for churches of the commemorative martyrium type; in the West, the martyrium was to be in-corporated in the basilica, the area of the chancel and the apse thus being adapted to new developments which ended by modifying the whole of the construction. To this type of Latin Cross church was opposed the Greek or Syrian formula, which adhered tenaciously to the central plan. The question of the roofing was not always as decisive in Italy as it was in the remainder of the West, for the temptation to have recourse to the decorated ceiling, and to treat the walls freely, not as architectonic members, but as areas of decoration, was always present for the heirs of Rome and Ravenna.

Sculpture. A large number of ivory diptychs, known as 'consular diptychs', were used, bound like books, like that of Stilicho (c. 400, Monza, Cath. Trea-sure), or employed for liturgical purposes, like that of Boethius (Mus. Cristiano, Brescia). The first (Monza) is of an exquisite Hellenistic elegance in its slender form and delicate execution; the second is confused and heavy, but both have an absolutely Roman character.[31] The same comparison can be made between the *Symmachos diptych* (*c.* 400; left side: Paris, Musée de Cluny; right side: London, V. and A.) which evokes on a Hellenistic model a scene of sacrifice, and the two squat and ungainly allegories of Rome and Constantinople (*c.* 450; Vienna, Kunsthistorisches Mus.).

The evolution of sculpture in Ravenna was more complex; there the sarcophagi constitute a particularly complete series.[32] Conceived as small temples, they were not, as was the case with pagan ones, placed against the wall, but treated as independent buildings and placed on a base. The sculpture is less charged with meaning, contains fewer figures, and the relief is lower.

One of the oldest (fifth century) is a true aedicule, with acroteria, its façades decorated with arches between niches, carrying a representation of the *traditio legis*—the transmission of the Law. The design is clear and lucid (church of S. Francesco). Later types fully demonstrate a desire for even greater simplification, and a more abstract quality: the *S. Rinaldo sarcophagus* (first half of the fifth century; cathedral), with its curved lid, has a symbolical composition of Christ between two apostles flanked by palm trees; the Lamb between palms set against a plain ground is all that appears on the sarcophagus from the mausoleum of Galla Placidia (second half of the fifth century). But the types with figures continued long after the inception of the symbolic reliefs. It was only at the end of the sixth century that the lambs and the peacocks completely ousted the human figure. In the seventh and eighth centuries, they reappear with the Cross in reliefs of a distressing platitude; for example, those on the tomb of Archbishop Felix (S. Apollinare in Classe).

The most striking feature of Italian art in Ravenna is to be found in the superb *transennae* (pierced stone or marble screens before a tomb or in the chancel) and the capitals in the churches. The lithe Corinthian type found in S. Apollinare Nuovo may be contrasted with the more linear, twisted forms in S. Apollinare in Classe. The most powerful are those of Byzantine type in S. Vitale, which are like truncated pyramids, base uppermost, with their surface covered with branches of foliage, cut with a drill, curved and pierced to obtain the rarest effects of light. The *transennae*, more solid at the date of S. Apollinare Nuovo, are more transparent and lacy in S. Vitale; in their abstract pattern of convolutions, almost like embroidery, a bird or a palm motif appears as if by a miracle, to make them into one of the most outstanding points of contact between the delicacy of the Antique and the new aspiration towards a non-figurative art. In certain cases—for instance, in the sarcophagus of Barbatianus, confessor of Galla Placidia, carved about 540—the vigorous adaptation of an Eastern form of decoration can still be seen.

The ambo of the cathedral, erected in the time of Bishop Agnellus (556–69), is divided into compartments which contain symbolical animals carved in a tame and flat relief. The almost contemporary throne of Maximianus (Episcopal palace) is an imported work, displaying a clever mixture of Hellenistic and Syrian motifs, eastern and even Egyptian, which gratified the Byzantine love of syncretism. Through the carving of its ivory panels, this strange object attains a kind of luminosity.

Carolingian Italy: Eighth to Tenth Centuries

Three events were to convulse Italy during the course of the seventh century, and separate it definitely from the Antique world. In 568, three years after the death of Justinian, the Lombards left Pannonia, where he had held them, and descended upon the peninsula. Driven back, Byzantium held only the Exarchate. These barbarians, the most uncivilized yet seen in Italy, were converted; Gregory the Great (540–604), who took a long view of the defence of Rome, initiated a grand policy for driving them back. Nevertheless, the Lombards penetrated everywhere, and adapted themselves to their new land; finally, the alliance between the Holy See and the Franks, Pepin and Charlemagne, led, in 773, to the elimination of the Lombards politically.

The balance of the Mediterranean world was already undermined by Moslem invasions. Italy had not been spared these; in 638 Omar had entered Jerusalem; Egypt, Africa were conquered; Sicily fell, Rome was attacked, and profaned in 646. The Byzantines retook Taranto in 680, but not Sicily. In this manner, after the establishment of the Lombards in the North, pressure from the Saracens in the South completed the political and cultural break-up of the country. The defence of Christianity against Islam, undertaken by the Carolingians, the elimination of the Lombards, the consolidation of the Roman See, now transformed into the Papal States (756), are the more important in that the Carolingian effort was, or strove to be, a return to Antique order and culture. It was therefore of direct concern to Italy, at least in the more thinking, intellectual and artistic circles. But this revival was not a renaissance; its model was as much Byzantium as Rome, and also it coincided with the tremendous crisis of iconoclasm which was to lead to the total rupture between the two halves of the Christian world. By 726, Leo III the Isaurian had begun his destruction of images; Rome protested. The persecution was to revive with unparalleled violence during the ninth century. Like the Syrian or Egyptian monks alarmed by Islam, the painters and sculptors of the Balkans sought refuge in the West. This influx affected chiefly Rome and the cities of the Adriatic, like Venice, which was then in the early years of its existence. The iconoclastic fury of Constantine V Copronymus who, in 754, condemned the 'criminal art of painting', was answered in Rome by the decision to multiply the images in the sanctuaries; it was then that S. Maria in Cosmedin, SS. Nereo ed Achilleo, and later S. Prassede, were rebuilt and enlarged. The Greek art of mosaic seemed to have found a refuge in Italy. This movement, which led to a return to the syncretism of the fifth century, coincided with a more ardent veneration of relics. This is the period of multiple translations of relics, and the birth of pious legends about works of art which were given extreme veneration, such as the supposed portrait of Christ by S. Luke and the angels in S. Maria Maggiore, and the *Holy Face* at Lucca.

But, by the end of the ninth century, the divisions in the Empire resulted in a complete check to the Carolingian enterprise; one effect of this had been the resuscitation of certain learned tendencies in sculpture and painting in face of more popular ideas. Italian art and culture were long to remain, at least in the valley of the Po, in contact with civilizations north of the Alps, while the southern provinces and the maritime cities shared after their own fashion in Byzantine or Islamic civilization.

This kind of accumulation of interests and tendencies, this conflict of styles, brought about the abandonment of some things and the acquisition of others. The seventh century was a poor one; the tenth century was likewise. Between the two, from about 730 to about 840, the welcome given by the Roman church to Eastern artists, and the interest taken by the Carolingians in everything connected with great art and classic form, created conditions so favourable that this period constitutes one of the major formative periods of the history of Italian art.

THE MINOR ARTS

Like the Merovingians, the Lombards were goldsmiths. It was through their jewels, their handicrafts, their ornamental motifs that they communicated not only with the East, but with the even more distant Asia. The cups and dishes of the treasure of Canoscio (Umbria) of the fifth and sixth centuries, with their decoration of strap-work and jewelled crosses, the fibulae, plaques and necklace of Castel Trosino (Ascoli), and of Nocera Umbra (Umbria) of the sixth and seventh centuries, covered with tightly packed motifs, knots, and cloisonné enamels that make the whole surface shimmer, are examples. But what the Italianized Barbarian artist preferred above all else was to insert pure Antique elements into a composition according to his own taste, for example the Roman cameos among the cabochon stones in the book cover of Queen Theodolinda (seventh century, Monza), or to transpose an Antique figure scene, like that in the copper frontal of Agilulf of 600 (Bargello), where the figures are somewhat agitated and confused, though it is possible that the Lombard artisan wished to adopt the naïve and rustic style of the Roman provinces.

In any case, it was through their attachment to these popular and uncouth forms, to the exclusion of aulic art, that the Lombards played their decisive role. A strange Campanian marble of the ninth century (Cathedral, Aversa) shows a monster with a mane composed of small circles, coiling above a highly stylized horseman; the relief is flat, and telling only in its outline. It may be a S. George, but he is here represented by a sculptor who was a stranger to everything that relief sculpture meant for Rome and for Byzantium. This genuinely barbaric accent is rare in Italy, where the dominant forms were arabesques, and interlacings used over a whole surface, with the curves and the palmettes which create almost a kind of vertiginous calligraphy. Numerous chancel plaques in which appear Byzantine or simple Early Christian motifs, or even Sassanid monsters, bear witness to this, as for example at Cividale in Friuli. The font and the altar known as the Ratchis altar are among the coarsest examples, and at the other extreme is the fine sarcophagus of Teodota, a concubine of King Cuniperto (720), where two peacocks facing each other among interlacings translate into low relief the finest Ravennate inspiration (Pavia Mus.). But as a general rule the decorative motifs lose their established, architectural, identifiable character, to arouse, by means of strapwork and interlacings, a sense of uncertain movement and to produce an astonished wonderment through the profusion and the glitter of gems.

The equilibrium which still survived in Rome during the fifth century was lost in this way; the precious character of the object, increased by every possible means, triumphed over the quality of the image, and the material and ornament destroyed the feeling for the monumental. In a few generations there ensued the disintegration of that unity of architectonic space and coloured surface, of that continuous and luminous structure of decoration, which transformed the whole building into a 'treasure house' and which, in Ravenna particularly, was the last phase of Antique art. Architecture was to be no more than a massive and bare form; precious beauty became the attribute of the object. This is a significant shift in the very basis of artistic sensibility.

It was facilitated by the growing divorce between the art of the cities and that of provincial centres, parallel to the difference between learned Latin and the vernacular; the indifference shown to classical canons of proportions in figures has its counterpart in the abandonment of classical norms in the written word and in literary forms. It is essential to realize that this process of relaxation licensed borrowing from every source, and the acceptance of the most unusual motifs. The disorder of the period accelerated their circulation, and it is frequently impossible to distinguish the precise nature of the influences. An amusing example of this is provided by the *Hen with Seven Chicks* in silver gilt, with sapphire and garnet eyes (basilica of S. Giovanni, Monza) which is generally said to have been a gift of Theodolinda—an original of the seventh century, a copy of the eleventh, Moslem work of the tenth? Every hypothesis has been advanced.

ARCHITECTURE

In architecture, the difficulties of dating and attribution are exceedingly great. They are made worse by the many reworkings which, though they attest to the historical vitality of the buildings, conceal their origins. Without forming any kind of school, the great centres of the earlier periods—Rome, Milan, Ravenna—retain a certain authority, even if only in the sense that certain monuments were mechanically repeated, others enlarged.

Rome. In Rome itself, the basilical form underwent characteristic changes. It was in the 'Byzantine' sense that S. Lorenzo-fuori-le-mura was reworked under Pelagius II (579–90); similarly, on the nave of S. Agnese-fuori-le-mura (625–38) a gallery was erected with elegant columns that echo the movement of the arcades and articulate the building in height, so as to enhance the effect of the mosaics, where the rich idol-like saint seems to soar through the apse. At S. Giorgio in Velabro, in the time of Leo II (682–83), the nave was, on the contrary, made narrower, and the materials were mixed. In the Quattro Santi Coronati, which was reworked in the ninth century and again in the twelfth, the wide nave springing from two terminal pilasters begins a new development.

In the eighth century, Hadrian I (772–95) restored the Aurelian walls and began to raise the defensive towers, as well as the Lateran palace. S. Maria in Cosmedin, enlarged at the end of the eighth century under this pope, was given two lesser apsidal forms on either side of the high apse, as in Byzantine basilicas. Another novelty was the enormous piers which, here and there, interrupt the perspective of columns which thus lose their expressive value as supports. The wide nave and large arches of S. Maria in Domnica, built under Pascal I (817–24) make the traditional basilical structure much heavier and spread out. The exactly contemporary S. Prassede has a transept and is flanked by a perfectly Greek chapel with a dome decorated with a mosaic of the Second Coming; this chapel of S. Zeno, added to the right aisle by Pope Pascal is, with its hemispherical dome articulated above a cube, a perfect example of the canonical forms of Byzantine monumental symbolism as they were expressed at Edessa and Salonica.

As far as can be judged from texts, from hymns and manuscript illuminations, church interiors, with rich draperies between the columns, silk hangings in the choir, suspended chalices, and numberless silver lamps, were never more sumptuous than during this period, when all Christianity with its artists and relics flowed towards the capital.

Northern Provinces: Lombard Art. North of Ravenna, in the solitudes where reigned the malaria of the lagoons, the Benedictines raised during the course of the ninth century the Abbey of Pomposa, reproducing faithfully the type of S. Apollinare in Classe and perhaps even using capitals and marble from the basilica. The exterior fixes the type of decoration with long arcadings which was to be one of the legacies of Ravenna to Early Romanesque architecture; the baptistery, rebuilt at this time, stresses this element. At the same time, the circular sanctuary of S. Donato at Zara in Dalmatia imitated, but in a barer, more static, colder form, the structure of S. Vitale; but it does not seem necessarily to be derived directly from the other. During the eighth century, the forms of S. Apollinare may be perceived underlying those of S. Maria in Valle in Cividale (Friuli), of which it is an elegant reduced version.

The Lombards seem to have been ambitious builders. The records mention the buildings of Liutprand (712–44). What is chiefly known of his is an edict, completing older laws, in which are mentioned the 'Maestri Comacini' whose privileges were defined. It has been suggested that this term describes the engineer, working *'cum machina'*, and it is perhaps reasonable to see in it a reference to the enterprising architects of Como, rather than, paradoxically, an allusion to the distant Commagene (on the Euphrates). But the mention is valuable, even if the history of this class of artisans is obscure. It helps to sum up the evolution which permanently separated the lessons of Rome and Ravenna.

This explains the importance of a building like S. Pietro at Toscanella, near Viterbo; a text of 739 mentions a certain *Rodbertu magister comacinus*. The alteration of the Roman model can be sensed everywhere. The walls of the nave are massive, borne by very wide arches, punctuated by heavy denticulations, resting on columns with crude capitals; in the choir cruciform piers make their appearance. There is no feeling of suppleness, but only of massive weight, without light or lightness. The basilical form is to be found again, impoverished in the same way at S. Salvatore in Brescia (753) and at S. Pietro at Agliate. With these masters, the aulic art of Ravenna loses its sumptuous character, its decoration and its wonderful world of figures; the classic art of Rome loses its elegance and its precision. The only qualities sought after were bareness and strength.

The Lombards had made Pavia and Monza their favourite cities but Milan regained its importance with the eighth century. From Charlemagne onwards it became the tradition to be crowned King of Italy in Milan, and Emperor of the West in Rome. Interesting churches were not lacking in the ninth century; basilicas with nave and two aisles, apses with niches, bands and arcadings in the walls, as at S. Vincenzo in Prato at Milan; centrally planned buildings of multifoil design recalling S. Lorenzo, with three small apses and eight niches, like the Oratory of Ansperto (861–81) later absorbed under the name of Cappella della Pietà into S. Satiro; four columns support the dome of this enchanting building which was later completed by Bramante with an enclosing wall. But the principal work of the period was the new basilica of S. Ambrogio; the transformation of the fourth century sanctuary took place under Archbishop Angilbert II (824–59). The apse was enlarged by flanking it with smaller parallel apses, according to the new formula; on the outside appeared the crowning arcaded galleries. The atrium was enlarged by Bishop Anspert. Here Lombard art develops the more organic, more powerful style demanded by the new Carolingian culture. The enlarged choir of the basilica seems made for the golden altar of Vuolvinius, a masterpiece of Carolingian goldsmiths' work, dated by the mention of Angilbert (835).

The building activity of the Lombards established an innovation to which the classical regions of Italy offered but little resistance, but from the ruin of the ancient systems some characteristics survived which were decisive for the future. Among these were the use of three apses in the chevet of a church; outside they were ornamented with niches, the walls of the nave were decorated with arcadings, sometimes single, sometimes interlaced, sometimes with two or three arches carried on corbels. In this manner there arose a new unity, a new rhythm, founded less on ornament than on the basic motives furnished by the walls and the masses. A new system arose from this: a Roman nave, a Byzantine apse, Ravennate bands, which become Lombard bands, and to these elements were later added the square or round tower of the campanile, cylindrical in the Marches and Tuscany, square and broken up into storeys by means of decorative bands framing the bays in the area of the Paduan plain. The tower of S. Satiro in Milan, prototype of a long series of such towers, is usually dated in the ninth century.

But the interest in centrally planned compositions did not diminish. Buttresses strengthen the walls crowned by galleries in the baptistery of Biella, built on a quatrefoil plan (tenth century), and in that at Galliano, near Cantù (beginning of the eleventh century). The experience of the Lombard centres at the end of the eleventh century was to be virtually unique in Europe. The role of initiator played by the Benedictine abbot William of Volpiano in Normandy and Burgundy after the year 1000 is a striking proof of this. And with this the threshold of Romanesque art is crossed.

Byzantine Centres: Venice and the Southern Provinces. Two regions lay outside the scope of this slow evolution of a new order and, in the period which followed, they sometimes successfully contradicted it; these areas were the lagoons of the Adriatic, and the South.

Defended against the Lombards by the lagoons, a small enclave closely linked with the Byzantines of Ravenna grew up during the sixth and seventh centuries. Torcello was then the chief among the islands. The now vanished baptistery, a rotunda with eight columns, has been dated at 640; the cathedral, on a basilical plan, with a crypt, was founded in the seventh century, and was rebuilt and consecrated in 1008. It was long believed that the first basilica of S. Mark's, founded in 829 on an island already possessing a political organization headed by a Doge, was of the same type. Giustiniano Partecipazio had recently had the new S. Zaccaria built by Greek workshops; two travellers had brought the relic of the saint from Alexandria. The church was, therefore, a martyrium, and there are many reasons for believing that from the start it was of Greek cross form.[33]

In the South, witnesses to the authority of Byzantine forms can be found in S. Sofia in Benevento (eighth century), built on an octagonal plan, with a brilliant and complicated arrangement of columns inside, and S. Maria delle cinque torri (. . . . . of the five towers) at Cassino (eighth century) built on a Greek cross plan, with three apses, four corner towers and one over the crossing. S. Benedict of Nursia, a Campanian, had founded the monastery of Monte Cassino in 528, and had given a new structure, founded on obedience and work, to monastic institutions. The movement originated in the East; its enormous success is very characteristic of the dramatic and troubled period from the sixth to the eleventh century, and the Benedictine rule was to give a more active character to the monasticism of the West.

These southern centres had a twofold importance: during the Pre-Romanesque period it was not so much the buildings that counted as the products of the *scriptoria*, whence issued manuscripts of cardinal importance. These *scriptoria* were based on Greek culture, and each burst of activity in Monte Cassino corresponds to a new influx of Eastern artists and models.

PAINTING

The interior of nearly all the Roman basilicas of the sixth to the ninth century was made splendid by gold ground mosaics, placed on the triumphal arch and the spherical vault of the apse. That of S. Agnese, an immense golden bowl from which soars the figure of the crimson cheeked saint in her jewelled robes, goes back to the seventh century. So does the row of rigid and glittering saints in glory in the Oratory of S. Venanzio, built under John IV (640–42). Even during the eighth century, the tradition was not lost; in the Vatican, the Oratory of Pope John VII (705–7) was decorated with mosaics of which a few fragments survive—an *Adoration of the Magi* in S. Maria in Cosmedin —and it is clear that to the delicacy of colour is joined a genuinely expressive draughtsmanship.

It is not possible to consider all the works of the period as simply products of Greek art in Italy; in S. Maria Antiqua, for example, beside the panels in the nave (of the seventh and eighth centuries), which look like icons translated into fresco, there exist compositions in which the solidity of the forms recalls Roman art. The crucified Christ, clothed in the tunic or *colobium*, is framed by figures which are set in a wide rocky space. The angel of the Annunciation is a heavy figure, in weighty draperies, which suggests the painting of the late Imperial period. At the beginning of the ninth century, the decoration of S. Maria in Domnica, or of S. Prassede, is Eastern in type; angels and saints are set in a gold ground, their nimbuses dwarfing their faces and in the sky are thin little figures.

Painting in fresco continued to display a definite inheritance from Roman forms side by side with Greek ones. In S. Clemente (lower church 869), the *Ascension* bears comparison with Carolingian illumination, while on the triumphal arch is a vision of the Heavenly Jerusalem, glittering with precious stones, towards which move saints and confessors wearing golden crowns and, below them, the elect robed in white. Similarly, in S. Zeno, the angels on the blue globes, the medallions of saints, the Deësis with the throne, is of the same style as that of Eastern illuminations, like those of S. Gregory of Nazianzus (Bib. Nat. Paris).

The problem has assumed a new importance with the recent discovery of the remarkable frescoes in the little church of Castelseprio, north of Milan. These consist of a New Testament cycle running from the Annunciation to the Presentation in the Temple which implies a knowledge of the rich iconographical tradition of Byzantium and the Christian East, with a setting, landscapes, and poses showing a remarkable classical vitality. A medallion with a half-length figure of Christ, and the scene of the *Hetoimasia* (angels surrounding the throne of the Saviour), with an accurately calculated optical effect, are placed on the triumphal arch and dominate the whole. On the lower part of the short wall is a painted decoration of niches, with curtains under a deep Greek-key ornament. The date and origin of this cycle are highly controversial; they compel the reconsideration of all painting in the two halves of the Roman Empire, from the sixth to the tenth century. If executed about 650–700, the frescoes could only be the work of artists from the Syrian-Palestinian coast who had fled before the Arab invasion, and would thus be one of the many exotic introductions into the current of Italian art during the course of the seventh and eighth centuries. If they belong to the tenth century, and more exactly to the years between 925–50, the classical characteristics of their style could only be explained by a Greek origin, in which case they represent an instance of Greek renaissance in the heart of Lombardy, which would seem to confirm the analogies between Castelseprio and two Constantinopolitan manuscripts, the *Paris Psalter* and the *Joshua Roll* (Vatican). In either case, the exceptional series in Castelseprio has nothing to do with Italian art.

It must however be mentioned because neither hypothesis can be advanced without fear of contradiction in view of the absence of any painting, or mosaic, or illumination which can be definitely dated in Byzantium and the Near Eastern centres from the sixth to the end of the ninth century. The famous manuscripts to which the series has been attached are themselves subject to the difficulties of chronology common in these three centuries. There exist in Italy, however, certain vestiges—needless to say, strongly influenced by the ideas prevailing in the East and in Christian Greece—that indicate the existence, from the seventh to the ninth century, of a style of painting sufficiently widespread for the examples at Castelseprio not to be definitely of foreign origin. Monumental painting during the ninth century is far from limited to Roman examples; a cycle containing a crucified Christ, a Virgin and saints, treated with a fulness and liveliness far more Carolingian than Byzantine, is to be found in a very damaged condition in the Oratory of S. Lorenzo in S. Vincenzo del Volturno, positively dated, by the mention of Abbot Epiphanius, between 843 and 864. Illumination also offers a synopsis of the whole period during which the replacement of a descriptive manner by a firm contour and brilliant, expressive colour took place. Irish strap-work decoration was known and disseminated in Italy, for example at Bobbio which was founded by S. Columba; hence, during the eighth century, the manuscripts of S. Gregory with animal initials and very bright colour (Ambrosiana). During the ninth century these appeared in the Terence manuscripts, for example, scenes 'in the Antique manner', very probably copied from older manuscripts, but possessing remarkable vivacity and 'Impressionist' liveliness of handling.

This passing revival did not affect other styles; they were all current at the same time. The Benedictine *scriptoria* of Monte Cassino, Nonantola, and Subiaco spread, during the ninth and tenth centuries, works which were sometimes of pure Greek type, sometimes inspired by Coptic or Syrian models, and above all, the liturgical Exultet rolls for Easter Sunday, in which the watercolour figures, in flat areas of colour, make a running commentary on the music and the text with liveliness reminiscent almost of popular art.

SCULPTURE

Carved chancel plaques frequently reflect Byzantine models transmitted by ivories (Rome, S. Maria in Trastevere, and Bobbio), or by oriental stuffs (Oratory of S. Aspreno, Naples), in which their gryphons, peacocks, palmettes, and leafy branches are to be found.

The stucco figures at Cividale, and above all the *ciborium* at S. Ambrogio in Milan (tenth century), correspond to the strengthening of the style imitating the Antique, characteristic of Carolingian ideas in art. But the most remarkable example is the famous golden altar signed *Vuolvinius magister phaber*: this coffer represents on each face in inset panels scenes from the Gospels or the lives of the saints, particularly S. Ambrose on the back; on the front is Jesus among the Apostles and the symbols of the Evangelists. The setting is composed entirely of richly worked bands of gold and silver gilt, decorated with a profusion of jewels and pearls. This huge massive and glittering chased chest, which contains a wealth of images and symbols, is a kind of compendium of Carolingian art.[34] The figurative quality of the Antique world can be found in the chasings of the reliquary, and barbaric luxury is adapted to the strict composition of the panels.

II

Roman and Byzantine Art

(Eleventh and Twelfth Centuries)

Introduction: The Papacy and the Empire

After the break-up of the Carolingian Empire, the crown of Italy finally devolved in 962 upon Otto the Great, King of Germania, who founded the Holy Roman Empire by harnessing, for three centuries —which correspond to the Romanesque period—the peninsula to the German princes' dream of empire. Italy became the battlefield and the main bone of contention of these unlooked-for heirs of Rome; in the year 1000, the Eternal City had no more fervent admirer than the young Otto III, son of a Greek princess and pupil of the monk Gerbert of Auvergne, who later became pope as Sylvester II.[35]

Italy thus underwent the effects of the medieval dream of a universal empire, which her past history supported, but which she opposed with all her strength; first through the feudal overlords who, during the breakdown of the political structure, had installed themselves in the Apennines, in Tuscany, and even in Rome itself, then through the hostility of the growing city republics. The communes of the Lombard plain—Brescia, Milan, Piacenza, Como, Parma, Lodi—often barred the route to the Emperors, and in 1158, Frederick Barbarossa had to crush them with great ferocity. Finally, there was the attitude of the Papacy which, upheld by the Cluniac Benedictines, from 1073 onwards openly sustained against the Empire a struggle destined to ensure in the West the distinction of the two 'Swords'—the Spiritual and the Temporal Powers. Until the triumph of Innocent III (1198–1216), after the disappearance of Barbarossa (1190), the Germans were ceaselessly upon the roads and in the fortresses of the Apennines; Italy was in the centre of Western European affairs; its architecture, sculpture—particularly in the North and in the twelfth century—bore ample witness to the constant contacts with Provence and countries north of the Alps, maintained, incidentally, by the great number of pilgrimages like that of S. Nicolas of Bari (discovery of the relics in 1087) or the numerous sanctuaries of S. Michael.

The great initiative of the Papacy, the Crusades, launched by Urban II in 1095, not only contributed to the participation of Italy and its ports in the life

51

Principality of Benevento

Principality of Salerno

of the West during two centuries, but also had the effect of restoring it to the centre of Mediterranean politics, and its links with the Balkans and with Asia Minor were never closer.[36] There were never enough powerful feudal lords in Italy—except in the Norman South—for Italians to play much of a role in the Holy Land, but the maritime republics of the Tyrrhenian and Adriatic coasts were then very active. Before Venice, who derived so great an advantage from the tragic and shameful taking of Constantinople by the Fourth Crusade (1204), there was Genoa which fought against the Saracens from the tenth century onwards, the powerful Pisa which conquered Sardinia, Corsica, and the Balearics, and warred with Genoa during the thirteenth century, and in the South Amalfi, which was only dominated by the Normans in 1131, and by Pisa in 1137, and on the Adriatic coast, Bari, the residence in 999 of the Byzantine Governor of Apulia.

Politically, Byzantium was in decline. The Greek schism completed in 1056 marked its final separation from the West. The Exarchate of Ravenna had broken away in the eighth century; in the South she kept Apulia, the country round Otranto, and Calabria. It was a weak domination, resented by the Apulians who seem early on to have appealed to the Norman adventurers drawn into Italy by pilgrimages. They were recorded there in 1021, and in 1030 in Apulia and the Terra di Lavoro.

By 1057, Robert Guiscard had conquered a duchy—Benevento—for himself. The Normans were henceforward to count in the balance of Italian affairs, and in 1054 they intervened in Rome, summoned by Gregory VII against the Emperor. In 1060, they entered Sicily, and in about twenty years, Roger, the brother of Robert, who became Count in 1072, had cleared it of Saracens. For two centuries Sicily and the Kingdom of Naples were to be a staging point in the centre of the Mediterranean for a vast movement of expansion of the West towards the East, and a meeting point for all cultures.

From the marriage of Constance, the daughter of Roger II, with Henry, the heir of Barbarossa, there finally sprang, after a short struggle between the Norman dynasty and the Swabian princes, one of the most interesting paradoxes of the Middle Ages—the coronation of a Hohenstaufen as King of Sicily in 1194 and, a few years later, the emergence of a German emperor whose court was to be in Palermo and whose heart was more attached to the South than to the Germanic world: this was the fabulous Frederick II.

The new culture inaugurated by this period will be studied later, but this remarkable conclusion is quite sufficient on its own to preclude Romanesque Italy, from the tenth to the thirteenth century, from being considered merely as the point of contact of two distinct countries, one adhering to the Holy Roman Empire and participating in Western culture, particularly through Lombardy, and the other—the Kingdom of Naples and Sicily—an adherent of a purely Mediterranean and Byzantine culture.

The facts are more complicated; through her continual contact with the ancient and new styles of the whole of the Christian world, Italy participated in the elaboration of new forms in the West. On the other hand, she also provided a point of contact between them and those of the East, provoking the most curious compromises. Here and there, pure and quite straightforward insertions of foreign types may be discerned, and these are very important buildings: S. Mark's in Venice of Constantinopolitan derivation, the Abbey of S. Antimo of Auvergnese origin, La Martorana or S. Cataldo in Palermo based on a Greek type, or, more complex still, S. Nicolo in Bari which combines a Norman façade and an Eastern apse—but such clear-cut cases are fairly rare. Very soon, among the works which derive from these particular examples, it is possible to distinguish characteristic inflexions, weaknesses but also corrections, which become marks of originality.

Above all, during this same unstable period of the eleventh and twelfth centuries, there appeared in Italy some buildings of a completely novel and powerful character which can be compared, line for line, with the imported monuments, so as to form a proper idea of activity in the country: these were the new basilica of S. Ambrogio in Milan of the eleventh century, the group of cathedral buildings at Pisa, the baptistery in Parma, S. Miniato in Florence, and, in certain respects, Monreale and Cefalù.

The competition between all these forms was of immense benefit to Italian art; to omit, as it is sometimes tempting to do, this long medieval parenthesis, and to make true Italian art begin in 1300, is to misunderstand the original reactions which permitted it to develop.

Romanesque Italy is, in short, the reduced image of the entire Christian world where Western formulae, so powerfully defined on the Rhône, in Burgundy, in Aquitaine or in Normandy, everywhere encountered those of the Greek world. This last had, during the tenth and eleventh centuries, its highest moment of richness and brilliance—what is in fact meant by the term 'Macedonian Renaissance'. Once again there was a Constantinopolitan art, an aulic art, at once secular and religious. In architecture there was achieved the final form of the canonical types of domed, centrally planned buildings, while in France the basilical system with an apse was developed and perfected. None of this was lost on the peninsula. In sculpture, Antique models were not disregarded. By means of ivory caskets, for example, the images of heroes, of emperors, were spread not only in Italy but through all Christianity. Hercules, for instance, is to be found on the façade of S. Mark's. The whole search for great art involves turning towards Greece; the bronze doors at Amalfi (1066), like those at Monte Cassino, both great works, were commissioned in Constantinople. The more popular current represented by monastic painting is chiefly active in the provinces; its initiators were the Syrians, the Copts, the Africans, whom the Moslem conquest had scattered across Europe. This art had its own evolution; in such fixed institutions as the workshops of Monte Cassino it became rigid, but spread abroad in secular centres, as during the thirteenth century in Tuscany, it ended by revivifying the ancient, Early Christian foundation, once common to both cultures. This is the point at which Italian painting detaches itself from its Byzantine sources; this is the moment when Cimabue, Cavallini and Giotto arose at the end of the thirteenth century.

Contacts with the West were no less constant. The Lombard masters crossed the Alps; the Norman abbots intervened in the provinces of the South. Later, Cluniacs and Cistercians ensured definite exchanges, and the revival of monumental sculpture was followed in Italy with as much attention as in the whole of Christendom, and with a kind of impatience to share in it.

1. North of the Apennines during the Eleventh and Twelfth Centuries

ARCHITECTURE

As in the whole of Europe, including the East, the eleventh century in Italy was a period of fine architecture; but the richness of interior decoration remained so strictly linked with the building, with mosaics in the apses, amboes, inlays, and on the outside with the sculptures on the façade, that it is difficult to separate techniques so exactly adapted one for the other.

Lombardy demonstrated its vitality with the second rebuilding of the basilica of S. Ambrogio (the first went back to the Carolingian period) which, about 1100, was a notable contribution to the rise of the whole of Romanesque architecture. The atrium was reworked in the spirit of the ninth century and, in the façade, three large windows were cut. By enlarging the choir into three apses, it was possible to construct the whole church with a nave and two aisles without adding a transept. The really radical alteration was in the roof; by a decision which does honour to the Lombard master builders, a ribbed vault was decided upon for the whole building, with square bays in the nave, and with two corresponding squares in each bay in the aisle. In consequence, the supports were strengthened by using large cruciform piers with applied half-columns supporting the arches both across and parallel to the nave. The aisles are two-storeyed, and smaller piers are inserted between the massive piers to link each of the main bays with the smaller bays at the side. The forms are squat, the masses are heavy, the system is powerfully conceived; it would be interesting to know if this arrangement—which, in any case, is early for a solution of this type—is before 1098, the date of the discovery of the bodies of SS. Gervasio and Protasio, or after 1117, the year of the devastating earthquake. There is no question of an anticipation of Gothic; the pointed ribs of S. Ambrogio constitute a coherent architectonic device, but one adapted to a Romanesque system—that is, designed to stress the mass and the importance of the walls, not to do away with them. On the outside, buttresses, varying in size and alternating like the piers inside, complete the structure. There are no windows in the body of the nave. It is only through the large windows in the façade and from the semidome over the apse that enough light enters to enable the spectator to pick out the numerous effects of modelling underlined by the harmony of the piers, the profiles of the arches, the reliefs of the capitals, the string course marking the position of the gallery, stressed by its machicolated ornament, which, incidentally, is to be found also in the little cornice of the entrance porch under the gable of the façade.

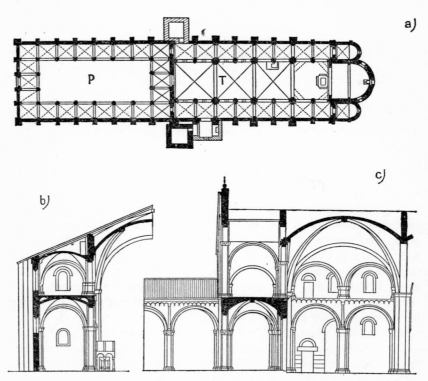

FIG. 7. *Basilica of S. Ambrogio, Milan (after Dehio and Bezold). a, Plan: P, Portico; T, Square Bays; b, Section at aisles; c, Section on the axis of the first bay, showing the ribs of the nave vaulting and the arches of the aisles.*

The collapse of the third bay in 1196 caused the dome to be rebuilt on squinch arches, with a drum surrounded on the outside by open loggias. Between 1128 and 1144 the canons had also built on the left a campanile divided into rhythmically proportioned storeys and bays by horizontal machicolated ornament and deep vertical bands.

S. Ambrogio does not present a typical plan, because of the conditions imposed by the preceding building, the absence of a transept and the peculiarity of the lighting. But this incontestably Lombard creation shows what the new

Romanesque solutions were to consist of. The continuous, and, so to speak, unlimited space of the Early Christian basilica was to be controlled, cut up, defined by sculpture and ornament; the main stress lines of the building were to be emphasized so as to create a greater effect of tension, drama, and effects of light and shadow, depth and relief, which were to replace the earlier chromatic effects. But these points were rarely so well understood in the rest of Italy.

The immediate successor to S. Ambrogio was south of Milan, at Pavia, where, at the beginning of the century, S. Michele Maggiore followed rather freely the ideas set out in S. Ambrogio. The architect added a transept roofed with semi-circular vaulting which is a radical departure from his model; he raised the height of the choir, sought to give an impression of slenderness, and tended to combine useful elements with the decoration. In the façade, divided into three vertical sections by moulded buttresses, the main innovation is the gallery of blind arcades under the gable, while the reliefs inserted here and there in the form of bands give the wall a kind of superficial sculptural effect, without any attempt to give them structural significance. The portals, particularly the right one, include archivolts crammed with branch and foliage motifs, with, in the lintel, a medallion of Christ between two saints. To this church are linked almost all the churches of Pavia built during the twelfth century: San Pietro in Ciel d'Oro (1132), where S. Augustine, Boethius and Liutprand are buried, has a façade with a single gable, decorated by majolica plaques, and originally had a portico; other churches in the group are S. Teodoro and S. Lazzaro, and the now destroyed cathedrals of S. Stefano and S. Maria del Popolo.

North of Milan, at Como, S. Abbondio (1013–95) is another original variant; it has a nave and four aisles, with an open timber roof, and a series of arcades borne on columns, with a vast choir of a Burgundian type. The façade reproduces the system of storeys, with an elegant arrangement of bands and pilasters. A whole group can be linked with this example: S. Giacomo at Como, and, at Isola Comacina, S. Eufemia, now destroyed. The plan of S. Fedele, also at Como, bears witness to the tardy interest of the Lombard architects in central plans: four bays prolong a trefoil composition which has a dome borne on squinches; the complexity of the interior space is increased by the storeys which further enlarge the wide bays. On the outside, buttresses emphasize the articulations of the building, and on the apse a spacious loggia, borne on slender colonettes, looks as if it would open up the wall to admit the light. The mitre-shaped main portal is framed in rather crude bas-reliefs. This church is not without links with the trefoil-shaped church of S. Maria im Kapitol at Cologne. Did master builders from the Rhineland work here? Or does the coincidence merely indicate that, starting from the same models, Lombardy and the Rhineland tended to evolve in the same direction?

The appearance of Romanesque churches in Italy is never deceptive, but it tends more to the repetition of certain characteristics than to adhere to a rigid system. The basilical plan generally omits the transept—though at Piacenza there is a vast transept surrounded by aisles; it includes a nave and two aisles (as in the cathedrals of Modena, Piacenza and Parma); the apse has small galleries like loggias (S. Maria Maggiore in Bergamo, Modena cathedral); between the buttresses which are used to create a picturesque effect, the façade has reliefs embedded in it. Modena cathedral is the work of a *mirabilis artifex*, one Lanfranco, who began work in 1099, although the building was not finished until 1184, and the Gothic vaulting was added in the fifteenth century. Nevertheless, the façade is very harmonious, with its high arches which isolate, in the upper storey, groups of three bays forming a gallery; this motif is used again on the south side, enriched with much sculpture and with a superb portal; it is also used inside. Lanfranco's cathedral was, in fact, originally, one of the few fortified churches in Italy.[37] Similar elements are to be found at Cremona, and Ferrara (1135). But the most revealing group is at Parma; the Latin cross church has ribbed vaulting in its longitudinal rectangular bays, and three storeys one of which is a continuous gallery like a triforium. This sense of scale is reflected also in the façade, flanked with two towers (one unfinished); two horizontal rows of triple arcades, and the usual gallery under the gable, enliven and hollow out the wall, as if to bring the lightened construction into closer relation with the atmosphere.

French characteristics are relatively numerous in this church, the most complete in Emilia. It is built on a rectangular open site, and completed on the right by a baptistery; the desire to create an urban enclave worthy of an independent commune, which is patent here, is now to be met with increasing frequency. Of all the circular or polygonal baptisteries of Northern Italy—at Biella, Asti, Cremona—Parma is the largest; the buttresses on the corners are linked by large arches, and on the upper storeys by small loggias with columns and architraves, giving an effect of lightness which compensates for the mass of the structure. After 1260, its dome was covered with important frescoes, which will be discussed later. With this work, built in 1196, the name of Benedetto Antelami is associated, which leads to the consideration of the part played by sculptors in the buildings of the twelfth century.

SCULPTURE

In sculpture there was the same rush of new activity as in architecture; the one led to the other. But in France the very structure of the building controlled the sculptor, fixed the places where his work was to be sited (capitals, tympana, keystones), and circumscribed his effects; Italian architects had less authority. Eleventh- and twelfth-century reliefs were all for liturgical purposes, but two currents may be distinguished. The first, of coarser type, always uses animal motifs with strap-work, in the abstract and bewildering old Lombard style. This was very prevalent during the eleventh century, and may be seen in Como, in Milan, and in Pavia, where it inspired the fabulous beasts on the façade of S. Michele, so oddly placed, like fossils, amid the stone courses. This kind of motif had been travelling for a long time all through Europe, after being widespread in Asia; the Lombards were not the only ones to be attracted by it.

It continued for long, but was crossed about 1100 with another current, more knowledgeable and ambitious, which first appears in the stucco reliefs on the ciborium of S. Ambrogio (tenth century) and was strengthened, during the eleventh century, by the revival in sculpture in southern France, where the desire to create completely plastic effects was once again to the fore. One master sums up this style—Wiligelmo, who, in 1099, was the companion of Lanfranco in Modena. He signed the reliefs on the façade, in which stories from Genesis are unfolded in a frieze. A background of arches accentuates the relief of the squat, virile forms in which may easily be distinguished affinities with scenes on Rhenish ivories, and with work produced in Toulouse, though the energy of, for instance, the figure of Abel is of a very positive originality.

The workshop of Wiligelmo can be followed at Pieve di Quarantoli, at Cremona, at Piacenza, and finally at the Abbey of Nonantola (1121). But one of the results of its success was that it created, around 1130, an atmosphere receptive to influence from Burgundy—for instance, the Atlas figure from the church at Portile (Este, Mus.)—and from Provence—the rail round the crypt of Modena cathedral is inspired by S. Trophime in Arles. The scene was set for the appearance of those at once daring, archaic and energetically modelled compositions—the Modena metopes. Thus the way was also prepared for the resurgence of Lombard sculpture, a century after Wiligelmo, in the work of Benedetto Antelami, who is usually placed between 1177 and 1223, and between Parma and Vercelli. In 1178 he signed the *Deposition* in Parma cathedral where, against a background of niello-inlaid foliage patterns, stand small figures tightly enveloped in finely pleated draperies, in a thoroughly Byzantine composition. In 1196, he signed his name on the architrave of the baptistery portal, of which he may have been the author; for the first time, in fact, the sculptor is dominated by the architectural framework, and exactly fills a tympanum with a *Madonna and Child*, between the *Magi* and *Joseph's Dream*, with a *Tree of Jesse* cleverly fitted into the archivolt. Antelami's figures are serious, massive, tightly knit, but their hardness is enlivened in such free-standing figures as the one known as *Spring*; his style becomes more varied in the many scenes from the Gospels (for instance, the *Flight into Egypt*) and allegories (*Balaam's Ass*) which the baptistery workshop produced in quantity, in imitation of French cathedrals. After 1200 the influence of this forceful and knowledgeable artist is to be found everywhere north of the Apennines; it explains the vivacious and rather playful manner of an anonymous sculptor who, about 1220, carved the six high-relief panels of the *Months* for the now destroyed south portal of Ferrara cathedral. He uses short figures of quite Roman type to set out his scenes from everyday life.

PAINTING

It seems possible to discern the Carolingian antecedents of Lombardy in the frescoes of S. Vincenzo at Galliano (1007). In the *Christ between Jeremiah and Ezekiel*, in the apse, the motifs adapted from marine animals in the border, and the blue shadows in the scenes in the nave are derived from the world of Ottonian illumination. In the scene of the *Apocalypse* in S. Pietro al Monte at Civate (twelfth century) the agitated angels look back to the major influence of Byzantine art, revived in Venice. The thirteenth century was weak; sometimes a surprising liveliness of colour is to be found in the deep and delicate reds and blues, as in the *S. Agatha icon* in Cremona, and there are important remains in the upper valleys of the Adige and in Trentino. The most striking witness is the dome of the Parma baptistery executed about 1260. At the top is a star-scattered awning, below it are five concentric zones in which are placed the Evangelist symbols and the Apostles, the Deësis and the Prophets, the life of the Baptist, and finally, in the lunettes and pendentives, curious allegories (Seasons, Virtues, etc.). There is certainly a great deal of the purely conventional in this huge work, but, in the third zone, that of the Deësis, which is the most vigorously handled, the faces of the Virgin, and of Balaam, have an exceptional gravity and graphic force, in which may be sensed the revival of a classical inspiration similar to that which occurred, at about the same time, in Roman centres.

2. Verona and Venice

ARCHITECTURE

The transformations in the Benedictine Abbey of Pomposa, near the lagoon of Comacchio, are particularly interesting. It was given a triple arched atrium, with the wall covered in terracotta ornaments, and with majolica dishes inserted as decorations, and an immense nine-storey campanile built in 1063, in which the window openings increase in number as the structure rises, so as to lighten progressively the high square silhouette. It is the outside of the church that attracts attention.

Further south, in the Marches, a tendency to swither between Byzantine and Lombard forms may be discerned. Hence works without any strong indication of style, such as the churches of Ancona: for instance, S. Ciriaco, a basilical church transformed into a Greek cross structure with a polygonal dome during the thirteenth century, and with a portal at the top of a flight of steps, and S. Maria della Piazza, with a Lombard portal by Maestro Filippo (1210).

The frontier between Lombard art and those forms which had originated in Ravenna, and which were now spreading outwards from Venice, passed through Verona after the year 1000. S. Lorenzo was built about 1100 on a Benedictine plan which included a transept with orientated apses, and with two other apses flanking the choir; vaulting, the alternation of piers and columns, and an upper storey are found there, but the effect is dominated by the two-colour striping of brick and stone, which runs horizontally round the church. But such complete examples as this are rare. The façade and apse of the cathedral, rebuilt during the fourteenth century, also make use of the effective device of long and narrow bands, the last ornamental stage of the Lombard type of band. These are to be found also in the façade of S. Zeno, built mainly between 1120 and 1138. Modena provided the original model, but the treatment of the façade is of unsurpassable delicacy; a gallery of small blind arches is placed half-way up, on either side of the porch which is borne on columns; two triangular-shaped buttresses support the raised central portion, under the pediment separated by a narrow band borne on modillions. The interior is not vaulted, but has a trefoil-shaped timber roof, and the alternating piers with their high engaged columns enable the wall to develop a calm spaciousness. A campanile constructed in stripes of different colours soars on the right of the church, and the grand design and the warm coloured stone make the whole into a building of unsurpassed beauty. It also owes much of its charm to the reliefs in the façade, inserted like pieces of decoration round the porch; they were carved by a certain Niccolo, whose art derives from that of Wiligelmo in Modena, and who had already signed work in the cathedral at Ferrara (1135). Niccolo lacked the temperament of his master; his silhouettes are softer, but he has a feeling for the decorative arrangement of his material: the *Creation of Eve* takes place against a geometrical background, among the Old Testament scenes on the right; the Gospel scenes, on the left, are by a certain Guglielmo, who arranges his figures rather charmingly in the arches of a portico. The bronze doors, where lively silhouettes of unsurpassed energy stand out against the plain background, in panels framed in arches and bands, are a later work (end of the eleventh century) and of German origin; they are connected with the art of Hildesheim and the Rhenish bronze-founders, but they had a posterity in Italy, with the doors by Bonannus at Pisa and Monreale.

In Venice, the end of the eleventh century was taken up with the reconstruction of the S. Mark's of 829, which, as has been seen, was probably already on a Greek cross plan. Too small, and damaged by a fire, it was replaced on the same plan by a sumptuous sanctuary of which the prototype was the church of the Holy Apostles (or Apostolium) in Constantinople, the work of Anthemius of Tralles, the architect of S. Sophia, and like that famous work, an exceptional building. The vast church of the Holy Apostles, built between 536 and 546, was the development of an ancient martyrium in Greek cross form, with five domes on cylindrical drums, a narthex and an atrium with a portico. But the model for S. Mark's was the building renovated about 930–40 by the architect Belona during the reign of Constantine VII, which explains certain disparities between S. Mark's and the Apostolium, in so far as it can be reconstructed nowadays. Serving both as the tomb of S. Luke and an Imperial mausoleum, this church was felt to provide the noblest model for the tomb of S. Mark and the sepulchres of the Doges.[38]

The new basilica was consecrated in 1094. The four arms of the cross, surmounted by domes pierced by windows, create a light and airy composition around the central dome, and the overall revetment of mosaics tends to increase the feeling of immaterial and magical splendour. The enormous corner piers are pierced by arches on two storeys, and this, with the columns of the bays which, until the fire of 1145, bore an upper storey (the matroneum), gives a lively animation to the interior space.

Outside, the enveloping architecture is enlivened with a multitude of aediculae, niches, colonettes and even little cupolas which are distributed along a continuous arcade forming an atrium. The façade proper was reworked after 1419, but it retained its little circular pediments, and its two storeys, with the clusters of small columns which, at ground level, frame the portals and the tympana decorated with mosaics.

S. Mark's became one of the richest treasuries of Christianity, and its campanile with its wide vertical bands, rebuilt in 1904 after it fell down, provided an entirely Western accompaniment. This is far from being a mere accessory detail, since, as it revived—through its Greek model—the type of the Early Christian sanctuary with an atrium and a quadriporticus, the basilica of S. Mark's was never treated as an isolated building, but as the centre of a vast urban composition, and it was the Piazza which became the natural development of it during the Renaissance.

The other churches of the lagoons are already rather eclectic arrangements; Murano cathedral reproduces the Greek cross of S. Mark's, and has a fine pavement (1140), and on the outside of the apse two storeys of loggias, probably of Lombard inspiration, but separated by a frieze and with Eastern mouldings. Torcello cathedral, rebuilt in 1008, had copied the basilical type; the little S. Fosca, of a century later, has the Greek cross plan with only the dome missing; its entrance loggias have high stilted arches derived from Saracen examples.

DECORATION AND MOSAICS

S. Mark's was thus a centre in Venice for all kinds of valuable elements, and, from the twelfth century, was constantly active both in sculpture and painting. The creators of the *S. Demetrius* were closer in spirit to Eastern models, while the artists of the *Dream of S. Mark* (in the central portal) were closer to the Lombard style. The architects of this portal had already shown, in the *Labours of the Months* where the figures spring from huge leaves, a typically Gothic nervous energy (second half of the thirteenth century).

The Pala d'Oro, the enormous altarpiece in goldsmiths' work, consisting of eighty enamel plaques, which fills the choir, was a labour lasting for more than four centuries. The principal enamels are fifth-century Byzantine, but the altarpiece was frequently reworked—in 1105, and again in 1200, until the final, Gothic, setting was given to it by the Sienese goldsmith Gianpaolo Bonisegna in 1345.

The famous mosaics in S. Mark's were themselves completed, and continually restored, right up to the nineteenth century. The Byzantine world of the Golden Age had elaborated during the tenth century a 'classic' system of mosaic decoration,[39] adapted to the central plan. It was characterized by the hierarchic arrangement of the sacred themes, which, at Hosios Lukas for instance, was very strict. There, the Christ in Majesty adored by Angels, the Virgin and Child, the Saints, all had their set place, and the same applied in the Hagia Sophia in Kiev (before 1054). It was also marked by a feeling for the overall effect which transformed the whole church into a huge icon. At the moment when in the Balkans competition from fresco painting was beginning to arise, decoration underwent new developments in Italy, in the former Byzantine provinces in the North and South, in Venice and Sicily.

In S. Mark's, the general arrangement agrees with these principles. In the apse is a *Christ in Majesty*, and the great themes of the dogma are divided between the other domes. In the older parts, the virtuosity of rhythm and colour is remarkable; the second dome, with the *Ascension* and the *Passion*, is more confused. In the atrium, the stories from Genesis are the work of Venetian artists who sometimes composed clumsily, but sometimes with a freshness of narrative and charming detail that recalls Romanesque art. There are, finally, the recently discovered frescoes in S. Zan Degolà (S. Giovanni Decollato), some contemporary with S. Mark's (end of the eleventh century)—for instance the figures of the Apostles under a balcony on which appears a half-length effigy of S. Helena—and some clearly later—for instance, the *Annunciation* in which the two figures occupy the extreme edges of an arched space—but obviously Greek in execution and spirit.[40]

3. Tuscany

ARCHITECTURE

S. Miniato bears the name of a third-century saint, and a martyrium probably commemorated him from Carolingian times on the lovely site on the hill which dominates the Florentine plain. The new foundation of 1018 was not finished until the second half of the century. The recasting of the basilical system is interesting: powerful transverse diaphragm arches, supported by cruciform piers, punctuate the nave and break the rhythm of the columns of the nave arcades; a crypt roofed with groin vaulting fills the apse, and the very large choir is raised very high. The warm light is due to the translucent marble plates in the windows which increase the effect of the green and white marble revetment. The pavement, with inlaid black and white motifs—for instance, the wheel of the Zodiac—is like a carpet of lace translated into stone, and is dated 1207. The mosaic in the semi-dome of the apse dates from 1297, but, like the later ambo and a great deal of the reworked marble decoration, it fits perfectly into the decorative ensemble in which Florentine taste seems to have crystallized with complete harmony. This is what makes the twelfth-century façade so interesting. The internal divisions are reflected in its design, and the frieze clearly divides the storeys. The five arches at ground level repeat those of the apse, with large panels clearly set into their frames; the upper storey, probably somewhat later, includes a mosaic and rather more detailed motifs. What counts here is the insistence on the abstract design, which concentrates all the effect on the plane of the wall, where line and colour meet with restraint and clarity.

This lovely and balanced building, which alone sums up all Florentine Romanesque architecture, was accompanied by others such as S. Pier Scheraggio (later swallowed up in the Uffizi), the Collegiata at Empoli, after 1093, and chiefly the Badia at Fiesole, on the hill opposite S. Miniato. The arches of its façade, and its windows with triangular pediments, have poly-chrome frames which, by contrast with S. Miniato, exaggerate the more abstract effect. In short, Florence subjects Lombard structures to the calm dignity of polychrome walls, but these forms did not gain general currency in Tuscany. The church of SS. Apostoli, mentioned as early as 1075, may once have had a polychrome revetment too; the green marble columns of the nave suggest this. But the interest of the building and the significance it had during the fifteenth century for men like Brunelleschi in fact derives from the precision of its outlines, and the bareness of its forms, in which the curve of the arches, the empty shell of the apse, and the clarity of its proportions make their full impact. These are Romanesque not Early Christian qualities; the architectonic structure is present.

In the Tuscan hills, there were many monastic houses directly influenced by French monasteries such as Cluny. There is no other explanation for the choir with radiating chapels, grafted, incidentally, on to a nave with an open timber roof, at S. Antimo (twelfth century). Umbrian churches like S. Eufe-mia, Spoleto, with a gallery similar to that of S. Lorenzo at Verona, and the cathedral at Todi derive from Lombard examples. Spoleto Cathedral—the work of Giovanni da Gubbio (1140)—like those at Narni and Assisi, follows the model of the Tuscan basilica, stressing the masses and striving to achieve in the façades an almost pictorial effect, or one like goldsmiths' work, by means of numerous panels of sculpture.

The basilica of S. Francesco at Assisi was begun in 1228; inspired by Gothic models, it introduced into Umbria quite another kind of architecture which became general among the many monastic foundations in the rest of Tuscany. Hence the difficulty of placing quite a number of important churches begun

1 Rome, St. Peter's: façade of Old St. Peter's (*left*),
detail from the fresco of the *Fire in the Borgo* (1517)
by Raphael and his pupils

2 Rome, Sta. Costanza: interior

3 Rome, Sta. Sabina: interior

4 Ravenna, S. Vitale: interior

5 Ravenna, Mausoleum of Galla Placidia: interior

6 (a) Ravenna, S. Vitale: detail from the mosaic of *Theodora and her ladies*, in the presbytery

(b) Torcello, Cathedral: relief with peacocks

7 Castelseprio: detail from the fresco of the *Nativity*

8 (a) Campania: barbaric relief

(b) Milan, S. Ambrogio: Golden Altar frontal

during the Romanesque period but rapidly influenced by the Gothic style, which, as will be seen, never attained the same vigour in Italy as it did in France. Thus, the cathedral in Siena, begun at the end of the eleventh century, and of which the dome was finished in 1264, had two precedents worthy of mention; the little cathedral at Sovana, with its nave and two aisles covered with groined vaulting in the Lombard style and its dome over the choir; and the new Cistercian churches, like S. Galgano. The many transformations of the plan, the enlargement of the ambulatory, the shifting of the position of the façade, mark the need for significant changes, at the moment when the city became one of the great centres of Italian painting.

PAINTING AND MOSAIC

If Pisan and Lucchese painting is principally confined to crucifixes, that of Florence and Siena consists chiefly of Madonnas. In these too, Greek icons, particularly those of the Odegitria type—the Virgin, seated upon a throne held by angels, and indicating the Child—are mainly a pretext for a minute rendering of drapery folds and ornaments.

The celebrated altarpiece by Guido (Siena, Pal. Pubblico) is not a very good witness, for the date of 1221 is probably that on an older panel, repeated a half-century later. It is only after 1250 that are found the often delicately coloured works of graceful liveliness, like the *Altar Front of S. Peter* (Siena Mus.); but the most remarkable artist of the Pisan region was the so-called S. Martino Master who painted, between 1260 and 1270, a fine panel of the *Maestà* (the Virgin enthroned) framed by a dozen narrative scenes (Pisa Mus.) of a matchlessly vivid movement and beauty of colour, which make him an immediate precursor of Duccio. An attempt has recently been made to identify him with a certain Ranieri d'Ugolino.[41]

Florentine crucifixes are closer to Pisan art than to that of Lucca. Three dating from the beginning of the century have been preserved at S. Gimignano. Beneath the stucco *Virgin* in S. Maria Maggiore in Florence are painted compartments—the *Annunciation,* the *Holy Women at the Sepulchre*—of a most delicate quality. With Coppo di Marcovaldo, whose works can be dated between 1260 and 1274 (the *Madonna of the Servites* in Orvieto, and the one in Siena), form is, on the contrary, more strictly defined. The Magdalen Master (between 1250 and 1270) is interesting in this sense, through his use of red and browns on a gold ground, but his style is much more that of a popular art.

But the great event in the painting of this whole period was the decoration of the baptistery in Florence. The Florentines had always believed it to be an Early Christian foundation, and it must be admitted that its octagonal form and its internal trabeated architecture support this tradition. The development of the galleried upper storey, the high arches which frame it outside and the attic with its geometrical panels crowning the whole exterior structure, the octagonal dome, built of light materials and reinforced by arches masked by the covering, belong to the Romanesque period. It was consecrated in 1059, but the exterior revetment of white and green marble, the angle pilasters in horizontal stripes of coloured marble, the geometrically designed windows, the strict arrangement of the decoration, must be twelfth-century work. It was not until 1202 that the rectangular sanctuary, or *scarsella*, replaced the apse.

The building, a very old church, possessed baptismal baths that Dante evoked in his Inferno, by comparing them with the holes in which the Simoniacs were burnt

(ne) men ampi né maggiori
che que' che son nel mio bel san Giovanni
fatti per luogo de' battezzatori

(Inferno, XIX, 16–18)

The 'bel San Giovanni' is a sort of compendium of Roman, Early Christian and Byzantine types, but it retains the essence, the size, and the dignity rather than the detailed arrangement of its models (see pl. 18).

The mosaics of the *scarsella*, or absidal niche, of 1225 are, like that on the façade of S. Miniato, attributed to a Jacopo Francescano, who came probably from Venice. The decoration of the dome occupied the whole thirteenth century, and the first geniuses of Tuscan painting, one of whom was Cimabue, were certainly trained in this workshop through which passed, successively over the years, the most varied masters. The master responsible for the *Hell* is similar, in his use of tortured forms, to the manner of Giunta and Coppo; the one who executed the *Story of Joseph* heralds Giotto. The composition is strictly hierarchical, according to the Eastern model, and the enormous mass of motifs is arranged in superimposed zones. At the top are the hierarchies of angels; above the *scarsella*, occupying three zones, is the Christ Pantocrator in a circular glory, seated on the throne of the world; on his left, a hideous Hell, and on his right the Heavenly Jerusalem, which recalls Western Romanesque types. The other sections of the dome are divided into four storeys each with a cycle from the Old Testament, separated by superimposed columns which rise and converge towards the celestial baldaquin of the vault, where real space is dissolved in the brilliance of the gold. The whole scheme may appear rather confused, but it has a decorative power perfectly adapted to the structure of the baptistery, and for a long time it must have been for Tuscan artists, including Giotto, an unequalled iconographical repertory. One of the last artists to work there, the master of the *Stories of Joseph* and *the Baptist,* displayed an inventiveness and modernity which bring him close to Mosan enamellers.[42]

4. Rome: architecture, inlays, mosaics

The century which witnessed the struggle between the Papacy and the Empire created a ghastly void in Rome itself. The great basilicas fell into a ruinous state, and money as well as artists was lacking to restore them. In 1045, there was begging throughout Christendom for S. Peter's, and in 1050 for S. Paolo-fuori-le-mura. In 1084, at the height of the quarrel between Gregory VII and Henry IV, the Imperial troops were driven out of Rome by Robert Guiscard, but the Norman mercenaries destroyed several quarters of the city with their churches, among which was the old S. Sylvester dedicated by Stephen II to S. Denis. S. Clemente, and the Quattro Santi Coronati were burnt, and only very slowly were they rebuilt. 'Rome, thou art nought but a ruin, yet nothing can be compared to thee', wrote the French bishop, Hildebert de Lavardin, in 1107 in a romantic poem.

Paschal II (1099–1118) and his successors, at the moment when the triumph of the Holy See was foreseeable, strove to restore to Rome her monumental dignity. On the site of the old S. Clemente a second church (1108) was built with an atrium and a portico; its nave and two aisles reveal a remarkable effort to imitate the Early Christian style, with its precious and shimmering atmosphere. As in S. Maria in Cosmedin, piers faced with pilasters interrupt the sequence of its sixteen columns; the rail round the choir, the *transennae*, and the pavement of marble inlaid with a geometrical pattern are true reconstructions; above all, in the mosaics, the motif of vine branches found in the Lateran baptistery is reproduced with great energy.

For these reasons, Rome knew nothing of Romanesque art proper, just as, with one exception, she knew nothing of Gothic. The populace, who went to hear the harangues of Arnold of Brescia, only had Antiquity in mind. After S. Clemente, in 1119 S. Maria in Cosmedin was rebuilt with an atrium, nave and two aisles, ambones, and always with a timber roof; in 1140, S. Maria in Trastevere where, with a sort of Neoclassical purism, the architrave borne on columns was reintroduced. This had already been done in recent, small churches such as S. Crisogono and was to be used again at S. Lorenzo-fuori-le-mura. In these buildings also the entrance portico reappeared; and again, they were completed in the apse and on the façade with mosaics, and at ground level by rich marble pavements. This is as far removed as possible from contemporary architecture in the North, conceived all in masses, walls, vaulting, and sculptured decoration, and where the traditional polychrome effects of variegated stone, and not fresco or stained glass, were used to give colour to the sanctuaries. But in Rome builders no longer looked to Antique models, but sought the collaboration of the Greeks; the popes called on the Monte Cassino workshop, recently reorganized as a result of the arrival of Byzantine artists, who thus brought back to Rome the techniques of former days.

In fact, it seems that no mosaics were made in Rome between those in the Carolingian church of S. Marco, done under Gregory IV (827–44), and those in S. Clemente, which were followed by the ones in S. Maria in Trastevere and S. Maria Nuova (1161). In the apse these traditionalist artisans set the frieze of lambs, above it the group of saints framing Christ and the Virgin, and at the top a sort of multicoloured pavilion. One modern feature has been noted: near the prophets who flank the triumphal arch are birds in cages which symbolize the imprisonment of Christ in the Incarnation—probably an iconographical novelty. Notably, in the apse of S. Maria in Trastevere, the *Coronation of the Virgin*, more elegant than hieratic, must derive from a similar stained-glass window given by Abbot Suger to the first cathedral of Paris.[43]

Once again flourishing in Rome, the art of mosaic was to assume there an even greater significance than before. The Benedictine workshop of Monte Cassino was in complete decadence when, in the middle of the thirteenth century, the Roman school began to form masters independent of purely Byzantine forms (which may be seen at the end of the twelfth century at the Abbey of Grottaferrata) and eager to equal the great compositions of the West.

The same comment can be made with regard to the art of the marble workers. The rich geometrical and polychromatic ornaments to be found in S. Maria in Trastevere, and in about fifteen other churches such as S. Clemente, and the Ara Coeli, built or reworked during the twelfth century, came from Monte Cassino. Like carpets underfoot are the great wheels of marble with complicated strapwork, the multicoloured forms which probably recalled the glittering floors of Byzantine palaces, but there is also in them something of their Roman heritage. There were several dynasties of artisans, generically known as 'Cosmati'; several masters are known to have borne the name Cosma, and there were others called Vassalletto, from 1150 onwards. But through the development of inlay work on façades, and in the use of twisted columns, which determined the new decorative motifs, this technique ended by becoming one among the various forms of ornament of Romanesque art. Vassalletto's cloister at the Lateran, so graceful, so ordered, and so luminous, is one of the late masterpieces of this art (1225–35), as is the atrium of S. Lorenzo-fuori-le-mura (about 1220), and the paschal candle of S. Paolo-fuori-le-mura. Some among these ornamental artists became architects, like Jacopo di Lorenzo and his son Cosma, and improved on the classical forms in the façade at Civita Castellana, between Rome and Viterbo (1210). The entire composition—an Ionic portico with an entablature and a large central arch, and with light inlays in the friezes—clearly presages the clarity of the Pazzi Chapel.

The same taste for inserting coloured and bright elements into architecture is displayed in campaniles. They sprang up in great number in Rome during the twelfth century, and are consistently decorated with marble crosses, with white modillions against the brickwork, and with little tabernacles. The success of these high towers of somewhat feudal appearance is the only important trace of Lombard and northern influence in Rome, the only obvious concession made by Rome to Romanesque art. All types may be found. The archaic kind had two plain openings on each storey (S. Maria in Cappella, 1290), or with intermediary colonette (S. Bartolomeo in Isola, about 1110); the more developed type, with three openings and two colonettes, had pilaster framings (S. Maria in Cosmedin, S. Pudenziana, twelfth century), derived perhaps from the circular bell-tower of S. Apollinare Nuovo in

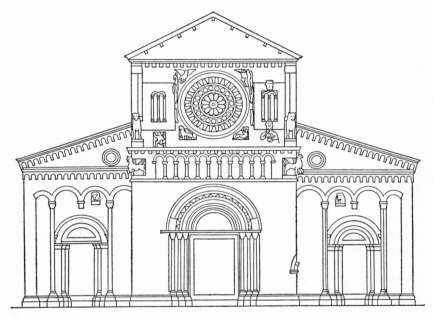

FIG. 8. *S. Pietro, Toscanella. Façade (after Dehio and Bezold): Arches, gallery, and reliefs round the rose window.*

Ravenna. Finally, there was the more open type with four bays and three colonettes (S. Francesca Romana, S. Alessio) which have wrongly been thought to derive from Old St. Peter's.[44]

The originality of the many Romanesque basilicas of the province of Lazio was due to a harmonious assimilation of Lombard forms. These are evident in S. Maria di Castello at Tarquinia (begun in 1121), with nave and two aisles and groin vaulting, and in the churches of Viterbo—S. Sisto and S. Giovanni in Zoccoli. Contact with the Roman basilical tradition created, in a clearly defined site, the fine monumental group of Tuscania or Toscanella: S. Maria Maggiore (twelfth century), which has a rich façade with a rose window and an arcade, an interior enlivened by an ornamental ambo, and frescoes of the *Last Judgement* on the triumphal arch; and S. Pietro, which, rebuilt at the end of the eleventh century, had its eighth-century nave completed by an elegant façade (over-restored in 1870) with three portals of the Pisan type, a delicate arcade, and numerous re-uses of older material.

Mural painting in all the Roman regions also had an interesting revival, of which the most important instance is certainly the fresco decoration of the church at Anagni (in the crypt, between 1231 and 1251). Three workshops have been distinguished, in which were revived both the Roman tradition of the grandiose (the old men of the *Apocalypse*) and liveliness of drawing.

5. The Maritime Republics: Pisa and Bari

Pisa only became a truly Tuscan city and dependent on Florence in 1405. Her great period was from the eleventh century to the defeat of Meloria (1284), when she lost her dominion over the Tyrrhenian Sea.

The victorious expeditions against Reggio Calabria (1005) and Palermo (1063), are inscribed on the façade of the cathedral, where the architect did not omit to sign his own name, Buschetos, adding proudly *non habet exemplum niveo de marmore templum*—this is a church without compare. Contrary to Tuscan habit, the Pisans were proud of making innovations, and there is in fact nothing comparable to this unusual masterpiece, absolutely different from Tuscan custom and not without connections with S. Demetrius in Salonica, and to the scenographic grouping of the cathedral (1063–thirteenth century), the baptistery (1153–fourteenth century), the campanile or Leaning Tower (1173–1350), and the Campo Santo (1278–fifteenth century). The three elements of the church are treated as monumentally separate forms, though assembled in a composition at once solemn and full of charm.

Consecrated in 1118, the cathedral was continued in the second half of the century by a master of the name of Rainaldo who lengthened the nave and designed a new façade. An excellent choice was made among traditional motives. The Latin cross plan is stressed by a strongly projecting aisled transept; there are four aisles flanking the nave, but the apse corresponds only to the nave; there is an oval dome over the crossing; the nave, bordered by sixty-eight heavy Corinthian columns, with arches over, has open tribunes of which the pilasters, with alternating horizontal stripes of black and white, introduce a note of colour; the pointed triumphal arch increases the impression of height. The outside walls are treated with great care, since the building was to be seen from all four sides, like a reliquary. There are strips, blind arcades, framed windows which are combined over three storeys strongly stressed by cornices; roses and polychrome inlays accentuate the precious effect. On the façade all is subordinated to the effect created by the four rows of arcades like loggias one above the other, hollowing out the wall behind slender columns, giving it a vibrant quality and an equally remarkable unity. High arches

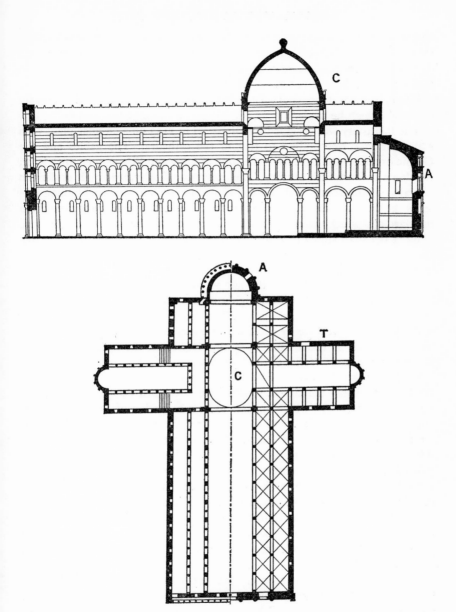

FIG. 9. *Pisa Cathedral. Plan and longitudinal section (after Dehio and Bezold):* A, *Apse; T, Transept and side aisles; C, Elliptical dome. The right side of the plan shows the ground plan, the left side shows the small columns in the interior and on the exterior (façade and apse).*

stretch across the lower storey, with three doorways. The principal one is framed by two columns covered with an Antique type of leaf ornament; its bronze doors (1180) were destroyed in a fire in 1596. Those of the transept survive, by the same sculptor, Bonannus of Pisa, whose delicacy, driven to the point of an actual thinness of form, can also be seen in the doors of Monreale in Sicily, which he made.[45]

The baptistery was designed by Diotisalvi in 1153 on a circular plan, with a decoration corresponding to that of the cathedral. The gables and the Gothic ornamentation are from the time of Giovanni Pisano, a century later; the dome is fourteenth century. The campanile, finally, was built under the direction of Bonannus, and is a cylindrical tower, surrounded by open galleries on the same model as the cathedral. The tilt, which it acquired from the start, and which was not allowed to interfere with the work (the bells were hung at the sixth storey in 1301, and somewhat later the seventh storey was built, at a height of 55 metres), gives the impression that the galleries climb in a spiral.

This unusual decoration is the last word in the Pisan picturesque style which replaced the wall with a highly ornamental surface that made great play with effects of light and shadow. The decoration of the baptistery doors seems to have been done by Sicilian workmen; it associates vigorous foliage motifs with panels taken from Byzantine ivories on the lintel, more freely handled in the jambs (series of the *Months*).

Almost all the churches of Pisa adhered to these models: examples are S. Frediano, probably the closest to Buschetos's original design, and S. Paolo a Ripa d'Arno (1148), which appears to be a reduction of Rainaldo's work. The influence of Pisan art, with Buschetos's cathedral as a paradox of Mediterranean art on the fringe of Tuscany, worked chiefly on the islands and ports dominated by Pisa. For instance, in Sardinia, the basilica of S. Gavino at Porto Torres, of the beginning of the twelfth century, with its double apse, its alternate piers in the nave and its striking decoration, derives from Pisa, and, in turn, gave rise to a whole group, among which are S. Maria at Silanus and S. Giusta at Oristano. Pisan influence also worked in the interior of Tuscany, in the Val di Nievole, and chiefly at Lucca and Pistoia, and even beyond Florence, since Pisan arcades are to be found on the façade of the Pieve at Arezzo.

The architects of Pistoia soon created an even more colouristic version of the Pisan style. In 1156, at S. Andrea, the horizontal stripes in two colours are combined with large and small arches, not without changing their value. In S. Giovanni Fuorcivitas (end of the twelfth century), the green and white stripes acquire in this way an intense and luminous quality.

At Lucca, the naves of S. Alessandro and S. Frediano (1147) are directly inspired from Pisa, but not the façades, which have small columns bearing an architrave. At the cathedral of S. Martino, built by Guidetto in 1204, a deep portico opens out under the galleries of the façade, and at S. Michele in Foro the high arches of the lower storey appear in the upper storeys with thinner colonnettes, with a more fanciful effect due to the height of the pediment.

SCULPTURE

The artists of Pisa, a maritime city, liked to scatter over choir-screens and walls coloured marble ornamented with roses and with Eastern designs. But in the middle of the twelfth century, Pisan masters had to reckon with the effect of Lombard sculpture. Their first important master was Guglielmo, who designed between 1159 and 1162 a large ambo (dismantled and removed in 1312 to Cagliari in Sardinia). Motifs derived from sarcophagi are to be found in the narrative friezes which have a mosaic background. The fine architraves, with squat Roman figures, are generally considered to have been inspired by it; examples are those of Gruamonte (more flattened), at S. Giovanni Fuorcivitas in Pistoia, and those by Biduino (1130) at S. Cassiano at Pisa, and the Misericordia at Lucca. The art of these monumental ambones is continued in S. Gennaro, Capannori (1162), and at S. Michele at Groppoli, near Pistoia (1194).

After 1200, sculpture reveals a better knowledge of Lombard work, in particular at S. Martino, Lucca, where Guidetto decorated the portico with heavy foliage motifs, and where, around the central portal, a follower of Antelami carved with exemplary energy *Scenes from the Life of S. Martin*; in the right tympanum the group of S. Regulus bending before the executioner is silhouetted against the bare ground. Also during the thirteenth century, an unknown Lucchese sculptor carved the celebrated group of *S. Martin dividing his cloak*, fixed by consoles to the cathedral façade. The horseman in his voluminous chlamys could be derived from a classical model, while the trembling figure of the tall beggar is a more personal invention.

But it was principally through the reliefs on its great pulpits that Pisa and its surrounding territory made so great a contribution to medieval sculpture, in the middle of the thirteenth century. In rapid succession, at S. Bartolomeo in Pantano at Pistoia in 1250, a master from Como, Guido Bigarelli, took up Guglielmo's ideas, but arranged his reliefs, with their smooth backgrounds, in rigid compartments; at Barga, near Pisa, a workshop deriving from Guidetto, which also worked at S. Michele in Lucca, erected towards the middle of the century a pulpit with figures and inlays; in 1260, Nicola Pisano erected the great pulpit in the baptistery, and this marked the beginning of a new era.

PAINTING

In Italy, marble revetments, mosaics, and, more rarely, frescoes, are the true painting of the Romanesque period, completed in liturgical furnishings by panels of goldsmiths' work, enamels and ivories. It was from this double monumental or precious repertory, delimited by Byzantium and by Carolingian creations, that slowly, during the course of the twelfth century, an independent art of painting was to evolve. This production, closely tied to religious needs, had its original starting place in Pisa and Lucca; it began with the large crucifixes, painted on wood or stretched leather, destined for the iconostasis or the triumphal arch, with the peculiarity that scenes from the lives of the saints, the Gospels or the Passion, frame the body of Christ, or the panels destined for tabernacles, altar fronts, and later the altarpieces which evolved from them. There is no question of the iconographical form known as the *'Christus patiens'*, with its sometimes grimacing expression of pain and the S twist of Christ's body (Pisa cathedral crucifix, now Mus.) being a freer and more Western type than the traditional *'Christus triumphans'* (Sarzana cathedral, and crucifix by Alberto Sotio in Spoleto cathedral, 1187).

At Lucca, Berlinghiero, recorded in 1228, was the head of a workshop which adopted the type of stiff, calm Christ, with open eyes, surrounded by a small number of figures, Mary and S. John, treated in a linear style with fair tenseness. His son, Bonaventura, painted a picture signed and dated 1235, of *S. Francis* (Pescia church); it contains six small pictures on either side of the saint, who is treated in an impersonal, frontal style, and as if stretched out upon the gold ground, with the aspect of a superhuman and Eastern ascetic. But the small scenes are quite different. While they retain a wonderful calligraphic quality, and an almost Eastern fantasy, as in the scene of the *Sermon to the Birds*, they display a strong feeling for the concrete character of things— flowers, animals. This is one of the works always advanced as evidence for the transformation of art under the impact of the new Franciscan sensibility. Similarly, in Pisa, Giunta Pisano (active 1241–54), more familiar with Byzantine expressionism, exploited the 'pathetic' type, eminently suitable for a public stirred by the Franciscan legend. As in the visions of Bonaventura, this is a spectacle evoking grief and fear, but the body of this *'Christus patiens'* is a kind of schematic arch in which the projection of the ribs and the contorted features are accentuated by the shadows and the points of light. The crucifix painted in 1236 for Brother Elias is lost; but there is another in S. Maria degli Angeli in Assisi, of *c.* 1241.

Giunta's style in Pisa is in contrast to that of Enrico de Tedice, mentioned in 1254; his is more vivacious and lively, if he is in fact the painter of the *Madonna* with scenes from the life of Christ in the Bargello. Giunta's art developed particularly in Umbria, with the so-called S. Francis Master. But his activity coincided with the beginning of the decoration of the baptistery, and this concentration of Tuscan and Roman masters led rapidly to the surpassing of Romano-Byzantine formulae.

Apulia. The cathedral of Troia (1093–1125) was directly imitated, at least in the lower storey, decorated with high arches with lozenge forms inscribed within them, from Pisa. The pediment with its rose window is of a different style. The façade of S. Maria at Siponto, which in turn imitates Troia, remained unfinished. The Pisan element is allied to a Lombard scheme in churches like those at Trani and the cathedral at Zara.

Passing from the great ports of the Tyrrhenian Sea to those on the Adriatic involves hardly any change in character. Bari, during the twelfth century, played somewhat the same role as Pisa. A fine limestone comparable to the Carrara used in Pisa permitted very sharp cutting in the profiles and in the reliefs. S. Niccolo at Bari, an important pilgrimage church, around which the town was planned, was built from 1087 onwards, consecrated in 1105, and finished in 1132. A nave and two rib-vaulted aisles, a transept, three storeys, two towers on the façade, unfinished however, provide so many Romanesque, probably Lombard, characteristics. A kind of curtain wall encloses the volumes of the apse in a geometrical frame of Eastern type. This much admired and much visited building had great influence, and explains certain features of Bari cathedral, and those of Bitonto, Ruvo, Trani—this last more slender and flanked by a campanile.

A centre of communications, during the age of the pilgrimages and the Crusades, Bari was quickly affected by Lombard sculpture, as is proved by the four figures which support the niello-inlaid pulpit, and by the lion supports of the columns of the portal.

At Trani cathedral (S. Niccolo) the door jambs are decorated with tightly packed, seething figures like those at Souillac; at Ruvo, monsters and smiling heads have a more Antique quality. Other excellent examples are the doors at Troia, built up of compartments containing a great variety of motifs, and those at Trani, by Barisano, where the panels in low relief are framed in chased bands, of an entirely Moslem laciness. The same forms—flat cornices with heavy nails at the corners, a low relief tending to the linear, and ornamental fantasies—are found on the doors at Ravello and Monreale.

6. Southern Italy and Sicily

ARCHITECTURE

The church of the Abbey of Monte Cassino had a nave and two aisles, a non-projecting transept, and a triple apse; it was consecrated in 1071. It was the work of Abbot Desiderio, who imported for its decoration painters and mosaicists from Constantinople, and had his liturgical furnishings made there. One of the monks at the Abbey, Leone d'Ostia, recorded all the works undertaken by Desiderius to make his monastery 'the wonder of the West'. At this date, there was only one model for ambitious and art-loving abbots—Byzantium.

The Abbey was destroyed in the earthquake of 1349. But its design is known from buildings in the group deriving from it. First, S. Angelo in Formis, near Capua, built by the same Desiderius, where some frescoes survive and where in the atrium are pointed arches of an Arab type. The basilical design also appears at Carinola and at Sessa Aurunca (1113), at Salerno cathedral (1085) with its fine quadriporticus, at Capua cathedral (*c.* 1120), and at Caserta Vecchia, only consecrated in 1153. The stilted arches and the vaulting indicate borrowings from Arab art. The centre of this style is the small town of Amalfi, where the cloister, built in 1103 by Giulio di Stefano, is remarkable for the thinness of the columns and for the interlaced arches, in threes, a motif developed in S. Domenico at Salerno. This motif, which draws out the arches into lozenge shapes, creating an interplay of voids and solids like a *transenna*, is one of the most charming discoveries of the Middle Ages.

It reappears, during the thirteenth century, in the bell towers of Amalfi and Gaeta. There are other Moslem features in the buildings of Ravello—in the cathedral of 1036, in S. Maria a Gradillo, and in the Palazzo Rufolo which, with typical vaulting, has a courtyard with such a profusion of decoration as to recall certain Venetian façades. To these buildings it is necessary to make the mental restoration of their original colour and luxurious decoration, for which Monte Cassino had provided the model; for instance, mosaic pavements in churches like S. Demetrio Corone, near Cosenza, ambones with geometrical motifs at Salerno, and particularly at the cathedral in Ravello, with its two ramps and its large symmetrical motifs (another larger ambo, more classical and heavier in style, was placed facing the earlier one in 1272), episcopal thrones in wood, more commonly in marble, like that in Bari, of which the lions or the elephants supporting them attest to their origin in the aulic art of Persia.[46] To these must be added the ivory *paliotti*, like that in Salerno, and the bronze doors of Benevento (beginning of the thirteenth century) where the same themes are to be found expressed in forceful modelling.

The architecture of the Tyrrhenian coast was thus coloured sometimes rather furtively, sometimes strikingly, by Arab influence. The implantation of Byzantine ideas was much stronger in southern Calabria and in Apulia. The cathedral of Canosa, in Apulia, is of Latin cross plan, but with five domes; it is flanked by the extraordinary tomb of Bohemond (d. 1111) which is a sort of cube with Pisan type arches and a Moslem dome on an octagon, all in perfect harmony. The same elements are to be found at the cathedral of Molfetta. Later, but outstanding for the assimilation of Romanesque with Arab forms in the dome, is S. Cataldo at Lecce, founded by Tancred in 1180, but reworked in 1716, in the centre of a district of magnificent stone.

Calabria had provided a refuge for Basilian monks during the persecutions, hence the many small churches on a square or Greek cross plan, corresponding to what during the ninth century was the classic type of the Byzantine church. The best known are S. Marco at Rossano, and the Cattolica at Stilo, with three apses and five domes. But, after the middle of the eleventh century, the infiltration of Northern ideas began to affect these little constructions.

The huge unfinished church of the Trinità, near Venosa, on the foothills of the Apennines, above the Gulf of Manfredonia, was probably founded by Robert Guiscard and the monk Berenger in 1063. The plan adopted was that of the Cluniacs, and contained—this is exceptional in Italy—an ambulatory with radiating chapels. The cathedrals of Acerenza and Aversa, which imitate this feature, derive it probably from the example of Venosa.[47] There were other similar foundations, in particular S. Eufemia, and it is possible that these churches, like the odd Roccelletta at Squillace which has a large choir, similar to that of churches in Normandy of the type of Bernay, derive from these now destroyed Calabrian monasteries, rather than from Sicilian models such as Cefalù.

The influences from Normandy in plan and in vaulting contribute to the astonishing complexity of these edifices. But the imported Romanesque features must not be allowed to conceal the fact that the wall treatment is quite different and the decoration unknown elsewhere. At S. Giovanni at Stilo (before 1101) and in the north side of S. Pietro at Agro (1117) in Calabria, are to be found interlaced arches which have nothing Romanesque nor Byzantine about them; more than probably Saracen in origin, they were called to a noble future, since they are to be found at Durham at the same time. Their presence at Amalfi has already been commented upon.

In Sicily, where the Arabs succeeded the Byzantines (827) and the Normans succeeded the Arabs (1072), this complexity of architectural forms reaches a climax. Antique buildings had been used by Christians—at Syracuse, the cathedral was the Greek temple of Minerva adapted in the seventh century (reconstructed after 1693); at Agrigento, S. Biagio and S. Maria dei Greci were established in the remains of chapels originally dedicated to Demeter and Minerva. During the eleventh century, the dominant note is of a Southern richness, which is the opposite pole to the synthesis of the Lombard masters, who concentrated upon the structure. Arab buildings were numerous mainly in the western part of the island; a Moslem traveller of the tenth century described the fine Palermitan stone, and it is recorded that Robert Guiscard, shocked at the contrast between the poor S. Maria delle Grotte at Palermo and the Saracen monuments, immediately had the church rebuilt in ashlar and marble.

The oldest Norman foundation, dated by a Charter given by Count Roger in 1081, is the church of Troina in the centre of the island, a fairly modest affair with nave and two aisles and a transept, rebuilt during the sixteenth century. It was built of stone and mortar, like the near-by priory of Piazza Armerina (beginning of the twelfth century). In the west, where the Arabs predominated, mainly ashlar was used; in the east, where there were numerous monastic communities of Eastern origin, brick was the main material.[48]

The first cathedral of Catania, which was also one of Roger's foundations, before the end of the century, was most important. The apse, strongly built of blocks of lava, partially resisted the earthquakes of 1169 and 1693 which caused the church to be rebuilt. The ancient columns were incorporated into

77

Vaccarini's enormous eighteenth-century pilasters. The development of the transept is remarkable and indicates a Norman scheme, with two features which give it a peculiar aspect. The small apses are separated from the transept by a hollow wall with communication with the larger apse through passages; there is therefore a separate and self-contained choir—an Eastern element. Secondly, the transept is extended, at either end, by square members which serve as bases for the towers; these flank the building, which probably had a curtain wall around it, and must therefore originally have been a fortified church. Only Modena cathedral displays this military character, which can easily be explained in Catania by the still precarious position of the conquerors at the end of the eleventh century.

In the western region there is a whole group of churches of the twelfth century which abandon the use of timber roofs in favour of the Saracen type of dome, and tend to reduce apses to niches in the thickness of the wall. S. Trinità at Delia, near Castelvetrano, with a small dome over a Greek cross, presents an arrangement of clearly articulated volumes, underlined by arches and strips. This exotic profile is also found at S. Giovanni degli Eremiti (1132), of an extreme geometric rigour, with a Latin cross incorporating the remains of an Arab mosque, and at S. Cataldo (before 1160), which on the outside is a parallelepiped surmounted by three stilted domes. The most remarkable of these Greco-Arab combinations is that of the Martorana, or church of the Admiral, founded by George of Antioch before 1143; it has a square plan with a quadriporticus, columns and pointed arches, supporting niches, the whole decorated with Byzantine mosaics. This masterpiece also has a campanile with two storeys with large arches and two other with colonnettes which give it the appearance of being open against the sky.

With the transformation of the County into a Kingdom, grandiose foundations were not lacking. These were, principally, the cathedrals of Cefalù (1131-48), Palermo (1185), begun by Roger II, and Monreale (1172-89), the work of William II. All these had a large choir and a basilical type of nave, and an outstanding decoration. At Cefalù, the church is built upon a platform up against the cliff; it is huge (74 metres long), and its sixteen columns carry stilted arches which give it great height. Steps separate the transept from the nave, and the choir from the transept; the open timber roof was painted in the thirteenth century. The apse is covered with mosaics (of 1148). Outside, projecting buttresses, linked by arches, accentuate the masses, and two massive towers flank the façade which is hollowed out to form a portico, reconstructed during the fifteenth century.

The only surviving twelfth-century part of Palermo cathedral is the huge apse; the rest disappeared in the fifteenth-century reworkings (the right portal), and principally during Fuga's eighteenth-century rebuilding (the dome, and the façade). The most satisfying building, the one in which every possible use has been made of earlier experiments, is the cathedral at Monreale. The fusion of elements as diverse as possible results in an absolutely novel composition, packed with significance. An atrium precedes the Norman façade, with its two square towers; the interior, a hundred metres long, consists of a nave and two aisles ending in three apses; the nave—three times wider than the aisles, an untraditional feature—has arches on Corinthian columns some of which are Antique, with capitals in the Byzantine taste; the sharply pointed arches impart a feeling of energy increased by the glittering figures, one row above the other, of the mosaics.

The exterior emphasizes the movement and articulation of the masses, in the spirit of Western Romanesque, but the upper storeys with their interlaced arches, decorated with discs and geometrical forms, belong to the repertory of the Moslem world, with its continuous surface animation. A magnificent cloister, with sharply pointed arches, slender coupled columns, and projecting buttresses at the corners, completes an ensemble of dazzling liveliness and beauty. On the upper part of the twin Corinthian capitals are scenes with a Byzantine iconography, treated with an ease and an incisiveness in the relief which frequently recalls the art of Languedoc.

The Cappella Palatina, consecrated in 1140, is a kind of treasure house of the Arabian Nights changed into a sanctuary. Its plan, incidentally, is a smaller version of the one used for the cathedral; a nave and two aisles, with flat ceilings, and an apse with a dome. Nothing can compare with the profusion and the fascinating brilliance of its decoration, completed in stages over the century; the ceiling is composed of Arab stalactites, the floor of a polychrome pavement, the walls are covered with inlays, the dome and the soffits with mosaics. The quintessence of the decorative repertory of the Christian and Saracen East is here expressed with a wealth of foliage motifs and spirals; a superb marble candlestick even includes small Hellenistic figures.

MOSAIC

In Sicily, even more than in Venice, the Byzantine masters of mosaic had to adapt their art to unfamiliar structures, and conform to new requirements, which led to numerous deviations from the Greek canons. At Cefalù, the motifs proper to the dome are used for the shell of the apse (1148), where the huge Christ Pantocrator stresses the central axis, and the circle of saints and apostles is arranged in hierarchic bands in the choir (1270). This scheme was used again in the Cappella Palatina, decorated in two distinct phases (1140 and 1160) and at Monreale.

Even at the Martorana, where all is Greek, the mosaic betrays provincial changes. In the vaulting of the transepts, the figures of the apostles have the overlarge proportions of the figures at Cefalù, from which they are derived. The necessary corrections to the perspective, which at the bottom crushes the figures of the angels, are not applied. At Monreale, the composition is the largest of the twelfth century; the entire choir is treated as a unit, and some figures are even set in profile. To this 'simultaneous' type of vision is opposed the narrative sequence in the nave, fatal in a basilical plan. Recourse to landscape backgrounds—silhouettes of mountains, dwarf trees—was a means of linking the Old Testament scenes and of giving an animated accompaniment to the story (*History of Jacob*, for instance), which is a complete novelty. Despite these innovations, and its very late date, the Monreale decoration is the closest, in its drawing and in the purity of its colour, to the grand Byzantine style. This is because here, as in Venice, the stylistic evolution, which seemed to be undergoing a progressive withdrawal from Greek forms, was interrupted by the arrival, from time to time, of Constantinopolitan artists, who reimposed a return to the norm. Byzantine pressure was increased, not relaxed, around 1200.

SECULAR BUILDINGS

The palace of the Norman kings stresses even more strongly the contrast between the geometry of the exterior—cubic, and enclosed in arcades—and the enchanting rooms within, covered with mosaics, as for instance at Palermo, in the so-called Torre Pisana in the palace of Roger II, in the Zisa (from the Arab *aziz* meaning brilliant) of William I, with its steps, niches, stalactites and fountains, in the Cuba of William II (1180). Roger II's room in the royal palace is decorated with hunting scenes, palm trees and countercharged animals which, in the exquisite brilliance of the mosaics, recalls oriental illumination. The castles of Trani and Bari in Apulia are, with fewer Islamic features, massive fortresses, with angle towers, in which Imperial power was made manifest. They remained within the local Romanesque tradition, but, with Frederick II, they were distinctly modified in the direction of Gothic.

9 Milan, S. Ambrogio: atrium and façade

10 Verona, S. Zeno, bronze doors:
(a) detail of the *Baptism of Christ*
(b) detail of a section of the doors

11 Modena, Cathedral: façade

12 (a) Ravenna,
    Campanile of
      S. Apollinare Nuovo
  (b) Viterbo,
      Campanile of S. Francesco

13 Florence, S. Miniato: pavement

15 Pisa: the Baptistery, the Cathedral and the Campanile

16 Antelami: *The Descent from the Cross*, relief

(Parma Cathedral)

III

The Thirteenth Century

Introduction: Frederick II and S. Francis

The second half of the thirteenth century saw, in every province of Italy, profound stylistic changes which were accompanied by regressive and even archaic elements. There was no question of a stylistic revolution, such as would have marked the decisive alignment of Italy with the Gothic world at its zenith; nor was there yet a return to antique naturalism such as marked the dawn of the Renaissance. Both these explanations have seemed plausible, each in its turn, and they still have a certain charm. The first is based on the astonishing success of the Franciscan preaching movement (S. Francis died in 1226) and on the importance of the Assisi painters around 1290; the second takes its rise from the extraordinary action of the Emperor Frederick II Hohenstaufen, King of Naples and Sicily (d. 1250), and from the leading part played by the sculptural workshops in the Ghibelline city of Pisa after 1270. One fresh impulse, quasi-mystic in origin, comes from Cimabue and Giotto, and starts from Umbria; the other impulse is classic and comes from Apulia and through Giovanni Pisano.

In actual fact it is not possible to make a neat pattern of these events by isolating any one of these movements. The whole course of this evolution is peculiar to Italy and its very complexity is its main fascination, for Italian Ghibelline culture was a new synthesis, in the long run a 'progressive' one, of Byzantine influences, Gothic novelties and a certain basic Romanesque undercurrent. Franciscan preaching—the importance of which has probably been exaggerated—started an artistic movement which led to a similar synthesis in religious art. It is the final convergence of these two strands which is so interesting to define, in order to understand the first complete affirmation of Italian art. They must be related to the elements of Romanesque art, the final expression of which is to be found in the great works of Pisa and Assisi, and, later on, at Siena and Florence. They punctuate the thirteenth century and were largely inspired by the two artistic 'invasions' of the century—the influence of Byzantium and the introduction of French Gothic.

THE THIRTEENTH CENTURY

REACTION TO BYZANTINE INFLUENCE: PAINTING

The capture of Constantinople by the Crusaders in 1204 marked the triumph of Venetian policy and led Italian commerce towards the East. For its counterpart it has the great influx into Italy of Greek decorators and mosaic workers. Naturally enough, they went to Venice and it was from there that they were summoned, in the first decade of the century, by Pope Innocent III to work on the mosaics of St. Peter's, and by his successor Honorius III to work at S. Paolo-fuori-le-mura. Little enough is left of these works but there are contemporary panels, such as the poorly preserved *Madonna* in S. Maria in Trastevere, which still show the oppressive monotony of the Greek formula. About the middle of the century, in the chapel of S. Silvestro in Quattro Santi Coronati, there is a fresco which may properly be called Byzantine and is of an extreme banality. Nevertheless, it was at Rome and through the imitation of Byzantine work—albeit of a higher quality—that the very fertile school of mosaicists and fresco painters was to be formed at the end of the century. This school can be summed up in the names of Pietro Cavallini and Jacopo Torriti.

The mosaic of *Christ Pantocrator* executed by Torriti and Fra Jacopo da Camerino in the apse of S. John Lateran (1291), shows how he had already mastered the Greek style at its highest level. In the apse of Sta Maria Maggiore (1295), Torriti designed the famous *Coronation of the Virgin*, which does not lack any of the Byzantine subtleties in tone, or in the treatment of the ivory throne; but the study of antique examples can be recognized in the addition of such motifs as the winged genii, the river-gods, and the majestic fall of the draperies.

In the work of Pietro Cavallini the dominance of traditional means, the impulse to use antique models, and the grand style, tinged with rhetoric, can all be seen decisively affirmed. In the mosaics of the *Life of the Virgin* in S. Maria in Trastevere (*c.* 1291) the iconography is Byzantine but the composition is simpler and more flexible in arrangement, as in the *Adoration of the Magi*. The highest point of this evolution is probably reached in the frescoes in S. Cecilia in Trastevere, still more powerful and of an undeniable majesty, but unfortunately much damaged. The Byzantine types of the figures are surpassed in force and majesty in a classical interpretation, solemn and at the same time forward-looking. The great image of the Christ of the Last Judgement is no longer the Christ Pantocrator but now has the penetrating grandeur of Giotto's figures. The frescoes in the atrium of St. Peter's, representing the Evangelists with SS. Peter and Paul, have disappeared, but they were inspired by the same sentiments of modernity and antiquity and by the same return to basic principles.

The pontificate of Boniface VIII and the Jubilee of 1300 thus witnessed a return to Roman grandeur in monumental art. The Italian masters had something to set against the force of Byzantine compositions, and this achievement was the more important in that Rome had become a centre of attraction. A well-known Florentine sculptor, Arnolfo, worked in S. Paolo and in S. Cecilia at the same time as Cavallini. A kind of alliance seems thus to have come about, at the end of the century, between the Roman masters and the Tuscans. The new workshops at Assisi had become a fresh meeting-place for them, fired by the same ambition to excel Byzantine art.

In fact the problem was very much the same all over Italy. We have already examined it at Pisa, Lucca, and in the whole of Tuscany. In the case of Giunta it was little more than a modest adaptation of the Greek tradition to the painting of crucifixes and icons. For a few sensitive works, such as those of Margaritone, there were innumerable arid and derivative ones and even the work of Coppo di Marcovaldo shows that the Greek style retained its hold into the second half of the century. The icons derived from the *Theotokos* in the Baptistery confirm this dreary impression, yet the decoration of the Baptistery led the Florentines to an awareness, in their turn, of the urgent need for an original and sincere interpretation. According to tradition it was Cimabue who took the lead in this salutary initiative. He collaborated on the *Scenes from the Life of Joseph* in the Baptistery; in 1272 he was in Rome; he worked at Assisi, in the Lower Church (*Madonna of S. Francis*), and then in the transept of the Upper Church, where some scenes from the Apocalypse and the Life of the Virgin are attributed to him. His career is not without obscurities and his most secure works are the *Crucifix* from Sta Croce, the *Sta Trinità Madonna* (Uffizi), whose dark robe is strewn with accents of luminous gold under-lining the folds, and the mosaic of *S. John* in the apse of Pisa Cathedral. All of them show the same attempt as that made by the Romans to master the traditional forms of the '*maniera greca*' and to inspire them with gravity and restraint.

This is the reason why the new style is not to be judged as a simple turning away from what they considered as a merely worn-out art, for the innovators were rather trying to extract the essentials of the older style with the utmost possible clarity. In speaking of Giunta or Cimabue it has often been remarked that they correspond to the new emotional sensibility of the Franciscans, devoted to Christ and His Mother with a new tenderness and pathos. This is, partly at least, a misunderstanding.

The most pathetic themes, such as the Pietà or the Imago Pietatis, the harrowing scenes of the Passion, or those which were intended to canalize the emotions of the devout—Mary embracing the Crucifix—or those motifs which were intended to soften the rude manners of the times, such as the shepherds adoring the Child—all these had already appeared in Byzantine art of the twelfth century, both in icons and in illuminated Mss, and had been further spread by paintings such as those in the rock churches of Cappadocia. It is true to say, therefore, that the success of these themes in Italy was, initially at any rate, but one more aspect of Byzantine influence.[49]

This tendency towards more human imagery and greater sentiment had,

83

however, its parallel in the West. The sculpture of the great French cathedrals, which by then had been making its influence felt for nearly a century, was also an art charged with human feeling. Italy became acquainted with this great art after a short delay, so that some motifs arrived simultaneously from both directions, from France through Lombardy and from the Greek East through Apulia and Campania. Tuscany and Umbria, situated at the cross-roads and with a long experience of Romanesque, were the two provinces most eager to react, around 1300, to these new external stimuli and they were in a position to do so in the most original way. In any case, these new influences were not as irreconcilable as some would have us believe, for the Gothic style of the Ile-de-France itself was a response to the wave of Byzantine influence that spread throughout the whole of the West about 1200.[50] The situation of Italy was thus no different from that of Europe as a whole, but she was able to make a synthesis and to continue and extend these ideas.

THE REACTION TO GOTHIC: ARCHITECTURE

Gothic architecture entered Italy through what seemed the least favourable regions—Latium and Sicily. The abbey of Fossanova, in the mountains of the Volscian country above the Pontine marshes, was consecrated by Innocent III in 1208. The layout of the monastery (church, chapter-house, cloisters, cells . . .) reproduces exactly the Cistercian formula, as, for example, it may be seen at Fontenay. The church, begun in 1187, is high, light, and bare: except at the crossing the vaults are groined, over oblong bays, and separated by transverse pointed arches which rest on colonnettes set against the piers. A thin string-course separates the upper part with the windows from the arcade below. The apse has a flat end wall and is flanked by square chapels; in the end wall is a rose window, corresponding to the one on the entrance façade. These Burgundian forms, with their extreme simplicity setting off the structure and the geometry of the main masses, were a great novelty in Italy. At this date there were not yet any flying buttresses on the exterior, but there are buttressing strips. The only anomalies are the octagonal bell tower over the crossing, and particularly the small triangular pediment inserted into the relieving arch of the façade over the portal, corresponding to the gable above the rose window. There is also a Cosmatesque mosaic over the lintel.

A few years later, in 1217, Honorius III consecrated another Burgundian abbey at Casamari, in the mountains east of Rome. In the middle of the century the French type recurs all over Lazio: at San Martino al Cimino, near Viterbo, where there are alternating piers but still no flying buttresses; Piperno at Terracina in the north; and in the south Sta Maria d'Arabona, near Chieti. All these spread the influence of Casamari as far as the Abruzzi, Subiaco and Sulmona. The ever-increasing number of deviations from the norm indicate that the masters in charge of the work were now all Italians. The important fact, however, is the resistance of Rome, which, as we have seen, was then the stronghold of a conscious return to the ideal of the Constantinian basilica and to Byzantine decoration. The only Gothic church in Rome, S. Maria sopra Minerva, begun in 1280, is a sister-church to S. Maria Novella in Florence (1278); the Minerva reproduces the principal characteristics of S. Maria Novella and its exotic style is explained by the fact that it is the church of the Dominicans, and from the beginning the Mendicant Orders had taken over the Gothic style from the Cistercians.

Until at least 1250 Gothic buildings looked strange in the Italian landscape and their presence is to be explained in terms of the links between monasteries which were not confined to Lazio but were equally valid in Tuscany and the North. Sant' Andrea at Vercelli was founded in 1219 by the canons of S. Victor at Paris and its first abbot was a Frenchman. They were, however, manifestly content with general instructions as to the style. Here we find the plan, the square towers framing the façade, the lantern tower over the crossing, the flying buttresses, the small pinnacles on the gables, and, in the interior, the strong ribs prolonged by colonnettes which belong to the Ile-de-France style, but the treatment is timid, lacking in rigour and self-assurance: it would be no more than an awkward and archaizing Gothic, were it not that these weaknesses express a Romanesque and Lombard point of view that refuses to do away with the wall and prefers contrasts of light which do not prejudice the structural effects to nervous, linear effects of line and volume. In particular, this can be seen on the great façade which is given animation by a double gallery on the upper part and three portals on the lower, the whole being divided by important buttresses.

From the twelfth century the Cistercians of Northern Europe had established some 'Clairvaux' (*Chiaravalle* in Italian) to the north of the Apennines; one in the Marches, one near Piacenza, and the third near Milan. These abbeys were completed at the beginning of the thirteenth century in the same ambiguous Gothic as the monastery of Milan, consecrated in 1221. Its piers are heavy and cylindrical, the ribs retain the rectangular profile of Lombard ogival arches, and the lantern tower is a Romanesque belltower. The builders of the Abbey of S. Galgano, near Siena (now in ruins), were clearer in their minds about what they wanted; they must have come from Casamari and they understood the value of space in Gothic art, both in articulation and in expressive members. Nevertheless, the execution betrays the work of local artisans, especially in the use of brick in the upper parts. In 1237, the Cistercians of Settimo, at the gates of Florence, created a spacious and lively complex in their refectory and guest-chambers, basing themselves on S. Galgano. Gothic penetrated into Tuscany and its action was to be decisive and fecund: in Siena there is evidence for a certain link between the Abbey of S. Galgano and the cathedral, and in Florence the style was brought in by the Mendicant Orders. What was still needed was the climate of inventive daring which grew up in the second half of the century around such strong personalities as Giovanni Pisano in Siena and Arnolfo di Cambio in Florence.

THE THIRTEENTH CENTURY

Half a century ago it was generally believed that the Gothic style did not reach Southern Italy until the end of the Swabian period, through a Cypriot architect with the French name of Philippe Chinard. On the contrary, it is certain that, about the second decade of the thirteenth century, the Cistercians had introduced into the Hohenstaufen lands that strict and elegant style of which a good example is the unfinished basilica at Morpurgo, near Lentini (in the province of Syracuse). It was begun about 1225 on a strictly Cistercian plan—that of Clairvaux itself—which had become a kind of standard international plan, as is shown by the Sketchbook of Villard de Honnecourt. In 1224 Frederick II summoned lay-brothers from all the Cistercian abbeys of Sicily, Apulia and the Terra di Lavoro, to construct *castra et domicilia*. In fact there are precise correspondences in the profiles of the mouldings, the disposition of the masses and the geometry of the plans between the Cistercian architectural types and the castles of the Swabian period at Syracuse (Castello Maniace), at Augusta, and at Catania (Castello Ursino), which will be mentioned later.

Gothic was more than an architectural setting, it was a whole *milieu*. In most of the buildings cited above the decoration, and especially the capitals, clearly show the vitality and freedom of the French style. Starting from the forms of Romanesque and the style of the Antelami, the transition to the robust precision of Gothic can be seen, from 1210, in the charming series of the *Months* at Ferrara, or (to choose another example from Central Italy) in the similar but more animated set on the Pieve at Arezzo. In woodcarving, the *Deposition* at Tivoli, of the beginning of the thirteenth century, is a very frigid set-piece; while the serenity of the *Madonnas* at Vico (Lazio) and Mentorella (c. 1220), who clutch the Child to them, is a serenity still lifeless and inert.

Only dominating personalities could break away from this situation and bring new life to Italian sculpture. This was to be the task of Nicola Pisano and his son Giovanni, from 1270 onwards; but their works must be seen, like the southern castles, in the framework of the 'Ghibelline Renaissance'.

ILLUMINATION

The complexity of Italian art in the thirteenth century can be seen very clearly in the modest art of manuscript illumination, always open to fertilizing contacts and thus a particularly good pointer to general developments. Exultet Rolls continued to be produced in monastic *scriptoria* at Benevento, Monte Cassino and Subiaco. They were illustrated by vignettes, often of a popular and stereotyped kind, and the same taste for rather bright colours can be found in secular treatises, such as those produced for the Medical School at Salerno. There, once again, the Greek influence was far from weakening during the course of the century. It was stronger than ever in Lombardy, where it encountered a taste for Gothic scenes, set on an ornamental ground with enamel-like colours; the fusion is evident in the *Sacramentary of S. Benedetto in Polirone* (Mantua, Bibl. Comunale). There are often occasions when it is almost impossible to determine the exact provenance of one of these manuscripts. Sometimes the Gothic origin is revealed by the elegance and distinction of the drawing, as in the *De arte venandi*, illuminated for Manfred between 1258 and 1266; sometimes the hieratic element reveals Byzantium, as in the *Epistles* illuminated by Giovanni di Gaibana in 1259 (Padua Cathedral), which perhaps comes from a Venetian workshop.

The centre which seems to have been the most receptive and also the first to express an authentically Italian reaction was Bologna, the international University city, where there was a heavy demand for books. The output was for a long time of a disconcerting variety: it includes the famous Statutes of the Carpenters' Guild (*Falegnami*) dating from 1248, a manuscript of 1264–65 which has vigorous, Romanesque figures heightened with watercolour, and Statutes of 1270, which are stiffer and rather Byzantine in their colour scheme.

Dante mentions two painters who had made their names in the Parisian art of illuminating—Oderisi da Gubbio and Franco Bolognese. Their works are not certainly identifiable. It may be that the hand of Oderisi, more solid and more detailed, is to be recognized in the initial letters of a *Digest of Justinian* (Turin, Bibl. Naz.), where the grotesques and a charming stylization of vegetable forms indicate a study of Gothic examples. The same hand may be responsible for a Dominican Gradual which has beautiful filigree framework (Modena, Bibl.). Nevertheless, the type of the medallions, the layout of the compositions, even the enamel-like glitter of the tones reveal someone familiar with Greek art. In Franco's work the accent lies rather on greater plasticity and contrast (Modena, Bibl.) and leads into the fourteenth century.

THE FRANCISCAN MOVEMENT

The appearance of S. Francis of Assisi (1182–1226) marked the re-awakening of the Italian conscience; the endless political rivalry between the Papacy and the Empire was bound to produce a reaction in the direction of mysticism, the need for a more personal and intimate religion. The astonishing sermons of Assisi brought this to Italy, most particularly to the provinces of Umbria and Tuscany, where, little by little, all the spiritual fruits of the Middle Ages were accumulated.

The tender and deliberately popular genius of S. Francis, his almost familiar love for the Divine Persons, profoundly disturbed religious sentiment in Italy and at the same time rejuvenated it. Yet in this he was comparable with those Byzantine monks of the twelfth century who had themselves left their monasteries to address themselves to the crowd and to infuse piety and devotion into everyday life. The spontaneous defiance of the Saint towards the artificial framework of devotion, his independence from the clergy, his desire for a simpler Christianity all recall certain aspects of the Albigensian heretics of Languedoc. There was a moment when the Church regarded the Franciscan movement with some suspicion, since the new ideas seemed to undermine its own structure. A good deal has been written about the love of Nature, the Canticle of the Sun, and the Sermon to the Birds, all of it intended to present the son of the draper of Assisi as the initiator of a new 'naturalism' in art. In fact the sermons of the *poverello* had a profound effect upon the hearts of men; but it was only his legend which, much later and through the agency of S. Bonaventura, had any effect upon art. The vision of S. Francis himself has all the purity and elegance of the most exquisite of Gothic images. The *Speculum perfectionis* reveals one of his secrets: 'The blessed Francis strove above all to maintain, outside the times of prayer and the Divine Office, a continuous spiritual joy, both inward and outward. This he also prized greatly among the brethren, and he often reproached them for manifestations of sadness or melancholy. . . .' His new Order was to overcome the cruel *accidia* which afflicted monastic life by a humble and generous acceptance of the conditions of human life and of nature. In his intentions his Brothers were to be the great instrument of the *laetitia spiritualis* which was to spread out over the earth. Fra Jacopone da Todi (1306), 'jongleur de Dieu', the inspired troubadour, gave poetic form to the new spirituality. At first, however, the Franciscan movement could exert only a restrictive action on art. The severe portrait at Subiaco may be a direct recollection of the Saint from before 1228, but the great altarpiece of 1235 by Berlinghieri presents an aloof, impersonal monk, still very much in conformity with Eastern ideals. The testament of the Saint lays down that 'temples of large dimensions and richly ornamented' are to be avoided, and this was the reason why the Chapter of Narbonne, in 1260, limited both dimensions and decoration and confined itself in practice to the adoption of the Cistercian model, then considered to be the most sober and austere in Christendom. In the middle of the century, a generation after the death of the Saint, his new Order thus adopted the most simple architecture and the most modest iconography. Franciscanism deviated towards a spectacular art and a kind of picturesque luxury when the workshop at Assisi became the meeting-place of those masters from Rome and from Florence who were then anxious to give proof of their powers. This new direction was in keeping with the whole trend of Italian art at that time and it cannot properly be called Franciscan art, for the same principles were adopted by the Dominicans in their turn.[51]

The Basilica at Assisi was founded in 1228 by Pope Gregory IX; from 1230 the body of S. Francis was enshrined in the crypt of the Lower Church. The whole complex of buildings lies at the end of a ridge of Monte Subiaso and consists of a series of superimposed buildings carried on immense arcades. It was the work of Brother Elias, who deflected the Order towards power and pomp. The Lower Church has a long narthex preceding a single nave, and consists of four square bays crossed by heavily accented ribs and with round arches. In the next century these bays were flanked by chapels built between the buttresses. The squat and vigorous masses of the buildings are relieved by the richness of the thirteenth- and fourteenth-century paintings which cover them entirely.

The Upper Church, consecrated in 1253, has a single nave and likewise offers the spectacle of an accomplished Gothic structure, entirely articulated by pointed vaults and the play of ribs and colonnettes. It was on these walls, below the lancet windows, that the Life of the Saint was painted, under the direction of Giotto, at the end of the century. The façade is divided into two bands, in the Roman manner, crowned by a triangular gable and decorated with a rose window and a double-bay portal. The sides of the building have alternating half-cylindrical buttresses and windows, introducing a play of colour in pink and white, which, together with the absence of flying buttresses, gives the architecture its Italian character.

The chronology of the frescoes executed at Assisi is far from being established; the numerous additions made in the course of the fourteenth and fifteenth centuries and the perpetual restorations have made this a very difficult problem, perhaps an insoluble one, but the importance of this great workshop between 1280 and 1300 was immense. It was Cimabue, in association with a Roman master such as Torriti, who, from 1277 onwards, began the decorations in a new style, of which we have already seen the point of departure in Byzantine art as well as its marked 'Western' inflexion. In the Lower Church the *Madonna with S. Francis*, and in the Upper Church the frescoes of the apse and transept are to be referred to this determined master, whose development was both rapid and sure. The breadth of handling of the *Crucifixion*, with its skilfully grouped crowd and the impressive vortex of the angels, leaves no doubt on this point. The perceptive image of S. Francis in S. Maria degli Angeli at Assisi is also attributed to him.

All the major figures in the art of Central Italy of the generation of 1300 seem to have been to Assisi and to have come under Cimabue's influence; it is even possible that they included Duccio, the founder of the Sienese school.[52] It was there that Giotto, whose style had been formed in Rome, was to be summoned about 1296, by Fra Giovanni di Muro, General of the Franciscans. The whole series of twenty-eight frescoes of the *Life of S. Francis* can hardly be by his own hand, but it is likely that he was provided with many assistants in order to get the work done and achieve a modern cycle for the decoration of the sanctuary before the great Jubilee of 1300. In any case, it is a new chapter in the history of Italian art.

In short, what happened at the end of the thirteenth century to the Franciscan reform was similar to what happened three centuries later to the equally reforming spirit of the Council of Trent: a recall to spirituality, to the dignity of Christian art and concern for simplicity and austerity in buildings during the first phase. Piety was strengthened, mystics multiplied, and the Church, once more reinvigorated, canonized the new Saints who had done it honour. From then onwards, to pay proper homage to these great men, to proclaim the

Faith more clearly, to rally the faithful, and to struggle against laxity and heresy, more and more luxury and magnificence of decoration began to creep back into churches; human nature was once more triumphant. The fresco cycles at Assisi must be seen in this perspective, which explains their place at the turning-point of the Italian Middle Ages. The 'Spirituals', the most rigorous interpreters of the Rule, more and more bridled by the official Establishment, never ceased to protest against this deviation; in 1310 Ubertino da Casale denounced the costly luxury of Sta Croce.

THE SOUTHERN 'RENAISSANCE'

In the heart of Apulia, about sixteen miles from Barletta, there is a sort of Belvedere crowned by the Castel del Monte, built about 1240 for Frederick II.

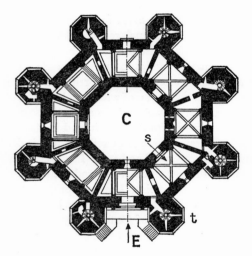

FIG. 10. *Castel del Monte (Apulia). Plan (after A. Cecchelli): E, Entrance; C, Inner courtyard; t, Eight octagonal angle towers; s, Groin vaulted room.*

It is a regular octagon of unforgettable nobility. The angles have octagonal towers about 80 feet high and the internal court is also an octagon: the rooms are therefore vaulted trapezoidal shapes in each of the eight sides, on both floor-levels. The portal is in the form of a Triumphal Arch.

The arrangement of the waterpipes is very notable and has led to a suggestion of Arab experts, but the problem is larger than that. Similar refinements of comfort can be found in the castle at Syracuse, Castello Maniace, and in the one at Catania, Castello Ursino, both of which were also constructed about 1240, on a square plan with circular angle towers containing perfect radiating vaults. All these buildings have a strict geometrical layout. There is a striking likeness between these castles and the clear articulation, the vaulting of the chapter-houses and the monastic lodgings and workrooms of the Cistercians. We have already seen how, from 1224, Frederick II confided important works to the Cistercians, the propagators of Gothic in Italy; nevertheless, there was no question of religious buildings, and Castel del Monte, in spite of its severe exterior, was very much a pleasure house, a princely residence with columns of pink marble and *cipollino*, delicate window mouldings, Arabian-style mosaics next to crocket capitals—in short, a Gothic frame for an Arabian Nights atmosphere. There are similarities of plan and layout between these princely dwellings and the castles of Syria or the *castra* of the Ummayids, all earlier than the eleventh century, which Frederick had had occasion to see on his visits to Asia Minor, and which, moreover, had already been more or less adapted by the Moslem architects in the time of the Normans.

From all this we may conclude that if at the time of the Normans the Cluniac type of plan had been altered or adapted in the course of execution by Arab workmen, then equally, in the Swabian period, plans which had originated with the Saracens were realized and interpreted in detail by workmen who had come from the Cistercians.[53] It is this original synthesis of contemporary techniques, in the service of an ideal properly called Imperial, which marks the importance of the reign of Frederick II.

But it was in sculpture above all that it manifested its vitality. The Gate at Capua (1234–39), which marked the entry to the Kingdom, consisted of two octagonal towers and a round arch flanked by lions and surmounted by busts of the members of the Prince's entourage and embellished by allegorical statues, such as the immense *Capua fidelis*. Only the head of this statue survives and it is 32 inches high. The planning is broad in conception and one is tempted to think of it in terms of Neoclassicism, but it is necessary to be more precise. The Gate carried a statue of the Emperor in tunic and chlamys, which certainly indicates a desire to return to the Roman style; but this Latinity is always strongly coloured by Byzantine majesty and it is as well to recall that the famous marble colossus (?Theodosius), more than 16 feet high, which is characteristic of the frozen style of the orientalized later Empire, was brought back during this century by the Venetians and installed at Barletta.

Simply from the point of view of Italy, the creations of the Emperor Frederick II seem to be a prelude to the Renaissance of the fifteenth century. But, seen in the perspective of the Middle Ages in the Mediterranean as a whole, as well as of Romanesque Italy, the work of the Emperor seems at first like an adaptation of the pagan and courtly Renaissance of Byzantium in a society at once Italian and cosmopolitan, the peculiarities of which went beyond the bounds of the West though not beyond those of the age.

No better proof of this can be found than the decoration of the right-hand pulpit at Ravello and the bust which comes from it, signed by Niccolo da Foggia and dated 1272. This allegory of the Church shows us a sort of glowering Empress, laden with jewels, beneath an overcharged crown, in which everything is hieratic and Byzantine.

Such were the sources of Nicola Pisano (born towards 1220), the last great Romanesque sculptor, in whom we can see the transition to a new age, a transition such as we have already seen at work in Rome, Tuscany and Umbria. An Apulian by origin, and therefore mainly a Southerner, he was born at Pisa where, in 1260, he carved the *Pulpit* in the Baptistery. The inspiration for this work is composite; lion-supports from Romanesque porches, Gothic arches and trefoils, Byzantine motifs, and bas-reliefs imitated directly from Antique sarcophagi. By his dense but ordered composition, and the arrangement of draperies to vary the effects of light the artist attains to unheard-of power of effect, but his genius is still trammelled by reminiscence and the complex but sometimes faltering culture of his age seems to flow through him. The great step was made in the *Pulpit* at Siena cathedral (1266–68) where Nicola was surrounded by four helpers on whom the future was to depend: his son Giovanni, the Florentine Arnolfo, Donato and Lapo; but it was towards the strongest and most expressive Gothic and not towards a form of Neoclassicism that the advances were to be made.

The end of the thirteenth century thus constitutes, in all spheres, a decisive turning-point in Italian art; in 1268 the new masters declare themselves with the *Siena Pulpit*; in 1272 Cimabue was in Rome along with Torriti and Cavallini; about 1280–85, Arnolfo, the pupil of Nicola Pisano, was also in Rome where he came in contact with the mosaicists and sculptors; from 1280 the Assisi workshops were open. On account of this extraordinary activity attention is concentrated on central Italy. The activities of the Masters there were to reach out into all the provinces and Italy, come to maturity under their example, was able to join actively in the Gothic art of the West.

IV

Gothic Art

(Fourteenth to Fifteenth Centuries)

Introduction: Historical background.
The political position of the City States

In 1300 there took place one of the most memorable Jubilees of the Roman Church. The Imperial power, humbled after the destruction of the Hohenstaufens by Charles of Anjou, had suffered a long interregnum. The brief foray of Henry of Luxembourg, crowned Emperor in Aix in January 1309, who descended upon Italy to be consecrated in Rome in 1312, revived the hopes of the Ghibelline clans of the peninsula, in particular in Siena and Pisa, but his death in 1313 saw them quickly and totally overthrown. No more foreign troops crossed the Alps to enforce their rule upon Italy until the wars at the end of the fifteenth century.

The power of the Guelfs thus became stronger. Boniface VIII attracted to Rome all the most lively minds of the Christian world; all over Italy new buildings were undertaken, in particular at Assisi, so as to show restored churches to the pilgrims coming to celebrate the Holy Year. The Pope seemed to be realizing the theocratic dream conceived a century earlier by Innocent III. But in fact the power of the papacy was also to dwindle rapidly away. Philip the Fair, King of France, began his struggle with the Holy See in 1300; the Archbishop of Bordeaux, elected Pope as Clement V, came under the control of the French crown and from 1309 to 1376 there followed the exile in Avignon, prolonged from 1378 to 1417 by the crisis of the great Schism. For Italy, the Trecento is a century without either Emperor or Pope. Never, since the days of Constantine, had such a position as the simultaneous eclipse of both the 'swords' existed. The peninsula was entirely left to itself politically, and every variety of individual power found expression, within the framework of an emphasis on urban, industrial and banking development.

The power of the French throne which, under Louis IX, had reached into the Holy Land and as far as Tunis, and had gained a foothold in Southern Italy, appeared to have replaced the Empire, but it was not in fact destined to exercise any serious dominion over the peninsula. Driven out of Sicily in

1282, by the intrigues of the Aragonese princes who reigned in the Tyrrhenian seas, the Angevins maintained a brilliant court, particularly during the reign of Robert (1309–43), rather than a focus of political power. This state of affairs was to last until 1435, when King René, the last claimant, was ousted by Alfonso of Aragon. During the fourteenth century, the struggle against England paralysed French power. But the Kings of Naples were the instruments of a cultural influence the more interesting in that contacts with Tuscany, leap-frogging over Rome, were very close. On the other hand, Siena was in close contact with Avignon, where her painters went to work, and, finally, Lombardy was closely linked to the art of Northern France and the Rhineland.

The Byzantine Empire was unable to prevent itself from becoming a field of exploitation for the Latins, and in particular for the nobles and merchants of Italy. The commercial empires of Genoa and Venice were established by means of a chain of trading posts and depots all over the Eastern Mediterranean. Euboea belonged to Venice from 1204 onwards, and remained hers until 1470; Gazaria (the Crimea) belonged to the Genoese. Italian principalities replaced French fiefs in the Balkans, and in Florence even, for a few months of 1343, there was an overlord who had become Duke of Athens. The development of painting on both sides of the Adriatic in the Serbian provinces, which, after the fine work done at Mileseva (1235) and Sopocani (1265), was still continued at Gracanica (c. 1320), and in the Greek provinces, for instance at Mistra in the fourteenth century, can be explained automatically through the continued contacts.

During the fourteenth century, the internecine warfare within the cities, and between them, reached its highest level, and also its most fruitful phase. This is the great creative age of the City States: Florence, Siena, Bologna, Verona, Milan were all shaped by a series of monumental compositions on which immense sums were spent. What the building of cathedrals and the planning of towns had been during the thirteenth century for France, where it was more carefully studied than elsewhere, it became in turn for Italy during the fourteenth century. This was the age of cities, but no longer that of communes and small republics.

The exception was Florence, where all efforts to establish a dictatorship were successively defeated, only to end, after the popular revolt of the 'ciompi' (1378), in government by the big guilds and the bourgeoisie. In many of the provinces, feudal overlords established their dominion: the Visconti in Milan, the Scaligeri in Verona. In Venice even, there was in 1355 the bid by Marino Faliero to establish a kind of personal régime. The political struggles were many and violent, but these conflicts, though often spectacular, did not impede an evolution which, on the social plane, made the links between centres and classes more flexible, and on the political plane, confirmed the growth of four or five large centres of business and culture such as Milan, Venice and Florence. The North assumed a new importance, in face of the

South which remained the great victim of feudal anarchy and the absence of organization. Yet in Rome, abandoned by the Popes and left prey to every kind of disorder, came in 1374 the strange and memorable epic of Cola di Rienzo, who revived the Antique tribunes in an endeavour to bring about a political re-awakening in Italy and a restoration in the capital of the republic of Cicero, to the plaudits of Petrarch and other men of letters. His brief moment of power in the Capitol and at the Lateran, followed by the inflated proclamations addressed to the temporal rulers of the day which precipitated his fall, was the last phase of the medieval dream of a return to an imperium based on Rome; only in the culture of the age did it survive as an idea and there it served to nourish poetry and symbolism, as is testified by Petrarch himself. In Florence in 1300, the year of the Jubilee, Dante was prior of the Guilds; in 1302 he was banished by a return to power of the Black Guelfs under Charles of Valois. He passed the remainder of his life as an exile among the Ghibelline nobles of the North, in Verona and in Ravenna, composing the most complete, the most elaborately conceived poem of the whole Middle Ages, a poem in the form of a vision which took place precisely during the Holy Week of 1300. The first grand affirmation of vernacular Italian is a masterpiece at once profoundly Gothic, medieval in its interior articulation, and deliberately national, expressing all the passions, the experiences, the sensations even, of an Italian conscious of the beauty of his country and his classical heritage. What has been called the 'humanism' of Dante is a compromise between the Antique world and the nature of the Christian world; he therefore helps to make comprehensible the transformation of Italian culture, which is expressed no less strongly by the sculptures of Giovanni Pisano and Arnolfo, and the fresco cycles of Giotto. These are the geniuses who strove to raise Italian poetry and art to a new and incontestable level by the personal use they made of all that had been achieved up to now.

The following generation was weaker and more sensitive. In literature, it was Boccaccio and Petrarch who, in close contact with French culture, adopted the forms of chivalry—one with clear and laughing ease, the other with tenderness and aestheticism—which fascinated Italian poets and storytellers. It was also a tormented generation which hovered between the pleasures of the world and the introspective withdrawal which promotes study; it was this curiosity about men and things, alternating with historical reveries and the melancholy of the 'studio', that little by little gave to Italian writers the human and personal note that is especially theirs. Gothic preciosity, and a genius for the expression of feeling is what we find in Simone Martini and Bernardo Daddi. The parallelism remains constant.

About 1300, therefore, came the emergence of the dominant personalities whose activity was to be reflected through the whole peninsula and give it, for the first time, a real unity. About 1340 appeared those less powerful but more sensitive personalities who refined the style and adapted it completely to the nuances of Italian sensibility. These are the two aspects of Italian Gothic—

the first heroic, the second elegant and chivalrous—which are overlooked by the traditional interpretation which places the beginning of the Renaissance in 1300 with Dante, the Pisani, and Giotto. At the end of the fourteenth century, Filippo Villani declared that with Cimabue painting began its return to Nature, and Giotto was a master on a par with the Antique. The humanists —Boccaccio, then Landino, and Poliziano—having adopted Giotto, his name was naturally placed at the head of the artistic genealogies imitated from Pliny, in which, from the fifteenth century onwards, the course of modern art was set out. The historian Vasari inherited this point of view and gave full expression to it in his monumental work, his *Lives* of the artists (1550). But the style of Niccolo Pisano derives from a strongly Byzantine interpretation of sarcophagi; the meaning that Antiquity had for Torriti and Cavallini, Giotto's direct precursors, has already been noted. It was through a double contact with Byzantine and Gothic that the memory of the works of Antiquity was evoked. The aim around 1300 was not to revive Antiquity but to gather together all the elements of a grand and complete art. The works created in the same period by so many outstanding personalities finally restored to Italy the consciousness of her own genius, and this emergence of a truly national culture dominated the future to the point that the two successive phenomena represented by the development of Italian art in 1300 and the anti-Gothic 'humanist' reforms of the fifteenth and sixteenth centuries may be considered as a single unit.

To this gross simplification, the classic fiction adds yet another; it only takes account of the major line, that of the Tuscans, running from Giotto to Masaccio, and from Masaccio to Leonardo and Michelangelo. But everything cannot be explained through the influence of the Florentine masters; to Giotto and to the 'Giottismo' derived from him must be opposed, during the fourteenth century, the originality of the Sienese; the art of Duccio and Simone Martini. If Giotto is more Roman, then Duccio is more Byzantine, and the two forces were equally active right up to the fifteenth century. Moreover, in the North, from Bologna to Verona, although painters took Tuscan style into account, they almost always either adapted it, or ran counter to it, as best suited them.

Finally, there was no continuous progress. The masters of 1300 were far from having been understood and followed by succeeding generations. Those that expressed themselves after 1300–40, for instance, Andrea Pisano in his Baptistery Doors, or Orcagna in his tabernacle, rapidly developed a taste for thin forms, angular and truly Gothic, directly inspired from Western examples. The heirs of Giovanni Pisano and Giotto retained only the semblance of their grandiose forms, and inclined in the direction of French fashions. It is possible, however, that the confusion provoked by the terrible epidemic of the middle of the century, the Black Death of 1348, which emptied half of the cities of Tuscany, tended temporarily to weaken and to disturb the creative power of the artists. But finally, at the end of the century, closely linked with

17 Florence: interior of the Baptistery

18 Ruvo, Apulia: main door of the Cathedral

19 Mosaics: (a) Rome, apse of S. Clemente
(b) Venice, S. Mark's: *The Building of the Tower of Babel*

20 Monreale, Sicily: cloister

21 Toscanella, Lazio: façade of S. Maria Maggiore

22 (a) Ferrara, Cathedral: relief of the
Month of March

(b) Pisa, Cathedral: Pulpit by Giovanni Pisano,
detail of column figure of *Hercules*,
typifying Fortitude

(a) Margaritone
d'Arezzo:
S. Francis
(*Siena Pinac.*)

(b) Pisa, Baptistery:
Pulpit by
Nicola Pisano,
detail of
angle figure of
Hercules,
typifying Fortitude

61 Castel del Monte, Apulia

Avignon, Prague and Paris, Italy took part through Siena and Lombardy in the development of the most highly refined aspect of Gothic, the 'international style', in which her contribution was of the greatest importance. Later, and in parallel manner, the great cathedral building operations were developed. Generally speaking, architecture, sculpture and painting were more 'Gothic' in 1400 than they were in 1300, and the Tuscans, who had held the initiative until 1340, ceded it to the Northern centres of Lombardy and Verona. This evolution governs the study of the century.

1. Tuscany: Architecture (1280–1340)

The Gothic system of building, introduced at the beginning of the thirteenth century by the Cistercians and taken over by the Franciscans, exists in Italy in its simplest form rather than in its fullest monumental and decorative form. The Order of Preachers, founded by St. Dominic during the papacy of Innocent III, was anxious to build new churches suitable for sermons, and they therefore based their buildings on these Gothic models. In a great number of Italian cities of the thirteenth and fourteenth century, a Franciscan and a Dominican house were built at opposite ends of the town.

The sanctuary at Assisi, one of the most celebrated in Italy, was the major influence in Umbria; the church of Sta Chiara at Assisi (1265), those of Gualdo, Terni, Narni and S. Francesco at Perugia testify to this. But in Perugia, the church of S. Domenico, founded in 1305—and rebuilt by Maderna in the seventeenth century—introduced the variation in Gothic church building which consisted of making the aisles of the same height as the nave,* a variant form found in the south of France, which increases the interior space. This form was used in the cathedral at Perugia, and at S. Fortunato at Todi, which took over a century and a half to build.

While the glory of the Saint of Assisi created in Umbria an important artistic centre lasting several decades, the church of S. Domenico was being built in Bologna to house the tomb of the founder of the Order of Preachers (1233). The Dominicans adopted the same Gothic formulae, and simplified them in the same way that the Franciscans did. The church with its flat choir flanked by two chapels, its nave and two aisles constructed with rectangular vaults, was rebuilt after 1298. The church of S. Francesco, founded in 1236, consecrated in 1251, and finished in 1270, is closer to the French type: the nave, with sexpartite bays, is markedly higher than the aisles, and is supported on the outside by flying buttresses. Nine radiating chapels are grafted on to the ambulatory. The gabled façade is, however, of a Lombard type.

The same type of choir with nine radiating chapels, each built on a square plan, is to be found in the Santo at Padua, founded in 1232 to serve as the tomb of S. Anthony. But here the building develops in a series of domes

*Known in English as a hall-church.

imitated from those of S. Mark's picked out with Lombard motifs on the façade and the exterior walls.

In Southern Italy, the collapse of the Hohenstaufen dynasty after the death of Frederick II, and the arrival of Charles of Anjou, brother of Louis IX of France, gave Naples a paramount importance for several generations, and established a kind of official character for French Gothic. Master masons were summoned from France: a certain Tibaud of Saumur built the abbey of Realvalle; the church of the Franciscans, S. Lorenzo, was finished with an ambulatory and with flying buttresses; S. Domenico Maggiore, built from 1239 onwards, with a nave and two aisles, and a transept, had an entirely Gothic character. Most of these buildings have now disappeared, but the portal of S. Eligio, and the surviving parts of S. Maria Donna Regina, remain as evidence.

Northern Italy was, in short, rather slow in developing something original out of the existing forms of Gothic architecture. In the South, apart from the Neapolitan essays, they were principally exploited in secular architecture and in tomb sculpture. It was in Tuscany that Gothic finally evolved, between 1250 and 1300, coherent and lasting forms. Two centres must be considered, which each contributed something to the complete adaptation of Gothic to Italian needs: Siena and Florence.

Siena. Siena supplanted Pisa and Lucca, in neither of which was Gothic thoroughly understood. In the Campo Santo at Pisa, built after 1278 by a certain Giovanni di Simone,[54] the rectangular cloister is constructed from semi-circular arches even though these are subdivided into pointed bays with headings of trefoils and quatrefoils. The delightful little chapel of S. Maria della Spina, on the banks of the Arno, is hardly more than a highly decorative shrine (1322–35). At Lucca cathedral, built from 1308 onwards, the Gothic decorative elements do not affect the structure, which remains Romanesque. In Siena, on the other hand, the vicissitudes of the cathedral's building history are evidence of a determination to achieve a complete and original solution. A Romanesque structure with a nave and two aisles stood on the highest point of the town from the beginning of the thirteenth century, but the present building owes much to the work undertaken during the second third of the century, under the direction of monks from the Abbey of S. Galgano, such as the 'Fra Vernaccio' mentioned in a document of 1259.

In 1264, efforts were made to complete the hexagonal dome. The builders made full use of the Gothic pier composed of clustered columns, developed the transept, and gave a rectangular form to the apse, all of which suggest Cistercian example; but the design of the semi-circular arches, the systematic arrangement of alternate layers of black and white marble, the setting of the dome upon squinch arches, and the decoration of the drum with a kind of continuous *loggia*, rule out any question of the original cathedral being merely an adaptation of Gothic style, Cistercian or other.

It was with the façade, commissioned in 1284 from Giovanni Pisano, that the transformation of the building into something much more complicated took place. The deep portals were surmounted by triangular gables and flanked with small towers with carefully articulated corner buttresses garnished with pinnacles. But Giovanni Pisano did not frame the doorways with sculpture as was the Western custom; the sculpture was placed on the first storey. He kept to the colour effects, and made use of marble inlays, proper to Tuscan Romanesque, but with a new and almost excessive exuberance in the forms, which is his personal and energetic interpretation of Gothic.

It became the starting point for a series of transformations that kept Sienese architects busy for more than a century. In 1317, the apse was continued at the east end as a projection and under its foundations was inserted the baptistery, which is a vaulted crypt with Gothic ribs, clothed with a brilliant marble revetment.

The Sienese Gothic dream had come to life, and grew rapidly. From 1339 onwards, Lando di Pietro was commissioned to convert the cathedral into a vast and exceedingly high building, the original nave of which was to become a transept. A start was made on the work, but it became impossible to complete it. The outline of the construction, and a characteristic doorway, alone have survived. From 1357 the vaulting of the nave was raised, and Giovanni Pisano's façade was completed by Giovanni di Cecco, from 1366 onwards.

These works took into account what was happening at Orvieto. The cathedral there was begun about 1288 on the plan of a Roman basilica, with a nave and two aisles, semi-circular arches in the nave arcade, and an open timber roof; at the end of the century bays with pointed arches were added to the choir and transepts. The Sienese Lorenzo Maitani designed the façade after 1309, and supervised the building of it until his death in 1330, though it was not completed until 1369. The three huge portals under their gables, with archivolts and deep splays, recall Western examples; the arcaded gallery and the tall pinnacles which surmount the buttresses confirm this impression. The serene grandeur of proportion is increased by the square motif which frames the central rose window, but the profusion of inlays, the reliefs on the buttresses, the mosaics—generally reworked—transform it into a brilliantly coloured screen without architectonic significance.

In Siena, the cathedral easily eclipses the churches built by the Mendicant Orders. S. Francesco, to the east of the town, was building from 1326 onwards; S. Domenico, to the west, was built between 1293 and 1391. Both have a single nave, huge proportions and a stark exterior, very different from Florence.

Florence. Towards the middle of the thirteenth century the two great Orders began to build their churches on a new plan. About 1246, the Dominicans—then ruled by Pietro da Verona, the future S. Peter Martyr—began building their monastery of S. Maria Novella; in 1250 S. Annunziata for the Servites, Ognissanti for the Umiliati, and then Sto Spirito for the Augustinians were built. A little later Sta Croce was begun for the Franciscans, S. Trinità for the monks of Vallombrosa, and finally the Carmine for the Carmelites.

In S. Maria Novella the basilical plan with a rectangular east end comprising choir and four square chapels, the bays vaulted with pointed arches, the piers with clustered columns, demonstrate clearly the Cistercian starting point. But during the course of construction, between 1279 when the nave was started, and 1348 when the campanile was built, there took place that process

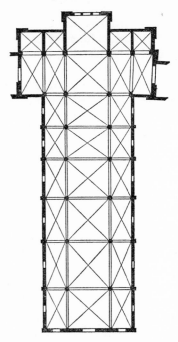

Fig. 11. *S. Maria Novella, Florence (Dominican Church). Plan (after Dehio and Bezold): Rib vaulting; Five chapels on the axis of the nave; Flat chevet.*

of evolution which resulted in the creation of Florentine Gothic. For not only does the relationship between the bays of the nave and those of the aisles tend to create a unified interior space of the type of Romanesque buildings, but the elevation which presents only one high and wide arch under a single round window has nothing particularly Gothic about it in the Western sense of the term, and the decoration of the capitals and the string courses is often of the same type as that found in Tuscan and Pisan churches of the twelfth century. For a long time situated outside the city, the Dominican monastery, with cloisters and large halls, eventually became one of its principal monuments. The façade remained unfinished until it was completed by Leon

Battista Alberti in 1470 with a harmonious design of panels in inlays of green and white marble.

The church of S. Trinità, in which the work of Niccolo Pisano may be recognized, was built between 1258 and 1280, on a similar plan and system, but it demonstrates even more clearly the characteristic desire to achieve a harmonious compromise between Gothic verticality and the more familiar horizontal forms. The church of the Franciscans, Sta Croce, begun in 1252, rebuilt in 1295, was not consecrated until 1443 by Cardinal Bessarion. The

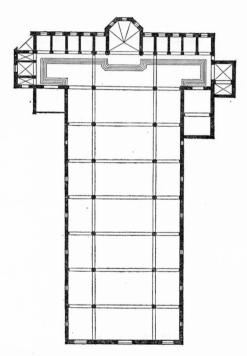

FIG. 12. *Sta Croce, Florence (Franciscan Church). Plan (after Dehio and Bezold): Open timber roof; Eleven chapels on nave axis; Flat chevet.*

interior is of a truly classic serenity; the octagonal piers and the relation of the central bays to those of the aisles create a flowing sense of space. But by using an open timber roof, which brutally interrupts the rising movement of the architectural members, and in completing the gallery running round the church above the main arcades, the builder risked spoiling the relations between the nave and the tall windows, and the luminous effect of the polygonal apse flanked by its eight chapels. He displayed a great deal of skill in giving the building a coherence which is no longer founded on principles admitted in Romanesque and Gothic, but on a relationship between open forms and solid wall, depth and surfaces.

It has already been observed that the general proportions of the church are those of the old basilica of St. Peter in Rome: at St. Peter's the nave was 89 metres long, against 90·5 at Sta Croce, 21 metres wide at St. Peter's against 19·5 at Sta Croce, 38 metres high at St. Peter's against 34 at Sta Croce. Together with S. Maria Novella, which supplied a Gothic model, the Early Christian churches of Rome played a decisive role in the conception of Sta Croce, which is why it may be considered as an example of the 'proto-Renaissance' of the thirteenth century. The builder seems to have conceived a kind of compendium of all known architecture, and created a building typical of Christian art. Hence the interest the Florentines showed in this unprecedented edifice, which Vasari believed could be attributed to the greatest builder of the period, Arnolfo di Cambio. An unfortunate modern façade, which pretends to be a reconstruction, was added in 1863 by Matas.

The most singular building of the period is, however, the new cathedral of Florence. The old church of S. Reparata, restored during the eleventh century, was considered to be too small and unworthy of such a city. In 1294 it was decided to replace it with an edifice in the grand style. The direction of the undertaking was confided to Arnolfo, who received in 1300 a general exemption from taxes, as being a master-worker whose 'industry, experience and talent' were to give to Florence 'the most beautiful and most noble church in Tuscany'. Insofar as it is possible to guess at the projected façade, it was to be of a Gothic strongly influenced by local Tuscan ideas, with wide horizontal inlaid bands, side doors smaller than the principal entrance, and statues in niches.[55] But Arnolfo was dead by 1302, and the works which had been undertaken at such great expense were suspended. This explains the slow progress and the alterations made in a project the original scheme of which has proved very difficult to reconstruct. During the first half of the fourteenth century, work was continued only for two short periods—from 1318 to 1323, on the façade, and from 1331 to 1333, when it was decided to build the campanile which was entrusted to Giotto. In 1355, the campanile having been built, attention was once more directed to the main building and a new model by Talenti was agreed on. This modified chiefly the height of the church, and projected an attic storey on the side elevations. Under his direction the main structure of the façade above Arnolfo's storey was built. He retired in 1365. During the next half century, a commission directed the works with modifications introduced as they went along; for instance, round windows were substituted for the Gothic windows planned for the upper storey. The third, final, and decisive phase came with the enlargement of the choir and Brunelleschi's construction of the dome (1420–36), which will be discussed later. Many projects were made for the decoration of the façade until the mediocre neo-Gothic construction of de Fabris was finally erected (1875–87).

The church projected by Arnolfo had a nave and two aisles (as it has now), the nave being somewhat wider and the whole lower than now; the nave had only four bays, and a gallery ran over the arcades. On to this large basilica was to be grafted a three-lobed choir, of the same shape as the present one, but much smaller, and the five chapels in each lobe gave it a five-sided figuration; the whole was to be surmounted by an enormous octagonal dome, which required on the nave side two strong piers to complete the structure necessary for its support. The general form, the gallery round the nave, and the decorative system, were inspired by the example of Siena, but the eclectic influence of Sta Croce was important; a methodical reinterpretation of old Tuscan buildings in the light of the works of Antiquity, studied by Arnolfo during his stay in Rome, may be divined. The art of the Cosmati enabled him to re-establish the old art of marble inlays, and this may possibly be the origin of the small twisted columns placed like mullions in the windows of the aisles. The octagonal dome which was to crown the edifice appeared like a transposition of that of the baptistery, also octagonal and oval in profile. But the powerfully modelled mass which was thus to be raised over the choir, the exceptional amplitude which it gave to the interior space, the siting of rectangular chapels within the apses, and certain of the decorative details, strongly suggest the precedent of Roman works—for example, the palace of Augustus on the Palatine. Arnolfo must, therefore, be credited with the ambition to create an architecture which would be not only typically Tuscan, but also Roman and Italian. And it was this that posterity kept well in mind.

Secular Buildings. This was the age of Dante. Florence was then a fortified city, enclosed in a wall with 150 bastions, of which the Torre di S. Niccolo survives as a reminder. Already the town was considered as sumptuous— 'beautiful houses, filled with works of the useful arts, as nowhere else in Italy' (Dino Compagni). Many of these houses, with their interior courtyards, opened on to arcades, had wooden cornices, and were crowned with feudal-looking crenellations (Palazzo Spini). Rows of houses with projecting upper storeys allowed the streets to be properly planned, for example in the Via Maggio (1295), and numerous and precise rules imposed certain aesthetic obligations on builders.[56] In 1325, it was laid down, *pro majore pulcritudine civitatis*, that all houses should be built of stone up to a height of four *braccie* (about 8 feet).

Palaces. In 1293, as a result of their political success, the new priors of the guilds decided to begin on an important municipal building.[57] This was the Palazzo Vecchio, the seat of the Signoria, which, in the civic field, was as clear and characteristic as the cathedral of Florence among the churches of Tuscany. The main part of the existing structure was built between 1299 and 1304, more than probably under the supervision of Arnolfo, and its fortress appearance is emphasized by the lower storey being almost without openings and heavily rusticated. The upper storeys are stressed by a string course on which are placed two-light windows under a visible relieving arch. The building is surmounted by a huge sentry walk projecting over machicolations forming a whole storey with windows, and is topped by a high tower with smooth walls built between 1310 and 1314. The summit of the tower also has a storey borne on machicolations, and ends in a structure like a baldaquin designed to carry the bells of the city. It is the most typical image of Florentine energy and clarity. In 1306, a small square was planned in front of the palace, but this was much enlarged in 1319.

After the triumph of the Guelfs and the merchant middle class, the Bargello, or Palazzo del Podestà, destined originally for the 'Captain of the people'—the Capitano del Popolo—was begun in the thirteenth century and

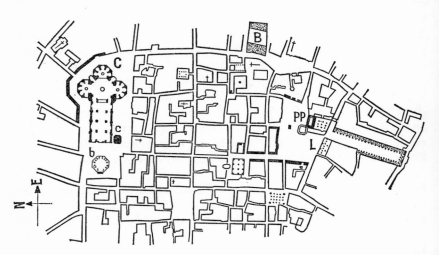

FIG. 13. *Florence. The two main building centres during the fourteenth and fifteenth centuries (after W. Braunfels): C, Cathedral (with b Baptistery, and c Campanile); B, Bargello; O, Or San Michele; PP, Palazzo Vecchio (Palazzo dei Priori); L, Loggia.*

finished in the middle of the fourteenth century, in a style which recalls at the same time the Castello di Poppi, and a simplified version of the Palazzo Vecchio. In the courtyard a beautiful and celebrated staircase leads up to an elegant loggia; its essential feature is the great vaulted hall of 1254, rebuilt by the cathedral architects after a fire, between 1332 and 1345.

This would indicate that after 1330 all the official architectural undertakings, whether secular or religious, were under the direction of those responsible for the cathedral. This is typical of the care taken by the Florentines to increase and unify the buildings of their city; if the work on the cathedral was so frequently interrupted it was because teams of workmen were sent to labour on other civic enterprises. This new feeling for the city as an entity was stressed in the decision taken in April 1334 to confide to Giotto, *accipiendus in patria velut magnus magister*, fortifications and bridges, as well as the cathedral and municipal buildings.[58]

One of the results was the campanile, which, echoing the old tower of S. Reparata, not destroyed until 1357, was an original contribution to the architectural planning of the city. With it was defined the modern axis of the city which was to be completed in the course of the fourteenth century by Or San Michele, a building with both a municipal and religious purpose, projected in 1297, and officially decided upon in 1336, and the Loggia, near the Palazzo Vecchio, later known as the Loggia dei Lanzi.

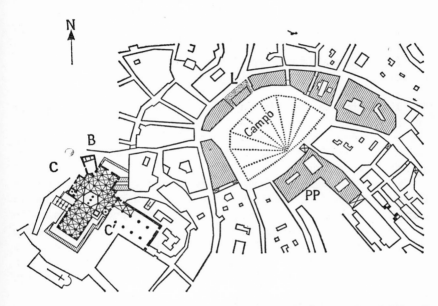

FIG. 14. *Siena. The two main building centres during the fourteenth and fifteenth centuries (after W. Braunfels): C, Cathedral; B, Baptistery; C′, New nave, left unfinished; PP, Palazzo Pubblico; L, Loggia (or Mercanzia); Campo, Main Square.*

Siena in the fourteenth century displayed the same features. The centre of the city was considered as the 'masterpiece' of which the municipal buildings were the elements; there was the same tendency to sacrifice the burdensome works on the cathedral to the municipal building works; town planning functions were vested in the principal architect or in the building commission.[59] The centre of Siena has remained more typical than that of Florence, which was too extensively rebuilt during the nineteenth century, and the streets converge upon the semi-circular *campo* which has preserved its old form of a theatre. The town hall was built from 1288 onwards by the Oligarchy of the

Nine; it consists of a central block and two wings (the latter were increased in height during the seventeenth century). The interruption of work on the cathedral in 1297, which provoked the departure of Giovanni Pisano, was probably due to the desire to speed up the building of the Palazzo Communale.

The three-light windows within a sharply pointed arch, the arcading of the ground floor, the lightness of the crowning members, and above all the immensely tall and thin 'Torre del Mangia' (begun later, in 1325, and finished between 1338 and 1348, on a design by Lippo Memmi), gave it a more charming, more picturesque air, less virile than that of Florence.

The harmony of the great square was established by the building of the Palazzo Sansedoni, which runs along one side, with windows in the same style as, and elements built to conform to, the Palazzo Pubblico opposite. In the town itself, palaces of the same kind were built, such as the Tolomei, and the Chigi-Saracini. The gateways of Siena were equally carefully designed— the Porta Ovile and the Porta Pispini—as were the city fountains.

The Fountains. In town planning schemes, great fountains were of major importance. They were placed at cross-roads, or most carefully sited in squares, and they may the more easily be considered as symbols of the vitality and power of the city since they were indispensable to its hygiene and to the growth of its population.

The great fountain at Perugia, built by Niccolo Pisano and finished by Giovanni, portrays in the reliefs on its two tiers of basins all the splendour of human civilization. In Siena, where the greatest importance was attached to the wells, Giovanni was working in 1295 along the same lines. What the Fonte Branda had been to the preceding century, the Fonte Nuova, treated as a Gothic portico, was in the fourteenth century. The Fonte Gaia was opened in 1343. A statue of the Virgin was placed upon it in the middle of the century. For it Jacopo della Quercia created the great ensemble (1409–19) which was intended to add to the glory of the city.

Bridges. The utilitarian, civic and religious character of fountains is also proper to bridges. Those of Florence date for the most part from the thirteenth century—for instance, the Ponte alla Carraia (1220), the Ponte delle Grazie (1237), and the Ponte S. Trinità (1252). A terrible flood in 1333 destroyed three of them, hence the rapid construction of the new Ponte alla Carraia (1346) and of the Ponte Vecchio (1345) with shops on it, now the only surviving old bridge in the city. The bridges had a chapel on them, like that, for instance, of S. Michael on the Ponte S. Trinità, or sometimes the huts of hermits, as for example on the Ponte delle Grazie, against the first pier of which lived, from 1390 to 1396, the holy Apollonia.

In Pisa, in 1322, there was the chapel of S. Maria del Ponte Vecchio; in 1323, the church of S. Maria della Spina (raised to its present position of the Arno quayside in 1871) was built. Like a reliquary on the river's edge it was designed to enable the passerby and the traveller to solicit the grace that crossing a bridge always seems to demand.

Central and Northern Italy. In all Central Italy, from the mid-thirteenth century onwards, renewed activity took place in secular architecture. There was for instance, the town hall of Massa Marittima, a high frontage with crenellated walls (before 1250), and the Palazzo degli Anziani in Pisa (1286). To the houses with projecting façades must be added those with towers, and with decorated openings. These were the years when the towns acquired their characteristic silhouette, still to be seen in the towers of San Gimignano, near Siena. But the fourteenth century was to be the great epoch of the town halls. The most notable examples in Umbria were the Palazzo del Capitano (1296), the Palazzo dei Priori at Todi, the more massive Palazzo Communale in Perugia (1292–97), the work of Servadio and Benvenuto, later completed by a series of three-light Gothic windows and a monumental doorway. In Orvieto, the papal palace may be compared with that at Viterbo from 1266 onwards with its large main hall and loggia. A little later, at Gubbio, came the Palazzo dei Consoli built by Angelo da Orvieto in 1333 on high arches which give it a cliff-like appearance, and the Palazzo del Podestà begun in 1349.

In Northern Italy, there was the Palazzo della Ragione in Padua, dating from 1219; its main hall, or *salone*, was built from 1306 onwards with a boat-shaped roof (restored during the eighteenth century). At Cremona, the Torrazzo, a high bell-tower (1250) with a later octagonal top, flanks a Gothic cathedral with nave, aisles and triple apse, the transepts of which were enlarged in the fourteenth century. In Venice the Doge's Palace, combined with the basilica of S. Mark's into a town-centre unmatched elsewhere, was continued throughout the fourteenth century, and chiefly about 1400, to which date must be deferred the examination of this major masterpiece of Italian Gothic.

2. Tuscany: Sculpture (1280–1330)

The great building programme undertaken in Tuscany during the second third of the thirteenth century was matched by the large number of important works in sculpture. They were intimately connected; the dominant figures were those of the Pisano workshop, Niccolo, Giovanni and Arnolfo, who were at the same time sculptors, decorators and architects. It is possible that Niccolo may even have inspired, as Vasari maintains, the church of Sta Trinità in Florence (from 1258–80), and the Castello at Prato which derives from southern types. Giovanni Pisano, who was a *capo d'opera* even more versatile and advanced in his ideas, created the choir at Massa Marittima (1287), the façade of Siena cathedral (after 1284), and the choir at Prato (1317), where the motif of small square chapels flanking a large apse reappears. Finally, Arnolfo rounded off the century by his work as a sculptor, and by his very important works in architecture.

Nevertheless, in Italy, sculpture remained dissociated from architecture; with two or three exceptions, the large, carefully organized façades were never completed. Architecture sought to achieve a spaciousness which ruled out complexity, and sculpture not governed by the requirements of the wall tended to restrict itself within a framework of columns or arches; it returned to the group and the individual statue. Niccolo Pisano died about 1280, after having trained the school of sculptors which was to dominate the period. He taught them that through the amplitude and grandeur of their figures sculpture could once more attain, as in Antiquity, to the greatest splendour of expression. The Middle Ages were not to see more ambitious and determined artists, and their very full careers show that they were immediately understood and followed. Their style spread rapidly through all the provinces of Italy, but it is only in confrontation with the Gothic masterpieces of the thirteenth century, French and German, that the full measure of their abilities may be judged. Their originality is immediately apparent in the programme of their works.

The Great Pulpits. After the pulpit of Pisa Baptistery (1255–60), where Niccolo introduced the powerful style derived from Antique reliefs, and the Siena cathedral pulpit (1266–68), where Gothic tendencies are in the ascendant, similar large works were undertaken in Tuscany: for S. Giovanni Fuorcivitas in Pistoia (1270), the pulpit by Fra Guglielmo, raised on two columns and attached to the wall; for S. Andrea in the same town (1301), Giovanni Pisano's hexagonal pulpit, with statues between the reliefs; and, finally, the one in Pisa Cathedral (1302–12) on a circular base with ten columns and a central pier surrounded by statues. The upper portion contains nine reliefs separated by figures of prophets, while sibyls are placed on the columns between the arches.

The Grand Tombs. The *Shrine of S. Dominic*—the *Arca*—erected between 1264 and 1267 in the church dedicated to him in Bologna, was by Niccolo and a group of assistants, among whom was the young Arnolfo. The sarcophagus was originally borne by caryatides or groups of figures back to back (now scattered); those of three angels in deacon's dress (Bargello) may very well be the work of Arnolfo himself.[60] Like the *monument to Cardinal Annibaldi* (d. 1276) in the Lateran, from which a frieze of figures against a mosaic background survives, the *tomb of Cardinal de Braye* at Orvieto, which Arnolfo executed towards 1282, was a large wall tomb placed under an arch, with the dead man resting upon a sarcophagus decorated with columns, above a base with panels of Cosmati work. In 1312, Giovanni made the *monument to Margaret of Luxemburg*. The final form of the great Gothic wall tomb is due to a Sienese disciple of Giovanni, Tino da Camaino; *the monument to the*

Emperor Henry VII at Pisa (1313) consisted of a figure lying upon a sarcophagus accompanied by a procession of dignitaries whose place in the original composition is now unknown. The *tomb of Cardinal Petroni* at Siena (1317-28) develops Arnolfo's type at Orvieto, by decorating its several storeys with statues. Finally, the four-storey *tomb of Bishop Orso* in Florence cathedral (1321-23), the parts of which have been broken up, presents a complex and skilfully designed arrangement: the dead man, sleeping, is seated above a sarcophagus borne by carefully executed caryatides. This model was imported into Naples by Tino; with a certain monotony of inspiration he made in turn the *tombs of Mary of Hungary* in S. Maria Donna Regina (1324-25), *Charles of Calabria* and *Mary of Valois* in S. Chiara.

Arnolfo. His career, begun in Siena and Bologna in the shadow of Niccolo Pisano, was pursued in Rome in the service of Charles of Anjou, whose bust he carved in a forthright style (Conservatori, Rome). He also worked on a fountain in Perugia, destroyed in 1308, of which several strange 'thirsty figures' survive. In Rome at S. Paolo-fuori-le-mura (1285) and S. Cecilia (1293) he erected large square tabernacles (*ciboria*) with pointed arches, surmounted by pinnacles, with statues on the angles, and in them is to be found the deliberate geometric clarity which was to be the guiding inspiration of his work as an architect. Arnolfo therefore belonged to the workshops of Siena and Rome, but his work had a distinctive, sometimes even archaistic note, and a personal quality of restraint.

The *capo maestro* of Florence cathedral also directed the workshop responsible for the façade. A few most interesting fragments have survived, such as an enthroned *Virgin between two saints*, where the statement of the volumes borders on heaviness, and above all a reclining *Virgin of the Nativity*, with its smooth planes and a freedom of execution as remarkable as its feeling of tender calm (Mus. dell' Opera).[61] Through his early death Florence lost the chance of having in sculpture a school equal to that which Giotto already represented in painting. For thirty years all work was suspended, and masters from outside, like Tino da Camaino, had to be called in. The Pisan masters dominated Tuscany without any opposition, and in fact for thirty years all turned on Giovanni Pisano.

Giovanni Pisano. A visit to France between 1270 and 1275 is sometimes postulated for Giovanni, and this would help to explain his abilities as an architect and his considerable plastic experience, since he was patently as much conscious of French cathedral sculpture as he was of the Antique reliefs which Vasari considered sufficient to explain his genius. The first work attributed to him—the half length *Madonna* in Pisa—is, in the arrangement of the drapery, the cutting of the face, the mouth, etc., the Gothic interpretation of a Byzantine type well known from the tenth century onwards through ivories and mosaics, but not appearing in Pisan painting until after

1271. Niccolo had left the Perugia fountain to be finished by his son (about 1277); the delicate reliefs of the *Months* and the *Liberal Arts* are, in great measure, the work of the father; the small allegorical statues which crown them are by Giovanni.[62]

From 1284 to 1299, the sculptor divided his attention between the façades of Siena cathedral and the Pisa Baptistery. For the latter, he carved the busts and statues which surmounted it outside, of which there are now in the museum twenty-one pieces 4 ft. 6 in. high, very weatherbeaten, but with an astonishing force in the pose, the movement and the drapery. For Siena, he carved a host of figures nearly 6 ft. high, placed on the gables and the buttresses. The programme was too ambitious; after Giovanni left Siena it was not until 1377 that the upper storey was finished. But he was the only Italian *capo maestro* able to conceive a programmed façade, on the model of the northern cathedrals. For him, Siena may have represented a rejoinder to Rheims. Few sculptors have so vigorously explored the possibilities of their art from classic tension to Gothic inflexion, and in his last statues, to an almost Baroque movement; in turn, he evokes Sluyter and Rodin.

Giovanni thus tended towards increasingly violent effects, and increasingly complex compositions, similar to battle scenes, as if he were driven to compress a suffering and gesticulating world within bounds which he felt to be too narrow for it. A comparison between the reliefs on the *Pistoia pulpit* (1301), which have deeply undercut, undulating forms that writhe tormentedly, with the ordered panels by Niccolo, of forty years earlier—for instance, those of the *Adoration of the Magi* and the *Crucifixion* on the Pisa Baptistery pulpit—makes clear how much Giovanni's troubled and turbulent temperament introduced of the pathetic, the powerful and sometimes even of the tortured. The atlantes writhe grimacingly, the lions seem to roar under the weight of the columns. In him there appears a genius who, like Michelangelo later, could not establish a school; he advanced alone, without encountering any rivals of his own stature.

On the *Pisa cathedral pulpit* Giovanni worked for twelve years. He placed his portrait at the feet of S. John, and two Latin inscriptions proclaim insistently his glory and his torments. In this work he reached the highest point of his inspiration. The pulpit, over 12 ft. high, is circular instead of hexagonal in plan, and as complex and crowded as a cathedral. Smooth columns rising from the base and figures replacing columns alternate on the outside, and further support is provided under the centre of the floor of the pulpit by pyramids of figures. There is an agitation, a conglomeration of forms, an almost excessive richness of handling, in the nine reliefs of the upper part. The *Massacre of the Innocents*, the *Passion*, the *Last Judgement*, seem to cram into an area about to burst open, elongated figures, expressive faces, heads of barbarians, shepherds, ruffians, angels or saints, in which survive his recollections of antique sarcophagi and of French lintels—a treasure house of universal sculpture.

Giovanni also made the *Madonna* in the Arena Chapel in Padua, the beautiful marble of the *Madonna of the Girdle* at Prato, the remains of the group placed in 1312 on the tympanum of the Porta S. Ranieri in Pisa cathedral, to celebrate the entry of the Emperor Henry VII into his faithful city (now in the Mus. dell' Opera, Pisa). This shows angels, the Emperor and the city of Pisa surrounding a seated figure of the Virgin, with drapery of a noble and splendid amplitude, of a type hitherto unknown in Tuscan sculpture.

Pupils. The numerous following which grew up in Giovanni Pisano's shadow spread whatever was imitable in his art – the most important were Giovanni di Balduccio, in the Northern provinces, and Tino da Camaino in Tuscany itself and in Naples. The latter, of Sienese origin, was the most talented if not the most attractive. If he lacked the nervous sensibility of his master, he began by retaining his feeling for weight and bulk. In 1351, in the *Tomb of the Emperor Henry VII* and in the figures of his councillors he treated the forms largely, massively, as he did in 1317, in the *tomb of Cardinal Petroni* in Siena cathedral, where massive angels hold aside the hangings of the baldaquin. In the *tomb of Cardinal Orso*, erected during a stay in Florence (1321–23), Tino created a noble example in the figure of the sleeping cardinal in his embroidered vestments; the angels of the baldaquin (V. and A., London) are of an extreme delicacy, and a supple and almost pictorial modelling. After leaving Pisa, Tino's art loses its strength. In Naples, where he became the official sculptor of the Angevin court, he created nothing new. His royal tombs there are high catafalques in which occasionally some details, like the small figures of children on the sarcophagus of *Mary of Hungary*, relieve the monotony of the execution.

Goro di Gregorio, in the *Arca di S. Cerbone* at Massa Marittima (1324), and in the *tomb of Guidotto de' Tabiati* in Messina (1333) uses Giovanni's art to achieve a kind of easy narrative style. Agostino di Giovanni and Agnolo di Ventura in the *Guido Tarlati tomb* in Arezzo (1330) arrange small narrative panels separated by statuettes in four rows, and set the portion with the figure of the dead man up on consoles above them. All vigour of effect is lost.

The death of Giovanni Pisano in about 1317, the dispersal of the workshop, the definite departure of Tino for the court of Anjou in 1323–24, left a void which the labours of the only notable Sienese, Lorenzo Maitani, could not fill. In effect, sculpture moved further and further away from the grand style of the Pisani. Adopting the elegant aspects of Western Gothic, the taste of the day was expressed in an art of applied ornament, thin reliefs with festoons, delicate garlands and profiled figures. The finest example of this is the decoration of the façade of Orvieto cathedral, directed by Maitani until 1330. The scenes of the *Genesis*, the *Tree of Jesse*, and the *Last Judgement* in five zones, which occupy respectively the four divisions of the lower storey, display a style composed of tendril ornament and medallions, of striated reliefs in which appear charming details of stylized landscape of a typically Sienese pictorial taste. Several different hands must be distinguished, before and after

Maitani himself, in this tender and delicate decoration which for so long occupied the time, and whittled away the energy, of the Sienese workshops. Nothing could better demonstrate the divorce between architecture and sculpture from which, at bottom, Giovanni also suffered. The initiative was now with the goldsmiths and engravers, something clearly demonstrated when Andrea Pisano was commissioned in 1330 to work on the doors of the Baptistery in Florence. But this represents a new phase of Italian Gothic.

3. Tuscany: Painting (1280–1340)

Giotto. The emergence of Giotto is less mysterious, since to the name of his master Cimabue, vouched for by tradition and by Dante himself, may be added those of Cavallini and Torriti and to the Florentine heritage the stimulating fields of Rome and Assisi. For a gifted child from the Mugello, whether or not he was a shepherd boy as the legend has it, the mosaic workshops of the Baptistery in Florence around 1280 were the finest training ground available, and the plastic freedom which heralds Giotto may be found already in the works of the master of the *Story of Joseph*. Cimabue himself systematically developed his own gifts by moving from Florence to Rome and from Rome to Assisi; through the painter-mosaicists, it was the ultimate form of the Neoclassical and Romanized Byzantine that imposed itself upon the new generation, with more stable forms, a more open drawing, a search for dignity, solemnity, obtained through strong, compact compositions with the minimum of the conventional. This lesson was valid for sculptors as well as painters. But the only work which testifies to Giotto's activity in Rome, the *Navicella* (*The Ship of the Church*), a mosaic in St. Peter's generally dated 1300,[63] impressed itself upon posterity for quite other reasons: it was the vivacity of the expressions, the language of physiognomy, which was so much admired, as even Alberti in the fifteenth century records.

To the monumental quality of his style must be added therefore a quality of expressiveness which, as has been seen, had also interested certain Byzantine artists, but which, at the beginning of the thirteenth century, was the great strength of French Gothic. Giotto, like Giovanni Pisano, was moved by the necessity of broadening the whole basis of figurative style, taking into account all known experience. This will to achieve, and the capacity to do so, appeared quite clearly as early as Assisi, in the frescoes of the Upper Church, which there is good reason to assign to the last years of the thirteenth century, and in part at least to Giotto. This great series, the first devoted to the *Legend of S. Francis*, episode by episode according to the new official texts of the Order, and in particular to the life written by S. Bonaventura, betrays the presence of artists formed in Rome, but in the present state of knowledge Giotto's exact role cannot be established.[64] In what can be attributed to him, such as the *Dream of Innocent III*, and the *Death of the Knight of Celano*, the

25 Orvieto: Cathedral façade

26 Siena: aerial view of the Cathedral

27 S. Galgano, near Siena: ruins of the Abbey

28 Florence: interior of Sta. Croce

29 (a) Giotto: *Faith*

(*Padua, Scrovegni Chapel*)

(b) Tino da Camaino: *Charity*

(*Florence, Bardini Mus.*)

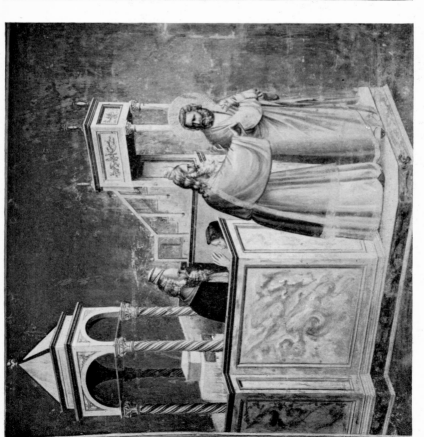

30 (a) Giotto: *Joachim expelled from the Temple*
(*Padua, Scrovegni Chapel*)

(b) Duccio: detail of the *Three Marys at the Sepulchre from the Maestà*

31 Ambrogio Lorenzetti: *View of a town*

(Siena, Pinac.)

32 Sano di Pietro: *S. Bernardino*

(*Siena, Palazzo Pubblico*)

architectural elements, the baldaquins, the temple façades, play an outstanding part; the figures tend to be cut off from one another, the silhouette stands out and assumes the role of a plastic entity, the colour of which emphasizes its solidity.

If it is clear that the work at Assisi, devoted to the glorification of the '*poverello*' in his historic aspect, was decisive for the younger artists, too much stress must not be placed on Giotto's 'Franciscanism'. Twenty years later, in the *Bardi Chapel* of Sta Croce (between 1317 and 1323), the painter created his definitive version of the Legend of Assisi; this was a solemn, powerful evocation, but a perfectly balanced one, and in a sense, a 'bourgeois' one. Giotto is primarily the exponent of a style.

The frescoes in the *Scrovegni Chapel* in the Arena at Padua (1303–5) reveal, in fact, a great decisiveness in the sense of strict composition. This building, a plain hall with a barrel vaulted roof, with six windows on the south side, was conceived as a vehicle for paintings; it may even be that Giotto himself was its architect. The painting, here, takes over the whole wall and organizes it to suit itself; above a base with niches in grisaille are three tiers of panels separated by painted bands; the panels contain scenes from the *Life of Christ* and the *Life of the Virgin*, conforming strictly to the most received iconography. Giotto is not an innovator in this field. On the triumphal arch is an *Annunciation* in two parts, and below it curious 'abstract' compositions showing empty rooms; on the inside of the façade is a *Last Judgement*.

In each narrative panel reigns an admirable order to which all is conducive: the sobriety of the setting (an architectural motif or bulky rocks); the welding together of groups of figures into massive blocks so as to isolate the eloquent gesture; the use of drapery to envelop the bodies so that they look like free-standing statues edged with long folds, thus isolating the essentials; and finally original invention, such as the look exchanged between Jesus and Judas in the scene in the Garden of Olives, which authoritatively concentrates all the attention on the most powerful dramatic moment.

The Byzantine formula, which limited itself to decorating the surface of the wall, thus appears as a useless restriction: the figure must appear to move itself. Space is opened up, but not of a kind which would merely tunnel out an illusionistic distance behind the figures; it is a kind of narrow and taut stage in which massive and cubic forms can play their parts. The figures thus acquire a right to movement and mimicry, not to exhibit their own character, nor to reveal their own individuality, but solely to show their complete participation in the action. Giotto's 'naturalism' imposes on itself a two-fold limitation: neither the space nor the expression is rendered for itself alone; space is only the condition of an adequately solid and convincing development of the forms, expression becomes an intensely dramatic effect.

With Giovanni Pisano, the piling up of figures and the multiplicity of scenes shown in one relief are the rule. The opposite is true of Giotto; the composition is arranged in a frieze, on a shallow plane, parallel to that of the wall, and without any allowance being made for the position of the spectator's eye. This narrative frieze is not continuous, as in Antique reliefs, but broken up as many times as is necessary for complete clarity. Giotto's art is entirely one of cutting and organization. A good example of this is to be found in the way in which the Story of Joachim is set out. If, in his grandeur and his energy, he corresponds to Giovanni Pisano's 'heroic' style, his art is informed by a totally different kind of tension: his awe-inspiring solemnity is derived from reason rather than passion. The clearest proof of this is in the scene of the *Lamentation*, at once vehement yet restrained.

'Naturalism' is therefore not Giotto's last word, since he aspires to the conquest of a complete style comparable to that of the great cathedral sculpture of three-quarters of a century earlier, or to Dantesque poetry which distils and surpasses through its discipline the material richness of Western lyricism.[65] In scholastic terms, the painter tends to make explicit in each character its own motivating force, grasping it according to what S. Thomas Aquinas defined as the 'principle of individuation'. Each character plays its part with complete consistency to its original, hence the artist's evident pleasure in deploying all his forces in action. But, at the same time, he obeys a higher principle to which all must finally be subordinated: this law, which directs and reconciles all individual energies, is that of a cosmos designed to make manifest a divine plan through certain moments of action, of drama. Hence his prodigious gifts of objectivity, of quiet statement, found only in the greatest Western artists such as Piero or Poussin.

Giotto was, therefore, the perfect master of his style when he returned to Florence. Among the altarpieces and crucifixes attributed to him in these years, there survive principally the *Madonna in Majesty* painted for the church of the Ognissanti (now in the Uffizi) which is the complete opposite of the suave and inert thirteenth-century representations. This new intellectual quality enabled Giotto to attract all the most positive minds in the new Florence. In the chapel of the Bargello he painted a *Last Judgement*, finished in 1337, but now indecipherable. In Sta Croce he was responsible for several chapels, from which the damaging nineteenth-century restorations have recently been removed: the Bardi chapel, with its cycle of the *Life of S. Francis* (*c.* 1320); the Peruzzi chapel, with its scenes from the *Lives of SS. John the Evangelist and the Baptist*,[66] of somewhat later date. The great gift for balance, and that facility for the sublime which characterize Giotto are here displayed in their highest degree. The *Burial of S. Francis* in the first series, the *scenes on Patmos* in the second, are truly classical through their assurance and ponderation. Moreover, certain effects of light in the scene of the *Stigmatization*, and in the architecture in the *Herod's Banquet*, display a virtuosity capable of further development. This is confirmed by the lovely golden colour, the more delicate drawing of certain works of this period, such as the dismembered *polyptych of S. Stephen* (of which two fragments are in the Museum at Chaalis, and one in the Horne Mus., Florence).

Giotto's fame extended far beyond Tuscany. His labours made him rich. Invited in 1330 to the court of King Robert of Anjou in Naples, he painted there works which are now lost, and in particular, a Gallery of Famous Men in the Palazzo del Uovo. The Visconti invited him to Milan. It was probably with the aim of keeping Giotto in Florence that the Council voted the important decree of 1334, which made him responsible for all the official work in architecture and town planning. But he was to die in 1337, leaving a well-established workshop.

It was conjectured that he returned to Assisi, and Vasari attributed to him the celebrated decoration of the Lower Church, and in particular, the choir with the *Allegory of the Virtues of S. Francis*, the *Marriage with Poverty*, and the *Apotheosis of the Saint carrying the standard of Christ*. These are the work of a follower who, about 1330, weakened Giotto's style with concessions to the picturesque, and above all, richness of colour. Other Florentine painters were not as advanced as Giotto: for instance, the unidentified painter of the panel of *S. Cecilia* in the Uffizi, who had worked in the Upper Church at Assisi, and whose style is drier, stiffer, and sometimes of a more popular aspect than Giotto's.

Like Giovanni Pisano, Giotto remains, basically, an isolated figure. After him all was changed, but no one could continue his style; his successors, who collaborated with him and finished some of his works, were never able to do more than quarry from their master's prestige. In any case, their history is confused. About 1400, a chronicler cited three main heirs of Giotto: Taddeo Gaddi, who for 24 years worked in the studio, and was the painter of the *Baroncelli chapel* in Sta Croce (1332–38), a masterpiece of the kind of elegant and minor form of 'Giottesque', which was to be the dominant style of the century; Maso di Banco, probably identical with the Giottino of whom Vasari speaks (and quite distinct from the real Tommaso di Stefano, called 'Giottino', who is recorded at the Vatican in 1369) and to whom is due the *S. Sylvester cycle* in the second Bardi chapel in Sta Croce, where the intensity of the colour reinforces the feeling for plasticity; and finally a certain Stefano, whose painting Vasari declared to be remarkable for its 'new unity' of colour and drawing, for one tone melting into another. Recently efforts have been made to reinstate this painter, who would throw a great deal of light on the transformation which took place about 1330–40. He may perhaps be identified in Assisi, in the *Crucifixion* in the Chapter-house, the *Coronation of the Virgin* (Lower Church), and the *Virgin with four Saints* (Sta Chiara). But his gift of delicate luminosity did not in fact have much repercussion in Florence, even in subsequent years.[67]

To these earliest 'Giottesques' must be added Bernardo Daddi, known from the *Madonna with SS. Matthew and Nicolas* (1328, Uffizi), and the frescoes in the *Pulci Beraldi chapel* in Sta Croce, higher in colour, richer in decorative detail, than Giotto's works.

Giotto's followers are therefore to be found principally in Sta Croce and at

Assisi; for a half century it was these two centres that saw the main developments of Tuscan painting. The progressive weakening of Giotto's heroic style in his pupils, their much prettier, more anecdotal, often more colourful manner, cannot be explained without considering the influence of contemporary Sienese masters. The effect of this was the greater in that the two schools were in constant contact during the first half of the fourteenth century, all the great Sienese painters having worked in Assisi, and sometimes even in Florence. The difference in the character of the two centres is all the more interesting in that it was defined in the course of a rapid parallel development.

But, apart from the influence of Giotto, mention must be made of Naples, where Cavallini, like Giotto, worked at the beginning of the century. The huge nuns' choir of S. Maria Donna Regina contains many frescoes (*Last Judgement, Passion, Lives of SS. Catherine, Elizabeth*, etc.) where Cavallini may perhaps have had a hand in the splendid choirs of angels, and where, above all, his studio may have played a part. Of Giotto's decorations in the castle of Charles of Anjou nothing survives, but the advent of two such masters gave an impulse to the local school which worked in the middle of the century in the church of S. Maria Incoronata, built by Queen Joanna about 1350, with the paintings on the vaulting of the *Triumph of the Church* and the *Sacraments*, which Vasari believed to be by Giotto, and which are now attributed to a certain Oderisio, and in S. Lorenzo where Charles of Anjou placed the altarpiece by Simone Martini. For half a century, Naples was to be the meeting point of international currents—French, Provençal, then Aragonese and finally Flemish—but it was only in the middle of the fifteenth century that their significance was understood.

Duccio. In 1285, the confraternity of S. Maria Novella (the *Laudesi*) of Florence commissioned a large altarpiece of the Virgin from a painter who must be identified, after many arguments, as the Sienese Duccio. The panel was the *Rucellai Madonna*, for long attributed to Cimabue. The common origins of the two masters are obvious, as is their intention to extend the confines of Byzantine convention to reach Gothic sensibility. But Duccio's chosen path was that of the precious effect, with jewel-like colours, elegant and finely outlined motifs, and multicoloured light.

Nothing is known of Duccio's early training (he was born about 1250) before this important work, but from his known works it is obvious that he had studied the sculpture of Niccolo Pisano, and particularly the Siena pulpit, several motifs from which he adopted in his *Nativity*, and his *Adoration of the Magi*. During the years following the *Rucellai Madonna*, the attributions made to him include the cartoons for the circular stained-glass windows of Siena cathedral (1287–88), and the *Madonna of the Franciscans* (1290, Siena Pinac). He seems to have been a turbulent character much in the public eye. In 1308, in a meticulously detailed contract, the city of Siena commissioned from him a large double-sided altarpiece of the Virgin, or *Maestà*, with, on one face, the Madonna enthroned amid saints, and on the other a carefully composed narrative of the Life of Christ extending over more than forty panels. Virtually the whole of the work still exists in the Museo dell' Opera in Siena, except for a few panels from the predella and the pinnacles; the two faces of the altarpiece have merely been separated. The *Maestà* was carried to the cathedral on the 9th June, 1311, amid scenes of popular rejoicing, which associated the traditional devotion of the Sienese to the Virgin with their pride in owning a masterpiece which was, after a fashion, a civic possession.

It was within the dense, subtle and intensely colouristic world of the grandest Byzantine tradition that Duccio achieved his general remodelling of painting. The simple and efficacious composition that Giotto used on a monumental scale is found, on a gold ground and on, so to speak, a minor note, in the panels of the *Maestà*. The *Three Marys at the Sepulchre*, the disciples in the *Apparition of Christ*, those in the *Garden of Olives*, form balanced and clear groups; the story takes place against the same architectural or rocky setting as in Giotto; the figures are unified in the action, but, for example in the *Incredulity of S. Thomas*—a small panel from one of the pinnacles—the tension of the event, the inward drama, is revealed by the nervous quality of the line and the play of the silhouettes. Often, as in *Christ before Caiaphas*, the full and sculptural forms recall Niccolo Pisano. In short, this art is like that of the Pisani, of a high order of narrative power; Duccio knows how to give the sensation of the action and the event.

But this intensity of expression, this gift of concentrating a sacred episode from the Gospels into a scene which brings it vividly to life, counts less in Duccio than the transfiguration of the scene by means of colour. Never had it attained to such precise subtlety: the archangel seated upon the salmon pink tomb of Christ wears red sandals, and the veils of the three Marys are in turn violet, green, red, against the iron grey rocks, flecked with points of light. Each panel is a network of saturated colours that envelops the scene in a quivering web of light. The articulations of pictorial space are but a structure to support a space of fluid enchantment, in which a gold ground stresses its fairyland quality. Duccio thus preserves the essential secret of Byzantine beauty with an art concerned with creating a coherent plastic world, and with the most beautiful use of line—an art made more expressive by Gothic. This example was decisive; he created an exquisite and complete style, which was continued by Segna di Bonaventura (active from 1298 to 1326) his pupil, and perhaps his nephew, and above all by the vigorous Ugolino di Nerio (active until at least 1339, *Sta Croce Polyptych*, London and Berlin).

Duccio died soon after completing his masterpiece, about 1318; there remained in Siena an even more brilliant painter, more accomplished and able to ensure the full development of the school. What Giotto had been for Florence, as an initiator and model, and what he had created in the way of firmness of style and confidence in the future, was achieved in Siena by the twin personality of Duccio-Simone. In fact, Duccio and Giotto began from the same premises—the Byzantine style of Cimabue—but from it they drew almost completely opposite conclusions: Giotto by stressing the Antique qualities of composition and mass; Duccio by concentrating on colour. The opposition is accentuated with Giotto and Simone, who, from 1320 onwards, became, as it were, the team leaders of Tuscan painting. Both have their place in the prevailing Gothic, Giotto in the monumental aspect of the great image makers of the thirteenth century, Simone in the elegant field of the miniaturists of the fourteenth century.

Simone Martini. Simone's career begins with the *Maestà* painted in 1315 for the Palazzo Pubblico in Siena. Nothing is known of what preceded this lovely and subtle masterpiece in which Duccio's manner, very thoroughly understood, is exploited with more decorative purpose. The Madonna is enthroned under a huge, scintillating and embroidered canopy. For the same building, Simone also painted the equestrian portrait (1328) of the condottiere *Guidoriccio da Fogliano*, who gained an important victory for Siena. His decorative silhouette stands out against an arid landscape dotted about with the small grey shapes of stockaded strongholds. Simone's art is that of the secular commemoration, the legend of chivalry. What finally gives him his place was his activity at Assisi and the role he played outside Tuscany. He was a painter with an international influence. About 1330, probably, he decorated the *chapel of S. Martin* in the lower Church at Assisi. In a carefully-organized spatial composition, the somewhat conventionally charming figures are grouped in rather more confusion than would have been the case with Giotto; the details of stuffs and the genre interest play a more important part, and stress the Gothic elements in Sienese art. As a Ghibelline city, Siena had close relations with Naples, Milan and France. From 1317 onwards, Simone was in contact with the Angevin Court, and may perhaps have visited Naples to paint the great panel in honour of S. Louis of Toulouse. In 1339 he went to Avignon, where he painted the *Madonna des Doms*, a *S. George* (now lost), and the *Christ returning to His family* (1342, Liverpool), and he died there in 1344. To the map of the exchanges which, during the second half of the century, established such close links between Siena and the development of Western art in Bohemia, Catalonia and Flanders, Simone added an important route towards the Rhône. But Sienese Gothic reached its most expressive achievement in his *Annunciation* (Uffizi) of 1333, with wings containing the figures of two saints by his assistant Lippo Memmi. All the details of this marvellous work—the delicately diapered gold ground, the blue-cloaked Virgin shrinking from the angel in a white dalmatic, the tender curves that envelop them, the vase with its four lilies—increase to the highest point possible, by means of exquisite precision of detail and beauty of colour, the empire which Sienese art aims at exerting over the hearts and sensibilities of its audience. The same trend is perceptible in the harmonious and melancholy poems of Petrarch, for whom Simone, who became his friend in Avignon, summed up in a celebrated miniature which heads a manuscript of Virgil, the vision of the Antique world as it appeared to an educated literary man of the Trecento.

Pietro and Ambrogio Lorenzetti. With the younger Pietro Lorenzetti, the great creative period of Tuscan art ends in Siena itself on a dominant Florentine and Giottesque note. With him the great workshops at Assisi came to an end. There he decorated the right arm of the transept of the Lower Church with five large scenes, one of which was the *Crucifixion*. In them— notably in the *Deposition*—the tragic themes of Byzantine painting can still be seen, but the feeling is now more violent, the forms fuller, and the draperies more dignified.

The exact date of this most expressive work is unknown, but its affinities with the polyptych of the *Virgin and Saints* at Arezzo, dated 1320, suggest a date between 1320 and 1330. The *altarpiece of the Carmine* (1329) includes a predella with scenes from the lives of the Hermits, of the remarkable variety of richness of effect which also characterizes the *Birth of the Virgin* (1342), with its use of tiled flooring, draperies and stuffs, at once solid in form and high in colour. Unfortunately, the development of this powerful painter is difficult to reconstruct. In 1335, he was working with his brother Ambrogio on frescoes in the Hospital at Siena; after having been, about 1320, in the ambience of Giotto, about 1340 he came under the influence of his brother, who, recorded in the Florentine Guild from 1327 to 1334, was moving away from Giottesque solidity to develop a more emotional, flowery, and Gothic art. Ambrogio's development seems quite steady in the *Madonna* at Vico l'Abate (1319), the two surviving frescoes in S. Francesco in Siena, the triptych of *S. Nicholas of Bari* (Uffizi) of about 1332, the polyptych of the *Crucifixion* (Frankfurt), the frescoes at Montesiepi, and the *S. Michael* from Badia a Rofeno (now Asciano), an astonishing, undulating, Gothic warrior in gilded armour, trampling on a monster composed of writhing interlacings. The *Madonna with six angels and four Saints* (Siena Pinac.) of about 1340, illustrates clearly the growing sacrifice of strength of colour. But it would be mistaken to see merely a trend towards fantasy; never was his observation of reality sharper. The celebrated decoration in the Palazzo Pubblico, the two *Allegories of Good and Bad Government* (painted between 1337 and 1339), are proof of this. This is the finest surviving example of the type of civic art of which Giotto provided a model in the Bargello. The lessons of the Arena Chapel can still be perceived in the weight of the allegorical figures that dominate the composition, and in the care taken with the balance of the groups, but Ambrogio is carried away by the desire to particularize, to multiply precise details of an amusing or poetic character requiring the audacity of a brilliant narrator and the ability of a cartographer. The result is merely a conglomeration from which an aspect of the city, or a corner of a vineyard, or a realistic group, could be singled out. Two small panoramic views by

Ambrogio (in the Siena Pinac.), survive from a series of Sienese cities and are of the same date: one represents a fortified harbour, perhaps Talamone, the other Montepulciano. These highly detailed representations may once have formed part of a piece of furniture; they demonstrate the artist's ability to render realistic images in a delicate play of colour under a high horizon which transposes them into enchanting visions.

The contact between the Duccesque delicacy of handling continued by Ugolino di Nerio and the new robustness of Pietro was what probably provoked the interesting evolution of a master of the first half of the century, provisionally named (by Berenson) for this reason Ugolino Lorenzetti. He painted the charming *Nativity* (Fogg. Mus., Cambridge, Mass.) and the large *altarpiece of S. Peter* at Ovile (Siena), and is identifiable as Bartolommeo Bulgarini.

4. Tuscany (1350-1410)

Florentine Architecture. In Florence, successive Signorie developed and enriched the civic building works started at the beginning of the century; instances of this are the hall inside the Bargello, and the Mercatanzia on the Piazza della Signoria. The loggia projected in 1323 was not built until 1376, on the plans of Benci di Cione and Francesco Talenti, who were the architects of the cathedral. The piers of this open hall are in fact derived from the type used in the cathedral. Throughout this period the civic buildings show adaptations from the methods of construction worked out in the principal ecclesiastical undertaking, and often at its expense.

But, in the development of an architecture suitable for prestige building, the stress comes to lie on the decorative effect. There came therefore a moment when the need for beauty triumphed over the utilitarian concept, and this, operating within the Gothic framework, led to the spectacular compositions of the later period. This may be observed in the Loggia dei Lanzi, where the three high, round arches introduce a new note of elegance and clarity. It was already to be observed in Or San Michele, where work was completed at about the moment when the Loggia stressed the new town centre.

The Duke of Athens was expelled in 1343. A few years later, the Priori, now returned to control, had placed 'at the four corners of the palace of the People of Florence' four stone lions (*marzocchi*) which were to be the symbols of the city. This civic pride and sense of corporateness was to be seen in the most important building of the century—Or San Michele. As early as 1336, the Signoria had planned to replace the old grain market with a *templum in statura et forma palatii*, on the street—now the main axis of the city—which links the cathedral to the Palazzo della Signoria. This 'temple' would serve the double purpose of a church dedicated to the Virgin and an improved granary. Eighty thousand gold pieces were set aside for its construction, and it was to become the most original building in Florence, designed for the glory of the Guilds which ensured the prosperity of the town. The architect of this masterpiece is unknown: Vasari mentions Taddeo Gaddi, which would suggest that the project originated during Giotto's (d. 1337) period of office as city architect, when Taddeo would perhaps have been working under him. In 1343 a chapel dedicated to S. Anne was added, in honour of the liberation of the city. After 1349, Neri di Fioravanti and Benci di Cione were in charge of the work. The year 1352 saw the beginning of the elaborate tabernacle by Orcagna, for the lower hall, and to him has sometimes been attributed the design of the whole. Work on the outside decoration, on the marble revetment, on the niches set into it all round, continued long after 1380. Built on an almost square plan, Or San Michele is an unusual building in the form of a cube, each side having three large Gothic arches, and on the upper storey tall, two-light windows.

Work proceeded slowly at the cathedral; the great labour of the façade and the upper storey was in process of completion (1355–65) by Talenti, who had persuaded the Board of Works to agree to modifications to Arnolfo's original plan, particularly in the lower ranges of the exterior elevations on the side. But the really important matter was the drum of the dome. Before the completion of the fourth bay, the Board of Works opened a new competition in 1367; nine plans were submitted, and that of Andrea Orcagna and three collaborators was accepted. It increased still further the size of the cathedral, and above all raised the dome on a large drum, which had for result the sacrifice of Arnolfo's concept of a building dominated by the effect of its interior space in favour of a new concept which paid more attention to the effect of large architectural masses within the city.

Florentine Sculpture (1330–60). After the death of Arnolfo, there was no sculptor of eminence in Florence. Tino da Camaino was invited to make the tomb of Bishop Orso, who died in 1321, and a decade later Andrea da Pontadera, better known as Andrea Pisano, came to the city to make the bronze doors for the Baptistery, which were completed in 1336. These consist of twenty-eight panels separated by a framework ornamented with nail heads and rosettes; each panel encloses a quatrefoil such as are found in the portals at Lyons, Rouen, and many other cathedrals, and within this Gothic cartouche the scene is set out as elegantly and clearly as possible. These reliefs are therefore in no sense the continuation of the art of Giovanni Pisano; they constitute an entirely new point of departure.

The scheme is the same as that of Bonannus at Pisa (1180), the doors being divided into square panels with a border. The principal theme, scenes from the Life of the Baptist, is directly borrowed from the mosaics in the Baptistery itself, as is borne out by a comparison of the motifs, for example those of *S. John leaving for the desert*, or *S. John preaching*, but the sculptor has subjected them to the same process of elaboration as that used by Giotto in the Peruzzi Chapel of Sta Croce. Andrea thus returns to 'Byzantine' sources and gives them a modern elaboration; a disciple of Giotto in his mode of composition, in his clear drawing and feeling for mass, he in fact restates in bronze sculpture what the painter had stated in painting. The result, however, is more mannered, pretty rather than monumental. This is because Andrea was governed by a concept of the precious which is only to be explained by models of Parisian Gothic transmitted through miniatures and ivories.

The elongation of his figures, the shallow Y-shaped folds in the drapery, the calm softness, the polish, ensure for Andrea his eminent place among the exponents of minor Gothic in Florence.[68]

According to Vasari, Andrea was already a man of sixty when the doors were commissioned. On the death of Giotto, he was charged with the execution and decoration of the second and third storeys of the campanile (1337–1343), and it was he, on the models of the master apparently, who finished on the first storey, and elaborated on the second, the series of hexagonal medallions with the *Allegories of the Mechanical Arts*, completed on the second storey by the *Planets*, the *Virtues*, the *Sacraments* and the *Liberal Arts*. This is an encyclopaedia of human labour and knowledge, comparable to the famous fresco in the Spanish Chapel, and it occupies a front rank in the history of Italian encyclopaedic cycles adapted from the important Western programmes.[69]

The plastic vigour of certain medallions—notably the *Hunt*, and *Agriculture*—suggest the influence of Giotto. The tender imagination of Andrea may be perceived in the series from Genesis; others are the work of the studio, which, under the general direction of Talenti from 1343 to 1357 continued the decoration, in particular on the second storey, where the figures of the *Planets* and the *Sacraments* are silhouetted elegantly against the bare field of the hexagon. The most vigorous are the work of Arnoldi, whose rather archaizing harshness may be explained by his Lombard origins, from whom was commissioned in 1361 the lunette for the Loggia del Bigallo. In the high niches of the third storey were placed, in the fifteenth century, the figures of the *Prophets*.

Until his death in 1349 Andrea was also the *capo maestro* at Orvieto, where he continued the polychrome façade above the delicate sculpture decoration by Lorenzo Maitani. His son Nino succeeded him in this employment; he was principally the creator of supple, smiling, Madonnas holding the Child on one projecting hip in the French fashion, with which he crowned his tombs—for instance the Cavalcanti tomb in S. Maria Novella, and that of the Doge Marco Cornaro in Venice (1368). A typical work, the *Madonna feeding the Child*, in S. Maria della Spina in Pisa, is at once derived from a composition by Ambrogio Lorenzetti and from Ile de France Madonnas known from ivory and alabaster statuettes.

The most complete figure of the middle of the century is that of Andrea di

Cione, better known as Orcagna, who was an architect, sculptor, painter and goldsmith. Between 1352 and 1359 he created the polychrome tabernacle for Or San Michele, with, on the back, a large, crowded and exuberant relief representing the *Dormition and Assumption of the Virgin*—a Gothic tympanum brilliantly confined within the limits of goldsmiths' work. The comparison of the tabernacle itself, with its gables, pinnacles and precious effects, with the tabernacles of Arnolfo of seventy years earlier, gives the measure of his more complex, more refined taste, though one attracted by coherent and solid form. It is therefore not surprising to find him intervening in the great debate of 1367 on the final model of the cathedral, and advocating the exterior effect —that the beauty of the shape of the choir end of the cathedral would be significant for the town itself. To his activity as a painter is due the fine *polyptych of the Strozzi Chapel* in S. Maria Novella, painted between 1354–57, in which the theme of Christ in Majesty, forgotten since 1300, reappears. The work is powerful, though strained, the draperies ample but their colour dry; the rather inert faces suggest a revival of hieratic feeling in the midst of the powerful Giottesque formula. To his brother Nardo di Cione must be attributed, according to Ghiberti, the three frescoes in the same chapel—adept representations of *Hell*, *Purgatory*, and *Paradise*, of a facile dryness alike in both tender and horrific parts.

Painting, from 1350 to 1375: Crisis. If one of the stories of Sacchetti is to be credited, the Florentine painters of about 1360 believed that their art was in full decadence. The first generation of Giotto's pupils did not have the strong personality of Simone in Siena; with the Lorenzetti, a kind of equilibrium seemed to have emerged for the whole of Tuscan art, but both the brothers disappeared in 1348, victims probably of the Black Death. Sienese painting was cut off short. Lippo Vanni may be placed among the followers of Ambrogio; brilliant as a miniaturist (*Antiphonal*, Mus. dell' Opera, Siena, 1345) he tended towards Simone in his polyptychs and frescoes (1372, Cloister of S. Domenico, Siena). The followers of Simone were numerous: Andrea Vanni, who painted a *S. Catherine of Siena* (Pinac.), Luca di Tommè, and Bartolo di Maestro Fredi, the author of frescoes in the Collegiata at S. Gimignano (1367). In the same church also worked a little known but original painter, Barna, who expressed himself in a manner at once constrained and violent; his *Way to Calvary*, for instance, is of an exceptional harshness. Thus, certain painters may be seen renouncing harmony in favour of an expressive ugliness; other, like Orcagna, return to pre-Giottesque compositions.

In 1365, the Prior of S. Maria Novella, Fra Zanobi dei Guasconi, commissioned from the painter, Andrea di Bonaiuto, the decoration of a chapel built on a square plan, with cross vaulting, recently erected in the cloister (1350–55), and now known as the *Spanish Chapel*. The iconographic scheme, the spirit of which may be discovered in the 'Mirror of Penitence' by Jacopo Passavanti, sets out the glorification of the Dominican Order both in theology and in the life of the Church, on side walls left entirely free for the purpose, with a *Crucifixion* on the end wall, and *Allegories of the Church* in the four divisions of the vault. This array of encyclopaedic images is treated with animation, but with more dogmatic ardour than delicacy. The array of the Arts, enthroned on the left-hand wall beneath S. Thomas, the vision of the *domini canes* (the Dominicans, the Hounds of God) coursing amid the crowd outside the cathedral of Florence, on the right-hand one, are admirable inventions of a proselytizing spirit which was a very real force during the general crisis in morals and religious institutions.

In fact, there was in Tuscany, at the middle of the century, a great development of programme painting, no longer of the narrative type as in the time of Giotto, but didactic and accompanied by a return to the forms of the more solemn styles of the thirteenth century. The fresco of the *Last Judgement* in the Strozzi Chapel, the vanished frescoes of Orcagna in the nave of Sta Croce, and the celebrated composition attributed to Traini in the Campo Santo of Pisa, the so-called *Triumph of Death*, all date from about 1350. The meeting of the *Three Quick and Three Dead* is the vision described by the holy hermit to awaken the worldly to the knowledge of salvation, with the pleasures of the world shown in a garden of love worthy of the Decameron, and with an imaginative power, a colour and a draughtsmanship which make it one of the most brilliant creations of the fourteenth century. Its destruction, in July 1944, resulted in the discovery upon the wall—as in other parts of the badly damaged Campo Santo—of fine preparatory drawings, or *sinopie*.[70] The only other work known by Traini is an *altarpiece of S. Dominic* (1344–45); the Pisa fresco was in a style far superior to that of Orcagna, and in the best tradition of the beginning of the century. But its theme shows once again the obsession with death and the exhortations to penitence which were widespread in Tuscany during the years dominated by the plague.

This scourge, which appeared in 1340, and in 1348 carried off two thirds of the inhabitants of Siena, and half those of Florence, was to reappear in 1363 and 1374. It is necessary to bear this in mind the better to understand the spiritual crisis, the activity of the religious orders, the character of the commissions, and even the weakening of the art.[71] The sacred characters, abandoning their humanity, once more affirm their hierarchic position; Orcagna's great altarpiece for the Strozzi Chapel is the most significant expression of this. The iconographic innovations which proliferate during this period (the Madonna of Humility, the Glory of the Trinity, the Coronation of the Virgin shown taking place in Heaven) indicate a rapid adaptation to the requirements of an uneasy piety and to ecclesiastical directives.

The general demand for pictures is shown by the great number of small pictures designed for private devotional purposes. The principal types of these had appeared at the end of the thirteenth century; they were fixed by the middle of the fourteenth century, and continued to be in demand until the sixteenth and seventeenth centuries. The Man of Sorrows, the Pietà, the

TUSCANY

Madonna of Humility—these devotional images often owe their popularity to an example by a great master. The new concept of the Madonna, which associates the exaltation of Christian virtue with the touching representation of motherhood, was possibly 'invented' by Simone Martini and hall-marked by Sienese emotion. In its most triumphal form, it reappears in the *Madonna of Foligno* by Raphael.

The origin of these small pictures is often to be found in a detail of large scale compositions; they isolate the most moving moment and concentrate attention on the pathetic pose and expression. Chiefly, they were the Pietà and the Madonna of Humility, well suited to intensify the direct relationship between the spectator and the sacred figure.[72] The third quarter of the century thus witnessed a simultaneous transformation in the character of the large scale fresco cycle and the small devotional picture.

Painting in 1400: Revival in Siena. By the end of the century new tendencies were beginning to appear. Spinello Aretino, born in 1347, who was commissioned to paint the frescoes in the sacristy of S. Miniato, and then those of the history of Siena in the Palazzo Pubblico there (1407), displays a conscious though often frigid return to Giotto. Gherardo Starnina, whom Vasari names as the master of Masolino, and who is best known for his connection with Castile, was perhaps the author of the *Thebaid* in the Campo Santo, in Pisa. Agnolo Gaddi, the son of Taddeo, painted frescoes of the *Legend of the True Cross* in Sta Croce about 1390—the same picturesque and legendary theme that inspired Piero della Francesca at Arezzo—and decorated Prato cathedral; Agnolo understood how to compose his scenes, and tell his story with naïve details. Of a similar character was the Antonio Veneziano who came from the north, and who worked with Taddeo, before he completed with realist vigour the *Scenes from the Life of S. Raniero*, Patron Saint of Pisa, in the Campo Santo there, which had been begun by Andrea Bonaiuti. The lesson of Giotto, filtered through that of Simone, as being the essential of Gothic narrative art, had a renewed lease of life about 1400, within the framework of a style whose chief characteristics are determined by the so-called International Gothic emanating from Paris, Cologne, and Prague.

At this moment, a pupil of the Gaddi, Cennino Cennini, described the art of painting according to workshop practice in a treatise which achieved great popularity among the Nazarenes and the Pre-Raphaelites of the nineteenth century. Cennino was a pupil of Agnolo Gaddi, the son of Taddeo who was held at the font by Giotto and who remained his pupil for twenty-four years. He proposed to describe 'the most complete art that any man ever possessed' and which had led painting 'from Greek to the Latin manner'—to the art of Giotto. But what he in fact provided were the studio recipes, the manner in which colours were ground, the manufacture of bole, the tinting of paper, the preparation of grounds for panels and for fresco, the proper methods for painting different kinds of beards and draperies, all precious information on the workshop practice of the times. For art itself, the counsels were quite straightforward: choose the best master, copy his drawings, and naturally, respect nature; for instance 'If you want to acquire a good style for mountains, so that they look natural, take some large stones, rugged, and full of cracks, and copy them. . . .'

A host of artisans were trained according to these precepts; all over Italy each town contained several flourishing workshops. There was one field in which the Italians, particularly the Tuscans, were past masters, and this was mural painting—work in fresco which required the readiness of hand and the established patterns that Cennini codified. This resulted in a process of 'going one better'; the novelties of Western art were scanned, so as to produce a more brilliant version. Around 1400, Tuscany, led by Siena, was in touch with the whole of the West, and the final development of the fourteenth century comes at the same time as that commonly called International Gothic.

The pre-eminence of Siena was established in Florence with Lorenzo Monaco, a Camaldolese monk, who came to the convent of S. Maria degli Angeli, and worked for the Carmine and then for S. Maria Novella (1404–25). The Sta Trinità *Annunciation* is exquisite, the *Adoration of the Magi* (Uffizi) is all elegance and fantasy, but there is strength in the *Coronation of the Virgin* of 1413 (Uffizi).

Lorenzo Monaco's charm owes much to his work as a miniaturist, and his ability to extend to a whole composition the brilliance of colour proper to illumination finally seduced Florentine artists: Fra Angelico was to be his pupil. This taste for the exquisite, for vibrant silhouettes, for lively colour, explains the popularity in Tuscany of the Northern master who summed up all the Gothic preciousness, with its love of jewels and pearls: Gentile da Fabriano. His *Adoration of the Magi* (1423) painted for Sta Trinità, conceived as a strange and overloaded evocation of the scene, marks the summit of this art, as delightful as it was inconsistent. In Florence it was soon to meet with increasing hostility, but sufficient survived to affect the style of Masolino when he decorated the Baptistery of Castiglione d'Olona in 1435. But the place which was really to become the refuge of Gothic sensibility, with its imaginary landscapes, its elongated and weightless figures, its compositions set out on gold ground or against flowery trellises, its graceful or grimacing saints draped in exquisite stuffs, was Siena. In the middle of the fifteenth century, Ghiberti records that the Sienese considered their major artist to have been Simone Martini, and his *Annunciation* was still being imitated. In an almost unaltered atmosphere of devotion and simplicity appeared two major exponents of Sienese grace: Stefano di Giovanni, or Sassetta, a master of sinuous figures, who painted the series of the *Life of S. Francis* (panels in London, Chantilly, and the Berenson coll., Settignano) between 1437 and 1444, and Giovanni di Paolo, who set his *Madonna of Humility* (c. 1436) in a flowery orchard, curved about her, and surmounted by an imaginary Tuscan landscape all divided into rectangular fields.

Manuscripts and 'Tavolette di Biccherna'. The art of the anonymous painter of the charming manuscript of S. George (Vatican), illuminated for Cardinal Stefaneschi about 1330, derives from the ultimate manner of Simone Martini with such a lively Gothic accent that it has been argued that the painter could have been French. To the same hand must be given the panel of the *Madonna with Saints and Angels* (Louvre), of a more Florentine character. The superb frontispiece of the *Caleffo dell' assunta* (Siena Archives), of about 1334–36 is by Niccolo di Ser Sozzo Tegliacci (?–1363), who was less assured and inspired in the polyptych painted in 1361 with the rather inadequate help of Luca di Tommè (Siena, Pinac).

Dante's *Divine Comedy* inspired a great number of illustrated manuscripts, but with very deceptive results: Inferno is usually more fully treated than Paradiso and Purgatorio. Generally, the scenes are limited to small details, scattered weakly over a ground sown with flowers or stars (MSS. 1387, Bib. Naz. Florence). The Northern miniatures, like those of the Venice Manuscript, the illustration of which is attributed to the School of Altichiero (*c.* 1400), make the figures larger, but the result is still mean.

Florentine miniatures have more charm in their evocations of rather down to earth and precise views of daily life and its setting, as in the *Biadaiolo* illustrated between 1320 and 1330 (Bib. Laur., Florence). Liturgical works received rather laborious ornament of a Giottesque character, in particular those emanating from the workshop of Jacopo del Casentino (*Antiphonal*, mid-fourteenth century, Basilica of Impruneta). The end of the century saw the pre-eminence in this type of work of the scriptorium of the Camaldolese convent: a certain robustness, born of a distant Giottesque inheritance, is added to the Sienese elegance which comes more and more to the fore with Don Simone Camaldolese, and above all with Lorenzo Monaco (*Choral*, 1409, Bib. Laur.).

One of the finest collections of Sienese miniatures is to be found in the records of the Biccherna. The name goes back to the middle of the thirteenth century and is derived from the seat of the magistracy of the city finances, which, being renewed every six months, involved a regular compilation of registers. These were given wooden covers which—at least after 1259—were decorated with paintings: scribes, magistrates, scenes of commerce, allegories in the spirit of Lorenzetti's (those of 1340 and 1344), religious scenes such as the Circumcision in the manner of Lippo Vanni (1357), or political ones, such as the coronation of the Emperor Sigismund, by a pupil of Giovanni di Paolo (1433). The series, which was continued until the eighteenth century, and which was often the work of men close to the really great names, bears admirable witness to Siena's fidelity to a style of painting in bold colours, red and gold, in which everyday life is heightened to the symbolic level.[73]

Sculpture and Goldsmiths' Work: Florence. Filarete, in about 1460, was to condemn Gothic art by declaring it to be an architecture of goldsmiths and painters. The decline from the monumental to the precious which characterizes the fourteenth century, and such an example of this as Orcagna, fully justifies this judgement. Sienese and Florentine goldsmiths were then, in fact, at the height of their powers and the same style sufficed for the chapel and the reliquary.

The *Bolsena Corporal*, made by Ugolino di Vieri (1338) is in the form of a miniature cathedral with three gables, decorated with narrative scenes on enamel plaques. Another reliquary, that of S. Juvenal also at Orvieto, is hexagonal in form; a third, that of the *Skull* of S. *Saviniano*, in gilded copper, with rock crystal and enamels, the work of Viva and Ugolino (second half of the fourteenth century) is also important for its precious architecture.

Tuscan goldsmiths' work took over the programmes of sculpture: between 1368 and 1372 two silver-gilt life-sized busts of SS. Peter and Paul, of an impressively powerful energy, were made; the same applies to the bust of S. *Zanobi*, decorated with translucent enamels, the invention of which is attributed to Italian workshops.

Among its specialities are, finally, the altar-front (or *paliotto*) and the complete retable—that in S. Giacomo in Pistoia is by Andrea d'Ognabene (thirteenth century), completed in the fourteenth century. Works of this kind are to be found even in the provinces, and the cathedral of Ascoli Piceno possesses a *paliotto* with twenty-seven scenes, framed in foliage, which is a provincial imitation of Tuscan types.

But the great *Silver Altar* of the Baptistery in Florence, commissioned in 1366, must be considered on quite another plane. Only eight panels were completed by Betto di Geri and Leonardo di Giovanni, the four side reliefs being added a century later by Antonio Pollaiuolo, Verrocchio and two other goldsmiths. It is a large aedicule in which, under a heavy horizontal cornice, pilasters decorated with niches and gables frame the large panels of scenes from the Life of S. John the Baptist. This compendium of Florentine virtuosity and minuteness blends without excessive contrast the soft and highly detailed Gothic manner with the more lively qualities of the Quattrocento.

After Orcagna, Gothic traditionalism reigned supreme in Florentine monumental sculpture, particularly in the workshops at the cathedral, where the façade was being ornamented with statues by Giovanni d'Ambrogio and others, on Agnolo Gaddi's designs. At Or San Michele the niches were awaiting the large statues which the Guilds had undertaken to provide; the upper storeys of the campanile were also awaiting the large statues of the Prophets. It was not until 1415–30 that new personalities came on the scene. The best established workshop was that of Niccolo Lamberti, who made a S. *Luke* for the Guild of Judges at Or San Michele (1403–6), and a S. *Mark* for the cathedral. Later he went, with Piero di Niccolo, to Venice, where he worked on the façade of S. Mark's, leaving in Florence Nanni di Bartolo

33 Andrea Pisano:
(a) detail from the bronze doors of the Baptistery, Florence
(b) *Hercules*, relief from the east wall of the Campanile, Florence

34 Jacopo della Quercia: Tomb of Ilaria del Carretto

(*Lucca Cathedral*)

35 Giovanni di Paolo: *The Presentation of the Virgin*
(*Siena, Pinac.*)

36 Jacobello del Fiore: *S. Michael*

(*Venice, Accademia*)

37 Niccolò di Ser Sozzo Tegliacci: *The Assumption*
(frontispiece to records of the City of Siena)

38 Lorenzo Monaco: *The Journey of the Magi*

(*Drawing, Berlin Print Room*)

39 Naples, S. Chiara: Tomb of King Robert

40 Milan: east end of the Cathedral

known as Rosso. Rosso collaborated with Donatello on the campanile in about 1420, on the figures of the *Baptist* and *Obadiah*—but he introduced nothing new, any more than did Bernardo Ciuffagni, another later collaborator of Donatello, in his *Joshua* for the cathedral. The only consistent artistic figure is that of Nanni di Banco, who, born in 1380, belonged to the younger generation, but was to die about 1420. He was never able to give the full measure of his abilities, and his work remains rather hesitant. The youthful *Isaiah*, with a square face and solid drapery, made for the cathedral, and the bronze group of the *Quattro Santi Coronati*, made for Or San Michele (*c.* 1410) are impressive and full of the influence of the Antique, but their style is not unified. In the tympanum of the *Porta della Mandorla*, at the cathedral (1414–21) which represents the Madonna in a Glory supported by angels, the handling is more brittle and the interest is dispersed over the whole surface.

Siena. The same tendencies may be observed in Sienese sculpture. The memory of Nino Pisano and of his smiling and graceful Madonnas remained vivid, in particular in polychromed wood sculpture. Numerous churches acquired groups of the Madonna and Child, or of the Annunciation, with large blue or red veils, and stereotyped gestures. These are modest works, and part of a continuous workshop tradition: for instance, the *Annunciation* (1369–70) at Montalcino, the work of a Maestro Angelo, and that by Domenico di Niccolo (*c.* 1400), also at Montalcino. This prolific sculptor, born in 1363, worked until 1450; he owed his fame and his nickname of 'Maestro del Coro' to the marquetry decoration of the chapel in the Palazzo Pubblico, executed between 1415 and 1428. Work of the same type by his pupil Mattia di Nanni, in the same palace (1425), gives an idea of the poor design of these *intarsie.* Among the workshops, so numerous in Siena at the beginning of the fifteenth century, may be mentioned that of Giovanni di Turino, goldsmith and sculptor, born about 1385, who worked on the Font in the Baptistery (1417).

Jacopo della Quercia must be considered apart. In 1407, when in his thirties, he was in Lucca, where he carved, with Francesco da Valdambrino, the tomb of Ilaria del Carretto. The dead woman, in long, smooth draperies, lies on a sarcophagus of which the only ornament is a heavy garland borne by putti, superbly dignified in effect. This energy, which was not a Sienese characteristic, recalls something of the art of Giovanni Pisano; it may also be seen in the remains of the *Fonte Gaia*, the monumental fountain of the town (now in the Palazzo Pubblico) on which he worked from 1409 to 1419, in the figure of *Justice* and the group of the *Madonna and Child* where his style may be seen developing with fullness of invention and a new energy. But a decade has now been reached when sculpture in Florence as in Siena, and even in the West as a whole, was undergoing rapid changes.

The Competition of 1401. In 1401 there took place in Florence a great competition for the second pair of doors for the Baptistery, in which all the major Tuscan sculptors took part: Jacopo della Quercia (born in 1374), Francesco da Valdambrino (1363–1435), both Sienese, and the Florentine goldsmiths Brunelleschi (born in 1377), Lorenzo Ghiberti (born in 1378), and Simone da Colle Valdelsa and Niccolo d'Arezzo who were older men. The young Donatello, born in 1386, could also have competed. This general confrontation must be considered as the end of the Gothic age, to which all the masters in fact belonged. The purpose was to furnish a companion to Andrea Pisano's doors of seventy years earlier, and the winner, Ghiberti, carefully imitated their design, their quatrefoil framing of the scenes, and refined still further Andrea's graceful style and delicate silhouettes. In this work was perpetuated and strengthened the delicate—classic, in fact—tradition of the Middle Ages in Italy. Ghiberti's work was not finished until 1425; the third set of doors were in an altogether different spirit.

The subject set for the competitors in 1401 was the Sacrifice of Abraham. Della Quercia's version seemed, so it was said, to be lacking in delicacy; the energetic style, composed of strong contrasts and clear relief, which was reborn in him, and which was thus rejected, was nevertheless inherent in Gothic, though it had been weakened during the course of the century. The version by Brunelleschi, whose failure was to turn him away from sculpture, displayed an exaggerated relief, a variety of accent, and an absence of movement and action resulting in the heaping up of his figures one on the other. This was the type which was later taken up by Donatello. It was not therefore in a kind of artistic vacuum that the great labours of the next century were to be undertaken, but, on the contrary, in a climate of activity and competition which was maintained during the first half of the Quattrocento, when the main task was to finish the works started in the previous century, and not to make a break with the past.

5. Northern Italy (1350–1430)

PAINTING

Cimabue and Duccio were Tuscan painters; Giotto and Simone Martini were Italian, even Western, painters. Giotto worked at Padua and at Naples, Simone at Naples and at Avignon. Their art awakened interest everywhere, and helped by the rise of the City-States, provincial schools were able to make their own adaptations of the new style. Their development more often than not depended on the arrival of a master and became rigidly fixed afterwards. But the really remarkable thing was the vigour with which, all over an Italy which seemed suddenly aware of having discovered the kind of art that suited her, frescoes of pious subjects were painted on the walls of churches, and civic allegories or galleries of illustrious men covered those of town halls.

The Marches and Emilia. It was probably from the workshop at Assisi that what is now called the Riminese School evolved. After 1307, the *dossale* (or *paliotto*) by Giuliano da Rimini with a *Madonna and Saints* (Gardner Mus., Boston), and, slightly later, the small panels by Giovanni da Rimini (Palazzo Venezia, Rome) give some idea of this original development of the Giottesque, with expressive exaggerations and Byzantine archaisms, this mixture being even clearer in the little anonymous triptych of S. Chiara (1330, Rimini). Pietro da Rimini, who probably painted the large and tortured *Crucifix* at Urbania, decorated between 1330 and 1340 the vaulting of S. Chiara in Ravenna, and a little later, Baronzio, who painted the heavy altarpiece dated 1345 (Urbino), executed frescoes in S. Maria in Porto Fuori, near Ravenna (destroyed 1944). Both endowed their saints with the powerful facial expressions and rather sharp gestures which for long remained a fashion in the Marches. The link with Florentine centres was established in the middle of the fourteenth century with Allegretto Nuzi of Fabriano (recorded 1346— died 1376), whose style is parallel with that of Nardo di Cione and has a highly personal delicacy, for instance in the *S. Ursula Triptych* (1365, Vatican). Sweetness is the main characteristic of the brothers Lorenzo and Jacopo Salimbeni da Sanseverino, at Urbino in 1416; the anecdotic and everyday detail inspired by French art and by illumination was their main concern.

The most interesting centre for this tendency towards realism was Bologna. The illustration of law books and the classics, so brilliant in 1300, never declined during the course of the century. In the second half, Niccolo di Giacomo, influenced by Orcagna, worked in a broad style which can be followed from the 1354 Decretals to the register of the Monte di Pietà of 1394.

But in painting proper between 1330 and 1350 appeared the complex figure of Vitale da Bologna, who is restrained and close to Daddi in his frescoes in Udine, but develops in his *S. George* into a vigorous and almost popular artist. This seems to be the dominant characteristic of the whole school. After 1350 appeared a certain Dalmasio, who also worked in Tuscany, and Jacopino di Francesco de' Bavosi, who painted the intense figures in the *Coronation with four Saints* (Bologna, Pinac.). After 1400, Giovanni da Modena, gifted in the expression of energetic details and even grimacing faces (*Bolognini chapel* in S. Petronio, Bologna), and the group of artists active in the S. Petronio workshop, round off the physiognomy of this uneven and impulsive school, rich in minor artists of originality, but equally distant from the dignity of Giotto and the good breeding of Siena.

Also to be linked with this movement is the lively Tommaso da Modena, who covered the walls of the chapter house of S. Niccolo at Treviso (1351) with forty fairly detailed portraits of Dominican saints. Though Barnaba da Modena (active between 1360 and about 1380) worked in a more restrained manner, yet one still affected by Emilian portentousness, he remained faithful to Sienese and Byzantine charm.

Lombardy. The Alpine regions were naturally dominated by French and Rhenish Gothic. The churches of Piedmont, like S. Andrea at Vercelli, abbeys like Vezzolano, castles like Montiglio, were decorated about the middle of the century in a manner which was easily surpassed in vigour and in pointedness by the frescoes by Giacomino Jacquerio of Turin in S. Antonio at Ranverso. The charming and truly chivalresque series in the castle of La Manta (*the Nine Worthies and the nine Noble Dames,* the *Fountain of Youth* of 1420) must be attributed to Jacques Iverny of Avignon and belong to French art, as do the contemporary figures of saints in the castle at Fenis, near Aosta.

In Milan, where Giotto himself had perhaps been, the development was very slow. Giovanni da Milano is the link with Central Italy. In 1365 he decorated the Rinuccini chapel in Sta Croce in Florence; in 1369 he was summoned to Rome by Urban V. It is difficult to say if Giovanni arrived from Lombardy already primed with knowledge about the style current in Florence and Siena, or if he reacted to their art on the spot and developed the qualities of imagination and chiaroscuro which were then absent in Florence. His Lombard work is even more difficult to reconstruct. Many painters, most of whom are anonymous, in whom Tuscan sobriety is softened by contact with the oversweet style of Paris, decorated abbeys like that of Viboldone in Milan. The interesting result of these contacts is that which is known as 'Lombard Style'—that is, the style of Lombard illumination. This was to become one of the most important Italian contributions to the International style of 1400.

The oldest animal studies known are those in a sketchbook by Giovannino de' Grassi (at Bergamo, *c.* 1360), and after him, by Michelino da Besozzo (*c.* 1400), master of a fine and precise hand, like that of oriental calligraphers. His drawings of dogs, of birds, are of a precision, a clarity of contour, which suggest the natural history specimen; they are as exact and pure as emblems. It is the type which is to be grasped, as in contemporary portrait, and these studies are not conceived as impressions of animal life, but as details to be incorporated verbatim into the larger compositions of miniatures or paintings.[74]

The variety of these animals, the care with which their poses, their plumage or their claws are particularized, are an indication of a new curiosity, and this remains an aspect of Gothic liveliness of mind. Realist details of this type enlivened religious and secular compositions of an even more fairytale character. The success of the Lombard Style spread, about 1400, through the whole of the West, in France and in England, and led to the compilation of Herbals (series of medicinal plants, or *Tacuinum sanitatis*), and to complete zoological plates, though the majority of Italian artists saw in the new material little more than another curiosity to be displayed in accessories and ornaments.

They left to Flemish painters the elaboration of the idea of a natural space inhabited by flowers and birds—that is, landscape. But this final development implies a knowledge of Italian ideas. The great landscape of the *Ager senensis* by the Lorenzetti with its logical richness and wealth of detail, which long remained unequalled, was one of the sources of the Limbourg brothers' inspiration, and the frescoes of the Torre Aquila at Trento, executed about 1390–1400, are based on the theme of the 'months', which was to be their great glory.[75]

Padua and Verona. With Giovanni da Milano, the more solid representatives of monumental painting, derived from the Tuscans, are Altichiero and Giusto de' Menabuoi. About 1380, the latter completely covered the interior of the Baptistery at Padua with frescoes. The imitation of the Arena frescoes is patent in the series of the Old Testament; the *Crucifixion* and, on the vault, the *Paradise*, are of a descriptive fullness the more interesting in that the painter still follows the framework of a Byzantine composition.

In Verona, a painter recorded in 1360, Turone, knew nothing of Giotto; in his *Last Judgement* in the choir of S. Anastasia he merely softens the Byzantine form. In his panel, or *ancona*, composed of many scenes, from the convent of S. Caterina (Verona Mus.) he shows an awareness of German illumination.

The real founder of the local school was Altichiero who, according to Vasari, worked in the Scaligeri palace before settling in Padua, where he collaborated with Avanzo in the chapel of S. Felice in the Santo (1379), and in the Oratory of S. George, which they decorated with vigorous frescoes from the life of S. Lucy. Some frescoes by him, of the *Madonna and four Saints*, have been discovered in the Castelvecchio at Verona. Without the example of Giotto, which he knew from Padua, his style would have lacked firmness; within the strong framework derived from the Tuscan master, he inclines towards a typically Northern realism, a charming inflexion of line, but a profusion of forms.

There was thus established in Verona, towards the end of the fourteenth century, a tender and almost romantic style, encouraged by the spread of Northern ideas, and, in particular, by close contacts with the School of Cologne. The valley of the Adige opened up the road to the North and its fairytale, imaginative art. With Stefano da Zevio, a pupil of Altichiero, space loses all depth as in miniatures and tapestries, and charming motifs such as angels, flowers, fantasy animals, painted in subtle colours, surround the figures as in the *Madonna with the Rose Hedge* painted for the convent of S. Domenico, now in Verona Museum.

These Paradise gardens, full of doves and with wonderful buildings against the horizon, and calling for costumes of exquisite richness, were never more in vogue than in the early years of the fifteenth century. But this precious style no longer served only for pious works, but lent itself also to profane subjects; to princely portraits, for instance, and more often still, it mingled the two in delicate compositions whose principal exponents were Gentile da Fabriano (1360–1428), and Pisanello (before 1395 until after 1450). Gentile, who knew the works of Altichiero, Tommaso da Modena, Parisian Gothic artists and, through Venice, the ultimate splendours of Byzantine illumination and icons, added to the sinuousness of his line the most seductive qualities of delicate blue colour and tender shadows. After 1408, he painted frescoes (now lost) in Venice, and when on his way to Rome to decorate the Lateran stopped at Florence, where he left the two works which permit the fullest evaluation of his charming style: the *Adoration of the Magi* painted for Sta Trinità (1423, Uffizi) and the *Quaratesi Altar* (1425, central panel Royal Collection). In Verona he left an artist of genius to continue his ideas. This was Pisanello, who far surpassed Gentile's highly mannered court art, and elevated this calligraphy of minutely detailed forms into the basis of a great style. The *Madonna of the Quail* (Verona) where the gold and purple brocade, the diaphanous veils, hold the attention as much as the delicate outlines of the birds, is attributed to his Veronese period. The *Angel of the Annunciation* painted above the Brenzoni Tomb in S. Fermo Maggiore is also a kind of pageboy-cum-bird, with eyed wings. In S. Anastasia about 1436, Pisanello decorated the arch over the Pellegrini chapel with a fresco of *S. George delivering the Princess of Trebizond*, which is his masterpiece. It contains everything—fantastic architecture, the fantasies of an intrepid realism and curiosity, a gibbet, Mongols, the rump of a carthorse and, with exquisite exactness of profile, a noble lady dressed in the height of fashionable elegance. From 1420–30 Pisanello was to be the finest representative of Gothic preciosity. But the formal system he used was so carefully worked out and so coherent that it permitted him to assimilate a host of new observations, and there was an aspect of Pisanello which cannot be understood except from the standpoint of new generations of the fifteenth century.

Venice. In Venice, the almost permanent workshop at S. Mark's for long took the place of painting. The first known painter, Paolo Veneziano, whose *Coronation of the Virgin* (Washington) dates from 1324, and who died about 1360, was, as might be expected, orientated towards Constantinople, where he may perhaps have spent some time. He derives from the frescoes and icons of the period of the Paleologi, though with an increasingly Gothic vividness of line and elegance.[76] The *Dormition of the Virgin* (Vicenza), a fragment from a polyptych of 1333, is of a truly Greek delicacy and severe purity, the golden haloes balanced like discs between figures of an exiguous and uniform modelling. The exterior paintings on the *Pala d'Oro* (1345) and the polyptych of the *Coronation of the Virgin* (Venice, Accad.), show a closer approach to the plasticity of the Trecento. Paolo Veneziano was a prolific artist who established a style that dominated his period. Thus it was that Venice again turned strongly towards Byzantium, at a moment when the rest of Italy was undergoing a powerful Gothic influence. This involved ignoring completely what Giotto had achieved in Padua, and not all Paolo's pupils could bring themselves to do that. Lorenzo Veneziano, who is documented from 1356 to 1379, remained, despite his travels, well within the refined, luxurious, golden formulae established by his master, as for instance in the *Vicenza polyptych* (1366). With Guariento, trained in Padua and active there between 1335 and 1365, a more ambitious art penetrated into Venice, when he was summoned in 1368 to decorate the Sala del Gran Consiglio. But it was, basically, only another form of provincialism.

Niccolo di Pietro, recorded from 1394 to 1430, displays, in his *Coronation of the Virgin* (Rovigo), a knowledge of Rhenish art. But Venetian painting still had to be brought into line with the flamboyant, exotic and precious character of its surroundings, and for this purpose nothing could have been more suitable than the International Gothic of Cologne and Siena. Gentile da Fabriano, who came to Venice in 1408, played the vital role; he was the master of Jacopo Bellini, and after him of Jacobello del Fiore (d. 1439), a weak but sensitive artist, and of the exquisite Michele Giambono (d. 1462), whose rich luxuriance, and great subtlety in the use of gold and ornament, seem, for instance in his *S. Crysogono* in S. Trovaso, to have been slowly distilled from all the essences of the Orient.

Gentile established himself as a portrait painter, and he possessed the gift of delineating character, as may be seen in the *triptych with Madonna and Donor* in Berlin. One other aspect of his art is a now closed book, for his frescoes in the Doge's Palace in Venice (1409), the Broletto in Brescia, and the Lateran (begun 1427–28, continued by Pisanello, and destroyed in the seventeenth century) have all disappeared. Well-documented works (already discussed) executed during his stay in Florence, between 1423 and 1427, gave a powerful impulse to the Gothic revival of the beginning of the century. In Venice itself, the seductive quality of this style had a deep effect, and its hold did not weaken upon Jacopo Bellini, nor on Antonio Vivarini, until the middle of the fifteenth century.

SCULPTURE

The inspiration of Gothic travelled into Milan, by about the middle of the century, through the circle of Giovanni Pisano. Giovanni di Balducci, who made, about 1335, the tomb of Castruccio Castracani at Sarzana, completed in 1339 the reliquary (or *arca*) of S. Peter Martyr in S. Eustorgio in Milan. Eight pilasters with statues attached to them, eight narrative reliefs, eight crowning figures, are superimposed in this bulky and placid object which is surmounted by a stone baldaquin, also crowned with statuettes, under which are a Madonna and two saints. Giovanni also worked on the Visconti tomb in S. Gottardo (now in the Castello Sforzesco), and on the reliquary of S. Augustine in S. Pietro in Ciel d'Oro in Pavia (*c.* 1370), which is an even heavier version of his own major work.

His was the principal workshop of the Po valley. Bonino, who came from the small town of Campione, near Lugano, spread his formulae, and the sculptors who worked in his orbit were called the *Maestri Campionesi*. They were particularly successful in Verona, where in 1375 Bonino made the monument to Cansignorio della Scala, a curious structure in which the sarcophagus is enclosed within a huge baldaquin surmounted by an equestrian statue, and Gothic tabernacles with allegorical statues surround the whole, giving it a complicated and romantic aspect. A charm deriving from its evocation of chivalry emanates from the rather crude monument to the Cangrande della Scala (d. 1329) who befriended Dante; seated on his heavily caparisoned horse, he stretches out his sword, true image of the feudal Ghibelline.

In Bergamo, about 1375, Giovanni da Campione worked on the portal of the transept of S. Maria Maggiore, where, twenty years later, he placed a statue of *S. Alexander*. In Milan itself, Giacomo da Campione carved a *Christ in Glory* for the south sacristy of the cathedral.

Another notable family was that of the Masegne in Venice. Based on a Byzantine style brought up to date by Nino Pisano's elegant Gothic (Cornaro tomb), they developed a style which was perfectly suited to complicated constructions, in which marble polyptychs enclose statues of varying sizes within pierced niches. Examples of this may be found in their rood-screen in S. Mark's (1394) and in the altar of S. Francesco in Bologna (1388–92). Pier Paolo, who died in 1403, is perhaps the most delicate, and Jacobello, who died in 1409, the most powerful among them. The two tendencies, which could also be seen in Florentine and Sienese Gothic sculpture, seem also to be present here. Contacts with Tuscany were by no means weakened; Niccolo Lamberti went to Venice and worked either on the façade of S. Mark's or on the Doge's Palace. The *Judgement of Solomon*, a relief on the corner of the palace, is attributed to Nanni di Bartolo, another Florentine associated with Lamberti.

ARCHITECTURE

Towards the end of the fourteenth century, there developed all over Northern Italy and parallel with the Flamboyant Gothic of Western Europe, a more slender style of greater movement and picturesqueness. Thus, for instance, the great university city of Bologna, where the local school of painting had been re-established, wanted a cathedral worthy of its fame. In 1390 the building of S. Petronio was begun, on plans furnished by Antonio di Vincenzo (1340–1402), who had a very active career. The building dragged on interminably and the vaulting was only finished in the seventeenth century.[77] The project was inspired by the cathedral in Florence, where Antonio lived in 1393, but in more Gothic form, with a nave and two aisles sharply differentiated, and amplified by side chapels built between the buttresses, so that the profile of the building became a triangle—a design accentuated in Milan. The octagonal piers are a development of those of Florence cathedral, though not unconnnected, in the arrangement of pilasters and capitals, with local precedents supplied by S. Francesco. By foregoing the dome, the building was given a less classical, less imposing character, which was accentuated by the lightness of the ribs and the profiles.

Lombardy. The Campione masters are to be found working as architects in Lombardy, just as the dalle Masegne are in Venice. This identity of sculptor and master builder was common in Italy at the time of Arnolfo and Giovanni Pisano; it became all the more significant when a certain mannerism of effect became general. Giovanni di Balducci had decorated the façade of the

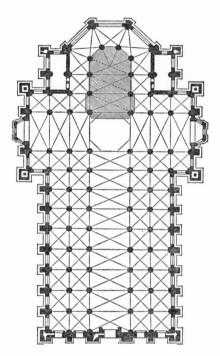

FIG. 15. *Milan Cathedral. Plan (after Dehio and Bezold): Nave and four aisles; Large transept; Polygonal apse.*

now vanished church of S. Maria di Brera with polychromed revetments; Matteo da Campione used this idea again in the cathedral of Monza (1390–1396). The campanile of S. Gottardo (1333) in Milan used, in its upper colonnade and projections, certain Romanesque forms, but to the one in Cremona, the Torrazzo of 1250, an octagonal crowning member, similar to a Gothic reliquary, was added at the end of the century.

Between 1280 and 1380 little seemed to be happening in Lombardy. The relatively true Gothic of S. Andrea at Vercelli (1219) provoked no response. The current practice in the fourteenth century was to graft a Romanesque façade on to a Tuscan structure, as may be seen in the churches of Piacenza, or in S. Maria del Carmine in Pavia. The political revival of the province was due to Gian Galeazzo Visconti (1347–1402), who extended its confines as far as the borders of Venice and Tuscany, and his last years were marked by important foundations, the completion of which occupied all the succeeding century: Milan cathedral, started in 1386, the Certosa of Pavia begun in 1396, the cathedral of Monza, where the reconstruction was started in the same year, as well as the cathedral of Como, but each of these buildings was conceived in a different spirit, and the men in charge were clearly lacking in tradition and experience. The chequered history of Milan cathedral is evidence of their uncertainty, which finally ended in the decision to consult foreign builders and in theoretical discussions made the more learned by their desire to erect a building which should have no precedent in Italy.[78]

The minutes of the discussions between the local engineers and the French and German specialists, who were successively called in, have survived. They make the approach to technical problems quite clear. In 1391, the mathematician Stornacolo had proposed a design based on equilateral triangles enclosed one within the other, which should carry the outermost points of the elevation to a height of 84 *braccia*, but the insistence of the Milanese upon achieving an effect of unusual width was already remarkable. All the Milanese architects of the Renaissance worked on the building, and the construction was extremely laborious. When it was consecrated in 1577 the work was unfinished, the apse and the spire were completed later, to say nothing of the façade, which was planned by Tibaldi in 1567, and finally designed in 1858.

The cathedral was planned with a nave and four aisles, with a strongly projecting transept, and an ambulatory. The side vaulting is sufficiently high to enable the interior space to be opened up with a fine breadth of effect. The piers with their clustered colonnettes are imposing, but at the height of the capitals they are decorated with a coronet of niches and statues which breaks the rising movement. The interior is immense, 148 metres by 87 metres (roughly 480 feet by 270 feet). But despite the effort to push Gothic forms as far as possible, there is a kind of chill about the edifice, above all in its exterior, where the host of small pinnacles and statues do not seem to be imposed by the rather stiff articulations of the building. But the artists of the Quattrocento believed they had built, and Italy long believed herself to possess, a masterpiece of Flamboyant style superior to all the French cathedrals.

The Certosa of Pavia was also built slowly during the course of the century; the monastery was finished first, about 1450, by Bernardo da Venezia and Giacomo da Campione, then after this date the church was built by the Solari, and finally by Amadeo. The delay permitted the façade to be developed as a sort of reliquary in goldsmiths' work, after a fashion which was to become general during the Renaissance.

Venice. The end of the fourteenth century saw an even more striking development of Gothic in Venice. The whole town became a masterpiece in a new
form of Flamboyant. The church of the Franciscans, S. Maria dei Frari, in the
western part of the town, took a century to build from 1340 onwards; it has a
nave and two aisles, a transept, and a deep apse with six chapels opening off it.
The ribs of the vaulting rest on powerful cylindrical piers and the façade is of
an extreme simplicity. The brick bell-tower dates from the end of the fourteenth century. The Dominican church, SS. Giovanni e Paolo (better known
locally as S. Zanipolo), on the east of the city, is as spacious, and has slowly
been decorated with numerous grand tombs. Though the church was built
in 1390, the façade is still unfinished. In the northern quarter of the city, S.
Alvise, built in 1388, has a single nave, and the Madonna dell' Orto, built
during the fifteenth century, has a nave separated from the two aisles by
columns; both stress the Gothic in their decoration, the profile of the arches
and the type of the capitals.

In 1419, the façade of S. Mark's had to be rebuilt because of a fire, and, as
has been mentioned, the Tuscan and Lombard workshops indulged, in the
upper part, in a profusion of pinnacles and Gothic aediculae, which then seemed
to be the last word in decoration. But it is in the palace façades that this late
Gothic, brought to the highest pitch of ornament and amplified by the colour
and light usual on the lagoons, is displayed at its most brilliant. The most
beautiful effects of the whole late Middle Ages were achieved by means of this
profound and easily predictable harmony between the Flamboyant, which
denied the existence of the wall, and metamorphosed all the surfaces into
arabesques and rhythms, and the Roman and Byzantine tendency to abolish
space so as to create a lively and scintillating setting. The luxurious aspect of
maritime cities like Pisa and Amalfi, where the oriental type of interlaced
arches were to be found, is, so to speak, revived in Venice and enriched by
two centuries of varied conquests. The Fondaco dei Turchi, which, with its
double colonnade, goes back to the thirteenth century, the Palazzo Pisani
near S. Maria Formosa, and the Palazzo Foscari, on the Grand Canal, still
present ancient features. The building which best expresses the flowering of
the more theatrical aspects of Venetian Gothic is the Ca' d'Oro, built between
1421 and 1440 for Marco Contarini, by Marco d'Amadio and the Bon
family. The entrance and the double colonnade were originally picked out in
gilding; the courtyard contains an exterior staircase and a wellhead by
Bartolommeo Bon (1427).

At the Doge's Palace, designed at the beginning of the fourteenth century,
the front part was built during the second half of the century, and the wing
on the Piazzetta was built when the old Romanesque palace was demolished
in 1424, to permit of the building being constructed all in one style. The
eastern wing dates from the end of the fifteenth century. There again, the
new was in perfect harmony with the traditions of the high Middle Ages. The
seeming paradox of placing over a colonnade surmounted by an open gallery

an upper storey the plane surface of which is pierced irregularly by windows—that is, a solid surface over voids—is explained above all by the relationship of colour. The arrangement forms, in effect, a pink and ochreous band over a double black dominant. And upon this dazzling harmony is placed the elaborate balcony for the Doges, grafted by the Dalle Masegne on to the façade facing the sea, as if to add the final touch of goldsmiths' work to a composition which is mainly the creation of a painter's mind.

Part Two

THE RENAISSANCE

I

The Quattrocento

Introduction: Italian Culture and the Northern World

At the beginning of the fifteenth century, Italy became more and more involved in the politics of the West. After the end of the Great Schism (1378–1417) which followed the Avignonese exile, the Papacy returned to Rome for good. The restoration of the city was undertaken under Nicolas V, who inspired the brilliant and rewarding jubilee of 1450. This was followed by the methodical creation by Sixtus IV of a great State, which was to play a major role in the wars and intrigues of the end of the century. Around Rome numerous lesser powers were established and confirmed. Florence was dominated by the banker Cosimo de' Medici, master of the Signoria after his return from exile in 1434 until his death in 1464, then by Piero (1464–69) and the fabulous Lorenzo (1469–92) whose adroit policy brought peace to Italy. In Milan the Visconti, who became dukes in 1395, were ousted in 1447 by a great *condottiere*, Francesco Sforza. Under Lionello d'Este and then Borso, Ferrara became one of the most important creative centres of the period. In Naples, where René of Anjou made a fruitless attempt at reconquest in 1438, the house of Aragon, under Ferdinand in 1458, established itself even more strongly than before. The Venetian empire, gathering together the remains of Byzantium, extended itself as far as Crete and Cyprus (acquired in 1489).

Never had diplomacy been more intelligent and adroit, nor the role of Chancery secretaries more interesting, as is demonstrated by the reflections of a Machiavelli, compared with those of the Frenchman Commines. Violence finally became rather rare, though it was atrocious when used, as is demonstrated by the way in which Lorenzo de' Medici crushed the revolt in Volterra, and the Pazzi conspiracy (1478). War became the business of mercenary captains who could become the heads of states, either as fantastic tyrants, like Sigismondo Malatesta of Rimini (1417–68), or as just and enlightened princes, like his enemy Federigo da Montefeltro who made Urbino into one of the most interesting and original centres of the period (1444–82).

The taste for splendour, for beauty on a monumental scale, for luxury in clothes and dwellings, was very strong in all these courts, where festivities, foundations, and artistic commissions proliferated, and this spirit of

143

Geneva
DUCHY OF
Chambéry
SAVOY
PRINC. OF PIEDMONT
Turin
Saluzzo
Tenda
Nice
Monaco
Ajaccio
Bonifacio
Alghero
Bosa
Sassari
KINGDOM
OF
SARDINIA
(to Aragon)
Cagliari
Calvi
Nonza
Biguglia
CORSICA
Adda
DY. OF
MILAN
Pavia
Placenza
Sesia
Brescia
Crema
Verona
Asola
Mantua
Territory of Genoa
Genoa
Savona
Carrara
Lucca
Pisa
Volterra
Piombino
Monte Oliveto
REP. OF
SIENA
Elba
Orbetello
Parma
Reggio
Modena
Fornovo
Vicenza
Venice
Ferrara
Bologna
Faenza
Forlì
Cesena
Ravenna
Rimini
Florence
Sansepolero
Arezzo
Siena
Cortona
Pienza
Perugia
Orvieto
Viterbo
Rome
Palestrina
Terracina
Gaeta
Naples
Ischia
Capri
Amalfi
CADORE
FRIULI
Aquileia
Fiume
ISTRIA
Pesaro
Fano
Senigalia
Urbino
Ancona
Loreto
Assisi
Spoleto
Abbruzzi
Pescara
Chieti
Subiaco
Pty. of
Benevento
Foggia
Malfi
Basilicata
Potenza
Outer Calabria
Policastro
Save
Drave
OTTOMAN
EMPIRE
Zara
DALMATIA
REP. OF RAGUSA
Adriatic
Sea
KINGDOM
Capitanata
Vieste
Manfredonia
Bari
Taranto
Otranto
Rossano
Crotone
Squillace
NAPLES
Lower Calabria
Tyrrhenian
Sea
Palermo
Messina
Reggio
Trapani
Marsala
KINGDOM OF
SICILY
(to Aragon)
Girgenti
Catania
Syracuse
Ionian
Sea

Papal States
Republic of Florence

competition enabled Italy to make great progress in the arts. The armies of the great European powers—France, the Hapsburgs, Spain—which met in the peninsula from the invasion (1494) of Charles VIII of France onwards were to be astonished, and enchanted, by the brilliance of Italian culture. The fifteenth century, in this sense, truly represented the first Renaissance.

But the crisis in the Church was becoming more obtrusive. Projects for reform were mooted at the Council of Constance (1414–17), then at the Council of Basle (1431–49), and the monarchical organization of the Church was questioned. With Pius II Piccolomini, the humanist who was Pope from 1458 to 1464, the union between the Holy See and the modern currents of art and culture became closer. The important fact also explains the growing denunciations of the Roman hierarchy and its 'scandals' by Italian and then by foreign preachers. Efforts to achieve, in extremis, the union of the Greek and Latin churches, only resulted in an illusory agreement at the Council of Florence (1439), and finally failed, despite the pressure exercised by the Infidels who conquered Constantinople in 1453, and were only held at the gates of Hungary by the energy of a great prince, Matthias Corvinus, a powerful adherent of Italian culture (1458–90).

The problems of the Papal States, their defence and growth, disproportionately preoccupied Sixtus IV and Alexander VI. It would be wrong to conclude from this that there was any weakening of Christian faith. Never had the religious orders been more active; in Siena, the Franciscan S. Bernardino (1380–1444), in Florence, the Dominican S. Antoninus (1389–1459) who became the city's archbishop, exerted an immense authority and, under Savonarola, Florence was to experience a revival of religious enthusiasm. Confraternities became numerous in all cities; even prominent citizens joined them. Nobles, merchants, bankers, all preoccupied with worldly pleasures and luxuries, were extremely troubled about their salvation. In Florence, the humanists of the old school, pupils of Salutati, Niccolo Niccoli and Leonardo Bruni, were primarily moralists and men of learning who discovered among the Stoics of Antiquity arguments in favour of a wise and noble life. The feeling that there existed a profound harmony between the teaching of the past and Christianity was to become the mainspring of the more ambitious movement initiated by the Neo-Platonic humanists of the time of Lorenzo de' Medici. The Platonic Theology (Platonica Theologica) of Marsilio Ficino (1482) is a new apologia; the mysticism of Plato and Plotinus rejoins naturally, through S. Augustine and Dante, the Christian ideal. Great poets were the equal of theologians. As a result of the enthusiasm which raised the most varied minds to a higher level, moral conflicts began to appear, and the love of beauty played a new role in speculative philosophy. During the thirty years up to the death of Lorenzo (1492) and the French invasion (1494), there reigned in Italian circles an exalted confidence in man and the completeness of his powers, without which it would be difficult to understand the appearance of classic art. At the same time, literature in the vernacular,

THE QUATTROCENTO

encouraged by such potentates as Lorenzo de' Medici, himself a poet, and Alfonso of Aragon, achieved a delicate beauty with Poliziano's unfinished stanzas for the *Giostra* (composed between 1475 and 1478), and Sannazaro's *Arcadia* (composed between 1481 and 1496) gave a new depth to the mythological pastoral. But the love of Antique fable did not preclude a marked return to the epic themes of chivalry in the *Orlando Innamorato* by Boiardo in Ferrara (1476–84), and of burlesque type in the *Morgante* by Luigi Pulci published in 1483 in Florence, to say nothing of the popular vein of storytellers such as Burchiello or Lorenzo.

The fifteenth century was a period of anxiety and not of scepticism, of intrepid curiosity and not of revolt. The new authority of secular culture did not work for the deliberate ruin of tradition but to exploit all fields more vigorously: it was the humanist Lorenzo Valla who wrote the epitaph for Fra Angelico. The aspiration towards a renewal of mankind implies a moral reform and a re-orientation of knowledge. Philological and scientific problems dominated Northern circles, with the great teacher Guarino of Verona (1370–1460), the traveller and archaeologist Ciriaco d'Ancona (1391–1455), learned epigraphists such as Felice Feliciano of Verona, storyteller, poet and friend of Mantegna,[79] and the Paduan mathematicians and astrologers. In Tuscany, where philology penetrated deeply with Poggio and the Greeks of the 'Studio', Leon Battista Alberti (1404–72) added to it a new dimension by writing definitive treatises on Painting (1435), on Sculpture, and, finally, a great work on Architecture (written 1450–72, published 1485) very freely adapted from Vitruvius. The commentaries of Ghiberti appeared at a moment when Filarete and Francesco di Giorgio were writing their manuals on architecture, and Piero della Francesca was preparing a treatise on perspective. The fifteenth century was the age of theoretical speculation: in Mantua with Mantegna; in Florence in the workshop of Verrocchio, from whence came Leonardo; in Milan with Bramante. The artist-theorists constructed on original foundations a discipline, rivalling that of the old 'liberal arts', in which the best of the lessons of Antiquity were assimilated.

The example of countries north of the Alps, condemned under the polemical description 'Gothic', soon ceased to have any effect in architecture, but retained its importance in sculpture and painting. To the Italian revival corresponded, in fact, the remarkable movement which began in Burgundy with Sluter (Champmol, 1411) and in Flanders with Jan van Eyck (the *Ghent Altar* 1425–32), which turned its back on the softness of International Gothic and found an echo in Italy. Thus the Florentine Ghiberti praised a Cologne sculptor 'di excellentissimo ingegno' (whom an anonymous Florentine, author of a series of illustrious artists, defined later as 'called Gusmin'), who died in the time of Martin V (d. 1431): he was, Ghiberti said, a goldsmith, author of elegant statuettes worthy of Greek art, with particularly expressive heads.[80]

Van Eyck was known and appreciated: the Cardinal Ottaviano owned his

146

Women bathing, which was singled out for praise by the humanist Fazio (1456), chronicler of the Neapolitan court, which here played a vital role. Three memorable visits throw into still sharper relief these interconnections of about 1450: that of Rogier van der Weyden to Ferrara, that of Petrus Christus, who made known van Eyck, to Milan,[81] and that of Jean Fouquet to Rome. The results of these exchanges are to be seen in the work of Antonello, through whom the history of Italian painting is linked up with the development of Western art as a whole. About 1480, other points of contact were established through the arrival of Joos van Wassenhoven (Justus of Ghent) at Urbino, the arrival of the Portinari altar by van der Goes in Florence, and the knowledge of the works of Memling, which were important for the workshops of Verrocchio and Ghirlandaio.

It is in its general development that the two facets of the Italian problem must be understood. Firstly, between 1420 and 1440, Italy, through Florence, established a reaction against the flowery and courtly style of 1400, and showed itself able to counterbalance, in every field, the art of the northern countries. Secondly, the later generations consolidated, gave new inflexions to, occasionally even contradicted, the new basis of this great movement, but the advances they made sufficed to assure to the peninsula a clearly dominant position in the West.

Art in Florence (1420–1470)

PUBLIC COMMISSIONS

At the beginning of the century, the Florentines were less concerned with the Palazzo della Signoria and its surroundings than with the second centre being created round the cathedral (still unfinished, despite the decisions taken in 1368), the campanile of which the niches were still partly empty and the lower medallions incomplete, and the Baptistery which still had only one set of bronze doors, those by Andrea Pisano of 1336. Between about 1400 and 1425, a great deal of work was undertaken round these main points. The new era opens in 1401 with the competition for the second set of Baptistery doors. For twenty years, from 1403 to 1424, they gave Ghiberti the control of an important workshop, through which passed many artists, bronze workers, and decorators, or even future painters such as Uccello. Through the third pair of doors (1425–52) Ghiberti would have retained until his death a sort of official primacy had it not been contested and finally wrested from him by the alliance of his two rivals, Brunelleschi and Donatello. At the cathedral itself the small workshop concerned with the Porta della Mandorla, on the north side of the cathedral, is of particular interest; it began operations in the last years of the fourteenth century, and the artists who worked there achieved a more energetic kind of stone carving than Ghiberti's delicate art. They probably included Jacopo della Quercia for the small figures on either side,[82] Nanni di Banco in the upper tympanum (1414–21) in which a certain prettiness is discernible, and the young Donatello, who executed a vigorous small figure of a prophet for it (1406–8).

During these years, a good many statues were commissioned, either for the duomo or the campanile. Nanni di Banco's *Isaiah* (1408), and the first *David* by Donatello (1408–16), with the drapery strongly accented on the hip, were destined for the cathedral, as were the *Prophets* executed by Bernardo Ciuffagni with Nanni di Bartolo (Il Rosso), who collaborated with Donatello. The last named carved a splendidly majestic *S. John the Evangelist* (1415), and finally, on the campanile, where he was assisted by Il Rosso who did the *Abraham* (1421), he filled the niches with the powerful figures of *Jeremiah* (1423–26) and *Habakkuk* (1427–36).

At the same time, the work on Or San Michele was brought nearer to completion. Nanni di Banco composed in a hemicycle the group of togate figures of the *Quattro Santi Coronati* (c. 1410), and competition stimulated the sculptors. In 1414, the figure of the *Baptist* by Ghiberti was placed in a typically Gothic niche; in 1419, his *S. Matthew* was placed in a more regular setting, with the pointed arch surmounted by a triangular pediment. Donatello, who had already created his *S. George* (1416), and *S. Mark* (1421), provided, with his customary decisiveness, the final formula: in 1423, the setting for *S. Louis of Toulouse* (later replaced by Verrocchio's group of the *Incredulity of S. Thomas*) was conceived as a small façade framed by two Corinthian pilasters, with a shell niche, flanked by Ionic colonettes.

Though the completion of the cathedral dome was the stage most gratifying to Florentine pride, the new era opened by bringing to a lucid conclusion the projects conceived by the fourteenth-century masters.

It was Brunelleschi who solved the problem bequeathed by the Gothic master builders. Amid continuous intrigues and arguments he contrived, in 1420, to get his project adopted, and the work on it done between 1421 and 1436, thanks, according to his anonymous biographer, to the insistence and support of his friends, who were men of culture—that is, humanists. It is a commonplace that the plans submitted in 1366 and accepted in 1367 defined the shape of the dome; it appears that these specifications merely excluded the use of visible buttresses and recommended the use of invisible chains. This was what Brunelleschi achieved, by giving the dome and its octagonal drum a new shape. It is not necessary to accept that the new model of 1419 was the combined work of Brunelleschi, Ghiberti and Battista d'Antonio, as might be supposed from the confused discussions recorded by the documents. From 1423 onwards, Brunelleschi was designated as *inventor et gubernator cupolae*, and this cannot only be due to the triumph of a faction. Ghiberti was retained on the work, but Brunelleschi, particularly after 1426 when the preliminary works were being finished, had the supreme control.

The dome was conceived as a double shell, the walls of which were joined together by transverse links, supported by powerful internal frameworks, some in brick, and the others—probably the most important—in stone, to which correspond the strongly projecting ribs of the outer shell. It was thus possible to dispense with cumbersome scaffolding, though wooden supports seem to have been used to test the resistance of the materials as the dome went up. Among the practical inventions which accompanied the work may be mentioned the use of a lifting tackle.[83] Thus the dome was raised without

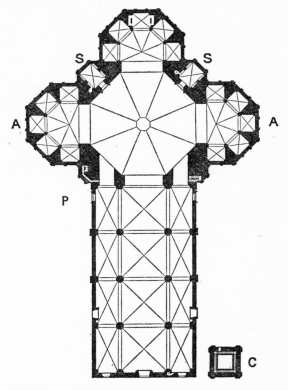

FIG. 16. *Florence Cathedral. Plan (after Dehio and Bezold): A, Apsidal structures; S, Sacristies; C, Campanile; P, Porta della Mandorla.*

exterior supports and without visible framework, to the amazement of the people.

Arnolfo's project had been much enlarged; the final diameter of the dome was $138\frac{1}{2}$ feet, and its relation to the choir which it surmounts, and the nave, had been modified. Outside, the ovoid profile, its eight facets separated by heavy ribs, is on a scale which is no longer Gothic, yet it has nothing in common with the massive domes of Roman architecture. Its volume, broken up at the base by the drum with its huge circular windows—the *oculi*—develops with suppleness on rigid ribs, in a new blend of energy and elegance. But the great beauty of the dome is in its setting within the townscape, not only in the attraction which it exercises on the perspective of streets and roofs, but in the manner in which it completes, in the centre of the vast bowl of the Arno valley, between the hills of Fiesole and the slopes of Galuzzo, the harmony between the architecture and its site. When Brunelleschi died, the lantern was still lacking, but he had designed it, pierced with arches and provided with

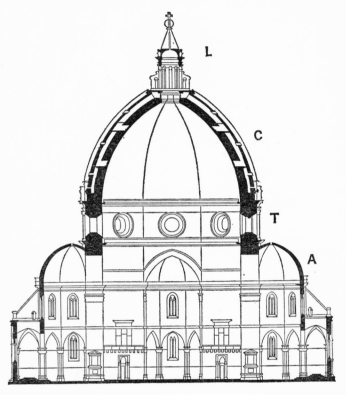

FIG. 17. *Florence Cathedral. Section at the transept (after Dehio and Bezold): A, Apsidal structures; T, Drum with oculi; C, Dome with double shell; L, Lantern.*

complicated members, and it was executed by his successors. The gallery at the base of the dome was begun in 1507 by Baccio d'Agnolo, but was so criticized that it was never completed. Also remaining unfinished was the façade of the cathedral. A fair number of statues commissioned between 1410 and 1420 were destined for it. After the dome was built, the question became more pressing, but local passions did not make for simple solutions. When, in 1490, Lorenzo de' Medici opened a competition for this façade, more than forty contestants appeared: painters, goldsmiths, carpenters, smiths, heralds, musicians. Everyone wanted to have a say; the experts hesitated; nothing was done.

This atmosphere peculiar to Florence, where, as Donatello said, criticism was more valuable than praise, explains the appearance of literary publicity. About 1450, Ghiberti began to make public the memoirs in which he justified his art, as that of true culture, and asserted his participation in all the important works of the period. The humanist clan maintained the glory of Brunelleschi by amusing and popular tales, such as the one about the fat carpenter, and by a life, attributed to Antonio Manetti, which placed the architect at the beginning of a new age. Landino, in his preface to his edition of the *Divine Comedy* (1481) stressed the eminent position of Brunelleschi among the artists who had oriented the new culture.

ARCHITECTURE

The Initiator: Brunelleschi. Filippo Brunelleschi (1377–1446), who started life as a goldsmith and sculptor, only became an architect when he was nearing his forties, but he then had, for the next quarter of a century, an exceedingly active career which completely altered the history of Western architecture. At the cathedral in Florence he solved brilliantly the problem set by the Tuscan master builders of the fourteenth century; everywhere else he was an innovator. In 1418, in the Palazzo di Parte Guelfa, he used a fourteenth-century base as a podium for an upper storey enclosed within pilasters, and pierced regularly by high round-headed windows surmounted by *oculi*. The Foundling Hospital of the Innocenti of 1419 has a clear internal arrangement with two cloisters, and the façade consists of an arcade nine bays long. Carried on thin columns of perfect regularity (the curve of each arch surmounts a square of which the columns represent the sides), this arcade is framed between two pilasters set one either end, and a long horizontal band running above the arches. Circular medallions decorate the spandrels. Thus nothing escapes the overall design. A clear arrangement is perceived in which everything is resolved into measurable relationships. The volumes are used as a surface articulation, the surfaces as an extension of the lines governing the composition, their tension creating a new kind of beauty. This is fully set out in his two masterpieces in S. Lorenzo—the old Sacristy begun in 1420, and the church proper begun in 1421 and finished about 1429. The sacristy is conceived as an hemispherical dome, supported on a cube by four pendentives, which is a rationalization of the Gothic arrangement of a square room with a vaulted crossing, as in the Spanish chapel at S. Maria Novella (1350). The architectural elements are made extremely precise—the arches are clearly defined, the angle pilasters support entablatures which clearly stress the horizontals—and, as remarkable and rich in consequences, the lesser elements, such as the frames of doors and windows, are inserted into the composition with great strictness, introducing a human scale and repeating motifs

151

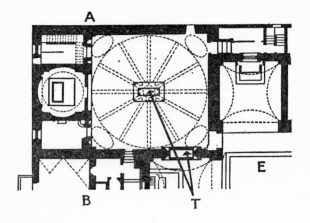

FIG. 18. *Old Sacristy, S. Lorenzo, by Brunelleschi: a, Plan with projection of dome: E, Transept of the church; T, Medici tombs; b, Elevation taken at AB (excluding the dome), showing the 'harmonic' relation of the parts.*

analogous to the principal elements, and creating in each of the sections in which they are placed harmonic sub-divisions attuned to the overall effect. But in order to create this series of harmonies, founded on a uniform module, it is necessary to dispose of simple elements, themselves subject to strict rules and as rational as possible. Hence the methodical use made by Brunelleschi of well-known elements of Antique architecture: columns, fluted pilasters, pediments, cornices with varied profiles. It was during a long visit to Rome, made with Donatello after his defeat in the 1401 competition, that the architect undertook a methodical study of the resources of Roman art, and built up a repertory of the forms which he needed.

But it would be wrong to attribute to him an archaeological purpose. Not only did the religious and secular building programmes have nothing in common with Roman temples or baths, but the use he made of these forms has nothing Roman, nothing Antique, about it. The purity and elegance of Brunelleschi retains, in this sense, something of Gothic, through its insistence on lines, the articulation of surfaces, the clearly defined ribs, the tension between the main members, the minimizing of solid surfaces. But the types of his forms are derived from Romanesque art, from examples of the eleventh and twelfth centuries, such as SS. Apostoli, S. Pietro in Scheraggio, S. Miniato (churches which are sometimes grouped under the name Tuscan Proto-Renaissance)—such forms as the hemispherical arch, the regular alignment of colonnades, the geometrical clarity of volumes, the panelling of surfaces. The first great example of this orientation was the basilica of S. Lorenzo, with nave and two aisles, a wide transept and a flat choir; a wooden ceiling, as in early Christian churches, covers the central space, a dome is raised over the crossing. The return to fourth- and fifth-century forms, which was one of the main preoccupations of the Quattrocento, was achieved by using the precedents set by Romanesque art, which, rightly or wrongly, was always thought of as belonging to 'good architecture'. Only it was no longer the massive walls, the infrequent openings of Romanesque art which counted, but the ideal of a uniform space, limited and ordered, in contrast to the fragmentation, the extending of forms, the interminglings of Gothic. Brunelleschi accentuates unity, coherence, serenity. Thus, the lovely sequence of arches borne on Corinthian columns is framed by pilasters at its extremities, and by a continuous horizontal entablature which supports the windows.

Brunelleschi's assurance came from his having evolved, through perspective, a means of reducing a building to a mathematical rule, and of isolating the interplay of relationships of which it is constituted, and which forms its 'idea'. He had translated the squares round the Baptistery and the Palazzo Vecchio into perspective models; in the same way, each new building was defined with a hitherto unsought clarity. As a subtle organism, an harmonious body, the building presents itself unequivocally as a work of art. The charming little Pazzi Chapel next to Sta Croce (1429–46), with its unfinished façade (a triangular pediment is missing), shows the artist's virtuosity in the use of slender Corinthian columns on the exterior, with dark, *pietra serena* pilasters of the same order inside to stress the main lines and the tense framework of the building. A smaller edition of the volumes of the Old Sacristy of S. Lorenzo is provided by the entrance portico, the Corinthian columns, the elegant framings, with their wealth of ornament which even use terracotta medallions to increase the effect. Thus in Brunelleschi's discretion and control flowers a grace which gives an Attic character to the new style.

Brunelleschi also found time to work on other novel ideas. At the same time as he developed the basilical plan, he considered re-establishing the central plan so dear to Early Christian art.

The church of S. Maria degli Angeli (after 1434), which unfortunately remained unfinished, was octagonal in form and with sixteen sides on the exterior, delimited by articulated pilasters; deep niches were hollowed out of the walls, creating a double apse within each cell-like chapel. An umbrella-shaped dome, like that of the S. Lorenzo sacristy, surmounted this composition full of movement, which, had it been completed, would certainly have been full of striking contrasts of light and shadow. Sometimes another visit to Rome, in 1433, has been suggested as an explanation of his interest in the niches and shafts of light which gave new importance to the wall. But allowances must also be made for a particularly rich imagination, which had been able to free itself from the rather dry simplifications of his early years and now tended to greater expressiveness, paying less and less heed to outside considerations in its wholehearted pursuit of an entirely renewed art, and imposing a sublime concept of art upon a subjugated client.

In his old age Brunelleschi enlarged enormously the domain of pure architecture. The treatment of the lantern of the dome, as a kind of gigantic reliquary, designed to hang in the sky, is evidence of this.[84] In 1444, two years before his death, he began Sto Spirito, on the south bank of the Arno. It is a reworking of his solution given in S. Lorenzo of the problem of basilical form. The dome of the cathedral had meanwhile been finished; the architect wanted to take up again the problem of inserting a transept with a luminous dome into a nave, in a more coherent manner. In Sto Spirito, the transept, built to a proportion of three squares against the four squares of the nave, is linked to it by the continuous line of the arcades, which unify the interior space by multiplying the perspective effects. More remarkable still is the hollowing of the outer wall of the chapels into deep niches, so that a kind of undulation seems to be affecting the wall, which ceases to be a surface contained within fixed members, but achieves an importance of its own, so that the rather even light for the first time suggests a moving and pathetic quality.

Brunelleschi left many unfinished works. His detractors proclaimed that he had conceived more than he had achieved, and his defenders that he had been betrayed by his helpers in the Palazzo di Parte Guelfa, in S. Maria degli Angeli, and at Sto Spirito where, in 1487, arguments were still going on as to how many openings should be made in the façade. But the very violence of the arguments testifies to the speed with which these problems became important.

Popularization: Michelozzo. From a utilitarian standpoint, it was necessary to adapt Brunelleschi's art to the needs of Florentine society. In 1440 he had made for the Pitti palace a project which was still very much a fourteenth-century design, with its air of severity, its rustication, its two storeys of seven windows each; yet it was modern in the vigour of its string courses, the design of the openings, and its horizontal extension. But the project for a palace, presented at the same time to Cosimo de' Medici for the Via Larga, had frightened the patron. Michelozzo (1396–1472) was a friend of Donatello, and had collaborated with him in the niche at Or San Michele (*c.* 1423), in which the triangular pediment borne on pilasters marked a new development in decoration. He had built the little church of Bosco ai Frati in the Mugello, and in Venice had built the Library of S. Giorgio Maggiore (1433, destroyed in 1614) which was paid for by Cosimo de' Medici, whom he had followed into exile. The same patron commissioned (between 1437 to 1452) the rather bold arcaded cloisters, the church with a five-sided apse, and the library of the convent of S. Marco, with its three-aisled design of pleasing clarity. And it was Michelozzo who built, from 1444 to 1459, the huge palace of the Medici (later almost doubled in size to become the Riccardi palace in the seventeenth century). Here is to be found the Florentine type of palace-block, but the rustication, extremely pronounced in the lower storey, becomes less so in the upper storeys, the top one being smooth, while an enormous overhanging cornice surmounts the whole, and gives it an irresistible energy of form. The twin-light windows within a single round arched frame enliven its feudal-looking walls. All the novelty is in the interior courtyard, treated as a square colonnade with three arches on each side, and repeating, above a wide frieze ornamented with medallions à *l'antique* (1452) the same type of windows as in the exterior. This is a kind of lay cloister—an adaptation of the peaceable arrangement of the convent to the needs of town houses. The happy interpretation of volumes is enhanced by the axial perspective which, from the entrance doorway, led across the courtyard and out into the garden at the back, luminous and full of flowers. Michelozzo thus brought to perfection the very type of building essential to the times. The feudal villas in the environs of Florence, acquired by the Medici family, were renovated and modernized by him: Careggi from 1434 onwards; Cafaggiolo in 1451. These villas crowned by battlements look like little fortresses; Michelozzo retained this aspect, kept the machicolations which served as a cornice, and opened in the plane of the walls a balanced system of windows borne on consoles. This sober arrangement admitted only of very slight alterations; on the main façade at Careggi, for instance, a loggia carried by an arcade was added (remade in the sixteenth century). The robust interiors had the courtyard elements added to them, and at Careggi these were cleverly inserted into the defensive walls.

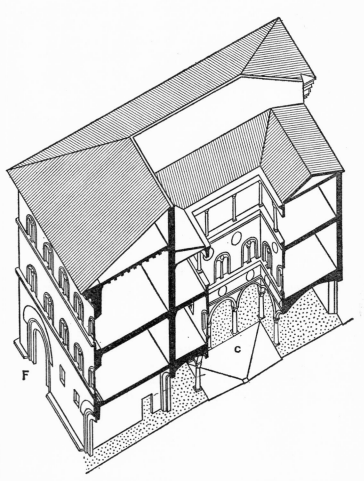

FIG. 19. *Medici Palace, Florence, by Michelozzo. Axonometric section (after O. Morisani): F, Façade on the Via Larga; C, Square courtyard with nine arcades.*

Finally, Michelozzo was asked to construct a similar courtyard in the Palazzo Vecchio (1454; altered in 1565), with a matching upper storey. These villas appear to lack elegance, but their charm lay in their views, their terraces, their plantations, and the shade which surrounded them. The success of this new type of town house was certain among wealthy Florentines. A chronicler records that between 1450 and 1478, more than thirty palaces were built in Florence. All the rich families then had a large town house, and comfortable villas frequently connected with their farms, which were often real manors, in the middle of parklands, like that described by Giovanni Rucellai in his *Zibaldone* (1464).

Michelozzo's more intimate works, like the chapel of the Noviciate in Sta Croce (1445), the Crucifix Tabernacle in S. Miniato (1447-48), and that in the Annunziata (1448-64) executed by Pagno di Lapo Portigiani, betray his desire to achieve, through an abundant and complex decoration of garlands, medallions and friezes, a Donatellesque vigour which results in an overloading of the structure. Brunelleschi's style was thus not always understood nor carefully followed during the middle of the century.

From about 1460 onwards, Michelozzo, together with another Florentine sculptor-architect, Averlino (Filarete) was to introduce the modern style into Lombardy. In Milan, the two-storied Casa Portinari, of which the gateway liberally decorated with motifs is now in the Castello Sforzesco, was built on a pattern repeated in the Palazzo dei Rettori in Ragusa (1464), and above all in the Portinari Chapel in S. Eustorgio in Milan (1462), which is a Lombard reinterpretation, tending towards greater height and more ornament, of the S. Lorenzo sacristy.

Systems: Alberti. With Brunelleschi architecture became the leading art; with Alberti it was closely associated with all the problems of humanist culture. Leon Battista Alberti (1404-72) belonged to a great and rich Florentine family, but he was born in Genoa, lived as a child in Venice, studied law and the sciences in Padua, and did not return to Florence until he was thirty, and then only for short visits, since he was attached to the Papal Court. Alberti was not an architect, but a writer, a theorist, proficient, however, in all techniques. It was only after he was forty that he became interested in monumental art, creating a few works as an accompaniment to the preparation of his great treatise *De re aedificatoria* (written after about 1450 and printed after 1485), which is a modern recasting, a critical interpretation, of Vitruvius.

Alberti succeeded in raising architecture to the rank of the liberal arts by frankly separating the conception, or the mental labour, from the execution, by opening a gulf between the artist and the workman or head of a workshop. The essential is in the idea and in the plans which make it understood. It may be doubted if Alberti ever did more than give directions and sketches; Bernardo Rossellino built for him the Palazzo Rucellai in Florence (*c.* 1446-1451); Matteo de' Pasti executed the Tempio at Rimini (*c.* 1447). To this

THE QUATTROCENTO

intellectualization, which was an essential step, is linked an effort at complete rationalization. Everything in a building may be calculated and analysed; beauty is the absolute value of an aesthetic organism, of which no part may be modified. This beauty illumines the human soul with pure joy, creates an ineffable accord between man and the universe. By mathematical calculation, by the play of proportions, or, in terms borrowed from the *Timaeus* of Plato, by mediates, the harmonious edifice evokes a particular happiness which confirms the beneficent power of its practical design. *Venustas* absorbs *firmitas* and *utilitas*: architecture is functional, in the highest sense of the word. This striking proposition laid its impress on the whole period.

Alberti escaped, through his knowledge of all Italy, from the limitation of a purely Florentine culture. In his synthesis, the basic yardstick became for the first time the Antique, of which only the vaguest or most fanciful idea could be conceived in Florence. From this resulted, first, a variety of programmes which could only flatter the craving for expansion and creation of the period: villas, monumental perspectives, palaces, churches, aqueducts, fortresses —architecture was everywhere, subjecting all nature to the will of man, and disorder to beauty. But the exact knowledge of Roman remains and texts led to precisely the opposite of the Brunelleschian style. The major element is now the wall and the engaged members which enliven it without breaking it up, and hence the Antique Orders which were to be superimposed in a façade—Doric in the lower storey, then Ionic, then Corinthian. Another result is to forbid the placing of an arch upon a column, since this was only meant to carry an entablature. To the multiplication of programmes corresponded a stricter discipline in the parts.

Alberti's evolution revealed the strength of his critical thought; in twenty years he adopted, correcting himself in the process, several attitudes towards Antique architecture. In Florence, he made free use of elements of Roman decoration, mixing them with Gothic; in Rimini, he strove for objective precision, and attempted a façade which should be a proper reconstruction; but this was a voluntary limitation from which, in Mantua, he freed himself in order to compose new forms 'in the Antique manner'.

The Palazzo Rucellai (*c.* 1446–51), with its three orders of pilasters delicately linked by the windows, and its three horizontal string courses creating an harmonious alignment, remained a lone work in Florence. In the narrow Via della Vigna it is impossible to see it properly in perspective, but the loggia in the Antique manner opposite to it accentuated its originality. Giovanni Rucellai also financed the modernization of the façade of S. Maria Novella. Alberti preserved the medieval niches, framed them by extending the coloured marble panels, enclosed the main door in a large arch resting upon pilasters, and later, in 1470, added a triangular pediment between two large volutes, which is a remarkable anticipation of Counter-Reformation types. Alberti's original feature lay in the harmonious proportions, based on 2:3, and in the unsurpassable balance of the composition.

At about the same time, Alberti was directing the reconstruction of the church of Rimini, a thirteenth-century Franciscan foundation, which, by the addition of side chapels of irregular shape during the fourteenth century, had become the mausoleum of the Malatesta. The rebuilding began in 1447, with the interior chapels opened into the nave with large arcades. In 1450, the commemorative medal by Matteo de' Pasti, who was responsible for the actual work, revealed the great project for a new exterior conceived by Alberti—a façade based on an Antique triumphal arch, with a curvilinear pediment broken into by a window; beyond the nave was a short transept and an enlarged apse surmounted by a huge dome. The side walls were to be

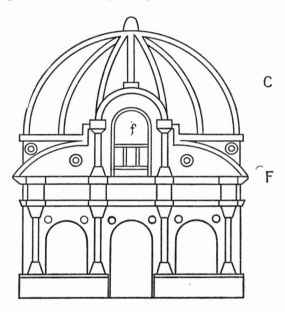

FIG. 20. *S. Francesco, Rimini* (*Tempio Malatestiano*). *Reconstruction of Alberti's projected façade* (*after the medal of 1450 and the diagram by R. Wittkower*): *F, Triumphal façade; C, Dome* (*not built*); *f, Unfinished window.*

decorated with niches, under a general revetment of marble, taken from antique monuments, and in large measure stripped from Ravenna. Work was interrupted by the death of Sigismondo in 1468, and the 'temple' remained without its pediment. The apse was rebuilt, without regard for the plan, in the sixteenth century, and a campanile was erected in 1501. Almost totally destroyed in 1945, the Tempio had been rebuilt by 1950.[85]

Alberti's directives are astonishing documents. The architect wanted, in the interior, clear proportions—the nave is three times longer than its width, as in S. Andrea in Mantua—and a strong alternation of voids and solids. On the exterior, he opposed Matteo de' Pasti's tendency to combine with freedom decorative motifs of vegetable origin with those of a purely formal kind; 'remember what I have told you, this façade must be a work on its own'— that is, without any connection between it and the forms of the Gothic chapels. Thrusts were to be taken into consideration, the roof lightened by using wood, but the façade was to preserve 'the scale and proportions of the pilasters. To modify them, would force all this music out of tune'.

In his great Mantuan compositions, S. Sebastiano (1460; not consecrated until 1529), built on a central plan, raised over a crypt by an arcaded basement, and S. Andrea (1470 onwards) on a basilical plan (the dome is by Juvara), Alberti's designs underwent considerable modification. In both cases, the composition was to create an imposing effect, the walls were given a powerful rhythm. In S. Andrea, the side elevation of the chapels reproduces the principal motif of the façade, and the whole structure establishes the same grandiose rhythm of a B a—two pilasters raised on bases, an arch, two pilasters again. The interior space is at the same time a block of strict relations, without any empirical residue, without obscurity, and a life-enhancing experience on a heroic plane.

Alberti thus established at the heart of the Quattrocento rules and models from which no later architect could escape. Some of his solutions had to wait to be properly understood and developed. It was not in Florence that Alberti's personal tastes were fully reflected. In Urbino, a city which offered more scope, things could be carried much further. Rome was only at the beginning of the road; there, Alberti's ideas were put into practice by Bernardo Rossellino, who had directed the building of the Palazzo Rucellai in Florence. Agostino di Duccio (1418–81), the sculptor of so much of the Tempio in Rimini, developed Alberti's monumental forms in S. Bernardino in Perugia (1457–61), of which the façade consists of a large triumphal arch surmounted by a pediment, though the side pilasters decorated with niches, and the mass of the building covered with terracotta decoration, dissipates the effect. This tendency to soften his rigorous expression is equally apparent in Florence, about 1460, parallel to the anti-Donatellesque reaction in sculpture. The masterpiece of Florentine preciosity was the funerary chapel of the young Cardinal of Portugal in S. Miniato, constructed in 1460 by Antonio Rossellino, Bernardo's brother, with the help of Luca della Robbia who did the glazed terracotta parts, and several painters. The Greek cross plan, with short arms, under a little dome, is in the Albertian spirit, but the revetments of polychrome marble, the delicacy of the reliefs on the tomb, dissolve the structure by their appeal to an heightened sensibility. The Tuscan taste for polychromed architecture, so powerfully manifested in the Romanesque period, remained unaffected by the reforms of Brunelleschi and Alberti.

SCULPTURE

Donatello. Like his friend Brunelleschi, Donatello was an exceptional per-

41 Venice, Doge's Palace

43. Venice. Fondaco dei Turchi (state before restoration)

43 Poggio a Cajano: view of the Villa, painted in the 17th century

(Florence, Uffizi)

44 Alberti: façade of the Tempio Malatestiano, Rimini

45 Urbino: view of the Ducal Palace

46 Brunelleschi: interior of the Pazzi Chapel

(*Florence, Sta. Croce*)

47 Brunelleschi: interior of Sto. Spirito, Florence

48 (a) Masaccio: detail from the fresco of the *Tribute Money* in the Brancacci Chapel

(*Florence, Carmine*)

(b) Donatello: detail of *S. John the Evangelist*

(*Florence, Campanile*)

(c) Piero della Francesca: detail of *S. John the Evangelist*

(*New York, Frick Coll.*)

sonality, a tireless and demanding genius. During his long career (1386–1466) he affected all forms of art, influencing painters like Mantegna, decorators and architects as much as sculptors. The works of his youth reveal his disdain for the merely formal beauty, the drapery and the calligraphy then in fashion. The marble *S. George* destined for Or San Michele (1416) still has a Gothic twist and balance in the figure, but facile serenity is surpassed by a new insistence on energy, strongly marked in the pose and in the character of the head (Bargello). The *Baptist* (1415), made for the cathedral (now Mus. dell' Opera) which personifies inspiration, expresses at the same time a new tension. Donatello's art renews the vocabulary of gesture and rejects the conventional use of drapery. His *Habakkuk* for the campanile (1427–36; Mus. dell' Opera) is a haggard, bony creature, with a shapeless garment thrown across his shoulder and tumbling heavily between his legs. This figure is compounded of daring aggressiveness and power; the linear patterns of Gothic have vanished. What one is faced with is the enlargement of life, an all-embracing grasp of interesting natural characteristics. In this sense, he was impelled by a powerful realism, but one guided by a desire for expression geared ever more closely to his astounding technique.

About 1440, he created the lovely bronze *David* (now in the Bargello), nude, wearing a laurel-wreathed hat which makes the wittiest contrast with the decapitated giant's enormous helmet. The highly strung, graceful young body in its nonchalant contrapposto is clothed in light, the gleaming skin is full of reflections, the patches of chasing contrasting with the smooth areas of the torso and the legs.

In his bronze panels, like that of the *Feast of Herod* in the Siena font (1421–1427), Donatello introduced an architectural background which permitted a diminution of depth in which the greatest variety of relief was exploited. In his *Annunciation* in Sta Croce (1433), carved with great care and skill in stone, he framed an interplay of extremely sensitive gestures within a complex tabernacle surmounted by putti, the background being filled with the most delicate decorative motifs. Here again, the artist amplifies and diversifies the links between figure and ornament.

In the *Cantoria*, or choir balcony of the cathedral (1433–39, Mus. dell' Opera), the richness of handling begins in the architecture with its heavy consoles and with the background of gold mosaic, against which are poised the wild groups of childish dancers. Some among them are derived from Antique motifs, Roman, but originating in Praxiteles, which could still be seen in the churches of Ravenna. These Antique examples opened up new paths in Donatello's imagination; with him, the putto, an infant figure full of vitality and fantasy, becomes an essential character in sculpture, considered as a subject in itself, as a little laughing faun, placed as an emblem-bearer before pilasters, as a garland-carrier upon cornices and friezes, and moving on from thence into painting.[86]

He worked from 1443 to 1452 in Padua, which he discovered to be more sensitive to the Antique than Florence, and there, using the equestrian statue of Marcus Aurelius in Rome as a model, he created the first monumental work of the period, the figure of *Gattamelata* (1444-47). The pedestal presents some of the characteristics of a tomb; the horse, with complete mastery over technical difficulties, achieves the programme adumbrated in the *condottiere* painted by Uccello in Florence cathedral. In contrast, the armour is of an extraordinary richness in the chasing. It was also in Padua that Donatello executed the large assemblage of reliefs and statues destined for the high altar of the basilica of S. Antonio (1447-50). The seated *Virgin* evokes an archaic type and looks like an idol; the reliefs, the subjects of which are the miracles of the saint, introduce with a new largeness of effect grand motifs of architectural perspective—vaults, colonnades and so on—filled with agitated crowds. The relations between the planes are carried to an almost unbearable degree in his late bronzes—for instance, in the *Crucifixion* (Bargello) where the most daring kind of *rilievo schiacciato*, or flattened-out relief is used, and in the more tumultuous and even confused, but unbelievably pathetic, reliefs made for the ambones in S. Lorenzo (1461 onwards), in which the hand of his collaborators, Bellano of Padua and Bertoldo, is discernible. The old artist's originality remained as vivid as ever in the quavering figure of the *Magdalen* (Baptistery), in the little *Judith*, a dreamy heroine and counterpart of a sort to the *David*, destined to crown a fountain (hence the Bacchic putti on the base), and which became, in 1495, the emblem of Republican virtue.

His works remained unequalled in their intensity of expression; they exploit all the resources of the medium, but by their authoritative use of form they always avoid the dissolution of plastic values. This variety itself, where the energy of Giovanni Pisano and the Dionysiac reliefs of Antiquity find a meeting place, surpasses all categorizing of styles. Dominating from on high, he was not always appreciated without some reserve in Florence.

Ghiberti. During Donatello's long absence in Padua, another current had regained the ascendancy; in 1452, the final pair of doors of the Baptistery successfully established, at the mid-century, a suave and serene style which perfectly suited Florence. Their author, Ghiberti, compiled his Commentaries, in which he appeared as the representative of modern art, informed by the Antique, but his understanding of optics and his use of Pliny remained medieval. For him, the Renaissance existed, but it had achieved its aims in the fourteenth century, with Giotto and Duccio, whom he places very high, and with Nino Pisano. He accepted the role of a continuator of a balanced and harmonious art, whose Gothic nature struck him less in that he enriched it regularly by tiny borrowings from the Antique, calling everything that appealed to him 'Greek'. What he displayed was a continuous development on an already fixed base, which is fully evidenced by the tender inflexion, and the criss-cross drapery of his *Baptist*, and his *S. Matthew* (1422) on Or San Michele.

The comparison between the two great works of this sculptor-goldsmith shows the steady development of the formula. The second pair of Baptistery doors, commissioned from Ghiberti after the 1401 competition, were divided, like those by Andrea Pisano of three-quarters of a century earlier, into 24 quatrefoil panels, containing scenes from the life of Christ: *Christ walking on the waters and supporting S. Peter* recalled Giotto's treatment of the theme; the exquisite group of figures in the *Flagellation* had the sinuous grace of Parisian ivories.

In the third pair of doors, Ghiberti adopted a quite different scheme. The two leaves of the doors are divided into sixteen large bronze panels; the rather ingenuous floral frame was replaced by bands with niches (containing prophets) separated by medallions from which emerge heads that were intended to furnish a note of the surprising. The large scenes, drawn from the Old Testament, present a host of details charmingly combined with all the resources of very varied relief: landscapes with decorative trees, architecture—like the Temple at Jerusalem—very delicately incised, the figures in the foreground standing out in a continuous and smooth modelling. The harmonious draperies, the supple compositions, characteristic of Ghiberti, lose their effect in this 'pictorial' fusion; the minuteness of the drawing clashes with the glitter of the gilded parts, which recent cleaning has revealed in their original brightness. Not only does Ghiberti succeed in grafting the new upon the old; he maintains the unity of style between the goldsmith and the sculptor.

Luca della Robbia. This workshop played a very important rôle, its insistence on delicacy in detail and effect being reinforced by the work of Luca della Robbia (1400–82). Trained as a goldsmith, he created in the *Cantoria*, executed between 1431 and 1438 at the same time as Donatello's, an interpretation filled at once with animation and restraint, truth and idealization, of the labours of choirboys on the theme of the psalm *Laudate Dominum*. On the campanile, he completed the five hexagonal medallions which still wanted (1437–39), following those by Andrea Pisano, with a pure and melodious invention, and for the sacristy of the cathedral he made the bronze doors (1445–69), in which he was helped by Michelozzo and Maso di Bartolomeo. The clarity of the scenes represented, and the supple modelling of the figures, explain why Luca finally adopted the techniques of polychromed and glazed terracotta to popularize his reliefs. In the pediments of large numbers of tabernacles and niches, often in the form of medallions or *tondi*, encircled with thick garlands of flowers, are to be found Luca's tranquil Madonnas draped in blue and white mantles, and his mild and benevolent saints. This art, made use of by the best artists and becoming very popular, became a source of livelihood for a whole tribe of della Robbias including Andrea (1435–1525), who made more complicated and spectacular altarpieces (at La Verna and Assisi, for instance), and Giovanni (1446–1527), whose effects are more elaborate, but in less good taste (*Madonna and Saints*, Arezzo).

Stimulated by the fire of Donatello's inspiration and by Ghiberti's discipline, Florentine sculpture developed, towards the middle of the century, a number of types of decorative composition. Little attention, however, was paid to large scale figure sculpture.

From Donatello's reliefs are derived, though without his lively and direct approach, the works of Agostino di Duccio (1418–81), Alberti's assistant in Rimini, who was responsible for many decorative reliefs in which are evoked the gods of fable, the Liberal Arts, the prophets, and which are peopled by round-headed, compact-bodied putti. The sculptor, fond of complicated groups, never hesitated before iconographical oddities, and multiplied his figures—his *Virtues*, for instance, hermetic and treated rather as heraldic emblems—with a roundness of form which finally became more graphic than plastic. The marble *Madonna* made for Florence cathedral (now in the Bargello), has more feeling and inclines towards Desiderio; he is uneven in the façade of S. Bernardino in Perugia, which he treated as if it were a shrine decorated with goldsmiths' work, dominated by a serene *Annunciation*, while inside, framed in pilasters, are figures personifying Franciscan Virtues, much coarser in style.

Tombs. Bernardo Rossellino (1409–64), who had been Alberti's assistant, created in the *tomb of Leonardo Bruni* in Sta Croce (1444) the lasting form of the Florentine monument, modified to suit the latest taste. In this work of major importance, the old Gothic type of monument in tiers was remodelled with a new emphasis on clarity. Under a wide arch supported on fluted pilasters set into the surface of the wall was placed a sarcophagus of antique form, with two Ghibertian type angels bearing inscriptions; this was surmounted by a catafalque bearing a figure of the dead man, and in the tympanum was a lunette with a tender image of the Madonna. Some years later, one of Donatello's pupils particularly sensitive to delicacies of modelling, Desiderio da Settignano (1428–64), created a more ornamented version, more colouristic in the delicacies of the relief, though less direct in their draughtsmanship, in the *tomb of Carlo Marsuppini* (1455). Desiderio's feeling is of the tenderest, and he concentrates on the most delicate type of female beauty in his very low '*schiacciato*' reliefs which etch the contour of a *Madonna* (Bargello) with a taut line, and his limpid portrait heads have the same quality. His marble tabernacle in S. Lorenzo, inspired by Donatello, is a model of careful design in rich effects. With Desiderio, Florentine taste reached an unsurpassable peak. The third monument which reveals the rapid evolution of sculpture towards pictorial effects is the *tomb of the Cardinal of Portugal* (1459) by Antonio Rossellino in the chapel at S. Miniato which has already been mentioned. The arch is now a niche, shadowed by a curtain; on the wall coloured by a rich alabaster plaque, the *tondo* of the Virgin above and the catafalque below seem to float in a tender harmony where death is forgotten. Antonio repeated this charming formula at Monteoliveto in Naples, but in the bust of *Matteo Palmieri* (Bargello) his handling is much firmer and more direct. After Desiderio's premature death, the dominant figure about 1470 was Mino da Fiesole (*c.* 1430–84), a rather uncertain artist, attracted by Ghiberti's flowing line, and affected by the softness which followed upon the energy of the preceding generation. He carved busts of *Piero de' Medici* (1453) and *Diotisalvi Neroni* (1464), and in his *tomb of Leonardo Salutati* (Fiesole, 1464), he endeavoured to escape from Rossellino's formula by placing a simple bust of the dead man on a coloured marble ground above the sarcophagus, and in the *tomb of Conte Ugo* in the Badia by simplifying the structure so as to enable the tender figures in his relief of *Charity* to make their full impact. In Rome, he returned to the Bruni type in his *tomb of Paul II* (1475: now dismembered). In the minutely detailed panels of the Prato cathedral pulpit he succumbed eventually to the influence of Ghiberti.

In the work of Benedetto da Majano, after 1470, eclecticism was the dominant feature. But a strong reaction was building up in the powerful workshops of the painter-sculptors Antonio Pollaiuolo and Verrocchio, which were to revive realism and re-establish the important bronze commission.

PAINTING

Masaccio. In the midst of the success of Lorenzo Monaco's soft style, and of Gentile da Fabriano's enchanting works decked with little naturalistic details, the youthful Masaccio (1401–28) suddenly appeared with a new kind of painting, sternly hostile to all the charms and sweetnesses of Gothic. The old sources indicate that Masaccio was a pupil of Masolino (1389–*c.* 1435), and it was with him that in February, 1426, Masaccio was to be found decorating the Brancacci chapel in the Carmine in Florence. It would appear that the older man came under the influence of the younger painter whom he was training, and it would be unreasonable to separate them radically, as if Masaccio had come out against his master. The latter tended, as did Ghiberti in sculpture, to modify Gothic formulae, and to keep them as a framework for innovations which exploited, principally, perspective and modelling. Masolino's drawing became more vigorous from his earliest *Madonna* (1423: Bremen) to the one in the church of S. Stefano in Empoli. In 1427, interrupting his work in the Carmine, he went to Hungary, summoned there by Pippo Spano, and was in Rome in 1429, where he continued Masaccio's work at S. Clemente. Later he worked at S. Fortunato in Todi, and lastly in Castiglione d'Olona, where his frescoes in the baptistery—the *Feast of Herod*, the *Baptism of Christ*, etc.—show that a pictorial space, developed through the use of sharply retreating perspectives of arcades and the use of solidly conceived figures, did not appear to him as incompatible with the softer and more insinuating qualities of the Trecento. Rather did they become for him an ingredient of extra piquancy—friezes of putti, and the famous panorama

THE QUATTROCENTO

of Rome suggested other means of enriching a picture, as well as an awakened sensitivity.

Masaccio began from another standpoint: his affinity with Donatello was such that the painter probably received his first impulse from the sculptor. Like Donatello, he created powerful figures, of a solid humanity, indifferent to gracefulness, but robust and of heroic cast. His almost exclusive insistence on modelling and on tonal values revived the deep seriousness, the plastic severity, of Giotto. This lesson, administered with unprecedented energy, had an immense force. But it is not always easy to distinguish between Masaccio and his master, and the chronology of the decisive years is a difficult matter.[87] In the *Madonna with S. Anne:* (between 1420–24, Uffizi) which revives a Trecento canonical form, the Madonna and Child, of a larger and more certain modelling of form, are by Masaccio, as well as one of the angels on the right, where, for once, he sought to create a figure of real charm. In the now dismembered *Pisa polyptych* (1426; central *Madonna and Child*, London; side panels in Pisa, Berlin, etc.), Masaccio concentrates on the light which falls upon the large masses; the stunted figures of the saints have the imperious quality of Donatello's Prophets on the campanile. In the chapel in the Carmine decorated with scenes from the Life of S. Peter, the *S. Peter preaching* and the *Healing of Tabitha and the lame man,* set in the famous perspective view of a Florentine square, and the *Adam and Eve in Paradise* on the entrance arch are by Masolino; what he did, in fact, was to set his figures in a more coherent space. Masaccio may be recognized in the fuller volumes, in the greater authority he gives his figures, and the authenticity of their types. As in Dante's Inferno, humanity can recognize itself in these evocations. In the *Tribute Money,* the sombre figures of the Apostles, in their heavy garments, stand gesticulating with slow dignity against a bare rocky landscape, like those of Giotto. In the fresco of the *S. Peter giving alms,* the rude, peasant quality of the figures is accentuated by a dramatic use of light, of a poignant intensity. Masaccio composed his scene by choosing a vanishing point at the same level as the spectator, so that all the heads, near and far, are at about the same level (the law of isocephaly), but he uses this only to obtain from his figures meaningful interchanges of glance, not picturesque effects.

In Rome, where he died suddenly in full strength at the age of twenty-seven, Masaccio must have worked on the *Crucifixion* in the chapel of S. Catherine in S. Clemente, and on a polyptych for S. Maria Maggiore, of which a fragment has recently come to light (London, National Gallery). In Florence, he had painted in S. Maria Novella a *Trinity* placed within an arch seen in plunging perspective, between two strongly characterized donors seen in the same light and the same pictorial space as the Virgin and S. John. Painting, having become the most severe means of expression, confers upon its creations a grandeur, a gravity, and a dignity composed of concentration, self-knowledge, and self-control, of which the West had not seen the like for many years.

ART IN FLORENCE

Between 1430 and 1440 there was a kind of void in Florence. Masaccio was dead, Masolino never came back. The financial crisis which followed the war with Milan, and the institution of the *catasto* (a form of graduated tax), limited commissions. No new masters appeared. The only commissions were those of the Medici, after their definite return from exile in 1434, and the only activity was in the works on the cathedral. At this juncture Leon Battista Alberti published his thesis *De Pictura* (1436), in which, on a model derived from rhetoric, he discussed the principles of art (perspective), the work itself (drawing, modelling, colour), and the artist (the laws of composition). This memorable treatise adumbrates, not so much a memorandum on the new style, as a programme for the art of the future: it is an anticipation born of Alberti's humanist culture, which discerned in Pliny the formulae of a great art to be rediscovered, and his confidence in a few enlightened minds.

The Colourful Style: Domenico Veneziano, Fra Angelico and Fra Filippo. In 1438, Domenico Veneziano (*c.* 1410–61), a painter of Venetian origin, who had long been outside his own country, wrote to Piero de' Medici and offered to equal the works of Fra Angelico and Fra Filippo, whom he cites specifically as the two most eminent painters in Florence. It would be of the greatest interest to know whether Domenico had already been in Florence, and if his influence had already penetrated there. For, as a master of oil painting, he possessed a gift for light pale colour of a springtime freshness, and was able to manage, not only effects of light and shade, but also his pictorial space in a manner which created a counterbalance to the authority of Masaccio.

From 1439 to 1445 Domenico Veneziano worked in the choir of S. Egidio, having the young Piero della Francesca as his assistant. Andrea del Castagno also painted there, between 1451 and 1453, three scenes from the Life of the Virgin. The loss of these works means that there are now no proper grounds for comparison. The known works by Domenico range from the cluttered space of the Gothicizing *Adoration of the Magi* (1435, Berlin), to the lovely clarity in perspective and lighting of the *S. Lucy Altarpiece* (*c.* 1445, Uffizi), and to the fresco fragment, *SS. Francis and John the Baptist* (*c.* 1460, Sta Croce), an essay in rather sharp realism of a striking character.

Thus, between 1420 and 1440, painting had discovered, one after another, the fundamentals of a new purpose. The reactions that took place were most interesting: the use of a harmonious and properly designed space was common to all, but the alignment into two camps of the artists who specialized in delicate colour—Fra Angelico (*c.* 1400–55) and Fra Filippo Lippi (1406–69) —and those who specialized in drawing—Paolo Uccello (1397–1475), and Andrea del Castagno (1423–57)—gave a new direction to the whole of Florentine painting.

Under the protection of Cosimo de' Medici, the reformed Dominicans, or Observantists, of Fiesole and of the convent of S. Marco in Florence, had for their prior a painter, soon known by the name Angelico, whose first known work was the altarpiece at Fiesole of 1428. The taste for gold, for pearly colour, for brocades, which was to be displayed to the full in his panel of the *Madonna and Angels* (Vatican), did not exclude a good grasp of space composition and a clear sense of modelling, apparent in his *Linaioli Madonna* (1433, S. Marco Mus.), and in his *Deposition* (S. Marco Mus.). Through the clarity of his arrangement of the parts, Fra Angelico is an adherent of the new style. His taste for moderately antique architectural forms, *à la* Michelozzo, adds a faint humanist note, as in the Cortona *Annunciation*, and, principally, in the *chapel of Nicholas V* (1447, Vatican). But the heavy damasks, and the gold haloes, were reappearing, and after 1435 memories of the archaic compositions of the early Trecento are multiplied, as if by a kind of conservative reflex, readily understandable in a man in religious orders. This frame of mind is perceptible, principally, in the decoration of the convent of S. Marco, somewhat spoilt by the great number of pious and tame helpers. The year 1440 or thereabouts saw him creating the chief altarpiece, the predella of which (now dispersed) is far more pleasing, the so-called *Analena altarpiece*, that at Bosco ai Frati, and finally, the important *Coronation of the Virgin* (Louvre), which combines in the creation of a kind of celestial idyll the splendours both of perspective and Domenico's tender colour, though without opening up any new ground.

A Carmelite brother, though little suited to the cloister, Filippo Lippi also was attracted first to Masaccio, and then to the exponents of colour, and was a ceaseless searcher after new ideas. The *Trivulzio lunette* (1434, Castello Sforzesco), shows him able to set animated figures against a background of blue sky; in the *Barbadori altarpiece* (Louvre), he took a particular delight in filling with lively and decorative figures and objects a space which is none too well defined; and in the *Coronation of the Virgin* (1441–47, Uffizi), he permitted himself an excessive, though well-controlled, overcrowding of the picture surface by the variety of his colour and his skill in soft blended modelling. His use of detail is lively, shot through with effects of colour that permeate the composition with tender feeling. The lovely, rich, green, brown and golden colouring of the *Nativity* (Uffizi) is of a lyricism that less scrupulous followers were to hasten to counterfeit. But, shortly afterwards, as if he felt that he should take a sterner grip on himself, Lippi endeavoured in the choir of Prato cathedral to illustrate scenes from the lives of S. Stephen and S. John the Baptist (1452–64) with compositions having a much stricter perspective framework. In the apse of Spoleto cathedral, where the work was finished by his assistant Fra Diamante, he returned to a more exuberant imaginativeness to express his vision of Paradise.

The Draughtsmen: Uccello, Castagno. With Uccello, the shattering effect of the innovations of Masolino and Masaccio reacted upon workshop practice. Uccello is first recorded in Ghiberti's shop; in 1425 he was invited to work on mosaics at S. Mark's in Venice; from 1443–45 he produced the cartoons for the stained-glass windows (of which, one, the *Resurrection*, is of an admirable compression) of Florence cathedral. Before becoming a master of perspective, Uccello was a competent journeyman, and taking everything into account, his career as a painter appears to have been slow and uncertain. In 1436, he was commissioned to paint on the north wall of the cathedral a fresco of an equestrian figure to serve as a monument to the *condottiere* Sir John Hawkwood. The enormous horse in this *tour de force* of illusionism, painted entirely in grisaille, stands out against the dark background. About 1439–40, in the cloister of S. Miniato, Uccello painted some scenes from monastic life of which nothing remains but a few shadowy traces of dizzying perspective. The large compositions in the Chiostro Verde of S. Maria Novella, the *Creation* and, above all, the *Deluge*, with its fleeing perspective and its extraordinary drowned figures, date from 1445–50, and the three *Scenes from the Battle of S. Romano* (between 1456 and 1460; Uffizi, Louvre, and London Nat. Gall.) are late works, of dazzling fantasy. They allow no reason for seeing in Uccello a successor to Masaccio nor the father of perspective, but rather a backward-looking, uneven and strange poetic talent, which fits in with the legend created round him, who reduced the world to a geometric exercise, and was endlessly addicted to researches in descriptive mathematics, introducing everywhere that oddly faceted object called a *mazocchio*, and on occasion painting red fields and blue villages. But the truth is rather more complex, since there is also in Uccello an artist with a charming Gothic gift of narrative, clearly to be seen in the backgrounds of his large pieces which contain many motifs like those in Franco-Flemish tapestries. In this sense, there are exquisite details in his *Night Hunt* (Oxford), and in the *Nativity* (Karlsruhe) which is sometimes attributed to him. The sinuous Gothic line is united to brilliant deployment of the geometrical spiral; the tender silhouette gives place to a mechanical schematism. Formal extremes reach their furthest point, and it was eventually to decorators, engravers, and *intarsia* and inlay workers that Uccello's art was the most useful.

Andrea del Castagno, a much younger man who died prematurely, had understood the need for a return to Masaccio's powerfulness. After a short stay in Venice, where he painted the chapel of S. Tarasio in S. Zaccaria (1442), he decorated the refectory of the convent of S. Apollonia (1445–50) with huge spectacular compositions, the most monumental and the most strictly composed in all Florentine art. His effects are astounding; his heroic vigour transmutes all, but Masaccio's grave humanity is lost. In the Villa Carducci at Legnaia (*c.* 1450) he displayed in his series of *Famous Men and Women* dominated by a superbly drawn *Eve*, his power as a decorator, but at the cost of dry and arid colour. The cornices with their putti are those of a sculptor, as is the effigy of Niccolò da Tolentino in the cathedral, painted twenty years after the one by Uccello and next to it in the church. The harshness of his drawing sometimes detracts from his resolute style, but under his bronze draperies, and in his grimacing angels which look as if they were carved from boxwood, there lies the obsessive plastic tension essential to the Quattrocento. Something of this may be seen in the Master of the Carrand triptych (Bargello) who has been identified with Giovanni di Francesco (d. 1459).

The Turning Point of 1460: Antonio Pollaiuolo. After the departure in 1440 of Piero della Francesca, who was never to return and was to be of no further account in Florence, a void becomes noticeable about 1460–70 with the disappearance of all the masters of the great generation. There remained, of course, prolific and amusing illustrators, who bring to mind the popularizers of the romances of chivalry: Pesellino (1422–97), who painted cassone panels, and Benozzo Gozzoli, one of Angelico's assistants, who created in 1459, in the chapel of the Medici palace, the most charming of all fairy stories in which an exotic cavalcade winding its way through a Tuscany in flower seems to be an enlargement of miniatures painted in 1400. The same impression is gained in front of the *Life of S. Augustine* in San Gimignano (1463–65), and the Biblical scenes in the Campo Santo in Pisa (1468–84), which have the same disarming garrulity, and in which Gozzoli incorporated without scruple—as without any malicious intention—the many discoveries of his great contemporaries. The taste for extravagant costumes, for fantastic armour, equally the fashion at the Burgundian court, penetrated into Florence at this moment: the illustrated Chronicle, made about 1460 by a goldsmith who was probably Maso Finiguerra himself, presents a repertory of almost wild romantic pictures. But the grand style is less in favour with the new generation. Domenico Veneziano's only descendant, Alesso Baldovinetti (1425–99), who, by the exquisite serenity of his colour (for instance, in the Louvre *Madonna*), the solemn breadth of his vision (as in the *Nativity* in the forecourt of the Annunziata, 1450), the variety of his compositions in the chapel in S. Miniato, is a kind of Piero but less active and less inventive, and for some unknown reason quite uninfluential, though equally pure and delicate. Castagno's descendant was Antonio del Pollaiuolo (*c.* 1432–98), sculptor in bronze and painter, who for a time oriented Florentine art towards violent movement, frenetic gestures, and an agitation of mind and body. His most unrestrained work, the designs for embroidery made for the cathedral (1466–80, Mus. dell' Opera), is filled with sudden and nervous agitation. Antonio completed, by his pursuit of anatomy, and by his precision in modelling, a half-century of experiments in the truthful rendering of the human body, but he represents the opposite pole from Masaccio's impassive humanity. The harmony of perspective space is also compromised by another of Pollaiuolo's passions—that for the Flemish type of landscape, which invaded the panels of the *Labours of Hercules* (1460, et seq., small versions in the Uffizi) and the *Rape of Dejanira* (Yale University). Had it not been for the numerous and successful artifices of composition, the over-detailed human figure silhouetted against an immense panorama would, for the first time, have created the impression of an oppressive naturalism, as for instance, in the *Martyrdom of S. Sebastian* (1475, London). It was against this that Florentine painting at the end of the century was to react.

The Spread of the New Style. The diffusion of the forms elaborated in Tuscany was rapid and decisive. The effect of Michelozzo in Milan, and of Donatello and Uccello in Padua is proof of this. Florentine ascendancy was so powerful that for several decades Venice itself applied to Tuscany for painters (Castagno), sculptors (Donatello) and even for mosaicists (Uccello). This hegemony could not last. In Milan, Padua, Ferrara, Urbino, workshops which had benefited from the passage of Tuscan masters developed original styles which the Florentines had to compete with and to take into account. About 1460–70, Florence was no longer quite the centre of the artistic map, since motifs from compositional types of Florentine origin were being elaborated everywhere. Some of these must be mentioned.

In painting, where Florence was less dominant, Fra Angelico had produced in his *S. Marco altarpiece* (1437–41; based perhaps on a lost Masaccio model) the first *Sacra Conversazione*—a gathering of male and female saints around the Virgin set in a corner of Paradise. In sculpture, the typical forms of Florentine origin, which became dominant, were numerous: the highly characterized and individualized bust, such as the *S. Rossore* by Donatello (1427, Pisa, Mus.), or those in the sober, delicate style of Desiderio; the monumental tomb, with the sarcophagus framed by pilasters, created by Rossellino; the melodious *tondo* in the style of della Robbia.[88] As noteworthy are the equestrian statue revived by Donatello, and the free-standing statue on a column, which he also had established in 1428–31 with the *Dovizia* (now lost) in the Mercato Vecchio.[89]

It was above all in the definition of types of compositions, between the monumental and the decorative, that the Florentine workshops had their profound effect. Everywhere they promoted the development of common solutions by the widespread use of the aedicule with pediment, cornice, and pilasters, adapted to the monumental niche by Donatello and Michelozzo (1423), then to the framing of a bas-relief, as in Donatello's *Annunciation* of about 1434, to the design of an altarpiece for the Linaiuoli by Ghiberti in 1434, or to the tabernacle by Desiderio in S. Lorenzo of 1461. This structure, which was suitable for small compositions, could also be used for windows, gateways, façades, and tombs. Hence, it appeared naturally in painting, where many grand and elaborate constructions, thrones, niches, etc., were used, beginning with the fresco of the *Trinity* by Masaccio, and the *S. Lucy altar* by Domenico Veneziano where the arches are still of a Gothic type.

Finally, engraving became important in Florence, principally in the form of engraving on metal as spread by the goldsmith Maso Finiguerra (1426–64). He is generally credited with the *niellos* (engraved silver plaques, with a filling of black enamel in the chased lines of the engraving; from impressions made with sulphur, prints on paper could be made) such as the *Coronation of the Virgin* of about 1455. It was chiefly Antonio Pollaiuolo who achieved the most remarkable clarity, and in his engravings was able to create a line which both characterized and enveloped the form. His famous *Battle of Nude Men* (c. 1470) impressed Mantegna and through him Dürer.

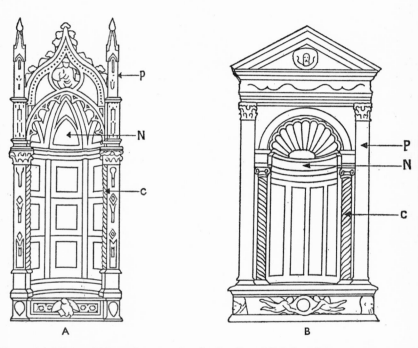

A   B

Fig. 21a. *Niches (fifteenth century). A: Or San Michele, niche on the west of the south façade (built in 1411 in fourteenth-century style); N, Semi-dome with ribs; p, Pinnacles; c, Twisted column; B: Or San Michele, central niche on west façade, or Parte Guelfa niche (built between 1418 and 1425 by Michelozzo for Donatello); N, Semi-dome with shell; P, Fluted Corinthian pilaster; c, Twisted Ionic column.*

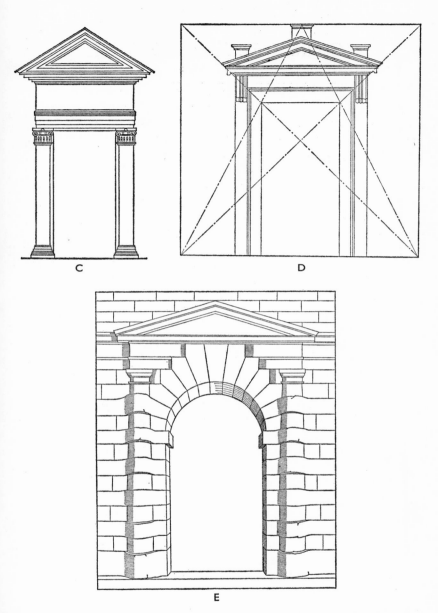

C

D

E

FIG. 21b. *Niches (fifteenth and sixteenth centuries). C: S. Maria del Calcinaio, Cortona, by Francesco di Giorgio. Design of a lower window. D: Serlio, Libri d'architettura, I, fol. 16 verso. Design of a door inscribed with a square three times the size of the opening. E: Serlio, id., IV, fol. 12 recto. Door with rusticated order, banded columns and projecting springer.*

173

Central Italy: Siena and Umbria

SCULPTURE AND ARCHITECTURE

In 1419 Jacopo della Quercia finished his *Fonte Gaia* in Siena. In 1417, Ghiberti was called there to work on the monumental Font in the Baptistery (1416–24), a six-sided basin with six niches and six statues at the corners, which later also required Donatello's collaboration. This composite masterpiece was not to be followed by other important undertakings during the century. The sole apparent preoccupation of the cathedral's workshop was to finish the pictorial inlaid marble pavement begun in 1372 with framings in which black and white squares alternate in the same manner as the layers of the piers. Several sibyls were made, in 1422, in the front part of the nave, historical scenes in 1426 by Paolo di Martino, and in 1434 by Domenico di Bartolo; the most remarkable are the huge *Liberation of Bethulia*, inlaid in black on white, made in 1473 by Federighi on a design by Urbano di Cortona, and the *Massacre of the Innocents*, made in 1481 by Matteo di Giovanni Bartoli, where the terrible scene is set within a delicate and curious arcade.

Siena offered no school of architecture to compare with that of Florence; the city was neither enlarged nor rebuilt during the fifteenth century. Francesco di Giorgio, architect, sculptor and painter, dominated the whole period. He is also an instance of the importance of woodcarving, a traditional technique in Siena, where Virgins of the type created by Nino Pisano continued to be carved and painted by Francesco di Valdambrino (?–1435) and Domenico di Niccolo, also called dei Cori (1363–1453). Antonio Federighi, the *capo maestro* of the cathedrals of Siena and Orvieto (from 1451–56), worked on the holy water stoup of the cathedral, designed, though without adhering openly to the Florentine manner, the arcades of the papal loggia (1460), and carved several rather heavy statues such as his *S. Nicolas of Bari*.

Vecchietta (1412–80) was a much more interesting artist, a lively painter who was inspired in his sculpture by Donatello's ideas. His bronzes are still well within the Gothic tradition; for instance, his *Resurrection* (1472, Frick Coll., New York), and the *Resurrected Christ* (1476, Siena Hospital). His wood carvings—the large relief of the *Assumption* (S. Giorgio di Montemerano), the *S. Antony Abbot* (1475, Narni), the *S. Bernardino* (Bargello)—lack the delicacy of his bronzes. Among his pupils were Neroccio di Bartolomeo Landi (c. 1440–1500) who made the touching *S. Catherine* (1474, Siena), and Francesco di Giorgio Martini (1439–1502), who began by sharing a workshop with Neroccio before going to Urbino and the Marches. A harsh *S. John the Baptist* in a blue tunic (1464, Fogliano), and a *S. Christopher* with a fine peasant head (Louvre) are attributed to him as a wood carver, and among his bronzes are the sensitively handled relief of the *Flagellation* (Perugia), and two vigorous candelabra-bearing angels (1497, Siena). His bronze reliefs of the *Deposition* (Venice, Carmine) and the *Flagellation* (Perugia) contain a large number of slightly indicated figures in front of the architecture which surrounds them. To these subtleties are opposed accents of great vigour: Francesco di Giorgio is a complex personality. He created moreover, in Siena and Urbino, cartoons for a large number of decorative works such as *intarsie* and inlaid pavements, like that in the chapel of S. Catherine in S. Domenico, Siena.

His main works in architecture are in Umbria and in the Marches, for in the years around 1460–80, there was a continual interchange between Siena, Perugia and Urbino, which by-passed Florence. Francesco di Giorgio's share in the palace at Urbino will be discussed later; he also designed the fortress of Sassocorvaro near Pesaro, one courtyard of the palace in Ancona (1486), and the town hall at Jesi (1486). He wrote a treatise on architecture, highly scientific and technical, particularly important for his study of fortifications and sieges. The originality of his mind and the quality of his culture both appear in the church of the Calcinaio, sited on the slopes above Cortona (1485). This huge Latin cross building, with an octagonal dome over the crossing, stresses the geometrical purity of the volumes and reduces the decorative elements to pilasters and string courses. All the windows outside and inside are crowned by large projecting triangular pediments. The abstract quality of the work is Albertian; the use of dark *pietra serena* in the framings, the pilasters and the cornices, underlines the composing elements and gives to the interior a kind of Brunelleschian elasticity, but the unusual height of the vaulting, the slender height of the dome on high pendentives, gives it an almost fantastic quality. The architect has exercised control over his imagination, but has nevertheless introduced into the classic harmony something of Sienese imaginative exuberance.

The last interesting personality among the sculptor-architects of Siena is Giacomo Cozzarelli (1455–1515), who designed the dome of the Osservanza in Siena, and built the palace of Pandolfo Petrucci '*il Magnifico*' on the Florentine type (c. 1505). Among his statues in wood are a *S. Vincent Ferrer* (Siena, Sto Spirito), and a *S. Nicolas of Tolentino* (Siena, S. Agostino), which already display the increasing weakness which reached its peak with the decorated figures by Lorenzo di Marino ('Il Marrina': 1476–1534).

PAINTING

For a long time, Siena remained the 'Gothic conscience' of Florence. Long after 1440, Simone Martini was still considered there as a model to imitate. The gold ground was still used there as late as 1490 by Matteo di Giovanni, and Sienese painting continued to be an art of panel painting, of pious icons created in an atmosphere of rather highly strung devotion, maintained at high pitch by sermons. The most gifted adapted Florentine novelties according to

175

their fancy. Sassetta (*c.* 1392–1450), who derived from Taddeo di Bartolo, belonged, from the imaginative and colouristic point of view, to the Trecento; in 1423–26 he painted the charming *Linaioli triptych* in the International Gothic style, but, about 1432, the retable of the *Madonna of the Snows* (Contini-Bonacossi Coll.) shows him ready to adopt perspective devices in an ordered setting. Soon after, about 1434, the *polyptych of S. Domenico* at Cortona breaks with this broader type of construction to return to linear drawing, and the curious set of the *Life of S. Antony* (*c.* 1436) has an amusingly miniaturist calligraphy. Giovanni di Paolo (1403–83), an artist whose imagination was much less pure and tender, used consciously distorted naturalistic motifs, for instance, in his *Paradise* (New York). The predella, with the history of the Baptist, painted twenty years later, is the most astonishing example of Sienese poetic fantasy, which borrowed from contemporary innovations only those elements which were capable of enriching the system of thin forms, upright composition and arbitrary colour of the Trecento (*c.* 1445, London, Chicago).

While the workshop of Sano di Pietro (1406–81) supplied a clumsy vulgarization of the imaginative art of Sassetta, and the anonymous painter of the altarpiece in the Osservanza (near Siena, 1436) still retains the ornamental richness of the fourteenth century, Domenico di Bartolo (*c.* 1400–*c.* 1450) more receptive of Florentine innovations, is much closer to Fra Filippo Lippi —for instance, in his *Madonna of Siena* (1433), and his frescoes in the Spedale della Scala (1441–44). The two most interesting figures are the two painter-sculptors, Vecchietta and his pupil, the versatile Francesco di Giorgio. Vecchietta was a follower of Donatello in sculpture, but having worked at Castiglione d'Olona, alongside Masolino, where he strove to be 'hypermodern' in his drawing, he introduced an inventive and turgid note in his frescoes in the old sacristy of the Hospital in Siena (1436–49), and added to Sienese painting a note of rather brusque energy in his *Madonna with Six Saints* (1457, Uffizi), and his *Assumption* in Pienza (1461), with its rather sculptural handling. His real heir is Matteo di Giovanni (1435–95) of Borgo San Sepolcro. A prolific artist, full of ornamental devices, he dominates the second half of the century; as early as his small *Crucifixion* (1460, Asciano) and above all in his *Assumption* (*c.* 1475, London), he affects a Pollaiuolesque agitation of movement, joined, in his strange *Massacre of the Innocents* (1482, Siena, S. Agostino), to influences from the North. Close to him may be placed Benvenuto di Giovanni dal Guasta (1436–1518), who also derived from Vecchietta. His unequal style reflects both the charm of Gozzoli and the sharpness of Northern masters known to him through Liberale da Verona towards 1475–85; *Madonna with Saints* (1483, Siena, S. Domenico) is an example.

Francesco di Giorgio deserves attention also as a painter; he was the only Sienese who was able to assimilate Florentine severity and blend it with Sienese poetry. His little *Annunciation* (*c.* 1470, Siena Pinac.) is of a lovely

49 (a) Arnolfo di Cambio:
Madonna and Child

(*Florence, Mus.
dell' Opera*)

(b) Donatello:
Madonna and Child

(*Padua, the Santo*)

50 Fra Filippo Lippi: *Madonna and Child*

(*Florence, Pitti Palace*)

51 Antonio Rossellino: Tomb of the Cardinal of Portugal

(Florence, S. Miniato)

52 Antonio Pollaiuolo: *Dancers*

(*Florence, Villa la Gallina*)

53 Antonio Pollaiuolo: detail from the background of the
Martyrdom of S. Sebastian

(*London, National Gallery*)

54 Francesco di Giorgio: *Annunciation*

(*Siena. Pinac.*)

55 Rome, St. Peter's: detail from the bronze doors by Filarete

harmony of pink and blue tones, in rhythmic diagonals which create an unreal perspective. Small secular subjects such as the *Chess players* (1485, New York), full of blonde and graceful figures, seem to come from *cassoni*. The *Nativity* (*c.* 1490, Siena, S. Domenico) is, despite the oddness of the design and the undulation of the lines, so energetically conceived that it has passed for a work by Signorelli. The links between Siena and Umbria were then constant and permitted the development of a kind of fantasy Renaissance, quite different from Florentine gravity.

In Perugia Domenico Veneziano, Fra Angelico, Gozzoli and Domenico di Bartolo all worked at different times, and an interesting local school arose from their influence: Benedetto Bonfigli (1420?–96) who created, on rather monotonous types, scenes of a frequently remarkable delicacy of detail bathed in a golden light, as in the triptych (1467) by him in the Perugia museum; Giovanni de' Boccati (between 1445–80) who worked also in the Marches and was in Urbino about 1455, and who painted an elegant and highly decorative altarpiece in 1447, and the *Madonna of the Orchestra* (both in the Perugia Mus.). He had a pleasant collaborator in Caporali (1420–1505) and a rather unequal succession in Fiorenzo di Lorenzo (*c.* 1440–1522) who somewhat resembles Gozzoli, and Niccolo da Foligno (*c.* 1430–1502) who painted small devotional works of rather poignant type, and with movement and breadth.

This concern for spacious setting, for pale colour, spread broadly over fanciful buildings, but with vivid details and strikingly sharp faces, flourished in Umbria. It later inspired Perugino's earliest manner, in the series—as delightful as theatrical decors—for the Life of *S. Bernardino* (1473, Perugia Mus.) in which Bonfigli and Pintoricchio also collaborated. It is among these Sienese-influenced Umbrian painters that must be sought the painter of the dazzling and enigmatic little compositions with strongly coloured architectural settings, known as the Barberini panels (*c.* 1470 *Birth of the Virgin*, New York, Met. Mus., and *Presentation*, Boston Mus.). The fresco of *the Charge to S. Peter* in the Sistine Chapel which Perugino painted in 1481, is the only one in the whole chapel breathing this open quality of Umbrian space, which cannot be explained without taking into account the influence of Piero della Francesca.

Piero della Francesca (*c.* 1416–92). This 'Monarch of painting', to use Pacioli's expression, represented, for the generation of the mid-century, a personality as complete as that of Giotto, and, like him, of a national and not provincial character. His genius overflowed the limits of Tuscany, and above all of Florence, to which, after his early years when he was able to study the work of Masaccio and to work with Domenico Veneziano in S. Egidio (1438–1440), he never returned. His career was passed among the small cities of his native country, Borgo San Sepolcro and Arezzo, and in the princely courts of the further side of the Apennines, Ferrara first, then Urbino. The demands of a modern intellect and an indeterminable inherited archaism, a Virgilian love of nature and the experience of courtly ceremonial, equally nourished the richness of an art which seems born of itself alone and of a perfect naturalness. The *Baptism of Christ* (London) and the *S. Jerome* (Venice) have their figures silhouetted against a background of the patchy landscape of a

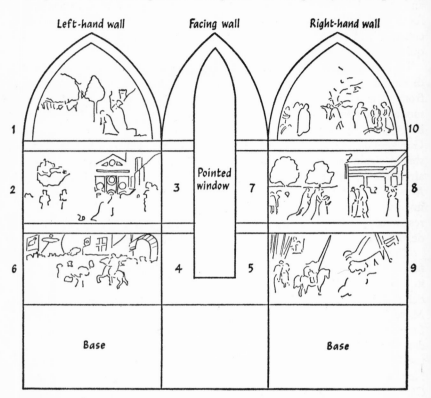

FIG. 22. *S. Francesco, Arezzo. Fresco cycle of the Legend of the True Cross, by Piero della Francesca. Plan (amended from the ed. by A. Skira): 1, Return of the Cross to Jerusalem; 2, Discovery and proof of the True Cross; 3, Torture of the Jew, who reveals the site of the Cross; 4, Annunciation; 5, Dream of Constantine; 6, Victory of Heraclius over Chosroes; 7, Transfer of the wood to the Temple; 8, Meeting of Solomon and the Queen of Sheba before the bridge made of the wood of the tree; 9, Victory of Constantine over Maxentius; 10, Death of Adam at the foot of the tree.*

Tuscan valley, brilliant with pallid light, and with the blond colouring of Domenico Veneziano.

The lovely little panel of the *Flagellation* (Urbino) shows a strict perspective system, though one less rigid than Uccello's, irradiated by this clear effulgence of light. A feeling for interval and for intense colour are both pushed to the limits of their expressive power, in such a way that the whole universe is transformed into a perfect and luminous framework, in which there are no imperfections, and in which humanity is immutably fixed in its courses.

In the impersonal quality of this art lies its nobility, though it is qualified by two features; a rustic, even peasant, aspect which is patent even in the types Piero used; another, courtly, heroic, legendary, which explains the epic nature of his art. This is particularly clear in the cycle of the *Legend of the True Cross* in S. Francesco at Arezzo (1452–59), possibly through a reflection of the crusade projected during the pontificate of Pius II. Complex battles, filled with helmets, trumpets and standards, where the sky and the woods are reflected in the curve of rivers, hillsides crowned by towers, horses pulled up before the pure colonnade of a temple—there is no aspect of human life which is not magnified by an appropriate setting. The solidity of his forms and the grandeur of his types can be seen reaching their fullest development in the most astonishing variety. The *Monterchi Madonna* and the *Resurrection* in Borgo San Sepolcro are of the same date. The polyptych painted for S. Agostino in Borgo, between 1454 and 1469, and now dismembered (parts in Lisbon, London, New York, Frick Coll., Milan, Poldi-Pezzoli, etc.), consisted of several figures of saints of a grave Masaccesque bearing, but with an air of having been compounded of jewel dust and rare colours.

In a diptych painted in about 1465 (Uffizi), he represented Federigo da Montefeltro and his wife Battista Sforza in profile against a wide and luminous landscape. In the *Sinigallia Madonna* the interior surrounding the figures, in which the light is broken up into a myriad facets, bears comparison—and suggests contact—with Flemish works. Finally, he created, in about 1475, the *Brera altarpiece*, in which the Virgin is seated amid a circle of saints, in an apse from which hangs an ostrich egg, representing at once the medieval symbol of Immaculate Conception and the intellectual symbol of the cosmos.[90] This rather unusual feature shows that Piero could recreate as he wished the grandest iconography, just as he wrote for Federigo of Urbino a detailed treatise on perspective for the use of painters, and a work on geometric regular bodies in which he placed his art on a mathematical basis.

While Lorenzo da Viterbo (1446–70) is very close to Piero in his *Scenes from the Life of the Virgin* in Viterbo (1464), Piero's echo is quickly stilled in the art of the next generation; he counted for little in Florence, and for equally little in Venice, during the first period of Giovanni Bellini's career. In Umbria itself, he failed to make a lasting impact, though he confirmed in their taste for calm spaciousness such artists as Perugino. Unexpectedly, he inspired the 'heroic' tension of Signorelli's figures cast as if from bronze; without his example it would not be possible to explain the way in which the figures of Melozzo da Forli (1438–94) tend to develop on every axis of the space they inhabit. Melozzo introduced into Rome, with his fresco of *Sixtus IV inaugurating the Vatican Library* (1474), the rhythm of large painted architectural settings, and, in the apse of SS. Apostoli (1480), sharply foreshortened angels, around an *Ascension* (now transferred to the Quirinal Palace). The same virtuosity in perspective was displayed at Loreto, in the Cappella del Tesoro (after 1477) and in S. Biagio at Forlì, where he was assisted by Palmezzano (1456–1517). But around 1470 the centre of all these brilliant interpretations of space, deriving ultimately from Piero, was at Urbino.

Urbino

The stronghold of Urbino dominates the last passes over the Apennines on the road to the Adriatic. During the middle of the century, the city became one of the most remarkable in the Marches, thanks to the energy of the Montefeltro dynasty, and the ability of the most outstanding *condottiere* prince of the century, Federigo, who, after having fought against his neighbour, the Malatesta Lord of Rimini, consolidated his power by becoming the Gonfalonier of the Papal forces (1474) and Duke of Urbino.

During a whole generation, Federigo practised an enlightened patronage and as a result Urbino became one of the most brilliant centres of modern culture. The Florentine bookseller, Vespasiano da Bisticci, was commissioned to establish an outstanding library; Landino and Ficino were appreciated by the Duke. Alberti lived there. Above all, Urbino, close to the valleys of Borgo San Sepolcro and Arezzo, attracted Piero della Francesca, who was resident on several vitally important occasions, before 1450 and again towards 1465. It would seem reasonable to connect with the first of these visits the project to transform the old Gothic palace of the Montefeltros, decorated with frescoes by Boccati, into a spacious Renaissance residence, and with the second visit, the lucid conception of the courtyard and the upper storeys, which Federigo employed the Dalmatian Luciano Laurana (1420–79) to build for him. Both through its being so much in advance of its time, and through its enormous size, the palace at Urbino is of prime importance.[91] The interior courtyard, the monumental staircase, the doorways, have a clarity of outline, a simplicity of design, a volumetric severity which is perfectly in accord with the purism of Alberti and Piero; it can even be considered as their most eloquent masterpiece. Since the castle overhung a cliff, a façade with three superimposed loggias was built facing the landscape, between the towers, and the various storeys were planned accordingly. Florentine decorators were already active in Urbino; Francesco di Simone Ferrucci (1437–93) of Perugia (where he made a well-balanced ciborium) came to work in the Marches, and Domenico Rosselli carved fine cornices and composed friezes of putti. The most important sculptor was Francesco Laurana, a compatriot of the architect (?–1502), who had already been in Naples where he carved the exquisite *Madonna* in S. Maria Materdomini, and who, influenced by Piero, created the clear and impressive bust of Battista Sforza.

In a later phase, in which Francesco di Giorgio figured, new developments were undertaken on the southern side of the castle, with the Cortile di Pasquino, and with new rooms, hanging gardens, and clever planning, which can be attributed to the architectural imagination of the Sienese. The interior is not less remarkable than the outside; under the *studiolo*, or 'Cabinet', of Federigo, entirely covered in *intarsie* on the lower part of the walls and with the gallery of illustrious men in the upper part, were built twin sanctuaries, decorated by the Lombardo-Venetian workshop of Ambrogio Barocci, one dedicated to the Muses, with a decor by Giovanni Santi (1435–94), the father of Raphael, the other to the Holy Ghost. This *intarsia* decoration (1475) consisted of forms of every kind—illusionistic niches containing statues, a loggia with a landscape view, cupboards with still-life objects in them, and was executed by the Florentine workshop of Baccio Pontelli, after cartoons derived from Botticelli and Francesco di Giorgio. The series of the twenty-eight doctors and poets, from Homer to Petrarch, grouped in pairs, was by several hands (1476–77): the Fleming, Justus of Ghent, the Spaniard Berruguete, Melozzo da Forlì, and perhaps even the young Bramante (born in 1444). The last named was brought up in the artistic environment provided by Urbino, and was inspired by Melozzo in the colossal and vigorously modelled figures of men at arms in the Casa Panigarola (1479) in Milan, before devoting himself to architecture. The huge castle of Urbino remained unfinished at the death of Federigo (1482), whose military engineer Francesco di Giorgio had become.

Other works can be connected with his residence in Urbino: the convent of S. Chiara (before 1472) and above all the church of S. Bernardino (between 1482–91) which has been thought to be by Bramante, built on a spur opposite the town to serve as a tomb for Federigo. The single nave is joined to a trilobed central plan which recalls those of Early Christian martyria, surmounted by a dome which is enclosed in a flattened double cylindrical drum. Horizontal string-courses delimit the storeys, as in Cortona. The whole building bears witness to the freshness of imagination which underlies the originality and charm of Urbino as an artistic centre in the second third of the century.

DECORATIVE ARTS

Majolica. By the fourteenth century, Italian earthenware (or *majolica*), which had been developed principally in the region of Faenza in Romagna, had already adopted the technique of tin based coloured glazes, to which a second firing gave their characteristic varnished appearance. After imitating at first Islamic pottery, which had come via the Island of Majorca—hence the name—the pottery of the Apennines had its most interesting period of development between 1440 and 1530. Factories multiplied on the Tuscan slopes of Cafaggiolo, and above all on the northern slopes at Deruta, Gubbio and Urbino.

Among the products were tiled pavements, like that in the Marsili Chapel in S. Petronio, Bologna, of 1487, useful wares, and luxury wares such as marriage plates and ceremonial vases with suitable mythological and heraldic decoration. The blue and yellow borders of plates were generally covered with garlands or ornamental motifs, the centre having an historiated decoration; grotesques on a dark ground appeared at the end of the century. In the set on the theme of Orpheus by Pellipario (Mus. Correr, Venice), Umbrian taste triumphed in the quieter shapes and colours.[92]

Rome and Southern Italy

ROME:

TOWN PLANNING, ARCHITECTURE, DECORATION

The city which Donatello and Brunelleschi visited at the beginning of the century was an inchoate place, fully justifying Platina's remark that Martin V found in 1420 'a town so full of destruction that it no longer had the look of a city'. As the town was repopulated and reorganized, under the impulsion of successive popes, so it became the centre of archaeology. Until then the trade in antiquities flourished principally in Padua and Venice. This double movement characterized the evolution of the pontifical city during the course of the century. After the mid-century, the palace of the Conservatori, on the Capitol, became a kind of museum; under Julius II, the antiquities were rearranged in the Belvedere of the Vatican. Pietro Barbo, the future Pope Paul II, hunted for antiquities in competition with the Medici, and had the sarcophagus of S. Agnes placed in the Palazzo Venezia.

After his final victory over the Antipope Felix, and after the Council of Basle in 1449, Nicolas V (1447-55) determined to revive the prestige of the seat of Christianity by means of a vast rebuilding project which included great buildings like St. Peter's, the forty pilgrimage churches, and whole quarters such as the Borgo which lay between the Vatican, Castel S. Angelo and the Tiber.[93] The portico of the basilica was to give on to a big flight of coloured marble steps, and an open space of five hundred *braccie*, extended by three parallel roads bordered with arcades, was to stretch out in front of it. This was no innovation but something deriving from a medieval Roman tradition. The scheme clearly anticipated what was to be achieved in the next century. The execution was confided, according to the Florentine Gianozzo Manetti, to Bernardo Rossellino, which indicates the Albertian inspiration of the project. When he was attached to the papal court Alberti had written a memorandum on the measurement of the ruins; his treatise on architecture was written during the middle of the century.

For a long time the Tuscans were in the forefront of all these projects. Bernardo Rossellino was, most probably, the architect of the Palazzo Venezia

(1455 et seq.), built for the great collector Cardinal Pietro Barbo; the outside is similar to that of one of the Florentine fortress-palaces of the fourteenth century, with only the large rectangular windows of the first storey to introduce a lighter note. The courtyard consists of two large arcades with pilasters and engaged columns, imitated from nearby Roman models like the Theatre of Marcellus. But Rossellino's principal work was the large town planning scheme commissioned by Pope Pius II Piccolomini for the modernization of his native village of Corsignano,[94] renamed Pienza, in the mountains to the south of Siena. In his *Commentarii*, the pope has left a record of the programme which was followed: the refaced Gothic cathedral with a polygonal choir, stained-glass windows, and a nave and two aisles of equal height on the hall-church pattern, was framed by the pontifical palace on the right and the bishop's palace on the left, placed obliquely so as to achieve a kind of theatrical set. The second palace has a regular façade, with flat rustication, directly inspired by the Palazzo Rucellai in its string-courses and pilasters, a courtyard and three storeys of colonnaded loggias one above the other opening on to the valley and the terraces.

In Roman churches, like S. Agostino (*c*. 1480), the façade is Florentine and Albertian. The interior, with heavy articulated piers and vaulting, is less developed. The replanning of the city was also a matter of great concern to Sixtus IV, who was surnamed *Restaurator Urbis*; it was he who laid out on the left bank of the Tiber, in front of the Castel S. Angelo, the great fan of roads which Julius II completed with the Via Giulia running along the Tiber, lower down. This was the starting point for the steady shift of the city eastwards, as far as the Corso and the neighbourhood of the Capitol, where the Palazzo Venezia, originally conceived as a country house, eventually came to stand at one of the main cross-roads of the city.

Among the new buildings were the hospital of Sto Spirito, with its colonnades and its rounded arches, on the Esquiline, and the very important Palazzo Riario (later the Cancelleria) built by the pope's nephew between 1489–95. Its strongly rhythmical three-storey façade, and its huge inner courtyard with two loggias one above the other, is extremely interesting. The regular courses of the stone, the flat pilasters delimiting the bays, still bear witness to the principles of the Palazzo Rucellai, but in the type of the windows, in the delicacy with which the elements are put together, there is a touch of Laurana and the art of Urbino, and, finally, in the breadth with which its very size is handled, there is a conscious adaptation to Rome of all these stylistic features. There is nothing to justify the old attribution to Bramante, but it can be explained by the need to attribute so clear and grand a building to a great master. It may perhaps have been the work of a Lombard, Andrea Bregno, whose activity in Rome is marked by his numerous tombs.

For its monumental sculpture, Rome, during the last half-century, had turned chiefly to Florence, but towards the end of the century competition from Lombards was beginning to be felt. In 1432–33 Donatello had stayed for

THE QUATTROCENTO

a short time in the Vatican, and a rather confused *tabernacle* in St. Peter's can be attributed to this date. The bronze doors of St. Peter's, with their strikingly archaeological profusion of detail, had been cast by Filarete between 1439 and 1445, and later Mino da Fiesole executed in Rome the *tomb of Paul II* (1475) on the Florentine model, and that of *Francesco Tornabuoni* (S. Maria sopra Minerva) on Desiderio's sarcophagus design. The Vatican also attracted one of the masters of the new generation: Antonio Pollaiuolo. In Rome he created his masterpiece, the *tomb of Sixtus IV* (1484–94), which, through its style, its iconography, its siting in a mausoleum-chapel in the apse of St. Peter's, was the epitome of the ambitions of a great modern pontiff.[95] On a sort of bronze coffer, like a slab of stone lifted above the level of the floor, the reclining figure is surrounded by allegories of the Liberal Arts and the Virtues, expressed with great clarity, and separated by thick branches of acanthus. The *tomb of Innocent VIII* (1492–98), which returns to the form of the wall tomb, has the sarcophagus with the reclining figure of the dead man surmounted by the statue of the living pope.

Andrea Bregno (*c.* 1418–1506) was influenced chiefly by Mino; his structurally simplified tombs are covered all over the pilasters, and in the panels, by a wealth of small motifs with candelabra and branches, characteristic of Northern taste. The *tombs of Cardinal Coca* in S. Maria sopra Minerva, and of *Cardinal Cristoforo della Rovere* in S. Maria del Popolo are examples of this. Among local workshops, the most important was that working round Paolo Romano on the *Ciborium of Sixtus IV*.

If Rome was not an original centre in painting, any more than it was in sculpture, it is necessary to stress the way in which it nourished the inspiration of both with classical examples. Fra Angelico worked in the Vatican in 1440, and again during the pontificate of Nicolas V, when he painted the charming chapel of S. Lawrence and S. Stephen (1447–50), with its straightforward and sustained perspective effects. But these visits no more gave rise to a Roman school than did the passage of Jean Fouquet, invited to paint the portrait of Pope Eugenius IV, and that of Piero in 1459. Sixtus IV turned successively to the Umbrians and the Florentines; Melozzo to his greater glory created the scene of the *Inauguration of the Vatican Library* (*c.* 1477), in which the firmness of the profile portraits corresponds, in a spirit closer to Mantegna than to Piero, to the solemnity of the space; he also decorated the apse of SS. Apostoli (*c.* 1480).

The most fortunate venture of Sixtus IV was to summon to Rome in 1481 to decorate the new chapel in the Vatican which bears his name, a complete team of all the Florentine painters who counted for anything, from Botticelli to Signorelli; this commission will be discussed later. In the period that followed, the Umbrians rose to first place, with Pinturicchio's work in the Appartamento Borgia (1492–94), where the facile drawing and picturesque colour are enriched by numberless charming decorative details, stuccoes, gilding, flutings, which create a lively and amusing decoration, strange alike

in the style as it is in the nature of the imagery, which mixes the bull Apis
with the ceremonies of the Church in honour of the Borgias.

NAPLES AND SICILY:
ARCHITECTURE AND PAINTING

Southern Italy, where Spain was dominant, and Naples, which had passed to
the Aragonese princes despite the final efforts of the Angevins, were linked to
Spain and played an interesting role in the distribution of styles which characterizes the middle of the century in the Western Mediterranean. Between
1450–60 three great artists who passed through Naples formed their style
there: Antonello da Messina, and the two sculptors Francesco Laurana and
Niccolo dell' Arca.

The palace of the duke of S. Stefano in Taormina presents a typical Sicilian
form of picturesque Arabo-Gothic; the Palazzo Bellano at Syracuse, with its
pointed arches and courtyard with exterior staircase, is of a more developed
type. In Palermo, a composite portico with a triangular pediment was added
to the cathedral on the south side. In 1463, the Genoese Gagini, who worked
on the portal of S. Agostino, arrived in the island; after him, his son Antonello
spread a Lombard type of decoration. The only master who showed a proper
understanding of order and the balance of masses as they had been discovered
in Central Italy was Matteo Carnelivari, who built the Aiutamicristo, Pietra-
tagliata and Abbatellis palaces (c. 1488). He also built S. Maria della Catena
in Palermo, which presents on the façade an atrium with three high arches
surmounted by a heavy screen wall. In short, all these provincial centres were
slowly permeated by decorative forms on which they never really concentrated their attention until the advent of Baroque.

In the Abruzzi, in S. Maria di Collemaggio at Aquila, and at Soleto, the
richness of decorative sculpture continued the Gothic Trecento. Similarly, at
Carinola, and Fondi, small towns near Naples, the taste for a colouristic
decoration long survived, even in a building of Tuscan inspiration such as the
Palazzo Cuomo. The entrance gateway of the Castel Nuovo in Naples, commissioned by Alfonso V from the Catalan workshop of Sagrera from 1452 onwards, and finished in 1466 by Pietro da Milano, offers further evidence of
this trend.[96] It includes a frieze by Francesco Laurana and is surmounted by
a loggia crowned with niches. Inserted between two round towers, the superimposed storeys recall the high structures of the Angevin tombs in S. Chiara,
and perhaps also the Porta Imperiale at Capua of 1234. This impression is not
lessened by the quality of the detail, nor by the faithful imitation of the antique in the lower archway flanked by its coupled columns. The same triumphal spirit informs the Porta Capuana (1485), more regular and classical,
commissioned from Giuliano da Maiano, the builder of the villa at Poggio
Reale, by Ferdinand II, to whom Giuliano da Sangallo was to send plans for
a symmetrical and harmonious palace.

185

THE QUATTROCENTO

There again, complete developments only came during the seventeenth century. What did take root in Naples, and flourished exceedingly during the sixteenth century, were the groups of monumental sculpture—the *Nativity*, the *Deposition*—almost unknown in Tuscany, but much in favour in Burgundy and Languedoc. Following Niccola da Bari, Guido Mazzoni imported them into Monteoliveto in Naples. They consist of *tableaux vivants* in painted wood or terracotta, which lend themselves to every kind of naïve and picturesque development, always of a theatrical kind, suited to Southern piety.

It is principally in painting that links between Southern Italy, Spain, France and Flanders appear most clearly. The strongest evidence of this attraction is in the writings of Bartolommeo Fazio, a humanist at the court of Alfonso of Aragon, who in 1455 mentioned among the most illustrious men of the time Jan van Eyck (Johannes Gallicus) as *pictorum princeps*, and Rogier van der Weyden (Rogerius Gallicus), while he ignored Masaccio. A great anonymous master, in whom have been detected in turn Catalan, Valencian, Parisian and Lombard influences, sums up the complexity of this Western Mediterranean style developed around the large ports; this was the painter of the *Triumph of Death* in the Sclafani palace in Palermo (1440–50), in which may be seen the masterpiece of the late International Gothic style, with its tapestry-like composition lacking depth, its elegant silhouettes, its peculiar macabre quality. The most recent hypotheses suggest a Catalan or a Ferrarese painter.[97] This alone would be sufficient to prevent Sicily being considered as an inactive region at the time when the young Antonello was emerging.

While Antonello was finally entirely successful in marrying the new style of Italian painting to Northern art, his southern origins account for many things. In Palermo, a Sienese, Niccolo di Magio, a very minor artist, worked among painters from the Marches and Spanish masters like Pedro Serra. In Messina, it is not the isolation of the local school that complicates the position, but the presence of almost every style of Western art, and it is therefore the more difficult to speak of a Sicilian school. One of the most interesting artists, the Master of the Piazza Armerina Crucifix, who has a feeling for the monumental, may have come from Provence and be linked with Quarton. After the departure of Antonello, the very eclectic Antonello da Saliba and the dry Antonino Giuffré and Riccardo Quartararo, who were still working after 1500, were not of much account.

Antonello himself (1430–79) reflects the international culture acquired in Messina and Naples, where he was in contact with Colantonio, whom a humanist declared to have been his teacher, and whose activity may be dated between 1440 and 1470 (a Franciscan altarpiece of *S. Laurence*, of which a *S. Jerome* in Naples formed part, is by him). Antonello's *S. Jerome in his study* (London) is at once spacious and minute in detail, Flemish yet with perspective effects. From his earliest works onwards—the *Crucifixion* in Sibiu (Rumania), the *Three Angels* and the *S. Jerome* in Reggio Calabria—he displays his knowledge of the resources of Flemish art, and more precisely his knowledge of Jan van Eyck. But, at the same time, he had acquired, probably through Laurana, solidity in the posing of his figures (for instance, the *Salvator Mundi*, 1465, in London), and an authority in composition, above all in his portraits (1475, Louvre, Turin), which recall those of Piero. To the rendering of textures, and to the rich outer envelope of things, Antonello gave a new attention, but in his *Annunciations* (Munich, Palermo, Syracuse) it is the nobility of the gesture, the calm way in which the Virgin's blue mantle falls, the clarity of the composition, which dominate, along with novel adventures in perspective, such as the raised hand seen edge on.

With him, the Italo-Flemish accord, reached in the South, travelled up the peninsula and was to lead to very positive results, but Antonello discovered strong confirmation all along his route. The decisive moment came between the *polyptych of S. Gregorio* in Messina, dated 1473, with its corpulent saints, and its airy and placid feeling, the *S. Sebastian* in Dresden, in which the nude martyr is set peaceably on an infinitely extended paved floor, and the *S. Cassiano altarpiece* painted in 1476 in Venice (of which the fragments are now in Vienna). The architecture which envelops the group of saints both diversifies and distributes the light powerfully, and there reigns in this composition an atmosphere of saturated colour which in Venice was to be the point of departure for a new world.

Northern Italy

In 1450 Northern Italy was still very much cut up into separate units; each city, anxious to be in the forefront of development, strongly entrenched in its own tradition, reacted differently to the new ideas emanating from Tuscany. Sometimes their adaptation depended upon a master temporarily working there, sometimes upon the work of an accommodating popularizer.

In Genoa, for instance, the chapel of S. John in the cathedral, built by Domenico Gagini from 1456 onwards—before he left for Sicily—contains panels framed in branch motifs between bands with little niches inset into them, imitating Ghiberti, though the effect is closer to Romanesque art. Matteo Civitali (1436–1501) of Lucca, who had built in the manner of Alberti the small chapel of the Holy Face at Lucca, decorated Gagini's Chapel of S. John about 1480 with statues of *Adam, Eve* and the *Prophets*, in a more modern but rather feeble style.

In painting the situation is more interesting. From Genoa comes the curious picture of the *Annunciation* (*c.* 1480, Louvre) which is attributed to Carlo Braccesco; its novel colour scheme of yellow and black, its grave forms, its fine sense of space, mark the artist out as unusual. In Savona, which gained a certain prominence towards the end of the century through Cardinal Giuliano della Rovere, an artist from Nice, Louis Brea (1443–1523), adopted for his large altarpieces the Lombard manner of Foppa. It was chiefly around Nice that this Ligurian art reached its best development.

Milan. The power of Lombardy, a rival politically to Florence, was growing rapidly when Francesco Sforza, who controlled Milan in 1450, began attracting Tuscan artists to his city. The Portinari Chapel, imitated structurally from the Old Sacristy of San Lorenzo in Florence, was built at S. Eustorgio (1462) on plans supplied by Michelozzo, but with small towers on the corners which rather change its appearance.

With the sculptor-architect Filarete, Sforza came in contact with the most romantic, most exuberant of the disciples of Brunelleschi. His Treatise on architecture, written about 1464, is a kind of confused romance in which his opposition to Gothic art is frankly declared, but in which the many examples of ideal architecture which illustrate his thesis are each more fantastic and overloaded than the next: imaginary cities, buildings piled up like pyramids, façades with turrets and pediments, decorated with a plethora of branches, medallions and statues, and picked out with much colour. It was, in fact, a new metamorphosis of Gothic taste accomplished under the cover of Antique-inspired motifs. Filarete built the great central tower of the Castello in Milan in diminishing storeys (1451–54), and the Ospedale Maggiore (1456–65) with large courtyards of a Brunelleschian type, and his most typical feature is the placing of two storeys of open loggias borne on slender columns one above the other in the small courtyards. Soon after, bays with pointed arches were added to the façade. The lesson of Filarete is particularly valid for Bergamo: the project for a façade for the cathedral, of 1457, is a good example of 'Renaissance Gothic', or, if the term be preferred, of the Romantic architecture of the Renaissance which was to be current for two generations in Lombardy. The façade, surmounted by a huge pediment, is ornamented with polychrome elements borrowed from Lombard Gothic, and with a heavy gallery with pilasters and niches. Above this there is a huge dome on a polygonal drum, flanked by two towers with small balustrades edging the storeys.

The astonishing Colleoni Chapel, built somewhat later (1470–75) beside the church, is a kind of manifesto of the new school led by Giovanni Antonio Amadeo (1447–1522)—elements broken up, piled-up decoration, the insertion of Antique medallions into all the empty spaces, candelabra motifs under the pilasters and so on. An architecture of strongly accented members is covered all over with reliefs and small ornaments which make it look like the most modish kind of art. At the Certosa of Pavia, Amadeo decorated the small cloister; he returned in 1481 as director of the works and from 1491–98, with Benedetto Briosco, set about the façade. The gallery in the middle clearly divides the façade into two storeys; the upper part, finished in the sixteenth century, lacks its crowning member. The first storey is well articulated, but all the architectural elements are smothered with an almost intolerable conglomeration of statuettes, reliefs, and inlays; the basement is encrusted with giant medals, and even the mullions of the windows are carved into candelabra shapes.

The same florid effects can be found in civic architecture, for example in the entrance gate of the Palazzo Stanga in Cremona (now in the Louvre), and in works by Ambrogio Bonosci, such as the courtyard of the Palazzo Fodri. A tendency towards greater simplicity can also occasionally be discovered in Cremona towards the end of the century, for instance in the two-storey façade articulated with superimposed pilasters of the Palazzo Raimondi (1496). This was because in civil as in ecclesiastical architecture the ideas of Bramante, who came from Urbino to enter the service of Ludovico Sforza in 1481, were already creating a trend towards greater monumentality. The best example is in the cathedral of Pavia, the model for which was worked out jointly by Bramante, Amadeo and others, in 1488.

Until the arrival of Leonardo in 1482, Lombard painting consists almost only of Vincenzo Foppa (1427–1515). He used dark colours for his adaptations of Florentine models, for instance in his Madonnas, and in the *polyptych of S. Maria delle Grazie* (now in the Brera). Before 1468, he had decorated the Portinari chapel in S. Eustorgio, with a feeling for open space in both the landscape and the architectural elements. The ordered volumes and serene perspectives of Piero lost something of their clarity, but the search for monumentality inspired by the example of Mantegna and the Ferrarese governed the whole school. A huge work, the *altarpiece of S. Martino* at Treviglio, painted between 1485 and 1500 by Butinone (active 1484–1507) and Zenale (1436–1526) represents the last word in this brittle detailed style, cluttered with lines and tightly packed motifs. By contrast, Bergognone (1450–1523) softened the forms and tended to much quieter compositions, as for instance in his *Crucifixion* and his *S. Ambrose* in the Certosa at Pavia (1490).

Verona, Padua and Mantua: Mantegna. Padua and Verona, conquered in turn by Venice, thus found themselves linked with a new culture. But during an appreciable part of the century it was rather they who introduced new ideas into the city on the lagoons. It was only much later that windows with multiple lights and porticoes, such as are found in the Loggia del Consiglio in Padua, and the Loggia in Verona built by Fra Giocondo, were borrowed from Venetian architecture.

With the success of Gentile da Fabriano, who decorated the Doge's Palace, followed by that of Pisanello, International Gothic spread throughout North Italy the delicate Virgins, the fantastic costumes, the heraldic bestiary which characterized it, but the passage of artists like Masolino, on his way to Hungary or to Castiglione d'Olona in Lombardy, provided a stimulus of another sort. In fact, Pisanello was more interested in silhouette than in space in for instance his *SS. George and Antony* (London), but in his medals and drawings he repeated motifs from Antique sarcophagi and thus enriched a profoundly

Gothic system of forms with new elements in rather the same way that Ghiberti did.

Several years later, in 1457, the abbot of S. Zeno commissioned from Mantegna a large triptych which was destined to destroy the charming, fairy-tale world of Gothic. The altarpiece took the form of a building with four columns, carrying a heavy lintel and a rounded pediment, projected backwards in perspective. This aedicule imitates the construction which, in the Santo in Padua, Donatello had used to frame his statues of saints. Padua, which possesses no important buildings, had early marked out a path for itself by getting Donatello (1443-52) and Uccello, who decorated the Casa Vitaliani with now vanished 'giants' to work in the city. Donatello had an over-emphatic follower in Bellano (1430-96), whose reliefs in the Santo are his most important work, but Squarcione (1394-1474) was a notable collector in a province which, from the end of the Trecento onwards, was a centre for the trade in antiques, and also for the first medals struck in imitation of Roman coins by Venetian goldsmiths. Squarcione's workshop, where there reigned an almost 'surrealist' imagination stimulated by the romanticism of Antiquity, was the training ground of a whole group of young artists filled with fervour for Roman grandeur, passionately interested in archaeology, and clever at ornamenting their compositions with steles, inscriptions and classical arms.

The most gifted among them, Niccolo Pizzolo and Andrea Mantegna (1431-1506), began in 1449 the decoration of the Ovetari chapel in the church of the Eremitani. Pizzolo died at the end of 1453. In 1454, the youthful Andrea undertook the second series of *Scenes from the Life of S. James*, in the Ovetari Chapel, with the powerful representations of the trial and martyrdom of the saint, in which were manifested a hard, firm style, and an unprecedented power of draughtsmanship. Neither Tuscany nor Umbria had ever seen the like. His grasp of perspective enabled him to create extraordinary angles of vision, his drawing gave to forms the consistency of metal, nature was reduced to an architectonic system of blocks and flagged pavings, humanity was garbed in Roman fashion and lived among the buildings of Antiquity. The *S. Zeno altarpiece* reveals Donatello as the great source of Mantegna's art; he possessed the sculptor's enthusiasm, but not his delicacy. The predella is in three panels, the *Crucifixion* (Louvre), the *Agony in the Garden* (in situ), and the *Resurrection* (Tours), which form a sequence of harsh landscapes with red rocks. His iron drawing is linked to a cold, dry and conventional sense of colour, which towards the end of the century tended to become monochrome, as, for instance, in the cartoons of the *Triumph of Caesar* (1482-92, Hampton Court), and the *Dead Christ* (Brera). In Mantua, Mantegna worked for the Gonzagas, and decorated for them the *Camera degli Sposi* (1474) with an illusionistic dome in which he silhouetted against the sky laughing servant maids, a peacock, and putti, and with frescoes which show the throne room during the ceremonial meeting of the Marquis with his Cardinal son. These scenes from the life of the court were framed with pilasters and with ornamental bands that articulate the whole of the vaulting, with medallions of the Caesars and representations of Hercules. The framing of the doors, the windows, the fireplaces even, form part of the decoration, and this celebrated room became the frequently imitated model for state apartments with history scenes, in which were glorified a prince and his house, in a conventional heroic and mythological style.

For the *studiolo* of Isabella d'Este, also conceived as a unified decoration in which the pictures were framed in *intarsie*, Mantegna painted the *Parnassus* (Louvre) in a slightly facetious spirit, to face his own heavily didactic allegory of the *Triumph of Virtue* (also Louvre). The ensemble was completed after 1506 by Costa's laborious pictures of *Comus*, and the *Garden of Harmony*, Perugino's *Combat of Venus and Chastity*, and for Alfonso d'Este's *studiolo* Bellini painted his amusing *Feast of the Gods*, very different indeed from Mantegna's harsh style. The taste for solid forms, hard as coral and rock crystal, reached its height in the archaeological display which surrounds the Aigueperse *S. Sebastian* (1485, Louvre) and is still paramount in the large verdant apse of the *Madonna della Vittoria* (1496, Louvre) in which Francesco Gonzaga kneels before the Virgin.

A friend of humanists like Felice Feliciano, Mantegna recreated the Gospels and the lives of the saints in terms of Roman history. He worked out an enormous repertory of Antique medallions, military processions, tritons, monsters, which from now on form part of the common language of painting. His treatment of space was dry and sonorous; all natural forms, including those of the human body, are enclosed in stone and metal contours. This severe style, this rigorous domination of nature, corresponded to a deep-seated ideal of the period. A whole system of ornament was based upon it, with tube-like folds in the draperies, and garlands and motifs taken from Antique reliefs. Finally, this style found a new and decisive outlet in engraving, of which Mantegna was so great a master; compositions like his *Triumph of Julius Caesar*, the *Bacchanal*, and the *Tritons*, enjoyed a wide currency.

Mantegna's art represents a kind of necessary summing-up of the tastes and researches of the fifteenth century in the north of Italy. His influence was to be felt in Verona, in 1460–70, on an artist like Domenico Morone, and on the local school of illumination; in Padua he influenced the curious Bernardo Parenzano (1437–1531), and he exercised a more general effect in Venice through the Bellinis and Bartolomeo Vivarini. In the Marches, he influenced Carlo Crivelli (1430–95), a Venetian who filled his picture surfaces with floral brocades, highly worked decorative motifs and enormous garlands, and, finally, he had an impact on Ferrarese art.

Bologna and Ferrara. In Bologna, the Bentivoglio had employed a Florentine, Pagno di Lapo, to build a palace, now vanished, with an exterior colonnade and a terracotta decoration, as was admired in the northern provinces. The Palazzo Sanuti (later Bevilacqua) of 1481 has a spacious courtyard decorated with a terracotta frieze between the two storeys, and on the outside a curious rustication of regularly cut blocks. The same feature is found on a grand scale in the Palazzo dei Diamanti in Ferrara, the work of Biagio Rossetti, who also built the Constabili palace and the church of S. Francesco (1494).

Niccolo da Bari (1440–94), also known as Niccolo dell' Arca from his work on the *arca* or tomb of S. Dominic in Bologna for which he made a cover ornamented with statuettes (1469–73), was an Apulian who worked in Naples about 1480. Later (1485), in S. Maria della Vita in Bologna, he treated the great Burgundian theme of the dramatic *Pietà*, which reaches the heights of frenzied emotion in the figures of the despairing mourners in tumultuous draperies, and the agonized S. John. This violence contrasts with the *Tartagni tomb* (1477), a variant of the Florentine model by Francesco Ferrucci. In painting, Marco Zoppo (1433–78) is a close follower of Mantegna's hard style, but the major Ferrarese artists who received many important commissions in Bologna dominated local painting. Francesco del Cossa lived there after 1470 and after 1473 painted the *Griffoni altarpiece* in S. Petronio. The great Ercole de' Roberti decorated the Garganelli chapel in S. Petronio with frescoes (now vanished), but the lively style of these painters was lost at the end of the century through the adherence of Lorenzo Costa and the Bolognese to the mild and soft style of Francia.

Ferrara, which was a fief of the house of Este from 1208 onwards, and remained theirs until 1597 when it was absorbed into the papal states, only became an artistic centre at the end of the reign of Niccolo III (1393–1441), and during those of Lionello (1441–50) and Borso (1450–70). In a few years a taste for luxury and culture developed there; in the city itself and in its environs the Estes multiplied castles and villas built with a refinement, and sharp sense of enjoyment which left behind the phrase 'delizie estensi', but all these have now vanished. The most important, for the Italian and Flemish masterpieces which it contained, for the allegorical paintings with which it was decorated, and on which Cosme Tura was still working in 1460, and for its *intarsie*, was the retreat of Belfiore. Ercole I (1471–1505) imposed on Ferrara a controlling plan of a geometrical type, and the Roverella and Diamanti palaces thus stand on rectilinear avenues. A square tower with its storeys edged by pilasters was added to the cathedral. The most admired artists were continually invited there. Pisanello was there in 1432 on his return from Rome, and again between 1438–48. The portraits which he painted of members of the princely house (Louvre) show that it was really the sumptuous quality of International Gothic that Ferrara enjoyed. Jacopo Bellini was there in 1441, and it was an event of the greatest importance that Rogier van der Weyden, in Italy for the 1450 Jubilee, brought the Flemish technique with him in 1449 and painted for Lionello a triptych, which included the Uffizi *Entombment*, and also the portrait of a member of the Este family. Ferrarese interest in dazzling surfaces, and in sharpness of drawing, was thus much encouraged. In the same year, Mantegna was summoned to paint Lionello, and if he failed to return to Ferrara during the succeeding years, the example of his severe art, in which a romantic kind of culture is expressed, dominated the brilliant generation which was given its expression by the artists who worked in the Palazzo Schifanoia ('Begone dull care'), whose enlargement and decoration was carried out under Duke Borso. In 1449 Piero della Francesca painted frescoes in the municipal palace of the Corte Vecchia, and in a chapel for the Augustinians; the former were destroyed when the building was rebuilt in 1480, and the latter were early ruined by damp. But in illumination, a great Ferrarese speciality, and in particular in the huge Bible of Borso d'Este, something of the influence of Piero may be detected,[98] although in monumental painting it tended to enlarge rather than to discipline the imaginative resources of the younger generation of painters. They retained his power but not his harmony, his tension but not his serenity.

In short, the result of this welter of cross-currents was a tense and strained workshop, elaborate to the point of the fantastic, relishing disturbing effects of relief and brilliant colours with strident reds and greens. It possessed an original poetic feeling quite distinct from that of the Florentines, despite its debt to Donatello, and it competed with the more highly wrought manner of the Vivarini and Crivelli in Venice. Three great names adorn it: Tura (c. 1430–1495), Cossa (c. 1436–c. 1478), and Ercole de' Roberti (c. 1450–96). Cosme Tura had been a pupil of Squarcione in Padua, and his particular reactions to new Florentine ideas stem from that source. He studied Donatello and knew Mantegna, before he returned to Ferrara to pursue his own strange and implacable style; this consisted of a peculiar severity in contour and in the tight folds of drapery, a completely unreal use of colour which still further heightens these effects, a predilection for excessive and minutely detailed ornamental forms, shells, garlands, dragons, twisted motifs like steel shavings, and, naturally, armour and all the elements of a fantastic and semi-heraldic antiquity.

The *Roverella altarpiece* of 1470–74 (the centre panel is in London, the lunette in the Louvre, and the medallions are dispersed in various collections), was a huge composition built up in stages which bore witness to Tura's monumental powers; the drapery folds like veins, the symmetrically piled up groups of spiky and metallic forms, even the Hebrew inscriptions, meet fearlessly the demands made by ornament. The very large and fine *Annunciation* originally formed, with the *S. George* now separated from it, the organ doors of the cathedral (1489).

The complicated detail of the draperies, and the quantity of ornament in these static and sculptural compositions give parts of them an air of agitation and a tenseness which hovers on the brink of the grimacing, but it would be wrong to impute to Tura the harshness of style of a provincial artist; the little *Madonna*, and the *Pietà* in Venice (c. 1474, Correr) display the beauty of his enamel-like colour and the subtle highlighting which he could achieve.

It is not always easy to work out the links between Tura, who established the school and gave it its particular character, and the two younger men who developed his style: Francesco del Cossa with his much more ponderous energy and 'terribilità', and Ercole de' Roberti, with his exaggerated elegance and imagination. That they collaborated on the Palazzo Schifanoia is not to be doubted. The decoration of this palace, enlarged by Duke Borso after 1458 and where work continued until 1478, is the most impressive survivor of Ferrarese luxury and extravagance. The façade is ornamented with panels of marble and has a large entrance portal; in the principal state apartment is the most complete array of 'astrological demonology' to survive from the fifteenth century. This saloon of the 'months' was based on the theme of the powers which govern human life, each month being presided over by an astral 'lord' prefigured in an explicit 'storia'. One band is reserved to the appropriate sign of the Zodiac and the corresponding 'decans'; the lower band shows how the life of the court and that of the country proceeded under the reign of the Estes, in luxury and order and in accord with the laws of nature. This represents the last word in humanist pseudo-science, according to the treatises of Pietro d'Abano and Manilius. A brilliant and formidable style pervades these powerful compositions: the most beautiful, the most implacably conceived is the *Allegory of September*, in which the astonishing chariot of Lechery surges forward; there are reasons for believing that Ercole de' Roberti executed this section, and that it was his first and among his most startling works. On the east wall, the months of March, April, and May are by Cossa, partly after Tura's designs; December may be by one of their pupils, Galasso.

The *Annunciation*, with its heavily moulded architecture, in the large altarpiece of the Osservanza (now in Dresden) is a very typical work by Cossa, as is the grave *S. Petronio*, under an arch inhabited by angels, which once formed the central portion of the Griffoni polyptych in Bologna. But the unusual predella in the form of a frieze, with the miracles of S. Vincent Ferrer (Vatican) taking place among wildly exuberant architectural motifs and *contre-jour* effects of light, and the small figures of saints (of which there are two in the Louvre), are to be attributed to Ercole. These additions may have been made in 1475–77.

The authorship of the strange masterpiece the former *S. Lazzaro altarpiece* from Ferrara (now Berlin) is much more difficult to determine. The decoration of the throne, the animals, the landscape in the background, appear to be by Ercole; the rather stolid figures seem to be by Cossa. The *Madonna Enthroned* (Brera) painted about 1480, displays more balance and control, as if the artist wished following Bellini, to rejoin, the main current of painting. The surer composition prevents the relief of the decorative parts from obtruding, yet they are full of fantasy. The solids and the voids correspond in an original yet serene manner, so that the whole is bathed in intense light. His small works have the same brilliance and inspiration. His style was much imitated, and his influence was particularly strong on illuminators, among whom Guglielmo Giraldi was the best.

At the end of the century, Lorenzo Costa's (1460–1535) development is characterized by his infection with Umbrian softness, which can be seen in his *Bentivoglio Madonna* (S. Giacomo Maggiore, Bologna). The panels of the *Argonauts* (Padua), like the pictures painted for the *Studiolo* in Mantua, betray the effects of a new humanist culture. When Costa underwent the influence of Francia in Bologna, there remained in Ferrara nothing but uneven artists like Aspertini, who were already hovering on the brink of Mannerism.

DECORATIVE ARTS

Illumination. Only after 1500 did the development of books and engravings which took place during the second half of the century bring about the ruin of illumination; between 1460 and 1490 there took place, by a phenomenon common to the whole of Western Europe, a final flowering of the art of the illuminated manuscript.

In Florence, after the great activity of monastic scriptoria stimulated by Lorenzo Monaco and Fra Angelico at S. Marco, and continued by Zanobi Strozzi (who executed nineteen manuscripts for S. Marco between 1446 and 1453), the illustration of great liturgical books and humanist works stimulated by the bookseller Vespasiano da Bisticci, occupied several specialist workshops. Francesco d'Antonio, who made Lorenzo de' Medici's *Book of Hours* (1485) spread the use of characteristic framings of white branch motifs. An unidentified master illustrated, with all the sprightliness of a *cassone* painter, the *Virgil* in the Biblioteca Riccardiana (*c.* 1460). Gherardo (1445–97) and Monte (1448–1528) di Giovanni del Fora, who are close to Ghirlandaio in style, were very prolific illuminators; the *Dydimus* (Morgan Library, New York) is one of their masterpieces, with the *Psalter of Matthias Corvinus* (1490). Attavante (1452–1517) is a less important master who was inspired by contemporary work, as in the *Breviary of Thomas James, Bishop of Dol* (Lyons Library).

It was in the north that illumination reached its highest development. In Ferrara, Borso d'Este commissioned his *Bible*, with its six hundred pages in two volumes, from a team led by Taddeo Crivelli and Franco de' Rossi, between 1455 and 1461; the borders imitate jewelled, flowered fields, the many medallions are inspired by the most popular compositions of Pisanello, Cosme Tura, and perhaps even of Piero della Francesca. All the artists who worked on it did not have quite the same brilliance in colour; the most luminous, and closest to Roberti, is probably Guglielmo Giraldi, who also drew inspiration from Piero della Francesca in the bewitching frontispiece to his *Aulus Gellius* (Ambrosiana, Milan), and is on a par with the Umbrians in his celebrated illustrations to Dante (1475, Vatican).

The last word, however, is with the Veronese and Paduan tradition, whence came Liberale da Verona, an artist of great originality who worked as a painter at Viterbo and in Rome[99] and as an illuminator in Siena between 1467 and 1476, where, for the Monteoliveto monastery and the cathedral, he illuminated wonderful *Antiphonals* (Chiusi, and the Piccolomini Library, Siena), together with Girolamo da Cremona, who also executed a *Breviary*, more Paduan in style, in 1474 (Florence, Bargello). In style, his works are full of bizarre and poetic effects; the voluminous forms of his ornamental capitals are filled with dolphins, fluted folds, candelabras, and this fantastic framework envelops figures with wiry outlines and minutely detailed landscapes, Mantegnesque in feeling, often seen in the most dizzy perspective.

Marquetry. One of the most remarkable results of the evolution of taste is perhaps to be found in the transformation of *intarsia* work; nothing exploited more fully the decorative quality, in addition to the theoretical aspects, of the new type of construction by perspective.[100] Wood inlays, used extensively in choir stalls and sacristy furniture, included, at Orvieto for example (*c.* 1330), rich mouldings and rosettes inlaid into the panels. In Siena, Domenico de' Cori at the beginning of the century translated the articles of the Creed into marquetry panels for the chapel in the Palazzo Pubblico (1415–28); the work lacks a suitable style, the foreshortenings are clumsy, and the treatment of space inadequate.

The success of perspective permitted a very different type of ornament, the development of which has been attributed to Brunelleschi. He painted panoramas of city views, with paved squares, and with very strongly stressed lines of recession in the architecture, which could easily be treated as a system of regular patterns and then be cut up into sections like a puzzle. Uccello studied this sort of construction and those based on geometric bodies such as cubes, polyhedrons, scrolls, and spirals, from which were evolved a whole new repertory, probably used from 1440 onwards in Florence, in the decoration done by Antonio Manetti in the cathedral sacristy, brought up to date by the Maiano brothers in 1463. But it was in the northern provinces that these ideas developed into important compositions in the grand style.

From 1461–65, the Lendinara brothers, who were connected with Piero della Francesca, made admirable abstract decorations in the stalls of Modena cathedral; between 1462–69 in the Santo in Padua, they executed fine perspective scenes. In Urbino, in 1474, was built the *studiolo* with an *intarsia* decoration by the Florentine workshop of Baccio Pontelli, based on designs partly by Francesco di Giorgio. A third stage may be distinguished at the end of the century, with the celebrated *intarsia*-maker Fra Giovanni da Verona who did the stalls and the lectern in S. Maria in Organo in Verona (1500), where all the panels consist of perspective *vedute* remarkable for their poetic quality, and who also worked at Monteoliveto (1505), and in the Vatican in the Stanza della Segnatura (1511–12). Parallel to this, in Siena, the woodworker Barili executed, in 1500, very beautiful panels (now in S. Quirico

B   School of Belbello da Pavia: *An illustration to Plutarch's 'Lives'*
(*London, British Museum, Add. MS.* 22318)

d'Orcia) which were perhaps an imitation of a now vanished model. Decadence soon followed with the panels by Capodiferro in Bergamo (1524–30), full of puerile details, based on cartoons by Lorenzo Lotto.

Venice

ARCHITECTURE AND SCULPTURE

Given the particular conditions of a city built upon many islands which meant that perspective vistas were rare and panoramic views staggering, given the originality of traditional forms that had never been abandoned, and the development, as late as it was brilliant, of Venetian Gothic (the Ca d'Oro was not begun until 1421), it would be reasonable to suppose that Venice for long lay outside current artistic trends. Nothing of the sort happened. From the beginning of the century onwards, the maritime state also became a power on Terra Firma, and was in contact with Padua and Verona, whence it drew artists. But this was accompanied by a period of exhaustion in the artistic workshops of Venice proper, and during two or three decades foreign artists, in particular Florentines and then Lombards, were to be called in to work in the city. The Tuscans arrived about 1415–25: the sculptor Pietro Lamberti; Uccello, invited to work as a mosaicist; a little later, Donatello, Lippi, Castagno. About the middle of the century, work was resumed on the still unfinished Doge's Palace: the *Porta della Carta* by Giovanni and Bartolomeo Bon (1440–43), the colonnade and *Arca dei Foscari*, begun by them in 1457 and finished by Rizzo, have the regular pilasters, mouldings and niches combined with elaborate capitals, the pinnacles with statues, which are the hallmarks of Lombard style. Antonio Rizzo (before 1465–98) of Verona had, in fact, worked at the Certosa of Pavia before settling in Venice, where he was joined by the brothers, Antonio and Paolo Bregno. In his *Doge Niccolo Tron monument* (1473 onwards, Frari) Rizzo introduced a clever variant of the Florentine wall-tomb, extended to five 'storeys' and surmounted by a semicircular pediment; in it, niches and statues are multiplied, as in Gothic monuments. But the figure style is rather severe, the volumes straightforward, even geometric in handling, like the *Adam and Eve* on the *Arca dei Foscari*. In the eastern side of the courtyard of the Doge's Palace, which he completed after the fire in 1483, he used an arcade of pointed arches, but the big flight of steps facing the *Arca dei Foscari*—the *Scala dei Giganti*, dominated by the two statues, which is the grand ceremonial staircase of the palace—was treated with deliberate solemnity, and kept free of all superfluous decoration. The work was completed and the wings finished by Pietro Lombardo (c. 1450–1515) in the ornamental yet delicate style, strongly modelled yet controlled, which in fact goes to make up the Lombardo-Venetian decorative style. Pietro Lombardo was also a sculptor whose Madonna reliefs, for instance, were worked with a delicacy and grace akin to that of Bellini. In Venice proper, he designed very light tombs, though with stronger contrasts and more hollowed out than the Tuscan type, such as those of Pasquale Malipiero and Pietro Mocenigo of *c.* 1480 (both in SS. Giovanni e Paolo).

The Lombard style is supreme in the charming façade, covered in inlays and colour contrasts, of the Palazzo Dario by Pietro Lombardo, and in the reliquary-like church of S. Maria dei Miracoli (1481–89). Every possible device is used to concentrate the effect on the surface, and to destroy both depth and the solidity of the wall. The arches are decorated with coloured marble and gem-studded crosses; a huge semi-circular pediment decorated with rosettes, after the manner of Bramante, crowns an apparently weightless wall; inside, these are matched by a semi-circular, coffered, timber barrel-vault, which is borne by extremely ornate walls, and a very elaborate staircase leads to a raised choir.

The Bergamasque architect Mauro Coducci (1440–1504) rationalized these very Venetian fantasies into an architectonic framework based on Alberti's principles; this may be seen in the façade of the Scuola di S. Marco (1485–95) in two orders, the Palazzo Corner with its rusticated basement storey articulated rhythmically by pilasters Albertian in style, and the Palazzo Vendramin Calergi (1481–1509) where the principal storey, repeated twice, presents large two-light windows inserted exactly between columns which are coupled in the two end bays. Medallions and Gothic tracery sometimes appear in the detail, but the feeling for reform is evident. The same applies to S. Michele in Isola (1469–79) with nave and two aisles, and the façade of S. Zaccaria (1483), where the nave had recently been built by Gambello. There is a clear division into storeys by means of the string-course, and, in the case of S. Zaccaria, further division into compartments and bays, with, above, a curvilinear pediment linked to the sides by semi-circles which give a characteristic profile to buildings in this group, which also include S. Maria Formosa and S. Giovanni Crisostomo. It is of interest that Coducci, in this last church (1497 onwards), created a variant of the Byzantine Greek Cross type enclosed within a square, with domes, but without marble revetments or mosaics. The design of the Torre dell' Orologio and the Procuratie Vecchie, begun about 1480 along one side of the Piazza di S. Marco, are also attributed to Coducci, but their execution belongs to the sixteenth century, and presents problems which will be discussed later.

PAINTING

About 1430–40 the passage of Florentine artists—Uccello, Lippi, and probably Masolino—began to have an effect upon the more up-to-date Venetian workshops, particularly on that of Antonio Vivarini of Murano (*c.* 1415–80). For long he signed his works with a certain Giovanni d'Alemagna (d. 1450); the thrones of his *Madonnas* (Padua and Milan) are given a florid architecture.

When, later, he collaborated with his brother Bartolomeo (1432–99) who was trained by him, the example of Mantegna, which had already had its effect on them in Padua, reached an extreme in the triptychs in S. Maria Formosa (1473), and in S. Giovanni in Bragora (1478), in the bony saints, and the complicated drapery, painted in brilliant colour, recalling the vivid tones of stained glass and enamel.

The plastic force of the Florentines transmitted by Mantegna underwent, in the Byzantine-Gothic climate of Venice, a curious reinterpretation (and one better understood nowadays) expressed in cramped outlines, and the brilliant and ample colour of the Murano school. Its full development ran to elaborately denticulated forms, and minute novelties of ornament such as jewels and marble flowers in shapes similar to those of poppy heads; these are to be found particularly in the works of Carlo Crivelli (1430–95), who left Venice in 1457 and whose principal works were executed in the cities of the Marches (*Ascoli Piceno polyptych* (1475); the *Annunciation with S. Emidius* (1493, London); the *Coronation of the Virgin* (1493, Brera), with its surrealist architecture). His brother Vittore (*c.* 1440–1501) and an Austrian called Pietro Alemanno (*d.* 1498) continued his highly-flavoured and over-decorated style, but without his sharpness, which owes much to the influence of the Ferrarese.

The Bellini workshop followed quite another path; the career of Jacopo (*c.* 1400–70) makes this clear. His Madonnas, his portraits, derive from the Veronese, from Gentile da Fabriano, but in his Sketchbooks (Louvre, British Museum) he worked out space compositions by means of an exaggerated perspective with enormously distant vanishing points, making an almost visionary use of these effects. His sons Gentile (1429–1507) and Giovanni (*c.* 1430–1516), whose sister Nicolosia married Mantegna in 1454, dominated the two main directions which were taken by Venetian painting, apart from the Murano group.

Gentile, a sharply perceptive portraitist in his profile portrait of *Lorenzo Giustiniani* (1465, Accademia) went to Constantinople to paint Mahomet II and his court (1480, London); he is a chronicler in pictures, a story teller, for whom descriptive means suffice. But he has the gift of grouping his crowds in open squares in an even light; his vision of space is both decorative and balanced, as may be seen in his *Procession of Relics in the Piazza S. Marco* (1496, Accademia), and in his even more picturesque *S. Mark of Alexandria preaching* (Brera), which is a naïve representation of 'oriental splendour'. This narrative artist, who established in Venice a fashion for large compositions, of a type which Mansueti (*c.* 1485–1527) painted with such simplicity, stimulated the development of Vittore Carpaccio (*c.* 1455–1526), who painted large series such as the *Life of the Virgin* and *S. Stephen* for the Scuola degli Albanesi, and above all the *Life of S. Ursula* (1490–1515, Accademia), which may be compared with Memling's almost contemporary work, but treated as an adventure story with complicated town views (of Cologne and Rome, for

THE QUATTROCENTO

instance) and with delightful interiors (*the Dream of the Saint*). In the panels of *S. George*, *S. Jerome* and *S. Tryphon*, made for the Confraternity of S. Giorgio degli Schiavoni (1502) he reached his highest point, in the intensity of the landscape, the golden light, and the magical oriental fairytale quality of the image. But he also indulges in minute detail worthy of Crivelli, and a love of mineral forms, which explain the hermetic aspect of a work like the *Lamentation* in New York. The unity of the warm colour in his last works shows the influence of Giovanni Bellini.

All Venetian painting came to be concentrated in the long and splendid career of Giovanni Bellini (*c.* 1430–1516). Like the rest of the Venetian artists, he accepted Mantegna's grand vision of space, but, through the example of Antonello (1475), transformed it into a concept of space apprehended through light. By his consideration of the harmonious relation of tonal values, by concentrating on the creation of a tonal atmosphere, Giambellino brought all the painters of Venice, including those who were not interested in his vision of serenity, to the degree of maturity which rendered possible both the reforms of Giorgione and Titian and the resistance of Lotto.

A most important part of his *oeuvre* has been lost, since the histories which he painted in the Sala del Gran Consiglio in 1480 were burnt in 1577, but the curve of his development is inscribed in the succession of great altarpieces: the four triptychs (*c.* 1464, Accademia) painted for the church of the Carità, Mantegnesque in feeling, but with a deliberate insistence on perspective and on supple contours; the *S. Vincent Ferrer altar* (*c.* 1465, SS. Giovanni e Paolo) where the raking light from below dominates all the effects; the great *Pesaro altar* (*c.* 1475) where he reacted to the spatial amplitude of Piero della Francesca, followed by the *S. Giobbe altarpiece* (1486–87, Accademia) where the perspective provides an easy framework for the play of light reflected from the gold mosaic apse; and culminating in the triptych of the *Frari Madonna* (1488) and the great *S. Zaccaria altarpiece* (1505) which explore all the resources of harmonies of values, light and dark, and of diffused tonalities.

There is an equally clear development in the half-length Madonna and Child groups, starting with the *Potenziani Madonna* (Lehmann Coll., New York), with firm contours and garlands, and the *Enthroned Madonna with a sleeping Child* (*c.* 1470, Accademia), to the more luminous compositions dating from the 1490's (Bergamo, Louvre). The effect of Antonello can clearly be seen in the blues of the Madonna, and quantities of imitations have exaggerated the suavity of the theme. It is in the rendering of landscape, or, more precisely, in the harmonizing of the figure with the natural world seen as form and light, that Giovanni Bellini is incomparable: in the *Christ Blessing*, where the figure is arrayed in a shining robe (Louvre), and the Brera *Pietà* (*c.* 1470), the sky streaked with clouds plays a subtle role. Soon after (1480) in the *S. Francis* (Frick Coll., New York), qualified by the connoisseur Michiel as 'admirably studied', in the *Transfiguration* (Naples), with its subtle shafts of light, in the lovely *Allegory* (Uffizi), where a river flows between golden hills, in front of a luminous marble terrace, Bellini achieves that balance between space and tonal values from which rises the serene, pure and hitherto unheard paean in praise of the visible world. This inward splendour is so precious to him that he again exalted it in the panel of *S. John Chrysostom* (1513) where the saint is silhouetted against the sky, and in the strange *Feast of the Gods* (*c.* 1510, Washington), of which Titian finished the landscape. This is one of the artist's few secular compositions, with the exception of about ten or so portraits—nearly all in three-quarter view and against a background of sky—and the lost decorations. He had no need to move outside the traditional framework of Christian art to endow earthly beauty with a limpidity and a moving charm, and to create the thrill of something new.

The concentration of pictorial means thus obtained soon imposed itself throughout Venice. An artist dominated by pastoral feeling, Giambattista Cima da Conegliano (1459–1517), adopted Bellini's vision in his altarpiece of 1489 (Vicenza Mus.), his *Baptism* in S. Giovanni in Bragora in Venice of 1494, and his masterpiece, the *Carmini Madonna* in Venice of 1510. The Bergamasque Andrea Previtali (1470–1528), the Venetian Marco Basaiti (1470–1530) cannot be distinguished except as Bellini followers, and this applies also to Vittore Belliniano (*d.* 1529), and to a minor and more eclectic master like Lattanzio da Rimini (*d.* 1524).

Towards the end of the century, there were very few artistic personalities able to resist the attractions of the new style. The heir of the Vivarini, Alvise (1445–1503), had nevertheless turned towards the manner of Antonello, towards effects of relief and precise edges of forms, as in the *S. Clare* with a red missal (*c.* 1480, Accademia), and the little *S. Antony* standing out against the sky, as sharp as the lily he holds between his two fingers (*c.* 1480, Correr). This sharpness in his volumes, these cold tones, which were destined to be swept away by Giorgione, had their effect on an artist like Pennacchi, and provided a starting point for Lorenzo Lotto. Bartolomeo Montagna (1450–1523), who came from Brescia and was trained in Venice, and who founded a local school in Vicenza, also began from Antonello's clarity of form, though in his altarpiece from S. Michele in Vicenza (1499, Brera), and in his *Cartigliano polyptych* (*c.* 1509), he accentuates the envelope of light and shadow. Despite his brown and leaden colour, he did not escape the influence of Bellini, any more than did his pupils Giovanni Buonconsiglio (Marescalco; 1470–1535) and Cima.

The only artists who did diverge from the Bellinesque were those stimulated by Venice, but who turned towards Northern art, such as Boccaccio Boccaccino (1467–1525) in Cremona, and the unusual Jacopo de' Barbari (*c.* 1450–*c.* 1510), whom Dürer congratulated himself on finding in Venice, and who worked for northern courts. It was he who, with copperplate engravings inspired by Mantegna's, and on themes from Antique fables, engraved in 1500 the famous view of Venice which proclaimed the grandeur and beauty of the city.

PRINTING

This heading underlines the important development reached in Venice by engraving, particularly by woodcuts, and by book publication. As early as 1469, Johann and Vindelin de Spira (Speyer) introduced printing techniques, and in a few years there were in Venice as many printing workshops as there were in the whole of the rest of Italy, perhaps more than a hundred. The first known incunable is the 1469 Pliny; works with fine margins (Appian, 1477) and with illustrations (*Supplementum chronicarum* by B. Benali, 1486) were produced in quantity, and included celebrated editions such as Petrarch's *Trionfi* (1488), the *Novelle* of Masuccio (1492), *Herodotus* (1494), charmingly imitated from illumination, and the important *Hypnerotomachia Polifili* (1499), the woodcuts of which created a new, simple and delicate style, derived from the Antique, and are the glory of Aldus Manutius.

Florence and Italy at the end of the Century

The Tuscans long continued to hold the first place in architecture. In 1508, Luca Pacioli could still write in his treatise *De divina proportione:* 'Whoever wishes to build in Italy today, goes to Florence for an architect'. Thus Giuliano da Majano (1432–90) spent a good part of his career outside Florence, in Arezzo, Siena, and Naples, where he designed the Porta Capuana (1485), and created a new type of villa at Poggio Reale (1487). While he was working on the cathedral of Faenza (after 1476), he was also working with his brother Benedetto (1442–97) on the huge pilgrimage church of Loreto, where one by one all the principal architects were called in. Giuliano da Sangallo (c. 1443–1516), who was the head of a family of important architects, also worked in Loreto, Naples in the service of Ferdinando of Aragon, Savona for Cardinal Giuliano della Rovere, the future Pope Julius II (1495), and finally in Rome. The principles of Florentine art penetrated everywhere, but these contacts with the most varied regions of the peninsula and with foreign countries introduced reciprocal currents bringing a wealth of new ideas which tended to overwhelm even the strongest personalities.

Collections of examples were compiled; each workshop composed repertories of Antique details, types of Antique and modern plans. Among these sketchbooks of studies made at the end of the century, the most remarkable is the Barberini Codex (Vat. Lat. Barb. 4424) compiled by Giuliano da Sangallo, where may be found, alongside detailed drawings of Roman ruins, many plans, like that of the rotunda of S. Costanza, and even that of S. Sophia in Constantinople, copied from the sketchbook of a traveller.

Giuliano da Majano, sculptor in wood and architect, built the chapel of S. Fina in S. Gimignano, the cathedral in Faenza (1476–86), partly following Sto Spirito by the use of a curious combination of the vaulting borne on pilasters and basilical colonnades, the Palazzo Spannocchi in Siena (1473), and perhaps also the more elegant Palazzo Antinori in Florence. His brother Benedetto placed a long loggia of seven arcades in front of S. Maria delle Grazie in Arezzo, disregarding the façade, but carefully planning the square around the church.

With Cronaca (Simone del Pollaiuolo, 1457–1508), he was the builder of the most important Florentine palace of the end of the century, the Palazzo Strozzi (1489 onwards). The wooden model, which still exists, shows with what care the work was thought out, and how surprisingly faithful it was to the severe examples of Brunelleschi and Michelozzo, with a strongly stressed regularity and symmetry. The windows are the same as those in the Medici palace, but the ground floor is more severe still, and the cornice even more impressive. Inside, the vast proportions of the palace allowed the courtyard to be planned with singular nobility, with the walls of the first floor solid above the colonnade, and surmounted by an open loggia.

Sangallo, trained as a sculptor in wood, was the only all-round artist that

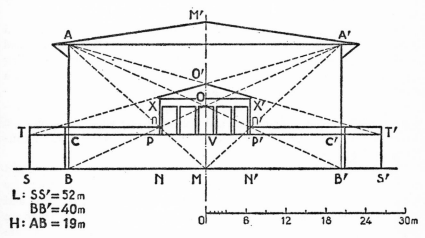

FIG. 23. *Villa at Poggio a Caiano, by Giuliano da Sangallo. Scheme for the façade (after a drawing by the author). The combination of the diagonals of the façade and the basement determines the position of the central portico.*

Florence could offer who could compete with Francesco di Giorgio as a military engineer. In 1483, Giuliano planned the fortifications of Ostia, which were executed by Baccio Pontelli, those of Poggio Imperiale in 1488, and of Borgo San Sepolcro in 1502. He was a parallel figure to Bramante, whom he met at Loreto (1499; dome) and whom he had the honour to succeed at the Vatican in 1514. In fact, his works alternated between Rome, where he began his large sketchbook (Barberini Codex) in 1465, and Tuscany. To him are due two decisive creations: Lorenzo de' Medici's villa at Poggio a Caiano, and the church of the Madonna delle Carceri. The first, situated between Florence and Prato (*c.* 1480–85), no longer has anything in common with Michelozzo's villas, and presages, in some respects, Palladio. The house is placed on an eminence in the midst of gardens, orchards and park, and is built on a square plan, raised on a kind of colonnaded pedestal; it presents a façade unified by a continuous projecting roof, with the central feature of a five-bay Ionic portico, crowned by a terracotta frieze below the pediment. This portico, built into the wall, is a portent of the future; it leads into an enormous saloon with a coffered ceiling which occupies the centre of the villa.

Giuliano's second memorable design was that of the church of S. Maria delle Carceri in Prato (1485–91), where the Greek cross plan reappeared for the first time in an independent building. The architect has perfectly understood and conveyed the quality of a structure in which the space is concentrated. The stressing of the various elements in *pietra serena* is Brunelleschian in type, but the pilasters and pediments of the façades are composed on a massive, Albertian model. His qualities of learning and imagination can also be seen fully in the sacristy of Sto Spirito, a two-storey octagon, built by Cronaca with a coffered barrel-vaulted vestibule.

The design of the Palazzo Gondi (1490–94) is less original, less sculptural than that of the Strozzi, built in the same years. But Giuliano, who had worked out a façade for Sto Spirito, displayed late in life (1515) an astonishing inventiveness in his projects for the façade of S. Lorenzo. They appear as a Baroque-tending conclusion to his Roman period, with architectonic elements, cornices and their supporting members, treated like living and suffering figures, stretching themselves and balancing one another. Niches are inserted everywhere, figures decorate the acroteria and create a rather confused tension which inspired Michelangelo.

The elder Sangallo is the exact contemporary of Francesco di Giorgio and Bramante. At the end of the century he was the most outstanding personality in Florentine architecture; Francesco was the first architect in Siena and Urbino; Bramante was the most influential architect in Lombardy. There are differences of temperament and taste between them; Giuliano's forms were more heavily, but more strictly, dependent on the Antique; Francesco's were more inventive, more unusual; Bramante's, finally, more ambiguous. But Giuliano's church in Prato, Francesco's in Cortona, and Bramante's S. Maria presso S. Satiro in Milan, all built about 1485, had remarkable analogies; unitv was beginning to emerge.

Bramante, who was trained in Urbino, went to Bergamo, where he worked as a painter in about 1477, and moved to Milan in about 1480 to work for Ludovico Sforza until 1499. His first work was S. Maria presso S. Satiro (1482–86), where a Latin cross form with a dome over the transept is inserted into a space limited by the street outside, so that a false perspective replaces the apse. The contiguous octagonal baptistery has a ground floor with niches and a storey richly decorated with medallions and friezes, surmounted by a dome. The whole has rather the character of a *tour-de-force*, through the insertion of a complicated decorative scheme within an Early Christian system.

At Pavia, the projected cathedral (1488 onwards) was worked out by Bramante together with several other artists—Amadeo, Francesco di Giorgio —and in the choir presents an interesting adaptation of the central plan. The fashion for this form was general. It constituted, however, a revolution which was justified from every standpoint. For the whole century, the sacristy of S. Lorenzo became a tantalizing model of a geometrically pure building: its square plan permitted an easy transition to the cube, with a regular arrangement of a small hemispherical dome. In S. Maria delle Carceri this resulted in a domed Greek cross plan which created a homogeneous central space, completely freeing the interior volume. S. Maria degli Angeli (Brunelleschi) and the rotunda of the Annunziata (Michelozzo-Alberti) had shown the search for the pure central plan based on an octagon.

This represented a return to an Early Christian type, the memory of which was still strong in the octagon of the baptistery of S. Maria presso S. Satiro (1482) and in the choir of S. Maria delle Grazie (1492–98, Milan). It certainly represents a systematic process of thought; for it was to be followed by the masterly church of the Incoronata at Lodi (1488–94), built under the direction of Giovanni Battaggio, where a clear and harmonious sense of volume is manifested, and by the most beautiful works of the succeeding epoch. It is patent that here a deeper knowledge of Antique central plans is in evidence; baptisteries with circular colonnades like S. Costanza, and buildings with niches like the temple of Venus Pudica, which were included in the Sangallo sketchbook. Thus, from all directions, the elements of a new kind of approach appeared; there was a move away from the intuitive character of Brunelleschi's classicism, through a more profound archaeological inquiry, and possibly also a stronger sense of the symbolical meaning in Alberti's sense, and, therefore, of the adaptation of these Antique examples to religious buildings. The best witnesses to the attraction exercised by these researches are the studies and theoretical musings on the central plan in Leonardo's Ms. B. (*c.* 1490, Institut, Paris).[101]

SCULPTURE

In about 1470 may be placed the turning point in Florentine culture which corresponds to the advent of Lorenzo, to the success of the Academy at Careggi, where reigned, besides Platonism, an unconditional glorification of the artist, a cult of an astrological and effulgent cosmos, and the hope of a humanist and religious Golden Age. In this climate, at once poetic, naturalist and abstract, participated those workshops which enjoyed the favour of the Medicean circles, and wherein were practised all the arts, under the dominating influence of sculpture.

Benedetto da Majano, who specialized in marble sculpture, followed the lead of Rossellino in his tabernacle at the Badia of Arezzo (1475), and in his *bust of Filippo Strozzi* (Louvre); his pulpit in Sta Croce, of a rather Donatellesque form, includes statues of Virtues in rose marble niches, and, between fluted columns, large narrative panels, inspired by painting. His *S. Fina altar* in S. Gimignano (1475) and his door to the Sala del Giglio in the Palazzo Vecchio (1480) are more solemn; his *Madonna* on the Strozzi Tomb in S. Maria Novella (1491) is colder.

More original and more varied, Antonio Pollaiuolo (*c.* 1432–98) designed, as has already been mentioned, a series of embroidered panels for the Baptistery (1466–*c.* 1480, Mus. dell' Opera). Towards 1460, he painted some energetic *Labours of Hercules* for the Medici palace. The same Hercules theme decorated the breastplate of a youthful warrior, whose lively and haughty bust is perhaps his most masterly piece of sculpture (Bargello). The small statuette of *Hercules crushing Antaeus* (Bargello), also made for the Medici, is rather loose in composition, but has a most forceful silhouette. This violent theme was fashionable; exaggerated muscularity suited Antonio's interest in anatomy, and in his sculpture and painting alike his strongly characterized types reveal the born exponent of naturalism. But when he came to make his papal tombs in the Vatican twenty years later he brought to them other qualities still: great variety in the workmanship, a barely restrained exuberance of virtuosity, but in the reliefs themselves—in the figures of the Liberal Arts for instance—a much less harsh invention. This eager interest in detail is replaced by an astonishing sobriety in a work which there is good reason to attribute to Sangallo: the *Sassetti tomb* in Sta Trinità (*c.* 1486). It consists of a sarcophagus of dark porphyry, crested with Antique bucrania, and placed in a niche which recalls the arcosolium of the catacombs.

In 1476, the commission for a tomb for Cardinal Forteguerri at Pistoia, which had been given to the Pollaiuoli, was finally, on the intervention of Lorenzo, transferred to Verrocchio and his workshop. The two firms were often rivals thus. Verrocchio (1435–88) owed much of his fame to his numerous Medici commissions—armour, and the monuments to Piero and Giovanni (1472). Between the highly worked surface of Pollaiuolo's bronzes and the smooth marble of the followers of Desiderio, Verrocchio went his own highly original way in his great compositions such as the *Incredulity of S. Thomas*, made for Or San Michele (1461–62), full of contrasts, of twisted drapery and large diagonal lines, and in his small works, such as the *Putto with a Dolphin*, destined for Careggi, and now in the Palazzo Vecchio.

He adopted and adapted the most famous motifs of Donatello; in his bronze *David* (1476, Bargello) the figure is given quite another poise and a more dreamy, almost indifferent, look; in his last work, the equestrian statue of Bartolomeo Colleoni, commissioned in 1481 by the Venetians, and cast after his death by Alessandro Leopardi, all is held in movement. An acute psychological subtlety is expressed in the terracotta of the *Resurrection* (*c.* 1470), and in the reliefs of *Alexander and Darius* (*c.* 1480). But he also knew how to rely on decoration alone to create a supreme effect. In his Medici tomb in S. Lorenzo (1472), the sarcophagus with its massive acanthus feet is inserted into a net made of thick bronze ropes, with a superb inscription in the Antique manner in lieu of any kind of imagery. It is easy to understand Leonardo's interest in a master both so intelligent and so deliberate in the creation of forms of surprising novelty and completeness; he was in Verrocchio's workshop from 1469 to 1475.

With the Pollaiuoli in Rome, and Verrocchio in Venice, the only sculptor to represent in Florence the vigorous art of Donatello in face of the eclectic style of the Majano brothers, was Bertoldo (c. 1420–91). One of Donatello's assistants on the S. Lorenzo ambones, he endeavoured to maintain his master's strength in tumultuous compositions, in which the multitude of hollows and bosses catch the light—for instance, in the Louvre *Madonna* and the Bargello *Battle of the Centaurs*. He was powerfully influenced by Antique motifs, which he retranscribed into bronze statuettes and small plaques (the *Orpheus* series, for instance). Lorenzo, with whom he was closely connected, chose him as keeper of his collections; these were rich in sculpture, precious stones, cameos, vases and other objects, and were regularly increased by the purchase of Antiques, so much so that the principal pieces of sculpture were placed in the garden of a summerhouse, near the Convent of S. Marco about 1489–90. A few privileged people were permitted a free access, including Granacci and the young Michelangelo. This circumstance which, in any case, could not have lasted for very long (Bertoldo died in 1491 and Lorenzo in 1492), gave birth a half-century later to the legend of the art school, the 'academy' of Lorenzo.[102] But this academic role that, in the long run, Florence would have played, Rome was able to ensure to the next century on a grand scale.

Sculpture at the turn of the century is therefore as full of contrasts in types as in its modelling. Thus Giovan Francesco Rustici (1475–1554), a friend of Leonardo, created with large effects of light and shadow, the group of three figures on the northern façade of the Baptistery (1506–11). Andrea Contucci, better known as Sansovino (1470–1529), the most important sculptor, used an impeccable but restrained stylization in his *Font* at Volterra (1501), and in his *Christ and the Baptist* in bronze over the east door of the baptistery in Florence (1502–5). Then he went to Rome. There he made the two tombs of Ascanio Sforza (1505) and Girolamo della Rovere (1507) in S. Maria del Popolo, of a Florentine type but softened by reliefs, and amplified by statues in niches and on the crowning members. There is much less newness of inspiration in Benedetto da Rovezzano (1474–1522) who made the *Soderini tomb* in the Carmine, in Andrea Ferrucci (1465–1526) or in Baccio da Montelupo (1469–1535) who made statues for the cathedral. All the genius seems to have been reserved for their contemporary, Michelangelo.

THE QUATTROCENTO

PAINTING

In Florence the everyday production of altarpieces, and coffers or *cassoni*, was the province of ordinary workshops in which the current styles were vulgarized—for instance, that of Neri di Bicci (1415–91), who has left a fascinating day-book of his commissions.

The choice of Tuscan and Umbrian masters summoned to Rome in 1481 gives an interesting picture of the position. The Pollaiuoli, who were not fresco painters and who were important chiefly as sculptors, were not invited. For similar reasons, the Verrocchio shop was not represented either; Leonardo had just left Florence for Milan. Along with the most original genius, Botticelli, were summoned the facile illustrators, Ghirlandaio and Cosimo Rosselli, and with them Umbrians, among whom the hard style of Signorelli and the soft manner of Perugino were already in opposition.

Antonio Pollaiuolo, as a painter, sacrificed beauty of tone to strength of contour; for him line has an evocative power of its own, and he sought for a harmony between figure and landscape, to which he accorded a new and extremely important place, in imitation of Flemish painters, as, for instance, in the London *S. Sebastian* (1475). His brother Piero (1443–96) helped him in the *S. Miniato altarpiece* (1467), and painted the *Coronation of the Virgin* (1483, S. Gimignano), which is lacking in vigour; he is stronger in profile portraits, where, following the examples of Uccello and Piero, the Florentine abstract taste for figures silhouetted against a luminous background is given full rein. The growing interest of the younger generation for the Flemish manner is expressed in Domenico Ghirlandaio (1449–94); his rather sad colour, his placid drawing, descriptive, without accent, his feeling for the clear distribution of his figures made of him a kind of softened Castagno, conscious of Baldovinetti and Verrocchio, but deliberately eclectic. Still rather naïve in his *Life of S. Fina* in S. Gimignano (1475), after his journey to Rome to work in the Sistine chapel (1481) and his contact with the realism of the Portinari altar of Hugo van der Goes, which had arrived in Florence by 1482, Ghirlandaio undertook frescoes in which contemporary figures treated as portraits were mingled with the sacred events. Of this type are the *S. Francis cycle* in Sta Trinità (1483–86), and the *S. John the Baptist cycle* in S. Maria Novella (1485–90), both with an uneven anecdotic and decorative quality, and a more personal feeling for space, which also informs the *Adoration of the Magi* in the Spedale degli Innocenti (1488). The very popular workshop was continued after the death of Domenico by his brother Davide, who, in about 1490, became a specialist in mosaic decoration, which was fashionable in the time of Lorenzo. Domenico trained many of the younger artists, among whom were Granacci and Michelangelo.

The only workshop to have a profound effect on the whole of Italian art was that of Verrocchio. His activity as a painter was limited to the period 1470–80, and the best example is the *Baptism of Christ* (Uffizi) which he perhaps left unfinished, and where Leonardo introduced in his turn the original detail of the two angels on the left, the modelling of which form such a contrast to the dry contours of the neighbouring figures. Verrocchio's style is most fully expressed in the *Madonna with SS. Donato and John the Baptist* (1478–85, Pistoia) a novel composition of simple breadth, fully modelled, brilliant, and with a delicate landscape background, in which he was assisted by Lorenzo di Credi (1459–1537), his principal pupil with Leonardo. Among the *Madonnas* (London, Berlin) attributed to him it is difficult to decide what is by his own hand, and how much of him passed into the art of Leonardo. Of the harsh style of the Pollaiuoli he retained only the feeling for harmonious contour, but this is no more than one of the factors in the luminous exchanges within the delicate modelling, which imitates both the sheen of bronze and the flutter of draperies. This stressing of relief effects occurs in harmonious, balanced compositions, classical in taste. Towards 1500, Ugolino Verino in *De Illustratione urbis Florentiae*, declared that Verrocchio was the master of 'all those whose name flies among the cities of Italy'.

Once Leonardo had gone, Botticelli (1445–1510) remained the major Florentine painter of the end of the century, but without any effect outside Tuscany. He expresses to the highest degree sensitiveness of contour and, by his opposition to pure landscape and atmospheric effects, he represents a sort of counterweight to the influence of Verrocchio. He develops from Filippo Lippi, but his earliest works have a sonority close to that of Verrocchio, for instance the *Fortezza* (1470, Uffizi), the little *Judith and Holophernes* (Uffizi), of splendid sonority. The line, which was the mainstay of the naturalism of the Pollaiuoli, becomes lighter, purer, and develops a more poetic liberty in the *Madonnas*, often painted in large medallions, or *tondi*, with elegant and, so to speak, musical compositions; examples are the *Madonna with the candelabra* (Berlin), the *Madonna of the Magnificat* (1485, Uffizi), the *Madonna with the Pomegranate* (1487, Uffizi), to which must be added the large *Coronation of the Virgin*, of an extreme suavity, and built up of tense lines, culminating in the rainbow garland of angels (1489, Uffizi).

Sensitive to the inspired and voluptuous ambience of the Medici Palace and of Careggi, Botticelli became the painter of new, allegorical, poetic mythologies, very different from the Roman processions of Mantegna, of a more distilled sensibility than the bacchanals and pagan feasts of the next century. These are those sacred visions, those decorative dreams of candid humanism, the *Primavera* (1478, Uffizi), like a tapestry garden, the *Birth of Venus* (1486, Uffizi), which appears to illustrate lines from Poliziano's *Giostra*, *Minerva and the Centaur* (*c.* 1488, Uffizi), a truly Medicean and Florentine allegory in which the goddess looks like a saint dressed in a ballet costume; this learned series was completed by an astonishing 'literary' reconstruction of the *Calumny of Apelles* (1495, Uffizi), tight in drawing, pushed to the limits of the bizarre, but limpid and lively.

About 1480, Botticelli's art was described as 'virile, of perfect proportions'.

He had recently painted for Ognissanti (1480) a *S. Augustine*, seen in his study in perspective, like an Antonello da Messina or a Flemish work, but outstanding in firmness of drawing, brilliance, and intensity of physiognomy. The frescoes in the Sistine Chapel, *Moses and the daughters of Jethro*, and *Corah, Dathan and Abiram*, are more uncertain in composition but extremely rich in detail. In the frescoes from the Villa Lemmi (1485–86, Louvre), which were nuptial allegories, the contours are exquisite, the colours pale, the space limited. His great compositions tend towards an increasing irrealism. The elongation of forms presages Mannerism, as, for instance, in his S. Barnaba altarpiece of the *Virgin with six Saints* (1485–86, Uffizi) and in the *Coronation of the Virgin* from S. Marco, already mentioned, where the crown of angels conceals his indifference to perspective, and is in harmony with the gestures of the protagonists. The internal tension results in a simplification of the drawing, as in the *Pietà* (1498, Munich). This orientation of the artist towards archaicism, and his rejection of 'modern' principles in favour of the creation of works of an exacerbated and visionary piety, are attributed to the preachings of Savonarola. Works showing these tendencies include the *Crucifixion* (Fogg Mus., Cambridge, Mass.), and the *Mystic Nativity* (1500, London). But the truth is more complex. Botticelli was in the forefront of a general return to Gothic expressionism, which had been manifested in the luxurious art of Ferrara, in the overworked and over-gilded manner of Crivelli, and in Florence in the harshness of the Pollaiuoli. But in Botticelli the movement finally worked itself out and finished in sentimental irreality.

The evidence for its not being merely a fortuitous feature of his late style is provided by the beauty of the drawings (originally intended to be finished in watercolour, but left as drawings) made to illustrate Dante's *Divine Comedy*. In the Inferno, the figures are tense and animated; in the Purgatorio, they gain in intensity; in the Paradiso there reigns a marvellous sense of a space limited only by the fine line of a circle, in which the ecstatic faces alone count. The same evolution characterizes the neurotic, impulsive but often cluttered art of Filippino Lippi (1457–1504), the son of Fra Filippo, and Botticelli's pupil. His youthful works, once collected under the name of a fictitious artist known as 'Amico di Sandro', exaggerated the delicacy and the aesthetic leanings of Botticelli; in the *Vision of S. Bernard* (1480, Badia) these are given rein in the landscape. In 1484, Filippino was considered worthy to complete the unfinished frescoes by Masaccio and Masolino in the Carmine—*the Scenes from the life of S. Peter*. In the Carafa Chapel in S. Maria sopra Minerva in Rome (1489–90) where he treated the *Triumph of S. Thomas Aquinas* as a kind of theatrical performance, and again, and principally, in the *Scenes from the lives of SS. John and Philip* in the Strozzi Chapel in S. Maria Novella (1498–1502), he displayed an extraordinary vivacity, which indicated both his romantic view of history and the hardening of a draughtsmanship still lively and inspired though incapable of discipline within the composition. This is the exact opposite of the weighty art of the Umbrians. This impression of crisis in ideas and style appears no less clearly in Piero di Cosimo (1462–1521), an eclectic artist, inspired by Verrocchio and Ghirlandaio, who went out of his way to find strange themes of recondite mythology such as *Prometheus* (Strasbourg), or primitive savagery in *The Hunt* (New York) and *Lapiths and Centaurs* (London). All these painters, soon forgotten in the sixteenth century, were only rediscovered to have some interest a generation later, during the age of Mannerism. This was not to be the case with their contemporary Leonardo.

The historians who, following Vasari, have placed at the beginning of the sixteenth century and have defined as starting with Leonardo the break with the 'hard and sharp style' (Vasari) have falsified the truth. The phenomenon was fairly general during the last decades of the fifteenth century, when the two styles impinged on each other, somewhat confusedly, and does not derive only from Leonardo. Venetian art pursued on its own a remarkable evolution, already implicit in the second period of Giovanni Bellini, and stressed more sharply at the end of the century, when a move was made away from the sharpness and plasticity of Mantegna towards the melting tones and atmosphere which directly prepared the way for Giorgione. The same thing happened in Ferrara, and at Mantua with Francia and Costa. It was principally in Umbria that the conflict between the dry manner, that is, the sculptural manner, and the more relaxed and softer one, came to a head in the opposition between Signorelli and Perugino, who were both, to a certain point, the heirs of Piero della Francesca. Luca Signorelli (1450–1523) began with Piero, but from his *Flagellation* (c. 1475, Brera) onwards, he introduced a note of violence, a dramatic use of light, which became more noticeable still in the Sacristy of Loreto where he worked, c. 1480. He tended, therefore, towards strongly symmetrical compositions, with stereotyped contrasts of colour and physiognomy, as in the group of *Saints around the Madonna* (1484, Perugia), and in the evocation of a pagan myth in the *Pan* (c. 1490, destroyed during the war) painted for Lorenzo de' Medici. Signorelli restored the links between Umbria and Florence. But he ended his days at Cortona, at the head of a mediocre provincial workshop. He was, before Michelangelo, whom he inspired, the last great Tuscan fresco-painter of the fifteenth century. In the Sistine Chapel in his *Scenes from the Life of Moses*, at Monteoliveto Maggiore (1497 and later) in his *Legend of S. Benedict*, and above all in the *chapel of S. Brizio* in Orvieto cathedral (1499–1504) where his use of anatomy, his ability in foreshortening, his feeling for pathos, triumph in his unforgettable rendering of the *Antichrist* (who is almost certainly Savonarola[103]) and the *Last Judgement*. One link between this search for the forceful and the softer manner pursued in Siena at the same time as in Perugia is provided by the Camaldolese monk Bartolomeo della Gatta (1448–1502/3). His two pictures of *S. Roch* (1479, Arezzo) are close to Signorelli, but in 1481 he collaborated with Perugino in the Sistine Chapel, and softened his style in the Castiglione Fiorentino *Madonna* (1486–87), to turn later still to Melozzo and even to Piero.

THE QUATTROCENTO

Perugino (1445–1523) also worked in the Sistine Chapel, and created a grand historical cycle consisting of heroes, sibyls, the *Nativity*, and the *Transfiguration*, in the Sala del Cambio in Perugia. He also was in contact with Florence, and for a time worked for Isabella d'Este (his *Love and Chastity*, 1505, Louvre, was painted for her). But he sacrificed everything to the softness of his types, to their calm, smiling features, and his symmetrical compositions are harmonious but flaccid—as, for instance, in his *Madonna with the Baptist and S. Sebastian* (1493, Uffizi). From Piero he derived and successfully exploited a full and convincing space construction, clearly organized and with a wonderful feeling for spaciousness and atmosphere, in which a building set in the middle of the picture space is perfectly suited to the action (as in his *Charge to S. Peter* in the Sistine), or in which a tender landscape provides a delicate backcloth (*Apollo and Marsyas*, Louvre). It was this perfect harmony, this freshness, that attracted Raphael. But at the same time Leonardo was maturing a style which took no account of these serene solutions, since he was transforming the very basis of the problem.

Leonardo da Vinci (1452–1519). As a pupil of Verrocchio, he inherited all the aspirations of the Florentine Quattrocento in all the arts, and must be considered from this standpoint.

He was born in the little town of Vinci in 1452, and was placed by his father with Verrocchio in 1469; he established himself as one of the younger masters in Florence from 1475–81, went to Milan to work for Ludovico Sforza until 1499, and divided his last years between Florence (1500–6), Milan (1507–13), Rome (1513–15) and Touraine in France (1515–19). He produced little; no building, for all that he was concerned with architecture; little or nothing survives of his sculptures done in Verrocchio's workshop, nothing at all of his Milanese masterpiece; finally, as a painter, there are at most ten or so works, of which several remained unfinished. One of his masterpieces, the *Last Supper*, is an over-restored ruin; another, the *Battle of Anghiari*, had disappeared by the sixteenth century. But the artist's significance is immense, his personality fascinating, and both can be explained through the high quality of his works and the abundance of his writings.

As a young man, in Verrocchio's shop, Leonardo had modelled heads of women and children with lovely curly hair and smiling expressions; this was an inheritance from Desiderio, but with an added enchantment. The regal Virgin set in a flowery meadow and the elaboration of all the details in the important *Annunciation* (c. 1475, Uffizi) show the extent to which Florentine aesthetic quality and subtlety of feeling had affected the young artist; the same impression of impeccable form and style, allied to a Verrocchiesque feeling for modelling, is given by his portrait of *Ginevra de' Benci* (c. 1478, Liechtenstein coll.). The *Adoration of the Kings* (Uffizi), commissioned in 1481 by the monks of S. Donato a Scopeto and abandoned by Leonardo when he left Florence, shows the conclusions at which he had finally arrived, and his solution to the problem of his generation: the gradation of chiaroscuro, the 'sfumato' which dissolves the contour and the plastic mass into a new, more suggestive, more poetic reality, which would give full play to discoveries in the expressiveness of physiognomies. At the same time, a complete and carefully ordered composition was worked out, with the central group contained within a triangle.

In Milan, Leonardo continued to work on painters' problems. In the *Virgin of the Rocks* (version begun in 1483, Louvre; replica by Leonardo and his assistants, London) the artifice of the construction is more obvious, but so is its transfiguration by light; in the *Last Supper* (1495–97) in the refectory of S. Maria delle Grazie, a traditional theme treated with the maximum of order and symmetry is charged with the maximum of emotion. He also tackled the problems of sculpture, and architecture, the first with the equestrian statue of Francesco Sforza, of which only the giant model was slowly elaborated, with carefully worked out contrasts, and the second with his participation in the arguments on the spire of the cathedral, and on the cathedral at Pavia (1488); as a result of his contacts with Bramante, he multiplied his studies of architectural forms capable of conveying the greatest impression of equilibrium with the richest handling of volumes. He decorated with interlaced trellises of foliage the Sala delle Asse in the Castello Sforzesco (1498; remade during the nineteenth century), planned numerous festivities, and took part in the life of the court surrounding Cecilia Gallerani, Ludovico's mistress, who gathered around her a following of poets and musicians. He composed rebuses on gallant themes in 1497. Finally, with Bramante, with the mathematician Pacioli, with Lombard engineers and technicians who occasionally met, under the patronage of Ludovico, in learned disputations, he was able to develop in every direction the knowledge he had acquired in Florence.

In Florence, he had conceived the unity of all the arts and its application to all aspects of nature and life. In 1481, his letter to Ludovico set forth his universal powers. In Milan, the artist conceived the necessity of giving a foundation in theory to his doctrine by means of a series of treatises on perspective, on anatomy, on mechanics, and so on, which would constitute a new *organon*. He undertook to this end an enormous labour of investigation and inquiry, projecting from time to time partial publication (1489 and 1508) which in fact never took place. His ideas remained unknown until the publication of his manuscripts; only the treatise on painting, in a version put together in the sixteenth century, appeared in 1651. This enormous compilation led Leonardo to elaborate a complete cosmology, which should associate earth and sky in the pulsations of a universe animated by irradiating light, and shot through by the incessant struggle of the elements, above all by water, the basic constituent of the blue atmosphere which envelops the distances of pictures. The summit of all theory was to be found in the painted work of art, which rendered possible the only complete worship of beauty and richness of form; the only practical achievement was to be obtained from mechanics,

THE QUATTROCENTO

which subdued and organized nature, and provided unforeseen enjoyments as much for the pleasures of life as for the most material necessities.

By the time the Milanese duchy fell in 1499, Leonardo was celebrated all over Italy; Isabella d'Este in Mantua sought to attract him to her court and commissioned her portrait from him, for which the admirable cartoon, which alone was executed, survives (Louvre). The return of the exile to Florence in 1500 marked the beginning of a new era. The cartoon for the *S. Anne* (1501, London), and the slightly later picture (Louvre), in which the lively group is composed in a pyramid against a landscape of glaciers, the portrait of *Mona Lisa* (Louvre), where never before had painting attained to such subtlety in modelling, to such tonal unity through chiaroscuro, and finally, the involved *Battle of Anghiari*, in the great hall of the Palazzo Vecchio (1503–5), achieved a universal fame. His finished works became more rare, his projects more numerous; with the *Bacchus* (*c.* 1506, Louvre), much altered, must be mentioned the lost *Leda*, known from drawings and painted between 1504–8, and two *Madonnas* painted for Louis XII of France (*c.* 1506–1510, now lost). In Milan, Leonardo returned to the equestrian statue with the *Trivulzio monument* (1511–12); in France, he resumed his architectural projects for royal buildings (1516). But these last years saw, above all, the multiplication of those astonishing drawings in which man, analysed through anatomy, and nature examined in cataclysms of storm and flood, seem as if ready to yield the same secret.

II

The Cinquecento: Rome

Introduction: Historical Background: the Triumph of the Papacy

Lorenzo de' Medici died (1492) before the full development of the High Renaissance, and his patronage was immediately followed by Savonarola's revolution when art once again tended to be circumscribed by the piety of Dominican tradition. The re-awakening in 1500, under the Gonfalonier Soderini, was remarkable. Suddenly Florence again became for a few years the artistic centre of Italy. Botticelli, Filippino Lippi, Lorenzo di Credi, Davide Ghirlandaio were all then working there. The position of leader was still vacant; but everything pointed to its being taken by Leonardo da Vinci, who had painted the *Last Supper* in Milan (1495–97), and who reappeared in Florence in 1501 with an enormous prestige. His cartoon of the *Madonna and Child with S. Anne* excited general interest. He was commissioned to execute some engineering works and a gigantic decoration in the Sala del Consiglio of the Palazzo Vecchio. He was working on it in April 1505, but his preparations were faulty and the colour ran; by the end of the year he had abandoned the work. His plans to deflect the course of the Arno also failed. Though he returned to Milan in 1506 full of the bitterness of failure, he had exercised a decisive influence on the young Raphael (in Florence from 1504–8) who finally broke away from his Umbrian insipidity, and he had stimulated Michelangelo who was commissioned to paint the fresco of the *Battle of Cascina* in the Palazzo Vecchio next to Leonardo's *Battle of Anghiari*.[104] In 1506 Michelangelo, and in 1508 Raphael, went to Rome, summoned by Julius II (1503–13). In a few months all was changed. The immense works projected by enlightened patrons drew all the most famous artists to Rome, which in one decade regained a unique importance, and restored, under the aegis of the Church, that imperial significance accorded to it by the history of the humanists.

The movement created by Julius II grew under his successor, Cardinal Giovanni de' Medici, son of Lorenzo, who became Pope Leo X (1513–21). Through the personality of Raphael, the Roman School acquired a decisive importance, and Florence tended to become subordinate to it. The return of the Medici in 1515, and the government by legates during the pontificate of Clement VII (1523–34), the posthumous son of Giuliano de' Medici, led, notwithstanding the republican revolt of 1527, to the restoration of Medici power, and to the Grand-Duchy of Tuscany set up under Cosimo I, in 1537.

This waning of Florence was only one of the consequences of the enormous growth of Rome and of the Papal States during the intrigues of the Wars in Italy. The conquests of Charles VIII of France (Milan 1494; Naples 1495), renewed by his successor Louis XII (1501), were lost in 1512 through the efforts of the coalition engineered by Julius II. With the arrival of Francis I of France (1515), a series of treaties confirmed the French position in Milan, that of the Papacy in Central Italy, and of Spain in Naples. During the next thirty years the peninsula became the closed field of Imperial power through the Spanish domination resulting from the election of the Emperor Charles V (1519). Clement VII's efforts to rid himself of two invaders at the same time ended in the disaster of the Sack of Rome (1527), amid scenes of unbelievable horror, by the Imperial troops under the Constable of Bourbon. This terrible blow to Papal power marks the end of the Renaissance, or, at least, of that period when the optimism of Rome, humanist confidence, and the patronage of the arts went hand in hand. The vandalism of the Lutheran soldiery also made it clear that the period of the Religious Wars had in fact begun.

The relaxation of morals that made it possible for thousands of courtesans to occupy high social positions, the learned diversions, and the rising vogue of the theatre (the celebrated *Calandria* by Bibbiena with settings by Peruzzi was presented in 1514), the growth of libraries and above all the multiplication of humanist salons among the rich patricians (Paolo Giovio, Colocci, the banker Agostino Chigi, Castiglione, the arbiter of taste), provided Rome with an unparalleled atmosphere of culture and pleasure. Erasmus himself wrote, 'I cannot conceive a greater happiness than to return to Rome' (1515).

This Roman culture was more passionately concerned with archaeology than any other. For several decades Rome had been the great centre for antiquities and the meeting place for collectors. In this city filled with ruins, where vast empty districts were laid out in lovely gardens, finds succeeded finds: the Apollo of Antium, the Laocoon on the Esquiline, immediately identified by Giuliano da Sangallo and Michelangelo in 1506, and the celebrated 'torso' all entered the papal collections for which the palace of the Belvedere was planned. The trade in forgeries was considerable; it was in Rome that in 1496 Michelangelo's *Sleeping Cupid* was sold as an antique to the Cardinal di San Giorgio. One of the best archaeologists of the day, the architect Fra Giocondo of Verona, supervised the excavations and all new building so as to prevent too much destruction of interesting remains. In 1515, Raphael succeeded him in this function as Curator of Antiquities. He began a huge survey, parts of which may have been used by the increasing number of engravers specializing in scenes of ruins, and which finally bore fruit in the plans of Pirro Ligorio (1561) and Lafreri (1577). At no other time were antique marbles and inscriptions given so much learned consideration by masters so well able to interpret their significance. But despite Julius II's appetite for power, and Leo X's epicureanism, it would be wrong to accuse the Renaissance of displaying unprecedented cynicism. Though no serious reforms were made within the Church by the Lateran Council which met from 1512 to 1517, it gave the theologians an opportunity to express their views, and the inaugural discourse of Cardinal Egidio of Viterbo is an adequate reminder that the general development of humanism encouraged a deeper understanding of religious philosophy, and led the thoughtful to adopt a questioning attitude, with a remarkable perception of the drama of the moment and of its universal significance.[105]

ARCHITECTURE

Despite the efforts of Sixtus IV, Rome, at the beginning of the century, consisted largely of ruins, gardens and unhealthy marshland with a sordid and densely populated centre dominated by the Castel Sant'Angelo and was very unsafe, since the ancient cellars in the hillsides had become the haunts of bandits. His nephew Julius II undertook considerable works for the amelioration of public health at the Vatican, along the banks of the Tiber, in the Arenula district and round the Capitol. He rebuilt the walls of the city between the Porta Salaria and the Porta Pia, remade many roads, encouraged cardinals to repair streets and restore churches, and in 1512 could commend himself in an inscription placed in Via Sto Spirito for having ornamented the city *pro majestate imperii*. After 1527 and during the stringencies of the Counter-Reformation, town planning projects in Rome were brought to a halt, and it was only at the end of the century that the modernization of the city was again embarked upon.

Bramante. It was Bramante whom the Pope entrusted with the supervision of these works. He had arrived in Rome in 1499, with an architectural experience that enabled him to sum up the progress made during the preceding age; he had learned to appreciate the beauty of a delicate and sensitive use of geometry at Urbino, the clarity of Alberti's works in Mantua, and the grand treatment of masses studied by Leonardo at the time of the meetings in Pavia.

His first two works demonstrate the extremes of his range: in the Tempietto at S. Pietro in Montorio (1502) a circular portico in the strictest Doric order is united to a cylindrical core surmounted by a small dome. The complete project, known from an engraving by Serlio, called for a surrounding circular cloister with niches which were to echo those of the central structure, and serve as a sheath for this miniature temple, so like a Vitruvian model. The little cloister of S. Maria della Pace (1504) conceals under its apparent simplicity a very complex arrangement of supporting members: the cruciform piers are repeated on the first storey, where they alternate with columns carrying the entablature, of a type close to the second cloister of S. Ambrogio (Milan). As he became better known, Bramante was probably also consulted for the Palazzo della Cancelleria, then building on the Corso, for the design of the interior courtyard in three storeys is of a clarity quite new in Rome.

Bramante developed a completely monumental style with the Palazzo dei Tribunali (not completed, and known from Peruzzi's drawings) which was to consist of a large courtyard with an arcade, towers at the corners, and, flanking the Via Giulia, a lower storey composed of huge rusticated blocks. The House of Raphael in the Borgo (destroyed in the seventeenth century) presented a rusticated lower storey inspired by the Forum of Trajan, surmounted by a row of five high windows each between two Doric columns.

In church design, Bramante was interested chiefly in the development of the central plan: SS. Celso e Giuliano (rebuilt in the eighteenth century) consisted of a square from which projected rounded apses, repeated on a smaller scale at the four corners. The plan of S. Biagio della Pagnotta (within the Palazzo dei Tribunali) is even simpler. The use of massive and simple volumes in the choir of S. Maria del Popolo sums up this development.

The effects which Bramante had sought to obtain in his Milanese work through the careful cutting of the ornament, and the use of decorative devices and false perspective, were progressively eliminated. Their place was taken by a precise planning of the volumes of a building which more fully expressed his intentions through the contrast of light and shadow obtained from the alternation of voids and solids.

At the Vatican, Bramante proposed as early as 1503 to redesign the whole hillside so as to link the Belvedere of Innocent VIII to the basilica and the papal palace. A drawing by G. A. Dosio (Uffizi) and a later fresco in the Castel Sant'Angelo reveal that the architect had conceived the intervening space as a continuous edifice, closed on the east by a wall with rhythmical bays hollowed out into a huge absidal exedra (finished in the form of a niche by Pirro Ligorio in 1562), where the gigantic fir cone brought from the atrium of Old St. Peter's was eventually placed. The lie of the ground enabled terraces, flights of steps and exedrae to be built on the pattern of the temple at Praeneste, so as to restore antique Rome and serve as an archaeological garden.[106] But this superb design was destroyed by the construction of the library (1588), and then by the Chiaramonti wing (1617). Bramante's rapidly completed design was the first radical transformation of a Roman site for centuries. It created a striking impression, but it was only the accompaniment to another even more sensational work—the new basilica of St. Peter's, which was not to be completed until the next century. Bramante's original conception was, therefore, subjected to the most unfortunate alterations; but its significance cannot be overrated, in that it expresses the desire for monumental clarity which he inspired around him, and is an example of the vitality displayed in all his projects, and the amazing optimism of all his undertakings. In a pamphlet entitled 'Scimia' (Monkey) written by an opponent, he is made to

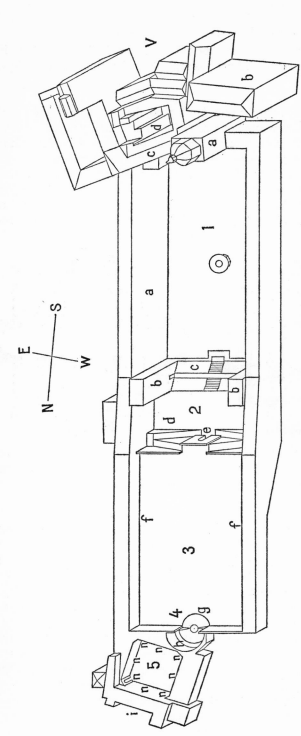

Fig. 24. *Courtyard of the Belvedere of the Vatican. Reconstruction of Bramante's project, 1503–4 (after J. S. Ackermann): V, Vatican Palace—a, Torre Borgia; b, Sistine Chapel; c, Palace of Nicolas V (library, Appartamento Borgia and Stanze); d, Cortile del Papagallo. 1, Lower courtyard—a, arcades; b, towers. 2, Middle courtyard—c, grand stairway; d, ramps; e, nymphaeum. 3, Upper courtyard—f, loggie. 4, Exhedrae—g, circular stair; h, hemicycle. 5, Courtyard of the Statues—i, Villa of Innocent VIII; n, niches.*

declare that he intended to destroy the old Paradise to build a more beautiful one, and if his architecture did not please S. Peter, he would address himself to Pluto.

The New St. Peter's. On the 18th April, 1506, the first stone of a pier of the dome was laid. The new basilica was to be on a central plan, and this was immediately laid out on the ground. The tomb of the pontiff, which Michelangelo was to design, was to be placed in the new church, which would thus also become the mausoleum of Julius II. The replacement of Constantine's basilica had been planned ever since the pontificate of Nicolas V, and Bernardo Rossellino had laid the foundations of a new choir. From the beginning of his reign Julius II had decided to proceed with the rebuilding.

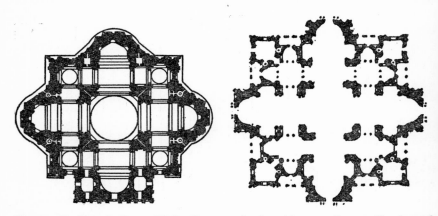

FIG. 25. *St. Peter's, Rome. Right: Bramante's plans (four small centrally planned structures within the arms of the major Greek cross). Left: Michelangelo's plan (reinforcement of the supports of the dome, the principal piers, and the exterior wall).*

He applied first to Giuliano da Sangallo, then to Bramante, who destroyed the major part of the transept and the apse, and added a temporary choir to the remainder which won for him the unkind nickname of '*maestro ruinante*'. The original plan, known from the medal by Caradosso and from a drawing by Antonio da Sangallo, was of an extreme complexity: a Greek cross with projecting apses, with four smaller domed subsidiary spaces on a similar plan, and towers at the corners. A recension by Bramante and Peruzzi added ambulatories to the apses and detached the towers, according to a form which recalled S. Lorenzo in Milan. This reference to Early Christian types is of importance: it is one of the reasons for the prestige of the central plan. The other is its mathematical perfection; that is, the multiplicity of simple relationships to be obtained within a single major unit, and the combination of shapes most proper to represent the Cosmos, the Divine Labour. The shape of the new church is therefore not without the highest religious significance, in which humanism and the Christian order were united. The recourse to antique Roman examples also was important, since the Pantheon supplied the form of the interior pilasters.

Julius II died in 1513, and Bramante in 1514. The undertaking was not so far advanced that it was essential to adhere entirely to the first project. Arguments started and for thirty years two opposing trends contended. From 1514 to 1518, Raphael and Giuliano da Sangallo worked out a plan which reverted to a rectangular Latin cross scheme while retaining the central plan at the crossing. The difficulties were such that Antonio da Sangallo worked out a compromise plan (1534–46) which retained the Greek cross, but added to it an apsidal bay, and re-introduced the towers. All unity of design was lost; the main block of the building was subdivided into three unequal storeys, and the dome, built up of two drums of disparate size, was crowned by a lantern surrounded by colonettes.

In 1547, Michelangelo, brought in to clear up the confusion, declared that 'to depart from Bramante was to depart from the truth'; he returned to the Greek cross plan rigorously simplified in the domed corner elements; he enveloped the whole with an exterior wall clearly articulated by giant pilasters running through all the storeys and supporting a cornice which embraced the whole structure. Everywhere unity was restored. He concentrated above all on the dome and designed for it a drum simply articulated by windows between coupled columns. The dome itself was erected later (1586–93) by Giacomo della Porta, who elongated its shape, possibly deriving the new form from a variant design by Michelangelo. After 1607, Maderna transformed the nave by the addition of bays and side chapels which finally resulted in the victory of the Latin cross plan. The consecration took place on the 18th November, 1626, after 120 years, during which time there had been twenty popes, and ten architects had been employed on the works. The building is therefore a basically composite one: the final intervention of Bernini, who created the oval *piazza* and the colonnades of the forecourt, enveloped in a Baroque setting the work which might have been the highest expression of the Renaissance. To pass beneath the enormous portico and walk up the nave is to move backwards through time; only under the dome, in the central viewpoint, can the meaning of the gigantic project of Bramante and Julius II still be grasped.

The Sangalli. Some ambiguity remained in Bramante's style: the size of the space he conceived was not always matched by an equal precision in the parts, and the feeling remains that his art is due to successive changes of intention. Nothing of this kind is observable in Michelangelo, but since he did not concentrate on architecture until after 1546, it was left to the Sangalli—Antonio (1455–1534) already mentioned, and his nephew Antonio da Sangallo the Younger (1473–1546)—to express during the first third of the century the new ideal of Roman architecture. With them the huge work accomplished by Giuliano da Sangallo finally reaches maturity. It was furthered also by the publication of books which, quite different from the 'notebooks' of Giuliano, were based upon a methodological and systematic coherence; these were the

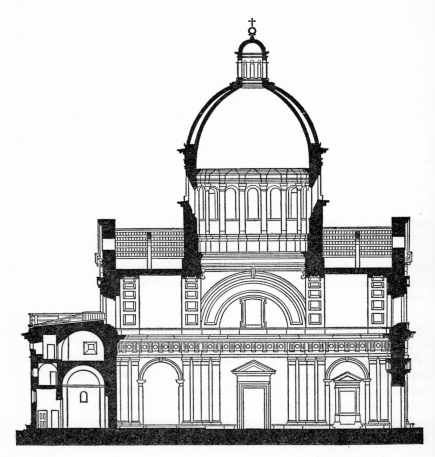

FIG. 26. *S. Biagio, Montepulciano, by Antonio da Sangallo. Section along the central axis (after Stegmann and Geymüller). Central plan extended by the sacristy, on the left; Doric order; high drum under the dome, buttressed by four barrel vaulted arches.*

great editions of Vitruvius of 1486 (Rome), 1496 (Florence), 1511 by Fra Giocondo (Verona), and 1521 by Cesariano (Como), with admirable engraved plates and measured drawings of the buildings, and by the treatises, inspired by Alberti's (which appeared in 1485), of which Serlio's was the most popular. Serlio's first four Books (1537) contained, under the title

Regole generali di architettura, sections on geometry, perspective, Roman archaeology and the study of the five orders; the following Books contained the plans of temples, towns and country houses, and, in the last incomplete Book, a study of restorations.

A   B

FIG. 27. *Sixteenth-century bays. A, Palazzo della Cancelleria, Rome. Uniform ground floor with simple bay; upper storeys with coupled pilasters framing the bays. B, Palazzo Farnese, Rome. Window treated as an aedicule with a pediment as the only rhythmic element.*

For the Piazza dell' Annunziata in Florence Giuliano's brother Antonio the Elder designed an arcade repeating Brunelleschi's Innocenti, but his career took him to Rome, and it was there that he received the commission for the votive church of S. Biagio (1519–26) on an arresting hill site on the slopes of Montepulciano in Tuscany. In this masterpiece of harmony he was inspired both by the precedent of the Madonna delle Carceri at Prato and by

223

Bramante's plan for St. Peter's. The plan is a Greek cross; two detached towers were intended though only one was built. The interior, exceptionally noble in form, is articulated by a Doric order, incorporating secondary motifs of windows, openings, pediments and string-courses repeated exactly on the outside, where the formal masses are articulated, with strictness and clarity, in front of the immense panorama of the Val di Chiana and Val d'Orcia.

Antonio was as successful in his designs for the Cervini and Contucci palaces in Montepulciano, the Palazzo Del Monte at Monte San Savino, where he developed the theme of an upper storey with rhythmically arranged windows and pediments resting on a rusticated basement, of the type that Bramante had originated in the House of Raphael.

His nephew, Antonio, who was to specialize in military architecture in Florence (Fortezza da Basso, 1534), Rome and Perugia (Rocca Paolina, 1540), had first designed in Rome the church of the Madonna di Loreto (1507) in front of Trajan's Forum, and palaces clearly articulated by pilaster orders, such as the Palazzo Palma (originally Baldassini) and the Banco di Sto Spirito. But his most memorable work was the palace which was commissioned (1511) by Cardinal Alessandro Farnese, the future Paul III, at the end of the Via Giulia, behind the Campo de' Fiori. The original project was enlarged in 1541. The form of the palace returned to the Florentine type in three storeys with orders on the façade and with those parts built of travertine projecting beyond the brickwork; the interior court was inspired by the theatre of Marcellus, with its windows boldly set between intercolumniations. In 1547, after a competition, Michelangelo became responsible for the cornice of the façade, which was given a greater projection. Michelangelo also added the third courtyard storey which, with its clusters of pilasters, adds an unexpected note of tension to the building. The loggia in the rear was built by Giacomo della Porta.

A different interpretation of the classical themes adumbrated by Bramante is to be found in the more picturesque compositions Raphael and Peruzzi conceived rather from the aspect of decoration. But this can only be considered in relation to their whole œuvre.

RAPHAEL

Between the summer and the autumn of 1508, a young artist, Raphael Sanzio (1483–1520), was summoned to Rome on the recommendation of Bramante. Born, like the architect, in Urbino, the son of a knowledgeable painter, he retained from the gentle culture of the Montefeltro, which was to receive its complete expression in the treatise *Il Cortegiano* (The Courtier) by Baldassare Castiglione (written after 1514 and published in 1528), a receptivity and sense of balance so perfect as to belong only to a superior order of being.

Raphael appeared at the moment when the strained style of Melozzo was

57 Urbino, Ducal Palace: detail of intarsia work
in the Studiolo of Federigo of Montefeltro

58 Agostino di Duccio: façade of S. Bernardino, Perugia

59 (a) Anonymous engraving
of a grotesque helmet,
c. 1475

*(Washington, National Gallery,
Print Room)*

(b) Botticelli: drawing
for the first Canto of
Dante's "Paradiso"

60 (a) Majolica bowl
      made at Cafaggiolo,
      16th century
   (b) Torch holder
      in wrought iron,
      Florentine,
      beginning of the
      16th century

   (*Florence, Bargello*)

61 Angel heads:
(a) Fra Angelico, *Annunciation*
(*Cortona Mus.*)
(b) Botticini, *Tobias and the Angel*
(*Florence, Uffizi*)
(c) Michelangelo, *Madonna and Child*
(*London, National Gallery*)
(d) Filippino Lippi
(*Strasbourg Mus.*)

62 Cosimo Tura: detail from the fresco *The Triumph of Vulcan*

(*Ferrara, Palazzo Schifanoia*)

63 Mantegna: *Lodovico II Gonzaga and his court*
(*Mantua, Ducal Palace, Camera degli Sposi*)

64 Gentile Bellini: *The Doge Andrea Vendramin*

(*Rotterdam, Boymans-Van Beuningen Mus.*)

giving way to the tender and balanced vision of Perugino. From his earliest efforts—the *Three Graces* (Chantilly) and the *Knight's Dream* (London)—this influence determined his pale colour, his open landscapes, the calm of his compositions. He collaborated with Perugino in the Cambio in Perugia, in the *Assumption* (Uffizi), with Pinturicchio in the Piccolomini library in Siena. He was already a past master in facial expression. The masterpiece and the final work of this period is the *Marriage of the Virgin* (Brera, 1504). The design is Peruginesque: the temple in the centre is pierced by doorways which allow the light to shine through and the perspective creates a wonderful sense of space, but all is brought to a pitch of concentration, of elegance, of suavity without affectation, of which Perugino was never capable. It is in highly worked drawings of this period (*Head of a Saint*, Lille; *S. George*, Uffizi; *Head of an angel* and *Head of S. John*, London; *Female heads*, Oxford) that it is possible to see the extent to which Raphael's genius represents a constant striving towards perfection; he shatters the conventional to seize upon the convincing movement, the most expressive gesture.

This disposition led him to study and to assimilate with extraordinary ease the manner of all his seniors. As free of complacency as he was of envy, he provides an example of a life devoted to learning, but what he achieved in this way through the pervasive force of his genius rapidly led to the misunderstandings of academism. His brief life sufficed for him to sum up all available experience and adapt it to his grand design. It could in fact be described in terms of each master who appears in turn to dominate the successive stages of his development. Thus the pupil of Perugino turned in 1504 towards Leonardo and he adopted the *sfumato* technique in the *Madonna del Granduca* (1505, Pitti), the pyramidal composition in the *Belle Jardinière* (1507, Louvre) the union between figure and landscape with *Maddalena Doni* (1505, Pitti). Then the virile force of Michelangelo attracted him after the second of the *Stanze* (1511); from him he derived the use of the nude and of dramatic movement, which he assimilated perfectly in the *Fire in the Borgo* and the *Sibyls* of S. Maria della Pace (1514). Then certain aspects of the tonal painting of the Venetians—of Titian himself—of which he was aware through Sebastiano del Piombo, were also incorporated into the new style. This is discernible in the *Mass of Bolsena* in the second *Stanza* (1511–14), in the *Madonna with the Fish* (1513, Prado), and in some portraits. But it is precisely in this ability to submit himself to the influence of others so as to increase his knowledge, in this almost impersonal sense of values, that the classic character of the artist is revealed, as is confirmed by the admirable drawings he never ceased to produce, such as the silverpoint *Madonna* (British Museum), and the *Man standing* (Louvre). In this development two decisive factors intervened. From 1509 onwards, his enormous success—which even overshadowed Michelangelo—led Raphael to speed up his production. But after 1515, this success and the charming personality of the painter attracted numerous collaborators and this involved a risky compromise: the execution of the greater part of his works was left entirely to his pupils and it is only possible to speak of the studio.

The works in the Vatican are his main claim to fame. The three *Stanze* of the Vatican—rooms in the apartments taken over by Julius II, above the abandoned Appartamento Borgia—for which he had been summoned to Rome, were painted in fresco on all four walls. The first, called the *Stanza della Segnatura* (1509–11) has four 'histories': the *Disputa*, or 'Disputation concerning the Holy Sacrament', the *School of Athens*, the *Parnassus*, and the *Decretals*, which illustrate, not as has been said a system for the arrangement of a library, but the four aspects of a Christian and Platonic doctrine which recognizes Truth in its two aspects: revealed and natural, the Beautiful and the Good. Each scene is the occasion for an original composition where 'famous men' are arranged in a perspective at once harmonious and reasonable. The *Stanza d'Eliodoro* (1511–14) illustrated the great theme of the trials of the Papacy: *Heliodorus driven from the Temple*, *The Liberation of S. Peter*, *Leo I halting Attila before Rome*, the *Miracle of Bolsena*. The allusions to the troubled reign of Julius II are obvious, and in the third scene (the last to be executed), the pontiff has the features of Leo X. The brilliance of the grouping of the figures is heightened by an increasing interest in tonal richness and grand effects of light, evident in the *Liberation of S. Peter*. The third *Stanza* was painted from 1514–17 with further histories with a contemporary significance: the *Fire in the Borgo*, the *Battle of Ostia*, the *Coronation of Charlemagne*, the *Oath of Leo III*. Here only the compositions as a whole are by Raphael; the execution and numerous details are by a pupil, Giulio Romano, assisted by Luca Penni in the backgrounds.

Raphael was the born interpreter of the humanist longing for a serene and reconciled humanity, towards a fusion of wisdom and the forms of beauty, which found an echo in Rome. His art, the product of a rich and conscious culture, is leavened by contact with the antiquity of the Golden Age which is itself a repertory of clear forms and expressive gestures. The Ancients and the Moderns are closely associated and on Parnassus, around the Muses, the poet's crown is distributed equally to the ideal portrait of Homer and to Sannazaro and Tebaldeo, the painter's friends.[107] Raphael's essentially classical character was never better displayed than in his sublime representations of the Virgin, from the suave Florentine compositions to the *Madonna of Foligno* (1511–12, Vatican), the perfect *Madonna della Sedia* (1514–15, Pitti) and, above all, the *Sistine Madonna* (1513, Dresden). This last work develops, through the marvellous and perfectly sustained sequence of its forms and line, a composition of an Early Christian type in a kind of luminous and tender Epiphany, animated by the moving figure of the old Pope in ecstasy and the meditative reserve of S. Barbara. The fusion of Christian and Antique iconography is as perfect in the *Vision of Ezekiel* (*c.* 1518, Pitti), of which the cartoon alone is by the master. A more and more subtle feeling for profound emotion appears in the *S. Cecilia* (1514–15, Bologna), and in the *Trans-*

figuration (1517–20, Vatican), of which the composition and certain details are Raphael's final expression, while the agitation of the over-emphatic lower part betrays the hands of Giulio Romano and Penni.

The world of fable is treated with as light a touch in the decorations in the Farnesina where, for his patron and friend, Agostino Chigi, he painted a fresco of *Galatea* (1514) and designed the incidents of the *History of Psyche* (1518) for the garden loggia, of which the execution in a reddish tonality is by Giulio Romano and Penni. But the field in which his insight and knowledge of psychology were most happily associated with his keen perception of tonal values and restrained colour, is that of the portrait. Examples are *Angelo* and *Maddalena Doni* (1505, Pitti), the *Cardinal Ippolito d'Este* (1510, Prado), *Inghirami* (1514, Pitti), the poets *Navagero and Beazzano* (1516, Doria Gallery), *Leo X* (1517, Pitti), and the superb *Castiglione* (*c.* 1515, Louvre), in an admirable state of preservation, where in a harmony of greys the artist establishes not only the likeness, but the sitter's tender and gentle personality. Raphael seems to have portrayed himself as well as his sitters in these portraits of literary men which reflect not only a character but an ideal, the vocation of poetry and love. The artist's authority thus created a kind of portrait which approximated to a symbol.

With the *Sala di Costantino* (1517–25) finished after Raphael's death by Giulio Romano and the studio, the decoration of the *Stanze* ends in a wearisome rhetoric which was to weigh heavily on the future. They were to become doubtful models for Italian palaces, as were the decorations of the *Loggie* of the Cortile di S. Damaso, which must be considered with the architecture. But a third work, destined to have the greatest repercussions of all, was the group of the great cartoons for the tapestries of the *Acts of the Apostles* (1515–16, Victoria and Albert Museum, London, on loan from the Royal Collection), which were woven in Brussels. The huge tapestries (Vatican) fail to convey all the originality and suppleness of the compositions. Raphael here brought together for the last time, in the *Feed my sheep* and the *Death of Ananias,* the essential characteristics of his inspiration—spatial breadth and the power to express it fully, richness of contrasts, and that sense of perfect balance maintained to the extreme limit, which escapes affectation through the constant vigilance of the intellect. Ingres attempted to define it by saying, 'The conception appears to be simple, and, as in the works of God, all seems to be a pure effort of will-power.'

Raphael's art was passionately admired by knowledgeable connoisseurs, by humanists, like Bibbiena, Leo X's cardinal who wanted to give him his niece in marriage, poets such as Tebaldeo or Ariosto. At his death, he was spoken of as a 'divine' being, without whom the world had lost its savour. What was chiefly prized in him was that *mirabil giudizio* which governed the composition, and more particularly, his feeling for *costumi,* that is, his power to express character through drawing. During the academic age, the accent was to be placed on *convenienza*—decorum—but during the period which followed 1520, the pupils of the studio grossly exploited the discoveries of their master, and it was the cleverness of the invention and the use of the Antique which were the most imitated.

Raphael as an Architect, and Baldassare Peruzzi. As Bramante's relative, Raphael succeeded him (1514) as architect of St. Peter's. He provided a new plan which, by introducing narrow ambulatories and multiplying the side chapels, could only have resulted in plunging the building into a sort of chequered gloom. At S. Eligio degli Orefici (1509), a pure and diminutive church on a central plan, the 'cosmic' quality of the motif is underlined by the inscription: *astra Deus, nos templa damus; tu sidera pande*, which makes of the vault a replica of the celestial dome. In the Chigi Chapel in S. Maria del Popolo (finished in 1520), also on a Greek cross plan, the surfaces are enlivened by the polychrome revetment, and the dome by a mosaic decoration in which figure the planets and the angels that govern them.

Raphael's tendency in architecture was therefore to stress the surface aspects, to accentuate the pictorial quality in structures based on geometric harmony; in the Palazzo Branconio dell' Aquila in Rome (which disappeared in the building of the Piazza in front of the Vatican) the façade above the five strongly marked arches was covered with balconies, stuccoes, panels and niches. The effect, known from a drawing by Parmigianino (Uffizi), is of an extremely animated picturesqueness. The more serene Palazzo Pandolfini, built in Florence by Gian Francesco da Sangallo to Raphael's design (1516–1520) offers a symmetrical arrangement of perfect regularity; the two storeys are equal, the pediments of the aedicular windows, Doric on the ground floor, Corinthian on the first floor, are alternately triangular and segmental; the only two elements of contrast are the strict rustication of the portico and the cornice above a monumental inscription with the prelate's name. The much lighter interior court offers a pleasing contrast to the façade, which must have been designed by Raphael himself.

Like Bramante, Raphael had the opportunity to 'tame nature' in his designs for a vast villa for Giuliano de' Medici, the future Clement VII, on the slopes of Monte Mario. The combination of architecture and landscape, arranged in terraces and gardens, was to be used as an example by Serlio in his *Regole*. The building, badly damaged in 1527, was rebuilt by the daughter of Charles V (hence the name Villa Madama). The huge domed vaults, ornamented with coffering and stuccoes, the loggia walls hollowed out into niches, openly proclaim their classical source in the ruins of the *domus aurea*—Nero's Golden House. What was wanted was an appropriate setting for learned leisure, pleasure and relaxation. It was in somewhat the same spirit and from the same sources that the *Loggie* of the Vatican—twelve vaulted bays forming a gallery open on to the courtyard of the façade and separated by pilasters with *graffitti*—were conceived (1517–19). Each section is ornamented with a different decorative composition; set in the network of tracery or in bands, often of fake architecture, are panels after the designs of the master illustrating the Book of Genesis, which constitute what has been called Raphael's Bible. The great variety of the motifs suggests a number of executants; but they also represent the definitive adaptation to classical decoration of the thousand fantasies of *amorini*, animals, and imaginary plants which are called 'grotesques' (from the caverns or grottoes of the Esquiline Hill), and in which Giovanni da Udine specialized. Another charming example is on the floor above, in Bibbiena's sybaritic bathroom (1516–17).

Raphael's principal architect collaborator was Baldassare Peruzzi (1481–1536) of Siena, who came to Rome in 1503, and who also worked with Bramante, though his architectural taste was closer to the imaginative quality of a Francesco di Giorgio. His masterpiece, the Villa Farnesina facing the Palazzo Farnese, built for Agostino Chigi (1508–11), crowned by an antique frieze supported by pilasters, and with two projecting wings, is not merely a 'folly' in the Neoclassic taste. With its stucco facings, its astrological decorations (depicting the horoscope of the owner), it is a compendium of all the most fashionable ornaments of the time. By means of the huge *trompe-l'oeil* decoration simulating a view seen through a colonnade, an intimate relationship is established between the architecture and the gardens; the leafy bowers and the verdant arcades repeat the architectural decor, in the same way that the fake balcony painted on the wall introduces the landscape and the distant panorama into the house.[108] This is the last word in a refinement which restored the setting of the Imperial antiquity to the great Romans of the day.

At Carpi, near Modena, Peruzzi provided the plans for the cathedral, and those of the new Sagra (1514–22). He also furnished for the festivities of Leo X the decors and theatrical settings where his gifts in perspective and his taste for large-scale architectural fantasies were brilliantly employed. In 1520, he was even entrusted with the direction of the works on St. Peter's; he built palaces derived from Raphael's prototypes, and in Siena directed the municipal architecture, inspired villas, and built the villa at Belcaro with a grand loggia. It is difficult to consider his last work, the Palazzo Massimi in Rome (after 1532), as a classical work, since it is more of an example of the rise of Mannerism. His virtuosity develops an element of capriciousness at the expense of architectonic considerations: the external wall follows the curve of the street and becomes a screen which does not explain what lies behind it; the courtyard gives the impression of being the realization of a fake perspective, like that in the Farnesina, rather than the internal articulation of the structure. The Doric columns support an enormous entablature, pierced with ill-conceived openings which become a row of false windows; the entrance gallery has unexpected effects of light and shadow. This overstressing of what should properly remain small nuances and surprising details reveals, by the importance given to it, the passionate and almost romantic character which is blended with the use of classical forms after the third decade of the century. The links between Peruzzi and Raphael help to explain why, after the dispersal of 1527, Giulio Romano and his associates turned, in architecture as in painting, in a direction totally opposed to that of their master; the more they claimed kinship with him, the more they make clear the unique qualities of his grace and precision.

MICHELANGELO

In Florence, between 1501 and 1506, Michelangelo Buonarroti (1475–1564) although the younger by twenty-three years, had competed with Leonardo in the decoration of the Sala del Consiglio; chroniclers speak of a violent antipathy between the artists. To purely personal reasons of character may be added the opposition of their ideas. Leonardo mistrusted the learned, adumbrated a general theory of knowledge which ignored religious issues, and raised painting—as a magic mirror of the effects of nature—above all the arts. On the other hand humanism and poetry were part of Michelangelo's being; if he wrote, it was not treatises on method, but poems; the anguish of human destiny so obsessed him that he could conceive no other purpose for art than to express it, through the subtlety of the human body, and the variety of man's passions, and their sublimation through faith. Lastly, for him the major art was sculpture, which liberated a latent form, given life by the intuition or 'Idea' of the artist from the confines of matter.

In Rome, where he was summoned by Julius II, Michelangelo soon clashed with the Bramante-Raphael clique. In 1506, he fled from Rome on the eve of the day when the first stone of the New St. Peter's was laid. He was as mistrustful and irascible as the younger man was affable, as self-willed as Raphael was accommodating. Later Michelangelo accused his rival of being the cause of all the difficulties that he met with in Rome, adding 'All he knows, he got from me'. He considered himself the more powerful and complete artist, abandoning grace to Raphael and reserving *terribilità* for himself. Classic art was born of the struggle between these two great branches of inspiration, which gave a complex organization to, and conferred an unprecedented value on, the visual forms elaborated by the Quattrocento.

The Early Works. Michelangelo's early years are rather obscure. He was for a time in the studio of Ghirlandaio, who taught him painting; he then entered the household of Lorenzo de' Medici, and a famous anecdote makes him discuss with the ruler of Florence an Antique faun's head in the garden-museum of San Marco, where he was the pupil of Bertoldo. But this academic schooling is only a fiction; Michelangelo had but two teachers—Donatello and the Antique. One of his biographers states that it was through his slightly older friend Granacci (1469–1543) that he obtained permission to work in the garden at S. Marco to escape from the boredom of Ghirlandaio's workshop, that there he knew the old Bertoldo, who died in 1491, and carved the *Battle of the Centaurs* (*c.* 1490, Bargello), so full of movement, and the

MICHELANGELO

Madonna della Scala (*c.* 1490–92, Casa Buonarroti), in a low relief style derived from Donatello.

At the same time as it gave him the opportunity to develop freely, Lorenzo's favour also brought the young artist into personal contact with Antique sculpture—rare in Florence—and with the humanism of Poliziano, Landino and Pico della Mirandola. This humanism postulated a concordance between pagan wisdom and Christian theology which gave a new balance to culture, raised the creative artist and the poet to the summit of the hierarchy of the inspired, and considered the harmonious play of the faculties, reason and intuition as necessary to human activity. Michelangelo, who later found himself in contact with the heirs of the Accademia de' Careggi, such as Diacceto, always adhered to these doctrines in his happier moments, but he experienced very early, and with increasing violence, the anguish and anxiety which they describe as one of the conditions of genius, or 'saturnism'.

In 1494, panic stricken, he fled to Venice, and then stayed in Bologna, where he meditated on the lessons to be learned from Jacopo della Quercia and carved an angel and two small statues for the Shrine of Saint Dominic. After returning to Florence, where he heard Savonarola prophesying doom, he went to Rome where for Jacopo Galli he made the *Drunken Bacchus* (1496–97, Bargello), with its multiple viewpoints, a clever, but rather facile work. The first masterpiece of this period was the *Pietà*, commissioned by a French cardinal in 1498, finished in 1499, and placed in St. Peter's: a theme of Gothic pathos, treated with an admirable assurance and skill in a pyramidal composition, in which the lovely, nude figure of the dead Christ stands out against the great masses of drapery.

Back in Florence in 1501, Michelangelo spent four years of immense activity; from him were commissioned the huge marble *David* (finally placed in 1504 in front of the Palazzo Vecchio), twelve apostles for the cathedral (only the *S. Matthew*, 1503, now in the Accademia, was begun, and the powerful twisting movement of the figure is only roughly sketched out) and, finally, the big fresco of the *Battle of Cascina*, which was to be the counterpart to Leonardo's *Battle of Anghiari* in the Palazzo Vecchio. During the same period, he made the large medallions or *tondi*, two in sculpture (the *Pitti Madonna*, 1505, Bargello, and the *Madonna with the bird*, *c.* 1505–6, R.A., London) which betray Leonardo's influence in the graduation from finish to unfinish, and one painted, the *Holy Family*, or *Doni tondo* (1503, Uffizi), where the strongly knit group creates a kind of undulating block, with figures from a pagan pastoral scene in the background. In all these works, Michelangelo adapts Leonardo's motifs, but gives them a striking plastic clarity. He had already discovered the power of gigantic forms; the *David*, cut from a block that no one had yet been able to use, is a triumph of the subtleties of anatomy and *contrapposto* (the weight of the body falls on the right side), suggesting tension underlying repose. The variety of nude figures in violent movement also became the main subject of the fresco for the

231

Signoria. In 1505, when he went to Rome to begin work on the tomb of Julius II, the artist was the master of a repertory and a style which already gripped the attention of the artistic world.

The Tomb of Julius II and the Sistine Ceiling, 1505–10. In March 1505, Julius II commissioned his tomb. One year later the project was suspended, and the forms which haunted the imagination of the artist were realized in an immense composition which was entrusted to him in May 1508—the decoration of the barrel vault of the Sistine Chapel. Instead of the sculptured monument, which he was never to bring to a satisfactory conclusion, it was, paradoxically, a gigantic fresco, 130 feet by over 40 feet, which was to be the complete expression of Michelangelo's genius. It was finished in October 1512.

The contrast between this formidable, taut framework of arches and bands which stress the structure of the vault, and the calm arrangements of Masaccio and Piero is the same as that between the anatomically developed marble *David* and the one by Donatello. Twelve prophets and sibyls—type figures expressing inspiration—accompanied by their genii, seem to bear the architecture upon their enormous shoulders; in form and gesture these figures are the highest achievement of Florentine linear draughtsmanship amplified by Roman monumentality. Below the seers, a lower zone contained in the lunettes and spandrels of the windows shows in monochrome the miseries of the sinful humanity of the Old Testament. Above the seers, on the flat part of the vault, in a zone delimited by a massive cornice in sharp perspective, appearing like celestial visions, are nine panels alternately large and small, illustrating the Book of Genesis. These nine histories are surmounted by the astonishing figures of the seated, nude adolescents—the *ignudi* who sum up in their expressive poses all the emotions of the soul. Thus the vocation of the soul and the stages of its progress are fully illustrated, and the tripartite arrangement set out in a fairly explicit symbolic language the great images of faith.

In this vast work are to be found a host of motifs from the Quattrocento and even from earlier periods blended into forms inspired by the Antique; it is the sum of two centuries of art, of Greek plastic beauty, and of Roman triumphal order. The delicate and airy tones of the colour accentuate the ethereal quality of the visions. There is no part that is not significant, no useless makeweight; even ornamental details such as the putti have been thought out anew to become the genii of inspiration. The strict framework expresses the disciplined style and the classical intent; from the architectural elements to the sublime scenes on the vault, such as the *Creation of Adam*, runs a powerful current of life animating the whole, attesting both to the artist's power to represent a higher world and to his aspiration towards the divine.

In its first version, the tomb of Julius II was a design of the same kind. It was worked out with such care that Michelangelo could truly claim that its failure was the tragedy of his life. From 1505 to 1545 there were six projects, whose variations are difficult to work out; the final wall-tomb in S. Pietro in Vincoli is only the last version, and the feeblest.

According to the project of 1505, the tomb was to be a free-standing edifice, a kind of *arcus quadrifrons* embellished with niches and allegorical statues on the lower storey, surmounted by statues on a flat upper storey, and crowned

FIG. 28. *Michelangelo: Tomb of Julius II. Reconstruction of the first project of 1505, made by Ch. de Tolnay with drawing by D. Fossard. Ground floor: the Slaves placed in front of the herm caryatides, framing the* Victories *placed in front of the niches. Upper storey:* S. Paul *and* Moses. *Top storey: Cosmic allegories and the sarcophagus.*

by a pyramidally composed top portion supporting the sarcophagus. This arrangement in storeys symbolically blended the apotheosis of the pontiff with the exaltation of the Church. In 1513, on the death of the pope, a new contract made with his heirs provided for a huge wall tomb which preserved the arrangement of the façade of the free-standing monument, but with an arch encompassing it and the figure of the Virgin. The surviving drawing (Berlin) identifies two of the *Slaves* (Louvre), carved towards 1513 as part of the original lower storey; on the upper storey, a figure of Moses was planned for the right side, and a S. Paul for the left. All these figures expressive of intense pathos, still retained the plastic clarity of style of the Sistine ceiling, though with a more anguished expression in the *Slaves*, which derived from the Laocoon, and a more vehement one in the *Moses*, held only within the classical order by its powerful structure.

In 1516, all that remained of the original project was a two-storey façade, which was again whittled down in 1522; finally it was decided to use those pieces of sculpture already carved. *Rachel* (or Faith), and *Lea* (or Charity), stand on either side of the figure of *Moses*, surmounted by a Sibyl and a Prophet by Raffaello da Montelupo, and by the sarcophagus and the figure of the Virgin. The composition is less hieratic, and animated by a kind of movement towards the much calmer upper storey. Nothing remains of the Antique triumph. These changes in his greatest work correspond to the artist's own spiritual evolution, for after the period when he was dominated by the pursuit of beauty and platonic love, he arrived at a state of pious austerity which cut him off entirely from the Renaissance.

Without being surrounded by pupils, as Raphael was, Michelangelo had his admirers and eager collectors of his drawings, and the style of the Sistine Ceiling evoked many imitators among painters.

Thus, around the superb unfinished *Manchester Madonna* and *Entombment* (both National Gallery, London), it has been possible to group six pictures, among which may be singled out a pretty but rather complicated *tondo* of the *Virgin and Child and S. John* (Vienna); they are by one of the artists working in the Michelangelo circle, such as Jacopo l'Indaco (1476–1526), or maybe the Antonio Mini to whom the artist gave his *Leda*, and they must have been painted between 1515 and 1530.[109]

The Return to Florence; 1520–34; the Medici Chapel. Although he was a childhood friend of Leo X, during his papacy Michelangelo avoided Rome where Raphael reigned supreme, spending years supervising the quarrying of marble at Carrara and Pietrasanta (1516–19). He returned to Florence to offer to make a tomb for Dante without payment (1519), and remained in Florence after 1520, despite the entreaties of his Roman friends. In 1527–29 he took charge of the fortifications at S. Miniato during the republican revolt, crushed in 1530.

All the works of this period turn on the Medici church of S. Lorenzo. In 1515, the pope commissioned him to make a façade for the church; this was his first work in architecture. He wanted this composition to be the 'mirror of all Italy', but it was never executed and is known only from drawings (Casa Buonarroti, Florence), from which can be seen how great was Michelangelo's debt to Giuliano da Sangallo. The rhythm of the bays framed by superposed pilasters is treated with a tension which seems to make the façade quiver; he proposed to people it with statues and considered it as a plastic entity divorced from the edifice itself. From 1520 to 1527, and again from 1530 to 1534, he worked on another addition to S. Lorenzo—the arrangement as a funerary chapel of an old chapel placed symmetrically to Brunelleschi's sacristy. He imposed on the architecture a strongly marked rhythm, in which the rising forms are contained by horizontal bands, and set face to face the tombs of the dukes, Giuliano (Active Life) and Lorenzo (Contemplative Life), framed in allegories which further underlined the stages of existence and its progressive transfiguration. The dukes are enthroned above their sarcophagi, on which recline the figures of *Day* and *Night*, *Dawn* and *Dusk*, symbols of the transient world. The designs prove that it was intended to have figures of the River-gods of the Underworld below the figures symbolizing Time, and on either side of Giuliano statues representing Heaven and Earth like those projected for the tomb of Julius II. Over the whole, frescoes were to represent the Resurrection and the Brazen Serpent. Thus, all the arts together were to compose a huge symphony of death and the resurrection of the soul, with, as an ideal centre, a figure of the Virgin resigned to her destiny. All the decorative motifs—trophies, dolphins, candelabra, masks—are borrowed from the repertory of sarcophagi and Antique funerary altars. The reclining allegories derive from Roman reliefs, the dukes are captains in armour; but never has the authority of the composition and of the modern style been more evident. It seems that the artist sought to heighten the contrasts between the finished parts and those left in a rough-hewn state, like the face of *Day*, as if to remain forever at the point of crystallizing the image, and, moreover, the elongation of the figures, the increasing flexibility of the forms in the last ones to be executed, the *Giuliano* and the *Virgin*, indicate a certain leaning towards Mannerism, consonant with Florentine tendencies.

In 1524, Michelangelo was commissioned to build the Laurenziana Library, and to give it an entrance loggia, or hall. He provided the designs and directed the work until 1527, and again after the troubles, from 1530 until 1534. After his departure the work remained unfinished. In 1560 Ammanati executed, after Michelangelo's designs, the curious staircase which, like a viscous substance, seems to flow down to the ground. Michelangelo deliberately sought to create an impression of strangeness: the hall is high, and the divisions adopted provide him with the most unusual proportions; instead of bearing the weight of the structure above them the columns are embedded in the wall and paired by volutes which appear to emerge as excrescences from the walls, and between these constricted masses are suspended windows like tabernacles. From all these contradictory effects, which for long remained enigmatic, it has been possible to define his style as classical in its elements, but anticlassical in spirit—the architectural counterpart of Mannerist painting.

Michelangelo in Rome: 1534–64. Dominated from 1532 onwards by his passionate friendship for a young Roman nobleman, Tomaso Cavalieri,

FIG. 29. *Staircase of the Laurenziana Library, by Michelangelo. Section (after Stegmann and Geymüller): f, blind windows; c, columns set into the wall, above projecting consoles.*

Michelangelo returned to Rome for good after the death of his father in 1534. The cult of earthly beauty, which he conceived as rising painfully above the normal condition of humanity, inspired him to produce grief-stricken poems, and very carefully worked out drawings, delicately modelled, on mythological themes such as Ganymede, Tityus and Phaeton.

The new pontiff, Paul III, forced Michelangelo once again to return to painting by commissioning from him the *Last Judgement* which covers the end wall of the Sistine Chapel (55 feet by 40 feet, 1536–41). The composition is conceived as a huge spiralling movement, rising on the left, descending on the right, around a Christ standing in the centre, like a Hercules or an angry Jupiter; in contrast to the ceiling painting of which it is the complement, there is no architectural framework to contain it. The scene is unfolded without any regard for the requirements of the place itself, in a visionary space controlled solely by spiritual attractions and repulsions. This breaking of all bounds involves also a renunciation of human scale; it is the movement of the whole which is paramount. The brown bodies twist and turn against an intense matt blue, above the sinister reds of the mouth of Hell. If, fifteen years earlier, Michelangelo had provided the materials for Florentine Mannerism, here he anticipates all the developments of Baroque and its unlimited space, but with a personal note of despair, and a stormy atmosphere which belongs to him alone; the Christ is the personification of anger, only the Virgin shows pity, and the artist's own portrait is hidden in the flayed skin of S. Bartholomew. The wealth of the detail is amazing; many figures are reworked from the sketches which remained unused in the *Battle of Cascina* (1504), others come from Antique sculpture (Niobe, Apollo), but the order and the joyful power of the ceiling has been rejected, and the classical style forgotten.

The exclusive use of nude figures for religious scenes and allegories was no longer understood, and Aretino, in an open letter (1545) sharply criticized the work on this score. Michelangelo's last work in painting was the decoration of the Pauline Chapel, between the Sistine Chapel and the Sala Regia. There he painted the *Conversion of S. Paul* and the *Crucifixion of S. Peter*; almost abstract in style, in the midst of disorganized groups, the convulsed features of S. Paul and the glowering face of S. Peter stand out. The style has become heavy, the masses compact. When he no longer had the physical energy necessary for fresco painting, Michelangelo composed black chalk drawings of a more and more massive and withdrawn character; for instance the *Crucifixions* (*c.* 1550, Windsor; *c.* 1550–56, Oxford, London).

In this groping style, where the forms are only roughed out and the gestures minimized, he expressed a tragic faith and profound urge to penitence, nourished through his contacts with members of Roman groups of 'spirituals', which included the circles round Ochino and Pole, for instance, and the Catholic reformers gathered round Vittoria Colonna. The artist addressed some of his most exalted poems to this noble lady, whom he knew from about 1536 or 1538, and whose affection brightened his advancing years until her death in 1547. Their colloquies on art were given literary form by Francisco de Ollanda. It was at this time that Vasari's *Lives* (1550) and the one written by Condivi (1553) finally established his fame, and the Florentine Academy, with Benedetto Varchi, made him the object of a very real cult. His imitators were legion.

In 1546, at seventy years of age, under Paul III, Michelangelo finally became an architect. Once again, he continued the Sangalli, taking over from Antonio the Younger, both at St. Peter's and at the Palazzo Farnese, where his intervention was to be marked, as has already been mentioned, by an increased insistence on dignity, and by a grandeur of concept verging on the colossal. At St. Peter's, for which he made four models between 1546 and 1561, his main effort was directed towards the crowning of the masses by the dome, sometimes conceived as spherical, sometimes, as at Florence, conceived as ovoid and ribbed—although he was not to bring it to completion. At the Farnese, it is the powerful cornice and the upper storey that bear the imprint of Michelangelo. In the same way his other works tended to augment the qualities of pathos and of the superhuman in architecture, to make it live and work before one's eyes, as in the project for the construction of S. Maria degli Angeli in the Baths of Diocletian (1561), the impressive Porta Pia (c. 1560), the designs for S. Giovanni dei Fiorentini (1550–59) and, above all, the layout of the Capitoline square made famous by the engravings of Duperac (1567–68–1569). Michelangelo flanked the trapezoidal space with two palaces whose flat roofs are supported by an order of giant Corinthian pilasters which enclose the two storeys consisting of an open portico on the ground floor and a *piano nobile* with windows, whose segmental pediments are accented by a rounded shell form. The relationship between the enormous entablature and the giant pilasters is echoed in the lower storey, where the architrave of the portico rests on angle columns; the horizontals and the verticals cut harshly across each other. The feeling of the whole is rigid, eternal, characteristically Roman.[110]

Rarely now did Michelangelo return to sculpture, and then only for himself. His last works were groups of the *Pietà*: the one in Florence cathedral, worked out on a pyramidal design and still constructed with tremendous solidity (before 1550–55); the *Palestrina Pietà* (id. Florence, Accad.), where the sagging body of Christ is supported by two more composite figures, which has sometimes caused the group to be attributed to an imitator; and finally the unfinished *Rondanini Pietà* (1555–64, Milan, Castello), where the standing Virgin struggles to support the collapsing body of the dead Christ (it was roughed out in 1555, and the sculptor was still working on it at his death). In this work, entirely unlike any other, Christ and his Mother melt one into the other in a stricken mass. Far beyond the classic beauty of the Apollo-like *Risen Christ* in S. Maria sopra Minerva (1514–20, Rome), the complex elegance of the Medici Chapel, and the presage of the Baroque contained in the emotion which pervades the *Rachel* on the Julius Tomb, these rough-hewn forms express the tireless passion of an old man who finally rejected all concepts of beauty and even physical reality itself, to express more clearly a purely spiritual state.

In every technique Michelangelo ranged through every style, and two centuries of Italian art may be instanced in his works. He lay in state on the 19th February, 1564, in SS. Apostoli in Rome, and his remains were carried by stealth to Florence and buried on the 12th March, in Sta Croce. On the 14th July, 1564, his official funeral was held in S. Lorenzo; this was one of the first public functions of the recently founded Academy of Fine Arts, and characteristic of a society permeated by public bodies, which the solitary and rebellious genius of Michelangelo would never have endured.

The Classical Style in Lombardy and Tuscany

The achievements in Rome during the first twenty years of the century in architecture, sculpture and painting, intimately associated in such typical works as those in the Vatican, the Farnesina, and similar works, sundered Italy completely from the remainder of the West, and conferred upon her the prestige of a classical style of which there was no equivalent in France, the Netherlands or Germany. Its most general characteristic is the need to create absolute clarity in the forms, in the whole as in its details: Michelangelo's *Pietà* in St. Peter's, the *School of Athens*, the church of S. Biagio at Montepulciano are entirely successful examples of the complete articulation of plan, of the control of the elements which ensure order, and despite the abundance and diversity of the parts, of the essential unity of the work of art. This determination to compose order, which is expressed by the stability of the forms, by the clear establishment of the position they occupy in space, by the differentiation of types, has for its ultimate principle the satisfaction of the understanding of solid form seen and comprehended as a whole. (Wölfflin.)

The experiments of the Quattrocento had in great measure prepared the way, but the new Roman classicism surpassed them in two ways: it reacted sharply against the tyranny of detail, the multiplication of mere objects, and against the peculiar loss of concentration which characterized the last generation of the fifteenth century, particularly in the north of Italy, and even in Tuscany, with, for example, Filippino Lippi. What Raphael achieved, starting from Perugino, with his harmonious feeling for space, and by Michelangelo with his concentration on the splendours of the human body, is the transition to the type, the diminution of the individual and that movement towards the sublimation of particular forms in one great essential form, described in Raphael's famous letter to Castiglione, and by Michelangelo in his poems as a reflection of heavenly beauty. The success of this effort explains the assurance with which these great masters regarded the art of Antiquity. They all acknowledged it without being dominated by it. There is not one

THE CINQUECENTO: ROME

Renaissance church which is a copy of a Roman building; not one work which imitates a Greek or Roman composition, and yet nearly all breathe the same air of dignity, clarity and idealization. There again is a break with the fifteenth century, which studied ancient reliefs so as to nourish its taste for realism. In Rome the Antique became the basis for an immense 'heroization' of man,[111] which had a philosophical background implicit in the symbolism of the central plan, in the hierarchy of the great compositions, and in the balance which it holds between the cosmos and man.[112]

Roman art was thus able to become through its form, as much as through its content, the expression of a new culture. But it would never have had so profound an effect had it not also developed an entirely new decorative language in character with the pervasive quality of its humanism. The relation between a building and its ornament is strictly defined by the repetition of motifs and the constant nature of its elements. A creative work such as the Sistine Ceiling is at the same time a plastic vision and a new kind of decoration; Raphael framed his compositions with swags, grotesques, pilasters, etc. from which was to be derived a new kind of ornament. It was not so much new pictures as a new formal world which was to come into being in Italy after the finishing of the Sistine Ceiling and the *Stanze*.

Leonardo had stimulated all these new ideas without providing the same answers; from Florence, where he had first so deeply stirred and then so profoundly disappointed everyone's expectation, he returned to Milan. At the beginning of the pontificate of Leo X he went to Rome (1513–15) but did not stay long enough to achieve anything. In Lombardy as in Tuscany there resulted from his influence developments which must be considered in parallel with Roman classicism.

NORTHERN ITALY

Bramante's forms—octagonal domes raised on multiple drums, choirs articulated in great masses—were adopted in S. Maria della Croce, near Crema, by Battagio and Montanaro (c. 1500), and at Como cathedral (1491–1519). In the same spirit, Bramantino (c. 1465–1536), a notable painter and architect, gave to the Trivulzio chapel in S. Nazzaro (1518) a cubic base stressed by angle pilasters, and ornamented it with a portico on the façade, treated in a pure, but rather elongated, Doric. In Umbria, architects from Milan and Lazio worked, under the influence of Bramante, at Todi, near Foligno, where in 1508, Cola di Matteuccio da Caprarola began the building of a centrally planned church of the same type as that at Montepulciano, with a Greek cross and three polygonal apses, of which one (the north) was semi-circular, with Corinthian pilasters. He was succeeded by Ambrogio da Milano, but the dome was not built until 1607. At Parma, Bernardino Zaccagni, with the help of his son Gianfrancesco, built towards 1520 the Bramantesque Greek cross church of the Steccata.

240

65 Giovanni Bellini: *Agony in the Garden*

(London, National Gallery)

66 Carlo Crivelli: *Madonna and Child with symbols of the Passion*

(*Verona, Mus. Civico*)

67 Butinone and Zenale: detail from the *S. Martino altar*

(*Treviglio*)

68 *The Baptism of Christ:*
(a) from the Pala d'Oro
(Venice, S. Mark's)

(b) Masolino da Panicale; frescoes in the Baptistery, Castiglione d'Olona

69 (a) Masaccio:
S. Peter
baptizing
the neophytes

(Florence,
Brancacci
Chapel,
Carmine)

69 (b) Verrocchio
and
assistants:
The
Baptism of Christ

(Florence, Uffizi)

70 Piero della Francesca: *The Baptism of Christ*

(*London, National Gallery*)

71 Tintoretto: *The Baptism of Christ*

(Venice, Scuola di S. Rocco)

72 Leonardo da Vinci: *Leda and the Swan*, drawing

(Rotterdam, Boymans-Van Beuningen Mus.)

In painting, Leonardo's pupils and imitators are numerous and of very uneven quality. Ambrogio da Predis (1472–c. 1506), dark in colour, and Boltraffio (1467–1516), in a brighter range, painted Madonnas and portraits, in which they endeavoured to preserve their master's softness and the ambiguous expression of his female faces. Giampetrino, Marco d'Oggiono, Cesare da Sesto, repeat Leonardesque motifs, while adopting—above all the last named—Raphaelesque types and Flemish landscape. Andrea Solario (c. 1460–1524), who worked in France, at Blois where a *Last Supper* probably by him has been discovered, and at Gaillon, has a more personal development of which the *Madonna and Child with the green cushion* (Louvre) is an example.

Bramantino added to the massive figures of his master Bramante a romantic feeling for ruins and an animation which though entirely individual tends somewhat to grimaces (*Nativity*, Brera). Bernardino Luini (c. 1480–1532) is a stronger personality. He had links with Foppa and Borgognone, resisted strong chiaroscuro through his preference for tender greys and cold blues, sometimes used chalky tones and calm Umbrian horizons in his wall decorations, and developed into a good fresco painter in the Villa Pelucca at Monza (1522–25). His Leonardesque characteristics are weak and are limited to the glance and to the smile. In his big decorations (Saronno, 1525–27; Lugano, 1529), Luini uses complicated imagery and reveals affinities with Germanic art, expressed from 1515 onwards in the *Annunciation* (Brera) where he copied in grisaille Dürer's print of Adam and Eve from the 'Little Passion'. The Piedmontese style of Gaudenzio Ferrari (1470–1546) who was both a sculptor (at Varallo Sesia) and a painter (at Vercelli, 1530, and Saronno, 1535), is equally inspired by German and Umbrian elements, which coalesce in a prolix and overloaded narrative.

Sodoma (1477–1549), who was of equal importance in Tuscany and Piedmont, is a more attractive painter. His continuation of the *Scenes of the Life of S. Benedict* at Monte Oliveto (1505–8) was exuberant and worldly, and he then settled in Siena whence the friendship of Peruzzi drew him to Rome. He adapted Leonardo's subtlety and delicate modelling as readily to the voluptuous decoration of the bedroom of the Farnesina (*Alexander and Roxana*, 1511–12) as to the evocation of the ecstasies of S. Catherine (chapel in S. Domenico, Siena, c. 1526).

Domenico Beccafumi (1486–1551) derives both from Sodoma and Raphael, and he combines colour and chiaroscuro in strange twilit scenes shot through with fires, but the decorations in the Palazzo Pubblico in Siena (1537) and his allegories and *trompe l'oeil* scenes betray the influence of the Mannerism then coming from Tuscany. In his bronzes (angel-candelabra in the cathedral), the form is more stable, but loaded with accents of light and reflections.

Florence: Painting. In Florence, the first decades of the century were dominated not by Michelangelo, but by the example of Leonardo and Raphael.

Gian Francesco da Sangallo built the Palazzo Pandolfini (1516–20) on Raphael's design; Baccio d'Agnolo (1462–1543) at the Palazzo Bartolini in the Piazza della Trinità (1520–29) adapted the Roman type of façade, and in the Palazzo Rosselli del Turco returned to the robust Florentine type. In painting, the two dominant figures were those of Fra Bartolommeo and Andrea del Sarto.

Fra Bartolommeo (1475–1517), drawn by Savonarola into the Dominican Order, was one of the first to attempt a synthesis of the new trends, in a style softened by the use of *sfumato* handling, but based always on a large spacious setting in which his figures could be composed in balanced groups.

Although he possessed an innate gift for creating simple and typical figures he did not immediately achieve unity (*Last Judgement*, 1499, S. Marco); he attained it with his *Vision of S. Bernard* (1506, Accademia), and his *Madonna and Saints* (1508, Lucca), which conform to a Peruginesque pattern, and he reached a kind of grandeur, through amplitude of composition and delicacy of light, in the *Marriage of S. Catherine* (1512, Pitti). He is coherent, flexible, but cold in the *Christ* (1517, Pitti), more vivid in the *Pietà* (1517, Pitti). His companion Albertinelli (1474–1515) in his *Visitation* (1503, Accademia) and *Annunciation* (1510, *ibid.*), achieved a mild and tame conventional, classical sensibility. His *Tabernacle* (1500, Milan) shows that he was familiar with Flemish painting.

In the shadow of Leonardo and Raphael, Andrea del Sarto (1486–1531) expressed, in a kind of dreamy passivity, the Florentine version of the classic style. 'Faultless works' Vasari called them in 1550; it would perhaps be better to say masterpieces of perfectly assorted parts, of calm balance, but with the inquietude of ineradicable melancholy. In this respect, the *Madonna delle Arpie*, which Wölfflin called the 'most noble Madonna in Florence', is informed by a kind of pensive self-consciousness which distinguishes it from Raphael's *Sistine Madonna*.

For the atrium of the Annunziata Andrea painted frescoes of *Scenes of the Life of S. Filippo Benizzi* (1509–10), set in clear and placid landscapes, and a *Birth of the Virgin* (1514) whose skilful management suggests that he might have become a more sensitive Ghirlandaio. For the little cloister of the discalced Carmelites, or Scalzi, he composed monochrome *Scenes of the Life of the Baptist* (ten panels, of which two were completed by Franciabigio, and four allegories of Virtues) from 1512–24. All is of a touching and truthful intimacy, of an impressive ability and without any taint of mere routine in the compositions. There is the same rather melancholy ease, but with more lively colour harmonies, in the *Charity* (1518, Louvre). The *Madonna del Sacco* (1525), painted in a lunette of the Annunziata and seen in sharp perspective, is a brilliantly off-centre composition, planned on a double pyramid, which sets off admirably the face and gesture of the Virgin, while leaving in a transparent half-shadow the silhouetted figure of S. Joseph on the left. Limpid landscapes and delicate contrasts also dominate the *Holy Family* (Borghese, Rome), but the conscious arrangement of draperies favourable to a display of virtuosity in modelling and reflected light replaces the earlier tenderness of feeling. In the saints in the *Assumption* (*c.* 1530, Pitti), Andrea finally succumbs to this abuse. His artistic personality is wonderfully expressed in the modelling of his portraits (*self-portrait*, Pitti). His influence was paramount on convinced eclectics: Giuliano Bugiardini (1475–1554), Bachiacca (1495–1552), who painted the *Story of Joseph* in the Borghese, which finally won over Michelangelo, and above all Franciabigio (1482–1525) who was his assistant at the Scalzi; close to Andrea in his portraits of dark and dreamy young men (Louvre), he is closer to Raphael in the *Holy Family* (Vienna).

Sculpture: Andrea and Jacopo Sansovino. Andrea Sansovino (*c.* 1470–1529) worked in the spirit of Bramante at Monte San Savino and followed him at Loreto (1509); but it was in Tuscany that, under the influence of Leonardo, he proved, as a sculptor, his qualities of delicate idealization, in works such as the baptismal font at Volterra (1501), and the *Baptism of Christ* (1502–5, Florence, Baptistery). His important compositions at S. Maria del Popolo in Rome (the *tomb of Ascanio Sforza*, 1505, and the *tomb of Girolamo della Rovere*, 1507) combine with great clarity the Roman type of niche with the heavy decorated arches of the Lombards. His taste is purer, the motifs he introduced more carefully studied, and more easily handled in the set of bas-reliefs of the *Life of the Virgin* made for the Santa Casa at Loreto, where the idealization becomes one with the invention.

In Florence, no one personality replaced Sansovino, though sculptors were always numerous, for instance Gian Francesco Rustici, Andrea Ferrucci (both already mentioned) from whom were commissioned for the cathedral the Apostles which Michelangelo had abandoned, and Benedetto da Rovezzano (1474–1552), whose *tomb of S. Giovanni Gualberto* (1503–13, fragments in the Bargello) lacks coherence. After 1525 Michelangelo's collaborators in the Medici Chapel appear: Il Tribolo (1500–50), to whom we owe the fountains of the Villa Petraia and the Villa di Castello, and Giovan Angelo Montorsoli (1507–63) creator of the Orion fountain at Messina. But they have more of the fantasy of Mannerism than the classical spirit, as it was defined in Rome and conceived in terms of Raphael by Il Lorenzetto (1490–1541), in the figures of *Elijah* and *Jonah* carved by him for the Chigi Chapel.

Andrea Sansovino's only heir, capable of infusing into his frigid serenity something of Raphael's teaching, was his pupil Jacopo Tatti, who adopted his master's name. Jacopo Sansovino (1486–1570) was a Florentine, a friend of Andrea del Sarto, and as sensitive as he was to the most delicate aspects of modelling. He went to Rome in 1506, studied Raphael and enlarged the boundaries of his art, as may be seen in the *Bacchus* (1511–14, Bargello); then he was fired by the example of Michelangelo and his massive energy, in, e.g. *S. James* (S. Maria di Monserrato). His career underwent an abrupt change in 1527; he left Rome after the Sack and went to Venice where, with an unusual talent for adaptability, he became the great architect of the city, and a sculptor who can be associated with Titian. This eclectic, who had experienced all that Florence and Rome had to teach, blended their lessons with those of Venetian art. He was the complement to Sebastiano del Piombo who, formed by Giorgione, went to Rome in 1511, and there gained a reputation first with Raphael and then with Michelangelo, from whom he separated in 1532. It was towards 1530 that Venice reacted to the influences of Roman art; there, the classical moment of the Renaissance bore a very different aspect.

Small Bronzes. It was principally in Florence and in Padua that the art of the statuette and the bronze plaquette was developed during the second half of the fifteenth century. This taste for 'Antique' knick-knacks is characteristic of humanist collectors and the artists who served them. The art of bronze working was revived by Ghiberti and even more by Donatello, and the lesser Florentine masters formed in their workshops created small pieces which enjoyed immediate popular success. In Florence, from 1480 onwards were produced statuettes of *Hercules* from the Pollaiuolo workshop (Vienna), and works by Bertoldo di Giovanni (*Bellerophon*, Vienna; *Hercules*, pvt. coll. New York), and allegorical plaquettes modelled with nervous intensity (Victoria and Albert Mus., London). The most celebrated medallist was Niccolo Spinelli (1430–c. 1510) who left invaluable portraits of Lorenzo de' Medici, Pico della Mirandola, Ficino and Poliziano, and of other celebrated men and pontiffs.[113]

Most of these small works, where the imitation of the Antique easily slipped into the creation of pastiches, were produced in Northern Italy. Superb models already existed in the splendid medals created by Pisanello in Ferrara (*John VIII Paleologus*, 1438; *Lionello d'Este*, 1443) and those by Matteo de' Pasti (*Malatesta and Isotta*, 1446). Donatello's residence in Padua was decisive in importance, and from his vigorous art derived, besides such anonymous works as the small version of the *Spinario* (Vienna), the realistic animals, the lamp bearers, the little fauns by Andrea Riccio (1470–1532), whose small self-portrait bust, with curly hair, is an original and lively example. At the same period was produced the curious mirror 18 cms. in diameter, representing allegories of nature (the *Martelli mirror*, V. and A. Mus., London) which was once attributed to Donatello. A large number of plaquettes of gorgons, tritons, bacchantes, and pagan gods (coll. in Santa Barbara, California) were produced in these centres, either Florentine or Paduan.

A more sophisticated taste which sometimes even foreshadows Neo-classicism and the French Empire style, is discernible at the beginning of the sixteenth century in Ferrara, with Antico (c. 1460–1528), whose busts, and statuettes of Venus, are very delicately modelled (Vienna); no less elegant was the still anonymous Northern artist who signed himself Moderno, to whom we owe plaquettes filled with figures set against architectural backgrounds and treated with extreme refinement and delicacy, such as that of the *Flagellation*.

THE CLASSICAL STYLE IN LOMBARDY AND TUSCANY

With Sansovino, and another Florentine who came to Venice about 1530, Danese Cattaneo (1509–75), the statuettes of tall, slim figures in the now obligatory *contrapposto* already tend towards Mannerism. It was in Florence, in the middle of the century, with Cellini, Bandinelli and Vincenzo Danti, that the elegant bronze figurine with studied gestures was to reach its highest point of perfection.

III

The Cinquecento: Venice

Introduction: Historical Background: Exoticism and Humanism

In 1495, Philippe de Commines described the Grand Canal as 'the most beautiful street in all the world, and the one with the most beautiful houses in it'. The brilliance of its civil ceremonies, transformed into naval pageants, and the solemnity of its religious feasts, were the wonder of all—of the Milanese Pietro Casola; of the German F. Faber, who was astonished by the slave traffic, 'forbidden since the time of Pope Zacharias', by the elaborate dresses of the women at religious festivals, by the coming and going of innumerable ships in the harbour; and of Dürer whose stay in Venice in 1505–6 was a continued delight to him.

Sooner or later the discovery of the Cape route and of the New World was bound to threaten the extraordinary prosperity of the city. In 1504, an important citizen was already alarmed by the fact that from Syria arrived 'empty galleys, and, what had never yet been seen, completely without spices', but the now fatal decline of Venetian power was masked in the sixteenth century, and even in the seventeenth century, by the paramount role that the city played in European politics and culture.

The Venetian colonial empire in the Eastern Mediterranean receded step by step before the Ottoman advance, which became irresistible after the fall of Constantinople in 1453. Euboea was evacuated in 1470, and there remained only a few strongholds in Morea. Cyprus, given to Venice in 1489 by Catarina Cornaro, was only a temporary compensation (until 1570) for the loss of their trading stations. The incursions of the Turks began through the Adriatic and the Balkans; in 1501 an invasion through the north was feared, and Leonardo, who was passing through Venice, suggested fortifying the banks of the Isonzo. The Turkish menace provided the French with a counterpoise to the power of the Empire (Vienna was besieged by the Turks in 1529) and with a means of bringing pressure to bear on Venice. The wars of Italy once over, the alliance of the two forces of Venice and the Empire (well illustrated by Titian's success at the Court of Charles V) permitted Don Juan of Austria to win the decisive victory at Lepanto (1571), in which the Venetian fleet played a most important part.

INTRODUCTION: HISTORICAL BACKGROUND

From the beginning of the fifteenth century onwards, Venice was a main-land power, and became part of the Italian world through her conquests, which included Padua, Verona, and Brescia, and bordered on the Milanese states. She was also deeply involved in the Franco-Imperial wars; in 1509, the League of Cambrai between the pope and the King of France ensured her defeat and the loss of her mainland provinces which she was later to regain by negotiation. Venice was paramount in the refinements of diplomacy; the Grand Council had always required its ambassadors to make the most precise reports, which remained models of their type, and prove how keen an interest the Venetians took in the manners, customs and characters of others.

In this voluptuous, wealthy and cosmopolitan city open to exotic influences and quick to use them for decorative purposes, humanism was bound to develop quite differently from the intellectualism of Florence or the pomp of Rome. In 1468, Cardinal Bessarion gave his manuscripts to the city (now in the Biblioteca Marciana); printing workshops like that of Aldus Manutius specialized in Greek texts; Byzantine learning and literature found an exceptionally favourable ground. Ermolao Barbaro (d. 1493) represented the new Aristotelianism which developed from it and was continued at Padua by Pomponazzi's philosophy of nature. Pacioli's treatise on applied mathematics was published in Venice (*De divina proportione*, 1509). The fusion of these somewhat esoteric theories with religious and 'Pythagorean' symbolism was undertaken in the peculiar work by Francesco Giorgi on the *Harmony of the World* (1525). To this very strong current of Venetian learning was opposed a more facile literature, better adapted to the tastes of the nobility, which found a lively and utterly immoral exponent in Aretino (1492–1556), who settled in Venice in 1527, to become the fashionable author of licentious dialogues (1532), plays and books of devotion, and to be famous all over Italy for his huge correspondence, his brilliance as a critic and his friendship with Titian.

The *cognoscenti* met in gardens like those at Murano, celebrated in a Latin poem by Castaldi, and which Bembo loved. The famous plan of Venice engraved in 1500 by Jacopo de' Barbari shows the large number of these green enclaves behind the palaces. At the same time, the taste for villas was encouraged, like that described by Crescenzio's treatise on agriculture (1495) and, around Bassano and Padua, the building began of houses with loggias or frescoes on the outside. These manors on the mainland formed the natural setting where—around beautiful women, not elegant noblemen as in Florence—the Venetian men of letters could pursue their discussions on love and culture. The *Asolane* of Pietro Bembo (1505) are a case in point, and the masterpiece of the genre.

In these villas, as in the palaces of the lagoon, on which the Venetian patricians lavished the greatest care, painting plays an essentially decorative role. It becomes part of the walls and ceilings, and sculpture is placed in niches and on balustrades. This unity of the whole building is as striking in the churches,

247

where the altarpieces agree in scale and tonality with the architecture. This harmony was never spoilt, even by the modernization of the decoration or the rapid evolution of style, and the essential character of the city—that of a continuous façade—was made more complex and varied but was never fundamentally altered. In succession to Rome, which was temporarily relegated to second place, Venice achieved towards 1530 a tremendous artistic development, and opened a new phase in classic art, followed immediately by a Mannerist crisis which will have to be considered separately.

ARCHITECTURE AND SCULPTURE

It has been said that the architecture of Venice consists only of exteriors: the open space, or the canal bordered by façades, forms a monumental unit of which the buildings are only the outer walls. The palace front, with its ornate galleries, its arched storeys full of shadows, continues in large measure the Antique type of villa with portico.[114] In the fifteenth century, this picturesque formula reached its highest point of richness and elegance by combining the exotic tradition of the Levant with Gothic forms. With the introduction by the Lombardi of the classic orders of the Florentine type, this exuberance seemed to subside. But Sansovino's introduction of more imaginative and ornamental forms derived from Raphael and Peruzzi led, towards 1530, to a decisive change of development, in which Venice restored to classical architecture its great decorative function, through the multiplication of voids, and the animation of the surfaces, and reverted to ideas worthy of the fourth and fifth centuries or of the greatness of Byzantine art, thus adding an inflexion proper to the monumental concepts of the Renaissance.

Sansovino, who took refuge in Venice in 1527, succeeded Bon as 'chief of the Procurators of S. Mark'—that is, as principal civic architect. In his first works, such as the Palazzo Corner (1532) he was obviously still careful to maintain a strict rhythm; his rigid design arranged the rustication of the ground floor in regular courses and used coupled columns; instead of many small arcaded openings, two superimposed orders, noble and deep, ornament the façade. The mass remains dignified, but it is broken up and given colour by the variety of shapes. This adaptation to Venetian atmosphere is even clearer in the works around S. Mark's, where Sansovino restored two domes and began the definitive replanning of the square.

If there is in Venice one façade conceived as a background to an open space, it is that of S. Mark's. In the middle of the thirteenth century, the plan had been fixed by filling in a canal which crossed the square, by paving it with bricks, by creating a framework of houses with low arcades and embattled parapets, and by raising the façade of the church itself. The idea of a huge four-sided arcaded space, which the atrium of the basilica would close one side, crystallized when the fifty arches of the Procuratie Vecchie, of a strict severity of rhythm, were begun on the northern side of the square. The *loggetta* of the Campanile (1537–40), with cornices and projections, triumphal arches, bas-reliefs, and four bronze statues in niches, was the first modern addition, brilliantly suited to the luminosity of the square. The great Library (1536–54), which was finished by Scamozzi, is an even more complex and brilliant example of Sansovino's splendidly decorative style. It is built in two superimposed arcaded orders, Doric below and Ionic above, and given the richest decorative treatment, with festoons, sculpture, and a very deep

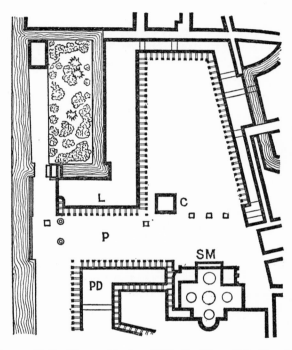

FIG. 30. *Piazza di S. Marco, Venice. SM, Basilica; PD, Doge's palace; C, Campanile; L, Library; P, Piazzetta.*

entablature surmounted by a balustrade crowned with statues. This masterpiece was flanked by the far less successful Mint (*La Zecca*, 1537–45) constructed of exaggeratedly rusticated blocks, and made even heavier by the later addition of an extra storey.

The building of the Library, along one side of the Piazzetta and on the corner of the main square of S. Mark's, made it necessary to extend a monumental façade right down the south side of the square, facing the Procuratie Vecchie. This was done in 1586 by Scamozzi, who raised the Procuratie Nuove to three storeys and retained the lively rhythm of the Library by following Sansovino's precedent of one large window filling each bay of the arcade. His work was finally completed by Longhena (1640), and the short

249

base of the quadrilateral—the Napoleon wing—was built in 1810. But the monumental conception was due to Sansovino who, during forty years, exercised much the same sovereignty in Venetian architecture as Titian did in painting. He created, in the Villa Garzoni at Candiana, near Padua (c. 1555), the pattern of the patrician villa designed for festivities, with a noble courtyard formed by a single storey colonnade decorated with statues, but despite the surface variety the forms are clearly related to each other, and there is neither the complication of Mannerism nor the subtle movement of Baroque. In sculpture the position is much less clear; in the statues on the *loggetta*, in the bronze *Evangelists* on the sanctuary balustrade of the high altar of S. Mark's, in his reliefs on the sacristy door, Sansovino's delicate elaboration of detail in the draperies, and his effort to create an airy shimmering effect, detracts from the solidity of his forms.

In 1534, S. Francesco della Vigna was begun on Sansovino's plan, which called for a nave and two aisles and a dome, but was later reduced to a single nave plan. The question of the proportions of the building having arisen, Doge Andrea Gritti requested a humanist known for his 'Pythagorean' speculations on numbers, Francesco Giorgi, to provide a memorandum on the subject. His measurements, founded on the musical scale and on the proportions that best express universal harmony, beginning with the series 1:2:3:4:8:9:27, were adopted by a commission which included Titian, Serlio and another humanist, Fortunio Spira.[115] This demonstrates the interest which was attached to the mathematical purity of buildings, perceptible only to connoisseurs, and the kinship between the arts. The presence of Serlio, who was then preparing the publication of his first four books (1537), is a good indication of the original work which was being done in Venice on the ideas of harmonic beauty. Finally, thirty years later (1568–1572) Palladio designed a great façade for S. Francesco della Vigna which takes account of the table of proportions established by Giorgi, and this establishes a link between Palladian aesthetics and the learned speculations which accompanied the adoption of classical forms in Venice.

Serlio is as important for Venice, where he lived from 1531 to 1541, as he was for Rome. His architecture, founded on the use of the Orders, yet picturesque and full of variety, with niches and voids which ensured, as he put it, 'a finer effect', is an extension of the style of Raphael and Peruzzi. It suited Venetian taste: a system of harmonic relationships—the geometry of which suggests an analogy with music—is overlaid with effects of light and shadow and variety of plane, which treat the surface of the building in a painterly way. Venice was receptive to all these new ideas, which were beginning to appear in treatises. For instance, the patrician Alvise Corner (Luigi Cornaro) wrote a commentary in which, while he recommended using the Antique Orders, he yet insisted on comfort and on modern ideas of beauty. But it was in Padua, with the help of Falconetto, that he was able to carry out his ideas.

Even before Sansovino arrived in Venice, a knowledge of the new architecture derived from Bramante and Raphael was spread in Venetian territories by Giovanni Maria Falconetto (1468–1535), whose native Verona was also the home of the humanist epigrapher Felice Feliciano, and of Fra Giocondo, the great archaeologist and editor of Vitruvius. He had a varied career, studied the Antique in Rome, and towards 1520 attracted the notice of Alvise Cornaro who sent him to work in Padua. In the palace near the Santo, he built (1524) the loggia whose very simple design displays to the best advantage the delicate reliefs and the alternation of pediments on the first storey, and the *odeon*, or music room, with its carefully calculated acoustics, and with, inside, stuccoes and paintings in the 'Pompeian' style of Giovanni da Udine. He also built the entrance, in the form of a Roman triumphal arch, of the Villa Cornaro (Este), a bishop's residence, with a colonnaded façade raised on a rusticated base in the midst of fields, and in Padua the Porta S. Giovanni (1528) and the Porta Savonarola (1530). It is the simplicity of style and clarity of form in these works, that prevents them from slipping into the growing Mannerism. With Sanmicheli, who was a pupil of both Falconetto and Sansovino, this step is often taken.

Michele Sanmicheli (1484–1559) was trained in Rome, where he learned much from Antonio da Sangallo the Younger. In S. Domenico in Orvieto he built the Cappella Petrucci (1518–21), then, in his native Verona to which he returned in 1527, palaces which were symptomatic of the break-up of classical ideals. In the Palazzo Pompei, with its rusticated ground floor and loggia on the first storey, a kind of affectation is perceptible. The Palazzo Bevilacqua shows a striving after originality which shatters the symmetry and creates discords in the composition; an enormous cornice and balcony runs between the two storeys and, on the *piano nobile* (the first floor) there is a cumbersome alternation of large and small arches, of columns with straight and spiral fluting, of triangular and segmental pediments, in an almost incongruous rhythm. An altogether Venetian vivacity, using every possible means of animating the planes and defining one against another, was later used in the Cappella Pellegrini at San Bernardino in Verona.

In Venice proper, the Palazzo Cornaro Mocenigo (1543) is in a more restrained style; in the Grimani palace he uses the arrangement in two storeys articulated by a coupled order in the side bays—a motif which already had been used in, for example, the Palazzo Vendramin, but he also obtains an effect of variety and depth by the extra height given to the ground floor, by the compression of the bays, and by introducing two rectangular bays, like minor motifs, on either side of the central arched bay. Sanmicheli built many villas, now nearly all destroyed; imposing remains survive of the Villa della Torre at Fumane (Verona), and of the arch of the Villa Bevilacqua. He was also the military architect of the Republic and fortified Verona (1528–38), Padua (1534), the Lido and the islands of the Lagoon.

The revolution in the treatment of light came during the first ten years of the century. This decisive reform in tonal painting, which had been adopted by Giovanni Bellini, was to be increased in scope by Titian, so as to involve every kind of painting, and to be adequate for all needs. Bellini did not die until 1516, when Titian was about twenty-five years old; Giorgione, born about 1477, died prematurely in 1510. There was therefore a moment, from 1505–10, into which—as in Rome—all the experience of the past and the present was concentrated. Bellini's last manner, in the *pala* of S. Giovanni Grisostomo (1513), is redolent of 'Giorgionismo' and the unfinished works of Giorgione, and those of Bellini, were completed by Titian; the assessment of the currents of influence and the attributions remain therefore a somewhat delicate affair. But this fact alone proves how profound were the repercussions of Giorgione's art and its seductive power.

Florentine historians, such as Vasari, have postulated a contact between Giorgione and Leonardo, who passed through Venice in 1501. The *sfumato*, the subtle haze in which Leonardo bathes his contours, certainly struck an answering chord in the sensibility of a young and cultured artist like Giorgione, but instead of softening a form comprehended with the feeling for plasticity inherent in the Florentines, the Venetian use of chiaroscuro was to become the final stage in the creation of an effulgence of light, in which the modulation of form had for long been expressed through tonal values. From 1480–90 onwards Bellini's works are lit with a strange brilliance; towards 1500, Carpaccio evoked with his *Two courtesans* (Correr Museum) the moist colours of 'Venice under the Sirocco' (Palluchini), and Montagna or Buonconsiglio introduced the oblique light, the leaden tones of stormy weather. All seems to lead to a new understanding of visual effects and themes, and to the discovery of a simple basic idea which would satisfy the changing tastes and the humanist culture of the Venetians.

Giorgione and 'Giorgionismo'. Giorgione's short and dazzling career (*c.* 1477–1510) is shrouded in mystery. He was born at Castelfranco, between Vicenza and Treviso. Since he painted, according to Vasari, a portrait of Caterina Cornaro, it may be supposed that he frequented the garden at Asolo, visited by Bembo and his friends. Giorgione was immediately admired in these circles. He was barely dead when Isabella d'Este inquired after a painting of a nocturnal scene which she thought was among his effects. His works were, therefore, greatly sought after, and in the difficult analysis of 'Giorgionismo' the most reliable information is found in the notes made by the connoisseur Marcantonio Michiel, from 1525 to 1545, on Venetian Collections.

The first phase is that of the *Castelfranco Madonna* (1504) or *Madonna and child between SS. Francis and Liberale*, and the *Judith* in the Hermitage

(Leningrad). The structure of the altarpiece seems a little awkward, but it enables the Virgin to be placed against a landscape bathed in light, and allows an oblique light coming from the left to envelop the two saints standing on the pavement beneath the throne. The composition is strictly balanced, as is the case in all Giorgione's pictures. It is arranged in planes parallel to the picture plane, and the forms are seen frontally. It responds to a typically classic feeling for space, but this arrangement is of less importance than the unification of the picture by means of the light, and it is the light pervading the whole canvas whose qualities Giorgione was to accentuate. Maturity was reached with the *Tempesta* (Venice Accad.), which Michiel describes as a '*paesetto*'. The painter makes his enigmatic narrative theme the occasion for a landscape, with a naturally haphazard arrangement of bushes and ruins, drained of colour by the sharp glare of a sudden flash of lightning, so that he has taken for his object the most transitory of effects, and instead of arranging the parts of his picture on a stable plan, has set them in a Nature controlled by the atmosphere and poised in the most fleeting instant of time. This is the great change. In 1508, some questions about the payment for the frescoes which Giorgione executed on the Fondaco dei Tedeschi on the Grand Canal allow their date to be precisely determined. Their subject matter was unknown even in their own day; an esoteric quality, due probably to his friendships with humanists, is discernible in the painter, and this would explain the strange yet human quality of the *Three Philosophers* (Vienna) which dates from the same time, and the X-rays of which reveal the Three Magi deep in their astrological calculations in the shadow of a dark rock. The picture is at once large in feeling and quite simple. The *Sleeping Venus* (Dresden) achieves grandeur through the dignity of its presentation, the amber colour of the nude, the diffused light and the breadth of the landscape. As if to explain the note of voluptuous and tender reverie, legend has it that Giorgione divided his time between love and music.

In his last work, the *Concert Champêtre* (Louvre), all the elements of the tender melancholy dear to the painter's heart—feminine beauty, music, twilight, and a pastoral landscape—are blended in an intimate and poignant unison. Titian may have had a hand here; the question is posed even more sharply in the *Concert* (Pitti) where, between two less incisively rendered heads, the face of the Augustinian renders with remarkable intensity the effort to express the inward feeling of a whole art.

Tradition still attributes many portraits to Giorgione, including one of himself as *David* (Brunswick). The restrained portraits, in warm tones against dark grounds, which proliferate towards 1510 are due to his influence. A work such as the *Old Woman* (Accademia, Venice) shows a striking realism, probably of a Northern origin, like some of the paintings of fantasy subjects still ascribed to him. All these novel, voluptuous, or subtle ideas were well off the beaten track; they were a stimulus both to the imagination and the practice of painters. According to Vasari, Giorgione painted directly in full colour, without any preliminary drawing. This was a new development of considerable importance, and one quickly taken up.

Among the pupils and followers of Giorgione, Titian was the most outstanding; but Palma Vecchio, Savoldo and Sebastiano del Piombo were all closely connected with 'Giorgionismo'. Palma (1480–1528) was born at Bergamo; he developed a type of opulent, blonde beauty, resplendent in gold and purple satin (*Portrait of a woman*, Poldi-Pezzoli, Milan), which only retains from the influence of Giorgione and Titian a facile effect of sumptuosity, although in his altarpiece in S. Maria Formosa, Venice (*S. Barbara and other Saints*), his brilliant colour and smooth technique are very impressive. In another Bergamasque, Cariani (*c.* 1485–1547), Palma's style becomes much heavier.

Savoldo (born near Brescia *c.* 1480–after 1548), turns towards a purely Northern, Lombard and even at times slightly Flemish taste, and in him the influence of Giorgione tends towards effects of light and of dark shadows, amid rich and warm harmonies of colour (*Altarpiece*, Brera; *Tobias and the Angel*, Borghese).

Sebastiano Luciani (1485–1547), who became Fra Sebastiano del Piombo when he was given the sinecure of the Papal Seal at the Vatican in 1531, shifted from Bellini to Giorgione, and was particularly attracted by the latter's 'grand manner', that of the Fondaco dei Tedeschi, which inspired his own magnificent figures set in niches—the *S. Louis of Toulouse*, and *S. Sinibald* in S. Bartolommeo a Rialto (1508). This grandiose and monumental art reveals a great deal of the period, but it also explains why Sebastiano was attracted to Rome in 1511; he could assimilate neither the grace of Raphael (his *Portrait of a Woman*, Uffizi, remains heavy) nor the strength of Michelangelo—the Viterbo *Pietà*, despite the striking quality of its sombre colouring, is hardly more than a rather heavy-handed night-piece.

Titian. Titian's long life (*c.* 1485–1576) guaranteed to sixteenth-century Venetian painting the continuity which, during the fifteenth century, it had owed to Giovanni Bellini. By holding the Venetian School to its development of the treatment of light, he rapidly extended his influence over Northern Italy, as far as Emilia, and finally as far as Rome. Through his fame the Italy of the Renaissance, where all was interrelated, achieved a hitherto unknown unity of culture and taste. At the same time, Titian disciplined the imagination of the Venetians and, towards 1530–40, when Sansovino, Serlio, and even Aretino, were introducing Roman ideas, he brought it to a classic completeness which balanced the contributions from Central Italy.

Titian became a national, and even an international, power. Deeply attached to Venice, where he was loaded with honours, he yet had remunerative and flattering relations with all the princes and prelates of his day—Alfonso d'Este at Ferrara (1516), Federigo Gonzaga at Mantua (1524), Francis I of France, the Emperor Charles V, who made him a Count Palatine

C   Titian: *Nymph and Shepherd*
(Vienna: Kunsthistorisches Museum)

(1534) and invited him to Augsburg in 1548 and 1551, Pope Paul III who gave him a magnificent welcome in Rome (1545). With his extraordinary vitality, and through the faithful disciples who surrounded him, he organized a studio which transformed all the characters and scenes of mythology and Holy Writ into an epic pageant, though one dominated and inspired by a controlling 'chromatic alchemy' (Lomazzo). Several thousand canvases may be grouped around this studio. His genius develops in several stages, in which may be seen successively the emergence of his personality (until 1520), the highest point of his classicism (1520–35), his awareness of Mannerism (1535–45), and, after this crisis, the flexible, but increasingly monumental style, which, after 1560, culminated in an unexpected deepening of spirituality and pathos.

On his arrival from Cadore, whose landscape of alpine foothills inspired him as often as did the Lagoons, Titian immediately followed the style of Bellini (*Jacopo Pesaro presented to S. Peter*, 1503, Antwerp) and Giorgione, with whom he worked, in 1508, on the Fondaco dei Tedeschi. But the decorations which he did in the Scuola del Santo at Padua (1511) reveal, despite their shortcomings, their differences in temperament. Titian builds up the forms, defines the space, and seeks to balance large areas of colour rather than to merge them into a 'musical' continuity. His instinct, therefore, is for the enrichment of a picture, for sumptuous adornment, and for the responses common to man, to nature and to the spirit. Rather than in the *Concert* (Pitti) where he only finished a Giorgione, it is in the *Sacred and Profane Love* (1514, Borghese), on the theme of the two forms of beauty derived from platonic humanism, that this new horizon is opened. The orientation is equally clear in the *Flora* (*c.* 1516, Uffizi), with her ample forms and serious beauty, and in the clear and unambiguous portraits, all in greys and whites, such as the Louvre *Man with a Glove* or the Uffizi *Knight of Malta*.

The monumental quality, linked to a classic simplification, a balanced arrangement of the colour masses, and a heroic conception of life, was achieved in a series of religious masterpieces, such as the *Assumption* (1518, Frari), or the *Madonna appearing to SS. Francis and Alvise* (1520, Ancona), where the Virgin hovers above a moving, twilit landscape with its forms outlined by the chiaroscuro. Next came the great *Pesaro Madonna* (1526, Frari) where the masterly composition ranges the donors in a foreground parallel to the picture plane, the saints and the Madonna in a strongly oblique middle ground, with enormous columns in the background to pull the whole composition together above the swirling banner on the left. The *Entombment* (*c.* 1525, Louvre) achieves by quite different means a composition as dense as that of the finest Raphaels, and the huge *Presentation of the Virgin* (*c.* 1534–1538, Accademia) established the authority of the artist, who was willing to risk even the commonplace in the treatment of the architectural parts, and in the distribution and the lighting of his crowds of figures. In the same years that saw the production of these religious works, a series of magnificent mythologies commissioned by the Duke of Ferrara (1523), showed the painter's ease and imperturbable lucidity in compositions based on the loves, the hunts and the adventures of pagan fable, set in the radiant light of day. Among them were the Prado *Worship of Venus* and *Bacchanal*, the *Bacchus and Ariadne* in London, and later the *Venus of Urbino* (1538, Uffizi), a lighthearted and unequivocal version of Giorgione's theme. This frank and passionate love of worldly beauty and happiness is also expressed in the variety and elegance of the innumerable portraits of this period—*Tommaso Mosti* (1526, Pitti) all in greys, *Ippolito Riminaldi* (*c.* 1540, Pitti) dressed in black, with green eyes in a golden face, *La Bella* (*c.* 1536, Pitti) in her gorgeous brocades, and later, but in the same style, the enormous *Aretino* (1545, Pitti) in a dark velvet gown, with his assured and insolent look.

Already, in the *Bacchus and Ariadne*, the painter was abandoning Antique forms and was tending towards a much freer use of colour, towards the elongation of forms to be seen in the *S. John Baptist* (Accademia), towards an abuse of foreshortening, as in the ceiling painted for Sto Spirito, now in the Sacristy of the Salute. This is the moment when the ferment of Mannerism can be seen working most strongly in the colour of an artist like Schiavone. The *Christ crowned with thorns* (1542, Louvre) contains gigantic and tormented figures which are the aftermath of Titian's visit to Rome. During these years of crisis, his colour changed its character. Faced with the sensual and virtuoso painters of the mid-century, the ageing master retaliated by increasing the intensity of his colour, and by casting a unifying veil of sparkling light over the landscape, the draperies, or the even more broadly brushed in forms. This may be seen most clearly in portraits where a sombre fire glows beneath brocades of muted colour—the *Paul II Farnese* (1546, Naples), the *Emperor Charles V on horseback* (1548, Prado), and later, the portrait of the collector *Jacopo Strada* (1567, Vienna). It can also be felt in compositions whose colour is more and more glittering in effect or bathed in golden light, as the *Crucifixion* (1557, Ancona), the *Annunciation* (1565, S. Salvatore, Venice) where the light turns the subject into high drama, in the sombre *Martyrdom of S. Lawrence* (*c.* 1570, Gesuiti). In his old age, Titian displays the vitality of a Michelangelo, but instead of retiring from the world in angry piety he achieved the serenity and universality of a Raphael. The painter's last works remind one of Shakespeare—a learned, conscious art, somewhat showy and theatrical, with a strongly impersonal quality, is increasingly pervaded by a highly personal emotion and pathos, of which the scintillating *Lucrezia* (Accademia), and the *Crowning with Thorns* (Munich) are the poignant expression. In the strange *Pietà* (Accademia) which he destined for his own tomb, and which was finished by Palma Giovane, a huge niche dominates a scene impressive alike for its pathos and serenity.

Titian's following was numerous—too numerous even. His all-powerful influence worked on Palma Vecchio, and on the Veronese painter, Bonifazio de' Pitati (1487-1553), in his lively and colourful *fêtes champêtres*.

73 Faces:
(a) Pisanello: *Lionello d'Este*
(*Bergamo, Accademia Carrara*)
(b) Antonello da Messina: *Unknown Man*
(*Turin, Palazzo Madama*)
(c) Raphael: *Madonna of the Goldfinch*
(*Florence, Uffizi*)
(d) Leonardo da Vinci: *S. Anne*
(*Paris, Louvre*)

74 Bramante: the Tempietto, Rome

75 (a) Rome,
      Palazzo della
      Cancelleria
   (b) Florence,
      Palazzo
      Rucellai

76 Raphael: *The Liberation of S. Peter*

(*Detail from the fresco in the Sala d'Eliodoro, Vatican*)

77 Michelangelo: *The Libyan Sibyl*

(*Detail from the Sistine Ceiling, Vatican*)

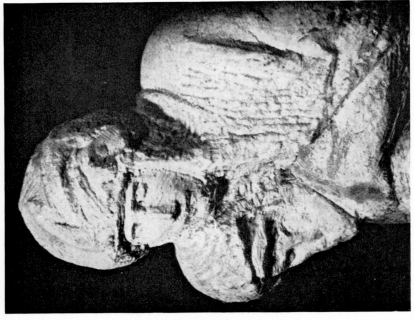

78 Michelangelo: (a) Detail from the *Pietà*
(Rome, S. Peter's)

(b) Detail from the *Rondanini Pietà*

79 Rome, St. Peter's and the Vatican

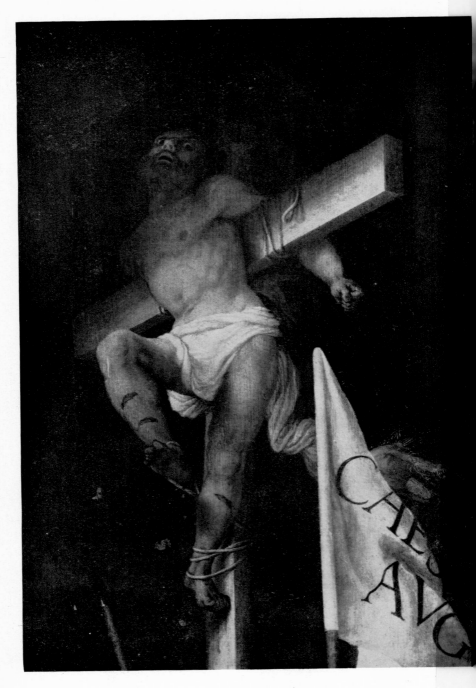

80 Lorenzo Lotto: detail from the *Crucifixion*

(*Monte S. Giusto*)

PAINTING

The Trevisan Paris Bordone (1500–71), is a more typical example. His tendency to exaggerate his tonal effects, and the rich style of his portraits, can be seen in *The Nurse* (Pitti); then, under the impact of Titian's style, his colour became frigid, and he produced nothing but mannered and discordant mythologies and religious scenes (*Paradise*, Venice Accad.).

Pordenone, too (*c.* 1483–1539) is an interesting case. Of Lombard origin, born in Friuli, he worked at Ferrara, Trieste (chapel in the Cathedral, 1520) Cremona, 1522, Piacenza, 1528–32, and finally in Venice, 1535–38, in a robust and impetuous style not uninfluenced by Giorgione's use of chiaroscuro (*Madonna*, 1516, Susegana). But he remained unaffected by Titian's serene unity in his immense combinations of architecture and drapery which presage Tintoretto. He is the link with the Mannerists.

Lorenzo Lotto. Opposition to Titian rapidly died out in Venice. Only the anarchical, confused, strangely dreamy temperament of Lorenzo Lotto (*c.* 1480–1556), still rooted in more popular imagery, enabled him, at the price of bitter disappointments, constant worry and even a kind of exile far from Venice, to resist Titian's overpowering influence.

Although he was a Venetian, Lotto worked outside his native city and in small towns like Recanati (*Madonna with seven Saints*, 1508) or in minor chapels like the Oratory at Trescore, near Bergamo (1524). His account book, which he started in 1538 at Ancona, is the record of his setbacks. He ended his days as a lay brother at Loreto. In 1529 he received a Venetian commission for the altarpiece of *S. Nicolas in Glory* for the Carmini; in the group of heavenly figures he can be seen striving his utmost, while the sombre grey landscape, treated with the greatest freedom, has a disjointed and overloaded quality, hardly to Venetian taste. His lack of success may be as much explained by his inability to acquiesce to the taste of the day as by his rejection of Titian's example. With him may be seen what Longhi calls the 'revolt punished'.

It was to Alvise Vivarini and Giovanni Bellini that he was most directly attached. His many portraits—*Bernardo dei Rossi* (1505, Naples) on a green ground, the *Young Man in a Cap* (1526, Castello, Milan), the superb *Andrea Odoni* (1526, Hampton Court) in tender greys and greens, the *Bearded man* before a landscape background (*c.* 1530, Borghese)—show clearly this derivation and his attachment to the cool colour and clear light that particularize all the irregularities of his sitters' features. And if he eventually tended towards a grave and nostalgic feeling of intimacy and towards a definitely softer handling (*Old Man*, 1542, Brera), it was without making any use of the fashionable new ideas on the treatment of drapery or forcefulness of presentation, just as he never really understood classical composition. In 1509 he went to Rome, where for a time he imitated Raphael without any very great success (*Deposition*, 1512, Jesi).

Uneven and unquiet, often bizarre, and sometimes platitudinous in his

257   T

landscapes (*S. Jerome*, 1506, Louvre; *Susannah*, 1517, pvt. coll., Florence), Lotto accentuates the seething clutter and lack of control which betray his affinities with German art. These are obvious in his altarpiece of *S. Bernardino* (1521, Bergamo) where the little angel is a brother of Grünewald's, and even more so in the huge *Crucifixion* (1531, Monte San Giusto) with its confused groups shot through by shafts of light, its tangle of diagonals, and its grimacing thieves, as in a Dürer or an Altdorfer. Certain trivial details underline these affinities—the cat frightened by the Angel in the *Annunciation* (1527, Recanati), the oxen dragging the martyr in the strange and moving predella of the *S. Lucy altar* (1531, Jesi)—which must also be understood as aspects of popular imagery. Lotto lived more and more among the poor, and from them derived a far keener understanding of the crowd than that of any other painter, as for instance in his *S. Anthony giving alms* (1542, SS. Giovanni e Paolo, Venice). By these means he escapes the simplification, the ordered composition and the larger-than-life quality proper to classic art. He remains a marginal figure, and the peculiarities of his temperament transmute his confused *oeuvre* into a forerunner of the Baroque.

Lotto's case is not unique. Among the other disoriented artists thrust out of Venice by Titian's ascendancy was Romanino (1484–1566), who came from Brescia to Venice in 1513 (*Madonna enthroned*, Padua). Under even stronger German influences than those working on Lotto, he turned, with his Cremona frescoes (1520) and those at Pisogno (1534), to a violent and wayward expressionism. And among the many minor artists of Lombardy who veered towards Venice, Altobello Melone also worked with brio in the cathedral at Cremona (*Massacre of the Innocents*, 1517).

MOSAICS AND GLASS

Just as Venetian architecture became familiar, through Roman types of decoration, with the ideas of late Antiquity, so is it tempting to consider the stress placed, in painting, on brilliance and on the radiance of colour, as something suggested by mosaic decoration. This is confirmed by the new interest shown in mosaic in the circles round Titian himself after almost a century of neglect during which Florence had taken the place of Venice in this particular technique.[116] Titian in his youth had been a pupil of a member of the Zuccati family, which restored the mosaics in the ceilings of S. Mark's. With his encouragement, the workshops of the basilica were re-established; a *Judgement of Solomon* by Vincenzo Bianchini (1532–48), was executed on the façade, on the vaulting of the vestibule were worked a *S. Mark in ecstasy* by the Zuccati family on a cartoon by Lotto, and an *Apocalypse* (1512–64) in the interior above the entrance. There were also portable mosaic portraits, made with coloured glass *tesserae*, like that of Bencho by the Zuccati (1542, Bargello). Neither the Florentine Vasari, nor the Venetian Ridolfi, who praise these works, sees in them anything anachronistic, nor do they protest against the violence they do to style.

FERRARA AND PARMA

Murano, in the Lagoon, probably had a tradition of workshops from Roman times—like Aquileia. From the thirteenth century onwards it was the centre where the techniques of glass making (a Byzantine importation) were practised on an industrial scale; its products were to be found all over Italy. At the end of the fifteenth century, the Alexandrian technique of flashed glass, which allowed an engraved design to appear either in clear glass or in colour, reappeared in Venice. The glass of the fifteenth century still had much in common with the enamel of inset miniatures, as in the Barovier goblet in Murano, or the dish with a woman's profile head and vine leaves (Trento Mus.).

In the sixteenth century, the shapes were more elongated and elegant, the material became more transparent, and 'crystal' was evolved. The production was very varied; dishes, cups, bottles encrusted with coats of arms (Bologna, Mus. Civico), and even liturgical lamps were made. Blown glass was valued for its thinness, and other sorts for diamond point engraving, for instance in dishes and ampullae with spouts and a lacy filament decoration (Florence, Pitti). In the seventeenth century came brilliant developments—winged chalices, figurines, and even cups and jars imitating porcelain. Another Venetian speciality, from the sixteenth century onwards, was the plating of glass and the production of mirrors in wide bevelled frames. The invention of drawn glass for mirrors, at Saint Gobain (1693) and the competition of Bohemia and Germany in crystal making eventually destroyed the Venetian supremacy.

FERRARA AND PARMA: CORREGGIO

Titian's prestige stood high at the courts of Ferrara and Mantua; from 1516 onwards, Alfonso d'Este commissioned his series of Venus Myths; in 1525 the Gonzagas also entered into negotiation with the painter, and, in 1536–38 he painted the gallery of the twelve Caesars in the Reggia at Mantua. These Courts, with that of Urbino, were the most brilliant and cultured in Italy.

In 1501, at the age of twenty, Lucrezia Borgia married Alfonso d'Este, to become Duchess of Ferrara. Until her death in 1519, she held court to a group of poets, including Pietro Bembo and Ariosto, who composed his *Orlando furioso* there. At Mantua Isabella d'Este, the wife of Francesco Gonzaga, dominated the intellectual life of the Court. She was a passionate collector who maintained agents in all the towns with a trade in antiques, and prided herself upon laying down for painters, who, like Giovanni Bellini and Leonardo, often evaded the commissions, the most precise programmes for their works. It was for her and for her brother Alfonso that the first of the Bacchanals based upon the Antique texts of Philostratus and Catullus[117] were painted at Mantua. Federigo, Isabella's son, having, like his uncle, a taste for mythological *erotica*, commissioned from Correggio, in consultation with his friend Aretino, the pictures of *Io*, *Leda* and *Danäe*. Stimulated by this current, it was at Ferrara, as an offshoot of Venice, that several remarkable painters were to appear, and further away still, at Parma, one of the most revealing artists of the period—Correggio.

From Costa to Dosso Dossi. After Ercole de' Roberti, the Ferrarese School, which was a kind of Gothic outpost, experienced the softening effect of Umbrian style through the work of Francia (1450–1517) and Amico Aspertini (1475–1522). A typical work of this group is the Oratory of S. Cecilia in Bologna, near S. Giacomo Maggiore (1506) where Lorenzo di Costa (1460–1535) also worked. The latter thus passed from the manner of Cosme Tura (*S. Sebastian*, *c.* 1485, Dresden) to a broader manner, closer to Bellini (*S. Petronio altarpiece*, 1492, Bologna; *Portrait of Giovanni II Bentivoglio*, 1490, Uffizi) and finally, in his works executed for the Gonzagas of Mantua, where he succeeded Mantegna in 1506, to the style of Perugino and Francia. For Isabella's *studiolo* he painted literary allegories like the *Gateway of Comus* and the *Garden of Harmony* (both in the Louvre) in praise of *honesta voluptas*. Twenty years later, Correggio completed the series, begun with Mantegna's *Parnassus* (1497), with the *Allegories of Pleasure and Vice* (*c.* 1525, Louvre).

Garofalo (1481–1559) who had been in Rome, painted some decorations for Lucrezia in 1506; after another visit to Rome in 1515, when he was in contact with Raphael, he returned to Ferrara in 1517 to paint decorations for the reigning family on mythological themes (such as the *Diana and Endymion*, Dresden), numerous serene altarpieces (the *Trotti altarpiece*, 1523, Ferrara pinac.), and large allegorical compositions such as the *Two Testaments* for the Refectory of the Augustinians (Ferrara pinac.). His pupil Girolamo da Carpi (1501–56) is more Mannerist and involved. There is less originality still in the work of Ortolano (before 1487–after 1524), who became obsessed by Raphael (*Deposition*, Borghese), and who strove to adapt to his own use the more complex manner of the dominant artistic personality of Ferrara—Dosso Dossi.

According to Vasari, Dosso Dossi (*c.* 1480–1542) was a contemporary of Ariosto; the remark has a certain critical value since it places the painter on the same level as romantic, courtly and passionate poetry. He first appeared in 1512 in Mantua, and in 1517 was in Ferrara. He had already been in contact with the art of Giorgione and Titian and he was able—perhaps better than anyone in Emilia—to appreciate its soft modelling of form, and the scope it offered for the painting of nudes and landscapes. He went to Rome and met Raphael, but all his working life was passed in Ferrara, where he decorated the palace of the Estes. A large part of his output was in portraits (*Warrior*, Uffizi), and in hauntingly poetic mythologies, where nude figures are in the most complete harmony with the sky, the distant views and the atmosphere (*Apollo and Daphne*, Borghese; *Antiope*, Marquess of Northampton). Dossi also used strange and exotic themes (*Group of men and women*, Uffizi; *Circe*,

Borghese) in which rich draperies scintillate, and a heavy atmosphere renders the faces pallid and the light overcast. His religious pictures do not always have this intensity, (*S. Jerome*, Vienna; *S. Michael*, Dresden). Towards 1525 Paolo Giovio praised the painter's tender style, saying that he was among the first to introduce a wealth of lovely detail into his landscapes, so as to make a new genre of it. The Court of Ferrara was rich in Flemish pictures; the art of Dossi, with that of the Venetians, is the Italian reply to Bosch and Patinier.[118]

Parma: Correggio. Leonardo, Raphael, or Titian were, to a certain extent, predictable: there was no reason to foresee in Emilia, and more particularly in Parma, the innovating and voluptuous temperament of Correggio (before 1489–1534). His artistic education was probably got in Mantua, from the works of Mantegna, whose frescoes and ceilings stimulated him to amazing feats of foreshortening, and with Costa, who professed a tame synthesis of Umbrian softness with a smattering of Venetian colour and handling. But all this Correggio had assimilated and surpassed by the time he painted his *Madonna of S. Francis* (1515, Dresden). He maintained his contacts with Mantua since he painted a series of mythologies for Ercole d'Este, but unlike Dosso, he never went to Venice, though like Garofalo, he appears to have gone to Rome in 1519.[119] In the last analysis, his secret was a kind of withdrawal upon himself in his provincial fastness, so that at a time when artists everywhere else were provoking excitement and curiosity by their experiments, he was able not only to absorb the results of their discoveries but to achieve a delicious thrill of novelty which neither Leonardo nor Raphael, Titian nor Michelangelo, had succeeded in expressing. And this nuance of the 'suave and tender', as it was phrased by Stendhal (who, like all connoisseurs formed in the taste of the eighteenth century, placed Correggio beyond compare) was indispensable in Italy, coming as it did at a moment when all was being subjected to a vigorous process of over-emphasis. Between Rome and Venice, which represented the twin poles of classical control over sensibility, there yet remained to be developed the vein of a beauty based on charm and pure sentiment which Correggio's virtuosity was to develop to the point where it became a direct precursor of the Baroque.

In 1519, in Parma, he was commissioned to paint an umbrella-shaped ceiling in a room in the convent of S. Paolo. He composed it in the form of sixteen compartments decorated with garlands of fruit, each section resting on a monochrome lunette and containing an oval medallion in which are seen figures of putti silhouetted against the sky. The putti represent the ages of man and the figures in the lunettes mythological divinities. This 'invention', at once learned and full of fancy, reminds one of a fantasy Farnesina or Sistine ceiling. From then on, Correggio settled in Parma. In the apse of S. Giovanni Evangelista he painted a *Coronation of the Virgin* (1520–24, fragment in the Mus.) and the cupola with the *Ascension*, which expands in a vision of golden light, so that the Apostles, ranged in a circle and seen in the sharpest fore-shortening, become mauve silhouettes amid the shot silk and mother of pearl tones of the clouds which surmount the rising figure of Christ. The memory of Mantegna's *Camera degli Sposi* is still strong; but all the contours are more melting, and all is a more shimmering and delicious illusion of Heaven. Correggio next composed for the octagonal dome of the cathedral the enormous decoration of the *Assumption of the Virgin* (1526–30). The silhouettes of the Apostles, scattered among the large circular windows and the lower part of each segment of the dome, fill the drum of the dome and set the scale for the foreshortening. In concentric circles of multicoloured brilliance, the groups of angels and the elect spiral upwards among the clouds, in a luminous, inexhaustible, limitless, fluid world, where order melts into ecstasy. It is more powerful than Raphael, more subtle than Michelangelo, yet 'this miraculous gift of movement is thrown away with the most foolish prodigality' (Berenson).

It was towards 1530 that he painted for the Reggia in Mantua the paired subjects of *Io* and *Ganymede* (Dresden), *Leda* (Berlin) and *Danäe* (Borghese) which are less evocative of love than of the thrills of voluptuousness. The forms are clearly defined, and sufficiently enveloped in chiaroscuro to allow the colour—iridescent in the flesh tones, tender yet heady at the same time—to create the most powerful effect.

In his altarpieces—the *Marriage of S. Catherine* (Louvre), *Madonna and Saints* (Dresden)—the softness of the feeling, the agitation in the forms and the lack of restraint in the details, would be embarrassing were they not offset by the new surge given to the composition, where he tends, more and more, to use diagonals, which, in turn, create a sense of instability. The masterpieces resulting from this new approach, which ran directly counter to all the classic rules so far established, and which, in their novelty of effect, offered immense scope for future development, were the *Madonna of S. Jerome* (or '*Il Giorno*': Parma) and the *Nativity* (or '*La Notte*': Dresden) where the formal cohesion is broken up by the diagonals, and the colour is either absorbed into the brilliance of the light, or drowned in nocturnal splendour. Nothing survives here of the intellectual clarity of Leonardo, or of Raphael's serenity, or Michelangelo's virility, or even of the unity and warmth of Venetian colour, which could originally have stimulated Correggio's art, but in this return to surface effects and to purely physical enjoyment is displayed so much passion and so fine an inventiveness that Correggio never had time to create his own form of Mannerism. It may be thought that as a master he used debatable means to further dangerous ends, but he remains a great master none the less, and he served to counterbalance the conscious nobility of thought and the elevated aims of the more spectacular artists among his contemporaries.

IV

Mannerism

Introduction: The Historical Background.
The Council of Trent, court life, the academies

To popular imagination, the Sack of Rome seemed to have the force of a divine judgement. The Eternal City was no longer sacrosanct. The extent of the change in the position of Italy, and the political consequences of the Hapsburg domination, were quickly grasped. After the reconquest of Florence and the Peace of Cambrai (1529), Charles V, now all-powerful, had himself crowned Emperor at Bologna in February, 1530. Under the Spanish yoke, which now and then, ineffectually, such popes as Paul IV (1555–59) tried to shake, the peninsula assumed a new political configuration. By the time of the Treaty of Cateau-Cambrésis (1559), which marked the end of the wars in Italy, all the Italian states that had become principalities or duchies were either dominated by Spain (Naples), or allied to her (the Gonzagas, the Estes), or without danger for her (Tuscany under Cosimo I (1537–74), Francesco (1574–87) and Ferdinando (1587–1609)). The only states that retained real independence were Savoy, under Emmanuel-Philibert (d. 1580), and Venice.

The religious crisis was as profound. Luther's revolt at Wittemberg (1517) had little immediate effect; but soon the news from Germany revealed the horror in which a part of the Christian world held the papacy, and evoked a wave of penitence and discipline, which was apparent before 1527 in the religious orders. The Theatines were founded in 1524, the Suore Angeliche in 1535, the Barnabites in 1533, the Brothers of Charity and the Jesuits in 1540, and, by 1580, the latter already had fourteen provinces, a hundred and forty-four colleges and five thousand members. The Counter-Reformation was under way; the Inquisition was re-established in 1542, ecclesiastical censorship in 1543, and the first Index of prohibited books appeared in 1557. Soon all Italy was shaken by the crisis. A policy of reform was adumbrated under Paul III (1534–49), who admired Erasmus and protected Sadoleto, Contarini and Reginald Pole, but with Cardinal Carafa, who became Paul IV (1555–59), the reaction was sharp. It became more flexible under Pius IV (1559–65), who was helped by Carlo Borromeo, Archbishop of Milan,

263

a man of deep asceticism and inflexible spirit and a model for a new kind of saint combining energy with mysticism. The interminable Council of Trent, which opened in 1545 and closed in 1563, imposed a new catechism, and a Roman breviary and missal that considerably simplified the liturgy; while it recognized, as a move against Protestantism, the cult of images (twenty-fifth session), its recommendations brought about a purification and a control of religious art, of which Veronese was a victim in 1573, when he was arraigned for the buffooneries introduced into his *Christ in the House of Levi* (Accademia, Venice). The freedom of medieval iconography, and of the apocryphal books, was condemned by Molanus in 1570, and the forms created by the great generation of 1510–20 were preferred, though without anything of the optimism, the vitality or the strength that inspired them.

From 1525–30 onwards, in the more sensitive artistic circles such as Florence, things reached the point where both medieval simplicity and the eager confidence in humanity and its resources felt in 1500 were equally shaken, and an art evolved which was obscure and full of contradictions, redolent with the inquietude of Leonardo and the torment of Michelangelo, and which used the classical forms in a spirit which was alien to them. Each master elaborated his own private formula of irrational and capricious effects, often reviving ideas current at the end of the fifteenth century which had been stifled after 1500, and betraying a new affinity with the intensity and confusion of German art.[120] This is the fundamental aspect of what has come to be called *Mannerism*.

In the middle of the century, the affectations of artists who believed that they could achieve an 'ideal style' by improving on the art of the great exemplars were praised as 'manner'. It is to this somewhat decadent 'manner' that the more vigorous seventeenth century opposed 'nature'. There was, in fact, a kind of weariness, a need for a pause in development, which encouraged a mixture of styles, and, in the end, led to an eclecticism expressed in the dream of the perfect picture. In 1590, the Milanese Lomazzo in his *Idea del Tempio della Pittura*, wrote that to represent Adam and Eve 'the drawing of Adam should be entrusted to Michelangelo, the colouring to Titian, adapting suitable proportions and expression from Raphael; the drawing of Eve to Raphael, and the colouring to Antonio da Correggio. These would then be the two finest pictures in the world'.

This recipe was only the result of the arguments and conflicts caused in the arts by the spread of academies. In 1541, Grand Duke Cosimo I founded the literary *Accademia fiorentina*, where Benedetto Varchi in 1546 lectured on a sonnet of Michelangelo's on the artist's ideal; in 1562, the *Accademia del Disegno* was founded. These institutions, which soon spread everywhere, had for their aims the establishment in art, as in everything else, of a code of the best usage; hence unending quarrels among savants and artists, who aspired to the role of official representatives of thought or art.

The funeral of Michelangelo in 1564 was a typical example of these newly established artistic customs. His remains were transported secretly to Florence, and the Grand Duke commissioned the newly founded 'Accademia del Disegno' to arrange the ceremony with the most resounding pomp. The committee included Vasari and Bronzino, Cellini and Ammannati; Cellini was virtually excluded, which resulted in his hatred of Vasari to which he gave free rein in his Memoirs. One of the reasons for this enmity was the importance given in the allegorical figures of the catafalque to Painting (placed on the right) over Sculpture (placed on the left), before Architecture and Poetry.[121] Such was the atmosphere of the times, with its official cele-brations, its academic rivalries and its artifices. In the mid-century appeared Vasari's monumental work (1550, enlarged, after a great success, in 1568), compounded of general theory, guide book and compendium of information, and based on a framework of the theories of genius and the relationship between art and nature achieved through the study of the masters, that of the primacy of drawing over colour, and, finally, that of the 'modern manner' born with Giotto, made clear by Masaccio, expressed by Leonardo, consum-mated by Michelangelo: it is the last word in all Florentine criticism.

The first discordant note was that sounded by Ludovico Dolce in his 'dialogue' entitled *L'Aretino* (1557), where Raphael is rated above Michel-angelo for drawing, and Titian above them both for colour. Vasari took this Venetian reaction into account in his second edition in 1568, and during the whole of the sixteenth century, and well into the next, Venetian authors carried on a lively controversy with the Florentines, on historical grounds (e.g. that the mosaics in S. Mark's were earlier than those in the Baptistery in Florence) and on aesthetic grounds (the greater importance of colour over drawing). A new phase in Mannerist thought is to be found in the theoretical treatises of Armenini (1587), and Lomazzo (*Trattato*, 1587, *Tempio della Pittura*, 1590) which link the neo-platonic philosophy of the Idea (or mental image) with contemporary naturalism.

Florentines like Vasari or Borghini (1584), Venetians like Paolo Pino (1548) or Dolce (1557), Lombards like Lomazzo—all the authors of treatises agree upon the co-existence of rational rules and the necessity for the inspiration or idea which is beyond reason.[122] In the books on architecture which multiply after Serlio's, and above all with Palladio (1570) and Vignola (1562), there is a tendency towards a formal codification which is made more evident by their preoccupation with archaeology and mathematics. The taste for the fable and its imagery, which had furnished the material for so many masterpieces by Raphael and Titian, ended in a multiplication of learned treatises and books on iconology which fed the humanist culture of the second half of the century with abstruse symbolism.[123] Thus was formed a decorative art, overloaded with programmes, allegories and emblems, enamoured of the obscure and the astonishing.

All these complications, this overweighting of artistic culture corresponds to the evolution of the small Italian social unit into one of the seigneurial type, to the appearance of court life, in the solemn and stiff form which Italy had never known before the Spanish occupation. This is the last element in Mannerist art; it provided people with a mask, and their life with a refined and conventional façade. The term 'manner' fits it all the better in that, in contemporary literature, it is normally used for an elegant and modish bearing, and is associated with the adjectives *pellegrino* (foreign—that is, distinguished) and *affettato* (fashionable, chic, apparently casual).[124]

The aggrandizement achieved through setting and behaviour having developed into the purest affectation, this official and over-elaborated art turned as a kind of compensation to rusticity, to the life of the people, and in the development of the *capriccio* to a relaxing of formal etiquette. Hence the success of all the paradoxical and fantastic, even buffoon and exaggerated, forms, that occasionally reached—as in Rosso—to the point of a provocative Satanism.

The vogue for miniature pictures and portraits is typical of Mannerist taste: twenty-four miniature portraits by Bronzino were in the Cabinet of the Palazzo Vecchio, and it is said that Catherine de' Medici, Queen of France, left a collection of three hundred and forty-one. Giulio Clovio, miniaturist and collector, 'an excellent painter of small things' according to Vasari, achieved the tour-de-force of reducing to miniature proportions the famous compositions of Raphael and Michelangelo. Despite the rigid outward appearances of the Counter-Reformation and court life, there was a profound Mannerist emancipation, one remarkable result of which was the new interest in still-life and landscape, in which Italian taste was stimulated not only by Northern example but also by its own needs. A vase of flowers with some large lemons and a small lizard is known from a later copy to have been painted about 1538 by Giovanni da Udine, one of Raphael's assistants; a generation later, the display of fruit and flowers is already sufficiently commonplace for it to be treated as an amusing and paradoxical *capriccio* in the famous compositions of Arcimboldo.[125] The same may be said for landscape. Two other pupils of Raphael, Maturino Fiorentino and Polidoro da Caravaggio, painted in about 1540 lively examples in a style composed of small touches of colour which recall the antique paintings in S. Silvestro al Quirinale, and from Dosso Dossi to Niccolo dell' Abate and the Bassanos the new genre developed in a fascinating way. One of the most interesting facets of the period is the tendency towards 'realism of detail within the basic irrealism of the whole' (G. Briganti). Not less revealing is the invention of the 'rustic' style, after the examples created in Mantua, with its affectations of heaviness, its grottoes and surprises, and the huge humorous fantasies that abound in all gardens. This period of complicated culture and official rhetoric made every appeal to instinct.[126] It could even be characterized by the dissociation of elements that, at moments of strength and certainty, immediately amalgamate: form and content. The middle of the sixteenth century witnessed the extremes of 'formalism' in works artificially derived from already existing styles, and of an intellectualism that attached more and more complex symbolic values to the image, to ornament and to architecture.

All these traits, which complemented one another without forming a unity, characterized a period that cannot be considered as faithful to that which preceded it, and which did not directly prepare the one that followed. If the term Mannerism is retained it must be extended to cover all the arts including architecture, which was one of its greatest fields.[127] But it is impossible to explain completely this series of phenomena, without stressing the real revolution that took place in the first decades of the sixteenth century with the spread of engravings. After Mantegna and the Florentines, the technique of the reproductive print was perfected by Marcantonio Raimondi (c. 1475–1534); after much copying of Dürer, he entered the service of Raphael towards 1510 and was to spread the great compositions of his master all over Europe. Later, the Dutchman Cornelis Cort received a kind of exclusive right for the reproduction of Titian's models (1556). Between these two dates, the course of artistic life was profoundly transformed by this new practice, which ensured the same rapid diffusion to the discoveries of geniuses that the printed book gave to ideas, and submitted them to immediate imitation or condemnation. Engravings are at once more manageable, easier to disseminate, and their execution more detailed than that of drawings, for which reason it was natural for the great masters to control their production.

Without engraving, the potent influence of the great masters, and the parallel evolution of different schools, would never have been possible; nor, in particular, would the absorption of Northern art and the knowledge of Dürer, which, from 1515–20 onwards, affected painters as different as Luini, Lotto and Pontormo. The nervous tension of the Mannerists was aggravated by the rapid circulation of these sheets, which maintained curiosity abroad; taken for the equivalent of drawings, they furnished ever more and more motifs to artists already inclined to elaborate. Certain disorders of the period clearly derived from this, as did also the internationalization of formulae. By 1540–50, Italian Mannerism was current all over Europe. Taking these elements into consideration, three periods may be distinguished in the development of Italian Mannerist art:

1. From c. 1515 to 1535–40: in Tuscany, the anti-classical orientation, and Beccafumi, Pontormo and Rosso; in the Po valley, the antithetical development given to 'manner' by Giulio Romano and Parmigianino.
2. From 1540 to 1570: the growth of academism, with the triumph of court art (Bronzino, Vasari), and the disciples of Michelangelo (Daniele da Volterra); Venice followed her own road with Palladio, Veronese and Vittoria, and Rome with Vignola.
3. From 1570 to c. 1610: while Lombardy (Luca Cambiaso), Venice (Tintoretto), and Central Italy (Baroccio), all developed more personal styles owing much to their interests in light effects, the Carracci, from Bologna to Rome, organized a new era of coherent classicism.

Mantua and Parma: 'Romanism' and Purism

The *Palazzo del Tè* in Mantua, built at great speed by Giulio Romano (*c.* 1499–1546), is one of the key buildings of the period. After he had completed the Stanze of Raphael, he entered before 1527 the service of the son of Isabella d'Este, the prodigal Federigo Gonzaga, who in 1530 had been raised to the rank of Duke by the Emperor Charles V. The palace, built on a former island and standing in the midst of paddocks, contained the huge stables destined for the horses which were one of the glories—and one of the principal sources of revenue—of the family. Later, under Francesco IV, a whole room, the work of B. Pagni (1525–70) was devoted to their glory, with their likenesses covering the walls.

The palace is a sort of recapitulation of the energetic classicism worked out by Alberti and the Roman architects. Its Doric derives perhaps from the Palazzo Farnese; the garden front develops the motif of arcades carried on columns, with a short architrave on either side, which had already been used by Bramante and was to become the Palladian motif. All these elements are given a new tension, by being so drawn out in width as to distort the proportions, and by a powerful feeling of plasticity due to the use of a massive rustic order of heavy blocks of stone, creating violent contrasts, which are increased by the use of colour, particularly in the metopes. This intentional heaviness is to be found again in the Cathedral at Mantua (rebuilt in 1545), an immense pastiche of early Christian basilicas, with nave and four aisles, separated by columns and coffered ceilings with stucco decorations.

At the Palazzo del Tè, Giulio Romano wanted to create a complete scheme of decoration, and show himself the heir of both Raphael and Michelangelo. The rooms were decorated from top to bottom; carpets covered the marble paving, the doors were inlaid with marquetry work, and from above the marble dado rose the frescoes that covered the vaulted ceilings. Two series may be distinguished: in the first, on the North, around the *Sala di Psiche* (1521–31), all is devoted to the praise of pleasure, and on the south side the rooms round the *Sala dei Giganti* are devoted to themes of power and triumph (1530–35). These two symmetrical ensembles concentrate with striking energy and virtuosity the two principal sources of Mannerist decoration: on the one hand, foreshortenings and illusionism that make the walls and ceiling run together as a continuous decoration, eroticism and somewhat explicit love scenes, Venus and Adonis, Psyche—all the voluptuous paganism that was so to delight Ingres; on the other hand, monstrous battles and cataclysms which, as if to make the interior match the rough stones of the exterior rustic order, cover the room from top to bottom with heaps of rock tumbling on to horrible giants, and with the illusionism strengthened by the fireplace being so placed that the fire would be at the point where the Cyclops are crushed under Etna. Other rooms develop such flights of symbolism as a Gonzaga Olympus, or images borrowed from the astrological cosmos—for example, the *Sala dei Venti*.[128]

The Palazzo del Tè had an immense influence, in Italy as well as abroad. Serlio cited it as a 'veritable model of architecture and painting of our times'; it inspired some works in Vicenza, and Palladio made use of some of its arrangements. The decorations were completed by graceful stuccoes and friezes (*Triumph of Sigismond*) by an assistant of Giulio Romano's, of great international importance. This was Primaticcio (1504–70), who in 1531 entered the service of Francis I at Fontainebleau, where he found the Florentine Rosso already employed. His frames, his stucco ceilings and his allegories in the Pavilion of Pomona and the apartments of the King and Queen made the French palace into the second capital of European Mannerism. Primaticcio, who was also known as Boulogne and the Abbé de Saint-Martin, was sent to Rome in 1540; he brought back 125 statues and antique copies for the French Royal collections. From his second period in France date the *Grotto of the Pines*, the *Ballroom* and the *Gallery of Ulysses* with its fifty-eight panels and its seventy-six small compositions in the ceiling; for these Primaticcio left a number of drawings in which the vigour of Giulio Romano and Rosso is refined and lightened by a mixture of Parmigianinesque elegance and fantasy (*Dance of the hours*, Frankfort-on-Main). From about the time of his second journey to Italy (1563) may be dated the *Ulysses and Penelope* (pvt. coll., New York) and *Helen fainting* (pvt. coll., London), compositions remarkable for their subtlety and their equilibrium.

About 1542, Niccolo dell' Abbate (1509–71) became Primaticcio's collaborator at Fontainebleau; later, he went back to Italy to produce landscape compositions with green and blue distances, rich in detail and effects of light, and inhabited by strangely elongated figures (*Moses saved from the waters*, Louvre). Before this, he had mostly painted pure fresco decorations in Bologna of a romantic *fête galante* type (e.g. in the Palazzo Poggi, now the University, and the Palazzo Zucchini-Solimei).

Most of the painters of this generation elongated their figures and placed them with extreme wilfulness within the picture space so as to create striking contrasts between them, with, for example, a huge, sharply foreshortened, silhouetted figure in the front of the picture. These tricks were due to Parmigianino (1503–40), one of the most remarkable Mannerists of Emilia, a painter whose refinement was totally opposed to the violence of the Roman artists, and who was the rather decadent heir of Correggio and Raphael. While Giulio Romano developed in the Palazzo del Tè a decor in which the excessive vitality becomes somewhat paradoxical, Correggio's pupil proposed, on the contrary, to achieve a quintessence of grace and charm by eliminating every trace of realism from the figures in his compositions. His style ('*maniera*') was imitated by a host of painters, because he brought to it 'a light of delicious grace' (Vasari). With Parmigianino, painting endeavours even more than it did with Raphael, to attain that 'grace beyond the reach of art', which emanates from the inward image—the *disegno interno* of the theorists—fused with the forms of nature. This Mannerist attitude was to govern the whole personality of the painter—neurotic, unstable, elegant. From 1522 to 1524, he competed with Correggio; then, to see the works of the great masters, he suddenly went to Rome, where his youth caused a sensation. Driven out by the German occupation, he went to Bologna, and in 1531 returned to Parma, before the death of Correggio, whom he survived by only six years. Full of charm, and eager to believe that the 'spirit of Raphael had gone into him', Parmigianino also had a taste for the unusual, for the kind of private symbolism expressed in his *Self portrait in a convex mirror* (1524, Vienna), and, even more, in the alchemy in which he consumed his time and strength, to the point of becoming on the eve of his death 'strange and melancholic'.

Parmigianino's development is interesting because of the romantic life history which is mingled with his formal work. The *Marriage of S. Catherine* (c. 1521, S. Maria, Bardi), already shows him setting in a contrived light graceful figures whose hands and hair are treated with conscious art. The tender and lively quality of the painter's genius is clear in two series of frescoes which may be dated from his first period in Parma: *two chapels in S. Giovanni Evangelista* (1522) where large figures shimmer in the diffused light of deep niches, and the lunettes of the *Story of Actaeon* in a ceiling of the Castello at Fontanellato (1523), accompanied by a decoration of putti.

Many admirable drawings, full of half-suggested forms and strong accents, survive from these works and lead on to the subtle elaboration of his three main works: the *Vision of S. Jerome* (1527, London; drawings in the British Museum), the *Madonna with the Rose* (1530, Dresden) and the *Madonna del collo lungo* (. . . 'with the long neck,' 1535, Pitti; drawings in the British Museum and Parma). Here all is brought to a pitch of the highest perfection: the refinement of the type used for the Virgin, seen full face, with downcast eyes, her hair beautifully dressed, and with the style of a great lady with slender limbs, her well-cared-for hands holding a lively or indolent putto; the play of the 'serpentine figure' (like an S) in which the principal silhouette is inscribed and on which all the other forms are based; the spatial devices in which sudden plunges into depth suggest in the middle distance strong foreshortenings and an impossible distance; the colour, pale and filmy, enveloping these strange inventions in a caressing haze. The immodest *Amor* (1534, Vienna), incredibly popular, represents the secular side of this style.

Parmigianino revealed another side of his unusual genius in the church of the Steccata (1531–39). Reversing the terms of his commission, he covered the vaults of the ceiling with huge painted cofferings containing decorative rosettes, with, in the lower ranges, female figures of an extraordinary elegance bearing vases, and he introduced a few biblical figures—Adam, Moses— in the subordinate parts. Finally, he was an admirable portrait painter, who knew how to present a whole generation, as neurotic as he was, with pictures of itself which it found quite irresistible: the *Man with an antique relief*

(Strafford coll.), the *Young prelate* (Borghese), *Malatesta Baglioni* (Vienna), the *Bearded man wearing a beret* (Uffizi) which is perhaps a self-portrait, and several delicious female figures, the *Lady in a turban*, or so-called 'Turkish slave' (Parma), *Anthea* (Naples), where the nature of the forms corresponds exactly to the sophistication sought by the artist. Parmigianino was admired from Venice to Naples, thanks to engravings which, during the rest of the century, disseminated his compositions and his types, so that it is difficult to overestimate his importance. Coming hard upon Correggio's voluptuous art and adding to his new manner a coldness and an unprecedented intellectual abstraction, Parmigianino gives the impression of having created the new form of beauty, and without him Mannerism would never have been either as conscious or as elaborate.

Under his direct influence was Girolamo Bedoli (*c.* 1500–69), who also took the name of Mazzola (Parmigianino's real name), and who continued to decorate Parma (Cathedral, 1538–44, Steccata, 1550) with frescoes and altarpieces.

Tuscany

ARCHITECTURE: SCULPTURE: GARDENS

By the middle of the century, the new duchy (1537), where no trace of the old civic spirit survived, was conspicuous for its princely pomp. The public buildings were enlarged and given new suites of state rooms, parks and gardens were developed round them, and, finally, the custom was established of grandiose settings for official ceremonies.

Baccio Bandinelli's group of *Hercules and Cacus* was placed in front of the Palazzo Vecchio in 1534, to counterbalance the Republican *David*. From 1540 onwards, the Palazzo della Signoria and its environs were radically altered, first by Vasari and then by Buontalenti; the buildings were enlarged along the Via della Ninna, which was spanned by an arch (1565–71) to link the buildings on one side of the road with the new offices of the Uffizi on the other. There was also considerable reworking of the interiors: new apartments were made on the first floor beyond the *Sala dei Cinque Cento*—the *suite of Leo X*, with its frescoes of the life of Cosimo, and on the floor above the *suite of the Elements* (from 1550 onwards) filled with allegories, elaborated by Vasari, which he described in detail in his *Ragionamenti*, and on the west side, along the Via della Ninna, the richly decorated apartments of Eleonora of Toledo. In contrast with the solemnity of these state rooms, Vasari introduced a note at once more intimate and more sophisticated with the precious and elaborate decoration of the little *studiolo*, or private study, of Grand Duke Francesco (1573). In his rebuilding of the centre of Florence, Vasari inserted (1560 onwards) between the Palazzo Vecchio and the Loggia dei Lanzi the long passage containing the administrative offices known as the *Uffizi*. The perspective effect ends in a feature that appears to be a loggia, and which is silhouetted like a screen against the sky, with the Arno beyond; its open gallery is not given a strict rhythm, and the colonnade on the ground floor is surmounted by an overstressed mezzanine. The architect himself explained the difficulties he had to overcome in using for it the Doric order which the Grand Duke preferred. The window frames and their pediments project strongly and the dark stonework contrasts with the walls, creating an effect borrowed from the Laurenziana Library. The strongly projecting cornices stress the rather stage-set quality of the whole design. On the Arno side, it is prolonged by the corridor which links the Uffizi to the Pitti Palace, reworked by Ammannati. The pilastered *loggie* which Vasari built in the Piazza Grande at Arezzo are much simpler (1573).

Bernardo Buontalenti (c. 1536–1608), who specialized in ceremonial displays and fireworks, introduced an element of fantasy and caprice into villa architecture, sometimes by means of curves, as in the *Villa at Pratolino* (1569–81, now demolished), sometimes, as at the *Villa at Artimino* (c. 1580), by the insertion of a large flight of steps (built from his designs) into a bare façade with a flat, unaccented portico, following the precedent set at Poggio a Cajano. Buontalenti also executed the *façade of S. Trinità* (1593) with its two disproportionate storeys, and made a project for the Cathedral façade (1587) which was too regular in design: the superposed orders were to be separated by panels which destroyed the effect. Such a work as this indicates clearly the direction that Mannerism had taken in Florence, and the force of the example set by the Laurenziana Staircase. While the classical vocabulary is still used, its elements are no longer balanced or kept distinct.

The most interesting man is Ammannati (1511–92), who was trained in Rome with Vignola, and, with the Jesuit *Collegio Romano* (1578), provided the most extreme example of feigned austerity of style. In Florence, he displayed a freer imagination in the *courtyard* of the Pitti Palace (1558–75), which repeats the motif of Brunelleschi's rustication on the façade, but he combined it, as had been done at Mantua, with superposed orders in alternate bays, so that the applied columns seem to swell under bands of stone. The wider central bay, stressed by the Palladian motif, owes its importance to the ground floor entrance on to the gardens, and the brilliant light striking through the opening adds to the contrast and accentuates the theatrical effect. The *S. Trinità bridge* (1567–69; destroyed in 1944, but now rebuilt as it was before) is his masterpiece. Based, perhaps, on a design by Michelangelo, this delicately curved structure is borne on three long arches, and if the Pitti in some ways suggests Giulio Romano, this work recalls the elegance of Parmigianino. The statues of the *Seasons* which decorated the bridge were by Francavilla, Landini and Caccini, whose pupil, Pietro Bernini, was the father of the great Baroque sculptor.

Baccio Bandinelli (1488–1560), a pupil of Rustici, was the only sculptor to practise a balanced style with a straightforward axial composition and a

81 (a) Cima da
Conegliano:
detail from the
*Madonna
with the Orange*

(*Venice, Accademia*)

(b) Mansueti:
detail from the
*Miracle of the
Holy Cross*

(*Venice, Accademia*)

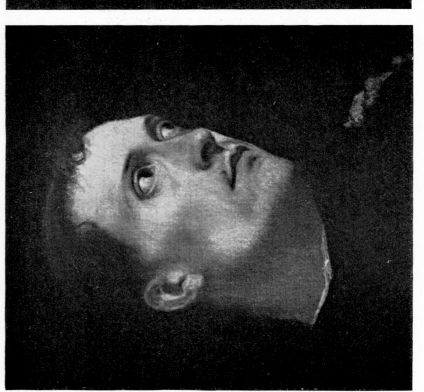

82 Titian: (a) Detail from the *Concert*
(Florence, Pitti Palace)
(b) *Self-portrait*

83 Titian: (a) *The Pesaro Madonna*
(*Venice, S. Maria dei Frari*)
(b) *The Deposition*
(*Venice, Accademia*)

84 (a) Andrea Riccio: *Arion*
(*Paris, Louvre*)

(b) Moderno: *Madonna and Child with Saints*, small relief

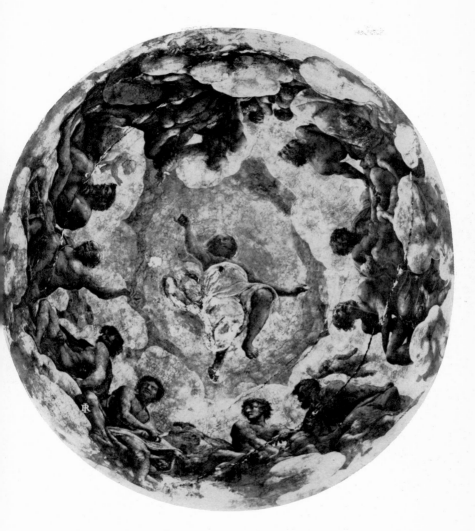

85 Correggio: the dome of S. Giovanni Evangelista, Parma

86 Pontormo: *Christ before Pilate*

(*Certosa di Val d'Ema, near Florence*)

87 Bronzino: *Don Garcia de' Medici*

(*Madrid, Prado*)

88 (a) Benvenuto Cellini:
detail from the base
of the *Perseus*

(*Florence, Loggia dei Lanzi*)

88 (b) Giovanni da Bologna:
Venus
(*Florence, Boboli Gardens*)

simple silhouette. The example of Michelangelo kept him within the limits, though not within the spirit, of classical form, as is shown by his *Hercules and Cacus* (1534, Piazza della Signoria), his *Bacchus* (1547, Pitti) with its affected silhouette, and the rather elegant *reliefs* for the Cathedral (1547, Museo del Opera). Vincenzo de' Rossi (1525–87) followed Bandinelli's style in his *Labours of Hercules* (Pal. Vecchio). Vincenzo Danti (1530–76) displays both elegance and the grand manner in his allegory of *Honour* (Pal. Vecchio), and the *Decollation of S. John the Baptist* over the south door of the Baptistery has a certain largeness of effect, despite its too carefully calculated symmetry.

The most striking figure is Benvenuto Cellini (1500–71), goldsmith, bronzeworker, swashbuckler, adventurer, who tried his fortune at the French Court in 1540, and then returned to Florence c. 1545, to work for Grand Duke Cosimo; he was also the author of astonishing memoirs in which he revenges himself upon his many enemies, among them Vasari whom he detested and belittled. He is a typical figure of the period with his perpetual fever of excitement, his virtuosity and his ambition. He revelled in its bizarre and convoluted taste (*Salt-cellar of Francis I*, Vienna, a masterpiece of extraordinary complexity), in its delicacy and formalism (*Nymph of Fontainebleau*, Louvre), and in the *Perseus* (1533, Loggia dei Lanzi) he displayed his great virtuosity in anatomy. The accomplishment of the modelling is fully set off by the variety of accent and subtlety of treatment of the bas-reliefs and the meticulously executed statuettes which decorate the base. It demonstrates all the possibilities of bronze and suffers only in that he does not distinguish sufficiently between monumental form and decoration, confusing sculpture with goldsmiths' work. His vitality, his richness of invention, infinitely surpass the frigidity of his rival Bandinelli. The *marble crucifix* (Escorial) executed in competition with Bandinelli, has a nobility of truth absent from the mean execution of the *bust of Cosimo I* (1548, Bargello).

Giovanni da Bologna (1529–1608) belongs to the next generation, more disciplined and more academic. Born at Douai in Northern France, trained in Antwerp, he came via Rome to Florence in 1553, to become a sculptor-architect. With his international background he was able, in the pervasive intellectual climate of Florence, to evolve a style full of subtle, daring and slightly decadent grace. As an architect, he built the *chapel of S. Antonino* at S. Marco, with its over-elaborate reliefs (1581–89), and transformed into an over-ornate *funerary chapel* one of the chapels in the choir of the Annunziata (1598). More voluptuous and sensitive than Cellini, he made the elegant bronze *Venus* of the Petraia (1567), and the one with the modest gesture in the Boboli gardens (1570), which, with her heavy thighs and small head, has every accent of modernity despite her Grecian air. In the *Charity* (1581, Genoa University) and in the elongated, sharp-formed little *Mercury* (1564, Bargello), the bronze is enlivened by a host of small accents of light which together form a continuous and flowing line. Despite his efforts to be classical, the equestrian monument to *Cosimo I* (1581, Piazza della Signoria), and *Ferdinando I*, finished by Pietro Tacca (1608, Piazza dell' Annunziata) are overladen, and there are too many carefully contrived twists of movement in the *Rape of the Sabines* (1583, Loggia dei Lanzi), where the multiplicity of axes results in the sculpture losing all sense of direction. The *Apennino*, a colossal rock of spongy stone in the gardens at Pratolino, carved into the form of a River-God, is one of those Mannerist frolics, at once comic and astonishing, which suited the Florentine brand of intellectualism. It raises the whole question of the Florentine side of one of the most typical features of the period: the architecture of fêtes and gardens.

Fountains and Parks. The Mannerist sculpture of Florence can be summed up by its monumental fountains. In the Piazza della Signoria is the large fountain of *Neptune* (1563–76) by Ammannati, with the marble giant in the centre (known as the *Biancone*) too closely imitating Michelangelo's *David*, and, fanning out round him, the fauns and voluptuous naiads of Pietro Tacca and Giovanni da Bologna; it is the ultimate touch of fantasy in the austere square. These compositions in several stages, where the statues are arranged elegantly round stretches of water, are more in place in the princely gardens where the majority of sculptors collaborated with the Grand Duke's inventive engineer of forests and waters, Niccolo Tribolo (*c.* 1500–50). About 1540, at the Villa di Castello near Rifredo, he built terraces on the hillside with bronze animals in the niches and grottoes of the upper part, and, lower down on a stretch of level ground, a fountain of *Hercules and Cacus* and various other statues, all by Giovanni da Bologna. This sculptor also specialized in compositions of Venus standing over a large marble basin, with great elegance in their silhouette. At the Villa Petraia (1545–46), and above all in the Boboli Gardens, each statue designed to be seen in the open was given an appropriate setting; Giovanni da Bologna's *Venus at the bath* is in the lower grotto, and at the top, in the midst of greenery, is his enormous *Abundance*. The Boboli Gardens were laid out in ilex, bays and cypresses on the hillside behind the Pitti Palace, to which they are linked by a large amphitheatre reconstructed in the eighteenth century. The general design, due to Tribolo (1550), consists of a system of perspectives and sloping paths, leading from copse to copse, as if from one room of surprises to another. Each of these green 'rooms' contains a statue, such as the *Ocean* by Giovanni da Bologna, encircled by three great rivers seated upon their urns, at the central point of a little island, or at a still higher level, the *Neptune* by S. Lorenzi (1565). Another attraction was the aviary garden, set in avenues which were completed in the following century. At the entrance, Buontalenti arranged a large grotto with a *Venus* by Giovanni da Bologna, and the grotto of the Menaboni with humorous paintings. With these huge, ambiguous compositions, the like of which were only to be found in Lazio and, to a lesser extent, around Venice, must be associated the fantasies of the ephemeral architecture which was then of unprecedented importance in Florence.

TUSCANY

Ceremonial Entries and Funerals. Fêtes in the streets and ceremonial architecture, extending a fifteenth-century practice, were revived in 1511 with the Carnival of Death designed by Piero di Cosimo; the extraordinary 'Entry of Leo X into Florence'[129] took place in 1515 and remained the model of the genre, with its decorated arches thrown across the streets, its false façade of the Cathedral designed by Jacopo Sansovino and painted in *trompe-l'oeil* by Andrea del Sarto, and its temples and pyramids in the squares. In the second half of the century, these carefully planned elaborations decked out with symbols were the subject of books which increased their scope; such, for instance, as those in 1565 for the marriage of Francesco de' Medici with Joanna of Austria (Vasari), and in 1589 for that of Ferdinando I and Christina of Lorraine (Buontalenti). They always consisted of temples and fantastic little buildings to be used by dancers and musicians. But the buildings of the city itself could also be adapted for these shows: in 1585, a *Teatro Mediceo* was organized by Buontalenti on the first floor of the Uffizi, and in 1579, the internal court of the Pitti, only just completed, served as a setting for a theatrical performance and a mock naval battle, known from the engravings of Gualterotti. These devices are interesting in that they pave the way for the development, during the seventeenth century, of the Baroque stage setting, which regularly made use of such motifs. But these fêtes were also the occasion for processions, 'triumphs', tableaux of allegories, in carefully designed costumes inspired by the treatises of Giraldi (1548) and Cartari (1556) on the ancient gods; innumerable drawings done for these purposes show how much the fantastic imagination of the sixteenth century created in the way of new repertories.[130]

State funerals were also affected by this kind of display, and their arrangement is known from books such as the *Descrizione della pompa funerale* of Cosimo de' Medici (1571), when giant skeletons holding scythes were erected along the arches of the Cathedral. A memorial service for the Queen of Spain, engraved by Callot, took place in 1612; in 1634 the *Esequie del Serenissimo Principe Francesco de' Medici* was published, with engravings by Stefano della Bella, who was one of the last artists faithful to the spirit and style of Tuscan Mannerism.

PAINTING

The mild and reticent form of classicism practised by Andrea del Sarto and Fra Bartolommeo soon led to a reaction in Florence. This was led by Pontormo (1494–1555), using as a model Michelangelo's limpid painting, which set particular emphasis on contour. While still in Andrea's studio Pontormo composed the *Visitation* in the narthex of the Annunziata. Through the patronage of Cardinal Ottaviano he was commissioned to decorate the great hall of Poggio a Cajano, which had remained unfinished at the death of Lorenzo de' Medici. As the principal artist working there, Pontormo under-

275

took the side lunettes, which contained huge circular windows; he only had time to paint one—the east one—on the theme of *Vertumnus and Pomona*, a lovely pastoral scene, in which a contour of intensely vivid and sensitive power silhouettes the figures of the peasants, which are painted in delicately modulated tones of grey and almost colourless yellows.

In the great cloister of the Certosa at Galuzzo, near Florence, Pontormo painted a cycle of the *Passion*, now very damaged, where his use of dense crowds, and elongated, contorted silhouettes, are directly inspired by Dürer's prints, for which Vasari accused him of having betrayed the Tuscan manner for a German one, the modern for the Gothic. In the superb *Deposition* in S. Felicità in Florence (1526), painted in cool and lightly fused tones of mauve, pink and green, the undulating forms are arranged in a composition carefully built up in sinuous tiers of figures. In his portraits, Pontormo uses pale colour, a simple arrangement of the figure on the canvas as so to show off the hands, and his wonderful draughtsmanship immediately establishes the sitter's personality (the *Lute player*, Guicciardini coll., Florence; *Two men*, ibid.; *Young man*, Castello Sforzesco, Milan).

But his tension led him to tread dangerous paths; in the *Visitation* at Carmignano (c. 1530) he isolated the central group by means of a use of violent foreshortening, and turned four figures—instead of the traditional two—into a block of pallid yellow and red draperies. The portrait of *Alessandro de' Medici* (c. 1525, Lucca), was the forerunner of a series of tall, slender and haunted figures. It was at this moment that he was encouraged by Michelangelo, who gave him cartoons to work up. After 1545, Pontormo was commissioned to decorate the choir of S. Lorenzo; he worked on it for ten years and on the themes of the *Fall*, the *Deluge* and the *Resurrection* he produced the most astonishing tangle of nude figures arranged in sweeping movements entirely deficient in monumental effect; this work was not successful and the frescoes were whitewashed in 1738. Balance had become impossible.

Rosso (1494–1540) was the other neurotic artist, though of lesser genius, who was an able representative of Florentine Mannerism. A born revolutionary, he wandered from one Florentine studio to another and to the subtleties of Andrea del Sarto and Fra Bartolommeo he opposed masses of drapery organized in facets of strident colour, and his haggard or tormented faces contrast with their insight into mind and feeling. His *Madonna with four Saints* (1518, Uffizi) was rejected by the commissioner. The large *Deposition* (1521, Volterra), muddled and strange, is a witness to the survival of fifteenth-century patterns—those of Filippino, for example. *Moses defending the daughters of Jethro* (c. 1523, Uffizi) reveals the influence of Michelangelo in its cold, pale colour, and in the vigorous drawing of the avalanche of bodies in the foreground, but the *Marriage of the Virgin* (1523, S. Lorenzo) has the colourless figures strangely silhouetted in an unreal light. His stay in Rome, from 1523 to 1527, when he was unable to work, finally unbalanced his neurotic genius. In Arezzo, where he next worked, he tended more and more to the elongated silhouette, but his composition became increasingly classical (*Carrying of the Cross*, Arezzo, 1528–29).

From Arezzo he went to France in 1530. In the *Gallery of Francis I* at Fontainebleau, he triumphed in frescoes and stuccoes, thanks to his unusual powers of imagination, which reached their fullest expression in the mythological scenes and their frames, and are displayed in the many related drawings, sketchy, brusque, and full of half-statements, while engravings ensured the widespread success of his inventions. The *Pietà* (1537–40, Louvre) painted in iridescent colour, but overcrowded with figures of an exaggerated elegance, is, without doubt, the masterpiece of this capricious and emotional art.

With the second generation, Florentine painting abandons emotional expression and adapts itself to a limited circle preoccupied with ceremonial; with Pontormo's pupil Bronzino (1503–72) it becomes a court art. Dry and disagreeable in his *Pietà* (Berlin), and in the decoration in the *chapel of Eleonora of Toledo* in the Palazzo Vecchio (1555–64), he achieved a revealing success thanks to the contrived impassivity of his portraits. In one of his famous portraits of *Eleonora of Toledo and Don Garcia* (Turin), the striped curtains, of an irreproachable but trifling realism, create a setting for the dress, the hands, the regal bearing of the sitter. The violet dress worn by *Lucrezia Pucci* (Uffizi), and the brocade dress in another Uffizi portrait, have become the centre of the picture. This attention to costume recalls Flemish minuteness, but the effect is concentrated on the immobility of the pose, the ivory of the flesh tones, the transparence of the unreal setting in which the figures are seen. There is more character in the little *Maria de' Medici*, with the laughing eyes (Uffizi), and in the portraits of artists, with affected gestures and an eloquent setting (Lisbon, Berlin). Here may be seen the development of the most official kind of art, the kind advocated by academies.

Salviati (1510–63), portrait painter and brilliant draughtsman, pupil of Andrea del Sarto (*Charity*, Uffizi), sets the type with his *allegories* in the Audience Chamber of the Palazzo Vecchio (1544). And it is in the Palazzo Vecchio, as has been seen, that Florentine academism, under Vasari and his many pupils, reached its highest point in dull and cluttered compositions, based on historical subjects, on allegorical moralities, on a culture derived from handbooks of mythology, such as that by Cartari, which finally ended by stultifying Florentine taste for good. For a century, Tuscany still supplied decorative painters: Cigoli (1559–1613), who as an architect was the last to work on the courtyard of the Palazzo Nonfinito begun by Buontalenti, and who also painted large frescoes; Alessandro Allori (1535–1607) and his son Cristofano (1577–1621), faithful to the traditions of fresco, the younger man working in the lighter colours of the Bolognese School; or Passignano (1558–c. 1638), who preferred a more sombre palette.

After them came the innumerable history painters of the seventeenth century, like Francesco Furini (1604–49), who worked at the Pitti Palace with Giovanni da San Giovanni (1592–1636). The latter, fairly close to the Carracci, showed himself capable of brilliant inventiveness in the refectory of the Badia at Fiesole, in the chapel of the Palazzo Rospigliosi in Pistoia, and in the church of the Quattro Santi Coronati in Rome. A link was thus established between the last of the Mannerists and the prodigious Pietro da Cortona who, from 1637 to 1647, was painting ceilings in the Pitti Palace.

Venice

It was not until about 1535 that the Mannerist crisis disturbing Central Italy began to be felt in Venice. When it finally arrived there, it was associated with the prestige of Michelangelo's manner, and it came at a moment when Titian and Sansovino appeared to have achieved a harmony—and one peculiar to Venice—between picturesque vision and classical composition.

In architecture, where the use of the orders had become general through the precepts of Serlio and the grand though rather formal art of Sanmicheli, the same effort of imagination permitted both the absorption of a Mannerism obsessed with the pretty or the violent, and the gratification of the perennial Venetian desire for harmony. This was Palladio's achievement.

ARCHITECTURE: PALLADIO

Palladio was born in Padua in 1508 and died in Vicenza in 1580. He owed his lovely name to the humanist Trissino, who was his patron and the inspirer of his early works. Until about 1540 he was a humble mason; the friendship of Trissino, the journeys to Rome made in 1541, 1547, 1549 and 1559—supplemented by the small guide to the *Antiquities of Rome* (1554) and the study of Vitruvius, whom he illustrated in Barbaro's 1556 edition—a constant meditation on the absolute beauty of forms, corresponding to Venetian musical aesthetics, enabled him to express the aspirations of his age in the most original and sensitive way. His immensely successful treatise, *I Quattro Libri* (1570), throws the clearest light on his genius. The first book deals with general principles according to Vitruvius, the second with houses and in particular with villas, the third with towns and the last with temples. Throughout, the Antique is held up as the model, or rather, as the key to an harmonic language, but his concern with the modern needs to which it must be adapted is no less strong. Palladio's strength lay in his discovery of the precise and perfect balance between knowledge and taste. His career took him constantly from Vicenza to Venice, from the Lagoons to the *terra firma*, and towards the end of his life he spent most of his time in Vicenza.

His starting point seems to have been more with Falconetto and his purism than with Sansovino and Sanmicheli, but only a better knowledge of his early works could throw a proper light on this. His main aim was to react against the abuses caused by decoration, by narrowing the limits of the problem, by defining according to good Antique usage the value of the various elements—pediments, columns, arches, cornices—and by using an appropriate motif in preference to a haphazard accumulation. This search for a sensitive and carefully thought out effect, which is one of the elements of Mannerist refinement, is linked to two far greater preoccupations: an internal harmony of the parts which will establish the edifice as a thing of beauty in itself, deliberately created (and in this Palladio shows himself heir to the

FIG. 31. *S. Francesco della Vigna, Venice, by Palladio. Design of the façade (after R. Wittkower) showing the two interlocked temple fronts.*

thought and even to the philosophy of Alberti); the importance of the relationship between this harmonious, geometric form with the place, the atmosphere, and nature itself, according to Venetian tradition. Villas are raised on a base which blends them with their setting, and their colonnades create effects of light and shadow comparable to those of the palaces on the Lagoons. But Palladio must not be considered merely in relation to purely Venetian problems, but in relation to the Renaissance as a whole.

One of the great problems was to ally the Antique temple and the church. Alberti, in his façades, ranged from the triumphal arch motif (Rimini) to the portico (S. Sebastiano, Mantua), and finally combined the two (S. Andrea, Mantua). A different solution was reached by Peruzzi at Carpi (1515), where a giant order corresponds to the full height of the nave, and a smaller order to the aisles. This harmonious arrangement occurs in Palladio who adapts a true prostyle temple front in *S. Francesco della Vigna* (1562), and again

MANNERISM

later at S. Giorgio Maggiore (1566–80). The central element is made to project slightly, and has its columns raised on high bases, though according to a drawing in London this was not the original plan for S. Giorgio and its triangular pediment. The aisles had their own quite distinct system. Thus two separate façades interpenetrate, or rather, two arrangements of classical façades are blended one into the other, like two harmonic systems whose fusion tends to obscure the framework. This is the Mannerist aspect of the solution, which became even more marked in the *Redentore* (1577–80). It was due to Trissino that Palladio was commissioned in 1545 to build the *Palazzo della Ragione* in Vicenza, the model for which was accepted in 1548. Like Alberti at Rimini, he had to case an old building in a new shell. This consisted of two storeys of arcades borne on Doric and Ionic orders, with the columns doubled at the angles, and it uses for the first time the motif of the triple opening contrived by supporting the arcades on two columns flanked on either side by a short entablature, a motif which Serlio had published in his fourth book. Each of the galleries is as vigorously conceived as a façade in the Palazzo del Tè, but the dominant impression is that of a profound and peerless sense of rhythm.

It was Vicenza, rather than Venice, that Palladio remodelled with his palace façades, each more imaginative than the next. The massive rustication of the *Palazzo Thiene* (1556), with different orders mingled in its almost rough-hewn surface, reveals his use of Mannerist architectural forms, better known after his visit to Rome in 1554. There is, in the *Palazzo Chiericati* (1566), a relationship both of pure form and picturesque effect in the alternation of the light wall areas with the dark porticoes, which transpose the features of a courtyard with those of a façade, and there is the same tendency towards an unclassical use of classical elements in the *Palazzo Valmarana* 1566 where the colossal order is used with a diminished projection in the large pilasters. Finally, in the *Loggia del Capitano* (1571) in Vicenza, the giant Corinthian order seems to compress the three-bayed edifice; the window balconies rest on consoles which have slipped down as far as the haunches of the arcades below, which, in their lightness and delicacy, contrast with the columns of the main order. It is as if the Capitoline Palace had been translated into a more colourful and Palladian language, while it also remains a more than usually tense and dramatic expression of the artist's vigour. He built many villas for wealthy Venetian clients, drawn to the countryside of the Brenta, the Monte Berico and the Marche di Treviso, and he insisted upon a strict ordering of the twin aspects of the building. The symmetrical plan—already used by Giangiorgio Trissino in his *Villa Cricoli* (Vicenza, 1535–38)—became the rule. From the first of these works onwards, the *Villa Godi Porto* (Lonedo, *c.* 1540), he stresses the symmetry of four rooms of equal volume proportioned to a central hall; the *Villa Thiene* (Cicogna, *c.* 1550), the strict disposition of the *Villa Malcontenta* (1560), and the geometric perfection of the *Villa Rotonda* (1550 onwards) all conform to this type. This sense of order is expressed in the arrangement of the rooms on a central axis, the placing of the flights of steps, the addition of service quarters and wings, but it also expresses something far more subtle, only made explicit by the publication of the plans with the dimensions of the rooms calculated in feet in the *Quattro Libri*. That is, the architect's concern with harmonic relationships that conform to the musical and cosmic scales, and the methodical selection of basic proportions and the use of a constant module: variations on 2:3 in the *Villa Godi Porto*, on 1:1 and 1:2:3 in the *Villa Thiene*, on the numerical relationships corresponding to the major and minor keys in the *Villa Barbaro* at Maser.

Palladio's second success was in his placing of these houses, so that they blend into the landscape, and are integrated into their setting like a building contrived by a humanist painter—for the villa is to inherit the temple front, which was not to be used in its most complete form in church architecture. The *Villa Godi Porto* still consists of a recessed central block borne on arcades with a flight of steps in front, set between two much simpler blocks. Later at the *Malcontenta*, at Mira, Palladio conceived the villa with a temple front, opening on to a pedimented hexastyle portico and raised on a high basement. In the *Villa Badoer* (Rovigo, 1549) the steps were placed in front of the portico, increasing the contrasts of light and shade. The façade is four times repeated in the astonishing *Villa Capra* near Vicenza (known as the Villa Rotonda), which clearly demonstrates how much the design is governed by the site of the house, here placed at the top of a hill from which it appears to emerge. At the *Villa Barbaro* (*c.* 1560, Maser), the giant order sets off the central block, which is extended by two wings, each with a freely designed pediment, linked to the main block by an arcade, and the whole is silhouetted against the broad screen of the wooded hillside. The plan of the *Villa Trissino* at Meledo added to these elegant features differences in level and the immense symmetrical colonnade, curved at first and then straight, inspired perhaps by antique Thermae and on a scale that presages the Baroque.

The villa-palace, organized in clearly defined blocks, with superposed orders and therefore several storeys, is the grandest and largest version of the form: examples are the *Villa Pisani* at Montagnana (*c.* 1566) with its motif of four columns grafted on to the façade itself, and the *Villa Cornaro* (1565) at Piombino Dese, where the temple front motif is reversed to become an interior portico. A variant, which shows Palladio's freedom of invention, is the two-storey façade of the loggia of the *Villa Santa Sofia* at Valpolicella, which is enlivened by a rusticated Corinthian giant order.

The Palladian villas also sum up a half-century of experience in the importance of their interior decoration. In 1510, on the outside walls of the Fondaco dei Tedeschi, Giorgione had painted secular allegories set in niches; the basic idea survived. By 1540, Giulio Romano had exploited every variety of illusionism in his enormous rooms, and Raphael followers such as Giovanni da Udine had spread the use of grotesque ornament. These types of decoration

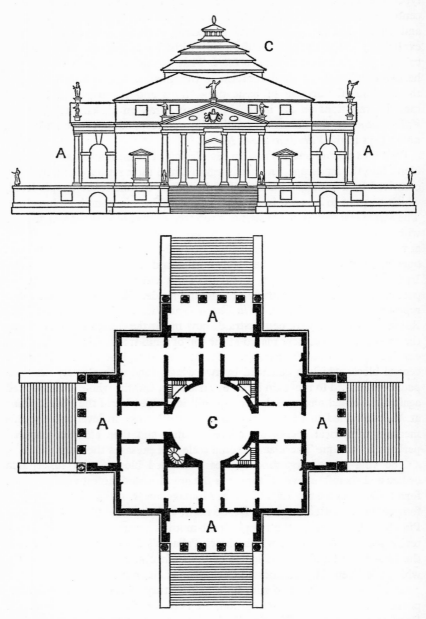

FIG. 32. *Villa Rotonda, Vicenza, by Palladio. Elevation and plan (after Scamozzi): C, Dome; A, Wings.*

were used in all the villas, but never with more charm and assurance than in the *Villa Barbaro* at Maser, where, thanks to Veronese's lively imagination, the learned themes of the ceilings are continued in amusingly illusionistic views set between columns, so that the landscapes introduce nature and the open sky into the house with much more ease and gaiety than in the Farnesina. The usual decorators for the interiors of Palladio's houses were Zelotti at the *Villa Godi Porto* and the *Villa Emo*, and G. A. Fasolo who painted the scenes with giants, and tender, agreeable scenes such as the *Concert of ladies* at the *Villa Coldogno* near Vicenza (*c.* 1570).

Palladio's last works, like the *Villa Trissino* and the church of the *Redentore* in Venice (1580), where the triple lobed choir opens on to the nave through a triumphal arch screen so that the effects of filtering light seem to increase its size, accentuate the colouristic quality of his vision. The summit of his *œuvre* is, in a sense, the little *Teatro Olimpico* begun in 1580, with its semi-elliptical auditorium, made of wood ornamented with stuccoes and crowned by a colonnade with niches and statues. On the stage is a permanent architectural setting, with three entrances and five streets running back in sharp perspective. It was in Rome that the most determined efforts had been made to treat stage scenery in the spirit of contemporary architecture; the performance of *La Calandria*, a light comedy by Bibbiena staged with settings by Peruzzi in 1514, was the first example of the kind of decor which consists of hollowing out the stage in a centralized perspective setting round which are grouped various types of buildings. Serlio, following Vitruvius, designed three types of setting—for tragedies, comedies and satyric plays—one based on the architecture of classical antiquity, one using ordinary modern types of buildings, and the third with a rural setting (Book II, 1537). In Venice, for one of the comedies of Aretino (1541), in Florence for the marriage festivities of Francesco de' Medici (1565) designed by Vasari, and for those for the marriage of Cesare d'Este (1585) designed by Buontalenti, this formula—which was to reappear during the Baroque—was fully developed.

Palladio's arrangement, which re-introduced the continuous setting with multiple entrances, was a learned reaction against the fantasies of theatrical perspective. From the experiments made in the *Basilica* at Vicenza (1561–62), and in Venice in the *Convent of La Carità* (1565), the Vicenza setting was evolved. Only the roof of the auditorium was not built as the architect wished; the stage itself was executed by Scamozzi (1552–1616) after Palladio's drawings, as a sort of ideal city inspired by the streets of Vicenza, and which became 'Thebes' for the memorable presentation of Œdipus Rex in 1585.[131] With its monumental proscenium and the corresponding gallery which encircles the tiers of the auditorium, Palladio's creation has an extraordinary unity and force. This arrangement is also found at Sabbioneta, near Mantua, where Scamozzi again used the form in a colonnade crowned by statues (1588–89), and the model was also used with two tiers of galleries by G. B. Aleotti in the Teatro Farnese in Parma (1618).

MANNERISM

Scamozzi finished a number of Palladio's works, including the *Villa Rotonda* (1606). He also built, along the south side of the Piazza San Marco in Venice, the *Procuratie Nuove* in a style freely adapted from that of Sansovino's Library, and he set out his theory in a weighty treatise (1615).

With Palladio must be considered the sculptor Alessandro Vittoria (1525–1608), the only other notable artist of the period. He made bust portraits of all the most famous Venetians of the day, such as *Antonio Grimani* (Chapel of S. Sebastiano), and *Pietro Zen* (terracotta, Seminary). The carefully disposed accents in the relief catch the light to create a pictorial effect, which is also found in his religious statues, such as the *S. Jerome* (altar on the right, Frari). As a decorative artist, Vittoria derives from Giovanni da Udine, adapts himself admirably to the setting, and uses all techniques so as to increase the charm of the effect, in, for instance, chimneypieces, statues and masks at the Villa Barbaro, Maser (1566), and the *Seasons* in the high niches of the Villa Pisani, Montagnana (1565–76).

PAINTING: TINTORETTO, BASSANO, VERONESE

The pressure of the new aesthetic ideas emanating from Mantua, Parma and Florence affected Venetian painters by about 1540. Titian responded by transforming his palette, and by enlarging the bounds of his ambition, which led him to Rome and Augsburg to become the painter of popes and emperors. Some minor artists were permanently affected by it.

Schiavone (*c.* 1522–63) derived from Giorgione and Titian, but found his true bent through the engravings of Parmigianino. His Mannerism is all movement and caprice; the clothes of his nymphs and the curves of the landscape turn into curls (*Bacchus and the Nymphs*, pvt. coll., Venice); his colour, here fluid and here impasted, dissolves and coagulates too arbitrarily (*Adoration of the Magi*, Carmine, Venice). Following in his footsteps was Pietro di Mareschalchi of Feltre (1520–*c.* 1580), who derived his taste for shot colours from Bassano, and his love of complex colour schemes from Tintoretto. His masterpieces, a *Banquet of Herod* (1576, Dresden) and a *Pietà with SS. Clare and Scholastica* (*c.* 1570, Bassanello), assure him a notable place through their blend of precision and softness. But to Venice was reserved the privilege of extracting an altogether new beauty from the general Mannerism, and of 'Venetianizing foreign art' (M. Florisoone).

The coldness of Veronese, who derived from Mantua and Parma, corresponds, in a certain degree, to the rigorous purity of Palladio. A passionate imagination and an unparalleled rapidity of execution, both increased by contact with the works of Michelangelo, led to the spiritual exaltation of Tintoretto's effects of light. The need to express nature with truth and directness was what inspired Bassano.

Tintoretto (1518–94). He was born in Venice itself, of lowly origins, and

284

was imbued with feelings which were at once strong, simple and popular. Work was his only passion, and he was eager to get all commissions for himself. He lived only in Venice—except for a journey to Rome, supposed to have taken place about 1545, and a visit to Mantua in 1580—and did not want to know anywhere else. Rather than from Titian, he seems to have derived from Pordenone's troubled style and from Schiavone's irrealism, which liberated him from the bonds of descriptive accuracy. From 1539 onwards, he was an independent master, but only a visit to Rome can account for the sudden outburst in his dramatic sense, and the explosions of light which run through the *Last Supper* (1547, S. Marcuola) and, above all, through the *Miracle of the Slave* (1548, Accademia).

His mastery of his art was such that he was able to heap up, or to string out, his figures in violent explosions of light and shadow, and to arrange them along the many axes of the varied, fantastic and obedient space which he created for them. Thus he could approach Titian (*Genesis*, Scuola di S. Trinità, and *Adam and Eve*, Accademia) and even Veronese, as in the memorable *Suzanna and the Elders* (c. 1560, Vienna) where the huge, weightless, and translucent nude is silhouetted against the depth and richness of the garden. He sets out to achieve a vehement narrative force, using three forms of Mannerism which in his hands revealed for the first time all their formidable resources: the elongated, sinuous figure which, far from merging into the golden haze, often increases its effectiveness; a composition in depth and full of daring foreshortenings; and a light—intense, variable and transfiguring—that ranges like a spotlight over the quivering, nerve-racked, and infinitely pathetic world of his creation.

Three pictures of the *Miracles of S. Mark* date from 1562 to 1566; in the *Discovery of the Saint's Body* (Brera) all the figures sway in a dark and glittering perspective. In 1564, Tintoretto contrived to get the commission for the huge *cycle of the Scuola di S. Rocco*, which he finished only in 1587, and which represents the summit of his art, his visionary Bible. He began on the upper storey, with a *Way to Calvary* constructed in a series of zigzags, a *Christ before Pilate* in which Christ's white garment explodes like a lightning flash amid the sombre tones and strange reds, and a *Crucifixion* of fantastic size, filled with a swirling mass of figures, almost unbearable to look at. In the lower room, the most lyrical pieces, the ones richest in invention, are those, like the *Flight into Egypt*, set in a landscape shadowed by palmtrees, or the *Magdalen*, painted between 1585 and 1587. The strange turmoil in which the painter sets all the incidents of the Bible story enabled him to introduce countless bizarre or naturalistic details: the Virgin's chair in the *Annunciation*, or the straw of the crib, for instance. Never before had an artist's handling achieved more freedom or more power. Other works of his are to be found in all the churches of Venice, in particular S. Maria dell' Orto and S. Rocco.

In the Doge's Palace, Tintoretto had also obtained huge commissions—in the Sala del Collegio and in the Sala del Senato, where is his celebrated ceiling of *Venice, Queen of the Seas.* In 1588 he painted for the Sala del Maggior Consiglio, rebuilt after 1577, the great *Paradise,* said to be the largest picture in the world, and in which the strain of the immense undertaking and the rather uncertain exaggeration of scale are only too noticeable in the schematic nature of the over-accentuated forms. Tintoretto's portraits are as numerous as his decorative works: *Alvise Barbaro* (Pitti), *Jacopo Soranzo* (Milan), are examples, but even more do they conform to a set type; that is, they are constructed upon a contrast between the dark background, the costume and the face, and this circumscribes the artist's impetuosity. Nowhere is this more striking than in the charcoal sketches, full of hatchings and small rounded forms, where in the energetically modelled reeling and swaying figures one can feel the artist working from small models, in the raking light of lamps, with a rather mechanical but still unsurpassed energy.

Jacopo Bassano (*c.* 1510–92). Bassano is not among the greatest masters. He worked first with Bonifazio de' Pitati, and lived in a provincial circle— one might even say a village circle—dominated by Pordenone, Paris Bordone and the Emilians (the *Good Samaritan,* Rome, Capitoline Mus.). But a knowledge of Roman and Flemish Mannerism and the example of Titian and Tintoretto raised and stimulated him, and, helped by his large family (including Francesco (*c.* 1549–92) and Leandro (1557–1622), a notable portrait painter) he created a new formula whose spread marked the end of the century in Italy and in Europe. The *Adoration of the Shepherds* (1568, formerly S. Giuseppe, now Mus., Bassano), the *Baptism of S. Lucilla* (1580, Bassano), are important examples of this type. He also painted innumerable genre scenes of a rather monotonous composition and colour, in which all the occupations appear. The pastoral scene has invaded painting or, more accurately, the sombre sky, the leafy horizons, the cattle and the peasants have become essential participants in all religious and secular scenes, which they accompany and express in a kind of overcast and chaotic dreamworld. With a sometimes surprising licence he dwells on the glitter of metals, the reflected light of the moon or the last remnants of daylight, creating a limited nocturnal world, basically naturalistic, with an often remarkable richness of contrasts, from the skies to the objects themselves, as in the *Crucifixion* (1562, S. Teonisto, Treviso). This realism and this obvious passion for light effects necessarily had prolonged repercussions; they prepared the way for the genre scene, but the 'Bassanesque' manner of Leandro, for example, is more often boring and heavy, as in his *Suzanna and the Elders* (pvt. coll., Venice), with its too dark greenery and conventional glitter.

Veronese (1528–88). If Tintoretto's personality and his astonishing cycle of S. Rocco mark in Venetian painting the exaggeration of an almost excessive Mannerism, the art of Veronese, which also has a grand decorative purpose, is successful by quite different means, and by its conscious return to the style of 1530—a chilly atmosphere, and a shadowless daylight in which forms are silhouetted and derive the maximum effect from their contours. This orientation can be explained by his origins: a pupil of Antonio Badile of Verona, Veronese was plunged into a centre still faithful to Bellini and Mantegna, with ramifications towards the Parma of Correggio and the Mantua of Giulio Romano. This is clearly indicated by his first works: the *Bevilacqua Altarpiece* (1548, Verona) and the *Gonzaga Altarpiece* (1552, Mantua). In 1553, on his arrival in Venice, where he remained until his death, he worked on the ceiling of the Sala dei Dieci in the Doge's Palace, with Ponchino's Mannerist team, and he adopted the sharply foreshortened perspective and the lively contrasts which characterized their style (*Venice receiving Juno's gifts*), though modified by his own taste for a more limpid and transparent light. But this reaction against Venetian 'tone' in favour of a clearer light and one without an enveloping atmosphere, has nothing archaizing about it, because it takes into account the palpitating and radiant spaciousness of Venice itself. The *decoration of S. Sebastiano* (1555–56), Veronese's favourite church, and the one he was buried in, the ceiling medallions with the story of Esther and the scenes of S. Sebastian on the walls, are of a striking breadth and inventiveness. His great gift was to provide a worldly society, overwhelmed by the torments and grandiloquences of Tintoretto, with the sumptuous, balanced and superficial decor it demanded. If Tintoretto represents the ultimate conjunction of Michelangelo and Titian, Veronese is more the fruit of a union between an attenuated Correggesque style from Emilia and the style of Raphael seen through the eyes of Giulio Romano and the Veronese G. B. Caroto (1480–1555). The finest moment of this happy progress is reached in the decoration of the *Villa Barbaro* at Maser (1556 onwards), where wide landscapes seen between painted colonnades, amusing *trompe-l'oeil* scenes set on fake balconies, and comic fantasies suited to fashionable taste, succeed each other in the Palladian setting to which they are perfectly attuned. The memory of Pompeian decoration is evident in the playfulness and variety which are here given free rein, and Veronese's visit to Rome in 1560 therefore assumes a certain importance.

To the convolutions and the plunges into depth which seemed indispensable to 'inspired' art, Veronese opposes large architectural constructions seen frontally, articulated by porticoes and colonnades in crowded compositions, full of figures and details, but clearly organized, and frequently obeying the compositional rule of the triangle within a rectangle. Of this kind are the *Family of Darius before Alexander* (*c.* 1565, London) and the sequence of banquets: the *Marriage of Cana* (1563, Louvre) and the *Christ in the House of Levi*, designed for the refectory of the convent of SS. Giovanni e Paolo (1573, Accademia), crowded with all ranks of Venetian society, from nobles to cooks, in a bewildering accumulation of gestures and materials. This is the canvas which drew down upon the artist the strictures of the Inquisition; he extricated himself by claiming the licence allowed to poets and fools. The love of light, of pleasure, of the elegant picturesque, of the charm of pale, rich tones against a blue sky, is so strong that it often slips into artificiality, as in the *Annunciation* (*c.* 1556, Uffizi) or into erotic sophistication, as in the little panel of *Mars and Venus* (Turin).

The *S. Menna* and the *S. John the Baptist*, destined for the organ case of S. Giminiano in Venice (*c.* 1565, Modena) belong to a moment when he was experiencing the influence of Giorgione. After 1570, and the pictures of the Cucina series (Dresden), his colour loses its limpidity and reflects the nocturnal world of Tintoretto and the Bassanos: such are the *Vision of S. Helena* (London), and the *Agony in the Garden* (Brera), where he is less rich, less heavy, and more pathetic, as he is in the little *Calvary* against a background of clouds, with the yellow draperies of one of the mourners (Louvre). In 1575, he painted in the Sala del Collegio in the Doge's Palace a decoration which is set into the gilded cofferings; in 1578, he was commissioned to paint the *Triumph of Venice* for the Sala del Maggior Consiglio. Thus all the great Venetian masters were in turn brought in to decorate the city's principal building. Among the crowd of painters who surrounded Tintoretto and Veronese, must be mentioned Palma Giovane (1544–1628) and his narrative *cycle in the Oratorio dei Crociferi* in Venice (*c.* 1590).

But fatigue began to be felt in Venice, until now more watchful even than Florence, more active than Rome. The seventeenth century, with only a few exceptions, was to be a period of rest when it was considered adequate to defer to the great examples which were having such far-reaching effects all over Europe. One of the most curious features of Venetian art is the survival of the workshops of the *Madonnieri*—Madonna painters—who provided works of popular imagery. It was from these artisan ranks that emerged a young Cretan—Theotocopoulos (1545–1614)—who adapted the themes of Byzantine iconography to a freer style of handling (*polyptych*, Modena). He may have worked under Titian, knew Bassano and his raking light, Tintoretto and his febrile energy, then went to Rome when he was twenty-five and finally, after 1576, settled in Toledo, where—as El Greco—he became the protagonist of a new Art.[132]

Late Forms: Rome, Bologna, Milan

In the second half of the century, new forms of Mannerism emerged, in which Rome again was paramount. The followers and imitators of Michelangelo multiplied in painting as in architecture; the progressive decline of Florence left more and more scope for the Lombards and the Emilians. The upshot was a development along two lines; the first consisted of an art more daring than confident, elaborated by numerous confused practitioners; the second consisted of a small learned reaction which sought to meet the needs of the age according to reasonable standards through a conscious eclecticism.

89 (a) Niccolo
dell' Abbate:
Abundance, drawing

89 (b) Beccafumi:
Saturn, drawing

90 Giulio Romano: the Palazzo del Te, Mantua
(a) entrance façade
(b) garden front

91 Giulio Romano: *Venus and Adonis surprised at the bath*

(*Mantua, Palazzo del Te*)

92 Palladio: the Loggia del Capitan, Vicenza

93 Villas by Palladio:

(a) La Malcontenta,
    Mira,
    near Venice
(b) Villa Rotonda,
    near Vicenza

94 Palladio: the façade of S. Francesco della Vigna, Venice

95 Tintoretto: *The Discovery of the Body of S. Mark*

(*Milan, Brera*)

96 Veronese: *Mars and Venus*

(*Turin, Galleria Sabauda*)

LATE FORMS: ROME, BOLOGNA, MILAN

An academic doctrine was worked out to be transmitted to the next generation.

Rome: Decoration, Architecture and Vignola. Towards the middle of the century, there was a return to the large painted decoration, an evolution in the art of designing villas, and finally, a clearly worked out doctrine in architecture. The *Deposition*, painted in 1541 by the Tuscan Daniele da Volterra (1509–66), created a sensation among the younger artists. An involved composition, derived from Michelangelo, it also contained passages of great delicacy derived from Raphael. Rome was again invaded by Florentine artists: examples of this are Jacopino del Conte (1510–92), and Salviati (1510–63), who delighted in all the technical devices—great plunges into depth, cut-off and repoussoir figures in the foreground—designed to create everywhere a feeling of undulation, sinuosity and spiralling movement.

One of the most remarkable of the Mannerist chapels in Rome is the *Oratory of S. Giovanni Decollato* (the Baptist), built by the Florentines in about 1495 in the Velabro. It was not until about 1545 that the room devoted to the history of the Baptist was decorated by Zuccaro, Vasari and Jacopino del Conte with panels containing the most incongruous dynamic effects. The *Oratory of S. Lucia del Gonfalone*, near the Via Giulia in Rome, is a similar but later work, containing histrionic decorations by Federigo Zuccaro, Raffaellino Motta and others.[133]

The fashion for painted cycles was becoming the vogue in palaces. Contemporary with Vasari's allusive decorations in the Palazzo Vecchio in Florence (1558), there were in Rome those of Jacopo Zucchi in the Palazzo Ruspoli built by Ammannati, of which the author analyses the wildly confused allegories in a detailed programme, and the similar cycle laid down by Annibale Caro (letter of 1562) for the painter Taddeo Zuccaro in the Palazzo Farnese at Caprarola.[134] Zuccaro (1529–61) was an important representative of erudite and pompous art, who worked both at Caprarola and in the Sala Regia of the Vatican with his brother Federigo. The activity of the latter (1542–1609) fills the end of the century and extends to Venice (Sala del Maggior Consiglio), Milan, England and Spain. His house in the Via Gregoriana in Rome is remarkable for the Mannerist device of framing windows and doors in gigantic and monstrous masks. But, still more important, in 1593 Federigo was the president of an Academy of Design, which had for a generation (1577) replaced the old Guild of S. Luke. The movement, so characteristic already in Florence, in favour of theoretical discussions on the arts, like the contemporary regimentation of literature, was also successful in Rome. The new organization centred attention once again on Rome, which could offer examples of all the most vital forms. A treatise by Romano Alberti (1599) gives a résumé of the discussions of 1593 on the dignity of the arts; Zuccaro's treatise, *L'Idea* (1607), was to crown this teaching by showing the best method of avoiding decadence, and the role of the internal image (or

289   v

idea) in its relation to the exterior image (or form) which is to be physically realized. Above all, huge fields of activity were proposed for the artist. The end of the century saw—thanks to this academic movement—an ever-increasing influx of artists to the papal city. A rival academy had been founded in Bologna by the Carracci, but Annibale and Agostino went to work for the Farnese in Rome in 1595, and it was an opposition deriving more from the personalities of the Carracci than from any divergence of aim. In architecture, too, Rome again led the way.

As in the Veneto and in Tuscany, the programme decorations were adapted to the building, and the villa or country house imposed on the surrounding landscape a design intended to set it off to the best advantage. The master-piece of this kind of natural scene-setting is the *Villa d'Este*, planned by Pirro Ligorio (1550). Set on the edge of the plateau of Tivoli, it offers splendid views over Rome and Lazio, and one of its attractions was an amusing imitation of Roman buildings in the avenue called *La Rometta*. But further surprises were planned in the way the cypress walks descended to a roundabout, in the fountains and pools in the midst of sombre verdure, in the water-organ. Pirro Ligorio (1510–83), who had worked at S. Giovanni Decollato, also built a casino in the Vatican grounds for Pius IV, called the *Villa Pia* (1560); with the help of Rocco da Montefiascone, he covered its first storey with reliefs in stucco, set garlands upon the panels, festooned the edges of the niches, to create an amusing 'folly'. The *Villa Medici*, built after 1564 on the edge of the Pincian Hill by Nanni Lippi for Cardinal Ricci, and most probably enlarged after 1578 by Ammannati for Cardinal de' Medici, is larger but in the same spirit. Its terraces and portico open on to a symmetri-cally planted garden, and it lends itself admirably to the display of antique marbles.[135] More complex and more brilliant still, the *Villa Giulia*, built on the slopes of the Parioli hills near the Ponte Milvio for Pope Julius III, was designed to use all the resources of the most grandiose scenic effects. It was a turning point. All the major Mannerists—Ammannati, who described the work in a letter dated May, 1555, Vignola, Fontana, and Vasari—worked there, while Taddeo Zuccaro decorated the ceilings with numerous stuccoes.[136] All the new motifs were combined there; a courtyard in hemicycle, a deep nymphaeum, a screen façade concealing a second courtyard, terraces and statues (Fig. 33). The principal façade is almost entirely by Vignola; this is the most restrained part of the work.

Vignola (1507–73) had made his début at Bologna, before coming to Rome, with the peculiar façade of the Palazzo dei Banchi, in square, harsh proportions and complicated detail. In Rome, in 1554, he designed the church of S. *Andrea in Via Flaminia* in severely simplified and superposed masses, with perfectly plain walls and heavy cornices. He was a member of the Mannerist group responsible for the Villa Giulia, but left there to work from 1559 to 1564 on the pentagonal *Villa Farnese* at Caprarola. He took as his starting point the plan conceived by Antonio da Sangallo and Peruzzi for a fortress, but he managed to arrange the interior as a palace, through the creation of a circular courtyard containing two storeys of high arcades, treated with a strong sense of rhythm. The exterior is entirely his; the façade with its insistent horizontals rises over two huge semi-circular flights of steps which

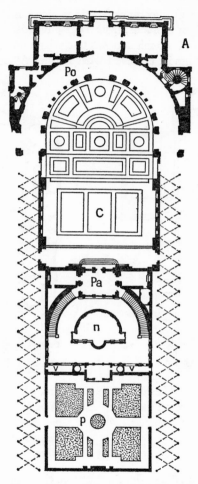

FIG. 33. *Villa Giulia, Rome. Plan of the whole (after Letarouilly): A, Casino, or main block; Po, Portico; C, Main courtyard; Pa, Garden pavilion; n, Nymphaeum; v, Screen; p, Formal garden.*

virtually dominate the architectural mass on the hill above. The gardens and the buildings in it are, for the most part, later. This vast work led Vignola to design simply, to use the antique, and to discipline his imagination. Again for the Farnese, Vignola worked on the gardens of the Palatine, of which the two-storey monumental rusticated gateway long remained famous, and started on a palace at Piacenza which was never finished.

His fame rests on his book *Regole delli cinque ordini d'architettura* (1562; Rules for the Five Orders of Architecture), a manual inspired by Vitruvius, and whose good sense and clarity ensured its popularity for the next three centuries. The reasoned selection of antique elements, checked against the ruins, came at the right moment; from 1542 onwards, a Vitruvian Academy had existed in Rome, whose mission was to clarify a doctrine which was often confused in detail. Vignola achieved what neither Alberti, nor Peruzzi, nor

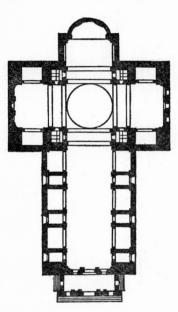

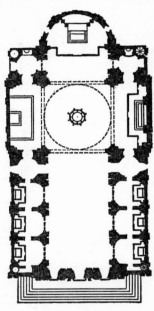

FIG. 35. *S. Andrea, Mantua, by Alberti. Plan: large transept; aisles cut up into side chapels.*

FIG. 36. *The Gesù, Rome, by Vignola. Plan: transept proportioned to the dome; side chapels opening into one another.*

even Serlio, had been able to do. At the same time, his example supported a resistance to the aesthetic licence of Mannerism, and responded to the severity increasingly required by the post-Tridentine church. Vignola was thus inspired to return to those Albertian solutions which were generally nearest to antiquity; this is borne out by the Gesù in 1568. This building, which he did not live to finish, was completed by the façade built by della Porta (1575) and a good deal altered by the seventeenth-century decorations; the ceiling was painted by Baciccia (1672–79) and the S. Ignatius Altar is by Padre Pozzo (1700). But its appearance is of major importance.[137]

The plan seems to be inspired by some of the rough designs made by Michelangelo for S. Giovanni dei Fiorentini (1560), themselves connected with S. Andrea in Mantua. In all three cases, there is a large barrel-vaulted nave, with side chapels opening on to it. The idea of a huge central space in the crossing could also have been derived from the *tepidarium* of the baths of Caracalla. In 1554, Michelangelo offered to provide, gratis, as an act of private devotion, the plans of a mother church for the Jesuit Order. After Vignola had been chosen, a letter of Cardinal Alessandro Farnese showed how precisely the Jesuits had stated their requirements: a church with a single vaulted nave for preaching, and the architect had to work within those terms of reference. The light diffused in the crossing by the dome left in shadow the side chapels which seem to shade off softly the outlines of the building; the enormous semi-circular barrel vault resting on a long and strongly marked entablature treated as a cornice is supported by the fluted pilasters which give a true Roman solemnity to the whole. But it is of interest to note that this arrangement corresponds to the simplified type which a Ferrarese architect, G. Tristano (1515–74), who became a Jesuit in 1555, endeavoured to perfect for the churches of the Order.[138]

During the following years, the Lombards became increasingly important in Rome, particularly in architecture. Giacomo della Porta (1540–1602) was employed to finish Michelangelo's work at the Capitol (1564–94), the Palazzo Farnese (where he inserted the big loggia overlooking the Tiber), and at St. Peter's where he completed the dome (1586–93) and changed its shape. He completed the Gesù by adding a façade with doubled pilasters in the lower storey, with one pediment forced inside another crammed on to the central axis, and with large volutes at the side which were not on Vignola's much simpler plan. This arrangement retains a good deal of the frigid complexity of Mannerism. In *S. Luigi dei Francesi* (1589) the five bays of the flat façade have only slight points of emphasis; the volutes have been eliminated and the rather constrained effect is the exact expression of the sixteenth-century Counter-Reformation spirit. The same effects are to be found in the *Villa Aldobrandini* at Frascati (1598–1604), where in the midst of a luxuriant garden and above a cleverly constructed ramp, rises a bastion-like wall with unexpected openings, oddly crowned by the great mass of the villa and its raised central feature, with, at its extreme edges, the two widely separated portions of a broken pediment. Della Porta followed Pirro Ligorio at the *Palazzo della Sapienza*, to work on the curved courtyard (1579), and added to the amenities of Rome with several fountains.

Domenico Fontana (1543–1607), who came from Como, worked early in his career in Rome at the Quirinal Palace, and at S. Maria Maggiore, where he built—for Cardinal Montalto, later Pope Sixtus V—the centrally planned *Chapel of the Holy Sacrament*, with an incredible luxury of colour in the polychrome marbles and the reliefs on the walls. His façade for the Lateran palace, derived from that of the Palazzo Farnese, is flanked by a two-storey

MANNERISM

loggia which stresses the monumental effect. He was Sixtus V's great town planner. This pope (1585–90), who reorganized the administration of the Holy See by creating the nine Congregations, wanted to give Rome its definitive appearance. Fontana conceived the plan by which six roads were to leave S. Maria Maggiore in six straight axes each ending in an open place containing an obelisk; two of these were the Via Merulana leading to the Lateran, and another narrow but vital highway leading by the Via Sistina to the Trinità dei Monti. Caligula's obelisk was raised with great difficulty on the square in front of St. Peter's (1586), and the statues of the Dioscuri were finally set up on the Quirinal Hill (or Monte Cavallo). Among his other works were the bringing of an abundance of water into the city by the aqueduct twenty kilometres long called the Acqua Felice, and the multiplying of monumental fountains. In a very few years was accomplished a vast undertaking, similar to that which had been conceived in the time of Julius II; imagination on a grand scale, and the art of planning a great whole, had reappeared in Rome.

Fontana, accused of peculations, finally went to Naples, where he built the Royal Palace, with a long façade in three orders which was completed in the eighteenth century. With him, we are on the threshold of the Baroque.

Lombardy and Emilia. Of the three principal architects of Milan, only one, Alessi, was purely a builder; the second, Leoni, was a sculptor, and the third, Tibaldi, was a painter.

Galeazzo Alessi (1512–72) came from Perugia. He derived from the Michelangelo of the Roman period, and first worked in Genoa where he built *S. Maria di Carignano* (1552) on a central plan with five domes and with the rather heavy façade indecisively framed by campaniles, the *Villa delle Peschiere* in two storeys crowned by a huge balustrade, and the *Palazzo Cambiaso*. Then he went to Milan, where his most important work is the *Palazzo Marino* (1558), with a courtyard curiously articulated by arches framed by two columns from which, amid a profusion of ornament, the upper storey rises. His *facade of S. Maria presso S. Celso* (1568–72) is also very Lombard in feeling.

Leone Leoni, born at Arezzo (1509–90), came to Milan, where he was the designer of the *Casa degli Omenoni* (1573) which has huge caryatides replacing the pilasters and transforming the façade into a sculptured relief. A specialist in bronze and large-scale portraiture, he made the *Vespasiano Gonzaga as an antique warrior* (Sabbioneta) and the *tomb of Giovanni-Giacomo de' Medici* (1560–62, Milan cathedral); his heroes in complicated armour wear expressions of affected dignity. He worked for the Emperor Charles V, whose *portrait in armour* he made (Prado), and he spent the rest of his life in Spain, where his son Pompeo (1533–1608) continued in his footsteps as a sculptor in bronze.

The most important of the three artists was Pellegrino Tibaldi. He was born at Valsolda in Lombardy in 1527, and from 1549 to 1553 was in Rome, where his contacts were with the second generation of Mannerists—those too strongly influenced by the Sistine Chapel and the frescoes of the Cappella Paolina, and dominated by Daniele da Volterra, with whom he worked at S. Trinità dei Monti. He built and decorated a palace for Cardinal Poggi in Bologna, with scenes from the story of Ulysses, notable for their extreme use of foreshortening, and built in a severe Doric a *chapel of S. James the Greater* which he ornamented with spirited and imaginative allegories. He also worked at Loreto, at Ancona, and in numerous palaces in Bologna.

In 1567, having become the chief architect of Milan cathedral, he modernized the interior, the choir with its caryatides, and the high altar. He built *S. Fedele*, with its heavy pedimented cornice, and the courtyard of the archbishop's palace, with its double rusticated storey. He also worked in Lombardy, at Novara (*S. Gaudenzio*, the *Palazzo Bellini*), at Varallo, at Pavia (*Seminary*), and during the next twenty years displayed an enormous activity in architecture and engineering, in a style as emphatic as it was confused. In 1577, his fame was such that he was summoned to Spain, where he worked with Federigo Zuccaro and Luca Cambiaso in the Escorial (*Library*, *Great Cloister*) as the past master of virtuoso painting in pallid colour.

In Northern Italy, particularly in Lombardy, there was a parallel development of two pictorial trends, one of which tended to emphasize effects of chiaroscuro and realistic detail, the other stressing softness in modelling and form. With Perino del Vaga (1500–47), a Florentine and a brilliant collaborator of Raphael on the Loggie, the new motifs were introduced into Genoa, with heroic captains (Palazzo Doria), mythologies, and decorations of galleys. Another interesting figure from the preceding half-century, who had spread the Mannerist conception of colour as far as Genoa, was the Sienese Domenico Beccafumi (1481–1551). He was in Rome from 1519 to 1541, and in Genoa in 1537; from 1530 onwards he composed figures whose colour changed from the bottom upwards, creating a vision in iridescent colour shot through with reddish gleams, the best examples of which are in Siena.

In the next generation, rich—too rich even—in experience, Luca Cambiaso stands out (1527–85). He displays a wide range of culture acquired in Florence and Rome. Lombard influence working on the heightened perception of light derived from Correggio explains his daring nocturnal scenes, such as the Brera *Nativity*. He can also be large and simple, as in the *Palazzo Parodi* in Genoa, and in the *Presentation* (S. Lorenzo); sometimes he displays a forceful simplicity as in the *Entombment* (S. Maria di Carignano) and the *Charity* (Berlin). His drawings contain interesting experiments in 'geometrization'. He died in Spain, where he worked in the Escorial. Another artist, so greatly appreciated by the Hapsburgs that he became the court painter to Maximilian II and Rudolf II, was the Lombard Giuseppe Arcimboldo (1530–93), who conceived the original idea of transforming still-life objects into allegorical fantasies. Amusing accumulations of fruits and vegetables, fishes and shells, fruits and appropriate flowers, are composed into figures symbolizing *Earth* or *Water*, in the series of the Elements, or a Season (*Spring*, 1563, Vienna; *a jar of vegetables*, undated, Cremona). From this came an interest in freaks of nature coupled with a typically Mannerist conceit, which delighted Lomazzo.

Several Lombard cities had interesting groups of artists. Savoldo was important in Venice, where he eventually settled, but it was in Brescia that he left his nocturnal compositions and his almost monochrome landscapes (*Magdalen*, London; *S. Matthew*, New York). Romanino (1485–c. 1566) moved from a Giorgionesque manner (*S. Francis*, Brescia) to an art full of daylight and varied colour (*Life of Christ*, 1520, Cremona Cathedral), and then returned to painting pictures steeped in shadow (*Chapel of Corpus Domini*, and *S. Matthew*, in S. Giovanni Evangelista, Brescia). Moretto (1498–1554) used Venetian colour with the Lombard taste for strongly modelled forms (*Nativity, Christ and the Angel*, Brescia Museum). Like Romanino and Moretto—and to an even greater degree than them—Giovanni Battista Moroni (c. 1530–78) was a solid, unemotional and carefully naturalistic portrait painter, particularly gifted in his handling of greys (*Gentleman*, 1554, Ambrosiana; *A. Navagero*, 1565, Brera), and interested equally in sitters of quality and artisans (the *Tailor*, London). In another entirely provincial centre—Cremona—the Campi family of five painters, who owed much to Giulio Romano, are worth noting: Giulio (1502–72), whose vigorous painting adorns the dome of the church of S. Sigismondo, his brother Antonio (d. 1591), who was an architect and historian, and the third brother Vincenzo (d. 1591), who was mainly a portrait painter. The last two participated in the decoration of the Cathedral. In works such as the *Death of the Virgin* by Antonio (1577, S. Marco, Milan), and the *S. Matthew* by Vincenzo (1580, S. Francesco, Pavia), there is a firmness and decision which has suggested them as precursors of Caravaggio, who, born in 1563, developed in mounting rebellion against all current forms during the last period of the century.

Quite another orientation, which fell in with the piety of the Counter-Reformation, is that of the Umbrian Federigo Baroccio (1528/35–1612), who worked from Pesaro to Rome, from Milan (1592–1608) to Genoa. He blended a natural tenderness with the sensual undertones of Correggio and the complicated light effects of the Lombards. In the *Deposition* (1569, Perugia), he unwisely allowed himself to be inspired by Rosso, but in his *Madonna del Popolo* (1579, for the parish church of Arezzo, now Uffizi) and the *Rest on the Flight* (Vatican), he developed charming colour effects, with delicate harmonies and languid compositions. The *Circumcision* (Louvre), with its tender harmonies of grey-blue and pink, fascinated many painters during the course of the next centuries; the grace of the *Annunciation* (Vatican) can be felt almost as an anticipation of Greuze. Technically brilliant in his management

296

of soft blendings of colour, Baroccio was the first artist of Central Italy to use pastel in his studies, which can occasionally remind one of Degas (Berlin). Another artist in Bologna worth mentioning was Bartolommeo Passarotti (1529–92), a friend of Vignola and associated with Zuccaro in Rome. His *Purification* (Bologna, Pinac.) shows him to be a specialist in delicate transitions of colour but rather laboured tone. He founded an academy that, to a certain extent, served as the starting point for the Carracci academy, though theirs eventually ended by being in opposition to it.

The Carracci. Baroccio's career lay in the Emilia where the triumvirate of the Carracci reigned. They appeared in the second half of the century and their achievement has to be judged in the perspective of Mannerism, of which it represents—like the teaching of Vignola—the culmination, the clarification and eventually the extinction. Their action was through the Academy of the Incamminati ('those who sought the right way'), which, from 1585 to 1595, made their reputation in Bologna. It is a gross simplification to consider only the sonnet by Agostino in honour of Niccolo dell' Abbate (d. 1571), which was quite probably written much later by Malvasia to codify eclecticism.[139] It recommends to the future painter the power of Michelangelo, the naturalness of Titian, the purity of Correggio, the harmony of Raphael, the propriety of Tibaldi, the inventiveness of Primaticcio, and the grace of Parmigianino. The choice, even, of these examples underlines the polemical value of the poem; the author wished to oppose a reaction of taste to the ill-disciplined and inflated art of Vasari or Zuccaro, and sweet reason and good sense to the surfeit of the exuberant and the prodigious. On a more general level, the academic reform linked up with the teaching of the Council of Trent, which wanted to see a controlled and moderate art, leaning more to truth of expression than to beauty of form; in short, a more clearly didactic and literary art. At the same time—and this is significant—the Carracci were the first to practise caricature, of which there are some amusing examples in the Munich Print Room: the new seriousness is matched by a new humour.

The industrious Ludovico (1555–1619) was for long a waverer between Venice, Florence, Parma and Mantua; he organized the School which was to establish a good manner in painting, just as a good usage in language was also being established at that time. He worked above all in Bologna—the *Palazzo Fava*, the *Palazzo Sampieri*, the *Chapel of S. Domenico*, the *courtyard of S. Michele in Bosco*, etc.—and to these large, carefully worked out decorations correspond altarpieces in which he sought to express sweetness and breadth, such as the *Madonna of the Scalzi* (1588, Bologna, Pinac.) and the *Madonna of S. Hyacinth* (1594, Louvre).

His cousin, Agostino (1577–1602), the best educated of the three, was a celebrated engraver (more than 350 plates), and was in this way a diffuser of Venetian models. He left a few works, such as the *Communion of S. Jerome* (Bologna, Pinac.), accomplished but cold, and, above all, he collaborated in the great decorations of his younger brother Annibale (1560–1609), the overwhelming genius of the family. Annibale, a pupil of Tibaldi, was considered too much of a realist in his decorations on the theme of the Argonauts in the *Palazzo Fava* (1584), but he triumphed with those of the *Palazzo Magnani*, based on the story of Romulus and Remus, with massive framings of atlantides and garlands, and with those in the *Palazzo Sampieri* (1593–94). The Roman element, derived from Raphael, overshadowed more and more the Venetian and Correggesque elements. In Rome, in 1595, he received from Mgr. Agucchi the programme for the decorations in the Palazzo Farnese—the stories of Hercules in the *Camerino* (or small saloon), and mythologies drawn from Ovid in the *Gallery*. Here all the elements are distributed with the most consummate ability—medallions, nudes, panels, garlands, friezes, deriving from the Sistine Ceiling and the Palazzo Poggi in Bologna, and of which the development may be followed in numerous drawings in the Louvre. This wonderful work, of astounding vitality, swamped its narrow academic programme; its breadth, its essentially decorative quality, create something entirely new and complete, and the clarity of its blonde and pale colour—as far as possible removed from the exaggerations of the specialists in light effects—was a decisive example for the new age that was making ready.

Towards 1600, the two brothers separated; Annibale, who also left some lovely landscapes, like the *Christ and the Woman of Samaria* (Vienna), of a remarkable intensity, died in 1609. Thus declined the Mannerism of the Renaissance. Some of its practitioners were still to be found in the courts of the seventeenth century—in Florence, where there was a taste for the Lorrainer Callot, and even in Rome, where Giuseppe Cesari, the Cavaliere d'Arpino (1568–1640), is, in fact, a last survival. Summoned by Gregory XIII, towards 1585, he retained in his works in the Lateran, at his patron's Frascati villa, and at the Palazzo dei Conservatori, the loaded pomposity and the customs of the middle of the century. But the factitious eclecticism of a few workshops could no longer indefinitely prolong the imitation of the great masters; the reaction in favour of effects of light, encouraged by the Lombards, prepared the violent revolution of Caravaggio, and the temperament of Annibale Carracci opened a direct road to the Baroque.[140] The last generation of the century thus fulfilled its mission. What brought the Renaissance to a close was not only Vignola's *Regole delli cinque ordini* (1562), but also Scamozzi's *Idea dell' architettura universale* (1615); it was not only Vasari's great *Lives of the most excellent painters, architects and sculptors* . . . (1st ed. 1550, 2nd ed. 1568) but also Lomazzo's *Idea del Tempio della pittura* (1590); it was not only the Accademia del Disegno that Vasari and the Grand Duke founded in 1562, but also the Accademia degli Incamminati in Bologna in 1585; it was not the mythological ceiling by Jacopo Zucchi in the Palazzo Ruspoli (1568), but the Farnese Gallery (1597–98).

LATE FORMS: ROME, BOLOGNA, MILAN

FURNITURE

The most characteristic piece of furniture of the fifteenth century was the *cassone* or painted coffer, a linen chest and object of display, whose evolution is interesting. Originally, it was simply a rectangular chest of carved or gilded wood, ornamented on three faces, often with panels painted on stucco; during the sixteenth century, it adopted more and more an architectonic form, recalling sarcophagi, with feet, cornices and bands, and with the sides sometimes rounded. It was covered with escutcheons arranged as a decoration, sometimes with arabesques and garlands, sometimes with scenes with figures, the first above all in Tuscany, the latter in Venice and Rome (Berlin; Bardini coll., Florence). To the delicate work of the wood sculptor is often joined marquetry work, which, above all in the first half of the century, inserted abstract motifs, or even architectural perspectives, on the front of the chest (Contini coll., Florence). From 1550 onwards, the woodwork displays greater opulence, with engraved or turned motifs, and a repertory astonishingly replete with Mannerist ornaments of figures and abstract forms. The lighter and more elegant forms can be found only in the large wardrobes, or clothes presses in two storeys, framed with pilasters, or sometimes simply composed of moulded panels to show off the surface of the wood itself—walnut or chestnut.

In some sidetables a Doric order was adapted, with sober panels and with series of triglyphs crowning them (Palazzo Davanzati, Florence).

Three-legged tables with a hexagonal top supported by volutes and lions' claws are frequent in Tuscany in the middle of the century. Tables with six very strongly moulded legs linked together by a circular framework are to be found in Bologna towards 1600. The commonest type has two side-members, joined together with a stretcher which on occasion can be made to support balusters. This permits of an arrangement of fairly simple mouldings, but more often than not, and above all in Tuscany and Emilia, it is loaded with volutes, foliage and arabesques, which multiply the decorative profiles (example in the Palazzo Davanzati). Sometimes, a more sober taste may be seen in the simple band decoration of a heavy console (pvt. coll., Milan). For chairs, Northern Italy seems to have favoured the little seat with two panel-supports, covered with dolphin-masks (Bardini coll., Florence), the X-type, with decorations of inlays, also being popular. The four-legged type with a straight back appeared, first with very ornamented stretchers (Figdor coll., Vienna), sometimes with large cartouches (Mus. Naz., Florence), then towards 1600 with leather, velvet and bands covering the seat or the back. The second great period of Italian furniture was to be the eighteenth century.

Part Three

THE BAROQUE

Historical Background: The Church: the 'Market of Kingdoms'

In Italy, the crisis of Mannerism comes in between the Renaissance and the Baroque in the same way that the crisis of the thirteenth century came between Romanesque and Gothic. The analogy is made valid by the pressure of two opposing cultural aspirations. The first, a purely secular one, arose from the growth of court life and princely luxury, from the concentration upon Rome, which (as in the days of Frederick II) bulked so large during the period of the great families of princes and cardinals—Medici, Gonzaga and Farnese; the other, born of the religious crisis, was a serious and spiritual aspiration towards moral reform and strictness of faith, which gave to the noblest aspects of the Counter-Reformation something in common with the rise of the Mendicant Orders. The Jesuits proposed to reform completely and to control all the culture which emanated from Humanism. But this trend in favour of austerity coincided with the last phase of Mannerism, when, for example, the column was banished from façades and replaced by pilasters so as to avoid ostentation. This attitude changed towards 1610–20; by 1620–30, great new works had established a new art everywhere, one full of strength, variety and splendour, which, like Trecento Gothic, involved a collaboration between all the arts and restored to the first rank of importance the town planning projects which had always been of basic concern in Italy. This was the 'Baroque', so called, like 'Gothic' during the fifteenth century, as a term of abuse by opponents who condemned its forms as irregular and absurd.

For two centuries Italy had been outside the current of European affairs. From 1600 onwards, her political configuration showed the results of the iron grip of the Spanish Hapsburgs, who controlled the country from Milan in the north to the twin kingdoms of Naples and sicily in the South, and dominated the territories in between—the duchies of Parma, Mantua, Tuscany, and the Papal States—either by matrimonial alliances or by policing operations made necessary by the brigandage and the poverty of the depopulated countryside. The only states which escaped this bondage were the Duchy of Savoy (which stretched as far as Annecy and Nice), and the two maritime republics of Genoa (with Corsica) and Venice (with her possessions in Istria, Dalmatia and the Ionian Islands), both then completely decadent. This painful feeling of frustration went on increasing until the wars of the French Revolution. In 1713, Milan, Naples and Sardinia passed to the Austrian crown. Consolation-prize principalities, which ensured

Duchy of Tuscany

Italy's domination by foreigners, were established as a feature of European negotiations; and this unhappy land was described by A. Scorel as the great 'market of kingdoms'. She thus remained cut off from the general development of the Europe of the Enlightenment, despite the noble work of certain historians, philosophers and jurists such as Vico in Naples (*Scienza nuova*, 1730) and Beccaria in Milan (*Delli delitti e delle pene*, 1764), who protested against her shameful passivity.

Despite the efforts made by enlightened princes, such as Leopold, the brother of the Emperor Joseph II, in Tuscany, and Charles VII in Naples, to abate the exorbitant privileges of the clergy, there was to be no real emancipation in Italy before the advent of Napoleon.

The power with which all others had to contend was that of the Roman Church: at the end of the sixteenth century, she had everywhere reconquered her initiative, above all in Bavaria and Bohemia. In Europe, with the Thirty Years War (1618–48), she triumphed through the retreat of Protestant Germany, and in the rest of the world—the Far East and America—through her powerful missions. The French monarchy, which struggled against Jansenism (1660–68) and drove out the Protestants (1679–85), furthered the greatness of the Holy See up to the moment of the struggle over the *régale* (the claim of the French Crown to the temporalities of a vacant bishopric) which came to a head in 1673, and the Gallican crisis, which established the limits of pontifical power. During the succeeding century, the power of the Papacy declined rapidly in Europe; the evolution of new customs and ideas was such that after a variety of campaigns, scandals and manifestations hostile to the Jesuits all over Europe, Pope Clement XIV decreed the dissolution of the Society in 1773.

Apart from the Capuchins, founded in 1525, and separated from the Franciscans in 1619, there were fewer new Orders and fewer new saints in Italy than in France or Spain. Up to the middle of the seventeenth century, the Jesuits were all-powerful. A militant, missionary and teaching order, they numbered a thousand at the death of S. Ignatius Loyola in 1556, 10,000 in 1608, and nearly 18,000 by 1679. But it is an over-simplification to make the Jesuits responsible for Baroque art. They played their part in it, and in general, encouraged it in their anxiety to make the Church as 'modern' as possible, by following the latest movements of taste.[141]

The dictates of austerity, the mistrust of artists, and the codifications in art slackened after about 1600. With Cardinal Bellarmin appeared a spirit of greater confidence; with Paul V (1605–21) the lugubrious Vatican of the time of Pius V in 1570 was far behind; Urban VIII (1623–44), Innocent X (1644–1655), and, even more, Alexander VII (1655–67), a member of the Chigi family, were pontiffs of great wealth and luxury. Since Pius V, papal tombs had been ornamented only with simple reliefs; the new pontiffs required tombs of great size and magnificence. This militant optimism which brought in its train the exaltation of saints, the cult of martyrs, the desire to teach dogma through art, led to the spectacular style which was to dominate the age, and which, it is now obvious, was far from being as devoid of religious thought as was believed in the nineteenth century.[142] But it is also certain that through its admiration of its own marvellous facility, luxury and illusionism, the Baroque allowed itself to be carried away by a sort of triumphant abandon which often prevented it from discerning the mediocre from the grandiose, and as a result it placed religious art in an equivocal position from which it did not escape unscathed.[143] Medieval naïvety and simplicity were lost; they survived only in provincial works, in popular folk tradition, of which the most typical is the Neapolitan Christmas crib. The 'official' style, common to palace and church, endeavoured to be learned and aristocratic, as was the art of the Renaissance, but without the latter's purity and intellectual clarity. It sought to create an emotional state of mind where, as in a dream, all would become possible: heroism—whence the revolting scenes of martyrdom which Pomarancio painted in S. Stefano Rotondo—emotion before the supernatural aspects of death and transfiguration, and voluptuous ecstasy. According to an idea common to Calderon and Bossuet, and widespread in the poetry of the period, human life, taken in its totality, must be conceived as an illusory adventure, a dream. The art of the theatre is the one most fitted to exploit illusionism, but Italy knew nothing of classical comedy and tragedy, preferring the more ambiguous and more sensual form of opera, which became the forcing house for many Baroque inventions. Its immediate result is to be seen in the design of church naves and princely reception rooms, which, with their balconies and boxes, were treated like façades, and in the scenographic construction of squares and towns, in which Italian genius surpassed itself.

These achievements encouraged the development of an immense culture, of which only a bare idea may be obtained from the learned treatises of Jesuits like Padre Kircher (1601–80) on archaeology, Padre Pozzo (1642–1709) on architecture, or works on symbolism which have recently been studied anew (C. Giarda, 1626).[144] Papal Rome once more resembled the Imperial Rome of the third and fourth centuries, by its taste for the exotic, which explains, for instance, the multiplication of obelisks in Rome and in many other towns—the elephant carrying an obelisk in the Piazza della Minerva in Rome (1677), repeated at Catania, is an amusing example of this.

The culture of the Baroque era is marked, finally, by the new spread of the natural sciences, the vogue for anatomical plates which, from the end of the sixteenth century, assured to all educated men a knowledge of the human body and its singularities. The engravings in such works as the *Epitome* of Vesalius (1543), or the *De corporis humani structura* of F. Platter (1583) or the plates of Julius Casserius (1627), are of an amazing vividness. In them the skeleton strikes fantastic poses and approaches the macabre imagination which inspired the surprising decoration of skulls and shin-bones arranged in garlands and rosettes in the chapel of the Capuchins in S. Maria della

HISTORICAL BACKGROUND

Concezione (1624) and later still, Ercole Lelli, the author of the anatomical models of the University of Bologna, in the two caryatides of flayed figures on the professorial throne of the anatomy theatre (c. 1750).

The taste for caricature is a final revealing characteristic. No longer is it merely irony, but buffoonery rather, and even a tendency to the monstrous, that fits in with the vitality of the age. Annibale and Ludovico Carracci had made fashionable the type of caricature portrait centred on the exaggeration of a single feature, and reconstructed a whole personality round a nose, a hat, or a stomach. Bernini was to become famous for these comic figures, which spared neither Cardinal Scipione Borghese nor the Pope, and Leone Ghezzi (1674–1755) was later to be celebrated for similar caricatures of the Roman clergy. As to the taste for deformity and the bizarre, it was fully expressed in decoration, in painted *capricci*, and even in the subtleties of architecture, with a liberty and a relish which were not to be found anywhere else.

These varied aspects of the period demand a new classification in works of art. Félibien, in his *Entretiens* (1685) only recognized two great schools: Venice and Lombardy on the one hand, and Florence and Rome on the other—North and Central Italy. There was room for a third, for Naples and the South, where the creative spark revived. But there are two other categories which must be distinguished in the art of each of these centres, for it was during the seventeenth century that the full results of the distinction between the various genres of paintings appeared. On the one hand was the grand Baroque design, which comprised the whole field of architecture and decoration, and on the other more modest yet not less intensive plane, the work of painters who are segregated within their particular genre. These are two fairly distinct planes which justify treatment under separate headings. Venice, however, must be considered apart: after a long period of inaction, she enjoyed during the eighteenth century a remarkable resurgence which left her, alone of all the Italian centres, in contact with European art as a whole.

I

Architecture and Decoration

Rome, The North and the South

The Baroque Style, which was current from about 1620 to about 1750, is most clearly defined in its architecture. From the starting point of a properly constituted vocabulary of Orders, profiles, and elements borrowed from the Antique, systematically codified in the treatises of Alberti (c. 1450) Vignola (1562) and Galli Bibiena (1711), Italian art achieved the complete swing of taste which gave to Baroque its full significance, enabling a distinction to be made between it and the academic formalism that preceded it and the Rococo elaboration by which it was succeeded.

During the first third of the sixteenth century monumental works were made to be seen from a frontal viewpoint, combined in clearly defined, harmonious relationships, and considered as volumes to be given a plastic articulation. The period known as Mannerism learned to contradict these effects while appearing to use the same formal language; turbulent and bizarre during the first half of the century, it was of a mannered frigidity during the academic reaction that followed.

About 1630, in Rome—which had recovered her capital position—three artists created works which, though far removed from the classical ideal, were yet the expression of a complete understanding of monumentality, and entirely opposed to oversubtle and constrained effects. In their larger members, the forms are dilated and multiplied, interlaced rather than set one within the other; 'undulations sweep through the body of the forms' (Wölfflin), and in them breadth of feeling, a quivering, nervous vitality, restless and unstable, finds expression. The physical space of the architecture is impinged upon by imaginary space, within the building as well as on the outside.

Bernini, who dominated the scene for a half-century, was the most vigorous exponent of this new ideal. With him, the 'symphonic' integration of all the arts is no mere catchphrase. Under his guidance, an amazing style of decorative painting was evolved for the great religious and secular buildings whose interior space it appeared to increase; sculpture in niches or at the end of perspective vistas stressed the focal point of the programme and, through the use of stucco, an ornament composed of full, rounded and coloured forms

ROME, THE NORTH AND THE SOUTH

lent itself to endless developments. Like all strong styles, the Baroque re-awakened the dormant energy of provincial centres; the end of the century witnessed a hitherto unknown development in Piedmont and Sicily, while other regions, such as Tuscany, showed the greatest reserve.

Italian Baroque declined at the beginning of the following century. In larger undertakings, it came up against competition from French art, which, in the guise of 'Grand Manner', had begun its career of conquest in Europe, and the springs of its invention began to dry up. It became stale, tended to dwindle into prettinesses derived from the art of Borromini, leading to Rococo, and finally encountered the opposition of the classicizing taste, entrenched in Tuscany and Venice, which was soon to bring about the philosophical, academic and archaeological reaction of Neoclassicism. But the fairly general decline during the eighteenth century never interrupted those works in which Italian genius is probably best expressed: the art of urban development, of the theatre and of garden layout—that is, the management of large empty spaces, of imaginary space and of natural space.

TOWN PLANNING AND THE THEATRE

Sixtus V's great work had many continuators. The majority of the great Roman undertakings were completed one after another, provided with façades where these were required, and included in an ordered space. First St. Peter's: Giacomo della Porta had built the dome by 1593; the nave and the façade were still those of Old St. Peter's, as may be seen from the sixteenth-century views. Paul V had them pulled down, and in 1607 held a competition (among the competitors were Domenico Fontana, Carlo Maderna and G. A. Dosio) which was won by Maderna. His project consisted of an extension of the nave by two bays, which would transform the central plan into a basilical plan more in sympathy with the programme of the Counter-Reformation, and in the provision in front of an enormous portico stretching beyond the nave and the side chapels on either side so as to extend the façade in width, give strong horizontal support to the dome, and carry an attic storey continuing that with which Michelangelo had crowned the apse. To compensate for this spreading effect, the Pope, in 1612, planned two campaniles: the works were carried on until 1621 and then again taken up without being concluded. Eight Corinthian columns and four pilasters articulate in a giant order this powerful but colourless ensemble. Something more was needed.

It was under Alexander VII Chigi that it was decided to expand this façade into a vast open space enclosed by a colonnade which should, according to a contemporary document, demonstrate by its very form that the Church enfolds all the faithful in her maternal embrace. Bernini built these two arms in the form of a double colonnade with four rows of Doric columns surmounted by 162 statues arranged in procession towards the basilica (1656–65).

The problem was to combine in one huge scheme the vista of the dome, proper consideration of the façade, and the siting of the oval space around the obelisk of Sixtus V. Bernini seems to have hesitated over closing the area with a third arm which would have echoed the façade, and which, in the end, was not executed. The two axes of the forecourt are of 240 and 340 metres; in the centre are two fountains and the obelisk of Heliopolis from Caligula's Circus, which was at the side of the basilica before Fontana erected it in front of St. Peter's in 1593. This, after a century and a half, was the Baroque completion of the masterpiece dreamed of by Bramante and Michelangelo.

In the same way, Bernini created lively scenic effects by placing fountains with sculpture in the vicinity of the principal palaces with which he was concerned. Thus, the famous *Triton* is close to the Palazzo Barberini (1640), and for Innocent X he made the astonishing *Fountain of the Rivers* in the Piazza Navona (1647–52), converting it into a kind of 'cour d'honneur in front of his palace' (Golzio). The base of the obelisk is a heap of theatrical-looking rocks inhabited by gigantic river gods in complicated poses. Never was more perfect expression given to the idea of filling a vast open space with an irregular mass, angular in shape and of continuously changing profile, and through the marriage of its strange forms with the jets of water tumbling into the basins, introducing an unexpected quality of movement into the decoration, which, moreover, echoes the design of an antique circus.

Bernini's projects achieved, in his last years, an exceptional visionary quality. He dreamed of moving Trajan's column next to that of Marcus Aurelius, and of flooding the Piazza Colonna so as to hold fêtes on it in front of the Chigi palace (recorded in Chantelou's Diary). His pupil Carlo Fontana also elaborated grandiose schemes of embellishment. One of them consisted of the demolition of the Borgo, between the Castel Sant'Angelo and the piazza in front of St. Peter's, so as to construct two huge colonnades (recorded in his book on the Baptistery of St. Peter's, 1697); the other, even stranger, was to erect a domed church with two campaniles inside the Colosseum (recorded in *Anfiteatro Flavio*).

The immense confidence of the Baroque artists explains the audacity with which they undertook hazardous remodellings, such as that by Borromini in the Lateran, where he joined the columns of the nave together in pairs so as to form huge piers which he then hollowed out with monumental niches, or the addition of Giuseppi Chiari's ceiling paintings (*c.* 1710) to the ancient basilica of S. Clemente. In many cases, they constructed façades which decorated the exterior but were in no way related to the character of the building itself. In the creation of splendid scenic effects, they took immense pains to direct the eye along a fine vista, as for instance in the Piazza del Popolo, at one of the main entrances to the city, where the junction of the Corso with the Via del Babuino and the Via Ripetta is set off by two apparently identical churches by Carlo Rainaldi—S. Maria di Monte Santo and S. Maria dei Miracoli (1662), both with regular pediments facing the square, and with octagonal domes and symmetrical bell-towers.

Further efforts in this direction were made during the second quarter of the eighteenth century. Galilei built the grandiose façade of the Lateran basilica (1735), and Fuga that of S. Maria Maggiore (1741–43). In 1732, Clement XII held a competition for the Fontana di Trevi, in which the French sculptor Bouchardon took part, and the completion of the project on a design by Nicola Salvi, which involved the erection of a triumphal arch and a palace façade behind the fountain, dragged on until 1765. The design of the Piazza S. Ignazio was even more significant. The church, begun in 1625, had a rather uninspired façade attributed to Algardi; the houses, designed as if they were the backcloth and wings of a stage set, were built on the side of the square facing the church by Filippo Raguzzini in 1725. In the same spirit, F. de Sanctis, working on a project by Specchi, designed the superb Spanish Steps (1728) which rise in harmoniously contrasted flights towards Giacomo della Porta's façade of S. Trinità dei Monti.

Although the example was set by Rome, many other towns acquired a more carefully planned architectural appearance. In the north this was particularly true of Turin which was developed according to the perpendicular axes of an antique military camp, and in the south, of Naples, where Domenico Fontana began the boulevards along the seafront, and above all of Sicily, where after the disastrous earthquakes of 1684 and 1693, several cities were rebuilt within a few years, according to designs remarkable both for their architectonic strictness, and their attention to scenographic effects. Catania thus became one of the great examples of Baroque rationalism, expressed in symmetry and chequerboard town plans, but the lesser known town of Noto, rebuilt on one of the slopes of the Monti Iblei on a fishbone plan which creates sharp perspective vistas in the upper part, is one of the masterpieces of illusionistic and theatrical town planning of the eighteenth century in Europe.

Garden Design. In garden layout, as in town planning, it was Domenico Fontana who led the way, in the Villa Montalto (now destroyed), built for Sixtus V on the Via Viminale, by creating extensive views closed by small buildings with waterfalls and statues. Giacomo della Porta had already done similar things in the Villa Aldobrandini at Frascati (1604), where the water tumbles cascade by cascade to the main basin in front of the villa—the first example of a new type of design in water gardens. A Dutchman from Utrecht, Vasanzio (d. 1621), specialized in pleasure houses (such as the Casino Borghese) and in water-gardens with huge and grandiose nymphaea, as in the Villa Mondragone, Frascati.

Pietro da Cortona's first work was the Villa del Pigneto (c. 1630; now destroyed) with a huge niche, a grotto, a fountain with tritons, all surrounded by flights of steps. Elements which were properly part of the garden were thus co-ordinated as part of a façade. The rather scattered effects of fantasy common in Mannerist gardens were conceived on a larger scale and were more carefully planned. Nothing is more expressive of Baroque taste than this kind of architecture of woods and leafy bowers, this ever-changing picture created by stretches of placid water, tamed into canals and shimmering basins, tumbling in cascades or gushing, foaming, from innumerable jets, these sculptures decorating a nymphaeum, these rusticated gateways, and marble statues standing at the end of long vistas. The living yet vividly theatrical garden is its most perfect achievement.

The villas of the sixteenth century were enlarged and completed in this way. At Bagnaia, Cardinal Riario's villa, finished on Vignola's designs in 1578, there are small cascades tumbling down to the Fountain of the Moors; the cascades of the Villa Torlonia at Frascati were designed by Maderna; and a little later, Algardi designed on the slopes of the Janiculum, the Villa Doria Pamphili, where the rustic, grotto-like nymphaea and the water garden were treated with much greater freedom. All over Italy, the types created during the sixteenth century were being enlarged and improved; in Genoa, on the example of Alessi and Montorsoli; in Lombardy, the Villa d'Este at Cernobbio, built in 1568 by Pellegrino Tibaldi, was transformed; at Isola Bella, the Borromeo family created during the middle of the seventeenth century an entirely novel garden composed of tiers of terraces, fantastic grottoes and a display of superb botanical beauty.

French fashions dominated during the first half of the eighteenth century. The gardens at Racconigi in Piedmont were laid out by Le Nôtre, and the park at Stupinigi was by Bernard. The Bourbons modelled themselves on their cousins at Versailles; at Colorno, near Parma, the Trianon was copied; in Naples, the villas at Portici and Capodimonte have their gardens arranged in bosquets and formal flowerbeds. But the supreme example of Baroque magnificence is the enormous park at Caserta, the Versailles of the Neapolitan Bourbons, where in the middle of the eighteenth century Vanvitelli, assisted by an army of engineers and sculptors, created the largest design for a palace and its setting of the whole period. An aqueduct (Carolino) brings the water to a hill in front of the immense palace; from there it tumbles in a huge cascade, at the foot of which, like tableaux vivants, statues of *Diana and her nymphs* and *Actaeon and his dogs* are placed among the rocks; further off is *Aeolus and the winds*, and *Venus and Adonis* appear among other settings of fountains, rockeries, and trees (by P. Solari and A. Brunelli). This was the last example of its kind, for villa gardens were already affected by the English Romantic type of garden (as at the Villa Borghese), and by Neoclassic taste (as in the Villa Carlotta at Tremezzo (1747) and the Villa Albani by Marchionni in Rome).

The Theatre. Theatres, which were so successful at the end of the sixteenth century, sprang up all over Italy; at the Court of Turin (*c.* 1618), in Milan, Ferrara, and Piacenza, where it is inside the Palazzo Communale (1644–46); in 1632 a theatre for three thousand spectators was inaugurated in the Palazzo

Barberini. But the focal point of experiment was the stage itself, upon which was concentrated an ever-growing volume of ideas for increasing perspective effects, of research into stage design, lighting and accessories, which definitely established illusionism.[145]

FIG. 36. *'Fantasy' altar by P. Pozzo. Elevation and plan (restored): on four pilasters arranged in an X design four angels support over the void a large crown over the Virgin.*

A very fine wash drawing by Bernini (Berlin) representing the setting of the sun, which looks like a romantic watercolour, is a design for the curtain of *Marina*, a spectacular opera staged in Rome in 1628, and played at sundown. It was in stage design that the sculptor achieved his greatest successes. But stage technique was already very highly developed. The *Pratica di*

313

fabbricare scene e macchine per teatri (Pesaro, 1637) by Niccolo Sabbatini, had worked out the principles of moveable stage scenery.[146] At the end of the century, the treatise by Padre Pozzo, *Prospettiva di pittori e architetti* (2 vols., Rome, 1693) supplied all the recipes for creating fantastic, illusionistic spaces, while retaining the principle of the central vanishing point, and he thus set out how, for great religious ceremonies, the altar can be made the focal point of a theatrical system of flats and fake perspectives (Fig. 36).

Italy owed her European pre-eminence in stage design to the Bolognese family of the Bibiena. Ferdinando Galli (1657–1743) was the head of this artistic dynasty, all of whose members specialized in theatrical scenery, the most fantastic flights of perspective, and architectural drawing. Himself a brilliant and learned designer, he used in his sets contrasts of light and shadow, of one space melting into another, and of diagonal vistas, while arranging his forms with an energy and freedom which echoed the art of the fugue. His brother, Francesco (1659–1739), built the Teatro Filarmonico in Verona, but the greatest genius among these Baroque scene designers was Giuseppe (1696–1756), an engraver of astonishing inventiveness, who freely mixed the arts of garden design, statuary and architecture in compositions of bewilderingly rich detail. Giovanni Maria (1739–69) returned to stricter architectural compositions, piling up arcades and colonnades in exciting effects of foreshortening, and the last, Carlo, made greater use of ruins.

The Bibiena family were virtuoso performers in monumental decoration. The Galliari family, who came from Andorno, and whose most interesting member is Fabrizio (1709–90), displayed a taste which leans more towards the picturesque and the exotic; his brother Bernardino, as well as designing for the stage and for fêtes, was a decorator of palaces who worked in Lombardy and, even more, in Piedmont. The Bolognese School also included Stefano Orlandi (1681–1760), whose scenes of palaces set amid parklands are full of feeling for atmosphere, and Giovanni Nicolo Servandoni (1695–1766) who, by his stronger feeling for dramatic effect, prepared the way for Piranesi. The inventions of these men, however, survived the Baroque and the end of the century still saw rustic scene settings with idyllic landscapes by P. Gonzaga (1751–1831) and Neoclassical ones by Antonio Basoli (1774–1848).

Rome: Architecture and Decoration

Lombard architects continued to dominate the scene during the early years of the century. After della Porta and Domenico Fontana, the most active among them was the Ticinese Carlo Maderna (1556–1629). His decisive part in the completion of St. Peter's has already been discussed. He tended to exaggerate somewhat his effects of plasticity, to model more vigorously those forms which the Mannerists had tended to flatten, and to seek for striking contrasts of light and shadow. He is among the earliest to use the language of

Baroque; it is only necessary to compare the two-storey façade of S. Susanna (1603) where the column is once more featured, with the façade of the Gesù, and that of S. Andrea della Valle (1608)—with its five bays at ground level, three at the first storey, and a very clear projection when seen from the side (remembering always that it was actually built in 1665 by Carlo Rainaldi on Maderna's designs)—with S. Luigi dei Francesi, to grasp the changes. But Maderna is devoid of imaginative power; he sticks to Vignola's types, which have academic sanction. The plan of S. Andrea is only an amplification of that of the Gesù, just as that of S. Maria della Vittoria (1608–20) is a reduction of it. Maderna worked at the Palazzo Marrei (1598–1618) which was decorated by Lanfranco and Albani among others, at the Palazzo Chigi on the Piazza Colonna, at the Palazzo Barberini, the design of which, with two projecting wings, was radically altered by Bernini when he took over in 1629. Nevertheless, the garden façade seems to be his work.

The Three Masters. It was not until the second quarter of the century that three strong personalities created Baroque architecture: a Florentine, a Neapolitan and a Lombard—Pietro da Cortona, Bernini and Borromini—who were respectively a painter, a sculptor and an architect. Pietro da Cortona (1596–1669), with the church of SS. Luca e Martina (1635) in the Forum in front of the arch of Septimius Severus, created the first example of a façade whose central portion swells forward as if it were compressed by the coupled pilasters at the sides; a second storey repeats the arrangement of the lower one under a cornice surmounted by flaming urns. The interior is as strongly independent: a return to a Greek cross plan, more plastic in form, white walls in which niches between heavy columns create deep shadows. Twenty years later, he built the façade of S. Maria della Pace (1656), for which a five-sided piazza was originally planned. This façade consists of a projecting semi-circular porch like an atrium, which seems as if thrust forward by the concave wings on either side; the upper storey shaped entirely by the elements of the orders and their projection and recession, has little other ornament. The same outward sobriety appears in S. Maria in Via Lata (1658–62) with its flat façade composed of high porticoes one over the other. Between these two works, Pietro da Cortona worked with great success as a painter on the ceilings of palaces and churches. Among the artists linked with him, and who occasionally were architects as well, Antonio Gherardi (1644–1702) built the curious Avila chapel in S. Maria in Trastevere (1680), where angels support the lantern of the dome with an illusionistic effect of strange originality in the sculpture.

'Thanks to Bernini, we have seen the three great arts in full possession of their ancient dignity' wrote Baldinucci in 1681, and he could find only Michelangelo to compare him to. In fact, Gian Lorenzo Bernini (1598–1680), surrounded by friends and poets like the Cavaliere Marin, with an international reputation such that he was invited to Paris in 1664 by Louis XIV, through his many talents and his role of scenic designer to a whole age, was an artistic personality on the scale of Rubens. But he was essentially a sculptor, and from the age of twenty onwards his statues brought him fame and papal patronage, for, with the exception of Innocent X, who preferred Algardi and Borromini, he received constant commissions from the popes. First it was for works of sculptural decoration, as in the *Baldacchino* in St. Peter's (1624), a real manifesto of the new style through the use of the twisted columns, the movement given to the entablature and the play of colour in the bronze, carried a stage further in the apse by the enormous Cathedra Petri (1657) of bronze heightened with gold and raised high upon clouds between the figures of the four Doctors of the Church.[147] In the building of the Palazzo Barberini, which Urban VIII entrusted to him in 1629, the central portion, which in many ways resembles a recasting of the Farnese

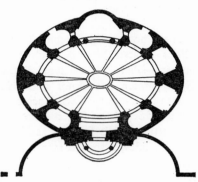

FIG. 37. *S. Andrea al Quirinale, Rome, by Bernini. Plan: Oval design, with oval and square chapels hollowed out of the wallspace. Oval dome.*

Palace, though with a more plastic and unquiet effect obtained by the open arcades of the entrance on the ground floor, is attributed to him. Bernini also provided designs for the Palazzo Montecitorio (*c.* 1650) which is not unlike the Palazzo Massimo, but of his plan only the rusticated pilasters at the sides remain.

During the pontificate of Alexander VII (1655–67) Bernini turned to architecture: the colonnades of St. Peter's and the Scala Regia, with its double file of Ionic columns, are his chief feats in perspective. He then showed his full ability in three centrally planned churches—the small church at Castelgandolfo (1658–62), a Greek cross with a dome, which is not unlike S. Eligio degli Orefici in Rome; the collegiate church at Ariccia (1662–64), a rotunda approached through a portico like an atrium, with colonnades on either side to extend the structure laterally so as to create a monumental effect; and S. Andrea al Quirinale (1658). In this last work—Bernini's favourite—a semicircular portico is attached to a façade with a pediment borne on pilasters; the oval plan is on the short axis, and inside the impression of smallness is counteracted by the fluted pilasters which bear the strongly modelled entablature, by the low dome, and the decoration of coloured marble. This is a brilliant exposition of the grandiose qualities of the Baroque, which uses all the resources of the Antique (i.e. the Pantheon) and the Renaissance to create a convincing and expressive whole. Bernini's projects for the Louvre, which were thwarted by the French architects, and those for the apse of S. Maria Maggiore (1669; not executed) were of a similar kind.

Bernini had many followers. Carlo Rainaldi (1611–91) built the two churches on the Piazza del Popolo, S. Maria in Campitelli (1657) with a shadowy nave increasing the dramatic effect of the brightly lit sanctuary, and a façade in travertine with massive projecting columns, and many other works, such as the apse of S. Maria Maggiore which is little more than a reduction, already academic in style, of a vast project by Bernini. Martino Longhi the Younger (1602–56) built the high altar of S. Carlo ai Catinari and the façade of SS. Vincenzo ed Anastasio (1650), with its six Corinthian columns close grouped round the entrance and repeated in the upper storey under a triple pediment. One of the most important architects of the day was Carlo Fontana (1638–1714) who came to Rome about 1650. He worked for the Chigi family, on a garden in the Quattro Fontane district, built S. Marcello al Corso (1683) with its concave façade, the domed Cappella Cibò in S. Maria del Popolo with its revetment of coloured marbles, erected the rather unimaginative portico in front of S. Cecilia in Trastevere (1702), and, after the death of his son Francesco, finished SS. Apostoli (1702–14) with a nave and two aisles, with profiles of an almost Rococo tension.

Algardi, Bernini's chief rival in sculpture, is generally credited with an important part of the Villa Doria Pamphili, on the Janiculum (1644), not only in its exterior decoration of rich stuccoes and reliefs, but also for the plan inspired by Palladio. He is also mentioned in connection with the church of S. Ignazio, with nave, two aisles and transept, but its interior is chiefly the work of the Jesuit, Orazio Grassi (1583–1654).

The only pure architect among the Baroque masters was Borromini (1619–1667), who came from Bissone near Lugano, and began by working in a humble capacity under Maderna. In his origins, his violent and hypochondriac character, and in the entirely singleminded forcefulness of his art, he may be likened to Caravaggio.

Some of the sculpture in St. Peter's, a few details of the Barberini palace, such as windows of rather Mannerist inspiration, are attributed to him. His first known work is the little church of S. Carlo alle Quattro Fontane (1638–1641) of which the façade (1667) was the last thing he built. The interior entablature, carried on slender Corinthian columns, follows the undulations of the elliptical plan, to create an impression at once strange and light, elegant and ambiguous, redolent of his whole art. At the monastery of S. Philip Neri (1637–43) where he built the façade, the oratory and the clock tower, all sorts of strange features (the concave plan, the entablature broken into by the

317

pediments of the window) contribute to the feeling of animation and strain. In contrast to Bernini's art of expanding forms, a kind of contraction now seems to be affecting architecture, achieved through the use of the elliptical plan, which instead of opening out the space cuts it up, by creating the odd impression that two independently centred volumes are interpenetrating, and by the substitution in the façade of a play of forces which contradict each other for the full articulation of pure Baroque.

The lovely S. Ivo della Sapienza (1642–50) represents perhaps the culminating point of this inspired research. The arrangement of the order and the arcades on either side of the courtyard are continued across the curved end which forms the façade of the church; above appear the convex forms of the upper storey, delimited by delicate cornices between which are tall windows. This upper storey, which appears to be a drum, in fact encloses the dome itself of which only the uppermost part appears above the upper cornice. Curved ribs stretch across the roof to support the tall and elegant lantern which is crowned by a strange spiral form. The lantern, whose proper function is to serve as a weight to stabilize the outward thrust of the dome, here seems to be a substitute for it. The richness of his ideas is expressed in the unfinished church of S. Andrea delle Fratte (1654–65), where the topmost portion of the campanile is treated like a precious jewel. The effect is one of density and tension, but in contrast to Bernini's art where all the elements are linked together easily, this is an architecture in which the forms are hardened and detached by outlining them with thin fillets of shadow and arabesques of light.

The same applies to the Collegio di Propaganda Fide, to the church of Santa Maria dei Sette Dolori (1662–66) where he even set pilasters diagonally, and to the ensemble of S. Agnese (1653–61) in Piazza Navona—finished after his death—where he used curved forms and silhouetted airy shapes against the sky. This is an art which never lends itself to the generalities of town planning, as does the true Baroque of Pietro da Cortona and Bernini; the building becomes an object in itself and burgeons in strange forms; the façade of S. Carlino, schematic yet cut up, makes this tendency clear. Milizia expressed all the disapproval of the Neoclassicists by calling it 'the knick-nackery of a fanciful furniture maker'. The strange Cappella Spada in S. Girolamo della Carità in the Via Monserrato proves that this virtuosity even went so far as an almost droll fantasy. But Borromini's exquisite sense of rhythm also enabled him to create a small work like S. Giovanni in Oleo (1658), a rotunda with a drum ornamented with a frieze with palmette motifs. In S. Giovanni Laterano, which Innocent X commissioned him to restore (1647–50), he retained the plan of the original building, but clothed it with a decoration of fluted pilasters and aediculae of great rhythmical clarity. One of his rare secular works was the enlargement of the Palazzo Falconieri in the Via Giulia, where he built a third storey consisting of a loggia with a triple Palladian motif.

FIG. 38. *S. Ivo della Sapienza, Rome, by Borromini. Section: F, Windows in the dome; T, Tribunes; L, Lantern with crowning spiral element.*

319

ARCHITECTURE AND DECORATION

Through Borromini's tense style and wonderful technical virtuosity, architecture arrived at a point of subtle disequilibrium. It becomes the expression of a dramatic tension whose often delightful solution makes a strongly recondite appeal. But when this tension becomes mere playfulness, then it also becomes the Rococo, which derived many of its ideas from Borromini. He also had a disciple worthy of himself in Guarini, who worked in Turin.

Painted Ceilings. The style of the High Renaissance had integrated painting and architecture by creating a frame for painting, or by suggesting a feigned structure in which the visions of the painter could be inserted—as is the case with Michelangelo and Correggio. Mannerism had used painted decoration to refute the rule of the architecture and the plane of the wall, by amusing illusionistic contradictions, as in Giulio Romano, Veronese, Zucchi. The essence of Baroque, here again, is to go to extremes and to put the cart before the horse: the whole building takes part in the pictorial illusion created by the ceiling, and becomes the means of a gigantic *trompe l'oeil* display, so that the spectator shall be swept off his feet and carried away.

Artists from Emilia played an important part in this aspect of the Baroque. Giovanni Lanfranco (*c.* 1580–1647) from Parma, who was an assistant of the Carracci in the Farnese Gallery, became slowly more and more audacious in his *di sotto in sù* perspective, as in the Palazzo Mattei (1615), the Casino Borghese (*Olympus*, 1616), and the dome of S. Andrea della Valle (1621–25), which derives from Correggio. After 1634 he was active in Naples, where he painted the *Paradise* in S. Gennaro (1643) instead of his rival Domenichino.

In Bologna, Il Dentone (1575–1632) concentrated on fictive architecture, developing the example of Veronese at Maser. In the ducal palace at Sassuolo near Modena, altered in 1634, his pupils Agostino Mitelli (1609–60) and Angelo Michele Colonna (1600–87) created in 1646 vast walls of *trompe l'oeil* decoration with colonnades, projecting cornices, tiers of balconies with figures and birds, that prepare the way for Tiepolo and Solimena.

In the second quarter of the century it was above all the illusionistic ceiling with immense airy perspectives in a luminous sky, developed from Mantegna's example in the palace at Mantua (1461–74), which was the fashion. In many palaces the centre of the ceiling was opened up in a huge medallion surrounded by columns and cornices in steep perspective, so that the opening is projected much further off, and in the 'sky' thus created were set classical allegories and mythological figures such as the *Aurora* by Guercino (1621) in the Casino Ludovisi in Rome, figures of heroes such as the *Apotheosis of Alexander* by Colonna and Mitelli (1638) in the Pitti palace, symbols of the planets such as the five rooms also in the Pitti (1637–47) by Pietro da Cortona and Ciro Ferri, triumphal allegories of great families such as the *Glory of the Barberini* in the great saloon of the Palazzo Barberini in Rome by Pietro da Cortona—the most famous of the genre. In 1640, Pietro da Cortona finished this great ceiling decoration; through the arrangement of the flights of figures, the

97 Bernini: Rome, *Fountain of the Four Rivers*
Piazza Navona

98 Vignola: Rome, Villa Giulia, Nymphaeum

99 Tivoli: Water organ in the gardens of the Villa d'Este

100 (a) Rome, window in the Casa Zuccari, Via Gregoria

100 (b) Catania, window in the Benedictine Monastery

101 Entablatures:
(a) Venice, S. Giorgio Maggiore
(b) Lecce, Basilica di Santa Croce

102 Borromini:
S. Carlo alle
Quattro Fontane

(a) façade
(b) interior

103 Guarini: interior of S. Lorenzo, Turin

104 Bernini: Baldacchino over the High Altar and the Confessio

(*Rome, St. Peter's*)

clouds and the figures in diagonal movement obscuring the construction of the room, all leads up, above the spectator's head, to the apotheosis in the centre of the sky where Providence surrounded by allegories greets the Barberini family, while the Virtues uphold a laurel garland which surrounds enormous bees, the emblem of the family.

Pietro da Cortona, with a large team of assistants, next decorated the rooms in the Pitti palace, where airy compositions in pale colours mix hyperbolic praise of the great with the fables of Olympus. Even more important was the work done from 1647 to 1651 in the dome of S. Maria in Vallicella in Rome, where a celestial glory, full of movement and light, and crowded with figures, floats overhead. These successful examples in a new genre of painting appear to overwhelm the more modest art of easel pictures, which explains the return to the 'classics' of an artist like Testa.

Despite the opposition of Andrea Sacchi (1599–1661) a pupil of Albani, who objected that these visual fêtes compromised both piety and sincerity, and who gave an example of simplicity in his *Sala della Divina Sapienza*, in the Palazzo Barberini, the fashion became widespread in religious art during the second half of the century. Northern Italy also added its quota of huge decorations to the pomp of the Roman prelates, and in its turn the Church rejoiced in this lavish and paradoxical art, sometimes inspired in invention, often rather unintelligible in meaning, and committed to depicting celestial jubilation and the effulgence of the divine as an exhausting aerial and light-filled fête. The eddying clouds of paradise separate the celestial hierarchies, the groups of saints, into vertiginous layers culminating in the Divine Glory, represented in a blaze of light. The incorporation of the coloured stucco figures, projecting irregularly beyond the cornices and over the supporting members of the building, increase the illusion. Artists particularly distinguished for this kind of work were Filippo Gherardi (1643–1704) of Lucca, at S. Pantaleone, and with G. Coli (1636–81) at S. Croce dei Lucchesi; and Baciccia (or Gauli, 1639–1709), who, on Bernini's advice, was chosen by the General of the Order, Padre Oliva, to paint the *Apotheosis of the Name of Jesus* in the mother-church of the Jesuits (1672–79). Instead of a structure of concentric clouds, a very strong oblique shaft of light, heightened by effects of chiaroscuro, carries the eye from the central radiant symbol to the enemies of the Faith, represented as hideous allegorical figures which tumble, defeated, towards the walls of the church. The limitless quality of the heavens has never been more forcibly presented than here. The *Adoration of the Lamb* in the apse of the Gesù, and the *Glory of S. Francis* in SS. Apostoli, are much more clearly enclosed within their framework, but both are also full of animation.

The art of the Jesuit Andrea Pozzo (1642–1709), born in Trento and trained in Lombardy, and whom the Padre Oliva brought to Rome in 1681, is very different. By means of astonishing and gigantic perspective elaborations of arcades and columns, fictive churches were raised upon the walls of the actual churches and prolonged into a limitless and distant sky. The figure, which Baciccia still respected, here becomes only an element lost among swirling creatures tossed like multicoloured flights of birds through heavenly colonnades. The *Glory of S. Ignatius*, in S. Ignazio in Rome (1685) is his masterpiece—at once entirely unreal and perfectly reasoned, each detail justified by curious theological, pedagogical or archaeological arguments. Pozzo elaborated his ideas in his remarkable treatise on perspective (1693) which gave him an immense influence over the wilder forms of Baroque in the North, stimulated by his own work in the Liechtenstein palace in Vienna. In Italy, his influence worked on Solimena in Naples, and more distantly on Tiepolo, though it had almost no effect in Rome where it was impossible to surpass his virtuosity.

Monumental Sculpture. Before the ascendancy of Bernini, Roman sculpture was principally in the hands of the Lombard Stefano Maderna (*c.* 1576–1636; apparently no connection of the architect) who made the recumbent figure, with delicate and sensitively handled drapery, of the patron saint of S. Cecilia in Trastevere in Rome, and the altar of the Cappella Paolina, and of Francesco Mochi (1580–1654), a Tuscan, whose *Angel of the Annunciation* in Orvieto (1610) opens the Baroque era of rich forms full of movement, designed to be seen from several viewpoints. After erecting two vigorous equestrian monuments in honour of the Farnese in Piacenza, Mochi went to Rome in 1629 and created a theatrical *Veronica*, with agitated draperies, in one of the niches under the dome of St. Peter's (1640).

Pietro Bernini (1562–1629), the father of Gian Lorenzo, was a Tuscan who worked first in Naples and then in Rome, where he created examples of the picturesque relief, very varied in modelling in the *Assumption* in S. Maria Maggiore (1609), and of the fantastic in the boat-shaped fountain in the Piazza di Spagna. The first works of Gian Lorenzo were executed with his father, who probably collaborated in the *Rape of Proserpine* (1621–22). This vigorous group, together with the *David* (1623) and the *Apollo and Daphne* (1622–25) all carved for Cardinal Scipione Borghese, were the first successful works by the son (Borghese, Rome).

If the *Rape of Proserpine* be compared with Giovanni da Bologna's *Rape of Sabines*, of a half-century earlier, it will be seen that Bernini, even if he broke out of the limits of the block and opened the composition up in every direction, refused to diffuse the interest in the way that the Mannerist sculptor was prepared to do. He returned to the principle of the single viewpoint and therefore of the single action, in order to create movement in his figures. The *David*, the only freestanding single figure in the sculptor's whole *oeuvre*, is caught in a twisting movement and an effort that turns it into a kind of continuous spiral. The *Apollo and Daphne* is a virtuoso composition which contrives to render all the tactile qualities of the various materials it represents—wood, flesh, hair, draperies—in forms so detailed as to be entirely broken up.

322

ROME: ARCHITECTURE AND DECORATION

Even more in famous busts like those of *Costanza Bonarelli* (*c.* 1635, Bargello), *Scipione Borghese* (1632, Borghese), and *Louis XIV* (1665, Versailles), Bernini sought to achieve his effects through an astonishing vividness of life and handling. In huge statues like the *S. Longinus* (1638, St. Peter's), the gesture and the drapery are the complement of the architecture. He is at his best in sculpture designed to be seen in an architectural setting. The *Tomb of Urban VIII* (1628–47), placed in a niche on the right of the apse in St. Peter's, consists of a sombre sarcophagus flanked by two allegories of *Justice* and *Charity* in white marble, and surmounted by a socle bearing the figure of the pope; the pyramid formed in this way is full of movement and colour. The *Tomb of Alexander VII* (1671–88) also in St. Peter's, conveys with even greater eloquence the vanity of earthly glory by the slipping movement of the marble drapery beneath the majestic figure of the pope. But nothing displays better the concordance between Bernini's powerful and undulating forms and their surrounding architecture than the *Baldacchino* and the *Cathedra Petri*, without which the enormous basilica would seem uninhabited.

In S. Maria della Vittoria, between 1644 and 1651, Bernini created the *Cornaro chapel* in the form of a small theatre with opera boxes and a decoration of coloured marbles; above the altar, on a shallow stage S. Theresa is seen in ecstasy, struck by an angel poised like an Eros. In this composition all is so full of audacity—its flamboyance, the mixture of the arts, the ambiguity of the religious representation—that this epitome of the most extreme form of Baroque must either revolt or deeply impress the spectator, as much by the richness of the means used as by the artist's brilliant exploitation of them. It was also in a kind of alcove that he placed the figure of the Blessed Ludovica Albertoni in S. Francesco a Ripa, swooning in a vision of divine Love, her hand on her breast, in a flutter of drapery.

His great open-air creations—the *Fountain of the Rivers* in Piazza Navona, and the *Triton* in Piazza Barberini, kneeling on magnificent shells—have already been mentioned among the great inventions of Roman town planning, which he inspired. To these must be added the statue of *Constantine* at the end of the atrium of St. Peter's; the rearing horse is seen against the diagonal of a huge simulated curtain, whose strong movement seems to have communicated itself to the whole construction.

Many of Bernini's works called for collaborators and followers. Thus, in the other niches of the main piers of the dome of St. Peter's were placed around his *S. Longinus* the huge *S. Andrew* by François Duquesnoy from Brussels (1597–1643), who also carved a tamer *S. Susanna* for S. Maria di Loreto in Rome, the *Veronica* by Mochi, already mentioned, and a *S. Helena* by Andrea Bolgi. Antonio Raggi (1624–86) decorated the dome of S. Andrea al Quirinale for Bernini with stuccoes, and composed statues and decorative sculpture for the Gesù. Domenico Guidi (1625–1701), who carved a statue of *Clement IX*, and a very deeply undercut relief of the *Holy Family* in S. Agnese, in his *Thiene tomb* in S. Andrea della Valle derives both from

323

Bernini and his rival Algardi, as does Ercole Ferrata (1610–86), who also carved reliefs for S. Agnese, and whose grandiloquent monuments rendered the grand style insipid.

In contrast to Bernini's energetic draughtsmanship, his tension, and his effects derived from mixed materials and the use of colour, Algardi (1602–54), already mentioned as an architect, is a much more sober and measured artist, conforming more to the ideal of the Bolognese painting of Guido Reni or Guercino. His style is often called classical, despite the fact that it is full of Baroque elements—for instance, in the billowing drapery of the bust of *Olympia Pamphili* or in the setting of his *Decollation of S. Paul* in S. Paolo, Bologna (1641).

Trained with Domenichino by Lodovico Carracci, Algardi came to Rome in 1628. He worked on restorations, made statues in stucco for S. Silvestro al Quirinale, then, after a quarrel with Bernini which lasted for the rest of his life, he gained the patronage of Innocent X, during whose pontificate he made the confused relief of *Attila repulsed by S. Leo* in St. Peter's (1646–50) and the white marble *Tomb of Leo XI* with its reticent and moderate figures (1645–50), which appears to be a deliberate correction of the type of the tomb of Urban VIII. His bronze *Innocent X* (Pal. dei Conservatori) is infused with the kind of energy that permeates his rival's *Urban VIII*. Nevertheless, the direction taken by Algardi is towards a formal balance, underlined by his return to High Renaissance types, as in the frieze motifs on the *Tomb of Innocent X* and the busts in the niches of the *Frangipani tombs* in S. Marcello.

The End of Roman Baroque in the Eighteenth Century. For almost a century Roman sculpture maintained the style established by the two masters, with hardly more change than the exaggeration of certain features—for example, the macabre effect of Bernini's skeletal figures on his papal tombs is pushed to the edge of absurdity by Giuseppe Mazzuili or Rusconi (1658–1728) in the *Paravicino tombs* in S. Francesco a Ripa. Bernini's use of polychrome becomes a mere *tableau vivant* device in the *S. Stanislas Kostka* by the French sculptor Pierre le Gros (1666–1719) in S. Andrea al Quirinale. The unhappier aspects of Baroque sculpture, treated without sufficient attention to the architecture, end in the specious declamations of Pietro Bracci (1700–73) in the tomb of *Maria Clementina Sobieska* in St. Peter's or of Bernardo Cametti in the *Muti tombs* in S. Marcello (1725). For futility and oddness nothing can surpass the tomb of *Maria Flaminia* (c. 1720) by Posi and Penna in S. Maria del Popolo—an eagle lifts a drapery from a medallion of the princess, gazed at by a lion at the base of the composition. But the solemnity of Roman Baroque inhibits its transition into Rococo. This is confirmed by the heavy decoration loaded on to his *S. Mark* by Michelangelo Slodtz, a French sculptor working in Rome from 1726 to 1746, who also carved a rather affected *S. Bruno* for St. Peter's (1740).

It was in St. Peter's, in the tombs in the side chapels, that Roman sculpture retained most vitality. In the *Tomb of Gregory XII* (1723), Rusconi combines the movement of Bernini with the design of Algardi; in the same spirit Filippo della Valle (1696–1770) designed the *Tomb of Innocent XII*, with a figure of *Temperance* which recalls Tuscan Mannerism; this is in the Corsini chapel in the Lateran, and the basilica was also decorated with statues by Bracci, and almost every other Roman sculptor of the mid-eighteenth century worked there.

Moreover, in architecture, an attachment to the principles of monumental Baroque may be discerned which, except in rare cases, precludes the emergence of Rococo forms. Gabriele Valvassori (b. 1683) in the Palazzo Doria (1720–34) remained faithful to the Borromini of the Propaganda Fide; Gregorini, architect of the Oratorio del SS. Sacramento in S. Maria in Via (c. 1730), with a curved façade, and of S. Croce in Gerusalemme (1743), was inspired by Borromini's Oratory of S. Philip Neri. Finally Sardi, whose career has been insufficiently studied, built the only really peculiar and decadent church of Roman Rococo: S. Maria Maddalena in Campo Marzio (1735 onwards) entirely covered with marble, stucco and gilded ornament, with a two-storey *cantoria* in gilded wood, and a façade which is an uneasy interpretation of Borromini.

The reaction and the return to the sixteenth century are already noticeable with the Florentine Alessandro Galilei (1691–1736), and are partly due to the fashion for Palladian architecture in England, where he also worked. His façade of S. Giovanni dei Fiorentini (1734) is inspired by that of the Gesù, and his Lateran façade (1735) which completed Borromini's remodelling of the church, recalls Michelangelo at the Capitol. Another Florentine, Ferdinando Fuga (1699–1782), after building the Palazzo della Consulta (1732–34) in a cold and dry style, S. Maria della Morte (1732) with more vigour in the forms, and the harsh Corsini palace (1736), developed more flexibility in the façade of S. Maria Maggiore (1741–43), which he treated as a loggia, although it has too many volutes, garlands and statues. A more interesting note was sounded by Carlo Marchionni, who, between 1743 and 1763, built the Villa Albani in an ordered richness for Winckelmann's collector-patron. But a work of great originality which really brought Baroque to an end was the Villa of the Knights of Malta (1765) built without any concessions to curved or broken lines, and the church connected with it, also loaded up to the vaulting with a decoration of learned stuccoes in Romantic profusion. This unexpected work was by the Venetian engraver Piranesi.

Piedmont and Central Italy

Lombardy, from whence came so many Baroque masters, never produced any very great architecture. Lorenzo Binaghi (c. 1556–1629) built the church of S. Alessandro in Milan in 1602 on a Greek cross plan, with an undulating pediment on the façade between two campaniles, which cannot have been

ARCHITECTURE AND DECORATION

without some interest for Borromini. The only notable artist, able to find work despite the extreme poverty of the duchy, was Richini (1583–1658), who built the entrance to the Ospedale Maggiore and the courtyard with the double order. Besides a number of minor palaces, he designed in 1651 the Palazzo di Brera (formerly the Jesuit college) with a cloister composed of arches borne on coupled columns, after the fashion of Tibaldi and Alessi. The strangest work of the period was the sanctuary of the Sacro Monte at Varese, which an architect called Bernascone began in 1604, and which was only finished long after his death. It consists of fifteen chapels dedicated in 1680 to the mysteries of the Rosary, each containing a theatrical tableau of large terracotta figures. The rather simple range of buildings is entered through a portico which serves as a setting for an *Annunciation* by C. Prestinaro, and a splendid piece of sham architecture serves as a background for twenty-two statues by F. Silva representing Jesus in the Temple among the Doctors. These strange little buildings are an extreme example of a popular and applied form of illusionism.

Genoa. Rubens was in Genoa between 1605 and 1607, and in 1622 published engravings of the principal palaces of the city. The works of Galeazzo Alessi and his followers thus gained a European importance, and during the rest of the century Genoa was a much visited centre. In 1602 the Via Balbi was planned, and in 1661 the link between this street and the Via Nuova. These essays in town planning were principally the work of Bartolomeo Bianco (d. 1651), the engineer to the Republic, who built the University, where the courtyard offers the same arrangement of a two-storey arcade carried on coupled columns as the Brera. His successor, G. A. Falcone, built the Palazzo Durazzo, which later became the Royal palace (c. 1650). The Palazzo Rosso (1672–77), built by Pier Antonio Corradi, is an impressive structure replete with every Baroque effect. Each of the major rooms has a dome, and the pendentives which link it to the arcades and side walls are so handled as to result in a skilfully modelled, almost square cornice from which the eye is directed along illusionistic colonnades to where allegorical figures disport themselves in the heavens—as, for instance, in the *Autumn* by Domenico Piola (1627–1703), a pupil of Cappellini, an imitator of Castiglione, and an important decorator of Genoese churches and palaces. More complicated still are the ceilings of the rooms dedicated to *Spring* and *Summer* by Gregorio Ferrari (1644–1726), a son-in-law of Piola, assisted by Haffner: enormous, lumpy, fleshy festoons and cornices support a whole population of caryatides, monsters, putti, descending on to the walls and climbing upwards towards dancing and flying allegorical figures. Ferrari worked in most of the secular and religious buildings of Genoa, in particular the churches of SS. Giacomo e Filippo, where the Apostles watch the Assumption from a loggia, and the Crociferi, where he painted a *Triumph of the Cross*.

The most remarkable of the Baroque churches of Genoa was the Annun-

326

ziata, finished in the middle of the century, with tall Corinthian fluted columns bearing garlanded arches of a particularly elegant and light-hearted character. The façade was a Neoclassic work with an Ionic portico by C. Barabino (1830). This charming building, terribly damaged in 1944, was entirely covered inside with gay and airy paintings filled with movement, by one of the Seminis and the brothers Gian Battista and Gian Andrea Carlone, whose decorative painting is often found in Genoa; they were completed by a *S. Gaetano in Glory* by Piola (1674). The nave of S. Siro was similarly covered between 1657 and 1670 with frescoes devoted to S. Peter (the sketches for which are in the Spinola Gallery); the cornices serve as a screen to push the scenes back into a limpid and luminous space.

The French sculptor Puget, who worked in Genoa between 1661 and 1667, executed a *S. Sebastian* for S. Maria in Carignano, and an *Immaculate Conception*. These works introduced a fashion for Berninesque sculpture, to the profit of Daniele Solari (1634–98) his assistant, and Filippo Parodi (1630–1702) who made a *S. John* for S. Maria in Carignano, as well as tombs in Venice. Among the members of the very active family of the Schiaffini, Francesco (*c.* 1691–1765) was a Bernini follower, as is shown by his *Rape of Proserpine*, and A. M. Maragliano (1664–1741) a wood carver, created an allegory of *Time*, very complicated in form (Palazzo Bianco).

Piedmont: Guarini and Juvara. After the departure of the French in 1562 Piedmont became, under Duke Charles Emmanuel I, a very vigorous small state whose capital, Turin, was modernized by the building of the Piazza S. Carlo by Castellamonte. His son Amedeo built, under Charles Emmanuel II, the Royal palace which was decorated by the Fleming Miel (d. 1663), before the arrival of the Frenchmen C. F. Beaumont and Charles Van Loo in the following century. Many minor architects such as Vittozzi (1539–1615) built interesting centrally planned churches—for instance, the SS. Trinità (1606). In Turin, there was a curious freedom in planning, which gives a very special flavour to the form which Baroque developed there, and explains the extreme quality of imagination found in her two greatest architects, Guarini and Juvara. Padre Guarino Guarini (1624–83) a member of the Theatine Order, a theologian and mathematician, was trained in Rome, where he was particularly affected by Borromini. He taught in Sicily, where his works in architecture have disappeared, and in Paris, before settling in Turin, after 1666, to use all the resources of his vast learning in the service of the Duke of Savoy. He built the Palazzo Carignano (1680) with its undulating façade and its elliptical entrance hall, and among his churches are the strange little chapel of the Holy Shroud (Cappella del SS. Sindone, 1668–94) in the Cathedral, entirely covered in black marble and crowned by a small dome with a pointed lantern, and S. Lorenzo dei Teatini (1668–87). In the latter, a domed, central, square space opens out into a world of illusion; the choir consists of an oval room with an ambulatory which penetrates the central square; the corners have exhedrae which recall Early Christian edifices, while the dome, pierced by eight windows and spanned by sixteen interlaced girder arches which bear the airy lantern, recalls the forms of one of the Cordovan mosques,[148] or, later in date, the dome of the crossing of Saragossa cathedral. Singularity of form contends with technical virtuosity in these constructions, in which his contemporaries, struck by their irregularity and their towering height, saw a return to Gothic. It is, in fact, a very learned Baroque, fascinated by geometrical paradoxes and deriving much of its effect from the very free use of the materials: bare brick in the palaces, and the most colourful marble in the interiors.

Filippo Juvara (1676–1736) was born in Messina, and after a Roman training in the studio of Carlo Fontana, became a member of the Accademia di S. Luca in 1716, and created the settings for musical tragedies in the Palazzo della Cancelleria. He entered the employ of Vittorio Amadeo II of Savoy in 1715. He was, above all, a master of large-scale compositions. The basilica of Superga, high above Turin, was built as a thank-offering for the liberation of the city in 1706, and as a mausoleum for the House of Savoy. It is a centrally planned church surmounted by a high dome, and with a huge portico, and is flanked by two graceful campaniles, in front of which the main structure projects; it also has a mosaic floor. In it may be discerned the use of a Roman repertory of forms, rather than the imaginative inspiration of Guarini. The façade of S. Cristina (1718) is more flexible, in its use of curved forms and a decoration of columns which are extended upwards by tall candelabra, while the interior of the Carmine (1732) is closer to Guarini, in its decoration of niches and small domes. The artist's imaginative variety was also given full rein in his numerous monumental altars: for instance, those of S. Teresa (1728), and the Visitation (1730), at Superga, which consist of statues grouped in the manner of Bernini.

Finally, Juvara created a style for Piedmontese palaces. The façade of the Palazzo Madama is in six bays, with a giant order under a finely detailed, balustraded flat roof (1718 onwards). Also by him are the two colonnaded buildings which close the Via del Carmine (1728), and, his last work and one of his most celebrated, the hunting lodge at Stupinigi (after 1729) set in the middle of the plain. The architect conceived this as constructed with two wings which are so arranged as to form three successive courts before joining the main block of the building. This is an oval central pavilion consisting of a huge central saloon rising through two storeys with a balcony running round the inside at first-floor level. The ceiling is decorated with frescoes by the Valeriani brothers, and niches and festoons are arranged so as to stress the rhythms of this superb ensemble. In all these various palaces, Juvara had proved his ability as a lavish decorator—for instance, in the main saloon of the royal palace, entirely decorated in white stucco. As an architect with an international reputation, Juvara was summoned to Madrid to make the plans for a new royal palace after the fire of 1734.

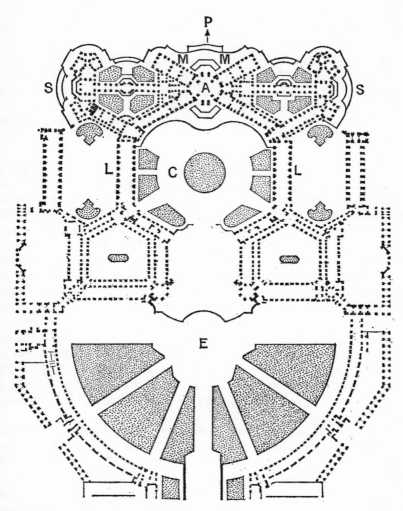

FIG. 39. *Hunting lodge at Stupinigi, Piedmont, by F. Juvara. Plan (after V. Golzio): A, Central pavilion; L, Side wings, varying in height; C, Main courtyard; S, Pavilions in the wings; M, X-shaped block; E, Entrance; P, Park.*

329

In Piedmont, his successors were Alfieri (1700–67), who built the theatre in Turin (1740) and S. Giovanni Battista at Carignano, and Bernardo Vittone (c. 1705–70), architect of the sanctuary of Vallinotto at Carignano, and of S. Maria di Piazza in Turin (1751), where an oval dome rests on two half-domed apses, and which is elaborately decorated with stuccoes. The building activities of the nobility, still considerable in the middle of the century, attracted decorators in the Rococo taste; next to the French painters, who followed the Van Loos to Turin, Pier Francesco Guala (1698–1757) must be noticed for his intense, and even rather heavy, version of Rococo. He was active in Lombardy as well, in the Palazzo Gozzani in Treviglio (1737), and in Milan where he painted the ceiling of S. Vincenzo di Paolo (1757).

Bologna. Emilian architecture was less important than painting, with the single exception of theatrical design, where the art of scene painting was restored and the models for the future established. In Parma, under Dukes Ottavio and Ranuccio I the Teatro Farnese in the Palazzo della Pilotta was built by G. B. Aleotti ('L'Argenta', 1546–1636), on the model of the theatre in Vicenza; the important treatise by Ferdinando Galli Bibiena (1657–1743) was published in Bologna in 1711. His brother Francesco, after working in Rome, Vienna, and Nancy, settled in Bologna in 1726, and in 1749 built the Teatro Filarmonico; his son Antonio (1700–74) built the charming municipal theatre. In fact, the models and the picturesque ideas—'idee'—of these famous scenic designers were exploited by all the Baroque architects, not only by those who specialized in theatrical design. But it should be noted that, from the projects of the sixteenth century onwards for the façade of S. Petronio (Varignara in 1518, Peruzzi in 1523, G. Rainaldi in 1626) Bologna had always retained a certain predilection for Gothic forms; for an opera in 1694, M. A. Chiarini designed a setting in pointed arches which anticipates the Romantic taste for the Middle Ages.[149]

Giovanni Ambrogio Magenta (1565–1635), a Lombard, was commissioned to build the cathedral of S. Pietro in Bologna with a barrel-vaulted roof (the façade was executed later by Torreggiani) and, in 1611, the church of S. Paolo, where the vaulting presents curious perspective effects. In 1687–1690, G. B. Bergonzoni built an interesting church on a Greek cross plan, reminiscent of Borromini: S. Maria della Vita, of which the dome was only built a century later.

In the eighteenth century, Carlo Francesco Dotti (c. 1670–1759) was the architect of the Madonna di S. Luca on the Monte Guardia which dominates Bologna; charming loggias surround the building, whose dome rises from cleverly contrived abutments. Brilliant perspective effects were achieved with the arch at Meloncello, but with heavy architectural elements à la Magenta. Torreggiani is more graceful in the Oratory of S. Filippo in the Palazzo Montanari. S. Paolo, where there are some brilliant Baroque paintings, such as the *Nativity* by Cavedoni (*c.* 1650), also contains the *Decollation of S. Paul* by Algardi, where the balance of mass contrasts with Baroque gestures. After he settled in Rome, Algardi expressed academic reserve in face of Bernini's brio. His pupil Camillo Mazza (1602–72) modelled the *Pietà* in terracotta for the Certosa, and the low relief plaques and tabernacles of his son Giuseppe were produced in great number for as far afield as Venice. Filippo Scandellari (1717–1801) specialized in clay figures for Christmas cribs.

Florence. After the great part she played in Mannerism, Tuscany contributed little to the Baroque. Exuberance and fancifulness were not her strong points, and she produced neither a Bernini nor a Guarini. Work was limited to additions like that of the Gaddi chapel in S. Maria Novella, by Giovanni Antonio Dosio (1533–1609), and the enormous and gaudy domed Cappella dei Principi at S. Lorenzo, all decorated in sombre marbles, by Matteo Nigetti (1560–1649). S. Gaetano, also by Nigetti, is a fairly graceful interpretation of Roman types, using the local *pietra serena* for the salient forms. Gherardo Silvani (1579–1675) who was also a sculptor, built or restored many palaces (Pallavicini, Castelli, Corsini) in the Tuscan tradition. Only the sculptor Giovan Battista Foggini (1625–1725) tried, in the Cappella Feroni in the Annunziata (1682) to adopt the more sumptuous type of Baroque. S. Ferdinando, Leghorn, a Latin cross church ornamented with stuccoes, is attributed to him (1717). During the eighteenth century, the Grand Duke's architects, Galilei and Fuga, worked in Rome where they exercised a moderating effect upon Baroque style, so that in Florence there remained only Ferdinando Ruggieri (1691–1741) to build the Palazzo Capponi and to begin the curious S. Firenze, whose façade is that of the Oratory of S. Philip Neri flanked by two churches; it was finished by Zanobi del Rosso (1724–98) in 1775. Only the attics above the segmental pediments owe anything to Rococo.

Among the great Roman decorators, first Pietro da Cortona and then Luca Giordano came to work in Florence. But there were also the pupils of Cristoforo Allori (1577–1621): Francesco Furini (*c.* 1600–46) and Giovanni da San Giovanni (1592–1636) who were influenced, as he was, by the light painting of the Carracci, and who covered the walls of the Pitti palace, and in particular those of the Museo degli Argenti, with historical frescoes. In sculpture, more originality was displayed. Pietro Tacca (1577–1640) showed, in his four Moors chained to the base of the monument to Ferdinand I at Leghorn, a Baroque vigour which is lacking in his rather amusingly Mannerist fountains, like that in the Piazza dell' Annunziata. Giovanni Caccini (d. 1613) worked on the statues of the Ponte Sta Trinità, and on the polychromed tabernacle of Sto Spirito. Giovan Battista Foggini, who knew Bernini and Algardi, was important for his portraits (monument to Galileo, 1737, Sta Croce), but his chief bent was for animated bas-reliefs such as those in his two Florentine decorative commissions, the Corsini chapel in the Carmine, and the Feroni chapel in the Annunziata, where he was assisted by the Bavarian Permoser. This was as far as Florence would go along the Baroque road; during the second half of the century, all her sculptors were absorbed by Neoclassicism.

The South: Naples, Apulia and Sicily

Naples. Domenico Fontana worked in Naples from 1594 onwards, and built the church of Gesù e Maria and the royal palace, supervised the construction of a new aqueduct as far as Torre Annunziata (without discovering Pompei) and the layout of the boulevards along the seafront. Francesco Grimaldi (1543–1630), who came from Lucca and had worked for a while under Maderna, built the first buildings in the new style—S. Paolo Maggiore (1590–1603), with an enormous nave and a façade flanked by a flight of steps, and the SS. Apostoli, also with a single nave, with a rich decoration imposed on a Renaissance structure. He also provided the plans for the Cappella del Tesoro in S. Gennaro (1608). Cosimo Fanzago (1593–1678), from Bergamo, a sculptor, painter, and skilful practitioner in stucco and marble, was more imaginative in the Certosa di S. Martino, where he built the large and light cloister (c. 1630), and in numerous chapels, in the royal palace (c. 1640), S. Maria Maggiore (c. 1660), and particularly at the Abbey of Monte Cassino, where the complete reworking of the church with nave and two aisles decorated with very lively frescoes by Luca Giordano, is attributed to him (after 1649; destroyed in 1944). He was also the architect of S. Giorgio Maggiore, where the plan is inspired by the Gesù, and of S. Maria l'Egittica at Pizzofalcone in Naples, with a peculiar central plan. Many of the old buildings, such as Sant'Angelo a Nilo, S. Restituta, were transformed by Arcangelo Guglielmelli, to whom is also due the library in the very rich and very sumptuously decorated convent of the Gerolomini (c. 1730). The façade of their church, founded at the beginning of the seventeenth century, was only built after 1780, in white marble, with two open bell-towers on either side of the central pediment, in imitation of S. Agnese; this was the last Baroque work in the city, and also the last work of Ferdinando Fuga.

Fuga, who came to Naples from Rome in 1751, had built at Resina the Villa la Favorita, a rather frigid two-storied edifice, and the huge Hospice for the Poor, with its façade over 1,000 feet long, which was not finished until 1829. In the kingdom of the Two Sicilies, two other names were dominant during the eighteenth century: Domenico Antonio Vaccaro (c. 1680–1750) and Luigi Vanvitelli (1700–73). The first, like Fanzago, was a sculptor and decorator; he was responsible for the octagonal church of Montecalvario and the charming cloister of the Clares with walls in majolica panels, copied from Portuguese *azulejos* by G. and O. Massa. His pupil, Ferdinando Sanfelice, a scenic designer who built a curious pyramidal tower in 1740, for a fête in honour of the queen, designed palaces of an extreme fantasy such as that of Serra di Cassano. But the one architect who really counted was Luigi Vanvitelli, the son of a Dutch painter. He had known Juvara, and had worked in Ancona (Hospital, Gesù, c. 1733, and 1745), Rome (interior of S. Maria degli Angeli, where he departed from Michelangelo's plan), and in 1751 was summoned to Naples by Charles III who wished to found a new capital. This new Versailles was Caserta. The exterior was worked on from 1752 to 1774, and was built in travertine and brick on a vast rectangular plan, about 600 feet long, and with a five-storey façade divided into two sections beneath a balustraded roof, and a colonnaded projecting central feature as at Versailles. The ensemble has a classic nobility, disturbed only by the unreasonable scale of the undertaking, which would have been even more emphasized by angle towers and a central cupola, its actual size being stressed by the Baroque decoration of the interiors, and the immense park which surrounds it. This last has already been discussed as one of the major masterpieces of Italian scenic decoration. A grand stairway with a central flight and two return flights on either side enabled a tremendous setting of marble arcades and colonnades to be constructed. In Naples itself, Vanvitelli built the double church of the Annunziata (1760, finished by his son Carlo)—a lower church, with, over it, a single nave church divided into three units, with a central dome and a continuous interplay of high columns—and the church of the Mission Fathers. This is Baroque both massive and full of movement.

Besides the manufacture of Christmas cribs with innumerable figures, which, during the eighteenth century, became an important industry,[150] Neapolitan sculpture consisted chiefly of fountains and urban decorations—for instance, the obelisk covered with escutcheons and small reliefs by Fanzago at S. Domenico (1658), and the obelisk of the Immaculate Virgin by Giuseppe di Fiore, of about 1750. The most important group of Baroque sculptures in which allegory is pushed to the extremes of illusionism is in the Sansevero chapel in S. Maria della Pietà (1749–66); it consists of a painted ceiling by F. Maria Rosso (1749), a recumbent shrouded figure of *Christ* by Giuseppe Sanmartino, and the ridiculous *Allegory of Deception* by Queirolo, all works which display a misused virtuosity. The sculptors who worked at Caserta have already been mentioned.

Grand-scale painting was employed, as in Rome, in all the principal religious and secular buildings. Two important masters worked this field: Luca Giordano (1632–1705) and Francesco Solimena (1657–1747). The first was born in Naples, and developed a genuine Baroque manner in a field where interest in effects of light tended to exclude large decorative schemes. Possessed of an almost intemperate virtuosity, which gained him the nickname of 'Luca fa presto' ('Luca, go quickly') he required huge schemes which would give full rein to his eclecticism. He worked in Florence (Riccardi palace, *Apotheosis of the Medici*), Venice, Rome, spent ten years in the Escorial (1693–1702), and was esteemed another Pietro da Cortona. In Naples, he left some astonishing works, such as the ceiling of the Abbey of Monte Cassino (1677, now destroyed), the *Expulsion of the Moneychangers from the Temple* in S. Filippo Neri (1684) with its large movement of light and shadow under colonnades, and the ceiling of the Cappella del Tesoro at S. Martino (1704), much lighter in effect.

So international a figure is essential to an understanding of Italian Baroque, because of the exuberance of his temperament, and his extraordinary talent for extracting from all styles that which could serve to heighten and dramatize the decor. Innumerable pictures, and almost all his frescoes, bear witness to his boundless inventiveness which ranged in the same work from the sublime to the down-to-earth, and to his spirited execution, which seems to overreach Tintoretto. From a starting point in Ribera (*Jacob's Dream*, Naples, Mus. Naz.) and in the exaggerated chiaroscuro of the Southern Caravaggesques, his style gained in lightness and animation through his contact with the Venetians (*Christ among the Doctors*, Rome, Gal. Naz.) and he was eventually converted, through the example of Pietro da Cortona and the ceiling of the Barberini Palace, to the spacious painting, filled with light clouds and airy distances, of which the ceiling in the Riccardi palace in Florence (1682) remains a good example, despite its artificialities.

The development of Luca Giordano shows the extent to which adherence to the systems of Roman Baroque, by encouraging illusionistic compositions teeming with figures, conceived as both a part and an extension of the architecture, demanded a performance quite contrary to that required by the more limited art of the easel picture. The precocious and self-confident talent of Francesco Solimena developed along the same paths, and in the huge melodramatic scenes which he painted for S. Paolo Maggiore (*Fall of Simon Magus, Conversion of S. Paul*, 1689-90) he enjoyed displaying his great ability in delicate tones, and in the elongation and twisting of his figures in every direction, and in the Sacristy of S. Domenico Maggiore (1709) and in the tumultuous fresco of the *Expulsion of Heliodorus* in the Gesù Nuovo (1725) he was able to compose spiralling groups of figures with the utmost assurance. With him, pictorial rhetoric achieved a particular passion of vision, shot through with shafts of stormy light in massive architectural constructions, which may not have been without some influence on Delacroix. Solimena manages everything with great ingenuity, but it is possible to evaluate his qualities of invention in the detail from his sketch of the *Massacre of the Giustiniani at Chios* (Naples). With his heir, Francesco de Mura (1696–1752), everything becomes much heavier, as, for instance, in SS. Severino e Sosio in Naples (1740).

Apulia. As during the Romanesque period, when the cities of the South reacted independently from each other to the styles emanating from Byzantine and Norman centres, Apulia and Sicily presented, under Spanish influence, entirely original versions of Mediterranean Baroque, at Galatone,

Martina Franca, and above all at Lecce. The beauty of the local stone, of a fine golden grain, easy to work, though unfortunately easily damaged by frost, facilitated the transformation of mural decoration into an interplay of reliefs, of goffering, of heavily undercut framing, which destroyed the sense of the structure. In each case, it is a question of purely local architects, among whom the most interesting, between 1650 and 1700, were Francesco Zimbalo and Giuseppe Cino. On the façades of secular buildings are balconies in the form of monsters, chimerae, centaurs, and on churches the upper storeys and the

FIG. 40. *Lecce, Apulia. Window in the Seminary, by G. Cino. Profusion of 'bead' motifs, and 'cartilage' forms. (After a photograph by the author.)*

cornices are covered with astounding carvings. The most astonishing is the church of Sta Croce, rebuilt after 1549, and finished in 1695, where six columns carry an entablature the frieze of which is borne on thirteen animal caryatides. Extraordinary inventions, veils of lace, are spread over the whole building, and the interior, with its nave and two aisles, looks like a Florentine church covered with garlands. The ensemble of the cathedral by G. Zimbalo (*c.* 1670–82), and the bishop's palace with lovely richly framed windows set between rusticated pilasters, suggests the influence of Spanish Plateresque, but in this late adaptation the building retains a considerable monumental regularity.

ARCHITECTURE AND DECORATION

After 1709, Borrominesque influence may be perceived in S. Matteo and in S. Chiara, where the curved façade has many projections.

Sicily. The position is different again in Sicily. Town planning and architecture there are more closely linked than elsewhere, and the architecture displays all the resources of decoration, and the Baroque seems to be trying to equal the dazzling effect of the mosaics and marbles of the twelfth century. At Syracuse, the cathedral was rebuilt so as to incorporate the Temple of Minerva; its façade is by Picherale (1728-57). But it was chiefly in Palermo, from the eleventh to the nineteenth century, from the Martorana to the Teatro Grande, that there is a long history of which the seventeenth century forms an interesting episode. It was then that the town took on its definitive form, with its two rectilinear axes linking the four monumental gateways of the sixteenth century (Nuova, 1535, on the East; Felice, after 1582, on the port, etc.) meeting on the Piazza dei Quattro Cantoni (1609) by G. Sasso, a Lombard, later amplified by four façades in the Roman taste. Giacomo Amato (1643-1732), who was trained in Rome, built the Pietà (1689), with projecting columns forming two heavy storeys, and the church of S. Salvatore (after 1682) with an elliptical dome. The church of S. Giuseppe dei Teatini, begun in 1612 by G. Besio, of Genoa, and the Jesuit House were transformed in the second half of the century, one by the frescoes and gilded stucco-work by Tancredi of Messina (1655-1725), the other by Serpotta's stuccoes. The Roanno chapel in Monreale cathedral was built by Angelo Italia and local sculptors who heaped up marble and stucco on the pavement, walls and ceiling so as to create a masterpiece of overloaded and dazzling decoration.

During the eighteenth century, Sicily had two celebrated architects, Juvara, born in Messina, who spent his working life in Turin, and Gian Battista Vaccarini (1702-68) of Palermo, though in fact the latter's masterpiece in Catania was the culmination of the work of a generation of architects, who around 1700, had undertaken the reconstruction of the cities on the east coast, destroyed by earthquake in 1693.[151] The rebuilt streets, disposed so as to create fine vistas, were given a rather exuberant decoration, as at the Palazzo Biscari, on the original façades on the Via dei Porta-Croci, and at the convent of the Benedictines. Vaccarini—a remarkable man, and a great theorist—who was trained by Carlo Fontana in Rome, imposed order on this exuberance. He rebuilt the cathedral with a strongly modelled façade, and a fountain, copied from Bernini, with an obelisk carried upon an elephant's back. At S. Agata, on an elliptical plan, encircled inside by a kind of frieze of arabesques (the dome was not finished until 1767), and in the Palazzo del Municipio, he was much closer to Borromini. The Valle and Sarravalle palaces are much more sober, but have a capricious arrangement of balconies which seem to hang a Rococo element on a much more severe structure.

Like Catania, Noto, on the slopes of the Monti Iblei, in the midst of almond

105 Mazzuoli: Tomb of Stefano Pallavicino

(Rome, S. Francesco a Ripa)

106 Catania, Palazzo Valle, central balcony

107 Annibale Carracci: detail of the fresco decoration in the Palazzo Farnese, Rome

108 Caravaggio: *David*

(*Rome, Galleria Borghese*)

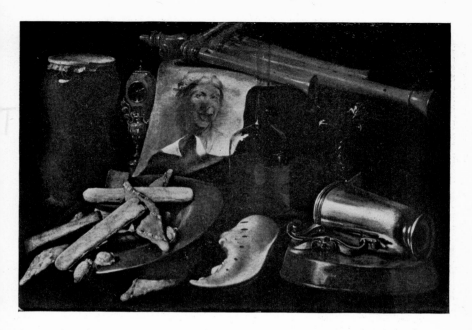

109 (a) Baschenis: *Still-life*
(*Milan, private collection*)
(b) Caravaggio: *Basket of fruit*
(*Milan, Ambrosiana Library*)

110 Padre Pozzo: *Apotheosis of S. Ignatius*

(*Rome, S. Ignazio*)

111 Castiglione: *Temporalis Eternitas*

(*From the Royal Collection, Windsor, by gracious permission of H.M. The Queen*)

112 G. B. Crespi: *Portrait of Fulvio Grati*

(*Bologna, private collection*)

trees, was rebuilt after 1693. As at Lecce, the local white *tufa* lends itself to highly detailed sculpture, and façades; balconies and window frames are of an exceptional richness and virtuosity, as in the convent of S. Salvatore (1706), and the Palazzo Nicolaci (*c.* 1730). It is the planning of the city itself which is its most remarkable feature, since it is conceived as a vast

FIG. 41. *Benedictine convent, Catania. Bay in the façade; pilasters in a colossal order, cornice with consoles, and highly ornamental windows with balconies.*

theatre, where the cornices of the buildings in the streets as they climb the hillside create increasingly steep perspective vistas. This exceptional ensemble, a masterpiece of Baroque stage design, was the work of local architects— Landolina, Nicolaci and Carnelivari. Despite Vaccarini's great reputation, it was to Fuga that the Bishop of Palermo applied for the restoration of the cathedral (1767). The interior was given pilasters and a barrel vault, the dome raised. At the same time, villa architecture became important and in Palermo itself, the courtyard of the Palazzo Bonagia was enriched by a staircase in red marble treated as a delicately pierced screen. Among the country houses of the island, the Villa Palagonia stands out as a supreme example of an architectural caprice.

At Bagheria, a few miles to the east of Palermo, Ferdinando Francesco Gravina, Prince of Palagonia had had a country villa built, about 1715, by a Neapolitan Dominican friar called Tommaso. His son, Ferdinando Francesco III, about 1746, added to it an astonishing decoration which, long before Goethe's visit in 1787, was famous all over the Europe of the Enlightenment as a vast 'folly'.[152] Around a central villa, transformed into a palace of illusion by a succession of trick decors, extended several oval courtyards impinging one on the other, with the surrounding wall crowned with statues, and forming seventy outbuildings. The whole was approached through a balustraded gallery about 700 feet long, flanked by thirty statuary groups, which made a kind of grotesque museum of funny figures—jesters, comic musicians, horned monsters in devil-like poses, animals from a crazy upside-down world, like lions sitting at table, and, even more significant, parodies of mythologies such as Achilles and Chiron among punchinellos, and a Harlequin Laocoon. At the southern entrance, statues of Wisdom and Folly facing each other underline the moral of this enormous and monumental piece of facetiousness.

Among the many workers in marble, stucco and polychromed decoration, the most outstanding was Giacomo Serpotta (1656–1732), who displayed an unusual virtuosity in the modelling of small scenes in bas-relief. The Oratorio di S. Lorenzo (1699–1706) and that of the Rosary in S. Domenico (c. 1710–17) are entirely decorated with his small figures and cartouches in a delicate relief, frequently with finely worked detail. It is a Rococo style, partly derived from Roman example, but developed in accordance with local taste, that already contains Neoclassical elements.

II

Painting: Rome, Bologna, Genoa and Naples

Full Baroque is the art of Bernini and Pietro da Cortona; it is a decorative and architectural system. But not every aspect of the development of Italian painting can be included in its framework. Aspirations towards the grandiose, and the display of allegorical or sacred themes, are an expressive feature of the taste of the time, and had repercussions everywhere, but there was also, as an opposing current, a constant effort to endow the painted representation with a specific organization, and to give depth to pure painting. This tendency was set on a firm basis in Rome, between 1600 and 1610, by two groups that had much in common and which it is often the custom—though an artificial one—to oppose one to the other. One proclaims itself as a break with almost everything that preceded it; this is the 'luminism' of Caravaggio. The other is the clear and idealized painting of the heirs of the Carracci. Both established themselves energetically, though not without impinging frequently upon each other, during the first half of the century, when the chief field was in Rome. In the second half of the century, the provincial schools, above all Naples, Bologna and Lombardy—to say nothing of Venice, until then something of an outsider—once more resume an important role. But the lesser genres attract more and more artists towards the painting of landscapes or town views (*vedute*), scenes of everyday life, or fantasies, through which Italian art found an answer to the art of the Netherlands. But the reactions emanating with increasing strength from Neoclassicism after 1750 were a sterner test. The community of feeling between the Neoclassicists of Milan, Venice and Rome, stimulated by example from London, Paris and Dresden, brought about a rapid drying up of Italian vitality.

Caravaggio and Luminism

There are few revolutions quite so clear-cut as Caravaggio's; it is the most remarkable in all Italian art. Violent temperaments such as his are not infrequent in the arts, but to the personal rebelliousness and pride were

339

added his stormy life and even the glimpse of the Roman underworld, which helps to explain his sudden rise and his dramatic end.

His career has but few fixed points. He was born in 1573, in Caravaggio, a village in Lombardy. He received his early training in Milan with Simone Peterzano, came to Rome towards 1588 and worked for the Cavaliere d'Arpino. Then, in 1590, the patronage of Cardinal del Monte obtained for him an important commission—that of the chapel of S. Matthew in S. Luigi dei Francesi—on which he worked for ten years. In 1603, the artist, who was already famous as far afield as the Netherlands, was imprisoned for libel, as a result of the complaints of Baglione. During the years that followed, his brushes with the police became more and more frequent—quarrels, fights, scandals, flight to Naples, Malta, and Sicily. After being wounded in a brawl, Caravaggio was returning to Rome when he died of malaria at Porto d'Ercole, on the 18th July 1610, when under thirty-seven years of age.

His *oeuvre* consists of about forty-odd works, and about a score of attributions or copies, the chronology of which is particularly difficult to establish. X-rays have revealed under the surfaces of the S. Luigi dei Francesi compositions, first sketches which now rule out any question of his having had a steady evolution towards strong contrasts of light and shadow, and of any classification of his works based on such a theory.[153] What really disgusted this formidable personality was Mannerist art, with its allegories and conventional forms; what his instinct demanded was 'natural painting'. The Venetians apart, there were in Lombardy examples of serious and simple painting, with Moretto and Moroni, and of popular painting with night effects, with Romanino and Campi. With absolute directness, Caravaggio created a brutal and straightforward vision of the incidents in the Bible or mythology that he elected to represent: Bacchus is an urchin with curly black hair, the *Calling of S. Matthew* takes place in a kind of guard room where a game of cards is in progress, the executioners of S. Peter are muscular bullies (1610, S. Maria del Popolo, Rome), and the grave diggers in the S. Lucy are clumsy peasants (1608, Syracuse); the Virgin who receives the devotions of the pilgrims is a serious Roman matron (1604-5, S. Agostino, Rome), and in the *Death of the Virgin* (1607, Louvre) she is a peasant woman in a smock whose body is swollen by decomposition.

This has been interpreted as the expression of a new kind of piety preparing, for example, the realist developments of Spanish art, and, linked with a revival of Franciscan thought, appearing as a new Biblia Pauperum. But there is also present in Caravaggio an urge to create a scandal, almost as a part of his particular genius, of his crude and violent vision, making no concessions, like that of Courbet. He exceeds the bounds of religious inspiration; the choice he makes of equivocal or vulgar types (*Bacchus*, Borghese) with which to represent the Apostles or *Eros* (Berlin), stimulated a taste for genre scenes and realist secular subjects, while at the same time it brought more humanity into religious art.

CARAVAGGIO AND LUMINISM

In this painting which sought to avoid preconceived notions, the essential lies, not in the atmosphere, the shimmer of the colour, or the graceful outline, which he dismissed as adventitious adjuncts, but in the solidity of the bodies, the massive and detailed reality of the object. His poetic discovery was to reveal this by the most powerful contrast—that of light and shadow, which bathes large parts of the figures in darkness and allows a face, the nape of a neck, a hand, to emerge from the gloom with an astounding relief, which the painter enjoyed accentuating in the *Narcissus* (an attributed work of *c.* 1595, Borghese) or the *David* (1605, Borghese). Hence the strong contrasts in the composition, which, to display to the full the sudden shafts of light, the sharp and telling edges, is made symmetrical or irregular, or balances forms in violent foreshortenings emphasized by strongly stressed foregrounds, as in the *Conversion of S. Paul* (S. Maria del Popolo). The abrupt placing of the subject within the picture space, the use of diagonals, were popularized by decorative painters, though with far less sense of drama; the strength of Caravaggism lies in its feeling for the nocturnal aspect of the world, in the quality of surprise gained from its hard light effects, and in the movement of large masses of form.

There is, therefore, no point in enclosing Caravaggism within the framework of the Baroque. It proceeds from quite different premises which, as the spread of the movement in Europe clearly demonstrates, can be allied equally with the severity of classicism (for example, Latour and Vermeer) or the fullness of a typically Baroque attitude, as in the early works of Rubens. Among the crowd of Tenebrists in Rome during the years 1610–20, there were Italians who tended towards classicism, like Orazio Gentileschi or Saraceni, and others who adopted the second course, like Serodine. Caravaggism was a generally liberating influence, and had repercussions everywhere. In some of the works of Orazio Borgianni (*c.* 1578–1616) such as the *Birth of the Virgin* (1613, Savona) is even to be found a first step towards the glittering and neurotic art of Magnasco, which could be described as the Rococo of Tenebrism.

Caravaggio disappeared from Rome certainly after 1607; therefore the success of his style would not perhaps have been so rapid and so widespread had it not been taken up and propagated by an excellent imitator, as much as follower, Bartolomeo Manfredi (*c.* 1580–1620) whose *Concert* (Uffizi) long passed as a genuine Caravaggio, and who was the teacher of the majority of the Northerners in Rome. He also painted a *Bacchus with a drinker* (Gal. Naz., Rome) where the drawing is not very distinguished. The new doctrine of Tenebrism was received by painters of every school, and even opponents of Caravaggio, such as Giovanni Baglione (1571–1644) the future historian, who endeavoured to paint in pale colour, or his pupil Salini (1570–1625), were affected by it.

Orazio Gentileschi (1565–1638) derived from Tuscan Mannerism. In Rome, he knew Reni and Caravaggio, whose style he treated in a very personal

PAINTING: ROME, BOLOGNA, GENOA AND NAPLES

way, modelling the forms delicately (*Flight into Egypt*, Louvre), softening the composition (*Lot and his daughters*, pvt. coll., London), developing the landscape (*S. Christopher*, Berlin). After having worked in the Marches, Genoa and Paris, he went to England and worked from 1626 onwards for the Court of Charles I. His daughter Artemisia (1597–after 1651) was talented and precocious, interested in odd effects of form seen in strong light and shadow, and retained a feeling for elegance even in scenes of extreme violence; she worked in Florence (*Judith*, Uffizi) and then in Naples, where she settled about 1630 and in contrast to Ribera represented a more delicate form of Caravaggism (*Judith*, Naples).

The Venetian Carlo Saraceni (*c*. 1585–*c*. 1625) approached the painting of light effects from a Giorgionesque starting point (*Rest on the Flight*, 1606, pvt. coll., Frascati) and was first interested in landscape, like the Rhinelander Elsheimer, and in genre scenes like the Frenchman Le Clerc. After 1610, his manner became broader (*S. Cecilia*, Gal. Naz., Rome), more sober and even lighter in colour as in the *S. Raymond preaching* (S. Adriano, Rome). Giovanni Serodine (1594–1631) came from Tessin on the Italo-Swiss border and only arrived in Rome about 1615. He brought to the Caravaggesque manner a liveliness of touch and an invigorating feeling for colour, for example, in the *Sons of Zebedee* (1617, Ascona), in the *S. Lawrence* (*c*. 1620, Valvisciolo near Sermoneta) and in the *S. Peter reading* (Züsi Coll., Lugano).

The group of militant Caravaggesques included many other painters from all over Europe. During at least ten years—1610–20—naturalism and tenebrism were in the forefront of men's minds, and they contributed to the rise of new genres. Caravaggio himself declared that 'it cost him as much effort to paint a good flower piece as a picture with figures in it', and the *Basket of fruit* (*c*. 1596, Ambrosiana, Milan)—whether or not it formed part of a larger work—proves this. History painting had to be saved from Academic conventionalism, just as still-life, or what was called the painting of 'cose naturali' (natural things), had to be allowed to escape from decorative programmes.[154]

His initiative was followed even by minor masters well outside the group, like Pietro Paolo Gobbi, known as Gobbo de' Frutti (active *c*. 1630–50), and Mario Nuzzi, known as Mario de' Fiori (1603–73) both of whom worked in Rome. In Naples, there was Paolo Porpora (1617–*c*. 1670), who was fond of painting still-life subjects of heavily laden tables in the Dutch fashion,[155] and his pupil, G. B. Ruoppolo (1620–85). In Northern Italy, at Bergamo, there was the astonishing Evaristo Baschenis (1617–77), who specialized in painting beautiful lutes and violins, with full and rounded bodies, and gilded scrolls emerging from the shadows, and huge kitchens heaving with meat and game.

Naples. After the middle of the fifteenth century, the only painters in Naples had been birds of passage. The arrival of Caravaggio brought about a revival in the Southern school, which began with G. B. Caracciolo, known also as

CARAVAGGIO AND LUMINISM

Battistello (*c.* 1570–1637), who was directly inspired by Caravaggio in his *Liberation of S. Peter* (*c.* 1612, Monte della Misericordia) and his *Miracle of S. Anthony* (*c.* 1607, S. Giorgio dei Genovesi). His compositions are rhythmically articulated with bands of light and shadow (*Washing of the Feet*, 1622, Certosa di S. Martino) but later become milder due, probably, to the influence of the Bolognese. The revival in the South is also evident in Andrea Vaccaro (1598–1670) who worked at the Certosa di S. Martino where he painted a *Massacre of the Innocents* with two huge executioners silhouetted as if in relief (Naples, Mus.) and in Massimo Stanzioni (1585–1656) a painter of numerous altarpieces such as the *Virgin appearing to the Carthusians*, a dramatic *Deposition* (Certosa di S. Martino) and sentimental figures like the *S. Agatha* (Naples, Mus.).

Reinforced by the presence of Artemisia Gentileschi after 1630, the movement gained in importance, and was also helped by several powerful artists among whom was the Spaniard Ribera (1588–1652), of Valencia, who came to Italy before 1610, settled there and became Court Painter in Naples. The aspect of Caravaggio which above all attracted Ribera was his realist violence, but he vulgarized it by conventional additions, like the little angels in his *Deposition* (Certosa di S. Martino), by anatomical crudeness in the *S. Sebastian* (Naples, Mus.) and by ponderous joviality in his *Drunken Silenus* (Naples, Mus.), which explains his lack of success with delicate subjects like *Venus and Adonis* (Rome, Corsini). As a portrait painter he is beyond compare (*Missionary*, Poldi-Pezzoli, Milan). In his use of vigorous diagonals like those of Tintoretto, he provided a clear impulse to the type of Baroque which finally found expression in Luca Giordano. Nevertheless, a Neapolitan, Aniello Falcone (1600–65), who possibly was a pupil of Ribera, used in landscapes and even in battle pieces an ordered composition and pale colour which place him right outside Ribera's orbit.

In the art of Mattia Preti (1613–99) Caravaggism becomes much more complex. He went to Venice and to Flanders, was interested in the works of the great decorators and displayed an elaborate learning which excludes facile interpretations. Basically, he is a Tenebrist, but he used more pallid colour in S. Andrea della Valle (1651), and composed a brilliant decorative ensemble for S. Pietro a Maiella in Naples. His allegories of the plague, inspired by the disasters of 1656, maintain, in the sketches in the Naples Museum, a grand sense of movement, but his most successful works were the three banqueting scenes of *Dives and Lazarus* (Corsini Gall., Rome), *Belshazzar*, and *Absalom* (both Naples, Mus.), where against a background of arches recalling something of Veronese, a fitful daylight gives flickering effects of colour, and, as in Tintoretto, lends drama to the scene where the expressions of fear and cruelty are forcefully delineated.

This is really a kind of Romanticism, and the same may be said of Luca Giordano, who has already been mentioned among the decorators in the following of Pietro da Cortona. He was an extraordinarily prolific artist, and

343

mixed, in his rather cluttered but richly diversified compositions, facets from Neapolitan, Roman and even Venetian style. With him Caravaggism is a long way away, and the same may be said of Salvator Rosa (1615–73), as much a poet and engraver as a painter, whose place is principally among the regenerators of landscape painting. After the second half of the seventeenth century, Tenebrism is no more than one of the elements of all current painting; it is no longer the property of any school, and true naturalism disappeared in the often subtle search for expression.[156]

But in Spain an artist like Luca Giordano was a decisive figure, and it is his followers, forming a group of Baroque artists specializing in the grander effects of light, stormy skies, and swirling figures after the fashion of the Neapolitans Solimena and Francesco de Mura, whom it is now customary to suggest as a formative influence on Goya, rather than Tiepolo's Baroque of pale colour and downy clouds. This Roman and Neapolitan Baroque which prolonged Ribera's influence and was acceptable in Spain, has as its last representatives Corrado Giaquinto (1699–1765), Giuseppe Paladino of Palermo, and the Roman Gregorio Guglielmi, who all worked on the decoration of the church of the Trinitari Calzati (1746–49) in the Via Condotti in Rome. Guglielmi also worked at Schönbrunn in a style at once daring in perspective and rigid in his treatment of form, reflecting nothing of Tiepolo's dominance.[157] Giaquinto, who executed the gay and glittering decorations in S. Croce in Gerusalemme, worked with Sebastiano Conca (1680–1764) in S. Lorenzo in Damaso. Conca's Roman career, from 1723 until his departure for Madrid in 1753, made him the leader among a small group of Southern painters, of dark and romantic feeling, among whom were Onofrio Avellino (d. 1714) and—though they retained more of their native Neapolitan style—Gaspare Traversi (d. 1768) and Giovanni Bonito (1707–1789), the latter a lively portraitist.

The Academic Movement

Mannerism had always oscillated between the more arbitrary aspects of fantasy and a conventionalized aestheticism. After Lomazzo, Zuccaro, and the academic theorists, the problem was to weld together the 'ideal image' which most fully represented the artist's inward inspiration—his 'idea'—with fidelity to the external characteristics of human beings and objects. What was needed was to give a new aspect to naturalism, taking a general view instead of concentrating on precise realization or deliberate abstractions. This is the point at which the Carracci's teaching intervened, and the import of this has already been discussed. By comparison with the arbitrary and intellectual nature of Mannerism, it represented a return to nature; but in face of the total naturalism of the followers of Caravaggio, it presents the starting point of a new idealism, full of pride in its responsibilities. This may be recognized in the concentration of means which brought about a return, in the cause of unity of action, to the frieze composition, and results in some instances in the elegant clarity of a Greek relief, and in others in a balanced distribution of pale tones in a brilliant light. It may also be recognized in the taste for lively narrative scenes, and in the predilection for allegories, Seasons, Virtues, the Ages of Man or Muses which are really variations on a single type, demonstrating the culture of the artist.[158] This art had three principal exponents divided between Rome and Bologna: Domenichino, Guido Reni, and Guercino.

Bologna and Rome: Domenichino, Guido and Guercino. Domenichino (1581–1641) was born in Bologna and was one of Annibale Carracci's assistants in the Farnese Gallery. With his *Martyrdom of S. Andrew* (1608, S. Gregorio al Celio, Rome), painted in fresco, appeared the rather dry drawing, clear and forced, imitated from Raphael, which characterized alike his qualities and his inadequacies. In the chapel of S. Nilo in the Abbey at Grottaferrata (1608–10), the narrative scenes have an archaizing quality, and this may also be said of the ceiling painting of the *Chariot of Apollo* (1613–14, Palazzo Costaguti, Rome), which may be compared with similar compositions by the other two leading artists of Romano-Bolognese Academism. In the huge pendentives of the dome of S. Andrea della Valle (1624–1628), Domenichino painted the Evangelists with a forcefulness of handling derived from Michelangelo, which is not to be found in the remaining parts of the church executed by him, nor in the *Virtues* in S. Carlo ai Catinari (1630), nor in S. Gennaro in Naples (1630–41). Two works ensured that the artist's fame endured until the eighteenth century: the *Last Communion of S. Jerome* (1614, Vatican) which is a more pathetic and more deeply studied version of Agostino Carracci's picture, and the *Hunt of Diana* (c. 1620, Borghese) already of a slightly Rococo taste and agitation, but set in a delicate landscape; it is the balance between horizon and planes which Poussin found interesting in Domenichino's art, and also the repertory of narrative gestures and poses which, for more than a century, established a complete and penetrating language for the expression of emotion.

It was, however, in the art of the Bolognese Guido Reni (1575–1642) that the classicizing reforms of the Carracci reached their apogee. He came to Rome with his friend Francesco Albani towards 1602, and there painted a *Charity* (c. 1604, Pitti) and a *Crucifixion of S. Peter* (1602–3, Vatican), in a Caravaggesque style, then frescoes in S. Gregorio al Celio; a martyrdom in the chapel of S. Andrew, and a superb corona of angel musicians in the chapel of S. Silvio. The *Samson Victorious* (c. 1610, Bologna) is proof of his growing interest in the nude, studied in Antique sculpture, and the *Massacre of the Innocents* (Bologna) displays a forcefully dramatic concentration, a clear composition quite different from Caravaggesque pathos, and an attitude which presages Poussin. Reni returned to Bologna in 1614. With the *Assumption*

(S. Ambrogio, Genoa), designed in two layers, begins his series of more confused works, packed with incident, of which another example is to be found in the *Madonna of the Rosary* (Bologna). What was to have the greatest success and to be reproduced to a point of satiety in church art were his devotional figures typical of the new, sentimental and facile piety, such as the *Mater Dolorosa* (pvt. coll., Bologna) and the *Ecce Homo* (Louvre), of distressing insipidity. His mythological compositions, though possessed of a flaccid smoothness, are happier in their arrangement and their colour, as, for instance, in the *Aurora* in the Casino Rospigliosi (*c.* 1613–14) and the four canvases commissioned by the Duke of Mantua about 1620 on the subject of Hercules (two in the Louvre) which have more vitality. The most elegant and famous of these works, where classical taste reaches a pitch of great purity and fulfilment, is the *Atalanta and Hippomenes* (1625, Naples).

Among the early followers of the Carracci, guided by Reni, were Francesco Albani (1578–1660), who worked at S. Maria della Pace (1611–14) and who composed in the Palazzo Verospi in the Corso graceful medallions with figures of *Days* and *Rivers* of a pastoral mythology. The same sources of inspiration reappear in even softer guise in the circular panels of the *Elements* where rosy *amorini* dance in tender landscapes (Turin). His altarpieces, pale in colour and rather soft in drawing, like the *Martyrdom of S. Andrew* in the church of the Servi in Bologna, suffer from his great facility, which is less obtrusive in the pretty autumn landscapes which form a setting for some of his mythological scenes, such as the *Hermaphrodite and Salmacis* (Louvre). He represents the delicate and nerveless extreme reached by the most highly refined trend of neo-Hellenism. This limpid and measured art is Andrea Sacchi's (1599–1661) starting point alike for his portraits and for his altarpieces (*S. Romuald*, Vatican).

A certain number of Guido's compositions, during the 1620s, such as the *Sacred and Profane Love* (Pisa), and the *Infant Bacchus* (Pitti), are a more placid elaboration of motifs current in Caravaggesque circles. The same tendency to use these new ideas may be discerned in minor masters such as Il Masteletta (1575–1655) who was principally a landscape painter, and Lionello Spada (1576–1622), and finally also appears in the art of the third major painter of the Academic persuasion, Guercino (1591–1666). Trained in Bologna, and too young to have been more than a pupil of Ludovico Carracci alone, he went to Venice in 1618, as is shown by his *S. William of Aquitaine* (1620, Bologna). His masterpiece, the *Aurora* in the Casino Ludovisi in Rome (1621; sole remaining portion of the villa destroyed in 1870), is, in its use of foreshortening, its contrasts of colour, the important role played in it by landscape, of an almost romantic grandeur when compared with the earlier work by Reni. The same impression is given by other delicate landscapes in the Casa Pannini at Cento (*c.* 1615; now in the Mus. in Cento). This is because Guercino was able to absorb the use of strong chiaroscuro, and the *S. Petronilla* (Capitoline Mus., Rome) painted at this period, displays

D   Guercino: *The Mystical Marriage of S. Catherine*
(*London, Denis Mahon Collection*)

in the gravediggers an imitation of Caravaggio's realism. The artist's last period, when he was working in Piacenza, was much heavier (*Prophets*, 1626, Cathedral), but in some of his half-figure compositions (*Hagar*, 1657, Brera) a Correggiesque sensibility is in evidence.

In the followers of these artists academism involved loss of all power of conviction. After Albani followed Carlo Cignani (1628–1719) whose pictures are over-emotionalized (*Joseph*, Dresden; *Madonna*, Uffizi), and Carlo Maratti (1625–1713), who developed the formula in innumerable altarpieces. Guido Reni had a large following, among whom were Guido Canlassi (1601–81) in Romagna, and Andrea Sirani (1610–70). But the success of the perspective painters, led by the decorators G. Curti and A. M. Colonna, caused the Bolognese School to develop 'quadratura' or scenic perspective painting, in which, after the Bibiena family, minor artists like V. M. Bigari (1692–1776) and M. A. Franceschini (1648–1729) achieved prominence.

It was from a very broad international culture that, towards the end of the century, the Bolognese Giuseppe Maria Crespi (1665–1747; also known as 'Lo Spagnolo') emerged. From the older Bolognese, such as Guercino, he retained all the subtlety of chiaroscuro, from Venetians their vibrant use of colour, and from the Dutch their love of genre subjects. He shows his ability as an inventive decorator in the *Gods* and the *Seasons* in the Palazzo Pepoli, Bologna (1692) and as a painter of religious pictures in scintillating colour, though treated with homely realism, as in his series of the *Seven Sacraments* in Dresden. Yet he tends above all to the painting of genre, ranging from portraits expressive of atmosphere (*Count Fulvio Grati*, pvt. coll., Bologna, and *Self-portrait*, Bologna), to delicate and monochromatic still-life subjects (*Bookshelves*, Conservatorio, Bologna), and scenes of popular life. In these, a heavy sky and a lowering atmosphere are enlivened with many flashes of light (*Peasant family*, Budapest; *Charity*, pvt. coll., Milan) and sometimes his landscapes achieve a dark and stormy quality as in the *Fair at Poggio a Caiano* (Uffizi). His mythological compositions have similar characteristics (*Nymphs and amorini*, Kress coll., New York).

Lombardy. A parallel development, though on a strictly provincial level, may be observed in Lombardy, where the tradition of faithful realism, in Moroni's sense of the term, survived to the point that during the seventeenth century his portraits were considered on a par with those produced in Venice, and his *Schoolmaster* was even taken for a Titian. Painting in Milan and Bergamo broke with Mannerism, and during the first half of the century turned towards Rome and Bologna; later it tended, in portrait as in genre scenes, towards a more generally European type of realism. The first generation is that of Giovanni Battista Crespi (1567–1632; also known as 'Il Cerano') and Morazzone (1571–1626) who were painters of sentiment and feeling in the sense implied by the Counter-Reformation and the aesthetics of Cardinal Borromeo, the first in his *Madonna with SS. Charles and Bruno* (Certosa of

Pavia), and the *Nativity* in Turin, the second in his harsh *S. Francis in ecstasy* (Milan, Castello) and his *Annunciation* (Lucca), where the colours melt into the chiaroscuro. Both collaborated with Procaccini (*c.* 1570–1625) in the involved and thoroughly Tenebrist *Martyrdom of SS. Rufina and Secunda* (Brera), and the series of ten pictures of the *Miracles of S. Charles Borromeo* for Milan cathedral.

The echo of Caravaggism is still stronger, after 1620, in the delicate and calculated style of Daniele Crespi of Busto Arsizio (1592–1630); in his *S. Charles at table* (S. Maria della Passione, Milan) the still-life on the table is sober and striking, and his *Dead friar* (Brera) could pass for a Velazquez. His pupil, Melchiorre Gilardini (d. 1675), is no less skilled in contrasts, and the same may be said for the pupils of Morazzone, like Tanzio da Varallo (1575–1635) whose figures are powerfully modelled in strong light, Francesco del Cairo (1598–1674), who also painted nocturnal scenes, and Carlo Francesco Nuvolone (1608–61), a tender portraitist, deriving from Guido Reni. The springs of inspiration of Evaristo Baschenis, who has already been mentioned among the Caravaggist practitioners of still-life, represent quite another aspect of Lombard painting.

At the end of the seventeenth century, Magnasco's position in Milan is not dissimilar to that of G. M. Crespi in Bologna; a highly strung sensibility, tending towards a very personal vision, and an interest in low life picturesque are expressed in abbreviated and glittering forms which are outside the range of local practice and are better considered in quite another light—that of landscape and the *capriccio*. But Magnasco is an artist apart. A Milanese by adoption, he returned to Genoa in 1735 at a moment when an art as different from his as it is possible to conceive was in favour in Lombardy. The representatives of this were Crivelloni (d. *c.* 1730) and Tavella (1668–1738), a landscape painter, both influenced by Dutch painting, and the trio of 'painters of reality': Giuseppe Ghislandi (also called Fra Galgario, 1655–1743), a Bergamasque portrait painter, straightforward, solid, and of a remarkable fluency in the handling of low tones (*Young man*, Poldi-Pezzoli, Milan; *Prelate*, pvt. coll., Milan; *Isabella Camozzi*, pvt. coll., Costa di Mezzate); Antonico Cifrondi of Chesone (1657–1730), more limited in his portraits of artisans; and finally Giacomo Ceruti, who worked between 1724 and 1738, and who appears almost a latterday Le Nain of the peasantry, the labouring poor and beggars (*Peasant leaning on his spade*, pvt. coll., Milan; *Child with a basket*, pvt. coll., Treviglio; *Family at the hearth*, pvt. coll., Brescia; *Two beggars*, pvt. coll., Brescia; *The washerwoman*, Brescia), and as an unaffected portraitist (*Old peasant woman with a wicker basket*, pvt. coll., Brescia; *Two sisters*, pvt. coll., Milan). His sober compositions, grey and restrained in colour, independent of current fashions and with a certain harshness even, record the sadness and unhappiness of the poor in the middle of the eighteenth century.

Landscape, 'Veduta' Painting and Capricci .

In Italy, the Baroque period did not witness anything like the development which took place in landscape in the Netherlands with Rembrandt and Ruysdael. The greatest interpreters of the Roman scene, and of the ideal landscapes, were the Italianized Frenchmen, Poussin and Claude Lorraine, who were guided in their chosen direction by the works of Annibale Carracci and Domenichino.[159] The grand allegorical style in decoration, the Caravagesque realist scene, the conscious balance of the Bolognese, all of which were dependent on strictly studio work, in no way advanced the independent study of nature and its more accidental quality. The architectonic and sculptural aspects of Italian Baroque only sanctioned as an offshoot of its large compositions a romantic and imaginative treatment of landscape, conceived as a form of artistic escapism. Hence the association of views of nature with all the more wayward aspects of the imagination.

From the picturesqueness of Naples in the first half of the century, full of conspirators and adventurers, there emerged a curious painter, given the name of Monsu Desiderio (c. 1590–1650) by the chronicler Bernardo de Dominici (1742), who records him as a painter of 'vistas and perspectives'. To him are attributed a certain number of panoramic views of Naples, one of which (Naples Mus.) has on the back the name Desiderius Barra of Metz, which suggests that the painter was a Lorrainer, in which case he is not the author of some engravings dated 1630–31 (Berlin Print Room), signed by Desiderius Pistoriensis. It has recently been discovered that the chief painter of these pictures of imagination, or 'capricci', related to topographical panoramas, was a François Denomé. These compositions are absolutely extraordinary. Church naves explode in a lurid glare which recalls Beccafumi (pvt. coll., London); saints are martyred in the moonlight in front of buildings with Corinthian colonnades and half-ruined Gothic windows (pvt. coll., Cleveland); fantastic ghostly cities offer a nocturnal version of the archaeological constructions of Antoine Caron; the flickering quality of statues and capitals is accentuated by an impasto like rivulets of gold or silver (Louvre, Mus. des Arts Décoratifs). This is the Mannerist taste for scenic effects and monumental vistas transmogrified into the bizarre and the fantastic, peopled in the scenes of horror with pygmies gesticulating like frightened insects.

These works also appear to have had considerable currency among foreign collectors, but they constitute a notable symptom which is underlined by the influence of Callot in Italy. Monsu Desiderio, alias Denomé, was a compatriot of Callot, whom, in any case, he would have known in Florence where Callot worked from 1619–29. In 1612, the latter engraved the collection of scenes of the commemorative ceremony in honour of the Queen of Spain; his pupil, Stefano della Bella (1610–64) engraved that of Francesco de'

Medici in the same style in 1634; he left sketches, costumes for balls, fantastic figures (which are reminders that he was a stage designer in Paris) and delicate and panoramic drawings of Roman tombs and views of Venice, *c.* 1661. To him may be likened the Genoese G. B. Castiglione (1610–65) who was attracted by Van Dyck in painting and Rembrandt in engraving; he went to Rome towards 1634 and returned there frequently. His scenes from the Bible and from fable are confused pastorals, closer to Bassano than to Poussin, but with a fascinating liveliness of handling in numerous drawings in which landscape plays a large part (*Oriental figures before a Herm,* Windsor). The discontinuous composition, the oddness of the forms, are also to be seen in his pictures of tombs, his scenes of Noah, Rebecca, and similar subjects. Salvator Rosa much admired him. His last works and *bozzetti* display a passionate movement and a nervosity of handling instinct with ecstasies and visions. The great vogue for his works came in the eighteenth century, with Sebastiano Ricci and Tiepolo, and with Marcola (1711–80) and Palmieri (1720–*c.* 1804) who directly imitated him, in a style expressive of pastoral and Rococo fantasy.

Salvator Rosa and Magnasco. Salvator Rosa (1615–73), who came from Naples, was an artist of full and agitated temperament, a great improvisor, the author of vehement 'Satires', who passed as an eccentric because, according to Passeri, he painted his landscape studies actually on the spot. He worked in Naples, Rome, and Florence, where he painted from 1641 to 1649 'landscapes, seascapes and battles' (*Forest scene with figures, Seascape with two towers, Cavalry battle,* Pitti). Because he conflicted with academic teaching, Salvator was accused of inadequate drawing, and of being satisfied with dabs of colour; his landscapes of rocks and trees, with screes, torrents, tattered trees, are the counterpart to the dramatic landscapes of Ruysdael. This tormented vision is the opposite of Poussin's, who describes an idealized and heroic nature. It is of a violently naturalistic character, as may be seen in his battle subjects where he insists on the confusion, the dust and the muddle of the struggle. Everything is seen on a small scale and in the disorder of actuality; for this reason, he is far from Caravaggism which isolated incidents and enlarged the figures. He opens a new register in painting, by establishing novel and immediately successful genres.

Salvator had a following: a Frenchman, Jacques Courtois, Italianized into Giacomo Cortese (1621–76), and Michelangelo Cerquozzi (1602–60), who were also painters of battle scenes. The second also acquired a name in another specialized line, the 'bambocciata', which was developed from the paintings of Pieter van der Laer, better known by his nickname of Bamboccio, a Dutchman who came to Rome in 1628, and created a fashion for burlesque scenes, with peasants in taverns and other clowns, adapting thus to Rome the manner of Ostade and Teniers. The Roman caricaturist P. L. Ghezzi (1674–1755), and Antonio Amorosi (1660–1736) may also be cited, and the latter's use of light is as remarkable as the vigour of his themes (*Peasant man and woman*, pvt. coll., London).

Alessandro Magnasco (1667–1749) derives both from Callot and Salvator Rosa. He made his career between Genoa, Florence (where he lived 1723–27) and Milan, working in landscape, genre subjects and the *capriccio*. Silhouettes of monks, beggars, ruffians treated with a summary and caricatural touch, are scattered in tormented valleys and rocky landscapes uniformly dark and shattered, and from these elements he created a new and at the same time impressionist and fantastic manner; he may be considered as the Watteau of the mob (*Carpenters*, pvt. coll., Venice; *Scene in a garden*, Palazzo Bianco, Genoa).[160] Often he sets in deep central perspectives banqueting scenes pullulating with monks or with minute soldiery (Bassano Mus.); some of his canvases describe torture scenes in a landscape of ruins (*Martyrdom of S. Erasmus*, pvt. coll., Milan). His poetic feeling for the *capriccio* is stressed in the *Punchinello with a guitar*, sketched in with the liveliest brush (pvt. coll., Venice); his feeling for the fantastic is evident in a picture of skeletons in the church of S. Maria di Campomorto at Siziano, near Pavia.

In the first quarter of the eighteenth century, Italian landscape, reinforced by French and Dutch example, thus arrived at a new stage of maturity. To Magnasco's torments was opposed the heritage from Poussin who, it must not be forgotten, enjoyed such prestige in Rome that in 1658 the Academy of S. Luke had considered electing him as their Principe.[161] To this current belong the often facile works of Andrea Locatelli (1660–1740), strongly inspired by Gaspar Poussin and, aided by the use of tumultuous accidentals derived from Salvator Rosa, those by Pietro Mulier, appropriately nicknamed 'Tempesta', who, after working in Rome, Genoa and Venice, settled in Milan where his milder aspects were continued by Carlo Antonio Tavella.

Views of Rome: Painters and Engravers. By about 1700 there was another fashionable craze: that for opera and imaginary architecture, of the kind created by Ferdinando and Francesco Bibiena. Given wide currency through engraving, at the same time as they were developed in the theatre, these compositions, which are a synthesis of Baroque aspirations in monumental art, stimulated speculation on invented space, in the same way that the perspectives of a Peruzzi had sharpened and defined the character of the classic space of the Renaissance. The part of painting proper which most closely corresponds to them is that of the Roman *vedutisti*, or painters of townscapes, and the ground was prepared by the architectural compositions of Viviano Codazzi, who worked first in Rome and then in Naples. As in Magnasco, minute figures are disposed in a lively space, but to his world of *lazzaroni* they opposed a perfect good taste: the golden light of Rome instead of the stormy skies of 'Savage Salvator', monumental perspectives, large open squares set sometimes with ruins amid trees, instead of stygian precipices. The most remarkable of the painters who evoked the beauty of Rome was Giovanni

PAINTING: ROME, BOLOGNA, GENOA AND NAPLES

Paolo Pannini (1691–1765), whose genre was pursued, though in a more Flemish manner, by the Frenchman Hubert Robert. His views of the Quirinal are examples of real open-air painting, with atmospheric vibration and effects of light and shadow (Quirinal palace). Trained by the Bibiena, he delighted in colonnades, in glimpses of architectural distances into which the crowd of figures introduce a myriad points of light (*Opening of the Porta Santa*, formerly London). His commemorative scenes of state entries and royal visits (Naples, Mus.) are less effective. This art was to reach its highest point in the works of the greatest of Italian etchers, Piranesi.

It was from sixteenth-century practice that the tradition of collections of engravings of Roman buildings derived, and it was continued in the seventeenth century by G. B. Falda in his books of engravings of buildings, fountains and gardens (*c.* 1670), and then by Giuseppe Vasi in his *Magnificenze di Roma antica e moderna* (1747–61). Vasi inspired the Venetian G. B. Piranesi (1720–78) who had studied as an architect—in 1765 he furnished plans for S. Maria del Priorato—and had settled in Rome in 1740. Of a powerful, violent, extrovert temperament, Piranesi devoted himself with passionate intensity to collections of huge etched plates—a series of *vedute* (incomplete: 137 plates, begun in 1746) in which the landscape of Rome was treated with the richest effect. The *Antichità Romane* (in four volumes, 1756) were the new, exciting and varied record of the remains of antiquity; the *Carceri* (*Prisons*: 1750, re-issued in 1760) are staggering fantasies, an architectural dream-world, where Bibiena's art is subjected to a hallucinatory invention. Piranesi sums up all the mutations of taste of the century: from the classical standpoint of scientific plates, with the subject seen frontally, to the Baroque of *di sotto in sù* perspective which endows the building, or ruin, with an extraordinary vitality, to the wildest flights of architectural imagination which had survived as a vital undercurrent during the whole period. This intrepid and learned genius possessed an exceptional technical ability; he effected an easy transition from the 'impressionist' style, allusive and diffused, of Venetian painters, to a hard manner, more in conformity with Roman practice, and designed to express fully the plastic quality of reliefs and antique marbles. Italian painting appears to abdicate, towards the middle of the eighteenth century, in favour of black and white which sums up all its preoccupation with chiaroscuro and perspective, space and atmosphere.

113 Giuseppe Ghislandi (Fra Galgario): *Portrait of a man*

(*Bergamo, private collection*)

114 Piranesi: *Sculpture in Arcade*

(Cambridge, Mass., Fogg Art Museum)

115 Tiepolo: *The Meeting of Antony and Cleopatra*
(*Venice, Palazzo Labia*)

116 Longhena: S. Maria della Salute, Venice

(a) Guardi:
  S. Maria della Salute
  (drawing)
(b) de Pisis:
  View of the Salute
  (painting)

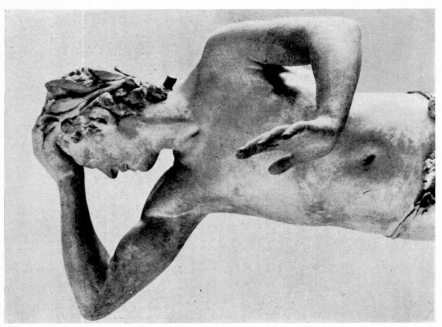

119 G. Calderini: Rome, Palace of Justice

120 Rome, Fountain in the garden of Palazzo del Grillo
(18th century)

III

Venice

There is little point in arguing whether or not Baroque penetrated into Venice; for many people, the city of the lagoons, with its architecture eternally reflected in the waters, its decor of glimpsed vistas, its enveloping atmosphere, embodies the very essence of the Baroque, which consists of a love of the fleeting appearance and of the spectacular. For the historian, the perspective is different. Venice did not react during the seventeenth century to the true Baroque style of 1630, which embraced architecture, painting and decoration; painting played a minor role, and architecture is summed up in the name of Longhena, who adhered strictly to local precedents. The Early Christian, Byzantine, and Gothic heritage of Venice released her from the obligation of being Baroque in the sense that Rome, Turin, or Catania were. She remained so obstinately faithful to her sixteenth-century style that, towards 1730, she was able to reassert her leadership in Italian art, and to restore its greatness. In architecture, she was the first city to advocate a return to Palladio—that is, to Neoclassicism; in painting, if she indulged in the final fling of Baroque with Tiepolo, it was in honour of Veronese and Titian, and she balanced this noble paradox by developing the most important aspect of the painting of the future, that of pure—or uncommitted—painting.

ARCHITECTURE AND SCULPTURE

Venetian buildings are always inspired by local precedents and if the creations of Baldassare Longhena (1598–1682) react against the dryness of Scamozzi, this reaction is not comparable with that of Bernini against Vignola. After some training as a sculptor which later enabled him, like Bernini, to furnish designs for an altar (S. Pietro di Castello) and for tombs (*cenotaph* in S. Giorgio Maggiore, *Pesaro monument* in the Frari), and after his apprenticeship to Scamozzi, Longhena showed his full ability in his masterpiece, S. Maria della Salute, begun in 1631 and finished in 1687. This is the most Baroque building in Venice, but no centrally planned Roman church offers a similar internal or external design. The octagonal plan is clearly expressed outside by successive façades, and stressed by the drum with its large windows surmounted by a smooth dome, without the projecting ribs usual in Florence or Rome; the deep choir, crowned by a smaller dome, is preceded by a transept with absidal ends in the Palladian manner, supporting small belfries. The internal planning, with the variety given by large arches and the arresting setting of the choir, recalls S. Vitale, Ravenna, in its complexity. The exterior is both plastic and picturesque through the richness of the motifs, and above all, through the admirable invention of the volutes, wound in tight spirals, and arranged as flying buttresses round the drum, to serve as bases for statues of saints, the principal entrance portico being a typically Palladian triumphal arch motif. The church of the Scalzi (1680–84), completed by Sardi's façade, has a single nave and is covered with a rich revetment of marble and gold to create a constantly animated surface. The façade of S. Maria dei Derelitti, with its orders and cornices heaped one upon the other, is one of the rare examples of Rococo licence, but it reveals Longhena's care in adding to the richness and scintillation of Venice. An imitator, Alessandro Tremignon, rather overdid the same effects of excessive decoration on the façade of San Moisè (1668). In the Palazzo Pesaro, Longhena's masterpiece in secular architecture, begun in 1679 and finished in 1710, with the top storey by G. Massari, the robustness of the orders rising above the rusticated base and the lively effect of the light and shadow caused by the deep arcades combine to create a grandiose result which transposes the style of Sansovino's Library on to the Grand Canal.

The only building of which the interior is Baroque in the Roman sense is the Jesuit church (the Gesuiti) by Domenico Rossi (1729) with a façade by G. B. Fattoretto: the walls imitate the undulation of a blue drapery, everything is covered in stucco and frescoes, and the high altar has a baldaquin with ten twisted columns. Like the art of Tiepolo, this displayed the last word in brilliance. Seventeenth-century architecture had paid little attention to Palladio, except for a few instances such as the Villa Pamphili by Algardi, on the Janiculum, built in 1644. The return to Palladio characterized Neoclassicism. It appeared earlier in Venice than elsewhere, with S. Vitale by Andrea Tirali (c. 1700), and S. Simeone Piccolo (c. 1730) by Scalfarotto (1700–64). G. Massari (c. 1686–1766) borrowed extensively from S. Giorgio Maggiore for his façade of the Gesuati (1725–36) on the Zattere, and his Palazzo Grassi-Stucky (1705–45), without loggias, and ornamented with pilasters instead of columns, stresses its simplicity by sheer contrast.

In villa architecture, there was a similar evolution. Carlo Borella, Francesco Muttoni (d. 1748) in the Villa Trissino, and Girolamo Frigimelica (1653–1732), designer of the Villa Pisani at Strà, represent the extremes of Baroque complexity matched inside by the luminous decorations of Tiepolo. But Muttoni himself in the Villa Cordellina at Montecchio (1735–60), and Massari in the Villa Lattes (1715) at Istrana, returned to a clarity in planning which extends even to the arrangement of the gardens—fit sites for Canova's Neoclassical sculpture.

During the seventeenth century Venetian sculpture had been supplied chiefly by Bolognese such as Giuseppe Mazza (1653–1741), who made the bronze reliefs in the church of SS. Giovanni e Paolo, by the Genoese family of Roccatagliata who carved the very deeply undercut *paliotto* of S. Moisè (1633), by the Flemish Giusto La Corte (*c.* 1660) who carved the altar of the Salute, and by the German Melchior Barthel, author of the statues on the Pesaro tomb in the Frari. The most interesting Venetian is Orazio Marinali (1643–1720), a pupil of La Corte, who carved character busts (coll. Querini Stampaglia) that treat the marble as if it were a malleable substance, and garden statues in a very full Baroque style (Villa Da Porto, Trissino). Giovanni Bonazza (active between 1695 and 1730), with his three sons, carved the marble reliefs in the chapel of the Rosary in SS. Giovanni e Paolo, in a manner complicated by exhaustive detail.

The pictorial taste in relief became less insistent and lighter in the panels by Giovanni Morlaiter (1699–1781) in the Gesuati, and in his models for evangelical scenes (pvt. coll., Venice). Giovanni Marchiori (1696–1778) achieved elegance in his *Sibyls* in the Scalzi, and a brilliant composition in his lunette in S. Rocco (1741).

Furniture. The field in which Venice occupied a predominant position was that of furniture; in interior decoration, she could hold her own with Versailles and Paris, with quantities of saloons decorated with stuccoes, panels, arabesques and medallions, like those at the Palazzo Widmann. The commodes, chairs, wall lights in curved forms which enliven these rooms and saloons are often of an unrivalled craftsmanship. One of the most remarkable of the master carpenters was Andrea Brustolon of Belluno (1662–1732) who designed the monumental chairs in the Palazzo Venier (now in the Ca' Rezzonico) in box and ebony. The middle of the eighteenth century saw the rise of the fashion for furniture, painted, lacquered or even decorated with vignettes (like those by the Remondini of Bassano).

PAINTING

The Seventeenth Century. While Venetian Renaissance painting had been assimilated all over Europe, and Roman Baroque, which owed so much to it, dominated in Italy, in Venice itself the seventeenth century was without any brilliance. Painters who tended towards Mannerism, like Palma Giovane (1546–1628) and numerous lesser masters like Padovanino, known from the treatises of Sansovino (1581, completed in the edition of 1663) and Boschini (1660) barely rise above the level of mediocrity. A clearer inspiration informs the three painters who understood Rubens. First there was Domenico Feti (1589–1624), Roman by birth, a pupil of Cigoli, and favourite painter of Cardinal Federigo Gonzaga who took him to Mantua, from whence he passed to Venice in 1621. His large frescoes in Mantua cathedral do not reflect his true bent, which was more inclined to genre scenes and to small formats.

His parables are treated in a Flemish spirit, with somewhat trivial natural or rustic details, in the style of Bassano, but with a feeling for lively and flickering colour—as, for instance in the *Miracle of the Loaves* (Mantua cathedral), and the *Lost Sheep* (Dresden). His *Melancholia* (Louvre) shows that he was aware of the seriousness of Roman luminism.

Giovanni Lys (*c.* 1595–1630), or 'Pan', who was born in Oldenburg, trained in Holland, passed through Paris and Rome, and settled in Venice where he concentrated on *bambocciate* (*Inn scene*, Cassel), struck out into the mythological nude (*Toilet of Venus*, Uffizi) and finally composed his *Vision of S. Jerome* (S. Niccolo da Tolentino) and *Sacrifice of Isaac* (Uffizi) in large and rather jerkily arranged blocks of warm colour.

Men from other parts of the country were beginning to flock into Venice to learn the Venetian manner, whence the third artist, a Genoese, Bernardo Strozzi (1581–1644), who was a Capuchin until he became a secular priest in 1610. After essays in the manner of Rubens and other Flemings (as in his *Cookmaid*, Palazzo Rosso, Genoa; and *Berenice*, Bologna) and Caravaggio, he settled in Venice in 1630. His palette became lighter through the influence of the works of Veronese, and richer in tones and effects, as in the *S. Lawrence distributing alms* (S. Niccolo da Tolentino), and the *Rape of Helen* (Posnan) which already reflects much of the spirit of the lighthearted *galanteries* of the Rococo.

Cut off from Mannerism by the ideas introduced by these forceful personalities, Venetian painting was to see evolve, towards the middle of the century, the painters of the transitional period: Andrea Celesti (1637–1706), who decorated S. Zaccaria, Sebastiano Mazzoni (1615–85), who introduced the dramatic lighting and the varied and lively brushwork of Rembrandt (*Death of Cleopatra*, Accademia, Rovigo), and, above all, Francesco Maffei (*c.* 1620–60) of Vicenza, who worked in Brescia (*Miracle of S. Anthony*, S. Francesco, *Translation of relics*, Cathedral), in Vicenza (*Visitation*, *Flight into Egypt*, *Assumption*, Oratory of the Zitelle; *Trinity with Sarah and Abraham*, church of the Servites), and in Padua (*Adoration of the Magi*, S. Tomè). A personality more amusing than impressive, Maffei knew Greco and Callot, and his evocations of martyrdoms, and his biblical scenes (*Sacrifice of Melchisedech*, Foscari Coll., Venice) have a bizarre and fantastic quality not lacking in poetry, while remaining highly artificial, as in the over-clever, theatrical setting in two planes, light in the background and dark in front, of his *Martyrdom of the Franciscans at Nagasaki* (S. Francesco, Schio). Maffei's art retains a certain distinction through the amazing variety of his handling, his fat and lively impasto creating the most delicious effects in small and scintillating pictures which are like soft and spicy delicacies.

Neither Gregorio Lazzarini (1655–1730), with his large and carefully drawn composition in S. Pietro di Castello, nor the decorator Fumiani, provided Venice with the necessary stimulant, but rather those who were influenced by the forceful art of Luca Giordano, and, even more, by Sebastiano Ricci (1659-1734). A pupil of Mazzoni, working in Central Italy (his decorations in the Pitti and the Palazzo Marucelli in Florence in 1706 were already Rococo), he concentrated on the grandest pieces of scene setting, lightening his colours bit by bit (*Madonna and nine saints*, S. Giorgio Maggiore) and using the ultra-sensitive tonalities with numerous reflections of Magnasco (*Susanna before Daniel*, Turin; *Ecstasy of S. Theresa*, S. Mark's, Venice). G. A. Pellegrini (1675-1741) was as knowledgeable an artist, and one, moreover, in contact with French painters; he developed into a past master of soft and tender handling, in which he was equalled by the international specialist of pastel portraits, Rosalba Carriera (1675-1757), whose works are full of airy and delicate transitions.

But Venice at this point, by re-assuming the leading position in painting which was increasingly her due internationally, opened up several new paths which it would be most unwise to confuse. The rupture which developed, in favour of the Rococo, between the fashionable intimacy and naturalism on one hand and the rhetoric of the Baroque on the other, enabled some painters to go as far as they could towards developing a pure painting, without Antique or religious themes, of a kind that presaged the nineteenth century. Some, whose champion was to be Canaletto, relied on the visual refinement and the objective formula of the *vedutisti*, others, linked rather with Maffei and Sebastiano Ricci, retained a more nervous brushwork, and a search for more imaginative qualities, and their leader was to be Guardi. In any case, their art was not held to be the most important; they were only considered as minor masters. The Academy, founded in 1755, was dominated by the representatives of the Grand Manner, restored to fashion by Piazzetta, which alone seemed able to revive the glories of the Renaissance, and its most important protagonist was Tiepolo.

Baroque Decoration. In those two most potent creators of atmosphere, wall decorations and ceilings painted with figure subjects, both of which had been practised in Florence and Rome during the previous century, an original style evolved rather late in Venice. After the end of the century, Giovanni Antonio Fumiani (1650-1710) undertook in S. Pantaleone immense illusionistic decorations in which the martyr-saint, who died under Diocletian, passes amid a crowd of soldiers and horses over impossible staircases rising straight up, and under enormous porticoes.

This feeling for ceiling composition was not lost on Piazzetta (1682-1754) and he developed it, without the aid of architecture, purely by means of contrasts of light in his *Glory of S. Dominic* (1777; Chapel of the Rosary, SS. Giovanni e Paolo, Venice). This was not, however, achieved without effort. It was chiefly in Bologna, with Crespi, that he learned to handle masses in a painterly fashion, and to merge his colours in a brilliant alternation of light and shadow. As a result of this tardy discovery of Caravaggism, Piazzetta was still using sombre tones in his *Christ* (Accademia); his visions

357

have a spasmodic rhythm, full of accents of light and sparkle, to represent the miraculous events dear to the piety of the Baroque, such as the *Virgin appearing to S. Philip Neri* (*c.* 1725, Church of the Fava, Venice), and the *Ecstasy of S. Francis* (1732, Vicenza). His colour becomes somewhat lighter in the *Assumption* (1735, Lille) perhaps through the influence of Tiepolo. The altarpiece in the Gesuati (1739) is more sober in colour, and the vision takes place in full daylight, but in the years that follow his inspiration becomes more rigid, the dark tones return, and despite his use of genre themes, as in the *Indovina* (*The Riddle*; 1740, Accademia), or modernized treatment of Biblical themes, such as the *Rebecca* in the Brera, his reddish shadows lack subtlety. The energetic twisting movement of his forms, the passages of brown and pale tones which melt together to create a quivering atmosphere, his clever choice of the moment of maximum psychological impact (as in the *Judith and Holofernes*, Accademia di S. Luca, Rome), awakened in Venice a renewed interest in a narrative and detailed treatment of great dramatic scenes. It is significant that Piazzetta should have illustrated Tasso's 'Gerusalemme Liberata' (1745).

The strange Dalmatian, Federigo Bencovich (1677–1753), who lived in Venice, worked also in Würzburg, and died in Gorizia, reflected this trend of Venetian art in his *Juno* ceiling (1707, Palazzo Foschi, Forlì), but also displayed a more strained energy in his *Blessed Pietro Gambacorti* in S. Sebastiano, Venice.

The supreme virtue of Tiepolo (1696–1770) was that he subordinated these rather narrow imaginative sources and this strident expression to the exigencies of an International Baroque, devoted to extravagant architectural devices and limitless settings, and established these ideas in Venice by reviving with a sometimes rather irritating contrivance, but also with prodigious brio, the full, glowing day and splendid, light-filled atmosphere of Veronese. Following in the footsteps of Bencovich and Piazzetta, Tiepolo began by adopting the dark manner which, as far as Naples with Solimena, had swept all before it in Italy. The shaft of light and the double diagonals of the *Martyrdom of S. Bartholomew*, in S. Stae (1721), are not new, but at least the artist has related his effects to the setting; one of the secrets of his enduring and often irresistible power lies in this clever calculation of his effects, so that the discovery of the painting on crossing the threshold, or on looking upwards, gives the spectator a kind of shock. In the Palazzo Sandi Porto (1725) and the Ca' Dolfin (1725–28, fragments in Vienna), he painted cycles which displayed the readiness of his invention. In 1731, for the *Triumph of the Arts* in the Palazzo Archinti (destroyed in 1944), and in the Palazzo Dugnani, both in Milan, he designed his first ceilings with history subjects seen through the frame of huge illusionistic cartouches. Here his colour was already light, and the background architecture was treated in tones of mauve, pink and blue. After assimilating the Venetian repertory of the seventeenth century, he turned with success to the painters in cool shot colours, combining, according to the kind of ceiling to be painted, sky effects, fictive architecture and crowds of moving figures. By the time he came to paint the *Scenes from the Life of the Baptist* (1732–33) in the Colleone chapel in Bergamo, his palette was completely light in tone.

For the next quarter of a century, Tiepolo covered the walls of palaces and the ceilings of churches with the luminous voids of his enormous frescoes, where the figures, in delicate tones of amber, lilac and green, eddy lightly yet still remain readable as figures. As a result of their even luminosity, these compositions introduce a Baroque animation, but without the Roman tendency towards over-inflated and overwhelming effects. After the ceiling in the Gesuati (1737–39), Tiepolo, now acknowledged as the most celebrated of the *virtuosi*, painted in the Palazzo Clerici in Milan a seething ceiling filled with every hue on the theme of the *Course of the Sun through the World* (1740), and in Venice in the churches of the Scalzi (a fresco destroyed in 1915) and the Pietà (1754). On two walls of the high central saloon of the Palazzo Labia he created his greatest masterpiece in fabulous scenic effects of a Shakespearian fantasy and solemnity, and of true Baroque brio, with the *Feast of Antony and Cleopatra* and the *Embarkation of Cleopatra*—works of an unsurpassably romantic audacity.

In 1756, Tiepolo was elected President of the newly founded Academy in Venice. He was not only supreme in his own city, but was summoned to Würzburg to decorate the Residenz of the Prince-Bishop (1751–53) with *Scenes from the Life of Frederick Barbarossa*. There he was accompanied by his son Gian Domenico, who became more and more his collaborator in these enormous cycles from which all sincerity has now disappeared in favour of a lavish display, and where the repetition of types and gestures becomes more and more noticeable. The artifice is very visible in the *Reception of Henri III of France* in the saloon of the Villa Contarini (1755–46, now in the Musée Jacquemart-André in Paris). Almost as if he were aware of this decline, Tiepolo gave a more personal quality, as well as greater sentimentality, to scenes drawn from Homer, Virgil, Ariosto and Tasso, in the Villa Valmarana (1755) completed by contemporary scenes by his son in the entrance pavilion. In 1762, he was summoned to Madrid, where he decorated the Royal palace, and where he eventually died.

Tiepolo's great fame is patent from 1736 onwards, when he was praised in an ambassador's letter for his 'infinite fire, brilliant colour, and astonishing speed'. In considering the vast quantity of his works, the question is whether the effort expended upon the creation of dazzling effects representing the ultimate in Baroque theatrical taste and culminating in an expatiation on the romantic aspects of pathos and sentiment (as in the *S. Lucy*, SS. Apostoli, Venice, 1748), should not be relegated to a different plane, and interest be concentrated rather upon the very numerous and naturally more spontaneous studies and sketches which are much more sensitive in handling. These often have more vivacity, while demonstrating at the same time his extraordinary assurance as a theatrical costumier and designer, as for instance in the *Martyrdom of S. Giovanni Vescovo* (1743, Bergamo cathedral) where the effect of the white satin tunic is already fully worked out in the sketch.

Tiepolo's art, of a brilliance and an expansiveness far in excess of that of his French Rococo contemporaries, looks back to the luminous beauty of the sixteenth-century masters, but after the artifices of the Baroque, this return to the past is only the supreme and final effort of a whole civilization all the resources of which have now been thoroughly worked out. Tiepolo sums up a complete repertory, and his narrative zest, his feeling for infinite space, give for the last time the impression of a great entity in which everything subserves the representation in visual terms of religious and secular themes. But when he recalls Titian in the *Danäe* (1736, Stockholm), or directly imitates Veronese in the *Decollation of the Baptist* (1733, Stockholm), or when he brings all his authority to bear on portraiture, as in his rendering of the *Procurator Querini* (1746, Venice), he betrays beneath the brio a vanity, a somewhat grimacing assurance, an excessive volubility, which are disconcerting. The great number of his drawings show how thoroughly he studied his subjects, and also his deeply felt emotion, in which poetry was blended with romance. They make doubly clear the menacing nature of the Neoclassicists' reaction, which rejected both facile emotional content, and what was believed to be meretricious means.

In this sense it is true that 'the arrogant scepticism of Tiepolo cost Italian painting too dear' (R. Longhi). The evocative and lively substance of his art could not survive the crumbling of its setting and the disappearance of its bedizened models. His collaborators, Fabio Canal (1703–67), Jacopo Guarana (1720–1808), and his own son Gian Domenico (1727–1804), whose scenes from the Villa at Mirano have been reconstituted in the Ca' Rezzonico, do not amount to very much.

Landscapes and Capriccios. It was as a result of his failure in large-scale decorative painting (Palazzo Saguda, Bologna, 1734) that Pietro Longhi (1702–85), a pupil in Bologna of Crespi, turned towards homely subjects, genre painting, and scenes from everyday life. He tried to put all Venice into his little pictures; the chronicle of small events (*The Negro Ambassador, The Rhinoceros*, Ca' Rezzonico), little comedies of manners (*The Geography Lesson*, Padua; *The dancing Lesson*, Accademia, Venice), more rarely outdoor scenes (*Duck hunting*, Mus. Querini Stampalia) and portraits (*Gentleman*, Treviso). They are Molière in painting, or rather, as has often been remarked, a sharp and bantering Goldoni. The documentary value of this pictorial daily record has been perhaps overstressed at the expense of the artist's qualities as a painter, since he has a feeling for interiors, and his colour is sensitively handled and applied with skill. But care must be taken not to over-estimate these small and clever glimpses of the past, which are composed with little effort from puppet-like figures, and which, in their search for wit, pinpoint the manners and customs of an insignificant society. Longhi's art is nothing like Hogarth's, and has even less to do with Chardin's.

His son Alessandro (1733–1813) was almost exclusively a portrait painter, in the showy and sometimes pompous manner of his time; good examples are *Alvise Mocenigo* (1756, Venice Accad.), *Goldoni* (Correr), and *Cimarosa* (Leichtenstein Coll.). Amateurs of pastoral landscape, of a tame and placid Arcadia in the French taste, could satisfy their needs with F. Zuccarelli (1702–88), in Venice from 1732 onwards, and Giuseppe Zais (1719–84), who were both members of the Venetian Academy. But the full impact of Venetian taste took quite another form—one which was adumbrated at the beginning of the century by Luca Carlevaris of Udine (1665–1731), who had painted landscapes of Rome, and a Dutchman, Gaspare Van Vittel, Italianized as Vanvitelli (1653–1736) who, to a certain extent, staked out Venice's claim to a special kind of marine painting.[162] The artist who best understood this development and realized its possibilities to the full was Antonio Canal, or Canaletto (1697–1768). He was in Rome in 1719, and again in about 1740; he derived from Pannini, and, with a rigid and inflexible feeling for light, and a crystalline definition of volumes, he exploited fully his predecessors' essays in perspective. (*Grand Canal seen from the Rialto*, Uffizi, and Liechtenstein Coll.). With more pervasive charm, he distilled the essence of his pictures in his series of etchings. Traditional topography had found its master; his success was international and he was imitated everywhere.

In London, at the invitation of the collector Consul Smith from 1746 to *c*. 1756, Canaletto came under Dutch influence and was inspired by the very detailed manner of J. van der Heyden. He developed a much more artificial style; for Algarotti, he composed extremely frigid ideal views (Parma gall.) made up from famous monuments. His nephew, Bernardo Bellotto (1720–80), carried the formula of the luminous landscape into all the Courts of Europe—Vienna, Munich, Dresden, and finally Warsaw, which became in his hands just so many Venices or ideal cities.

Like Pellegrini, who impinged on French Rococo, Jacopo Amigoni (*c*. 1682–1752), who worked in London, Paris and Madrid as a brilliant composer of decorations (*Venus and Adonis*, Venice Accad.), and Gianbattista Pittoni (1687–1767) whose colour is fat and vibrant (*Miracle of the Loaves*, Accad.), link Venetian painting to European art. In this art there survives a poetic impulse, a taste for glitter and thin fillets of light, which receives its fullest development with Francesco Guardi (1712–93). Trained by his brother, a minor painter, he began by painting ceiling figures which floated airily (*Four Allegories*, *c*. 1750–53, now in Palazzo Labia) or compositions built up in small touches and luminous patches of colour of astonishing intensity (*Tobias*, 1750, Church of the Archangel Raphael, Venice); all that is most characteristic in the artist is already present in these works, and his lively squiggles of paint, filmy lights and shadows that flicker through his ruins, his landscapes, his small figures, reach the highest flights of visionary poetry.

VENICE

This spirited handling, which Magnasco also used, is applied straightforwardly to *veduta* painting—to the landscapes of the lagoon (*Isola di S. Giorgio*, Accad. Carrara, Bergamo; *The Piazzetta*, Ca' d'Oro; *The Regattas*, pvt. coll., Paris), with its tenements, and the portrayal of things worn away by weather, which assume a pre-Romantic and nostalgic quality far removed from Canaletto. This painting, in which all appears to be sacrificed to the handling, and which returns over and over to the same motifs, seems to be the direct forerunner of Impressionism. In reality, it is far too dependent on tonal values, on greys, on silvery tones, and above all too far above pure sensation (*View of the Lagoon*, Poldi Pezzoli, Milan) for it to be possible to interpret it entirely in this sense. This is confirmed by the quality, at once sharp and precise, of the preparatory drawings, which far exceed brief notes, and contain real scales of tones; this strict construction of hatchings and blots, which expresses the Venetian habit of evoking the structure beneath the vibration of forms, later becomes much weaker in Migliara (1785–1837).

Engravings. From the start of the eighteenth century, in Venice as in Rome, the collections of descriptive plates proliferated, inspired by Venetian painters who were often themselves engravers. What Marco Ricci had done for views of Rome in his twenty-two etchings of 1730, Marieschi did with his vistas of Venice (1741) and G. F. Costa with his fashionable *Delizie del Fiume Brenta*. These engravings developed consistently not only topography and architecture, but also the light; their whites, their greys, their shadows achieve, despite their monotone, a calmness and freshness which is to be found in the engravings of Canaletto.

The thirty-five *Capriccios and Fantasies* of G. B. Tiepolo derive from another source of inspiration which becomes the forerunner of the demonic fantasies of Goya's *Caprichos*. The masses of ruins, of ragged trees with tombs and owls derived from Castiglione, are inhabited only by wizards, tatterdemalions, ghosts and grinning masks. The light technique of the artist, as easy in engraving as in painting, avoiding cross-hatching and heavy shadows, contradicts somewhat these strange imaginings, born of a varied and lighthearted culture. The engravings of Gian Domenico Tiepolo (*Road to Calvary*, *Flight into Egypt*) imitate Rembrandt too closely.

Through the skill of the *vedutisti*, Venetian engraving ensured the prestige of Venetian painting throughout Europe—in London, Dresden, Warsaw, and elsewhere. But it was with Piranesi, who became a Roman, that the twin aspects of Venetian art, its realism and its visionary quality, were most fully realized in a study of the diffusion of light upon the grandiose and complicated world of Rome. Comment has already been made upon the archaeological exactness and the nostalgic feeling he displays, which have nothing at all to do with the Baroque, and in fact were among the factors that turned Italy, and all Europe, away from Baroque art.

Part Four

MODERN ART

Introduction

The Drama of the Unification of Italy

The history of modern Italy consists of immense hopes, often expressed in enthusiastic and emphatic language, alternating with the most cruel disillusionments. Political unification, followed by the creation of a modern state, has cost the Italians more in suffering, in illusions, and in bitterness than any other European people. The literature of the nineteenth century is dominated by the '*Letters of Jacopo Ortis*' (1802), a masterpiece of impassioned despair by Ugo Foscolo; '*My Prisons*', by Silvio Pellico (d. 1854), a victim of the Austrian rule; the great historical novel of '*The Betrothed*' by Manzoni (1827—republished 1840), which evokes popular life in Lombardy about 1630; and the superb, grief-stricken *canzoni*, the pathetic 'nocturnes' of Leopardi (d. 1837). The polemical and revengeful literature of the Risorgimento, the high-flown works of the Fascist period, with its intemperate nationalism, have never effaced this sadness, this often poignant background of grave humanity, full of solicitude for the humble, which is perhaps today the most original aspect of Italian genius. But this inspiration is often, if not constantly, masked by two rather deceptive characteristics, which afford immense pleasure to the vivacious Southern temperament: eloquence and virtuosity, the display of the spectacular and the verve of the brilliant executant. The nineteenth century unfortunately abused both.

During the two centuries which followed the Renaissance, from 1550 to 1750, the peninsula was a passive stake in the rivalry between France and the Hapsburgs. By the Treaty of Aix-la-Chapelle (1748) the partition of Italy into small principalities was even more firmly established. From 1750 to 1950 the country achieved its unity as a modern State by a long and arduous succession of stages. For a long time she was supported by France against the Austrian Empire, then, when unity had been achieved (1871), she entered at various times into alliances with Central European powers (the Triple Alliance, 1882; the Rome-Berlin Axis, 1937) in order to achieve an Imperial development in the Mediterranean and in Africa at the expense of the Western Powers.

The emancipation of Italy is thus closely linked with the history of European crises. Napoleon's victorious campaign in Lombardy (1796) resulted in the first effort at a simplification of the political map, first into three (1805) and later into two States (1809), attached to the French Empire. The dismemberment and the absolutism being re-established by the Congress of

Vienna (1815), it was against Austria that the patriotic and liberal revolts of 1848 took place, followed by the victorious campaigns of Victor Emmanuel II, King of Piedmont, and of Napoleon III (1859). The rapid annexations which followed were due to the action of a national *condottiere*, Garibaldi; they led to the creation of the Kingdom of Italy (1861), completed by the occupation of Rome (1870), which was only recognized in 1929 by the Holy See. In 1918, at the end of the First World War, the conquest of the provinces of Trentino and Istria completed Italy's northern frontier, though this was considerably modified in 1945 on the Adriatic.

After having experienced the narrow absolutism of the petty States described by Stendhal, Italy evolved a constitutional régime which, from 1849 onwards, ensured the dominance of the Piedmontese monarchy, and was maintained after 1871. From 1922 to 1943, she underwent the experience of an authoritarian régime, nationalist and democratic, which, within the framework of the monarchy of Victor Emmanuel III, resulted in a more thorough administrative and economic unification. When this régime collapsed in 1943, with Italy's defeat in the Second World War, it dragged the monarchy down with it, and in 1947 a plebiscite established a republic.

This process of political, social and economic maturing absorbed the energies of several generations, after the long sleep of the absolutist period. It shows the difficulty experienced by a Mediterranean country, without a properly constituted middle class, without *bourgeoisie*, in entering the framework of the modern world. The educated classes had the sensation of being an anomaly: in 1843, the Abbé Gioberti set out in a famous work 'the moral and civil primacy of the Italians', which revived the old Guelf dream of a unity created by the Holy See; eighty years later, the neo-Caesarism of Gabriele d'Annunzio and the Fascist intellectuals was an equally high-flown illusion, and a much more dangerous one.

Without taking these circumstances into account, it would be difficult to explain why, after guiding and fertilizing European art during eight centuries, Italy, during the course of the last two centuries, became only a rather backward provincial centre. To the general weakening of the aristocracy, and the indifference of the Church, which no longer played the magnificent role of former days, must be added the absence of those great *bourgeois* patrons who elsewhere ensured the formation of fine collections of the modern works, the kernel of the great museums of the twentieth century. This *élite* of cultivated merchants and travellers, which had formerly assured the primacy of Tuscany, Lombardy, Venice, over the other countries of the West during the fourteenth and fifteenth centuries, is, with very few exceptions, just what was lacking in Italy during the nineteenth. The decay was common to both public and the critics, above all during the second half of the nineteenth century.[163] Three typical phenomena will make this clear.

Firstly, the success of Neoclassical doctrines, which by their excessive rationalism, their system of 'ideal beauty', their simplification of history,

366

came as a dangerous reinforcement to academic sterility: Italy, the land of treatises on art, became the victim of their rigidity. The two dominant figures for architecture were Milizia in Rome and Temanza in Venice, supplemented by historians like the Abbé Lanzi for painting (1789), and Count Cicognara for sculpture (1813–18). Milizia (1725–98), an Apulian who lived in Rome from 1761 onwards, published there his *Vite de' piu celebri architetti* (1768; Lives of the most famous architects), a declaration of war on Jesuit Baroque, and he maintained in his polemical writings the necessity for order and for a strict attention to means and functions. Temanza (1705–89) supported these ideals, using as arguments Vitruvius and Palladio, who was re-edited by the learned Vicentines, Bertotti-Scamozzi (1776) and Arnaldi (*Idea di un Teatro*, 1762). But these speculations extended far beyond the problems of contemporary art. The most striking example of this over-theoretical fecundity is supplied by studies of ideal cities, which were numer-ous during the eighteenth century, just as they had been during the Renaissance. An extraordinary theorist, the Venetian Jesuit Carlo Lodoli (1690–1761), imagined a rational city where the materials and the geometric forms were in strict correspondence with the purpose of the buildings; this first definition of 'functionalism' was spread by the work of A. Memmo—*Elementi dell' architettura lodoliana o sia l'arte del fabbricare con solidità scientifica e con eleganza non capricciosa*, Rome (1786),[164] which pushes as far as did Ledoux in France the idea that beauty is based on geometric rationalism. But the most coherent effort to conform to these abstract views was to be made during the French empire, and after a few remarkable essays in replannings in Milan and Rome, in Genoa and in Trieste, nothing further was evolved to counteract the monotony of arcaded façades and the complacency of the academies.

All over Europe, a passionate curiosity about the picturesque periods of history or the Orient (as in Delacroix, for example), coupled with an irre-sistible enlargement of the bounds of the imagination and of feeling for land-scape (as in Constable), and an interest in the contemporary scene (as in Géricault), compensated for, and soon surpassed, the academic fiction and its bogus antiquity. Only Italy, humiliated and a prey to conspiracies and dreams, played no part in this movement, and knew only the inferior products of romantic anecdote, with which she appeared to be satisfied, perhaps because she had already experienced, with Tiepolo, Piranesi and Guardi, its most exceptional forms. The absence of a generous and full Romanticism is, without doubt, one of the deep reasons for the general weakening in the arts.[165] The effects of this were to be felt until the end of the century, for, after the great upsurge of the Risorgimento and the unification of the country, the feeling for 'modernity' which elsewhere provoked a new flowering in the arts, was also lacking in power and subtlety under the *bourgeois* monarchy. There again, no movement appeared which could have held a balance be-tween the ponderous creations of official art, and above all, of official

INTRODUCTION

architecture, and the small genre scenes which satisfied even the most advanced artists, the *Macchiaioli* of Tuscany and the over-brilliant virtuosi of Naples or Venice. The compensatory counterplay between North and South no longer existed. The artists of the *avant-garde* took advantage of their position, and nothing came of the small revolution of the early twentieth century. It was only after 1945 that society in Italy, facing new tasks, could offer new scope for the age-old taste and ingenuity which, during the last two centuries, had regularly set its sights either too high or too low.

One last fact illustrates this demoralization, and helps to explain the gaps in Italian art during this critical period: the exile of many artists. During the eighteenth century, foreign commissions are the price paid for the primacy of the Italians, above all in architecture and decoration, and this remained so until about 1850. Brenna (1745–1820) and Corazzi (1792–1877) went to Warsaw; Rinaldi (1709–97) built the Marble Palace in Moscow (1768–85) and Quarenghi (1744–1817) the architect to the Czars, built the Alexander Palace with its vast colonnade at Tzarkoie-Selo (1792–96); the reconstruction of Moscow, after 1812, was directed by Domenico Gilardi (1788–1845). There was a graver symptom in the nomad life of painters like Pasini (1826–1899) who went to the court of the Shah of Persia, and like Fontanesi (1818–1882), who lived in Switzerland, Paris and Tokyo where he taught (1877–78), and above all in the regular exodus of artists to Paris, where de Nittis and Boldini gained a cosmopolitan prestige which the Italian public did not offer to its most favoured virtuosi. And in our time, the flight of Modigliani to France in 1906, the ebb-tide of the *avant-garde*, with its illusions and its promise, and the tendency, which is not yet at an end, for the most adventurous, if not the best, to become part of the École de Paris, is in accord with this trend.

I

Neoclassicism

1750-1830

From about 1750 to 1850 the whole of Europe went through a crisis in doctrine and learning, linked to a reaction in favour of purism, and then to a craze for the antique, which created a whole style in civilization and art. The *Ruines de la Grèce* of Le Roi (1758), and the *Geschichte der Kunst des Altertums* of Winckelmann (1764), which finally established the return to Hellenism, were paralleled by the magnificent collections of plates by Piranesi. Not only are they the most splendid Romantic evocation of the ruins of antiquity, but they also represent a repertory of useful forms, in the series called *Vasi e Candelabri* (1768–78), which was to become the grammar of the Empire style. The ground was prepared for secondary aspects of the new style, like the Egyptian taste which flourished in England and France after 1790, by the fantasies of Roman decorators. Decoration with hieroglyphics, cynocephales, sphinxes, and exotic ruins appeared in the English café on the Piazza di Spagna, and Piranesi published the motifs used in it in his *Diverse maniere d'adornare i cammini* (1769).

Italy, therefore, took part in the antiquarianizing trend through engravers, by the publication of the excavations at Herculaneum (1757–92), and through the teaching of theorists. She played a double role through Roman archaeology, and through the esteem enjoyed, with Palladio in architecture, by the elegant sculpture of Mannerism, the serene examples of Raphael, and the tenderness of Correggio. But creative initiative was beyond her. The neo-Palladian movement was strengthened in England by Robert Adam; the three most original architects of the end of the eighteenth century were an Englishman, John Soane, a Frenchman, Ledoux, and a Prussian, Frederick Gilly. The French painter David succeeded Mengs, and another Frenchman, Prud'hon, created the new 'Correggiosity'. Italy possessed the thin, sharp talent of Canova, who avoided being too deeply involved with Paris, and the minor talents of the architect Piermarini and the painter Appiani in Milan.

ARCHITECTURE

Giuseppe Piermarini of Foligno (1734–1808), an assistant of Vanvitelli at

NEOCLASSICISM

Caserta, was one of the first architects of the peninsula to use on a grand scale the formula of a classicism based on Vignola, Palladio and the Antique. Summoned to Milan in 1769, he entirely reconstructed the Royal Palace (finished in 1778), and applied to it a giant order of Ionic pilasters; he also reworked the Palazzo Belgioioso (1777) with still greater severity, giving it a rusticated basement, a colossal order, and a terrace, and enlivened it with a slightly projecting pediment, sections of re-entrant wall and French ornaments on the windows. This is still the spirit of Caserta. The Opera House of La Scala (1776–78) is even more remarkable; the exterior has a rusticated portico with engaged half-columns on the first floor, which offers a strong contrast to the richness of the interior, with its six tiers of boxes and its extremely wide stage. Piermarini's villas are regular and lacking in animation; for instance, Cusani at Desio, Perego at Cremnago; Monza, however, has a rectangular plan, a central block with three bays projecting into the garden with all its effect concentrated on the stressing of the horizontals, emphasized by the cornices and the vertical pilasters. There is more delicacy and charm in the royal villa in Milan itself, built by a Viennese pupil of Piermarini's, L. Pollak (1751–1806) in 1799, in gardens laid out in 1782 by his master in the north-eastern quarter of the town (now the Museum of Modern Art).

Simone Cantoni (1736–1818) turned from the weakened Baroque of the ducal palace in Genoa (1778) to Neoclassical affectation in the Palazzo Serbelloni (1794, Milan), the façade of which has two storeys within an Ionic portico crowned by a large pediment. But development along the lines of the Greco-Roman style of the French Empire became much more rapid when, under the Napoleonic occupation, Milan became the capital of the Kingdom of Italy (1805–14). It is from this period that date the triumphal Arch of Peace (Arco del Sempione) by Luigi Cagnola, built in marble with a bronze quadriga in honour of Napoleon (1806) and later rededicated to the Emperor Francis I of Austria (1838), the arch of the Porta Nuova (1810) by Giuseppe Zanoia, the town planning schemes of the Foro Bonoparte with its arcades (1801) projected for the surroundings of the Castello Sforzesco by Giovanni Antonio Antolini, those also of the Stadium or Arena (1806–7) built by Luigi Canonica, and the grandiose town plan of 1807, from which are derived the arterial roads that have governed the later development of the city.

The same effort towards large-scale systematization, in the framework of antique-style colonnades and with the help of rectilinear street plans, is to be found in Turin, where F. Bonsignore (1763–1843) erected, on the model of the Pantheon, the church of Gran Madre di Dio (1818–31). Similarly in Genoa, Carlo Barabino (1768–1835) created by means of terraced ramps links between the three towns one above the other on the hillside, constructed a façade with a projecting portico for the church of the Annunziata (1830), and built the Carlo Felice theatre with a hexastyle Corinthian portico.

In Venetia, Neoclassical taste, for which the ground was prepared by

Temanza and the Palladians, was given delicate expression by Giovantonio Selva (1751–1819) in the theatre of La Fenice, of which the façade survived the fire in 1792 and the auditorium was rebuilt after the fire of 1836, and the church of the Nome di Gesù (1813–34) finished by A. Diedo; the Paduan Giuseppe Iappelli (1783–1852) designed the symmetrical Caffè Pedrocchi (1816–31) at Padua, numerous villas and suites of furniture in a very pure style (Villa Cittadella, 1813, Vigodarzere). Town planning played an important role in the creation of public gardens in Venice (1807) and in the construction of the Napoleonic wing of the Procuratie Nuove by Giuseppe Maria Soli.

A whole city was modernized: this was Trieste, which was reunited to the kingdom from 1809 to 1814. Matteo Pertsch (1780–1854) built numerous houses, the Rotonda dei Pancera, and the Palazzo Carciotti (1802–6), with a fine terraced façade facing the sea: Antonio Mollari built the Borsa, with a Doric portico (1802–6); Pietro Nobile built the church of S. Antonio with an Ionic portico (designed in 1808, finished in 1849), which, based on the Pantheon, formed the closing vista at the end of the canal.

The two aspects of the Neoclassical movement, one correcting Rococo through its restraint and delicacy, the other becoming the Empire style— that is, striving for the grandiose—were both to be found in Emilia, in Parma, with the ducal villa of Colorno built by the French architect Petitot, and, under Napoleon's Empress Marie-Louise, the rather solemn ducal theatre (1829), with its bare walls above an Ionic atrium, by Niccolo Bettoli (1780– 1854). In Tuscany, Gaspare Maria Paoletti (1728–1813), who was the master of two generations of architects, designed, in an entirely Roman manner, the Doric Terme Leopoldine (1773) at Montecatini, and also made alterations in the Pitti palace (Sala Bianca, 1780), which were completed by his pupil Giuseppe Cacialli (1770–1828) when he decorated new rooms in a Roman style. Paoletti also reworked the villa of Poggio Imperiale, finished by another pupil, Pasquale Poccianti (1774–1848) and by Cacialli, who were inspired in the north façade by the Medicean villa of Poggio a Caiano. Poccianti also worked for a half-century on the aqueducts of Leghorn and on its reservoirs, treating them in a grand manner, imitating antique exedrae, as for example in the Cisternone. The enlargement of the old fortified port, founded by the Medici dukes (1577), resulted in a large town-planning project centred on the elliptical Piazza Voltone (or Carlo-Alberto). Another pupil of Paoletti, Gaetano Baccani (1792–1867), restored the nave of Florence cathedral (1842) and built the campanile of Sta Croce.

In Rome, where the classical reaction had begun and the doctrines of Neoclassicism had been formulated, it was principally in the Vatican and in the replanning of the city that the new ideas had their greatest currency. Michel Angelo Simonetti (1724–81) built under Pius VI the rooms of the Museo Pio Clementino (1774); for Pius VII, Raffaello Stern (1771–1820) designed a new wing consisting of a room over 225 feet long, parallel to the library wing, with marble and alabaster columns and a mosaic floor to complete the old museum (1817-21). It was at this period that all the western part of the Vatican was converted into a museum, with the creation of the picture gallery (1815), the Etruscan museum (1836) and the Museum of Egyptology (1839).

The major architect of Papal and Napoleonic Rome was Giuseppe Valadier (1762-1839), a pupil of his father Luigi, and a learned architect who restored the Arch of Titus and designed a number of façades—for instance, those of S. Pantaleone (1806), S. Andrea delle Fratte (1821), and S. Rocco (1833). He was also responsible for important restorations undertaken in the Papal States, notably those at Rimini and in the cathedral at Urbino where the dome collapsed in 1789. Valadier made very interesting surveys (which have made it possible to attribute the building to Francesco di Giorgio), but he rebuilt the structure entirely and it was completed by a rather tame façade by C. Morigia.[166]

Together with the Apollo and the Valle Theatres, his main work was the definitive planning of the Piazza del Popolo, which, until the middle of the nineteenth century, had always been the major entrance to Rome from the north. The siting of the twin churches by Carlo Rainaldi (1662) on either side of the Corso defined the junction of the three main roads which led into the centre of Rome. Valadier's plans were ready in 1794, and after being revised by Gisors and Berthault, were carried out between 1816 and 1820.[167] Inspired by the achievements of Gabriel in Paris and of the Woods in Bath, he endeavoured to unite the square and the open spaces of the Pincio by creating, above the square, a terrace giving wide views over the city, which should be linked to the square below with ramped roads suitable for carriages. Hemicycles with paved walks closed the sides and completed the design of the square.

On the 16th July, 1823, the basilica of S. Paolo-fuori-le-mura was destroyed by fire. The reconstruction was chiefly the work of Luigi Poletti (1792-1869) of Modena. The interior was rebuilt by 1854, and the exterior portico with its colossal order of four large columns followed his design, as does the mediocre campanile. He was professor at the Academy of S. Luke and trained most of the architects of the period.

In Naples, where the memory of Vanvitelli survived during the whole century, the main works of note are those by Antonio Niccolini (1772-1850), who built the Teatro di S. Carlo, with an Ionic storey carried on a huge rusticated colonnade, and attempted a large-scale replanning in front of the royal palace, which had been begun during the reign of Napoleon's brother-in-law Joachim Murat by Leopoldo Laperuta, in imitation of the colonnades of St. Peter's. This was completed by the church of S. Francesco di Paulo, designed by Pietro Bianchi (1787-1849), with a hexastyle portico (1817-46) grafted on to a rotunda on the example of the Pantheon. Similar neoclassic effects were introduced at Bari by Niccolini (the Teatro Piccini), in Palermo

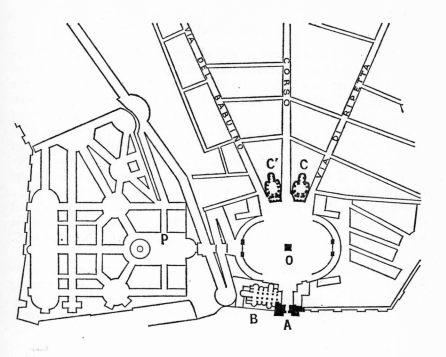

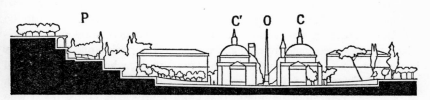

FIG. 42. *Piazza del Popolo, Rome, by G. Valadier, 1816. Plan and section (after W. Armstrong): A, Porta del Popolo (ancient gateway on the Via Flaminia); B, S. Maria del Popolo, rebuilt under Sixtus IV; C and C', S. Maria dei Miracoli and S. Maria in Montesanto (1662) by C. Rainaldi; O, Flaminian obelisk, placed there in 1589; P, Ramps and terraces of the Pincio.*

373

NEOCLASSICISM

by G. Venanzio Marvuglia and by Domenico Lo Faso Pietrasanta (1783–1863), an archaeologist who designed the loggia on the Piazza Umberto I (1845).

PAINTING

The Neoclassic reaction was started in Rome by Marco Benefial (1684–1764), from whom stems a narrow academism, dominated by Maratta, and without any connection with the art of the Carracci. It was developed by the French painter Subleyras (1699–1749), by Pompeo Battoni, a Tuscan who became a Roman (1708–87), and painted pictures offering a mild mixture of Correggio and Raphael, and grand manner portraits (for instance, his *Soderini* in Tarquinia), and principally by the German Anton Rafael Mengs (1728–79) the friend of Winckelmann and a convert to Catholicism (1746). His *Parnassus* in the Villa Albani, and his specious allegories (in the Sala dei Papiri in the Vatican) caused him to be proclaimed the greatest painter of the period. As a result of his many pamphlets, which aimed at a restoration of the cult of Raphael and the doctrines of Bellori, he was considered as the reformer of a corrupt age. In 1771 he was the head of the Academy of S. Luke.

The arrival of David in 1774, and his return in 1784, gave a new urgency to these archaeological and moralizing speculations; this was the age of a strong and controlled rhythm, of bas-reliefs corrected in the process of their translation into engravings, of melodrama derived from antique subjects. But it was to the foreigners, ever more numerous in Rome, that this intellectual, highly conscious art appealed—to the Germans, disciples of Mengs, friends of Goethe and later of Overbeck, who in the Villa Massimi on the Pincio practised a naïve Pre-Raphaelitism, to the English, and finally to the members of the French Academy in Rome. Italy played an important role in the new European style, but apart from Canova, venerated by the Academy of S. Luke, Italian artists hardly existed. The only Italian artists that can be fitted into the Neoclassic framework are the feeble Gaspare Landi (1756–1830), who painted tame frescoes and more lively portraits (one of Canova is in the Borghese), Vincenzo Camuccini (1771–1844), who painted historical subjects in Naples and Genoa, was esteemed by the Popes, and painted a work for S. Paolo-fuori-le-mura (1835) and F. Agricola (1795–1857) who painted historical and religious subjects.

In Florence, Pietro Benvenuti (1769–1844) who showed something of David's eloquence in his *Triumph of Judith* (1803, Arezzo), shone at the court of Napoleon's sister Eliza Baciocchi (of which a group portrait, 1813, is at Versailles), spread academic doctrine, and decorated the dome of the Cappella dei Principi at S. Lorenzo (1827–36). Luigi Sabatelli (1772–1830) was trained by him in decorative painting, and worked in a heavy and inflated style in the Capponi and Pitti palaces.

Only in Lombardy was there anything of interest, with Andrea Appiani

NEOCLASSICISM

(1754–1817), who decorated the dome of S. Maria presso S. Celso (1795), the Villa at Monza (1789), and who, in Milan, executed well-managed frescoes to the glory of the Napoleonic Empire, and a large monochrome frieze in the Sala dei Cariatidi (1807) in the royal palace. Under the affectations of his classical style there still survive the elegance and ability which put his *Apollo and Daphne* (1811) far above the dismal confections of Mengs. His friend, Giuseppe Bossi (1777–1815), decorated in the academic style the Villa Melzi at Bellagio. Outside contemporary trends, the Bergamasque Vincenzo Bonomini (1756–1839) covered the rooms of his native province with Pompeian motifs, idyllic and strange, creating, for instance, in the church of S. Grata at Bergamo a series of macabre scenes derived from the dancing skeletons of the medieval Dance of Death.

SCULPTURE

By about 1780, the mannered, precious and graceful elements, which broke up the Baroque, had gained the upper hand, for instance, in Tuscany with Innocenzo Spinazzi (1720–95) who carved the *Veiled Faith* in S. Maria Maddalena dei Pazzi and the tomb of Machiavelli in Sta Croce, where a soberly designed sarcophagus supports a small allegorical figure (1787).

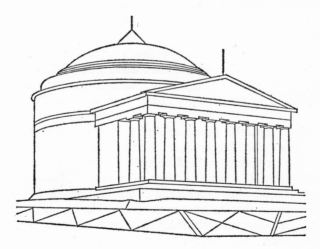

FIG. 43. *Temple at Possagno, by A. Canova, P. Bosio, and G. A. Selva,* 1819–30.

The full meaning of the style only became clear through the genius of Antonio Canova (1757–1821). Born at Possagno, near Bassano and the Monte Grappa, he was Venetian by birth. A museum founded in the house where he was born, full of models and plaster statuettes, and the odd little Doric temple built from his own designs (1819–30) with a Greek Doric Portico and a Roman structure behind, reveal very adequately the two sources of his art, which was the purest, and in many respects, the highest expression of Neoclassicism. He was formed in Venice as a pupil of Morlaiter and Marchiori, in a rarefied aesthetic atmosphere among casts of antique sculpture. Hellenistic sources of inspiration were for him as for his contemporary the poet André Chenier, a new form of elegance, a more subtle aspect of Rococo feeling—hence his *Daedalus and Icarus* (Mus. Correr) with its overworked modelling and its superfluity of directional axes.

Residence in Rome in 1779 brought him in contact not only with Winckelmann and Mengs and their severe ideology, but also with Rome itself and its insistence on the monumental. His full promise was revealed in 1780, with *Apollo crowned with laurels*. His tombs of the Popes Clement XIII (1787-91) in St. Peter's, and Clement XIV (1787) in SS. Apostoli, ran counter in their regularity of line, the repose of the figures, the serenity of the invention, to all the movement and energy of Baroque, though they preserve the breadth and the oratorical insistence of the traditional papal tomb. From delicacy the artist passed to eloquence, from Greek to Roman; *Hercules* (Rome, Mus. d'Arte Moderna) and *Hector* (Venice, Palazzo Treves) are the examples which illustrate this evolution, which was even clearer when Canova became the favourite sculptor of Napoleon and his family—examples of this period are the bust of the First Consul (1803), the project for an equestrian statue, Napoleon as the Apollonian hero, nude and holding a figure of Victory (bronze, 1810, Brera; marble, 1811, Apsley House, London). In 1811 the direction of the museums of Rome was offered to him, but he refused it.[168] His portraits of the members of Napoleon's circle—Madame Laetizia now at Chatsworth, Cardinal Fesch, Alexandrine, wife of Lucien Bonaparte, the Baron Daru—draw their inspiration from antique figures and are carved with great delicacy in a highly polished marble, with the refinement of alabaster.

It was principally in his figures of Pauline Borghese, daringly represented as *Venus Victrix* (1804, Borghese) and *Venus coming from the bath* (1805-12, Palazzo Pitti), that Canova realized his most subtle and harmonious compositions. Their controlled sensuality, their sober and learned appearance show his sense of uncluttered design, his use of the fine line that Ingres also used, but with a coldness that becomes even more evident in groups such as the *Cupid and Psyche* (Louvre), which look like huge trinkets. His *Three Graces* (Leningrad) inspired Ugo Foscolo to a poem, whose Romantic inspiration is redolent of soft and flowery Hellenism.

There is far greater breadth in the tombs of his last period—those of the Stuarts (1821, Rome, St. Peter's), Alfieri (1810, Florence, Sta Croce) in white marble with the allegorical figure symbolizing Italy leaning on a sarcophagus raised on a high oval base, Maria-Christina of Austria (church of the Augustinians, Vienna), where against a pyramidal mausoleum are portrayed the figures of the funeral procession—but there is much less clarity in his monument to Carlo III (1820, Piazza del Plebiscito, Naples).

Canova's most faithful disciples were the Dane Thorwaldsen (1770–1844), a competent craftsman who endeavoured to achieve powerful effects and who passed forty years in Rome, where he made the tomb of Pope Pius VII (1831, St. Peter's), and Pietro Tenerani (1789–1869), who made the rather dreary tomb of Pius VIII (St. Peter's).

II

Eclecticism

1840-1910

The twentieth century was dominated in Italy, as elsewhere, by the development of archaeology and history, but instead of the Gothic revival, Florence and Rome went through a phase of a return to Botticelli or Carpaccio—that is, of a revival of the Quattrocento. The craze for realism and for a mean naturalism prevailed over the examples of Caravaggio and the Bolognese, while religious art rehashed the more dreary examples of the seventeenth century. In architecture, where the confusion was even greater, a development ranging from the imitation of the High Renaissance to that of even earlier periods may be seen, only to be followed, towards the end of the century—under pressure from styles imported from Paris and Vienna—by a return to the freedom of Baroque.

ARCHITECTURE

In his history of design (*Storia dell' Arte del Disegno*, 1852–56), the learned Pietro Selvatico rejected the idea of only one kind of beauty, or of one absolute style, in favour of an historical interpretation which should justify each period, but he arrived at the dangerous conclusion that each of the styles of the past should be adapted to a modern use: the basilica for churches, Renaissance forms for houses, Roman types for cemeteries, Arabic fantasies for cafés, and so on. The divergence between ideas promulgated from the standpoint of learned culture and what actually happened was even graver in the case of Boito (1834–1914), an interesting historian, defender of a 'national architecture' and creator of distressing pastiches such as the Home of Rest for Musicians in Milan, and the Palazzo delle Debiti (Loans Office) in Padua, built in an Islamic style. Thus, the new features of modern society —banks, arcades of shops, stations—subjected to these arbitrary schemes, were not permitted to develop in a manner worthy of the past, except occasionally in their mass and their decorative richness; and on the other hand the modernization of ancient cities where remarkable buildings from the past ought to have been preserved, was not achieved without blunders.

378

ECLECTICISM

Restoration and Modernization. The vainglory of municipalities, and the pseudo-science of archaeologists, can alone explain the misplaced zeal with which it was thought possible to complete, if not to restore, the works of the past. In Italy, this was a movement parallel to the 'ideal' restorations of Viollet-le-Duc in France. In Florence, in particular, Sta Croce was given an arbitrary façade, with three triangular gables in a pastiche of the Trecento, by Matas (1863). As in the fifteenth century, a competition enabled Emilio de Fabris and Luigi del Moro to build a façade for the cathedral in Florence, something which the Renaissance had never been able to achieve. It was inspired by the side walls of the cathedral with its coloured marble, white, red and green, and its three portals are unhappily overloaded with mediocre mosaics and finished off with bronze doors. As typical of the spirit of false preservation is the medieval town arranged in Turin, on the banks of the Po, with genuine fragments, copies and free imitations of the Piedmontese castles of Fenis and Issogno, this conglomeration being the work of A. d'Andrade in 1884.

It was in Rome that the most spectacular transformations took place, after 1870. 'Twenty-two years of government by the house of Savoy have done more for the destruction of Rome than all the invasions of the Goths and Vandals' wrote an eye-witness in 1892. The archaic city, *bourgeois* and provincial, still surrounded by the desert created all round her by malaria, was the 'second Rome' of Pius IX and had scarcely begun to change when it became necessary to install there all the paraphernalia of a capital.[169] The new main roads of the Via Nazionale (1867) and the prolongation of the Corso Vittorio Emanuele, were cut through the city in the most graceless manner, and without being atoned for by any architecturally worthwhile buildings, except the old ones already there. The Piazza del Esedra, at the edge of the Baths of Diocletian, was planned in 1888 with buildings in a hemicycle by Gaetano Koch (1849–1910), and with the enormous Art-Nouveau fountain of Rutelli (1901) whose bronze nymphs are even more showy than those by Ammannati. The archaeological promenade through the Forum was laid out in 1897.

The new centre of Florence was not less ambitious, grandiloquent and eventually disastrous. The old quarter of the Calimala was destroyed from 1887 onwards to make way for a large square with an equestrian statue of Victor Emmanuel II by Zocchi (1890) in front of a triumphal arch on an unreasonable scale dominating a heavy arcade (1893). The happier replanning of the hill of S. Miniato, with ramps constructed on embankments above the Porta S. Niccolo, the circular boulevard planted with trees running from the Viale dei Colli, the loggia and belvedere on the Piazza Michelangelo, were for the most part the work of Giovanni Poggi (1811–1901).

Innovations and Projects. Italy had its share of those innovating minds, conscious of the new role of the engineer and impatient for the chance to

379

build large and impressive structures, such as the Eiffel tower, built in Paris in 1889, to the glory of the new technology. One of these was an engineer from Novara, Alessandro Antonelli (1798–1888), who erected on the cathedral of his native city, S. Gaudenzio, the extraordinary dome (planned as early as 1840) raised on two drums fitted one above the other, and finished with a three-tiered spire. Somewhat later, he achieved a similar soaring effect with his Mole Antonelliana in Turin, not by extending traditional elements but by using one cast-iron structure to support another dome with four concave faces (84 metres high), surmounted by a spire formed like a telescope to a total height of 168 metres. But these inventions, which remind one of some of Filarete's projects, were interesting experiments without any future and cannot be considered as the forerunners of a modern architecture keyed to the industrialization of society. Another architect of note was Ernesto Basile (1857–1932) of Palermo, who built a gallery for machines at the Palermo Exhibition in 1892, the rather freely designed Villa Igea in 1898, the Villa Basile (1904) and the Villa Farsini (1906) in Palermo, where his straightforward handling and control of the masses contrasts with the flaccidity of contemporary architecture in general. His new façade for the seat of the Parliament at Montecitorio in Rome shows an abuse of the organic motifs of Art Nouveau.

Theatres, Stock Exchanges, Cemeteries. Among public buildings the theatre maintained during the nineteenth century the important position it had enjoyed during the eighteenth. Under the Hapsburgs, the Scala in Milan, and under the Napoleonic Empire, the San Carlo in Naples, had set the pattern for those huge auditoria of which the exterior consists of a façade with a loggia supported by a rusticated portico. Thus, following in the footsteps of Piermarini and Niccolini, the pattern was repeated by Barabino at Genoa (1828), Bettoli at Parma (1829); Poletti, who rebuilt S. Paolo-fuori-le-mura in Rome, also built new theatres at Fano (1845–63), Terni (1849), and Rimini (1857), all with the exterior form of temples with porticoes. Another specialist in theatres was Aleandri of Sanseverino (1795–1885), who multiplied examples with two storeys of which one was an Ionic colonnade, as for instance at Ascoli and Spoleto (1845–84).

It was principally in the southern cities that this type of building was revived during the nineteenth century. The pseudo-Pompeian style was used in the Politeama Garibaldi at Palermo (1847), by Almeyda (1834–1901), and an elaborate Corinthian order for a hexastyle portico as in the Grand Theatre (1875–97) by G. B. Basile (1825–91) and his son Ernesto, who also built similar theatres at Agrigento and Marsala. At Catania, finally, the Teatro Bellini was built from 1873 to 1890, by Andrea Scala and by Sada, with strong influence from Garnier's Paris Opera House in the arrangement of the storeys, the acroteria and the overhanging groups of sculpture.

There was also a certain uniformity in Stock Exchanges and Banks, of a generally classicizing, though not antique, style. The Cassa di Risparmio (Savings Bank) in Milan is a pastiche in the rusticated style of Michelozzo by G. Balzaretti (1801–74); the Banca Commerciale by Luca Beltrami (1854–1933) has a façade with a colossal order of eighteenth-century type. In Bologna the Cassa di Risparmio was built (1868–76) in a delicate design faithful to the Quattrocento by Giuseppe Mengoni (1829–77); Arturo Cipolla (1823–74) made a fine design for the Banca d'Italia (1862–64). In Florence, the same architect built the Banca d'Italia (1869) in well-understood pseudo-Cinquecento style, and the central market (1874) with an equally capable understanding of cast iron.

There must also be noticed, chiefly in the North, the development of cemeteries. At Genoa, Barabino had designed the first monumental cemetery at Staglieno, grafting two colonnades on to a central rotunda, with a portico projecting in front. The scheme was completed by G. B. Regaseo (1799–1872) with terraces on the hillside. At Brescia, Rodolfo Vantini (1791–1856) designed a real town plan for the cemetery, with a central rotunda, long rectilinear wings and, on the main axis, a beacon similar to an antique column (1815–49).

Arcades, Monuments. Projects for open spaces of the type of the Antique forum had been revived during the Napoleonic Empire, particularly in Milan. Under the monarchy the tendency was to design arcades. After winning a competition, Mengoni built the Galleria Vittorio Emanuele (1863) in Milan; the triumphal arch façade facing the cathedral square derives both from the Antique and from Lombard Renaissance forms, which dominate in the interior, under the glass dome of the crossing, to the point of evoking the Certosa of Pavia. In Naples, the Galleria Umberto I (1887–90), designed by Rocco and decorated by Ernesto di Mauro, expresses a rather more sensitive eclecticism of Renaissance forms.

In Rome, as in Venice, the idea caught on of closing a view with a pastiche; this is the case with the Palazzo Sforza-Asarini by Piacentini (1846–98), the Palazzo Marignoli by Podesti (1857–1909) the Palazzo Margarita and Palazzo Lavaggi-Pacelli by Koch, which are all modelled on Sangallo or Peruzzi.

In 1883, as a result of a competition, the Perugian architect Gugliemo Calderini (1837–1916) was commissioned to build the new Palace of Justice in Rome (1888–1910). This heavy and overloaded building, which weighs down upon a whole district of the city, has at least the merit of seeking to recapitulate all the Antique and Baroque architectural forms by placing on a rusticated base storeys with heavily decorated windows, enormous string-courses and cornices, and in the centre, an entrance arch affording a view of the interior courtyard.

The inflated pomp, the vigorous, though confused and empty, imagination of the period are summed up in the monument to Vittorio Emanuele II,

ECLECTICISM

the major part of which was built by Giuseppe Sacconi (1854–1905) on the northern slope of the Capitol, facing the Corso. He won first prize in a national competition in 1884, and his conception of an enormous circular colonnade in Brescia white marble, backing on to the hill and raised up on a complicated podium, clearly appealed to contemporary taste. Its colour clashes with all the tonalities of Roman building and soil; its enormous mass does not fit any of the long vistas it commands; unassimilable and infinitely complicated, it remains, despite good detail in sculpture, the symbol of the failure of nineteenth-century Italy to create a style to match its glories.[170]

SCULPTURE

The technical ability and the breadth of culture of the Florentine Lorenzo Bartolini (1777–1850) only enabled him to oscillate between a richly detailed naturalism, the result of his facility, and a cold and over-calculated composition as, for instance, in his *Charity* (1824, Pitti), *Confidence in God* (1835, Poldi Pezzoli, Milan) and the Demidoff Monument (1830, Florence, on the Lungarno Serristori).

Giovanni Dupré (1817–82) a Sienese, carved a *Dying Abel* (Mus. d'Arte Moderna, Florence) and a *Sappho* (Mus. d'Arte Moderna, Rome) displaying an affected realism. Pandering to the taste for anecdote ruined the style of Adriano Cecione (1838–86), critic and leading light among the Florentine *Macchiaioli*, who prided himself on his free and varied handling; examples are his *Child with a Cock* (Mus. d'Arte Moderna, Florence) and *Mother and Child* (Mus. d'Arte Moderna, Rome). Giulio Monteverde (1837–1917) had a showy and descriptive vigour, to be seen in his figure of *Jenner* (1873, Genoa, Palazzo Bianco).

It was in the South that this taste for realism of detail declined most rapidly into the poorest kind of cleverly executed genre scenes, in which Achille d'Orsi (1845–1929) with his *Proximus tuus* (Mus. d'Arte Moderna, Rome), and particularly Vincenzo Gemito (1852–1929), with his *Fisherman* (1878, Mus. Nazionale, Florence), his *Water seller* (Mus. d'Arte Moderna, Rome) and his celebrated busts, were especially successful.

Luigi Borro (1826–86) gave to his busts and monuments (*Daniele Manin*, 1875, Venice) a rich and complicated air, and Carlo Marochetti (1805–67) endowed his figure of Emmanuel-Philibert (1828, Turin) with the romantic picturesqueness of a storybook chivalry. Vincenzo Velo (1822–91) had little of this quality in his sentimental compositions, such as the *Dying Napoleon* (1867, Versailles), and *Victims of work* (Mus. d'Arte Moderna, Rome). There is less energy and more humour in the works of del Zotto (1841–1918), who made the statue of Goldoni in Venice, and G. D. Grandi (1843–94), who retained from the influence of Carpeaux a taste for very detailed working, with crumpled drapery and a multitude of sharply lit details, though in a disjointed style as, for instance, in his *Lara's Page* (1872, Mus. d'Arte Moderna,

Milan) and his monument to the Five Days (1893, Milan). The most interesting figure is that of the Piedmontese Medardo Rosso (1858–1927), who not only introduced an Impressionist feeling for light into his busts, through his lively handling, but also a kind of Lombard chiaroscuro, or *sfumato*, which appears to envelop the soft forms in mist; through his tender sentiment and a certain feeling for intimacy, he is very close to the painter Eugène Carrière. The Baroque taste for fluency in the use of materials, deprived of all monumental feeling and pervaded by mannerisms, dominates the work of the extraordinary Andrea Lippi (1888–1916).

PAINTING

As in all the other arts, what was missing in nineteenth-century Italian painting was an artist of full and dominating genius. The local schools were now no more than provincial schools, reacting with longer or shorter delays to the foreign movements which seem sometimes to have robbed Italy of her inheritance and her own true gifts. The limpid beauty of the Antique and of the Tuscan Mannerists such as Bronzino had been taken over by Ingres, the epic quality of the Renaissance by Delacroix. All that survived in Italy was the Romanticism of small subject pictures and monotony in portraiture. Only Francesco Hayez (1791–1882), a Venetian who settled in Milan, and was the painter of small artificial costume-history pieces in the style of Delaroche (*Sicilian Vespers*, Mus. d'Arte Moderna, Rome; *The Kiss*, Brera) and more sensitive portraits against plain backgrounds (*Signor d'Adda Falco*, pvt. coll., Venice; *Manzoni, Cavour*, Brera) is worth noticing; the Raphaelesque revival among German and French artists found no echo in his limited art.

The finest interpreters of Italian landscape were not Italians: in Rome the field belongs to Corot, in Venice to Turner, Monet and Dufy. The nobility of the Primitives inspired the affected but intense school of the Ruskin followers—the minor pre-Raphaelite painter-poets who invaded Florence. In the field of elegant, worldly painting, the energetic reforms introduced by Manet were lacking; in that of analysis all that was represented by the perceptive and modern taste of a half-Italian such as Degas was not to be found in Italy; neither was the passion for the southern sun which filled Van Gogh, nor the archaizing nobility of Gauguin—all manifestations which, like Impressionism, were almost unknown in Milan and Florence. They were satisfied with Fortuny and Meissonier.

The most notable landscape painter, linked with Corot but chiefly influenced by Ravier, is the strange Fontanesi (1818–82), an Emilian with a passion for travel, whose case is typical of the profound disturbance of artists involved in the wars of liberation, dissatisfied and ready to exile themselves, even as far as Japan. His works show his feeling for chiaroscuro and quivering atmosphere (*Morning*, Turin; *Pasturage*, Mus. d'Arte Moderna, Rome).

ECLECTICISM

The two main centres of the Age of Baroque in Italy lost all their importance, and nothing survived of the tradition but the over-inflated or mean aspects. In Rome, where there remained an undercurrent of intellectual life, a derivative academism reigned with Podesti (1800–95), Fracassini (1838–1868) and Mariani (1828–1901), who decorated the Basilica of S. Lorenzo, and with Maccari (1841–1919) who painted history pieces in the Senate. The efforts of the movement called 'In arte libertas', founded by Nino Costa (1826–93), in an attempt to revive the Roman inspiration of Poussin, under the aegis of Corot, were also unsuccessful. With Sartorio (1860–1932) the literary painting of the Baroque impinged on symbolism.

In Venice, severed from Italy from 1815 until 1863, nothing came of the splendid development of the eighteenth century, which instead became part of the general European tradition. After 1874, Federico Zandomeneghi (1841–1917), a sensitive colourist, settled in Paris—that is, close to Renoir and Degas. The family anecdotes of Giacomo Favretto (1849–87), with their gay and glittering colour and their facile, over-detailed treatment, (*The Wedding*, 1879, pvt. coll., Milan) may in some ways recall Longhi, but his vivacity is spoiled by his concentration on detail, which was to become even greater, after a visit to Paris in 1879, through his admiration for the meretricious charm of Fortuny, as for instance, in his *Flower Girl* (1881, pvt. coll., Milan). With him in Venice reigned the boring painter of religious subjects, Luigi Nono (1850–1918), competent and studied, as in his *Refugium peccatorum*, 1882, and *The child's funeral*, 1885. With these two extremes of empty solemnity and exaggerated liveliness in small pictures, Italian painting faces the double pitfall of two temptations to decadence, from which there is no escape even in the three most active centres, which were then, as during certain moments of Gothic, Milan, Florence and Naples.

Milan. Francesco Hayez having become the leading painter in Milan at the expense of the Neoclassical Appiani, grand scenes of patriotic history were rendered by popular painters such as Bertini (1825–98), Casnedi, Faruffini (1831–69), who were addicted to historical subjects, and to a certain showiness of colour, also used by the Induno brothers, Domenico (1815–78) painter of *The Model* (1864, Milan) and Girolamo (1827–90).

Giovanni Carnevali, nicknamed 'Il Piccio' (1806–73), a rather neurotic artist, akin to Lotto, whom he studied, was particularly interested in effects of light and mist; he avoided defining his forms, concentrated on rendering atmospheric vibrations and became a hazy and shadowy Correggio, as in his *Bacchus and Ariadne* (pvt. coll., Milan), and *Lot and his daughters* (pvt. coll., Crema). His technique, with rich impasto and glazes, often has a sentimental note as in his *Self-portrait*, Bergamo, and sometimes directly imitates Guido Reni (as in his *Madonna*, 1863, pvt. coll., Milan). The revival of Lombard chiaroscuro, and the genuine or false subtleties of the followers of Leonardo, came with Tranquillo Cremona (1837–78), whose sensual effusions and blurred tones are the vehicle for his sentimental subject matter in works such as *Mother love* (Mus. d'Arte Moderna, Milan), *The Ivy* (Mus. Civico, Turin) the *Visit to the College* (pvt. coll., Genoa), and in his soft and lifelike portraits which presage so strongly the manner of Carrière, such as *Madame Duchamps*, 1865 (Mus. Civico, Milan). Around Cremona were grouped the emancipated artists, nicknamed the '*scapigliati*' ('the dishevelled'), chief among whom was Ranzoni (1849–89) who also exploited the romantic portrait with tender half-smiles seen in shadowed softness, for instance in *Princess Troubetskoy* (pvt. coll., Crema), *Madame Luvoni* (pvt. coll., Padua), *Sick Girl* (Mus. Civico, Milan). A third period is marked by the work of Grubicy de Dragon (1851–1920), a Hungarian by origin, who was a friend of Seurat and Signac, and who adapted the scientific technique of Divisionism while proclaiming, in an article published in 1896, that the technique had been anticipated by Ranzoni in 1870. The first exhibition of Divisionist painting in Italy took place in 1891, and the technique, which increases the luminosity of the picture while breaking up the continuity of its appearance, was thoroughly understood by three painters; Pelizza da Volpedo (1868–1907) in his landscapes such as the *Procession* (1895, pvt. coll., Milan) and *April Morning* (1904, pvt. coll., Florence), and Previati (1852–1920) who exploited it so as to give an aura of mystery to his religious and allegorical subjects, such as *Children in a field* (Mus. d'Arte Moderna, Florence), *Madonna with a Lily* (Mus. d'Arte Moderna, Milan), and Giovanni Segantini (1859–99), an artist passionately interested in country life and a close observer of the changing light on alpine landscapes (at Brianza). He achieved his earliest success in Milan in 1885 with his serious works, comparable to those of Troyon or the Dutch Old Masters, with their heavily accentuated shadows. After his contact with Grubicy about 1886, his palette became much lighter, and his brush-strokes more broken, more suited to the landscape of the Engadine (*Hay Harvest*, 1899, Segantini Mus., St. Moritz) and this also appears in his portraits, such as that of C. Rota (1897, Milan, Hospital). He also experimented with symbolical themes, inspired by the Pre-Raphaelites or by Böcklin, in his triptych of *Nature, Life and Death*.

Less ambitious, and endowed with a lively technique in broad brushwork, C. Tallone (1853–1919) was a celebrated portraitist in Lombardy (for example, *Bernasconi*, pvt. coll., Busto Arsizio; *Reffelle*, pvt. coll., Padua), and in Piedmont the painters most to be noted for their rendering of light and brilliance, with the temperamental Fontanesi, were, among the realists of integrity, Vittorio Avondo (1836–1910) and the highly detailed but sensitive painter Enrico Reycend (1855–1928) of Turin.

Florence. The Romantic Movement appeared in Tuscany with J. Bezzuoli (1784–1855), who painted the *Entry of Charles VIII into Florence* (Mus. d'Arte Moderna, Florence), and with his pupils Ciseri (1821–91) and E. Pollastrini (1817–74), who painted large and frigid religious and historical compositions. It was against this lifeless kind of art that, about 1855, arose the movement which emanated from the Caffè Michelangelo, where the ideals of the Risorgimento, the liberation from academism so as to evolve a more expressive handling, the love of truth and of the modern, were combined to express the vital aspirations of Italian art. A picture by the painter and critic Adriano Cecioni (1836–86) called *Caffè Michelangelo* (1861, pvt. coll., Milan) shows the chief members of the group in a small and sombre sketch, *à la* Daumier. Cecioni was the appointed apologist of the group and later accepted and welcomed, as an honorific title, the denigratory name of the 'daubers' (*macchiaioli*) which had been given to it. In the arguments about it, one of the main contentions of its advocates was the need to substitute an impression for a description, to base the composition on a striking arrangement of patches of colour, and to bring painting back to its essential principles: colour, and tonal values and relations. This aesthetic of simplification applied to contemporary subjects produced remarkable results during the two decades 1860–80.

But all the painters involved were soon to realize how much they could benefit from a study of Quattrocento artists, since they were sometimes inspired by Quattrocento use of overlapping areas of clearly defined forms. The principal artists in the group were Fattori, Lega and Signorini, and with them may be considered one or two other interesting men such as Vito d'Ancona of Pesaro (1825–84) who worked in Paris between 1868–79, and who painted female figures with brilliantly coloured details silhouetted in the light of the interiors (*Conversation*, pvt. coll., Milan) or gardens (*Lady with a sunshade*, pvt. coll., Milan) in which they are posed, G. Abbati (1836–68), a sober painter of landscapes in solidly constructed tonal values, R. Sernesi (1838–66) who rendered with delicacy the long lines of the Tuscan Apennines, O. Borrani (1833–1903), extremely precise in his handling of light, V. Cabianca (1837–1902), who specialized in effects of sunlight and moonlight, and C. Banti (1824–1904) less positive, more atmospheric, a '*macchiaiolo de luxe*'.

Giovanni Fattori (1825–1908), of Leghorn, who was a member of the 'Action party' in 1868, owed his success to large military pictures like the *Battle of Magenta* (1862). It was in his small, very direct and spontaneous works that, encouraged by Nino Costa and unaffected by programmes of ideas, he created with the simplest means the most lovely examples of the beauty and light beloved of the *macchiaioli*. The *Rotunda at Palmieri* (1866, Mus. d'Arte Moderna, Florence) long and narrow in format, extends sky, sea and horizon into even bands against which the figures are clearly silhouetted. The little picture of French soldiers (1859, pvt. coll., Crema) where strong blue stands out against the yellow ground and a grey wall, provides a fine instance of the lively and satisfying notation of colour that is used with less success in the *White Wall* (pvt. coll., Valdagno). This regular use of the figure silhouetted in a frieze and handled with large, flat, dense brushstrokes is to be found in his *Red barrels* (1867, pvt. coll., Milan), *Lega on the rocks* (1867, id.) and, in darker vein, in his *Farms in the Maremma* (date unknown) or *Net-menders* (1850, pvt. coll., Florence). This insistence on the same motifs indicates the painter's limits, but his love of strong light, reflected off matt or chalky surfaces, gives a quality of originality to the patch of colour which, according to Cecioni's own words, 'forms the basis' for the picture. His portraits are more accomplished, some in an open air setting (*Diego Martelli*, 1867, pvt. coll., Milan), others, such as *Argia* (date unknown, pvt. coll., Turin), or *The Red Shawl* (no date, pvt. coll., Milan) recalling the teachings of Bezzuoli, though with less artifice.

Silvestro Lega (1826–95), a friend of Borrani and Fattori, never had any success and lived in miserable poverty. His scenes of everyday life, such as *Curiosity* (1870, pvt. coll., Milan), which shows a woman peeping through a green shutter, *The Visit* (*c.* 1875, Mus. d'Arte Moderna, Rome), and *La Ciociara* (no date, pvt. coll., Florence), which contains a great sweep of striped red woollen stuff, have a strange, sombre quality little enlivened by their colour contrasts. On the other hand, Telemaco Signorini (1835–1901) controversialist, memorialist of the Caffè Michelangelo, follower of Garibaldi, was a more brilliant and even agitated personality, in touch with the latest artistic currents in Paris and London, and working in a vein of heightened realism, as in his *Lunatic Asylum* (1865, Mus. d'Arte Moderna, Rome), which earned him the praise of Degas. His *Sunny day at La Spezia* (1860, pvt. coll., Rome) is a magnificent composition in impasted lights and sombre motifs, and his *Mercato Vecchio* (pvt. coll., Padua) and his *Porta Adriana at Ravenna* (Mus. d'Arte Moderna, Rome) show considerable variety in the handling of tones and patches of colour, solidly put together with great directness. In his landscapes the arrangement of his subjects on the canvas is often lively, as for instance in his *Riomaggiore* (1890, pvt. coll., Valdagno) with its rows of yellow terraces, and his genre scenes, such as *The Sleeping Children* (1896, pvt. coll., Genoa) occasionally recall Toulouse Lautrec. In other words, Signorini fairly quickly freed himself from the true '*macchia*', or the individual touch of colour, and from its provincial limitations and theorizing. He turned rather towards Manet, though he could never resist the diffusion of concentration which compromised his talent. The connection with the '*macchiaioli*' of Giovanni Boldini of Ferrara (1842–1931) was even shorter still. From 1869 onwards, after visiting London and passing through Paris in 1872, he became the modish painter of the fashionable world, producing a large number of agitated and voluptuous portraits, tense and strained in handling and of a superficial elegance.

Naples. The memory of Salvator Rosa and other landscape painters, revived by the Dutchman A. Pitloo (1790–1837), encouraged the birth of a School of Posillipo, composed of artists determined to escape from the boredom of history painting. Giacinto Gigante (1806–76) becomes thus a kind of southern

Corot, somewhat inclined towards Turner. Filippo Palizzi (1818–99) continues in this placid and carefully composed vein, with cattle grazing in the sunlight. But in Naples it was impossible to stay within the limits of a reticent poetry without brilliance or bravura. Two artists are typical of the ensuing agitated and transiently attractive style, by which the Southern imagination is all too quickly carried away; Domenico Morelli (1823–1901), a great traveller, who knew the Nazarene painters, and also Delacroix, had an immense success with his very spectacular religious paintings, full of shafts of light and deep shadows dear to the seventeenth century, and with carefully thought out compositions such as the *Embalming* (Mus. d'Arte Moderna, Rome). With Eduardo Dalbono (1841–1915), the play of light on the bay and the animation of his scenes of everyday life in the city bring the art of the *Vedutisti* up to date. Bernardo Celentano (1835–60) painted chiefly historical pieces such as his *Council of Ten* (Mus. d'Arte Moderna, Rome), and Michele Cammarano (1851–1920) specialized almost in the battle scenes of the Risorgimento, large and very realistic. In Gioachino Toma (1836–91) there is an exceptional simplicity of tone, with the use of large areas of shadow, and compositions in greys and pallid colours, as in his *Signora Sanfelice in Prison* (1874, Erba coll.).

The arguments over the latest artistic developments were as energetic and keenly pursued in Naples as they were in Florence. In 1868, in his remarks on the exhibition known as the '*Promotrice*' a well-informed commentator, V. Imbriani, argued against the premises of academic theory, and advocated the necessity of seeking the pictorial idea indispensable to a picture in an organic and complete intuition, which he was already calling the '*macchia*' or patch of colour,[171] and which justified far more than a merely local movement.

But the example of Giuseppe de Nittis, a leading practitioner of showy bravura execution, who had an enormous success in Paris as a fashionable portrait painter, came down heavily in favour of a meretricious superficiality of manner, which, with the glitter of the Spaniard Fortuny and his 'empty brilliance', gave full scope to the Southern love of loud and motley colour effects. Francesco Michetti (1851–1929) recreated amid a welter of fanciful costumes the animated and often pathetic scenes of Neapolitan life (*Ex-voto*, 1883, Mus. d'Arte Moderna, Rome) and countryside (*Jorio's Daughter*, 1895, Berlin), in a d'Annunziesque drama. Another neurotic, Antonio Mancini (1852–1930), a friend of Gemito and influenced by Morelli and Cremona, achieved in his highly inventive pictures of urchins, of Neapolitan street sellers, a typical level of meretricious brushwork and over-emphasized colour.

III

The Twentieth Century

T he twentieth century saw the end of the long stagnation which prevented Italy from playing her full part in the development of modern art. Nevertheless, the movements born between 1910 and 1920, though significant upheavals, were without any staying power, and unable to affect all the arts; Italy had nothing equivalent, for instance, to the impulse given in every branch of art by the *'Esprit Nouveau'* of Ozenfant and Jeanneret between 1920 and 1925. Despite the effort to absorb the impact of machinery and to break down forms, Futurism had no useful extension into architecture (the efforts of Sant'Elia are quite apart) or decoration. The most notable artists were those able to fall back on themselves and cut themselves off in time.

Since 1948, large exhibitions in Milan, Venice and Rome have enabled Italian culture to form a clearer idea both of its own resources and foreign example. Matured by their long wait, the younger generation has quickly demonstrated Italian ability to create a valid modern architecture, and its speed in assimilating the most recent forms of painting, which have been encouraged since 1948 by the new Biennale exhibitions in Venice. There are now numerous sculptors, and the work done by decorators is increasingly successful and quickly becomes widely known through such manifestations as the Triennale exhibitions in Milan.

PAINTING AND SCULPTURE

Futurism and 'La Pittura Metafisica' (1909–20). At the beginning of the century, the artificial compositions of Previati in Milan, those of Armando Spadini (1883–1925) in Florence, which reflect something of French Impressionism, or of Adolfo de Carolis (1874–1928), who provided a pompous decoration for the Palazzo del Podestà in Bologna, were the productions of an art reduced to apathy and inanition. The reaction of the younger generation came violently with the movement known as Futurism, but this was, in fact, only one aspect of a general movement, hostile to traditional forms in art, of which Fauvism and Cubism in Paris, and the Blaue Reiter in Munich were also part. The most important centre of Futurism was, in fact, Paris. Marinetti's Manifesto was published in Le Figaro of 20th February 1909, and was a literary, political and poetical declaration in honour of 'feverish insomnia, the perilous leap, and the knock-out blow'. In 1910, the five painters of the movement, Boccioni, Carrà, Russolo, Balla, and Severini, composed a manifesto made clearer by the catalogue of the exhibition held in Paris in 1912, which proclaimed the necessity of war, and demanded the total banning of the nude in painting for ten years.[172] The contemplation of the female form was to be replaced by that of the machine, which was now to be reckoned the true modern muse. Ugo Boccioni (1882–1916), the imaginative inspirer of the movement, and author of a treatise on the new painting and sculpture (1914), limited himself to expressing the 'force lines' which he considered as the objective expression of the objects or scenes he represented, as in his *Dynamism of a footballer* (1911, pvt. coll., Rome), treated in multiple facets, or his *Lancers charging* (1915, pvt. coll., Milan), in powerfully articulated grey lines. But he also sought to express states of mind, through an interplay of short lines, woven into oblique and spiral movements, as in the series of the *Farewells* (1911, pvt. coll., New York) and he was particularly successful in 'unanimist' visions, such as the *Forces of a street* (1911, pvt. coll., Schinznach). This is a heightened rendering of modern life, obsessed with cars, railways, jazz, and expressing itself in an ideal of pictorial 'dynamism' that was to associate atmospheric vibrations in the Impressionist sense with the simultaneous facets of the objects envisaged by cubism, which was reproached with being static, dead. The Italian school thus sketched out a novel development, in which Gino Severini (born 1893) was to produce lively masterpieces (*Dance of the pan-pan at Monico*, 1912, pvt. coll.) or dynamic hieroglyphics such as the *Bal Tabarin* (1912, Museum of Modern Art), Giacomo Balla (born in 1874) increasingly abstract arcs of vibration, Carlo Carrà (born in 1881) more confused inventions. Futurism tended to escape altogether from painting through symbolic images, or through the composition of representations of movement intended to evoke an emotional response. Hence Boccioni's stimulating effect on sculpture which became irrealist in order to render the essence of movement, and which renounced the noble traditional materials (bronze, marble) in favour of hitherto unheard-of elements such as 'a valve which opened and shut to create a rhythm as beautiful but infinitely newer than that of an eyelid'.

Of all the *avant garde* European movements, Futurism was the only one with a nationalist and warlike character; with the First World War, in which Boccioni was killed very early on, and then with Fascism, it disappeared on its own. But already another revolutionary inspiration had appeared, that of the 'Metaphysical Painting' which, from 1910–20, was at once the complement and the ruin of Futurism. Giorgio de Chirico (born in 1888) lived in Greece and then in Munich, where he knew the symbolical and rhetorical painting of Böcklin, and where he became filled with idealistic philosophies; towards 1910 he was in Paris. He began to paint, in deliberately muted tones, grey and without formal qualities, pictures of imaginary cities, destined to symbolize the 'states of mind' of dreams—in fact, surrealist compositions. This was to deny, in favour of a revolution of the imagination, all the formalist revolutions which had followed upon each other. A summary design, and fantastic perspectives, sufficed to create the essential, which was an allegory redolent of inexpressible nostalgias (*Melancholy and Mystery of a street*, 1914, pvt. coll., New York) or of a decided nihilism.

Chirico and Carrà, already tired of Futurism, were thrown together in Ferrara by their military service from 1916 to 1918, and there worked side by side to create a more complete art, obviously influenced by the red brick palace of the Estes, by the striking paintings of Ercole de' Roberti, and the frescoes of the Palazzo Schifanoia. This was the moment of the *Disquieting Muses* (1917, pvt. coll., Milan), where tailor's dummies replace the statues in fantasy squares, and of the series where derisory objects invade the human figure, such as *Hector and Andromache* (1917, id.). Carrà painted fewer pictures of this type (*Penelope*, 1917, pvt. coll., Milan) and after 1919 the publication of his book on the '*Pittura Metafisica*' led to a rupture with Chirico. In 1925, Chirico, repudiating the 'lay mystery' and occult symbols, started painting pseudo-antique scenes with ruins and decorative horses, and nudes inspired by Renoir, while Carrà turned to sombre and monotone landscapes, gloomy in feeling.

The only painter who really developed during the 'metaphysical' period was Giorgio Morandi (born 1890), the most remarkable of the living masters. Naturally austere, strict, and insisting passionately on simplicity, he produces only still-life subjects, and these represent for him a subtle, unique communion with the dream and the invisible; for instance, in *Bottles and counter* (1916, pvt. coll., Milan). In 1918, the accent became more insistent and the forms harder (*Still life with a loaf and a leaf*, pvt. coll., Milan) under the influence of Chirico and Carrà, and then his work resumed the subtle and inevitable course which gives his white tones such a finely introspective quality. This marks yet another interruption in the flow of Futuristic pretentiousness and excitableness, but the asceticism of Morandi also involves the rejection of all academic reactions in the sense of Chirico and Carrà.

Morandi is a great example; he represents the innate awareness of an art which must rediscover a valid basis in seriousness and formal values. The problem had already presented itself in Bologna in the seventeenth century.

The tragic story of Amedeo Modigliani (1884–1920) who came to Paris in 1906 to lead the life of a 'peintre maudit', and brought his own unique contribution to all the latest movements, demonstrates, by an entirely contradictory energy and elegance, the resources of Italian genius. Among the cubist painters who concentrated on the hieratic, he alone handled distortion with harmonious continuity and was able with complete authority to give back to line its proper function—that which it had had for Botticelli and Pontormo (*Portrait of Soutine*, 1917, pvt. coll., Paris; *Marie*, 1918, pvt. coll.,

Paris), particularly in his nudes, with their exquisite undulation and their completely new inflexions, simplifications and discoveries.

The Contemporaries. Just as there was no Manet in Italian painting during the nineteenth century, so there was no Matisse in the twentieth. The revolutions of 1910–20 were at once too violent and too short lived; before political events deflected artists towards an art derived from academic tradition, the impetus was lost. A few painters escaped stagnation by identifying themselves with French movements: the Lombard Arturo Tosi (born in 1871) with his rather broad landscapes; the Mantuan Pio Semeghini (born in 1878), who added an individual softness and tenderness to a palette based on some of the Fauves and on Bonnard; the Ferrarese Filippo de Pisis (1896–1956) with his brief summary line, applied over colours with interesting contrasts, particularly in views of Venice and Paris; and, from certain standpoints, the Florentine Massimo Campigli (born in 1895) who, from the archaic forms of Roman portraits and the ochre tones of antique frescoes, has evolved his highly personal style.

Mario Sironi's unequal production shows the need to develop in a similar direction a vision which was strongly affected at the beginning by Futurism.

In the group called the '*Novecento*' (twentieth century), the ambition to work on a grand scale, the unconditional glorification of their Roman and classical past, fell in with the ideas of the Fascist régime and its programme of exaggerated nationalism. This tendency towards the rhetorical was opposed by the rather disordered fantasy of Luigi Bonichi, known as Scipione (1904–33), who created a vision of Rome in reds and dark greens. All over Italy there arose many discreet focuses of opposition, always in the form of small groups, as for instance the Six in Turin (from 1928) with Francesco Menzio (born in 1899) and Paulucci, who clung to a Post-Impressionist liberty. Abstract painters, such as the Florentine Magnelli (born in 1888), who sought the quintessence of Giottesque frescoes, and Mauro Reggiani (born in 1897), who concentrated on pure construction, like Atanasio Soldati (born in 1896), found themselves, after 1932, in the Galleria del Miglione in Milan, with the young initiators of a strictly political and social realism, such as Renato Guttuso (born in 1912). In 1933 the Roman Group was formed with Giuseppe Capogrossi (b. 1900), Cagli (b. 1910) and Cavalli (b. 1904), who were concerned with both balance and transposition.

The end of the war led to an acute and often heart-searching recognition of the inadequacies of Italian painting in comparison with European movements: the 'New front in the arts', founded in 1946 in Venice, with Guttuso, Birolli (b. 1906), Santomaso (b. 1907), and Vedova (b. 1919), declared itself as dedicated to a broader, more modern inspiration. Italian tradition was to supply a ground-base.

Sculpture was not further advanced by the researches of Boccioni than it was by the experiments of Modigliani. During the long period 1920–45,

121 Fattori: *Woman with a sunshade*

(*Milan, private collection*)

122 M. Sirone: *The Cyclist*

(*Rome, private collection*)

123 Marino Marini: *Small horseman*

(*New York, Curt Valentin collection*)

124 G. de Chirico: *Turin Melancholy*

125 Modigliani: *Portrait of Jean Cocteau*

(*Paris, Mme Paul Guillaume*)

126 Morandi: *Jugs*

(*Milan, Jesi collection*)

127 Manzù: *Head of a woman*

(*private collection*)

128 (a) Nervi: Rome,
Palazzo dello
Sport

128 (b) Ricci:
Villa
near Florence

where there is little to remark upon, were formed artists like Giacomo Manzù (b. 1908) of Bergamo, a clever portraitist, and above all Marino Marini (b. 1901), one of the most powerful living sculptors, who attains in deliberately limited themes (nudes, heads, but chiefly horses and riders), executed in bronze, an ironic strangeness and an uncommon firmness of plastic feeling. To them must be added Umberto Mastroianni (b. 1910), cubist in inspiration, the Bolognese Minguzzi (b. 1911), unquiet in spirit and with an exasperated sensibility, the mannerist accent of Marcello Mascherini (b. 1906) of Udine, and, among experimenters in abstract forms and smooth masses, Andrea Viani (b. 1906).

ARCHITECTURE

In 1914, Antonio Sant' Elia (1888–1916) published the manifesto of a Futurist architecture, deliberately modern in its forms, functional, rationalized, without unnecessary decoration, which was to be, at the same time, a pure art form.[173] From this sensational programme the essential analytical basis was lacking and his visions of tremendous cities partook more of the order of a last survival of Italian scenic decoration of seventeenth- and eighteenth-century type rather than of constructive suggestions. Thus, the stimulus of the Futurist revolt did not produce, in this field either, any real effect, and the 'future' lay elsewhere.[174]

In 1926, the formation of the Group of Seven, led by Giuseppe Terragni (1904–43) and A. Libera, centred on Milan, and based on the exhibitions of 'rational architecture' held in 1928 and 1931, appears to have started a movement of reappraisal, desperately needed not only by the northern industrial centres, but by all the rapidly expanding cities, including Rome itself. This movement, sustained for a short while by the Fascist régime, resulted in works of a breadth and clarity of conception as interesting as the new railway station in Florence by Michelucci (plans passed in 1933) and the new city of Sabaudia (1934). Pier Luigi Nervi (b. 1891) built the Stadium in Florence (1932) and the brilliant aeroplane hangars with their reinforced concrete shells, destroyed during the Second World War (Orbetello, 1938) where the constructional elements were interlaced. But this effort encountered the double opposition of the academism of the archaeologists such as Giovannoni, and the opportunism of Marcello Piacentini, the author of a specious 'Architettura d'oggi' (Rome, 1930). The latter, with his associates, set out to create a pseudo-Roman 'Fascist style', of a 'virile' bareness, based on simplified structures where the repetition of pilasters with sharp edges and giant columns created a kind of monumental obsessiveness. Piacentini created an example of this in his Tower at Brescia (1932), and it was realized in its most painful form in the Italian pavilion by Michele Busiri Vici at the New York Exhibition of 1939, and, with the greatest emptiness, in the enormous tower with arcaded storeys by Bernardo Lapadula, which is the only surviving part of the world exhibition planned for Rome in 1942. For fifteen years this style carried the day all through Italy, particularly in Rome, where the Palazzo Littorio may still be cited as an example. It also dominated new town planning schemes such as the Foro Mussolini (now the Foro Italica) by E. del Debbio (1933), and the University City in Rome, as well as the new towns in the Pontine Marshes, Littoria and Pontinia.

Developments in Italy may be judged from examples provided by the new railway stations in two major cities. In Milan, the central station by Stacchini (1931) is of colossal proportions, swollen by the enormous mass of the façade, by vast flights of steps and over-large exits, forming a distressing and inconvenient building. The Fascist régime had planned in 1937 an even more spectacular effort for the new Stazione Termini in Rome, which was to contain both the main-line services and the administrative offices. When the construction stopped in 1942, the parallelopipedic lateral portions, uniformly bare and pierced by severe arcades, were already built along the service roads, but the façade, for which an immense colonnade supporting a portico inevitably too light and completely unnecessary, had been planned, was not yet started. In 1947, a national competition enabled Messrs. Montuori and Calini and a group of architects to create, out of the very unpromising starting points of the problem, a superb modern façade (1952). Its horizontal rhythm, stressed with brio in the office building, and the profile with the double curve of the hall and the station proper, testify to an intelligent adaptation to modern means, and to their sense of the monumental.

It is principally in Northern Italy that rationalist architecture has had, before and since the war, the most important development. The suburb of Como called Novocomum by Terragni, finished in 1929, is a strict and imperious adaptation of the cubic, structural architecture of the Weimar Bauhaus and its teacher Gropius. The Casa del Partito at Como (1932–33), with its roof-garden, and its interior proportions calculated on the Golden Section, and the house belonging to Giuliano Frigerio (1939–40) also at Como, more complex and restless in the articulation of its parts, are of the same kind. In Turin, from 1927 onwards, there was the modernization of the Fiat factories by Matteo Trucco, and the work by the brilliant and controversial Giuseppe Pagano (1896–1945) who denounced the rhetoric of conventional architecture and insisted that more attention should be paid to the psychological and social effects of urban organization. He himself built, with his group, the Bocconi University in Milan (1936–42) in a very different spirit from that of the Roman one. After the war, the tendency towards light materials, and supple planning, suitable for constructions of fifteen to twenty storeys, provided that the position of the block is adapted to the site and surrounded by verdure, had its greatest successes in Lombardy. The new quarter of the Piazza Fiume in Milan and the office building of the Montecatini Company at Porta Nuova in Milan, all in aluminium and glass (1951), are remarkably successful undertakings.

THE TWENTIETH CENTURY

In Florence, a Tuscan Group formed round G. Michelucci had created, before the war, a fine example of imaginative good taste with the charming railway station finished in 1936. Their ease in handling coloured materials and in using the open type of plan in ingenious arrangements is particularly noticeable in the works of the young Leonardo Ricci. But the ability of the Tuscan architects has been but indifferently used in the reconstruction of the centre of Florence and in the Via Por Santa Maria.

In Venice, which for a century had remained almost entirely outside the architectural development of the rest of Italy, the problem of inserting a modern building into an old and privileged organism took on a particularly topical character, with controversy aroused by Frank Lloyd Wright's project to build a house in marble and glass in memory of the young architect Masieri, on the corner of the Grand Canal and the Rio Nuovo. This building, particularly restrained and harmonious, would have added to the panorama of the Grand Canal one more accent amid the multiple styles which are displayed there, with that strict adaptation to surroundings and atmosphere which Venice has always imposed.[175] Moreover, the basic struggle in Italian architecture surely lies between respect for a traditional formula and the desire for the modern, and, as is clearly established by the contrast between the 'monumental style' of Fascism, and the general spirit of reconstruction of 1945, between the anachronistic temptation to design enormous, massive, isolated buildings, spectacular but lifeless, and the spontaneous Italian preference for adapting the building to the site, and to the human and local setting. The triumph of this last trend makes it possible to link the happiest among the more recent undertakings with what took place during the Communes and the Age of Baroque, when the design of the site, the public square, always took precedence. This orientation also explains why the younger generation of architects has revealed itself the most able in Europe to assimilate and develop with brio the trends initiated by Scandinavian and American schools, which advocate, instead of geometrical masses, buildings articulated according to the particular needs of man, to the siting of each group, and to local requirements.

NOTES

1 Valuable remarks by F. Rodolico, *Le Pietre delle Città d'Italia*, Florence 1953.
2 K. Swoboda, *Römische und romanische Paläste*, Vienna 1924.
3 R. Wittkower, *Brunelleschi and 'Proportion in Perspective'*, in *J.W.C.I.*, XVI, 1953.
4 Jean Seznec, *La survivance des dieux antiques*, Fr. ed. London 1940; Eng. trans. New York 1953.
5 J. Carcopino, *Études d'histoire chrétienne*, Paris 1953, p. 104.
6 F. Heer, *Die 'Renaissance'—Ideologie im frühen Mittelalter*, in *Mitteil. des Instituts f. österreichische Geschichtsforschung*, LVII, 1949, 1 and 2.
7 Pöseler, *Der Rückgriff der römische Spätdugentomalerei auf die christliche Spätantike*, in *Beiträge z. Kunst des Mittelalters* (Brühl Congress) Berlin, 1950.
8 W. Braunfels, *Mittelalterliche Baukunst in der Toskana*, Berlin 1953, ch. iv.
9 E. Panofsky, *Renaissance and Renascences*, in *Kenyon Review*, 1944, p. 201.
10 E. H. Swift, *Roman Sources of Christian Art*, New York 1951, and the qualifying remarks of P. Lemerle, in *Revue des Études byzantines*, 10, 1952.
11 Principally since A. Riegl, *Spätrömische Kunstindustrie*, Vienna 1901; Ital. trans. Florence 1953.
12 In the sense used by H. Wölfflin, *Kunstgeschichtliche Grundbegriffe*, Munich 1917 (Eng. trans. *Principles of Art History*, London 1932), limited to the sixteenth and seventeenth centuries.
13 E. Panofsky, *Early Netherlandish Painting*, Princeton 1953, Introduction: The polarization of European painting.
14 J. B. Ward Perkins, *The Italian Element in Late Roman and Early Mediaeval Architecture*, in *Proceedings of the British Academy*, 33, 1947.
15 P. L'Orange, *Studien zur Geschichte des spätantiken Porträts*, Oslo 1933.
16 H. von Schönebeck, *Die christliche Sarkophag-Plastik unter Konstantin*, in *Mitt. des deutschen arch. Inst.*, Rome 1936.
17 A. Grabar, *Plotin et les origines de l'esthétique médiévale*, in *C.A.*, I., 1945.
18 G. Wilpert, *Le Pitture delle Catacombe romane*, Rome 1903.
19 F. Gerke, *Die christliche Sarkophage der vorkonstantinischen Zeit*, Berlin 1940.
20 R. Krautheimer, *Corpus basilicarum christianarum Romae*, Vatican City 1937.
21 L. Bréhier, *Les Origines de la basilique chrétienne*, in *Bulletin monumental*, LXXXVI, 1927; S. Lang, *A few suggestions . . .*, in *Rivista di archeologia cristiana*, XXV, 1949, and XXX, 1954.
22 R. Delbrueck, *Notes on the wooden doors of Santa Sabina*, in *A.B.* XXXIV, 1952, 2.
23 A. Grabar, *Martyrium. Recherches sur le culte des reliques et l'art chrétien antique*, 2 vols., Paris 1943–48; id., *Christian Architecture, East and West*, in *Archaeology*, II, 1948.
24 R. Krautheimer, *Introduction to an 'Iconography' of mediaeval Architecture*, in *J.W.C.I.*, V., 1942.
25 A. Calderini, G. Chierici, C. Cecchelli, *La Basilica di San Lorenzo Maggiore in Milano*, Milan, 1952.
26 M. Salmi, in *Mélanges A. della Seta*, Rome 1950.
27 W. Seston, *C.A.*, I, 1945; and P. Courcelle, *id.*, III, 1948.
28 E. Dyggve, *Ravennatum Palatium sacrum*, Copenhagen 1941.
29 Cecchelli, in *Felix Ravenna*, III, 1950; S. Bettini, in *Arte del primo Millennio*, Milan 1952.
30 M. Mazzoli, *La Basilica di Sant' Apollinare in Classe*, in *Studi di antichità cristiana*, XXI, Vatican, 1954.
31 W. F. Volbach, *Elfenbeinarbeiten der Spätantike und des frühen Mittelalters*, 2nd ed., Mainz 1952.
32 M. Lawrence, *The Sarcophagi of Ravenna*, New York, 1945.
33 F. Forlati, *Il Primo San Marco*, in *Arte Veneta*, IV, 1950.
34 V. Elbern, *Der karolingische Goldaltar von Mailand*, Bonn, 1952.
35 H. Focillon, *l'An mil*, Paris, 1952.
36 Y. Renouard, *Les hommes d'affaires italiens du Moyen Age*, Paris, 1949; the author's remarks also cover the following period.

NOTES

37 L. Olschki, *La Cattedrale di Modena e il suo rilievo artoriano*, in *Archivum romanicum*, 1935.
38 S. Bettini, *L'Architettura di San Marco*, Padua, 1946.
39 O. Demus, *Byzantine Mosaic Decoration*, London, 1948.
40 G. Fiocco, *Arte Veneta*, V, 1951.
41 E. Garrison, *B.M.*, August, 1947.
42 R. Longhi, *Giudizio sul Dugento*, Pr., II, 1948.
43 E. Mâle, *Rome et ses vieilles églises*, Paris, 1942, p. 207. English trans. London, 1960.
44 Serafini, *Torre Campanarie*, Rome, 1927; E. Mâle, *op. cit.*, p. 222.
45 A. Boeckler, *Die Bronzetüren des Bonanus von Pisa und des Barisanus von Trani*, Berlin, 1949.
46 A. Grabar, *Trônes épiscopaux des XIe et XIIe siècles en Italie méridionale*, in *Wallraf-Richartz Jahrbuch*, XVI, 1955.
47 R. Bordenache, *La SS. Trinità di Venosa*, in *Ephemeris Dacoromana*, VII, 1937.
48 S. Bottari, in *R.d.A.*, III, 1953, I.
49 L. Bréhier, *L'Art chrétien*, 2nd ed., Paris, 1928.
50 W. Koehler, *Byzantine Art in the West*, in *Dumbarton Oaks Papers*, 1941.
51 The oversimplification by H. Thode, *Franz von Assisi und die Anfänge der Kunst der Renaissance in Italien*, 2nd ed. Berlin, 1904, has been revised by H. B. Gutman, *The Rebirth of the Fine Arts and Franciscan Thought*, in *Franciscan Studies*, V, 1945, and by A. Jullian, *Le Franciscanisme et l'art italien*, in *Phoebus*, I, 1946, Nos. 3–4.
52 According to R. Longhi, *Pr.*, II, 1948, 37.
53 S. Bottari, *Intorno alle origine dell' architettura sveva*, in *Palladio*, I, 1951. Krönig, *Staufische Baukunst in Unteritalien*, in the collected essays: *Kunst des Mittelalters*, Berlin, 1950.
54 And not by Giovanni Pisano, as has long been said. See P. Bacci, in *Dedalo*, I, 1920.
55 M. Weinberger, *The First Façade of the cathedral of Florence*, in *J.W.C.I.*, IV, 1940–41.
56 P. Moschella, *Le case a sporti in Firenze*, in *Palladio*, VI, 1942.
57 A. Lenzi, *Palazzo Vecchio*, Florence, 1912.
58 W. Paatz, *Die Gestalt Giottos im Spiegel einer zeitgenössischen Urkunde*, in *Festschrift C. G. Heise*, Berlin, 1950.
59 W. Braunfels, *Mittelalterliche Stadtbaukunst der Toskana*, Berlin, 1953.
60 J. Pope-Hennessy, in *B.M.*, XCIII, 1951, p. 347.
61 P. Bettini, in *Riv. A.*, 1950.
62 G. Nicco Fasola, *La Fonte Maggiore di Perugia*, Rome, 1951.
63 G. Vitzthum, *Zu Giottos Navicella*, in *Italienische Studien*, Leipzig, 1929; W. Paeseler, in *R.J.*, V, 1941, pp. 49–162.
64 R. Offner, in *B.M.*, XXXI, 1939, gives reasons against an attribution to Giotto; F. Jewett Mather, in *A.B.*, XXX, 1943, states the reasons in favour, which had been stated already (1927) and developed (1951) by P. Toesca. The textual sources have been investigated by P. Murray, in *J.W.C.I.*, XVI, 1953.
65 E. Rosenthal, *Giotto in der mittelalterlichen Geistesentwicklung*, Augsburg, 1924.
66 A. Venturi, B. Berenson, J. Gy. Wilde (*J.K.G.*, VII, 1930) date them 1320, and even 1330, after the Bardi chapel; C. Brandi (*Le Arti*, I, 1938), Baumgart, R. Longhi, on the contrary, date them before the Bardi chapel, that is, before 1319.
67 R. Longhi, *Stefano fiorentino*, in *Pa.*, 13 January 1951.
68 I. Falk and J. Lanyi, *The Genesis of Andrea Pisano's Bronze Doors*, in *A.B.*, XXV, 1943.
69 J. von Schlosser, *Giustos Fresken in Padua und die Vorläufer der Stanza della Segnatura*, in *J.W.*, Vienna, XVII, 1896.
70 P. Sampaolesi, *Le Sinopie del Campo Santo di Pisa*, in *B.A.*, XXXIV, 1948.
71 M. Meiss, *Painting in Florence and Siena after the Black Death*, Princeton, 1951.
72 E. Panofsky, *Imago Pietatis*, in *Festschrift für Max Friedländer*, Leipzig, 1927.
73 E. Carli, *Le Tavolette di Biccherna*, Florence, 1950. French trans., 1953.
74 O. Pächt, in *J.W.C.I.*, XIII, 1950, p. 17.
75 B. Kurth, *Ein Freskenzyklus zu Adlerturm zu Trient*, in *Jahrbuch der zentral Kommission für Denkmalpflege*, Vienna, V, 1911.
76 V. Lasareff, *Maestro Paolo e la pittura veneziana del suo tempo*, in *Arte Veneta*, VIII, 1954.
77 A. Gatti, *La Basilica Petroniana*, Bologna, 1913; see below, Third Part, I, III.

NOTES

78 James Ackermann, *Gothic Theory of Architecture at the Cathedral of Milan*, in A.B., XXXI, 1949.
79 G. Fiocco, F. *Feliciano amico degli artisti*, in *Archivio Veneto Tridentino*, IX, 1926.
80 G. Swarzenski, *Der kölner Meister bei Ghiberti*, in *Warburg Vorträge*, 1926-27, has identified the carved altar (Rimini) and an altarpiece with alabaster statuettes (Isola Bella). R. Krautheimer, in *A.B.*, XXIX, 1947.
81 G. Bazin, in *R.d.A.*, II, 1952, December.
82 G. Brunetti, *Jacopo della Quercia and the Porta della Mandorla*, in the *Art Quarterly*, IV, 1952.
83 F. D. Prager, *Brunelleschi's inventions and the renewal of Roman masonry work*, in *Osiris*, IX, 1950, pp. 457-554.
84 L. H. Heydenreich, *Spätwerke Brunelleschis*, in *J.B.*, 1931.
85 G. Ravaioli, M. Salmi, in *Studi Malatestiani*, Faenza, 1952.
86 A. Chastel, *Di mano dell' antico Prassitele*, in *Mélanges Lucien Febvre*, Paris, 1954.
87 U. Procacci, *Sulla cronologia delle opere di Masaccio e di Masolino tra il 1425 e il 1428*, in *R.A.*, XXVIII, 1953.
88 M. Hauptman, *Der Tondo*, Frankfort am Main, 1937.
89 W. Haftmann, *Das italienische Säulenmonument*, Leipzig, 1939.
90 M. Meiss, *Ovum Struthionis, Symbol and Allusion in Piero della Francesca's Montefeltro altarpiece*, in *Studies in Art and Literature for Belle da Costa Greene*, ed. D. Miner, Princeton, 1954.
91 P. Rotondi, *Il palazzo ducale di Urbino*, 2 vols., Urbino, 1953.
92 G. Polidori, in *Studi artistici Urbinati*, II, 1953.
93 T. Magnuson, *The project of Nicolas V for rebuilding the Borgo Leonino in Rome*, in *A.B.*, XXXVI, 1954.
94 L. H. Heydenreich, *Pius II als Bauherr von Pienza*, in *Z.K.G.*, VI, 1937.
95 L. D. Ettlinger, *Pollaiuolo's Tomb of Pope Sixtus IV*, in *J.W.C.I.*, XVI, 1953, 3-4.
96 R. Causa, *Sagrera, Laurana e l'Arco di Castelnuovo*, in *Pa.*, No. 55, July, 1954.
97 S. Bottari, *La Pittura del Quattrocento in Sicilia*, Messina, 1954.
98 M. Salmi, *La Bibbia di Borso d'Este e Piero della Francesca*, in *Rinascita*, Nos. 32-33, 1943.
99 F. Zeri, in *B.M.*, April, 1951; R. Longhi, in *Pa.*, No. 65, May, 1955.
100 A. Chastel, *Marqueterie et perspective au XVe siècle*, in *R.d.A.*, III, 1953, 3.
101 C. Baroni, *Elementi stilistici fiorentini negli studi vinciani di architettura a cupola*, in *Atti, I, Congresso Storia dell' Architettura*, Florence, 1938.
102 A. Chastel, *Vasari et la légende médicéenne: l'école du jardin de Saint-Marc*, in *Studi Vasariani*, Florence, 1952.
103 A. Chastel, *L'Apocalypse en 1500*, in *Humanisme et Renaissance*, XIV, 1952.
104 J. Wilde, *The Hall of the Great Council of Florence*, in *J.W.C.I.*, VII, 1944.
105 A. Chastel, *L'Antéchrist à la Renaissance*, in *Actes du congrès international d'études humanistes* (1952), Rome, 1953.
106 J. S. Ackermann, *J.W.C.I.*, XIV, 1951, and *The Cortile del Belvedere*, in *Studi e Documenti*, III, The Vatican, 1954.
107 D. Redig de Campos, *Dei ritratti di A. Tebaldeo e di altri nel Parnasso di R.*, in *Archivio Società Romana Storia Patria*, LXXV, 1952.
108 E. Gerlini, *Giardino e architettura nella Farnesina*, Rome, 1942.
109 F. Zeri, *Il maestro della Madonna di Manchester*, in *Pa.*, No. 43, July 1953.
110 Pio Pecchiai, *Il Campidoglio nel Cinquecento*, Rome, 1954.
111 G. Weise, *Renaissance und Antike*, Tübingen, 1953.
112 A. Chastel, *Le Platonisme et les arts à la Renaissance*, Congrès G. Budé (1952), Paris, 1953.
113 L. Goldscheider, *Unknown Renaissance Portraits*, London, 1952.
114 G. Fiocco, *La casa veneziana antica*, in *Atti Accademia Naz. Lincei*, 1949.
115 R. Wittkower, *Architectural principles in the age of Humanism*, ch. IV, London, 1949.
116 A. Chastel, *La mosaique à Venise et à Florence au XVe siècle*, in *Arte Veneta*, XII, 1954.
117 E. Wind, *Bellini's Feast of the Gods*, Cambridge (Mass.), 1948. E. Battisti, *Disegni inediti di Tiziano e lo studio d'Alfonso d'Este*, in *Commentari*, V, 1954, 3. J. Walker, *Bellini and Titian at Ferrara*, London, 1956.
118 E. H. Gombrich, in *G.B.A.*, July, 1953.
119 This journey was maintained by Mengs, and later by A. Venturi, but contested by C. Ricci.
120 W. Friedländer, *Die Entstehung des antiklassichen Stiles in der italienischen*

NOTES

Malerei um 1520, in *Repertorium für Kunstwissenschaft*, XLVI, 1925. English trans., *Mannerism and anti-Mannerism in Italian painting*, New York, 1957.

121 P. Calamandrei, *Sulle relazioni tra Giorgio Vasari e Benvenuto Cellini*, in *Studi Vasariani*, Florence, 1952.

122 E. Panofsky, *Idea*, Leipzig, 1924; Italian translation, Florence, 1952.

123 J. Seznec, *La survivance des dieux antiques*, London 1940; English translation, New York, 1953.

124 G. Weise, *Maniera und Pellegrino, zwei Lieblingswörter der italienischer Literatur der Zeit des Manierismus*, in *Romanistisches Jahrbuch*, III, 1950.

125 C. Sterling, *La nature morte de l'Antiquité à nos jours*, Paris, 1952.

126 E. Kris, *Der Stil rustique*, in *J.W.*, Vienna, 1926.

127 N. Pevsner, *The Architecture of Mannerism*, in *The Mint*, London, 1946.

128 F. Hartt and E. H. Gombrich, in *J.W.C.I.*, XIII, 1950.

129 J. Pope-Hennessy, in *B.M.*, Jan., 1959.

130 J. Seznec, *La mascarade des dieux à Florence en 1565*, in *Mélanges d'archéologie et d'histoire*, 1935.

131 L. Magagnato, *Teatri italiani del Cinquecento*, Venice, 1954.

132 R. Pallucchini, *La Vicenda italiana del Greco*, in *Pa.*, No. 45, September, 1953.

133 B. Zeri, *Pittura e controriforma*, Turin, 1957.

134 F. Saxl, *Antike Götter in der Spätrenaissance: ein Freskenzyklus und ein Discorso des Jacopo Zucchi*, in *Studien Bibl. Warburg*, VIII, Leipzig, 1927.

135 F. Boyer, *La construction de la Villa Médicis*, in *Revue de l'Art Ancien et Moderne*, LI, (Jan.–Feb., 1927).

136 J. Coolidge, *The Villa Giulia, a study of Central Italian architecture in the mid-sixteenth century*, in *A.B.*, XXV, pp. 177–225, 1943.

137 Pio Pecchiai, *Il Gesù di Roma*, Rome, 1952.

138 P. Pirri, *Giovanni Tristano e i primordi dell'architettura gesuitica*, Rome, 1955.

139 D. Mahon, in *J.W.C.I.*, XVI, 1953.

140 W. Friedländer, *Der antimanieristiche Stil um 1590*, in *Vorträge der Bibl. Warburg*, VIII (1928–29), Leipzig, 1930. English trans. see note 120.

141 C. Galassi Paluzzi, *Storia segreta dello stile dei Gesuiti*, Rome, 1951.

142 E. Mâle, *L'art religieux après le concile de Trente*, Paris, 1932.

143 P. Roques and P. P. Régamey, *La signification du baroque*, in *Maison-Dieu*, No. 26, 1951–52.

144 E. Gombrich, *Icones symbolicae*, in *J.W.C.I.*, XI, 1948.

145 G. Schöne, *Die Entwicklung der Perspektivbühne bis Galli-Bibiena nach den Perspektivbüchern*, Leipzig, 1933.

146 French translation by Neuchâtel, 1942, with a preface by L. Jouvet.

147 R. Battaglia, *La Cattedra berniniana di San Pietro*, Rome, 1953.

148 S. Giedion, *Space, Time, and Architecture*, 3rd ed., Cambridge (Mass.), 1954.

149 R. Bernheimer, *Gothic survival and revival in Bologna*, in *A.B.*, XXXVI, 1954, 4.

150 *Scultura nel presepe napoletano del Settecento* (catalogue by Bruno Molajoli), Naples, 1950.

151 A. Chastel, *Notes sur le baroque méridional*, in *Revue des sciences humaines*, Lille, No. 55–56, July, 1949.

152 K. Lohmeyer, *Palagonisches Barock*, Frankfort am Main, 1943.

153 L. Venturi, *Studi radiografici sul Caravaggio*, in *Atti Accad. Naz. Lincei (Sc. Morali)*, 1952.

154 C. Sterling, *La nature morte de l'Antiquité à nos jours*, Paris, 1952.

155 R. Causa, *Paolo Porpora e il primo tempo della 'natura morta' napoletana*, in *Pa.*, No. 15, March, 1951.

156 N. Pevsner, *Die Wandlung um 1650 in der italienischen Malerei*, in *W.J.K.G.*, 1932.

157 R. Longhi, *Il Goya romano e la cultura di via Condotti*, in *Pa.*, No. 53, May, 1954.

158 D. Mahon, *Studies in Seicento Art and Theory*, London, 1947.

159 K. Gerstenberg, *Die ideale Landschaftsmalerei*, Halle, 1923.

160 S. Sitwell, *The Hunters and the Hunted*, London, 1948, pp. 264 et seq.

161 P. du Colombier, *L'An mil six cent cinquante-huit*, in *Bulletin de la Société Poussin*, I, 1947.

162 H. Voss, *Studien zur venezianischen Vedutenmalerei des 18. Jahrhunderts*, in *Rep. K.W.*, 1926.

163 R. Longhi, preface to *Storia dell'Impressionismo*, Florence, 1952, the Italian edition of J. Rewald's *History of Impressionism*, New York, 1946.

164 M. Petrocchi, *Razionalismo architettonico e razionalismo storiografico*, Rome, 1947.

165 L. Vitali, preface to the *Lettere dei Macchiaioli*, Turin, 1953.

400

NOTES

166 P. Rotondi, in *Studi artistici urbinati*, I, 1949.
167 R. Pierce and T. Ashby, in *Town Planning Review*, XI, 1924; F. Boyer, in *Revue de l'Art*, May, 1932.
168 F. Boyer, *Autour de Canova et de Napoléon*, in *Rev. études ital.*, 1937; *Nouveaux documents sur Canova et Napoléon*, in *Rev. études ital.*, 1948.
169 Silvio Negro, *Seconda Roma 1850–1870*, Milan, 1943.
170 M. Venturoli, *La patria di marmo*, Pisa, 1957.
171 B. Croce, *La critica e la storia delle arti figurative*, 2nd ed., Bari, 1946.
172 F. T. Marinetti, U. Boccioni, etc., *Fondazione e manifesto del futurismo*, Venice, 1950.
173 In the collected essays: *Dopo Sant' Elia*, Milan, 1935.
174 B. Zevi, *Storia dell'architettura moderna*, Turin, 1950, ch. V.
175 S. Bettini, *Venezia e Wright* in *Metron*, Nos. 49–50, January, 1954.

Bibliography

Monographs on artists are given in the biographical index, monographs on places in the topographical index. Works marked ** are indispensable, * important.

I. General

(1) GENERAL SURVEYS

Encyclopaedias. **Venturi, A., *Storia dell'arte italiana* (23 vols., Milan, 1901–41). *Storia dell'arte classica e italiana* (Turin): II. Lavagnino, E., *L'Arte medioevale* (1936); III. Ancona, P. d', and Gengaro, M. L., *Umanesimo e Rinascimento* (3rd ed., 1948); IV. Golzio, V., *Il Seicento e il Settecento* (1950). **Toesca, P., *Storia dell'arte italiana*: I. *Dalle origini cristiane alla fine del secolo XIII* (Turin, 1927); II. *Il Trecento* (Turin 1951). *L'Enciclopedia monografica illustrata* (in 25 parts by G. C. Argan, S. Bettini, G. Delogu, M. Pittaluga, on all the arts, Florence, 1936 on).

Manuals. Ojetti, U., and Dami, L., *Atlante di Storia dell'arte italiana* (2 vols., Milan, 1925–1934). Ancona, P. d', Cattaneo, L., Wittgens, F., *L'Arte italiana* (3 vols., Florence, 1930; new ed., 1953). Paribeni, R., Mariani, V., Serra, B., *L'Arte italiana* (2 vols., Turin, 1934). Bottari, S., *Storia dell'arte* (2 vols., Milan, 1943). Pittaluga, M., *id.* (3 vols., 6th ed., Florence, 1948). Salmi, M., *L'Arte italiana* (3 vols., Florence, 1944; new ed., 1953). Mottini, G. E., *Storia dell'arte italiana* (2 vols., Milan). Costantini, V., *Storia dell' arte italiana* (4 vols., Milan). Alazard, J., *L'Art italien* (Paris): I. *Des origines à la fin du XIVe siècle* (1949); II. *XVe siècle* (1951); III. *XVIe siècle* (1955). Carli, E. and dell'Acqua, G. A., *Profilo dell'arte italiana* (2 vols., Bergamo, 1955).

Histories of Art. *Histoire de l'Art*, published under the direction of André Michel, Paris, 1905–29, chapters by C. Enlart, E. Bertaux, A. Pératé. *The Pelican History of Art* (under direction of N. Pevsner) forthcoming vols. by C. Seymour, L. H. Heydenreich. *Histoire mondiale de l'art* (under direction of G. Salles and A. Malraux) vols. by A. Grabar, A. Chastel, R. Longhi, R. Wittkower, C. Sterling (after 1958).

Photographs and Inventories. Series *Attraverso l'Italia*, published by the T.C.I. (12 vols., Milan, 1930 on). *L'Italia artistica*, published in Bergamo; 140 vols. so far. *Il Fiore dei Musei e Monumenti d'Italia*, published in Milan; 18 vols. so far.
Collection published by the Ministero dell'Istruzione Pubblica: *Cataloghi delle cose d'arte e d'antichità*, Rome, 10 vols. so far: Aosta, Pisa, Urbino, Fiesole, Treviso, Assisi, Vercelli, Cividale, Brescia. *Inventari degli Oggetti d'arte*

BIBLIOGRAPHY

d'Italia, Rome, 10 vols. for the provinces of: Bergamo (1931), Calabria (1933), Parma (1934), Aquila (1934), Pola (1935), Mantua (1935), Padua (1936), Ancona and Ascoli Piceno (1936), Sondrio (1938). **Albums.** Numerous titles include: *Italie*, text and photographs by M. Hürlimann (Atlantis, 1952). Bruhns, L., *Die Kunst der Stadt Rom* (Vienna, 1951). Decker, H., *Venedig* (*ibid.* 1952). Borsig and Bandinelli, *Die Toskana* (*ibid.* 1953).

Sources. **Vasari, G., *Le Vite de' più eccellenti architetti, pittori e scultori italiani, da Cimabue insino a' tempi nostri* (2 vols., Florence, 1550; 2nd ed., (enlarged), 3 vols., Florence, 1568; ed. Milanesi, 9 vols., 1878 ff.; ed. C. L. Ragghianti, 4 vols., 1945–1949; English ed. by G. de Vere, 10 vols., London, 1912–15). Borghini, R., *Il Riposo* (Florence, 1584). Baglione, G., *Le Vite dei Pittori, scultori ed architetti* (from 1572 to 1642) (Rome, 1642; new ed., 1925). Passeri, G. B., *Vite de' pittori, scultori ed architetti che hanno lavorato in Roma* (from 1641 to 1673) (Rome, 1772; new ed. by J. Hess, 'Römische Forschungen', XI, 1934). Pascoli, L., *Vite* . . . , (Rome, 1730–1736; new ed., 1933). Bellori, G. P., *Le Vite* . . ., (Rome, 1672, complete edition by A. Bertini-Calosso, Rome, 1931). Baldinucci, F., *Notizie de' Professori del disegno* . . . (6 vols., Florence, 1681–1728; new ed., 14 vols., Milan, 1808–1812). **Schlosser, J. von, *Die Kunstliteratur* (Vienna, 1924); Italian ed.: Schlosser, J. and Magnino, A., *La Letteratura artistica* (Florence, 1932; new ed., 1956). (These works contain bibliography of source material.)

Documents. Bottari, C. G., *Lettere pittoriche* (Rome, 1745–1783; new ed., Milan, 1822–1825). Milanesi, G., *Lettere d'artisti italiani dei sec. XIV e XV* (Rome, 1869). Milanesi, G., *Documenti inediti dell'arte toscana dal XII al XVI s.* in 'Il Buonarotti', S. III, vol. II (1884). Gaye, G., *Carteggio inedito* (XIV–XVI) (Florence, 1839–1840; reprinted Turin, 1961). Müntz, E., *Les Arts à la Cour des Papes* (XV–XVIth c.) (3 vols., 1878–1882).

(2) PERIODS

Middle Ages. Marucchi, O., *Le Catacombe romane* (Rome, 1934). Galassi, G., *Roma e Bisanzio* (2 vols.; new ed., Rome, 1953). Colasanti, A., *L'Arte bizantina in Italia* (Milan 1913). *Riegl, A., *Spätrömische Kunstindustrie* (Vienna, 1906; Italian translation, Florence, 1953). Picton, H., *Die longobardische Kunst in Italien* (Augsburg, 1931). Francovich, G. de, *Arte carolinghia ed ottoniana in Italia* ('R.J.', VII, 1943–1944).
*Collective work: *Arte del primo millennio* (1950 Congress), by E. Arslan (1953); *Studi sull'alto medioevo* (2nd Congress, 1952) (Spoleto, 1953). Gabelentz, H. von der, *Die kirchliche Kunst im italienischen Mittelalter* (Strasbourg, 1907). Vitzthum, G., and Volbach, W. F., *Die Malerei und Plastik des Mittelalters in Italien* (Potsdam, 1926). Schmarsow, A., *Italienische Kunst im Zeitalter Dantes* (Augsburg, 1928). Weise, G., *Die geistige Welt der Gotik und ihre Bedeutung für Italien* (Halle a/Saale, 1939). Paatz, W., *Italien und die künstlerischen Bewegungen der Gotik und der Ren.* ('R.J.', V (1941), pp. 163–222). Lavagnino, E., *Storia dell'arte medioevale italiana* (Turin, 1936).

BIBLIOGRAPHY

Renaissance. Müntz, E., *Histoire de l'art en Italie pendant la Renaissance* (3 vols., 1888–1894). **Wölfflin, H., *Die klassische Kunst* (Munich, 1899; English ed., *Classic Art*, London, 1952). *Dvorak, M., *Geschichte der italienischen Kunst im Zeitalter der Renaissance* (2 vols., Munich, 1927–1929). Bode, W., *Die Kunst der Frührenaissance in Italien* (Berlin, 1923). Schubring, P., *Die Kunst der Hochrenaissance in Italien* (Berlin, 1926). Hoffmann, H., *Hochrenaissance, Manierismus, Frühbarock, die italienische Kunst des 16. Jahrhundert* (Zürich, Leipzig, 1938). Frey, D., *Gotik und Renaissance* (Augsburg, 1929). Stokes, A., *The Quattrocento* (London, 1932). Johansen, P., *Renaissance* (Leipzig, 1936). *Paatz, W., *Die Kunst der Renaissance in Italien* (Stuttgart, 1953; 2nd. ed. 1954). Chastel, A., *Art et Humanisme à Florence au temps de Laurent le Magnifique* (Paris, 1959). Landsberger, F., *Die künstlerischen Probleme der Renaissance* (Halle a/S., 1922).

Modern. Wölfflin, H., *Renaissance und Barock* (Munich, 1888). **Wölfflin, H., *Kunstgeschichtliche Grundbegriffe* (Munich, 1915; English ed., *Principles of Art History*, London, 1932). Weisbach, W., *Der Barock als Kunst der Gegenreformation* (Berlin, 1921). Marangoni, M., *Arte barocca* (Florence, 1927). Ojetti, U., Barbantini, N., *Il Settecento italiano* (2 vols., Milan, 1932). Wittkower, R., *Art and Architecture in Italy*, 1600–1750 (Pelican History of Art, London, 1958). Brizio, A. M., *Ottocento, Novecento* (Turin, 1939). Willard, A. R., *History of Modern Italian Art* (London, 2nd. ed., 1902). *Archivio del Futurismo* (Rome, 1958). Callari, L., *Storia dell'arte contemporanea italiana* (Rome, 1909). Carrieri, R., *Pittura, scultura d'avanguardia* (Milan, 1950).

(3) PROVINCES

North. Ricci, C., *L'Arte nell'Italia settentrionale* (3 parts, Bergamo, 1910). *Malaguzzi-Valeri, *La Corte di Lodovico il Moro* (3 vols., Milan, 1913–1937). Jacini, C., *Il Viaggio del Po* (6 vols., Milan, 1937–1951). Arslan, E., *La Pittura e la Scultura veronese dal sec. VIII al sec. XVII* (Milan, 1943). Serra, L., *L'Arte nelle Marche* (2 vols., Pesaro, 1929; Rome, 1934).

Tuscany and Umbria. Rohault de Fleury, G., *La Toscane au Moyen Âge* (2 vols., 1873). Tarchi, U., *L'Arte medioevale nell'Umbria e nella Sabina* (Milan, 1940); *L'Arte del Rinascimento* (ibid., 1954).

Rome. Hermanin, G., *L'Arte in Roma del secolo VIII al XIV* (Bologna, 1945). Rinaldis, A. de, *L'Arte in Roma dal Seicento al Novecento* (Bologna, 1949). Fokker, T. H., *Roman Baroque Art* (2 vols., Oxford, 1938). Lavagnino, E., etc., *Altari Barocchi in Roma*, Rome, 1959.

South. Nicolini, F., *L'Arte napoletana del Rinascimento* (Naples, 1925). Frangipane, A., *L'Arte in Calabria* (Messina, 1927). Rinaldis, A. de, *Naples angevine* (1927). **Bertaux, E., *L'Art dans l'Italie méridionale* (1903). Hourticq, L., *L'Italie du Sud* (1938). Uneli, F., *L'Arte in Sicilia dal sec. XII al sec. XIV* (Milan, 1929). Arata, G. U. and Biasi, G., *Arte Sarda* (Milan, 1935). Scano, D., *Storia dell'Arte in Sardegna* (XI to XIVth c.) (Cagliari, 1907). Bologna, F., *Opere d'arte nel Salernitano dal XII al XVIII sec.* (Naples, 1955).

405

BIBLIOGRAPHY

Dalmatia. Dudan, A., *La Dalmazia nell'arte italiana* (Milan, 1921–1922).

(4) DICTIONARIES

In addition to the indispensable Thieme, U., and Becker, F., *Allgemeines Lexikon* . . . (37 vols., Leipzig, 1907–1950), and *L'Enciclopedia italiana di scienze, lettere ed arti* (Fondazione G. Treccani, 36 vols., Rome, 1924–1942; relevant articles), Gubernatis, A. de, *Dizionario degli artisti d'Italia* (Florence, 1906).

(5) EXHIBITIONS

A Commemorative Catalogue of the Exhibition of Italian Art, London (1930), by Lord Balniel, K. Clark and E. Modigliani (2 vols., Oxford, 1931). *Exposition de l'art italien, de Cimabué à Tiepolo*, Petit Palais (1935). *L'Art italien des XIXe et XXe siècles (ibid)*. *Mostra Medicea* (Florence, 1939). *Kunstschätze der Lombardei* (Zürich, 1948–1949). *XXth Century Italian Art*, by James T. Soby and A. H. Barr (New York, 1951). *Mostra nazionale della pittura e scultura futuristica* (Bologna, 1951). Baroni, C. and Alberto dell' Acqua, G., *Tesori d'arte in Lombardia* (Milan, 1952).

II. History of Various Arts

A. ARCHITECTURE

(1) TREATISES

Alberti, L. B., *De re aedificatoria libri X* (Florence, 1485; ed. M. Theuer, Leipzig, 1912; English ed. by G. Leoni, 1726, and new ed. by J. Rykwert, London, 1955). Filarete, *Trattato d'architettura (c. 1460; ed. W. von Œttingen (Q.S.F.K.G.), Vienna, 1896; new ed. by E. Kauffmann coming shortly). Martini, F. di Giorgio, *Trattato d'architettura civile e militare (c. 1480; ed. C. Promis and C. Saluzzo, Turin, 1841). Serlio, S., *Regole generali di architettura* (Venice, 1537; complete ed. 1584; English ed. (partial only), 1611). Vignola, *Regole delli cinque ordini* . . . (?, 1562; full edition, Rome, 1602). Palladio, *I quattro libri dell' architettura* (Venice, 1570; English ed. by G. Leoni, London, 1715, and later eds.). Scamozzi, V., *Idea dell'architettura universale* (Venice, 1615). Milizia, F., *Vita dei più celebri architetti* (Rome, 1768).

(2) GENERAL HISTORIES

Rodolico, F., *Le Pietre delle Città d'Italia* (Florence, 1953). Kirchmayer, M., *L'Architettura italiana* (3rd ed., Turin, 1950). *Burckhardt, J. and Lübke, W., *Geschichte der neueren Baukunst, I, Italien* (Stuttgart, 1867; 3rd. ed., 1891). Gromort, G., *Italian Renaissance Architecture* (Paris, 1922). Ricci, C., *L'Architettura del Cinquecento in Italia* (Turin, 1923). Frey, D., *Architet-*

BIBLIOGRAPHY

tura della Rinascenza da Brunelleschi a Michelangelo (Rome, 1924). Anderson, W. J., *The Architecture of the Renaissance in Italy* (5th ed. revised by A. Stratton, London, 1927). Scott, G., *The Architecture of Humanism* (London, 1924, new ed. 1961). *Giovannoni, G., *Saggi sull'architettura del Rinascimento* (Milan, 1931). Bumpus, F., *The Cathedrals and Churches of Italy* (London, 1926). Chierici, G., *Il Palazzo italiano dal secolo XI al secolo XIX* (3 vols., Milan, 1952–1957).

(3) PERIODS

Medieval. Mengozzi, O., *La Città italiana nell'alto medioevo* (Florence, 1931). *Verzone, P., *L'Architettura religiosa dell'alto medioevo nell'Italia settentrionale* (Milan, 1942). Capitani d'Arzago, A. de, *Architettura dei secoli quarto e quinto in alta Italia* (n.p., n.d. (1945)). Thümmler, H., *Die Baukunst des XI. Jh. in Italien* ('R.J.', 1939). Ricci, C., *L'Architettura romanica in Italia* (Stuttgart, 1925). Enlart, C., *Les Origines françaises de l'Architecture gothique en Italie* (1894).

Renaissance. Frankl, P., *Die Renaissance Architektur in Italien* (Leipzig, 1912). Durm, G., *Die Baukunst der Renaissance in Italien* (Leipzig, 1914). Baum, J., *Baukunst und dekorative Plastik der Frührenaissance in Italien* (Stuttgart, 1920). Willich, H. and Zucker, P., *Die Baukunst der Renaissance in Italien* (Potsdam, 1914). *Wittkower, R., *Architectural Principles in the Age of Humanism* (London, 1949; new ed., 1952). Lotz, W., *Die ovalen Kirchenräume des Cinquecento* ('R.J.', XVII, 1955).

Baroque. Delogu, G., *L'Architettura italiana del Seicento e del Settecento* (Florence, 1935). Ricci, C., *Architettura barocca in Italia* (Stuttgart, 1922).

Modern. Melani, A., *L'Architettura nel sec. XIX* (Milan, 1899). Piacentini, M., *Architettura d'oggi* (Rome, 1930). *Esposizione italiana di architettura razionale* (Rome, 1928 and 1931). Pica, A. D., *Nuova architettura italiana* (Milan, 1936). Persico, E., *Profezia dell'architettura* (Milan, 1945). Marangoni, M., Venturi, L., Argan, C. A., etc., *Dopo Sant'Elia* (Milan, 1935). Veronesi, G., *Difficoltà politiche dell'architettura in Italia, 1920–1940* (Milan, 1950). SERIES: *Ricerche sulle dimore rurali in Italia.* 'La casa rurale'. (13 vols., Florence, 1938–1954).

(4) PROVINCES

SERIES: 'Architettura delle regioni d'Italia': Martelli, G., *Architettura del medioevo in Calabria* (Rome, 1953). Delogu, R., *L'Architettura del medioevo in Sardegna* (Rome, 1953).

Northern Italy. Haupt, A., *Architettura dei Palazzi dell'Italia settentrionale e della Toscana dal sec. XIII al sec. XVII* (3 vols., Milan, Rome, n.d. (1930 on)). Ressa, A., *L'Architettura religiosa in Piemonte nei secoli XVII e XVIII* (in 'Torino', July 1941). Meyer, A., *Oberitalienische Frührenaissance* (2 vols., Berlin, 1897–1900).

407

BIBLIOGRAPHY

Lombardy. Rivoira, G. T., *Le Origini dell'architettura lombarda* (Milan, 1908; English ed., *Lombardic Architecture*, 2 vols., London, 1910). Kingsley Porter, A., *Lombard Architecture* (New Haven, 1917). Terrasse, C., *L'Architecture lombarde de la Renaissance*, 1450–1525 (Paris, 1926). Baroni, C., *L'Architettura lombarda da Bramante a Richini* (Milan, 1941). *Calderini, A., Chierici, G., Cecchelli, C., *La Basilica di S. Lorenzo Maggiore in Milano* (Milan, n.d. (1952)). Calderini, A., *Le Basiliche dell'età ambrosiana in Milano* (Milan, 1942).

Veneto. Paoletti, P., *L'Architettura e la Scultura del Rinascimento in Venezia* (2 vols., Venice, 1893–1897). Mazzotti, G., *Le Ville venete* (2nd ed., Treviso, 1953; English ed., London, 1958).

Liguria. Labò, M., *Studi d'architettura genovese, XVII s.* ('L'Arte', 1921, 1922).

Emilia. Malaguzzi Valeri, F., *L'Architettura a Bologna nel Rinascimento* (Rocca S. Casciano, 1899). Sighinolfi, L., *L'Architettura bentivolesca in Bologna* (Bologna, 1909). Salmi, M., *L'Abbazia di Pomposa* (Rome, 1936). Quintavalle, A. G., *I Castelli del Parmense* (Parma, 1955).

Tuscany. Salmi, M., *Architettura romanica in Toscana* (Milan, 1927). Horn, W., *Romanesque Churches in Florence* ('A.B.', XXV, 1943, 112–131). Behne, A., *Inkrustationsstil in Toskana* (Berlin, 1912). Beenken, *Die florentiner Inkrustationen in der Architektur des XI. Jh.* ('Z.B.K.', LX, 1926–1927). Demus, O., *Der toskanische Inkrustationsstil und Venedig* ('Alte und neue Kunst', I, 1954). Paatz, W., *Werden und Wesen der Trecento-Architektur in Toskana* (Burg b. M., 1937). **Paatz, W., *Die Kirchen von Florenz* (with E. Paatz) (6 vols., Frankfurt a/M., 1942–1954). Stegmann, C. and Geymüller, H. von, *Die Architektur der Renaissance in Toskana* (10 vols., Munich, 1890–1908; abbreviated English ed., 2 vols., New York, n.d.). Patzak, B., *Paläste und Villen in Toskana* (2 vols., Leipzig, 1913). Eberlein, H. D., *Villas of Florence and Tuscany* (New York, 1922). Reymond, M., *Brunelleschi et l'architecture de la Renaissance italienne au XVe siècle* (1912). Lensi Orlandi, G., *Le Ville di Firenze di qua d'Arno* (Florence, 1954).

Rome. Huelsen, Ch., *Le Chiese di Roma nel medioevo, cataloghi ed appunti* (Florence, 1927). *Armellini, M., *Le Chiese di Roma del secolo IV al XIX. New ed. by C. Cecchelli (2 vols., Rome, 1952). Zimmermann, F. X., *Die Kirchen Roms* (Munich, 1935). Mâle, E., *Rome et ses vieilles églises* (1942; English ed. by D. Buxton, 1960). Tomei, P., *L'Architettura di Roma nel Quattrocento* (Rome, 1942). Magnuson, T., *Studies in Roman Quattrocento Architecture* (Stockholm and Rome, 1958). Hutton, E., *The Cosmati, the Roman Marble Masters of the XIIth and XIIIth century* (London, 1950). Frey, D., *Beiträge zur Römischen Barock-Architektur* ('W.J.K.G.', 1924). Muñoz, A., *Roma barocca* (2nd ed., Milan, 1928). Angelis d'Ossat, G. de, *L'Architettura in Roma negli ultimi tre decenni del sec. XIX* ('Ann. R. Acc. S. Luca', 1942). Sapori, F., *Architettura in Roma, 1901–1951* (Rome, 1953).

Campania. Pane, R., *L'Architettura del Rinascimento a Napoli* (Naples, 1937).

BIBLIOGRAPHY

Pane, R., *Architettura dell'età barocca a Napoli* (Naples, 1935). Chierici, G., *Architettura religiosa a Napoli nei sec. XVII e XVIII* ('Palladio', I, 1937).

Southern Provinces and Sicily. Gavini, I. C., *Storia dell'Architettura in Abruzzo* (2 vols., Milan, Rome, n.d.). Haseloff, A., *Die Bauten der Hohenstauen in Unter-Italien* (Leipzig, 1920). Shearer, C., *The Renaissance of Architecture in Southern Italy* (Cambridge, 1935). Calandra, E., *Breve Storia dell'Architettura in Sicilia* (Bari, 1938). Agnello, G., *L'Architettura sveva in Sicilia* (Rome, 1935; new ed.). Stefano, G. di, *L'Architettura gotico-sveva in Sicilia* (Palermo, 1935). Schwarz, H. M., *Die Baukunst Kalabriens und Siziliens im Zeitalter der Normannen* ('R.J.', 1942–1944). Bottari, S., *L'Architettura della Contea* (Catania, 1948). Fichera, F., *G. B. Vaccarini e l'architettura del Settecento in Sicilia* (Rome, 1934). Caronia Roberti, S., *Il Barocco in Palermo* (Palermo, 1933).

Sardinia. Muraro, M., *Monumenti medioevali di Sardegna* (Florence, 1953).

(5) THEATRE AND STAGE DESIGN

Bacchelli, R. and Longhi, R., *Teatro e immagini del Settecento italiano* (Turin, 1954). Hyatt Mayor, A., Viale, M., Della Corte, A. and Bragaglia, A. G., *Tempi e aspetti della scenografia* (Turin, 1954). Prampolini, E., *Lineamenti di scenografia italiana* (Rome, 1950). Mariani, V., *Storia della scenografia* (Rome, 1927). Kernodle, G. R., *From Art to Theater, Form and Convention in the Renaissance* (Chicago, 1943). Tintelnot, H., *Barock Theater und barocke Kunst* (Berlin, 1939). Magagnato, L., *Teatri italiani del Cinquecento* (Venice, 1954).

(6) TOWN PLANNING

Various authors: *Esperienze urbanistiche in Italia.* Pref.: A. Olivetti (Rome, 1952). Brigante Colonna, G., *Il Volto di Roma* (Rome, 1937). Giovannoni, G., *Vecchie Città ed Edilizia nuova* (Rome, 1931). 'Urbanistica', Nos. 15–16, Oct. 1954.
GENERAL WORKS: Lavedan, P., *Histoire de l'urbanisme* (3 vols., Paris, 1926, 1952: relevant chapters). Gantner, J., *Grundformen der europäischen Stadt* (Vienna, 1928; *id.*).

B. PAINTING

(1) TREATISES

Cennini, C., *Il Libro dell'arte* (*c.* 1390) (Florence, 1859; English ed. by D. V. Thompson, Yale, 1933). Alberti, L. B., *Della Pittura* (1435) (ed. L. Mallè, Florence, 1950; English ed. by J. R. Spencer, London, 1956). Piero della Francesca, *De Prospectiva pingendi* (*c.* 1480) (ed. G. Nicco Fasola, Florence, 1942). Pino, P., *Dialogo della pittura* (1550), ed. R. Pallucchini (Venice, 1946).

(2) GENERAL HISTORIES OF PAINTING

Lanzi, Abbé L., *Storia pittorica della Italia* (Bassano, 1789; English ed. by T. Roscoe, 6 vols., London, 1828). *Crowe, J. A. and Cavalcaselle, G. B., *A New*

BIBLIOGRAPHY

History of Painting in Italy (2 vols., London, 1864–1866; revised ed. by L. Douglas, T. Borenius and others, 6 vols., London, 1903–14; Italian ed., 11 vols., Florence, 1886–1908). Marle, R. van, *The Development of the Italian Schools of Painting* (19 vols., The Hague, 1923 on). *Storia della pittura italiana* (12 vols., by P. Ducati, S. Bettini, L. Coletti, G. Fiocco, C. Gamba, R. Pallucchini, U. Nebbia, G. Lorenzetti, E. Somarè: Novara, 1942 on). Venturi, L., *Italian Painting* (3 vols., Geneva-Paris, 1950–1951–1952). Galetti, U. and Camesasca, E., *Enciclopedia della pittura italiana* (2 vols., Milan, n.d.). Coll. 'Zehn Jahrhunderte italienische Malerei', Zürich; XIVth and XVth c. have already appeared. *Storia della pittura*, in 5 vols. by M. Borda, V. Mariani, P. d'Ancona and U. Nebbia, Milan, in course of publication; Vol. IV, *La Pittura dell'Ottocento*, by P. d'Ancona (1954) has already appeared.

(3) PERIODS

Middle Ages. *Bognetti, G. P., Chierici, G., Capitani d'Arzago, A. de, *Santa Maria di Castelseprio* (Milan, 1948). Weitzmann, K., *The Fresco Cycle of S. Maria di Castelseprio*, Princeton, 1951 (reviewed by Meyer Schapiro ('A.B.', XXXIV, 1952, 2)). *Oertel, R., *Die Frühzeit der italienischen Malerei* (Stuttgart, 1953). Ladner, G., *Die italienische Malerei im XI. Jahrhundert* (in 'J.W.', 1931). Garrison, E. B., *Italian Romanesque Panel Painting* (Florence, 1949); *Studies in the History of Medieval Italian Painting* (Vol. I, Florence, 1954). Hautecoeur, L., *Les Primitifs italiens* (Paris, 1931). *Ancona, P. d', *Les Primitifs italiens du XIe au XIIIe siècle* (Paris, 1935). Longhi, R., *Giudizio sul Duecento* ('Proporzioni', II, 1948). Coletti, L., *I Primitivi* (3 vols., Novara, 1942–1947).

Renaissance. Escher, K., *Die Malerei des 14 bis 16 Jh. in Mittel und Unteritalien* (Handbuch der Kunstwissenschaft, Berlin, 1922) and Bercken, E. v. der and Mayer, A. L., *Die Malerei des 15 und 16 Jh. in Oberitalien* (*id.*, Berlin, 1947). Berenson, B., *Italian Painters of the Renaissance* (1st ed.: I, 1894; II, 1896; III, 1897; IV, 1907; new ed. London, 1952) and lists: *Italian Pictures of the Renaissance* (Oxford, 1932; Italian translation, Milan, 1936); *The Venetian School* (2 vols., London, 1957). Goering, M., *Italian Painting of the XVIth Century* (London, 1936). Schmeckebier, L., *A Handbook of Italian Renaissance Painting* (New York, 1938). Gould, C., *An Introduction to Italian Renaissance Painting* (London, 1957).

Mannerism. Pevsner, N., *Die italienische Malerei vom Ende der Renaissance bis zum ausgehenden Rokoko*. Handbuch der Kunstwissenschaft, Berlin, 1928. Briganti, G., *Il Manierismo e Pellegrino Tibaldi* (Rome, 1945). Zeri, F., *Pittura e Controriforma* (Turin, 1957).

Baroque. Baldinucci, F., *Dal Baroccio a Salvator Rosa* (selection by F. Battelli from Vols. V and VI of the *Notizie*, Florence, 1914). Delogu, G., *La pittura italiana del Seicento* (Florence, 1931). Moschini, V., *La pittura italiana del Settecento* (Florence, 1931). Ojetti, U., *Il Settecento italiano* (Milan, 1932). Fogolari, G., *La pittura del Settecento italiano* (Milan, 1932).

BIBLIOGRAPHY

Goering, M., *Italienische Malerei des siebz. und achtz. Jahrh.* (Berlin, 1936). Lorenzetti, G., *La pittura italiana del Settecento* (Novara, 1942). *Mostra di Caravaggio e dei Caravaggeschi* (R. Longhi) (Milan, 1951).

Modern. Ojetti, U., *La pittua italiana dell'Ottocento* (Rome, 1929). Somarè, E., *Storia dei pittori italiani dell'Ottocento* (Milan, 1928). Scheiwiller, G., *Art italien moderne* (Paris, 1930). Carrà, C., *Il Rinnovamento delle arti in Italia* (Milan, 1945). Nebbia, U., *La Pittura del Novecento* (Milan, 1946). Ghiringhelli, *Pittura moderna italiana* (Turin, 1949). Zervos, C., Buzzi, P., *Un demi-siècle d'art italien* ('Cahiers d'Art', 1950). Apollonio, U., *Pittura italiana moderna* (Venice, 1951).

(4) PROVINCES

Piedmont. Weber, S., *Piemontische Malerschule* (Strasbourg, 1911). Brizio, A. M., *La Pittura in Piemonte dall'età romanica al Cinquecento* (Turin, 1942). Dragone, A. and J., *I Paesisti piemontesi dell'Ottocento* (Turin, 1947).

Lombardy. *Toesca, P., *La Pittura e la Miniatura nella Lombardia* (Milan, 1912). Baroni, C., *La Pittura lombarda del Quattrocento* (Messina, 1952). Costantini, V., *La Pittura lombarda dal XIV al XVI sec.* (Milan, 1922). *I Pittori della Realtà in Lombardia* (catalogue by R. Cipriani and G. Testori, Milan, 1953).

Liguria. Grosso, O., Bonzi, M., Mercenaro, C., *Pittori genovesi del Seicento e del Settecento* (Genoa, 1938). Morassi, A., *Mostra della Pittura del Seicento e Settecento in Liguria* (Milan, 1942); *Capolavori della pittura a Genova* (Exhibition of 1946; Milan-Florence, 1952). Delogu, G., *Pittori minori liguri, lombardi, piemontesi, del Seicento e del Settecento* (Venice, 1931).

Verona. Vavalà, E., *La Pittura veronese del Trecento* (Verona, 1926). Arslan, E., *La pittura e la scultura veronese dal sec. VIII al sec. XVII.* (Milan, 1943).

Venice. Ridolfi, C., *Le Meraviglie dell'arte* (Venice, 1648; ed. D. von Hadeln, Berlin, 1914). Boschini, M., *Le Ricche Minere della pittura veneziana* (Venice, 1674). Molmenti, P., *La Peinture vénitienne* (Florence, 1904). Fiocco, G., *Venetian Painting of the Seicento and Settecento* (Florence and Paris, 1929). Mather, F. J., *Venetian Painters* (New York, 1936). *Cinque Secoli di pittura veneta* (catalogue by R. Pallucchini, Venice, 1945). *Longhi, R., *Viatico per cinque secoli di pittura veneziana* (Florence, 1946). Pallucchini, R., *La Pittura veneziana del Settecento* (2 vols., Bologna, 1952). Valsecchi, M., *La Pittura veneziana* (Milan, 1954; French translation, Paris, 1954). Godfrey, F. M., *Early Venetian Painters, 1415–1495* (London, 1954). Levey, M., *Painting in XVIII century Venice* (London, 1959). *Il Seicento Veneziano*, exhibition catalogue (preface: P. Zampetti), Venice, Doge's Palace, 1959.

Emilia. Venturi, A., *La Pittura del Quattrocento nell'Emilia* (Verona, 1931). BOLOGNA: Longhi, R., *Momenti della pittura bolognese* ('L'Archiginnasio', 1935). Baldani, R., *La Pittura a Bologna nel sec. XIV* (1909). Longhi, R., *La Mostra del Trecento bolognese*, 'Pa.', No. 5 (May, 1950). FERRARA: Gardner, E. A., *The Painters of the School of Ferrara* (London, 1911).

BIBLIOGRAPHY

Pittura ferrarese del Rinascimento (catalogue, Florence, 1933). *Longhi, R., *Officina ferrarese* (new ed., Florence, 1956). Ortolani, S., *Cosmè Tura, F. del Cossa, Ercole de' Roberti* (Milan, n.d.: 1941). Nicolson, B., *The Painters of Ferrara* (London, 1952).

Romagna. Buscaroli, R., *La Pittura romagnola del Quattrocento* (Faenza, 1931). *Mostra di Melozzo e del Quattrocento romagnolo* (catalogue by C. Gnudi and L. Becherucci, Forlì, 1938). *Mostra dei pittori riminesi* (catalogue by C. Brandi, Rimini, 1935).

Marches. *Pittura veneta nelle Marche* (catalogue by P. Zampetti, Ancona, 1950).

Tuscany. *Meiss, M., *Painting in Florence and Siena after the Black Death* (Princeton, 1951).
FLORENCE: Colnaghi, D. E., *A Dictionary of Florentine Painters from the XIIIth to the XVIIth c.* (London, 1928). Offner, R., *A Critical and Historical Corpus of Florentine Painting* (New York, 1931 on). Khvoshinsky and Salmi, M., *Pittori toscani* (2 vols., Rome, 1914). Borsook, E., *The mural painters of Tuscany* (London, 1960). Toesca, P., *Florentine Painting of the Trecento* (Florence and Paris, 1929). Salmi, M., *I Mosaici del bel S. Giovanni e la pittura nel secolo XIII* ('Dedalo', 1931). Clark, K., *Florentine Painting, XVth c.* (London, 1945). Antal, F., *Florentine Painting and its Social Background* (London, 1947). Offner, R., *Studies in Florentine Painting* (New York, 1927). Beccherucci, L., *Manieristi toscani* (Bergamo, 2nd ed., 1949). Cazzullo, M. P., *La Scuola toscana dei Macchiaioli* (Florence, 1948). Franchi, A., *I Macchiaioli toscani* (Florence, 1942).
SIENA: Edgell, G. H., *A History of Sienese Painting* (New York, 1932). Weigelt, C., *Sienese Painting of the Trecento* (New York, n.d.: 1930). Cecchi, E., *Trecentisti senesi* (Rome, 1928; English ed., London, 1931). Pope-Hennessy, J., *Sienese Quattrocento Painting* (London, 1947). Brandi, C., *Quattrocentisti senesi* (Milan, 1950). Carli, E., *Sienese Painting* (London, 1956).

Umbria. Bombe, W., *Geschichte der Peruginer Malerei* (Berlin, 1912). Gnoli, U., *Pittori e miniatori nell'Umbria* (Spoleto, 1923).

Rome. *Wilpert, G., *Die römischen Mosaiken und Malereien der kirchlichen Bauten vom 4 bis 13 Jh.* (Freiburg, 1917). Marle, R. van, *La Peinture romaine au Moyen Âge* (Strasbourg, 1921). Busuioceanu, A., *Pietro Cavallini e la Pittura romana del Duecento e del Trecento* ('Ephemeris Daco-Romana', Rome, 1925). Waterhouse, E. K., *Baroque Painting in Rome* (London, 1937).

Campania. Rolfs, W., *Geschichte der Malerei Neapels* (Leipzig, 1910). Rinaldis, A. de, *Neapolitan Painting of the Seicento* (Florence and Paris, 1929). Morisani, O., *Pittura del Trecento in Napoli* (Naples, 1947).

Sicily. Demus, O., *The Mosaics of Norman Sicily* (New York, 1950). *Antonello e la pittura del 400 in Sicilia* (catalogue by G. Vigni, 2nd ed., Venice, 1953). Bottari, S., *La Pittura del Quattrocento in Sicilia* (Messina, 1954); *La Cultura figurativa in Sicilia* (Messina, 1954).

BIBLIOGRAPHY

(5) COLLECTIONS

Luzio, A., *La Galleria dei Gonzaga venduta all'Inghilterra nel 1627–1628* (Milan, 1913). Berenson, B., *Venetian Painting in America* (New York, 1926). Offner, R., *Italian Primitives at Yale University* (New Haven, 1927). Acchiardi, P. d', *I Quadri Primitivi della Pinacoteca Vaticana* (Rome, 1929). Venturi, L., *Pitture italiane in America* (Milan, 1931; English ed., New York, 1933). Richter, J. P., *The Cannon Collection of Italian Paintings* (Princeton, 1936). Meiss, M., *Italian Primitives at Konopiste* ('A.B.', XXVIII (1946), I). Pallucchini, R., *I Dipinti della Galleria Estense di Modena* (Florence, 1945). Zeri, F., *La Galleria Spada in Roma* (Florence, 1952). Somarè, E., *I Maestri italiani dell'Ottocento nella raccolta Marzotto* (Milan, 1937). *Davies, Martin, *The Earlier Italian Schools* (National Gallery, London, 1951). Russoli, F., *La Pinacoteca Poldi-Pezzoli* (catalogue, Milan, 1955). Redig de Campos, D., *Itinerario pittorico dei Musei vaticani* (2nd ed., Rome, 1952). Della Chiesa, A. Ottino, *L'Accademia Carrara* (Bergamo, 1955). Moschini Marconi, S., *Accademia di Venezia, Sec. XIV e XV* (Rome, 1955). *Italian Paintings and Drawings at 56 Prince's Gate (Catalogue of the Seilern Collection)* (2 vols., London, 1959).

(6) EXHIBITIONS

Valentiner, W. R., *A Catalogue of Early Italian Paintings exhibited at the Duveen Galleries* (New York, 1924; New York, 1926). *Pittura italiana del Duecento e del Trecento* (1937; catalogue G. Sinibaldi and G. Brunetti, Florence, 1943). *La Pittura italiana del Seicento e del Settecento alla mostra di Palazzo Pitti*, by U. Ojetti, L. Dami, N. Tarchiani (Florence, Milan, 1946). *Exhibitions of Italian XIXth Century Paintings* (catalogue by E. Somarè, New York, 1949). Morandotti, A. and Briganti, G., *I Bamboccianti* (catalogue, Rome, 1950). *Il Lavoro nella pittura italiana di oggi* (Verzocchi coll., Milan, 1950). *Premio Parigi* (1950, catalogue). *Biennali of modern art* (catalogues from 1948).

(7) PROBLEMS OF CRITICISM

Morelli, G. (I. Lermoliev), *Die Galerien Borghese und Doria Pamphili in Rom* (Leipzig, 1890; English ed., 1892); *Die Galerien zu München und Dresden (ibid.,* 1891; English ed. 2nd impression, 1907); *Die Galerie zu Berlin (ibid.,* 1893); *Italian Masters in German Galleries* (Munich, Dresden, Berlin) (London, 1883). Venturi, A., *Studi dal vero* (Milan, 1917). Berenson, B., *The Study and Criticism of Italian Art* (3 vols., London, 1901, 1902, 1916). *Quadri senza casa* ('Dedalo', 1930, 1931, 1932). *Atti del seminario di storia dell'arte* (1953), under direction of C. L. Ragghianti, Annali della Scuola normale, Pisa, 1954.

(8) DRAWINGS

*Meder, J., *Die Handzeichnung* (Vienna, 1919). Dussler, L., *Italienische Meisterzeichnungen* (Frankfurt-am-Main, 1938). *Tolnay, Ch. de, *History and Technique of Old Master Drawings, a Handbook* (New York, 1943).

BIBLIOGRAPHY

Degenhart, B., *Zur Graphologie der Handzeichnung* ('Jahrbuch Bibl. Herziana', I, 1937). Degenhart, B., *Italienische Zeichnungen des frühen 15 Jh.* (Basle, 1949). *Italienische Zeichner der Gegenwart* (Berlin, 1956). Ivanoff, N., *I disegni italiani del Seicento* (Venice, n.d.).

Venice. Parker, K. T., *North Italian Drawings of the Quattrocento* (London, 1927). Hadeln, D. von, *Venezianische Zeichnungen des Quattrocento* (Berlin, 1925); *V. Z. der Hochrenaissance (ibid).*; *V. Z. der Spätrenaissance* (1926). Tietze, H. and Tietze-Conrat, E., *The Drawings of the Venetian Painters in the XVth and XVIth Centuries.* (New York, 1944).

Florence etc. Schendel, A. van, *Le Dessin en Lombardie* (Brussels, 1938). Ede, H. S., *Florentine Drawings of the Quattrocento* (London, 1926). **Berenson, B., *The Drawings of the Florentine Painters* (2nd ed., 3 vols., Chicago, 1938). Fischel, O., *Die Zeichnungen der Umbrer* (Berlin, 1917).

Catalogues. British Museum. *Italian Drawings. The XIVth and XVth Centuries*, by A. E. Popham and P. Pouncey (2 vols., London, 1950; suppl. 1952). Windsor Castle. *The Italian Drawings of the XVth and XVIth Centuries*, by A. E. Popham and J. Wilde (London, 1949); *Bolognese Drawings*, by O. Kurz (*id.*, 1955).

Exhibitions. Popham, A. E., *Italian Drawings exhibited at the Royal Academy, Burlington House*, London, 1930 (Oxford, 1931).

(9) MINIATURES

**Ancona, P. d', *La Miniature italienne du Xe au XVIe siècle* (Brussels, 1925). Toesca, P., *Monumenti e studi* (Milan, 1930).

General works. *Wickhoff, F., Dvorak, M. and Hermann, J. *Beschreibendes Verzeichnis der illuminierten Handschriften in Österreich*, continued by J. Hermann, VI vols. (Leipzig, 1928–1933). *Tesori delle Biblioteche d'Italia*, vol. I by M. Salmi and D. Fava: *Emilia e Romagna* (Milan, 1931). *I Manoscritti miniati delle Biblioteche italiane, I. La Biblioteca estense di Moderna*, Vol. I, by D. Fava and M. Salmi (Florence, 1950). **Ancona, P. d', *La Miniatura fiorentina* (XIth-XVIth c.) (2 vols., Florence, 1914). Vagaggini, S., *La Miniatura fiorentina nei secoli XIV e XV* (Milan, Florence, 1952).

Exhibitions. *Miniatures de la Renaissance* (catalogue by L. Michelini Tocci, Vatican, 1950). *Trésors des bibliothèques d'Italie* (IVth-XVIth c.) (Bibliothèque nationale, Paris, 1950). *Italian Manuscripts in the Pierpont Morgan Library* (catalogue by M. Harrsen and G. K. Boyce; New York, 1953). *Mostra storica nazionale della miniatura* (Palazzo Venezia, Rome) (catalogue, Florence, 1953).

(10) BOOKS

Sander, M., *Le Livre à Figures italien depuis 1467 jusqu' à 1530* (Milan, n.d.). Essling, Prince d', *Les Livres à Figures vénitiens de la fin du XVe siècle et du commencement du XVIe* (Florence, Paris, 1907–1914).

BIBLIOGRAPHY

Exposition du livre italien, 1926 (catalogue, Paris, 1926). Gregori, L. de', *La Stampa a Roma nel secolo XV* (Rome, 1933). Rava, C. E., *Arte dell'illustrazione nel libro italiano del Rinascimento* (Milan, 1945). *Marinis, T. de, *La Biblioteca napoletana dei Re d'Aragona* (4 vols., Milan, 1947–1952). Aeschlimann, E., *Bibliografia del libro d'arte italiano*, 1940–1952 (Rome, 1952).

(11) ENGRAVING

**Hind, A. M., *Early Italian Engraving: a critical catalogue* (in two parts; London, 1938–1948, 7 vols.). Delaborde, H., *La Gravure en Italie avant Marc Antoine* (n.e.; 1883), Kristeller, P., *Early Florentine Woodcuts* (New York, 1917). Goldsmith Phillips, J., *Early Florentine Designers and Engravers* (Cambridge, Mass., 1955). Blum, A., *Les Nielles du Quattrocento* (Paris, 1950). Pittaluga, M., *L'Incisione italiana del Cinquecento* (Milan, n.d.; 1930). Servolini, L., *La Xilografia a chiaroscuro italiana nel sec. XVI, XVII, XVIII* (Lecce, 1930). Giglioli, O. H., *Incisori toscani del Seicento* (Florence, 1942); *Incisori toscani del Settecento* (Florence, 1943). Pittaluga, M., *Acquafortisti veneziani del Settecento* (Florence, 1953). Vitali, L., *L'Incisione italiana moderna* (Milan, 1934). Ozzola, L., *La Litografia italiana* (Rome, 1923).

General Works. Vesme, A. de, *Le Peintre-Graveur italien* (XVIIth c. on) (Milan, 1906). Servolini, L., *L'Incisione italiana di cinque secoli* (Mlan, 1952). Pelluccioni, A., *Dizionario degli artisti incisori italiani* (Carpi, 1949).

Collections. Witt, A. de, *La Collezione delle Stampe della Galleria degli Uffizi* (Rome, 1938).

C. SCULPTURE

(1) TREATISES

Gauricus, P., *De Sculptura* (Florence, 1504; ed. Brockhaus, Leipzig, 1886). Cellini, B., *Due Trattati* (Florence, 1568; ed. C. Milanesi, Florence, 1857; English ed. by C. Ashbee, London, 1898).

(2) PERIODS

Waters, W. G., *Italian Sculptors* (London, 1911). Knapp, F., *Italienische Plastik vom 15 bis 18 Jh.* (Munich, 1923).

Middle Ages. Mayer, A. L., *Mittelalterliche Plastik in Italien* (Munich, 1923). Haseloff, A., *La Scultura preromanica in Italia* (Bologna, 1930). Jullian, R., *L'Éveil de la Sculpture italienne* (Paris, 1945). Francovich, G. de', *Benedetto Antelami, architetto e scultore e l'arte del tempo suo* (2 vols., Milan, 1952). Crichton, C. H. B., *Romanesque Sculpture in Italy* (London, 1954). *Pope-Hennessy, J., *Italian Gothic Sculpture* (London, 1955).

Renaissance. Freemann, L. J., *Italian Sculpture of the Renaissance* (New York, 1927). Schubring, P., *Die italienische Plastik des Quattrocento* (Berlin, 1919). Maclagan, E., *Italian Sculpture of the Renaissance* (Cambridge, 1925). Valentiner, W. R., *Studies of Italian Renaissance Sculpture* (London, 1950).

BIBLIOGRAPHY

Becherucci, L., *La Scultura italiana del Cinquecento* (Florence, 1934). Pope-Hennessy, J., *Italian Renaissance Sculpture* (London, 1958).

Modern. Neri, D., *Scultori francescani del Seicento in Italia* (Pistoia, 1952). Brinckmann, A. E., *Barock Skulptur* (Berlin, 1921). *Brinckmann, A. E., *Barock Bozzetti: Italienische Bildhauer* (4 vols., Frankfurt, 1923). Sapori, F., *Scultura italiana moderna* (Rome, 1949). Vigezzi, S., *La Scultura italiana dell'Ottocento* (Milan, 1932).

(3) PROVINCES

Northern Italy. Zimmermann, M. G., *Oberitalische Plastik im frühen und hohen Mittelalter* (Leipzig, 1897). Mayer, A. G., *Lombardische Denkmäler des 14 Jh.* (Stuttgart, 1893). Gabelenz, H. von der, *Mittelalterliche Plastik in Venedig* (Leipzig, 1903). *Planiscig, L., *Venezianische Bildhauer der Renaissance* (Vienna, 1921). Baroni, C., *La Scultura gotica lombarda* (Milan, 1944). Vigezzi, S., *La Scultura lombarda nel Cinquecento* (Milan, 1929). Vigezzi, S., *La Scultura in Milano* (Milan, 1932).

Tuscany. *Salmi, M., *La Scultura romanica in Toscana* (Rome, 1934). *Reymond, M., *La Sculpture florentine* (4 vols., Florence, 1897–1900). *Bode, W. von, *Florentiner Bildhauer der Renaissance* (Berlin, 1910; English ed., London, 1928). Gore, W. O., *Florentine Sculptors of the Fifteenth Century* (London, 1930). Galassi, G., *La Scultura fiorentina del Quattrocento* (Milan, 1949). Schubring, P., *Die Plastik Sienas im Quattrocento* (Berlin, 1907). Carli, E., *La Scultura lignea senese* (Milan, Florence, 1951).

Rome. Bessone-Aurelj, A. M., *I Marmorari romani* (Milan, 1925). Grisebach, A., *Römische Porträtbüsten der Gegenreformation* (Rom. Forsch. Bibl. Herziana, XIII) (Rome, 1936).

Southern Italy. Sheppard, C. D., *A Chronology of Romanesque Sculpture in Campania*, 'A.B.', XXXII (1950), 4. Wackernagel, M., *Die Plastik des 11 und 12 Jh. in Apulien* (Leipzig, 1911). Marzo, G. di, *I Gagini e la scultura in Sicilia* (Palermo, 1883); *Sculture lignee nella Campania* (Naples, 1950).

(4) MEDALLIONS AND PLAQUES

Heiss, A., *Les Médailleurs de la Renaissance* (Paris, 1883). Molinier, E., *Les Plaquettes* (2 vols., Paris, 1881). Armand, A., *Les Médailleurs italiens du XVe et du XVIe siècle* (3 vols., Paris, 1883–1887). Foville, J. de, *Pisanello et les médailleurs italiens* (Paris, 1909). Hill, G. F., *A Corpus of Italian Medals of the Renaissance before Cellini* (Oxford, 1930). Habich, G., *Die Medaillen der italienischen Renaissance* (Stuttgart, 1923). Middeldorf, U. and Goetz, O., *Medals and plaquettes from the S. Morgenroth Collection* (Chicago, 1943).

(5) BRONZES

Bode, W. von, *Die italienischen Bronzestatuetten der Renaissance* (3 vols., Berlin, 1907–1912). Planiscig, L., *Die Bronze Plastiken* (Vienna, Kunsthistorisches Museum, 1924). Bode, W. von., *Bronzes of the Renaissance and*

BIBLIOGRAPHY

subsequent Periods. Collection of J. Pierpont Morgan (2 vols., Paris, 1910). Ricci, S. de, *Renaissance Bronzes: the Gustave Dreyfus collection* (Oxford, 1931). Planiscig, L., *Piccoli Bronzi italiani del Rinascimento* (Milan, 1930).

(6) GEMS

**Kris, E., *Meister und Meisterwerke der Steinschneidekunst in der italienischen Renaissance* (2 vols., Vienna, 1929). Eickler, F. and Kris, E., *Die Kameen des Altertums, des Mittelalters und der neueren Zeit* (Vienna, 1926). Righetti, R., *Incisori di gemme e camei in Roma dal Rinascimento all'Ottocento* (Rome, n.d.).

(7) COLLECTIONS

Pope-Hennessy, J., *Italian Gothic Sculpture in the Victoria and Albert Museum* (London, 1952). Swarzenski, G., *Some Aspects of Italian Quattrocento Sculpture in the National Gallery (Washington)* (GBA, Oct. and Nov. 1943). Galleria Borghese, *Sculture dal sec. XVI al XIX*, by I. Faldi (Rome, 1954).

D. DECORATIVE ARTS

ARMOUR

Gelli, I. and Moretti, G., *Gli Armaioli milanesi* (Milan, 1903). Lensi, A., *Il Museo Stibbert*, catalogue, *Sala armi* (Florence, 1918).

BOOKBINDING

Rossi, F., *Mostra storica della legatura artistica in Palazzo Pitti* (Florence, 1922). Fava, D. and Pastorello, E., *Cento belle legature italiane* (Congresso delle Bibl. e di Bibliogr., 1929) (Rome, 1933).

CABINET MAKING

Ferrari, G., *Il Legno nell'arte italiana* (2nd ed., Milan, n.d.: 1926). Morazzoni, G., *Il mobilio italiano* (Florence, 1940). Marangoni, G., *Storia dell' arredamento* (2 vols., Milan, 1952). Odom, W., *A History of Italian Furniture* (2 vols., New York, 1918–1920). Bode, W. von, *Die italienischen Hausmöbel der Renaissance* (Leipzig, 1920). Ponti, C., *Il Mobile moderno in Italia* (Milan, 1930). Tinti, M., *Il Mobilio fiorentino* (Milan, 1929). Morazzoni, G., *Il Mobile veneziano del 700* (Milan, 1927).

DECORATIVE ELEMENTS IN ARCHITECTURE

Ferrari, G., *Lo Stucco nell'arte italiana* (Milan, 1910). Morazzoni, G., *Stucchi italiani, Maestri genovesi del XVI al XIX* (Milan, 1950). Ferrari, G., *La Terracotta e pavimenti di laterizio nell' arte italiana* (Milan, 1928).

FABRICS AND FASHIONS

Ricci, E., *Ricami antichi e moderni* (Florence, n.d.). Ricci, E., *Antiche trine italiane* (Bergamo, 1908). Flörke, H., *Die Moden der italienischen Renaissance*

BIBLIOGRAPHY

vom 1300 bis 1500 (Munich, 1917). Vocino, M., *Storia del costume* (Rome, 1952). Santangelo, A., *Tessuti d'arte italiani* (Milan, 1959).

FRAMES

Guggenheim, M., *Le Cornici italiane dalla metà del sec. XV allo scorcio del XVI* (Milan, 1897). Bach, E., *Florentinische und venezianische Bilderrahmen aus der Zeit der Gotik und der Renaissance* (Munich, 1902).

FURNITURE

Bode, W. von, *Die italienischen Hausmöbel der Renaissance* (2nd ed., Leipzig, 1920). Schiaparelli, A., *La Casa fiorentina* (Florence, 1908). Schottmüller, F., *Wohnungskultur und Möbel der italienischen Renaissance* (Stuttgart, 1921). Marangoni, G. and Clementi, A., *Storia dell'Arredamento* (4 vols., Milan; I and II, new ed. 1952; III, 1952; IV, 1954). Pedrini, A., *Il Mobilio, gli ambienti e le decorazioni del Rinascimento in Italia* (Florence). Terny de Gregory, W., *Vecchi mobili italiani* (Milan, 1953).

GARDENS

Gromort, G., *Jardins d'Italie* (148 plates and 25 diagrams; Paris, 1922). *Dami, L., *Il Giardino italiano* (Milan, 1924). Shepherd, J. C. and Jellicoe, G. A., *Italian Gardens of the Renaissance* (New York, 1925; new ed., London, 1953). Gothein, M. L., *Geschichte der Gartenkunst* (Jena, 1914); chapters on Italy. *Mostra del giardino italiano* (Florence, 1931).

GENERAL WORKS

Marangoni, G., *Enciclopedia delle moderne arti decorative italiane* (Milan, 1928).

GLASS

Levi, C., *L'Arte del vetro in Murano nel Rinascimento* (Venice, 1895). Zecchin, V., *Sulla storia dell'arte vetraria muranese* (Venice, 1952). Taddei, G., *L'Arte del vetro in Firenze e nel suo dominio* (Florence, 1954).

GOLDSMITH'S WORK

Accascina, M., *L'Oreficeria italiana* (Florence, 1933). Churchill, S. I. A. and Bunt, C. G., *The Goldsmiths of Italy: their Guilds, Statues and Work* (London, 1926). Morassi, A., *Antica Oreficeria italiana* (Milan, 1936).

IRONWORK

Ferrari, G., *Il Ferro nell'arte italiana* (3rd ed., Milan, 1927). Pedrini, A. and Ricci, C., *Il Ferro battuto, sbalzato e cesellato nell'arte italiana* (Milan, 1929). Pedrini, A., *Il Ferro battuto, sbalzato e cesellato in Italia* (XI–XVIII) (Turin, n.d.; 1951). Pettorelli, A., *Il Bronzo e il rame nellarte decorativa italiana* (Milan, n.d.).

BIBLIOGRAPHY

MAIOLICA

Ferrari, G., *La Terracotta e pavimenti in laterizio nell'arte italiana*, intro. by C. Ricci (Milan, 1928). Piccolpasso, C., *I Tre Libri del vasajo* (MSS., 1548; ed. Rome, 1857; English translation by A. van der Put and B. Rackham, London, 1934). Molinier, E., *La Céramique italienne au XVe s.* (Paris, 1888). Ballardini, G., *Corpus della majolica italiana* (I, Rome, 1933, II, 1938). Chompret, J., *Répertoire de la majolique* (2 vols., Paris, 1948). Morazzoni, G., *Le Porcellane italiane* (Milan, Rome, 1935). Minghetti, A., *Ceramisti* (Milan, 1939). Rackham, B., *Italian Maiolica* (London, 1952). Lane, A., *Italian Porcelain* (London, 1954).

COLLECTIONS: Rackham, B., *Catalogue of Italian Majolica in the Victoria and Albert Museum* (2 vols., London, 1940).

MARQUETRY

Jackson, F. H., *Intarsia and Marquetry* (London, 1903). Arcangeli, F., *Tarsie* (2nd ed., Rome, 1943). Chastel, A., *Marqueterie et perspective au XVe siècle* ('R.A.', III, 1953).

STAINED GLASS

Straelen, H. van, *Studien zur florentinischen Glasmalerei des Trecento und Quattrocento* (1938). Monneret de Villard, U., *Le Vetrate dipinte del duomo di Milano* (Milan, 1918). Scherill, C. H., *A Stained Glass Tour in Italy* (London, 1913). Carli, E., *Vetrata duccesca* (Florence, 1946). Marchini, G., *Vetrate italiane* (Rome, 1955; English ed., New York, 1956).

TAPESTRY

Conti, G., *Ricerche storiche sull' arte dell' arazzo in Firenze* (Florence, 1875). Campori, G., *L'Arazzena estense* (Modena, 1876). Urbani, G. M., *Degli arazzi in Venezia* (Venice, 1878). Braghirolli, V., *Sulle manifatture di arazzi in Mantova* (Mantua, 1879). Minieri Riccio, C., *La Real Fabbrica degli arazzi nellà città di Napoli* (Naples, 1879). Relevant chapters in Göbel, H., *Wandteppiche* (3 vols., Berlin, 1934).

III. SPECIAL STUDIES

(1) GUIDES AND GUIDE-BOOKS

**Burckhardt, J., *Der Cicerone* (Basle, 1860; French ed. by W. Bode, 1925; English ed., London, n.d.). Rumohr, K. F. von, *Italienische Forschungen* (Berlin, 1827; new ed. by J. Schlosser, Frankfurt-am-Main, 1920). Taine, H., *Voyage en Italie* (2 vols., Paris, 1865, 18th ed., 1930). Brinckmann, A. E., *Giotto bis Juvara, ewige Werte italienischer Kunst* (Hamburg, 1940). Möller van den Bruck, *Die italienische Schönheit* (3rd ed., Berlin, 1930).

BIBLIOGRAPHY

(2) ART AND CULTURE

Ruskin, J., *The Stones of Venice* (3 vols., Oxford, 1851–1858; abridged ed., 1879); *Mornings in Florence* (Oxford, 1875). Thode, H., *Franz von Assisi und die Anfänge der Kunst der Renaissance in Italien* (Berlin, 1885; French translation, n.d.); *Michelangelo und das Ende der Renaissance* (2 vols., Berlin, 1903). Sizeranne, R. de la, *Les Masques et les Visages à Florence et au Louvre* (1927). Burckhardt, J., *Die Kultur der Renaissance in Italien* (Stuttgart, 1860; English ed. by S. C. G. Middlemore, London, 1944, and many other eds.). Walser, E., *Gesammelte Studien zur Geistesgeschichte der Renaissance* (intro. W. Kalgi) (Basle, 1932). Hoogewerf, G. J., *Via Margutta* (Rome, 1953). Müntz, E., *Les Précurseurs de la Renaissance* (1882; Italian translation, Florence, 1902). Wackernagel, M., *Der Lebensraum des Künstlers in der florentinischen Renaissance* (Leipzig, n.d.; 1938). Weise, G., *Renaissance und Antike* (Tübingen, 1953). Chastel, A., *Art et Religion dans la Renaissance italienne* ('Humanisme et Renaissance', VI, 1945). Paatz, W., *Renaissance oder Renovatio* (in the collection *Beiträge zur Kunst des Mittelalters*, Berlin, 1950). Ferguson, W. K., *The Renaissance in Historical Thought* (Boston, 1948). Battisti, E., *Rinascimento e Barocco* (Turin, 1960). Praz, M., *Gusto neoclassico* (Florence, 1940).

(3) ICONOLOGY AND THEMES[1]

Ripa, C., *Iconologia* (Rome, 1610), and treatises on mythography studied in: Seznec, J., *La Survivance des Dieux antiques* (London, 1939; English ed., New York, 1953). Panofsky, E., *Studies in Iconology* (New York, 1939). Volkmann, L., *Bildinschriften der Renaissance* (Leipzig, 1923). Warburg, A., *Gesammelte Schriften* (2 vols., Leipzig, 1932). Cartwright, J., *Christ and His Mother in Italian Art* (London, 1897). Venturi, A., *La Madonna nell'arte* (French translation, 1900). Mazzoni, P., *La Leggenda della Croce nell'arte italiana* (Florence, 1914). Gerke, F., *Christus in der spätantiken Plastik* (Berlin, 3rd. ed., Mainz, 1948). Prampolini, G., *L'Annunciazione nei pittori primitivi italiani* (Milan, 1939). Schorr, D. C., *The Presentation* ('The Art Bulletin', XXVIII (1946), pp. 17–32). Kaftal, G., *Iconography of the Saints in Tuscan Painting* (Florence, 1952). Ancona, P. d', *L'Uomo e le sue opere nelle figurazioni italiane del medioevo* (Florence, 1923). Schiavo, A., *La Donna nella scultura italiana dal XII al XVIII secolo* (Rome, 1950). Howe, W. N., *Animal Life in Italian Paintings* (London, 1912). Wedgewood Kennedy, R. *The Renaissance Painters' Garden* (New York, 1948). Polidori Calamandrei, E., *Le Vesti delle Donne nel Quattrocento* (Florence, 1924).

(4) TYPES

Avery, M., *The Exultet Rolls of South Italy* (Princeton, 1936). *Schubring, P., *Cassoni. Truhen und Truhenbilder der italienischen Frührenaissance* (2 vols.; new ed., Leipzig, 1923). Alazard, J., *Essai sur l'Évolution du Portrait à Florence* (Paris, 1924). *Mostra del Ritratto storico napoletano* (catalogue by G. Doria and F. Bologna; Naples, 1954). Sandberg-Vavalà, E., *La Croce dipinta*

---

[1] In addition to the general works by E. Mâle, L. Bréhier, K. Künstle, R. van Marle, L. Réau.

BIBLIOGRAPHY

italiana (Verona, 1929). Buscaroli, R., *La Pittura di paesaggio* (Bologna, 1935)· Kallab, W., *Toskanische Landschaftsmalerei* ('J.W.', XXI, 1900). Haumann, I., *Das oberitalienische Landschaftsbild des Settecento* (Strasbourg, 1927). Zucchini, *Paesaggi e rovine nella pittura bolognese del Settecento* (Bologna, 1951). *Caricatura italiana dell' Ottocento e del Novecento*; pref. by C. A. Petrucci (Rome, 1954).

FOUNTAINS: Colasanti, A., *Le Fontane d'Italia* (Milan, 1928). Wiles, B. H., *The Fountains of Florentine Sculptors* (Cambridge, 1933).

TOMBS: Ferrari, G., *La Tomba nell'arte italiana* (Milan, n.d.; 1909). *Burger, F., *Geschichte des florentinischen Grabmals bis zur Michelangelo* (Strasbourg, 1904). Schubring, P., *Das italienische Grabmal der Frührenaissance* (Berlin, 1904).

(5) THEORY

Texts. To the treatises of Alberti and Leonardo mentioned above, should be added: Pacioli, L., *De Divina Proportione* (Venice, 1509); Lomazzo, *Trattato dell'arte della Pittura* (Milan, 1584) (English ed. by R. Haydocke, *A Tracte containing the Artes of curious Painting . . .* , Oxford, 1598); *Idea del Tempio della Pittura* (Milan, 1590); Milizia, F., *Dizionario delle belle arti del disegno* (Bassano, 1787).

Studies. Venturi, L., *History of Art Criticism* (New York, 1936). Blunt, A., *Artistic Theory in Italy, 1450–1600* (London, 1940). *Mahon, D., *Studies in Seicento Art and Theory* (London, 1947). Lee, R. W., *Ut pictura poesis* ('A.B.', XXX, 1940). Chastel, A., *Marsile Ficin et l'Art* (Geneva, 1954). Panofsky, E., *Meaning in the Visual Arts*, Ch. 5: *The first page of Giorgio Vasari's Libro* (New York, 1955).

(6) COLLECTIONS

Burckhardt, J., *Die Sammler* (1893) in *Collected Works* (Berlin, XIII, 1930). *Müntz, E., *Les Collections des Médicis au XVe siècle* (1888). Levi, C. A., *Le Collezioni veneziane* (Venice, 1900).

(7) PARTICULAR PROBLEMS

Flanders and Italy. Hoogewerf, G. J., *Vlaamsche Kunst en italiaansche Renaissance* (Antwerp, n.d.: 1934). Baumgart, F., 'M.J.', XIII (1944). Tervarent, G. de, *Instances of Flemish Influence in Italian Art, B.M.*, LXXXV (1944). Van Puyvelde, L., *La Peinture Flamande à Rome* (Brussels, 1950). *I Fiamminghi e l'Italia* (catalogue, Venice and Bruges, 1951). Isarlov, G., *Caravage et le Caravagisme européen*, only Vol. II has appeared (Aix-en-Provence, 1941). *Artists in Seventeenth Century Rome* (catalogue by D. Sutton and D. Mahon, London, 1955).

Germany and Italy. Wölfflin, H., *Italien und das deutsche Formgefühl* (Munich, 1931, Eng. ed., *The Sense of Form in Art*, New York, 1958). Longhi, R., *Arte italiana e arte tedesca* (Florence, 1941).

BIBLIOGRAPHY

The East and Italy. Soulier, G., *Les Influences orientales dans la Peinture toscane* (Paris, 1924). Pouzyna, I. V., *La Chine, l'Italie et les Débuts de la Renaissance* (Paris, 1935).

(8) RESTORATIONS

Zampetti, P., *Antichi dipinti restaurati* (Urbino, 1951). Brandi, C., *L'Institut de restauration de Rome* ('G.B.A.', Jan. 1954). Perogalli, C., *Monumenti e metodi di valorizzazione* (Milan, 1954). Lavagnino, E., *Fifty War-damaged Monuments of Italy* (Rome, 1946).

Biographical Notes

Abbati (Filippo), Naples 1836–Florence 1868. Florentine landscape painter. See p. 386.

Agnolo di Ventura, beginning of 14th c. Tuscan sculptor, author (with Agostino di Giovanni) of the Tarlati monument at Arezzo (1330). See p. 111.

Agostino di Giovanni, beginning of 14th c. Tuscan sculptor, pupil of Giovanni Pisano. See p. 111.

Bibl.: Cohn-Goerke, *Scultori senesi del Trecento,* 'Riv. A.', 1939.

Agricola, Urbino 1795–Rome 1857. Roman painter of the Accademia di S. Luca (Rome). See p. 374.

Albani (Francesco), Bologna 1578–1660. Bolognese painter, trained in the academy of the Carracci, influenced by Reni, active in Rome and Bologna. See pp. 345, 346.

Bibl.: A. Boschetto, *Per la conoscenza di Fr. A.,* 'Pr.' II 1948.

Alberti (Leon Battista), Genoa 1404–Rome 1472. All-round humanist and artist, born in exile, trained in Venice and Padua; employed at the Papal court as an abbreviator. Author of treatises in Italian and Latin on painting and sculpture, and on architecture, in which field he did his most important work, after 1446, in Rimini, Florence and Mantua. See pp. 146, 151, 160, 180, 187, 205.

Bibl.: C. Ricci, *L.B.A.,* Turin 1917. G. Mancini, *Vita di L.B.A.,* Florence 1911. P. H. Michel, *La Pensée de L.B.A.,* Paris 1930. M. L. Gengaro, *L.B.A. teorico ed architetto del Rinascimento,* Milan 1939.

Albertinelli (Mariotto), Florence 1474–1515. Pupil of Cosimo Rosselli, like Fra Bartolommeo, with whom he collaborated. See p. 242.

Bibl.: Bodmer, 'Dedalo' 1930.

Aleandri (Ireneo), Sanseverino 1795–1885. Theatre architect. See p. 380.

Aleotti, see **Argenta.**

Alessi (Galeazzo), Perugia *c.* 1512–72. Architect influenced by Michelangelo. Worked in Genoa and Milan. See p. 294.

Alfieri (Benedetto), Rome 1700–67. Piedmontese architect who succeeded Juvara in Turin and the Marches. See p. 330.

Algardi (Alessandro), Bolgona 1602–Rome 1654. Sculptor, decorator, architect, rival of Bernini, trained in Bologna, active in Venice, Mantua, Rome, Naples. See pp. 317, 324, 354.

Bibl.: H. Posse, *A.A.,* 'J.B.' 1905. O. Pollak, *A.A. als Architekt,* 'Zeitschrift für Geschichte der Architektur' 1910–11.

Allori (Cristofano), Florence 1577–1621. Florentine painter and decorator, pupil of his father Alessandro Allori. See p. 331.

Altichiero Verona, *fl.* 1369–84. Veronese painter, pupil of Turone, influenced to a great extent by the works of Giotto in Padua. See p. 133.

Bibl.: P. Schubring, *A.,* Leipzig 1898. L. Coletti, *Studi sulla pittura del Trecento a Padova,* 'Riv. A.' 1930–31. L. Bronstein, *A.,* Paris 1932.

Amadeo (Giovanni Antonio), Milan 1447–1522. Lombard architect and sculptor, follower of Filarete, worked on Colleoni chapel at Bergamo (1470–1475) and on façade of Certosa di Pavia (1491–98). See pp. 188, 189.

Bibl.: Malaguzzi-Valeri, *G. A. Amadeo*, Bergamo 1904. G. A. dell'Acqua, *Mantegazza ed Amadeo*, 'Pr.', II (1948).

Amato (Giacomo), 1643–1732. Sicilian architect, trained in Rome, active in Palermo. See p. 336.

Amigoni (Jacopo), Naples *c.* 1682–Madrid 1752. Widely travelled Venetian painter, court painter to Ferdinand VI. Many influences can be discerned in his work, notably those of S. Ricci and French decorative genre. See p. 361.
Bibl.: S. Ortolani, *J.A.*, 'Emporium' 1938.

Ammannati (Bartolommeo), Florence 1511–92. Tuscan sculptor and architect, pupil of Bandinelli and typical Mannerist. Worked in Venice, Padua and Rome before returning to the Florence of Cosimo II; 1558–70: court of Palazzo Pitti; 1563–76: Neptune fountain (Piazza della Signoria, Florence); 1578: Jesuits' College, Rome. See pp. 272, 274.
Bibl.: L. Biagi, *Di B.A.*, 'L'Arte' 1923. E. Vodoz, *Studien zum B.A.*, Burg 1942.

Amorosi (Antonio Mercurio), 1660–1736. Roman painter, specialized in unusual light effects. See p. 350.

Andrea del Castagno, see **Castagno.**

Andrea di Cione, see **Orcagna.**

Andrea del Sarto, see **Sarto.**

Andrea d'Ognabene, 13th c. Tuscan goldsmith, author of silver altarpiece of S. Giacomo, Pistoia. See p. 128.

Angelico (Fra Giovanni), Vicchio di Mugello 1387–Rome 1455. Florentine painter, Dominican friar in Fiesole (1407), then prior of monastery of S. Marco. His art is descended from Lorenzo Monaco and the illuminators of Sta Maria degli Angeli. Author of numerous altarpieces. Around 1430 adopted certain principles of the art of Masaccio: frescoes in monastery of S. Marco, and in chapel of Nicholas V in Rome. See pp. 167–70, 177, 184.
Bibl.: F. Schottmüller, *Fra A.*, 2nd ed., Stuttgart 1924. P. Muratoff, *F.A.*, Italian edition Rome 1930. G. Bazin, *F.A.*, Paris 1949. M. Salmi, 'Com.' 1950. J. Pope-Hennessy, *Fra A.*, London 1952.

Angelo (da Orvieto), 14th c. Architect of the Palazzo dei Consoli at Gubbio (1333). See p. 107.

Angelo, 14th c. Sienese wood sculptor, author of Montalcino *Annunciation* (1369–70). See p. 129.
Bibl.: E. Carli, *Scultura lignea*, op. cit., p. 113.

Antelami (Benedetto), *c.* 1150–*c.* 1225. Lombard sculptor and architect, outstanding figure in the history of Italian Romanesque art: *Deposition* (Parma Cathedral, 1178), works in Parma Baptistery (1196) and at Sant'Andrea di Vercelli (*c.* 1223). See pp. 59, 86.
Bibl.: R. Jullian, *La Sculpture romane* . . . , Paris 1945. G. de Francovich, *B.A., Architetto e Scultore* . . . , 2 vols. Milan-Florence 1952.

Antico (Pier Giacomo Ilario **Bonacolsi**, called l'), *c.* 1460–1528. Mantuan medallist. See p. 244.

Antolini (Giovanni Antonio), Castelbolognese 1754–Milan 1842. Architect, provided designs for Foro Bonaparte in Milan. See p. 370.

Antonelli (Alessandro), Novara 1798–Milan 1888. Architect from Novara, author of the cupola of S. Gaudenzio and of the 'Mole Antonelliana' (Turin). See p. 380.

Antonello da Messina, Messina *c.* 1430–79. Painter trained in the circle of Colantonio in Naples, where a number of Flemish influences had penetrated; active in Messina (1456–73) and Venice (1475–76) where he had a decisive influence on Giovanni Bellini; style characterized by static composition, fine quality of paint and unity of tone. See pp. 185, 186, 210.

Bibl.: J. Lauts, *A. da Messina*, Vienna 1940. Cat. of Antonello da Messina exhib. (Messina) by G. Vigni and G. Carandente, Venice 1953.

Antonio Veneziano, 14th c. Florentine painter, worked with Taddeo Gaddi and in the Campo Santo, Pisa. See p. 125.

Antonio di Vincenzo, *c.* 1340–1402. Bolognese architect, various secular works, also directed building operations at S. Petronio. See p. 136.

Bibl.: A. Gatti, *A. di V.*, 'Arch. Stor. Arte' 1894.

Appiani (Andrea), Milan 1754–1817. Lombard painter, neo-classical portraitist friendly with David, worked in Parma, Bologna and Rome. See pp. 374, 375.

Bibl.: E. Somaré, *A.A.*, Milan 1928.

Arcimboldo (Giuseppe), Milan 1530–93. Lombard Mannerist painter specializing in 'capricci' of fruit and flowers, still lifes and portraits; in service of the Hapsburgs *c.* 1560–87. See pp. 295–6.

Bibl.: B. Geiger, *I Dipinti ghiribizzosi di G.A.*, Florence 1953. F. C. Legrand and F. Sluys, *G.A. et les Arcimboldesques*, Paris 1954.

Argenta (Aleotti Battista, called), 1546–1636. Parmesan architect, built theatre of Parma, worked in Ferrara and Parma. See p. 330.

Arnoldi, sculptor, mid-14th c., possibly of Lombard origin, active in Florence 1351–61. See p. 122.

Arnolfo di Cambio, Florence *c.* 1250–1302. Florentine architect and sculptor, in charge of work on Sta Maria dei Fiori; between 1299 and 1302 directed building of Palazzo Vecchio. Has left a number of vigorous statues originally designed for former façade of cathedral. See pp. 83, 85, 92, 102–3, 107–9.

Bibl.: J. Keller, *Der Bildhauer A.*, 'J.B.' 1934–35. V. Mariani, *A.*, Rome 1943.

Arpino, see **Cesari.**

Aspertini (Amico), Bologna *c.* 1475–1522. Painter of the Ferrarese school, passed through Rome and Florence, active in Ferrara *c.* 1505, in Bologna *c.* 1510–15. Worked with L. Costa and Francia in oratory of Sta Cecilia. See pp. 195, 260.

Bibl.: R. Longhi, *Ampliamenti all'Officina Ferrarese*, Suppl. 'C.A.', IV (1940).

Attavante (Gabriello **di Vante,** called), Castelfiorentino 1452–Florence 1517. Florentine miniaturist, connected with Ghirlandaio and Leonardo whose compositions he sometimes adapts in manuscripts executed for Lorenzo de' Medici, Mathias Corvinus and foreigners like the Bishop of Dol. See p. 195.

Bibl.: A. de Hevesy, *La Bibliothèque du roi Mathias Corvin*, Paris 1923.

Avanzo, Verona *c.* 1350–*c.* 1400. Worked with Altichiero (1379). See p. 133.

Avellino (Onofrio), Naples 1674–Rome 1741. Neapolitan painter, pupil of Giordano and Solimena, settled in Rome (1715). See p. 344.

Averlino (Antonio), see **Filarete.**

Avondo (Vittorio), Turin 1836–1910. Piedmontese naturalist and landscape painter. See p. 385.

Baccani (Gaetano), 1792–1867. Tuscan architect, restored nave of Florence Cathedral and erected campanile of Sta Croce. See p. 371.

Baccio d'Agnolo, Florence 1462–1543. Florentine architect, author of façade of Palazzo Bartolini. See pp. 150, 242.

Bacchiacca (Francesco **Ubertini,** called), Florence 1495–1557. Florentine painter, pupil of Perugino, assistant to Franciabigio, came in contact with

Giulio Romano in Rome, author of frescoes in Palazzo Vecchio, Florence. See p. 243.

Baciccia (Giovanni Battista **Gaulli**, called), Genoa 1639–1709. Painter and decorator. Excels in indirect lighting and chiaroscuro effects. Author of *Glorification of the name of Jesus* in Gesù (Rome), 1672–79. See pp. 292, 321.

Baglione (Giovanni), Rome, 1571–1644, painter and historian, antagonist of Caravaggio, author of a collection of lives of Roman artists from 1572 to 1642. See p. 341.

Baldovinetti (Alesso), Florence 1425–99. Painter and mosaicist whose style develops from that of Veneziano and Castagno; between 1466 and 1473 painted frescoes in chapel of Cardinal of Portugal in S. Miniato. Author of Louvre *Madonna* (1450). Preserves without advancing on it in any way the calm, clear and monumental style of Piero della Francesca. See pp. 170, 208.
Bibl.: R. W. Kennedy, *A.B.*, New Haven 1938.

Balduccio (Giovanni di), born in Pisa, mid-14th c. Milanese sculptor, author of the tomb of St. Peter Martyr in S. Eustorgio, Milan (1339) and of S. Augustine in S. Pietro in Ciel d'Oro, Pavia (1370). See pp. 135–6, 137.
Bibl.: G. Biscaro, *G.D.B.* . . . , 'Arch. St. Lomb.' XXXV 1908. C. Baroni, *Scultura gotica lombarda*, Milan 1944. C. Gnudi, 'Belle Arte' I (1946–48).

Balla (Giacomo), 1871. Futurist painter, one of the originators of the movement. Signed manifesto of 1910. See p. 390.

Balzaretti (Giuseppe), 1801–74. Architect of the Milan savings bank. See p. 381.

Bamboccio (Pieter **van Laer**, called), Haarlem *c.* 1595–Rome 1642. Dutch painter who went to Rome in 1628 and introduced painting of genre scenes. See p. 350.

Bandinelli (Baccio), Florence 1493–1560. Florentine sculptor dominated by Michelangelo, pupil of Rustici, author of *Hercules and Cacus*, Piazza della Signoria, and of reliefs for Sta Maria dei Fiori. See pp. 271, 272–3.

Banti (Cristiano), Pisa 1824–Florence 1904. Florentine painter and landscapist, connected with 'Macchiaioli'. See p. 386.

Barabino (Carlo), 1768–1835. Architect, author of staircases of Teatro Carlo Felice and of cemetery of Genoa. See pp. 327, 370, 380, 381.

Barbari (Jacopo de' **Barbari**, called Walch Jacob), 1440–1516. Venetian painter and engraver, author of the large map of Venice (1500). Entered service of Emperor Maximilian. Very few documented works. See pp. 201, 247.
Bibl.: A. de Hevesy, *J. de B.*, Paris-Brussels 1929. L. Servolini, *J. de B.*, Padua 1944.

Barili (Antonio), Siena 1453–1516. Architect and worker in *intarsia*, author of choir of S. Giovanni chapel in Cathedral (now in S. Quirico d'Orcia), 1482–1502. See pp. 196–7.

Barisani (Tommaso), see **Tommaso da Modena.**

Barisano, 13th c. Modelled and cast doors of cathedrals of Trani, Ravello, Monreale. See p. 75.
Bibl.: see **Bonannus.**

Barna, second half 14th c. Sienese painter of tender style, author of frescoes in Collegiata of S. Gimignano. See p. 123.

Barnaba da Modena, active between 1360–80, Emilian painter. See p. 131.

Barocci (Ambrogio), mid-15th c. Lombard architect and sculptor, worked on Castello of Urbino (1475). See p. 181.

Baroccio (Federigo), Urbino 1528–1612. Painter influenced by Correggio,

active in Pesaro, Rome, Milan and Genoa. Worked in pastel. His tender, muted and flexible style became widely known. See pp. 267, 296–7.
Bibl.: A. Schmarsow, *F. B. Zeichnungen*, Leipzig 1909. H. Olsen, *F.B.*, Uppsala 1956.

Baronzio (Giovanni), active *c.* 1350. Painter in tempera and fresco active in Ravenna and Rimini. See p. 131.

Barthel (Melchior), Dresden 1625–72. German sculptor active in Venice; author of the statues on the Pesaro tomb in the Frari. See p. 355.

Bartolo (Domenico di), *c.* 1400–47. Sienese painter influenced by Lippi. See pp. 174, 176, 177.

Bartolini (Lorenzo), 1777–1850. Florentine sculptor. See p. 382.

Bartolo di Fredi, 14th c. Sienese painter, frescoes in Collegiata of S. Gimignano, 1367. See p. 123.

Bartolommeo (Fra), della Porta, Soffignano 1475–Florence 1517. Pupil of Cosimo Rosselli, entered Dominican order under influence of Savonarola, returned to painting in 1507; 1508 in Venice, 1516 in Rome. His style reflects the influence of Leonardo, Raphael and even Michelangelo. See pp. 242, 275.
Bibl.: H. von der Gabelentz, *F.B. und die florentinische Renaissance*, 2 vols., Leipzig 1932.

Bartolommeo della Gatta (Pietro di Antonio Dei), Florence 1448–Arezzo 1502 or 3, painter and illuminator, in contact with the heirs of Piero della Francesca, Melozzo, Signorelli, Perugino. See p. 211.

Bartolommeo Veneto, 16th c. Venetian painter, mentioned in 1502 and 1530, pupil of Giovanni Bellini and in contact with Northern art.

Basaiti (Marco), 1470–after 1530. Pupil and assistant of Alvise Vivarini. See p. 201.

Baschenis (Evaristo), Bergamo 1617–77. Bergamasque painter specializing in still lifes of lutes, violins, meat and game. See pp. 342, 348.
Bibl.: L. Angelini, *E.B., pittore bergamasco*, Bergamo 1943. Cat. of exhib. *Pittori della Realtà in Lombardia*, Milan 1953.

Basile (Ernesto), 1857–1932. Architect from Palermo; author of villas and of the theatres in Agrigento and Marsala. See p. 380.

Basoli (Antonio), 1774–1848. Architect and scenic designer. Worked in Bologna and Rome. See p. 314.

Bassano (Jacopo **da Ponte**, called), Bassano 1510 or 1518–92. Son of Francesco, came to Venice *c.* 1534. Four of his sons became painters. Attracted by Mannerism, underwent the influence of Pordenone and Parmigianino. Painter of rustic subjects and night scenes. See p. 286.
Bibl.: W. Arslan, *J.B.*, Bologna 1931. S. Bettini, *L'Arte di J.B.*, Bologna 1932. *J.B.*, exhibition catalogue (preface: P. Zampetti), Venice, Doge's Palace 1957.

Bastiani (Lazzaro), mentioned 1449 to 1512. Venetian painter.
Bibl.: L. Collobi, *L.B.*, 'C.A.' 1939.

Battagio (Giovanni) of Lodi, late 15th c.–early 16th c. Lombard architect, author of the Incoronata of Lodi (1488–94), of Sta Maria della Croce, near Crema (*c.* 1500), and of Como cathedral (1491–1519). See p. 205.

Battoni (Pompeo), Lucca 1708–Rome 1787. Roman painter, mainly of portraits, obedient to the tradition of Raphael and Correggio. See p. 374.
Bibl.: L. Cochetti, 'Com.' III 1952.

Bazzi (Giovanantonio), see **Sodoma.**

Beccafumi (Domenico **di Pace**, called), Valdibiena 1484–Siena 1551. Sienese painter influenced by Fra Bartolommeo in Florence. Master of tonal gradations and fantastic light effects; *c.* 1519 stayed in Rome where he

assimilated the style of Michelangelo; 1537 in Genoa. See pp. 241, 267, 295.
Bibl.: L. Dami, *D.B.*, 'Boll. Arte' 1919. Gibellino-Krasceninnikowa, *D.B.*, Siena 1933. D. Sanminiatelli, 'B.M.', Feb. 1955.

Bedoli-Mazzola (Girolamo), Parma *c.* 1500–69, Emilian painter, pupil and follower of Parmigianino. See p. 271.

Bella (Stefano della), see **Stefano**.

Bellano (Bartolommeo), Padua 1430–96. Northern sculptor, pupil of Donatello. Active in Padua where he made the Santo reliefs. See pp. 162, 190.
Bibl.: W. Bode, *B.B. da Padova*, 'Arc. Stor. Arte', IV 1891.

Bellini (Jacopo), Venice *c.* 1400–70. Venetian painter best known for his albums of drawings and perspective studies. See pp. 135, 192, 199.
Bibl.: V. Goloubew, *Les Dessins de J.B.*, 2 vols., Paris 1912.

Bellini (Gentile), Venice *c.* 1429–1507. Venetian painter, elder son of Jacopo and brother of Giovanni. Official portraitist of the Doge. Influenced by Mantegna. See p. 199.
Bibl.: E. Cammaerts, *Les B.*, Paris 1912.

Bellini (Giovanni), or **Giambellino**, Venice *c.* 1430–1516. Venetian painter, son of Jacopo, brother of Gentile, brother-in-law of Mantegna, author of numerous altarpieces and Madonnas for which he evolved what became for many years a standard pattern, and in which can be traced the development of the new manner. Originally a strict draughtsman, he was converted by the colour harmonies of Antonello da Messina (1475) and bequeathed to painting a greater unity of tone. See pp. 199–201, 211, 252, 257, 259.
Bibl.: C. Gamba, *G.B.*, French translation Paris n.d. L. Dussler, *G.B.*, Frankfurt 1935. Cat. of exhib. *G.B.* (by R. Pallucchini), Venice 1949. G. Fiocco, *I Disegni di G.B.*, 'Arte Veneta' 1949.

Belliniano (Vittore **Matteo**, called), mentioned *c.* 1507–29. Venetian painter. Follower and assistant of Giovanni Bellini; archaic in style. See p. 201.

Bellotto (Bernardo), Venice 1720–Warsaw 1780. Pupil of Canaletto. Travelled widely abroad, settled in Warsaw. See p. 361.
Bibl.: H. A. Fritzsche, *B.B.*, Burg 1936.

Beltrami (Luca), 1854–1933. Milanese architect. See p. 381.

Bencovich, 1677–1753. Decorative painter of Dalmatian origin, worked in Venice in style of Piazzetta. See p. 358.

Benedetto da Rovezzano, 1474–1522. Florentine sculptor, adopted a sort of turgid classicism in the (unfinished) tomb of S. Giovanni Gualberto (Bargello, 1503–15). See pp. 207, 243.

Benefial (Marco), Rome 1684–1764. Roman neo-classical painter. See p. 374.

Benvenuti (Pietro), Arezzo 1769–Florence 1844. Florentine painter influenced by David, spread doctrines of academism. See p. 374.

Benvenuto, 13th c. Umbrian architect. Palazzo Communale of Perugia. See p. 107.

Benvenuto di Giovanni dal Guasta, Siena 1436–1518, pupil of Vecchietta, influenced by Gozzoli and Northern artists. See p. 176.

Bergognone (Ambrogio **da Fossano**, called), Fossano 1450–1523. Lombard painter. See pp. 189, 241.
Bibl.: F. Malaguzzi-Valeri, *I Seguaci del B.*, 'Ras. A.' 1905. N. Aprà, *Il B.*, Milan 1952.

Bergonzoni (Giovanni Battista), Bologna 1624–92. Bolognese architect. See p. 330.

Berlinghieri (Berlinghiero), first half 13th c. Tuscan painter, had large studio. See p. 74.

Berlinghieri (Bonaventura), first half 13th c. Tuscan painter, author of Pescia *St. Francis* (1235). See p. 88.

Bernardo da Venezia, late 14th c.–early 15th c. Lombard architect and sculptor, worked on Castello of Pavia, Milan cathedral and began Certosa of Pavia (c. 1400). See p. 138.

Bernascone (Giovanni), first half 17th c. Lombard architect. Began sanctuary of Sacro Monte at Varese. See p. 326.

Bernini (Gian Lorenzo), Naples 1598–Rome 1680. Internationally famous Roman architect, sculptor and painter; taught by his father, erected baldacchino in St. Peter's in 1624. 1655–67, during pontificate of Alexander VII, constructed portico of St. Peter's. Author of tombs of Urban VIII and Alexander VII (St. Peter's) and of numerous busts and statues, Bernini should be considered the founder of monumental and decorative Baroque. See pp. 315, 316, 322–5.

Bibl.: M. Reymond, *Le Bernin*, Paris, n.d. H. Brauer and R. Wittkower, *Die Zeichnungen des G.L.B.*, Berlin 1931. L. Grassi, *Disegni del B.*, Bergamo 1944, and *B. pittore*, Rome 1945. F. Baldinucci, *Vita di B.*, Rome 1947. R. Pane, *B. architetto*, Venice 1953. R. Wittkower, *B.*, London 1955.

Bernini (Pietro), Sesto Fiorentino 1562–Rome 1629. Tuscan sculptor, father of Gian Lorenzo, went to Naples, then Rome. See pp. 272, 322.

Bertini (Giuseppe), Milan 1825–98. Lombard painter, author of historical subjects. See p. 384.

Bertoldo (G.), *c.* 1440–Poggio a Caiano 1491. Florentine sculptor, worked with Donatello, author of plaques and Madonnas, curator of Lorenzo de' Medici's collections. See pp. 162, 230, 244.

Bibl.: W. Bode, *B. und Lorenzo dei Medici*, Freiburg 1925.

Besio (G.), 17th c. Genoese architect. Active in Sicily (1612). See p. 336.

Betto di Geri, 14th c. Florentine goldsmith, author of great silver altar in Florence Baptistery (1366). See p. 128.

Bettoli (Niccolò), Modena 1780–1854. Parmesan architect, author of Teatro Reale. See pp. 371, 380.

Bezzuoli (Jacopo), Florence 1784–1855. Florentine Romantic painter. See p. 385.

Bianchi (Pietro), 1787–1849. Neapolitan architect, author of church of S. Francesco di Paola, Naples. See p. 372.

Bianco (Bartolommeo), called **Il Baccio.** Como late 16th c.–Genoa 1651. Architect who came to Genoa with his father Cipriano, author of important Genoese palaces. See p. 326.

Bibiena, see **Galli.**

Biduino, first half of 12th c. Tuscan sculptor born in Bidogno, active in Pisa, Lucca and Pistoia. See p. 73.

Bigarelli (Guido), mid-13th c. Lombard sculptor born in Como. Active in Pisa, Pistoia; in 1250 carved pulpit of S. Bartolommeo in Pantano, Pistoia. See p. 73.

Bigari (Vittorio Maria), Bologna 1692–1776. History painter. See p. 347.

Binaghi (Lorenzo), *c.* 1556–1628. Lombard architect. Built church of S. Alessandro, Milan. See p. 325.

Birolli (Renato), 1906. Painter; one of founders of 'Fronte nuovo delle arti' (1945). See p. 392.

Bibl.: C. Maltese, 'Com.' I (1950).

Boccaccino (Boccaccio), Cremona 1467–1525. Painter from Cremona, pupil of Alvise Vivarini in Venice, decorated nave of Cremona Cathedral (1508–18). See p. 201.

Boccati (Giovanni de'), mentioned between 1445 and 1480. Painter active in Umbria and the Marches. See p. 180.

Boccioni (Umberto), 1882–1916. Painter who inspired Futurist movement (1910); author of a treatise on the new painting and sculpture (1914). See p. 390.
Bibl.: C. Carrà, *U.B.*, Milan 1910. F. Pastonchi, 'Cahiers d'Art' 1950. G. C. Argan, *B.*, Rome 1953.

Boito (Camillo), 1834–1914. Italian architect, historian and apologist of a 'national architecture', author of the *Casa di riposo dei Musicisti*, Milan. See p. 378.

Boldini (Giovanni), Ferrara 1842–Paris 1931. Florentine painter who, after a journey to London in 1869 and visits to Paris (after 1872) became the fashionable painter of fashionable society. See p. 387.
Bibl.: E. Somaré, *Ricordi di G.B.*, Milan 1941.

Bolgi (Andrea), 17th c. Roman sculptor, pupil of Bernini. See p. 323.

Bologna (Giovanni da, or Jean de Boulogne), Douai 1529–1608. Architect and sculptor trained in Antwerp, went to Rome, settled in Florence in 1553 where, thanks to the tenuous proportions and voluptuous attitudes of his figures, he became one of the major exponents of Mannerism. See pp. 273–4.
Bibl.: P. Patrizi, *J.B.*, V., n.d.

Boltraffio (Giovanni), Milan 1467–1516. Lombard painter, pupil and imitator of Leonardo. See p. 241.
Bibl.: W. Suida, *Leonardo und sein Kreis*, Munich 1929.

Bon (Bartolommeo), the Elder, ?–1464. Venetian architect and sculptor, worked on Doge's Palace and Ca' d'Oro. See p. 139.

Bon (Giovanni), ?–1442. Architect and sculptor, whose work is associated with that of his brother Bartolommeo the Elder. See p. 197.

Bonaiuto (Andrea di), called Andrea da Firenze, mentioned 1343–77. Florentine painter, close to Orcagna and Nardo di Cione. See pp. 123, 125.

Bonannus, mentioned between 1174 and 1180. Pisan bronze sculptor and *maestro dell' opera*, active in Pisa and Siena. See pp. 61, 72, 121.
Bibl.: A. Boeckler, *Die Bronzetüren des Bonannus von Pisa u. des Barisanus von Trani*, Berlin 1951.

Bonazza (Giovanni), active between 1695 and 1730. Venetian sculptor, author of reliefs in Cappella del Rosario, SS. Giovanni e Paolo. See p. 355.

Bonfigli (Benedetto), Perugia 1420–96. Painter of school of Perugia, influenced by Benozzo Gozzoli and Domenico di Bartolo. See p. 177.
Bibl.: Scalvanti, in 'Ras. d'Arte' 1902.

Bonifazio de' Pitati, 1487–1553. Veronese painter, pupil of Palma Vecchio, influenced by Giorgione and Titian. Worked largely in Venice. See p. 256.

Bonino da Campione, Lugano, second half of 14th c. Architect, pupil of Giovanni di Balduccio, school of 'Campionesi' masters. See pp. 135–6.

Boninsegna (Gianpaolo), Milan 13th c. Venetian goldsmith, made setting for *Pala d'Oro* (1345). See p. 63.

Bonito (Giovanni), Naples 1707–89. Neapolitan portraitist. See p. 344.
Bibl.: R. Longhi, 'Vita artistica' 1927.

Bonomini (Vincenzo), 1756–1839. Neo-classical painter from Bergamo, author of a cycle of macabre paintings at Sta Grata in Bergamo. See p. 375.
Bibl.: R. Bassi-Rathgeb, *V.B.*, Bergamo 1942.

Bonsignore (F.), 1763–1844. Architect from Turin. See p. 370.

BIOGRAPHICAL NOTES

Bonvicino (Alessandro), see **Moretto.**

Bordone (Paris), Treviso 1500–71. Venetian painter, pupil and imitator of Titian. See p. 257.

Borgianni (Orazio), Rome c. 1578–1616. Caravaggiesque painter. See p. 341.
Bibl.: R. Longhi, 'Vita artistica' 1927.

Borgognone, see **Bergognone.**

Borrani (Odoardo), Pisa 1833–Florence 1903. Florentine painter. See p. 386.
Bibl.: E. Cecchi, *O.B.*, 'Dedalo' VI 1926.

Borro (Luigi), 1826–86. Venetian sculptor. See p. 382.

Borromini (Francesco Catelli), Bissone 1599–Rome 1667. Architect of Lombard origin, worked with Maderna and settled in Rome, where he worked on cloister of S. Filippo Neri, S. Andrea delle Fratte, S. Ivo della Sapienza, Palazzo Falconieri, reconstructed interior of S. Giovanni in Laterano, and built S. Carlo alle Quattro Fontane. One of the most original masters of the Baroque. See pp. 315, 316, 317–20.
Bibl.: M. Guidi, *F.B.*, Rome 1922. E. Hempel, *F.B.*, Italian ed., Rome 1924. H. Sedlmayr, *Die Architektur B.*, Munich, 2nd ed., 1939. G. C. Argan, *B.* (Bibl. Mondadori), 1952.

Bossi (Giuseppe), Busto Arsizio 1777–Milan 1815. Lombard painter, friend of Appiani. Decorated Villa Melzi at Bellagio. See p. 375.
Bibl.: F. Samek, 'Belle Arti' 1947.

Botticelli (Alessandro **di Mariano, dei Filipepi,** called), Florence 1445–1510. Florentine painter, pupil of Filippo Lippi, author of numerous altarpieces in Florence. Painted for the Medici: *Adoration of Magi, Primavera, Birth of Venus.* Dominates Florentine art at the time when the frescoes of the Sistine chapel were being painted in Rome (1481–82). Disturbed by the moral crisis of the end of the century, drew a complete set of illustrations to Dante's *Divine Comedy.* See pp. 208–10, 215.
Bibl.: H. P. Horne, *B.*, London 1908. Y. Yashiro, *S.B.*, 3 vols., London 1925. J. Mesnil, *B.*, Paris 1938. E. H. Gombrich, 'J.W.C.I.' 1947. Y. Batard, *Les Dessins de B. pour la D.C.*, Paris 1953.

Braccesco (Carlo), late 15th c.–early 16th c. Painter of Milanese origin in Genoa between 1478–84; author of the Louvre *Annunciation* (c. 1480). See p. 187.
Bibl.: R. Longhi, *C.B.*, Florence 1942. G. V. Castelnuovi, *Un affresco del B.*, 'Emporium', April 1951. F. Zeri, 'B.M.', March 1955.

Bracci (Pietro), 1700–73. Roman Baroque sculptor, worked in the Lateran. See p. 324.

Bramante (Donato di Angelo), Monte Asdrualdo (Fermignano) 1444–Rome 1514. Painter and architect trained in Urbino; 1477–99 in Lombardy, first Bergamo, then Milan at the court of Ludovico il Moro. Monumental decorations and building at Sta Maria presso S. Satiro, the apse of Sta Maria delle Grazie, and the cathedral of Pavia. From 1500 on in Rome, cloister of Sta Maria della Pace (finished 1504), Tempietto of S. Pietro in Montorio, 1502. After 1505, reconstruction of the Vatican, provided main plan for new St. Peter's on which work was begun, and the court of the Belvedere, to which he gave a sense of volume that was completely new. See pp. 203, 204, 205, 213, 217–21.
Bibl.: M. Reymond, *B.*, Paris 1911. Vasari, *Vita di B.* (ed. Natali), Florence 1914. F. Malaguzzi-Valeri, *D.B.*, Rome 1924. C. Baroni, *B.*, Bergamo 1944. P. Rotondi, *B. pittore*, 'Emporium', March 1951.

BIOGRAPHICAL NOTES

Bramantino (Bartolommeo **Suardi**, called), *c.* 1465-1536. Lombard painter and architect. See p. 241.
Bibl.: W. Suida, *Bramante pittore e il Bramantino*, Milan 1955.
Brea (Louis), Nice 1443-1523. Painter born in Nice, active in Nice and Liguria; in contact with various artistic trends, worked with Foppa. Altarpieces in Nice, Savona, Taggia, Montalto, Genoa. See p. 187.
Bibl.: A. Labande, *Les Bréa*, Nice 1937.
Bregno (Andrea), *c.* 1418-1506. Lombard sculptor active in Rome; numerous tombs in Sta Maria sopra Minerva and Sta Maria del Popolo. See pp. 183, 184.
Bibl.: E. Lavagnino, *A.B. e la sua bottega*, 'L'Arte' 1924.
Brenna (Vincenzo), 1745-1820. Italian architect, travelled to Warsaw. See p. 368.
Briosco (Andrea), see **Riccio**.
Briosco (Benedetto), second half of 15th c. Milanese architect and sculptor, worked with Amadeo on façade of Certosa of Pavia. See p. 188.
Bronzino (Angelo **di Cosimo**), Monticelli 1507–Florence 1572. Florentine painter, pupil of Pontormo; in Florence decorated chapel of Eleanor of Toledo (Palazzo Vecchio) and painted numerous portraits (Uffizi). See pp. 265, 266, 267, 277.
Bibl.: A. MacComb, *A.B.*, Cambridge 1928. H. C. Smyth, *B's earliest works*, 'A.B.', XXXI 1944.
Brunelleschi (F.), Florence 1377–Rome 1446. Goldsmith, sculptor, became architect at age of forty, worked in Florence in service of Cosimo de' Medici. Revolutionized architecture after long stay in Rome by discreet use of antique motifs in buildings of early Christian and Romanesque pattern: cupola of Sta Maria del Fiore (1420), Old Sacristy of S. Lorenzo, S. Spirito, begun in 1444, Sta Maria degli Angeli (unfinished). See pp. 130, 151-5.
Bibl.: C. von Fabriczy, *F.B.*, Stuttgart 1892. H. Folnesics, *B.*, Vienna 1915. M. Reymond, *B. et l'architecture de la Renaissance italienne au XVe siècle*, Paris 1911. L. H. Heydenreich, *B.'s Spätwerke*, 'J.B.' 1931.
Brustolon (Andrea), Belluno 1662–1732. Venetian wood sculptor and cabinet maker, author of chairs in Palazzo Venier. See p. 355.
Bibl.: G. Biasuz and A. Lacchin, *A.B.*, Venice 1928.
Bugiardini (Giuliano), Florence 1475–1554. Florentine painter, pupil of Domenico Ghirlandaio. In Rome at beginning of 16th c.; in 1530 in Bologna. See p. 243.
Bibl.: A. Podestà, 'Emporium', June 1940.
Buonarroti, see **Michelangelo**.
Buonconsiglio (Giovanni), called Il Marescalco, Vicenza 1470–Venice 1535. Painter from Vicenza, pupil of Montagna, influenced by Ferrarese and Lombard painting. See pp. 201, 252.
Buontalenti (Bernardo), Florence *c.* 1536–1608. Florentine architect, enlarged Palazzo Vecchio. Specialized in court fêtes and firework displays. See pp. 271, 272.
Buscheto, 11th c. Architect of Pisa cathedral, 1063. See pp. 70, 72.
Busiri-Vici (Michele), 20th c. Architect of Italian pavilion at New York World's Fair of 1939. See p. 393.
Butinone (Bernardo), *c.* 1450–1507. Milanese painter, author of S. Martino altar at Treviglio, based on Mantegna's S. Zeno altar. See p. 189.
Bibl.: M. Salmi, *B.B.*, 'Dedalo' X 1929.
Cabianca (Vincenzo), Verona 1837–Rome 1902. Florentine painter. See p. 386.

Bibl.: Carlo Montani, *V.C.*, Rome 1925.
Caccini (Giovanni), Florence, *d.* 1613. Florentine sculptor. See pp. 272, 331.
Cacialli (Giuseppe), Florence 1770–1828. Tuscan architect, follower of Paoletti, worked on Palazzo Pitti. See p. 371.
Cagli, 1910. Painter of Roman group. See p. 392.
Cagnola (Luigi), 1762–1833. Milanese architect, built Arco del Trionfo della Pace ('del Sempione'), Milan 1806. See p. 370.
Bibl.: P. Mezzanotte, *Le Architetture di L.C.*, Milan 1930.
Cairo (Francesco del), 1598–1674. Lombard painter skilled in night effects. See p. 348.
Calderini (Guglielmo), 1837–1916. Architect of new Palace of Justice in Rome, 1888–1910. See p. 381.
Calini, 20th c. Architect, finished new Stazione Termini, Rome 1952. See p. 394.
Cambiaso (Luca), Moneglia (Liguria) 1527–Madrid 1585. Genoese painter and decorator, went to Florence and Rome. Pictures influenced by Lombard art and Correggio. Active in Genoa and Spain. See pp. 267, 295.
Bibl.: M. Labò, *L.C. Exhib.*, Milan 1927. B. Suida Manning, *The Nocturns of L.C.*, 'Art Quarterly' 1952–53.
Cametti (Bernardino), first half 18th c. Piedmontese sculptor, worked in Rome. See p. 324.
Cammarano (Michele), Naples 1851–1920. Neapolitan realist painter, specialized in battle scenes of Risorgimento. See p. 388.
Bibl.: M. Biancale, *M.C.*, Milan 1936.
Campagnola (Giulio), Padua 1482–after 1514. Paduan painter and engraver, influenced by Giorgione on whom he leans heavily in his landscape and figure compositions; four frescoes in Scuola del Santo, Padua, are attributed to him. See p. 254.
Bibl.: G. Fiocco, 'L'Arte' XVIII 1915. K. F. Sutes, 'Z.B.K.' LX 1926.
Campi. Cremonese family of painters influenced by Giulio Romano, consisting of Giulio, Cremona *c.* 1512–72, in contact with Venetians, and his brothers Antonio ?–1591, and Vincenzo ?–1591, portraitists and decorative painters, and their cousin Bernardino, 1522–*c.* 1584, more Mannerist in style. See p. 296.
Bibl.: R. Longhi, 'Pinacotheca', 1928–29. A. Fuerari, *La Pinacoteca di Cremona*, Cremona 1951.
Campigli (Massimo), Florence 1895. Florentine painter influenced by the School of Paris. See p. 392.
Bibl.: R. Franchi, *M.C.*, Milan 1944. M. Raynal, *C.*, Paris 1942.
Campione (*maestri campionesi*). Generic name of series of Lombard architects of 13th and 14th c. who worked at Modena, Cremona, Verona and Milan. See: Giovanni da Campione, Giacomo da Campione, Matteo da Campione. See pp. 136, 137, 138.
Camuccini (Vincenzo), Rome 1771–1844. Decorative painter, author of historical compositions and decoration of S. Paolo-fuori-le-Mura (1835). See p. 374.
Canal (Fabio), Venice 1703–67. Venetian painter, collaborator of Tiepolo. See p. 360.
Canaletto (Antonio **Canal**, called), Venice 1697–1768. Painter and engraver, mainly of '*vedute*'. His reputation, which was international, was particularly high in England. See pp. 357, 361.
Bibl.: R. Pallucchini, *C. et Guardi*, Paris 1942. V. Moschini, *C.*, Milan

1954. G. F. Guarnati, *Le Acquaforti di C.*, Venice 1945. D. von Hadeln, *Die Zeichnungen von A. Canal*, Vienna 1930.

Canlassi (Guido), 1601–81. Painter from Romagna. See p. 347.

Canonica (Luigi), 1764–1844. Milanese architect, originally from Ticino, author of Arena of Milan (1807). See p. 370.

Canova (Antonio), Possagno 1757–1822. Venetian sculptor; trained in studios of Morlaiter and Marchiori; in Rome influenced by Mengs and Winckelmann; his art is the purest expression of European Neoclassicism; tombs of Popes Clement XIII and Clement XIV. Napoleon's favourite sculptor. Active in Naples 1820; monument to Charles III. See pp. 369, 375, 376.

Bibl.: L. Coletti, *La fortuna di C.*, 'Boll. inst. Arch. Storia Arte' 1927. E. Bassi, *A.C.*, Bergamo 1943.

Canozzi da Lendinara, see **Lendinara.**

Cantoni (Simone), 1736–1818. Milanese architect, author of Ducal Palace in Genoa and Palazzo Serbelloni in Milan. See p. 370.

Capodiferro, 16th c. Lombard *intarsia* worker, active in Bergamo. See p. 197.

Bibl.: A. Pinetti, 'Bergamum', Sept. 1928.

Capogrossi (Giuseppe), Rome 1900. Roman painter (1933). See p. 392.

Caporali (Bartolommeo), *c.* 1420–1505. Umbrian painter, pupil of Gozzoli, influenced by Piero and Perugino. See p. 177.

Cappella (Francesco Daggiù, called), Venice 1714–Bergamo 1784. Decorative painter, pupil of Piazzetta, active in Bergamo (1749).

Bibl.: R. Pallucchini, *F.D. detto il C.* in 'Rivista di Venezia' 1932.

Caracciolo (Gian Battista), called Battistello, *c.* 1570–1637. Pupil of Caravaggio, founded Neapolitan group; works in Certosa di S. Matteo. See p. 342.

Bibl.: R. Longhi, *B.C.*, 'L'Arte' 1915.

Caravaggio (Michelangelo Merisi, called), Caravaggio 1573–Porto d'Ercole 1610. Of plebeian origin, trained in Milan in studio of Simone Peterzano. A strong personality, he dominates the first years of the 17th c. Active in Rome in 1588, in Naples, Malta and Sicily, where he fled in 1606. Difficult chronology: 1598–99, St. Matthew chapel in S. Luigi dei Francesi (Rome), *St. Matthew and the Angel* for S. Luigi dei Francesi (formerly Berlin); 1605–1606, *David* (Borghese); 1607, *Death of the Virgin* (Louvre). Caravaggio's naturalism had an immense influence. See pp. 296, 339–43.

Bibl.: L. Schudt, *C.*, Vienna 1942. R. Longhi, *Ultimi Studi sul C.*, 'Pr' 1943; *C.*, French ed. Paris 1952. Cat. of exhib. Milan 1951. L. Venturi, *C.*, Rome 1952. D. Mahon, 'B.M.' 1953. W. Friedlander, *C. Studies*, Princeton 1955.

Cariani (Giovanni de' Busi, called), Venice 1485–1547/48. Venetian painter influenced by Giorgione, Palma, Titian and Lotto. See p. 254.

Carlevaris (Luca), Udine 1663–1730. Venetian painter. After a stay in Rome, where he painted ruins, views and tempests, he settled in Venice *c.* 1700. See p. 361.

Bibl.: F. Mauroner, *L.C.*, Venice 1931. M. Pittaluga, *Acquafortisti veneziani del Settecento*, Florence 1953.

Carlone, name of a family of Genoese artists, which includes: Taddeo, Rovigo 1543–Genoa 1613, architect, painter and sculptor, author of the façade of the church of the Madonna, Savona, of numerous fountains, and of doorways in Genoa. Also his sons, Giovan Andrea, Genoa 1590–Milan 1630, painter, decorated the Annunziata di San Sisto; Giovan Battista, Genoa 1592–Turin 1677, painter, decorated the chapel of the ducal palace, worked in S. Siro, then worked for the Duke of Savoy. See p. 327.

BIOGRAPHICAL NOTES

Carnelivari (Matteo), 15th c. Architect from southern Italy. See p. 185.
Bibl.: S. Scardella, *L'Architettura di M.C.*, Palermo 1936.
Carnevali (Giovanni), called Il Piccio, Valtavaglia 1806–73. Romantic painter. See p. 384.
Bibl.: C. Carrà, *G.C.*, Bergamo 1946.
Carpaccio (Vittore), first mentioned 1486–died 1525. Venetian painter, pupil of Gentile Bellini, and like him skilled in large narrative pictures. Cycles of S. Giorgio degli Schiavoni (1502–10), and S. Ursula (1490–96, Accademia, Venice). See pp. 199–200.
Bibl.: Ludwig and Molmenti, *V.C.*, Milan 1906. G. Fiocco, *C.*, Rome 1931. V. Moschini, *C. la légende de sainte Ursule*, Milan 1948. G. Perocco, *C.* (coll. Rizzoli, Milan, 1960).
Carrà (Carlo), Alessandria 1888. Painter connected with Futurist movement in 1910, worked in Ferrara with G. de Chirico (1916–18). After 1925 concentrated on landscape. See pp. 390, 391.
Bibl.: R. Longhi, *C.C.*, 2nd ed. Milan 1945. P. Torriano, *C.C.*, Milan 1942. G. Raimondi, *La Congiuntura metafisica*, 'Pa.' no. 19 July 1951.
Carracci. Bolognese family of painters who founded the successful academy of 'Incamminati' (1585–95) to counteract Mannerism. Profound influence on 17th c. **Annibale**, Bologna 1560–Rome 1609. Painter, pupil of Tibaldi, sweeping decorative manner. Decisive example for new era. Active in Bologna and Rome, where he decorated Palazzo Farnese, 1595. **Lodovico**, 1555–Bologna 1619. Decorative painter from Emilia, founded a school. Active in Bologna, where he decorated Palazzo Fava. **Agostino**, Bologna 1557–Parma 1602. Cousin of Lodovico. Engraver, popularized Venetian compositions. Collaborated with brother Annibale. See pp. 297–8, 307.
Bibl.: A. Venturi, *C. e la loro scuola*, Milan 1895. H. Tietze, *A.C.'s Galerie im Palazzo Farnese*, Vienna 1907. A. Foratti, *L.C. nella teoria e nella pratica*, Città di Castello 1913. R. Peyre, *Les C.*, Paris 1924. R. Wittkower, *The Drawings of the C.* (at Windsor), London 1952.
Carriera (Rosalba), Venice 1675–1757. Painter of miniatures and pastels, pupil of Giuseppe Diamarini. Went to Paris and Vienna, where she became the most fashionable portraitist of the 18th c. See p. 357.
Bibl.: E. V. Hörschelmann, *R.C.*, Leipzig 1908.
Casnedi (Raffaello), Como 1822–Milan 1892. 19th c. Lombard master, painted historical scenes. See p. 384.
Castagno (Andrea **di Bartolo di Bargilla**, called Andrea del), San Martino a Corella 1423–Florence 1457. Went to Florence in 1440. Florentine painter of spectacular compositions and powerful plasticity; active in Florence, Venice. 1446, *Famous Men* series in La Legnaia, Florence. See pp. 168–70.
Bibl.: Mario Salmi, *Uccello, Castagno, Domenico Veneziano*, French translation, Paris 1939. G. M. Richter, *C.*, New York 1943. Cat. of *Mostra dei quattro maestri*, Florence 1954.
Castiglione (Giovanni Benedetto da), called Grechetto, Genoa 1610–Mantua 1665. Lombard painter, influenced by Van Dyck in his paintings and Rembrandt in his etchings. In 1634, paid the first of many visits to Rome; author of Bible scenes and pastorals. See p. 350.
Bibl.: A. Blunt, *The Drawings of G.B.C. and Stefano della Bella* (Windsor), London 1954.
Catena (Vincenzo), late 15th c. Venetian painter, pupil and imitator of Giovanni Bellini.
Bibl.: G. Robertson, *V.C.*, Edinburgh 1954.

BIOGRAPHICAL NOTES

Cattaneo (Danese), 1509–72. Florentine sculptor, settled in Venice. See p. 245.

Cavalli (Emmanuele), Lucera 1904. Painter of Roman group, settled in Florence. See p. 392.

Cavalli (Vitale), see **Vitale da Bologna.**

Cavallini (Pietro), c. 1250– c. 1330. Roman mosaicist, author of cycle of Sta Maria in Trastevere. Went to Naples at beginning of 14th c. See pp. 82, 92, 96, 112, 116.

Bibl.: Busuioceanu, *P.C. e la Pittura romana*, 'Ephemeris Daco-Romana' 1925.

Cavedoni (Giacomono), Sassuolo 1577–c. 1650. Bolognese painter and sculptor. See p. 331.

Cecioni (Adriano), Florence 1838–86. Florentine painter and sculptor, critic and leading spirit of Florentine 'Macchiaioli'. See pp. 382, 386.

Bibl.: E. Cecchi, *Nota su A.C.*, 'Vita artistica', 1927; *Opere e Scritti*, ed. E. Somaré, Milan 1932.

Celentano (Bernardo), 1835–60. Neapolitan history painter. See p. 388.

Celesti (Andrea), Venice 1637–1706. Venetian decorator. See p. 356.

Cellini (Benvenuto), Florence 1500–71. Florentine sculptor, goldsmith and medallist whose adventurous life is recounted in his famous *Memoirs*. 1523–30 Rome, 1540–45 at the French court, finally Florence and court of Cosimo I de' Medici. Author of Loggia dei Lanzi *Perseus*, bust of Cosimo I de' Medici, and Francis I's salt-cellar. See pp. 265, 273.

Bibl.: *Trattati dell'oreficeria e della scultura*, ed. G. Milanesi, Florence 1857. E. Plon, *B.C.*, Paris 1882–84. E. Molinier, *B.C.*, Paris 1894. H. Focillon, *B.C.*, Paris 1910. E. Camesasca, *Tutta l'opera del C.*, Milan 1955.

Cennino Cennini, Florence, late 14th c. Florentine, pupil of the Gaddi, author of treatise on the art of painting (c. 1390). See pp. 125–6.

Bibl.: *Trattato*, ed. G. Milanesi, Florence 1859; French translation by V. Mottez, Paris n.d.; English translation, New Haven 1932.

Cerquozzi (Michelangelo), 1602–60. Painter of battles and genre scenes, emulated manner of Salvator Rosa and Bamboccio. See p. 350.

Bibl.: G. Briganti, *Peter van Laer e C.*, 'Pr.' III 1950.

Ceruti (Giacomo), called Il Pitocchetto, first half 18th c. Lombard painter specializing in genre scenes, poor peasants, beggars. See p. 348.

Bibl.: Cat. of exhib. *Pittori Lombardi della Realtà*, Milan 1953.

Cesare da Sesto, 1477–1523. Lombard painter, pupil and imitator of Leonardo. See p. 241.

Cesari (Giuseppe), called Cavaliere d'Arpino, Rome 1568–1640. Roman painter who prolonged Mannerism. See p. 298.

Chiarini (Michelangelo), late 17th c.–early 18th c. Bolognese architect and scenic designer, collaborated with G. M. Crespi on many church decorations. See p. 330.

Chirico (G. di), born in 1888 at Volo (Greece). After visits to Munich and Paris (1910) composed imaginary townscapes and devoted himself to 'metaphysical painting', especially at Ferrara 1916–18. After 1920 reacted in favour of a classicizing style and produced polemics against the avant-garde. See p. 390.

Bibl.: W. Georges, *G. de C.*, Paris 1928. R. Carrieri, *G.C.*, Milan 1942. J. T. Soby, *G. di C.*, New York 1955.

Cifrondi (Antonico) (de Chesone), Chesone 1657–Brescia 1730. Brescian painter, author of portraits of artisans. See p. 348.

Bibl.: Cat. of exhib. *Pittori Lombardi della Realtà*, Milan 1953.

BIOGRAPHICAL NOTES

Cignani (Carlo), Bologna 1628–Forlì 1719. Bolognese painter, imitator of Albani. See p. 347.

Cigoli (Ludovico **Cardi**, called), Castelvecchio 1559–Rome 1613. Florentine painter and architect. See p. 277.
Bibl.: G. Battelli, *L.C. detto il C.*, Florence 1922.

Cima (Giambattista) (da Conegliano), Conegliano 1459–Venice 1517. Venetian painter, pupil of Montagna, influenced by Bellini. See p. 201.
Bibl.: R. Burckhardt, *C. da C.*, Leipzig 1905. L. Coletti, *Cima da Conegliano* Venice 1959.

Cimabue, Florence *c.* 1240–Pisa 1302. Florentine painter, trained in workshop of Florence Baptistery, active in Florence: *Scenes from Life of Joseph* (Baptistery); Assisi, Rome and Pisa, quoted by Dante as being most famous painter before Giotto. See pp. 81, 83, 89, 92, 96, 116, 118, 130.
Bibl.: R. Offner, *C.*, 'The Arts' 1934. A. Nicholson, *C.*, Princeton 1934. R. Salvini, *C.*, Rome 1946. *Postille a C.*, 'Riv. A.' 1950.

Cino (Giuseppe), 17th c. Architect, worked in Lecce. See p. 335.

Cione (Andrea di), see **Verrocchio.**

Cione (Benci di), second half 14th c. Florentine architect, connected with work on Florence Cathedral, directed work on Or San Michele. See pp. 120, 121.

Cipolla (Arturo), Naples 1823–Rome 1874. Architect of Banco d'Italia (Bologna) and Banco d'Italia (Florence). See p. 381.

Ciseri (Antonio), 1821–91. Tuscan painter of large pictures. See p. 385.

Ciuffagni (Bernardo), 1381–1457. Florentine sculptor, worked with Donatello; active in Florence and Rimini. See pp. 129, 148.

Civitali (Matteo), Lucca 1436–1501. Sculptor, architect, author of statues in Cappella del Volto Santo, Lucca (1480). See p. 187.
Bibl.: C. Yriarte, *M.C.*, Paris 1886.

Clovio (Giulio, **Jurai Glovicich**, called), Grizane (Croatia) 1498–Rome 1578. Miniaturist. See p. 266.
Bibl.: Mgr. Fourier-Bonnard, *Don G. C. miniaturiste*, Paris 1929.

Codazzi (Viviano), Bergamo 1603–72, *veduta* painter active in Rome and Naples. See p. 351.
Bibl.: R. Longhi, 'Par.' Nov. 1955.

Coducci (Mauro), Bergamo 1440–1504. Bergamasque architect, active in Venice; façade of Scuola di S. Marco (1485–95). See p. 198.

Cola di Matteuccio da Caprarola, active 1494–1518. Lombard architect, author of church at Todi (1508). See p. 240.

Colantonio, first half 15th c. Neapolitan painter, master of Antonello da Messina. See p. 186
Bibl.: C. Aru, *C. ovvero il maestro dell'Annunciazione di Aix*, 'Dedalo', 1931. G. Fiocco, *C. e Antonello*, 'Emporium' 1950. C. Grigioni, 'Arti figurative', III 1947.

Colonna (Angelo Michele), Varenna 1600–87. Decorative painter, pupil of Girolamo Curti, excelled in immense illusionist decorations. See pp. 320, 347.
Bibl.: S. de Vito Battaglia, *Note sur A.M.C.*, 'L'Arte' 1928.

Conca (Sebastiano), Gaeta 1680–Rome 1764. Neapolitan painter, pupil of Solimena, settled in Rome. See p. 344.

Coppo di Marcovaldo, Florence *c.* 1225–74. Florentine painter, author of *Madonna dei Servi*, Orvieto. See pp. 65, 66, 83.
Bibl.: P. Bacci, 'Arte' 1900. G. Coor-Achenbach, 'A.B.', XXVIII 1946.

Corazzi (Antonio), 1792–1877. Italian architect, went to Warsaw. See p. 368.

BIOGRAPHICAL NOTES

Corradi (Pier Antonio), 17th c. Genoese architect. See p. 326.

Correggio (Antonio **Allegri**, called), Correggio 1489–Parma 1534. Pupil of Mantegna in Mantua; in Parma 1520–30 gave new dimension of delicate illusionism to large-scale decorative painting. His religious compositions have matchless luminosity and grace. See pp. 260–2, 264, 270, 271.
Bibl.: S. de Vito Battaglia, *C.*, *Bibliografia*, Rome 1934. C. Ricci, *C.*, Rome 1930. P. Bianconi, *C.*, Milan 1953. E. Bodmer, *C. e gli Emiliani*, Novara 1943.

Cortona (Pietro **Berrettini**, called Pietro da), Cortona 1596–Rome 1669. Florentine architect and decorative painter, pupil of Baccio Ciarfi, whose reputation he made. Active in Rome and Florence. One of the three masters of Roman Baroque, both as architect (SS. Luca e Martino, 1635) and painter (Palazzo Barberini). See pp. 315, 320, 331.
Bibl.: H. Geisenheimer, *P. da C. e gli affreschi nel palazzo Pitti*, Florence 1909. A. Muñoz, *P. da C.*, Rome 1921.

Cosimo (Piero di), see **Piero di Cosimo**.

Cosmati. Generic name of Roman marble sculptors and decorators of 12th, 13th and 14th c. One Cosma, son of Jacopo di Lorenzo, worked on cloister of Sta Scholastica at Subiaco. See pp. 68, 84, 103, 108.
Bibl.: E. Hutton, *The Cosmati*, London 1950.

Cossa (Francesco del), Ferrara *c.* 1436–78. One of the three masters of the Ferrarese school; worked on decoration of Palazzo Schifanoia, author of Dresden *Annunciation*. See pp. 192, 193, 194.

Costa (Lorenzo), Ferrara 1460–Mantua 1535. Author of paintings in Studiolo of Isabella d'Este (Mantua), and master of the soft manner. See pp. 192, 195, 260, 261.

Costa (Gian Francesco), Venice 1711–73. Venetian engraver, author of *Delizie del Fiume Brenta*. See p. 362.

Costa (Nino), Rome 1826–93. Roman painter, founder of '*in arte Libertas*' movement. See p. 384.
Bibl: G. Cellini, *G.C.*, Bergamo 1933.

Courtois (Giacomo), called Cortese, 1621–96. French painter in manner of Salvator Rosa, specialized in battle scenes. See p. 350.

Cozzarelli (Giacomo), Siena 1455–1515. Architect and wood sculptor, built church of Sta Maria Maddalena at Porta Tufi (now destroyed) 1484. See p. 175.

Credi (Lorenzo di), Florence 1459–1537. Florentine painter, Verrocchio's most important pupil. See pp. 209, 215.

Cremona (Tranquillo), Pavia 1837–Milan 1878. Lombard intimist painter. See p. 384.
Bibl.: L. Beltrami, *Un maestro: T.C.*, Milan 1929. L. Naldini, 'Com.' V 1954.

Crespi (Giovanni Battista), called Cerano. Cerano 1576–Milan 1633. Brother of Daniele, studied in Rome and Venice; settled in Milan, court painter and director of works. See p. 347.

Crespi (Daniele da **Busto Arsizio**, called), 1592–1630. Lombard painter. See p. 348.

Crespi (Giuseppe Maria), called Lo Spagnolo, 1665–1747. Bolognese painter, portraits, decorative subjects and still lifes. See pp. 347, 348, 357.
Bibl.: H. Voss, *G.M.C.*, Rome 1921. Cat. of exhib. Bologna 1948, by F. Arcangeli and C. Gnudi.

Cresti (Domenico), called Passignano, 1558–1637. Florentine decorative painter. See p. 277.

BIOGRAPHICAL NOTES

Crivelli (Carlo), Venice c. 1430–c. 1495. Venetian painter. After trouble with the authorities in 1457 took refuge in the Marches, where he painted numerous altarpieces, rich and highly finished in style: *Virgin enthroned* (1482, Brera). See pp. 191, 193, 199, 210.
Bibl.: F. Drey, *C.C. und seine Schule*, Munich 1927. Cat. of exhib. *La Pittura nelle Marche*, Ancona 1950.

Crivelli (Taddeo), 15th c. Miniaturist of Lombard origin, settled in Ferrara, *Bible of Borso d'Este* (1455–61). See p. 195.

Crivelli (Vittore), brother of Carlo C., worked in the Marches, c. 1481–1501. See p. 199.

Crivellone (Angelo Mario **Crivelli**, called), ?–c. 1730. Lombard painter. See p. 348.

Cronaca (Simone **del Pollaiolo**, called), Florence 1457–1508. Architect of part of Palazzo Strozzi. See pp. 203, 204.
Bibl.: G. Marchini, *Il C.*, 'Riv. A.' 1941.

Curti (Girolamo), called Il Dentone, 1575–1632. Lombard decorative painter who specialized in illusionist architectural scenes, active in Modena. See pp. 320, 347.

Daddi (Bernardo), Florence. Giottesque painter, active between 1312 and 1345. Author of a signed triptych (Uffizi, 1328), and of frescoes in the Beraldi chapel in Sta Croce. See p. 115.

Dalbono (Edoardo), Naples 1841–1915. Neapolitan painter, mainly of landscapes. See p. 388.

Dalmasio (Lippo), second half 14th c. Bolognese painter, numerous frescoes, worked in Tuscany. See p. 131.
Bibl.: R. Longhi, *Mostra del Trecento bolognese*, 'Pa.', no. 5, May 1950.

D'Andrade (Alfredo), 19th c. Architect, designed *borgo medioevale*, Turin. See p. 379.

Danti (Vincenzo), Perugia 1530–76. Tuscan sculptor, author of *Beheading of St. John the Baptist*, south door of Florence Baptistery. See pp. 245, 273.

De Carolis (Adolfo), Piceno 1874–1928. Emilian decorative painter. See p. 389.

De Fabris (Emilio), 19th c. Florentine architect, built façade of Sta Maria del Fiore (1875–87). See p. 379.

De Ferrari (Gregorio), 1644–1726. Genoese decorative painter, son-in-law of Piola, author of ceiling paintings in Palazzo Rosso and of stucco and painted decoration in church of Assumption, Genoa. See p. 326.

Del Debbio, 20th c. Roman architect, author of Foro Mussolini (now Foro Italico) (1933). See p. 394.

Del Rosso (Zanobi), 1724–98. Florentine architect, finished S. Firenze (1775). See p. 331.

De Mura (Francesco), 1696–1752. Neapolitan decorative painter, influenced by Solimena. See p. 334.

De Nittis (Giuseppe), Barletta 1846–Paris 1884. Neapolitan painter who had successful career as fashionable artist in Paris. See p. 388.
Bibl.: L. Bénédite, *D.N.*, Paris 1926. E. Piceni, *G. de N.*, Milan 1934.

De Pisis (Filippo), Ferrara 1896–Milan 1956. Ferrarese painter who spent many years in Paris, settled in Venice; lively impressionist approach. See p. 392.
Bibl.: G. Raimondi, *D.P.*, Florence 1952.

Desiderio da Settignano, Settignano 1428–64. Florentine sculptor famous for *stiacciato* reliefs. Made the *Marsuppini tomb* (1455). See pp. 164, 165, 171, 184, 206, 212.

BIOGRAPHICAL NOTES

Bibl.: L. Planiscig, *D. da S.*, Vienna 1942.

Desiderio (Monsù), see **Monsu Desiderio.**

Diamante (Fra), mid-15th c. Assistant of Fra Filippo Lippi at Prato. See p. 168.

Diziani (Gasparo), Belluno 1689–Venice 1767. Pupil of Sebastiano Ricci. Bibl.: G. B. Baldissone, *G.D.*, 'Arch. Stor. di Belluno' 1941.

Domenico di Bartolo, see **Bartolo.**

Domenico di Niccolo, called de' Cori, Siena 1363–1453. Sienese sculptor and decorator, master of choir of Sta Caterina, Siena, author of *intarsia* decoration in chapel of Palazzo Pubblico: 1425–28. See pp. 129, 174, 196.

Bibl.: E. Carli, *Scultura lignea senese,* op. cit.

Domenico Veneziano, *c.* 1400–61. Painter of Venetian origin; called to Florence *c.* 1438, worked on choir of S. Egidio (1439–45, disappeared) with Piero della Francesca and Baldovinetti. See pp. 167, 170, 171, 177, 178.

Bibl.: G. Pudelko, *Studien über D.V.*, 'Mitteil. K. H. Inst. Florenz.' 1934. J. Pope-Hennessy, *The Early Style of D.V.*, 'B.M.', July 1951. *Cat. Mostra Quattro Maestri,* Florence 1954. R. Longhi, *Il maestro di Pratovecchio,* 'Pa', no. 35, Nov. 1952.

Domenichino (Domenico **Zampieri,** called), Bologna 1581–Naples 1641. Decorative painter born in Bologna, assistant to Annibale Carracci on Palazzo Farnese, represents Romano-Bolognese academism in his *Hunt of Diana* (1621) and his frescoes in choir of S. Andrea della Valle (1624). See p. 345.

Bibl.: L. Serra, *D.Z.*, Rome 1909.

Donatello, Florence 1386–1466. Florentine sculptor and outstanding personality whose influence extended to all forms of art; his career in Florence began with work for Or San Michele (*St. George,* 1416) and the cathedral (*Prophets,* for façade, *c.* 1410; *Prophets,* for campanile, 1433–39; *cantoria* 1433–39), was interrupted by journey to Rome (1432–33) and long stay in Padua (1443–53) where he executed statue of Gattamelata and altar of Santo. See pp. 130, 148, 160–4, 171, 190, 206, 207.

Bibl.: M. Reymond, *D.*, Florence 1917. H. Kauffmann, *D.*, Berlin 1935. R. Band, *D.'s Altar in Padova,* 'Mitt. K. H. Inst. Florenz' 1940. L. Planiscig, *D.*, Florence 1947. O. Morisani, *Studi su D.*, Venice 1953. H. W. Janson and J. Lanyi, *The Sculpture of D.*, Princeton 1957.

Donducci (Giovanni Andrea), see **Il Mastelletta.**

Dono (Paolo di), see **Uccello.**

D'Orsi (Achille), 1845–1929. Sculptor of genre scenes. See p. 382.

Dosio (Giovanni Antonio), Florence 1553–Rome 1609. Florentine architect. See pp. 218, 331.

Bibl.: C. Hülsen, *Das Skizzenbuch des G.D.*, Berlin 1933.

Dossi (Giovanni **Luteri,** called Dosso), Ferrara *c.* 1480–1542. Ferrarese painter, pupil of L. Costa, lived in Rome, then Venice where he knew Giorgione and Titian. Friendly with Ariosto, composed for court of Ferrara a number of mythological scenes and tapestry cartoons. Famous for his landscapes. See pp. 260–1, 266.

Bibl.: R. Longhi, 'Vita Artistica' 1927.

Dotti (Carlo Francesco), *c.* 1670–1759. Bolognese architect and sculptor. See pp. 330–1.

Duccio (Agostino di), 1418–81. Florentine architect and sculptor, Alberti's assistant in Rimini, author of marble *Madonna* for Florence Cathedral (Bargello) and of façade of S. Bernardino, Perugia. Specialized in low-relief plaques. See pp. 160, 164.

440

BIOGRAPHICAL NOTES

Bibl.: J. Pope-Hennessy, *The Virgin and Child*, London 1952. G. Brunetti, 'Belle Arti' 1952.

Duccio di Boninsegna, Siena *c.* 1250–1318. Vitally important Sienese painter. His style, which presents affinities with Byzantine art and Cimabue (in the *Rucellai Madonna*), evolves towards greater delicacy and concentration, which explains the fame of his masterpiece, the *Maestà* altarpiece. See pp. 89, 116–18.
Bibl.: C. H. Weigelt, *D. di B.*, Leipzig 1911. C. Brandi, *D.*, Florence 1951. E. Carli, *Vetrata Duccesca*, Florence 1946.

Dupré (Giovanni), 1827–82. Sienese sculptor. See p. 382.

Elia (Brother), first half 13th c. Franciscan monk, architect of basilica of Assisi. See p. 88.

Ercole dei Roberti, see **Roberti**.

Falca, see **Longhi** (Pietro).

Falcone (Aniello), Naples 1600–in France 1665. Neapolitan painter, famous for his battle subjects. See p. 343.

Falcone (Giovanni Angelo), Genoa ?–1657. Genoese architect, author of Palazzo Durazzo (Palazzo Reale). See p. 326.

Falconetto (Giovanni Maria), Verona 1468–1535. Venetian architect trained in Verona who introduced art of the High Renaissance into the Veneto. See p. 251.
Bibl.: G. Fiocco, *Le architetture di G.M.F.*, 'Dedalo' 1931.

Falda (Giambattista), Valduggia *c.* 1640–Rome 1678. Lombard engraver, author of albums devoted to Roman gardens and fountains. See p. 352.

Fanzago (Cosimo), 1593–1678. Bergamasque sculptor and decorative painter. Active in Naples, particularly in Certosa di S. Martino *c.* 1660, re-modelled church of Monte Cassino. See pp. 332, 333.
Bibl.: Fogaccia, *C.F.*, Bergamo 1945.

Faruffini (Federico), 1831–Perugia 1869. Milanese history painter. See p. 384.

Fasolo (Giovanni Antonio), Vicenza *c.* 1530–72. Painter and decorator from Vicenza, worked with Palladio. See p. 283.

Fattoretto (Giambattista), 18th c. Venetian architect, author of façade of church of Gesuiti and S. Stae, Venice. See p. 354.

Fattori (Giovanni), Leghorn 1825–Florence 1908. Painter, member of the avant-garde of 1868, author of large military scenes and smaller, more intimate works. See p. 386.
Bibl.: A. Soffici, *G.F.*, Rome 1921. E. Cecchi, *G.F.*, Rome 1933. R. Baldaccini, *F.*, Milan 1949. Francini Ciaranfi, *Incisioni di F.*, Bergamo 1944.

Favretto (Giacomo), Venice 1849–87. Venetian painter, visited Paris in 1879, revived genre painting by introducing new themes. See p. 384.
Bibl.: E. Somaré, *F.*, Milan 1943.

Federighi (Antonio), died 1490. Sienese sculptor, *maestro dell'opera* of cathedrals of Siena and Orvieto (1451–56). See p. 174.

Ferrari (Gaudenzio), Valdugia *c.* 1470–1546. Extremely productive Lombard sculptor and painter whose style derives from Raphael and Bramantino and is influenced by German art; frescoes and altarpieces (Vercelli, Saronno), important series of frescoes and statues at Sacro Monte di Varallo. *G.F.*, exhibition catalogue (preface: A. M. Brizio), 1956. See p. 241.

Ferrari (Gregorio), Porto-Maurizio 1644–Genoa 1720. Painter, decorator of Genoese churches and palaces: *Assumption* in the church of SS. Giacomo e Filippo, *Sala dell'Estate* in the Palazzo Rosso in collaboration with Haffner. See p. 326.

Ferrata (Ercole), 1610–86. Roman sculptor of academic style, author of statues and reliefs for Sant'Agnese. See p. 324.

Ferri (Ciro), Rome 1634–89. History painter, pupil of Pietro da Cortona, with whom he collaborated in Rome and Florence. See p. 320.

Ferrucci (Andrea), Florence 1465–1526. Florentine sculptor, influenced by Sansovino. See pp. 207, 243.

Ferrucci (Francesco di **Simone**), 1437–93. Florentine sculptor, active in Rimini and Urbino, author of Tartagni tomb in S. Domenico, Bologna (1477). See pp. 180, 192.

Feti (Domenico), Rome 1589–Venice 1624. Pupil of Cigoli, made painter to the court of Mantua. Active in Venice from 1621 on. See p. 355.
Bibl.: R. Oldenbourg, *D.F.*, Rome 1921. J. Wilde, *Zum Werke D.F.*, 'J.W.' 1936.

Fiammingo (Paolo **dei Franceschi**, called), Antwerp *c.* 1540–Venice 1596. Venetian decorative painter, pupil of Tintoretto, specialized in landscape.

Filarete (Antonio **Averlino**, called), Florence 1400–Milan 1469. Florentine architect and sculptor. Worked in Milan, on Castello (1451–54) and Ospedale Maggiore (1456–65). Decisive influence on Lombard architecture, author of a treatise on architecture (ed. by W. von Oettingen, Vienna 1890). See pp. 128, 146, 157, 184, 188.
Bibl.: M. Lazzaroni and E. Muñoz, *F. Scultore e Architetto del secolo XV*, Rome 1908. M. Salmi, *A.F. e l'architettura lombarda del primo Rinascimento*, 'Atti del 1° Cong. Naz. Storia dell'archit.' (1936), Florence 1938.

Finiguerra (Maso), Florence 1426–64. Florentine goldsmith, one of the first to practise engraving on metal. See p. 171.
Bibl.: S. Colvin, *A Florentine picture chronicle* . . . , London 1898.

Fiorenzo di Lorenzo, Florence *c.* 1440-1522. Florentine painter whose style derives from Benozzo Gozzoli and presents affinities with the Umbrian school. See p. 177.

Fiori (Giuseppe di), mid-18th c. Neapolitan sculptor. See p. 333.

Foggini (Giovan Battista), Florence 1653–1737. Florentine sculptor. Knew Bernini and Algardi. Active in Florence: monument to Galileo in Sta Croce (1737). See p. 331.
Bibl.: 'B.A.' V (1911).

Fontana (Domenico), Lombardy 1543–1607. Roman architect, worked for Pope Sixtus V in Rome, active in Naples from 1594 on. Author of large plan of Rome (1585–90). See pp. 293–4, 314, 332.

Fontana (Carlo), 1638–1714. Roman architect. Active in Rome, worked for Chigi family Designed Quattro Fontane garden and, in 1683, S. Marcello al Corso, with its curved façade. See pp. 317, 336.
Bibl.: Coudenhove-Erthal, *C.F.*, Vienna 1930.

Fontanesi (Antonio), Reggio Emilia 1818–Turin 1882. Landscape painter, friendly with Corot, influenced by Ravier. Exiled to Lugano in 1860. See pp. 383, 385.
Bibl.: C. Carrà, *A.F.*, Rome 1924. M. Bernardi, *Cinque Opere di A.F.*, Turin 1947.

Foppa (Vincenzo), Brescia 1427–1515. Lombard painter whose field of activity reached as far as Genoa and dominated the second half of the 15th c. in Lombardy: Portinari chapel in S. Eustorgio, Milan (before 1468). See pp. 188, 241.
Bibl.: C. I. Foulkes and R. Maiocchi, *V.F.*, London 1909. F. Wittgens, *F.*, Milan n.d. (1950).

BIOGRAPHICAL NOTES

Fracassini (Cesare), 1838–68. Roman decorative painter: basilica of S. Lorenzo. See p. 384.

Francavilla (Pietro), mid-16th c. Sculptor from Cambrai, collaborator of Giambologna, worked mostly in Genoa. See p. 272.

Francesca (Piero della), Borgo San Sepolcro c. 1410–92. Tuscan painter, pupil of Domenico Veneziano in Florence to which he did not return after 1440; active in princely courts of Ferrara (frescoes which no longer exist) and Urbino, and in Apennine towns such as Arezzo: *Story of the True Cross* in S. Francesco (1452–59). His style represents the most enlightened synthesis of conceptions of space and colour of the mid-15th c.; he left a treatise on perspective and a study of regular bodies which was put to use by his pupil Luca Pacioli. See pp. 146, 177–80, 184, 187, 189, 193, 195, 196, 211.

Bibl.: Edition of *De prospectiva pingendi*, by G. Nicco Fasola, Florence 1943. *De Corporibus regularibus*, by G. Mancini, 'Atti Acc. Lincei', Rome 1915. M. Salmi, *P. d. F. e il Palazzo di Urbino*, Florence 1945. R. Longhi, *P. d. F.*, 1st edition 1925, 2nd ed. Milan 1946; English edition London 1927. Kenneth Clark, *P. d. F.*, London 1952. H. Focillon (posthumous), *P. d. F.*, Paris 1952.

Francesco d'Antonio, 15th c. Florentine miniaturist, author of Lorenzo de' Medici's Book of Hours (1485). See p. 195.

Francesco di Giorgio Martini, see **Giorgio**.

Francesco di Valdambrino, Siena 1363–1435. Tuscan sculptor, took part in competition for second Baptistery door, Florence. See pp. 130, 174.

Francia (Francesco **Raibolini**, called), Bologna c. 1450–1517. Emilian painter, originally a goldsmith, assimilated Umbrian manner which he spread in Northern Italy: oratory of Sta Cecilia, Bologna (1506). See pp. 192, 195, 211, 260.

Franciabigio (Francesco **di Cristoforo**, called), Florence 1482–1525. Florentine painter, collaborated with Andrea del Sarto in Scalzi cloister. See p. 243.

Franco Bolognese, 13th c. Bolognese miniaturist mentioned by Dante. See p. 87.

Frigimelica (Girolamo), c. 1653–1732. Venetian architect, author of design for Villa Pisani at Stra. See p. 354.

Fuga (Ferdinando), 1719–80. Architect trained in Florence. Active in Rome, Florence and Palermo. Author of Palazzo della Consulta (1732) and of façade of Sta Maria Maggiore (c. 1750). Palazzo Corsini (1736). See pp. 325, 332, 337.

Fumiani (Giovanni Antonio), Venice 1643–1710. Venetian decorative painter, author of immense illusionist paintings at S. Pantaleone. See p. 357.

Furini (Francesco), Florence c. 1600–69. Florentine decorative painter influenced by the Carracci; author of frescoes in Palazzo Pitti. See pp. 278, 331.

Bibl.: L. V. Burckel, *F.F.*, 'J.W.' XXVIII 1908. A. Stanghellini, *F.F. Pittore*, 'Vita d'Arte' 1913. E. Berti Toesca, *F.F.*, Rome 1946.

Gaddi (Agnolo), ?–1396. Florentine painter, son of Taddeo, decorated Prato cathedral. See pp. 125, 128.

Bibl.: F. Salvini, *A.G.*, Florence 1935.

Gaddi (Taddeo), Florence c. 1310–c. 1366. Florentine painter and worker in mosaic and marble. Baroncelli chapel in Sta Croce. See pp. 115, 121, 125.

Gagini (Domenico), Palermo 1492. Genoese architect, active in Genoa (1456) and Palermo (1463). See pp. 185, 187.

Galilei (Alessandro), Florence 1691–Rome 1736. Florentine architect, mildly Baroque in style; façade of S. Giovanni in Laterano. See p. 325.
Bibl.: I. Toesca, 'English Miscellany', Rome II 1951.

Galli, called Bibiena, family of architects, scenic designers, painters and engravers from Bologna. **Giovanni Maria** (1625–65), pupil of Albani, took name of his birthplace (Bibiena), and had two sons, Ferdinando and Francesco, scenic designers like himself. **Ferdinando** (1657–1743), son of Giovanni Maria; with his four sons, forms head of dynasty; pupil of Cignani, Troili, Mavro Aldovrandini, painter and architect to Ranuccio Farnese in Parma, worked in Vienna; author of a treatise on civil architecture (1711). **Antonio** (1700–74), son of Ferdinando, author of Teatro Municipale of Bologna. **Giovanni Maria** (1739–69), concentrated on pure scenography. See pp. 227, 229, 314, 330, 351, 352.
Bibl.: A. Hyatt Mayor, *The Bibiena Family*, New York 1945.

Galliari, Piedmontese family of scene painters and decorators, which includes: **Fabrizio** (1709–90), Bolognese architect, pupil of the architect Nariani, decorated Bergamo cathedral. **Bernardino**, connected in 1750 with the Royal Palace in Turin, decorated the Palazzo Chivano. See p. 314.

Garofalo (Benvenuto **Tisi**, called), c. 1481–1559. Ferrarese painter, active in Rome and Ferrara, influenced by Dosso and Raphael. Author of allegorical compositions (Palazzo Costabili), open to Venetian influence but lacking in breadth. See p. 260.

Gemito (Vincenzo), Naples 1852–1929. Neapolitan sculptor; genre scenes. See p. 382.
Bibl.: C. Siviero, *G.*, Naples 1954.

Gentile da Fabriano, Fabriano c. 1360–1427. Umbrian painter, knew Altichiero and Gothic painters. Active in Venice, Florence and Rome. One of the most brilliant exponents of the International Gothic style in Italy. See pp. 126, 133–5, 189.
Bibl.: A. Colasanti, *G. da F.*, Bergamo 1909. B. Molaioli, *G. da F.*, Fabriano 1927. B. Molaioli, 'Boll. Ist. Arch. e Storia Arte', Rome 1929.

Gentileschi (Orazio), Pisa 1565–London 1638. Tuscan painter. Active in Rome where he knew Reni and Caravaggio, in Genoa and in England after 1626, where he worked as Court Painter to Charles I. See p. 341.
Bibl.: R. Longhi, 'L'Arte' 1916. G. Gamba, 'Dedalo' 1922. Rosenthal, 'G.B.A.' 1930. J. Hess, 'English Miscellany', Rome III 1952.

Gentileschi (Artemisia), Rome 1597–Naples, after 1651. Caravaggesque painter, daughter of Orazio. Active in Florence, settled in Naples in 1630. See pp. 342, 343.
Bibl.: R. Longhi, 'L'Arte' 1916. S. Ortolani, 'Emporium' April 1938.

Gherardi (Antonio), 1644–1702. Roman decorator, pupil of P. F. Mola and P. da Cortona, worked with the latter. See p. 315.

Gherardi (Filippo), Lucca 1643–1704. Painter and decorator associated with G. Coli, worked in Lucca, Venice and Rome. See p. 321.

Gherardo di Giovanni del Fora, 1445–97. Florentine miniaturist. See p. 195.

Ghezzi (Leone), 1674–1755. Roman caricaturist. See pp. 307, 350.

Ghiberti (Lorenzo), Florence 1378–1455. Florentine sculptor and goldsmith, won competition for Baptistery doors in 1401 and in 1425 received commission for third doors (*Porta del Paradiso*); also executed statues for Or San Michele. Having taken part in work on the cathedral, Ghiberti, who was anxious to leave behind him an up-to-date system of teaching, began c. 1450 his *Commentarii*, which he left unfinished. See pp. 130, 146, 148, 162–3, 187.

Bibl.: *Commentarii*, ed. by J. Schlosser, 2 vols., Vienna 1912. L. Planiscig, *L.G.*, Vienna 1940. J. Schlosser, *Leben u. Meinungen des florentinischen Bildners*, *L.G.*, Basle 1941. R. Krautheimer, *G.* Princeton 1956.

Ghirlandaio (Domenico), Florence 1449–94. Florentine painter, pupil of Baldovinetti, influenced by Verrocchio and Flemish art. Eclectic in taste, he worked in the Sistine Chapel (1481) and executed his greatest work in the choir of Sta Maria Novella (1485–90). See pp. 208, 215.

Bibl.: G. J. Davies, *G.*, London 1908. G. de Francovich, 'Dedalo' 1926 and 1930.

Ghirlandaio (Davide), 1452–1525. Florentine painter, directed workshop of brother Domenico after 1490. Specialized in mosaic. See pp. 208, 215.

Ghislandi (Giuseppe), called Fra Galgario, Bergamo 1665–1743. Portrait painter from Bergamo. See p. 348.

Bibl.: Catalogue of exhibition *Pittori lombardi della Realtà*, 1953.

Giacomo da Campione, 15th c. Lombard architect, worked on convent of Certosa di Pavia. See pp. 136, 138.

Giamberti, family name of **Sangallo**, q.v.

Giambono (Michele or Zambon), c. 1420–62. Venetian painter and mosaicist, influenced by Gentile da Fabriano and Pisanello. See p. 135.

Giampetrino, c. 1490–1540. Lombard painter, pupil and imitator of Leonardo. See p. 241.

Giaquinto da Molfetta (Corrado), Molfetta 1694–Naples 1765. Neapolitan painter and decorator, active in Rome from 1723 to 1753. See p. 344.

Gigante (Giacinto), Naples 1806–76. Neapolitan painter, interested in Corot and Turner. See p. 387.

Bibl.: E. Somaré, *Note su G.G.*, 'Il Frontispizio', Florence, October 1940.

Gilardi (Domenico), ? 1788–Lugano 1845. Italian architect, worked in Russia on reconstruction of Moscow after 1812. See p. 368.

Gilardini (Melchiorre), 1675. Lombard painter, pupil of Crespi. See p. 348.

Giordano (Luca), called Luca Fa Presto, Naples 1632–1705. Neapolitan painter and decorator, influenced by Caravaggio during his Neapolitan period. Active in Venice, Rome, Naples and Florence. See pp. 331, 332, 333–4, 343, 344.

Bibl.: O. Benesch, *L.G.*, Vienna 1923.

Giorgio (Francesco **di Giorgio Martini**), Siena 1439–1502. Architect, engineer, wood sculptor, painter, bronze worker and decorator. An outstanding personality, his influence was felt in Siena and Urbino where he was associated with the building of the Castello. Author of a treatise on architecture (c. 1480). See pp. 146, 174–5, 176–7, 181, 196.

Bibl.: *Trattato*, ed. by E. Promis, Turin 1881. S. Brinton, *F. di G. of Siena*, 2 vols., London 1934. A. S. Weller, *F. di G.*, Chicago 1943. R. Papini, *F. di G. architetto*, Florence 1946.

Giorgione (Giorgio **Barbarelli**, called), Castelfranco 1477–Venice 1510. Venetian painter, pupil of Giovanni Bellini, whom he follows closely in his Castelfranco altarpiece (1504). His feeling for 'atmosphere' becomes more marked in religious works such as *The Three Magi* (Vienna) and allegories such as the *Tempesta* (Venice, Accademia), and in his portraits and genre scenes which show the influence of Northern art. He died young and rapidly became a legendary figure. See pp. 200, 252–5.

Bibl.: Hourticq, *Le Problème de G.*, Paris 1931. G. M. Richter, *G. da C.*, Chicago 1937. G. Fiocco, *G.*, Milan 1942. Catalogue of Giorgione exhibition (by R. Pallucchini), Venice 1955.

BIOGRAPHICAL NOTES

Giottino, Florentine 14th c. painter, under whose name have been confused Giotto's pupil Maso di Banco (q.v.) and a later artist, Tommaso di Stefano, recorded at the Vatican in 1369. See p. 115.

Giotto di Bondone, Colle di Vespignano 1266–Florence 1337. Florentine painter, mosaic worker and *maestro dell'opera*, pupil of Cimabue and Pietro Cavallini. After a period of activity in Rome (*Navicella* in St. Peter's, 1295), he worked with immense authority in Assisi, Padua (Arena Chapel, 1305), and Naples (in the service of Robert of Anjou, 1330–33). Recalled to Florence in 1334, he was entrusted with the supervision of the construction of the cathedral, and built the campanile. See pp. 96, 104, 112–25, 132, 133, 134, 162, 163, 166.

Bibl.: F. Rintelen, *G. und G. Apokryphen,* Munich 1912; 2nd ed. 1923. I. B. Supino, *G.,* Florence 1920. C. Gnudi, *G.* (Eng. ed. London 1959). E. Rosenthal, *G.,* Munich 1924. C. Weigelt, *G.,* Leipzig 1925. T. Hetzer, *G.,* Frankfurt 1941. R. Oertel, 'Z.K.G.' 1941–42 and 1943–44. P. Toesca, *G.,* Turin 1941. R. Salvini, *G.,* Bibl., Rome 1938.

Giovanni d'Alemagna, brother-in-law and partner of Antonio Vivarini, with whom he worked in Venice until *c.* 1450, when he is thought to have died. See p. 198.

Giovanni d'Ambrogio, beginning 15th c. Florentine sculptor. See p. 128.

Giovanni da Campione, 16th c. Bergamasque architect and sculptor. See p. 136.

Giovanni de' Boccati, active between 1445 and 1460. Umbrian painter. See p. 177.

Giovanni di Cecco, 14th c. Architect and sculptor of Siena cathedral (1366) after Giovanni Pisano. See p. 99.

Giovanni di Gaitana, mid-13th c. Paduan miniaturist. See p. 87.

Giovannino de' Grassi, late 14th c. Lombard painter, sculptor and miniaturist, left albums of drawings typical of Gothic naturalism (Bergamo, 1360); worked in Milan cathedral (*Christ and the Woman of Samaria,* 1396). See p. 132.

Bibl.: O. Pächt, 'J.W.C.I.' XIII (1950).

Giovanni da Gubbio, 12th c. Umbrian architect, built cathedral of S. Ruffino, Assisi (1140). See p. 64.

Giovanni da Milano, mentioned between 1350 and 1359. Late Gothic Lombard painter, worked in Florence (Sta Croce) and Rome (Vatican). See pp. 132, 133.

Bibl.: M. Marabottini, *G. da M.,* Florence 1950.

Giovanni da Modena, late 14th–early 15th c. Emilian painter, author of St. John chapel in S. Petronio, Bologna (after 1390). See p. 131.

Giovanni di Paolo, Siena *c.* 1403–83. Sienese painter, influenced by T. di Bartolo, Sassetta and the Florentines, fantastic in style. See pp. 126–7, 176.

Bibl.: J. Pope-Hennessy, *G. di P.,* London 1937. C. Brandi, *G. di P.,* Florence 1947.

Giovanni da Rimini, recorded 1320–50. Painter and fresco worker active in Ravenna and Rimini. Author of small panels in the Doge's Palace, Venice (1330). See p. 131.

Giovanni da San Giovanni (Giovanni **Mannozzi,** called), S. Giovanni Valdarno 1592–Florence 1636. History painter, pupil of Rosselli and the Carracci. See pp. 278, 331.

Bibl.: O. H. Giglioli, *G. da S.G.,* Florence 1920; *Affreschi inediti di G. da S.G.,* 'Riv. A' 1929.

Giovanni di Simone, 13th c. Tuscan architect, author of Campo Santo, Pisa (1277). See p. 98.

Giovanni da Udine, Udine 1487–Rome 1564. Painter and stucco worker, collaborated with Raphael on Vatican Loggie (1517–20) and at Villa Madama (1520), with Giulio Romano at Mantua and Pierino del Vaga at Genoa; inventor of modern 'grotesques' and painter of still lifes. See pp. 229, 251, 266.
Bibl.: S. de Vito Battaglia, 'Arte' 1926.

Giovanni da Verona (Fra), Verona 1457–1525. Dominican monk and most important *intarsia* designer of 16th c., responsible for *intarsia* compositions in Sta Maria in Organo, Verona (1500) and Monte Oliveto, and one (now disappeared) in Stanza della Segnatura. See p. 196.
Bibl.: F. Lugano, *F.G. da V.*, Siena 1905.

Giraldi (Guglielmo), 15th c. Ferrarese miniaturist of unusual brilliance and delicacy who worked for Borso and Nicolo d'Este. See p. 195.
Bibl.: T. Liebart, 'L'Arte' 1911.

Girolamo da Carpi, 1501–56. Ferrarese painter, pupil of Garofalo, worked in an eclectic style close to his master's. See p. 260.

Girolamo da Cremona, active 1467–83. Northern painter and miniaturist trained by Mantegna and Liberale, author of the extraordinary Siena choir books. See p. 196.

Girolamo da Treviso, see **Pennacchi**.

Giuffrè (Antonino), 15th c.–died after 1510. Sicilian painter whose family was related to Antonello, the most important figure on the island at the end of the century. See p. 186.
Bibl.: Catalogue of Antonello exhibition, Venice 1953.

Giuliano da Rimini, recorded 1307–46. Painter from the Marches. Painted an altar dorsal for a brotherhood at Urbania (Gardner Coll. Boston). Active at Pomposa: frescoes for the refectory (1320). See p. 131.
Bibl.: Moschetti, *G. da R.* in the 'Boll. del Museo di Padova' 1931.

Giulio Romano (Giulio **Pippi**, called), *c.* 1499–1546. Architect, painter and decorator whose style derives from Raphael whom he assisted in the Vatican from 1517 onwards, from Bramante and from Michelangelo whose violent attitudes he exaggerates. After 1524 entered service of Federigo Gonzaga, built and decorated Palazzo del Te with combination of boldness and feeling that were to become hallmarks of Mannerism. See pp. 226, 227, 268–9, 272.
Bibl.: G. K. Loukomski, *J.R.*, Paris 1932. E. H. Gombrich, *G.R.*, 'J.W.' 1934 and 1935. F. Hartt, 'G.B.A.' 1947; *G.R.* (2 vols., Newhaven U.S.A., 1958).

Giulio di Stefano, 12th c. Sicilian architect, author of cloister of Amalfi. See p. 76.

Giunta Pisano (Giunta **Capitini**, called), mentioned between 1241 and 1254. Pisan painter, author of agonized images of crucified Christ in Byzantine manner; in 1241 worked at Assisi. See pp. 74, 83.
Bibl.: Van Marle, 'R.A.' 1920.

Giusto de' Menabuoi, late 14th c. Florentine painter, executed monumental frescoes in Padua Baptistery. See p. 133.
Bibl.: S. Bettini, *G. d. M.*, Padua 1944.

Gobbo (Pietro Paolo **Bongi**, called Gobbo de' Frutti or Gobbo de' Carracci), active *c.* 1630–48. Painter of still lifes or '*cose naturale*' and of tables piled with fruit. Emulator of Caravaggio. See p. 342.
Bibl.: E. Battisti, 'Com.' V 1954.

Gonzaga (Pietro), 1751–1831. Venetian decorator and scenic designer. See p. 314.

BIOGRAPHICAL NOTES

Goro di Gregorio, 1st half 13th c. Tuscan sculptor whose style derives from Giovanni Pisano; author of the *Arca* of S. Cerbone (Massa Marittima). See p. 111.

Gozzoli (Benozzo), Florence 1420–Pisa 1495. Florentine painter, apprenticed as goldsmith to Ghiberti, pupil of Fra Angelico with whom he decorated chapel of Nicholas V (Rome). After 1449 worked on his own as decorative painter of charm and vitality; church of S. Fortunato at Montefalco (1453–1459), Viterbo and Rome, Medici chapel (Florence 1459), S. Gimignano and Campo Santo, Pisa (after 1468). See pp. 170, 176, 177.

Bibl.: G. J. Hoogewerff, *B.G.*, Paris 1930. P. Toesca, *B.G. nella cappella dei Medici* 1954.

Grandi (Giuseppe Domenico), 1843–94. Sculptor influenced by Carpeaux. See p. 382.

Grassi (Orazio), 1583–1654. Roman painter, Jesuit, author of decorative paintings in church of S. Ignazio, Rome. See p. 317.

Bibl.: P. Levi, *Il fenomeno G.*, Leghorn 1910.

Greco (Domenicos **Theotokopoulos**, called El Greco), Crete 1541–Santo Domingo (Spain) 1614. Cretan painter who went to Spain, the early part of whose career belongs to Italy, 1560–76. Lived in Venice where he was influenced by Tintoretto; 1570–72, in Rome and Parma; the style for which he is known developed mainly during his second period in Venice. See p. 288.

Bibl.: R. Pallucchini, 'Pa.', no. 45, Sept. 1953.

Gregorini (Domenico), 1st half 18th c. Roman architect, author of oratory of Holy Sacrament in Sta Maria in Via (c. 1730) and of Sta Croce in Gerusalemme (1743). See p. 325.

Grimaldi (Francesco), Lucca 1543–1630. Neapolitan architect associated with Maderna, active in Naples. See p. 332.

Grubicy de Dragon (Vittore), Milan 1851–1920. Painter of Hungarian origin, friend of Seurat and Signac, apostle of 'divisionism' in Italy. See p. 385.

Guala (Pier Francesco), ? 1698–? 1757. Piedmontese decorative painter in Rococo style, active in Lombardy. See p. 330.

Bibl.: Catalogue of exhibition (by G. Testori), Ivrea 1954.

Guarana (Jacopo), 1720–1808. Venetian painter, pupil and collaborator of Tiepolo. See p. 360.

Guardi (Antonio), Venice 1698–1760. Founder member of Venetian academy in 1755, brother of Francesco Guardi, less distinguished in style. See p. 356.

Bibl.: F. de Maffeis, *A.G.*, Verona 1951.

Guardi (Francesco), Venice 1712–93. Worked with his brother Gian Antonio until 1760, influenced by Maffei, Magnasco, Canaletto, became outstanding painter of light effects in Venice in 18th c. See p. 361.

Bibl.: G. Fiocco, *G.*, Florence 1937. V. Moschini, *G.*, Milan 1954. A. Morassi, *Conclusioni su G.*, 'Emporium' 1951. J. Byam Shaw, *The Drawings of F.G.*, London 1949.

Guariento, 2nd half 14th c. Paduan painter whose style derives from Giotto, active in Venice and Padua; fresco of *Paradiso* in Sala del Gran Consiglio of Doge's Palace (now disappeared). See p. 135.

Bibl.: A. Moschetti, *G. Pittore padovano*, 'Atti Accad. Pad.' XL 1924. L. Coletti, 'Riv. A.' XII 1930.

Guarino Guarini, 1624–83. Piedmontese architect, Theatine, theologian and mathematician, trained in Rome, who taught in Sicily. Widely cosmopolitan in outlook, combined considerable virtuosity with free use of materials. In 1666 entered service of Duke of Savoy: Palazzo Carignan; in 1680, chapel of Holy Shroud in S. Giovanni Battista (Turin). See pp. 327–8.

Guercino (Francesco **Barbieri**, called), Cento 1591–Bologna 1666. Emilian painter trained in Bologna in school of Lodovico Carracci. Went to Venice, Rome (1621, Casino Ludovisi *Aurora*), Piacenza. Of all the academic masters Guercino has the clearest understanding of Caravaggiesque chiaroscuro. See pp. 346–7.

Bibl.: G. Cantalamessa, *Lo Stile del G.*, Rome 1925. D. Mahon, 'B.M.' 1937 and *Seicento Art and Theory*, London 1947.

Guglielmelli (Arcangelo), late 17th–early 18th c. Neapolitan architect and decorator, author of Baroque salon in library of Gerolomini convent. See p. 332.

Guglielmo, 12th c. Tuscan painter, author of Sarzana Crucifix, signed and dated 1138. See p. 74.

Guglielmo, 12th c. Lombard sculptor, worked on façade of S. Zeno, Verona, with Maestro Niccolo. See p. 61.

Guglielmo, 12th c. Tuscan sculptor, author of ambo (*c.* 1160) of Pisa cathedral, later transferred to Cagliari (1312). See p. 73.

Guidetto, early 13th c. Architect from Lucca. See pp. 72, 73.

Guidi (Domenico), 1625–1701. Roman sculptor, imitator of Bernini and his rival Algardi. See p. 323.

Guido Mazzoni, see **Mazzoni**.

Guttoso (Renato), Palermo 1912. In Venice became member of 'Fronte nuovo dell' arte' (1945). See p. 392.

Haffner (Antonio Maria), 1654–1732. Genoese decorative painter, worked as assistant to Ferrari and Piola in Palazzo Rossi. See p. 326.

Hayez (Francesco), Venice 1791–Milan 1882. Venetian painter, settled in Milan from 1818–82. Painted historical compositions and portraits. See pp. 383, 384.

Bibl.: G. Nicodemi, *Dipinti di F.H.*, Milan 1934.

Iappelli (Giacomo), 1783–1852. Paduan architect, author of designs for Caffè Pedrocchi (1831). See p. 371.

Induno (Domenico), Milan 1815–78. Milanese painter, pupil of Sabatelli and Hayez, played active part in Risorgimento. See p. 384.

Bibl.: G. Nicodemi, *D. e G. I.*, Milan 1945.

Induno (Girolamo), Milan 1827–90. Milanese painter, brother of Domenico. See p. 384.

Iverny (Jacques) of Avignon, 1st half 15th c. Painter and fresco painter from Avignon, worked in Piedmont on Castello della Mante (*c.* 1420–30); signed triptych in Turin. See p. 132.

Jacobello del Fiore, *c.* 1370–1439. Venetian painter in International Gothic manner. See p. 135.

Jacopino del Conte, Florence 1510–Rome 1598. Roman painter in Mannerist style. Trained at Florence under Andrea del Sarto. Active in Rome where he took part in decorating the Oratory of S. Giovanni Decollato in the Velabro (*c.* 1545). See p. 289.

Bibl.: F. Zeri, *J. del C.* in 'Proporzioni', Florence 1948.

Jacopino di Francesco, mid-14th c. Bolognese painter of unusual and powerful style. See p. 131.

Bibl.: R. Longhi, 'Pa.' no. 5 (May 1950).

Jacopo da Camerino, 13th c. Roman mosaicist, worked with Torriti on apse of S. Giovanni in Laterano (1291). See p. 82.

Jacopo Francescano, 13th c. Florentine mosaicist whose art is basically Venetian, active in Florence; apse mosaics, 1225 (Baptistery). See p. 66.

Jacopo dell'Indaco, 1476–1526. Assistant to Michelangelo. See p. 234.

Jacopo di Lorenzo, 13th c. Architect and worker in marble and mosaic; façade of Civita Castellana (1210). See p. 68.

Jacopo della Quercia, see **Quercia**.

Jacquerio (Giacomino), active early 15th c. Piedmontese painter, frescoes in S. Antonio di Ranverso. See p. 132.

Juvara (Filippo), Messina 1678–Madrid 1736. Architect, pupil of Carlo Fontana in Rome. In 1715 settled in Turin: basilica of the Superga (1731), church of the Carmine (1732) and Palazzo Madama (1718), royal villa of Stupinigi. Designer on a lavish scale, J. was summoned to Madrid. See pp. 328–30, 336.

Bibl.: A. Tellucini, *L'Arte dell'architetto F.J.*, Turin 1926. Various authors, *F. J.*, Turin 1937.

Koch (Gaetano), 1849–1910. Roman architect, author of Piazza dell'Esedra (Rome). See pp. 379, 381.

La Corte (Juste **Le Court**, called Giusto), Ypres 1627–Venice 1679. Flemish sculptor settled in Venice; altar of Sta Maria della Salute (1669). See p. 355.

Lamberti (Niccolo), *c.* 1370–Florence 1451. Florentine architect and sculptor, active in Florence and Venice where he worked on façade of S. Marco with his son Pietro di Niccolo (*c.* 1393–1435). See pp. 128, 136.

Bibl.: G. Fiocco, 'Dedalo', VIII (1927–28).

Landi (Gaspare), 1756–1830. Roman painter, author of rather superficial frescoes and of portraits (*Canova*, Borghese gallery). See p. 374.

Landini (Taddeo), Florence ?–Rome 1596. Florentine sculptor, author of *Seasons* on Ponte Sta Trinità. See p. 272.

Lando di Pietro, died in Siena 1340. Architect in service of Princes of Anjou, then architect of Siena cathedral which he tried to enlarge (1340–55). See p. 99.

Landollina, early 18th c. One of architects of Noto (Sicily). See p. 337.

Lanfranco, mentioned as 'Mirabilis Artifex', early 12th c. Architect of cathedral of Modena. See p. 59.

Bibl.: L. Olschki, *La Cathédrale de Modène*, 'Archivum Romanicum' 1935.

Lanfranco (Giovanni), Parma 1580–1647. Baroque decorative painter, assistant to Carracci, specialist in foreshortening and illusions of depth, active in Naples and Rome where he decorated Palazzo Mattei and Casino Borghese (*Olympus*). In Naples in 1634. See p. 320.

Lapadula (Bernardo), 20th c. Roman architect, author of tower of Universal Exhibition (1942). See p. 393.

Laperuta (Leopoldo), 19th c. Neapolitan architect, responsible for monumental reconstruction of Royal Palace (Naples). See p. 372.

Lattanzio da Rimini, documented 1495–1524. Painter from Romagna. Pupil of G. Bellini. Active at Rimini and Venice: *Sala del Gran Consiglio* (1495); church of the Crociferi (1499). See p. 201.

Bibl.: G. Fiocco, *L. da R.* in 'Boll. d'Arte' 1922–23.

Laurana (Francesco), mentioned between 1458 and 1502. Dalmatian architect and sculptor, active in Naples (arch of Alfonso of Aragon) and Urbino. Author of subtle portrait busts. Passed into service of King René in Provence. See pp. 180, 185, 187.

Bibl.: W. Rolfs, *F.L.*, Berlin 1907. R. Causa, 'Pa.', no. 55, July 1954.

Laurana (Luciano), *c.* 1420–79. Dalmatian architect, responsible for reconstruction of Castello of Urbino after 1468. See pp. 180, 183.
Bibl.: A. Colasanti, *L.L.*, Rome 1922.
Lazzarini (Gregorio), 1655–1730. Venetian decorative painter, author of huge composition in S. Pietro di Castello. See p. 356.
Lega (Silvestro), Modigliana 1826–Florence 1895. Florentine painter, friend of Borrani and Fattori, belonged to group of 'Macchiaioli' and painted scenes of everyday life. See p. 387.
Bibl.: M. Tinti, *S.L.*, Rome 1931. M. Valsecchi, *L.*, Milan 1950.
Le Gros (Pierre), 1666–1719. French Baroque sculptor, active in Rome. See p. 324.
Lendinara (Cristoforo **Canozzi** da), Lendinara *c.* 1420–after 1488. Painter and *intarsia* designer, collaborated with brother in studio of Belfiore, on cathedral of Modena and Santo, Padua. See p. 196.
Bibl.: G. Fiocco, 'L'Arte' 1913.
Lendinara (Lorenzo), Lendinara 1425–Padua 1477. Painter and *intarsia* designer associated with Piero della Francesca. See p. 196.
Leonardo di Giovanni, 14th c. Florentine goldsmith, worked on great silver altar for Florence Baptistery. See p. 128.
Leonardo da Vinci, Vinci 1452–Amboise 1519. Powerful Florentine genius whose activity covers every field: painter, sculptor, architect, engineer, has left notebooks filled with every kind of scientific observation, reflections on method and literary fragments which are no less remarkable. Pupil of Verrocchio (1469–75), left Florence in 1481–82, leaving behind him the unfinished *Adoration of the Magi*. Stayed in Milan at court of Lodovico il Moro, *c.* 1481–99. Various activities: *Virgin of the Rocks, Last Supper*, giant equestrian statue.
From 1500 to 1506, back in Florence: *Virgin and Child with St. Anne, Mona Lisa, Battle of Anghiari* (lost). Then returned to Milan, 1507–13, Rome 1513–15 at court of Giuliano de' Medici, and Amboise as guest of Francis I (1515–19).
Although his *oeuvre* is small, his influence was immense: researches into chiaroscuro, atmosphere and subtle effects. Summing up the theoretical writings of the 15th c., Leonardo compiled a scientific encyclopedia, which although fragmentary is nevertheless amazingly varied and precise. See pp. 205, 207, 209, 211, 212–14, 215.
Bibl.: Biographies: Vasari, *Vita di Leonardo*, ed. G. Poggi, Florence 1919. Paintings: H. Bodmer, *L.*, Berlin 1931. A. Malraux and J. Segnaire, *Tout l'oeuvre peint*, Paris 1950. Drawings: 'Publ. Commissione Vinciana', by A. Venturi, Rome 1928–36. Selection: A. E. Popham, *The Drawings of L. da V.*, London (2nd ed.) 1949. Studies: G. Séailles, *L. da V.*, Paris 1892. K. Clark, *L. da V.*, Cambridge (2nd ed.) 1952. L. H. Heydenreich, *L. da V.*, 2 vols., Basle and London 1953. Writings: J. P. Richter, *The Literary Works of L. da V.*, 2 vols., Oxford 1939. A. Chastel, *L. da V. par lui-meme*, Paris 1952. School: W. Suida. *L. u. sein Kreis*, Munich 1929.
Leoni (Leone), Arezzo 1509–Spain 1590. Tuscan architect and sculptor, active mainly in Milan. Specialized in bronzes and portrait busts. See p. 294.
Leoni (Pompeo), 1537–1608, son of Leone. Architect and sculptor, in Spain executed tombs and statues in bronze. See p. 294.
Leopardi (Alessandro), Venice *c.* 1466–*c.* 1522. Venetian architect, founder, sculptor. Made bronze cast of Verrocchio's statue of Colleoni and collaborated with A. Lombardo on the tomb of Cardinal Zen in St. Mark's. See p. 206.

Libera, 20th c. Architect, formed the 'Group of Seven' in Milan. See p. 393.

Liberale da Verona, *c.* 1445–1529. Powerful Veronese painter and miniaturist, author of frescoes in S. Fermo and Sta Anastasia, Verona, and of choir-books at Chiusi (1467–69) and at Siena (1470–76). See p. 196.
Bibl.: E. Carli, *L. da V.*, Milan 1953.

Ligorio (Pirro), Naples *c.* 1510–83. Roman architect and painter famous for stucco ornaments. See pp. 216, 218, 290, 293.

Lippi (Andrea), Pistoia 1888–1916. Tuscan sculptor and draughtsman, romantic and symbolist in feeling. See p. 383.

Lippi (Fra Filippo), Florence 1406–Spoleto 1469. Florentine painter and monk, pupil of Lorenzo Monaco but influenced by Masaccio, imparted new grace to original themes (*Coronation of the Virgin*). Frescoes in Prato cathedral (1452–65) and in choir of Spoleto cathedral (1467–69). See pp. 167–8, 176, 210.
Bibl.: M. Pittaluga, *F.L.*, Florence 1949.

Lippi (Filippino), Prato 1457–Florence 1504. Florentine painter, son of Fra Filippo and a Prato nun, worked with Botticelli: *Vision of St. Bernard* (Badia, 1480), finished Brancacci chapel, Carmine (1485–88), large tortured compositions in Rome: Caraffa chapel in Sta Maria sopra Minerva (1489–93) and Florence: Strozzi chapel in Sta Maria Novella (1495–1502). Numerous altarpieces, panels and *cassoni*, restless and complicated in style. See pp. 197, 198, 210, 215, 239.
Bibl.: U. Mengin, *Les deux Lippi*, Paris 1932. A. Scharf, *F.L.*, Vienna 1935.

Lippi (Nanni), 2nd half 16th c. Roman architect, author of Villa Medici. See p. 290.
Bibl.: A. Parronchi, 'Pa.', no. 37, Jan. 1953.

Locatelli (Andrea), 1660–1740. Roman painter, influenced by Poussin and Gaspard Dughet. See p. 351.

Lo Faso (Domenico L. F. Pietrasanta), 1783–1863. Sicilian architect and archaeologist. See p. 374.

Lombardo (Pietro), *c.* 1450–1515. Venetian architect and sculptor, responsible for a subtle yet richly ornamental style and delicate compositions: reliefs (Faenza) and tombs (SS. Giovanni e Paolo, Venice). See pp. 198, 248.

Lombardo (Antonio), late 15th–early 16th c. Son and assistant of Pietro. See p. 248.

Lombardo (Tullio), early 16th c. Sculptor, most consciously antique of whole group: recumbent figure of G. Guidarelli (Ravenna). See p. 248.
Bibl.: G. Mariacher, 'B.M.', Dec. 1954.

Longhena (Baldassare), 1598–1682. Venetian architect, pupil of Scamozzi. Trained as sculptor, author of Sta Maria della Salute (1631–87) and Palazzo Pesaro (1679). See pp. 299, 353, 354.
Bibl.: C. Semenzato, *L'Architettura di B. Longhena*, Padua 1954.

Longhi (Alessandro), Venice 1733–1813. Painter and engraver, pupil of Nogari, portraitist. See p. 361.

Longhi (Martino) the Younger, 1602–50. Roman Baroque architect and sculptor. See p. 317.
Bibl.: A. E. Brinckmann, 'L'Arte' XVI 1933.

Longhi (Pietro **Falca**, called), Venice 1702–85. Pupil of Balestra and Crespi, member of the Venetian academy, painter of Venetian society. See p. 360.
Bibl.: A. Ravà, *P.L.*, Florence 1923. O. Uzanne, *P.L.*, Paris 1924.

Lorenzetti (Ambrogio), active 1319–48. Painter, brother and pupil of

Pietro, more open to Florentine influence, frescoes in Palazzo Pubblico, Siena. See pp. 119–20, 122, 123, 133.
Bibl.: G. Sinibaldi, *I Lorenzetti*, Siena 1933. C. Volpe, 'Pa.', no. 13.
Lorenzetti (Pietro), active in Siena 1305–48. Painter trained by Duccio and Simone Martini, frescoes at Assisi, evolved monumental style which dominated Tuscan art. See pp. 119–20, 123, 133.
Bibl.: P. Bacci, *Dipinti di P.L.*, Siena 1930. E. Cecchi, *P.L.*, Milan 1930. C. Volpe, 'Pa.', no. 23.
Lorenzetto, 1490–1541. Florentine sculptor, executed two statues in Chigi chapel (Sta Maria del Popolo). See p. 243.
Lorenzi (Stoldo), 16th c. Florentine sculptor from Settignano. See p. 274.
Lorenzo di Mariano, called **Il Marrina**, Siena 1476–1534. Sienese sculptor. Pupil of Giovanni di Stefano. Active at Siena: high altar in marble, church of Fontegiusta. See p. 175.
Bibl.: M. Stoltz, *L. di M.*, 'A travers l'art italien', Paris 1949.
Lorenzo di Pietro, see **Vecchietta**.
Lorenzo Monaco, *c.* 1370–*c.* 1425. Florentine painter and monk, pupil of Agnolo Gaddi and the Sienese, active in Florence where he represents the sweetness of the late Gothic style. See pp. 126, 127, 165, 195.
Bibl.: O. Siren, *Don L.M.*, Strasbourg 1905. Pudelko, 'B.M.' 1938 and 1939.
Lorenzo Veneziano, mentioned between 1356 and 1379. Active in Venice, Padua, Bologna (1368). Follower of Paolo Veneziano, influenced by International Gothic, author of *Death of the Virgin* (Vicenza). See pp. 134–5.
Lorenzo da Viterbo, active *c.* 1446–70. Umbrian painter dominated by Benozzo Gozzoli. See p. 179.
Lotto (Lorenzo), Venice 1480–Loreto 1556. Venetian painter who, after a journey to Rome (1509), spent his life wandering from one place to another: Bergamo, 1526; Treviso, 1532; in the Marches, 1535. Became lay brother of the Santa Casa of Loreto in 1554. Interested in Flemish and German art, his cool colouring is the opposite of that of Titian. See pp. 197, 200, 201, 257–8, 267.
Bibl.: B. Berenson, *L.L.*, London (1st ed.) 1895, (2nd ed.) 1953. A. Banti and A. Boschetto, *L.L.*, Florence 1953. Catalogue of Lotto exhibition, Venice, 1953 (by P. Zampetti).
Luca di Tommè, active between 1355 and 1399. Sienese painter whose art derives from the Lorenzetti. See pp. 123, 127.
Lucchesino (Pietro **Testa**, called), Lucca 1611–Rome 1650. Tuscan painter who went to Rome to join Domenichino and study the antique. Attracted by the Baroque and Pietro da Cortona, his art was finally transformed by the example of Poussin. See p. 321.
Bibl.: A. Marabottini, 'Com.' V (1954).
Luciani (Sebastiano). See **Sebastiano del Piombo**.
Luini (Bernardino), *c.* 1481/2–1532. Lombard painter, pupil of Borgognone and Foppa, assimilated Umbrian manner and style of Leonardo. Last fresco painter of Northern Italy, interested in German art, slightly overloaded in his larger compositions. See pp. 241, 267.
Bibl.: Catalogue of exhibition, Como, 1953 (by A. Ottino della Chiesa).
Lyss (or **Liss**), called 'Pan'. Late 16th c.–1630. Painter of German origin and European outlook, settled in Venice. See p. 356.
Maccari (Cesare), 1841–1919. Roman decorative painter, author of historical scenes in Palazzo del Senatorio. See p. 384.

Maderna (Carlo), 1556–1629. Roman architect, influenced by Vignola, played decisive part in development of Baroque architecture, finished St. Peter's and built several palaces in Rome. See pp. 314, 315.

Bibl.: H. Egger, *C.M. Projekt für San Pietro in Vaticano*, Leipzig 1928.

Maderno (Stefano), *c.* 1571–1636. Lombard sculptor, settled in Rome, author of statue of Sta Cecilia. See p. 322.

Maffei (Francesco), Vicenza *c.* 1620–Padua 1660. Attracted by art of Veronese and Tintoretto, influenced by El Greco and Callot. Active in Venice and Brescia. See p. 356.

Bibl.: N. Ivanoff, *F.M.*, Padua 1942.

Magenta (Giovanni Ambrogio), 1565–1635. Lombard architect, active in Bologna. See p. 330.

Magnasco (Alessandro), Genoa 1667–1749. Milanese painter whose art derives from Callot and Salvator Rosa, active in Genoa, Florence and Milan. Genre scenes and fantasies of entirely personal inspiration. See pp. 341, 348, 351.

Bibl.: G. Beltrami, *M.*, Milan 1913. B. Geiger, *M.*, Bergamo 1949. Catalogue of exhibition, Bergamo 1949 (by A. Morassi).

Magnelli (Alberto), Florence 1888. Florentine abstract painter settled in Paris. See p. 392.

Maitani (Lorenzo), first half 14th c.–died 1330. Sienese sculptor, worked on façade of Orvieto cathedral, master of picturesque low relief. See pp. 99, 111–12.

Majano (Benedetto da), 1442–97. Florentine architect and sculptor whose works in marble are eclectic and influenced by Rossellino, collaborated with brother Giuliano on church of Loreto. Author of tabernacles and pulpits, and of altar of Sta Fina (S. Gimignano). See pp. 165, 202, 203, 207.

Bibl.: L. Dussler, *B. da M.*, 1924.

Majano (Giuliano da), 1432–90. Florentine architect and sculptor whose style derives from Brunelleschi, worked on cathedral of Faenza (1476). Active in Arezzo, Siena, Naples, where he built Porta Capuana (1485). Worked in Loreto with his brother, and with Cronaca on Palazzo Strozzi (after 1489), which is built on the Michelozzo pattern. See pp. 185, 202–3, 207.

Mancini (Antonio), Albano Laziale 1852–Rome 1930. Neapolitan painter, friend of Gemito, influenced by Morelli and Cremona; picturesque and vigorous style. See p. 388.

Bibl.: E. Cecchi, *A.M.*, Rome 1943.

Manetti (Antonio), 15th c. Florentine *intarsia* designer: works in sacristy of Florence cathedral (1440). See p. 196.

Manfredi (Bartolommeo), Ostiano, near Mantua *c.* 1580–Rome 1620. Follower of Caravaggio, played important part in diffusing his style after 1610. See p. 341.

Manozzi (Giovanni), see **Giovanni da San Giovanni.**

Mansueti (Giovanni), mentioned *c.* 1485–1527. Venetian painter, pupil and imitator of Gentile Bellini. See p. 199.

Mantegna (Andrea), Padua 1431–Mantua 1506. Mantuan painter and engraver, pupil of Squarcione, strongly influenced by Donatello and Uccello. From 1449 to 1454, worked in Eremitani church, and executed S. Zeno altar, Verona (1457–58). His rigid powerful style mellowed in service of Gonzaga: *Camera degli Sposi* (Ducal Palace), *studiolo* of Isabella d'Este (*Parnassus*, Louvre). He had a decisive influence on art of Venice, Padua, Ferrara; Mantegna engravings played important part in Dürer's artistic formation. See pp. 189–90, 193, 201, 211, 261, 267.

Bibl.: P. Kristeller, *A.M.*, London 1902. F. Knapp, *A.M.*, Berlin 1910. G. Fiocco, *A.M.*, Milan 1937. E. Tietze-Conrat, *A.M.*, London 1955.
Manzù (Giacomo), Bergamo 1908. Sculptor, bronzes and portraits. See p. 393.
Bibl.: B. Joppolo, *G.M.*, Milan 1946. A. Pacchioni, *G.M.*, Milan 1948.
Maragliano (Antonio Mario), 1664–1741. Genoese wood sculptor, fond of complicated volumes. See p. 327.
Maratta (Carlo), 1625–1733. Bolognese painter, author of numerous altarpieces. See pp. 347, 374.
Marco d'Amadio, 15th c. Venetian architect, built Ca' d'Oro (1421–40). See p. 139.
Marco d'Oggiono, ?–*c.* 1530. Lombard painter, pupil and imitator of Leonardo. See p. 241.
Marchionni (Carlo), 1702–86. Roman architect, author of the Villa Albani (1743–63). See p. 325.
Marchiori (Giovanni), 1696–1778. Venetian sculptor, author of *Sybils* in church of the Scalzi. See pp. 355, 376.
Marcola (Giambattista), Verona 1711–80. Veronese painter. Author of religious scenes, pastorals and lively *capricci*. See p. 350.
Marcovaldo (Coppo di), see **Coppo**.
Marescalchi (Pietro di), called Lo Spada, Feltre 1520–*c.* 1576/84. Venetian painter influenced by Bassano, Tintoretto and Schiavone. See p. 284.
Mariani (Cesare), 1828–1901. Roman painter, decorated basilica of S. Lorenzo. See p. 384.
Marieschi (Michele), Venice 1710–43. Venetian painter and engraver. See p. 362.
Bibl.: Mauroner, *M.M.*, 'Print Collector's Quarterly' 1940.
Marinali (Orazio), 1643–1720. Venetian sculptor, pupil of La Corte, author of portrait busts and garden statues. See p. 355.
Marini (Marino), Pistoia 1901. The most outstanding modern Italian sculptor. See p. 393.
Bibl.: R. Carrieri, *M.M.*, Milan 1948.
Marochetti (Carlo), 1805–67. Romantic sculptor. See p. 382.
Marrina, see **Lorenzo di Mariano**.
Martini (Simone), Siena 1284–Avignon 1344. Sienese painter who developed under influence of Duccio and gave extraordinary refinement to colour harmonies: *Maestà* (1315). Favourite artist of Robert of Anjou, visibly attracted by French Gothic art of the 'courtly' variety, ended his days in Avignon. Painter of international standing. See pp. 96, 116, 118–19, 123–5, 126, 127, 130, 175.
Bibl.: R. van Marle, *S.M.*, Strasbourg 1920. A. De Rinaldis, *S.M.*, Rome 1936. L. Coletti, 'Art Quarterly' XII 1949. Paccagnini, *S.M.*, London 1954.
Masaccio (Tommaso), Castello di Val d'Arno 1401–Rome 1428. Florentine painter, pupil of Masolino and Ghiberti, friend of Donatello and Brunelleschi; reacted strongly against facile elegance of Gothic, basing his art on volume, space and light. About 1427, with Masolino, decorated Brancacci chapel (Carmine, Florence). His short career, which ended in Rome, had a profound influence on Tuscan art. See pp. 165–7, 179, 210.
Bibl.: Documents: U. Procacci, 'Riv. Arte' 1935 and 1953. Studies: J. Mesnil, *M. et les débuts de la Renaissance*, The Hague 1927. M. Pittaluga, *M.*, Florence 1935. R. Longhi, *Fatti di Masolino e M.*, 'C.A.' 1940. M. Salmi, *M.*, 2nd ed. Milan 1948.
Mascherini (Marcello), Udine 1906. Sculptor from Trieste. See p. 393.

BIOGRAPHICAL NOTES

Masegne (the dalle), 14th c. Venetian family of architects and sculptors. 1394, rood-screen of St. Mark's, balcony facing sea of Doge's palace. Giovanni di **Rigozzi** dalle, worked on Loggia della Mercanzia, Bologna. Jacobello dalle, mentioned until 1409. Worked on rood-screen of St. Marks. Pier Paolo dalle, ?–1403. Worked on carved retable of S. Francesco, Bologna. See pp. 136, 137, 140.

Maso di Banco, first half 14th c. Florentine painter, pupil of Giotto. See p. 115.

Bibl.: L. Coletti, *Contributo al problema Maso Giottino*, 'Emporium', 1942.

Masolino da Panicale, Panicale 1383–1447. Pupil of Gherardo Starnina; influenced by Gentile da Fabriano, brought to Gothic tradition new awareness of space and volume. After 1425, in Florence, Brancacci chapel (Carmine); 1427, set out for Hungary; 1429, Rome, working in S. Clemente; died abroad. See pp. 126, 165–7, 198, 210.

Bibl.: P. Toesca, *M.*, Bergamo 1907.

Massari (Giorgio), 1686–1766. Venetian architect, collaborated on construction of façade of Gesuiti. See p. 354.

Mastelletta (Giovanni Andrea **Donducci,** called), 1575–1655. Bolognese landscape painter. See p. 346.

Mastroianni (Umberto), Frosinone 1910. Cubist sculptor, settled in Turin. See p. 393.

Matas (Niccolo), ?–Ancona 1872. Florentine architect, restored façade of Sta Croce, 1863. See p. 379.

Matteo da Campione, 14th c. Architect, *maestro dell'opera*, author of polychrome facing of cathedral of Monza (1390–96). See p. 137.

Matteo di Giovanni, Borgo S. Sepolcro, 1435–95. Sienese painter who reacted to Northern influence and dominated second half of century. See pp. 175, 176.

Mattia di Nanni, 14th c. Sienese sculptor and *intarsia* designer, worked on Palazzo Pubblico in 1425. See p. 129.

Maturino (Fiorentino), c. 1505–28. Tuscan decorative painter, imitator of Raphael. See p. 266.

Mauro (Ernesto di), 19th c. Italian decorator, worked on Galleria Umberto I, Naples. See p. 381.

Mazza (Camillo), Bologna 1602–72. Bolognese sculptor. See p. 331.

Mazza (Giuseppe), Bologna 1653–1741. Bolognese sculptor, son of Camillo, active in Bologna and Venice, specialized in reliefs, plaques, tabernacles. See pp. 331, 355.

Mazzoni (Guido), called Il Modanino, born in Modena–died in 1518. Sculptor who introduced wood sculpture and terracotta to Naples, author of monumental polychrome groups (*Holy Sepulchre* of Monte Oliveto, Naples), went to France in 1494. See p. 186.

Mazzoni (Sebastiano), Florence before 1615–Venice 1685. Tuscan painter, pupil of Allori, lively and fertile style. See p. 356.

Bibl.: L. Planiscig, in 'J.W.' 1916. R. Krautheimer, in 'Marburger Jahrbuch', V 1929.

Mazzucchelli (Pier Francesco), called Il Morazzone, Morazzone 1571–1626. Lombard painter influenced by Bolognese, went to Rome. See pp. 347-348.

Bibl.: Nicodemi, *P.F.M., detto il M.*, Rome 1927.

Mazzuoli, called Parmigianino, see **Parmigianino.**

Mazzuoli (Giuseppe), 18th c. Sculptor active in Rome. Author of Papal tombs in an exaggerated style derived from Bernini. See Plate No. 105.

BIOGRAPHICAL NOTES

Meldolla (Andrea), see **Schiavone.**

Melone (Altobello), end of 15th c.–beginning of 16th c. Painter of Cremona, close to Romanino. See p. 258.

Melozzo da Forlì, Forlì 1438–1518. Painter from Romagna whose style derives from Piero della Francesca, worked in Loreto, Urbino, Rome (apse of SS. Apostoli, c. 1480). See pp. 179–80, 181, 224.
Bibl.: R. Buscaroli, *M. da F. nei documenti,* Rome 1938. Catalogue of M. da Forlì exhibition, 1938.

Menabuoi (Giusto de'), see **Giusto.**

Mengoni (Giuseppe), 1827–77. Architect of Milan and Bologna. See p. 381.

Mengs (Anton Raphael), 1728–79. Bavarian painter, friend of Winckelmann. Converted to Catholicism, settled in Italy, became leader of neoclassical reaction and head of Accademia di S. Luca in 1771. See pp. 369, 374.
Bibl.: V. Christoffel, *Der Schriftliche Nachlass des R.M.,* Berlin 1918. K. Gerstenberg, *J. Winckelmann, A.R.M.,* Berlin 1929.

Menzio (Francesco), 1899. Piedmontese painter, inspiration behind 'Group of Six', Turin 1928. See p. 392.

Messina (Antonello da), see **Antonello.**

Michelangelo Buonarroti, Caprese (Casentino) 1475–Rome 1564. Florentine sculptor, painter and architect; the most powerful and complete personality of the 16th century. Passed through studio of Ghirlandaio, early training from Bertoldo, protected by Lorenzo de' Medici. In 1494, left Florence for Venice and Bologna; went to Rome in 1496. In 1502, in Florence, carved *David* (Piazza della Signoria) and worked on fresco for Palazzo Vecchio. Summoned to Rome in 1506 by Julius II; commissioned to execute his tomb, decorated immense ceiling of Sistine Chapel (1508–12), after 1515 returned to the tomb which was never finished (fragments placed in S. Pietro in Vincoli in 1545). 1520–34, back in Florence: Medici tomb in S. Lorenzo, architectural projects, republican activities. After 1534, settled in Rome, mentally more in tune with Catholic Counter-Reformation: *Judgement* fresco; architectural undertakings: St. Peter's, Piazza del Campidoglio. See pp. 207, 208, 215, 216, 221, 224, 230–9, 243, 254, 256, 261, 262, 265, 266, 267, 272, 274, 276.
Michelangelo transformed the plastic energy of Tuscan art into the language of the sublime; his art dominates the 16th century by virtue of the Mannerism which developed from his own tortured energy; his influence can be felt in 17th-century Baroque which looks back to his pathos and grandeur. He evolved from triumphant vigour to strength and feeling; the latter part of his life was mainly devoted to architecture and poetry.
Bibl.: Biographies: A. Condivi, *Vita di M.B.* 1553, ed. by P. D'Ancona, Florence 1928. Drawings: K. Frey, 3 vols., Berlin 1909–11 (and Supplement 1925). L. Goldscheider, London 1951. J. Wilde (Drawings in British Museum), London 1953. Studies: Ch. de Tolnay, *M.-A.,* Princeton, vol. I, 2nd ed. 1947; vol. II, 2nd ed. 1949; vol. III, 1948; vol. IV, 1954; V and VI in preparation; (in one vol.) *M.-A.,* Paris 1951. H. Thode, *M.-A.,* 3 vols., Berlin 1908–13, and *Kritische Untersuchungen,* 3 vols., *id.* Paintings: A. E. Popp, *Die Medici-Kapelle M.A.,* Munich 1922. Poems: ed. K. Frey, Berlin 1897. Letters: ed. G. Milanesi, Florence 1875. Bibl.: Steinmann-Wittkower, *M.-A., Bibl.* (1510–1926), Leipzig 1927. J. Ackerman, *The Architecture of Michelangelo* (2 vols., London 1961).

Michelino da Besozzo, c. 1400. Lombard miniaturist in the style of Gothic naturalism. See p. 132.

BIOGRAPHICAL NOTES

Bibl.: O. Pächt, 'J.W.C.I.' XIII (1950).
Michelozzo di Bartolommeo, Florence 1396–1472. Florentine architect, sculptor and decorator, trained under Ghiberti, then pupil and collaborator of Donatello; built monastery of S. Francesco (Bosco ai Frati), monastery of S. Marco (1437–52). From 1446–51 *maestro dell'opera* of Duomo. Author of Palazzo Medici, prototype for many noblemen's homes (1444–59), modernized number of Medici villas. In Milan built Palazzo del Banco Mediceo and Portinari chapel in S. Eustorgio. See pp. 155–7, 188.

Bibl.: F. Wolf, *M. di B.*, Strasbourg 1900. O. Morisani, *M.*, Turin 1951.
Michelucci, 20th c. Tuscan architect, author of the new railway station, Florence (1933). See pp. 393, 395.

Michetti (Francesco Paolo), Tocco Causaria 1851–Francavilla 1929. Neapolitan painter, author of lively popular scenes. See p. 388.

Bibl.: T. Sillani, *F.P.M.*, Milan 1932.
Migliara (Giovanni), 1785–1837. Venetian painter, follower of Guardi. See p. 362.

Bibl.: M. Pittaluga, 'Arte Veneta' 1954.
Minguzzi, 1911. Bolognese sculptor. See p. 393.
Mino da Fiesole, Fiesole 1429–84. Florentine sculptor, author of numerous busts and tombs, panels in Prato cathedral, tomb of Paul II in Rome; master of the delicate manner, soft modelling. See pp. 165, 184.

Bibl.: D'Angeli, *M.D.F.*, Florence 1905.
Mitelli (Agostino), 1609–60. Bolognese painter and decorator, pupil of Girolamo Curti. See p. 320.

Mocchi (Francesco), 1580–1654. Tuscan sculptor, active in Orvieto, Piacenza and Rome. Author of equestrian monuments for Farnese family in Piacenza. See p. 323.

Bibl.: L. Dami, *F.M.*, 'Dedalo' V 1924. A. Pettorelli, *F.M. e i gruppi equestri farnesiani*, Piacenza 1926.
Moderno, see under **Masters**, p. 482.

Modigliani (Amedeo), Leghorn 1884–Paris 1920. Tuscan painter who settled in Paris where his elegant and subtle line came as a revelation. Fragments of sculpture, severely controlled. See pp. 368, 391.

Bibl.: A. Salmon, *M.*, Paris 1926. G. Scheiwiller, *A.M.*, Milan 1927. M. Raynal, *M.*, Geneva 1951. E. Carli, *M.*, Rome 1952.
Mollari (Antonio), mentioned in Trieste 1802–06. Italian architect. See p. 371.

Monsù Desiderio, 17th c. (i.e. Monsieur, as of a foreigner), identified with D. Nomé, Metz 1593–after 1611, painter of fantasies, and not to be confused with Didier Barra of Metz who signed a view of Naples in 1647. See p. 349.

Bibl.: Catalogue of exhibition at Museum of Sarasota (Florida), by A. Scharf, 1950. Catalogue of exhibition at Gall. Obelisco (Rome), by G. Urbani, 1950. R. Causa, 'Pa', 75, March 1956.
Montagna (Bartolommeo), Brescia 1450–Vicenza 1523. Painter, founder of school of Vicenza, influenced by Antonello da Messina. See pp. 201, 252.

Montanaro (Giovanni Antonio), beginning of 16th c. Lombard architect, collaborated with Battagio on Sta Maria della Croce and Como cathedral. See p. 240.

Monteverde (Giulio), 1837–1917. Tuscan sculptor. See p. 382.
Monte di Giovanni del Fora, 1448–1529. Florentine miniaturist, brother of Gherardo. See p. 195.

Montorsoli (Fra Giovan Angelo), Florence c. 1507–63. Tuscan sculptor and architect of tremendous verve, connected with Michelangelo, worked in Messina, Bologna. See p. 243.

Morandi (Giorgio), Bologna 1890. Bolognese painter: after experimenting with metaphysical painting and landscape, devoted his energies with remarkable concentration to still lifes painted with restricted monochrome palette. See p. 391.

Bibl.: G. Marchiori, *G.M.*, Genoa 1945. C. Gnudi, *M.*, Florence 1946. M. Salmi, *G.M.*, Milan 1949. J. Berger and A. Arcangeli, 'Art News', Feb. 1955.

Morazzone, see **Mazzuchelli.**

Morelli (Domenico), Naples 1823–1901. Neapolitan painter in contact with the Nazarenes and Delacroix; author of spectacular religious pictures. See p. 388.

Bibl.: A. Conti, *D.M.*, Naples 1927.

Moretto (Alessandro **Bonvicino**, called), 1498–1555. Painter from Brescia who combined Venetian colouring with Lombard manner. See pp. 296, 340.

Morlaiter (Giovanni), 1699–1781. Venetian sculptor born in Alto Adige, author of panels for SS. Giovanni e Paolo. See pp. 355, 376.

Moro (Lodovico del), 19th c. Florentine architect, with de Fabris rebuilt façade of Sta Maria del Fiore. See p. 379.

Morone (Domenico), second half of 15th c. Painter from Verona whose art derives from Mantegna. See p. 191.

Bibl.: B. Berenson, *Metodo e Attribuzioni*, Florence 1947.

Moroni (Giovanni Battista), 1523–78. Lombard painter, weighty yet delicate portraits. See pp. 296, 340, 347.

Bibl.: G. Lendorff, *G.B.M.*, Winterthur 1933. D. Cubini, *M. Pittore*, Bergamo 1939. Cat. of exhibition *Pittori della Realtà in Lombardia*, Milan 1953.

Mulier (Pietro), see **Tempesta.**

Muttoni (Francesco) ?–1748. Venetian architect, author of Villa Trissino da Porto. See p. 354.

Nanni di Banco, 1373–1421. Florentine sculptor, pupil of Niccolo d'Arezzo. Influenced by Ghiberti, collaborated with Donatello, worked on Florence Cathedral and Or San Michele. See pp. 129, 148.

Bibl.: L. Planiscig, *N. di B.*, Florence 1946. P. Vaccarino, *N.*, Florence 1950.

Nanni di Bartolo, called Il Rosso, first half of 15th c. Florentine sculptor, collaborated with Donatello. See pp. 128–9, 136, 148.

Nardo di Cione, ?–1366. Florentine painter, brother of Orcagna. Author of frescoes in Strozzi chapel, Sta Maria Novella. See pp. 123, 131.

Neri di Bicci, 1415–91. Florentine painter, head of workshop of journeymen painters: *cassoni*, altarpieces. Left a diary. See p. 208.

Bibl.: *Le Ricordanze* (1453–75), ed. G. Poggi, 'Il Vasari', 1927, 1929, 1930.

Neri di Fioravanti, 14th c. Florentine architect, supervisor of building operations at Or San Michele (1349). See p. 121.

Neroccio di Bartolommeo Landi, Siena 1447–1500. Sienese painter and sculptor full of sensibility and feeling; author of statue of St. Catherine and tomb of T. Piccolomini; numerous altarpieces. See p. 174.

Nervi (P. L.), 1891. Florentine architect, author of Florence sports stadium (1932) and airport buildings at Orbetello (1938). See p. 393.

Niccolaci, mid-18th c. Sicilian architect. See p. 337.

Niccolini (Antonio), 1772–1850. Neapolitan architect, active in Naples: Teatro San Carlo. See p. 372.

BIOGRAPHICAL NOTES

Niccolo, 12th c. Lombard sculptor and *maestro dell'opera* who inscribed his name on cathedral of Ferrara (1135) and subsequently executed S. Zeno reliefs at Verona. His influence in Emilia was widespread: signed portico at Sagra di S. Michele (Turin). See p. 61.

Niccolo dell'Abbate, Modena 1509–Fontainebleau 1571. Emilian painter and faithful follower of Dosso and Parmigianino; sensitive to Flemish art, evolved delicately fantastic style in landscapes and portraits, settled in France in 1552, adopted manner of Primaticcio and played active part in school of Fontainebleau. See pp. 266, 269.
Bibl.: catalogue of Fontainebleau exhibition, Naples 1952.

Niccolo da Bari, called dell'Arca, Bari 1440–94. Sculptor from Apulia who worked in Bologna: *tomb of S. Dominic* in S. Domenico, and *Pietà* in Sta Maria della Vita. See pp. 185, 186, 192.
Bibl.: C. Gnudi, *N. d. A.*, Turin 1942.

Niccolo da Foggia, 13th c. Sculptor from Apulia; pulpit of Ravello (1272). See p. 91.

Niccolo da Foligno, *c.* 1430–1502. Painter from the Marches, author of restless religious paintings. See p. 177.
Bibl.: U. Gnoli, *Note varie su N. d. F.*, 'Emporium' Feb. 1909.

Niccolo di Giacomo, second half of 14th c. Bolognese miniaturist, influenced by Orcagna and the Tuscans. See p. 131.

Niccolo di Maestro Pietro di Paradisio, 1394–1430. Venetian painter influenced by Rhenish art. See p. 135.
Bibl.: E. Sandberg Vavalà, *Maestro Stefano und N. d. P.*, 'J.B.' 1930.

Niccolo di Magio, early 15th c. Sienese painter, worked in Palermo. See p. 186.
Bibl.: M. Accascina, *Pitture Senese nel museo di Palermo*, 'La Diana' 1930.

Niccolo di Ser Sozzo Tegliacci, Siena ?–1363. Sienese painter and miniaturist. Head of a school of illuminators. Designer of the Frontispiece to the *'Caleffo dell'Assunta'* (1334–36, Siena Arch.). See p. 127.
Bibl.: C. Brandi, *N. di Ser S.T.*, in 'l'Arte' 1932.

Niccolo Fiorentino (di Forzore **Spinelli,** called), Florence 1431–1514. Medallist at court of Lorenzo de' Medici; 150 pieces are attributed to him. See p. 244.
Bibl.: W. Bode, in 'J.B.' XXV 1904.

Nigetti (Matteo), 1560–1649. Florentine architect, author of S. Gaetano (Florence). See p. 331.
Bibl.: L. Berti, *M.N.*, 'Riv. A.' 1952.

Nittis (G. de), see **De Nittis.**

Nobile (Pietro), Campestri 1774–1845. Italian architect, active in Trieste. See p. 371.

Nono (Luigi), Fusina 1850–Venice 1918. Venetian painter, stiffly formal works. See p. 384.
Bibl.: P. Molinenti, *L.N.*, 'Il Secolo XX', 1919.

Nuvoloni (Carlo Francesco), 1608–66. Milanese painter and portraitist, called the Lombard Murillo. See p. 348.

Nuzi (Allegretto **da Fabriano,** called), mentioned 1346–73. Umbrian fresco painter, active in Rome and Rimini. See p. 131.

Nuzzi (Mario), called de' Fiori, 1603–73. Roman painter of still lifes whose art derived from Caravaggio. See p. 342.

Oderisi da Gubbio, 1240–99. Illuminator mentioned by Dante, active in Gubbio, Bologna and Rome. See p. 87.

Oderisio, mid-14th c. Painter to whom the frescoes of Sta Maria Incoronata at Naples have been attributed. See p. 116.

Orcagna (Andrea **di Cione**, called), active 1344–68. Florentine architect, sculptor, goldsmith and painter, author of Tabernacle of Or San Michele and of the polyptych in the Strozzi chapel (Sta Maria Novella). His style in sculpture derives from Andrea Pisano and in painting from Maso; dominates Tuscan Trecento Gothic. See pp. 96, 121, 123–4, 128, 131.
Bibl.: K. Steinweg, *A.O.*, Strasbourg 1929. H. D. Gronau, *A.O. und N. di Cione*, Berlin 1937.

Orlandi (Stefano), 1681–1760. Architect and decorator of Bolognese school, specialized in perspective effects. See p. 314.

Ortolano (Giovanni Battista Benvenuti, called) recorded 1512–24. Ferrarese painter. Pupil of Boccaccino and L. Costa. Painted altarpieces: *Pietà* (Modena), *Crucifixion* (Brera, Milan). See p. 260.

Padovanino (Alessandro **Varotari**, called Il), Padua 1588–1648. Painter influenced by Titian, Baroque in outlook. See p. 355.
Bibl.: Della Santa, *Il Pitt. A.V.*, Vicenza 1904.

Pagano (Giuseppe), 1896–1945. Architect of the Università Commerciale Bocconi, Milan (1936–42). See p. 394.

Pagni (Benedetto), known between 1525 and 1570. Painter and decorator in service of Francesco IV Gonzaga, author of Sala dei Cavalli in Palazzo del Te (Mantua). See p. 268.

Pagno di Lapo, 15th c. Florentine architect and sculptor whose style derives from Donatello, author of tabernacle in Annunziata (Florence) and of palace in Bologna. See pp. 157, 191.

Palizzi (Filippo), Vasto 1818–Naples 1899. Painter from the Abruzzi, member of the so-called 'Posilippo' school, brother of Giuseppe, Nicola and Francesco Paolo, all painters. See p. 388.
Bibl.: Sapori, *F.P.*, Turin 1918.

Palladio (Andrea **di Pietro**, called), Padua 1508–Vicenza 1580. Venetian architect, stonemason until 1540, studied in Rome, was adopted by poet and humanist Giangiorgio Trissino, became creator of villas on the mainland for the Venetian aristocracy. In these admirably varied designs, the classical portico is systematically adopted. At Vicenza he restored the basilica and refaced it with a portico featuring the arcade or Palladian motif (1549), built several palaces and began the Teatro Olimpico (1580) which was finished by Scamozzi. In Venice itself: churches of S. Giorgio Maggiore and Il Redentore. His art, which is of classical breadth, has nevertheless tenuous connections with Mannerism by virtue of its delicacy of detail, and, after 1575 (Loggia, Vicenza), massive effect. Palladio is the author of a treatise, *I Quattro Libri dell'Architettura*, Venice 1570, which was immensely influential in Italy and England. See pp. 204, 250, 265, 267, 269, 278–84.
Bibl.: G. K. Loukomski, *A.P.* 1924 (unreliable). G. Fiocco, *A.P. Padovano*, Padua 1933. R. Pane, *A.P.*, Turin 1948. On the villas: F. Bürger, *Die Villen des A.P.*, Leipzig 1909; G. C. Zorzi, *Studi palladiani*, 'A.V.' 1949, 1950, 1951. On the influence: G. C. Argan, *P. e la critica neo classica*, 'L'Arte' 1930; G. Zorzi, *I Disegni delle Antichità di Andrea Palladio* Venice 1958.

Palma Giovane (Jacopo **di Antonio Negretti**, called), Venice 1544–1628. Painter of large-scale works, son of Antonio and Giulia (granddaughter of Bonifazio Pitati), went to Rome before 1568. Influenced by art of Titian and Raphael. See pp. 256, 288, 355.

Palma Vecchio (Jacopo **Negretti**, called) Serinalta 1480–Venice 1528.

Venetian painter influenced by Giorgione and Titian, one of the best exponents of new Venetian manner. See pp. 254, 256.
Bibl.: G. Gombosi, *P.V.*, Berlin 1937.
Palmezzano (Marco), 1456–1517. Painter from the Marches, worked with Melozzo da Forlì. See p. 180.
Palmieri (Pietro), Bologna 1720–Turin 1804. Painter of *vedute* and *capricci*. See p. 350.
Panfilo, see **Nuvoloni.**
Pannini (Giovanni Paolo), Piacenza 1691–1765. Roman *veduta* painter trained by Bibiena in Bologna; his manner was continued by Hubert Robert. See p. 352.
Paoletti (Gaspare Maria), 1728-1813. Tuscan architect, master of two generations; author of Terme Leopoldine (Montecatini 1773). See p. 371.
Paolo di Martino, 15th c. Sienese decorator; pavement of Siena cathedral (1426). See p. 174.
Paolo Veneziano, called Maestro Paolo, Venice, late 13th c.–*c.* 1360. Painter closer to Byzantine art than to Giotto, as can be seen in his Dignano polyptych (1321). Outstanding personality; author of *Death of the Virgin* (Museum, Vicenza, 1933); collaborated with his brother and sons. See p. 134.
Bibl.: E. Sandberg Vavalà, *Maestro P.V.*, 'B.M.' 1930. V. Lasareff, 'Arte Veneta' 1954.
Parenzano (Bernardo), called Fra **Lorenzo,** 1437–1531. Painter from Squarcione's school. Active at Padua: frescoes showing the life of S. Benedict (S. Giustina Convent). See p. 191.
Parmigianino (Francesco **Mazzola,** called). Parma 1503–40. Emilian painter, pupil of Correggio, worked in Rome (1524) and in Parma where he decorated S. Giovanni Evangelista and painted portraits and religious pictures in a delicate and intense style which sometimes verges on the bizarre and which is his own particular form of Mannerism. See pp. 267, 269–71.
Bibl.: L. Frölich-Bum, *P. u. der Manierismus*, Vienna 1921. A. Quintavalle, *Il P.*, Milan 1948. S. J. Freedberg, *P.*, Cambridge, Mass. 1950.
Parodi (Filippo), 1630–1702. Baroque sculptor, active in Genoa and Venice. See p. 327.
Pasini (Alberto), 1826–97. Italian painter who went to court of Shah of Persia. See p. 368.
Passarotti (Bartolommeo), Bologna 1529–92. Emilian painter, friend of Vignola and Zuccaro in Rome, typical of the Mannerism against which the Caracci reacted.
Bibl.: H. Bodmer, *Correggio e gli Emiliani*, Novara 1943. See p. 297.
Passignano, see **Cresti.**
Paulucci (Enrico), Genoa 1901. Painter, settled in Turin. See p. 392.
Pelizza da Volpedo (Giuseppe), Volpedo 1868–1907. Milanese painter, landscapist, adept of 'divisionism'. See p. 385.
Bibl.: U. Ojetti, *Mostra di P. di V.*, Milan 1920.
Pellegrini (Giovanni Antonio), Venice 1675–1741. Widely travelled Venetian painter, pupil of P. Pagani; his 'loose' style owes much to Luca Giordano and Ricci. See pp. 357, 361.
Penna (Agostino), 18th c. Roman Baroque sculptor, collaborated with Posi on monument to Maria Flaminia Chigi, Sta Maria del Popolo. See p. 324.
Pennacchi (Girolamo), called Girolamo da Treviso or il Vecchio, *c.* 1455–1497. Venetian painter influenced by Mantegna and Alvise Vivarini. See p. 201.

BIOGRAPHICAL NOTES

Bibl.: E. Zocca, *G. da T. il V.*, 'Boll. Arte' 1932. G. Fiocco, *P.M.P.*, 'Riv. Ist. Arch. e storia Arte' 1929.

Penni (Luca), 1500-56. Painter decorator. Collaborated with Raphael and Giulio Romano in the Vatican *Stanze* and the Farnesina *Loggie*. See pp. 226, 227.

Perino del Vaga (**Buonaccorsi**, called), Florence 1500-47. Florentine painter, pupil of Raphael; decorated palaces in Genoa and Rome for Paul III. See p. 295.

Permoser (Balthasar), 1651-1732. Bavarian sculptor, worked in Carmine, Florence. See p. 332.

Pertsch (Matteo), 1780-1854. Architect and town planner from Trieste. See p. 371.

Perugino (Pietro **Vannucci**, called), Città delle Pieve 1445-Fontignano 1523. Umbrian painter, pupil of Verrocchio, worked in Perugia and in Rome on Sistine frescoes. Decorated Cambio in Perugia with assistance of the young Raphael, then Mantua in service of Isabella d'Este. His art shows freedom from stylistic effects and a harmonious treatment of space. See pp. 177, 179, 191, 208, 212, 225, 239.
Bibl.: F. Canuti, *Il P.*, 2 vols., Siena 1931.

Peruzzi (Baldassare), Siena 1481-1536. Sienese architect and painter, went to Rome in 1503 and worked with Bramante and Raphael. Author of various palaces, including Farnesina or Villa Chigi (1508-11) and Palazzo Massimo (1532). Also important as decorator and scenic designer (works in Rome and Siena). See pp. 216, 228-30, 241, 330.
Bibl.: W. W. Kent, *The Life and Works of B.P.*, London 1925.

Pesellino (Francesco), 1422-57. Florentine painter whose art derives from Filippo Lippi and Fra Angelico; his flowery Gothic manner was particularly well suited to the pictures and decorative panels (*cassoni*), usually secular in feeling, which he turned out in abundance. See p. 170.
Bibl.: W. Weisbach, *F.P.*, Berlin 1901.

Petitot (Alexandre), Lyons 1727-Parma 1801. French architect, active in Parma where he built the Palazzo del Governo. See p. 371.

Piacentini (Marcello), 20th c. Author of a neo-academic treatise entitled *Architettura d'Oggi* (1930). See pp. 381, 393.

Piazzetta (Giambattista), Venice 1683-1754. Venetian painter, son of a wood-sculptor, pupil of Crespi; his paintings, a great number of which are on pastoral themes, contain skilful manipulations of light and shade. See pp. 357, 358.
Bibl.: R. Pallucchini, *P.*, Rome 1942.

Picherale (Pompeo), 1670-1734. Sicilian architect. Active at Syracuse, where he constructed the façade for the cathedral (1728). See p. 336.

Piermarini (Giuseppe), Foligno 1734-1808. Italian architect, assistant of Vanvitelli at Caserta. Influenced by Vignola, Palladio and the antique, active in Milan where he altered the Palazzo Reale (1768-78), built the Scala (1776-1778) and several villas. See pp. 369, 370.
Bibl.: various authors, *G.P. Architetto*, Milan 1908.

Piero di Cosimo, 1462-1521. Florentine painter, pupil of Cosimo Rosselli, strongly influenced by Leonardo and Signorelli. Predilection for strange and fantastic themes. See pp. 211, 275.
Bibl.: F. Knapp, *P. di C.*, Halle 1899. R. Langton Douglas, *P. di C.*, Chicago 1946. E. Panofsky, *Studies in Iconology*, New York 1939.

Pietro Alemanno, ?-1498. Austrian painter, active in Northern Italy. See p. 199.

Pietro da Cortona, see **Cortona**.

Pietro da Milano, mid-15th c. Italian architect, built triumphal arch of Castel Nuovo (1466), Naples. See p. 185.

Pietro da Rimini, mid-14th c. Painter from the Marches, worked at Sta Chiara, Ravenna. See p. 131.

Pinturicchio (Bernardo di **Betto**, called Il), 1454–1513. Umbrian painter, pupil of Perugino, worked in Rome (Ara Coeli, 1490; Borgia apartments, Vatican, 1492–94), Spello (1501) and Siena (1503-08) in a facile ornamental style. See pp. 177, 184.
Bibl.: C. Ricci, *P.*, Perugia 1912.

Piola (Domenico), 1627–1703. Genoese decorator, pupil of Capellini, imitator of Castiglione, decorated churches and palaces in Genoa. See pp. 326, 327.

Piranesi (Giovanni Battista), Mogliano (Venice) 1720–Rome 1778. Architect and engraver. Except for building Sta Maria Aventina (1765), his entire career was confined to the publication, in Rome, of sets of engravings of unparalleled richness and imaginative power: *Vedute di Roma* (after 1748, 137 plates), *l'Antichità Romane*, 4 vols. (1756), *Carceri* (1744). His albums of decorative details gave a fillip to Neoclassical taste. See pp. 325, 352, 362, 369.
Bibl.: H. Focillon, *G.B.P.*, Paris 1918. A. M. Hind, *G.B.P.*, London 1922. A. Hyatt Mayor, *P.*, New York 1952. W. Körte, *G.B.P. als praktischer Architekt*, 'Z.K.G.' 1933. V. Mariani, *Studiando P.*, Rome 1938.

Pisanello (Antonio **di Puccio di Cerreto**, called), Pisa 1395–1450. Painter and medallist from Northern Italy, son of a Pisan father and a Veronese mother; grew up in Verona where he became pupil of Stefano da Zevio. Worked in Venice with Gentile Bellini (1425), in Mantua (1424), Rome (1431–33), Mantua and Milan (1440), Venice (1442), Ferrara and Naples; during his crowded career brought incisive character of Gothic draughtsmanship to its highest pitch and imparted new grace to the elegance of International Gothic. See pp. 133-4, 189.
Bibl.: G. F. Hill, *P.*, London 1903. B. Degenhart, *P.*, Vienna 1940. R. Brenzoni, *P.*, Milan 1952. J. Guiffrey, *Dessins de P. conservés au Musée du Louvre*, 4 vols., Paris 1911. B. Degenhart, 'Arte Veneta' 1949, 1953, 1954.

Pisano (Niccolo), *c.* 1225–87. Sculptor, native of Apulia where he was influenced by the Renaissance fostered by Frederick II. Settled in Tuscany, made pulpit in Pisa Baptistery (1260) and one in Siena Cathedral (1268) with four pupils: Giovanni, Arnolfo, Donato, Lapo: these works mark inception of grand style in Italian sculpture. Author, with his son Giovanni, of Fountain of Perugia (1273–80). See pp. 73, 86, 92, 96, 106, 107-10, 116, 117.
Bibl.: G. Swarzenski, *N.P.*, Frankfurt-am-Main 1926. G. H. and C. R. Crichton, *N.P. and the Revival of Sculpture in Italy*, Cambridge 1938. G. Nicco Fasola, *N.P.*, Rome 1941. W. R. Valentiner, in 'Art Quarterly', XV 1952.

Pisano (Giovanni), *c.* 1250–1314. Tuscan sculptor, son and pupil of Nicola. The most powerful personality in sculpture of the entire century, author of monumental pulpits of S. Andrea, Pistoia (1298) and of Pisa Cathedral (1302–1311). *Maestro dell'opera* of Siena Cathedral for which he designed a series of statues after the French manner (1294–98). Carved number of austere Madonnas: Arena Chapel, Padua, Prato Cathedral. See pp. 85, 86, 92, 99, 106, 107-15, 121, 129, 135, 137, 162.
Bibl.: A. Venturi, *G.P.*, 2 vols., Rome 1927. H. Keller, *G.P.*, Vienna 1942.

Pisano (Andrea **da Pontedera**, called), *c.* 1270–Orvieto 1348. Tuscan sculptor and goldsmith, came from Pisa to cast first bronze door of Baptistery

BIOGRAPHICAL NOTES

(1330–36). Author of delicate Gothic bas-reliefs on Campanile. 1347–48, *maestro dell'opera* of Orvieto Cathedral. See pp. 96, 111, 112, 121–2, 147, 163
Bibl.: I. Toesca, *Andrea e Nino Pisano*, Florence 1952.
Pisano (Nino), 14th c. Florentine sculptor, architect and goldsmith, son of Andrea. Collaborated with his father at Orvieto Cathedral, created a type of supple swaying Madonna that approximates to the French pattern. Author of *Cavalcanti monument*. See pp. 122, 129, 162, 174.
Bibl.: E. Carli, *Il Problema di N.P.*, 'L'Arte' 1934. I. Toesca, *Andrea e N.P.*, Florence 1952.
Pisano (Antonio), see **Pisanello.**
Pisis, see **De Pisis.**
Pitloo (Antonio), 1790–1837. Dutch landscape painter, spent some time in Naples. See p. 387.
Pittoni (Giambattista), Venice 1687–1767. Venetian painter, member of Academies of Bologna and Venice, typical exponent of 18th-century Baroque. See p. 361.
Bibl.: L. Coggiola, *G.B.P.*, Florence 1921.
Pizzolo (Niccolo), 1421–53. Paduan history painter and sculptor, pupil of Squarcione, worked on Eremitani chapel in Padua with Mantegna (1449–1453). See p. 190.
Poccianti (Pasquale), 1774–1852. Pupil of Paoletti, worked on aqueducts of Leghorn. See p. 371.
Podesti (Giulio), 1800–95. Roman academic painter. See p. 381.
Poggi (Giovanni), 1811–1901. Florentine architect and town planner, developed S. Miniato district (Florence). See p. 379.
Poletti (Luigi), Modena 1792–1869. Architect from Modena, pupil of Stern, active in Rome where he taught many of the architects of the period and restored S. Paolo fuori-le-Mura (after 1823). See p. 372.
Polidoro da Caravaggio, Caravaggio 1492–Messina 1543. Mannerist painter. Pupil of Raphael. Painted vivid landscapes in the style of *capricci.* Active in Rome, Naples and Sicily. See p. 266.
Pollaiuolo or Pollaiolo (Antonio **Benci,** called), Florence *c.* 1432–Rome 1498. Florentine painter, sculptor, bronze worker and medallist, remarkable for his anatomical studies. Collaborated with Ghiberti, followed Donatello and Uccello. The Hercules series (*c.* 1465), on which Piero collaborated, shows their style at its most distinctive. Summoned to Rome in 1490 to execute tomb of Pope Sixtus IV, then of Innocent VIII (1493–97). Outstanding engraver. See pp. 128, 165, 170–1, 172, 184, 206–7, 208, 210.
Bibl.: A. Sabatini, *A. e P.P.*, Florence 1944. S. Ortolani, *I P.*, Milan 1948. M. Cruttwell, *A.P.*, London 1907. C. Ragghianti, *A.P. e l'arte fiorentina*, 'C.A.', 1935–36.
Pollaiuolo (Piero), 1443–96. Painter and sculptor, brother of Antonio with whom he collaborated, usually in painting: author of *Virtues* (Uffizi) and *Coronation of the Virgin* (S. Agostino, San Gimignano, 1483). See p. 208.
Pollak (Leopoldo), 1751–1806. Viennese architect, pupil of Piermarini, author of Villa Reale, Milan (1799). See p. 370.
Pollastrini (Enrico), 1817–74. Florentine Romantic painter, author of large religious and historical compositions. See p. 385.
Pomarancio (Nicolo **Circignani,** called), late 16th c. Roman painter, decorator of S. Stefano Rotondo. See p. 306.
Ponte (Jacopo del), see **Bassano.**
Pontelli (Baccio), Florentine, directed *intarsia* workshop: *Studiolo* of Urbino (1474). See pp. 196, 203.

465   GG

Pontormo (Jacopo **Carucci**, called), Pontormo 1494–Florence 1557. Tuscan painter, pupil of Albertinelli and Andrea del Sarto whose delicacy he retains in a style of increasing pallor. His masterpieces are the peasant scenes in the villa at Poggio a Cajano (1520–22), and the *Passion* scenes in the Certosa of Galluzzo which show the influence of Dürer. Haunted by Michelangelo, he decorated the choir of S. Lorenzo (1546–57), painted over in the 18th century. Also painted sensitive portraits. The most extreme exponent of Florentine Mannerism. See pp. 267, 275, 276.

Bibl.: F. Mortimer Clapp, *J.C. da P.*, Newhaven 1916. C. Gamba, *Il P.*, Florence 1921. L. Becherucci, *Disegni del P.*, Bergamo 1943. C. de Tolnay, 'C.A.' 1950.

Pordenone (Giovanni Antonio **de' Sacchis** or **de' Codesanis**, called), Pordenone 1483/4–Ferrara 1539. Venetian painter active in Rome, Ferrara, Venice, Cremona and Piacenza. His style derives from Titian but is coloured by Mannerism. See p. 257.

Bibl.: F. Catalano, *P.*, Cremona 1940.

Porpora (Paolo), 1617–*c.* 1670. Neapolitan painter, specialized in studies of tables loaded with fruit in the Dutch manner. See p. 342.

Bibl.: R. Causa, *P.P. e il primo tempo della 'natura morta' napoletana*, 'Pa.', March 1951.

Porta (Giacomo della), 1540–1602. Roman architect. Finished Michelangelo's work on Campidoglio (1564–94), built loggia of Palazzo Farnese, dome of St. Peter's (1586–93), façade of Gesù (1575) and S. Luigi dei Francesi (1589). See pp. 221, 224, 393.

Posi (Paolo), 18th c. Roman Baroque architect. See p. 324.

Pozzo (Andrea), 1642–1709. Painter and decorator on an imposing scale. Jesuit lay brother in Rome in 1681. Author of extraordinary perspective effects: his *Glory of S. Ignatius*, S. Ignazio, Rome (1685) is a masterpiece of illusionism. Author of a treatise on perspective (1693). Active in Vienna and Naples. See pp. 292, 314, 321–2.

Predis (Ambrogio da), 1455–1508. Lombard painter, pupil and imitator of Leonardo. See p. 241.

Prestinaro (Cristoforo), 1579–1618. Lombard sculptor, specialist in terracotta. See p. 326.

Preti (Mattia), 1613–99. Painter and decorator known as *Il Cavaliere Calabrese*, influenced by Venetian and Flemish art and by Caravaggio. Active in Naples. See p. 343.

Bibl.: J. Alazard, 'R.A.A.M.' 1922. Rosenthal, 'G.B.A.' 1930. R. Longhi, *Recenti studi su Caravaggio*, 'Pr.' 1943.

Previati (Gaetano), Ferrara 1852–Lavagna 1920. Milanese painter, divisionist and later symbolist compositions. See pp. 385, 389.

Bibl.: V. Constantini, *G.P.*, Bergamo 1931.

Previtali (Andrea), Bergamo 1470–1528. Venetian painter, pupil of Giovanni Bellini, influenced, after 1502, by Palma and Lotto in his poetic approach to nature. See p. 201.

Primaticcio (Francesco), Bologna 1504–Paris 1570. Bolognese painter and decorator, pupil and assistant of Giulio Romano. Author of graceful stucco decoration and mythological compositions on which he had assistance of Niccolo dell'Abbate. In 1523 entered service of Francis I at Fontainebleau, where he found Rosso; in 1540 made journey to Rome on behalf of King of France. One of the greatest exponents of international Mannerism. See p. 269.

Bibl.: L. Dimier, *Le P.*, Paris 1900 and 1925. P. Barocchi, *Precisazioni su P.*, 'Com.' 1951.

BIOGRAPHICAL NOTES

Procaccini (Ercole), 1515–95. Bolognese painter. Arrived in Milan *c.* 1585. Father of **Giulio Cesare**, Bologna *c.* 1570–Milan 1626. Painter influenced by Correggio, important influence in Lombardy. See p. 348.

Puget (Pierre), Toulon 1622–94. French sculptor, spent six years (1661–1667) in Genoa where he founded an important school. See p. 327.
Bibl.: Lagrange, *P.P.*, 1865. M. Labò, *P.P.*, 'G.B.A.' 1925.

Quartararo (Riccardo), recorded 1484–1501. Sicilian painter. Active at Palermo and Naples where he collaborated with Costanzo de Moysis. Painted altarpieces. See p. 186.
Bibl.: S. Bottari, *T. da G. e R.Q.*, in 'Arte Antica e Moderna'. Bologna, no. 6 1959.

Queirolo (Francesco), Genoa 1704–Naples 1762. Genoese sculptor, author of *An Allegory of Deception*, Cappella Sansevero (Naples). See p. 333.

Quercia (Jacopo della), Siena 1374–1438. Sienese sculptor of vigorous style who took part in competition for Baptistery doors in Florence (1401), made several tombs in Lucca, the *Fonte Gaia* (destroyed) in Siena, and the impressive stone reliefs of the door of S. Petronio, Bologna (1425–38). See pp. 129, 130, 148, 174.
Bibl.: L. Biagi, *J. d. Q.: nuovi documenti e commenti*, Siena 1929; and *J. d. Q.*, Florence 1946. Nicco, *J. d. Q.*, Florence 1934.

Raggi (Antonio), 1624–86. Decorator and stucco worker: cupola of S. Andrea al Quirinale and Gesù (Rome). See p. 323.

Raguzzini (Filippo), first half of 18th c. Roman Baroque architect, author of Piazza S. Ignazio. See p. 311.

Raimondi (Marcantonio), 1475–1534. Engraver, copied Dürer, entered service of Raphael whose major compositions he popularized throughout Europe. See p. 267.

Rainaldi (Carlo), Rome 1611–91. Architect, decorator, son of Girolamo. Built churches in Piazza del Popolo, *c.* 1660 apse of Sta Maria Maggiore. See pp. 315, 317, 372.
Bibl.: E. Hempel, *C.R.*, Rome 1921. R. Wittkower, 'A. B.' 1937.

Ranzoni (Daniele), between 1849 and 1889. Milanese painter of delicate and Romantic style, member of '*scapigliati*' group. See p. 385.
Bibl.: Sarfatti, *D.R.*, Rome 1935.

Raffaello da Montelupo, *c.* 1505–70. Florentine sculptor, assistant of Sansovino at Loreto, follower of Michelangelo. See p. 234.

Raphael (Raffaello Sanzio), Urbino 1483–Rome 1520. Painter and architect, son of Giovanni Santi of Urbino, pupil of Perugino whose calm measured style he mastered as early as the *Sposalizio* of 1504. In Florence he succumbed to the influence of Leonardo (*Belle Jardinière*) and Fra Bartolommeo. Summoned to Rome in 1508, he had an outstanding success with the decoration of the Stanza della Segnatura (1509–11), d'Eliodoro (1511–14) and dell'Incendio (1514–17), the rest being the work of his pupils. His paintings in the Farnesina (1514–16) and the Vatican Loggie proved him to be a decorator of genius. The increasing richness of his style (which even began to show certain Venetian influences) is most marked in his portraits: Julius II, Leo X, Baldassare Castiglione. In 1516 Raphael provided cartoons for ten tapestries for the Sistine Chapel; after 1515 he succeeded Bramante as architect of St. Peter's and director of the Antiquities of Rome.
His influence was immense and was felt in every field of the arts; he incarnates the strength, clarity and sweetness of the classical style. See pp. 215, 224–30, 239, 241, 242, 243, 244, 255, 257–60, 261, 262, 264, 265, 266, 268, 269, 270.
Bibl.: V. Golzio, *R. nei documenti* . . . , Vatican 1936. T. Hetzer, *Gedanken*

von R. S. Form, Berlin 1932. C. Gamba, *R.*, Paris 1932. A. Venturi, *R.*, Rome 1920. W. Wauscher, *R. S. da Urbino*, English translation London 1926. S. Ortolani, *R.*, Bergamo 1942. O. Fischel, *R.*, 2 vols., London 1948; *R.'s Zeichnungen*, 8 vols., Berlin 1913–41. E. Camesasca, *R.* (coll. Rizzoli, 2 vols., Milan 1956). Hofmann, *R. als Architekt*, 2nd ed., 4 vols, Leipzig 1908–14. On the workshop: H. Dollmayr, *R. S. Werkstatt*, 'J.W.', XVI 1895. F. Hartt, *R. and Giulio Romano*, 'A.B.' 1944.

Regaseo (Gian Battista), 1799–1872. Architect, completed the cemetery of Genoa. See p. 381.

Reggiani (Mauro), Nonantola 1897. Milanese painter with passion for pure form. See p. 392.

Remondini da Bassano, 18th c. Venetian family of cabinet makers, famous for painted or lacquered furniture decorated with vignettes. See p. 355.

Reni (Guido), Calvenzano 1575–Bologna 1642. Bolognese painter who prolonged the example of the Carracci. Active in Rome with Albani. Back in Bologna, became master in use of light tonality. See pp. 345, 346.
Bibl.: F. Malaguzzi Valeri, *G.R.*, Florence 1928. Cat. Exhibition (by C. Gnudi), Bologna 1954. C. Gnudi and G. C. Cavalli, *G.R.*, Florence 1955.

Reycend (Enrico), 1855–1928. Painter from Turin. Impressionist landscapes. See p. 385.
Bibl.: R. Longhi, *Ricordi di E.R.*, 'Pa.', no. 27, March 1952.

Ribera (Giuseppe de), 1588–1655. Born in Valencia, called Lo Spagnoletto. Arrived in Italy before 1610, visited Rome and Parma, became painter to the court of Naples; remarkable portraitist, influenced by Caravaggio, evolved towards more Baroque forms. See p. 343.
Bibl.: A. Mayer, *G. de R.*, Leipzig 1923. H. Bedarida, *R.*, 'A travers l'art italien', Paris 1949.

Ricci (Sebastiano), Belluno 1659–Venice 1734. Widely travelled Venetian painter, pupil of Mazzoni and Crivelli, who worked in Bologna, Florence, Modena, Parma and Milan but did his most important work in Venice in a restless style influenced by Magnasco and Pietro da Cortona. See pp. 350, 357.
Bibl.: J. v. Derschau, *S.R.*, Heidelberg 1922.

Ricci (Marco), Belluno 1676–Venice 1729. Venetian painter, nephew of Sebastiano. Active in a number of Italian towns and in London. Author of Romantic landscapes and gouaches. See p. 362.

Ricci (Leonardo), Ancona 1918. Painter and architect, living in Florence. See p. 395.
Bibl.: L. Savioli, 'Architetti' III, 1952.

Ricciarelli (Daniele), see **Volterra**.

Riccio or Crispus (Andrea **Briosco**, called), Padua 1470–1532. Paduan bronze sculptor, author of Della Torre tomb in S. Fermo, Verona (fragments in Louvre) and of numerous statuettes and genre pieces. See p. 244.
Bibl.: L. Planiscig, *A.R.*, Vienna 1927.

Richini (Francesco Maria), 1583–1658. Baroque architect from Lombardy, active in Milan where he built the entrance and courtyard to the Ospedale Maggiore and Palazzo di Brera (formerly Jesuit college). See p. 326.

Rinaldi (Antonio), 1709–94. Italian architect, built the Marble Palace in Moscow, 1768–85. See p. 368.

Rizzo (Antonio), mentioned between 1465 and 1498. Sculptor and architect from Verona, worked on Certosa di Pavia (1465), then in Venice (Tron monument, 1473). See p. 197.
Bibl.: E. Arslan, *L'Œuvre de jeunesse d'A.R.*, 'G.B.A.', Sept. 1953.

BIOGRAPHICAL NOTES

Robbia (Luca della), Florence 1399–1482. Florentine sculptor, author of Cantoria of Florence Cathedral (1431–37), of medallions of Campanile (1437–1439) and of bronze door of the cathedral (1446). Invented form of glazed polychrome terracotta sculpture which had great success in Tuscany. See pp. 160, 163–4.
Bibl.: L. Planiscig, *L. d. R.*, Vienna 1940. P. Schubring, *L. d. R* (2nd. ed.), 1921.

Robbia (Andrea della), Florence 1435–1525. Florentine sculptor, nephew and pupil of Luca. Continued manufacture of faience decorative sculpture with assistance of his sons Girolamo and Giovanni. See p. 163.
Bibl.: W. v. Bode, *Die Familie della R.*, Berlin 1914. A. Marquand, *A. d. R. and his atelier*, 2 vols., London 1922.

Roberti (Ercole dei), Ferrara 1456–96. Ferrarese painter, pupil of Cosmè Tura with whom he worked on Palazzo Schifanoia. Named Court Painter in 1477, went to Bologna (1480–86) and worked on *Griffoni altarpiece* with Cossa. In 1492 accompanied Alfonso d'Este to Rome. Master of the original and incisive art of Ferrara: *Pala di San Lazzaro* (Berlin). See pp. 192–4, 260.
Bibl.: A. Venturi, *E. d. R.*, 'Arch. Stori. Arte' 1889. R. Longhi, *Officina Ferrarese*, op. cit. S. Ortolani, *Cosmè Tura, F. del Cossa, E.R.*, Milan 1941.

Roccatagliata (Niccolo), 17th c. Most important member of Genoese family of sculptors, active in Venice, author of very complicated *paliotto* of S. Moisè (Venice), 1633. See p. 355.

Rocco da Montefiascone, mid-16th c. Roman architect and decorator, worked as assistant to Pirro Ligorio at Villa Pia (1560). See p. 290.

Rocco (Ernesto), 19th c. Architect, designed Galleria Umberto I in Naples (1887–90). See p. 381.

Rodbertus Magister Comacinus, 8th c. Mentioned in a document of 739 as having worked on S. Pietro di Toscanella (Viterbo). See p. 46.

Romanino (Girolamo **di Romano**, called **Il**), 1484–1566. Lombard painter from Brescia, interested in Venetian art, particularly Giorgione, open to German influence. Worked in Cremona. See pp. 258, 296.

Romano (Giancristoforo), *c.* 1465–1512. Fashionable sculptor and medallist, active at the courts of Mantua, Urbino and Naples. Friend of the poet Bembo.
Bibl.: J. de Foville, 'G.B.A.' XXXIX (1908).

Romano (Giulio), see **Giulio Romano**.

Rosa (Salvator), 1615–73. Neapolitan painter and poet of fertile, spontaneous and unstable temperament. Active in Naples, Rome and Florence. Seascapes and battle pieces. See pp. 344, 350–1.
Bibl.: G. A. Cesareo, *Poesie e lettere inedite di S.R.*, Naples 1892. A. de Rinaldis, *Lettere inedite di S.R.*, Rome 1939.

Rosselli (Cosimo), Florence 1439–1507. Florentine painter, pupil of Neri di Bicci; author of frescoes in Annunziata (1476) and in Sistine Chapel. See p. 208.

Rossellino (Antonio), 1427–79. Florentine sculptor, brother of Bernardo, author of tomb of Cardinal of Portugal (1459) and of numerous busts and reliefs of the Madonna; flowing style. See pp. 160, 164, 171.
Bibl.: L. Planiscig, *B. und A.R.*, Vienna 1942.

Rossellino (Bernardo), 1409–64. Brother of Antonio, Florentine sculptor and architect influenced by Alberti on whom he relies heavily in his Palazzo Venezia (Rome 1455) and Palazzo Rucellai (1446–50); author of various town-planning projects for Pope Pius II Piccolomini, including plan and episcopal palace of Pienza. *Tomb of Leonardo Bruni* (Sta Croce 1444–50). See pp. 157, 160, 164, 182–3, 220.

BIOGRAPHICAL NOTES

Bibl.: L. H. Heydenreich, 'Z. f. Kg.' VI 1937.

Rossetti (Biagio), mentioned 1465–1516. Bolognese architect, redesigned Ferrara for Ercole I, built Palazzo dei Diamanti, Palazzo Costabile, church of S. Francesco (1494). See p. 192.

Rossi (Domenico), early 18th c. Venetian architect with Baroque leanings: church of Gesuiti in Venice (1729) and façade of S. Stae. See p. 354.

Rosso (Francesco Maria), mid-18th c. Neapolitan painter and decorator, painted vault of Cappella S. Severo in Sta Maria della Pietà (1749). See p. 333.

Rosso (Gian Battista **di Jacopo**, called **Il**), Florence 1494–1540. Painter and decorator, pupil in Florence of Andrea del Sarto and Pontormo; went to Rome which did not suit his nervous temperament. Summoned to France by Francis I, directed work on Grande Galerie of Fontainebleau where the combination of fresco and stucco became a model for the entire century. See pp. 276–7.

Bibl.: K. Kusenberg, *Le R.*, Paris 1931. P. Barocchi, *Il R.F.*, Rome 1950. S. Lougren, *R.F. à Fontainebleau*, 'Figura' (Stockholm), 1951, I.

Rosso (Medardo), 1858–1927. Piedmontese sculptor, master of flexible Impressionist style. See p. 383.

Bibl.: A. Soffici, *Il caso M.R.*, Florence 1909.

Rovezzano, see **Benedetto da Rovezzano**.

Ruggieri (Ferdinando), 1691–1741. Florentine architect of classicizing style, author of Palazzo Capponi, began S. Firenze. See p. 331.

Ruoppolo (Giovanni Battista), 1620–85. Neapolitan painter, pupil of Paolo Porpora, specialized in still lifes. See p. 342.

Rusconi (Camillo), 1658–1728. Baroque sculptor trained in Milan, settled in Rome where he relied heavily on Bernini for his Paravicini tomb in S. Francesco a Ripa; executed white marble tomb of Gregory XIII (1723) in controlled Baroque style. See pp. 324, 325.

Russolo (Luigi), Portogruaro 1885–Laveno 1947. Futurist painter, author of manifesto of March 1913 on '*l'arte dei rumori*'. See p. 390.

Rustici (Giovanni Francesco), Florence 1474–Tours 1554. Florentine painter, fellow-pupil of Michelangelo in studio of Ghirlandaio; strongly influenced by Leonardo, went to France in 1528. See pp. 207, 243, 272.

Rutelli (Mario), Palermo 1859–1941. Roman sculptor, author of statues decorating Guerrieri fountain in Piazza dell'Esedra, Rome (1885). See p. 379.

Sabatelli (Luigi), 1772–1830. Florentine painter and decorator, pupil of Benvenuti, author of decorations in Palazzo Capponi and Palazzo Pitti (Florence). See p. 374.

Sacchi (Andrea), 1599–1661. Bolognese painter, pupil of Albani; portraits, altarpieces, room in Palazzo Barberini. See pp. 321, 346.

Bibl.: H. Posse, *Der römische Maler A.S.*, Leipzig 1925.

Sacconi (Giuseppe), 1854–1905. Architect of monument to Victor Emmanuel II; winner of national competition, 1884. See p. 382.

Saliba (Antonello da), mentioned 1480–1535. Sicilian painter, son of wood-sculptor Giovanni and of a sister of Antonello da Messina; uneven talent, particularly after 1500. See p. 186.

Bibl.: Cat. of Antonello exhibition (by G. Vigni), Venice 1953.

Salimbeni (Jacopo), late 14th, early 15th c. Painter from the Marches, born in Sanseverino, with his brother Lorenzo exponent of pleasant unambitious Gothic style. See p. 131.

Salini (Tommaso, called Mao), Rome c. 1570–1625. Pupil of Baglione, painted flowers and still-life. See p. 341.

Salvator Rosa, see **Rosa**.

Salvi (Nicolo), 18th c. Roman architect, author of the Trevi Fountain (1735–44). See p. 311.

Salviati (Francesco **de' Rossi**, called Cecchino), 1510-63. Florentine painter whose art derives from Andrea del Sarto, worked in Palazzo Vecchio, 1544. See pp. 277, 289.

Sanfelice (Ferdinando), 18th c. Neapolitan decorator and scenic designer, pupil of Solimena. See p. 332.

Sangallo. Family of architects from Florence (15th c.) who later moved to Rome (16th c.), descended from Giuliano and Antonio Giamberti.

Bibl.: on the whole family: G. Clausse, *les San Gallo*, 3 vols., Paris 1900–1902 (unreliable).

Antonio Cordiani da Sangallo, called the Younger, 1473–1546. Nephew of Antonio the Elder, military architect in Rome and Orvieto. Began Palazzo Farnese in 1511 for Paul III and built several churches in Rome: S. Spirito in Sassia and Sta Maria di Loreto. Promulgated and popularized style of Bramante and Raphael. See pp. 221, 224, 251.

Antonio Giamberti da Sangallo, called the Elder, 1455–1534. Architect, trained by his brother Giuliano, demonstrates his ability to manipulate architectonic forms in a harmonious and classical manner in churches of Annunziata (Arezzo) and S. Biagio (Montepulciano, 1519–26). Follows Raphael's architectural style closely in palaces of Montepulciano and Monte San Savino. See pp. 220–4.

Francesco, called Il Margotta, 1494–1576. Florentine sculptor, son of Giuliano, author of statue of Paolo Giovio in cloister of S. Lorenzo.

Gian Francesco, 1482–1530. Architect, nephew of Giuliano, worked on Palazzo Pandolfini (1515–16). See pp. 228, 242.

Giuliano Giamberti, called Giuliano da Sangallo, 1445–1516. Florentine architect and sculptor, eldest of large family of architects who mark transition to classical style. After early works in wood sculpture (choir stalls in Medici chapel), he built in rapid succession the most representative works of the late 15th c.: centrally planned Madonna delle Carceri at Prato (1485–1491), villa at Poggio a Caiano (1480–85), sacristy of S. Spirito (1489–92) with Cronaca. He worked in Loreto, provided plans for palaces or fortresses in Naples, Savona and Rome, in service of future Pope Julius II. His designs for the façade of S. Lorenzo in Florence were to inspire Michelangelo. He compiled a collection of studies remarkable for their wealth of archaeological information. See pp. 185, 202–5, 216, 220–4, 234.

Bibl.: G. Marchini, *G. da S.G.*, Florence 1942. Drawings: Falb, *Il Taccuino Senese di G. da S.*, Siena 1892. C. Hülsen, *Il Libro di G. da S.* (Cod. Barb. 4424), Leipzig 1910. B. Degenhart, *Dante, Leonardo und Sangallo*, 'R.J.', XVII (1955).

Sanmartino (Giuseppe), 18th c. Neapolitan Baroque sculptor. See p. 333.

Sanmicheli (Michele), 1484–1559. Venetian architect, pupil of Falconetto and Sansovino, trained in Rome in studio of Antonio da Sangallo. Active in Rome: villas. Military architect: fortifications of Verona, Padua. See p. 251.

Bibl.: E. Langenskiöld, *M.S., The Architect of Verona*, Uppsala 1938. *M.S.*, exhibition catalogue (introduction: P. Gazzola), Verona 1960; London, R.I.B.A., 1961 (short introduction: P. Murray).

Sano di Pietro, 1406–81. Sienese painter, director of busy workshop which produced number of village Madonnas and altarpieces; for a period, *c.* 1445–50, under influence of Domenico di Bartolo, his style became more distinguished. See p. 175.

Bibl.: Truebner, *Die stilistische Entwicklung des S. di P.*, Strasbourg 1925. E. Gaillard, *Un peintre Siennois du XVe siècle: S. di P.*, Chambéry 1923.

Sansovino (Andrea **Contucci**, called), 1470–1529. Architect and sculptor whose style derives from Bramante. Worked as an architect in Rome and Loreto. Active in Tuscany (Siena), central Italy and Portugal, evolved towards subtle and deliberately effective style: *Baptism of Christ* (Florence Baptistery, after 1502), tombs in Sta Maria del Popolo (Rome). See pp. 207, 243–5.

Bibl.: G. H. Huntley, *A.S.*, Cambridge, Mass., 1935.

Sansovino (Jacopo **Tatti**, called), 1486–1570. Florentine sculptor and architect, strongly influenced by Andrea Sansovino, whose name he took, and by Andrea del Sarto; attracted by the art of Michelangelo and Raphael, he formed a delicate style on elements taken from these artists (*S. Jacopo*, Florence Cathedral 1511–18), which became stronger in Rome; from there he went to Venice (1527), where he built the Loggia of the Campanile, the Palazzo Corner and the Libreria. See pp. 243, 248–50, 254, 275.

Bibl.: H. Weihrauh, *Studien zum bild. Werk der J.S.*, Berlin 1935. G. Lorenzetti, *Itinerario sansoviniano a Venezia*, Venice 1929.

Sant'Elia (Antonio), 1888–1916. Architect, published manifesto of Futurist architecture in 1914. See p. 393.

Santi (Giovanni), 1435–94. Painter from the Marches, father of Raphael, active in Urbino. Author of rhymed chronicle. See p. 181.

Bibl.: H. Holtzinger, *F. di Montefeltro, Cronaca di G.S.*, Stuttgart 1893.

Santomaso (Giuseppe), Venice 1907. Venetian painter, founder of the '*Fronte nuovo delle arti*' (1945). See p. 392.

Sanzio (Raffaello), see **Raphael**.

Saraceni (Carlo), 1585–1625. Venetian painter, went to Rome where he became leader of Caravaggio's followers. See pp. 341, 342.

Bibl.: A. Porcella, *C.S.*, 'Riv. di Venezia' 1928.

Sardi (Giuseppe), 1630–99. Venetian architect, follower of Scamozzi, author of Sta Maria del Giglio and façade of Scalzi in Venice. See p. 354.

Sardi (Giuseppe), 18th c. Roman Baroque architect influenced by Borromini, author of Sta Maria Maddalena (Rome 1735). See p. 325.

Sarto (Andrea del), Florence 1486–1531. Florentine painter, pupil of Piero di Cosimo, influenced by Leonardo; preoccupied with balance of formal content and a sort of melancholy serenity in an atmosphere of gentle chiaroscuro: *Madonna delle Arpie*. Sometimes very close to Fra Bartolommeo and Raphael. His penetrating portraits represent one of the finest achievements of Florentine classicism. See pp. 242–3, 275.

Bibl.: I. Fraenkel, *A. d. S.*, Strasbourg 1935. A. J. Rusconi, *A. d. S.*, Bergamo 1935. H. Wagner, *A. d. S.*, Strasbourg 1951.

Sartorio (Giulio Aristide), Rome 1860–1932. Roman painter of pre-Raphaelite outlook. See p. 384.

Bibl.: A. Bertini-Calosso, Sartorio Exhibition, Rome 1933.

Sassetta (Stefano di Giovanni, called), Siena 1392–1450. Sienese painter, developed personal version of International Gothic incorporating Florentine concepts; follows logical line of development from Linaiuoli triptych (1423–1426) to altarpiece of St. Francis at Borgo San Sepolcro (1437–44) and polyptychs of S. Domenico at Cortona (1434) and the *Madonna of the Snows* (1432). See pp. 126, 176.

Bibl.: B. Berenson, *A Sienese Painter* . . . , 1st ed. 1909, new ed. Florence 1946. J. Pope-Hennessy, *S.*, London 1939. E. Carli, 'B.M.', May 1951.

Sasso (G.), Lombard architect. Active in Sicily. Designer of the *Piazza dei Quattro Cantoni* at Palermo (1609). See p. 336.

BIOGRAPHICAL NOTES

Savoldo (Gian Girolamo), Brescia *c.* 1480–1548. Lombard painter, registered in Venice in 1508, married Flemish woman in 1521. Very personal art, night scenes and almost monochrome landscapes. See pp. 254, 296.
Bibl.: W. Suida, *S.*, 'Pantheon' 1937. G. Nicco Fasola, *Lineamenti del S.*, 'L'Arte' II 1940. Creighton Gilbert, 'A.B.' XXVII 1945, and 'G.B.A.' 1953.

Scala (Andrea), 19th c. Architect from Catania, author of Teatro Bellini. See p. 380.

Scalfarotto (Giovanni), 1700–64. Venetian architect, author of S. Simeone Piccolo, *c.* 1730. See p. 354.

Scamozzi (Vincenzo), Vicenza 1552–1616. Venetian architect who finished a number of Palladio's works including Teatro Olimpico. Built Procuratie Nuove, inspired by Sansovino, in Venice in 1586. See pp. 249, 253.
Bibl.: R. Pallucchini, *V.S.*, 'L'Arte' 1936. F. Franco, *La Scuola scamozziana di 'stile severo' a Vicenza*, 'Palladio' 1937. F. Barbieri, *V.S.*, Vicenza 1952.

Scandellari (Filippo), 1717–1801. Bolognese sculptor who specialized in figures for cribs. See p. 331.

Schiaffini (Francesco), 1691–1765. Genoese sculptor, emulated Bernini. See p. 327.

Schiavone (Andrea **Meldolla**, called), Zara *c.* 1510–Venice 1563. Painter of Dalmatian origin who studied the engravings of Parmigianino. With his extremely fluid paint, he introduced Mannerism, properly speaking, into Venice *c.* 1540. See pp. 256, 284.
Bibl.: V. Moschini, *Capolavori di A.S.*, 'Emporium' 1943.

Scipione (Gino **Bonichi**, called), Macerata 1904–33. Painter of unusual aspects of Roman life. See p. 392.
Bibl.: U. Apollonio, *S.*, Venice 1945. C. Maltese, *S.*, 'Emporium' 1948.

Sebastiano del Piombo (Sebastiano **Luciani**, called), Venice *c.* 1485–Rome 1547. Venetian painter whose style derives from Giorgione; influenced by Michelangelo in Rome. See pp. 225, 244, 254.
Bibl.: L. Dussler, *S. d. P.*, Basle 1942. R. Pallucchini, *S. Viniziano*, Milan 1944.

Segantini (Giovanni), Arco 1858–Schafburg (Engadine) 1899. Milanese painter who concentrated on rural life and Alpine landscapes (at Brianza). See p. 385.
Bibl.: N. Barbantini, *S.*, Venice 1945.

Segna di Buonaventura, documented 1298–1326. Sienese painter. Pupil of Duccio. Painted altarpieces and crucifixes. See p. 117.

Selva (Giannantonio), 1751–1819. Venetian architect, author of Teatro della Fenice and church of Nome di Gesù. See p. 371.

Semeghini (Pio), born at Quistello, near Mantua, 1878, settled in Verona. See p. 392.
Bibl.: Cat. Exhibition (by L. Magagnato), Verona 1956.

Serlio (Sebastiano), Bologna 1475–France 1554. Architect, pupil of Peruzzi, writer on classical art, equally important in Rome and Venice where he lived from 1531–40. His treatise in 7 volumes was published and reprinted several times after 1537 (the 8th volume was never completed). Invited to France, built château of Ancy-le-Franc. See pp. 250, 254, 269.
Bibl.: G. C. Argan, *S.*, 'L'Arte' 1932. W. B. Dinsmoor, 'A.B.' 1942.

Sernesi (Raffaello), Florence 1838–Balzano 1866. Painter who specialized in delicate studies of the Tuscan Apennines. See p. 386.
Bibl.: E. Cecchi, *Ritratto di R.S.*, 'Vita artistica', II, 1927. F. Wittgens, *Dodici opere di R.S.*, Milan 1951.

BIOGRAPHICAL NOTES

Serodine (Giovanni), 1594–1631. Roman painter born in Ticino, went to Rome *c.* 1615. Follower of Caravaggio. See pp. 341, 342.
Bibl.: R. Longhi, *S.*, Florence 1954.

Serpotta (Giacomo), 1656–1732. Sicilian sculptor and stucco worker, reaches Rococo virtuosity in his bas-reliefs which occasionally show Roman influence. See pp. 336, 338.
Bibl.: F. Meli, *G.S.*, Palermo 1934.

Servandoni (Giovanni Nicolo), Florence 1695–Paris 1766. Architect, pupil in Piacenza of Giovanni Paolo Pannini (for painting) and in Rome of Giuseppe Rossi (for architecture). Went to Paris where he had a distinguished career as an architect (Saint-Sulpice). See p. 314.

Severini (Gino), Cortona 1893. Futurist painter who became more classical in style. See p. 390.
Bibl.: G. Severini, *Tutta la vita di un pittore*, Milan 1946. L. Courthion, *G.S.*, Milan 1930. L. Bardi, *G.S.*, Milan 1942. B. Wall, *G.S.*, London 1946.

Signorelli (Luca), Cortona 1450–1523. Umbrian painter, pupil and collaborator of Piero della Francesca but also influenced by anatomical precision of Pollaiuolo. Worked in the Marches, in Rome (Sistine Chapel, 1481), Florence (*Pan* for Lorenzo de' Medici, *c.* 1490). Painted numerous altarpieces. His dramatic lighting and powerful draughtsmanship show to best advantage at Monte Oliveto (1497) and above all in the Duomo of Orvieto (1499–1505). See pp. 179, 184, 208, 211.
Bibl.: L. Dussler, *S.*, Berlin 1927. Cat. Exhibition (Cortona–Florence) 1953. M. Salmi, *S.*, Rome 1953.

Signorini (Telemaco), Florence 1835–1901. Florentine painter, active follower of Garibaldi, brilliant polemicist in touch with artistic developments in Paris and London, painted sharply realist scenes. See p. 387.
Bibl.: E. Somarè, *T.S.*, Milan 1926. U. Ojetti, *T.S.*, Milan 1930.

Silvani (Gherardo), 1579–1675. Florentine architect and sculptor, author of Palazzo Castelli and Palazzo Corsini. See p. 331.

Simone Martini, see **Martini**.

Simone del Pollaiuolo, see **Cronaca**.

Simonetti (Michel Angelo), 1724–81. Roman architect, built rooms of Museo Pio Clementino (1774). See p. 371.

Sirani (Andrea), 1610–70. Roman painter, pupil of Guido Reni. See p. 347.

Sirani (Elisabetta), 1638–65. Woman painter, daughter of Andrea, worked in Bologna in wake of Carracci and Guercino.

Sironi (Mario), 20th c. Fertile painter formed by Futurist movement, unusual and original approach to urban landscape. See p. 392.
Bibl.: A. Sartoris, *M.S.*, Milan 1946. A. Pica, *M.S.*, Milan 1955.

Slodtz (Michel-Ange), French sculptor of Flemish origin, spent the years 1726–46 in Rome where he executed his statue of *St. Bruno* (St. Peter's, 1740). See p. 324.

Sodoma (Giovanni Antonio **Bazzi**, called), Vercelli 1477–Siena 1549. Lombard painter at first dominated by Leonardo, went to Siena where his manner became more complicated, worked at Monte Oliveto (1505–08). Then in Rome turned to Raphael (Farnesina) and returned to Siena where in 1526 he painted his *Ecstasy of St. Catherine* (S. Domenico). See p. 241.
Bibl.: L. Gielly, *S.*, Paris, n.d. E. Carli, Cat. Exhibition, Vercelli and Siena 1950.

Solari (Daniele), 1634–98. Genoese sculptor, assistant of Puget, influenced by Bernini. See p. 327.

Solario (Andrea), *c.* 1460–1524. Lombard painter, pupil of Leonardo,

474

went to France: in Blois, painted *Madonna with Green Cushion* (Louvre). See p. 241.

Soldati (Anastasio), 1896. Painter fascinated by pure form. See p. 392.

Soli (Giuseppe Maria), 1745–1823. Venetian architect, author of Procuratie nuove. See p. 371.

Solimena (Francesco), 1657–1747. Neapolitan decorator, specialized in powerful panoramas and architectural vistas. See pp. 320, 333–4, 344.

Sotio (Alberto), 12th c. Umbrian painter; in 1187 painted *Crucifix* in Spoleto Cathedral. See p. 74.

Spada (Lionello), 1576–1622. Bolognese painter, friend of Caravaggio; author of large-scale compositions full of strong contrasts. See p. 346.

Spadini (Armando), Florence 1883–1925. Florentine painter, tenuous connections with French Impressionism. See p. 389.

Spagnolo, see **Crespi.**

Spinazzi (Innocenzo), *c.* 1720–95. Roman sculptor, settled in Florence in 1784, author of Macchiavelli tomb in Sta Croce (1787). See p. 375.

Spinelli (Niccolo), see **Niccolo Fiorentino.**

Spinello Aretino, 14th c. Florentine fresco painter, free and eclectic style; sacristy of S. Miniato, Municipio of Siena, Campo Santo, Pisa. See p. 125.

Squarcione (Francesco), Padua 1397–1474. Paduan painter, master of Mantegna, collector, unstable and impulsive temperament. See pp. 190, 193.

Stacchini, born 1871. Architect of Stazione Centrale of Milan (1931). See p. 394.

Stanzioni (Massimo), called Il Cavaliere Massimo, 1585–1656. Neapolitan Caravaggiesque painter. See p. 343.

Bibl.: H. Schwanenberg, *Leben und Werk des M.S.*, Bonn 1937.

Starnina (Gherardo), late 14th c. Florentine painter, master of Masolino, in contact with Castille. See p. 125.

Bibl.: U. Procacci, 'Riv. A.' 1933.

Stefano, 14th c. Tuscan painter, pupil of Giotto. Author of *Coronation of Virgin* in lower church at Assisi, outstanding personality of the period 1330–1340. See p. 115.

Bibl.: R. Longhi, *S. fiorentino*, 'Pa.', no. 13, Jan. 1951.

Stefano della Bella, Florence 1610–64. Florentine painter and engraver whose art derives from Callot, went to Paris (1639–49). Author of theatre décors, costumes, fantastic figures, panoramic views. See pp. 275, 349.

Bibl.: A. Blunt, *G. B. Castiglione and S. d. B.*, London 1954.

Stefano di Giovanni, see **Sassetta.**

Stefano da Zevio, beginning of 15th c. Veronese painter, pupil of Altichiero; his paintings are illuminated manuscripts on a large scale: *Madonna of the Rose Bush.* See p. 133.

Stern (Raffaello), 1771–1820. Roman architect; during pontificate of Pius VII carried out large-scale alterations in the Vatican. See p. 371.

Strozzi (Bernardo), called Il Cappucino or Il Prete Genovese, Genoa 1581–Venice 1644. Venetian painter, Capuchin monk who left monastery and settled in Venice after 1630. Style influenced by Caravaggio. See p. 356.

Bibl.: G. Fiocco, *B.S.*, Rome 1921. P. Zampetti, 'Emporium' 1949.

Strozzi (Zanobi **di Benedetto**), Florence 1412–46. Florentine miniaturist, continued tradition under Fra Angelico at S. Marco (1446–53) and illustrated Chorals for the cathedral, *c.* 1470. See p. 195.

Bibl.: L. Collobi Ragghianti, *Z.S.*, 'C.A.' 1950.

Tacca (Pietro), 1577–1640. Florentine sculptor, active in Florence and Leghorn. See pp. 274, 331.

Taddeo di Bartolo, Siena 1362–1422. Sienese painter whose art derives from the Lorenzetti. See p. 176.

Talenti (Francesco), 1300–76. Florentine architect, *maestro dell'opera* of the cathedral, responsible for alterations to Arnolfo's plan and for final form of Campanile. See pp. 102, 120, 121, 122.
Bibl.: W. Paatz, *Werden u. Wesen der Trecento-Architektur*, Burg a. M. 1937.

Talenti (Fra Jacopo), first half 14th c. Author of Campanile of Sta Maria Novella. See p. 100.

Tallone, Savona 1853–Milan 1919. Popular Lombard portrait painter. See p. 385.
Bibl.: E. Somarè, *C.T.*, Milan 1946.

Tancredi da Messina, 1655–1725. Neapolitan fresco painter and stucco worker, decorated Jesuit church, Palermo. See p. 336.

Tanzio da Varallo, 1575–1635. Lombard painter. See p. 348.
Bibl.: *Tanzio da Varallo,* exhibition catalogue (preface: G. Testori), Turin 1959–60.

Tatti (Jacopo), see **Sansovino.**

Tavella (Carlo Antonio), 1668–1738. Milanese painter, specialized in landscapes and flowers. See pp. 348, 351.

Tempesta (Pietro **Mulier,** called), Haarlem 1637–Milan or Venice 1701. Landscape painter settled in Genoa, master of Tavella, specialized in storm scenes which earned him his nickname. See p. 351.

Tenerani (Pietro), 1789–1869. Roman sculptor, pupil of Canova, author of tomb of Pius VIII (St. Peter's). See p. 377.

Terragni (Giuseppe), 1904–43. North Italian architect, author of new district or 'Novocomum' of Como, completed in 1929. See pp. 393, 394.

Testa (Pietro), see **Lucchesino.**

Theotokopoulos, called El Greco, see **Greco.**

Thorwaldsen (Bertel), 1768–1844. Danish sculptor who spent 40 years in Rome. In 1831 executed tomb of Pius VII (St. Peter's). See p. 377.

Tibaldi (Pellegrino), called Il Pellegrini, Vasolda 1527–96. Lombard architect, painter and decorator. In Rome worked as engineer and architect: Trinità dei Monti. Summoned to Bologna, decorated Palazzo Poggi with clever use of foreshortening. Chief architect of Milan Cathedral (1567), then left for Spain and the service of Philip II, for whom he decorated library of Escorial (1587–91). See p. 294.
Bibl.: G. Briganti, *Il manierismo e P.T.*, Rome 1945.

Tiepolo (Giambattista), Venice 1696–Madrid 1770. Venetian painter and master of Baroque decoration in 18th century. Pupil of Lazzarini, influenced by Piazzetta, revived the clear colours and the breadth of Veronese in his fresco cycles on mythological (Udine, 1732–33; Villa Valmarana, 1737) or religious subjects (Scalzi, Venice). He was invited to Milan, Bergamo, Vicenza, to Venetian palaces (Labia) and the villas on the mainland; then abroad, to Würzburg (1750–53) with his sons, and finally to Madrid (after 1762). Lavish and heroic, the last of the great Baroque decorators, Tiepolo was also an original engraver fond of *capricci.* See pp. 350, 357, 358, 360, 362.
Bibl.: P. Molmenti, *G.B.T.*, Milan 1909. G. B. Morassi, *T.*, London 1955. Cat. Exhibition (by G. Lorenzetti), Venice 1951. Drawings: D. v. Hadeln, *The Drawings of G.B.T.*, 2 vols., Paris 1928. A. M. Brizio, 'O.M.D.', Sept. 1933. G. Lorenzetti, *Cahier des T. au musée Correr*, Paris 1946.

Tiepolo (Gian Domenico), Venice 1727–1804. Son of Giambattista, worked with his father. 1780, President of Venetian Academy. Specialized in genre subjects and engraved series of prints. See pp. 359, 360, 362.

Bibl.: E. Sack, *G.B. und G.D.T.*, Hamburg 1909.

Tino di Camaino, Siena c. 1295–c. 1337. Sienese sculptor, pupil of Giovanni Pisano, summoned to Pisa (tomb of Henry VII), to Florence (tomb of Bishop Orsi) and to Naples (tomb of Mary of Hungary). See pp. 108–9, 111, 121.

Bibl.: E. Carli, *T. di C. scultore*, Florence 1934. W. R. Valentiner, *T. di C.*, Paris 1935. O. Morisani, *T. di C. a Napoli*, Naples 1945.

Tintoretto (Jacopo **Robusti**, called), Venice 1518–94. Venetian painter, father of four more painters. Pupil of Titian, Paris Bordone and Pordenone, stimulated by Mannerism and the study of Michelangelo (strengthened by a visit to Rome in 1545–46), gave a completely new slant to Venetian painting. Diagonal compositions, startling perspective effects (*St. Mark rescuing the Slave*, 1548), floor-level lighting which drains the colour from the figures and wraps them in violent contrasts. Of almost unbelievable fecundity, he filled Venetian churches with his canvases and executed his most remarkable *tour de force* in the Scuola di San Rocco (1564–87). His style had a decisive influence on El Greco and the later phases of European Mannerism. See pp. 267, 284–6.

Bibl.: M. Pittaluga, *T.*, Bologna 1925. Cat. Exhibition (by R. Pallucchini), Venice 1937. L. Coletti, *T.*, Bergamo 1940. E. von der Bercken, *T.*, Munich 1942. R. Pallucchini, *La giovinezza di T.*, Milan 1950. Drawings: D. v. Hadeln, *Hz. des Tintoretto*, Berlin 1922. E. Tietze-Conrat, 'Die Graphische Künste', I 1936. H. Tietze, *T.*, London 1948.

Tirali (Andrea), c. 1660–1737. Venetian Baroque architect and sculptor (Valier monument, SS. Giovanni e Paolo), who later became more Neoclassical in feeling: S. Vitale, Venice, c. 1700, after Palladio. See p. 354.

Tisi (Benvenuto), see **Garofalo**.

Titian (Tiziano **Vecellio**, called), Pieve di Cadore c. 1485–Venice 1576. Venetian painter, pupil of Sebastiano Zuccato, then of the Bellini, profoundly influenced by Giorgione, several of whose works he finished after 1510. In a short time he rapidly dominated Venetian painting (*Sacred and Profane Love*, c. 1515; *Frari Assumption*, 1516–18) and was approached by the courts of Ferrara (after 1516) and Mantua (1524), became official portraitist to Charles V who made him a count palatine in 1534 and invited him to Augsburg in 1547. Innumerable religious and secular works. His immense success is due to his ability to widen the range of colour and light by atmospheric effects and to bring his compositions into line with the Roman taste of the High Renaissance. C. 1540, a slight Mannerist crisis is apparent in his work. His last period reaches heights of moving grandeur. He played an outstanding part in the history of Venetian art and its position with regard to the rest of Europe. See pp. 252–8, 264, 267.

Bibl.: V. Basch, *T.*, Paris 1927. W. Suida, *T.*, Paris 1935. H. Tietze, *T. Leben u. Werk*, Vienna 1936, London 1950. C. Gamba, *T.*, Novara 1941. L. Grassi, *T.*, Rome 1945.

Toma (Gioacchino), Galatina 1836–Naples 1891. Neapolitan painter, author of calm simple genre pictures. See p. 388.

Bibl.: M. Biancale, *G.T.*, Rome 1933. S. Ortolani, *G.T.*, Rome 1934. A. de Rinaldis, *G.T.*, Verona 1934.

Tommaso da Modena (**Barisani**, called), Modena 1325–76. Emilian painter active in the Veneto where he painted portraits of the Dominicans of

S. Niccolo at Treviso (1351); may have worked in Bohemia. See p. 131.
Bibl.: L. Coletti, *T. da M.*, Bologna 1933. B. Montuschi, *T. da M. miniatore*, 'Pa.' XVII 1951.

Tommaso da Napoli, beginning of 18th c. Franciscan monk, Neapolitan architect, built villa of Prince of Palagonia (Bagheria) *c.* 1715. See p. 338.

Torbido (Francesco), called Il Moro, Venice *c.* 1483–Verona 1562. Painter influenced by Giorgione and Giulio Romano: *Apostles* (1577, Cathedral of Verona).
Bibl.: D. Viana, *F.T. o il M.*, Verona 1933.

Torreggiani (Alfonso), 1682–1764. Bolognese architect, author of numerous palaces with restrained decoration (Palazzo Montanari). See p. 330.

Torriti (Jacopo), 13th c. Painter and mosaicist, worked on apse of S. Giovanni in Laterano (Rome), then apse of Sta Maria Maggiore (Rome, 1291–95). See pp. 88, 92, 96, 112.

Tosi (Arturo), 1871–1956. Lombard painter in contact with artistic developments in Paris. See p. 392.
Bibl.: G. C. Argan, *T.*, Florence 1942. G. Scheiwiller, *A.T.*, Milan 1942.

Traini (Francesco), *c.* 1350. Florentine painter: author of altarpiece of St. Dominic, Pisa, and possibly of the allegory of the *Triumph of Death* (Campo Santo, Pisa) which shows great qualities of imagination and draughtsmanship. See p. 124.
Bibl.: M. Meiss, 'A.B.' 1933.

Traversi (Gaspare), ?–1769. Neapolitan painter, close to Bonito. See p. 344.

Tremignon (Alessandro), 17th c. Venetian Baroque architect, imitator of Longhena. Author of façade of S. Moisè (*c.* 1668). See p. 354.

Tribolo (Nicolo **Pericoli**, called), *c.* 1500–50. Florentine sculptor and engineer, author of the ornamental terraces with niches and grottoes of the Villa Castello (1540) and of the arrangement of the Boboli Gardens. See pp. 243, 274.

Tristano (Giambattista), ? 1515–Rome 1575. Ferrarese architect. Entered the Jesuit Order and made several designs for Churches of the Society. See p. 293.
Bibl.: P. Pirri, *G.T. I primordi dell'Architettura Gesuistica*, Rome 1955.

Trucco (Mario), 20th c. Architect of the Fiat works in Turin, 1927. See p. 394.

Tura (Cosmè), Ferrara 1430–95. Ferrarese painter trained under Squarcione in Padua, and Donatello; met Piero della Francesca in Ferrara (1450). One of the three masters of the Palazzo Schifanoia (*c.* 1470); painted organ doors in Ferrara Cathedral and numerous official portraits. Developed unusually sharp style depending on powerful draughtsmanship. See pp. 192–4, 195, 260.
Bibl.: L. N. Cittadella, *Ricordi e Documenti*, Ferrara 1866. O. Hatzch, *C.T.*, 'Pantheon' 1940.

Turone, 14th c. Veronese painter mentioned in 1360. Decorated Sant'Anastasia. Style influenced by German miniaturists. See p. 133.

Ubertini (Francesco), see **Bachiacca**.

Uccello (Paolo **di Dono**, called), Florence 1397–1475. Florentine painter, decorator, mosaicist, pupil of Ghiberti. Mosaics in S. Marco, Venice (1425), author of cartoons for windows in Florence Cathedral (1445). Using naturalistic settings of Gothic origin, he imposes a geometrical approach on the visual world which is above all noticeable in his perspective exercises: equestrian monument (Florence Cathedral, 1436); battles (Uffizi, Louvre, London

1456–60); Old Testament cycle in cloister of Sta Maria Novella (*c.* 1445). See pp. 147, 162, 168–71, 190, 197, 198, 208.
Bibl.: W. Boeck, *P.U.*, Berlin 1939. M. Salmi, *P.U. Castagno, Veneziano*, 2nd ed., Milan 1938; and 'Com.' 1950 and 1954. J. Pope-Hennessy, *P.U.*, London 1950. Cat. Exhibition '*Quattro Maestri*', Florence 1954.

Ugolino di Nerio, documented *c.* 1317–27. Sienese painter, contemporary with Duccio. Painted a polyptych for Sta Croce in Florence (divided between Berlin, London, New York). See pp. 117, 120.

Ugolini di Vieri, documented *c.* 1380–85. Sienese goldsmith, carved shrines notable for their very ornate embossing: the Bolsena Corporal, S. Savinian. See p. 128.

Vaccarini (Gian Battista), Palermo 1702–68. Sicilian architect, pupil of Carlo Fontana; author of Palazzo Biscari and of Benedictine convent (Catania) during course of reconstruction of East coast towns. See pp. 336–8.

Vaccaro (Andrea), 1590–1670. Neapolitan painter, pupil of Caracciolo. See p. 343.

Vaccaro (Domenico Antonio), 1680–1750. Neapolitan sculptor and decorator; design for transformation of Sta Chiara (1724). See p. 332.

Valadier (Luigi), late 18th c. Roman architect, father of Giuseppe Valadier. See p. 372.

Valadier (Giuseppe), 1762–1839. Architect of pontifical Rome, pupil of his father Luigi; designed façade of S. Pantaleone. See pp. 372, 373.
Bibl.: I. Ciampi, *Vita di G.V.*, Rome 1870.

Valeriani, 18th c. Family of painter-decorators; most important members Domenico and Giuseppe, who worked at Stupinigi with Juvara and in palaces built by Alfieri. See p. 328.

Valle (Filippo della), 1696–1770. Roman Baroque sculptor, made tomb of Innocent XII. See p. 325.

Valvassori (Gabriele), Rome 1683–*c.* 1750. Roman architect, worked on Palazzo Doria from 1720–34. See p. 325.

Vanni (Andrea), mentioned between 1350 and 1375. Sienese painter whose art derives from Lorenzetti. See p. 123.

Vanni (Lippo), mid-14th c. Sienese painter and illuminator. See pp. 123, 127.

Vantini (Rodolfo), 1791–1856. Lombard architect, built cemetery of Brescia. See p. 381.

Vanvitelli (Gaspard **Van Wittel**), 1674–1736. Dutch painter active in Venice and Rome. Port scenes, precursor of Canaletto. See p. 361.

Vanvitelli (Luigi), 1700–73. Neapolitan architect, son of Van Wittel, knew Juvara. Active in Ancona, Rome, Naples, summoned to Naples by Charles III in 1751 to construct the sumptuous palace of Caserta. See pp. 332, 333.
Bibl.: F. Fichera, *L.V.*, Rome 1937.

Vanvitelli (Carlo), 18th c. Neapolitan architect, son of Luigi, continued latter's work at Caserta and completed Annunziata in Naples (1782). See p. 333.

Vasari (Giorgio), Arezzo 1511–Florence 1574. Tuscan architect, painter and decorator, responsible for monumental reorganization of Uffizi and Palazzo Vecchio under Grand-Duke Cosimo de' Medici. Best remembered for his *Vite*, the first modern history of art (1st ed. 1550, 2nd ed. 1568). See pp. 271, 272, 273, 289.

Vasi (Giuseppe), mid-18th c. Roman engraver, author of *Le Magnificenze di Roma Antica e Moderna*, 1747–61. See p. 352.

BIOGRAPHICAL NOTES

Vassalletto, 12th–13th c. Roman family of mosaicists and decorators on same level as Cosmati: S. Giovanni in Laterano, atrium of S. Lorenzo-fuori-le-Mura. To Pietro are attributed the paschal candelabra of S. Paolo-fuori-le-Mura and Anagni (1263). See pp. 6, 68.

Vecchietta (Lorenzo **di Pietro,** called), Siena 1392–1480. Sienese painter and sculptor; painter of uneven quality, restless compositions (frescoes in Siena Baptistery 1450); master of Matteo di Giovanni. Bronzes rely heavily on Donatello: *Resurrection* (private coll. New York, 1472), *Risen Christ* (Ospedale, Siena, 1476). Wood sculptures include *St. Anthony Abbot* (Narni), *St. Bernardino* (Bargello). See pp. 174, 176.
Bibl.: P. Vigoni *L. di P.*, Florence 1917.

Vedova (Emilio), Venice 1919. Painter of *Fronte nuovo delle arti*. See p. 392.

Vela (Vincenzo), 1822–91. Sculptor from Ticino, Romantic and realist subjects. See p. 382.

Veneziano (Domenico), see **Domenico.**

Veronese (Paolo **Caliari,** called), Verona 1528–Venice 1588. Venetian painter and decorator from a provincial centre, mainly influenced by Titian after his arrival in Venice in 1553 and Raphael after his journey to Rome in 1560. Worked in Doge's Palace from 1553 on, evolved type of clear, luminous and rhythmic composition, full of optimism and vitality. See pp. 286–7.
Bibl.: G. Fiocco, *V.*, Rome 1934. A. Orliac, *V.*, Paris 1939. Cat. Exhibition (by R. Pallucchini) Venice 1939. E. Tea, *P.V.*, Milan 1940. R. Pallucchini, *P.V.*, Bergamo 1946.

Verrocchio (Andrea **di Cione,** called), Florence 1435–Venice 1488. Florentine sculptor, goldsmith and painter. From 1465–80 headed most important workshop in Florence, with that of Pollaiuolo; executed busts, breast-plates, Medici tomb in S. Lorenzo (1472), bronze group of Christ and St. Thomas for Or San Michele (*c.* 1460–82). In 1482 went to Venice to make Colleone statue.
In painting, Verrocchio concentrates on technical researches and produces smooth well-lit effects: *Baptism of Christ*, with Leonardo (*c.* 1470–75), Pistoia altarpiece with L. di Credi (1478–82). His pupils include Leonardo, L. di Credi, Perugino. See pp. 128, 146, 147, 148, 165, 206, 208–9, 212.
Bibl.: M. Mackowski, *V.*, Leipzig 1901. M. Reymond, *V.*, Paris 1905. A. Bertini, 'L'Arte' 1935. L. Planiscig, *V.*, Vienna 1941.

Viani (Andrea), 1906. Sculptor of 'abstract forms'. See p. 393.

Vignola (Giacomo **Barozzi,** called), Vignola (near Modena) 1507–73. Roman architect trained in Bologna where he built Palazzo dei Banchi. In 1554 went to Rome, worked on Villa Giulia, then studied the antique and returned to a more rigorous style, providing the Counter-Reformation with the archetypal Gesù (1568–75), taken from elements in Alberti. Author of classic treatise on the orders (1562), which was widely read throughout the classical period. See pp. 290–3.
Bibl.: Various authors, *G. B. Vignola*, 1908. G. K. Loukomski, *V.*, Paris 1927.

Vincenzo (Antonio di), see **Antonio.**

Vincenzo dei Rossi, 1525–87. Florentine Mannerist sculptor, author of *Labours of Hercules* (Palazzo Vecchio); continued style of Baccio Bandinelli. See p. 273.

Vinci (Leonardo da), see **Leonardo.**

Vitale da Bologna (Cavalli, called), mentioned between 1330 and 1359. Emilian painter, decorated cathedrals of Udine and Pomposa. See p. 131.

BIOGRAPHICAL NOTES

Vittone (Bernardo), *c.* 1705–70. Architect from Turin where he succeeded Juvara, author of Sta Maria di Piazza (1751). See p. 330.

Vittoria (Alessandro), 1525–1608. Venetian sculptor, author of portrait busts and statues for churches, derives from Giovanni da Udine. Decorated Villa Barbaro at Maser and Villa Pisani at Montagnana. See p. 284.

Vittozzi (Ascanio), 1539–1615. Piedmontese architect. See p. 327.

Viva di Lando, second half 14th c. Sienese goldsmith, author of reliquary of skull of S. Saviniano (Orvieto), with Ugolini di Vieri. See p. 128.

Vivarini (Antonio), Murano 1415–Venice *c.* 1476–84. Venetian painter, brother of Bartolommeo and father of Alvise. Worked with Giovanni d'Alemagna until 1450. His style is characterized by richness of décor: *Coronation of the Virgin* (1444, Venice, S. Pantaleone). See pp. 193, 198.

Vivarini (Bartolommeo), Murano 1432–99. Brother, pupil and collaborator of Antonio Vivarini (1450–62); his rhythmic forms and bright colours played an important part in the development of Venetian painting: *Madonna* (1478). See pp. 191, 193, 199.

Vivarini (Alvise), Venice *c.* 1445–*c.* 1503. Venetian painter, son of Antonio, nephew of Bartolommeo, strongly influenced by Antonello da Messina whose solid manner he helped to popularize. See pp. 201, 257.

Volterra (Daniele **Ricciarelli**, called Daniele da), Volterra *c.*1509–Rome 1566. Tuscan painter dominated by Michelangelo. Worked in Rome and Tuscany. His *Deposition from the Cross* (1541) had a powerful effect on younger painters. See pp. 267, 289.

Vuolvinius, 9th c. Carolingian goldsmith, author of gold altar of S. Ambrogio, Milan. See p. 50.

Wiligelmo, 11th c. Lombard sculptor 1099. Worked with Lanfranco at Modena (signed reliefs on façade of cathedral). See pp. 59, 61.

Zaccagni (Bernardino), Parma 1455–*c.* 1530. Architect, with his son Gianfrancesco built the church of the Steccata in Parma. See p. 240.

Zais (Giuseppe), Formo di Canale 1709–Treviso 1784. Painter of open-air scenes influenced by Zuccarelli. See p. 361.

Zandomeneghi (Federico), Venice 1841–Paris 1917. Venetian painter, settled in Paris, 1874. See p. 384.
Bibl.: E. Picenzi, *F.Z.*, 1932.

Zanobi (di Lorenzo), 15th c. Florentine miniaturist. See p. 195.

Zanobi del Rosso, 1724–98. Florentine architect. Finished the Monastery of the Filippini (1775). See p. 331.

Zanoia (Giuseppe), 19th c. Milanese architect: arch of Porta Nuova. See p. 370.

Zelotti (Gian Battista), 16th c. Painter, decorator, worked with Veronese. See p. 283.

Zenale (Bernardino), 1436–1526. Lombard painter, harsh powerful style, with Butinone author of Treviglio altarpiece. See p. 189.

Zimbalo (Francesco), mainly active between 1660 and 1690. Architect from Apulia. Author of campanile of Cathedral of Sta Chiara at Lecce (1687). See p. 335.

Zocchi (Cesare), 19th c. Sculptor, author of equestrian statue of Victor-Emmanuel II (Florence, 1890). See p. 379.

Zoppo (Marco), 1433–78. Bolognese painter, harsh precise style, author of an album of drawings formerly attributed to Mantegna. See p. 192.
Bibl.: C. Dodgson, *A Book of Drawings attributed to Mantegna,* London 1923. G. Fiocco in 'Miscellanea Supino', Florence 1933.

BIOGRAPHICAL NOTES

Zotto (Antonio del), 1841–1918. Venetian sculptor, author of a statue of Goldoni. See p. 382.

Zuccarelli (Francesco), Pitigliano (Tuscany) 1702–88. Painter and engraver of landscapes, went to Venice *c.* 1732. Member of the Academy in 1763, pupil of Paolo Anesi. Visited Rome, Paris and London. See p. 361.
Bibl.: W. Arslan, 'Boll. Arte' 1934.

Zuccaro (Taddeo), 1529–66. Roman decorator, precious and pompous style. Worked at Caprarola. See pp. 289, 290.

Zuccaro (Federico), *c.* 1542–1609. Brother of Taddeo, painter and theorist, worked at Caprarola with his brother, frescoes on dome of Florence Cathedral with Vasari; his palace near Sta Trinità dei Monti was the headquarters of an Academy. Author of a treatise, 'Idea' (1607), 2nd ed. 1768. See p. 289.

Zucchi (Jacopo), 1541–90. Painter and decorator, collaborator of Vasari, worked on Palazzo Ruspoli built by Ammanati. See pp. 289, 298.

MASTERS

Magdalen Master, *c.* 1250. Florentine painter, gold, red and brown backgrounds of unusual quality. See p. 65.

Master of the Madonna of S. Martino (Rainieri d'Ugolino ?), active between 1260 and 1270. Pisan painter, author of *Maestà* (Museum of Pisa). Precursor of Duccio. See p. 65.
Bibl.: R. Salvini, 'Riv. Arte' 1950.

Master Filippo, 12th c. *Maestro dell'opera* who worked in Ancona, adapting Lombard formulae: Sta Maria della Piazza (1210). See p. 60.

Master of St. Francis, active between 1265 and 1275. Umbrian painter of the 13th c. Continued manner of Giunta. See p. 74.

Master of the St. George Codex, *c.* 1330. Sienese painter, follower of Simone Martini. See p. 127.

Moderno, early 16th c. Surname of Venetian bronze-caster who signed numerous plaques and classical statuettes: *Opus Moderni*. See p. 244.

Osservanza Master, early 15th c. Sienese master who takes his name from the polyptych in the church of the Osservanza in Siena. This serves as basis of reconstruction of an original personality, more 'Gothic' than Sassetta and earlier than Sano di Pietro. See p. 176.
Bibl.: A. Graziani, *Il maestro dell'O.*, 'Pr.', II 1948.

Topographical Index

Note. Buildings which have disappeared or been radically transformed are marked: *. Buildings damaged during the last war are marked: †, destroyed or very seriously damaged: ‡, rebuilt: §.

(After H. La Farge, *Lost Treasures of Europe*, New York 1946; French edition, *l'Europe blessée*, 1947.)

Acerenza (Lucania). *Cathedral* (13th c.), 77

Agliate (Lombardy). Church of *S. Pietro* (9th c.), characteristic work of the *maestri comacini*, 46.

Agrigento (Girgenti). Town in Sicily, founded in 582 B.C. by Greek colonizers. Many of its temples (5th c. B.C.) were transformed into churches in the Middle Ages, 77.

Captured by the Saracens in 828, by Ruggiero I in 1086, the town became the richest bishopric in Sicily and rival to Palermo.

Theatre, built by E. Basile (19th c.), 380.

Agro (San Pietro di) (Calabria). *Church*, founded in 1117, 77.

Amalfi. On the Gulf of Salerno, most important maritime republic of early Middle Ages, defeated by Pisa in 1195, 4, 53, 139.

Cathedral of Sant'Andrea (9th c.), restored in 19th c.; bronze doors brought from Constantinople in 1066, 13th c. campanile, 54, 76.

Cloister, work of Giulio di Stefano (1103), 76.

Bibl.: A. Schiavo, *Il Duomo di Amalfi*, 'Arte', 1939.

Anagni (Lazio), 40 km. from Rome, ancient church with frescoes from first half of 13th c., 70.

Ancona. Capital of the Marches and powerful maritime republic, became church possession in 16th c., 4, 255, 256, 257, 295.

†§*S. Ciriaco* (12th–13th c.), the cathedral; fine carved portico with 'protyrium' (13th c.), 60.

S. Domenico. Crucifixion by Titian (1557), 256.

Church of the Gesù, built by Vanvitelli (1733–45), 333.

S. Maria della Piazza (11th–13th c.). The portico is the work of 'Maestro Filippo' (1210), 60.

Ospedale, built by Vanvitelli, 333.

Prefettura, begun in 1417, continued with colonnaded courtyard and arcaded entry by Francesco di Giorgio (1486), 175.

Andria (Apulia). *Castel del Monte*, standing on a hill to the south of the town, built by Frederick II, *c.* 1240, 90, 91.

Aquila (Abruzzi), founded by Frederick II.

S. Maria di Collemaggio (1287), façade inlaid with marble and 14th-c. portico, 185.

S. Bernardino, founded in 1452 on the tomb of the famous Franciscan preacher.

Arabona (Abruzzi). *S. Maria*, Cistercian abbey founded in 1228, 84.

Ardara (Sardinia). *S. Maria del Regno*, 12th c. church, 72.

Bagnaia (Lazio), near Viterbo. *Villa Lante*, built by Vignola with splendid gardens for Cardinal Riario in 1578, 312.

Barga (Tuscany), north of Lucca. In the Romanesque church, pulpit by followers of Guido da Como (1256), 73.

Bari (Apulia). Flourishing sea-port since classical times, Saracen fortress in 9th c. and place of pilgrimage in Middle Ages, 4, 53, 75, 372.

Bibl.: F. Carabellese, *Bari*, Bergamo 1909. A. Vinaccia, *I monumenti medioevali di Terra di Bari*, Bari 1915.

S. Niccolo, basilica begun in 1087 to receive relics of the saint, completed in 1132 by King Ruggiero, 54, 75.

Episcopal throne of the end of the 11th c., 76

Bibl.: R. Krautheimer, *'W.J.K.G.'*, 1934.

S. Sabino (cathedral), founded in the 9th c., modified in 17th c., reconstructed according to original plans in 1904, 75.

Castello built by Frederick II (1232–40), 80.

Belcaro, near Siena. *Castello* built at the end of the 12th c., transformed and decorated by Peruzzi in 1536, 229

Bellagio, Lake Como. *Villa Melzi*, decorations by G. Bossi, 375.

Benevento (Campania). Capital of a Lombard duchy from 571 on. Papal state in 11th c., 47, 86.

Bibl.: A. Meomartini, *Benevento*, Bergamo 1909.

‡*Cathedral* (11th c.), rebuilt between 1179 and 1221, destroyed. 13th c. bronze doors, 76.

Sta Sofia, 8th c. octagonal church, 47.

Bergamo (Lombardy). Lombard duchy, independent city, then Venetian possession (1427), 4, 132, 200, 204, 254, 342, 347, 393

Cathedral. Design for façade dates from 1457. Church built in 1487, remodelled in 1689. Existing façade dates from 1886, 188.

Interior: *Martyrdom of the Bishop St. John of Bergamo*, by Tiepolo (1743), 360.

S. Bernardino, altarpiece by L. Lotto (1521), 258.

Sta Grata (18th c.), macabre panels by Bonomini, 375.

Sta Maria Maggiore (1137), campanile of 1436, 58.

Porches of transepts by G. da Campione in 1353 and 1375, 136.

Intarsia work of balustrade and choir by Capodiferro, partly from designs by Lotto (1524–30), 197.

Bibl.: P. Pesenti, *La Basilica di Sta. Maria Maggiore in Bergamo*, Bergamo 1953.

Cappella Colleone (1470–75), by G. A. Amadeo, who was also responsible for the tombs of B. Colleoni and his daughter (inside), 188. In the cupola and the apse, frescoes by Tiepolo (1732–33), 359.

Biella (Piedmont). 9th c. *Baptistery*, 47, 58.

Bitonto (Apulia), near Bari. 12th c. *Cathedral*, 75.

Bibl.: G. Mongiello, *La Cattedrale di Bitonto*, Caserta 1952.

Bologna. The Etruscan Felsina and Roman Bononia; played a considerable part in the Lombard League from 11th to 13th c. The University, famous for its lawyers, was founded in 1119, 20, 87, 94, 96, 131, 136, 191 ff., 226, 263, 295, 330 ff., 339–48.

Bibl.: C. Ricci and G. Zucchini, *Guida di Bologna*, Bologna 1930. A. Raule, *Architettura Bolognese*, Bologna 1952. I. B. Supino, *L'Arte nelle chiese di Bologna*, Bologna 1932. P. de Bouchaud, *Bologne*, Paris 1909.

TOPOGRAPHICAL INDEX

1. Churches

S. Domenico, built by the Dominicans in 1233 to receive tomb of their founder, 97, 297.

The 'Arca', by Nicola Pisano, was completed by Nicolo da Bari (1468–73) and Michelangelo (1494), 20, 108, 192, 231.

Bibl.: C. Gnudi, *L'Arca di San Domenico*, Florence 1949.

S. Francesco (1236–63). Façade inlaid with 'bacini' of majolica, 97, 136, 137

S. Giacomo Maggiore (1267), side portico of 1481. In the chapels, frescoes by L. Costa, Francia, A. Aspertini and Tibaldi, 195, 260, 295.

Sta Maria dei Servi (1383), painting by Albani, 346.

Sta Maria della Vita, built by G. B. Bergonzoni 1687–90, 192, 330.

S. Michele in Bosco (1494–1510). In the cloister, frescoes by L. Carracci, 297.

S. Paolo, built by G. A. Magenta (1605), 324, 330, 331.

S. Petronio, begun in 1390 by Antonio di Vincenzo, not completed until 1659. Central porch by Jacopo della Quercia (1425–38), 2, 131, 136, 182, 192, 260, 330.

Bibl.: A. Gatti, *La Basilica Petroniana*, Bologna 1913.

S. Pietro, cathedral, built by G. A. Magenta (1605), completed in 1747 by Torregiani's façade, 330.

Sto Stefano, a collection of small churches mostly between 11th and 13th c. on the site of older buildings, 32.

Oratorio di S. Filippo, by Torregiani, 330.

Sanctuary of the Madonna di S. Luca, built by C. F. Dotti (1723–57), 330.

Certosa. Pietà by C. Mazza, 331.

2. Secular Buildings

Banca d'Italia, built by A. Cipolla (1862–64), 381.

Cassa di Risparmio, by G. Mengoni (1868–76), 381.

Palazzo dei Banchi, by Vignola, 290.

‡*Palazzo dell'Archiginnasio*, by Antonio Morandi (1562–69), with Anatomy Theatre by A. Levanti, 1756 (now destroyed). The flayed figures, caryatids by Lelli, are to be restored, 306.

Palazzo Bevilacqua (1481), 191.

Palazzo Fava. A. and L. Carracci painted *The Argonauts* here in 1584, 297–8.

Palazzo Magnani. Story of Romulus and Remus by A. Carracci (1592), 298.

Palazzo Montanari, built by Torregiani (1752), 330.

Palazzo Pepoli, decorated by G. M. Crespi (1692), 347.

Palazzo del Podestà, built in 13th c., rebuilt in 1472, decorated by A. de Carolis (1911–28), 389.

Palazzo Saguda, decorated by Longhi (1734), 360.

Palazzo Sampieri. Paintings by L. and A. Carracci (1593–94), 297–8.

Palazzo Zucchini-Solimei (1544), decorated by Niccolo dell'Abate (*c.* 1550), 269.

University (formerly Palazzo Poggi), built and decorated by Tibaldi (1549), 269, 295, 298.

Teatro Filarmonico, built by Fr. Bibiena (1749), 330.

Teatro Communale, by A. Bibiena (1756), 330.

Arco del Meloncello, outside the city, by C. F. Dotti (1732), 330.

Borgo San Sepolcro (Tuscany), birthplace of Matteo di Giovanni and Piero della Francesca, fortified by G. da Sangallo in 1502, 171, 177, 179, 202.

486

Cathedral, begun in 1396, finished in 1519 (dome 1770). Façade and side porches (1498, 1507–09), 138, 240.

Bibl.: P. Gazzola, *La Cattedrale di Como,* Rome 1941.

Sant'Abbondio, consecrated in 1091, 57.

S. Fedele, 12th c., 57.

Casa Giuliano Frigiero, 1939–40, 394.

Casa del Partito, 1932–33, 394.

'*Novocomum*', new district built by Terragni (1929), 394.

Cortona (Tuscany), Etruscan foundation absorbed by Florence in 1411, 211.

S. Domenico (beginning of 15th c.). Triptychs by Fra Angelico and Sassetta (1434), 168, 176.

Outside the city: *Madonna del Calcinaio,* built by Francesco di Giorgio in 1485, 173, 175, 204.

Bibl.: G. Mancini, *Cortona, Montecchio Vesponi e Castiglione Fiorentino,* Bergamo 1909.

Cosenza (Calabria). In the environs: *S. Demetrio Corone,* church of S. Adriano: mosaic pavement, 76.

Crema (Lombardy), east of Milan. In the environs: *Sta Maria della Croce,* built by Battagio and Montanaro *c.* 1500, 240.

Cremona (Lombardy). Roman foundation (222 B.C.), independent city after 1334, fief of the Visconti of Milan, 189, 201.

Bibl.: E. Signori, *Cremona,* Bergamo 1928.

Cathedral, begun in 1190, continued by Giacomo Porrata of Como (1273). Porch with balcony by Giovanni di Balduccio of Pisa (1343), gable by Pietro Rho (1501), 58, 59, 258, 296.

Frescoes by G. Romanino, Pordenone, etc., 257, 296.

Portico by A. di Carrara (1491–1525) between the church and the *campanile* (1250), 107.

Baptistery built in 1160, remodelled in 16th c., 58.

Bibl.: A. Monteverdi, *Il duomo di Cremona, il battistero e il torrazzo,* Milan 1911.

Sant'Agata: icon (13th c.), 60.

Palazzo Fodri (late 15th c.), 189.

Palazzo Raimondi (16th c.), 189.

Palazzo Stanga c. 1500, 189.

In the environs: *S. Sigismondo* (late 16th c.). The cupola is decorated with frescoes by G. Campi, 296.

Desio (Lombardy), near Como. *Villa Cusani* by Piermarini, 370.

Empoli (Tuscany), on the Arno to the west of Florence.

‡§*Collegiata* (1093) and *campanile* destroyed in the war and rebuilt, 64.

†*S. Stefano, Madonna,* fresco by Masolino, 165.

Este (Veneto), south of Padua, birthplace of the Este family (11th c.), passed to Venice in 1405; *Villa Cornaro,* by Falconetto, 251.

Faenza (Romagna), near Forlì, famous for its production of majolica, 181.

Cathedral, built to plans of Giuliano da Majano (1476–86), 202.

Bibl.: A. Messeri and A. Calzi, *Faenza nella storia e nell'arte,* Faenza 1909.

Fano (Marches), on the Adriatic. In the *Palazzo della Ragione* (1299), Poletti built the *Teatro della Fortuna* (1845–63), 380.

Fanzolo (Veneto), near Castelfranco. *Villa Emo,* built by Palladio, decorated by Zelotti, 283.

Fenis (Val d'Aosta). §*Castello,* built in 1330 by A. de Challant (restored), contains early 15th c. fresco cycle, 132.

Ferrara (Emilia). Possession of the Este family (1208–1598). Court particularly brilliant in the 15th and early 16th c., 86, 143, 146, 171, 177, 192–5, 210, 211, 244, 254, 257, 259–61, 312.

Bibl.: G. Agnelli, *Ferrara e Pomposa*, Bergamo 1904.

Cathedral. Façade by Nicolo (1135), completed by arches at beginning of Trecento. Campanile begun in 1425, finished at end of 16th c., 58, 59, 61, 86, 193.

Bibl.: *La Cattedrale di Ferrara* (various), Ferrara 1935.

S. Francesco, built by B. Rossetti (1494–1515), 192.

Castello, begun in 1385 by Bartolino da Novara, finished in 1570, contains frescoes by D. Dossi, 261.

Palazzo Costabili, by B. Rossetti, 192.

Palazzo dei Diamanti, begun by B. Rossetti in 1492, finished in 1565, 192.

Palazzo Roverella, built in 1508, 192, 193.

Palazzo Schifanoia, begun in 1391, enlarged from 1458 to 1478 by B. Rosetti, contains cycle of frescoes by F. Cossa, Ercole de' Roberti, etc., 193–194, 391.

Bibl.: P. d'Ancona, *Les Mois de Schifanoia à Ferrare*, Milan 1954.

Teatro, built in 1606 by Aleotti, destroyed in 1679, 312.

Fiesole (Tuscany). Site of former Etruscan foundation, on a hill to the east of Florence.

La Badia, cathedral until 1028, rebuilt between 1459 and 1466; the Romanesque façade with its marble revetment has been preserved, 64, 278.

Monastery: in refectory, fresco by G. da S. Giovanni, 278.

Cathedral (1028), enlarged in 13th and 14th c. In the choir Salutati chapel and tomb of the Bishop by Mino da Fiesole (1464), 165

S. Domenico, built in 15th c., remodelled in 16th c. Altarpiece by Fra Angelico, 168.

Bibl.: G. del Basso, *Guida di Fiesole e dintorni*, Florence 1846. M. Lombardi, *Faesulae*, Rome 1941.

Florence. Etruscan colony, ruined by Sylla (82 B.C.), restored by Caesar (59 B.C.). City state from 11th c. on. Seat of the Medici under Cosimo, then Lorenzo (1470–92), then duchy and grand-duchy. Hapsburg possession in 1737; between 1865 and 1870 Florence was capital of the kingdom of Italy, 17, 18, 35, 64, 94 ff., 99–105, 120–3, 143–72, 195, 202–14, 215–16, 241–5, 259, 263 ff., 271–8, 307, 331–2, 350–1, 378 ff., 385–7

Bibl.: E. Gebhart, *Florence*, 6th ed. Paris 1935. M. Marangoni, *Firenze*, Novara 1930. G. Michelucci, L. Savioli, *Firenze, sviluppo e problemi urbanistici della città*, 'Urbanistica' (Turin), June 1953.

1. CHURCHES

S. Maria del Fiore (cathedral), begun by Arnolfo di Cambio (1296) on the site of Santa Reparata, continued by Talenti (1357–65). Brunelleschi constructed the dome from 1420–36, 20, 102, 109, 120, 121, 123, 124, 128, 136, 147, 148, 149, 150, 151, 161, 167, 231, 243.

Restored in 1842 by Baccani; de Fabris built façade between 1871 and 1887, 20, 371, 379.

Bibl.: P. Sanpaolesi, *La cupola di Santa Maria del Fiore*, Rome 1941. G. Poggi, *Il Duomo di Firenze, documenti sulla decorazione della chiesa e del campanile tratti dall'archivio dell'Opera*, Berlin 1909.

Porta della Mandorla, worked on by Nanni di Banco and Donatello, 129, 147.

Stained glass windows in the dome by Uccello, A. dal Castagno. Uccello:

equestrian figure of J. Hawkwood (1436) to which A. Castagno's figure of N. da Tolentino forms a pendant (1456), 162, 169–70.

(a) *Campanile*, by Giotto, A. Pisano and Talenti. Medallions by A. Pisano and his school and by L. della Robbia, statues by Donatello and his pupils, 105, 121, 123, 129, 148, 161.

(b) *Baptistery*. Possibly earliest building; dates from 5th c. Consecrated in 1059. 12th c. marble decoration. 13th c. mosaics. *Bronze doors:* S. (originally E.) door by A. Pisano (1336), N. (1403–24) and E. ('Porta del Paradiso') (1425–52) by Ghiberti. Sculpture by Sansovino, 1502, 20, 35, 65, 83, 112, 121 ff., 128 ff., 147, 153, 162–3, 206, 207, 238, 243, 273.

Bibl.: M. Salmi, *Il Battistero di Firenze*, Rome 1950.

(c) *Museo dell'Opera*, numerous sculptures executed for the cathedral or the Baptistery by Nanni di Banco, Donatello, L. della Robbia, A. Pollaiuolo, 161, 170, 206.

SS. Annunziata, founded in 1250; Michelozzo rebuilt it in the middle of the 15th c. and Alberti designed the centrally planned choir. In 1598, Giambologna decorated one of the chapels of the choir. Atrium with frescoes by Baldovinetti, A. del Sarto, Pontormo, 100, 157, 170, 205, 242, 275, 331, 332.

SS. Apostoli. Church founded in 11th c., remodelled in 15th and 16th c., 20, 64, 239.

Badia, founded in 10th c., enlarged in 1285, rebuilt in 1627. Porch by B. da Rovezzano (1495). Tomb of Count Ugo by Mino da Fiesole (1469–81), 165, 210.

Sta Croce, according to Vasari the work of Arnolfo di Cambio, begun in 1252, rebuilt in 1295, consecrated in 1443, 20, 90, 99, 100, 101, 102, 113–14, 124, 125, 132, 157, 167, 239, 331, 371, 376, 379.

Chapel of the Noviciate by Michelozzo (1445), 157.

(a) *Paintings*. Nave: frescoes by Orcagna (destroyed), 124.

Giotto, B. Daddi, T. and A. Gaddi, Maso di Banco decorated the chapels of the church, Spinello Aretino and L. di Nicolo the sacristy, 114.

(b) *Sculpture: Annunciation* by Donatello, 161, 171.

Pulpit by B. da Majano, 206.

Tombs of L. Bruni (Rossellino), Marsuppini (Desiderio), Galileo (Alfieri), 164, 331, 375, 382.

Bibl.: R. Schiamani, *La Basilica di Santa Croce*, Florence 1952.

(c) *Pazzi Chapel*, built in the adjacent cloister by Brunelleschi between 1429 and 1444, medallions by L. della Robbia, 68, 153.

S. Egidio, built in 1420,‡ frescoes in the choir by D. Veneziano and Piero della Francesca (1439–45) and by A. del Castagno (*Life of the Virgin*), 1451–1453, 167, 177.

Sta Felicità, rebuilt in 1736 by Ruggieri. *Descent from the Cross* by Pontormo (1528), 276.

S. Fernandino, built by G. B. Foggini (1717), 331.

S. Firenze, begun by Ruggieri and finished by Z. del Rosso in 1775, 331

S. Gaetano, built by M. Nigetti (1648), 331.

S. Lorenzo, by Brunelleschi, then Michelozzo; work completed in 1460, 151, 152, 154, 204, 205, 234, 238, 276, 374.

Tabernacle by D. da Settignano; ambones by Donatello, assisted by Bellano and Bertoldo, 162, 164, 171, 207.

(a) *Old Sacristy*, built by Brunelleschi (1420–29), medallions by Donatello, tomb of Giovanni and Piero de' Medici by Verrocchio (1472), 154, 188, 205, 206, 235.

Bibl.: W. and E. Paatz, *Die Kirchen von Florenz. Ein kunstgeschichtliches Handbuch*, Frankfurt, 6 vols, 1940–54.

2. SECULAR BUILDINGS

Banca d'Italia, by Cipolla, 1869, 380.

Boboli Gardens, designed by Tribolo (1550). Statues and fountains by Giambologna, P. Tacca, S. Lorenzi, 273, 274.

Fortezza da Basso, built to designs of A. da Sangallo the younger (1534), 224.

Fortress of S. Miniato, by Michelangelo (1530), 234.

Laurenziana Library. Vestibule and room by Michelangelo. Staircase by Ammanati, 235, 236, 272.

Loggia Rucellai, probably by Alberti (*c.* 1460), 158.

Loggia della Signoria (or *dei Lanzi*), begun in 1376 by Benci di Cione and F. Talenti, completed in 1381. Medallions by G. d'Ambrogio after designs by A. Gaddi (1384–89), 105, 120, 128.

In the loggia: *Perseus* by B. Cellini (1553), *Rape of the Sabines* by Giambologna (1583), 273, 274.

Mercatanzia (1359), 120.

Mercato Centrale, by Cipolla (1874), 381.

Palazzo Antinori, attributed to G. da Majano (*c.* 1465), 202.

Palazzo Bartolini, by Baccio d'Agnolo (1520–29), 242.

Palazzo Capponi, built by Ruggieri, decorated by L. Sabatelli, 331, 374.

Palazzo Castelli, by G. Silvani, 331.

Palazzo Corsini, by G. Silvani and A. Ferri (1648–56), 331.

Palazzo Gondi, by G. da Sangallo (1490–94), 204.

Palazzo Marocelli. Decorated by G. Ricci (1706), 357.

Palazzo Medici (later Riccardi), built by Michelozzo between 1444 and 1460. In the chapel, fresco by Benozzo Gozzoli (1459). In the long gallery (1670–88), *Apotheosis of the Medici* by L. Giordano (1683), 155, 156, 170, 209, 333, 334.

Palazzo Nonfinito, begun by Buontalenti in 1593, cortile completed by Cigoli, 277.

Palazzo Pallavicini, built in 17th c. by G. Silvani, 331.

Palazzo Pandolfini, built by G. F. da Sangallo according to designs by Raphael (1516–20), 228, 242

Palazzo di Parte Guelfa, built at beginning of 14th c., enlarged by Brunelleschi in 1418, 151, 154.

Bibl.: M. Salmi, *Il Palazzo della Parte Guelfa di Firenze e Filippo Brunelleschi*, 'Rinascimento' II 1951.

Palazzo Pitti, begun *c.* 1446 by Brunelleschi, remodelled and enlarged by Ammanati (1558–70) and Ruggieri, 155, 272, 274–5.

Princely residence from 1550 on, decorated late 16th c., 17th c., 18th c., and again in 19th c., 278, 320, 321, 331, 345, 371, 374, 376.

Bibl.: H. Geisenheimer, *Pietro da Cortona e gli affreschi del Palazzo Pitti*, Florence 1909.

Palazzo del Podestà ('Bargello'), built between 1245 and 1346; frescoes in the chapel attributed to Giotto, 104, 114, 119, 161, 163, 165, 196, 206, 207, 230, 231, 243, 323.

Palazzo Rosselli del Turco, by Baccio d'Agnolo (1517), 242.

Palazzo Rucellai, built by Rossellino after designs by Alberti (1446–51), 157, 158, 183.

Palazzo della Signoria (Palazzo Vecchio), presumably by Arnolfo di Cambio, enlarged by B. del Tasso and Vasari who was also responsible for most of the decoration of the interior (1560–72), 103, 104, 120, 153, 157, 206, 266, 271, 273, 277, 289.

In the Sala d'Udienza and Sala dei Gigli, doors and coffered ceilings by G. and B. Majano (1476–78), 206, 277.

Decoration of Sala del Gran Consiglio commissioned from Leonardo (1503–1505) and Michelangelo (1504), 10, 214, 215, 230, 231.

Courtyard remodelled by Michelozzi in 1454, carved decoration, 1565, 157.

In the entrance *Hercules and Cacus* by B. Bandinelli; *David*, copy of Michelangelo's statue, 231, 270, 273.

Bibl.: A. Lensi, *Palazzo Vecchio*, French translation, Florence 1912.

Palazzo Spini, built in 1289, 103.

Palazzo Strozzi, begun in 1489 by B. da Majano, finished by Cronaca in 1507, 202, 204.

Palazzo degli Uffizi, by Vasari (1560–80). Gallery on the Ponte Vecchio, towards Palazzo Pitti, remodelled by Ammanati, 271, 274, 277.

Piazza dell'Annunziata. Equestrian statue of Ferdinand I erected in 1608, and two fountains, by Giambologna and P. Tacca (1629), 223, 274, 331

Piazza della Signoria. Neptune Fountain (1563–76) by Ammanati, P. Tacca and Giambologna; statue of Grand-Duke Cosimo, 1581, 155, 271, 273, 274

Piazza Vittorio Veneto. Arcade and triumphal arch erected in 1893. Equestrian statue of Victor Emmanuel II by Zocchi (1890), 379.

Piazzale Michelangelo. Laid out by G. Poggi in 1877, 379.

*§*Ponte alla Carraia*, built in 1220, rebuilt in 1346 and again in 1952, 106.

Bibl.: A. Alinari, *La porta e il ponte alla Carraia di Firenze*, 'Arch. ital.', XCVII 1939.

‡**Ponte alla Grazie*, 1237, 106.

‡§*Ponte S. Trinità*, built in 1252, rebuilt in 1569 by Ammanati. The statues of the 'Seasons' by Francavilla, Landini, G. Caccini have partly been restored, 106, 272, 331.

Bibl.: C. L. Ragghianti, *Ponte a Santa Trinità*, Florence 1948.

Ponte Vecchio (1345), 106.

Spedale degli Innocenti. Portico by Brunelleschi (1419–21), medallions by L. della Robbia (1463), 151, 208.

Stadium, built by P. L. Nervi in 1932, 393.

Station, by Michelucci (1933), 393, 395.

3. VILLAS

Cafaggiolo, built by Michelozzo for Cosimo de' Medici in 1451, 155, 181.

Carducci-Pandolfini ('la Legnaia'), frescoes by A. del Castagno, some *in situ*, some at S. Apollonia, 169.

Bibl.: M. Salmi, *Gli affreschi di A. del C. ritrovati*, 'Boll. Arte' 1950.

Carreggi, Villa Medici, enlarged by Michelozzo in 1434, 155, 208.

Castello, acquired by the Medici in 1477, decorated with paintings (now destroyed) by Bronzino and Pontormo. 18th c. façade. Gardens designed by Tribolo *c.* 1540; in the grotto of the upper terrace bronze animals by Giambologna, 243, 274.

†*La Petraia*, old house rebuilt by Buontalenti for Grand-Duke Cosimo (1576–89). In the garden, *Venus* by Giambologna (1567), 243, 273, 274.

Poggio a Caiano, near Florence. *Villa* built for Lorenzo de' Medici by G. da Sangallo (1480–85); frescoes by A. del Sarto and Pontormo, 203, 272, 275, 371.

Poggio Imperiale, fortified by G. da Sangallo in 1488, remodelled in 16th c., then in 18th c. by Caccialli, Paoletti, Poccianti, 203, 271

Pratolino, north of Florence. *Villa* built by Buontalenti for Francesco de' Medici (1565–81), 272, 274.

Bibl.: G. Lensi Orlandi Cardini, *Le ville di Firenze*, Florence 1954. A. J. Rusconi, *Le Ville Medicee*, Rome 1938.

Fontanellato (Rocca di), near Parma. *Castello: Story of Actaeon* by Parmigianino, 1523, 270.

Forlì (Romagna). City state, feudal possession, then part of Papal States.

†*S. Mercuriale*, begun in 1176, altered several times. Triptych by Palmezzano, from church of S. Biagio,‡ 180.

Palazzo Foschi. Ceiling by F. Bencovich (1757), 358.

Fossanova (Lazio), near Sezze; the oldest Cistercian monastery in Italy. *Church* begun in 1187, consecrated in 1208, 84.

Frascati (Lazio), south-east of Rome, centre of the region of 'Castelli Romani'.

Bibl.: S. Kambo, *Il Tuscolo e Frascati*, Bergamo n.d.

Villa Aldobrandini, built by G. della Porta (1598–1604), decorated by G. Cesari, 293, 298, 311.

Villa Mondragone, by Vasanzio, 311.

Villa Torlonia, 16th c. Fountains and cascades by Maderna, 312.

Galliano (Lombardy), near Cantù. *S. Vincenzo*, frescoes dating from 1007, 47, 60.

Baptistery, 11th c., 47.

Galluzzo (Tuscany), south of Florence. *Certosa* founded by N. Acciaiuoli in 1342, later enlarged and remodelled.

Chiostro grande, with medallions by the della Robbia, frescoes by Pontormo, 276.

Genoa. Former Ligurian capital; after 11th c. powerful maritime republic rivalling Pisa and Venice; annexed to Piedmont in 1815, 53, 94, 303, 326, 339, 351, 367.

1. CHURCHES

Cathedral of S. Lorenzo, consecrated in 1118, completed in 16th c.; dome by Alessi (1567), Chapel of S. Giovanni Battista by D. Cagini (1456–78). Chapels decorated by Cambiaso, L. de Ferrari, F. Baroca, G. B. Castello, 187, 295.

‡*Annunziata*, early 17th c. Barabino added the classical portico in 1830. In the chapels, paintings by G. Assereto, D. Piola, L. Cambiaso, 327, 370.

Chiesa degli Crociferi. Triumph of the Cross, painting by G. de Ferrari, 326.

Sant'Ambrogio, built by Tibaldi in 1597. *Assumption* by Guido Reni, 346.

SS. Giacomo e Filippo. Assumption, fresco by G. de Ferrari, 326.

S. Maria di Carignano, begun by Alessi in 1552, finished in 1700. Statues in niches by P. Puget (*St. Sebastian*), F. Parodi (*St. John*), *Entombment* by Luca Cambiaso, 294, 295, 327.

S. Siro (cathedral until 985), burnt down in 15th c., rebuilt in 16th c. Façade by Barabino (1820), 327.

2. SECULAR BUILDINGS

Cemetery, by Barabino and Regaseo, 381.

Palazzo Bianco (17th–18th c.), 351, 382.

Palazzo Cambiaso, built by Alessi, 294.

Palazzo Doria, by Montorsoli (1528), with frescoes by P. del Vaga, 295.

Palazzo Ducale, by S. Cantoni (1778), 370.

Palazzo Parodi (1567), frescoes by L. Cambiaso, 295.

Palazzo Rosso, built by P. A. Corradi in 1677. Vaults with allegories of the *Seasons* by D. Piola, G. de Ferrari and Haffner, 326, 356.

‡*Palazzo Reale* (formerly *Palazzo Durazzo*), built by G. A. Falcone *c.* 1650, 326, 370.

Palazzo dell'Università (1634–1640), by B. Bianco; statues by Giambologna, 326.

Teatro Carlo Felice, built by Barabino in 1827, 370, 380.

Villa delle Peschiere, by Alessi, 294.

Bibl.: O. Grosso, *Genova e la Riviera Ligure,* Rome 1951.

Groppoli (Tuscany), near Pistoia. *S. Michele,* ambo of 1194, 73.

Grottaferrata (Lazio), near Rome. *Abbey* of Basilian monks founded by St. Nilus in 1004. Cardinal G. della Rovere, the future Pope Julius II, had it fortified after 1473, 68.

S. Maria, church consecrated in 1024, remodelled in 1190 and 1757. Late 12th c. frescoes and mosaics. In the chapel of St. Nilus, frescoes by Domenichino (1608–10), 345.

Gualdo Tadino (Umbria), near Foligno. *Cathedral* built in 1251.

S. Francesco, 13th c., 97.

Gubbio (Umbria), at foot of Monte Ingino, city state in 11th c., possession of the Montefeltre, della Rovere, then of the Church (1624), 181.

Palazzo dei Consoli, built in 1333 by Angelo da Orvieto, 167.

Palazzo del Podestà, by Gattaponi (1349), 107.

Bibl.: A. Colasanti, *Gubbio,* Bergamo 1905.

Istrana (Veneto). See **Treviso.**

Jesi (Marches), near Ancona. *Palazzo Pubblico,* by Francesco di Giorgio. *Pinacoteca,* paintings by Lotto, 175, 176, 257.

Lecce (Apulia). Norman possession (12th c.), became part of the kingdom of Naples in 1463, 'the Florence of Baroque art', 3.

S. Orenzo (now the cathedral), by G. Zimbalo (*c.* 1670–82), 335.

S. Cataldo, founded by Tancred in 1180, remodelled in 1716, 76.

Sta Chiara, 18th c., 336.

Sta Croce, built between 1549 and 1695, 335.

S. Matteo, 18th c., 336.

Palazzo del Vescovado, 17th c., 335.

Seminary, 18th c., 335.

Bibl.: G. Gigli, *Lecce e dintorni,* Bergamo 1911.

Leghorn (Livorno) (Tuscany). Annexed by Florence in 1421. Its period of major development as a port dates from the period of the Grand-Duchy (1577) and continued until the 19th c.

Monument to Ferdinand I. The statue of the Grand-Duke is by G. Bandini (1595), the four Moors by P. Tacca (1624), 331.

Aqueducts and Cisternone, the work of Poccianti (1842), 371.

Lodi (Lombardy), 36 km. from Milan on the Adda. *Incoronata* by Battagio (1488–94), 51, 205.

Bibl.: A. Terzagli, *l'Incoronata di Lodi,* 'Palladio' 1953.

Loreto (Marches), to the south of Ancona, place of pilgrimage.

Chiesa della Santa Casa, begun in early years of 15th c. G. da Majano, B. Pontelli, G. da Sangallo, Bramante, A. Sansovino and A. da Sangallo the younger took part in the building. Campanile by Vanvitelli (1750–54). The church contains the *Santa Casa,* a small Romanesque construction with no foundations, the outside of which is covered with marbles by A. Sansovino, A. and G. Lombardo, Raffaello da Montelupo, 9, 202, 203, 243, 295.

Sacristies frescoed by Melozzo da Forlì (1486) and L. Signorelli (1480, 181, 211.

Bibl.: A. Colasanti, *Loreto*, Bergamo 1910.

Lucca (Tuscany). Flourishing city state, independent until 1799, 43, 72 ff., 242, 276, 348.

Bibl.: A. Mancini, *Storia di Lucca*, Florence 1952.

S. Martino (cathedral), founded in 6th c., rebuilt in 1060. Façade by Guidetto da Como (1204), 72, 73, 97.

Tomb of Ilaria del Caretto by J. della Quercia (1407), 129.

'Tempietto del Volto Santo', built by M. Civitali in 1480, 187.

Bibl.: E. Ridolfi, *L'Arte in Lucca studiata nella sua cattedrale*, Lucca 1882.

Sant'Alessandro, 12th c., apse 13th c., 72.

S. Croce, by F. Gherardi and P. Coli, 321.

S. Frediano (1124–47). The font is by Matteo Civitali. Marble polyptych by J. della Quercia, 72.

S. Michele, begun in 1143, 13th c. façade, 72.

Misericordia, 12th c., by Biduino, 73.

Lugano (Ticino), on the lake, Swiss town since 1512. *S. Maria degli Angeli*, begun in 1499, frescoes by B. Luini (1529–32), 241, 342.

Manta (La) (*Castello* of Piedmont). Frescoes attrib. to J. Iverny, *Nine Heroes and Heroines of Antiquity* (1420), 132.

Bibl.: P. d'Ancona, *La Manta*, 'L'Arte' 1905.

Mantua (Lombardy), achieved great brilliance under Gonzaga rule (1326–1628), especially in time of Francesco II (1484–1519), husband of Isabella d'Este and their son Federigo II (1519–40), 214, 217, 254, 259, 284.

Cathedral, by Giulio Romano (1545). Façade of 1756, 268, 355.

Sant'Andrea, built by L. Fancelli after a design by Alberti, from 1470. Cupola by Juvara (1732–82), 159, 160, 279, 292, 293.

S. Sebastiano, begun in 1460 by Fancelli after design by Alberti, consecrated in 1529, 160, 279.

Palazzo Ducale, consists of several buildings of different periods; adapted and altered from 13th to 18th c. by Gonzaga, 259, 262, 287.

Frescoes in *'Camera degli Sposi'* by Mantegna (1474), 190, 261, 320.

Studiolo of Isabella d'Este: paintings by Mantegna, Perugino, 191, 195, 259–60.

Bibl.: N. Gianantoni, *Palazzo Ducale di Mantova*, Rome 1929.

Palazzo del Tè, built by Giulio Romano who also designed and executed most of interior. *Sala dei Cavalli* by B. Pagni. Stuccoes by Primaticcio, 268–269, 280.

Bibl.: G. and A. Pacchioni, *Mantova*, Bergamo, n.d.

Maser (Villa Barbaro), see **Asolo**.

Massa Marittima (Tuscany). *Cathedral*, begun early 13th c., finished by Giovanni Pisano (1287–1307). In 1324 Goro di Gregorio executed *arca di S. Cerbone*, 111.

Municipio, built in 1250, 107, 111.

Bibl.: L. Petrocchi, *Massa Marittima*, 'Arte e Storia', Florence 1900.

Messina (Sicily). Sicilian port on the Straits, occupied by Saracens in 893, first city to be occupied by the Normans. Almost totally destroyed by earthquake in 1908, 111, 186, 187.

Fountain of Orion, in front of the cathedral, by G. A. Montorsoli (1547–1551), 243.

Milan. Capital of Lombardy, Celtic foundation of Insubres; Roman in 222 B.C., one of the capitals of the Empire, centre of the Western Church with St. Ambrose. City state in Middle Ages, head of the Guelf faction, became possession of the Visconti (1330), then of the Sforza (1450) until French invasion (1494). Milan was subsequently occupied by the Spanish and the Austrians (1713–96 and 1814–59), 12, 24, 25, 33, 44, 45, 51, 58, 94, 132, 138, 143, 171, 208, 212–14, 303, 312, 339, 347–50, 367, 370, 384–5, 393.

Bibl.: P. Mezzanotte and G. C. Bascapè, *Milano nell'arte e nella storia*, Milan 1948. *Arte Lombarda dai Visconti agli Sforza*, exhibition catalogue (preface: R. Longhi), Milan, Palazzo Reale, 1958.

1. CHURCHES

Cathedral, founded in 1389, completed in 1856. After the '*maestri campionesi*' and the French and German architects who worked on the building in its early stages, it was continued in the 15th c. by Filippino degli Organi, G. Solari, G. A. Amadeo, then in the 16th c. by P. Tibaldi who drew up plans for the façade which was executed by Buzzi between 1616 and 1645. St. Charles Borromeo consecrated the cathedral in 1577, 20, 136, 137, 138, 294–295, 348.

Bibl.: C. Romussi, *Il duomo di Milano nella storia e nell'arte*, Milan 1908.

Sant'Alessandro, built by G. Binaghi in 1602, 325.

S. Ambrogio, basilica founded in 386, transformed in 9th c., rebuilt at the end of the 11th c., atrium 1150. A 9th c. campanile and one of the 12th c. frame the façade, 46, 50, 54, 55, 59, 217.

Outstanding gold and silver altar by 'Vuolvinus' (9th c.), 10th c. ciborium, 46, 59.

Bibl.: F. Reggiori, *La Basilica di Sant'Ambrogio a Milano*, Florence 1945.

S. Eustorgio, built in 12th and 13th c., transformed by construction of chapels (15th and 16th c.). *Cappella Portinari*, by Michelozzo (1462) contains frescoes by Vincenzo Foppa (1468), '*Arca*' of S. Peter Martyr by Giovanni di Balducci (1339), 135, 157, 188, 189.

S. Fedele, begun by P. Tibaldi (1569) and Bassi, finished in 19th c., 295.

SS. Gervasio e Protasio (or '*Basilica dei Martiri*'), built by St. Ambrose, rebuilt in 9th c., 33.

S. Gottardo (14th c.). The campanile dates from 1335, 135, 137.

S. Lorenzo, founded in 4th c., remodelled in 12th and 13th c., rebuilt at end of 16th c. by M. Bassi; *cappella di Sant'Aquilino* (4th c.), 30, 33, 34, 35, 38, 210.

Bibl.: A. Calderini, G. Chierici, C. Cecchelli, *La basilica di San Lorenzo Maggiore in Milano*, Milan 1952.

S. Marco, built in 14th c., remodelled internally in 17th c., façade 1871. *Madonna* by A. Campi (1577), 296.

S. Maria di Brera (destroyed), polychrome façade by G. di Balduccio, 137.

†*S. Maria delle Grazie*, begun by C. Solari in 1465, enlarged by Bramante who built the choir (1492–98), 189, 205, 212, 213, 215.

Bibl.: G. Portaluppi and A. Pica, *Le Grazie*, Rome 1937.

Sta Maria presso S. Celso, begun by Dolcebuono (1493), continued by Solari and C. Lombardo. Façade by Alessi (1568–72?). A. Appiani decorated the dome in 1795, 294, 375.

Sta Maria della Passione, begun in 1485. C. Solari built the dome, façade of 1692. Paintings by D. Crespi, 348.

S. Nazzaro Maggiore (12th c.), altered in 16th c., 33.

Palazzo Scalfani, built in 1330, became a hospital in 15th c. *Triumph of Death*, anonymous fresco painted *c.* 1450 (now transferred to Museum), 186.

Politeama Garibaldi, built by Almeyda (1874), 380.

Teatro Grande, built according to plans by G. Basile, from 1875 to 1897, 336, 380.

Villa Basile, built by E. Basile in 1904, 380.

Villa Farsini, by E. Basile (1906), 380.

Villa Igea, by E. Basile (1898), 380.

La Cuba, outside the city, *castello* built by William II in 1180, 80.

Bibl.: P. Lojacono, *L'Organismo costruttivo della Cuba alla luce degli ultimi scavi*, 'Palladio' 1953.

La Zisa, small 12th c. palace built by William I, 80.

Piazza delle Quattro Cantoni, designed by G. Sasso in 1609, 336.

Porta Nuova (1535), †*Porta Felice* (1582–1644), 336.

Parenzo (Istria), *basilica* and octagonal *baptistery* of the 6th c., 40.

Parma (Emilia). Roman colony, Ghibelline city, became possession of the Popes (1511), the Farnese (1545), the Bourbons (1749), of Marie-Louise (1815–47); united to House of Savoy in 1859, 51, 259 ff., 289 ff.

Cathedral (12th c.), porch by Giambono da Bissone (1281), *Deposition* by B. Antelami (1178), 58. In dome, *Assumption*, fresco by Correggio (1526–1536), 5, 262, 271.

Bibl.: C. Ricci, *Gli affreschi del Correggio nella cupola del duomo di Parma*, Rome n.d.

Baptistery (1196–1260), sculpture decoration by B. Antelami, frescoes of 1260, 6, 54, 58, 59, 60.

Bibl.: G. de Angelis d'Ossat, *Cronologia del B. di P.*, 'Palladio' III 1939.

S. Giovanni Evangelista, built by Zaccagni in 1510, 17th c. façade and campanile; in the dome, *Ascension*, fresco by Correggio (1520–24). Parmigianino decorated two chapels in 1522, 261, 270.

Sta Maria della Steccata, built by Zaccagni in 1530, frescoes by Parmigianino (1531–39), 240, 270, 271.

Bibl.: L. Testi, *Santa Maria della Steccata in Parma*, Florence 1922.

*Convento di S. Paolo: former refectory has vault painted by Correggio (1518—19), 261, 262.

Sta Maria, Bardi (near Parma), picture by Parmigianino, 270.

‡*Palazzo della Pilotta*, begun by the Farnese (1583–1622), never completed. On the second floor, G. B. Aleotti built the *Teatro Farnese* in 1619, 283, 330.

Bibl.: G. Lombardi, *Il Teatro farnesiano di Parma*, 'Arch. stor. per le provincie parmesani', N.S. IX 1909.

Teatro Regio, by Bettoli (1829), 371, 380.

Pavia (Lombardy), near Milan; capital of the Lombards (572–74), Ghibelline township, Visconti possession from 1364, 46, 57, 58.

Cathedral, begun in 1488 by C. Rocchi; Bramante and Amadeo worked there; 19th c. façade and dome, 189, 205, 213, 217.

Bibl.: I. F. Gianani, *Il D. di P.*, Pavia 1930.

*S. Francesco, altered in 18th c., painting by V. Campi (1580), 296.

*S. Lazzaro, 12th c., 57.

Sta Maria del Carmine, 14th–15th c., 138.

*Sta Maria del Popolo, 12th c., 57.

S. Michele, 12th c., 57, 59.

S. Pietro in Ciel d'Oro, consecrated in 1132. The *Arca* of S. Augustine (*c.* 1370) is the work of a follower of G. di Balduccio, 57, 135.

Baptistery of the Orthodox, built *c.* 450–52 by Bishop Neon, 35, 36.

Mausoleum of Galla Placidia, *c.* 450, 35, 40, 41.

Bibl.: G. Bovini, *Il cosidetto mausoleo di Galla Placidia in Ravenna*, Rome 1950.

Mausoleum of Theodoric, built in 526, 38.

Palace of the Exarchs, incorporated in 9th c. in church of *S. Salvatore a Calchi*, 38.

Realvalle (Campania). 13th c. abbey, 98.

Recanati (Marches), near Loreto. *Cathedral: Madonna and Saints* by L. Lotto (1508), 257, 258.

S. Maria dei Mercanti: Annunciation by L. Lotto (1527), 257.

Resina (Campania), between Naples and Pompeii, near Herculaneum. *Villa La Favorita* by Fuga (1751), 332.

Rimini. Umbrian, Gallic, then Roman colony; important cross-roads during Imperial period. Capital of the Pentapolis Marittima, fief of the Malatesta, Rimini became a possession of the Church in 1509, 131, 157–9, 180, 279, 372.

Sta Chiara, 14th c. Apse decorated with frescoes by Pietro da Rimini, 131.

‡§*S. Francesco*, founded in 13th c., entirely remodelled after plans by Alberti (1447–68), became '*Tempio Malatestiano*', 9, 157–9, 160.

Reliefs by Agostino di Duccio, 164.

Bibl.: C. Ricci, *Il T.M.*, Rimini 1925. D. Garattoni, *Il T.M.*, Bologna 1951.

Teatro Communale, built by L. Poletti (1857), 380.

Rome. *Urbs*, capital of the Roman Empire and seat of St. Peter, sacked many times in the course of the centuries, 13, 16, 18, 19.

In decline in 5th c., revived by Popes and Carolingians (8th–9th c.), 8, 16, 23, 24, 32, 36, 42–6.

Ruined in 11th c., in a state of anarchy in 13th c., abandoned by the Papacy in 14th c., 16, 51, 53, 68, 82, 92, 93–5.

Restored in 15th c., became capital of powerful State and centre of modern civilization, 15, 16, 17, 143, 179, 182 ff., 196, 207, 208, 215–39, 255, 263 ff., 285, 289–98, 308.

The influence of Rome, immense in the 17th and 18th c., declined abruptly after the Empire of Napoleon, 4, 18, 339–52, 357, 365 ff., 378, 384, 393.

Bibl.: L. Schudt, *Le Guide di Roma*, Vienna 1930. E. Bertaux, *Rome*, 3 vols., Paris 1931. E. Calvi, *Bibliografia generale di Roma*, Rome 1906–08. N. Denis and R. Boulet, *Romée*, Paris 1948.

I. CHURCHES

S. Adriano, formerly contained: *Preaching of St. Romuald* by C. Saraceni, 342.

Sant'Agnese in Agone, built by Borromini (1653–61), 318, 323–4, 332.

Sant'Agnese-fuori-le-mura, built 625–38 on the catacombs, 28, 29, 45, 48.

S. Agostino, built by G. da Pietra Santa (1479–83), interior remodelled by Vanvitelli (1750), *Madonna del Popolo* by Caravaggio (1604–05), 183, 340.

S. Alessio, interior remodelled in 1750. 12th and 13th c. frescoes in the crypt. Cosmatesque cloister, 69.

S. Andrea, built by Vignola in 1554, 290.

S. Andrea delle Fratte, built by Borromini (1654–65), façade by Valadier (1821), 318, 372.

S. Andrea al Quirinale, by Bernini (1653), *stucchi* in the cupola by A. Raggi, 316, 323, 324.

Bibl.: R. Villedieu, *Villa Médicis*, Rome 1953.
Villa Montalto (destroyed), built by Fontana for Sixtus V, 311.
Bibl.: L. Callari, *Le Ville de Roma*, Rome 1943.

8. VATICAN

St. Peter's. Original basilica founded by Constantine (324–44). In 1452 Nicholas V decided to reconstruct it, begun by B. Rossellino, afterwards entrusted by Julius II to Bramante, then to G. da Sangallo, 13, 27–30, 69, 184 ff., 203, 218, 220, 224, 232.

In 1547 Michelangelo intervened, then Vignola, G. della Porta, D. Fontana; Maderna lengthened the nave and constructed the façade (1607). Consecrated in 1626, 220, 221, 226, 229–30, 238, 293, 309, 314.

In the *atrium*, *'la Navicella'*, mosaic by Giotto, statue of Constantine by Bernini (1670), 82, 112, 323.

Sculptures: Michelangelo, *Pietà* (first chapel in right aisle), 231, 238.

Bernini: *Baldacchino* (1624), *Cathedra* (1657), *niches* in the piers of the dome, *tombs* of Urban VIII and Alexander VII, 316, 322, 323.

Sculptures by Algardi, Tenerani, Canova, 322, 323, 324, 376.

Piazza di S. Pietro. Bernini built the elliptical colonnade between 1656 and 1665. *Obelisk* of Caligula erected by D. Fontana in 1596, 294, 309–10.

Grottoes: tomb of Paul II, by Mino da Fiesole (1475), *tomb* of Sixtus IV by A. Pollaiuolo (1484–94), *ciborium* of Sixtus IV, 184, 206.

Bibl.: G. M. Pugno, *Storia della Basilica Vaticana*, Florence 1948. G. Smith, I. Giordani, *La Basilica di San Pietro in Vaticano*, Rome 1946. V. Mariani, *La facciata di San Pietro secondo Michelangelo*, Rome 1943.

Vatican Palace, residence of the Popes since the return from Avignon (1377). Original building, creation of Nicholas III (1272–80) constantly enlarged. *Chapel of Nicholas V* by Fra Angelico (1447–49), 131, 168, 218.

Sistine Chapel. Sixtus IV built the chapel which bears his name (1473–84), with mural decoration completed by Michelangelo's ceiling (1512), and later by his *Last Judgement* (1534–41), 177, 184, 210–14, 232–46, 294, 297.

Borgia Apartments for Alexander VI, decorated by Pinturicchio (1492–95), 184, 226.

Stanze by Raphael and his pupils (1510–20), 196, 225–6, 240.

Belvedere, built during pontificate of Innocent VIII (1484–92); joined to the Vatican by Bramante, 182, 216, 218, 219.

Loggie, on three storeys of the Cortile di S. Damaso (1512–18), decorated by Raphael and G. da Udine, 227, 228.

Cappella Paolina and *Sala Regia*, by A. da Sangallo (1540, finished in 1573), 23, 237, 289, 294, 322.

Casino, by P. Ligorio and R. di Montefiascone (1560), for Pius IV, 290.

Obelisk in the Piazza of St. Peter's and Library, by Fontana, 218, 294.

Scala Regia, by Bernini (1644), 316.

Small palace of Innocent VIII transformed into Museum. Pius VII added new wing (1817–21). Egyptian and Etruscan museums laid out in 1836, 218, 371, 374.

Bibl.: M. Pittaluga, *Die Sixtinische Kapelle*, Rome 1953. F. Hermanin, *L'Appartamento Borgia in Vaticano*, Vatican 1934. D. Redig de Campos, *Le Stanze di Raffaello*, Florence n.d. F. Baumgarten and B. Biagetti, *Gli Affreschi di Michelangelo e di E. Sabbatini e F. Zuccari nella cappella Paolina in Vaticano*, Vatican 1934. J. S. Ackerman, *The Cortile del Belvedere*, Vatican 1954.

Rossano (Calabria). *S. Marco*, 12th c., 76.

S. Fortunato, 13th–15th c. *Virgin and Child*, fresco by Masolino (1432), 97, 165.

S. Maria della Consolazione, built by Cola di Matteuccio da Caprarola (1508), then by Ambrogio da Milano, completed in 1607, 240.

Palazzo del Capitano (1296), 107.

Palazzo dei Priori, built 1293–1357, modified in 1514, 107.

Torcello, see **Venice**.

Toscanella (Lazio), near Viterbo. *S. Maria Maggiore*, 8th c., rebuilt in 12th c., 69.

S. Pietro, 12th c., 46, 69.

Trani (Apulia). Port on the Adriatic, used by the Greeks, the Normans and the Hohenstaufen.

Cathedral, begun in 1094, finished in 13th c., bronze doors by Barisano (c. 1179), 75.

Castello, built by Frederick II, 1233–49, 80.

Tremezzo (Lombardy), on Lake Como. *Villa Carlotta* (1747), 312.

Trento. Episcopal principality, seat of the œcumenical council (1545–63); Austrian territory in 1814, reverted to Italy in 1918.

Cathedral, built 1200 to 1505.

Castello del Buonconsiglio: Castelvecchio (13th–15th c.), with *Torre dell'Aquila*, frescoes of 1390–1400, and 'Magno Palazzo' built in 1536 (Museum), 133.

Bibl.: M. Sandona, *Il Castello del Buonconsiglio in Trento*, Trento 1954.

Trescore (Lombardy), near Bergamo; in the *Oratorio*, painting by L. Lotto (1524), 257.

Treviglio, near Brescia. *S. Martino*, Romanesque campanile, Baroque façade, polyptych by Butinone and Zenale (1485–1500), 189.

Palazzo Gozzani (1737): decorated by P. F. Guala, 330.

Treviso (Veneto). Roman foundation, prospered in Middle Ages, Venetian possession after 1388.

Bibl.: A. A. Michieli, *Storia di Treviso*, Florence 1938.

S. Niccolo (13th–14th c.). Frescoes by Tommaso da Modena in chapterhouse of neighbouring monastery (1351), 131.

S. Teonisto. Crucifixion, by J. Bassano (c. 1562), 286.

In the environs: *Villa Lattès*, by G. Massari (1715), 354.

Trieste (Venezia Giulia). Roman colony, city state in 1295, subject to Dukes of Austria in 1382. Reverted to Italy in 1918, Free State in 1947, Italian in 1954.

Bibl.: P. Rutteri, *Trieste, Spunti del suo passato*, Trieste 1954.

Sant'Antonio, by P. Nobile (finished in 1849), 371.

S. Giusto, cathedral, formed by the joining up, in the 14th c., of the church of the Assumption (5th c.) and the crypt of S. Giusto (6th c.). *Chapel:* picture by Pordenone (1520), 257.

Borsa, by Molari (1802–06), 371.

Palazzo Carciotti, by M. Pertsch (1802–06), 371.

Rotondo Pancera, by M. Pertsch, 371.

Trissino (Villa), see **Vicenza**.

Troia. Small town in Apulia. *Cathedral* (1095–1125): bronze doors by Oderisio di Benevento (1119), 74, 75.

Troina. Town in Sicily. *Cathedral*, founded by Count Ruggiero in 1081, rebuilt in 16th c., 77.

Bibl.: C. G. Canale, *La Cattedrale di Troina*, Palermo 1951.

Turin. Celtic foundation, possession of the Counts of Savoy since the 12th c., capital of duchy in 16th c. (1562). After the departure of the French, period of brilliance under Charles-Emmanuel I and II (17th c.). Capital of the new Kingdom of Italy in 1861, 277, 288, 311, 312, 348, 379.

Bibl.: P. Toesca, *Torino*, Bergamo 1911.

S. Giovanni Battista (cathedral), founded in 1498. The Cappella del Sacro Sindone was constructed by Guarini (1668–94), 327.

‡§*Chiesa del Carmine*, built by Juvara in 1732, 327.

S. Cristina, after plans by C. di Castellamonte. Façade by Juvara (1718), 328.

Gran Madre di Dio, by F. Bonsignore (1818–31), 370

S. Lorenzo dei Teatini, built by Guarini (1668–87), 327.

S. Maria di Piazza, by B. Vittone (1751), 330.

Sta Trinità, by Vittozzi (1606), was transformed by Juvara in 1718, 327.

Palazzo Carignano, by G. Guarini (1680), 327.

Palazzo Madama, built by A. di Castellamonte in the middle of the 17th c., altered by Juvara in 1718, 328.

†*Palazzo Reale*, by A. di Castellamonte (1658), transformed by Juvara. The ceiling of the Throne Room is by G. Miel, decorations by C. F. Beaumont, C. Van Loo, 327.

★Teatro Reale, built by Alfieri (1740), burnt down in 1936, 330.

Mole Antonelliana (1863), by A. Antonelli, 380.

Piazza S. Carlo, laid out by C. di Castellamonte, 327.

Equestrian monument to Emmanuel-Philibert by Marochetti (1828), 382.

Via del Carmine, arcades by Juvara (1728), 328.

Borgo medioevale, reconstruction carried out in 1884 by A. d'Andrade, 379.

Fiat works (1927), by Matteo Trucco, 394.

ENVIRONS

Stupinigi. Villa Reale, built by Juvara in 1729, surrounded by a park designed by Bernard, 312, 328, 329.

Superga. Basilica by Juvara (1731), 328.

Urbino (Marches), built on a hill dominating the valleys, fief of the Montefeltro since the 13th c., birthplace of Bramante and Raphael. Particularly brilliant during reign of Federigo (1444–82), passed to the della Rovere in 1508 and to the Church in 1631, 143, 147, 160, 171, 174, 178, 180 ff., 204, 217, 224, 259.

Cathedral, rebuilt by G. Valadier (1789–1801), façade by C. Morigia, 372.

S. Bernardino, by Francesco di Giorgio, 175, 181.

Bibl.: G. Giovannoni, *S. Bernardino di Urbino*, 'Belle Arti' I, 1946–47. P. Rotondi, *Quando fu costruita la chiesa di S. Bernardino in Urbino*, ibid.

S. Giovanni Battista, oratory decorated with frescoes by L. and J. Salimbeni (1416), 131.

Palazzo Ducale, enormous nexus of buildings around a first palace begun c. 1444; continued by L. Laurana after 1465, Fr. di Giorgio after 1475, 175, 180, 181, 196.

Bibl.: P. Rotondi, *Il Palazzo ducale di Urbino*, Urbino 1950.

Valvisciolo (Lazio), near Sermoneta. Cistercian *abbey* built in 1240. *St. Laurence* by G. Serodine (c. 1620), 342.

Varallo Sesia (Piedmont), near Novara.

Madonna delle Grazie, frescoes by G. Ferrari, 241.

S. Gaudenzio, polyptych by G. Ferrari, 241.

Sacro Monte, forty-five chapels grouped on a hill with statues and frescoes, notably by G. Ferrari, 241.

Varese (Lombardy), near Lake Maggiore. Sanctuary of *Sacro Monte* (1604–80). Fifteen chapels with the 'Mysteries of the Rosary', 326.

Bibl.: C. del Frate, *Santa Maria del Monte presso V.*, Varese 1933.

Venice. Founded in the 5th c. by the people of Venetia in flight from the Barbarians. Confederation of islands (end of 10th c.), powerful maritime republic whose dominion extended to the islands and coasts of the Adriatic and after the 4th crusade (1204) to the Eastern Mediterranean, 25, 53, 60–2, 82, 94.

In the peninsula dominion extended to Bergamo and the Po Valley; in conflict with Genoa and with the Slavs and Turks, Venice maintained its independence until 1797 when it was ceded to Austria; entered the Kingdom of Italy in 1866, 17, 139–40, 143, 191, 193, 197 ff., 231, 244, 246–59, 278–88, 303, 307, 339, 353–62, 365–8, 384, 395.

Bibl.: G. Lorenzetti, *Venezia e il suo estuario*, Milan 1927. V. Alinari, *Eglises et 'Scuole' de Venise*, Florence 1906. E. R. Trincanato, *Venezia minore*, Milan 1948. M. Brunetti, G. Lorenzetti, *Venezia nella storia e nell'arte*, Venice 1950. M. Muraro, *Venezia*, Florence 1953.

1. CHURCHES

S. Mark's, basilica founded in 829, reconstructed in 1063, consecrated in 1094. Late 13th c. sculptures. The façade was partly rebuilt in 1419 by Niccolo Lamberti, 47, 54, 61–3, 107, 128, 134, 136, 139, 357.

Mosaics, interior decoration of domes and façade, of vestibule and of interior continuously executed and restored from 12th to 19th c., 63, 169, 258.

The *Pala d'Oro*, a magnificent assemblage of gold, precious stones and enamels, some of which date from the 5th c., was executed in 1105. Rearranged in 1206, the final setting is the work of G. di Boninsegna (1345), 62, 134.

Rood-screen, statues of the *Virgin* and *Apostles*, by P. P. dalle Masegne (1394), bas-reliefs on sides by J. Sansovino (1537–44), statues of Evangelists and bas-reliefs (1549–69), 136, 250.

Campanile, rebuilt in 1912, copy of original building (12th c.) which collapsed in 1902, 129.

Loggetta by Sansovino (1537–40), 248.

Bibl.: L. Marangoni, *L'Architetto ignoto di San Marco*, Venice 1933. S. Bettini, *L'Architettura di San Marco*, Milan 1946. *Mosaichi antichi di San Marco a Venezia* (edited by S.B.), Bergamo 1944. O. Demus, *The Church of San Marco in Venice*, Dumbarton Oaks 1961.

Sant'Alvise, built in 1388; paintings by Tiepolo, 139.

SS. Apostoli. In the *Cornaro chapel*, Tiepolo's *Communion of St. Lucy* (1748), 359.

S. Bartolommeo a Rialto, reconstructed in 18th c. Organ doors, *Four Saints*, by S. del Piombo (1503), 254.

Chiesa della Fava, *The Virgin appearing to S. Philip Neri*, by G. B. Piazzetta, 358.

S. Francesco della Vigna, built by Sansovino (1534), the façade by Palladio, 250, 279.

S. Giorgio Maggiore, by Palladio (1580), completed by Scamozzi (1610). *Adoration of the Shepherds*, by Bassano (1591), 155, 280, 353, 354, 357.

Bibl.: F. Forlati, *Il restauro dell'abbazia di San Giorgio Maggiore di Venezia*, 'Palladio' 1953.

Gesuati, built by G. Massari (1725–36). Altarpiece by Piazzetta (1739). Ceiling by Tiepolo (1735–39), 355, 358, 359.

Gesuiti, built by D. Rossi in 1729, façade by G. B. Fassoretto, *Martyrdom of S. Lawrence* by Titian, 256, 354.

S. Giovanni Chrysostomo, built by Coducci in 1497, altarpiece by G. Bellini (1513), 108, 201, 252.

S. Giovanni Decollato ('San Zan Degolà'), late 11th c. frescoes, 63.

SS. Giovanni e Paolo (or 'Zanipolo'), church built by the Dominicans (13th c.), 139, 198, 200, 355.

Numerous *tombs* of doges, those of P. Mocenigo (*c.* 1480) and P. Malipiero (†1462) by P. Lombardo, 198, 355.

In the *chapels* paintings by G. Bellini, L. Lotto, Piazzetta, sculpture decoration by G. Bonazza, G. Mazza, etc., 258, 355, 357.

Bibl.: G. Fogolari, *I Frari e i SS. Giovanni e Paolo*, Milan 1949.

S. Giovanni in Bragora (1475). Paintings by Cima, B. and A. Vivarini, 199, 201.

S. Marcuola, begun by Massari (1736), unfinished. *Last Supper* by Tintoretto, 285.

S. Maria del Carmine, built in 14th c. Façade 16th c. Paintings by Cima, Lotto, Schiavone, 201, 257, 284.

Deposition, bas-relief by Francesco di Giorgio, 175.

S. Maria dei Derelitti, by Longhena, 354.

S. Maria Formosa, built by Coducci, 198, 199, 254.

S. Maria Gloriosa dei Frari, built by Franciscans in 1340–1443; campanile 1393, 134, 284.

Numerous *tombs*: Doge Fr. Foscari by A. Bregno, tomb of N. Tron by A. Rizzo (1473), of J. Pesaro by Longhena, 197, 353–4.

Paintings: *Pesaro Madonna* (1526) and *Assumption* (on high altar, 1518) by Titian; in sacristy: *Madonna* by Giov. Bellini (1488), 200, 255.

Bibl.: G. Fogolari, *op. cit. See* SS. Giovanni e Paolo, above.

S. Maria dei Miracoli, by P. Lombardi (*c.* 1489), 198.

S. Maria dell'Orto, 15th c., 139, 285.

Paintings by Tintoretto, 139, 285.

S. Maria della Salute, by Longhena (1631–87), altar by G. La Corte (*c.* 1660), 353–5.

In the *sacristy* (formerly Santo Spirito), paintings by Titian, 256.

S. Michele in Isola, by Coducci (1469–79), 198.

S. Moisè, façade by Tremignon (1668), *Paliotto*, by the Roccatagliata(1663), 354, 355.

S. Niccolo da Tolentino, paintings by J. Lyss and B. Strozzi, 356.

S. Pantaleone. *Martyrdom of the Saint*, by F. Fumiani, 357.

S. Pietro di Castello (1619), altar by Longhena, paintings by G. Lazzarini, 353, 356.

Archangelo Raffaele (1618). On the organ case, *Story of Tobias*, by Fr. Guardi (1750), 361.

Redentore, by Palladio (1580), 280, 283.

S. Rocco, built by B. Bon in 1495, reconstructed in 1725, façade 1771. *Story of S. Roch*, by Tintoretto, 285, 355.

S. Salvatore (1506–34), façade 18th c. *Annunciation* and *Transfiguration*, by Titian, 256.

S. Sebastiano (1505–48). On the ceiling: *Story of Esther* (1556), in the apse: *Story of S. Sebastian*, by Veronese, 284, 287, 358.

S. Simeone Piccolo, by Scalfarotto (*c.* 1730), 354.

S. Stae, façade by D. Rossi (1709); *Martyrdom of St. Bartholomew*, by Tiepolo (1721), 358.

Cappella Pellegrini: St. George, by Pisanello (1436), 134.
  Choir, fresco by Turone (1360), 133.
  S. Bernardino, Cappella Pellegrini, by Sanmicheli, 251.
  S. Fermo, formed by two superimposed churches, 11th and 12th c. *Annunciation*, fresco by Pisanello, 134.
  ‡*S. Lorenzo* (*c.* 1100), 61.
  Sta Maria in Organo, 15th c. Campanile of 1525, *intarsie* by Fra Giovanni di Verona (1500), 196.
  S. Zeno (1120–38), completed in 13th c. On the portal, reliefs by Maestro Niccolo and Guglielmo, bronze doors (work of German artists), 48, 61.
  Triptych by Mantegna (1457–59), 190.
  Arco del Cansignore della Scala (†1375) by Bonino da Campione, 136.
  Loggia del Consiglio, by Fra Giocondo (1470), 189.
  Castello (Museum), 133.
  Palazzo Bevilacqua and *Palazzo Pompei*, by M. Sanmicheli (*c.* 1530), 251, 287.
  Teatro Filarmonico, by Francesco Galli, 314.
  In the environs:
  *Bevilacqua. Villa by Sanmicheli (destroyed), triumphal arch entrance, 251.
  Fumane. Villa delle Torri, by Sanmicheli, 251.
  Pedemonte-Valpolicella. Villa Santa Sofia, by Palladio, 281.
Vezzolano (Piedmont), 20 km. from Turin.
  12th c. *abbey*, screen of 1189, mid-14th c. frescoes, 132.
Vicenza. Town of Venetia, Ligurian, then Roman colony, commune in Middle Ages, belonged to the Carrara, the Scaliger, the Visconti; annexed by Venice in 1404, became in 16th c. the 'Venice of the Terra Firma', 269, 278, 358.
  Bibl.: L. Magagnato, F. Barbieri and R. Cevese, *Guida di Vicenza*, Vicenza 1953.
  Church of the Servi: The Trinity with Sarah and Abraham, by Fr. Maffei, 356.
  Oratorio delle Zitelle, paintings by Maffei, 356.
  Loggia del Capitano, by Palladio (1571), 280.
  Palazzo Chiericati, by Palladio (1566), 280.
  Palazzo della Ragione or 'Basilica', Gothic building with classical façades by Palladio (1544–80), finished after his death (1614), 280, 283.
  Palazzo Thiene, by Palladio (1556), 280.
  Palazzo Valmarana, by Palladio (1566), 280.
  Teatro Olimpico, Palladio's last work (1580), finished by Scamozzi (1583), 283.
  Villa Capra. 'La Rotonda', begun by Palladio in 1550, finished by Scamozzi (1606), 280, 281, 282, 284.
  Villa Coldogno, by Palladio (*c.* 1570), interior decorated by Fasolo, 283.
  Villa Godi Porto, at Lonedo, built by Palladio (*c.* 1540), decorated by Zelotti, 280, 281, 283.
  Villa Thiene, unfinished work by Palladio (1550), 280–1.
  Villa Trissino, at Meledo, by Palladio, 280, 281, 283.
  Villa Trissino da Porto, built by F. Muttoni; in the garden, statues by O. Marinali (17th c.), 354, 355.
  Villa Valmarana, frescoes by Tiepolo (1757), 359.
  Bibl.: R. Pallucchini, *Gli Affreschi di Giambattista e Giandomenico Tiepolo alla Villa Valmarana di Vicenza*, Bergamo 1945. A. Morassi, *Tiepolo, la villa di Valmarana*, Milan 1946.

TOPOGRAPHICAL INDEX

INVENTORY 74

INVENTORY 1983